BIOGEOCHEMISTRY

FOURTH EDITION

BIOGEOCHEMISTRY
An Analysis of Global Change

FOURTH EDITION

WILLIAM H. SCHLESINGER
Duke University, Durham, NC, United States
Cary Institute of Ecosystem Studies, Millbrook, NY, United States

EMILY S. BERNHARDT
Department of Biology, Duke University, Durham, NC, United States

ELSEVIER

ACADEMIC PRESS
An imprint of Elsevier

Academic Press is an imprint of Elsevier
125 London Wall, London EC2Y 5AS, United Kingdom
525 B Street, Suite 1650, San Diego, CA 92101, United States
50 Hampshire Street, 5th Floor, Cambridge, MA 02139, United States
The Boulevard, Langford Lane, Kidlington, Oxford OX5 1GB, United Kingdom

Notices

Knowledge and best practice in this field are constantly changing. As new research and experience broaden our
understanding, changes in research methods, professional practices, or medical treatment may become
necessary.

Practitioners and researchers must always rely on their own experience and knowledge in evaluating and using
any information, methods, compounds, or experiments described herein. In using such information or methods
they should be mindful of their own safety and the safety of others, including parties for whom they have a
professional responsibility.

To the fullest extent of the law, neither the Publisher nor the authors, contributors, or editors, assume any liability
for any injury and/or damage to persons or property as a matter of products liability, negligence or otherwise,
or from any use or operation of any methods, products, instructions, or ideas contained in the material herein.

Library of Congress Cataloging-in-Publication Data

A catalog record for this book is available from the Library of Congress

British Library Cataloguing-in-Publication Data

A catalogue record for this book is available from the British Library

ISBN: 978-0-12-814608-8

For information on all Academic Press publications
visit our website at https://www.elsevier.com/books-and-journals

Publisher: Candice Janco
Acquisitions Editor: Amy Shapiro
Editorial Project Manager: Megan Ashdown
Production Project Manager: Bharatwaj Varatharajan
Cover Designer: Matthew Limbert

Typeset by SPi Global, India

Last digit is the print number: 10 9 8 7 6 5 4 3 2

Working together
to grow libraries in
developing countries

www.elsevier.com • www.bookaid.org

Dedication

To Planet Earth

WHS and ESB

Contents

Preface

This volume represents the latest update to our efforts to provide a text and reference book for the general field of biogeochemistry. Like the earlier editions, published in 1991, 1997, and 2013, this book focuses on the theme that biology controls the chemical conditions on the surface of the Earth and that without human stewardship, our rising impact on the planet will likely alter its conditions to the detriment of a sustainable future for the biosphere and the human enterprise.

The format of the book follows that of earlier editions. Eight chapters consider the biogeochemistry in different realms in space and time, whereas the last four chapters attempt to integrate this understanding to provide a global picture of how the Earth functions. Where needed, we have revised text, figures, and tables to reflect recent advances, and we have substantially updated the literature cited in response to an explosion of new published research in the past few years.

We hope these pages will excite students to join the field of biogeochemistry and contribute to the better management of planet Earth for future generations.

WHS and ESB
Millbrook, NY, United States
Durham, NC, United States

Acknowledgments

In the course of preparing four editions of this book, we have benefited from comments, references, data, and insights sent to us from colleagues worldwide, many of whom have been acknowledged in earlier editions. We would like to thank the many students enrolled in our Biogeochemistry course at Duke University over the last 40 years (1980–2019); they continue to challenge us to explain biogeochemistry more effectively. Here we especially thank various friends who have helped in the preparation of this edition, including Ron Amundson and Richard Phillips, who reviewed certain chapters, and Alex Glass, Dan Binkley, Evan DeLucia, Guy Dovrat, Adrien Finzi, Chris Geron, Jackie Gerson, Kevin Griffin, Pat Hatcher, Kirsten Hofmockel, Ben Houlton, Lu Hu, Stephen Jasinski, Heike Knicker, Ed Laws, Jochen Nuester, Sasha Reed, Bill Reiners, Phil Taylor, and Kevin Trenberth, who each offered helpful advice, data, figures, photographs, and reviews to help make this a better volume. At Duke, Jennifer Rocca helped with general editing and compiling of the literature cited, and Laura Turcotte offered a variety of administrative services to our effort.

Of course, any errors are our own, and we welcome comments from readers at any time—schlesingerw@caryinstitute.org or emily.bernhardt@duke.edu.

Processes and reactions

What is biogeochemistry?

Today life is found from the deepest ocean trenches to the heights of the atmosphere above Mt. Everest; from hot and saline deserts in Chile to the coldest snows of Antarctica; and from acid mine drainage, with pH < 1.0, to alkaline groundwaters with pH of > 12. More than 3.5 billion years of life on Earth has allowed the evolutionary process to fill nearly all habitats with species, large and small. And collectively these species have left their mark on the environment in the form of waste products, byproducts, and their own dead remains.

Look into any shovel of soil and you will see organic materials that are evidence of life—a sharp contrast to what we see on the barren surface of Mars. Any laboratory sample of the atmosphere will contain nearly 21% oxygen (O_2), an unusually high concentration given that the Earth harbors lots of organic materials, such as wood, that are readily consumed by fire. All evidence suggests that the oxygen in Earth's atmosphere is derived and maintained by the photosynthesis of green plants. In a very real sense, O_2 is the signature of life on Earth (Sagan et al., 1993; McKay, 2014).

The father of biogeochemistry—Valdimir Vernadsky—recognized that the influence of life on Earth is so pervasive that there is no pure science of geochemistry at the Earth's surface (Vernadsky, 1998). Indeed, many of the Earth's characteristics are hospitable to life today because of the current and historic abundance of life on this planet (Reiners, 1986). Granted some Earthly characteristics, such as its gravity, the seasons, and the radiation received from the Sun, are determined by the size and position of our planet in the solar system. But many other features, including climate, liquid water, free oxygen, and a nitrogen-rich atmosphere, are at least partially due to the presence of life. Life puts the *bio* in biogeochemistry.

At present, there is ample evidence that our species, *Homo sapiens*, is leaving unusual imprints on Earth's chemistry. The human combustion of fossil fuels is raising the concentration of carbon dioxide in our atmosphere to levels not seen in at least the past 5 million years (Stap et al., 2016). Our release of an unusual class of industrial compounds known as chlorofluorocarbons has depleted the concentration of ozone in the upper atmosphere, where it normally protects the Earth's surface from harmful levels of ultraviolet light (Rowland, 1989). In our effort to feed 7 billion people, we produce vast quantities of nitrogen and phosphorus fertilizers, resulting in the runoff of nutrients that pollute surface and coastal waters (Chapter 12). As a result of coal combustion and other human activities, the concentration of mercury in fish is much higher than a century ago (Monteiro and Furness, 1997), rendering many species unfit for regular human consumption. In pursuit of a comfortable life style, humans have fostered the proliferation of exotic chemicals at a rate that exceeds the growth of the human population (Fig. 1.1; Bernhardt et al., 2017b).

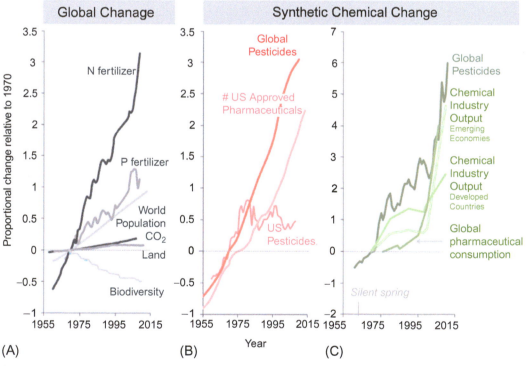

FIG. 1.1   (A) Trajectories for drivers of global environmental change; (B) increases in the diversity of US pharmaceuticals and the application of pesticides within the US and globally; (C) trends for the global trade value (in USD) of synthetic chemicals and for the pesticide and pharmaceutical chemical sectors individually. To allow comparison, all trends are shown relative to values reported in 1970 (Bernhardt et al., 2017b).

Certainly we are not the first species that has altered our chemical environment. It will be interesting to see if our creative innovation can be directed toward mitigating the environmental degradation that stems from changes in Earth's chemistry threatening to derail the improving quality of life for humanity.

Understanding the earth as a chemical system

Just as a laboratory chemist attempts to observe and understand the reactions in a closed test tube, biogeochemists try to understand the chemistry of nature, where the reactants are found in a complex mixture of solid, liquid, and gaseous phases. In most cases, biogeochemistry is a nightmare to a traditional laboratory chemist—the reactants are impure, their concentrations are low, and the temperature is variable. About all you can say about the Earth as a chemical system is that it is closed with respect to mass, save for a few meteors arriving and a few satellites leaving our planet. This closed chemical system is powered by the receipt of energy from the Sun, which has allowed the elaboration of life in many habitats (Falkowski et al., 2008).

Biogeochemists often build models for what controls Earth's surface chemistry and how Earth's chemistry may have changed through the ages. Unlike laboratory chemists, we have no replicate planets for experimentation, so our models must be tested and validated by inference. For instance, if our models suggest that the accumulation of organic materials in ocean sediments is associated with the deposition of gypsum ($CaSO_4 \cdot 2H_2O$), we must dig down through the sedimentary layers to see if this correlation occurs in the geologic record (Garrels and Lerman, 1981). Finding the correlation does not prove the model, but it adds a degree of validity to our understanding of how Earth works—its biogeochemistry. Models must be revised when observations are inconsistent with their predictions.

With sufficient empirical observations, mathematical models can be built to describe ecosystem function. Here, equations are used to express what controls the movement of energy and materials through organisms or individual compartments of an ecosystem, such as the soil. These models allow us to determine what processes control the productivity and biogeochemical cycling in ecosystems, and to identify areas where our understanding is incomplete. Models that are able to reproduce past dynamics allow us to explore the future behavior of ecosystems in response to perturbations that may lie outside the natural range of environmental variation.

In many cases, the Earth's conditions, such as the composition of the atmosphere, change only slowly from year to year, so we can use steady-state assumptions to model the activities of the biosphere, which we define as the sum of all the live and dead materials on Earth.[a] In a steady-state model of the atmosphere, the inputs and losses of gases from the biosphere are balanced each year; the individual molecules in the atmosphere change, but the total content of each stays relatively constant.

[a] Some workers use the term biosphere to refer to the regions or volume of Earth that harbor life. We prefer the definition used here, so that the oceans, atmosphere, and surface crust can be recognized separately. Our definition of the biosphere recognizes that it has mass, but also functional properties derived from the species that are present.

For some cyclic behaviors, such as the daily rotation of the Earth around its axis and its annual rotation about the Sun, it seems surprising that these phenomena were mysterious to philosophers and scientists throughout much of human history. These cycles affect the behavior of the biosphere, and they can be described by steady-state modeling. For instance, total plant photosynthesis exceeds respiration by decomposers during the summer. This results in a temporary storage of carbon in plant tissues and a seasonal decrease in atmospheric CO_2, which is lowest during August of each year in the Northern Hemisphere (Fig. 1.2). The annual cycle is completed during the winter months, when atmospheric CO_2 returns to higher levels, as decomposition continues when many plants are dormant or leafless. Certainly, it would be a mistake to model the activity of the biosphere by considering only the springtime period of rapid change, but a steady-state model can ignore the annual cycle if it uses a particular time each year as a baseline condition to examine changes over decades.

Steady-state models bring a degree of tidiness to our understanding of Earth's chemistry, especially when long periods of cyclic behavior control Earth's characteristics. Robert Berner

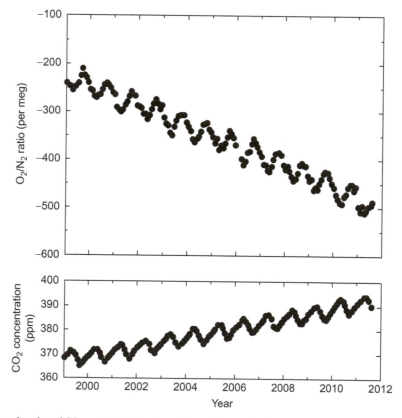

FIG. 1.2   Annual cycles of CO_2 and O_2 in the atmosphere. Changes in the concentration of O_2 are expressed relative to concentrations of nitrogen (N_2) in the same samples. Note that the peak of O_2 in the atmosphere corresponds to the minimum CO_2 in late summer, presumably due to the seasonal course of photosynthesis in the Northern Hemisphere. *From Ralph Keeling, unpublished data used by permission.*

and his coworkers at Yale University elucidated the components of a carbonate-silicate cycle that stabilizes Earth's climate and its atmospheric chemistry over long periods of geologic time (Berner and Lasaga, 1989). The model is based on the interaction of carbon dioxide with Earth's crust. When CO_2 in the atmosphere dissolves in rainwater to form carbonic acid (H_2CO_3), it reacts with the minerals exposed on land in the process known as rock weathering (Chapter 4). The products of rock weathering are carried by rivers to the sea (Fig. 1.3). In the oceans, limestone (calcium carbonate) and organic matter are deposited in marine sediments, which in time are carried by subduction into Earth's upper mantle. Here the sediments are metamorphosed; the calcium and silicon are converted back into the minerals of silicate rock, and the carbon is returned to the atmosphere as CO_2 in volcanic eruptions (Drewitt et al., 2019). On Earth, the entire oceanic crust appears to circulate through this pathway in <200 million years (Muller et al., 2008). The presence of life on Earth does not speed the turning of this cycle, but it may increase the amount of material moving in the various pathways by increasing the rate of rock weathering on land and the rate of carbonate deposition in the sea.

The carbonate-silicate model is a steady-state model, in the sense that it shows equal transfers of material along the flow-paths and no change in the mass of various compartments over time. In fact, such a model suggests a degree of self-regulation of the system, because any

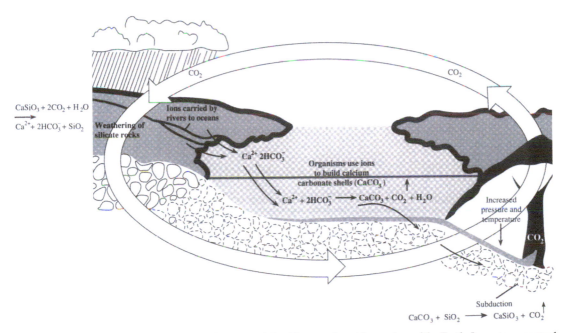

FIG. 1.3 The interaction between the carbonate and the silicate cycles at the surface of the Earth. Long-term control of atmospheric CO_2 is achieved by dissolution of CO_2 in surface waters and its participation in the weathering of rocks. This carbon is carried to the sea as bicarbonate (HCO_3^-), and it is eventually buried as part of carbonate sediments in the oceanic crust. CO_2 is released back to the atmosphere when these rocks undergo metamorphism at high temperature and pressures deep in the Earth. *Modified from Kasting et al. (1988).*

period of high CO_2 emissions from volcanoes should lead to greater rates of rock weathering, removing CO_2 from the atmosphere and restoring balance to the system.

Perturbations of the steady-state

While steady-state models are often useful, biogeochemists should be alert to situations when the assumption of a steady state may not be valid, such as during transient periods of non-linear behavior and rapid change. For instance, the storage of organic carbon on land increased strongly during the Carboniferous Period—about 300 million years ago, when most of the major deposits of coal were laid down. The unique conditions of the Carboniferous Period are poorly understood, and not apparently related to any cyclic behavior. Similarly, during the Eocene, 40 million years ago, high rates of volcanic activity may have upset steady-state conditions and resulted in a temporary increase in atmospheric CO_2 and a period of global warming (Owen and Rea, 1985).

During the Pleistocene epoch, changes in atmospheric CO_2 are best viewed in the context of cyclic changes during the past 800,000 years in a record obtained from the bubbles of air trapped in an ice core taken near Vostok, Antarctica (Fig. 1.4). During the entire 800,000-year record, the concentration of atmospheric CO_2 appears to have oscillated between high values (~280 ppm) during warm periods and lower values (~200 ppm) during glacial intervals. The glacial cycles are linked to small variations in Earth's orbit that alter the receipt of radiation from the Sun (Hays et al., 1976). During the peak of the last glacial epoch (20,000 years ago),

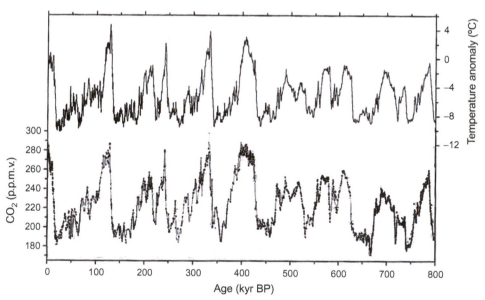

FIG. 1.4   An 800,000-year record of CO_2 and temperature, showing the minimum temperatures correspond to minimum CO_2 concentrations seen in cycles of ~120,000 periodicity, associated with Pleistocene glacial epochs. *Adapted from Lüthi et al. (2008).*

CO_2 ranged from 180 to 200 ppm in the atmosphere. Then CO_2 rose dramatically at the end of the last glacial (10,000 years ago) and was relatively stable at 280 ppm until the Industrial Revolution. The rapid increase in CO_2 at the end of the last glacial epoch may have amplified the global warming that melted the continental ice sheets (Shakun et al., 2012; Parrenin et al., 2013).

The recent increase in atmospheric CO_2 to today's value of more than 400 ppm has occurred at an exceedingly rapid rate, which carries the planet into a range of concentrations never before experienced during the evolution of modern human social and economic systems, beginning about 8000 years ago (Flückiger et al., 2002). Since the Industrial Revolution, humans have added more carbon dioxide to the atmosphere than the carbonate-silicate cycle or the ocean can absorb each year (Fig. 1.3). We have upset steady-state conditions on the planet. If the past is an accurate predictor of the future, higher atmospheric CO_2 will lead to global warming. But the current and ongoing changes in global climate must also be evaluated in the context of long-term cycles in climate with many possible causes (Crowley, 2000; Stott et al., 2000).

Because the atmosphere is well mixed, changes in its composition are perhaps our best evidence of human alteration of Earth's surface chemistry. Concern about global change is greatest when we see rapid increases in atmospheric content of constituents such as carbon dioxide, methane (CH_4), and nitrous oxide (N_2O), for which we see little or no precedent in the geologic record. These gases are produced by organisms, so changes in their global concentration must reflect massive changes in the composition or activity of the biosphere.

Humans have also changed other aspects of Earth's natural biogeochemistry. For example, when human activities increase the erosion of soil, we alter the natural rate of sediment delivery to the oceans and the deposition of sediments on the seafloor (Wilkinson and McElroy, 2007; Syvitski et al., 2005). As in the case of atmospheric CO_2, evidence for global changes in erosion induced by humans must be considered in the context of long-term oscillations in the rate of crustal exposure, weathering, and sedimentation due to changes in climate and sea level (Worsley and Davies, 1979; Peizhen et al., 2001).

Human extraction of fossil fuels and the mining of metal ores substantially enhance the rate at which materials are available to the biosphere, relative to background rates dependent on geologic uplift and surface weathering (Bertine and Goldberg, 1971; Sen and Peucker-Ehrenbrink, 2012). For example, the mining and industrial use of lead (Pb) has increased the transport of Pb in world rivers by about a factor of 10 compared to historical conditions (Martin and Meybeck, 1979). Recent changes in the content of lead in coastal sediments appear directly related to fluctuations in the use of Pb by humans, especially in leaded gasoline (Trefry et al., 1985)—trends superimposed on underlying natural and historical variations in the movement of Pb at Earth's surface (Marteel et al., 2008; Pearson et al., 2010; McConnell et al., 2018).

Recent estimates suggest that the global cycles of many metals have increased significantly and inadvertently due to human activities (Table 1.1). Some of these metals are released to the atmosphere and deposited in remote locations (Boutron et al., 1994; Preunkert et al., 2019; McConnell et al., 2018). For example, combustion of coal has raised the concentration of mercury (Hg) deposited in Greenland ice layers during the past 100 years (Weiss et al., 1971). Recognizing that the deposition of Hg in the Antarctic ice cap shows large variations over the past 34,000 years (Vandal et al., 1993), we must evaluate any recent increase in Hg deposition in the context of past cyclic changes in Hg transport through the atmosphere. Again, human-induced changes in the movement of materials through the atmosphere must be placed in the context of natural cycles in Earth system function (Nriagu, 1989).

TABLE 1.1 Movement of certain crustal elements through the atmosphere.

| Element | Continental dust | Sea spray | Volcanic emissions | Biomass burning | Volatilization | | Industrial particles | Fossil fuel combustion | Ratio anthropogenic: natural |
					Natural	Human induced			
Vanadium	155	0.52	7	5				287	1.71
Mercury	0.12	0.009	0.5	0.6	3.2	3.5		3.4	1.55
Lead	32	5	4.1	38			32	85	1.48
Copper	50	14	9	27			43	4	0.47
Zinc	100	51	10	147			88	5	0.30
Silver	2.3	0.01	0.01	1.2			0.44	0.05	0.14
Iron	55,000	200	8800	830			641	4200	0.07
Aluminum	96,000	810	4500	2125			397	5900	0.06

From Rauch and Pacyna (2009), except for vanadium (Schlesinger et al., 2017) and mercury (Selin, 2009; Sen and Peucker-Ehrenbrink, 2012). All data in 10^9 g/yr.

Globally, human impacts threaten the quantity and quality of freshwaters (Rodell et al., 2018). Collectively, about 20% of the easily accessible freshwater in rivers is usurped for human use (Jaramillo and Destouni, 2015). Throughout the world, groundwaters show ongoing depletion, a worrisome trend given that many groundwater resources are only replenished over thousands of years (Gleeson et al., 2012). Widespread construction of dams on major rivers has altered the hydrologic connection and sediment delivery between land and sea (Nilsson et al., 2005; Syvitski et al., 2005).

Other intentional human activities, such as the international trade of commodities affect global biogeochemistry. International shipment now approaches 2.5 billion tons/year. Foods and fertilizers carry a significant quantity of nitrogen and phosphorus, contributing to the redistribution of these elements between continents (Lassaletta et al., 2014). Widespread use of pesticides now accounts for accumulations of exotic chemicals in regions far from the point of application (Simonich and Hites, 1995).

Large-scale experiments

Biogeochemists frequently conduct large-scale experiments to assess the response of natural systems to human perturbation. Schindler (1974) added phosphorus to experimental lakes in Canada to show that it was the primary nutrient limiting algal growth in those ecosystems (Fig. 1.5). Bormann et al. (1974) deforested an entire watershed in New England to demonstrate the importance of vegetation in sequestering nutrients in ecosystems (Chapter 6). Several experiments have exposed replicated plots of forests, grasslands, and desert ecosystems to high CO_2 to simulate plant growth in the future environments on Earth (Chapter 5). And oceanographers have added Fe to large patches of the sea to ascertain whether it normally limits the growth of marine phytoplankton (Chapter 9). In many cases these large experiments and field campaigns are designed to test the predictions of models and to validate them.

Scales of endeavor

The science of biogeochemistry spans a huge range of space and time, spanning most of the geologic epochs of Earth's history (see inside back cover). Geologists study the chemical weathering of minerals in rocks and soils and document Earth's past by examining sedimentary cores taken from lakes, oceans, and continental ice packs. Microbiologists study the occurrence and activity of microbes in the wide variety of habitats they occupy on Earth. Atmospheric scientists provide details of reactions between gases and the radiative properties of the planet. Meanwhile, remote sensing from aircraft and satellites allows biogeochemists to see the Earth at the largest scale, measuring global photosynthesis (Running et al., 2004) and following the movement of desert dusts around the planet (Uno et al., 2009; Zhang et al., 2015b).

Meanwhile, molecular biologists contribute their understanding of the chemical structure and spatial configuration of biochemical molecules, explaining why some biochemical reactions occur more readily than others (Newman and Banfield, 2002). Increasingly, genomic

FIG. 1.5 An ecosystem-level experiment in which a lake was divided and one half (distant) fertilized with phosphorus, while the basin in the foreground acted as a control. The phosphorus-fertilized basin shows a bloom of nitrogen-fixing cyanobacteria. *From Schindler (1974); www.sciencemag.org/content/184/4139/897.short. Used with permission.*

sequencing allows biogeochemists to identify the microbes that are active in soils and sediments and what regulates their gene expression (Fierer et al., 2007). Physiologists measure variations in the activities of organisms, while ecosystem scientists measure the movement of materials and energy through well-defined units of the landscape.

Indeed the skills needed by the modern biogeochemist are so broad that many students find their entrance to this new field bewildering. But the fun of being a biogeochemist stems from the challenge of integrating new science from diverse disciplines. And luckily, there are a few basic rules that guide the journey.

Thermodynamics

Two basic laws of physical chemistry, the laws of thermodynamics, tell us that energy can be converted from one form to another and that chemical reactions should proceed spontaneously to yield the lowest state of free energy, G, in the environment. The lowest free energy of a chemical reaction represents its equilibrium, and it is found in a mix of chemical species that show maximum bond strength and maximum disorder among the components. In the face of these basic laws, living systems create non-equilibrium conditions; life uses energy to counteract reactions that might happen spontaneously to maximize disorder.

Even the simplest cell is an ordered system; a membrane separates an inside from an outside, and the inside contains a mix of very specialized molecules. Biological molecules are collections of compounds with relatively weak bonds. For instance, to break the covalent bonds between two carbon atoms requires 83 kcal/mole, versus 192 kcal/mole for each of the double bonds between carbon and oxygen in CO_2 (Davies, 1972; Morowitz, 1968). In living tissue most of the bonds between carbon (C), hydrogen (H), nitrogen (N), oxygen (O), phosphorus (P), and sulfur (S), the major biochemical elements, are reduced or "electron-rich" bonds that are relatively weak (Chapter 7). It is an apparent violation of the laws of thermodynamics that the weak bonds in the molecules of living organisms exist in the presence of a strong oxidizing agent in the form of O_2 in the atmosphere. Thermodynamics would predict a spontaneous reaction between these components to produce CO_2, H_2O, and NO_3^-—molecules with much stronger bonds. In fact, after the death of an organism, this is exactly what happens! Living organisms must continuously expend energy to counteract the basic laws of thermodynamics that would otherwise produce disordered systems with oxidized molecules and stronger bonds.

During photosynthesis, plants capture the energy in sunlight and convert the strong bonds between carbon and oxygen in CO_2 to the weak, reduced biochemical bonds in organic materials. As heterotrophic organisms, herbivores eat plants to extract this energy by capitalizing on the natural tendency for electrons to flow from reduced bonds back to oxidizing substances, such as O_2. Heterotrophs oxidize the carbon bonds in organic matter and convert the carbon back to CO_2. A variety of other metabolic pathways have evolved using transformations among other compounds (Chapters 2 and 7), but in every case metabolic energy is obtained from the flow of electrons between compounds in oxidized or reduced states. Metabolism is possible because living systems can sequester high concentrations of oxidized and reduced substances from their environment. Without membranes to compartmentalize living cells, thermodynamics would predict a uniform mix, and energy transformations, such as respiration, would be impossible.

Free oxygen appeared in Earth's surface environments sometime after the appearance of autotrophic, photosynthetic organisms (Chapter 2). Free O_2 is one of the most oxidizing substances known, and the movement of electrons from reduced substances to O_2 releases large amounts of free energy. Thus, large releases of free energy are found in aerobic metabolism, including the efficient metabolism of eukaryotic cells. The appearance of eukaryotic cells on Earth was not immediate; the fossil record suggests that they evolved about 1.5 billion years after the appearance of the simplest living cells (Knoll, 2015). Presumably the evolution of eukaryotic cells was possible only after the accumulation of sufficient O_2 in the environment to sustain aerobic metabolic systems. Aerobic metabolism in an atmosphere with free O_2 was essential to the evolution and dominance of multicellular (fleshy) organisms, carnivory, and fire (Belcher and McElwain, 2008; Sperling et al., 2013; Judson, 2017).

Although aerobic metabolism offered large amounts of energy that could allow the elaborate structure and activity of higher organisms, some humility is important. Eukaryotic cells may perform biochemistry faster and more efficiently, but the full range of known biochemical transformations in nature is found amongst the members of the prokaryotic kingdom.

Stoichiometry

A second organizing principle of biogeochemistry stems from the coupling of elements in the chemical structure of the molecules of which life is built—cellulose, protein, and the like.

Redfield's (1958) observation of consistent amounts of C, N, and P in phytoplankton biomass is now honored by a ratio that carries his name (Chapter 9). Reiners (1986) carried the concept of stoichiometric ratios in living matter to much of the biosphere, allowing us to predict the movement of one element in an ecosystem by measurements of another, and Sterner and Elser (2002) have presented stoichiometry as a major control on the structure and function of ecosystems.

Although an expected chemical stoichiometry in biomass allows us to predict the concentration of chemical elements in living matter, the expected ratio of elements in biomass is not absolute, such as the ratio of C to N in a reagent bottle of the amino acid alanine. For instance, a sample of phytoplankton will contain a mix of species that vary in individual N/P ratios, with the weighted average close to that postulated by Redfield (Klausmeier et al., 2004). And, of course, a large organism will contain a mix of metabolic compounds (largely protein) and structural components (e.g., wood or bone) that differ in elemental composition (Reiners, 1986; Elser et al., 2010). In some sense, organisms are what they eat, but decomposers can adjust their metabolism (Manzoni et al., 2008) and enzymatic production (Sinsabaugh et al., 2009) to feed on a wide range of substrates, even as they maintain a constant stoichiometry in their own biomass.

Often we speak as if a single element "limits" the productivity or activity of life in certain habitats. This is too simple. Sometimes several elements are in short supply, producing a condition of co-limitation of biotic activity (Kaspari and Powers, 2016). In many cases, organisms respond to the ratio between the available supply of the important elements for biochemistry (Zechmeister-Boltenstern et al., 2015). Sometimes the addition of one element will engender responses that increase the availability of other elements. And, sometimes shortages of a particular element for one component of an ecosystem, for example, plants, are accompanied by shortages of other elements for other components of the ecosystem (Kaspari and Powers, 2016). The growth of land plants is often determined from the nitrogen content of their leaves and the nitrogen availability in the soil (Chapter 6), whereas phosphorus availability explains much of the variation in algal productivity in lakes (Chapter 8). The population growth of some animals may be determined by sodium—an essential element that is found at a low concentration in potential food materials, relative to its concentration in body tissues.

In some cases, trace elements control the cycle of major elements, such as nitrogen, by their role as activators and cofactors of enzymatic synthesis and activity. When nitrogen supplies are low, signal transduction by P activates the genes for N fixation in bacteria (Stock et al., 1990). The enzyme for nitrogen fixation, nitrogenase, contains iron (Fe) and molybdenum (Mo). Over large areas of the oceans, Falkowski et al. (1998) suggest that iron, delivered to the surface waters by the wind erosion of desert soils, controls marine production, which is often limited by N-fixation. Similarly, when phosphorus supply is low, plants and microbes may produce alkaline phosphatase, containing zinc, to release P from dead materials (Shaked et al., 2006). Thus, the productivity of some ecosystems can be stimulated either by adding the limiting element itself or by adding a trace element that facilitates nutrient acquisition (Arrigo, 2005).

The chemical stoichiometry of life extends to metabolism, where certain elements must be available in the right proportions to drive energy-yielding reactions (Helton et al., 2015). The coupled biogeochemistry of elements in metabolism stems from the flow of electrons in the oxidation/reduction reactions that power all of life (Morowitz, 1968; Falkowski et al., 2008).

Oxidized ————————————→ Reduced

	H$_2$O/O$_2$	C	N	Fe	S
H$_2$O/O$_2$ (Oxidized)	X	Photosynthesis $CO_2 \rightarrow C$ $H_2O \rightarrow O_2$			
C	Respiration $C \rightarrow CO_2$ $O_2 \rightarrow H_2O$	X	Denitrification $C \rightarrow CO_2$ $NO_3 \rightarrow N_2$	Iron-Reducing Bacteria $C \rightarrow CO_2$ $Fe^{3+} \rightarrow Fe^{2+}$	Sulfate-Reduction $C \rightarrow CO_2$ $SO_4 \rightarrow H_2S$
N	Heterotrophic Nitrification $NH_4^+ \rightarrow NO_3^-$ $O_2 \rightarrow H_2O$	Chemoautotrophy Nitrification $NH_4^+ \rightarrow NO_3^-$ $CO_2 \rightarrow C$	Anammox $NH_4 + NO_2 \rightarrow N_2 + 2H_2O$	Feammox $NH_4^+ \rightarrow NO_2^-$ $Fe^{3+} \rightarrow Fe^{2+}$	?
Fe	$Fe \rightarrow Fe_2O_3$ (rust)	Iron Photosynthetic Bacteria $Fe^{2+} \rightarrow Fe^{3+}$ $CO_2 \rightarrow C$	$Fe^{2+} \rightarrow Fe^{3+}$ $NO_3 \rightarrow N_2/N_2O$	$Fe^{2+} \rightarrow Fe^{3+}$ $Fe^{3+} \rightarrow Fe^{2+}$	
S (Reduced)	Thiobacillus Thioxidans $S \rightarrow SO_4^{-2}$ $O_2 \rightarrow H_2O$ (Acid-mine Drainage)	Sulfur-based Photosynthesis and Chemoautotrophy $H_2S \rightarrow S/SO_4^{-2}$ $CO_2 \rightarrow C$	Thiobacillus denitrificans and Thioploca $S \rightarrow SO_4$ $NO_3 \rightarrow N_2/NH_4$	$SO_3^{-2} \rightarrow SO_4^{-2}$ $Fe^{3+} \rightarrow Fe^{2+}$	X

FIG. 1.6   A matrix showing how cellular metabolisms couple oxidation and reduction reactions. The cells in the matrix are occupied by organisms or a consortium of organisms that reduce the element at the top of the column, while oxidizing an element at the beginning of the row. *Modified from Schlesinger et al. (2011).*

Coupled metabolism is illustrated by a matrix, where each element in a column is reduced while the element in an intersecting row is oxidized (Fig. 1.6). All of Earth's metabolisms can be placed in the various cells of this matrix and in a few adjacent cells that would incorporate other trace metals. The matrix incorporates the range of metabolisms possible on Earth, should the right conditions exist (Bartlett, 1986). Coupled metabolisms link the biogeochemical cycles of elements at the global scale (Reinhold et al., 2019).

Lovelock's Gaia

In a provocative book, Gaia, published in 1979, James Lovelock focused scientific attention on the chemical conditions of the present-day Earth, especially in the atmosphere, that are extremely unusual and in disequilibrium with respect to thermodynamics. The 21% atmospheric content of O_2 is the most obvious result of living organisms, but other gases, including NH_3 and CH_4, are found at higher concentrations than one would expect in an O_2-rich atmosphere (Chapter 3). This level of O_2 in our atmosphere is maintained despite known reactions that should consume O_2 in reaction with crustal minerals and organic carbon. Further, Lovelock suggested that the albedo (reflectivity) of Earth must be regulated by the biosphere, because the planet has shown relatively small changes in surface temperature despite large fluctuations in the Sun's radiation during the history of life on Earth (Watson and Lovelock, 1983).

Lovelock suggested that the conditions of our planet are so unusual that they could only be expected to result from activities of the biosphere. Indeed, Gaia suggests that the biosphere evolved to regulate conditions within a range favorable for the continued persistence of life on Earth. In Lovelock's view, the planet functions as a kind of "superorganism," providing planetary homeostasis. Reflecting the vigor and excitement of a new scientific field, other workers have strongly disagreed—not denying that biotic factors have strongly influenced the conditions on Earth, but not accepting the hypothesis of purposeful self-regulation of the planet (Lenton, 1998; Tyrrell, 2013).

Like all models, Gaia remains as a provocative hypothesis, but the rapid pace at which humans are changing the biosphere should alarm us all. Some ecologists see the potential for critical transitions in ecosystem function; points, known as tipping points, beyond which human impacts would not allow the system to rebound to its prior state, even if the impacts ceased (Scheffer et al., 2009). Others have attempted to quantify these thresholds, so that we may recognize them in time (Rockström et al., 2009; Steffen et al., 2015). In all these endeavors, policy makers are desperate for biogeochemists to deliver a clear articulation of how the world works, the extent and impact of the human perturbation, and what to do about it.

Origins

Introduction

Six elements, H, C, N, O, P, and S, are the major constituents of living tissue and account for 95% of the mass of the biosphere. At least 25 other elements are known to be essential to at least one form of life, and it is possible that this list may grow slightly as we improve our understanding of the role of trace elements in biochemistry (da Silva and Williams, 2001).[a] In the periodic table (see inside front cover), nearly all the elements essential to life are found at atomic numbers lower than that of iodine, at 53. Even though living organisms affect the distribution and abundance of some of the heavier elements, the biosphere is built from the "light" elements (Deevey, 1970; Wackett et al., 2004).

Ultimately, the relative abundance of chemical elements in our galaxy and the subsequent concentration and redistribution of those elements on Earth's surface determined the

---

[a] Arsenic is known to be an essential trace element for some species, but recent reports of bacteria that are able to grow using arsenic as a substitute for phosphorus are largely discounted (Erb et al., 2012).

environment in which life arose and the arena for biochemistry today. In this chapter we will examine models that astrophysicists suggest for the origin of the elements. Then we will examine models for the formation of the solar system and the planets. There is good evidence that the conditions on the surface of the Earth changed greatly during the first billion years or so after its formation—before life arose. Early differentiation of the Earth, the cooling of its surface, and the composition of the earliest oceans determined the arena for the origins of life. Later changes caused by the evolution and proliferation of life strongly determined the conditions on our planet today. In this chapter, we will consider the origin of the major metabolic pathways that characterize life and affect Earth's biogeochemistry. The chapter ends with a discussion of the planetary evolution that has occurred on Earth compared to our near neighbors—Mars and Venus.

Origins of the elements

Any model for the origin of the chemical elements must account for their relative abundance in the Universe. Estimates of the cosmic abundance of elements are made by examining the spectral emission from the stars in distant galaxies as well as the emission from our Sun (Asplund et al., 2009). Analyses of meteorites also provide important information on the composition of the solar system (Fig. 2.1). Two points are obvious: (1) with three exceptions—lithium (Li), beryllium (Be), and boron (B)—the light elements, that is, those with an atomic number <30, are far more abundant than the heavy elements; (2) especially among the light elements, the even-numbered elements are more abundant than the odd-numbered elements of similar atomic weight.

A central theory of astrophysics is that the Universe began with a gigantic explosion, "the Big Bang," about 13.8 billion years ago (Mac Low, 2013). The Big Bang initiated the fusion of hypothetical fundamental particles, known as quarks, to form protons (^1H) and neutrons, and it allowed the fusion of protons and neutrons to form some simple atomic nuclei (^2H, ^3He, ^4He, and a small amount of ^7Li; Malaney and Fowler, 1988; Pagel, 1993). After the Big Bang, the Universe began to expand outward, so there was a rapid decline in the temperatures and pressures that would allow the formation of heavier elements by fusion in interstellar space. Moreover, the elements with atomic masses of 5 and 8 are unstable, so no fusion of the abundant initial products of the Big Bang (i.e., ^1H and ^4He) could yield an appreciable, persistent amount of a heavier element. Thus, the Big Bang can explain the origin of elements up to ^7Li, but the origin of heavier elements had to await the formation of stars in the Universe—about 1 billion years later.

The Big Bang cast its products of nucleosynthesis into the expanding space of the Universe. While the process is poorly understood, the gravitational attraction in local accumulations of hydrogen could initiate the agglomeration of more such materials in space, forming galaxies and stars. Stars "ignite, when they gather enough mass—largely hydrogen—and pressure to initiate the fusion of hydrogen to form helium (Mac Low, 2013).

As a star ages, the abundance of hydrogen (H) in the core declines as it is converted to helium (He) by fusion. As the heat from nuclear fusion decreases, the star begins to collapse inward under its own gravity. For these conditions, a model for the synthesis of heavier elements was first proposed by Burbidge et al. (1957), who outlined a series of pathways that

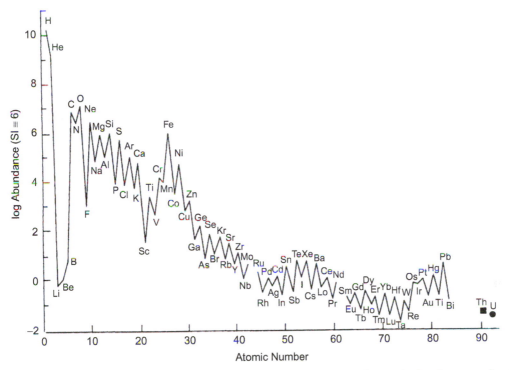

FIG. 2.1   The relative abundance of elements in the solar system, also known as the cosmic abundance, as a function of atomic number. Abundances are plotted logarithmically and scaled so that silicon (Si) = 1,000,000. *From a drawing in Brownlee (1992) based on the data of Anders and Grevesse (1989).*

could occur in the interior of massive stars during their evolution (Fowler, 1984; Wallerstein, 1988). When a star collapses, its internal temperature and pressure increase until He begins to be converted via fusion reactions to form carbon (C) in a two-step reaction known as the triple-alpha process. First,

$$^4\text{He} + {}^4\text{He} \rightleftharpoons {}^8\text{Be} \tag{2.1}$$

Then, while most ^{8}Be decays spontaneously back to helium, the momentary existence of small amounts of ^{8}Be under these conditions allows reaction with helium to produce carbon, which is stable:

$$^8\text{Be} + {}^4\text{He} \rightarrow {}^{12}\text{C} \tag{2.2}$$

The main product of this so-called helium "burning" reaction is ^{12}C, and the rate of this reaction determines the abundance of C in the Universe (Oberhummer et al., 2000). ^{16}O is built by the addition of ^{4}He to ^{12}C, and nitrogen by the successive addition of protons to ^{12}C. As the supply of helium begins to decline, a second phase of stellar collapse is followed by the initiation of a sequence of further fusion reactions in massive stars (Fowler, 1984). First, fusion of two ^{12}C forms ^{24}Mg (magnesium), some of which decays to ^{20}Ne (neon) by loss of an alpha

(^{4}He) particle. Subsequently, oxygen-burning produces ^{32}S, which forms an appreciable amount of ^{28}Si (silicon) by loss of an alpha particle (Woosley, 1986).

A variety of fusion reactions in massive stars are thought to be responsible for the synthesis—known as stellar nucleosynthesis—of the even-numbered elements up to iron (Fe) (Fowler, 1984; Trimble, 1997). Smaller stars, like our Sun, do not go through all these reactions and burn out along the way, becoming white dwarfs. The fusion reactions release energy[b] and produce increasingly stable nuclei (Friedlander et al., 1964). However, to make a nucleus heavier than Fe requires energy, so when a star's core is dominated by Fe, it can no longer burn. Fe accumulates, accounting for its anomalous peak in cosmic abundance (Fig. 2.1). Lack of further nuclear fusions leads to the catastrophic gravitational collapse and explosion of the star, which we recognize as a supernova. Heavier elements are apparently formed by the successive capture of neutrons by Fe, either deep in the interior of stable stars (s-process) or during the explosion of a supernova (r-process; Woosley and Phillips, 1988; Burrows, 2000; Cowan and Sneden, 2006). A supernova casts all portions of the star into space as hot gases (Chevalier and Sarazin, 1987). Heavy elements are also created during the collisions of neutron stars (Drout et al., 2017; Watson et al., 2019).

This model explains a number of observations about the abundance of the chemical elements in the Universe. First, the abundance of elements declines logarithmically with increasing mass beyond hydrogen and helium, the original building blocks of the Universe. However, as the Universe ages, more and more of the hydrogen will be converted to heavier elements during the evolution of stars. Astrophysicists can recognize younger, second-generation stars, such as our Sun, that have formed from the remnants of previous supernovas because they contain a higher abundance of iron and heavier elements than older, first-generation stars, in which the initial hydrogen-burning reactions are still predominant (Penzias, 1979). We should all be thankful for the fusion reactions in massive stars which have formed most of the chemical elements of life.

Second, because the first step in the formation of all the elements beyond lithium is the fusion of nuclei with an even number of atomic mass (e.g., ^{4}He, ^{12}C), the even-numbered light elements are relatively abundant in the cosmos. The odd-numbered light elements are formed by the addition of neutrons to nuclei in the interior of massive stars (s-process) and by the fission of heavier even-numbered nuclei. In most cases an odd-numbered nucleus is slightly less stable than its even-numbered "neighbors," so we should expect odd-numbered nuclei to be less abundant. For example, phosphorus (P) is formed by several nucleosynthesis reactions in massive stars (Woosley and Weaver, 1995) and dispersed to the Universe in supernova explosions (Koo et al., 2013). As an odd-numbered element, phosphorus is much less abundant than the adjacent elements in the periodic table, Si (silicon) and S (sulfur) (Fig. 2.1). It is interesting to speculate that the low cosmic abundance of P formed by nucleosynthesis may account for the fact that P is often in short supply for the biosphere on Earth today (Macia et al., 1997; Koo et al., 2013).

Finally, the low cosmic abundance of Li, Be, and B is due to the fact that the initial fusion reactions pass over nuclei with atomic masses of 5–8, forming ^{12}C, as shown in Eqs. (2.1), (2.2).

---

[b] When two light nuclei fuse, there is a slight loss of mass (m), which releases energy (E) in proportion to the speed of light (c), following Einstein's famous equation $E = mc^2$.

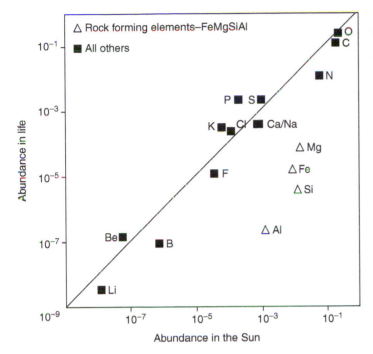

FIG. 2.2 Relative abundance of elements in living tissue versus abundance in the Sun. *From Langmuir and Broecker (2012), with data identified following Asplund et al. (2009).*

Apparently, most Li, Be, and B are formed by spallation—the fission of heavier elements that are hit by cosmic rays in interstellar space (Reeves, 1994).

This model for the origin and cosmic abundance of the elements sets some initial constraints for biogeochemistry. All things being equal, we might expect that the chemical environment in which life arose would approximate the cosmic abundance of elements. Thus, the evolution of biochemical molecules might be expected to capitalize on the light elements that were abundant in the primordial environment. Except for the rock-forming elements (Fe, Al, Si), the composition of living tissues is remarkably similar to that of the Sun (Fig. 2.2). Thus, it is of no great surprise that no element heavier than Fe is more than a trace constituent in living tissue and that among the light elements, no Li or Be, and only traces of B, are essential components of biochemistry (Wackett et al., 2004). The composition of life is remarkably similar to the composition of the Universe; as put by Fowler (1984), we are all "a little bit of stardust."

Origin of the solar system and the solid earth

The Milky Way galaxy is about 12.5 billion years old (Dauphas, 2005), indicating that the first stars and galaxies had formed within a billion years after the Big Bang (Cayrel et al., 2001). By comparison, as a second-generation star, our Sun appears to be only about 4.56 billion years old (Bonanno et al., 2002; Bouvier and Wadhwa, 2010; Wang et al., 2017a). Current models for the origin of the solar system suggest that the Sun and its planets formed from a cloud of interstellar gases and dust, likely including the remnants of a supernova

(Chevalier and Sarazin, 1987). This cloud of material would have the composition of the cosmic mix of elements (Fig. 2.1). As the Sun and the planets began to condense, each developed a gravitational field that helped capture materials that added to its initial mass. The mass concentrated in the Sun apparently allowed condensation to pressures that reinitiated the fusion of hydrogen to helium.

The planets of our solar system appear to have formed from the coalescing of dust to form small bodies, known as planetesimals, within the primitive solar cloud (Beckwith and Sargent, 1996). Then, collisions among the planetesimals would have formed the planets that we see today. The process is likely to have been fairly rapid. Several lines of evidence show planetesimals forming during the first million years of the solar system (Yin et al., 2002; Alexander et al., 2001; Wang et al., 2017a), and most stars appear to lose their disk of gases and dust within 400 million years after their formation (Habing et al., 1999). Recent observations suggest that a similar process is now occurring around another star in our galaxy, ß Pictoris (Lagage and Pantin, 1994; Lagrange et al., 2010), and Earth-size and larger planets have been detected around numerous other stars in our galaxy (Quintana et al., 2014; Gillon et al., 2017).

Overall, the original solar nebula is likely to have been composed of about 98% gaseous elements (H, He, and noble gases), 1.5% icy solids (H_2O, NH_3, and CH_4), and 0.5% rocky solid materials, but the composition of each planet was determined by its position relative to the Sun and the rate at which the planet grew (McSween, 1989). The "inner" planets (Mercury, Venus, Earth, and Mars) seemed to have formed in an area where the solar nebula was very hot, perhaps at a temperature close to 925°C (Boss, 1988). Venus, Earth, and Mars are all depleted in light elements compared to the cosmic abundances, and they are dominated by silicate minerals that condense at high temperatures (1500–2000 °C; Langmuir and Broecker, 2012, p. 94) and contain large amounts of FeO (McSween, 1989).

The mean density of Earth is about $5.5\,g/cm^3$. The high density of the inner planets contrasts with the lower average density of the larger, outer planets, known as gas giants, which

TABLE 2.1 Characteristics of the planets.

Planet	Equatorial radius (10^8cm)	Volume (10^{26} cm^3)	Mass (10^{27} g^3)	Density (gm/cm^3)	Corrected density[a] (gm/cm^3)
Mercury	2.44	0.61	0.33	5.43	5.40
Venus	6.05	9.29	4.87	5.24	4.30
Earth	6.38	10.83	5.97	5.52	4.20
Mars	3.38	1.63	0.64	3.93	3.70
Jupiter	71.49	14,313	1898.6	1.34	<1.3
Saturn	60.26	8271	568.4	0.69	<0.69
Uranus	25.56	683	87.0	1.28	<1.28
Neptune	24.76	625	102.40	1.64	<1.64

[a] Corrected density is the density that a planet would have in the absence of gravitational squeezing.
From Langmuir and Broecker (2012).
The mass of the Sun is 1.99×10^{33} g, i.e., $1000 \times$ the mass of Jupiter.

retained a greater fraction of lighter constituents from the initial solar cloud (Table 2.1). Jupiter contains much hydrogen and helium. The average density of Jupiter is $1.3 \, g/cm^3$, and its overall composition does not appear too different from the solar abundance of elements (Lunine, 1989; Niemann et al., 1996). Some astronomers have pointed out that the hydrogen-rich atmosphere on Jupiter is similar to the composition of "brown dwarfs"—stars that never "ignited" (Kulkarni, 1997).

The majority of the mass of the Earth seems likely to have accreted by about 4.5 billion years ago (bya)—within about 100 million years of the origin of the solar system (Allègre et al., 1995). The Earth may have captured planetesimals and meteorites with a wide range of composition (Dauphas, 2017). Kinetic energy generated during the collision of these planetesimals (Wetherill, 1985), as well as the heat generated from radioactive decay in its interior (Hanks and Anderson, 1969), would have heated the primitive Earth to the melting point of iron, nickel, and other metals, forming a magma ocean. These heavy elements, known as siderophiles,[c] were "smelted" from the materials arriving from space after which they sank to the interior of the Earth to form the core (Newsom and Sims, 1991; Wood et al., 2006; Stevenson, 2008; Norris and Wood, 2017).

As Earth cooled, lighter minerals progressively solidified to form a mantle dominated by perovskite ($MgSiO_3$; Hirose et al., 2017), with some complement of olivine ($FeMgSiO_4$), and a crust dominated by aluminosilicate minerals of lower density and the approximate composition of feldspar (Chapter 4). Thus, despite the abundance of iron in the cosmos and in the Earth as a whole, the crust of the Earth is largely composed of Si, Al, and O (Fig. 2.3). The aluminosilicate rocks of the crust "float" on the heavier semifluid rocks of the Fe-enriched mantle (Fig. 2.4; Bowring and Housh, 1995).

This model for accretion assumes that the Earth was not well endowed with volatiles (Albarède, 2009). Much of the Earth's inventory of "light" elements is likely to have been delivered to the planet as mineral constituents of the silicates in a special category of meteorites known as carbonaceous chondrites (Javoy, 1997; Sarafian et al., 2014). Even today, many silicate minerals in the Earth's mantle carry elements such as oxygen and hydrogen as part of their crystalline structure (Bell and Rossman, 1992; Meade et al., 1994). Of particular interest to biogeochemistry, carbonaceous chondrites typically contain from 0.5% to 3.6% C (carbon) and 0.01% to 0.28% N (nitrogen) (Anders and Owen, 1977), which may represent the original source of these elements for the biosphere. It is possible that carbonaceous chondrites were particularly abundant in a late veneer of materials delivered to the Earth's surface (Wetherill, 1994; Javoy, 1997; Alexander et al., 2001).

It is also likely that during its late accretion, Earth was impacted by a large body—known as Theia—which knocked a portion of the incipient planet into an orbit about it, forming the Moon (Lee et al., 1997). The Moon's age is estimated at about 4.51 billion years (Barboni et al., 2017). Earth's early history was probably dominated by frequent large impacts, but based on the age distribution of craters on the Moon, it is postulated that most of the large impacts occurred before 2.0 bya (Neukum, 1977; Cohen et al., 2000; Sleep, 2018). The Earth still receives a continuous flux of meteoritic material (Fig. 2.5), but the present-day receipt of extraterrestrial materials (8–$38 \times 10^9 \, g/yr$; Taylor et al., 1998; Love and Brownlee, 1993; Cziczo et al., 2001) is much too low to account for Earth's mass ($6 \times 10^{27} \, g$), even if it has continued for all of Earth's history.

---

[c]Technically siderophile means "loving" or showing an affinity for iron.

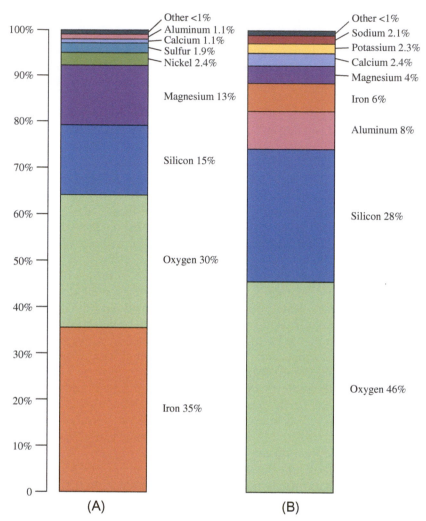

FIG. 2.3 Relative abundance of elements by weight in the whole Earth (A) and the Earth's crust (B). *From Frank Press and Raymond Siever 1986. Earth, fourth ed. W.H. Freeman and Company, Copyright 1986. Reprinted by permission.*

Several lines of evidence indicate that the primitive Earth was devoid of an atmosphere derived from the solar nebula—that is, a primary atmosphere. During its early history, the gravitational field on the small, accreting Earth would have been too weak to retain gaseous elements, and the incoming planetesimals were likely to have been too small and too hot to carry an envelope of volatiles. The impact of Theia is also likely to have blown away any volatiles that had accumulated in Earth's atmosphere by that time. Today, volcanic emissions of some inert (noble) gases, such as ^{3}He, ^{20}Ne (neon), and ^{36}Ar (argon), which are derived from the solar nebula, result from continuing degassing of primary volatiles dissolved in the magma of the deep mantle (Williams and Mukhopadhyay, 2019) or trapped in pockets (fluid

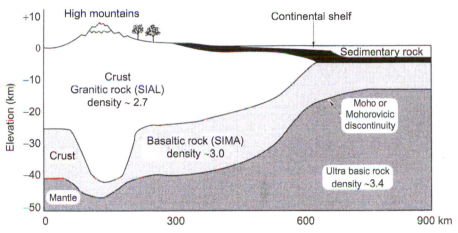

FIG. 2.4 A geologic profile of the Earth's surface. On land the crust is dominated by granitic rocks, largely composed of silicon and aluminum (Chapter 4). The oceanic crust is dominated by basaltic rocks with a large proportion of silicon and magnesium. Both granite and basalt have a lower density than the upper mantle, which contains ultrabasic rocks with the approximate composition of olivine ($FeMgSiO_4$). *From Howard and Mitchell (1985).*

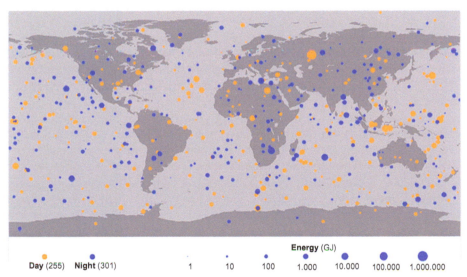

FIG. 2.5 Modern-day receipt of large (>1 m diameter) meteors (bolides) on Earth, as recorded by the NASA Near-Earth Object (NEO) observation program, 1994–2013. The large impact in Russia was the Chelyabinsk event in February 2013, releasing more than 1 million Gigajoules of energy.

inclusions) in later accreting chondrites (Lupton and Craig, 1981; Burnard et al., 1997). Otherwise, the Earth's atmosphere appears to be of secondary origin.

If a significant fraction of today's atmosphere is derived from the original solar cloud, we might expect that its gases would exist in proportion to their solar abundances (refer to Fig. 2.1). Here, ^{20}Ne is of particular interest because it is not produced by any known radioactive decay, it is too heavy to escape from Earth's gravity, and as an inert gas it is not likely to have been consumed in any reaction with crustal minerals (Walker, 1977).[d] Thus, the present-day abundance of ^{20}Ne in the atmosphere is likely to represent its primary abundance—that derived from the solar nebula. Assuming that other solar gases were delivered to the Earth in a similar manner, we can calculate the total mass of the primary atmosphere by multiplying the mass of ^{20}Ne in today's atmosphere by the ratio of each of the other gases to ^{20}Ne in the solar abundance. For example, the solar ratio of nitrogen to neon is 0.91 (Fig. 2.1). If the present-day atmospheric mass of neon, 6.5×10^{16} g, is all from primary sources, then $(0.91) \times (6.5 \times 10^{16}$ g$)$ should equal the mass of nitrogen that is also of primary origin. The product, 5.9×10^{16} g, is much less than the observed atmospheric mass of nitrogen, 39×10^{20} g. Thus, most of the nitrogen in today's atmosphere must be derived from other sources—likely from degassing of nitrogen held in minerals of the Earth's mantle.

Origin of the atmosphere and the oceans

The origin of the Earth's atmosphere is closely tied to the appearance and evolution of its crust, which differentiated from the mantle by melting and by density separation under the heat generated by large impacts and internal radioactive decay (Fanale, 1971; Stevenson, 1983). During melting, elements such as H, O, C, and N would have been released from the mantle as volcanic gases. Several lines of evidence point to the existence of a continental crust as early as 4.4 bya (Wilde et al., 2001; O'Neil et al., 2008; Greber et al., 2017), and its volume appears to have grown through Earth's history (Abbott et al., 2000; Hawkesworth and Kemp, 2006). Thus, the accumulation of a secondary atmosphere began early in Earth's history (Kunz et al., 1998).

Today, a variety of gases are released during volcanic eruptions at the Earth's surface. The emissions associated with the eruptions at mid-ocean ridges and island basalts, such as on Hawaii, offer a good indication of the composition of mantle degassing, since they are less likely to be contaminated by younger crustal materials that have been subducted (Charlou et al., 2002; Marty, 2012). Table 2.2 gives the composition of major gases emitted from various volcanoes. Characteristically, water vapor dominates the emissions, but small quantities of C, N, and S gases are also present (Tajika, 1998). Volcanic emissions, representing degassing of Earth's interior, are consistent with the observation that Earth's atmosphere is of secondary origin—largely derived from solid materials. The oxidation state of the upper mantle determined the volatiles that were degassed (Armstrong et al., 2019). The earliest atmosphere is likely to have been dominated by N_2, H_2O, and CO_2 (Schaefer and Fegley, 2010; Marty et al., 2013). Some elements, including H, can dissolve in magma so that a significant proportion of the Earth's inventory of water, carbon, and

---

[d] ^{20}Ne is one of the isotopes of neon. Isotopes of an element have the same number of protons in the nucleus, but differ in the number of neutrons, so they differ in atomic weight. Naturally occurring chemical elements are usually mixtures of isotopes, and their listed atomic weights are average values for the mixture. Most of the elements in the periodic table have two or more isotopes, with 254 stable (i.e. nonradioactive) isotopes for the first 80 elements.

TABLE 2.2   Composition of gases emitted by volcanoes, with a presumed mantle source.

Location	H_2O	H_2	CO_2	SO_2	H_2S	HCl
Tanzania, Oldoinyo Lengal	75.6		24.2	0.02		
Congo, Nyliragongo	70.5		23.7	4.55		0.26
Ethiopia, Erta' Ale	79.4	1.49	10.4	6.78	0.62	0.42
Antarctica, Erebus	57.9		36.4	1.4		0.69
Hawaii, Kilauea	37.1	0.49	48.9	11.87		0.08

All data are expressed in mol%.
Adapted from Oppenheimer (2003).

nitrogen may still reside in the mantle (Bell and Rossman, 1992; Murakami et al., 2002; Marty, 2012). The total extent of mantle degassing through geologic time is unknown, but perhaps is 50% based on the content of ^{40}Ar in Earth's mantle (Marty, 2012).

The isotopic ratio of argon gas on Earth (i.e., $^{40}Ar/^{36}Ar$) is suggestive of the proportion of our present atmosphere that is derived from mantle degassing. On Earth the isotope ^{40}Ar appears to be wholly the result of the radioactive decay of ^{40}K in the mantle (Farley and Neroda, 1998), whereas the isotope ^{36}Ar was delivered intact from the original solar nebula. Like ^{20}Ne, this noble element is too heavy to escape the gravity of the Earth.[e] Thus, the atmospheric content of ^{36}Ar should represent the proportion that is due to the residual primary atmosphere (i.e., the solar nebula), whereas the content of ^{40}Ar is indicative of the proportion due to crustal degassing. The ratio of $^{40}Ar/^{36}Ar$ on Earth is nearly 300, suggesting that 99.7% of the Ar in our present atmosphere is derived from the interior of the Earth. ^{40}Ar has been accumulating in the atmosphere since the Earth formed; one study suggests that the $^{40}Ar/^{36}Ar$ ratio was only 143 at 3.5 billion years ago (Pujol et al., 2013). In contrast, the Viking spacecraft measured a much higher ratio of 1900 for ^{40}Ar versus ^{36}Ar in the atmosphere on Mars (Mahaffy et al., 2013). This observation supports the emerging belief that Mars lost a large portion of its primary atmosphere, and that most of the atmosphere on Mars today is derived from degassing (Carr, 1987; Atreya et al., 2013).

The ratio of N_2 to ^{40}Ar in the Earth's mantle is close to that of today's atmosphere ($\sim$80)—implying a common source for both elements at the Earth's surface (Marty, 1995). Volcanic emissions and mantle degassing were undoubtedly greatest in early Earth history; the present-day flux of nitrogen from volcanoes (0.78–1.23×10^{11} g/yr; Sano et al., 2001; Tajika, 1998) is too low to account for the current inventory of N at the Earth's surface ($\sim$50 $\times 10^{20}$ g; Table 2.3) even if it has continued for all 4.5 billion years of Earth's history. Moreover, through geologic time, some nitrogen has also returned to the mantle by subduction (Zhang and Zindler, 1993).

It is possible that a late impact of comets contributed to the gaseous inventory on Earth (Chyba, 1990a). If so, the proportion must be small, because the isotopic ratio of H measured in the ices of the Hale Bopp and other comets does not match the ratio in the present inventory

---

[e] A small amount of ^{40}Ar has been destroyed by cosmic rays, producing ^{35}S in the atmosphere (Tanaka and Turekian, 1991).

TABLE 2.3   Total inventory of volatiles at the surface of the Earth[a].

Reservoir	H_2O	CO_2	C	O_2	N	S	Cl	Ar	Total (rounded)
Atmosphere (cf Table 3.1)	1.3	0.31	–	119	387	–	–	6.6	514
Oceans	135,000	19.3[b]	0.07	256[c]	2[d]	128[e]	2610	–	138,000
Land plants	0.1	–	0.06	–	0.0004	–	–	–	0.16
Soils	12	0.40[f,g]	0.15	–	0.0095	–	–	–	12.6
Freshwater (including ice and groundwater)	4850	–	–	–	–	–	–	–	4850
Sedimentary rocks	15,000[h]	30000[g]	1560	4745[i]	200[j]	744[k]	500[h]	–	52,750
Total (rounded)	155,000	30,000	1560	5120	590	872	3100	7	196,000
See also	Fig. 10.1	Fig. 11.1	Fig. 11.1	Fig. 2.8		Table 13.1			

[a] All data are expressed as 10^{19} g, with values derived from this text unless noted otherwise.
[b] Assumes the pool of inorganic C is in the form of HCO_3^-.
[c] Oxygen content of dissolved SO_4^{2-}.
[d] Dissolved N_2.
[e] S content of SO_4^{2-}.
[f] Desert soil carbonates.
[g] Assumes 60% of $CaCO_3$ is carbon and oxygen.
[h] Walker (1977).
[i] O_2 held in sedimentary Fe_2Q_3 and evaporites $CaSO_4$.
[j] Goldblatt et al. (2009).
[k] S content of $CaSO_4$ and FeS_2.

of water on Earth (Meier et al., 1998). Nevertheless, it is difficult to explain the presence of ice, organic matter, and nitrogen on the surface of the Moon, which has undergone only slight degassing, other than by delivery in a late arrival of comets impacting its surface (Thomas-Keprta et al., 2014; Füri et al., 2015; Li et al., 2018a). At least 22% of the xenon on Earth today traces to a cometary origin (Marty et al., 2017).

As long as the Earth was very hot, volatiles remained in the atmosphere, but when the surface temperature cooled to the condensation point of water, water could condense out of the primitive atmosphere to form the oceans. This must have been a rainstorm of true Biblical proportion! Several lines of evidence point to the existence of liquid water on the Earth's surface 4.3 bya (Mojzsis et al., 2001; Wilde et al., 2001). Although heat from large, late-arriving meteors may have caused temporary revaporization of some of the earliest oceans (Sleep et al., 1989; Abramov and Mojzsis, 2009), the geologic record suggests that liquid water has been present on the Earth's surface continuously for the past 3.8 billion years. Indeed, despite a few lingering meteor impacts, Earth may have harbored a temperate climate as early as 3.4 bya (Hren et al., 2009; Blake et al., 2010).

Various other gases would have quickly entered the primitive ocean as a result of their high solubility in water; for example:

$$CO_2 + H_2O \rightarrow H_2CO_3 \rightarrow H^+ + HCO_3^- \tag{2.3}$$

$$HCl + H_2O \rightarrow H_3O^+ + Cl^- \tag{2.4}$$

$$SO_2 + H_2O \rightarrow H_2SO_3 \tag{2.5}$$

These reactions removed a large proportion of reactive water-soluble gases from the atmosphere, as predicted by Henry's Law for the partitioning of gases between gaseous and dissolved phases:

$$S = kP \tag{2.6}$$

where S is the solubility of a gas in a liquid, k is the solubility constant, and P is the overlying pressure in the atmosphere. Under one atmosphere of partial pressure, the solubilities of CO_2, HCl, and SO_2 in water are 1.4, 700, and 94.1 g/L, respectively, at 25°C. When dissolved in water, all of these gases form acids, which would be neutralized by reaction with the surface minerals on Earth. Thus, after cooling, the Earth's earliest atmosphere is likely to have been dominated by N_2, which has relatively low solubility in water (0.018 g/L at 25°C).

Because many gases dissolve so readily in water, an estimate of the total extent of crustal degassing through geologic time must consider the mass of the atmosphere, the mass of oceans, and the mass of volatile elements that are now contained in sedimentary minerals, such as $CaCO_3$, which have been deposited from seawater (Li, 1972). By this accounting, the mass of the present-day atmosphere (5.14×10^{21} g; Trenberth and Guillemot, 1994) represents less than 1% of the total degassing of the Earth's mantle over geologic time (refer to Table 2.3). The oceans and various marine sediments contain nearly all of the remainder, and some volatiles have returned to the upper mantle by subduction of Earth's oceanic crust (Zhang and Zindler, 1993; Plank and Langmuir, 1998; Kerrick and Connolly, 2001).

Despite uncertainty about the exact composition of the Earth's earliest atmosphere, several lines of evidence suggest that by the time life arose, the atmosphere was dominated by N_2, CO_2, and H_2O (Holland, 1984), which were in equilibrium with the oceans, and by trace quantities of other gases from volcanic emissions that were continuing at that time (Hunten, 1993; Yamagata et al., 1991). There was certainly no O_2; the small concentrations produced by the photolysis of water in the upper atmosphere would rapidly be consumed in the oxidation of reduced gases and crustal minerals (Walker, 1977; Kasting and Walker, 1981).

During its early evolution as a star, the Sun's luminosity was as much as 30% lower than at present. We might expect that primitive Earth was colder than today, but the fossil record indicates a continuous presence of liquid water on the Earth's surface since 3.8 bya. One explanation is that the primitive atmosphere contained much higher concentrations of water vapor, CO_2, CH_4, and other greenhouse gases than today (Walker, 1985). These gases would trap outgoing infrared radiation and produce global warming through the "greenhouse" effect (refer to Fig. 3.2). In fact, even today the presence of water vapor and CO_2 in the atmosphere creates a significant greenhouse effect on Earth—about 75% due to water vapor and 25% from CO_2 (Lacis et al., 2010; Schmidt et al., 2010a). Without these gases, Earth's temperature would be about 33°C cooler, and the planet would be covered with ice (Ramanathan, 1988).

There are few direct indications of the composition of the earliest seawater. Like today's seawater, the Precambrian ocean is likely to have contained a substantial amount of chloride. HCl and Cl_2 emitted by volcanoes would dissolve in water, forming Cl^- (Eq. 2.4). The acids produced by the dissolution of these and other gases in water (Eqs. 2.3–2.5) would have reacted with minerals of the Earth's crust, releasing Na^+, Mg^{2+-}, and other cations by chemical weathering (Chapter 4). The earliest oceans may have had a pH of 6.5–7.0, reflecting the neutralization of the atmospheric acids by rock weathering (Halevy and Bachan, 2017). Carried by rivers, cations would accumulate in seawater until their concentrations increased to levels that would precipitate secondary minerals (Sleep, 2018). For instance, sedimentary accumulations of $CaCO_3$ of Precambrian age indicate that the primitive oceans had substantial concentrations of Ca^{2+} (Walker, 1983). Thus, it is likely that the dominant cations (Na, Mg, and Ca) and the dominant anion (Cl) in Precambrian seawater were similar to those in seawater today (Holland, 1984; Morse and Mackenzie, 1998). Only SO_4^{2-} seems to have been less concentrated in the Precambrian ocean than today (Grotzinger and Kasting, 1993; Habicht et al., 2002; Crowe et al., 2014).

Origin of life

Fundamental to all considerations of the origins of life are the characteristics of living systems as we know them today. To develop theories and a timeline' for the evolution of life on primitive Earth, we need to be able to recognize some or all of the universal traits of life in fossil sediments and in the products of laboratory synthesis of organic materials. These characteristics include the presence of a physical membrane, metabolic machinery for obtaining energy from the environment, and genetic material allowing heritability. These fundamental characteristics separate life from abiotic organic materials.

Even in the simplest cell, a surrounding or plasma membrane allows segregation of the building blocks of biochemistry, up to 30 elements, at concentrations and proportions that typically diverge substantially from the surrounding environment. Internal membranes, such as the mitochondrial membrane, allow separation of materials within cells, facilitating the capture of energy as electrons flow from electron-rich (reduced) to electron-poor (oxidized) substances that are obtained from the environment (Fig. 1.6). Autotrophic organisms produce their own organic materials by capturing energy from the Sun (photoautotrophy) or other external sources (chemoautotrophy); heterotrophic organisms consume the organic materials produced by others. Generic material allows these innovations to be repeatable and heritable so that organisms can grow and reproduce.

An initial constraint on the evolution of life was a lack of organic molecules on primitive Earth. In 1871, Darwin postulated that the interaction of sunlight with marine salts under a primitive atmosphere might have created these primordial organic building blocks.[f] Working with Harold

---

[f] From a letter from Charles Darwin to Joseph Hooker, 1871: "but if (and oh! what a big if!) we could conceive in some warm little pond, with all sorts of ammonia and phosphoric salts, light, heat, electricity, etc., present, that a protein compound was chemically formed ready to undergo still more complex changes, at the present day such matter would be instantly devoured or absorbed, which would not have been the case before living creatures were formed."

Urey in the early 1950s, Stanley Miller carried out an experiment by adding the probable constituents of the primitive atmosphere and oceans to a laboratory flask and subjecting the mix to an electric discharge to represent the effects of lightning. After several days, Miller found that simple, reduced organic molecules had been produced (Miller, 1953, 1957; Johnson et al., 2008b). This experiment, possibly simulating the conditions on early Earth, suggested that the organic constituents of living organisms could be produced abiotically.

This experiment has been repeated in many laboratories, and under a wide variety of conditions (Chang et al., 1983). Ultraviolet light can substitute for electrical discharges as an energy source; a high flux of ultraviolet light would be expected on primitive Earth in the absence of an ozone (O_3) shield in the stratosphere (Chapter 3). Additional energy for abiotic synthesis may have been derived from the impact of late-arriving meteors and comets passing through the atmosphere (Chyba and Sagan, 1992; McKay and Borucki, 1997) or from hydrothermal vents around volcanoes in the deep sea (Russell, 2006).

The mix of atmospheric constituents taken to best represent the primitive atmosphere is controversial. NH_3 and H_2 may have been important components of Earth's earliest atmosphere (Tian et al., 2005; Li and Keppler, 2014), and the yield of organic molecules is greatest in such highly reducing conditions. Nevertheless, an acceptable yield of simple organic molecules is found in experiments using mildly reducing atmospheres, composed of CO_2, H_2O, and N_2 (Pinto et al., 1980), which are the more probable conditions on the primitive Earth (Trail et al., 2011). The experiments are never successful when free O_2 is included; O_2 rapidly oxidizes the simple organic products before they can accumulate.

Interstellar dust particles and cometary ices contain a wide variety of simple organic molecules (Busemann et al., 2006; Sloan et al., 2009; Öberg et al., 2015), and various amino acids are found in carbonaceous chondrites and asteroids (Kvenvolden et al., 1970; Cooper et al., 2001; Herd et al., 2011; De Sanctis et al., 2017), suggesting that abiotic synthesis of organic molecules may be widespread in the galaxy (Orgel, 1994; Irvine, 1998). Significantly, it is possible that a small fraction of the organic molecules in chondrites and comets survived passage through the Earth's atmosphere, contributing to the initial inventory of organic molecules on Earth's surface (Anders, 1989; Chyba and Sagan, 1992). Even if the total mass received was small, exogenous sources of organic molecules are important, for they may have served as chemical templates, speeding the rate of abiotic synthesis and the assembly of organic molecules on Earth.

A wide variety of simple organic molecules have now been produced under abiotic conditions in the laboratory (Dickerson, 1978; Ruiz-Mirazo et al., 2014). In many cases hydrogen cyanide and formaldehyde are important initial products that polymerize to produce simple sugars such as ribose and more complex molecules such as amino acids and nucleotides. Even methionine, a sulfur-containing amino acid, has been synthesized abiotically (Van Trump and Miller, 1972). The volcanic gas carbonyl sulfide (COS) can catalyze the binding of amino acids to form polypeptides (Leman et al., 2004), and short chains of amino acids have been linked by condensation reactions involving phosphates (Rabinowitz et al., 1969; Lohrmann and Orgel, 1973). Phosphorylation of simple organic molecules by diamidophosphate appears to foster their assembly into more complex forms (Gibard et al., 2018). An early abiotic origin of organic polyphosphates speaks strongly for the origin of adenosine triphosphate (ATP) as the energizing reactant in virtually all biochemical reactions that we know today (Dickerson, 1978).

FIG. 2.6 The left-handed (L) and right-handed (D) forms, known as enantiomers, of the amino acid alanine. No rotation of these molecules allows them to be superimposed. Although both forms are found in the extraterrestrial organic matter of carbonaceous chondrites, all life on Earth incorporates only the L form in proteins. *From Chyba (1990).*

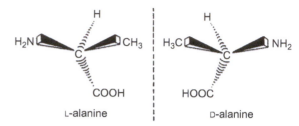

Clay minerals, with their surface charge and repeating crystalline structure, may have acted to concentrate simple, polar organic molecules from the primitive ocean, making assembly into more complicated forms, such as RNA and protein, more likely (Cairns-Smith, 1985; Ferris et al., 1996; Hanczyc et al., 2003; Bu et al., 2019). Metal ions such as zinc and copper can enhance the binding of nucleotides and amino acids to clays (Lawless and Levi, 1979; Huber and Wächtershäuser, 1998, 2006). It is interesting to speculate why nature incorporates only the "left-handed" forms of amino acids in proteins, when equal forms of L- and D-enantiomers are produced by abiotic synthesis (Fig. 2.6). Apparently, the light of stars is polarizing, creating an abundance of L-enantiomers during organic synthesis in the interstellar environment (Engel and Macko, 2001). Meteors may have carried a disproportionate abundance of L-enantiomers that acted as a template for organic molecules on Earth (Engel and Macko, 1997; Bailey et al., 1998; Pizzarello and Weber, 2004).

Recently, scientists studying the origins of life have focused on submarine hydrothermal vent systems, which today harbor a diversity of life forms, as the arena for Earth's earliest life (Kelley et al., 2002; Russell, 2006). Hydrothermal vents appear to support the abiotic synthesis of simple organic molecules, including formate, acetate (Lang et al., 2010), pyruvate (Cody et al., 2000), and amino acids (Huber and Wächtershäuser, 2006; Barge et al., 2019). Indeed, the energetics of amino acid synthesis is favorable in these environments (Amend and Shock, 1998; Ménez et al., 2018). An origin of life in the high temperature, extreme pH, and high salinity of these habitats may explain how life persists in such a wide range of extreme habitats today (Rothschild and Mancinelli, 2001; Marion et al., 2003).

Just as droplets of cooking oil form "beads" on the surface of water, it has long been known that some organic polymers will spontaneously form coacervates, which are colloidal droplets small enough to remain suspended in water. Coacervates are perhaps the simplest systems that might be said to be "bound," as if by a membrane, providing an inside and an outside. Yanagawa et al. (1988) describe several experiments in which protocellular structures with lipoprotein envelopes were constructed in the laboratory. In such structures, the concentration of substances will differ between the inside (hydrophobic) and the outside (hydrophilic) as a result of the differing solubility of substances in an organic medium and water, respectively. Mansy et al. (2008) show how primitive membranes may have allowed the transport of charged substances to the interior of protocells, allowing the evolution of heterotrophic metabolism. Huang et al. (2013) documented spontaneous assembly of protein-polymer conjugates into selectively permeable compartments of protocells, which are capable of encapsulating molecules and responding to stimuli with shifts in membrane porosity.

Some organic molecules produced in the laboratory will self-replicate, suggesting potential mechanisms that may have increased the initial yield of organic molecules from abiotic synthesis (Hong et al., 1992; Orgel, 1992; Lee et al., 1996). Other laboratories have produced simple organic structures, known as micelles, that will self-replicate their external framework (Bachmann et al., 1992). There is good reason to believe that the earliest genetic material controlling replication may not have been DNA but the related molecule, RNA, which can also perform catalytic activities (de Duve, 1995; Robertson and Miller, 1995).

Recent reports indicate some success in the abiotic synthesis of RNA precursors, which could subsequently support the abiotic synthesis of lengthy RNA molecules (Unrau and Bartel, 1998; Powner et al., 2009). Vesicles that form around clay particles are found to enhance the polymerization of RNA (Hanczyc et al., 2003). Recently Gibson et al. (2010) inserted synthetic DNA into bacteria, where it replaced the native DNA and began reproducing. This work brings us one step closer to the assembly of simple organic molecules into a complete self-replicating, metabolizing, and membrane-bound form that we might call life, with its origins in the laboratory.

A traditional view holds that life arose in the sea, and that biochemistry preferentially incorporated constituents that were abundant in seawater. For example, Banin and Navrot (1975) point out the striking correlation between the abundance of elements in today's biota and the solubility of elements in seawater. Elements with low ionic potential (i.e., ionic charge/ionic radius) are found as soluble cations (Na^+, K^+, Mg^{2+}, and Ca^{2+}) in seawater and as important components of biochemistry. Other elements, including C, N, and S, that form soluble oxyanions in seawater (HCO_3^-, NO_3^-, and SO_4^{2-}) are also abundant biochemical constituents. Molybdenum is much more abundant in biota than one might expect based on its crustal abundance; molybdenum forms the soluble molybdate ion (MoO_4^{2-}) in ocean water. In contrast, aluminum (Al) and silicon (Si) form insoluble hydroxides in seawater. They are found at low concentrations in living tissue, despite relatively high concentrations in the Earth's crust (Hutchinson, 1943). Indeed, many elements that are rare in seawater are familiar poisons to living systems (e.g., Be, As, Hg, Pb, and Cd).

Although phosphorus forms a soluble oxyanion, PO_4^{3-}, in seawater, it may never have been particularly abundant, owing to its tendency to bind to other minerals, especially iron oxides (Griffith et al., 1977). Some reactive phosphorus may have been delivered to the Earth by meteorites containing the iron-nickel mineral known as schreibersite, which could release P to seawater in the earliest oceans (Pasek et al., 2013). Unique properties of phosphorus may account for its major role in biochemistry, despite its relatively low geochemical abundance on Earth. With three ionized groups, phosphoric acid can link two nucleotides in DNA, with the third negative site acting to prevent hydrolysis and maintain the molecule within a cell membrane (Westheimer, 1987). These ionic properties also allow phosphorus to serve in intermediary metabolism and energy transfer in ATP.

In sum, if one begins with the cosmic abundance of elements as an initial constraint, and the partitioning of elements during the formation of the Earth as subsequent constraints, then solubility in water appears to be a final constraint in determining the relative abundance of elements in the geochemical arena in which life arose. Those elements that were abundant in seawater are important biochemical constituents. Phosphorus appears as an important exception—an important biochemical constituent that has been in short supply for much of the Earth's biosphere through geologic time.

Evolution of metabolic pathways

Although their interpretation is not without controversy (see Allwood et al., 2018; Javaux, 2019), the oldest fossils that are likely to represent Earth's earliest life are found in rocks deposited in shallow seas and hydrothermal vents 3.7 billion years old (Dodd et al., 2017; Hassenkam et al., 2017) or older (Tashiro et al., 2017). The earliest organisms on Earth may have resembled the methanogenic archaea that survive today in anoxic hydrothermal environments at pH ranging from 9 to 11 and temperatures above 90°C (Rasmussen, 2000; Huber et al., 1989; Kelley et al., 2005). Archaea are distinct from bacteria due to a lack of a muramic acid component in the cell wall and a distinct r-RNA sequence (Fox et al., 1980). Halophilic (salt-tolerant), acidophilic (acid-tolerant), and thermophilic (heat-tolerant) forms of archaea are also known (Fig. 2.7). Kashefi and Lovley (2003) describe iron-reducing archaea growing at 121°C near a deep sea hydrothermal vent of the North Pacific—a potential analog of one of the earliest habitats for life on Earth.

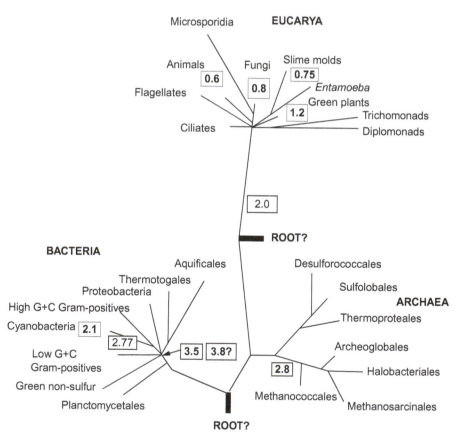

FIG. 2.7　Relationship of the three domains of the tree of life, with boxes showing the estimated time (billions of years ago) for the first appearance of various forms. *Modified from Javaux (2006).*

The most primitive metabolic pathway probably involved the production of methane by splitting simple organic molecules, such as acetate, that would have been present in the oceans from abiotic synthesis:

$$CH_3COOH \rightarrow CO_2 + CH_4 \qquad (2.7)$$

Organisms using this metabolism were scavengers of the products of abiotic synthesis and obligate heterotrophs, sometimes classified as chemoheterotrophs. The modern fermenting bacteria in the order Methanobacteriales may be our best present-day analogs.

Longer pathways of anaerobic metabolism, such as glycolysis, probably followed with increasing elaboration and specificity of enzyme systems. Oxidation of simple organic molecules in anaerobic respiration was coupled to the reduction of inorganic substrates from the environment. For example, sometime after the appearance of methanogenesis from acetate splitting, methanogenesis by CO_2 reduction,

$$CO_2 + 4H_2 \rightarrow CH_4 + 2H_2O \qquad (2.8)$$

probably arose among early heterotrophic microorganisms. Generally this reaction occurs in two steps: fermenting bacteria convert organic matter to acetate, H_2, and CO_2, and then archaea transform these to methane, following Eq. (2.8) (Wolin and Miller, 1987; Kral et al., 1998). Note that this methanogenic reaction is more complicated than that from acetate splitting and would require a more complex enzymatic catalysis.

Evidence for the first methanogens is found in rocks more than 3.5 billion years old (Ueno et al., 2006). Both pathways of methanogenesis are found among the fermenting bacteria that inhabit wetlands and coastal ocean sediments today (see Chapters 7 and 9). Without O_2 in the atmosphere, these early microbial metabolisms may have led to large accumulations of methane and an enhanced greenhouse effect on Earth (Catling et al., 2001).

Today, microbial communities performing methanogenesis by CO_2 reduction are also found deep in the Earth, where H_2 is available from geologic sources (Stevens and McKinley, 1995; Chapelle et al., 2002; Kietäväinen et al., 2017; Sherwood-Lollar et al., 2014; Worman et al., 2016). These microbial populations are functionally isolated from the rest of the biosphere, and indicate another potential habitat for early life on Earth (Sleep, 2018). Indeed, a vast elaboration of prokaryotes is found at great depths on land and in ocean sediments worldwide, where they persist with extremely low rates of metabolism (Krumholz et al., 1997; Fisk et al., 1998; Røy et al., 2012; Orsi et al., 2013). The total biomass of these microbes may contain 44×10^{15} g C—about 8% of the total living biomass on Earth today (Kallmeyer et al., 2012; Bar-On et al., 2018).

Before the advent of atmospheric O_2, the primitive oceans are likely to have contained low concentrations of available nitrogen—largely in the form of nitrate (NO_3^-; Kasting and Walker, 1981). Thus, the earliest organisms had limited supplies of nitrogen available for protein synthesis. The origin of nitrogen fixation (diazotrophy), in which certain bacteria break the inert, triple bond in N_2 and reduce the nitrogen to NH_3, dates to 3.2 billion years ago (Stüeken et al., 2015). Today this reaction is performed by bacteria that require strict local anaerobic conditions. The reaction,

$$N_2 + 8H^+ + 8e^- + 16ATP \rightarrow 2NH_3 + H_2 + 16ADP + 16P_i \qquad (2.9)$$

is catalyzed by the enzyme complex known as nitrogenase, which consists of two proteins incorporating iron and molybdenum in their molecular structure (Kim and Rees, 1992; Chan et al., 1993; Ćorić et al., 2015). The iron-based nitrogenase may have evolved first; the modern form of nitrogenase, containing molybdenum, may have appeared only 1.5–2.2 bya, having evolved from earlier forms in methanogenic bacteria (Boyd et al., 2011). In some species vanadium (V) can substitute for molybdenum (Zhang et al., 2016b). A cofactor, vitamin B_{12} that contains cobalt, is also essential (Palit et al., 1994; O'hara et al., 1988). Nitrogen fixation requires the expenditure of large amounts of energy; breaking the N_2 bond requires 226 kcal/mol (Davies, 1972). Modern nitrogen-fixing cyanobacteria couple nitrogen fixation to their photosynthetic reaction; other nitrogen-fixing organisms are frequently symbiotic with higher plants (Chapter 6). The evolution of nitrogen fixation may have exacerbated the limited availability of phosphorus to support the proliferation of life in the primitive oceans (Olson et al., 2016).

Photosynthesis: The origin of oxygen on earth

Despite various pathways of anaerobic metabolism, the opportunities for heterotrophic organisms must have been quite limited in a world where organic molecules were only available as a result of abiotic synthesis. Natural selection would strongly favor autotrophic systems that could supply their own reduced organic molecules for metabolism. Some of the earliest autotrophic metabolisms may have depended on H_2 (Schidlowski, 1983; Tice and Lowe, 2006; Canfield et al., 2006). Autotrophic cyanobacteria metabolizing hydrogen have also been isolated from the Earth's crust (Puente-Sánchez et al., 2018), namely:

$$2H_2 + CO_2 \rightarrow CH_2O + H_2O \tag{2.10}$$

We might also expect that another early photosynthetic reaction might have been based on the oxidation of the highly reduced gas, hydrogen sulfide (H_2S) (Schidlowski, 1983; Xiong et al., 2000). For H_2S, the reaction,

$$2H_2S + CO_2 \rightarrow CH_2O + 2S + H_2O \tag{2.11}$$

was probably performed by sulfur bacteria, not unlike the anaerobic forms of green and purple sulfur bacteria of today (Fig. 1.6). These bacteria could have been particularly abundant around shallow submarine volcanic emissions of reduced gases, including H_2S.

Several indirect lines of evidence suggest that some form of photosynthesis occurred in ancient seas of 3.8 bya. Photosynthesis produces organic carbon in which ^{13}C is depleted relative to its abundance in dissolved bicarbonate (HCO_3^-), and there are no other processes known to produce such strong fractionations between the stable isotopes of carbon.[g] Fossil organic matter with ^{13}C depletion is found in rocks from Greenland dating back to at least 3.8 bya (Mojzsis et al., 1996; Rosing, 1999; Schidlowski, 2001; Fig. 2.8). This discrimination,

---

[g]When two isotopes are available, here ^{12}C and ^{13}C, most biochemical pathways proceed more rapidly with the lighter isotope, which is often more abundant. This preferential use is known as mass-dependent fractionation, and it leaves metabolic products with a different ratio of isotopes than found in the surrounding environment. For instance, carbon in carbonaceous chondrites is enriched in ^{13}C (Engel et al., 1990; Herd et al., 2011); whereas carbon in the products of photosynthesis is depleted in ^{13}C.

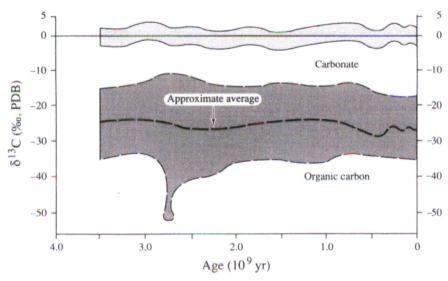

FIG. 2.8 The isotopic composition of carbon in fossil organic matter and marine carbonates through geologic time, showing the range (shaded) among specimens of each age. The isotopic composition is shown as the ratio of ^{13}C to ^{12}C, relative to the ratio in an arbitrary standard (PDB belemite), which is assigned a ratio of 0.0. Carbon in organic matter is 2.8% less rich in ^{13}C than the standard, and this depletion is expressed as $-28‰$ $\delta^{13}C$ (see Chapter 5). *From Schidlowski (1983).*

FIG. 2.9 Banded Iron Formation from the 3.25-billion-year-old Barberton Greenstone Belt, South Africa. *Collected by M.M. Tice (Texas A & M University); photo © 2010 by Lisa M. Dellwo.*

which is about -2.8% ($-28‰$) in the dominant form of present-day photosynthesis, is based on the slower diffusion of $^{13}CO_2$ relative to $^{12}CO_2$ and the greater affinity of the carbon-fixation enzyme, ribulose bisphosphate carboxylase, for the more abundant $^{12}CO_2$ (see Chapter 5). Some workers have suggested that the C-isotope depletion seen in these rocks is an artifact of metamorphism (van Zuilen et al., 2002), but other evidence for the existence of photosynthesis is also found in rocks beginning 3.8 bya, known as Banded Iron Formations ("BIF"; Fig. 2.9).

Under the anoxic conditions of primitive Earth, Fe^{2+} released during rock weathering and from submarine hydrothermal emissions would be soluble and accumulate in seawater. The deposition of Banded Iron Formation may indicate the first appearance of an iron-based photosynthesis, by anaerobic, Fe^{2+}-oxidizing bacteria (Kappler et al., 2005; Widdel et al., 1993; Li et al., 2013), namely:

$$4Fe^{2+} + HCO_3^- + 10H_2O \rightarrow CH_2O + 4Fe(OH)_3 + 7H^+ \tag{2.12}$$

Indeed, some workers believe that anoxygenic Fe-photosynthesis (Eq. 2.12) may have dominated primary production in the primitive ocean (Table 2.4).

Oxygen-evolving photosynthesis may have followed later, with the appearance of cyanobacteria, which are known oxygenic forms today (Soo et al., 2017), viz.

$$H_2O + CO_2 \rightarrow CH_2O + O_2 \uparrow \tag{2.13}$$

With the advent of oxygenic photosynthesis, O_2 would be available to oxidize Fe^{2+} and deposit Fe_2O_3 in the sediments of the primitive ocean, namely (Czaja et al., 2013):

$$4Fe^{2+} + O_2 + 4H_2O \rightarrow 2Fe_2O_3 \downarrow + 8H^+ \tag{2.14}$$

Thus, massive worldwide deposits of Fe_2O_3 in the Banded Iron Formations beginning 3.8 billion years ago are consistent with both Fe-based photosynthesis and the advent of oxygen-producing photosynthesis based on the photochemical splitting of water in sunlight. Note that these two pathways for the deposition of BIF (Eqs. 2.12, 2.14) are not mutually exclusive; both Fe-photosynthesizers and cyanobacteria can precipitate iron oxides in marine environments today (Trouwborst et al., 2007).

Despite the relatively large energy barrier inherent in the oxygen-evolving photosynthetic reaction, there must have been strong selection for photosynthesis based on the splitting of water (Schidlowski, 1983). Oxygenic photosynthesis yields considerably more energy than other forms of autotrophy (Judson, 2017) and water offered an inexhaustible substrate (Walker et al., 1983). There is strong evidence that photosynthetic microbes existed at least 3.4 bya (Tice and Lowe, 2004)—about 400 million years after the last great impacts of Earth's accretion. The deposition of Banded Iron Formation was widespread in the world's oceans and lasted for about 2 billion years (Johnson and Molnar, 2019). Most of the major deposits of iron ore in the United States (Minnesota), Australia, and South Africa are found in formations of this age (Meyer, 1985). Presumably the deposition of BIF ended when the Earth's primitive oceans were swept clear of Fe^{2+} and excess oxygen could diffuse to the atmosphere.

TABLE 2.4   Estimates of marine primary production about 3.5 billion years ago.

Process	Annual rate	See equation
H_2-based anoxygenic photosynthesis	0.35×10^{15} g C/yr	Eq. (2.10)
Sulfur-based anoxygenic photosynthesis	0.03	Eq. (2.11)
Fe-based anoxygenic photosynthesis	4.0	Eq. (2.12)
Present day, oxygenic photosynthesis	~50.0	Eq. (2.13); Chapter 5

Modified from Canfield et al. (2006).

Although rocks of 3 bya hold some evidence of free O_2 (Crowe et al., 2013; Planavsky et al., 2014a), many lines of evidence indicate that the Earth's atmosphere was largely anoxic until about 2.45–2.32 billion years ago (Farquhar et al., 2000, 2011; Bekker et al., 2004; Sessions et al., 2009). Most researchers attribute the delayed accumulation of oxygen solely to its reaction with reduced iron (Fe^{2+}) in seawater and the deposition of Fe_2O_3 in Banded Iron Formations (Cloud, 1973). Oxidation of other reduced species, perhaps sulfide (S^{2-}), may have also played a role, accounting for the slow buildup of SO_4^{2-} in Precambrian seawater (Walker and Brimblecombe, 1985; Habicht et al., 2002; Blättler et al., 2018). It is also possible that the early deposition of iron oxides in the Banded Iron Formation held phosphorus concentrations at low levels, slowing the proliferation of photosynthetic organisms (Bjerrum and Canfield, 2002; Reinhard et al., 2017) Several recent papers postulate an early evolution of aerobic respiration, closely coupled to local sites of O_2 production, which may have held the concentration of O_2 at low levels (Castresana and Saraste, 1995). Aerobic oxidation of methane may have kept oxygen at low levels until 2.7 bya (Konhauser et al., 2009).

Only when the oceans were swept clear of reduced substances, such as Fe^{2+}, S^{2-}, and CH_4, could excess O_2 accumulate in seawater and diffuse to the atmosphere. Thus, what is known as the Great Oxidation Event began about 2.4 bya, perhaps achieving 0.1–1% of the present level of O_2 in the atmosphere about 2.0 bya (Lyons et al., 2014; Canfield, 2014, p. 156).

Chemoautotrophy

Oxygen also enabled the evolution of several new biochemical pathways of critical significance to the global cycles of biogeochemistry (Raymond and Segrè, 2006). Two forms of aerobic biochemistry constitute chemoautotrophy. One based on sulfur or H_2S,

$$2S + 2H_2O + O_2 \rightarrow 2SO_4^{2-} + 4H^+ \tag{2.15}$$

is performed by various species of *Thiobacilli* (Ralph, 1979). The protons generated are coupled to energy-producing reactions, including the fixation of CO_2 into organic matter (refer to Fig. 1.6). On the primitive Earth, these organisms could capitalize on elemental sulfur deposited from anaerobic photosynthesis (Eq. 2.11), and today they are found in local environments where elemental sulfur or H_2S is present, including some deep-sea hydrothermal vents (Chapter 9), caves (Sarbu et al., 1996), wetlands (Chapter 7), and lake sediments (Chapter 8).

Also important are the chemoautotrophic reactions involving nitrogen transformations by *Nitrosomonas* and *Nitrobacter* bacteria,[h] respectively:

$$2NH_4^+ + 3O_2 \rightarrow 2NO_2^- + 2H_2O + 4H^+ \tag{2.16}$$

and

$$2NO_2^- + O_2 \rightarrow 2NO_3^- \tag{2.17}$$

---

[h] Recently two laboratories have independently characterized bacteria in the genus *Nitrospira* that can perform the complete nitrification reaction (van Kessel et al., 2015; Daims et al., 2015).

These reactions constitute nitrification, and the energy released is coupled to low rates of carbon fixation; thus, nitrifying bacteria are chemoautotrophs.[i] Using O_2 as a reactant, sulfide oxidation and ammonium oxidation (nitrification) provide indirect evidence for the presence of O_2 on Earth.

Anaerobic respiration in an aerobic world

With the appearance of SO_4^{2-} and NO_3^- as products of chemoautotrophic reactions, other metabolic pathways could evolve. The sulfate-reducing pathway, which depends on SO_4^{2-},

$$2CH_2O + 2H^+ + SO_4^{2-} \rightarrow H_2S + 2CO_2 + 2H_2O \tag{2.18}$$

is found in archaea dating to 2.4–2.7 bya, on the basis of the S-isotope ratios in preserved sediments (Cameron, 1982; Parnell et al., 2010). The late appearance of sulfate-reduction relative to photosynthesis may be related to the time needed to accumulate sufficient SO_4^{2-} in ocean waters, from the oxidation of sulfides, to make this an efficient means of metabolism (Habicht et al., 2002; Kah et al., 2004; Crowe et al., 2014). This biochemical pathway has been found in a group of thermophilic archaea isolated from the sediments of hydrothermal vent systems in the Mediterranean Sea, where a hot, anaerobic, and acidic microenvironment may resemble the conditions of primitive Earth (Stetter et al., 1987; Jorgensen et al., 1992; Elsgaard et al., 1994). In South Africa, simple microbial communities isolated at 2.8-km depth in the Earth's crust consist of sulfate-reducing archaea that fix nitrogen and are completely isolated from energy inputs from the Sun at the Earth's surface (Lin et al., 2006; Chivian et al., 2008).

Similarly, today an anaerobic, heterotrophic reaction called denitrification is performed by bacteria, commonly of the genus *Pseudomonas*, found in soils and wet sediments (Knowles, 1982), namely:

$$5CH_2O + 4H^+ + 4NO_3^- \rightarrow 5CO_2 + 2N_2 \uparrow + 7H_2O \tag{2.19}$$

The denitrifying reaction requires NO_3^- and its preferential use of $^{14}NO_3^-$ over $^{15}NO_3^-$ leaves the ocean enriched in $^{15}NO_3^-$. Rocks showing this enrichment are dated to at least 2.0 bya (Beaumont and Robert, 1999; Papineau et al., 2005) and perhaps earlier (Godfrey and Falkowski, 2009; Zerkle et al., 2017). At that time, nitrate must have been present as the product of nitrification reactions (Eqs. 2.16, 2.17), providing another indirect line of evidence for the presence of O_2 on Earth. Together, the photosynthesis, nitrification, and denitrification reactions couple the carbon and nitrogen cycles on Earth (Busigny et al., 2013).

Although the denitrification reaction requires anoxic environments, denitrifiers are facultatively aerobic—that is, switching to aerobic respiration when O_2 is present. This is consistent with several lines of evidence that suggest that denitrification may have appeared later than the strictly anaerobic pathways of methanogenesis and sulfate reduction (Betlach, 1982). Denitrification would have been efficient only after relatively high concentrations of NO_3^- had accumulated in the primitive ocean, which is likely to have contained low NO_3^- at the start (Kasting and Walker, 1981). Thus, the evolution of denitrification may have been

---

[i]Carbon fixation by these chemoautotrophs is rather limited, ~0.02 mol C/mol NH_4^+ oxidized to nitrate in modern forest soils (Norman et al., 2015). See also Sayavedra-Soto and Arp (2011).

delayed until sufficient O_2 was present in the environment to drive the nitrification reactions (Eqs. 2.16, 2.17). It is interesting to note that having evolved in a world dominated by O_2, the enzymes of today's denitrifying organisms are not destroyed, but merely inactivated, by O_2 (Bonin et al., 1989; McKenney et al., 1994).

The first O_2 that reached the atmosphere was probably immediately involved in the oxidation of reduced atmospheric gases and with exposed crustal minerals of the barren land (Holland et al., 1989; Kump et al., 2011). Chromium isotopes in sediments of 1.8–0.8 bya suggest that oxygen levels in the atmosphere may have been only 0.1% of today's levels (Planavsky et al., 2014b; Colwyn et al., 2019). Oxidation of reduced minerals, such as pyrite (FeS_2), would consume O_2 and transfer SO_4^{2-} and Fe_2O_3 to the oceans in riverflow (Konhauser et al., 2011). Deposits of Fe_2O_3 that are found in alternating layers with other sediments of terrestrial origin constitute Red Beds, which are found beginning at 2.0 bya and indicative of aerobic terrestrial weathering, also known as subaerial weathering (Van Houten, 1973). It is noteworthy that the earliest occurrence of Red Beds roughly coincides—with little overlap—with the latest deposition of Banded Iron Formation, further evidence that the oceans were swept clear of reduced Fe before O_2 began to diffuse to the atmosphere.

Canfield (1998) suggested that with O_2 in Earth's atmosphere, the oceans about 2 bya may have consisted of oxic surface waters and anoxic bottom waters, which later became oxic throughout the water column (Reinhard et al., 2009; Sperling et al., 2015). Analyses of submarine basalts suggests oxygen in the deep ocean at 540 mya (Stolper and Keller, 2018). In this model, the bottom waters may have contained substantial concentrations of sulfide, produced by sulfate-reducing bacteria using SO_4^{2-} that mixed down from the oxygenated layers above.

Oxygen began to accumulate to its present-day atmospheric level of 21% when the rate of O_2 production by photosynthesis exceeded its rate of consumption by the oxidation of reduced substances. Atmospheric oxygen may have reached 21% as early as the Silurian—about 430 mya (see inside back cover), and it is not likely to have fluctuated outside the range of 15–35% ever since (Berner and Canfield, 1989; Scott and Glasspool, 2006). What maintains the concentration at such stable levels? Walker (1980) examined all the oxidation/reduction reactions affecting atmospheric O_2, and suggested that the balance is due to the negative feedback between O_2 and the long-term net burial of organic matter in sedimentary rocks. When O_2 rises, less organic matter escapes decomposition, stemming a further rise in O_2. We will examine these processes in more detail in Chapters 3 and 11, but here it is interesting to note the significance of an atmosphere with 21% O_2. Lovelock (1979) points out that with <15% O_2, fires would not burn, and at >25% O_2, even wet organic matter would burn freely (Watson et al., 1978; Belcher and McElwain, 2008). Either scenario would result in a profoundly different world than that of today.

The release of O_2 by photosynthesis is perhaps the single most significant effect of life on the geochemistry of the Earth's surface (Raymond and Segrè, 2006; Judson, 2017). The accumulation of free O_2 in the atmosphere has established the oxidation state for most of the Earth's surface for the last 2 billion years. However, of all the oxygen ever evolved from photosynthesis, only about 2% resides in the atmosphere today; the remainder is buried in various oxidized sediments, including Banded Iron Formations and Red Beds (see Fig. 2.10 and refer to Table 2.3). The total inventory of free oxygen that has ever been released on the Earth's surface is, of course, balanced stoichiometrically by the storage of reduced carbon in the

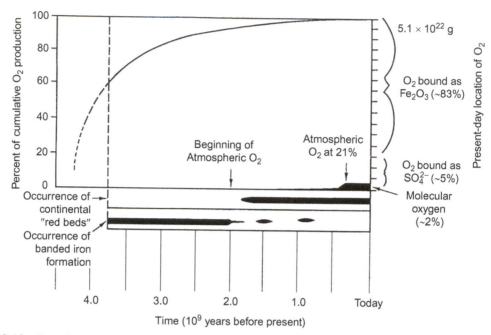

FIG. 2.10 Cumulative history of O_2 released by photosynthesis through geologic time. Of more than 5.1×10^{22} g of O_2 released, about 98% is contained in seawater and sedimentary rocks, beginning with the occurrence of Banded Iron Formation beginning at least 3.5 billion years ago (bya). Although O_2 was released to the atmosphere beginning about 2.0 bya, it was consumed in terrestrial weathering processes to form Red Beds, so that the accumulation of O_2 to present levels in the atmosphere was delayed to 400 mya. *Modified from Schidlowski (1980).*

Earth's crust, including coal, oil, and other reduced compounds of biogenic origin (e.g., sedimentary pyrite). The sedimentary storage of organic carbon is now estimated at 1.56×10^{22} g (Des Marais et al., 1992; see also Table 2.3), representing the cumulative net production of biogeochemistry since the origin of life.

The release of free oxygen as a byproduct of photosynthesis also dramatically altered the evolution of life on Earth. The pathways of anaerobic respiration by methanogenic bacteria (Eq. 2.7) and photosynthesis by sulfur bacteria (Eq. 2.11) are poisoned by O_2. These organisms generally lack catalase and have only low levels of superoxide dismutase—two enzymes that protect cellular structures from damage by highly oxidizing compounds such as O_2 (Fridovich, 1975). Oxygenic photosynthesis poisoned the environment for a wide variety of these species, and today, these metabolisms are confined to local anoxic environments in wetlands and sediments.

Eukaryotic metabolism is possible at O_2 levels that are about 1% of present levels (Berkner and Marshall, 1965; Chapman and Schopf, 1983). Fossil evidence of eukaryotic organisms is found in rocks formed 1.7–1.9 bya (Knoll, 1992; Fig. 2.5), and perhaps even as old as 2.1 billion years ago (Han and Runnegar, 1992). Large colonial organisms are reported from rocks 2.1 bya (El Albani et al., 2010). The rate of evolution of amino acid sequences among major groups

TABLE 2.5   Milestones in the deep history of the Earth.

Origin of the Universe	13.8 billion years ago
Origin of the Milky Way Galaxy	12.5
Origin of the Sun	4.56
Accretion of the Earth largely complete	4.5
(N.B. Impact of Theia at 4.51 bya forms the Moon)	
Liquid Water on Earth	4.3
Last of the great impacts	3.8
Earliest evidence of photosynthesis	
Depleted ^{13}C and Banded Iron Formations	3.8
Earliest evidence cellular structures	3.7
First evidence of cyanobacteria	2.7
First evidence of O_2 in the atmosphere	2.45
Evidence of seawater SO_4^{2-}, thus O_2	2.4
Evidence of denitrification, hence NO_3^- and O_2	2.3
Evidence of aerobic rock weathering (Red Beds)	2.0
First evidence of eukaryotes	2.0
End of Banded Iron Formation	1.8
Land plants	0.43
Genus *Homo*	0.002

of organisms suggests that prokaryotes and eukaryotes diverged 2.0–2.3 bya (Doolittle et al., 1996; Gold et al., 2017). All these dates are generally consistent with the end of deposition of the Banded Iron Formation and the presence of O_2 in the atmosphere as indicated by Red Beds (Table 2.5).

O_2 in the environment allowed eukaryotes to localize their heterotrophic respiration in mitochondria, providing an efficient means of metabolism and allowing a rapid proliferation of higher forms of life. Similarly, more efficient photosynthesis in the chloroplasts of eukaryotic plant cells presumably enhanced the production and further accumulation of atmospheric oxygen.

O_2 in the stratosphere is subject to photochemical reactions leading to the formation of ozone (Chapter 3). Today, stratospheric ozone provides an effective shield for much of the ultraviolet radiation from the Sun that would otherwise reach the Earth's surface and destroy most life (Chapter 3). Before the O_3 layer developed, the earliest colonists on land may have resembled the microbes and algae that inhabit desert rocks of today (e.g., Friedmann, 1982; Bell, 1993; Phoenix et al., 2006; Marlow et al., 2015; Fig. 2.11). Although there is some fossil evidence for the occurrence of extensive microbial communities on land during the

FIG. 2.11 Cyanobacteria inhabit the space beneath quartz stones in the Mojave Desert, California, where they photosynthesize on the light passing through these translucent rocks. *From Schlesinger et al. (2003); Photo © 2017, Lisa M. Dellwo.*

Precambrian (Horodyski and Knauth, 1994; Knauth and Kennedy, 2009; Watanabe et al., 2000; Strother et al., 2011), it is unlikely that higher organisms were able to colonize land abundantly until the ozone shield developed.

Multicellular organisms are found in ocean sediments dating to about 680 mya, but the colonization of land by higher plants was apparently delayed until the Silurian (Kenrick and Crane, 1997; Morris et al., 2018). A proliferation of plants on land followed the development of lignified, woody tissues (Lowry et al., 1980) and the origin of effective symbioses with mycorrhizal fungi that allow plants to obtain phosphorus from unavailable forms in the soil (Pirozynski and Malloch, 1975; Simon et al., 1993; Yuan et al., 2005; Chapter 6). Even primitive land plants are likely to have speeded the formation of clay minerals, which help preserve organic matter from degradation and thus further the accumulation of oxygen in the atmosphere (Kennedy et al., 2006; McMahon and Davies, 2018, Chapter 4). An oxygen-rich atmosphere was likely necessary for the evolution and proliferation of higher animals, especially carnivores (Sperling et al., 2013).

Comparative planetary history: Earth, Mars, and Venus

With the release of free O_2 to the atmosphere, life has profoundly affected the conditions on the surface of the Earth. But what might have been the conditions on Earth in the absence of life? Some indications are given by our neighboring planets Mars and Venus, which are the best replicates we have for the underlying geochemical arena on Earth. We are fairly confident that there has never been life on these planets, so their surface composition represents the cumulative effect of 4.5 billion years of abiotic processes. Our understanding of the atmosphere of Mars has improved markedly from the explorations by various robotic instruments

FIG. 2.12   The surface of Mars as seen from the *Viking 2 Lander* in 1976 (left) and *Curiosity* in 2018 (right).

that have landed on the surface of Mars, beginning with the Viking spacecraft in 1976 and continuing to the most recent lander, *Curiosity*, which arrived on 6 August 2012 (Fig. 2.12).

Mars

Table 2.6 compares a number of properties and conditions on Earth, Mars, and Venus. Two properties characterize the atmosphere of these planets: the total mass (or pressure) and the proportional abundance of constituents. The present atmosphere on Mars is only about 0.76% as massive as that on Earth (Hess et al., 1976). We should expect a less massive atmosphere on Mars than on Earth because the gravitational field is weaker on a smaller planet. Thus, Mars

TABLE 2.6   Some characteristics of the inner planets.

	Mars	Earth	Venus
Distance to the Sun (10^6 km)	228	150	108
Surface temperature (°C)	−71	15	464
Radius (km)	3380	6380	6050
Atmospheric pressure (bars)	0.007	1	92
Atmospheric mass (g)	2.4×10^{19}	5.3×10^{21}	5.3×10^{23}
Atmospheric composition (% wt.)			
CO_2	96	0.04	98
N_2	1.89	78	2
O_2	0.145	21	0
H_2O	0.10	1	0.05
$^{40}Ar/^{36}Ar$	1900	296	1

Compiled from various sources: Haberle (2013), Langmuir and Broecker (2012), Mahaffy et al. (2013), Nozette and Lewis (1982), Owen and Biemann (1976), and Wayne (1991).

probably captured a smaller allocation of the solar nebula during planetary formation, and we should expect that a small planet would have retained less internal heat to drive tectonic activity and outgassing of its mantle after its formation (Anders and Owen, 1977; Owen and Biemann, 1976). Estimates of the cumulative generation of magma are substantially lower for Mars ($0.17 \, km^3/yr$) than for Earth (26–$34 \, km^3/yr$) or Venus ($<19 \, km^3/yr$) (Greeley and Schneid, 1991). The lack of tectonic activity on Mars did not allow for the return of water trapped in subducted materials to return to the surface of the planet (Wade et al., 2017).

We should also expect that the surface temperature on Mars would be colder than that on Earth because the planet is much farther from the Sun. The average temperature on Mars, $-53 \, °C$ at the site of the Viking landing (Kieffer, 1976), ensures that water is frozen on most of the Martian surface at all seasons. Ice is found at both poles of Mars and in Martian soils in other areas (Titus et al., 2003; Mustard et al., 2001; Smith et al., 2009). In the absence of liquid water, we would expect that the atmosphere on Mars would be mostly dominated by CO_2, which readily dissolves in seawater on Earth (Eq. 2.4). Indeed, CO_2 constitutes a major proportion of the thin atmosphere of Mars, and the observed fluctuations of the ice cap at the south pole of Mars appear due to seasonal variations in the amount of CO_2 that is frozen out of its atmosphere and falls as snow (Leighton and Murray, 1966; James et al., 1992; Phillips et al., 2011a; Hayne et al., 2014).

Several attributes of Mars are anomalous. First, with most of the water and CO_2 now trapped on the surface, why is N_2 such a minor component of the atmosphere on Mars? Second, why do the surface conditions on Mars indicate a period when liquid water may have been present on its surface (Williams et al., 2013; Grotzinger et al., 2014, 2015). Could it be that a more massive early atmosphere may once have allowed a significant "greenhouse effect" on Mars and warmer surface temperatures than today (Pollack et al., 1987)? Such a scenario could explain an early sporadic occurrence of liquid water on Mars, but if it is correct, why did Mars lose its atmosphere and cool to its present average surface temperature of $-71°C$ (Haberle, 2013)?

Losses of atmospheric gases from Mars may have resulted from several processes. A thick atmosphere on Mars may have been lost to space as a result of catastrophic impacts during its early history (Carr, 1987; Melosh and Vickery, 1989) or by a process known as "sputtering" driven by solar wind (Kass and Yung, 1995; Jakosky et al., 2017). Impacts are consistent with the low abundance of noble gases on Mars relative to the concentrations in carbonaceous chondrites and the Sun (Hunten, 1993; Pepin, 2006). Catastrophic loss of an early atmosphere is also consistent with the observation that nearly the entire atmosphere on Mars is secondary, as evidenced by a $^{40}Ar/^{36}Ar$ ratio of 1900 (Mahaffy et al., 2013; Atreya et al., 2013), and a depletion of its $^{36}Ar/^{38}Ar$ ratio relative to other bodies in the solar system (Atreya et al., 2013). Mars shows some evidence of recent volcanic activity (Neukum et al., 2004), and a significant proportion of its thin atmosphere may have been derived <1 bya (Gillmann et al., 2011; Niles et al., 2010).

Loss of water from Mars may have also occurred as the water vapor in its atmosphere underwent photolysis by ultraviolet light. Observations of analogous processes on Earth are instructive. In the upper atmosphere on Earth, small amounts of water vapor are subject to photodisassociation, with the loss of H_2 to space. However, because the upper atmosphere is cold, little water vapor is present, and the process has been minor throughout Earth's history (Chapter 10). If this process were significant on Mars, we would expect that the loss of ^1H

would be more rapid than that of ^{2}H, leaving a greater proportion of ^{2}H$_2$O in the planetary inventory. Owen et al. (1988) found that the ratio of ^{2}H (deuterium) to ^{1}H on Mars is about 6 times greater than on Earth, suggesting that Mars may have once possessed a large inventory of water that has been lost to space (de Bergh, 1993; Krasnopolsky and Feldman, 2001; Villanueva et al., 2015). The relative abundance of ^{2}H on Mars has potentially increased through time (Greenwood et al., 2008; Mahaffy et al., 2015). Indeed, the enrichment of ^{2}H on Mars could imply an ocean that was once more than 100 m deep (Villanueva et al., 2015). Although a small amount of O_2 is found in the Martian atmosphere (Table 2.6), most of the oxygen produced from the photolysis of water has probably oxidized minerals of the crust, namely:

$$4FeO + O_2 \rightarrow 2Fe_2O_3 \tag{2.20}$$

giving Mars its reddish color (Fig. 2.12).

Nitrogen may have also been lost from Mars as N_2 underwent photodisassociation in the upper atmosphere, forming monomeric N. This process occurs on Earth as well, but even N is too heavy to escape the Earth's gravitational field and quickly recombines to form N_2. With its smaller size, Mars allows the loss of N. Relative to the Earth, a higher proportion of ^{15}N$_2$ in the Martian atmosphere is suggestive of this process, since the escape of ^{15}N would be slower than that of ^{14}N, which has a lower atomic weight (McElroy et al., 1976; Murty and Mohapatra, 1997; Wong et al., 2013). Both the Earth and Mars have a higher relative proportion of ^{15}N than the solar composition—Mars more so (Marty et al., 2011).

With losses of H_2O and N_2 to space, it is not surprising that the Martian atmosphere is dominated by CO_2. What is surprising is that the atmospheric mass is so low. As much as 3 bars of CO_2 may have been degassed from the interior of Mars, but only about 10% of that amount appears to be frozen in the polar ice caps and the soil (Kahn, 1985). Some CO_2 may have been lost to space (Kass and Yung, 1995), but during an earlier period of moist conditions, CO_2 may have also reacted with the crust of Mars, weathering rocks and forming carbonate minerals on its surface. A few studies have reported carbonate minerals on Mars (Ehlmann et al., 2008; Morris et al., 2010; Bultel et al., 2019), although little evidence of carbonates was reported from *Curiosity*.

In sum, various lines of evidence suggest that Mars had a higher inventory of volatiles early in its history, but most of the atmosphere has been lost in reactions with its crust or lost to space. The presence of water on Mars may have once offered an environment conducive to the evolution of life, especially in light of the relatively rapid appearance of life on Earth. Subsurface and hydrothermal environments are possible sites for early life on Mars (Squyres et al., 2008) and there are recent reports of liquid water beneath some Martian glaciers (Orosei et al., 2018). Some reports of high concentrations of methane in the Martian atmosphere (21 ppb; Webster et al., 2018) have not been seen by subsequent measurements (Korablev et al., 2019), suggesting that at best it is a short-lived species in the atmosphere. Fixed-N (i.e., NO_3; Stern et al., 2015), NH_3; Villanueva et al., 2013), and organic molecules (Leshin et al., 2013; Freissinet et al., 2015; Eigenbrode et al., 2018; Frantseva et al., 2018) are seen on the surface of Mars, but these may be of meteoritic origin. Organic molecules on the surface of Mars are likely to be oxidized by the Sun's ultraviolet light (Kminek and Bada, 2006). Thus, evidence for past life on Mars is rather scant (McKay et al., 1996), and there is no evidence of life on Mars today.

Through geologic time, the loss of water from Mars would remove a large component of greenhouse warming from the planet (Jakosky et al., 2018). The thin atmosphere that remains is dominated by CO_2, but it offers little greenhouse warming—raising the temperature of Mars only about 10°C over what might be seen if Mars had no atmosphere at all (Houghton, 1986). Our best estimate suggests that the volume of CO_2 now frozen in the polar ice caps (100 mbar) and soils (300 mbar) on Mars is insufficient to supply the 2 bars of atmospheric CO_2 that would be necessary for the greenhouse effect to raise the temperature of the planet above the freezing point of water (Pollack et al., 1987). Thus, it would be difficult to use planetary-level engineering to establish a large, self-sustained greenhouse effect on Mars, allowing humans to colonize the planet (McKay et al., 1991).

Venus

Unlike the Earth, the high surface temperature of 474°C on Venus ensures that its present inventory of volatiles resides entirely in its atmosphere. The atmospheric pressure on Venus is nearly 100 times greater than on Earth (Table 2.6). On Venus, the ratio of the mass of the atmosphere to the mass of the planet (1.09×10^{-4}) is only slightly less than the ratio of the total mass of volatiles on Earth (see Table 2.3) to the mass of the Earth (3.3×10^{-4}). These values suggest a similar degree of crustal degassing on these planets. Indeed, volcanism is observed on Venus today (Smrekar et al., 2010; Bondarenko et al., 2010), and the atmosphere on Venus is remarkably similar to the average composition of volatiles in carbonaceous chondrites (Pepin, 2006). Despite the evidence of outgassing, the $^{40}Ar/^{36}Ar$ ratio on Venus is close to 1.0, implying that, relative to the Earth, Venus may have also retained a greater fraction of gases from the solar nebula during its accretion (Pollack and Black, 1982).

The hot, high-pressure conditions have made it difficult to land spacecraft for the exploration of the surface of Venus. The massive atmosphere on Venus is dominated by CO_2, conferring a large greenhouse warming and surface temperatures well in excess of that predicted for a non reflective body at the same distance from the Sun (54°C; Houghton, 1986).[j] The relative abundance of CO_2 and N_2 in the atmosphere of Venus is roughly similar to that in the total inventory of volatiles on Earth (Oyama et al., 1979; Pollack and Black, 1982; Lecuyer et al., 2000). What is unusual about Venus is the low abundance of water in its atmosphere. Was Venus wet in the past?

The ratio of ^2H (deuterium) to ^1H on Venus is more than 100 times greater than that on Earth (Donahue et al., 1982; McElroy et al., 1982; de Bergh et al., 1991), suggesting that Venus, like Mars, may have possessed a large inventory of water in the past, but lost water through a process that differentiates between the isotopes of hydrogen. With the warm initial conditions on Venus, a large amount of the water vapor in the atmosphere may have been subject to photodisassociation, causing the planet to dry out through its history (Kasting et al., 1988;

[j]One of the widely cited reports of the Intergovernmental Panel on Climate Change (IPCC) indicates that the surface temperature on Venus in the absence of a greenhouse effect would be 47°C (Houghton et al., 1990). This is lower than the value given here because the IPCC report accounts for the reflectivity of the thick cloud layer on Venus, whereas our value considers the equilibrium temperature for a black body absorber in the orbit of Venus. See Lewis and Prinn (1984, p. 97).

Lecuyer et al., 2000). The oxygen released during the photodisassociation of water has probably reacted with crustal minerals (Donahue et al., 1982; McGill et al., 1983, p. 87).

At the surface temperatures found on Venus, little CO_2 can react with its crust (compare with Fig. 1.3), so high concentrations of CO_2 remain in the atmosphere (Nozette and Lewis, 1982). Various other gases, such as SO_2, that are found dissolved in seawater on Earth also reside as gases in the atmosphere on Venus (Vandaele et al., 2017). Continuing volcanic releases of CO_2 have accumulated in the atmosphere to produce a runaway greenhouse effect in which increasing temperatures allow an increasing potential for the atmosphere to hold CO_2 and other gases (Walker, 1977). Thus, the current temperature on Venus, 474°C, is much greater than we would predict if Venus had no atmosphere and is not conducive to life as we know it.

Moons and exoplanets

Increasing discoveries of planets around other stars and observations of organic molecules in interstellar dusts and meteorites beg for further exploration for the presence of extraterrestrial life in our galaxy and beyond (Quintana et al., 2014; Gillon et al., 2017). The massive atmospheres on the outer planets of our solar system are not conducive to life, but Europa, Ganymede, and the other moons of Jupiter and Saturn show evidence of subsurface oceans beneath surface ice (Bell, 2012; Nimmo and Pappalardo, 2016). Some of these habitats may be subjected to hydrothermal activity, providing a submarine habitat for the abiotic synthesis of organic materials and perhaps modest forms of metabolism (Gaidos et al., 1999; Marion et al., 2003). A small amount of O_2 is detected in the atmosphere of Europa (10^{-11} of that on Earth), where it may originate from a photolytic process similar to that leading to the loss of water from Mars and Venus (Hall et al., 1995).

The atmosphere of Saturn's moon Titan is dominated by nitrogen (96%) and methane (3%), with an $^{40}Ar/^{36}Ar$ ratio of 154, suggesting a secondary origin from outgassing (Niemann et al., 2005; Yung and DeMore, 1999, p. 202). Huge lakes of liquid methane are potentially stable on its surface (Stofan et al., 2007; Lorenz et al., 2008; Dhingra et al., 2019), and liquid methane falls from its atmosphere as rain (Hueso and Sánchez-Lavega, 2006; Turtle et al., 2011). On Saturn's moon, Enceladus, frozen water-ice covers much of the surface (Brown et al., 2006a), and geysers appear to spew water vapor into the atmosphere (Waite et al., 2006; Postberg et al., 2011). Recent analyses detect hydrogen in these plumes, likely from hydrothermal processes and potentially a precursor for the formation of methane by microbial activity (Eq. 2.8; Waite et al., 2017). Habitats with liquid water, hydrothermal activity, and reduced gases (especially methane) on planets and planetary moons are likely places to look for extraterrestrial life within and outside our solar system (Swain et al., 2008). Subglacial lakes in Antarctica are analogous terrestrial habitats on Earth (Priscu et al., 1998).

Certainly the most unusual characteristic of the Earth's atmosphere is the presence of large amounts of O_2, which is an unequivocal indication of life on this planet (Sagan et al., 1993; McKay, 2014). Having examined the conditions on Mars, Venus, and other bodies of our solar system, we can now offer some speculation on the conditions that might exist on a lifeless Earth. At a distance of 150×10^6 km from the Sun, the surface temperature on the Earth, assuming no reflectivity to incoming solar radiation, would be close to the freezing point of water (Houghton, 1986). Such cold conditions would seem to ensure that the atmosphere on the Earth has never contained much water vapor, so relatively little water has been lost to space as a result of photolysis in the upper atmosphere. Despite the small amount of H_2O in Earth's atmosphere, the atmosphere confers enough greenhouse warming to the planet to have maintained liquid oceans for most of its history. Thus, even on a lifeless Earth, most of the inventory of volatiles would reside in the oceans. The atmosphere on a lifeless Earth would be dominated by N_2, which is only slightly soluble in water. The size of Earth and its gravitational field ensure that photolysis of N_2 does not result in the loss of N from the planet. Moreover, the rate of fixation of nitrogen by lightning in an atmosphere without O_2 appears too low to transfer a significant portion of N_2 from the atmosphere to the oceans (Kasting and Walker, 1981; Chapter 12). Thus, the main effect of life has been to dilute the initial nitrogen-rich atmosphere on Earth with a large quantity of O_2 (Walker, 1984).

How long will the Earth be hospitable to life? Barring some unforeseen catastrophe, we can speculate that the biosphere will persist as long as our planet harbors liquid water on its surface. Eventually, however, a gradual increase in the Sun's luminosity will warm the Earth, causing a photolytic loss of water from the upper atmosphere, irreversible oxidation of the Earth's surface, and the demise of life—perhaps after another 2.5 billion years (Lovelock and Whitfield, 1982; Caldeira and Kasting, 1992). The Sun itself will burn out in 10 billion years. If we manage the planet well, studies of biogeochemistry have a long future.

Summary

In this chapter we have reviewed theories for the formation and differentiation of early Earth. In the process of planetary formation, certain elements were concentrated near its surface and only some elements were readily soluble in seawater. Thus, the environment in which life arose is a special mix taken from the geochemical abundance of elements that were available on Earth. Simple organic molecules can be produced by physical processes in the laboratory; presumably similar reactions occurred in high-energy habitats on primitive Earth.

Life may have arisen by the abiotic assembly of these constituents into simple forms, resembling the most primitive bacteria that we know of today. Essential to living systems is the processing of energy, which is likely to have begun with the heterotrophic consumption of molecules found in the environment. A persistent scarcity of such molecules is likely to have led to selection for the autotrophic production of energy by various pathways, including photosynthesis.

Autotrophic photosynthesis appears to be responsible for nearly all the production of O_2, which has accumulated in the Earth's atmosphere over the last 2 billion years. The major biogeochemical cycles on Earth are mediated by organisms whose metabolic activities couple the oxidation and reduction of substances isolated from the environment.

The Atmosphere

Introduction

There are several reasons to begin our treatment of biogeochemistry with a consideration of the atmosphere. The composition of the atmosphere has evolved as a result of the history of life on Earth (Chapter 2), and is now changing rapidly as a result of human activities. The atmosphere controls Earth's climate and ultimately determines the conditions in which we live—our supplies of food and water, our health, and our economy. Further, the atmosphere is relatively well mixed, so changes in its composition can be taken as a first index of changes in biogeochemical processes at the global level. The circulation of the atmosphere transports biogeochemical constituents between the oceans and land, contributing to the global cycles of chemical elements.

We begin our discussion with a brief consideration of the structure, circulation, and composition of the atmosphere. Then we examine reactions that occur among various gases, especially in the lower atmosphere. Many of these reactions remove constituents from the atmosphere, depositing them on the surface of the land and sea. In the face of constant losses, the composition of the atmosphere is maintained by biotic processes that supply gases to the atmosphere. We mention the sources of atmospheric gases here briefly, but they will be treated in more detail in later chapters of this book, especially as we examine the microbial reactions that occur in soils, wetlands, and ocean sediments. Finally, we discuss human impacts on the global atmosphere, as seen in ozone depletion and climate change.

Structure and circulation

The atmosphere is held on Earth's surface by the gravitational attraction of the Earth. At any altitude, the downward force (F) is related to the mass (M) of the atmosphere above that point:

$$F = M \times g, \tag{3.1}$$

where g is the acceleration due to gravity ($980 \, cm/s^2$ at sea level). Pressure (force per unit area) decreases with increasing altitude because the mass of the overlying atmosphere is smaller (Walker, 1977). Decline in atmospheric pressure (P in bars) with altitude (A in km) is approximated by the logarithmic relation:

$$\log P = -0.06 \, (A), \tag{3.2}$$

over the whole atmosphere (Fig. 3.1).

Although the chemical composition of the atmosphere is relatively uniform, when we visit high mountains, we often say that the atmosphere seems "thinner" than at sea level. The composition is the same, but the abundance of molecules in each volume of the atmosphere is greater at sea level, because it is compressed by the pressure of the overlying atmosphere. The lower atmosphere, the troposphere, contains about 80% of the atmospheric mass (Warneck, 2000). The lower density of the upper atmosphere is the reason that jet aircraft flying at high altitudes require cabin pressurization for their passengers.

Certain atmospheric constituents, such as ozone, aerosols, and clouds absorb and reflect portions of the radiation that the Earth receives from the Sun, so only about half of the Sun's radiation penetrates the atmosphere to be absorbed or reflected by the Earth's surface (Fig. 3.2). The overall reflectivity or albedo of the Earth, as measured by the CERES satellite (Clouds and the Earth's Radiant Energy System) and by "earthshine" received by the Moon, is about 29% (Goode et al., 2001; Kim and Ramanathan, 2012). Albedo is affected by the reflectivity of the Earth's surface, which differs among sea water, forests and ice cover. Particles in the atmosphere absorb or reflect about 5–10% of the incoming solar radiation. Although some regions with air pollution have greater albedo than pristine areas, the Earth's overall albedo has not changed appreciably during the past couple of decades (Palle et al., 2016). Greater albedo reduces the radiation reaching Earth's surface and leads to "global dimming."

The land and ocean surfaces reradiate long wave (heat) radiation to the atmosphere, so the atmosphere is heated from the bottom and is warmest at the Earth's surface (Fig. 3.1). Because warm air is less dense and rises, the troposphere is well mixed. The top of the troposphere extends to 8–17 km, varying seasonally and with latitude. The temperature of the upper

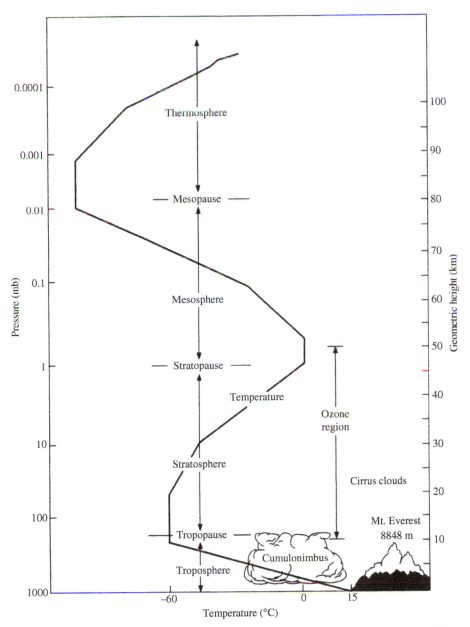

FIG. 3.1 Vertical structure and zonation of the atmosphere, showing the temperature profile to 100-km altitude. Note the logarithmic decline in pressure (left axis) as a function of altitude.

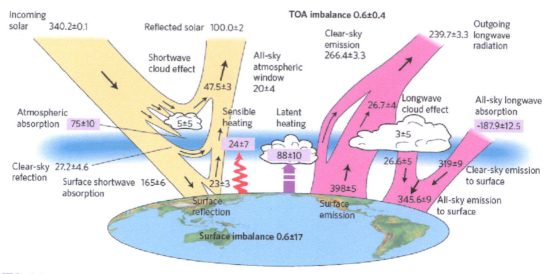

FIG. 3.2   The radiation budget for Earth, showing the proportional fate of the energy that the Earth receives from the Sun, about 340 W/m² largely in short wavelengths. About one third of this radiation is reflected back to space and the remainder is absorbed by the atmosphere (23%) or the surface (46%). Long-wave radiation (infra-red) is emitted from the Earth's surface, some of which is absorbed by atmospheric gases, warming the atmosphere—the greenhouse effect. The atmosphere emits long-wave radiation, so that the total energy received is balanced by the total energy emitted from the planet. *Source: Stephens et al. (2012).*

troposphere is about −60°C, which ensures that the atmosphere above 10 km contains only small amounts of water vapor and ice particles, especially over Antarctica.

Above the troposphere, the stratosphere is defined by the zone in which temperatures increase with altitude, extending to about 50 km (Fig. 3.1). The increase is largely due to the absorption of ultraviolet light by ozone. Vertical mixing in the stratosphere is limited, as is exchange across the boundary between the troposphere and the stratosphere, the tropopause. Thus, materials that enter the stratosphere remain there for long periods, allowing for high altitude transport around the globe.

The thermal mixing of the troposphere is largely responsible for the global circulation of the atmosphere, as well as local weather patterns (Fig. 3.3). The large annual receipt of solar energy at the equator causes warming of the atmosphere (sensible heat) and the evaporation of large amounts of water, carrying latent heat, from tropical oceans and rainforests. As this warm, moist air rises, it cools, producing a large amount of precipitation in equatorial regions. Having lost its moisture, the rising air masses are deflected towards the poles by the Coriolis forces. In a belt centered on approximately 30°N or S latitude, these dry air masses sink to the Earth's surface, undergoing compressional heating. Most of the world's major deserts are associated with the downward movement of hot, dry air at this latitude. These circulation patterns of rising, cooling air at the equator and sinking, warming air at 30°N and S latitude are known as Hadley Cells. A similar, but much weaker, circulation pattern is found at the poles, where cold air sinks and moves north or south along the Earth's surface to lower latitudes.

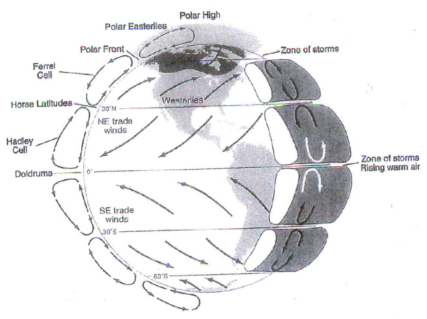

FIG. 3.3 Generalized pattern of global circulation showing surface patters, vertical patterns, and the origin of the Coriolis force. As air masses move across different latitudes, they are deflected by the Coriolis force, which arises because of the different speeds of the Earth's rotation at different latitudes. For instance, if you were riding on an air mass moving at a constant speed south from 30°N latitude, you would begin your journey seeing 1446 km of the Earth's surface pass to the east every hour. By the time your air mass reached the equator, 1670 km would be passing to the east each hour. While moving south at a constant velocity, you would find that you had traveled 214 km west of your expected trajectory. The Coriolis force means that all movements of air in the northern hemisphere are deflected to the right; those in the southern hemisphere are deflected to the left. *Modified from Berner and Berner (2012).*

Frictional drag between the polar cells and Hadley cells drives an indirect circulation in each hemisphere between 40° and 60° latitude, producing regional storm systems and the prevailing west winds that we experience in the temperate zone.[a]

The tropospheric air in each hemisphere mixes on a time scale of a few months (Warneck, 2000), allowing for regional transport of air pollutants that persist for more than a few days. For instance, in 1995 carbon monoxide (CO) from Canadian forest fires contributed to air pollutant loads in the eastern United States (Wotawa and Trainer, 2000). The eruption of the Eyjafjallajökull volcano in Iceland on 13–14 April 2010 produced a cloud of volcanic ash over Poland several days later (Pietruczuk et al., 2010; Langmann et al., 2012) and disrupted airplane travel over much of Europe for several weeks. Vertical mixing in the troposphere is driven by convection, especially in thunderstorms, so that much of the air in the upper troposphere is less than a week old (Brunner et al., 1998; Bertram et al., 2007). Each year, there is also complete mixing of tropospheric air between the Northern and the Southern

[a]NASA provides an excellent visualization of wind patterns across the Earth's surface at: https://youtu.be/w3SmRTh5wJ4.

Hemispheres across the intertropical convergence zone (ITCZ). If a gas shows a higher concentration in one hemisphere, we can infer that a large natural or human source must exist in that hemisphere, overwhelming the tendency for atmospheric mixing to equalize the concentrations (Fig. 3.4).

Exchange between the troposphere and the stratosphere is driven by several processes (Warneck, 2000). In the tropical Hadley cells, rising air masses carry some tropospheric air to the stratosphere (Holton et al., 1995; Fueglistaler et al., 2004). The strength of the updraft varies seasonally, as a result of variations in the radiation received from the Sun. When the height of the tropopause drops, tropospheric air is trapped in the stratosphere, or vice versa. There is also exchange across the tropopause due to large-scale wind movements, thunderstorms, and eddy diffusion.

Atmospheric scientists have examined the exchange of air mass between the troposphere and the stratosphere by following the fate of industrial pollutants released to the troposphere and radioactive contaminants released to the stratosphere in tests of atomic weapons during the 1950s and early 1960s (Warneck, 2000). In these considerations, the concept of mean residence time is useful. For any reservoir that is in steady state, mean residence time (MRT) is defined as:

$$MRT = Mass/flux,\qquad\qquad(3.3)$$

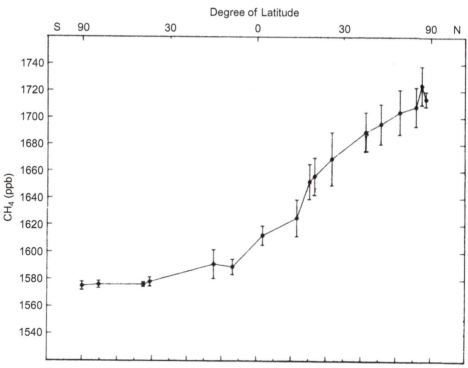

FIG. 3.4   The latitudinal variation in the mean concentration of methane (CH_4) in Earth's atmosphere. *From Steele et al. (1987). Used with permission of Reidel Publishing.*

where flux may be either the input or the loss from the reservoir.[b] Since the stratosphere is not well mixed vertically, the mean residence time of stratospheric air increases with altitude (Waugh and Hall, 2002). However, the return of stratospheric air to the troposphere, about 4×10^{17} kg/yr (Seo and Bowman, 2002), amounts to about 40% of the stratospheric mass each year, leading to an overall mean residence time of 2.6 years for stratospheric air. Thus, when a large volcano injects sulfur dioxide into the stratosphere, about half of it will remain after 2 years and about 5% will remain after 7.5 years.

Atmospheric composition

Gases

Table 3.1 gives the globally averaged concentration of some important gases in the atmosphere. Three gases—nitrogen, oxygen, and argon—make up 99% of the atmospheric mass of 5.14×10^{21} g (Trenberth and Guillemot, 1994). The mean residence times of these gases are much longer than the rate of atmospheric mixing. Because of their long residence times, the concentrations of N_2, O_2, and all noble gases (He, Ne, Ar, Kr, and Xe) are globally uniform and time-invariant.

Several hundred trace gases, including a wide variety of volatile organic compounds (VOCs), are also found in the atmosphere. The most abundant volatile organic compound from vegetation is isoprene, which is commonly emitted from many coniferous forest species (Guenther et al., 2000). For comparison, Table 3.2 shows the volatile emission of organic compounds from representative species in various world biomes. The various volatile organic compounds (VOCs) derived from vegetation include nonmethane hydrocarbons (NMHC) and oxygenated organic molecules, such as methanol (Park et al., 2013). Human activities also add a wide variety of trace organic gases to the atmosphere, including ethane and oxygenated organic gases, such as acetone and alcohols (Piccot et al., 1992; Huang et al., 2015). In urban areas, organic gases are released from a variety of products, including paints, pesticides, cleaning agents, and personal care products (McDonald et al., 2018).

Most trace gases are highly reactive and thus have short mean residence times, so it is not surprising that they are minor constituents in the atmosphere (Atkinson and Arey, 2003). The concentration of such gases varies in space and time. For instance, we expect high concentrations of certain pollutants (O_3, CO_2 and CO etc.) over cities (e.g., Kort et al., 2012; Pommier et al., 2013), high levels of ammonia (NH_3) near areas where cattle and pigs are kept (Leifer et al., 2017), and high concentrations of some reduced gases (methane and hydrogen sulfide) over swamps and other areas of anaerobic decomposition (e.g., Harriss et al., 1982; Steudler and Peterson, 1985). Remote sensing from satellites allows us to measure the spatial variation

---

[b] Assuming exponential decay of a tracer from a well-mixed reservoir that is in steady state, the fractional loss per year ($-k$) is equal to the reciprocal of the mean residence time in years (i.e., 1/MRT). The amount remaining in the reservoir at any time t (in years) as a fraction of the original content is equal to e^{-kt}, the half-life of the reservoir in years is 0.693/k, and 95% will have disappeared from the reservoir after 3/k years.

TABLE 3.1   Global average concentration of well mixed atmospheric constituents.[a]

Compounds	Formula	Concentration	Total mass (g)
Major constituents (%)			
Nitrogen	N_2	78.084	3.87×10^{21}
Oxygen	O_2	20.946	1.19×10^{21}
Argon	Ar	0.934	6.59×10^{19}
Parts-per-million constituents (ppm = 10^{-6} or $\mu L/L$)			
Carbon dioxide	CO_2	400	3.11×10^{18}
Neon	Ne	18.2	6.49×10^{16}
Helium	He	5.24	3.70×10^{15}
Methane	CH_4	1.83	5.19×10^{15}
Krypton	Kr	1.14	1.69×10^{16}
Parts-per-billion constituents (ppb = 10^{-9} or nL/L)			
Hydrogen	H_2	510	1.82×10^{14}
Nitrous oxide	N_2O	320	2.49×10^{13}
Xenon	Xe	87	2.02×10^{15}
Parts-per-trillion constituents (ppt = 10^{-12})			
Carbonyl sulfide	COS	500	5.30×10^{12}
Chlorofluorocarbons			
CFC 11	CCl_3F	280	6.79×10^{12}
CFC 12	CCl_2F_2	550	3.12×10^{13}
Methylchloride	CH_3Cl	620	5.53×10^{12}
Methylbromide	CH_3Br	11	1.84×10^{11}

[a] *Those with a mean residence time >1 year. Assuming a dry atmosphere with a molecular weight of 28.97, the overall mass of the atmosphere sums to 514×10^{19} g.*
Source: Updated from Trenberth and Guillemot (1994).

in the concentration of CO_2, CH_4, NH_3, NO_2, and other gases from space and to identify "hot spots" of emission (Hakkarainen et al., 2016; Jacob et al., 2016; Frankenberg et al., 2011; Warner et al., 2017; Van Damme et al., 2018; Griffin et al., 2019). Winds mix the concentrations of these gases to their average tropospheric background concentration within a short distance downwind of local sources. We can best perceive global changes in atmospheric composition, such as the current increase in CH_4, by making long-term measurements in remote locations.

Junge (1974) related geographic variations in the atmospheric concentration of various gases to their estimated mean residence time in the atmosphere (Fig. 3.5). Gases that have short mean residence times are highly variable from place to place, whereas those that have

TABLE 3.2 Biogenic volatile organic carbon (VOC) emission rates (μg-Cg/h) from leaves of woody species, adjusted to standard conditions of temperature and light using algorithms of Guenther et al. (1993).

Desert species	Isoprene	α-Pinene	β-Pinene	Camphene	D-Limonene	$\sum$Monoterpenes	Reference
Ambrosia dumosa	<0.1	1.6	3.0	0.06	2.0	7.9	Geron et al., 2006a
Chrysothamnus nauseosus	<0.1	0.28	0	0	0.21	0.65	Geron et al., 2006a
Hymenoclea salsola	<0.1	1.4	0.06	0.02	0.30	2.6	Geron et al., 2006a
Larrea tridentata	<0.1	0.37	0.12	0.44	0.74	2.0	Geron et al., 2006a
Boreal							
Abies balsamea	<0.1	0.61	1.9	0.51	0	3.4	Ortega et al., 2008
Picea glauca	14.9	0.25	0.19	0.07	0.44	1.4	Kempf et al., 1996
Pinus sylvestris	<0.1	0.34	0.02	0.02	0.03	0.8	Janson, 1993
Temperate							
Acer rubrum	<0.1	0.18	0.53	0.04	0.04	1.4	Ortega et al., 2008
Pinus taeda	<0.1	0.08	0.02	0.01	0.01	0.14	Ortega et al., 2008
Pinus ponderosa	<0.1	0.58	0.33	0.04	0.11	1.6	Helmig et al., 2013
Quercus rubra	67	0.28	0.10	0.05	0.20	1.7	Geron et al., 2001; Ortega et al., 2008
Tropical							
Apeiba tibourbou	0	1.0	0.43	0	0.14	3.6	Kuhn, 2002
Eucalyptus globules	56	1.0	0.4	0	0.14	3.6	Kuhn, 2002
Hymenaea courbaril	46	0	0	0	0	0	Kuhn, 2002
Cecropia sciadophylla	<0.1	6.5	2.2	0	0	155	Jardine et al., 2015a
Hevea brasiliensis	<0.1	2.6	2.1	0	0	30	Geron et al., 2006b
Azadirachta indica	<0.1	0	0.15	0.9	0.38	2.43	Singh et al., 2011

Continued

TABLE 3.2 Biogenic volatile organic carbon (VOC) emission rates (μg-C g/h) from leaves of woody species, adjusted to standard conditions of temperature and light using algorithms of Guenther et al. (1993)—cont'd

Desert species	Isoprene	α-Pinene	β-Pinene	Camphene	D-Limonene	∑Monoterpenes	Reference
Ficus religiosa	77	0	0	0	0	0	Varshney and Singh, 2003
Citrus limon	0.61	0.6	1.1	0	3.8	7.9	Varshney and Singh, 2003

Data compiled and provided by Chris Geron (US EPA).

long mean residence times relative to atmospheric mixing show relatively little spatial variation. For example, the average volume of water in the atmosphere is equivalent to about 13,000 km^3 at any time, or 24.6 mm above any point on the Earth's surface (Trenberth, 1998). The average daily precipitation would be about 2.73 mm if it were deposited evenly around the globe. Thus, the mean residence time for water vapor in the atmosphere is:

$$24.6 \, mm/2.73 \, mm \, day^{-1} = 9.1 \, days. \tag{3.4}$$

This is a short time compared to the circulation of the troposphere, so we should expect water vapor to show highly variable concentrations in space and time (Fig. 3.5). Most of the volatile organic compounds have short residence times in the atmosphere, so they are found at low concentrations in remote regions, such as Antarctica (Beyersdorf et al., 2010). This relationship between variation in concentration and residence time in the atmosphere extends to trace organic species (e.g., propane), which have residence times of a few days (Jobson et al., 1999).

The mean residence time for carbon dioxide in the atmosphere is about 3 years—only slightly longer than the mixing time for the atmosphere.[c] Owing to the seasonal uptake of CO_2 by plants, CO_2 shows a minor seasonal and latitudinal variation (±about 1%) in its global concentration of ~400 ppm (Figs. 1.1 and 3.6). In contrast, painstaking analyses are required to show any variation in the concentration of O_2 because the amount in the atmosphere is so large and its mean residence time, ~4000 years, is so much longer than the mixing time of the atmosphere (Keeling and Shertz, 1992).

Gases with mean residence times of <1 year in the troposphere do not persist long enough for appreciable mixing into the stratosphere. Indeed, one of the most valuable, but dangerous, industrial properties of chlorofluorocarbons (CFCs)[d] is that they are chemically inert and thus long-lived in the troposphere (Rowland, 1989). This allows chlorofluorocarbons to mix into the stratosphere, where they lead to the destruction of ozone by ultraviolet light (see Eqs. 3.54–3.57).

[c] For visualization of the mixing of CO_2 in Earth's atmosphere over an annual cycle, showing emissions from major industrial regions, see https://youtu.be/x1SgmFa0r04.

[d] CFCs are identified using code in which the units digit is the number of fluorine atoms, the tens digit the number of hydrogen atoms plus one, and the hundreds digit the number of carbon atoms minus one (cf. Table 3.1; Rowland (1989).

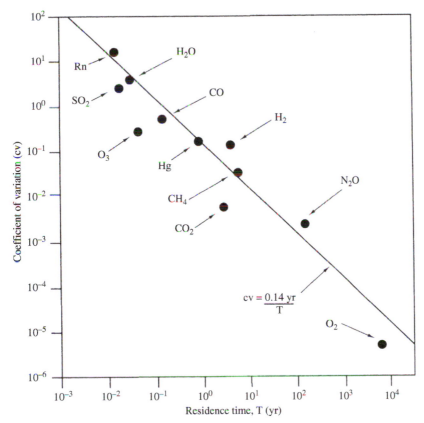

FIG. 3.5   Variability in the concentration of atmospheric gases, expressed as the coefficient of variation among measurements, as a function of their estimated mean residence times in the atmosphere. *Modified from Junge (1974), as updated by Slinn (1988).*

Aerosols

In addition to gaseous components, the atmosphere contains particles, known as aerosols, that arise from a variety of sources (Table 3.3). Increasingly, we recognize that the concentrations of particles in the atmosphere have important impacts on biogeochemistry, human health (Lelieveld et al., 2015), and climate (Mahowald, 2011). The total burden of aerosols in the atmosphere is expressed as Aerosol Optical Depth (AOD), which is now measured from space by satellites (Fig. 3.7).[e, f]

[e] Aerosol optical depth (AOD) is a dimensionless value that is the natural logarithm of the ratio of incident to transmitted light through the atmosphere.

[f] NASA visualization shows the global emission and circulation of aerosols, with soil dusts *(red)*, sea salt *(blue)*, smoke *(green)*, and sulfate particles *(white)* from industrial and volcanic sources. See: https://www.nasa.gov/multimedia/imagegallery/image_feature_2393.html.

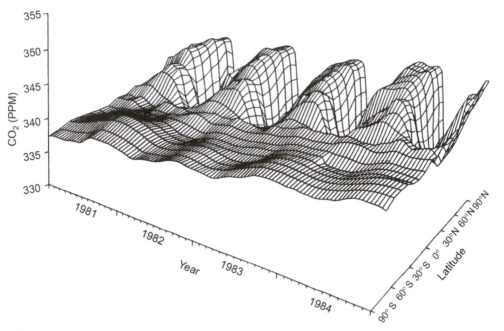

FIG. 3.6   Seasonal fluctuations in the concentration of atmospheric CO_2 (1981–1984), shown as a function of 10° latitudinal belts (Conway et al., 1988). Note the smaller amplitude of the fluctuations in the southern hemisphere, reaching peak concentrations during the northern hemisphere's winter.

Soil particles are dispersed by wind erosion (also known as deflation weathering or eolian transport) from arid and semiarid regions (Pye, 1987; Ravi et al., 2011). Particles with diameter <1.0 µm are held aloft by turbulent motion and are subject to long-range transport. Current estimates suggest that up to 2×10^{15} g/yr of soil particles enter the atmosphere from arid and barren agricultural soils (Zender et al., 2004), and about 20% of these particles are involved in long-range transport. While natural sources account for 81% of the global dust emissions (Chen et al., 2018a), the flux of soil dust has increased due to human cultivation, especially in semiarid lands (Tegen et al., 2004; Mulitza et al., 2010; Routson et al., 2019) and increases during droughts. Dust from the deserts of central Asia falls in the Pacific Ocean (Duce et al., 1980), where it contributes much of the iron needed by oceanic phytoplankton (Mahowald et al., 2005b, Chapter 9). Similarly, dust from the Sahara supplies nutrients to phytoplankton in the Atlantic Ocean (Talbot et al., 1986; Jickells et al., 2005) and phosphorus to Amazon rainforests (Swap et al., 1992; Okin et al., 2004; Yu et al., 2015). Some particulate pollution in the western United States is derived from the deserts of China (Yu et al., 2012), and emissions from arid lands in the Colorado Plateau affect the biogeochemistry in Rocky Mountain forests (Field et al., 2010). Dust from desert soils is monitored by several satellites, including NASA's MODIS satellite (Tanre et al., 2001; Kaufman et al., 2002; see Fig. 3.7). Typically, while it is in transit, soil dust warms the atmosphere over land and cools the atmosphere over the oceans, which have lower surface albedo (reflectivity) (Ackerman and Chung, 1992; Kellogg, 1992; Yang et al., 2009).

An enormous quantity of particles enters the atmosphere from the ocean as a result of tiny droplets that become airborne with the bursting of bubbles at the surface (MacIntyre, 1974; Wu, 1981). As the water evaporates from these bubbles, the salts crystallize to form seasalt

TABLE 3.3 The global production and atmospheric burden of aerosols from natural and human-derived sources.

	Mass emission 10^{12} g/yr	Mass burden Tg	Number produced per year	Number Burden
Carbonaceous aerosols				
Primary organic (0–2 μm)	95	1.2	–	310×10^{24}
Biomass burning	54	–	7×10^{27}	–
Fossil fuel	4	–	–	–
Biogenic	35	0.2	–	–
Black carbon (0–2 μm)	10	0.1	–	270×10^{24}
Open burning and biofuel	6	–	–	–
Fossil fuel	4.5	–	–	–
Secondary organic	28	0.8	–	–
Biogenic	25	0.7	–	–
Anthropogenic	3.5	0.08	–	–
Sulfates	200	2.8	2×10^{28}	–
Biogenic	57	1.2	–	–
Volcanic	21	0.2	–	–
Anthropogenic	122	1.4	–	–
Nitrates	18	0.49	–	–
Industrial dust, etc.	100	1.1	–	–
Sea salt				
$d < 1\,\mu m$	180	3.5	7.4×10^{26}	–
$d = 1\text{–}16\,\mu m$	9940	12	4.6×10^{26}	–
Total	10,130	15	1.2×10^{27}	27×10^{24}
Mineral (soil) dust				
$<1\,\mu m$	165	4.7	4.1×10^{25}	–
$1\text{–}2.5\,\mu m$	496	12.5	9.6×10^{25}	–
$2.5\text{–}10\,\mu m$	992	6	–	–
Total	1600	18 ± 5	1.4×10^{26}	11×10^{24}

From Andreae and Rosenfeld (2008).

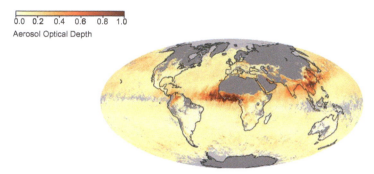

FIG. 3.7 Aerosols in Earth's atmosphere, measured as Aerosol Optical Depth (AOD) by the NASA MODIS satellite during March 2010. See Footnote e, p. 61 for the derivation of AOD. Note high amounts of aerosols exiting the southern Sahel region of Africa, blowing westward to the Amazon, and high concentrations of aerosols emitted from the deserts of China, blowing eastward across the Pacific Ocean. *From http://earthobservatory.nasa.gov/GlobalMaps/view. php?d1=MODAL2_M_AER_OD.*

aerosols, which carry the approximate chemical composition of seawater (Glass and Matteson, 1973; Möller, 1990). As in the case of soil dust, most seasalt aerosols are relatively large and settle from the atmosphere quickly, but a significant proportion remains in the atmosphere for global transport. Lewis and Schwartz (2004) compile global estimates of seasalt production, averaging 5×10^{15} g/yr, whereas Andreae and Rosenfeld, 2008; Table 3.3) estimate a total seasalt production of 10×10^{15} g/yr, which carries about 200×10^{12} g of chloride from sea to land (see Fig. 3.17 later).

Organic particles are produced from a wide variety of sources, including pollen, plant fragments, and bacteria (Després et al., 2012). Some plant aerosols contain significant quantities of potassium (Crozat, 1979; Pöhlker et al., 2012). Forest fires produce particles of charcoal that are carried throughout the troposphere, and small organic particles (soot) are produced by the condensation of volatile hydrocarbons from the smoke of forest fires (Hahn, 1980; Cachier et al., 1989). Forest fires in the Amazon are thought to release as much as 1×10^{13} g of particulate matter to the atmosphere each year (Kaufman et al., 1990). It is likely that the global production of aerosols from forest fires has increased markedly as a result of higher rates of biomass burning in the tropics (Andreae, 1991; Cahoon et al., 1992). Aerosols from these fires may affect regional patterns of rainfall (Cachier and Ducret, 1991) and global climate (Penner et al., 1992). Recent catastrophic fires in the western United States have contributed to the atmospheric burden of aerosols and to impacts on human health (McClure and Jaffe, 2018; Ford et al., 2018).

Periodically, volcanoes inject finely divided rock material—volcanic ash—into the atmosphere where it is deposited over large areas (Table 3.4), contributing to soil development in regions that are downwind from major eruptions (Watkins et al., 1978; Dahlgren et al., 1999; Zobel and Antos, 1991). Volcanic gases and ash that are transported to the stratosphere by violent eruptions undergo global transport, affecting climate for several years (Sigl et al., 2015).[g, h] During the past several centuries, aerosols from volcanoes have dwarfed those

[g]See https://arstechnica.com/science/2012/07/berkeley-earth-project-is-back-to-re-re-confirm-earth-is-warming/.

[h]Aerosols refract sunlight, so sunsets are particularly red in the years following major volcanic eruptions, as recorded inadvertently in landscape paintings in the 1800s (Zerefos et al., 2014).

TABLE 3.4   Composition of a volcanic ash sample collected during the eruption of Mt. St. Helens (Washington, USA) on 19 May 1980.

Constituent	Particulate sample	Average ash
Major elements (%)		
SiO_2	$\equiv 65.0$	65.0
Fe_2O_3	6.7	4.81
CaO	3.0	4.94
K_2O	2.0	1.47
TiO_2	0.42	0.69
MnO	0.054	0.077
P_2O_5	–	0.17
Trace elements (ppm)		
S	3220	940
CI	1190	660
Cu	61	36
Zn	34	53
Br	<8	~1
Rb	<17	32
Sr	285	460
Zr	142	170
Pb	36	8.7

Source: From Fruchter et al. (1980) and Hooper et al. (1980).

produced by incoming meteors, although the latter have been a potential source of plant nutrients, such as Fe, in the geologic past (Reiners and Turchyn, 2018).

Small particles, known as secondary aerosols, are also produced by reactions between gases in the atmosphere. For instance, when SO_2 is oxidized to sulfuric acid (H_2SO_4) in the atmosphere, particles rich in $(NH_4)_2SO_4$ may be produced by a subsequent reaction with atmospheric ammonia (NH_3; Behera and Sharma, 2011; Kirkby et al., 2011):

$$2NH_3 + H_2SO_4 \rightarrow (NH_4)_2SO_4. \tag{3.5}$$

(Ammonia is derived from a variety of sources, primarily associated with agricultural activities; Chapter 12.) Sulfate aerosols are also produced during the oxidation of dimethylsulfide released from the ocean (Chapter 9). Sulfate aerosols increase the albedo of the Earth's atmosphere, so estimates of the abundance of sulfate aerosols are an important component of global climate models (Kiehl and Briegleb, 1993; Mitchell et al., 1995).

Secondary aerosols are also produced from volatile organic compounds, such as isoprene, that are released from plants (Kavouras et al., 1998; Jimenez et al., 2009; Zhang et al., 2018b). Sulfuric acid vapor can oxidize volatile organic compounds to produce secondary aerosols that are important to cloud formation (Riccobono et al., 2014).

Finally, a wide variety of particles are produced from human industrial processes, especially the burning of coal (Hulett et al., 1980; Shaw, 1987). Globally, the release of particles during the combustion of fossil fuels rivals the mobilization of elements by rock weathering at the Earth's surface (Bertine and Goldberg, 1971, see Table 1.1). Overall, human activities probably account for about 10% of the burden of aerosols in today's atmosphere (Table 3.3). Fine particulate air pollution has significant human health effects (Samet et al., 2000; Pope et al., 2009; Lelieveld et al., 2015). Nanoparticles ($<0.3\,\mu m$), derived from a variety of natural and manufactured sources (Kumar et al., 2010; Hendren et al., 2011), are of particular concern.

Fortunately, the mass of industrial aerosols has declined in many developed countries where pollution controls have been instituted. One of the most widespread anthropogenic aerosols, particles of lead from automobile exhaust, has declined in global abundance over the past 40 years due to a reduction in the use of leaded gasoline (Boutron et al., 1991). In other regions, where air pollution is unregulated, concentrations of aerosols have increased (Streets et al., 2008; Dey and Di Girolamo, 2011), contributing to regional observations of global dimming. The largest changes are observed for East Asia and Africa (Mao et al., 2014; He et al., 2018).

Small particles ($<1.0\,\mu m$)[i] are much more numerous in the atmosphere than large particles, but it is the large particles that contribute the most to the total airborne mass (Warneck, 2000; Raes et al., 2000). The mass of aerosols declines with increasing altitude, from values ranging between 1 and $50\,\mu g/m^3$ near unpolluted regions of the Earth's surface. Although there is an inverse relation between the size of particles and their persistence in the atmosphere, the overall mean residence time for tropospheric aerosols is about 5 days (Warneck, 2000). Thus, aerosols are not uniform in their distribution in the atmosphere. As a result of their longer mean residence time, small particles have the greatest influence on Earth's climate, and annually they carry the largest mass of material through the atmosphere.

The composition of tropospheric aerosols varies greatly depending on the proximity of continental, maritime, or anthropogenic sources (Heintzenberg, 1989; Murphy et al., 1998). Over land, aerosols are often dominated by soil minerals and human pollutants (Shaw, 1987; Gillette et al., 1992). Over the ocean, the composition of aerosols is a mixture of contributions from silicate minerals of continental origin and seasalt from the ocean (Andreae et al., 1986). Various workers have used ratios among the elemental constituents of aerosols to deduce the relative contribution of different sources (e.g., Moyers et al., 1977; Rahn and Lowenthal, 1984).

Aerosols are important in reactions with atmospheric gases and as nuclei for the condensation of raindrops. The latter are known as cloud condensation nuclei, often abbreviated CCN. Raindrops are formed when water vapor begins to condense on aerosols $>0.1\,\mu m$ in diameter. As raindrops enlarge and fall to the ground, they collide with other particles and absorb atmospheric gases. Soil dusts often contain a large portion of insoluble material (Reheis and Kihl, 1995), but seasalt aerosols and those derived from pollution sources are readily soluble and contribute to the dissolved chemical content of rainwater. Reactions of

---

[i] The U.S. EPA designates small aerosols, those <2.5 mm, as $PM_{2.5}$.

atmospheric gases with aerosols or raindrops are known as heterogeneous or multiphase reactions (Ravishankara, 1997). Such reactions are responsible for the ultimate removal of many reactive gases from the atmosphere.

Biogeochemical reactions in the troposphere

Major constituents—Nitrogen

It is perhaps not surprising that the major constituents of the atmosphere, N_2, O_2, and Ar, have nearly uniform concentrations and long mean residence times in the atmosphere. Argon is inert and has accumulated in the Earth's atmosphere since the earliest degassing of its crust (Chapter 2). From a biogeochemical perspective, N_2 is practically inert; reactive N is found only in molecules such as NH_3 and NO. Collectively the reactive nitrogen gases are sometimes called "odd" nitrogen, because the molecules have an odd number of N atoms (versus N_2 or N_2O).[j] Despite its abundance in the atmosphere, N_2 is so inert that the rate of formation of odd, or reactive, nitrogen is the primary factor that limits the growth of plants on land (LeBauer and Treseder, 2008). Among atmospheric gases, only argon and the other noble gases are less reactive.

Conversion of N_2 to reactive compounds, N fixation, occurs in lightning bolts, but the estimated global production of NO by lightning ($<9 \times 10^{12}$ g N/yr; Chapter 12) is too low to account for a significant turnover of N_2 in the atmosphere. By far the most important source of fixed nitrogen for the biosphere derives from the bacteria that convert N_2 to NH_3 in the process of biological nitrogen fixation (Eq. 29.). The global rate of biological N fixation is poorly known because it must be extrapolated from small-scale measurements to the entire surface of the Earth (Chapters 6 and 9). Including human activities, global nitrogen fixation (land + marine) is not likely to exceed 430×10^{12} g N/yr, with the production of synthetic nitrogen fertilizer now accounting for about one-third of the total (Battye et al., 2017, Chapter 12).

The natural rate of nitrogen fixation would remove the pool of N_2 from the atmosphere in about 9 million years.[k] Fortunately, denitrification (Eq. 2.20) returns N_2 to the atmosphere. At present, we have little evidence that the rate of either N fixation or denitrification changes significantly in response to changes in the concentration of N_2 in the atmosphere. The biosphere is responsible for the maintenance of N_2 in Earth's atmosphere over geologic time, but it plays a minor role in stabilizing the concentration of atmospheric N_2 over shorter periods, since the pool of N_2 in the atmosphere is so large (Walker, 1984).

---

[j]The term "odd nitrogen" is not ideal. In practice it refers only to the various oxidized forms of nitrogen in the atmosphere, including N_2O_5, but not to NH_3, which also has an odd number of N atoms.

[k]Mass of N in the atmosphere (Table 3.1) divided by the global rate of biological nitrogen fixation (430×10^{12} g N/yr) gives a mean residence time of ~9,000,000 years for N_2 in the atmosphere. Thus, $k = 1.1 \times 10^{-7}$ and $3/k = 27$ million years.

Oxygen

In Chapter 2 we discussed the accumulation of O_2 in the atmosphere during the evolution of life on Earth. The atmosphere now contains only a small portion of the total O_2 that has been released by photosynthesis through geologic time (Fig. 2.8). However, the atmosphere contains much more O_2 than can be explained by the storage of carbon in land plants today. The instantaneous combustion of all the organic matter now stored on land would reduce the atmospheric oxygen content by only 0.45% (Chapter 5). The accumulation of O_2 in the atmosphere is the result of the long-term burial of reduced carbon in ocean sediments (Berner, 1982), which contain nearly all of the reduced, organic carbon on Earth (Table 2.3). The rate of burial is determined by the area and depth of the ocean floor that is subject to anoxic conditions (Walker, 1977; Hartnett et al., 1998). Because the area and depth vary inversely with the concentration of atmospheric O_2, the balance between the burial of organic matter and its oxidation maintains O_2 at a steady-state concentration of about 21% (see also Chapters 9 and 11).

A large amount of O_2 has been consumed in weathering of reduced crustal minerals, especially Fe and S, through geologic time (Fig. 2.8); the current rate of exposure of these minerals would consume all atmospheric oxygen in about 70 million years (Lenton, 2001; see Fig. 11.8). However, the rate of exposure is not likely to vary greatly in response to changes in atmospheric O_2, so weathering is not the major factor controlling O_2 in the atmosphere (Bolton et al., 2006). In sum, despite the potential reactivity of O_2, its rate of reaction with reduced compounds is rather slow, and O_2 is a stable component of the atmosphere. Oxygen has declined only 0.7% over the past 800,000 years, presumably due to lower rates of organic burial and weathering (Stolper et al., 2016). The mean residence time of O_2 in the atmosphere is on the order of 4000 years, largely determined by exchange with the biosphere (compare Figs. 3.5 and 11.8). As such O_2 is well mixed and uniform in the atmosphere. Annual photosynthesis and respiration cause seasonal variations in O_2 concentration of about $\pm 0.002\%$ (Fig. 1.2).

Carbon dioxide

Carbon dioxide is not reactive with other gases in the atmosphere. The concentration of CO_2 is affected by interactions with the Earth's surface, including the reactions of the carbonate-silicate cycle (Fig. 1.3), gas exchange with surface seawaters following Henry's Law (Eq. 2.4), consumption and production of CO_2 by photosynthesis and respiration by heterotrophs (Figs. 1.2 and 3.6). For the Earth's land surface, our best estimates of plant uptake (120×10^{15} g C/yr; Chapter 5) suggest a mean residence time of about 6 years before a hypothetical molecule of CO_2 in the atmosphere is captured by photosynthesis by land plants. The annual exchange of CO_2 with seawater, dominated by areas of cold, downwelling water and high productivity (Chapter 9), is about $1.5 \times$ as large as the annual uptake of CO_2 by land plants. Both plant and ocean uptake are likely to increase with increasing concentrations of atmospheric CO_2, potentially buffering fluctuations in its concentration (Chapters 5, 9, and 11). Following Eq. 3.3, the mean residence time for CO_2, determined by the total flux from the atmosphere (the sum of land and ocean uptake), is about 3 years, so CO_2 shows small seasonal and latitudinal variation in the atmosphere (Figs. 1.1 and 3.6).

The carbonate-silicate cycle (Fig. 1.3) also buffers the concentration of CO_2 in the atmosphere, but does not affect the concentration of atmospheric CO_2 significantly in periods of less than ~100,000 years (Hilley and Porder, 2008; Colbourn et al., 2015). We will compare the relative importance of these processes in more detail in Chapter 11, which examines the global carbon cycle. The current increase in atmospheric CO_2 is a non-steady-state condition, caused by the combustion of fossil fuels and destruction of land vegetation. CO_2 is released by these processes faster than it can be taken up by land vegetation and the sea. If these activities were to cease, atmospheric CO_2 would return to a steady state, and after several hundred years nearly all of the CO_2 released by humans would reside in the oceans (Laurmann, 1979). In the meantime, higher concentrations of CO_2 are likely to cause significant atmospheric warming through the "greenhouse effect" (refer to Fig. 3.2).

Trace biogenic gases

Volcanoes are the original source of volatiles in the Earth's atmosphere (Chapter 2) and a small continuing source of some of the reduced gases (H_2S, H_2, NH_3, CH_4) that are found in the atmosphere today (Table 2.2). However, in most cases, the concentrations of these gases in today's atmosphere are dominated by supply from the biosphere, particularly by microbial activity (Monson and Holland, 2001). Methane is largely produced by anaerobic decomposition in wetlands (Chapters 7 and 11), nitrogen oxides by fossil fuel combustion and soil microbial transformations (Chapters 6 and 12), carbon monoxide by combustion of biomass and fossil fuels (Chapters 5 and 11), and volatile hydrocarbons, especially isoprene, by vegetation and human industrial activities (Chapter 5). The production of trace gases containing N and S contributes to the global cycling of these elements, which is controlled by the biosphere (Crutzen, 1983). These and other trace gases are found at concentrations well in excess of what is predicted from equilibrium geochemistry in an atmosphere with 21% O_2 (Table 3.5).

Unlike major atmospheric constituents, many of the trace biogenic gases in the atmosphere are highly reactive, showing short mean residence times and variable concentrations in space and time (refer to Fig. 3.5). Concentrations of these gases in the atmosphere are determined by the balance between local sources and chemical reactions—known as sinks—that remove these gases from the atmosphere. Sinks are largely driven by oxidation reactions and the capture of the reaction products by rainfall. Currently the concentration of nearly all these constituents is increasing as a result of human activities, suggesting that humans are affecting biogeochemistry at the global level (Prinn, 2003).

Oxidation reactions in the atmosphere

Despite its abundance, O_2 does not directly oxidize reduced gases in the atmosphere. Instead, a small proportion of the oxygen is converted to the powerful atmospheric oxidants ozone (O_3) and hydroxyl radicals (OH) through a series of reactions driven by sunlight (Logan, 1985; Thompson, 1992). Ozone and OH are the primary gases that oxidize many of the trace gases to CO_2, HNO_3, and H_2SO_4.

It is important to understand the natural production, occurrence, and reactions of ozone in the atmosphere. Nearly daily we read seemingly contradictory reports of the harmful effects

TABLE 3.5   Some trace biogenic gases in the atmosphere.

Compound	Formula	Concentration (ppb) Expected[a]	Concentration (ppb) Actual[b]	Mean residence time	Percentage of sink due to OH
Carbon compounds					
Methane	CH_4	10^{-148}	1830	9 years	90
Carbon monoxide	CO	10^{-51}	45–250	60 days	80
Isoprene	$CH_2{=}C(CH_3){-}CH{=}CH_2$		0.2–10.0	<1 day	100
Nitrogen compounds					
Nitrous oxide	N_2O	10^{-22}	320	120 years	0
Nitric oxides	NO_x	10^{-13}	0.02–10.0	1 day	100
Ammonia	NH_3	10^{-63}	0.08–5.0	5 days	<2
Sulfur compounds					
Dimethylsulfide	$(CH_3)_2S$		0.004–0.06	1 day	50
Hydrogen sulfide	H_2S		<0.04	4 days	100
Carbonyl sulfide	COS	0	0.50	5 years	20
Sulfur dioxide	SO_2	0	0.02–0.10	3 days	50

[a] *Approximate values in equilibrium with an atmosphere containing 21% O_2 (Chameides and Davis, 1982).*
[b] *For short-lived gases, the value is the range expected in remote, unpolluted atmospheres.*

of ozone depletion in the stratosphere and harmful effects of ozone pollution in the troposphere. In each case, human activities are upsetting the natural concentrations of ozone that are critical to atmospheric biogeochemistry.

Ozone is produced by the reaction of sunlight with O_2 in the stratosphere, as described in the next section. Some of this ozone is transported to the Earth's surface by the mixing of stratospheric and tropospheric air (e.g., Hocking et al., 2007), where it contributes to the budget of tropospheric ozone. (Table 3.6). However, observations of high ozone concentrations in the smog of polluted cities (e.g., Los Angeles) alerted atmospheric chemists to reactions by which ozone is also produced in the troposphere (Warneck, 2000).

When NO_2 is present in the atmosphere, it is dissociated by sunlight (hv),

$$NO_2 + hv \rightarrow NO + O, \tag{3.6}$$

followed by a reaction producing ozone:

$$O + O_2 \rightarrow O_3. \tag{3.7}$$

This reaction sequence is an example of a homogeneous gas reaction, that is, a reaction between atmospheric constituents that are all in the gaseous phase. The net reaction is:

$$NO_2 + O_2 \rightleftarrows NO + O_3, \tag{3.8}$$

TABLE 3.6 Tropospheric ozone budget. All values in Tg (10^{12} g) of O_3/yr.

Sources	
Chemical production	4960
Downward transport from stratosphere	325
Total	5290
Sinks	
Chemical loss	4360
Dry deposition	908
Wet deposition	19
Total	5290

From Hu et al. (2017).

which is an equilibrium reaction, so high concentrations of NO tend to drive the reaction backward. Sunlight is essential to form ozone by these pathways, so they are known as photochemical reactions. At night, ozone is consumed by reactions with NO_2 to form nitric acid (Brown et al., 2006b).

Both NO_2 and NO, collectively known as NO_x, are found in polluted air, in which they are derived from industrial and automobile emissions.[1] Small concentrations of both of these constituents are also found in the natural atmosphere, where they are derived from forest fires, lightning discharges, and microbial processes in the soil (Chapters 6 and 12). Thus, the production of ozone from NO_2 has probably always occurred in the troposphere, and the present-day concentrations of tropospheric ozone have simply increased as industrial emissions have raised the concentration of NO_2 and other precursors to O_3 formation (Lelieveld et al., 2004; Cooper et al., 2010; Schneider and van der A, 2012). NO_x concentrations have increased over much of China (Fig. 3.8). As air pollution regulations have reduced emissions of NO_x in Europe and the U.S., O_3 concentrations have declined (Kim et al., 2006; Butler et al., 2011).[m]

Ozone is subject to further photochemical reaction in the troposphere,

$$O_3 + hv \rightarrow O_2 + O(^1D), \tag{3.9}$$

[1] NOx (pronounced "knocks") refers to the sum of NO and NO_2. NO_y is used to refer to the sum of NO_x plus all other oxidized forms of nitrogen—for example, HNO_3 and $CH_3C(O)O_2NO_2$ (peroxyacetyl nitrate or PAN).

[m] Reductions in NO_x emissions from power plants during the 2003 North American electrical blackout resulted in widespread reductions in O_3 levels (Marufu et al., 2004). NOx emissions were also much lower during the economic shutdown in major cities following the coronavirus outbreak (Bauwens et al., 2020).

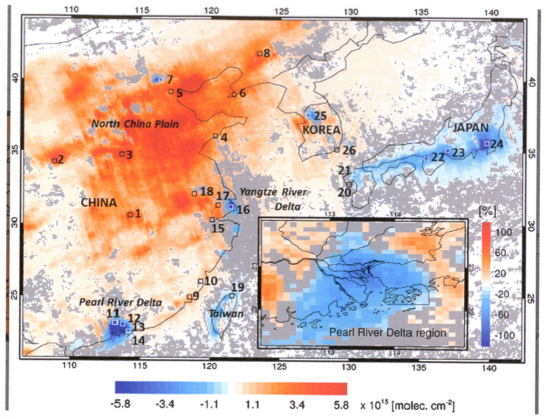

FIG. 3.8 Changes in NO$_x$ concentrations over East Asia from 2005 to 2014, as monitored by satellite. *From Duncan et al. (2016).*

where hv is ultraviolet light with wavelengths <318 nm and O(^{1}D) is an excited atom of oxygen. Reaction of O(^{1}D) with water yields hydroxyl radicals:

$$O\left(^1D\right) + H_2O \rightarrow 2OH. \tag{3.10}$$

The formation of hydroxyl radicals is strongly correlated with the amount of ultraviolet radiation (Rohrer and Berresheim, 2006). Hydroxyl radicals may further react to produce HO$_2$ and H$_2$O$_2$,

$$2OH + 2O_3 \rightarrow 2HO_2 + 2O_2 \tag{3.11}$$

$$2HO_2 \rightarrow H_2O_2 + O_2, \tag{3.12}$$

which are other short-lived oxidizing compounds in the atmosphere (Thompson, 1992; Crutzen et al., 1999).

Hydroxyl radicals exist with a mean concentration of about 1×10^6 molecules/cm^3 (Prinn et al., 1995; Wolfe et al., 2019). The highest concentrations occur in daylight (Platt et al., 1988;

Mount, 1992) and at tropical latitudes, where the concentration of water vapor is greatest (Hewitt and Harrison, 1985). The average OH radical persists for only a few seconds in the atmosphere, so concentrations of OH are highly variable. Local concentrations can be measured using beams of laser-derived light, which is absorbed as a function of the number of OH radicals in its path (Dorn et al., 1988; Mount et al., 1997).

Because of its short mean residence time, the global mean concentration of OH radicals must be estimated indirectly. For this purpose, atmospheric chemists have relied on methylchloroform (trichloroethane), a gas that is known to result only from human activity. Methylchloroform has a mean residence time of about 4.8 years (Prinn et al., 1995), so it is reasonably well mixed in the atmosphere. In the laboratory, it reacts with OH,

$$OH + CH_3CCl_3 \rightarrow H_2O + CH_2CCl_3, \tag{3.13}$$

and the rate constant, K, for the reaction is 0.85×10^{-14} cm^3 molecule^{-1} s^{-1} at 25 °C (Talukdar et al., 1992). Then, knowing the industrial production of CH_3CCl_3, its accumulation in the atmosphere and K, one can calculate the concentration of OH that must be present, namely,

$$OH = (Production - Accumulation)/K. \tag{3.14}$$

Hydroxyl radicals are the major source of oxidizing power in the troposphere. For example, in an unpolluted atmosphere, hydroxyl radicals destroy methane in a series of reactions,

$$CH_4 + OH \rightarrow CH_3 + H_2O \tag{3.15}$$

$$CH_3 + O_2 \rightarrow CH_3O_2 \tag{3.16}$$

$$CH_3O_2 \, HO_2 \rightarrow CH_3O_2H + O_2 \tag{3.17}$$

$$CH_3O_2H \rightarrow CH_3O + OH \tag{3.18}$$

$$CH_3O + O_2 \rightarrow CH_2O + HO_2, \tag{3.19}$$

for which the net reaction is:

$$CH_4 + O_2 \rightarrow CH_2O + H_2O \tag{3.20}$$

Note that the hydroxyl radical has acted as a catalyst to initiate the oxidation of CH_4 and its byproducts by O_2. Other volatile organic compounds are also oxidized through this pathway, which yields formaldehyde (CH_2O; Atkinson, 2000; Atkinson and Arey, 2003). Changes in the concentration of formaldehyde can be used as an index of the burden of organic gases in the atmosphere from natural sources and air pollution (Zhu et al., 2017).

The formaldehyde that is produced in these reactions is further oxidized to carbon monoxide,

$$CH_2O + OH + O_2 \rightarrow CO + H_2O + HO_2, \tag{3.21}$$

and CO is oxidized by OH to produce CO_2,

$$CO + OH \rightarrow H + CO_2 \tag{3.22}$$

$$H + O_2 \rightarrow HO_2 \tag{3.23}$$

$$HO_2 + O_3 \rightarrow OH + 2O_2. \tag{3.24}$$

for which the net reaction is

$$CO + O_3 \rightarrow CO_2 + O_2. \tag{3.25}$$

Thus, OH acts to scrub the atmosphere of a wide variety of reduced carbon gases, ultimately oxidizing their carbon atoms to carbon dioxide.

Hydroxyl radicals can also react with NO_2 and SO_2 in homogeneous gas reactions:

$$NO_2 + OH \rightarrow HNO_3 \tag{3.26}$$

$$SO_2 + OH \rightarrow SO_3 + HO_2 \tag{3.27}$$

and the latter reaction is followed by a heterogeneous reaction with raindrops:

$$SO_3 + H_2O \rightarrow H_2SO_4 \tag{3.28}$$

which removes sulfur dioxide from the atmosphere, causing acid rain. Sulfur dioxide is also oxidized by hydrogen peroxide, according to Chandler et al. (1988):

$$SO_2 + H_2O_2 \rightarrow H_2SO_4 \tag{3.29}$$

The reaction of OH with NO_2 is very fast, and it produces nitric acid that is removed from the atmosphere by a heterogeneous interaction with raindrops (Munger et al., 1998). The reactions with SO_2 are much slower, accounting for the long-distance transport of SO_2 as a pollutant in the atmosphere (Rodhe, 1981). Hydrogen sulfide (H_2S) and dimethylsulfide ($(CH_3)_2S$), released from anaerobic soils (Chapter 7) and the ocean surface (Chapter 9), are also removed by reactions with OH and other oxidizing compounds, leading to the deposition of H_2SO_4 (Toon et al., 1987). Thus, OH radicals cleanse the atmosphere of trace N and S gases by converting them to acid anions NO_3^- and SO_4^{2-} in the atmosphere.

The vast majority of OH radicals in the atmosphere are consumed in reactions with CO and CH_4. Although the concentration of methane is much higher than that of carbon monoxide in unpolluted atmospheres, the reaction of OH with CO is much faster. The speed of reaction of CO with OH accounts for the short mean residence time of CO in the atmosphere (Table 3.5). The mean residence time for methane is much longer, accounting for its more uniform distribution in the atmosphere (Fig. 3.5). One explanation for the current increase in methane in the atmosphere is that the anthropogenic release of CO consumes OH radicals previously available for the oxidation of methane (Khalil and Rasmussen, 1985; Rigby et al., 2017). Consistent with the relative distribution of sinks for OH, the concentration of OH in the atmosphere is slightly lower in the Northern hemisphere (Wolfe et al., 2019), but various global estimates suggest that OH concentrations have declined only slightly (or perhaps not at all) in recent years (Prinn et al., 1995, 2005; Montzka et al., 2011b). An increasing deposition of formaldehyde in the Greenland snowpack indicates the oxidation of methane has kept up with its increasing concentration in the atmosphere (Eq. 3.20; Staffelbach et al., 1991).

In unpolluted atmospheres, all these reactions consume OH. In "dirty" atmospheres, a different set of reactions pertains, in which there can be a net production of O_3, and thus OH, during the oxidation of reduced gases (Jenkin and Clemitshaw, 2000; Sillman, 1999). When the concentration of NO is >10 ppt, which we will define as a "dirty" atmosphere (Jacob and Wofsy, 1990), the oxidation of carbon monoxide begins by reaction with hydroxyl radical and proceeds as follows (Crutzen and Zimmermann, 1991):

$$CO + OH \rightarrow CO_2 + H \tag{3.30}$$

$$H + O_2 \rightarrow HO_2 \tag{3.31}$$

$$HO_2 + NO \rightarrow OH + NO_2 \tag{3.32}$$

$$NO_2 + h\upsilon \rightarrow NO + O \tag{3.33}$$

$$O + O_2 \rightarrow O_3. \tag{3.34}$$

The net reaction is:

$$CO + 2O_2 \rightarrow CO_2 + O_3. \tag{3.35}$$

Similarly, the oxidation of methane in the presence of high concentrations of NO proceeds through a large number of steps, yielding a net reaction of:

$$CH_4 + 4O_2 \rightarrow CH_2O + H_2O + 2O_3. \tag{3.36}$$

In both cases, NO acts as a catalyst leading to the oxidation of reduced gases by oxygen.

Fig. 3.9 shows the contrasting pathways of carbon monoxide oxidation in clean and dirty atmospheres. Crutzen (1988) points out that the oxidation of one molecule of CH_4 could consume up to 3.5 molecules of OH and 1.7 molecules of O_3 when the NO concentration is low, whereas it would yield a net gain of 0.5 OH and 3.7 O_3 in polluted environments (see also Wuebbles and Tamaresis, 1993). Although they were first discovered in urban areas, the reactions of dirty atmospheres are likely to be relatively widespread in nature. NO is produced naturally by soil microbes (Chapter 6) and forest fires, and concentrations of NO >10 ppt are present over most of the Earth's land surface (Chameides et al., 1992; Levy et al., 1999). In the presence of NO, oxidation of volatile hydrocarbons emitted from vegetation, and CO emitted from both vegetation and forest fires, can account for unexpectedly high concentrations of O_3 over rural areas of the southeastern United States (Fig. 3.10)[n] (Jacob et al., 1993; Kleinman et al., 1994; Kang et al., 2003) and in remote tropical regions (Crutzen et al., 1985; Zimmerman and Greenberg, 1988; Jacob and Wofsy, 1990; Andreae et al., 1994). In urban areas, where the concentration of NO_x is especially high due to industrial pollution, effective control of atmospheric O_3 levels may also depend on the regulation of volatile hydrocarbons (Chameides et al., 1988; Seinfeld, 1989). In rural areas, ozone formation is usually limited by the concentration of NO_x, especially during the growing season, when vegetation actively emits volatile organic compounds (Fig. 3.11; Aneja et al., 1996).

Understanding changes in the concentration of O_3 and OH in the atmosphere is critical to predicting future trends in the concentration of trace gases, such as CH_4, that can contribute to greenhouse warming. Some models predict an increase in O_3 (Isaksen and Hov, 1987; Hough and Derwent, 1990; Thompson, 1992; Prinn, 2003) in the atmosphere as a result of increasing human emissions of NO, creating dirty atmosphere conditions over much of the planet. These predictions are consistent with observations that the ozone concentrations in Europe have risen since the late 1800s (Volz and Kley, 1988; Marenco et al., 1994), though perhaps only

---

[n]The customary unit for the total number of ozone molecules in an atmospheric column, the Dobson Unit, is equivalent to 2.69×10^{16} molecules/cm^2 of the Earth's surface.

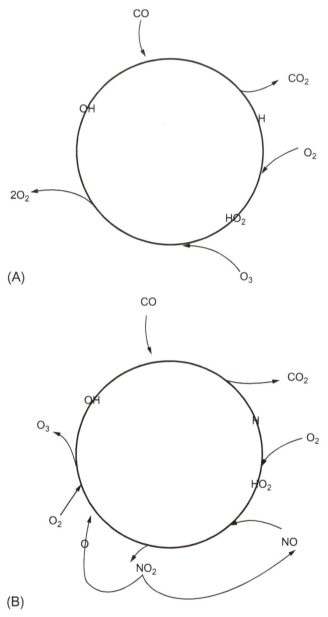

FIG. 3.9 Reaction chain for the oxidation of CO in clean (A) and dirty (B) atmospheric conditions.

40% or less globally (Yeung et al., 2019). Concentrations of H_2O_2, derived from OH (Eqs. 3.11 and 3.12), have increased in layers of Greenland ice deposited during the last 200 years, suggesting a greater oxidizing capacity in the northern hemisphere as a result of human activities (Fig. 3.12). The models are also consistent with indirect observations that the global

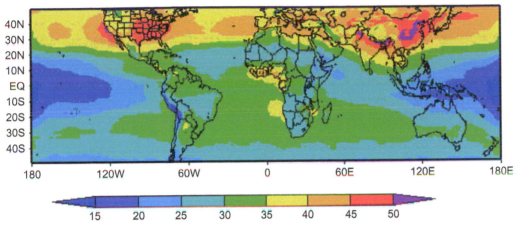

FIG. 3.10   Distribution of ozone in Earth's troposphere, for summer months, averaged over 1979–1991. Data are in Dobson Units, see footnote n, p. 75. Note high ozone concentrations over the eastern United States and China. *From Fishman et al. (2003). Used with permission of European Geosciences Union.*

concentration of OH has remained fairly stable in recent years, despite increasing emissions of reduced gases that should scrub OH from the atmosphere (Prinn et al., 1995, 2005; Montzka et al., 2011b). Oxidizing radicals in the atmosphere may increase in future warmer climates (Geng et al., 2017).

Several recent papers suggest additional pathways leading to the formation of OH (Li et al., 2008; Hofzumahaus et al., 2009), so that its production may not be restricted to the photochemical reactions outlined in Eqs. 3.8–3.10. Soil microbes that produce nitrite (NO_2^-) are a potential source of nitrous acid (HONO) in the atmosphere and OH radicals (Su et al., 2011).

Ozone has a mean lifetime of about 24 days in the troposphere (Hu et al., 2017). Some of the O_3 produced over the continents undergoes long-distance transport (Jacob et al., 1993; Parrish et al., 1993; Cooper et al., 2010; Brown-Steiner and Hess, 2011), resulting in the appearance of O_3 and its byproducts at considerable distances from their source (Fig. 3.10). In some rural areas, concentrations of O_3 from local production and transport from nearby cities inhibit the growth of agricultural crops and trees (Chameides et al., 1994). Other workers disagree, finding that local atmospheric conditions, rather than global changes in transport from polluted areas, determine the oxidizing capacity of the atmosphere over much of the planet (Oltmans and Levy, 1992; Ayers et al., 1992; Kang et al., 2003).

Atmospheric deposition

Elements of biogeochemical interest are deposited on the Earth's surface as a result of rainfall, dry deposition, and the direct absorption of gases from the atmosphere. The importance of each of these processes differs for different regions and for different elements (Gorham, 1961).

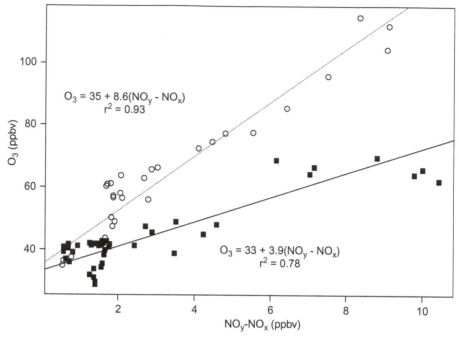

FIG. 3.11 Ambient O_3 vs. NO_y-NO_x concentrations in the atmosphere at Harvard Forest, northern Massachusetts (USA), 6–12 May 1990 (■) and 24–30 August 1992 (O). *From Hirsch et al. (1996). Used with permission of the American Geophysical Union.*

In many forests, a large fraction of the annual uptake and circulation of nutrient elements in vegetation may be derived from the atmosphere (Kennedy et al., 2002a; Avila et al., 1998; Miller et al., 1993). The atmosphere accounts for nearly all of the nitrogen and sulfur that circulates in terrestrial ecosystems, with rock weathering providing only smaller amounts (see Table 4.5 and Chapter 6). The chemical composition of rainfall has received great attention, as a result of widespread concern about dissolved constituents that lead to "acid rain."

Processes

The dissolved constituents in rainfall are often separated into two fractions. The rainout component consists of constituents derived from cloud processes, such as the nucleation of raindrops. The washout component is derived from below cloud level, by scavenging of aerosol particles and the dissolution of gases in raindrops as they fall (Brimblecombe and Dawson, 1984; Shimshock and de Pena, 1989). In one study, rainout contributed about 1/3 of the deposition of NO_3^- but 50–80% of the deposition of SO_4^{2-}, which typically forms cloud condensation nuclei (Aikawa et al., 2014). The dissolved content in both fractions represents the results of heterogeneous reactions between gases and raindrops in the atmosphere.

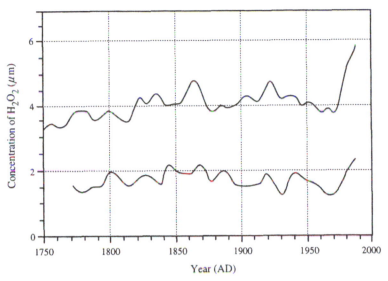

FIG. 3.12   Variation in the mean annual H_2O_2 concentration during the past 200 years as seen in two cores from the Greenland ice pack. *Modified with permission of Macmillan from Sigg and Neftel (1991).*

The relative contribution of rainout and washout varies depending on the length of the rainstorm. As washout cleanses the lower atmosphere, the content of dissolved materials in rainfall declines. Thus, the concentration of dissolved constituents in precipitation is inversely related to the rate of precipitation (Gatz and Dingle, 1971) and to the total volume that has fallen (Likens et al., 1984; Lesack and Melack, 1991; Minoura and Iwasaka, 1996). The concentration of dissolved constituents also varies inversely as a function of mean raindrop size (Georgii and Wötzel, 1970; Bator and Collett, 1997). This inverse relation explains why extremely high concentrations of dissolved constituents are found in fog waters (Weathers et al., 1986; Waldman et al., 1982; Clark et al., 1998; Elbert et al., 2000). Capture of fog and cloud water by vegetation is an important component of the deposition of nutrient elements from the atmosphere in some high-elevation and coastal ecosystems (Lovett et al., 1982; Waldman et al., 1985; Weathers et al., 2000; Templer et al., 2015).

The relative efficiency of scavenging by rainwater is often expressed as the washout ratio:

$$\text{Washout} = \frac{\text{Ionic concentration in rain (mg/L)}}{\text{Ionic concentration in air (mg/m}^3)}. \tag{3.37}$$

With units of m^3/L, this ratio gives an indication of the volume of atmosphere cleansed by each liter of rainfall as it falls. Large ratios are generally found for ions that are derived from relatively large aerosols or from highly water-soluble gases in the atmosphere. Snowfall is generally less efficient at scavenging than rainfall.

Whereas the deposition of nutrients by precipitation is often called wetfall, dryfall is the result of gravitational sedimentation of particles during periods without rain (Hidy, 1970; Wesely and Hicks, 2000). Dryfall of dusts downwind of arid lands is often spectacular; Liu et al. (1981)

reported $100/g/m^2$ /h of dustfall in Beijing, China, as a result of a single dust storm on 18 April 1980. In some regions, enormous deposits of wind-deposited soil, known as loess, were laid down during glacial periods, when large areas of semiarid land were subject to wind erosion (Pye, 1987; Simonson, 1995; Muhs et al., 2001). Today, various elements necessary for plant growth are released by chemical weathering of soil minerals in these deposits (Chapter 4).

The dryfall received in many areas contains a significant fraction that is easily dissolved by soil waters and immediately available for plant uptake. Despite the high rainfall found in the southeastern United States, Swank and Henderson (1976) reported that 19–64% of the total annual atmospheric deposition of ions such as Ca, Na, K, and Mg, and up to 89% of the deposition of P, was derived from dryfall. Dryfall inputs of P may assume special significance to plant growth in areas where the release of P from rock weathering is very small (Newman, 1995; Chadwick et al., 1999; Okin et al., 2004). Dry deposition contributes about 30–60% of the deposition of sulfur in New Hampshire (Likens et al., 1990; cf. Tanaka and Turekian, 1995). Similarly, 34% of the atmospheric inputs of nitrogen to Harvard Forest (Massachusetts) are derived from dry deposition (Munger et al., 1998). Organic nitrogen compounds deposited from the atmosphere are decomposed by soil microbes, providing additional plant nutrients (Neff et al., 2002a; Mace et al., 2003; Zhang et al., 2012c).

Dryfall is often measured in collectors that are designed to close during rainstorms. When open to the atmosphere, these instruments capture particles that are deposited vertically, known as sedimentation. In natural ecosystems, dryfall is also derived by the capture of particles on vegetation surfaces. When vegetation captures particles that are moving horizontally in the airstream, the process is known as impaction (Hidy, 1970). Impaction is a particularly important process in the capture of seasalt aerosols near the ocean (Art et al., 1974; Potts, 1978).

In addition to the uptake of CO_2 in photosynthesis, vegetation also absorbs N- and S-containing gases directly from the atmosphere (Hosker and Lindberg, 1982; Lindberg et al., 1986; Sparks et al., 2003; Turnipseed et al., 2006). Uptake of pollutant O_3, SO_2, and NO_2 by vegetation is particularly important in humid regions (McLaughlin and Taylor, 1981; Rondon and Granat, 1994), where plant stomata remain open for long periods. Lovett and Lindberg (1986, 1993) found that uptake of HNO_3 vapor accounted for 75% of the annual dry deposition of nitrogen (4.8 kg/ha) in a deciduous forest in Tennessee, where dry deposition was nearly half of the total annual deposition of nitrogen from the atmosphere. Globally, dry deposition of NO_2 and SO_2 comprises a significant fraction of the total deposition of N and S on land (Nowlan et al., 2014; Jaeglé et al., 2018). Vegetation can also be a source or a sink for atmospheric NH_3, depending on the ambient concentration in the atmosphere (Langford and Fehsenfeld, 1992; Sutton et al., 1993; Pryor et al., 2001), and plants can also remove volatile organic compounds from the air (Simonich and Hites, 1994).

The total capture of dry particles and gases by land plants is difficult to measure. When rainfall is collected inside a forest, it contains materials that have been deposited on the plant surfaces, but also large quantities of elements that are derived from the plants themselves (Parker, 1983, Chapter 6). Artificial collectors (surrogate surfaces) are often used to approximate the capture by vegetation (White and Turner, 1970; Vandenberg and Knoerr, 1985; Lindberg and Lovett, 1985). The capture on known surfaces can be compared to the airborne concentrations to calculate a deposition velocity (Sehmel, 1980):

$$\text{Deposition velocity} = \frac{\text{Rate of dryfall } (mg/cm^2/s)}{\text{Concentration in air } (mg/cm^3)}. \tag{3.38}$$

In units of cm/s, these velocities can be multiplied by the estimated surface area of vegetation (cm^2) and the concentration in the air to calculate total deposition for an ecosystem. For example, Lovett and Lindberg et al. (1986) used a deposition velocity of 2.0 cm/s to calculate a nitrogen deposition of 3.0 kg/ha/yr in a forest with a leaf area index[o] of 5.8 m^2/m^2 and an ambient concentration of 0.82 mg N m^3 in the form of nitric acid vapor. It is often unclear if deposition velocities measured using artificial surfaces apply to natural surfaces (e.g., bark), and accurate estimates of the surface area of vegetation are difficult (Whittaker and Woodwell, 1968). Clearly, further work on dry deposition is needed (Lovett, 1994; Petroff et al., 2008).

Atmospheric deposition on the surface of the sea is often estimated from collections of wetfall and dryfall on remote islands (Duce et al., 1991). The surface of the sea can also exchange gases with the atmosphere (Liss and Slater, 1974), often acting as a sink for atmospheric CO_2 (Sabine et al., 2004) and SO_2 (Beilke and Lamb, 1974) and as a source of NH_3 (Paulot et al., 2016) and dimethylsulfide (Chapter 9).

Regional patterns and trends

Regional patterns of rainfall chemistry in the United States reflect the relative importance of different constituent sources and deposition processes in different areas (Munger and Eisenreich, 1983). Coastal areas are dominated by atmospheric inputs from the sea, with large inputs of Na, Mg, Cl, and SO_4 that are the major constituents in the seasalt aerosol (Junge and Werby, 1958; Hedin et al., 1995). Areas of arid and semiarid land show high concentrations of soil-derived constituents, such as Ca, in rainfall (Fig. 3.13; Young et al., 1988; Sequeira, 1993; Gillette et al., 1992). Areas downwind of regional pollution show exceedingly low pH and high concentrations of SO_4^{2-} and NO_3^- (Schwartz, 1989; Ollinger et al., 1993), whereas agricultural areas have high deposition of NH_4^+ (Stephen and Aneja, 2008).

The ratio among ionic constituents in rainfall can be used to trace their origin. Except in unusual circumstances, nearly all the sodium (Na) in rainfall is derived from the ocean. When magnesium is found in a ratio of 0.12 with respect to Na—the ratio in seawater (refer to Table 9.1)—we may presume that most of the Mg is also of marine origin. In the southeastern United States, however, Mg/Na ratios in wetfall range from 0.29 to 0.76 (Swank and Henderson, 1976). Here the Mg content has increased relative to Na, presumably because the airflow that brings precipitation to this region has crossed the United States, picking up Mg from soil dust and other sources. Schlesinger et al. (1982) used this approach to deduce nonmarine sources of Ca and SO_4 in the rainfall in coastal California (Fig. 3.14).

Iron (Fe) and aluminum (Al) are largely derived from the soil, and ratios of various ions to these elements in soil can be used to predict their expected concentrations in rainfall when soil dust is a major source (Lawson and Winchester, 1979; Warneck, 2000). High concentrations of Al in dryfall on Hawaii were traced to springtime dust storms on the central plains of China

---

[o]Leaf area index (LAI) is a measure of the total area of leaves (m^2) over a certain area of the ground, usually 1 m^2.

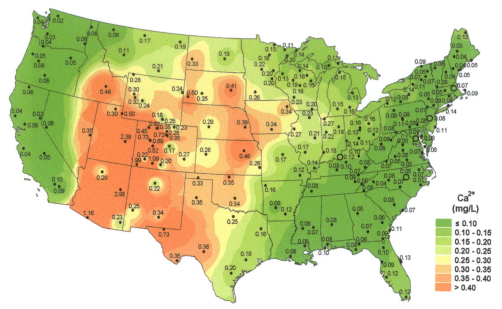

FIG. 3.13 Mean calcium concentration (mg/L) in wetfall precipitation in the United States for 2017. *From the National Atmospheric Deposition Program http://nadp.slh.wisc.edu/maplib/pdf/2017/Ca_conc_2017.pdf.*

(Parrington et al., 1983). Soil mineralogy in dusts from the 1930s' Great Plains dust bowl can be identified in layers of the Greenland ice pack (Donarummo et al., 2003). Windborne particles of soil and vegetation contribute significantly to the global transport of trace metals in the atmosphere (Nriagu, 1989, Table 1.1).

In many areas downwind of pollution, a strong correlation between H^+ and SO_4^{2-} is the result of the production of H_2SO_4 during the oxidation of SO_2 and its dissolution in rainfall (Eqs. 3.27 and 3.28; Cogbill and Likens, 1974; Irwin and Williams, 1988). Nitrate (NO_3^-) also contributes to the strong acid content in rainfall (HNO_3). These constituents depress the pH of rainfall below 5.6, which would be expected for water in equilibrium with atmospheric CO_2 (Galloway et al., 1976). In contrast, ammonia (NH_3) is a net source of alkalinity in rainwater, since its dissolution produces OH:

$$NH_3 + H_2O \rightarrow NH_4^+ + OH^-. \tag{3.39}$$

The pH of rainfall is determined by the concentration of strong acid anions that are not balanced by NH_4^+ and Ca^{2+} (from $CaCO_3$), namely (from Gorham et al., 1984),

$$H^+ = \left[NO_3^- + 2SO_4^{2-}\right] - \left[NH_4^+ + 2Ca^{2+}\right]. \tag{3.40}$$

In Kanpur, India, ammonia dominated the neutralization of acidity in rainfall during the wet season, while Ca played a similar role in the dry season, when more soil dust was present in the atmosphere (Shukla and Sharma, 2010). Low pH of rainfall in Europe is partially alleviated by Ca from the Sahara desert and NH_3 emissions from agriculture (Lajtha and Jones, 2013).

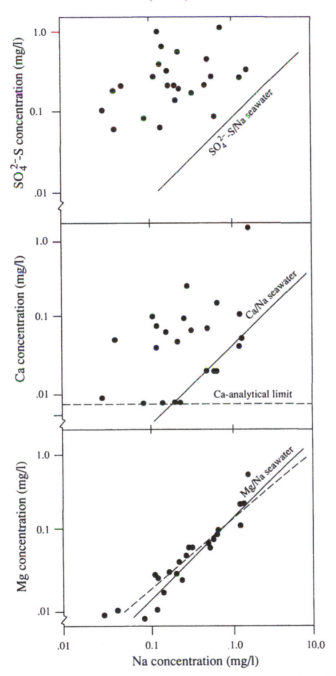

FIG. 3.14   Concentrations of SO_4, Ca, and Mg in wetfall precipitation near Santa Barbara, California (USA), plotted as a logarithmic function of Na concentration in the same samples (Schlesinger et al., 1982). The solid line represents the ratio of these ions to Na in seawater. Ca and SO_4 are enriched in wetfall relative to seawater, whereas Mg shows a correlation (dashed line) that is not significantly different from the ratio expected in seawater.

Globally, about 37% of the atmosphere's acidity is neutralized by NH_3 (Chapter 13), with a higher proportion in the Southern Hemisphere where there is less industrial pollution (Savoie et al., 1993). Across the United States, airborne concentrations of NH_3 and the proportional deposition of NH_4^+ have increased markedly in recent years, as air pollution regulations have lowered the emissions of NO_x (Kharol et al., 2018; Li et al., 2016a). Reduced forms of N (i.e., NH_4) are also important in China (Zhan et al., 2015). Monitoring of ammonia in the atmosphere has improved markedly with its monitoring on tall towers (Griffis et al., 2019) and the application of satellite remote sensing (Van Damme et al., 2015, 2018; Warner et al., 2017; Kharol et al., 2018).

In the eastern United States, the acidity of rainfall is often directly correlated to the concentration of SO_4^{2-}, which is related to pollutant emissions of SO_2 in areas upwind (Likens et al., 2005). Similar relationships are seen between emissions of NO_x and the NO_3^- content of rain (Butler et al., 2003, 2005). In the western United States, the relationship between acidity and the acid-forming anions is less clear because they have often reacted with soil aerosols containing $CaCO_3$ (Young et al., 1988).

In the past 100 years, increases in the concentration of NO_3 and SO_4 in the Greenland and Tibetan snowpacks have reflected the changes in the abundance of anthropogenic pollutants due to industrialization in the Northern Hemisphere (Mayewski et al., 1986, 1990; Thompson et al., 2000). There are no apparent changes in the deposition of these ions in the Southern Hemisphere as recorded by Antarctic ice (Langway et al., 1994). Similarly, the uppermost sediments in lakes and peat bogs of the Northern Hemisphere contain higher concentrations of many trace metals, presumably from industrial sources (Galloway and Likens, 1979; Swain et al., 1992; Allan et al., 2013).

Long-term records of precipitation chemistry are rare, but the collections at the Hubbard Brook Ecosystem in central New Hampshire and eastern Tennessee suggest a recent decline in the concentrations of SO_4 that may reflect improved control of emissions (Likens et al., 1984, 2002; cf. Kelly et al., 2002; Zbieranowski and Aherne, 2011; Lutz et al., 2012b). Improvements in air quality as a result of the implementation of the Clean Air Act in 1990 have resulted in significant decreases in the acidity of rainfall over the eastern United States and Canada (Figs. 3.15 and 13.2; Hedin et al., 1987; Lajtha and Jones, 2013). Similarly, the uppermost layers of ice in glaciers of the French Alps and Greenland show lower concentrations of SO_4^{2-} and NO_3^-, presumably due to control of pollutant emissions in upwind sources in recent years (Preunkert et al., 2001, 2003; Fischer et al., 1998).

Even with pollutant abatement, long-term records suggest that many natural ecosystems currently receive a greater input of N, S, and other elements of biogeochemical importance than before widespread emissions from human activities. Pollutant emissions have more than doubled the annual input of S-containing gases to the atmosphere globally (Chapter 13). Global deposition of inorganic nitrogen has increased about 8% between 1984 and 2016 (Ackerman et al., 2019), with most of the increase seen for NH_4^+ (Li et al., 2016a; Kharol et al., 2018). Excess deposition of nitrogen might be expected to enhance the growth of forests, but in combination with acidity, this fertilization effect may lead to deficiencies of P, Mg, and other plant nutrients (Chapters 4 and 6). Atmospheric deposition of nitrogen makes a significant contribution to the nutrient load and eutrophication of lakes (Bergstrom and Jansson, 2006), estuaries (Nixon et al., 1996; Latimer and Charpentier, 2010), and coastal waters (Paerl et al., 1999). The western North Atlantic Ocean receives about 20–40% of the sulfur

Sulfate ion concentration, 1985

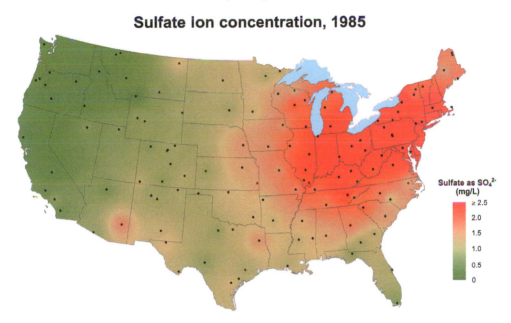

National Atmospheric Deposition Program/National Trends Network
http://nadp.isws.illinois.edu

(A)

Sulfate ion concentration, 2017

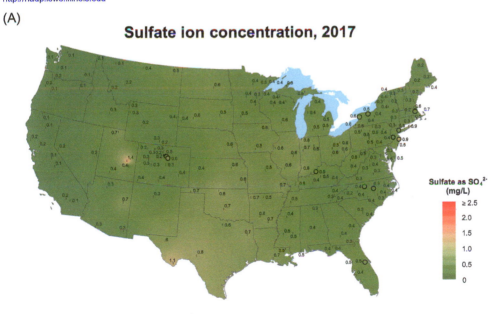

National Atmospheric Deposition Program/National Trends Network
http://nadp.slh.wisc.edu

(B)

FIG. 3.15 Sulfate (SO_4) concentration (mg/L) measured in samples of wetfall precipitation across the United States, showing the effect of the Clean Air Act in reducing SO_2 emissions and thus, SO_4 deposition between 1985 and 2017. *From the National Atmospheric Deposition Program.*

and nitrogen oxides emitted in eastern North America (Galloway and Whelpdale, 1987; Liang et al., 1998; Jaeglé et al., 2018), and increasing nitrogen deposition is recorded in the North Pacific Ocean (Kim et al., 2014). Although pollutant emissions have declined in North America and Europe, increasing emissions are seen from India and China (Lelieveld et al., 2001; Richter et al., 2005; Stern, 2006). The airborne concentrations of many air pollutants can now be measured using satellite technology (Richter et al., 2005; Clarisse et al., 2009; Martin, 2008; Yang et al., 2013a, 2014; Griffin et al., 2019).

Biogeochemical reactions in the stratosphere

Ozone

Ozone is produced in the stratosphere by the dissociation of oxygen atoms that are exposed to shortwave solar radiation. The reaction accounts for most of the absorption of ultraviolet sunlight (hv) at wavelengths of 180–240 nm and proceeds as follows:

$$O_2 + hv \rightarrow O + O \tag{3.41}$$

$$O + O_2 \rightarrow O_3. \tag{3.42}$$

Some ozone from the stratosphere mixes down into the troposphere, where the production of O_3 by these reactions is limited because there is less ultraviolet light. Most of the remaining ozone is destroyed by a variety of reactions in the stratosphere. Absorption of ultraviolet light at wavelengths between 200 and 320 nm destroys ozone:

$$O_3 + hv \rightarrow O_2 + O \tag{3.43}$$

$$O + O_3 \rightarrow O_2 + O_2. \tag{3.44}$$

This absorption warms the stratosphere (refer to Fig. 3.1) and protects the Earth's surface from the ultraviolet portion of the solar spectrum that is most damaging to living tissue (uvB). Stratospheric ozone is also destroyed by reaction with OH (Wennberg et al., 1994),

$$O_3 + OH \rightarrow HO_2 + O_2 \tag{3.45}$$

$$HO_2 + O_3 \rightarrow OH + 2O_2. \tag{3.46}$$

and by reactions stemming from the presence of nitrous oxide (N_2O), which mixes up from the troposphere. Tropospheric N_2O is produced in a variety of wkays (Chapters 6 and 12), but it is inert in the lower atmosphere. The only significant sink for N_2O is photolysis in the stratosphere. About 80% of the N_2O reaching the stratosphere is destroyed in a reaction producing N_2 (Warneck, 2000),

$$N_2O \rightarrow N_2 + O(^1D), \tag{3.47}$$

and about 20% in reactions with the $O(^1D)$ produced in Eq. 3.47, mainly

$$N_2O + O(^1D) \rightarrow N_2 + O_2, \tag{3.48}$$

but with a small amount forming NO:

$$N_2O + O(^1D) \rightarrow 2NO. \tag{3.49}$$

The nitric oxide (NO) produced in reaction 3.49 destroys ozone in a series of reactions,

$$NO + O_3 \rightarrow NO_2 + O_2 \tag{3.50}$$

$$O_3 \rightarrow O + O_2 \tag{3.51}$$

$$NO_2 + O \rightarrow NO + O_2. \tag{3.52}$$

for which the net reaction is:

$$2O_3 \rightarrow 2O_2. \tag{3.53}$$

Note that the mean residence time of NO in the troposphere is too short for an appreciable amount to reach the stratosphere, where it might contribute to the destruction of ozone. Nearly all the NO in the stratosphere is produced in the stratosphere from N_2O; only a small amount is contributed by high-altitude aircraft.

Eventually NO_2 is removed from the stratosphere by reacting with OH to produce nitric acid (Eq. 3.26), which mixes down to the troposphere and is removed by the heterogeneous reaction with raindrops.[P]

Finally, stratospheric ozone is destroyed by chlorine, which acts as a catalyst in the reaction:

$$Cl + O_3 \rightarrow ClO + O_2 \tag{3.54}$$

$$O_3 + hv \rightarrow O + O_2 \tag{3.55}$$

$$ClO + O \rightarrow Cl + O_2, \tag{3.56}$$

for a net reaction of:

$$2O_3 + hv \rightarrow 3O_2. \tag{3.57}$$

Although each Cl produced may cycle through these reactions and destroy many molecules of O_3, Cl is eventually converted to HCl and removed from the stratosphere by downward mixing and heterogeneous interaction with cloud drops in the troposphere (Rowland, 1989; Solomon, 1990).

The balance between ozone production (refer to Eqs. 3.41 and 3.42) and the various reactions that destroy ozone maintains a steady-state concentration of stratospheric O_3 with a peak of approximately 7×10^{18} molecules/m^3 at 30 km altitude (Warneck, 2000). Although the photochemical production of O_3 is greatest at the equator, the density of the ozone layer is normally thickest at the poles.

Since the mid-1980s, field measurements have indicated that the total density of ozone molecules in the atmospheric column has declined significantly over Antarctica, resulting

---

[P] Atmospheric chemists refer to this reaction as denitrification. It is not to be confused with the denitrification performed by certain bacteria, which remove NO_3 and produce N_2 in anaerobic soils and sediments (Chapter 7).

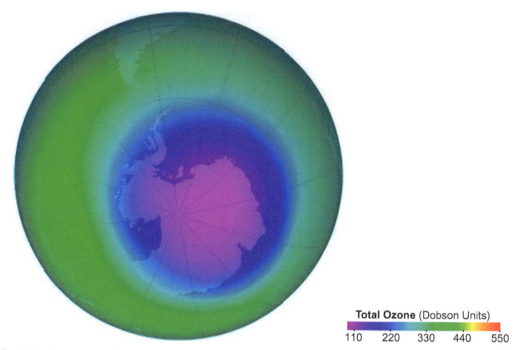

Total Ozone (Dobson Units)

110   220   330   440   550

FIG. 3.16   The average abundance of ozone in the atmosphere of the Southern Hemisphere during October 2006. The ozone "hole," seen in blue and purple, is actually the area where the abundance of ozone in the stratosphere is below 220 Dobson units—perhaps better described as a thinning rather than a hole. Data are in Dobson Units for the entire atmosphere, see Footnote n, p. 75. *From http://ozonewatch.gsfc.nasa.gov/monthly/monthly_2006-10.html.*

in an ozone "hole" over a large area (Farman et al., 1985; Fig. 3.16).[q] The decline, as much as 0.3%/yr, was unprecedented and represents a perturbation of global biogeochemistry. Destruction of ozone is likely to lead to an increased flux of ultraviolet radiation to the Earth's surface (Correll et al., 1992; Kerr and McElroy, 1993; McKenzie et al., 1999) and increased incidence of skin cancer and cataracts in humans (Norval et al., 2007). Greater uvB radiation at the Earth's surface is likely to reduce marine production in the upper water column of the Southern Ocean around Antarctica (Smith et al., 1992b; Meador et al., 2002; Arrigo et al., 2003). Ultraviolet radiation also causes deleterious effects on land plants (Caldwell and Flint, 1994; Day and Neale, 2002). Because previous, steady-state ozone concentrations were maintained in the face of natural photochemical reactions that produce and consume ozone, attention focused on how this balance might have been disrupted by human activities (Cicerone, 1987; Rowland, 1989; Hegglin et al., 2015).

Chlorofluorocarbons (freons), which were produced as aerosol propellants, refrigerants, and solvents, have no known natural source in the atmosphere (Prather, 1985). These compounds are chemically inert in the troposphere, so they eventually mix into the stratosphere

[q] The ozone hole is the area in which the abundance of O_3 in the atmosphere is less than 220 Dobson Units (see footnote n, p. 75).

where they are decomposed by photochemical reactions producing active chlorine (Molina and Rowland, 1974; Rowland, 1989, 1991):

$$CCl_2F_2 \rightarrow Cl + CClF_2, \tag{3.58}$$

which can destroy ozone by the reactions of Eqs. 3.54–3.56. These reactions are greatly enhanced in the presence of ice particles, which accounts for the first observations of the O_3 "hole" in the springtime over Antarctica (Farman et al., 1985; Solomon et al., 1986). In the absence of ice particles atmosphere, ClO reacts with NO_2 to form $ClONO_2$, an inactive compound that removes both gases from O_3 destruction. In the presence of ice clouds, $ClONO_2$ breaks down:

$$ClONO_2 + HCl \rightarrow Cl_2 + HNO_3 \tag{3.59}$$

$$Cl_2 + hv \rightarrow 2\,Cl, \tag{3.60}$$

producing active chlorine for ozone destruction (Molina et al., 1987; Solomon, 1990). Significantly, during the last 40 years, levels of active chlorine over Antarctica have increased in a mirror image to the loss of ozone from the stratosphere (Solomon, 1990). Stratospheric temperatures in the arctic are not as cold as over Antarctica, so there a fewer ice particles and a lower capacity for ozone destruction.

The relative importance of chlorofluorocarbons versus natural sources of chlorine in the stratosphere is apparent in a global budget for atmospheric chlorine (Fig. 3.17). Seasalt aerosols are the largest natural source of chlorine in the troposphere, but they have such a short mean residence time that they do not contribute Cl to the stratosphere. There is also no good reason to suspect that seasalt aerosols have increased in abundance in the last few decades.

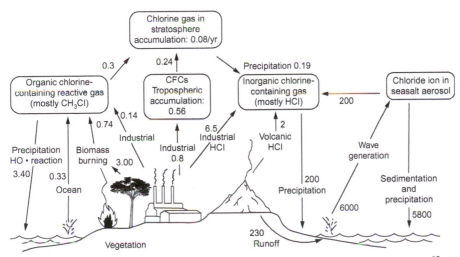

FIG. 3.17 A global budget for Cl in the troposphere and the stratosphere. All data are given in 10^{12} g Cl/yr. *Modified and updated from Möller (1990), Graedel and Crutzen (1993), Graedel and Keene (1995), with new data from McCulloch et al. (1999) and other sources listed in Table 3.7.*

Similarly, industrial emissions of HCl are rapidly removed from the troposphere by rainfall. Especially violent volcanic eruptions can inject gases directly into the stratosphere, sometimes adding to stratospheric Cl (Johnston, 1980; Mankin and Coffey, 1984; Cadoux et al., 2015). However, in most cases only a small amount of Cl reaches the stratosphere, because various processes remove HCl from the rising volcanic plume (Tabazadeh and Turco, 1993; Textor et al., 2003). After the Mount Pinatubo eruption, which released 4.5×10^{12} g of HCl, stratospheric Cl increased by <1% (Mankin et al., 1992).

The only significant natural source of Cl in the stratosphere stems from the production of methylchloride[r] by the ocean surface, by plants, especially tropical vegetation, and by forest fires (Table 3.7). Methylchloride concentrations have fluctuated during the Holocene (Verhulst et al., 2013), but there is no strong implication of an industrial source that might be indicated by an increase in the concentration of CH_3Cl in the Antarctic ice pack during the past 100 years (Butler et al., 1999; Saltzman et al., 2009), or by a significant difference in the concentration of methylchloride between the Northern and Southern Hemispheres (Beyersdorf et al., 2010). The current budget for CH_3Cl is slightly imbalanced, with sources exceeding sinks. Methylchloride has a mean residence time of about 1.3 years in the atmosphere, so a small portion mixes into the stratosphere.

In the global Cl budget, the relatively small industrial production of chlorofluorocarbons, which are inert in the troposphere, is a major source of Cl delivered to the stratosphere (Fig. 3.16; Russell et al., 1996). Increasing concentrations of these compounds have been strongly implicated in ozone destruction (Rowland, 1989; Butler et al., 1999). Happily, with the advent of the Montreal Protocol in 1987, which limits the use of these compounds worldwide,[s] there is already some evidence that the growth rate of these compounds in the atmosphere is slowing (Elkins et al., 1993; Montzka et al., 1996; Solomon et al., 2006) and that the ozone hole may be starting a slow recovery (Fig. 3.18; Solomon et al., 2016; Chipperfield et al., 2017).[t] Indeed, the main cause of continued human impacts on stratospheric ozone may stem from our continuing contributions to the rise in N_2O in Earth's atmosphere (Ravishankara et al., 2009; Chapter 12).

Similar reactions are possible with compounds containing bromine; in fact, Br compounds may be even more potent in the destruction of stratospheric O_3 than Cl (Wennberg et al., 1994). Industry is a source of methylbromide (CH_3Br), which has a long history of use as an agricultural fumigant (Yagi et al., 1995). Methylbromide is also released from the ocean's surface, vegetation, and biomass burning (Table 3.7). Sinks of CH_3Br include uptake by the oceans and soils and oxidation by OH radical. After rising throughout the Industrial Revolution (Saltzman et al., 2008; Khalil et al., 1993a), the atmospheric concentration of methylbromide appears to have declined in recent years (Yvon-Lewis et al., 2009). The global budget of CH_3Br and its mean residence time (about 0.8 years; Colman et al., 1998) in the atmosphere are poorly

---

[r] Also known as chloromethane.

[s] Other Cl-containing, ozone depleting substances were also banned by the Montreal Protocol, but there is some evidence of continuing clandestine emissions of CCl_4 (Liang et al., 2014) and CFC-11 (Montzka et al., 2018; Rigby et al., 2019).

[t] Unfortunately, some of the replacements for CFCs contribute significantly to the greenhouse warming of the Earth (Wuebbles et al., 2013; Rigby et al., 2014) and their use will be curtailed by a recent international agreement.

TABLE 3.7 Budgets of CH_3Cl and CH_3Br in the atmosphere (Tg/yr).

	CH_3Cl	CH_3Br	References
Sources			
Oceans	0.70	0.0015	Hu et al., 2012, 2013
Coastal vegetation	0.03–0.17	0.001–0.008	Rhew et al., 2014
			Hu et al., 2010
			Deventer et al., 2018
Upland vegetation	2.20		Verhulst et al., 2013
			Bahlmann et al., 2019
Freshwater wetlands	0.74	0.035	Hardacre and Heal, 2013
Biomass burning	0.73	0.034	Andreae, 2019
			Verhulst et al., 2013
Industrial uses	0.11–0.16	0.05	McCulloch et al., 1999
			Thompson et al., 2002
Total of sources (best estimates)	4.60	0.12	
Sinks			
Ocean uptake	0.37	0	Hu et al., 2013
Soil uptake	0.25	0.022–0.042	Shorter et al., 1995
			Serca et al., 1998
Reaction with OH	3.37	0.09	Thompson et al., 2002
Loss to stratosphere	0.28	0.006	Thompson et al., 2002
Total of sinks (best estimates)	4.27	0.24	

constrained—sinks exceed sources (Table 3.7). However, some CH_3Br persists long enough to reach the stratosphere, where it can lead to ozone destruction.

Among other halogen-containing gases, the lifetimes of bromoform ($CHBr_3$; Quack and Wallace, 2003; Stemmler et al., 2015) and methyliodide (CH_3I; Bell et al., 2002; Stemmler et al., 2014; Yokouchi et al., 2012), both produced by marine phytoplankton, and various inorganic fluoride compounds (e.g., CH_3F) are too short for appreciable mixing into the stratosphere. The observed increase of fluoride in the stratosphere appears solely due to the upward transport of chlorofluorocarbons, and it is an independent verification of their destruction in the stratosphere by ultraviolet light (Russell et al., 1996). However, F is ineffective as a catalyst for ozone destruction.

Satellite observations have greatly aided our understanding of changes in stratospheric ozone. The loss of ozone from the atmosphere has been monitored since 1979 when the first Total Ozone Mapping Spectrometer (TOMS) began records of the abundance of O_3 in a

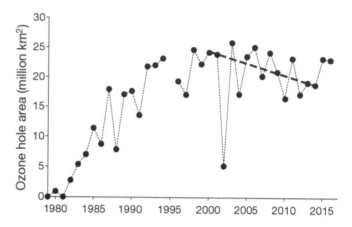

FIG. 3.18 Size of the ozone hole, defined as the area over which the ozone content is less than 220 Dobson Units, over Antarctica since the 1980s, showing its stabilization in recent years as a result of the Montreal Protocol, instituted in 1987. *Adapted and updated from Solomon et al. (2016).*

column extending from the bottom to the top of the atmosphere (Fig. 3.18). Both the area and the minimum column abundance of ozone appear to have stabilized in recent years, after strong O_3 losses at the beginning of the record.[u] Similar, though less extensive, ozone losses are reported for the Arctic (Solomon et al., 2007; Manney et al., 2011) resulting in greater exposure to uVB radiation in Europe (Petkov et al., 2014).

Stratospheric sulfur compounds

Sulfate aerosols in the stratosphere are important to the albedo of the Earth (Warneck, 2000). A layer of sulfate aerosols, known as the Junge layer, is found in the stratosphere at about 20 to 25 km altitude. Its origin is twofold. Large volcanic eruptions can inject SO_2 into the stratosphere, where it is oxidized to sulfate (Eqs. 3.27 and 3.28). Large eruptions have the potential to increase the abundance of stratospheric sulfate 100-fold (Arnold and Bührke, 1983; Hofmann and Rosen, 1983), and these sulfate aerosols persist in the stratosphere for several years, cooling the planet (McCormick et al., 1995; Briffa et al., 1998; Sigl et al., 2015). Following the eruption of Tambora in Indonesia, the year 1816 was known as the "year without a summer" in New England and much of Europe.

During periods without volcanic activity, the dominant source of stratospheric sulfate derives from carbonyl sulfide (COS)[v] that mixes up from the troposphere (Sheng et al., 2015). Showing an average concentration of about 500 parts per trillion, COS is the most abundant sulfur gas in the atmosphere (Table 3.1). The pool in the atmosphere contains about 2.8×10^{12} g S (Chin and Davis, 1995). Based on the global budget shown in Table 3.8, the mean residence time for COS in the atmosphere is ~5 years. Whereas most sulfur gases are so reactive that their mean residence time is much less than one year, the mean residence time of COS allows about one-third of the annual production to enter the stratosphere. Additional COS may be lofted to the stratosphere in the smoke plumes of large wildfires (Notholt et al., 2003).

---

[u] See http://ozonewatch.gsfc.nasa.gov/.

[v] Also abbreviated OCS.

TABLE 3.8 Global budget for carbonyl sulfide (COS) in the atmosphere.

Source or sink	COS (10^{12} g S /yr)	References
Sources		
Oceans	0.04[a]	Kettle et al., 2002
Anoxic soils	0.03	Kettle et al., 2002
Biomass burning	0.31–0.60	Andreae, 2019
		Stinecipher et al., 2019
Industrial	0.26	Campbell et al., 2015
Volcanoes	0.02	Chin and Davis, 1993
Oxidation of natural CS_2	0.14[b]	Kettle et al., 2002
		Campbell et al., 2015
Oxidation of DMS	0.15	Kettle et al., 2002
Total sources	0.95–1.25	
Sinks		
Vegetation uptake	0.24–0.74	Kettle et al., 2002
		Berry et al., 2013
Soil uptake (oxic)	0.13	Kettle et al., 2002
Oxidation by OH	0.09	Kettle et al., 2002
Stratospheric photolysis	0.02	Kettle et al., 2002
Total sinks	0.48–0.98	

[a] net.
[b] assumes that 44% of the global annual CS2 source is industrial.

The global budget for COS in the atmosphere has been revised repeatedly during the last few decades, and a current budget of COS shows a slight excess of sources over sinks. Several components of the budget, including net ocean uptake, are poorly constrained (Table 3.8). The concentration of COS in the atmosphere has increased during the Industrial Revolution (Aydin et al., 2002; Montzka et al., 2004), but it has declined slightly during the past couple of decades (Sturges et al., 2001; Rinsland et al., 2002), perhaps due to enhanced plant uptake (Campbell et al., 2017).

A large source of COS in the atmosphere stems from the oxidation of carbon disulfide (CS_2) emitted from anoxic soils and industrial sources. The oceans are an indirect source of COS, stemming from the oxidation of dimethylsulfide, emitted from phytoplankton (Chapters 9 and 13). In its global budget (Table 3.8), the small direct source of COS from the oceans is a net value, recognizing that COS is also taken up by seawater (Weiss et al., 1995). Wetland soils are a small source, and the global emission of COS from salt marshes is limited by the small extent of salt marsh vegetation (Aneja et al., 1979; Steudler and Peterson, 1985; Carroll et al., 1986). Other sources of COS include biomass burning (Nguyen et al., 1995; Andreae, 2019) and direct industrial emissions, including combustion of coal (Zumkehr et al., 2017;

Du et al., 2016). Anthropogenic sources may account for half of the annual sources of COS in the atmosphere (Zumkehr et al., 2018).

Some COS is oxidized in the troposphere via OH radicals, but the major tropospheric sink for COS, first reported by Goldan et al. (1988), appears to be uptake by vegetation and upland soils (Steinbacher et al., 2004; Kuhn et al., 1999; Simmons et al., 1999). COS shows seasonal fluctuations in the atmosphere, parallel to CO_2 and likely to reflect plant activity (Montzka et al., 2004, 2007; Campbell et al., 2008; Geng and Mu, 2006). Kesselmeier and Merk (1993) found that a variety of crop plants take up COS whenever the ambient concentration is greater than 150 ppt. Uptake by vegetation is now believed to account for half of the total annual destruction of COS globally (Table 3.8), and several workers have suggested that measurements of changes in the abundance of COS in the atmosphere may be useful as an indirect measure of changes in global photosynthesis (Blonquist et al., 2011; Chapter 5).

The small amount of COS that mixes into the stratosphere is destroyed by a photochemical reaction involving the OH radicals, producing SO_4 and contributing to the Junge layer. In fact, aside from the periodic eruptions of large volcanoes, COS appears to be the main source of SO_4 aerosols in the stratosphere (Hofmann and Rosen, 1983; Servant, 1986). Eventually, these aerosols are removed from the stratosphere by downward mixing of stratospheric air.

There is some evidence that these sulfate aerosols have increased in the stratosphere recent years (Hofmann, 1990). These aerosols affect the amount of solar radiation entering the troposphere, and they are an important component of the radiation budget of the Earth (Turco et al., 1980). Through direct and indirect (CS_2) sources, humans appear to make large contributions to the budget of COS (Table 3.8), and any increase in COS-derived aerosols in the stratosphere has potential consequences for predictive models of future global warming (Hofmann and Rosen, 1980). In return, global warming and an increasing flux of uvB radiation penetrating to the Earth's surface may enhance the production of COS in ocean waters (Najjar et al., 1995).

The effect of volcanic eruptions that loft SO_2 to the stratosphere is so dramatic that some workers have suggested organized programs to seed SO_4 aerosols in the stratosphere to combat global climate change (Crutzen, 2006). This form of geoengineering is not without critics, given its potential impacts on other aspects of Earth system function (Robock et al., 2009; Keith et al., 2016).

Models of the atmosphere and global climate

A large number of models have been developed to explain the physical properties and chemical reactions in the atmosphere. When these models attempt to predict the characteristics in a single column of the atmosphere, they are known as one-dimensional (1D) and radiative convective models. For example, Fig. 3.2 is a simple 1D-model for the greenhouse effect, which assumes that the behavior of the Earth's atmosphere can be approximated by average values applied to the entire surface. Two-dimensional (2D) models can be developed using the vertical dimension and a single horizontal dimension (e.g., latitude) to examine the change in atmospheric characteristics across a known distance of the Earth's surface (e.g., Brasseur and Hitchman, 1988; Hough and Derwent, 1990). On a regional scale, these are particularly useful in following the fate of pollution emissions (e.g., Rodhe, 1981; Asman and van Jaarsveld, 1992; Berge and Jakobsen, 1998). Three-dimensional (3D) models attempt to follow the fate of particular parcels of air as they move both horizontally and vertically in the atmosphere. These dynamic 3D models are known as general circulation models (GCMs) for the globe (Fig. 3.19).

Many models are constructed to include both chemical reactions and physical phenomena, such as the circulation of the atmosphere due to temperature differences. Chemical transformations are parameterized using the rate and equilibrium coefficients for the reactions that we have examined in this chapter. Because there are a large number of reactions, most of these models are quite complex (e.g., Logan et al., 1981; Isaksen and Hov, 1987; Lelieveld and Crutzen, 1990), but they give useful predictions of future atmospheric composition when the input of several constituents is changing simultaneously.

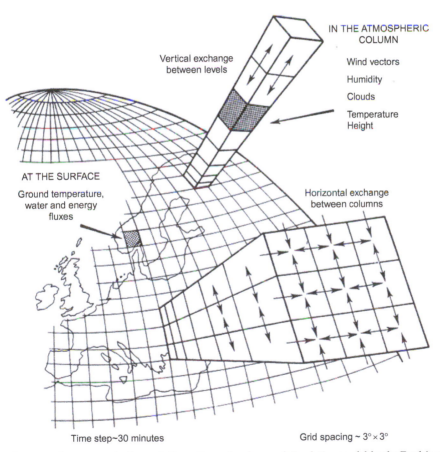

FIG. 3.19 Conceptual structure of a dynamic three-dimensional general circulation model for the Earth's atmosphere, indicating the variables that must be included for a global model to function properly. *From Henderson-Sellers and McGuffie (1987). Copyright ©1987. Reprinted by permission of John Wiley and Sons, Ltd.*

Nearly all climate models predict that a substantial warming of the atmosphere (2–4.5 °C) will accompany increasing concentrations of CO_2, N_2O, CH_4, and chlorofluorocarbons in the atmosphere (IPCC, 2013).[w] Largely stemming from fossil fuel combustion, the concentration of CO_2 is now higher than at any time in the past 5 million years (Stap et al., 2016). During the past 150 years, the concentrations of CH_4, and N_2O have risen above levels seen at any time during the past 10,000 years—spanning the entire history of human civilization (Flückiger et al., 2002). Atmospheric warming, resulting from the absorption of infrared (heat) radiation emitted from the Earth's surface, is known as the greenhouse effect or radiative forcing (Fig. 3.2). The predicted warming of future climate is greatest near the poles, where there is normally the greatest net loss of infrared radiation relative to incident sunlight (Manabe and Wetherald, 1980). Dramatic, recent declines in the extent of Arctic sea ice suggest that these predictions are already proving correct (Serreze et al., 2007; Notz and Stroeve, 2016; Chapter 10). For the same reason, future nighttime and wintertime temperatures worldwide are likely to show large changes relative to today's conditions. Presumably the oceans will warm more slowly than the atmosphere, but eventually warmer ocean waters will allow greater rates of evaporation, increasing the circulation of water in the global hydrologic cycle (Chapter 10). Water vapor also absorbs infrared radiation, so it is likely to further accelerate the potential greenhouse effect (Raval and Ramanathan, 1989; Rind et al., 1991; Soden et al., 2005; Willett et al., 2007). Thus, most models predict that higher concentrations of CO_2 and other trace gases in the atmosphere will make the Earth a warmer and more humid planet.

Incoming solar radiation delivers about 340 W/m² to the Earth (Fig. 3.2).[x] The natural greenhouse effect warms the planet about 33° C by trapping 153 W/m² of outgoing radiation (Ramanathan, 1988).[y] For the past 30 years or so, there has been a small increase in the Sun's luminosity (+0.12 to +0.16 W/m²) (IPCC, 2013; Foukal et al., 2006; Pinker et al., 2005), but the human impact on radiative forcing due to increasing concentrations of atmospheric trace gases currently adds about 2.3 W/m² to the natural greenhouse effect (IPCC, 2013), causing measurable changes in the spectral distribution of radiation leaving the Earth (Harries et al., 2001). Aerosols tend to cool the atmosphere, and increases in aerosols due to human activities are thought to reduce the global radiative forcing by about 1.2 W/m² (i.e., global dimming; IPCC, 2013; Bellouin et al., 2005; Mahowald, 2011). It is interesting to note that aerosol concentrations were higher (Patterson et al., 1999; Lambert et al., 2008) and CO_2 concentrations and temperatures were lower during the last glacial period (Fig. 1.4).

[w] Through its industrial use as a solvent, nitrogen trifluoride (NF_3) is potentially an important contributor to Earth's greenhouse effect, but it is not included in most assessments of climate change (Prather and Hsu, 2008; Weiss et al., 2008).

[x] The Sun's radiation, measured outside the Earth's atmosphere, delivers 1379 W/m², known as the solar constant (McElroy, 2002). A one-dimensional model for the Earth's radiation budget shows an annual input of ∼340 W/m² because only ¼ of the Earth's surface is exposed to sunlight at any given moment.

[y] The Earth's greenhouse effect is dominated by H_2O and CO_2. O_2 and N_2 provide only a trivial contribution to Earth's radiation balance, together adding 0.28 W/m² to the greenhouse effect (Höpfner et al., 2012).

Long-term records from tree rings and ice cores document substantial and continuing warming of the global climate, coincident with rising CO_2 (Mann et al., 1999; Thompson et al., 2000). Recent temperatures in Europe are greater than at any point in the past 500 years (Luterbacher et al., 2004). These changes in climate cannot be explained by natural phenomena alone (Crowley, 2000; Stott et al., 2000). How rapidly these changes in climate occur will be moderated by the thermal buffer capacity of the world's oceans, which can absorb enormous quantities of heat. Already, several long-term records suggest increases in the ocean's temperature worldwide (Barnett et al., 2005; Levitus et al., 2001).

Differential warming of the atmosphere and oceans will also change global patterns of precipitation and evapotranspiration (Manabe and Wetherald, 1986; Rind et al., 1990; Zhang et al., 2007), causing substantial changes in soil moisture of most areas outside the tropics. Arid regions, such as the southwestern United States, are especially likely to experience increased drought (Cook et al., 2004; Seager et al., 2007), consistent with recent trends in rainfall in this region (Milly et al., 2005), and discussed further in Chapter 10.

Climate change affects biogeochemistry (and vice versa) in a variety of ways. Land plants and ocean waters take up substantial quantities of CO_2 from the atmosphere, potentially slowing the rate of climate change (Chapters 5, 9, and 11). Clearing vegetation alters the albedo of the Earth's land surface, potentially altering radiative forcing (Forzieri et al., 2017). Warmer temperatures may increase the rate of decomposition of soil carbon now frozen in Arctic permafrost (Dorrepaal et al., 2009; Schuur et al., 2009), and trigger the release of methane now frozen in ocean sediments, resulting in positive feedbacks that further exacerbate global warming (Chapter 11). Changes in climate are likely to affect the distribution of many plants and animals, potentially causing extinctions (Thomas et al., 2004); they also impact a wide range of conditions affecting human health and economic activity (Hsiang et al., 2017).

Summary

In this chapter we have examined the physical structure, circulation, and composition of the atmosphere. Major constituents, such as N_2, are rather unreactive and have long mean residence times in the atmosphere. CO_2 is largely controlled by plant photosynthetic uptake and by its dissolution in waters on the surface of the Earth. The atmosphere contains a variety of minor constituents, many of which are reduced gases. These gases are highly reactive in homogeneous reactions with hydroxyl (OH) radicals and heterogeneous reactions with aerosols and cloud droplets, which scrub them from the atmosphere. Changes in the concentration of many trace gases are indicative of global change, perhaps leading to future climatic warming and higher surface flux of ultraviolet light. The oxidized products of trace gases are deposited in land and ocean ecosystems, resulting in inputs of N, S, and other elements of biogeochemical significance. Pollution of the atmosphere by the release of oxidized gases containing N and S as a result of human activities results in acid deposition in downwind ecosystems. The enhanced deposition of N and S represents altered biogeochemical cycling on a regional and global basis. Changes in stratospheric ozone and global climate are early warnings of the human impact on the atmosphere of our planet.

The Lithosphere

Introduction

Since early geologic time, the atmosphere has interacted with the exposed crust of the Earth, causing rock weathering. Many of the volcanic gases in the Earth's earliest atmosphere dissolved in water to form acids that could react with surface minerals (Chapter 2). Later, as oxygen accumulated in the atmosphere, rock weathering occurred as a result of the oxidation of reduced minerals, such as pyrite, that were exposed at the Earth's surface. At least since the advent of land plants, rock minerals have also been exposed to high concentrations of carbon dioxide in the soil as a result of the metabolic activities of soil microbes and plant roots. Today, carbonic acid (H_2CO_3), derived by reaction of CO_2 with soil water, is the major cause of rock weathering in most ecosystems. Higher levels of CO_2 in Earth's atmosphere can be expected to increase the rate of rock weathering globally. In recent years, humans have added large quantities of NO_x and SO_2 to the atmosphere, causing acid rain (Chapter 3) and increasing the rate of rock weathering in many areas (Cronan, 1980).

Biogeochemistry: An Analysis of Global Change, Fourth Edition
https://doi.org/10.1016/B978-0-12-814608-8.00004-9

Siever (1974) proposed a basic equation to summarize the close linkage between the Earth's atmosphere and its crust:

$$\text{Igneous Rocks} + \text{Acid Volatiles} = \text{Sedimentary Rocks} + \text{Salty Oceans} \tag{4.1}$$

This formula recognizes that through geologic time the primary minerals of the Earth's crust have been exposed to reactive, acid-forming C, N, and S gases of the atmosphere. The products of the reaction are carried to the oceans, where they accumulate as dissolved salts or in ocean sediments (Li, 1972). Large amounts of sedimentary rock have formed through geologic time; indeed, about two-thirds of the rocks now exposed on land are sedimentary rocks that have been uplifted by tectonic activity (Durr et al., 2005; Suchet et al., 2003; Hartmann and Moosdorf, 2012). Following uplift, sedimentary rocks are subject to further weathering reactions with acid volatiles, in accord with Siever's basic equation. Eventually, the geologic process of subduction carries sedimentary rocks to the deep Earth, where CO_2 is released and the solid constituents are converted back to primary minerals under great heat and pressure (Fig. 1.3; see also Siever, 1974).

This chapter reviews the basic types of rock weathering on land and the processes that drive these weathering reactions. Rock weathering is especially important to the bioavailability of elements that have no gaseous forms (e.g., Ca, K, Fe, and P; Table 4.1). Weathering is basic to soil fertility, biological diversity, and agricultural productivity (Huston, 1993). Reactions between soil waters and the solid materials in soil determine the availability of essential elements to biota and the losses of these elements in runoff. Conversely, land plants and soil microbes affect rock weathering and soil development, potentially regulating global biogeochemistry and the Earth's climate. In this chapter, we examine soil development in the major ecosystems on Earth. Finally, we will estimate the global rate of rock weathering in an attempt

TABLE 4.1   Approximate mean composition of earth's continental crust.

Constituent	Percentage composition
Si	28.8
Al	7.96
Fe	4.32
Ca	3.85
Na	2.36
Mg	2.20
K	2.14
Ti	0.40
P	0.076
Mn	0.072
S	0.070

Data from Wedepohl (1995) (see also Fig. 2.3).

to determine the annual new supply of biochemical elements on land and the total delivery of weathering products to rivers and the sea.

Rock weathering

Following geologic uplift and exposure at the Earth's surface, all rocks undergo weathering, a general term that encompasses a wide variety of processes that decompose rocks. Mechanical weathering is the fragmentation or loss of materials without a chemical reaction; in laboratory terminology it is equivalent to a physical change. Mechanical weathering is important in extreme and highly seasonal climates and in areas with much exposed rock. Wind abrasion is a form of mechanical weathering in arid environments, whereas the expansion of frozen water in rock crevices—often resulting in fractured rocks—is an important form of mechanical weathering in cold climates. Plant roots also fragment rock when they grow in crevices. Tectonic forces cause fractures in bedrock at considerable depth (St Clair et al., 2015). Through mining and road building humans have dramatically enhanced rates of mechanical weathering.

Where the products of mechanical weathering are not rapidly removed by erosion, thick soils develop, and the landscape is said to be "transport-limited" (e.g., Brosens et al. 2020). In other areas, finely divided rock and soil are removed by erosion—the loss of particulate solids from the ecosystem. The products of mechanical weathering may also be lost in catastrophic events such as landslides (Swanson et al., 1982). Indicative of the high rates of mechanical weathering and erosion at high elevations, rivers draining mountainous regions often carry exceptionally large sediment loads (Milliman and Syvitski, 1992).

Chemical weathering

Chemical weathering occurs when the minerals in rocks and soils react with acidic and oxidizing substances. Usually chemical weathering involves water, and some elemental constituents are released as dissolved ions that are available for uptake by biota or lost in stream waters. In many cases, mechanical weathering is important in exposing the minerals in rocks to chemical attack (e.g., Miller and Drever, 1977; Anbeek, 1993; Hilley et al., 2010). If the rate of mechanical weathering is slow, it may limit the rate of chemical weathering and lower the fertility of soils (Gabet and Mudd, 2009). Fractures in bedrock allow diffusion of CO_2 and O_2 and the percolation of ground water, allowing chemical weathering at depth and altering the flow and composition of groundwaters that subsequently reach the surface downslope (Kim et al., 2017). What soil scientists call the "critical zone" or regolith extends to a mean depth of 80 m (Xu and Liu, 2017), defining the zone of rock weathering and groundwater flow.

The rate of chemical weathering depends on the mineral composition of rocks. Igneous and metamorphic rocks contain primary minerals (e.g., olivine and plagioclase) that were formed under conditions of high temperature and pressure deep in the Earth. These primary silicate minerals are crystalline in structure, and they are found in two classes—the ferromagnesian or mafic series and the plagioclase or felsic series—depending on the crystal structure and the presence of magnesium versus aluminum in the crystal lattice (Fig. 4.1). Among these minerals, the rate of weathering tends to follow the reverse order of the sequence of mineral

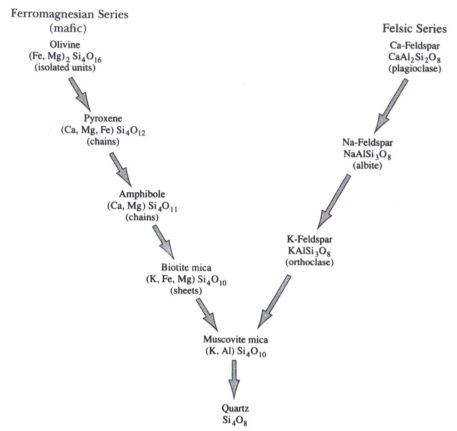

FIG. 4.1 Silicate minerals are divided into two classes, the ferromagnesian series and the felsic series, based on the presence of Mg or Al in the crystal structure. Among the ferromagnesian series, minerals that exist as isolated crystal units (e.g., olivine) are most susceptible to weathering, while those showing linkage of crystal units and a lower ratio of oxygen to silicon are more resistant. Among the felsic series, Ca-feldspar (plagioclase) is more susceptible to weathering than Na-feldspar (albite) and K-feldspar (orthoclase). Quartz is the most resistant of all. This weathering series is the reverse of the order in which these minerals are precipitated during the cooling of magma.

formation during the original cooling and crystallization of rock; minerals that condensed first are the most susceptible to weathering reactions (Goldich, 1938).

Minerals formed during rapid, early crystallization of magma at high temperatures contain few bonds that link the units of their crystalline structure. They also have frequent substitutions of various cations—for example, Ca, Na, K, Mg, and trace metals (Fe and Mn)—in their crystal lattice, distorting its shape and increasing its susceptibility to weathering. Thus, for example, olivine, which is formed under conditions of great heat and pressure deep within the Earth, is most likely to weather rapidly when exposed at the Earth's surface.

In rocks or soils of mixed composition, chemical weathering is concentrated on the relatively labile minerals, while other minerals may be unaffected (April et al., 1986; White et al., 1996, 2001). The dissolution and loss of some constituents may result in a reduction of the density of bedrock, with little collapse or loss of the initial rock volume—a process known as isovolumetric weathering. The product, "rotten" rock known as saprolite or grus, comprises the lower soil profile of many regions, especially in the southeastern United States (Gardner et al., 1978; Velbel, 1990; Stolt et al., 1992; Oh and Richter, 2005). In other cases, the removal of some constituents is accompanied by the collapse of the soil profile and an apparent increase in the volumetric concentration of the elements that remain (e.g., Zr, Ti, and Fe; Brimhall et al., 1991).

Quartz is very resistant to chemical weathering and often remains when other minerals are lost (Fig. 4.1). Quartz is a relatively simple silicate mineral consisting only of silicon and oxygen in tetrahedral crystals that are linked in three dimensions. In many cases, the sand fraction of soils is largely composed of quartz crystals that remain following the chemical weathering and loss of other constituents during soil development (Brimhall et al., 1991).

In addition to mineralogy, rock weathering also depends on climate (White and Blum, 1995; Gislason et al., 2009). Chemical weathering involves chemical reactions, so it is not surprising that it occurs most rapidly under conditions of higher temperature and precipitation (White et al., 1998; West et al., 2005; Dere et al., 2013). Chemical weathering is more rapid in tropical forests than in temperate forests, and more rapid in most forests than in grasslands or deserts. Chemical weathering proceeds, albeit at low rates, in cold Antarctic environments (Hodson et al., 2010). White and Blum (1995) show that the loss of Si in stream water, which is often a good index of chemical weathering, is directly related to precipitation and temperature over much of the Earth's surface (Fig. 4.2). With the ongoing global changes in climate, the likelihood of higher future temperatures and precipitation should increase the rate of chemical weathering worldwide (Cotton et al., 2013; Raymond, 2017).

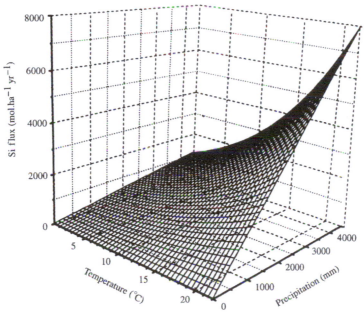

FIG. 4.2   Loss of silicon (SiO_2) in runoff as a function of mean annual temperature and precipitation in various areas of the world. *Modified from White and Blum (1995).*

The dominant form of chemical weathering is the carbonation reaction, driven by the formation of carbonic acid, H_2CO_3, in the soil solution:

$$H_2O + CO_2 \rightleftharpoons H^+ + HCO_3^- \rightleftharpoons H_2CO_3 \qquad (4.2)$$

Because plant roots and soil microbes release CO_2 to the soil, the concentration of H_2CO_3 in soil waters is often much greater than that in equilibrium with atmospheric CO_2 at 400 ppm (i.e., 0.04%) (Castelle and Galloway, 1990; Amundson and Davidson, 1990; Pinol et al., 1995). Buyanovsky and Wagner (1983) report seasonal CO_2 concentrations of greater than 7% in the soil beneath wheat fields in Missouri. Such high concentrations of CO_2 can extend to considerable depths in the soil profile, affecting the weathering of underlying rock in the critical zone (Sears and Langmuir, 1982; Richter and Markewitz, 1995). Wood and Petraitis (1984) found CO_2 concentrations of 1.0% at 36 m, which they link to the downward transport of organic materials that subsequently decompose at depth. Solomon and Cerling (1987) found that high concentrations of CO_2 accumulated in the soil under a mountain snowpack, potentially leading to significant weathering during the winter (see also Berner and Rao, 1997).

Plant growth is greatest in warm and wet climates. As a result of root growth and activity, these areas maintain the highest levels of soil CO_2 and the greatest rates of carbonation weathering (Johnson et al., 1977). Examining data from a variety of ecosystems, Brook et al. (1983) found that the average concentration of soil CO_2 varies as a function of actual evapotranspiration,[a] a composite measure of temperature and available soil moisture, at the site (Fig. 4.3). However, even in arid regions, rock weathering appears to be controlled

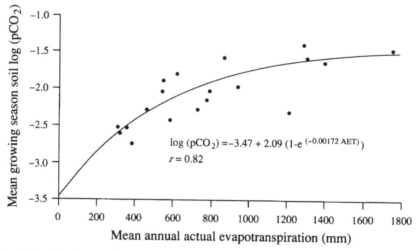

FIG. 4.3   The relationship between the mean concentration of CO_2 in the soil pore space and the actual evapotranspiration of the site for various ecosystems of the world. *From Brook et al. (1983).*

[a] The total of direct evaporation of moisture from the Earth's land surface and the uptake of soil water by plants which subsequently lose water to the atmosphere in a process known as transpiration (See Chapters 5 and 10).

by carbonation weathering (e.g., Routson et al., 1977). By maintaining high concentrations of CO_2 in the soil, plants and soil microbes control the process of rock weathering—a good example of *biogeochemistry* at the Earth's surface. Carbonation weathering transfers CO_2, an acid volatile, from the atmosphere to groundwater (Kessler and Harvey, 2001).

Carbonic acid attacks silicate rocks. For example, weathering of the Na-feldspar, albite, proceeds as:

$$2NaAlSi_3O_8 + 2H_2CO_3 + 9H_2O \rightarrow 2Na^+ + 2HCO_3^- + 4H_4SiO_4 + Al_2Si_2O_5(OH)_4 \qquad (4.3)$$

During this process, a primary mineral is converted to a secondary mineral, kaolinite, by the removal of Na and soluble silica. A sign that carbonation weathering has occurred is the observation that HCO_3^- is the dominant anion in runoff waters (Ohte and Tokuchi, 1999). Examining the depletion of Na in the critical zone, Dere et al. (2013) found weathering of feldspars increased exponentially as a function of temperature and linearly as a function of precipitation along a transect from New York to Puerto Rico. The formation of the secondary mineral, kaolinite, involves hydration with H^+ and water. The secondary mineral has a lower ratio of Si to Al as a result of the loss of some Si to stream waters. Because only some of the constituents of the primary mineral are released, this type of weathering reaction is known as an incongruent dissolution. Under conditions of high rainfall, as in the humid tropics, kaolinite may undergo a second incongruent dissolution to form another secondary mineral, gibbsite:

$$Al_2Si_2O_5(OH)_4 + 5H_2O \rightarrow 2H_4SiO_4 + Al_2O_3 \cdot 3H_2O \qquad (4.4)$$

Some weathering reactions involve congruent dissolutions. In moist climates, limestone undergoes a relatively rapid congruent dissolution during carbonation weathering:

$$CaCO_3 + H_2CO_3 \rightarrow Ca^{2+} + 2HCO_3^- \qquad (4.5)$$

Olivine ($FeMgSiO_4$) also undergoes congruent dissolution in water, releasing Fe, Mg, and Si (Grandstaff, 1986). The magnesium and silicon are lost in runoff waters, but usually the Fe reacts with oxygen, resulting in the precipitation of Fe_2O_3 in the soil profile. Similarly, pyrite (FeS_2) undergoes a congruent reaction during its oxidation:

$$4FeS_2 + 8H_2O + 15O_2 \rightarrow 2Fe_2O_3 + 16H^+ + 8SO_4^{2-} \qquad (4.6)$$

The H^+ produced in this reaction accounts for the acidity of runoff from many mining operations (Ross et al., 2018). As with the weathering of olivine, the Fe from weathered pyrite is subsequently precipitated as Fe_2O_3 in the soil profile or streambed (Bloomfield, 1972; Johnson et al., 1992). Often this reaction is mediated by the chemoautotrophic bacterium *Thiobacillus ferrooxidans* (Eq. 2.15; Temple and Colmer, 1951; Ralph, 1979; Schrenk et al., 1998). Weathering of minerals containing Fe^{2+} is a source of energy for chemolithotrophic microbes that oxidize it to Fe^{3+} leading to the breakdown of bedrock (Napieralski et al., 2019).

In addition to carbonic acid, organisms release a variety of organic acids to the soil solution that can be involved in the weathering of silicate minerals (Ugolini and Sletten, 1991). Many simple organic compounds, including acetic and citric acids, are released from plant roots (Smith, 1976; Tyler and Ström, 1995; Jones, 1998). Phenolic acids (i.e., tannins) are released during the decomposition of plant litter, and soil microbes produce a variety of organic acids during decomposition of plant remains.[b] Many fungi release oxalic acid that results in chemical weathering (Cromack et al., 1979; Welch and Ullman, 1993; Cama and Ganor, 2006; Li et al., 2016c). Organic acids from plant roots and microbes can weather biotite mica, releasing K (Boyle and Voigt, 1973; April and Keller, 1990). Soil microbes and plant roots appear to preferentially colonize and weather the surface of minerals that contain elements, such as phosphorus, that are otherwise in short supply for their growth and reproduction (Rogers et al., 1998; Banfield et al., 1999; Teodoro et al., 2019).

In addition to their contributions to total acidity, organic acids speed weathering reactions in the soil by combining with some weathering products, in a process called chelation.[c] When Fe and Al combine with organic acids, they are mobile and move to the lower soil profile in percolating water (Dahlgren and Walker, 1993; Lundström, 1993). When these elements are involved in chelation, their inorganic concentration in the soil solution remains low, the equilibrium between dissolved products and mineral forms is not achieved, and chemical weathering may continue unabated (Berggren and Mulder, 1995; Zhang and Bloom, 1999). Grandstaff (1986) found that additions of small concentrations of EDTA (an organic chelation agent) to weathering solutions increased the dissolution of olivine by 110 times over inorganic conditions. Organic acids increase the weathering of a variety of silicate minerals, including quartz, particularly when the soil solution is neutral or slightly acid (Tan, 1980; Bennett et al., 1988; Wogelius and Walther, 1991; Welch and Ullman, 1993).

Organic acids often dominate the acidity of the upper soil profile, while carbonic acid is important below (Ugolini et al., 1977b). In general, organic acids dominate the weathering processes in cool temperate forests where decomposition processes are slow and incomplete, whereas carbonic acid drives the chemical weathering in tropical forests where lower concentrations of organic acids remain after the decomposition of plant debris (Johnson et al., 1977; Ohte and Tokuchi, 1999).

There is much debate about changes in the rate of chemical weathering through geologic time, especially as a result of the evolution of vascular land plants (Berner and Kothavala, 2001; Drever, 1994). Greater rock weathering, beginning with the initial colonization of land by plants, is likely to have been responsible for a decline in atmospheric CO_2 300–400 million years ago, owing to the consumption of CO_2 in carbonation weathering (Knoll and James, 1987; Berner, 1997; Mora et al., 1996). Various workers have found evidence for higher rates of chemical weathering under areas of vegetation (Moulton et al., 2000), suggesting that plants add large amounts of CO_2 to the soil from root metabolism and from the decomposition of plant debris by microbes (Kelly et al., 1998; Perez-Fodich and Derry, 2019). This view holds that the process of photosynthesis acts to speed the transfer of an acid volatile, CO_2, from the atmosphere to the soil profile. Even before the advent of vascular land plants, the

[b] Based on their differential solubility in acid and alkaline solutions, the soluble organic acids were traditionally subdivided into fulvic and humic acids. While this categorization is still operationally useful, most soil scientists have abandoned these terms as inappropriate to describe our current understanding of the molecular structure of humus substances (Sutton and Sposito, 2005; DiDonato et al., 2016; Kögel-Knabner and Rumpel, 2018).

[c] Derived from the Greek word for claw, chelation applies when an organic molecule forms two or more separate covalent bonds with an ion, usually a metal, that keeps it in solution.

Earth's surface may have been covered with algae and lichens, producing relatively high levels of CO_2 in the soil (Keller and Wood, 1993; Retallack, 1997).

Other workers disagree, suggesting that tectonic uplift and erosion—processes that stimulate mechanical weathering—are the most important determinants of the rate of chemical weathering (Hilley and Porder, 2008; Dixon et al., 2012; Maher and Chamberlain, 2014). Still others argue that temperature and precipitation are the dominant factors controlling chemical weathering, with plants playing a lesser role (Gislason et al., 2009; Ohte and Tokuchi, 1999; Kump et al., 2000). In sum, a variety of factors—tectonics, mineralogy, climate, plants, and soil microbes—play important roles in determining the rate of rock weathering, and it is likely fruitless to attribute a predominant role to any one of them (Gaillardet et al., 1999; Gabet and Mudd, 2009; Anderson et al., 2002).

Because carbonation weathering consumes atmospheric CO_2, the rate of rock weathering on Earth has a major long-term effect on global climate. Taken alone, silicate weathering has the potential to remove CO_2 from the atmosphere and oceans in about 500,000 years (Moon et al., 2014; Colbourn et al., 2015; Chapter 11). Schwartzman and Volk (1989) suggest that without CO_2-enhanced chemical weathering, the temperature of the Earth would be too hot for all but the most primitive microbes. As atmospheric CO_2 increases, we can expect rates of rock weathering to increase. This weathering will be mediated directly by increasing soil CO_2 concentrations and indirectly by rising global temperatures (compare Fig. 4.2). Several experiments show that the amount of CO_2 in the soil profile increases when plants are grown at high CO_2 (Andrews and Schlesinger, 2001; Bernhardt et al., 2006; Williams et al., 2003; Karberg et al., 2005; Fig. 4.4). And in soil-warming experiments, greater microbial activity leads to higher concentrations of CO_2 in the soil pore space, and presumably greater carbonation weathering (Fig. 4.5).

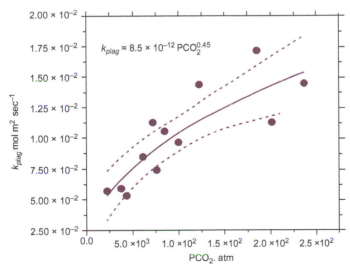

FIG. 4.4   Dissolution of Ca-feldspar (plagioclase) as a function of soil CO_2 concentrations in watersheds of the Sierra Nevada (California), subject to differential hydrothermal activity. *From Navarre-Sitchler and Thyne (2007).*

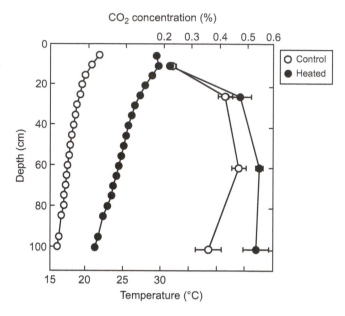

FIG. 4.5   Soil temperature and pore-space CO_2 concentration (mean ± S.E.) as a function of depth in control and experimentally heated (+5 °C) plots in a hardwood forest in Massachusetts. *From unpublished work of Megonigal.*

Secondary minerals

Secondary minerals are formed as byproducts of weathering at the Earth's surface. Usually the formation of secondary minerals begins near the site where primary minerals are being attacked, perhaps even originating as coatings on the crystal surfaces (Casey et al., 1993; Nugent et al., 1998). Although weathering leads to the loss of Si as a dissolved constituent in stream water (Fig. 4.2), some Si is often retained in the formation of secondary minerals (see Eq. 4.3).

Many types of secondary minerals can form in soils during chemical weathering. The secondary minerals in temperate forest soils are often dominated by layered silicate or "clay" minerals. These exist as small (<0.002-mm) particles that control the structural and chemical properties of soils. In general, two types of layers characterize the crystalline structure of secondary, aluminosilicate clay minerals—Si layers and layers dominated by Al, Fe, and Mg. These layers are held together by shared oxygen atoms. Clay minerals and the size of their crystal units are recognized by the number, order, and ratio of these layers (Birkeland, 1984). Moderately weathered soils are often dominated by secondary minerals such as montmorillonite and illite, which have a 2:1 ratio of Si- to Al-dominated layers. More strongly weathered soils, such as in the southeastern United States, are dominated by kaolinite clays with a 1:1 ratio of layers, reflecting a greater loss of Si.

Because secondary minerals may incorporate elements important to biochemistry, one cannot assume that the release of those elements from primary minerals by weathering leads to an immediate increase in the pool of ions available for uptake by plants. Magnesium is often fixed in the crystal lattice of montmorillonite, whereas illite contains K (Martin and Sparks, 1985; Harris et al., 1988). These are common secondary minerals in temperate soils. Similarly, although little nitrogen is contained in primary minerals, some 2:1 clay minerals incorporate N as ammonium (NH$_4$) in their crystal lattice (Holloway and Dahlgren, 2002). Ammonium contained in clay minerals can represent more than 10% of the total N in some soils (Stevenson, 1982; Smith et al., 1994; Johnson et al., 2012). The weathering of sedimentary rocks containing ancient clay minerals with "fixed" ammonium can release large quantities of nitrogen to stream waters (Holloway et al., 1998). Recognizing the widespread nitrogen limitation on land (Chapter 6), the release of nitrogen from rock weathering may play an important role in determining the availability of N for plant growth in some regions (Baethgen and Alley, 1987; Morford et al., 2011; Houlton et al., 2018).

In contrast to the loss of Si and other cations (e.g., Ca and Na) to runoff waters, Al and Fe are relatively insoluble in soils unless they are involved in chelation relations with organic acids (Huang, 1988; Alvarez et al., 1992; Allan and Roulet, 1994; Ross and Bartlett, 1996). In the absence of chelation reactions, these elements tend to accumulate in the soil as oxides (Perez-Fodich and Derry, 2019). Initially, free Fe accumulates in amorphous and poorly crystallized forms, known as ferrihydrite (Fe$_o$), which are often quantified by extraction in a weak oxalate solution (Shoji et al., 1993; Birkeland, 1984). With increasing time, most Fe is found in crystalline oxides and hydroxides, which are traditionally extracted using a reducing solution of citrate-dithionate (Chorover et al., 2004). Some of these mineral transformations involve bacteria and thus are biogeochemical in nature (Fassbinder et al., 1990). The formation of secondary Fe and Al minerals and the adsorption of chelated forms of these elements to clay minerals dramatically reduces the losses of Fe and Al to runoff waters (Ferro-Vázquez et al., 2014; Fuss et al., 2011).

Crystalline oxides and hydrous oxides of Fe (e.g., goethite and hematite) and Al (e.g., gibbsite and boehmite) are the dominant soil minerals in many tropical soils, where high temperatures and rainfall cause relatively rapid decomposition of plant debris and few organic acids remain to chelate and mobilize Fe and Al. Under these climatic conditions, the secondary clay minerals typical of temperate zone soils are subject to weathering, with the near-complete removal of Si, Ca, K, and other basic cations in stream water (Fig. 4.6). However, in an interesting example of the importance of biota in soil development, Lucas et al. (1993) show that kaolinite may persist in the upper horizons of some rainforest soils due to the plant uptake of Si from the lower soil profile and the return of Si to the soil surface in plant debris (see also, (Alexandre et al., 1997; Markewitz and Richter, 1998; Gerard et al., 2008; Conley, 2002). Similar plant "pumping" also maintains higher levels of K at the surface of many soils (Jobbágy and Jackson, 2004; Barre et al., 2009).

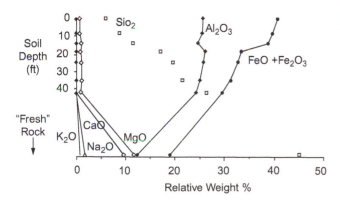

FIG. 4.6 Content of different rock-forming minerals (% mass) above basalt bedrock in Kauai, Hawaii. *From Bluth and Kump (1994).*

Soil chemical reactions

Following their release by weathering, the availability of essential biochemical elements to biota is controlled by a number of reactions that determine the equilibrium between concentrations in the soil solution and contents that are associated with soil minerals or organic materials. In contrast to the kinetics of weathering reactions, soil exchange reactions occur relatively rapidly (Furrer et al., 1989; Maher, 2010). Variations in the concentration of ions in runoff are determined by the time it takes the soil solution to reach chemical equilibrium with soil minerals versus the transit time of the runoff (Wlostowski et al., 2018).

Cation exchange capacity

The layered silicate clay minerals that dominate temperate zone soils possess a net negative charge that attracts and holds cations dissolved in the soil solution. This negative charge has several origins; most of it arises from ionic substitutions within silicate clays, especially in 2:1 clays known as phyllosilicates. For example, when Mg^{2+} substitutes for Al^{3+} in montmorillonite, there is an unsatisfied negative charge in the internal crystal lattice. This negative charge is permanent in the sense that it arises inside the crystal structure and cannot be neutralized by covalent bonding of cations from the soil solution. Permanent charge is expressed as a zone or "halo" of negative charge surrounding the surface of clay particles in the soil.

A second source of negative charge is found at the edges of clay particles, where hydroxide (-OH) groups are often exposed to the soil solution. Depending on the pH of the solution, H^+ may be more or less strongly bound to this group. In most cases, a considerable number of the H^+ are dissociated, leaving negative charges (-O-) that can attract and bind cations (e.g., Ca^{2+}, K^+, and NH_4^+). This cation exchange capacity is known as pH-dependent charge. The binding is reversible and exists in equilibrium with ionic concentrations in the soil solution.

In many temperate soils, a large amount of cation exchange capacity is also contributed by soil organic matter (Yuan et al., 1967). These are pH-dependent charges originating from the phenolic (-OH) and organic acid (-COOH) groups of soil humic materials. In some sandy soils, as in central Florida, and in many highly weathered soils, nearly all cation exchange is the result of soil organic matter (e.g., Daniels et al., 1987; Richter et al., 1994). Organic matter is also the major source of cation exchange in desert soils that contain relatively small amounts of secondary clay minerals as a result of limited chemical weathering.

The total negative charge in a soil is expressed as mEq/100 g or cmol(+)/kg of soil, which constitutes the cation exchange capacity (CEC). Exchange of cations occurs as a function of chemical mass balance with the soil solution. Elaborate models of ion exchange have been developed by soil chemists (Sposito, 1989). In general, cations are held on the exchange sites and displace one another in the sequence:

$$Al^{3+} > H^+ > Sr^{2+} > Ca^{2+} > Mg^{2+} > Rb^+ > K^+ > NH_4^+ > Na^+ > Li^+ \qquad (4.7)$$

(Sparks, 2002; Sahai, 2000). This sequence assumes equal molar concentrations in the initial soil solution and can be altered by the presence of large quantities of the more weakly held ions. Agricultural liming, for example, is an attempt to displace Al^{3+} ions from the exchange sites by "swamping" the soil solution with excess Ca^{2+}. In most cases, few cation exchange sites are actually occupied by H^+, which quickly weathers soil minerals, releasing Al and other cations into the soil solution and enhancing their potential for export to surface and groundwaters.

Cations other than Al^{3+} and H^+ are informally known as base cations, since they tend to form bases—for example, $Ca(OH)_2$—when they are released to the soil solution (Birkeland, 1984, p. 23). The percentage of the total cation exchange capacity occupied by base cations is termed base saturation. Both cation exchange capacity and base saturation increase during initial soil development on newly exposed parent materials. However, as the weathering of soil minerals continues, cation exchange capacity and base saturation decline (Bockheim, 1980; Huston, 2012). Temperate forest soils dominated by 2:1 clay minerals have greater cation exchange capacity than those dominated by 1:1 clay minerals such as kaolinite. Highly weathered soils in the humid tropics are dominated by aluminum hydroxide minerals, which offer essentially no cation exchange capacity at their natural soil pH.

Soil buffering

Cation exchange capacity acts to buffer the acidity of many temperate soils. When H^+ is added to the soil solution, it exchanges for cations, especially Ca^{2+}, on clay minerals and organic matter (Bache, 1984; James and Riha, 1986). Over a wide range of pH, temperate soils maintain a constant value (k) for the expression:

$$pH - \frac{1}{2}(pCa) = k \tag{4.8}$$

which is known as the lime potential. This expression suggests that when H^+ is added to the soil solution (lower pH), the concentration of Ca^{2+} increases in the soil solution (lower pCa), so that k remains constant. The ½ reflects the valence of Ca^{2+} versus H^+. As long as there is sufficient base saturation (e.g., >15%), buffering by CEC explains why the pH of many temperate soils may show relatively little change when they are exposed to acid rain (Federer and Hornbeck, 1985; David et al., 1991; Johnson et al., 1994; Likens et al., 1996).

In strongly acid soils, as in the humid tropics, there is little CEC to buffer the soil solution. These soils are buffered by various geochemical reactions involving aluminum (Fig. 4.7). Aluminum is not a base cation inasmuch as its release to the soil solution leads to the formation of H^+ when Al^{3+} is precipitated as aluminum hydroxide:

$$Al^{3+} + H_2O \rightleftarrows Al(OH)^{2+} + H^+ \tag{4.9}$$

$$Al(OH)^{2+} + H_2O \rightleftarrows Al(OH)_2^+ + H^+ \tag{4.10}$$

$$Al(OH)_2^+ + H_2O \rightleftarrows Al(OH)_3 + H^+ \tag{4.11}$$

These reactions account for the acidity of many soils in the humid tropics (Sanchez et al., 1982a). Note that the reactions are reversible, so that the soil solution is also buffered against additions of H^+ by the dissolution of aluminum hydroxide. The acid rain received by the northeastern United States appears to dissolve gibbsite [$Al(OH)_3$] from many forest soils, leading to high concentrations of Al^{3+} that are toxic to fish in lakes and streams at high elevations. As stream waters flow to lower elevations, H^+ is consumed in weathering reactions with various silicate minerals, stream water pH increases, and aluminum hydroxides are

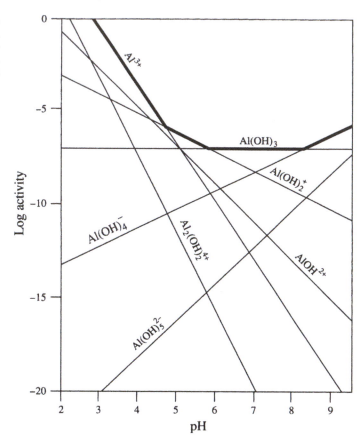

FIG. 4.7 The solubility of aluminum as a function of pH. For pH in the neutral range, gibbsite [Al(OH)$_3$] controls aluminum solubility, and there is little Al^{+3} in solution. Al^{+3} becomes more soluble at pH <4.7. *From Lindsay (1979).*

precipitated (Johnson et al., 1981). In other regions, dissolution of Al-organic complexes (rather than gibbsite) appears to control the concentration of dissolved Al^{3+} in the soil solution (Mulder and Stein, 1994; Allan and Roulet, 1994).

Anion adsorption capacity

In contrast to the permanent, negative charge in soils of the temperate zone, tropical soils dominated by oxides and hydrous oxides of iron and aluminum show variable charge, depending on soil pH (Uehara and Gillman, 1981; Sollins et al., 1988; Arai and Sparks, 2007). Under acid conditions these soils possess positive charge, as a result of the association of H$^+$ with the hydroxide groups on mineral surfaces (Fig. 4.8). With an experimental increase in pH, a soil sample is observed to pass through a zero point of charge (ZPC), where the number of cation and anion exchange sites is equal. These soils develop cation exchange capacity at high pH. For gibbsite, the ZPC occurs around pH 9.0, so significant anion adsorption capacity (AAC) is present in acid tropical soils in most field situations.

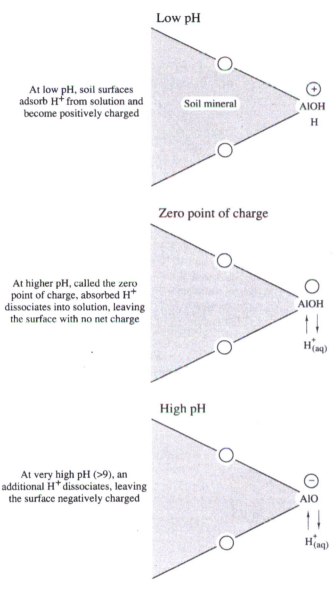

Low pH

At low pH, soil surfaces adsorb H$^+$ from solution and become positively charged

Soil mineral

$\oplus$
AlOH
H

Zero point of charge

At higher pH, called the zero point of charge, absorbed H$^+$ dissociates into solution, leaving the surface with no net charge

AlOH
$\uparrow\downarrow$
H$^+_{(aq)}$

High pH

At very high pH (>9), an additional H$^+$ dissociates, leaving the surface negatively charged

$\ominus$
AlO
$\uparrow\downarrow$
H$^+_{(aq)}$

FIG. 4.8 Variation in surface charge on iron and aluminum hydroxides as a function of pH of the soil solution. *From Johnson and Cole (1980).*

Anion adsorption capacity is also found on soil organic matter, but the ZPC of soil organic matter occurs at pH <2.0, so it offers little AAC in most conditions. The same is true of layered silicate minerals in temperate soils (Sposito, 1989; Polubesova et al., 1995). The ZPC of a bulk soil sample will depend on the relative mix of various minerals and organic matter (Chorover and Sposito, 1995). Tropical soils in Costa Rica show a ZPC at a pH of about 4.0, as a result of their mixture of soil organic matter and gibbsite (Sollins et al., 1988). Some anion adsorption capacity is found in temperate soils when iron and aluminum oxides and hydroxides occur in the lower soil profile (Johnson et al., 1981, 1986).

Anion adsorption capacity is typically greater on poorly crystalline forms of Fe and Al (oxalate-extractable), which have greater surface area than crystalline forms (dithionate-extractable) (Parfitt and Smart, 1978; Johnson et al., 1986; Chorover et al., 2004). Potential adsorption of various anions, including sulfate from acid rain, is positively correlated to the oxalate-extractable Al in a variety of soils (Harrison et al., 1989; Courchesne and Hendershot, 1989; Macdonald and Hart, 1990; Walbridge et al., 1991).

Anion adsorption follows the sequence:

$$PO_4^{3-} > SO_4^{2-} > Cl^- > NO_3^- \tag{4.12}$$

which accounts for the low availability of phosphorus in many tropical soils (Strahm and Harrison, 2007). Frequently anion exchange is described using a Langmuir model, in which the content of anions held on exchange sites is expressed as a function of the concentration in the solution (Travis and Etnier, 1981; Reuss and Johnson, 1986; Autry and Fitzgerald, 1993). Phosphorus, sulfate, and selenate (SeO_4^{2-}) are so strongly held that the binding is known as specific adsorption or ligand exchange, which is thought to replace -OH groups on the surface of the minerals (Fig. 4.9; Hingston et al., 1967; Guadalix and Pardo, 1991; Bhatti et al., 1998). Here, the adsorption of SO_4^{2-} from acid rain is associated with an increase in soil pH, a decline in apparent ZPC, and higher cation exchange capacity (e.g., Marcano-Martinez and McBride, 1989; David et al., 1991). All these anions are also involved in nonspecific adsorption, which is more readily reversible with changes in their concentration in the soil solution.

Anion adsorption capacity is inhibited by soil organic matter, especially organic anions, which tend to bind to the reactive surfaces of Fe and Al minerals (Johnson and Todd, 1983; Hue, 1991; Karltun and Gustafsson, 1993; Gu et al., 1995). Soils or soil layers that are rich in organic matter are less efficient in anion adsorption than those dominated solely by Fe and Al oxide and hydroxide minerals. Percolating waters often carry SO_4^{2-} from the upper, organic layers of the soil to lower depths, where it is captured by Fe and Al minerals (Dethier et al., 1988; Vance and David, 1991). Thus, anion adsorption capacity is determined by the effects of soil organic matter on a variety of soil properties: increasing AAC by inhibiting the crystallization of Fe and Al minerals, but reducing AAC by binding to the anion exchange sites (Johnson et al., 1986; Kaiser and Zech, 1998).

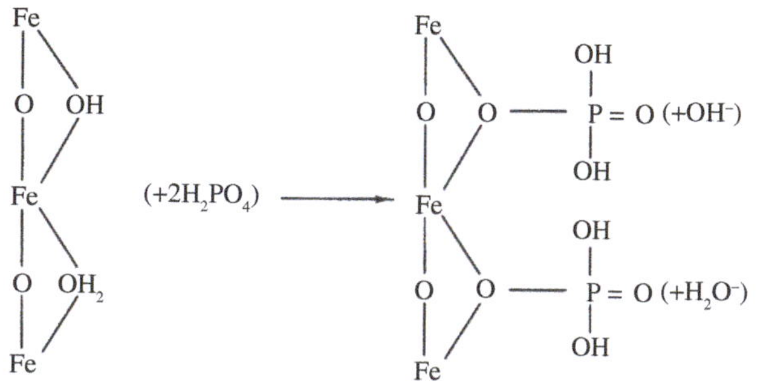

FIG. 4.9 The specific adsorption of phosphate by iron sesquioxides may release OH or H_2O to the soil solution. *From Binkley (1986).*

Phosphorus minerals

Phosphorus (P) deserves special attention, since it is often in limited supply for plants. The only primary mineral with significant phosphorus content is apatite, which can undergo carbonation weathering in a congruent reaction, releasing P:

$$Ca_5(PO_4)_3OH + 4H_2CO_3 \rightarrow 5Ca^{2+} + 3HPO_4^{2-} + 4HCO_3^- + H_2O \qquad (4.13)$$

Although this phosphorus may be taken up by plants (organic-P), a large proportion of the available P is involved in reactions with other soil minerals, leading to its precipitation in unavailable forms (Weihrauch and Opp, 2018).

As seen in Fig. 4.10, the maximum level of available phosphorus in the soil solution is found at a pH of about 7.0, although maximum plant uptake by roots is usually seen at lower values (Barrow, 2017). In acid soils, P availability is controlled by direct precipitation with iron and aluminum (Lindsay and Moreno, 1960; Arai and Sparks, 2007), whereas in alkaline soils phosphorus is often precipitated with calcium minerals (Cole and Olsen, 1959; Lajtha and Bloomer, 1988), and therefore may be deficient for optimal plant growth (Tyler, 1994). Binding to iron and aluminum oxides accounts for the low availability of phosphorus in many tropical soils (Sanchez et al., 1982a; Smeck, 1985; Agbenin, 2003). When phosphorus is captured in the interior of crystalline Fe and Al oxides, it is known as occluded P, which is essentially unavailable to biota.

Walker and Syers (1976) diagram the general evolution of phosphorus availability during the weathering of rocks containing apatite (Fig. 4.11). Initially there is limited P availability (e.g., Schlesinger et al., 1998; Darcy et al., 2018). However, apatite weathers rapidly, giving rise to phosphorus contained in various other forms and to a decline of total phosphorus in the system due to losses in runoff (Singleton and Lavkulich, 1987a; Filippelli and Souch, 1999; Selmants and Hart, 2010; Zhou et al., 2018). Non-occluded phosphorus includes forms that are held on the surface of soil minerals by a variety of reactions, including anion adsorption. With time, crystalline oxide minerals accumulate, and phosphorus accumulates in occluded forms. At the later stages of weathering and soil development, occluded and organic P dominate the forms of P remaining in the system (Cross and Schlesinger, 1995; Crews et al., 1995; Richardson et al., 2004; Yang and Post, 2011). At this stage, almost all available phosphorus may be found in organic forms in the upper soil profile, while phosphorus found at lower depths is largely bound with secondary minerals (Yanai, 1992; Werner et al., 2017). Plant growth may depend almost entirely on the release of phosphorus from dead organic matter, defining a biogeochemical cycle of phosphorus in the upper soil horizons (Wood et al., 1984; Roberts et al., 2015). Over longer periods, erosion and geologic uplift may rejuvenate sources of soil P by exposing new parent materials (Porder et al., 2007; Eger et al., 2018).

As seen for the weathering of silicate minerals, organic acids can influence the release of phosphorus during rock weathering. Jurinak et al. (1986) show how the production of oxalic acid (CH_2O_4) by plant roots can lead to the weathering of P from apatite (cf. Rosling et al., 2007). Oxalate production is directly related to soil phosphorus availability in sandy soils of Florida (Fig. 4.12). Organic acids can inhibit the crystallization of Al and Fe oxides, reducing the rate of phosphorus occlusion (Schwertmann, 1966; Kodama and Schnitzer, 1977, 1980), and allowing noncrystalline (amorphous) forms to dominate the P-adsorption capacity of the soil (Walbridge et al., 1991; Yuan and Lavkulich, 1994).

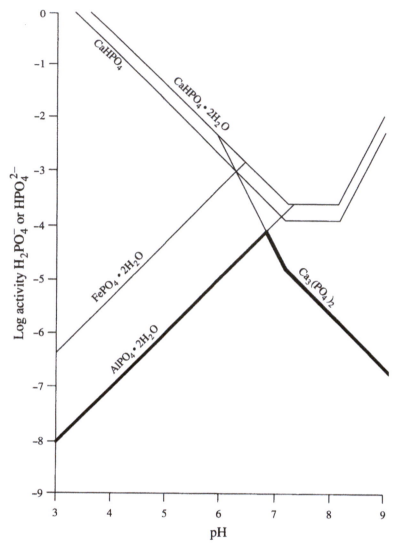

FIG. 4.10 The solubility of phosphorus in the soil solution as a function of pH. Precipitation with Al sets the upper limit on dissolved phosphate at low pH *(bold line)*; precipitation with Ca sets a similar limit at high pH. Phosphorus is most available at a pH of about 7.0. *Modified from Lindsay and Vlek (1977).*

In addition, phosphorus may be more available in the presence of organic acids, such as oxalate, which remove Fe and Ca from the soil solution by chelation and precipitation (Graustein et al., 1977; Welch et al., 2002; Wang et al., 2008). The production and release of oxalic acid by mycorrhizal fungi (Chapter 6) explains their importance to the phosphorus nutrition of higher plants (Bolan et al., 1984; Cromack et al., 1979) and the greater availability of

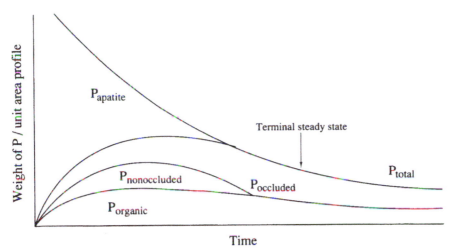

FIG. 4.11 Changes in the forms of phosphorus found during soil development on sand dunes in New Zealand. *Modified from Walker and Syers (1976).*

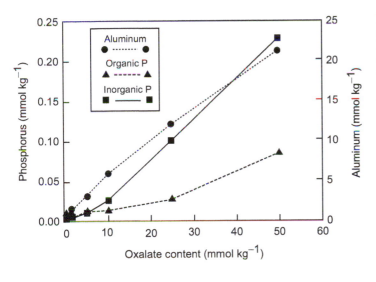

FIG. 4.12 Release of inorganic P, organic P, and Al from a spodic horizon of a pine forest in Florida, following a single oxalate addition at different levels. *From Fox and Comerford (1992). Used with permission of American Society of Agronomy.*

phosphorus under fungal mats (Fisher, 1972, 1977). Some workers believe that the mobilization of phosphorus by symbiotic fungi was a precursor to the successful establishment of plants on land (Chapter 2).

Soil development

Soils usually consist of a number of layers, or horizons, that collectively constitute the complete soil profile, or pedon. Rock weathering, water movement, and organic decomposition all influence the development of a soil profile under varying climatic conditions. Indeed, Jenny (1941, 1980) suggested that soil profile development could only be understood as a multivariate function:

$$\text{Soils} = (f) \text{ climate} + \text{biota} + \text{topography} + \text{parent material} + \text{time} \qquad (4.14)$$

Today, humans are one of the organisms that have dramatic effects on soil development in many regions (Amundson and Jenny, 1991; Richter, 2007). Recognition of the processes that occur in the different layers of soil is an essential part of understanding biogeochemical cycles on land. Chemical weathering occurs in the upper soil profile as well as in the underlying layers of fractured rock that constitute the critical zone. In this section, we consider soil development in forests, grasslands, and deserts.

Forests

In forests it is often easy to separate an organic layer, the forest floor or "O-horizon," from the underlying layers of mineral soil. The thickness and presence of the forest floor varies throughout the year, especially in regions where plant litterfall is strongly seasonal. In some tropical forests decomposition of fresh litter is so rapid that there is little surface litter (Olson, 1963; Vogt et al., 1986). On the other hand, slow decomposition in coniferous forests, especially in the boreal zone, results in the accumulation of a thick forest floor, known as a mor, that is sharply differentiated from the underlying soil (Romell, 1935). Much of the arctic zone is characterized by waterlogged soils, in which the entire rooting zone is composed of organic materials. Such peatland soils are known as Histosols. We will treat the special properties of waterlogged organic soils in Chapter 7.

The upper mineral soil is designated as the A-horizon. Soil water percolating through the forest floor often contains a variety of organic acids derived from the microbial decomposition of litter (Vance and David, 1991; Strobel, 2001). These organic acids dominate the weathering of soil minerals in the A-horizon. Solutions collected beneath the A-horizon carry cations and silicate, derived from weathering reactions (Table 4.2). Iron and Al are removed from the A-horizon by chelation with low-molecular-weight (LMW) organic acids that percolate downward from the forest floor (Antweiler and Drever, 1983; Driscoll et al., 1985; Zysset et al., 1999; Fuss et al., 2011). Some transport may occur in colloidal forms[d] (Ugolini et al., 1977a; Bazilevskaya et al., 2018). The removal of mineral components from the A-horizon is known as eluviation, whereas the downward transport of Fe and Al in conjunction with organic acids is known as podzolization (Chesworth and Macias-Vasquez, 1985; Lundström et al., 2000; Ferro-Vázquez et al., 2014).

Although it occurs throughout the world, podzolization is particularly intense in subarctic (boreal) and cool temperate forests (e.g., Ugolini et al., 1987; De Kimpe and Martel, 1976; Langley-Turnbaugh and Bockheim, 1998). Many of these areas are characterized by coniferous forests that produce litterfall that is rich in phenolic compounds and organic acids (Cronan and Aiken, 1985; Strobel, 2001). In these ecosystems, decomposition is slow and incomplete, and large quantities of organic acids are available to percolate from the forest floor

[d]These colloids are organo-mineral aggregations that are too small to settle from a solution by gravity.

TABLE 4.2 Chemical composition of precipitation, soil solutions, and groundwater in a 175-year-old *Abies amabilis* stand in Northern Washington.

Solution	pH	Total cations (mEq/L)	Soluble ions (mg/L)			Total (mg/L)	
			Fe	Si	Al	N	P
Precipitation							
Above canopy	5.8	0.03	<0.01	0.09	0.03	0.60	0.01
Below canopy	5.0	0.10	0.02	0.09	0.06	0.40	0.05
Forest floor	4.7	0.14	0.04	3.50	0.79	0.54	0.04
Soil							
15 cm E	4.6	0.12	0.04	3.55	0.50	0.41	0.02
30 cm B_s	5.0	0.08	0.01	3.87	0.27	0.20	0.02
60 cm B3	5.6	0.25	0.02	2.90	0.58	0.37	0.03
Groundwater	6.2	0.26	0.01	4.29	0.02	0.14	0.01

Data from Ugolini et al. (1977b). Copyright (1977) Williams and Wilkins.

into the underlying A-horizon. The pH of the soil solution is often as low as 4.0 (Dethier et al., 1988; Vance and David, 1991).

When the removal of Fe, Al, and organic matter is very strong, a whitish layer is easily recognized beneath the forest floor (Fig. 4.13). This horizon is sometimes designated as an A_e or E- (eluvial) horizon, which may consist entirely of quartz grains that are resistant to weathering and relatively insoluble under acid conditions (Pedro et al., 1978). These eluvial horizons reflect the importance of biota in soil development; in the absence of organic chelation, one would expect Fe and Al to accumulate as weathering products and for Si to dominate the losses from the upper soil profile.

FIG. 4.13 Soil profile, Nyanget, Svartberget Forest Research Station, north Sweden, showing the *E*- (whitish) and B-horizons. The scale is in decimeters. The role of biology in the process of podzolization is exemplified by the removal of Al and Fe minerals from the E-horizon in complexes with organic acids and their precipitation as oxides in the B-horizon. *From Lundström et al. (2000).*

During soil development, substances leached from the A- and E-horizons are deposited in the underlying B-horizons (Jersak et al., 1995; Langley-Turnbaugh and Bockheim, 1998). These are defined as the zone of deposition or the illuvial horizons, where secondary clay minerals accumulate. Clay minerals arrest the downward movement of dissolved organic compounds that are carrying Fe and Al (Greenland, 1971; Chesworth and Macias-Vasquez, 1985; Cronan and Aiken, 1985; Schulthess and Huang, 1991; Jansen et al., 2005). Typically, Fe oxides and hydrous-oxides precipitate first, and Al moves lower in the profile (Adams et al., 1980; Olsson and Melkerud, 1989; Law et al., 1991; Lundström et al., 2000; Ferro-Vázquez et al., 2014).

Strongly podzolized soils, Spodosols, are characterized by a dark spodic horizon-designated B_{hs} that is rich in Fe and organic matter. On fresh parent materials, a spodic horizon can develop in 350–1000 years (Singleton and Lavkulich, 1987b; Protz et al., 1984, 1988; Barrett and Schaetzl, 1992). In New England forests, the accumulation of organic matter in the spodic horizon appears to limit the loss of dissolved organic carbon in streams (McDowell and Wood, 1984). Most of the Fe and Al in the B-horizon is found in crystalline oxides (e.g., Fe_2O_3), but near root channels and pockets of buried organic matter, Fe^{3+} may be reduced to Fe^{2+} and leached away (Fimmen et al., 2008; Dubinsky et al., 2010; Fuss et al., 2011). Phosphorus can be mobilized from these areas of the B-horizon, which often appear grayish in the field (Chacon et al., 2006).

In warmer climates, decomposition is more rapid, smaller quantities of organic acids remain to percolate through the A-horizon, podzolization is less intense, and there is no sharply defined E-horizon (Pedro et al., 1978). In most areas of the tropics, decomposition is so complete that there is almost no soluble organic acid percolating through the soil profile (Johnson et al., 1977). In the absence of podzolization, Fe and Al are not removed from the upper soil profile by chelation; rather, they are precipitated as oxides and hydroxides that accumulate in the zone of active weathering. Unless they cycle through vegetation, cations and Si are lost to runoff. Podzolization is found in only a few tropical soils, usually those that develop on sandy parent materials (Bravard and Righi, 1989).

Soils in lowland tropical forests may be many meters in depth, because in many areas they have developed over millions of years without disturbances such as glaciation (Birkeland, 1984). In the absence of clear zones of eluviation and illuviation, the distinction of A- and B-horizons is difficult. Long periods of intense weathering have removed cations and silicon from the entire soil profile. The climate of most tropical regions includes high precipitation, and the solubility of Si increases with increasing temperature (Fig. 4.2). Many tropical soils are classified as Oxisols, on the basis of high contents of Fe and Al oxides throughout the soil profile (Richter and Babbar, 1991). Over large portions of lowland tropical rainforests, soils are acid (with Al buffering), low in base cations, and infertile with P deficiency (Sanchez et al., 1982b). Under extreme conditions, these soils are known informally as laterite.

A comparative index of soil formation and the degree of podzolization is seen in the ratio of Si to sesquioxides (Fe and Al) in the soil profile (Table 4.3). In boreal forest soils, Si is relatively immobile and Fe and Al are removed, which results in high values for this ratio in the A-horizon. The accumulation of secondary minerals such as montmorillonite in the moderately weathered soils of the glaciated portion of the United States yields Si/sesquioxide ratios of two to four as a result of the ratio of Si to Al in the crystal lattice. Silicon/sesquioxide ratios are lower in more highly weathered soils. During soil development in the southeastern

TABLE 4.3 Silicon/sesquioxide ($Al_2O_3 + Fe_2O_3$) ratios for the A- and B-horizons of some soils in different climatic regions.

Region	Number of sites	Mean Si/sesquioxide ratio		References
		A-horizon	B-horizon	
Boreal	1	12.0	8.1	Wright et al. (1959)
Boreal	1	9.3	6.7	Leahey (1947)
Cool-temperate	4	4.07	2.28	Mackney (1961)
Warm-temperate	6	3.77	3.15	Tan and Troth (1982)
Tropical	5	1.47	1.61	Tan and Troth (1982)

Note that the removal of Al and Fe results in high values in boreal and cool temperate soils, especially in the A-horizon. Lower values characterize tropical soils, where there is little differentiation between horizons as a result of the removal of Si from the entire profile in long periods of weathering.

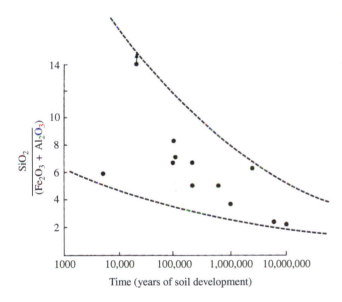

FIG. 4.14 Changes in Si-to-sesquioxide ratio during the long-term development of soils in the southeastern United States (Markewich and Pavich, 1991). *The data are recalculated from Markewich et al. (1989).*

United States, low ratios characterize older soils in which kaolinite (a 1:1 mineral) has accumulated as a secondary mineral (Fig. 4.14). Tropical Oxisols and Ultisols have very low values for this ratio in all horizons because they are dominated by kaolinite and iron and aluminum minerals.

Below the B-horizons, the C-horizon consists of coarsely fragmented soil material with little organic content. Carbonation weathering tends to predominate in the C-horizon, which may consist of a thick saprolite (Ugolini et al., 1977b). Weathering can extend to a depth of hundreds of meters in the soil profile, defining the critical zone (Richter and Markewitz, 1995), and it controls the chemistry of groundwater. For a rainforest in Puerto Rico, nearly all the Al was released by weathering in the upper soil profile, but 50% of the Si in stream water was derived from weathering in the C-horizon (Riebe et al., 2003). When the soil has developed from local materials, the C-horizon shows mineralogical similarity to the underlying parent rock. In contrast, when the parent materials have been deposited by transport, there may be little correspondence between the C-horizon and the underlying bedrock. Some of the world's richest agricultural soils are found on transported materials, such as glacial till, floodplain deposits, and volcanic ash.

The distribution of soils forms a continuous gradient over broad geographic regions. For example, the degree of podzolization varies beneath deciduous and coniferous forests in the same geographic region (Stanley and Ciolkosz, 1981; De Kimpe and Martel, 1976). Soil profile development on steep slopes is often incomplete as a result of landslides and other mechanical weathering events. Soils in floodplain areas and those that have received deposits of volcanic ash may have "buried" horizons (Dahlgren et al., 1999). In areas of recent deposition, soils with little or no profile development are known as Inceptisols and Entisols, respectively. In all cases one must remember that soil profile development is slow compared to changes in vegetation.

Soil profile development is also affected by human activities. In the Piedmont region of the southeastern United States, the forest floor often resides directly on top of the B-horizon. Here, the A-horizon was lost by erosion during past agricultural use. Past settlement often leaves a signature of human activity in the soil profile (e.g., Hejcman et al., 2013; Smejda et al., 2017), and military activity creates bomb-turbated soils (Hupy and Schaetzl, 2006).

In the northeastern United States, forest Spodosols have been exposed to decades of acid rain. At high elevations, the acidity of the solution percolating through the forest floor and A-horizon is not dominated by organic or carbonic acids, but by "strong" acids such as H_2SO_4 that have lowered soil pH (see Chapter 13). Similar patterns are reported for regions downwind of air pollution in China and Russia (Yang et al., 2015; Lapenis et al., 2004). At normal levels of acidity, aluminum is mobile only as an organic chelate, and it is precipitated in the B-horizon. Under the conditions of higher acidity, Al is mobile as Al^{3+}, a potentially toxic form that is carried through the lower profile to stream waters, with SO_4^{2-} as a balancing anion (Fig. 4.7; Johnson et al., 1972; Cronan, 1980; Reuss et al., 1987). Calcium, potassium and other cations are also lost from the soil profile (Likens, 2013). The overall rate of chemical weathering has increased (Cronan, 1980; April et al., 1986). Fortunately, with the implementation of the Clean Air Act in the early 1990s, the deposition of acidity and SO_4^{2-} and the stream water losses of Al^{3+} are beginning to decline (Palmer and Driscoll, 2002; Clow and Mast, 1999; Likens et al., 2002).

Grasslands

In contrast to soil development in forests, where precipitation greatly exceeds evapotranspiration and excess water is available for soil leaching and runoff, soil development in grasslands proceeds under conditions of relative drought. The products of chemical weathering are not rapidly leached from the soil profile, so soils remain near neutral pH with high base saturation. High contents of Ca and other cations tend to flocculate[e] clay minerals and organic

---

[e]Flocculation occurs when materials that are moving in water as colloids form insoluble aggregates, usually by the neutralization of hydrophyllic surface charges.

acids in the upper profile (Oades, 1988), limiting podzolization and the downward movement of weathering products. Often there is little development of a clay-rich or Bt-horizon in grassland soils until the upper profile has been leached free of Ca. Overall, the intensity of chemical weathering and podzolization in grassland soils is much less than in forests (Madsen and Nornberg, 1995).

Trends in the development of grassland soils are best seen by examining a transect across the midportion of the United States. Mean annual precipitation decreases westward from the tallgrass prairie near the Mississippi River valley to the shortgrass prairie at the base of the Rocky Mountains. Honeycutt et al. (1990) show that the thickness of the soil profile—measured by the depth to the peak content of clay—decreases from east to west along this gradient (Fig. 4.15). Along the same gradient, base saturation, pH, and the content of calcium increase (Ruhe, 1984; Gunal and Ransom, 2006). Similar trends are seen along a gradient of decreasing precipitation in tropical grasslands of eastern Africa (Scott, 1962). Soil pH increases abruptly when the mean annual precipitation is lower than potential evapotranspiration and there is little runoff (Slessarev et al., 2016). In the western Great Plains of the United states, the leaching of the soil profile is so limited that Ca precipitates as $CaCO_3$, which accumulates in the lower profile in calcic horizons, designated B_k and informally known as caliche.

The climatic regime of grasslands results in lower levels of plant growth than in forests, and smaller quantities of plant residues are added to the soil each year. Nevertheless, grassland soils contain large stores of organic matter because the limited availability of water also results in slower rates of decomposition (Chapter 5). Roots are a major source of organic matter in grassland soils, and there is evidence that plant matter added below ground decomposes more slowly than that added at the surface (Sokol and Bradford, 2019). Calcium also stabilizes the persistence of organic matter in grassland soils (Rowley et al., 2018). Most grassland soils in the temperate zone are classified as Mollisols, on the basis of a high content of organic carbon and high base saturation in the surface layers.

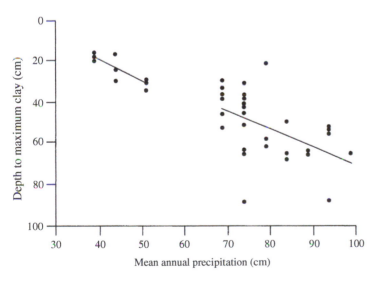

FIG. 4.15 Depth-to-peak content of clay in the soil profile, an index of weathering and soil development, decreases from east to west across the Great Plains of the United States as a function of the decrease in mean annual precipitation. *From Honeycutt et al. (1990). Used with permission of American Society of Agronomy.*

As seen for forest ecosystems, soil properties in grasslands vary locally as a result of differences in underlying parent materials and hillslope positions. Schimel et al. (1985) showed how the downslope movement of materials results in thin soils on hillslopes and accumulations of organic matter and nitrogen in local depressions. Similarly, Aguilar and Heil (1988) found that grassland soils derived from sandstone have lower contents of organic carbon, nitrogen, and total P than soils derived from fine-textured materials, such as shale, which have a higher content of clay minerals. The proportion of P contained in organic forms is greater for soils derived from sandstone. These differences in soil properties can strongly affect the local productivity of grasslands.

Deserts

The trends in soil development that are seen with increasing aridity in grasslands reach their extreme expression in deserts. Desert soils often contain clay minerals and horizons of clay accumulations derived from eolian dust, giving the appearance that substantial weathering has occurred (Singer, 1989; Reheis and Kihl, 1995). In fact, soil development in deserts occurs slowly because of limited weathering and infrequent leaching of the soil profile. In the face of limited weathering and runoff, many desert soils show a net gain of materials from atmospheric deposition (Ewing et al., 2006; Michalski et al., 2004a; Schlesinger et al., 2000b).

Despite the limited chemical weathering in deserts, the small amount of water percolating through these soils transports substances both vertically in the profile and horizontally across the landscape. As water is removed by plant uptake, soluble substances precipitate. Typically, well-developed desert soils contain $CaCO_3$ horizons that show progressive development and cementation through time (Gile et al., 1966). When desert soils contain distinct horizons, they are classified as Aridisols (Dregne, 1976).

Because $CaCO_3$ precipitates in desert soils in equilibrium with CO_2 derived from plant root respiration,

$$2CO_2 + 2H_2O \rightarrow 2H^+ + 2HCO_3^- \tag{4.15}$$

$$Ca^{2+} + 2HCO_3^- \rightarrow CaCO_3 \downarrow + H_2O + CO_2 \tag{4.16}$$

the carbonate carries a carbon isotopic signature that can be traced to photosynthesis (Chapter 5) and used as an index of past climate and vegetation (Quade et al., 1989). Seasonal variations in the activity of plant roots (Schlesinger, 1985) and soil microbes (Monger et al., 1991) affect $CaCO_3$ deposition and other aspects of soil development in deserts, despite the outward appearance that abiotic processes should be dominant. For instance, beneath desert shrubs, the deposition of $CaCO_3$ is inhibited by the presence of dissolved organic materials in the soil solution (Inskeep and Bloom, 1986; Reddy et al., 1990; Suarez et al., 1992), and calcic horizons are found deeper in the soil profile.

Depth to the $CaCO_3$ horizon in deserts shows a direct relation to mean annual rainfall and wetting of the soil profile (Arkley, 1963). Beneath the $CaCO_3$, one may find horizons in which $CaSO_4 2H_2O$ (gypsum) or $NaCl$ are dominant, reflecting the greater solubility and downward movement of these salts (Yaalon, 1965; Marion et al., 2008). Similar patterns are seen across the landscape, where Na, Cl, and SO_4 are carried to intermittent lakes in basin lows, while Ca remains in the upland soils of the adjacent piedmont (Drever and Smith, 1978; Eghbal et al., 1989; Amundson et al., 1989).

Despite sparse plant cover, much of the nutrient cycling in desert ecosystems is controlled by biota. With widespread root systems, desert shrubs accumulate nutrients from a large area and concentrate dead organic matter in the local area beneath their canopy. Most of the annual turnover of N, P, and other elements is controlled by biogeochemical processes in these "islands of fertility" (Schlesinger et al., 1996; Titus et al., 2002).

Models of soil development

The processes underlying soil profile development are conducive to simulation modeling. Models of soil chemistry include the weathering reactions described earlier in this chapter and equilibrium constants for the exchange of cations and anions between the soil solution and the mineral phases (Furrer et al., 1989). Depending on the time scale of the simulation, a model of soil chemistry usually routes daily or annual precipitation sequentially through the soil profile, where the solution achieves equilibrium with the soil minerals in each horizon. Removal of water from the soil profile is calculated from estimates of evaporation from the soil surface, plant uptake, and runoff to streams. Simulation models of soil processes have been constructed to predict soil profile development and to calculate losses of dissolved constituents in forested regions subject to acid rain (Reuss, 1980; Cosby et al., 1986; David et al., 1988; Perez-Fodich and Derry, 2019).

The long-term development of soils in arid regions is simulated in a model, CALDEP, developed by Marion et al. (1985), in which daily precipitation achieves equilibrium with carbonate biogeochemistry as it percolates through the soil profile. Plant root respiration is explicitly included in the calculation, and it varies seasonally and with depth in the profile. Plants also control the loss of water from the soil surface by evapotranspiration. The model suggests that the $CaCO_3$ horizon will be deeper in a soil profile developed from coarse textured parent materials, which allow greater percolation of water, and when plant root respiration varies seasonally, showing high values of soil CO_2 during the growing season.

Using current climatic conditions to parameterize precipitation and evaporation, the model was run to simulate 500 years of soil profile development (Fig. 4.16). The model predicted mean depths to the $CaCO_3$ that were much shallower than observed in a sample of 16 desert soils from Arizona. When the model was reparameterized using the cool, wet conditions that are thought to have been widespread in this region during the latest Pleistocene, the predicted depth of $CaCO_3$ closely matched that found in the field. These conditions produced greater percolation of soil moisture and lower rates of evaporation from the soil surface. Such models are only as good as the data used in the simulations, and rarely can models establish the importance of processes unequivocally. Nevertheless, models are useful for hypothesis development and for organizing research priorities. CALDEP suggests that most $CaCO_3$ horizons were formed during the Pleistocene, when deserts received more precipitation than today—consistent with the age of soil carbonates (Schlesinger, 1985). In the southwestern United States, most calcic horizons are >10,000 years old, and the $CaCO_3$ has accumulated at rates of $1.0–5.0 \, g/m^2/yr$ from the downward transport of Ca-rich minerals deposited from the atmosphere (Schlesinger, 1985; Capo and Chadwick, 1999; Van der Hoven and Quade, 2002).

4. The Lithosphere

FIG. 4.16 Depth to $CaCO_3$ in desert soils of Arizona, as a function of mean annual precipitation. The *dashed line* shows the prediction from the CALDEP model using current precipitation regimes. The *solid line* shows the best fit to actual data reported from the field. The *dotted line* shows the predictions when the model is run with postulated climatic data from the latest Pleistocene pluvial period. *Modified from Marion et al. (1985).*

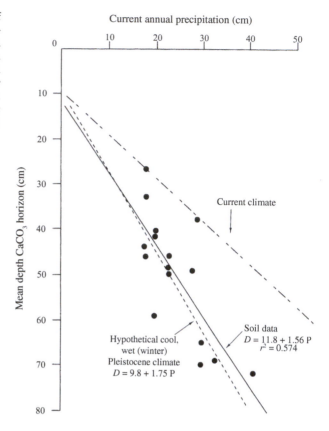

Weathering rates

Rock weathering and soil development are difficult to study because the processes occur slowly and the soil profile is nearly impossible to sample without disturbing many of the chemical reactions of interest. Nevertheless, estimates of weathering are needed to understand the biogeochemistry of local watersheds, where essential elements for biota are derived from the underlying rock (Chapter 6). Often we must infer weathering rates from what remains in the soil profile and what is lost to stream water (Likens, 2013). Estimates of the dissolved and suspended load of rivers allow us to calculate a global rate of chemical weathering, which supplies nutrient elements to land and marine biota.

Chemical weathering rates

Variations in the concentration of dissolved ions in streams can be related to the rate of discharge, the origin of their waters, and chemical weathering (Johnson et al., 1969; Maher, 2011). In a simple geochemical system, we might expect that stream water concentrations would be highest at periods of low flow, because most of the water would be derived from drainage of the soil profile where it would be in equilibrium with various rock weathering and ion-exchange reactions (Raymond, 2017). As stream flow increases, we might expect concentrations to decline as an increasing proportion of the flow is derived from precipitation and surface runoff, with little or no equilibration with the soil mineral phases. This simple geochemical model often explains the behavior of major ions in stream water (Ca, Mg, Na, Si, Cl, SO_4, and HCO_3), which are readily soluble in water and available in excess of biotic demand (Meyer et al., 1988).

During rainfall or seasonal flooding, the concentrations of dissolved ions in stream waters are often higher as the waters are rising than during the equivalent flows during the receding period (Whitfield and Schreier, 1981; McDiffett et al., 1989). The effect, known as hysteresis, is thought to result from an initial flushing of the highly concentrated waters that accumulate in the soil pores during periods of low flow. Not all ions show consistent hysteresis patterns, so to calculate the total annual loss of dissolved ions from a watershed, the stream flow discharge for each day must be multiplied by the concentration measured at that discharge and the products summed for all 365 days.

Even for elements that show lower concentrations at greater discharge, the total removal from the landscape is greatest during years of high stream flow (Fig. 4.17)—that is, the increase in flow predominates over the expected dilution of dissolved materials. In comparisons made over large geographic regions, concentrations are greatest in rivers that drain areas with limited runoff, but total transport is greater in rivers with greater discharge (e.g., Fig. 4.18; Bluth and Kump, 1994; Gaillardet et al., 1999).

Jennings (1983) shows a dilution of Ca and Mg with increasing discharge in areas of limestone in New Zealand, but the slope of the relationship changes slightly from summer to winter, reflecting a greater weathering of limestone by the more active respiration of roots during the summer (see also, Laudelout and Robert, 1994). When the mean residence time of the soil solution is short, plants exert control on stream water chemistry by the uptake of essential

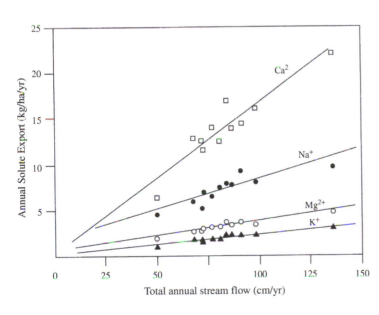

FIG. 4.17 Annual stream water loss of major cations as a function of the stream discharge in different years in the Hubbard Brook Experimental Forest in New Hampshire. *From Likens and Bormann (1995).*

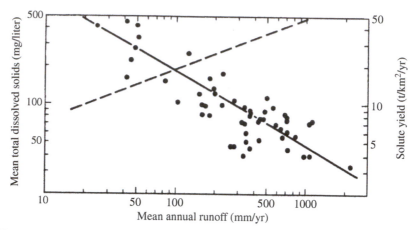

FIG. 4.18 Variation in the concentration of total dissolved solids (*solid line*) and the total annual transport of dissolved substances (*dashed line*, shown without data) for various streams in Kenya as a function of mean annual runoff. *From Dunne and Leopold (1978). Used with permission of Springer-Verlag.*

nutrient elements. In most areas, the concentrations of plant-essential nutrients, such as potassium and nitrate, show little relation to stream water discharge because they are regulated by microbial activity and root uptake (Lewis and Grant, 1979; Feller and Kimmins, 1979; McDowell and Asbury, 1994). For instance, in a deciduous forest in New Hampshire, the lowest potassium and nitrate concentrations in stream water are found during the low-flow periods of summer, when biotic demands are greatest (Johnson et al., 1969; Likens et al., 1994).

Plant uptake of essential elements and the retention of cations on soil minerals complicate the use of stream water concentrations to calculate the rate of chemical weathering over short periods (Taylor and Velbel, 1991; Gardner, 1990). However the eventual loss of cations to riverflow explains the decline in base saturation and pH during soil development (Bockheim, 1980). Losses of dissolved constituents from terrestrial ecosystems represent the products of chemical weathering and constitute chemical denudation of the landscape.

One of the best-known attempts to calculate the rate of chemical weathering began in 1963, when Gene Likens, Herbert Bormann, and Noye Johnson quantified the chemical budgets for the Hubbard Brook forest in New Hampshire (Likens, 2013). Here, a number of comparable watersheds are underlain by impermeable bedrock with no apparent subsurface drainage to groundwater. These workers reasoned that if the atmospheric inputs of chemical elements are subtracted from the stream water losses, the difference should reflect the annual release from rock weathering. They estimated the rate of rock weathering using the equation:

$$\text{Weathering} = \frac{(\text{Ca lost in stream water}) - (\text{Ca received in precipitation})}{(\text{Ca in parent material}) - (\text{Ca in residual material in soil})} \qquad (4.17)$$

The solution of this equation shows rather different amounts of weathering when the calculations are performed using various rock-forming elements (Table 4.4). The observed losses of calcium and sodium in stream water imply higher rates of weathering than what is

TABLE 4.4 Calculation of the rate of primary mineral weathering, using the stream water losses and mineral concentrations of cationic elements.

Element	Annual net loss (kg/ha/yr)	Concentration in rock (kg/kg of rock)	Concentration in soil (kg/kg of soil)	Calculated rock weathering (kg/ha/yr)
Ca	8.0	0.014	0.004	800
Na	4.6	0.016	0.010	770
K	0.1	0.029	0.024	20
Mg	1.8	0.011	0.001	180

Data from Johnson et al. (1968).

calculated using potassium and magnesium in the same equation. Johnson et al. (1968) suggest that the latter elements are accumulating in secondary minerals (illite and vermiculite) in the soil. In addition, trees may take up and store essential elements in long-lived tissues (e.g., wood growth), temporarily reducing the loss of some elements in stream water (Taylor and Velbel, 1991; see also Chapter 6).

In calculating weathering rates, it is often useful to examine the watershed budgets for Cl and Si. In nearly all cases, the atmospheric inputs of Si are trivial, so Si in streamwater is an index of net chemical weathering, which is its only source (e.g., Fig. 4.2). In contrast, the Cl content of rocks is normally very low. Nearly all Cl^- in streamwater is derived from the atmosphere, passing through the system relatively unimpeded by biotic activity or geochemical reactions (but see Lovett et al., 2005; Oberg et al., 2005). A balanced budget for Cl is often indicative of an accurate hydrologic budget for the watershed (Juang and Johnson, 1967; Svensson et al., 2012).

Release from rock weathering is the dominant source of Ca, Mg, K, Fe, and P in stream waters draining the Hubbard Brook Experimental Forest, whereas deposition from the atmosphere is the dominant input for Cl, S, and N, which have a small content in rocks (Table 4.5). In forests not subject to regional inputs of acid rain, the proportion of sulfur that is derived from the atmosphere is often somewhat lower than that at Hubbard Brook (e.g., Mitchell et al., 1986). A large number of watershed studies have been conducted, allowing similar calculations of weathering rates for a variety of ecosystems (Table 4.6; Henderson et al., 1978; Feller and Kimmins, 1979; Velbel, 1992; Likens, 2013).

A relative index of chemical weathering is calculated by summing the annual losses of various elements in stream water. In comparisons of ecosystems of the world, chemical denudation is found to increase with increasing runoff (Fig. 4.2). Total dissolved transport ranges from 47.6 to 372.7 kg/ha/yr among the watersheds in Table 4.6, and Alexander (1988) found that chemical denudation ranged from 20 to 200 kg/ha/yr among 18 undisturbed ecosystems in a variety of climatic regimes. Chemical weathering in southeastern Brazil was estimated to carry 280 kg/ha/yr (Fernandes et al., 2016). Often there is much regional variability in calculated weathering rates (Nezat et al., 2004; Schaller et al., 2010), but globally, rivers transport about 4×10^{15} g of dissolved substances to the oceans each year—an average of about 267 kg/ha/yr for the Earth's land surface (refer to Table 4.8 later in chapter).

TABLE 4.5   Inputs and outputs of elements from Hubbard Brook experimental forest.

	Inputs (%)		Output as
	Atmosphere	Weathering	percent of input
Ca	9	91	59
Mg	15	85	78
K	11	89	24
Fe	0	100	25
P	1	99	1
S	96	4	90
N	100	0	19
Na	22	78	98
Cl	100	0	74

Data from Likens et al. (1981).

TABLE 4.6   Net transport (export minus atmospheric deposition) of major ions, soluble silica, and suspended solids from various watersheds of forested ecosystems.

Watershed characteristics	Caura River, Venezuela	Gambia River, West Africa	Catoctin Mtns., Maryland	Hubbard Brook, New Hampshire
Size (km^2)	47,500	42,000	5.5	2
Precipitation (cm)	450	94	112	130
Vegetation	Tropical	Savanna	Temperate	Temperate
Net dissolved transport (kg/ha/yr)				
Na	19.4	3.9	7.3	5.9
K	13.6	1.4	14.1	1.5
Ca	14.2	4.0	11.9	11.7
Mg	5.7	2.0	15.6	2.7
HCO_3^-	124.0	20.3	78.1	7.7
Cl^-	−1.4	0.6	16.6	−1.6
SO_4^{2-}	1.5	0.4	21.2	14.8
SiO_2	195.7	15.0	56.1	37.7
Total transport (kg/ha/yr)	372.7	47.6	220.9	80.4

Modified from Lewis et al. (1987). Used with permission of Springer.

In most areas underlain by silicate rocks, the loss of elements in stream water relative to their concentration in bedrock follows the order:

$$Ca > Na > Mg > K > Si > Fe > Al \qquad (4.18)$$

but the order is affected by the specific composition of bedrock and the secondary minerals that are formed in the soil profile (Holland, 1978; Harden, 1988; Hudson, 1995). This general order reflects the tendency for Ca- and Na-silicates to weather easily and for the limited incorporation of Ca and Na in secondary minerals. The composition of individual streams may differ strongly from these patterns depending on local conditions. For instance, streams draining areas of carbonate terrain are dominated by Ca and HCO_3 (e.g., Laudelout and Robert, 1994), and where evaporate minerals are exposed, stream waters may contain high concentrations of Na, Cl, and SO_4 (e.g., Stallard and Edmond, 1983). In most cases, Fe and Al are retained in the lower soil profile as oxides and hydroxides, which are essentially immobile (Chesworth et al., 1981; Olsson and Melkerud, 2000; Lichter, 1998).

Because temperate forest soils are dominated by clay minerals with permanent negative charge, the loss of cations is determined by the availability of anions that "carry" cations through the soil profile to stream waters (Gorham et al., 1979; Johnson and Cole, 1980; Terman, 1977; Christ et al., 1999). In most cases, the dominant anion in soil water is bicarbonate (HCO_3^-); thus, the activities of plant roots and soil biota control chemical weathering and the composition of stream water. The high stream water concentration of HCO_3^- in the rainforest of Venezuela (Table 4.6) reflects the importance of carbonation weathering in tropical ecosystems (refer to Figs. 4.3 and 4.4). The total mobilization of cations and silicon in the Venezuelan study is also high, consistent with our expectations of rapid chemical weathering in tropical climates (Fig. 4.2). In Chapter 6, we will see how an increase in the availability of NO_3^-—a mobile anion—increases the loss of cations following disturbances, such as forest cutting (Fig. 6.15).

In both the northeastern United States and much of Europe, losses of base cations increased when the large amounts of H^+ and SO_4^{2-} were delivered to the soil by acid rain (Wright et al., 1994; Fernandez et al., 2003; Lapenis et al., 2004; Blake et al., 1999; Warby et al., 2009). Some of the cations were lost from the cation exchange capacity, while others were derived from increased rates of chemical weathering under the influence of strong acids (Miller et al., 1993; Likens et al., 1996). The cations move through the soil to stream water, with SO_4^{2-} acting as a balancing anion. After decades of acid rain in New England, the losses of cations from soils reduced cation concentrations in runoff to vanishingly low levels (Likens and Buso, 2012). With lower levels of acid input in recent years, these soils are now recovering (Lawrence et al., 2015). In contrast, in the more highly weathered soils of the southeastern United States, losses of cations from acid rain have been less dramatic because the soils in this region have greater anion adsorption capacity, so SO_4^{2-} is retained on soil minerals and cannot act as a mobile anion (Reuss and Johnson, 1986; Harrison et al., 1989; Cronan et al., 1990). As a historical index of anion adsorption in Tennessee, Johnson et al. (1981) found a lower content of SO_4^{2-} in the soils beneath a house built in 1890 (67 mg/kg) compared to that in adjacent field soils (195 mg/kg) that have been exposed to acid rain throughout the twentieth century.

In soils of the humid tropics, dominated by variable-charge minerals, one might expect that the abundance of mobile cations might determine the loss of anions from adsorption sites. Indeed, the loss of nitrate is retarded as water passes through experimental columns of tropical and volcanic soils (Wong et al., 1990; Bellini et al., 1996; Maeda et al., 2008; Strahm and Harrison, 2006), and the loss of NO_3^- from forests in Costa Rica is reduced by a high soil anion adsorption capacity (Matson et al., 1987; Reynolds-Vargas et al., 1994; Ryan et al., 2001). In many cases, the adsorption of SO_4^{2-} in tropical soils reduces its importance as a mobile anion in the stream waters and groundwaters draining these regions (Szikszay et al., 1990; Table 4.6).

The organic compounds in river water, especially the organic acids, are important in the dissolved transport of Fe and Al. Because, these metals form complexes with organic acids, they are carried at concentrations well in excess of the solubility of Fe and Al hydroxides in river water (Perdue et al., 1976). The importance of dissolved organic acids in the transport of metals to the sea is a good example of the influence of terrestrial biota over simple geochemical processes that might otherwise determine the movement of materials on the surface of the Earth.

The composition of "average" river water was calculated by Livingstone (1963) from measurements on a large number of rivers (Table 4.7; see also Table 9.1). Livingstone's estimate of total dissolved transport, 3.76×10^{15} g/yr, is largely confirmed by more recent work (e.g., Meybeck, 1979; see also Table 4.8). Nearly all the Ca, Mg, and K in river water is derived from rock weathering (Table 4.9). Weathering of carbonates is the dominant source for Ca, while silicates are the dominant source for Mg and K (Holland, 1978). Primary minerals in igneous rocks account for 30% of the dissolved constituents from chemical weathering delivered to the ocean—slightly less than the proportional exposure of igneous rocks at the Earth's surface (Durr et al., 2005). The chemical weathering of sedimentary rocks, especially carbonates, accounts for the remainder (Li, 1972; Blatt and Jones, 1975; Suchet et al., 2003).

Not all of the constituents in rivers are derived from rock weathering. Reflecting the importance of carbonation weathering, about two-thirds of the HCO_3^- in rivers is derived from

TABLE 4.7   Mean composition of dissolved ions in river waters of the world.

Continent	HCO_3^-	SO_4^{2-}	Cl^-	NO_3^-	Ca^{2+}	Mg^{2+}	Na^+	K^+	Fe	SiO_2	Sum
North America	68	20	8	1	21	5	9	1.4	0.16	9	142
South America	31	4.8	4.9	0.7	7.2	1.5	4	2	1.4	11.9	69
Europe	95	24	6.9	3.7	31.1	5.6	5.4	1.7	0.8	7.5	182
Asia	79	8.4	8.7	0.7	18.4	5.6	9.3		0.01	11.7	142
Africa	43	13.5	12.1	0.8	12.5	3.8	11		1.3	23.2	121
Australia	31.6	2.6	10	0.05	3.9	2.7	2.9	1.4	0.3	3.9	59
World	58.4	11.2	7.8	1	15	4.1	6.3	2.3	0.67	13.1	120
Anions[a]	0.958	0.233	0.220	0.017							1.428
Cations[a]					0.750	0.342	0.274	0.059			1.425

[a] *Millequivalents of strongly ionized components.*
Concentrations in mg/L.
Source: Livingstone (1963).

TABLE 4.8 Chemical and mechanical denudation of the continents.

Continent	Chemical denudation[a]		Mechanical denudation[b]		Ratio mechanical/chemical
	Total (10^{14} g/yr)	Per unit area (kg/ha/yr)	Total (10^{14} g/yr)	Per unit area (kg/ha/yr)	
N. America	7.0	330	14.6	840	2.1
S. America	5.5	280	17.9	1000	3.3
Asia	14.9	320	94.3	3040	6.3
Africa	7.1	240	5.3	350	0.7
Europe	4.6	420	2.3	500	0.5
Australia	0.2	20	0.6	280	3.0
Total	39.3	267	135.0	918	3.4

[a] From Garrels and MacKenzie (1971).
[b] From Milliman and Meade (1983).

TABLE 4.9 Sources of major elements in world river waters (in percent of actual concentrations).

Element	Weathering				Pollution
	Cyclic salt	Carbonates	Silicates	Evaporites	
Ca^{++}	0.03	65	18	8	9
HCO_3^-	<<1	61	37	0	2
Na^+	0.9	0	21	50	28
Cl^-	1.7	0	0	68	30
SO_4^{2-}[a]	0.2	0	0	29	28
Mg^{++}	0.11	36	54	<<1	8
K^+	0.04	0	87	5	7
H_4SiO_4	0	0	99+	0	0

[a] Sulfate is also derived from the weathering of pyrite.
From Berner and Berner (2012).

the atmosphere, either directly from CO_2 or indirectly via organic decomposition and root respiration that contribute CO_2 to the soil profile (Holland, 1978; Meybeck, 1987; Moosdorf et al., 2011). Because chemical weathering involves the reaction between atmospheric constituents and rock minerals, weathering of 100 kg of igneous rock results in 113 kg of sediments that are deposited in the ocean and about 2.5 kg of salts that are added to seawater (Li, 1972). Thus, a significant fraction of the transport of total dissolved substances in rivers is derived from the atmosphere and does not represent true chemical denudation of the continents

(Berner and Berner, 2012). A small fraction of the Na, Cl, and SO$_4$ in riverflow is also derived from the atmosphere as marine aerosols ("cyclic salts"; Chapter 3) that are deposited on land.[f]

The river transport of some dissolved ions has been increased by human activities, such as mining, which accelerate the natural rate of crustal exposure and rock weathering on Earth (Bertine and Goldberg, 1971; Martin and Meybeck, 1979; Raymond and Neung-Hwan, 2009). The widespread use of roadsalt as a deicer has dramatically increased the concentration of Cl in runoff waters (Kaushal et al., 2005), and losses of NO$_3$ from fertilizers account for its relatively high concentrations in the runoff from agricultural regions (Chapter 12). The flux of HCO$_3$ in the Mississippi River appears to have increased significantly during the past century as a result of a variety of human activities such as agricultural liming and rising CO$_2$ in Earth's atmosphere (Raymond et al., 2008; Raymond and Hamilton, 2018). Considering longer periods of time, it is likely that the chemical weathering and dissolved transport in rivers today is lower than immediately following the last continental glaciation about 18,000 years ago, when large amounts of finely ground materials were exposed to chemical weathering (Vance et al., 2009).

Gibbs (1970) used the concentrations of ions in major world rivers to examine the origins of the dissolved constituents in their waters. Rivers dominated by precipitation show low concentrations of dissolved substances, and a high ratio of Cl to the total of Cl + HCO$_3$, reflecting the importance of Cl from rainfall (Zone A in Fig. 4.19). Rivers in which the dissolved load is largely derived from chemical weathering show higher concentrations of dissolved substances, and HCO$_3$ is the dominant anion (Zone B), reflecting the importance of carbonation weathering in most soils. Rivers that pass through arid regions lose a significant amount of water to evaporation before reaching the ocean. These rivers (Zone C) show the greatest concentrations of dissolved ions and high ratios of Cl/(Cl + HCO$_3$), because HCO$_3$ has been removed by the chemical precipitation of minerals such as CaCO$_3$ in soils and stream sediments (Holland, 1978). In this scheme, seawater represents the endpoint of the evaporative concentration of river waters. These relationships are also seen when Na and Ca are used to scale the x axis in Fig. 4.19, with the relative concentration of Na as an index of rainfall and Ca as an index of chemical weathering (Gaillardet et al., 1999).

Mechanical weathering

In addition to chemical denudation, a large amount of material derived from mechanical weathering is eroded from land and carried in rivers as the particulate or suspended load (Jeandel and Oelkers, 2015). These materials have received less attention from biogeochemists because their elemental contents are not immediately available to biota. Overall, the mean rate of soil formation from mechanical weathering is about 0.1 mm/yr (Stockmann et al., 2014).

[f] The amount of Cl in rivers that is derived from the atmosphere has been the subject of some controversy. Some budgets (e.g., Table 4.9) suggest that only a small amount of Cl is "cyclic," based on the observations of Stallard and Edmond (1981, 1983) in the Amazon. Other workers have assumed that a larger fraction of the Cl in global riverflow is derived from the sea, with values ranging from 85% (Dobrovolsky, 1994, p. 83) to nearly 100% (Möller, 1990, cf. Fig. 3.17).

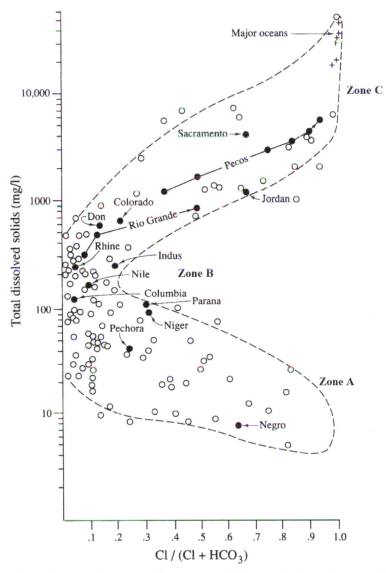

FIG. 4.19 Variations in the total dissolved solids in rivers and lakes as a function of the ratio of Cl/(Cl+HCO$_3$) in their waters. *From Gibbs (1970). Used with permission of the American Association for the Advancement of Science.*

The concentration of suspended sediment often shows a curvilinear relationship with stream flow, increasing exponentially at high flows (Parker and Troutman, 1989; Fig. 4.20). At low flows, suspended sediments are dominated by organic materials, but the proportion of organic matter declines as the amount of suspended sediment increases during high flows, when soil erosion is greatest (Meybeck, 1982; Ittekkot and Arain, 1986; Paolini, 1995). Long-term records show that the sediment transport during occasional extreme events often

FIG. 4.20   Concentration of particulate matter as a function of stream flow in the Hubbard Brook Experimental Forest of New Hampshire (United States). *From Bormann et al. (1974). Used with permission of the Ecological Society of America.*

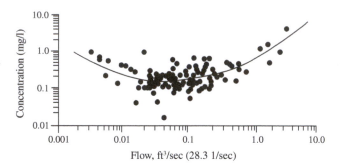

exceeds the total transport during long periods of more normal conditions (Van Sickle, 1981; Swanson et al., 1982). Large concentrations of suspended sediments are found during flash floods in deserts (Baker, 1977, Fisher and Minckley, 1978, Laronne and Reid, 1993). While the presence of vegetation appears to increase the rate of chemical weathering in soils, it retards the removal of materials by mechanical weathering (McMahon and Davies, 2018). In a world without plants, most hillslopes would be barren (Amundson et al., 2015).

Transport of suspended sediments in world rivers is affected by many factors, including elevation, topographic relief, and runoff from the watershed (Gaillardet et al., 1999). Large rates of sediment transport are found in mountainous regions, but areas of low topography dominate the Earth's land surface and contribute disproportionately to total sediment mobilization (Willenbring et al., 2013). Although rivers draining arid regions show high concentrations of suspended sediments, their total flow is limited, so the loss of soil materials per unit of landscape is rather low (Milliman and Meade, 1983). Rivers draining southern Asia carry 70% of the global transport of suspended sediments, 13.5×10^{15} g/yr (refer to Table 4.8). Large sediment loads in the rivers of China are derived from the erosion of massive deposits of wind-derived soils, loess, in their drainage basin. In contrast, despite carrying 20% of the world's riverflow, the Amazon River carries only about 600 to 900×10^{12} g of suspended sediment each year, about 5% of the world's total (Mikhailov, 2010; Wittmann et al., 2011). Most of the Amazon Basin is situated at low elevations with limited topographic relief, which accounts for its relatively low yield of suspended sediments (Meybeck, 1977).

Much of the sediment removed from uplands is deposited in stream channels and floodplains in the lower reaches of rivers (Trimble, 1977; Longmore et al., 1983; Gurtz et al., 1988). Thus, the sediment yield per unit area of watershed declines with increasing watershed area (Milliman and Meade, 1983). Despite large seasonal variations in volume, the daily sediment transport of the Amazon is rather constant as a result of storage of sediment in the floodplain during periods of rising waters and remobilization during falling waters (Meade et al., 1985).

Erosion increases when vegetation is removed (Bormann et al., 1974), and global land cleared by humans for agriculture, mining, and construction has dramatically increased the transport of suspended sediments in many parts of the world (Wilkinson and McElroy, 2007; Hooke, 2000; Haff, 2010; Vanmaercke et al., 2015). For instance, in the eastern United States, coal mining now vastly exceeds all natural processes that mobilize materials from the Earth's crust (Fig. 4.21). Worldwide, human activities have simultaneously nearly doubled the transport of suspended materials in rivers, but reduced the amount of the mobilized

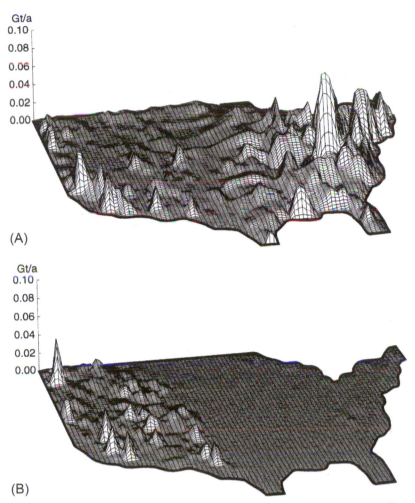

FIG. 4.21   Rate at which surficial materials are mobilized from the Earth's crust (Gt/yr = 10^{15} g/yr) in grid cells measuring $1 \times 1°$ latitude and longitude by humans (A) and rivers (B). The massive mobilization of crustal materials in the eastern United States is largely the result of surface mining for coal (Hooke, 1999).

material reaching the oceans owing to the construction of dams and reservoirs (Syvitski et al., 2005; Li et al., 2019a). Large amounts of suspended sediment are captured in lakes and floodplains and behind dams and other human structures (Dynesius and Nilsson, 1994; Vörösmarty et al., 2003; Walling and Fang, 2003).

Total denudation rates

The total denudation of land is dominated by the products of mechanical weathering, which exceeds chemical weathering by three to four times, worldwide (Table 4.8). The mean rate of total continental denudation is about 1000 kg/ha/yr, with approximately 75% carried in the suspended sediments in rivers (Alexander, 1988; Tamrazyan, 1989; Wakatsuki and Rasyidin, 1992; Gaillardet et al., 1999). The importance of mechanical weathering increases with elevation (e.g., Reiners et al., 2003; Gaillardet et al., 1999); differences in mean elevation among the continents explain much of the variation in mechanical weathering in Table 4.8.

Recent estimates of the total transport of suspended materials in all rivers of the world range from 12.6 to 13.5×10^{15} g/yr (Milliman and Meade, 1983; Syvitski et al., 2005). Assuming that the specific gravity of suspended sediment is 2.5 g/cm^3, these estimates are 4–5 times higher than estimates (1.27 km^3/yr) of the volume of deep ocean sediments derived from land (Howell and Murray, 1986) and much greater than the current mass (1.3×10^{15} g/yr; Plank, 2014) and volume (0.73 km^3/yr; Plank and Langmuir, 1998) of sedimentary materials being subducted into the mantle. Presumably, most sediment is deposited near the shore, in continental shelf deposits (Rea and Ruff, 1996). Gregor (1970) suggests that the global rate of sediment transport, an overall measure of mechanical weathering, may have been about four times greater before the land surface was colonized by plants (cf. Wilkinson and McElroy, 2007; McMahon and Davies, 2018). Indeed, today, especially high concentrations of suspended sediment are seen in rivers draining arid and semi-arid regions where vegetation is sparse (Milliman and Meade, 1983).

Because Fe, Al, and Mn are only slightly soluble in water, particulate and suspended sediments account for most of the removal of these elements from terrestrial ecosystems (Table 4.10; Benoit and Rozan, 1999). The numbers shown in the table are estimated by comparisons of the concentrations in bedrock to those in weathered residues (river particulates); both are normalized to Al. Suspended sediments are also enriched in phosphorus, owing to chemical reactions between dissolved P and various soil minerals (Avnimelech and McHenry, 1984; Sharpley, 1985). As a result of a variety of human activities, the river transport of many metals (e.g., Cu, Pb, and Zn) is now greater than the transport under preindustrial conditions (Table 4.11), but it is interesting to note that the concentration of

TABLE 4.10   Loss or gain of major elements during chemical weathering.

Element	wt%		% lost (−) or gained (+)
	Surficial rock	Weathered river particulate	
Al	6.93	10.6	0
Fe	3.59	5.75	0
Mn	0.072	0.25	+230
Na	1.42	0.27	−88
Ca	4.5	0.63	−91
Mg	1.64	0.63	−75
K	2.44	2.25	−40
Si	27.5	27.0	−36

Note: The changes in concentration are normalized to Al.
From Canfield (1997).

TABLE 4.11   Estimates of some elemental fluxes to the oceans in rivers (10^{12} g/yr).

	Ca	Na	Mg	Si	Fe	Cu	Pb	Zn
River particulate load	345	110	209	4430	733	1.55	2.3	5.4
River dissolved load	495	131	129	203	1.5	0.37	0.04	1.1
Total river load	840	241	338	4630	734	1.9	2.3	6.5
Theoretical load[a]	946	298	345	5780	754	0.67	0.33	2.6
Discrepancy	N.S.	N.S.	N.S.	N.S.	N.S.	+1.2	+2.0	+3.9
World mining production	–	–	–	–	–	4.4	3.0	3.9

[a] Based on weathering of average rock.
N.S., not significant.
From Martin and Meybeck (1979).

Pb in rivers has declined recently, presumably as a result of the decreased use of leaded gasoline in automobiles (Smith et al., 1987; Trefry et al., 1985).

Summary

In this chapter we have seen that the rates of weathering and soil development are strongly affected by biota, particularly through carbonation weathering and the production of organic acids in the soil profile (Kelly et al., 1998). It is logical to speculate that the rate of carbonation weathering was lower before the advent of land plants, when it depended solely on the downward diffusion of atmospheric CO_2 through the soil profile. However, at periods in Earth's early history, the concentration of atmospheric CO_2 was most certainly higher than today, presumably yielding high rates of carbonation weathering. Weathering is also driven by the availability of water. The high concentration of CO_2 on Venus (Table 2.6) is ineffective in weathering because the surface of the planet is dry (Nozette and Lewis, 1982).

By mining minerals from its crust and extracting buried fossil fuels from the Earth, humans have increased the global rates of chemical and mechanical weathering and added significant quantities of dissolved materials to global riverflow. Currently, the removal of fossil fuels from the Earth's crust mobilizes and oxidizes carbon at a rate (10×10^{15} g/yr), which exceeds the total of all natural chemical weathering (Table 4.8). Human exposure and erosion of soils in agricultural use have increased the global denudation due to mechanical weathering by a factor of about 2 (Syvitski et al., 2005), leading to increases in the rate of sediment accumulation in estuaries and river deltas (Chapter 8).

Chemical weathering is a source of essential elements for the biochemistry of life, but streamwater runoff removes these elements from the land surface. Chemical reactions among soil constituents and uptake by biota determine the rate of loss, but the inevitable removal of cations results in lower soil pH and base saturation through time (Bockheim, 1980; Huston, 2012). Phosphorus is particularly critical as a soil nutrient because it is not abundant in crustal rocks and is easily precipitated in unavailable forms in the soil. Old soils in highly weathered landscapes are composed of resistant, residual Fe and Al oxide minerals. In these soils, P is often deficient and even the small atmospheric inputs of P are important for plant growth (Chadwick et al., 1999; Chapter 6).

Introduction

Photosynthesis is the process that transfers carbon from its oxidized form, CO_2, to the reduced (organic) forms, known as carbohydrates that result in plant growth. Directly or indirectly, photosynthesis[a] provides the energy for all other forms of life in the biosphere, and the use of plant products for food, fuel, and shelter brings photosynthesis into our daily lives. The fossil fuels that power modern society are derived from photosynthesis in the geologic

---

[a] The energy derived from chemoautotrophy (Chapter 2) is tied to photosynthesis indirectly, since the O_2 used in chemoautotrophy is a product of photosynthesis that would not otherwise be found in significant quantities on Earth.

past (Dukes, 2003). Plant growth affects the composition of the atmosphere (Chapter 3), the development of soils (Chapter 4) and the losses of nutrients in drainage waters (Chapter 6), linking photosynthesis to other aspects of global biogeochemistry. Indeed, the presence of organic carbon in soils and sediments and O_2 in our atmosphere provides the striking contrast between *bio*geochemistry on Earth and the simple geochemistry that characterizes our neighboring planets (Chapter 2).

In this chapter we consider the measurement of net primary production—the rate of accumulation of organic carbon in land plants. A similar treatment of photosynthesis in the world's lakes and rivers is given in Chapter 8, and the oceans in Chapter 9. The rate of plant growth varies widely over the land surface. Deserts and continental ice masses may have little or no net primary production (NPP), while tropical rainforests can show annual production of $>1000\,g\;C/m^2$ (Malhi, 2012).

Various environmental factors affect the rate of net primary productivity on land and the total storage of organic carbon in plant tissues (biomass), dead plant parts (detritus), and soil organic matter. As any home gardener knows, light and water are important, but plant growth is also determined by the stock of available nutrients in the soil. These nutrients are ultimately derived from the atmosphere or from the underlying bedrock (Table 4.5). The overall storage of carbon on land is determined by the balance between primary production and decomposition, which returns carbon to the atmosphere as CO_2.

Photosynthesis

Containing a central atom of magnesium, the chlorophyll molecule is a prime example of how plants have incorporated an abundant product of rock weathering as an essential element in their biochemistry (Fig. 5.1). When photosynthetic pigments absorb sunlight, a few of the chlorophyll molecules are oxidized—passing an electron to a sequence of proteins that ultimately add the electron to a high-energy molecule, known as nicotinamide adenine dinucleotide phosphate ($NADP^+$), which is thus reduced to form NADPH. The chlorophyll molecule regains an electron from a water molecule, which is split by an enzyme containing manganese, calcium, and chlorine, in a complex three-dimensional structure (Yano et al., 2006; Cox et al., 2014; Suga et al., 2019). This reaction is the origin of O_2 in the Earth's atmosphere:

$$2H_2O \rightarrow 4H^+ + 4e^- + O_2 \uparrow \qquad (5.1)$$

In all cases, the photosynthetic pigments and proteins are embedded in a cell membrane, which allows protons (e.g., H^+ of Eq. 5.1) to build up to high concentrations on one side of the membrane and for this potential energy to be captured in a high-energy compound, adenosine triphosphate or ATP. In higher plants, the accumulation of protons occurs within the chloroplasts of leaf cells, whereas in photosynthetic bacteria, the reaction is conducted across the external cell membrane.

The high-energy compounds NADPH and ATP are then used by a suite of enzymes to reduce CO_2 and build carbohydrate molecules. The reaction begins with the enzyme ribulose

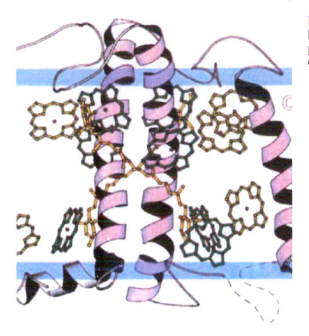

FIG. 5.1 Molecular structure of the light-harvesting complex of photosystem II, showing the position of the Mg-porphyrin groups. *From Kuhlbrandt et al. (1994).*

bisphosphate carboxylase, informally known as Rubisco, which adds CO_2 to the basic carbohydrate unit.[b] The overall reaction for photosynthesis is shown in the following equation,

$$CO_2 + H_2O \rightarrow CH_2O + O_2 \uparrow \tag{5.2}$$

but we should remember that the process occurs in two stages. First, the capture of light energy allows water molecules to be split and high-energy molecules to form. This drives the second reaction in which CO_2 is added to carbohydrates by Rubisco. Rubisco is likely the most abundant enzyme on Earth,[c] and the demand for nitrogen to synthesize Rubisco leads to nitrogen limitation of photosynthesis in many habitats (Chapter 6). The second reaction can be executed in the laboratory, given the necessary reactants and the provision

[b] For understanding global biogeochemistry, we focus on the photosynthesis of C3 plants, which account for the overwhelming proportion of plant biomass and net primary productivity on Earth. C3 plants are so named because the first product of the photosynthetic reaction is a carbohydrate containing three carbon atoms. However, some plant species, largely warm-climate grasses, conduct photosynthesis by another biochemical pathway, known as C4 photosynthesis (Ehleringer and Monson, 1993). C4 plants may account for up to 23% of global net primary production (Lloyd and Farquhar, 1994; Still et al., 2003), but their contribution to global biomass is small because most species are not woody. The overall photosynthetic reaction is identical to Eq. (5.2), but C4 plants have different water-use efficiency and a different C-isotopic fractionation in their tissues (average −12‰). The isotopic ratio of plant debris preserved in soils can be used to trace changes in the past distribution of C3 and C4 plants (e.g., Quade et al., 1989; Ambrose and Sikes, 1991).

[c] Recognizing that humans depend on green plants for their food, it is interesting to calculate that there are about 5 kg of Rubisco available in the Earth's vegetation to supply each of us with the products of photosynthesis (Phillips and Milo, 2009).

of Rubisco and other enzymes as catalysts, constituting a form of artificial photosynthesis (Schwander et al., 2016).

The efficiency of photosynthesis relative to the sunlight absorbed by chlorophyll is known as the *quantum yield efficiency*, which is normally close to 0.081 mol of CO_2 captured in carbohydrate per mole of photons absorbed, when soil water and other environmental factors are optimal (Singsaas et al., 2001). Photosynthesis can occur in markedly low-light environments, such as beneath snow and ice and under translucent rocks (Starr and Oberbauer, 2003; Schlesinger et al., 2003; Hancke et al., 2018); moonlight, however, is not enough (Raven and Cockell, 2006).

The carbon dioxide used in photosynthesis diffuses into plant leaves through pores, known as stomates,[d] which are generally found on the lower surface of broad-leaf plants. One factor that determines the rate of photosynthesis is the stomatal aperture, which plant physiologists express as stomatal conductance (g) in units of cm/s. Stomatal conductance is controlled primarily by the availability of light and water to the plant and the concentration of CO_2 inside the leaf, where it is consumed by photosynthesis. When well-watered plants are actively photosynthesizing, internal CO_2 is relatively low and stomates show maximum conductance. Under such conditions, the amount and activity of Rubisco may determine the rate of photosynthesis (Sharkey, 1985).

Water-use efficiency

There is a trade-off in photosynthesis; when plant stomates are open, allowing CO_2 to diffuse inward, O_2 and H_2O diffuse outward to the atmosphere. The loss of water through stomates, transpiration, is a major mechanism by which soil moisture is returned to the atmosphere (Chapter 10). In the Hubbard Brook Experimental Forest in New Hampshire (see Chapter 4), about 25% of the annual precipitation is lost by plant uptake and transpiration; stream flow increased by 26–40% when the forest was cut (Likens, 2013). Globally about 39% of the precipitation that falls on land is returned to the atmosphere by plants (Schlesinger and Jasechko, 2014). Because water for plant growth is often in short supply (Kramer, 1982; Green et al., 2019), these large losses of water by plants are somewhat surprising. One might expect natural selection for more efficient use of water by plants.

Plant physiologists express the loss of water relative to photosynthesis as water-use efficiency (WUE), namely,

$$\text{WUE} = \text{mmoles of } CO_2 \text{ fixed per moles of } H_2O \text{ lost} \tag{5.3}$$

For most plants, water-use efficiency typically ranges from 0.86 to 1.50 mmol/mol, depending on environmental conditions (Osmond et al., 1982). Water-use efficiency is higher at lower stomatal conductance and sensitive to the scale (e.g., leaf or ecosystem) at which it is measured (Medlyn et al., 2017).

As atmospheric concentrations of CO_2 rise, plants are able to acquire more CO_2 at lower stomatal conductance, thus increasing WUE (Bazzaz, 1990; Ceulemans and Mousseau, 1994). There is also some evidence that the number of stomates per unit of leaf surface has declined

[d] Also known as stomata.

as atmospheric CO_2 has risen during the Industrial Revolution (Woodward, 1987, 1993; Peñuelas and Matamala, 1990). The olive leaves preserved in King Tut's tomb (1327 BC) have a higher density of stomates than the leaves of the same species growing in Egypt today (Beerling and Chaloner, 1993).

Eq. (5.3) largely applies to short-term experiments in the laboratory. In nature, variations in water use among species in space and time confers some stability in regional WUE and plant productivity in response to shifts in water availability (Anderegg et al., 2018). For the biogeochemist, long-term average WUE can be estimated from the carbon isotope composition of plant tissues, especially tree rings. This method is based on the observation that the diffusion of $^{12}CO_2$ is more rapid than that of $^{13}CO_2$, which has a slightly higher molecular weight. Thus, over any time period, a greater proportion of $^{12}CO_2$ enters the leaf than $^{13}CO_2$. Inside the leaf, ribulose bisphosphate carboxylase also has a higher affinity for $^{12}CO_2$. As a result of these factors, plant tissue contains a lower proportion of $^{13}CO_2$ than the atmosphere by about 2% (=20‰) (O'Leary, 1988). The discrimination (*fractionation*) between carbon isotopes is expressed relative to an accepted standard as:

$$\delta^{13}C = \left[\frac{^{13}C/^{12}C_{sample} - {}^{13}C/^{12}C_{standard}}{^{13}C/^{12}C_{standard}} \right] \times 1000 \qquad (5.4)$$

using the units of parts per thousand parts (‰). Because atmospheric CO_2 shows an isotopic ratio of −8.0‰ versus the standard, most plant tissues show $\delta^{13}C$ of about −28‰—that is, (−8‰) + (−20‰). Sedimentary organic carbon with this isotopic signature is useful in determining the antiquity of photosynthesis as a biochemical process (Fig. 2.8).

The discrimination between $^{12}CO_2$ and $^{13}CO_2$ during photosynthesis is greatest (most negative $\delta^{13}C$) when stomatal conductance is high (Fig. 5.2). When stomates are partially or completely closed, nearly all of the CO_2 inside the leaf reacts with ribulose bisphosphate carboxylase, and there is less fractionation of the isotopes. Therefore, the isotopic ratio of plant tissue is directly related to the average stomatal conductance during its growth, providing a long-term index of water-use efficiency and environmental conditions (Farquhar et al., 1989; Cernusak et al., 2013). Significantly, $\delta^{13}C$ values of preserved plant materials and tree rings indicate that WUE of plants increased as the concentration of atmospheric CO_2 rose at the end of the last glacial period (Van de Water et al., 1994) and during the last several hundred years (Penuelas and Azcon-Bieto, 1992; Feng, 1999; Saurer et al., 2004; Watmough et al., 2001; Kohler et al., 2010). As the ^{13}C content of plant tissues has increased (higher or less negative $\delta^{13}C$), the ^{13}C content of atmospheric CO_2 has decreased, implying greater WUE for vegetation globally (Keenan et al., 2013; Keeling et al., 2017).

Nutrient-use efficiency

Over a broad range of plant species, the rate of photosynthesis is directly correlated to leaf nitrogen content when both are expressed on a mass basis (Reich et al., 1992, 1999; Atkinson et al., 2010; Fig. 5.3). Most leaf nitrogen is contained in enzymes; by itself, ribulose bisphosphate carboxylase usually accounts for 20–30% of leaf nitrogen (Evans, 1989). Seemann et al. (1987) found that photosynthetic potential is directly related to the content of ribulose bisphosphate carboxylase and leaf nitrogen in several species, suggesting that

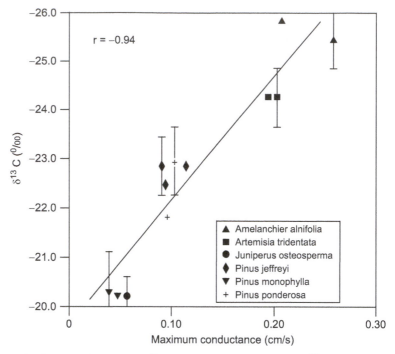

FIG. 5.2   Relationship between the content of ^{13}C in plant tissues (expressed as $\delta^{13}C$) and stomatal conductance for a variety of plant species in western Nevada. *Modified from DeLucia et al. (1988).*

the availability of nitrogen determines leaf enzyme content and, thus, the rate of photosynthesis in land plants. Leaf chlorophyll content is a good measure of photosynthetic capacity (Croft et al., 2017). In addition to nitrogen, leaf phosphorus content may be an important determinant of photosynthetic capacity in some species (Reich and Schoettle, 1988; DeLucia and Schlesinger, 1995), and adequate P often determines the relationship of photosynthesis to N (Raaimakers et al., 1995). Despite their central role in the biochemistry of photosynthesis, magnesium and manganese are seldom in short supply for plant growth.

Because most land plants grow under conditions of nitrogen deficiency, we might expect adjustments in nutrient use to maximize photosynthesis under varying conditions of soil fertility. The rate of photosynthesis per unit of leaf nitrogen—the slope of the line in Fig. 5.3—is one measure of nutrient-use efficiency (NUE) (Evans, 1989). Overall, the data of this figure would seem to indicate that most species have similar photosynthetic NUE, but subtle variations in NUE are seen among different types of plants (Reich et al., 1995) and among plants grown at different levels of fertility (Reich et al., 1994). For many plant species, when leaf nutrient content increases (by fertilization), NUE declines (Ingestad, 1979a; Lajtha and Whitford, 1989). Nutrient-use efficiency also appears inversely correlated to WUE across many species (Field et al., 1983; DeLucia and Schlesinger, 1991) and increases when plants are grown at high CO_2 (Springer et al., 2005).

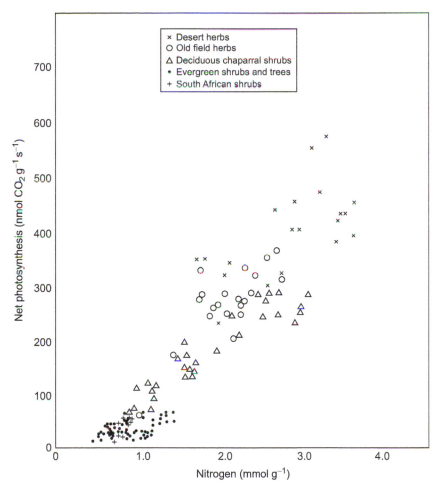

FIG. 5.3   Relationship between net photosynthesis and leaf nitrogen content among 21 species from different environments. *From Field and Mooney (1986).*

Respiration

Photosynthesis is usually measured by placing leaves or whole plants in closed chambers and measuring the uptake of CO_2 or release of O_2. The rates are a measure of *net* photosynthesis by the plant—that is, the fixation of carbon in excess of the simultaneous release of CO_2 by plant metabolism. Plant metabolism, known as respiration, is largely the result of mitochondrial activity in plant cells, and it is correlated to their nitrogen content, which is a good index of metabolic activity in most plant tissues (Fig. 5.4; see also Ryan, 1995; Vose and Ryan,

FIG. 5.4   Root respiration as a function of nitrogen content (%) in roots of loblolly and ponderosa pine. *From Griffin et al. (1997).*

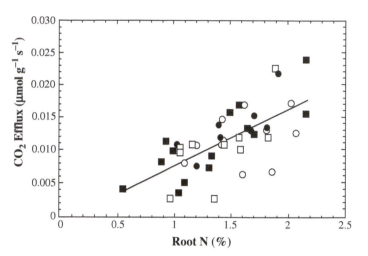

2002; Reich et al., 2006). Photosynthesis masks leaf respiration during the day, but respiration is a dynamic process that alters overall plant carbon balance (Tcherkez et al., 2017). For leaf tissues, mitochondrial respiration is lower during the day than at night,[e] but a process known as photorespiration also contributes to respiration during the day.[f] At similar temperature, respiration is higher for leaves from colder habitats, which typically have higher nitrogen content (Atkin et al., 2015).

In woody plants, a large fraction of the respiration is contributed by stems and roots owing to their large contribution to total plant biomass (Amthor, 1984; Ryan et al., 1994). In desert environments, trees show high total respiration owing to a greater allocation of tissue to sapwood (Callaway et al., 1994). For long-lived woody plants, maintenance respiration increases with stand age, consuming an increasing fraction of the gross photosynthesis and contributing to the reduction in the rate of plant growth with age (Kira and Shidei, 1967; DeLucia et al., 2007; Piao et al., 2010).

[e] Named for its discoverer, this is known as the Kok effect (Heskel et al., 2013; Wehr et al., 2016).

[f] Photorespiration is not simply total respiration during the day, but an increment to plant respiration that is observed in sunlight as a result of a competitive reaction of Rubisco with O_2, producing a toxic compound (phosphoglycolate or 2PG) that plants must expend energy to eliminate (Sharkey, 1988; Erb and Zarzycki, 2018). The reaction with O_2 is a function of the O_2/CO_2 ratio in the chloroplast, which is greater at high temperatures and during drought. Photorespiration can consume up to 30% of the C fixed in photosynthesis (Erb and Zarzycki, 2018). Plant physiologists are actively investigating ways to reduce or eliminate photorespiration in crop plants to increase agricultural yield (South et al., 2019). Although photorespiration has generally been regarded as detrimental to plant growth, there is some evidence that the process is important to nitrate assimilation in land plants (Rachmilevitch et al., 2004; Wujeska-Klause et al., 2019), internal phosphorus cycling (Ellsworth et al., 2015), and the protection of the photosynthetic mechanism at high light during periods of drought (Mahall and Schlesinger, 1982).

About one-half of the carbon fixation by photosynthesis is respired by plants, so the gross rate of photosynthesis is often twice the measured rate of carbon uptake (Farrar, 1985; Amthor, 1989). Although the rate of plant respiration shows a partial acclimation to higher temperatures (Atkin et al., 2015; Reich et al., 2016; Vico et al., 2019), plant respiration generally increases with increasing temperature, accounting for high rates of respiration in tropical forests and potentially higher rates of plant respiration with global warming (Ryan, 1991; Ryan et al., 1994, 1995). The increase in plant respiration in warmer conditions, versus the less sensitive response of photosynthesis, has important implications for global net primary production in the face of ongoing global warming (Dillaway and Kruger, 2010; Piao et al., 2010; Cai et al., 2010). Of course, as humans further impact the Earth, photosynthesis and respiration will also respond to the combination of ongoing changes in atmospheric CO_2 and moisture availability, as discussed in a later section.

Net primary production

The rate of plant growth, net primary production (NPP), measured by ecologists in the field is analogous to the rate of net photosynthesis measured by plant physiologists in the laboratory. For plants in nature, we say that:

$$\text{Gross primary production} - \text{plant respiration} = \text{net primary production} \qquad (5.5)$$

NPP is, however, not directly equivalent to plant growth as measured by foresters, ranchers, and farmers. Some fraction of NPP is lost to herbivores, fires, and in the death and loss of tissues, known collectively as litterfall. Foresters frequently call the NPP that remains the true increment, which adds to the accumulation of live biomass over many years. When mortality occurs during forest development, the true increment is the net increase in the mass of woody tissue in living plants, after subtracting the mass of individuals that die over the same interval (Clark et al., 2001).

The annual accumulation of organic matter per unit of land is a measure of NPP, often expressed in units of $g/m^2/yr$. Plant tissue typically contains about 45–50% carbon, so division by two is a convenient way to convert the accumulation of organic matter to carbon fixation (Reichle et al., 1973a). Net primary production can also be expressed in units of energy, by measurements of the caloric content of various plant tissues (Paine, 1971; Darling, 1976). Calories are particularly useful for expressing the efficiency of photosynthesis relative to the receipt of sunlight energy. Net primary production typically increases as a function of intercepted radiation (e.g., Runyon et al., 1994), but in forests, photosynthesis usually captures only about 1% of the total energy received from sunlight (Botkin and Malone, 1968; Reiners, 1972; Schulze et al., 2010). Even crop plants show maximum efficiencies of <5% of incident sunlight (Amthor, 2010).

Rain-use efficiency is a measure of water-use efficiency at the ecosystem level, and expressed as NPP per unit of precipitation. Normally this averages about $2g \ C/m^2$ per mm of rainfall (Sun et al., 2016). Rain-use efficiency is lower in barren, arid regions, where a larger proportion of the precipitation simply evaporates without contributing to plant growth (LeHouerou, 1984).

Measurement and allocation of NPP

Traditional methods for measuring NPP in forests and shrublands involve the harvest of vegetation and calculation of the annual growth of wood and the mass of foliage at the peak of annual leaf display (Clark et al., 2001). The harvest data of a few individuals are used to calculate the mass and increment in plants of varying size by virtue of tightly constrained relationships between diameter, height, mass, and growth of individual plants—known as plant allometry (Whittaker and Marks, 1975; Enquist et al., 2007; Niklas and Enquist, 2002). Independent estimates of the seasonal loss of plant parts can be obtained from collections of plant litterfall throughout the year.

In grasslands, where there is little or no true increment, estimates of NPP generally involve the difference between the mass of tissue harvested from small plots at the beginning and the end of the growing season (e.g., Wiegert and Evans, 1964; Lauenroth and Whitman, 1977; Singh et al., 1975). These estimates must be corrected for the consumption and loss of tissues during the same period. Similar approaches are used to measure the growth of roots by sequential coring throughout the growing season (Neill, 1992; Vogt et al., 1998; Makkonen and Helmisaari, 1999).

Allocation of net primary production varies with vegetation type and age. In forests, 25–35% of aboveground production is found in leaves (Whittaker et al., 1974), with this percentage tending to decrease with stand age. Allocation to foliage in shrublands is generally greater, ranging from 35% to 60% in desert and chaparral shrubs (Whittaker and Niering, 1975; Gray, 1982). In grassland communities, essentially all aboveground net primary production is found in photosynthetic tissues. Across a broad range of species, allocation of photosynthate to leaf growth relative to stem growth is 0.53 (Niklas and Enquist, 2002).

As seen in laboratory studies of photosynthesis, plant respiration accounts for about half of GPP in ecosystem studies (Waring et al., 1998; Fig. 5.5). As a result of their massive structure and high environmental temperatures, tropical forests may expend a greater percentage of their GPP in respiration (Ryan et al., 1994; Luyssaert et al., 2007; Piao et al., 2010; Malhi, 2012), leaving less for wood growth. Comparing plant communities in different regions, Jordan (1971) found that the proportional allocation of NPP to wood growth was greater in boreal forests than in tropical forests—that is, there is greater wood production per unit of foliage in boreal forests. Webb et al. (1983) found a logarithmic relationship between total aboveground NPP and foliage biomass for a variety of plant communities in North America, with some deserts showing exceptionally high values of this ratio (Fig. 5.6; also Flanagan and Adkinson, 2011). Compared to communities with abundant precipitation, desert shrublands show relatively low allocation of NPP to wood production (Jordan, 1971), perhaps as a result of a large allocation to roots (Wallace et al., 1974; Mokany et al., 2006).

Because measurements of roots are difficult, many studies of NPP include data only for the aboveground tissues. Root biomass ranges between 20% and 40% of total biomass in forest ecosystems (Vogt et al., 1996; Poorter et al., 2012, see also Table 6.5), and the annual growth of root tissues accounts for a significant fraction of the total NPP in most communities—averaging about 15–25% across a broad range of plant size (Pan et al., 2006; Niklas and Enquist, 2002; K. Niklas, personal communication, 2010; McCormack et al., 2015). Some NPP is lost in soluble organic compounds released from roots, perhaps as much as 20% in some

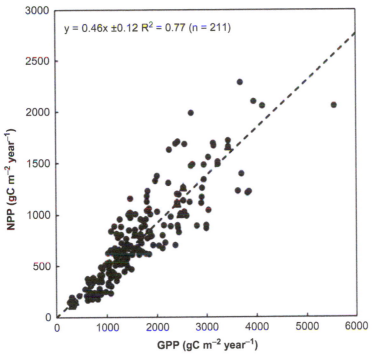

FIG. 5.5   Relationship between net primary production (NPP) and gross primary production (GPP) in different forest types. *From Collalti and Prentice (2019).*

circumstances (van Hees et al., 2005; Fahey et al., 2005).[g] These losses of organic carbon are known as rhizodeposition (Pausch and Kuzyakov, 2018). Trees allocate proportionally more photosynthate to root growth in low-fertility soils (Gower et al., 1992; Powers et al., 2005; Gill and Finzi, 2016), although the absolute amount of root growth is greatest on sites with high NPP (Raich and Nadelhoffer, 1989; Aerts and Chapin, 2000).[h]

Edwards and Harris (1977) reported that the growth and death of roots in a forest in Tennessee delivered 733 g C/m²/yr to the soil, whereas the aboveground production was 685 g C/m²/yr (Reichle et al., 1973a). Similarly, roots composed more than half of the NPP in coniferous forests in Washington (Table 5.1) and in the deciduous forest at Hubbard Brook (Fahey and Hughes, 1994). An even larger proportion of total NPP is allocated to root growth in many grassland ecosystems (Lauenroth and Whitman, 1977; Warembourg and Paul, 1977). Although there are strong relations between above- and belowground biomass (Cairns et al.,

[g] Some tree species appear to share carbohydrates among roots, perhaps by parasitism (Klein et al., 2016).

[h] Various workers use "total belowground allocation" to describe the total allocation of GPP to root and mycorrhizal respiration (Chapter 6), to leakage of root exudates, and to root production—only the latter has traditionally been considered a component of NPP (Raich and Nadelhoffer, 1989).

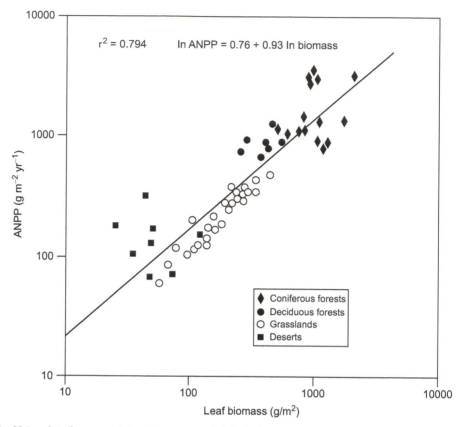

FIG. 5.6 Using data from a variety of ecosystems in North America, Webb et al. (1983) found a strong relation between annual aboveground net primary production and leaf biomass.

1997; Enquist and Niklas, 2002; Mokany et al., 2006; Cheng and Niklas, 2007), there are no obvious generalizations that allow us to predict the allocation of NPP to the *growth* of shoots and roots worldwide (Nadelhoffer and Raich, 1992; Gower et al., 1996; Litton et al., 2007).

Much of the plant photosynthate that is allocated belowground supports the growth of roots <2mm diameter, known as fine roots. In grasslands, the total length of fine roots may exceed 100km beneath each square meter of the soil's surface (Jackson et al., 1997). In most ecosystems, about half of these roots die each year (Gill and Jackson, 2000), consistent with observations of roots using transparent soil tubes, known as minirhizotrons (Strand et al., 2008; Eissenstat and Yanai, 1997). However, studies following the disappearance of carbon from isotopic-labeled root systems indicate considerable longevity of some fine roots—sometimes exceeding 5 years (Matamala et al., 2003; Gaudinski et al., 2001; Riley et al., 2009). It is likely that many root systems consist of a large fraction of roots with relatively short longevity and a smaller population that lasts several years (Joslin et al., 2006; Gaudinski et al., 2010; McCormack et al., 2015).

TABLE 5.1   Net primary production in 23- and 180-yr-old *Abies amabilis* forests in the Cascade Mountains, Washington.

Aboveground	23-yr-old		180-yr-old	
	g/m²/yr	% of total	g/m²/yr	% of total
Biomass increment				
Tree total	426		232	
Shrub stems	6		<1	
Total	432	18.37	232	9.33
Detritus production				
Litterfall	151		218	
Mortality	30			
Herb layer turnover	32		5	
Total	213	9.06	223	8.97
Total aboveground	**645**	**27.42**	**455**	**18.30**
Belowground				
Roots				
Fine (≤2 mm)	650	27.64	1290	51.87
Fibrous-textured	371		1196	
Mycorrhizal (host tissue)	79		94	
Coarse (>2 mm)	358		324	
Angiosperm fine root turnover	373		44	
Total root turnover	1381	58.72	1658	66.67
Mycorrhizal fungal component	326	13.86	374	15.04
Total belowground	1707	72.58	2032	81.70
Ecosystem total	**2352**		**2487**	

From Vogt et al. (1982).

Net ecosystem production and Eddy-covariance studies

As long as they are growing, plants allocate some NPP to the accumulation of biomass, with the remaining NPP passing to herbivores or decomposers, which convert organic carbon to CO_2 and return it to the atmosphere. We define net ecosystem production (NEP) as:

$$NEP = NPP - (R_h + R_d) \tag{5.6}$$

I. Processes and reactions

thus,

$$NEP = GPP - (R_p + R_h + R_d) \qquad (5.7)$$

where R_p, R_h, and R_d represent the respiration of plants, herbivores, and decomposers, respectively.[i] Except in unusual circumstances, NEP will be a partial fraction of NPP. In young forests, NEP may be 50% of NPP, but in older stands, when vegetation is not accumulating biomass, nearly all the NEP will be found in small increments to soil organic matter (Law et al., 2003; Pregitzer and Euskirchen, 2004). Nevertheless, old-growth forests continue to store carbon at rates that are significant to the global carbon cycle (Luyssaert et al., 2008; Lewis et al., 2009; Chapter 11). The annual increment to carbon storage in an ecosystem as a fraction of carbon uptake has been defined as C-use efficiency, which tends to be greater in ecosystems with favorable conditions for growth (Manzoni et al., 2018).

Since the early 1990s, ecosystem scientists have made indirect measurements of net ecosystem production of whole ecosystems by calculating the net uptake of CO_2 within a hypothetical column of the atmosphere with a small ground "footprint," typically $1 m^2$. Substantial theory underlies this approach. Namely, if there is no carbon exchange between the atmosphere and the biosphere, such as over the surface of a parking lot, we would expect the atmosphere to show a uniform concentration of CO_2 at all heights above the surface—with values close to the global average in the troposphere ($\sim$400 ppm; Table 3.1). In contrast, inside the canopy of a forest, photosynthesis will deplete CO_2 during the day, while CO_2 will be enriched in samples taken near the soil surface, where it is emitted by the activity of decomposers. These differences in CO_2 concentration with height persist in the face of winds that might transport fresh air from outside the ecosystem or mix air within the forest, resulting in a uniform concentration of CO_2. Thus, if we measure the concentration of CO_2 at various heights within the forest and the delivery of fresh air at each height, then we can estimate the carbon uptake, or CO_2 release, necessary to maintain the differences in CO_2 at different heights. Integrating over height and time, these measurements would indicate the net exchange of CO_2—net ecosystem production—in a forest or other types of vegetation.

Now consider the height-profile of carbon dioxide in the middle of a uniform expanse of vegetation. The air moving through the column above a hypothetical 1-m^2 ground footprint will arrive and leave with the same distribution of CO_2 with height. The only net exchange with the atmosphere will occur from above the canopy as a result of wind- and eddy-driven transport into or out of the vegetation. Eddy-covariance measurements of net carbon exchange, so named because they trace the simultaneous variation in CO_2 concentrations and vertical wind velocity, have been made in a large number of sites worldwide. The method requires a tower with wind speed and CO_2 analyzers at varying heights (Fig. 5.7), and works best in large areas of relatively uniform vegetation and flat topography, where the effects of turbulence are minimized (Baldocchi, 2003). At night, the outward flux of CO_2 would represent the total respiration in the ecosystem. GPP can be calculated from NEP using Eq. (5.7), with the assumption that the respiration flux measured during the night applies during the 24-h period.[j]

---

[i] When calculated for large areas, NEP is sometimes called net biome production (NBP) (Randerson et al., 2002).

[j] Net carbon uptake in eddy-covariance studies is sometimes known as net ecosystem exchange (NEE).

FIG. 5.7   An eddy-covariance (flux) tower in a deciduous forest in North Carolina. *Photo from G. Katul, Duke University.*

Eddy-covariance studies can provide simultaneous measurements of the net flux of several gases, such as CO_2 and H_2O, providing an estimate of the WUE of photosynthesis at the ecosystem level (Fig. 5.8).[k] Normally, GPP is directly related to transpiration (T), so a regression of ET on GPP, can indicate the proportion of evaporation (E), when GPP zero (Scott and Biederman, 2017). Eddy-covariance measures of carbonyl sulfide (COS), which is taken up and destroyed in plant leaves, can indicate GPP in forest ecosystems (Spielmann et al., 2019). COS measurements can also be used to partition ET into its components (Wehr et al., 2017), since the uptake of COS is directed related to T but unrelated to E (Chapter 3). Eddy-covariance has been used to estimate regional CH_4 flux in flooded tropical forests (Dalmagro et al., 2019) and Alaskan tundra (Taylor et al., 2018).

It is interesting to compare traditional harvest estimates to NEP obtained from eddy-covariance studies (Table 5.2). Barford et al. (2001) used eddy-covariance to estimate GPP of 1300 g $C/m^2/yr$ during an 8-year study of a deciduous woodland at Harvard Forest in Massachusetts. Total respiration was 1100 g $C/m^2/yr$, resulting in NEP of 200 g $C/m^2/yr$—a preliminary estimate of net carbon sequestration in wood and soil organic

[k] Developed by atmospheric scientists who were interested in the disappearance of CO_2 from the atmosphere, eddy-covariance studies assign plant uptake of CO_2 a negative value. Ecologists, using harvests to estimate increases in carbon storage in an ecosystem, express NEP as a positive value. Thus, NEP of -100 g $C/m^2/yr$ reported by an eddy-covariance study is equivalent to $+100$ g $C/m^2/yr$ reported by foresters, and both indicate net carbon storage in the ecosystem. In this book, we follow the latter convention and assign a positive value to all NPP and NEP estimates that indicate net carbon uptake by vegetation.

FIG. 5.8 Monthly gross primary production and evapotranspiration in various temperate deciduous forests, measured by eddy-covariance techniques. The slope of the line is an estimate of water-use efficiency, here equivalent to 1.4 mmol/mol (See Eq. 5.3). *From Law et al. (2002).*

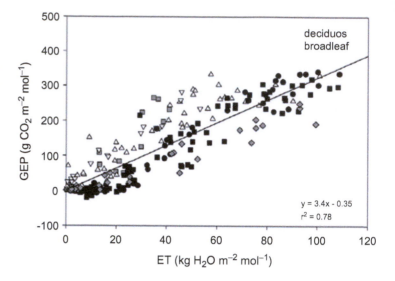

TABLE 5.2 Gross Primary Production (GPP), Net Primary Production (NPP) and Net Ecosystem Production (NEP) for some young, temperate and boreal forest ecosystems measured by harvest (H) and eddy-covariance (CV) methods.

Ecosystem type	Age	Method	GPP	NPP	NEP	Reference
Pinus sylvestris (Finland)	40	CV	1005		185	Kolari et al. (2004)
		H			228	
Picea rubens (Maine)	90	CV	1339		174	Hollinger et al. (2004)
Pinus taeda (North Carolina)	16	CV	2238		433	Juang et al. (2006), Hamilton et al. (2002),
		H		986	428	and McCarthy et al. (2010)
Pinus elliottii (Florida)	24	CV	2606		675	Clark et al. (2004)
		H			745	
Pinus ponderosa (Oregon)	56–89	CV	1208		324	Law et al. (2000, 2003)
		H		400	118	
Mixed deciduous (Massachusetts)	60	CV	1300		200	Barford et al. (2001)
		H			160	
Mixed deciduous (Michigan)	85	CV			151	Gough et al. (2008)
		H		654	153	

All data in g C/m²/yr.

matter. Independent measurements of NEP from traditional harvests indicate net carbon storage of $160 \, g \, C/m^2/yr$ showing agreement within 20–25%. In an experimental forest in Michigan, Gough et al. (2008) concluded that the differences between biometric (harvest) and meteorological (eddy-covariance) estimates of NPP were related to late-season photosynthesis which was allocated to storage rather than growth (cf. Babst et al., 2014). NPP measured by these techniques agreed within 1% when data were averaged over 5 years. Good agreement between eddy-covariance and harvest is seen in the few tropical forests studied (Malhi, 2012). Unfortunately, many investigators using eddy-covariance techniques have not simultaneously used traditional methods to validate the carbon accumulation at their field sites (Luyssaert et al., 2009), and the method is notoriously bad in regions of sparse vegetation (Ham and Heilman, 2003; Schlesinger, 2017).

Eddy-covariance studies of carbon uptake can be applied in a wide variety of situations, including ecosystem-level studies of the net carbon balance of cities. Satellite observations show an excess of $3.2 \, ppm \, CO_2$ in the atmosphere over the Los Angeles Basin (Kort et al., 2012). Despite harboring considerable forest cover, suburban areas in Baltimore show a net release of $361 \, g \, C/m/yr$ or $241 \, kg \, C/yr$ for each inhabitant (Crawford et al., 2011). Other world cities have even higher rates of net efflux, reflecting the balance between CO_2 uptake by vegetation vs. CO_2 release by fossil fuel combustion for heating and transport (Crawford et al., 2011; Bergeron and Strachan, 2011).

Remote sensing of primary production and biomass

Harvest measurements and eddy-covariance studies of NPP are labor intensive and necessarily applied only to small areas. Since the productivity of vegetation may vary greatly over the landscape, regional estimates of productivity by harvest are often prohibitively expensive. Since 1999, a satellite-based senor, known as the Moderate Resolution Imaging Spectroradiometer (MODIS),[1] has provided integrated estimates of GPP over large areas for studies of global change (Running et al., 2004). MODIS is the latest in a history of satellites, including LANDSAT and NOAA-AVHRR, which have provided estimates of global NPP using similar approaches (Box et al., 1989; Field et al., 1998). The history of these measurements and anticipation of future advances is reviewed by Ryu et al. (2019).

The basis of satellite measurements of GPP is the differential absorption of light by chlorophyll and other leaf pigments. Green plants look green because chlorophyll preferentially absorbs light in the blue and red portions of the solar spectrum, reflecting a large portion of the green light to our eyes. Despite its strong absorption of red light (760 nm), chlorophyll shows little absorption of infrared light at wavelengths of 800–1200 nm. Thus, to provide an index of the underlying "greenness" of the Earth's surface, satellites measure the surface reflectance in discrete portions of the visible and infrared spectrum (Fig. 5.9). By itself, the near-infrared reflectance of vegetation is well correlated to gross primary production (Badgley et al., 2019). Bare soil shows similar reflectance in the infrared and red wavebands,

[1] See https://modis.gsfc.nasa.gov.

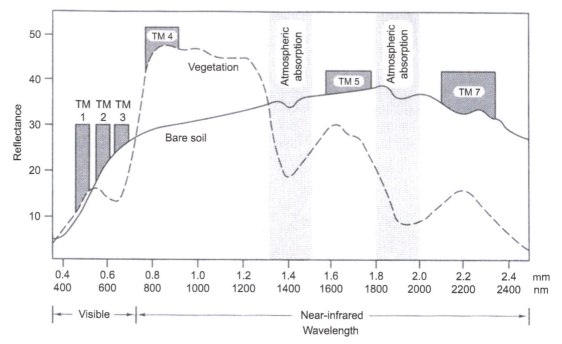

FIG. 5.9   A portion of the solar spectrum showing the typical reflectance from soil (—) and leaf (——) surfaces and the portions of the spectrum that are measured by the LANDSAT satellite.

whereas vegetation shows an infrared/red ratio $\gg 1.0$ as a result of the absorption of red light by chlorophyll. The normalized difference vegetation index (NDVI) is calculated as:

$$NDVI = (NIR - VIS)/(NIR + VIS) \tag{5.8}$$

where NIR is reflectance in the near-infrared and VIS is reflectance in the visible red wavebands, respectively. This index minimizes the effects of variations in background reflectance and emphasizes variations in the data that occur because of the density of green vegetation. NDVI allows global mapping of a greenness index for the Earth's land surface, and satellite measurements of greenness can provide estimates of NPP, assuming that greenness is directly related to leaf area[m] and that LAI is a good predictor of NPP (Gholz, 1982; Flanagan and Adkinson, 2011; Figs. 5.6 and 5.10).

The MODIS satellite has provided an estimate of GPP by assuming that:

$$GPP = \varepsilon \times NDVI \times PAR \tag{5.9}$$

where ε is a measured coefficient expressing the efficiency of conversion of sunlight energy into plant growth in various ecosystems (Field et al., 1995), and PAR is a measure of photosynthetically active radiation. The estimate is computed every 8 days for the Earth's land surface at 1-km spatial resolution (Running et al., 2004). Currently, the satellite measurements of

[m] See footnote o in Chapter 3. The term, *leaf area index* or LAI, has the units of m^2/m^2.

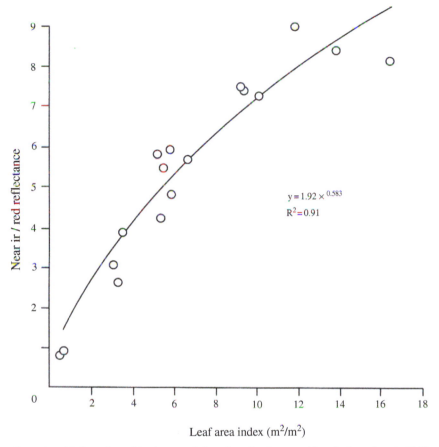

FIG. 5.10 The ratio of light reflected in the near infrared and red spectral bands (wavebands TM4 and TM 3 of LANDSAT (Fig. 5.9) is related to leaf area index in forest stands in the northwestern United States. *From Peterson et al. (1987).*

NDVI are coupled to independent measurements of surface climate conditions that affect ε. Direct satellite estimates of ε, using solar-induced fluorescence from chlorophyll, are well correlated with seasonal, regional and global patterns of plant productivity (Smith et al., 2018; Xiao et al., 2019; Mohammed et al., 2019). When actively photosynthesizing, plants give off (fluoresce) about 1% of incident light as red light that can be measured by satellite (Frankenberg and Berry, 2018). In most areas changes in the greenness of vegetation due to drought and harvest will have a direct relationship to changes in NPP (Zhou et al., 2014), although this relationship can be problematic in northern evergreen forests (Walther et al., 2016). Satellite measurements of leaf chlorophyll content also show promise in improving global estimates of plant production (Luo et al., 2019).

Remote sensing of biomass is more difficult than for LAI and NPP. Synthetic aperture radar (SAR) is used to measure vegetation biomass based on the absorption of microwave

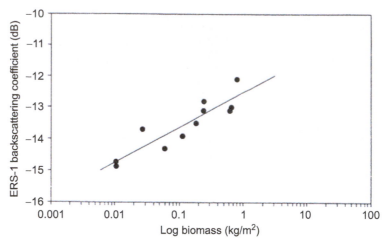

FIG. 5.11   The reflected microwave radiation (backscattering coefficient) measured by an airborne synthetic aperture radar (SAR) for stands of young loblolly pine (*Pinus taeda*) in central North Carolina. *Modified from Kasischke et al. (1994).*

radiation by the water held in woody biomass (Toan et al., 2011; Fig. 5.11). Biomass estimates have used radar or LiDAR to estimate forest height, which is often directly related to biomass (Treuhaft et al., 2004, 2010; Shugart et al., 2010; Quegan et al., 2019). Boudreau et al. (2008) used a combination of field measurements and aircraft and satellite LiDAR systems to estimate forest biomass from measurements of its height in Quebec. Similar techniques have been applied to tropical rainforests in Costa Rica and Brazil (Dubayah et al., 2010; Drake et al., 2003; Asner et al., 2010; Clark et al., 2011). Where compared to field inventories, the remote sensing estimates of forest biomass often agree to within 10% (e.g., Nelson et al., 2012; Baccini and Asner, 2013) although exceptions are marked (Mitchard et al., 2014). Using LIDAR, Hudak et al. (2012) used differences in biomass between repeat surveys only 6 years apart to estimate carbon gains in mature forests and carbon losses in harvested areas in Idaho. Similarly, Dalponte et al. (2019) used LIDAR to document an increase of 3.6%/yr in forest biomass over 5 years in the European Alps.

Global estimates of net primary production and biomass

Beer et al. (2010) estimate 123×10^{15} g C/yr of global GPP[n] based on 8 years of observations of the land surface by the MODIS satellite. Alternatively, Badgley et al. (2019) estimate GPP at 147×10^{15} g C/yr. Since one-half of the GPP is normally consumed by respiration, the indicated global terrestrial net primary production is about 60 to 70×10^{15} g C/yr—near the upper

---

[n] Many people have difficulty visualizing 10^{15} g C (=1 billion metric tons) of plant productivity. It is equivalent to the amount of carbon, obtained from atmospheric CO_2, which is contained in a block of wood roughly 1.2 km on each side.

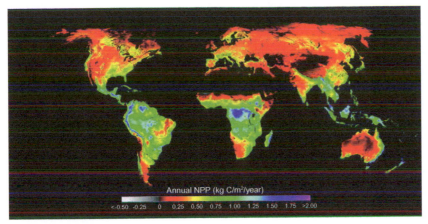

FIG. 5.12   Distribution of global net primary production (NPP) on land for 2002, computed from MODIS data. *From Running et al. (2004).*

end of the range of earlier estimates based on the modeling and aggregations of harvest data ($45\text{--}65 \times 10^{15}$ g C/yr) by various authors (Whittaker and Likens, 1973; Lieth, 1975; Field et al., 1998; Del Grosso et al., 2008; Ito, 2011). As expected, a global map of terrestrial NPP shows the highest values in tropical rainforests and the lowest values in areas of extreme desert and ice (Fig. 5.12). Repetitive measures by MODIS show the seasonal "green-up" of vegetation in the spring of each hemisphere.[°] Agricultural lands account for up to 10% of global NPP (Imhoff et al., 2004b), but only about 1% of the NPP from croplands is fed to humans or livestock (Wolf et al., 2015), and the rest is left in the field as crop residues or lost to spoilage during shipping and marketing.

Aggregations of data from harvest and eddy-covariance studies also suggest that the primary productivity of forests is greatest in the tropics and declines with increasing latitude to low values in boreal forests and shrub tundra (Table 5.3). In seasonal environments, photosynthetic rates often acclimate to changes in temperature (Lange et al., 1974; Gunderson et al., 2010). Thus, daily values for peak NPP are relatively similar in many ecosystems, and it is the length of the growing season, as determined by temperature and moisture, that determines annual NPP (Kerkhoff et al., 2005). Among European forests, net ecosystem production is lower in northern forests as a result of a greater effect of low temperatures on the length of the growing season, reducing GPP relative to ecosystem respiration (Valentini et al., 2000; Janssens et al., 2001; Van Dijk and Dolman, 2004). Along a gradient of decreasing precipitation, NPP declines from forests to grasslands, showing very low values in most deserts (Knapp and Smith, 2001). In all biomes, rain-use efficiency by vegetation is greatest during dry years, when it approaches a value of 0.21 g C/m^2/yr in aboveground NPP per mm of precipitation (WUE = 0.315 mmol/mol) across a broad range of ecosystems (see Eq. 5.3; Huxman et al., 2004).

[°] https://earthobservatory.nasa.gov/GlobalMaps/view.php?d1=MOD17A2_M_PSN.

TABLE 5.3   Biomass and net primary production in terrestrial ecosystems, with data compiled by Saugier et al. (2001), assuming a 50% carbon content in plant tissues.

Biome	Area (10^6 km^2)	NPP (g C/m^2/yr)	Total NPP (10^{15} C/yr)	Biomass (g C/m^2)	Total plant C pool (10^{15} g C)
Tropical forests	17.5	1250	20.6	19,400	320
Temperate forests	10.4	775	7.6	13,350	130
Boreal forests	13.7	190	2.4	4150	54
Mediterranean shrublands	2.8	500	1.3	6000	16
Tropical savannas/grasslands	27.6	540	14.0	2850	74
Temperate grasslands	15.0	375	5.3	375	6
Deserts	27.7	125	3.3	350	9
Arctic tundra	5.6	90	0.5	325	2
Crops	13.5	305	3.9	305	4
Ice	15.5				
Total	149.3		58.9		615

Evidence for the importance of temperature and moisture as controls of NPP is seen in regional comparisons of productivity, especially patterns along gradients of elevation. Whittaker (1975) found that net primary production declined with increasing elevation in the forested mountains of the eastern United States, presumably reflecting the influence of declining temperatures (i.e., a shorter growing season). In the southwestern United States, where precipitation is more limited, NPP tends to increase with elevation in communities ranging from desert shrublands to montane forests (Whittaker and Niering, 1975). Sala et al. (1988) show a direct relationship between NPP and precipitation within the grasslands of the central United States. Compilations of data from various world biomes show strong relations of NPP to temperature and mean annual precipitation (Scurlock and Olson, 2002; Schuur, 2003; Fig. 5.13). Overall, the availability of water seems to be the dominant control of carbon dioxide uptake by land vegetation (Jung et al., 2017; Humphrey et al., 2018; Green et al., 2019).

In forests of the northwestern United States, NPP and LAI are directly related to site water balance, which is the difference between precipitation inputs and losses of soil moisture by runoff and evapotranspiration during the growing season (Grier and Running, 1977; Gholz, 1982). Rosenzweig (1968) combined temperature and precipitation to calculate actual evapotranspiration, which shows a positive correlation to NPP in temperate zone ecosystems (cf. Webb et al., 1978; Hunt, 2017). Fisher et al. (2012) estimate soil nutrient deficiencies reduce global NPP by 16–28% over what might be determined by temperature and precipitation alone. The overall strength of the relationship between NPP, temperature and available moisture may partially derive from the influence of these variables on microbial processes that speed nutrient turnover in the soil (Chapter 6). In tropical rainforests, where both light

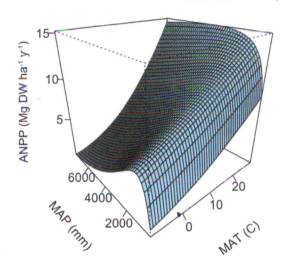

FIG. 5.13 Aboveground net primary production in world forests versus mean annual temperature (°C) and precipitation (mm). *From Taylor et al. (2017).*

and moisture are abundant, the relationship of NPP to these variables is weak, and local soil conditions determining fertility are potentially more important (Cleveland et al., 2011; Augusto et al., 2017).

Estimates of the total biomass of land plants range from 450 to 615×10^{15} g C, derived from the aggregation of harvest data worldwide (Olson et al., 1983; Saugier et al., 2001; Bar-On et al., 2018).[P] About 1/3 of global biomass resides in tropical rainforests (Avitabile et al., 2016). Based on a random sampling of land areas, total biomass in U.S. forests is about 18×10^{15} g C (Blackard et al., 2008), whereas vegetation in China and India contains about 13.6×10^{15} and 2.9×10^{15} g C, respectively (Tang et al., 2018; Kaul et al., 2011). By comparison, total biomass in the tropical forests of Brazil may be as large as 50×10^{15} g C (Nogueira et al., 2008, 2015).

The ratio of biomass/NPP is an estimate of the mean residence time for an atom of carbon in plant tissues (compare to Eq. 3.3). The global values yield an overall mean residence time of about 10 years, but this value varies from about 4 in deserts to >20 in some forests (see Table 5.3; compare to Fahey et al., 2005). For the entire United States, the mean residence time of C in vegetation is about 5 years—that is, biomass of 18×10^{15} g C divided by NPP of 3.5×10^{15} g C/yr, which is calculated from Xiao et al. (2010). Of course, we must remember that these are weighted averages. In forests some tissues, such as leaves, may last only a few months, while wood may last for centuries.

Estimates, such as those in Table 5.3, are calculated by classifying the land vegetation into a small number of categories and by assigning a mean value to the NPP and biomass of each category based on data from the widest possible number of field studies. The classification of vegetation is arbitrary, and estimates of the land area in each category often vary considerably (Golley, 1972). Moreover, the NPP data often do not reflect average values because ecologists

[P] A global database for forest biomass is maintained at https://GitHub.com/forc-db (Anderson-Teixeira et al., 2018).

often tend to select mature, well-developed stands for study. Random site selection often produces lower regional values (Botkin and Simpson, 1990; Botkin et al., 1993; Jenkins et al., 2001). Remote sensing estimates of NPP and biomass have the advantage of including the full range of variation seen in the field (Zhang and Kondragunta, 2006). MODIS offers continuous, realistic estimates of NPP over large scales, albeit with some loss of accuracy at local sites (Pan et al., 2006).

The fate of net primary production

As defined in Eq. (5.6), NEP would seem equivalent to the incremental accumulation of organic matter in an ecosystem—largely in wood growth and increments of soil organic matter. Even old forests that have stopped growing continue to store some organic matter in soils (Luyssaert et al., 2008; Law et al., 2003; Schlesinger, 1990; Zhou et al., 2006). Only a small amount of carbon from photosynthesis accumulates in ecosystems in other forms, including calcium oxalate and calcium carbonate in plant tissues (Stone and Boonkird, 1963; Braissant et al., 2004; Cailleau et al., 2004) and calcium carbonate in soils (Chapter 4). If it survives the appetite of herbivores, most of the remaining NPP passes to the decomposers, where it is respired to CO_2.[q] Of course, today humans are a dominant consumer of NPP, either directly in the consumption of food and forest products or indirectly in the destruction of vegetation for other purposes. Terrestrial vegetation is also subject to fire, with a yearly frequency in some grasslands and century-long return intervals for fire in some forests. Fires return carbon to the atmosphere, largely as CO_2, analogous to a large, generalist herbivore.

Fires are a normal part of the Earth's terrestrial ecosystems, especially in areas of tropical savanna (Cahoon et al., 1992). Past records show changes in the occurrence of fire through the Holocene, responding to changes in climate and human activity (Marlon et al., 2013; Nicewonger et al., 2018). In some periods, the occurrence of fire may have exceeded that of today (Ward et al., 2018); however, land-clearing by humans has likely increased the global extent of fire in recent years (Bowman et al., 2009; Mouillot and Field, 2005). Global estimates of CO_2 emissions from fires range average about 2.5×10^{15} g C/yr—about 4% of terrestrial NPP (Randerson et al., 2012). On average about 0.1–0.6% of the annual NPP of land vegetation is converted to charcoal each year (Santín et al., 2016; Jaffé et al., 2013; Wei et al., 2018). Fires also contribute to global sources of a variety of trace gases (Andreae, 2019; Jain et al., 2006; Chapters 3 and 6). Large fires in Kalimantan, Indonesia, in 1997 are estimated to have released between 0.81 and 2.57×10^{15} g C into the atmosphere in a single year (Page et al., 2002). Large fires have also resulted in major losses of carbon from boreal forests in recent years (Kasischke et al., 1995; Bond-Lamberty et al., 2007; Walker et al., 2019). Recent estimates show a modest decline in the area of forests burned

---

[q] The oxidative ratio (OR) is defined as the moles of O_2 emitted by a leaf or an ecosystem divided by the moles of carbon stored in organic matter. Given Eq. (5.2), it is not surprising that the OR calculated globally is close to 1.0 (Worrall et al., 2013; Battle et al., 2019), but it varies by plant species and other factors that affect the presence of other reduced compounds (e.g., NH_4^+) that accumulate at the expense of reduced carbon.

annually (Andela et al., 2017), but many ecologists anticipate a greater frequency of fires as a result of global climate change (Aragão et al., 2018).

Although herbivory may play a role in controlling forest productivity and nutrient cycling (Chapter 6), the consumption of plant tissues by herbivores is nearly always <20% of terrestrial NPP (e.g., Mispagel, 1978; McNaughton et al., 1989; Cyr and Pace, 1993; Cebrian and Lartigue, 2004). Higher values are nearly always associated with insect outbreaks (e.g., Kurz et al., 2008; Hicke et al., 2013) or managed grazing systems (Oesterheld et al., 1992). By consuming leaf area and root biomass, herbivores may have an indirect effect on NPP that is larger than their direct consumption (Reichle et al., 1973b; Llewellyn, 1975; Ingham and Detling, 1990). In Africa, bull elephants trash large amounts of vegetation, reducing biomass by much more than they consume. Globally, herbivores consume about 5% of terrestrial NPP (Whittaker and Likens, 1973). Respiration by decomposers consumes most of the rest (Street and Mcnickle, 2019; Chapter 11).

When volatile organic compounds (VOCs) are produced by plants and lost to the atmosphere, they represent a small portion of NPP that is oxidized by hydroxyl radicals outside of the ecosystem (Chapter 3; Kesselmeier et al., 2002). Plant emissions of isoprene and other organic compounds (Table 3.2) have been measured using chambers and eddy-covariance methods (Rinne et al., 2000). Globally, the emission of reduced carbon compounds from natural vegetation may exceed 1×10^{15} g C/yr, or about 2% of NPP (Guenther, 2002; Laothawornkitkul et al., 2009). This small fraction accounts for much of the CO and CH_4 in the atmosphere (Chapter 11). A small amount of carbon is lost from ecosystems in organic materials that are carried by streamwaters and groundwaters (Chapter 8) and respired to CO_2 outside the boundaries of the ecosystem, typically amounting to 1–10 g C/m^2/yr, or less than 1% of global terrestrial NPP (Schlesinger and Melack, 1981; Deirmendjian et al., 2017). Losses of volatile organic compounds to the atmosphere and organic carbon in streamwaters explain why NEP is not always directly equivalent to new incremental storage of organic matter in the ecosystem (Lovett et al., 2006; Chapin et al., 2006; Kindler et al., 2011).

Net primary production and global change

Since the beginning of civilization, humans have harvested the Earth's net primary production for food, fuel, and fiber. Indeed, the pages of this book were once part of a living tree. Cultivated lands and pasture now occupy about 40% of the world's land surface (Ellis et al., 2010; Goldewijk et al., 2010; Ramankutty and Foley, 1998; Sterling and Ducharne, 2008), and urban areas cover about 2% of the land that is not covered by ice (van Vliet et al., 2017). The cumulative losses of carbon from human impact on vegetation and soils are estimated to be as large as 357×10^{15} g C up to the beginning of the Industrial Revolution (Kaplan et al., 2011), and current annual emissions from forest destruction are estimated at 1.1×10^{15} g C/yr (Houghton et al., 2012).

Based on a compilation of government statistics, global forest cover in 2005 was $38.81 \pm 1.38 \times 10^6$ km^2, with annual deforestation of 0.73×10^6 km^2 and gains of 0.28×10^6 km^2 due to regrowth (Feng et al., 2016). Rates of tropical deforestation, largely for new cultivation, are estimated at 0.056–0.058×10^6 km^2/yr (Achard et al., 2002; DeFries et al., 2002; Keenan et al., 2013).[r] Directly and indirectly humans now use 11–24% of potential NPP on the land surface (Haberl et al., 2007; Imhoff et al., 2004b), with most of the carbon released to the atmosphere as CO_2. Ancillary human impacts on the land surface may raise our total appropriation of photosynthesis to 40% annually (Vitousek et al., 1986). These high values for the consumption of NPP by a single species do not bode well for the future of other species on the planet.

The human harvest of natural vegetation is not uniform across the planet. High rates of harvest in the tropics are balanced by the abandonment and regrowth of cultivated land elsewhere (Imhoff et al., 2004b). Recovering forests often have high rates of carbon accumulation (Poorter et al., 2016; Bonner et al., 2013) that attenuate with time (Law et al., 2003). In the southeastern U.S. coastal plain, young forests are storing 90,000 t C/yr (90×10^9 g/yr), which is equivalent to NEP of 100 g C/m^2/yr (Binford et al., 2006; Delcourt and Harris, 1980). Nevertheless, as a result of expanding urban areas, overall NPP in this region has declined by 0.4% (Milesi et al., 2003). Imhoff et al. (2004a) estimate a 1.6% loss of NPP due to urbanization across the United States, because tree growth in urban areas accounts for only small fractions of total forest biomass (3.6%) and NPP (0.7%) (Nowak et al., 2013). Recent estimates suggest an accumulation of carbon in vegetation and soils ranging from 0.32×10^{15} g C/yr for the U.S. (Lu et al., 2015) and 0.47×10^{15} g C/yr for North America (King et al., 2015).

In the Great Plains of the United States, cultivated lands have NPP about 10% above the level of native ecosystems in that region (Bradford et al., 2005), largely as a result of exogenous inputs to the landscape (Smith et al., 2014). Accounting for the CO_2 emissions associated with the production of fertilizer and pesticides, pumping of irrigation water, and cultivation itself, agricultural lands are a net source of CO_2 to the atmosphere (i.e., negative NEP; West et al., 2010). Of course, along with improved plant genetics, agricultural intensification is likely to be the only way to provide more food from less land to feed the world's growing human population (Balmford et al., 2018).

The rising concentration of carbon dioxide in the atmosphere from fossil fuel combustion and biomass burning increases the availability of a basic reactant for photosynthesis (Eq. 5.2). Early studies of plant response to high CO_2 showed an average 31% increase in growth with a doubling of CO_2 concentrations for woody plants in controlled experiments (Curtis and Wang, 1998; Wang et al., 2012). When it was noted that the growth responses were much lower in the absence of fertilization (Thomas et al., 1994; Hattenschwiler et al., 1997; Poorter and Pérez-Soba, 2001), investigators established large-scale, long-term experiments in a variety of ecosystems using Free-Air CO_2 Enrichment (FACE) technology (see Fig. 5.14 and Hendrey et al., 1999).

---

[r] Deforestation rates are notoriously subject to political and economic motivations and thus variable in different reports (see Mitchard, 2018). Country-specific statistics for the extent of forests are given in The Global Forest Resources Assessment of the Food and Agriculture Organization. www.fao.org/forestry/fra/fra2010/en/. The gross rate of forest cover loss is approximately 3 times larger than the net rate as a result of the regrowth of forests on some disturbed lands (Hansen et al., 2013). In the Amazon basin, the rate of deforestation declined significantly between 2004 and 2011 (Davidson et al., 2012; Keenan et al., 2013). For the United States, the gross rate of deforestation is about 1×10^4 km^2/yr (Masek et al., 2011), but as a result of reforestation and afforestation, there has been a net increase in forest area during the last decade.

FIG. 5.14 The Free-Air CO_2 Enrichment experiment in Duke Forest, central North Carolina. Each plot is 30 m in diameter and surrounded by 16 towers, which emit CO_2 so as to maintain a specified concentration in the cylindrical volume of the plot to the height of the forest canopy.

The results from various forest FACE experiments show an 18% increase in NPP with growth at +200 ppm CO_2—the atmospheric levels expected globally in 2050 (Norby et al., 2005; Norby and Zak, 2011). Crop plants show growth stimulations ranging from 12% to 14% in rice, wheat, and soybeans (Long et al., 2006). The change in photosynthesis in response to elevated CO_2 is attenuated by simultaneous changes in respiration in response to temperature, so that the ratio of photosynthesis to respiration is relatively homeostatic (Dusenge et al., 2019). In all species, growth at high CO_2 stimulates the proportional allocation of NPP to roots, which may increase the delivery of carbon to soils (Rogers et al., 1994; Jackson et al., 2009). High CO_2 improves water-use efficiency by vegetation, prolonging the period of growth in droughty climates (Battipaglia et al., 2013; Jung et al., 2017). Over time, adjustments in leaf area and hydraulic conductivity in tree stems can attenuate higher WUE in plants grown at high CO_2 (Tor-ngern et al., 2015). Nevertheless, there seems little doubt that the rise of CO_2 in Earth's atmosphere and thus the rate of anticipated climate change would be greater if it were not for CO_2 uptake in areas of undisturbed vegetation and the regrowth of vegetation on previously cleared lands.

Along with exposure to high CO_2, humans have increased the exposure of forests to ozone, acid rain, and various forms of reactive nitrogen, with variable effects on growth. Nitrogen deposition from the atmosphere, producing a number of changes in soil biogeochemistry (Chapter 6), is likely to increase carbon uptake and storage in forests through its effect as a plant fertilizer (Magnani et al., 2007; Thomas et al., 2010). On the other hand, ozone reduces the growth of most plants when concentrations exceed 100 ppb (Richardson et al., 1992; Gregg et al., 2003), although growth at high CO_2 partially compensates for the ozone effect (Reid and Fiscus, 1998; King et al., 2005; Poorter and Pérez-Soba, 2001; Peñuelas et al., 2012). Losses of photosynthetic efficiency from air pollutants are noted even among plants in high northern latitudes, where one might expect relatively low pollution loads (Odasz-Albrigtsen et al., 2000; Savva and Berninger, 2010).

Surprisingly, the growth increase due to high CO_2 is not seen in various inventories of forest growth (Groenendijk et al., 2015; Girardin et al., 2016; Bader et al., 2013; Silva and Anand, 2013). Tree-ring records are equivocal—some showing recent increases in growth (Soulé and Knapp, 2006), others showing little or no effect, perhaps due to concurrent drought (Barber et al., 2000; Gedalof and Berg, 2010; Andreu-Hayles et al., 2011; Peñuelas et al., 2012). Using satellite data to monitor NDVI (Eq. 5.8), several studies noted increases in global NPP during the 1980s and 1990s, largely through changes in high northern latitudes (Myneni et al., 1997), and perhaps globally (Nemani et al., 2003). Hember et al. (2019) report increases in the growth of boreal forests across Canada. Most of the change was attributed to changes in temperature, which determines the length of the growing season. Strangely, this effect seems to be reversed in the MODIS-derived NPP record from 2000 to 2009 which shows a 1% decline in global NPP attributed to increasing drought in the Southern Hemisphere (Fig. 5.15; Piao et al., 2011). Eddy-covariance studies in Europe show a severe reduction in NPP during 2003 due to heat and drought, so that the region became a source of CO_2 to the atmosphere (i.e., negative NEP; Ciais et al., 2005).

Certainly, global changes in NPP and biomass accompanied past climate changes associated with glacial intervals. At the last glacial maximum 19,000 years ago, carbon storage in land plants and soils was 30–50% lower than today (Bird et al., 1994; Beerling, 1999; Kohler and Fischer, 2004). NPP on the world's land surface was presumably depressed as well, because of lower plant cover; cold, dry climates; and low atmospheric

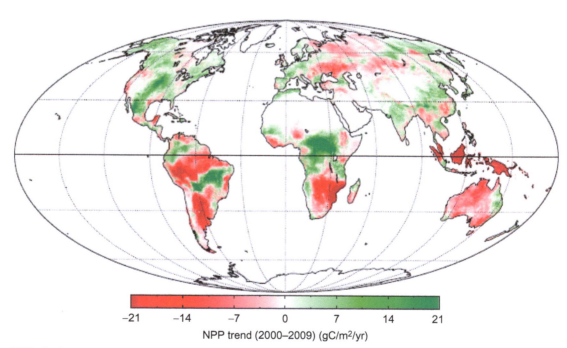

FIG. 5.15 Change in terrestrial net primary production (NPP) from 2000 to 2009 from MODIS. *From Zhao and Running (2010).*

CO_2 (Gerhart and Ward, 2010). At the last glacial maximum, Landais et al. (2007) estimate that terrestrial NPP was only 65–70% of today's value. With the future climate changes due to greenhouse warming, plant biomass may increase up to 10% over present-day conditions (Smith et al., 1992b). Although a transient period of drought may reduce terrestrial productivity during the next few decades (Rind et al., 1990; Smith and Shugart, 1993), overall, we might expect higher terrestrial NPP on a warmer, wetter world in the future (Wu et al., 2011). Changes in the distribution and productivity of vegetation are likely to have additional impacts on climate (Forzieri et al., 2017).

Detritus

The largest fraction of NPP is delivered to the soil as dead organic matter. Global patterns in the deposition of plant litterfall are similar to global patterns in NPP (Matthews, 1997). The deposition of litterfall declines with increasing latitude from tropical to boreal forests (Vogt et al., 1986; Lonsdale, 1988; Berg et al., 1999). Leaf tissues account for about 70% of litterfall in forests (O'Neill and De Angelis, 1981; Meentemeyer et al., 1982), but the deposition of woody litter tends to increase with forest age, and fallen logs may be a conspicuous component of the forest floor in old-growth forests (Lang and Forman, 1978; Harmon et al., 1986; McGarvey et al., 2015). In grassland ecosystems, where little of the aboveground production is contained in perennial tissues, the annual litterfall is nearly equal to annual net primary production. In most areas, the annual growth and death of fine roots contributes a large amount of detritus to the soil, which has been overlooked by studies that only consider aboveground litterfall (Vogt et al., 1986; Nadelhoffer and Raich, 1992).

Using actual evapotranspiration to predict global patterns of litterfall; Meentemeyer et al. (1982) estimated 27×10^{15} g C for the annual production of aboveground litterfall worldwide, similar to values ranging from 26 to 31.5×10^{15} g C/yr calculated by Shen et al. (2019) from 2347 field studies of litterfall in forests worldwide. Matthews (1997) indicates *total* detritus production of $\sim 50 \times 10^{15}$ g C/yr, suggesting that about half of global NPP occurs belowground. Her value is slightly smaller than most current estimates of terrestrial NPP ($\sim 60 \times 10^{15}$ g C/yr; Table 5.3), which is consistent with ancillary losses of organic carbon to fires (4%), herbivores (5%), the emission of volatile organic carbon compounds to the atmosphere (2%), and losses of dissolved organic carbon to streams and groundwater (0.5%), each as previously estimated.

The decomposition process

Most detritus, whether from litterfall or root turnover, is delivered to the upper layers of the soil where it is subject to the decomposition by microfauna, bacteria, and fungi (Swift et al., 1979; Schaefer, 1990). Decomposition leads to the release of CO_2, H_2O, and nutrient elements, and to the microbial production of organic compounds known as humus. Humus compounds accumulate in the lower soil profile (Chapter 4) and compose the bulk of soil organic matter (Schlesinger, 1977; Rumpel and Kogel-Knabner, 2011). The dynamics of the pool of carbon in soils is best viewed in two stages—processes leading to rapid turnover of the majority of litter at the surface and processes leading to the slower production, accumulation, and turnover of humus at depth.

The litterbag approach is widely used to study decomposition at the surface of the soil. Fresh litter is confined in mesh bags that are placed on the ground and collected for measurements at periodic intervals (Singh and Gupta, 1977). Simple models of decay are based on an exponential pattern of loss, where the fraction remaining after 1 year is given by:

$$X/Xo = e^{-k} \tag{5.10}$$

An alternative, the mass-balance approach, suggests that the annual decomposition of litter should equal the annual input of fresh debris so that the mass of detritus stays constant. Under these assumptions, a constant fraction, k, of the detrital mass decomposes, so that:

$$\text{litterfall} = k(\text{detrital mass}) \tag{5.11}$$

or

$$\text{litterfall}/\text{detrital mass} = k \tag{5.12}$$

When the detritus is in steady state, the values for k calculated from the litterbag and mass-balance approaches should be equivalent, and mean residence time for plant debris is $1/k$ (Olson, 1963; see also footnote b, Chapter 3). For a forest in the Pacific Northwest, Vogt et al. (1983) shows the importance of fine roots in the calculation of mean residence times by the mass-balance approach. When root turnover was included, the mean residence time for organic matter in the forest floor was 8.2–15.6 years, compared to 31.7–68.6 years calculated from aboveground litter alone.

With either approach, when decomposition rates are rapid, values for k are greater than 1.0, and there is little surface accumulation (e.g., in tropical rainforests; Cuevas and Medina, 1988; Gholz et al., 2000; Powers et al., 2009). In such systems, decomposition has the potential to respire more than the annual deposition of organic carbon in litterfall. In contrast, values for k are very small (e.g., 0.001) in some peatlands and boreal forest (Olson, 1963). Decomposition in grasslands shows a range of 0.20–0.60 in values for k (Vossbrinck et al., 1979; Seastedt, 1988), but values for deserts may be as high as 1.00 due to the action of termites (Schaefer and Whitford, 1981) and photooxidation of litter by ultraviolet light (Austin and Vivanco, 2006; Gallo et al., 2009). In many ecosystems, decomposition shows a rapid initial phase of decomposition, followed by a slower phase in which some material may persist for decades (Harmon et al., 2009). Two- or three-phase exponential models are often best to describe this pattern of decomposition and for the most accurate estimates of k (Minderman, 1968; Adair et al., 2008).

Decomposition rates vary as a function of temperature, moisture, and the chemical composition of the litter material. Across Europe, temperature was the major factor controlling litter decomposition (Portillo-Estrada et al., 2016). Microbial activity increases exponentially with increasing temperature (e.g., Edwards, 1975). For plant litter, this relation often shows a Q_{10} of ≥ 2.0, that is a doubling in activity per 10°C increase in temperature (Raich and Schlesinger, 1992; Kirschbaum, 1995; Katterer et al., 1998; Wang et al., 2019c). Van Cleve et al. (1981) found that the thickness of the forest floor in black spruce forests in Alaska was inversely related to the cumulative degree days favorable to decomposition each year. In contrast, soil

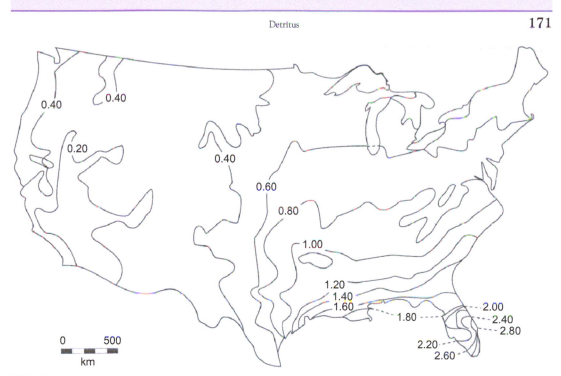

FIG. 5.16 Rates of decomposition of fresh litter in the United States predicted by a simulation model using actual evapotranspiration as the predictive variable. Isopleth values are the fractional loss rate (k) of mass from fresh litter during the first year of decomposition. *From Meentemeyer (1978a).*

moisture often limits the rate of decomposition in arid and semiarid regions (Strojan et al., 1987; Epstein et al., 2002; Amundson et al., 1989), and moisture assumes increasing importance when temperate forest soils are subject to experimental warming, which dries them (Peterjohn et al., 1994).

Meentemeyer (1978a) compiled data from various decomposition studies to relate surface decomposition to actual evapotranspiration, and used the resulting equation to predict regional patterns of decomposition (Fig. 5.16). His predictions are consistent with observations of surface litter in much of the United States (e.g., Lang and Forman, 1978). Actual evapotranspiration is also a good predictor of decomposition in Europe (Berg et al., 1993), but less successful in predicting the decomposition of fine roots (Silver and Miya, 2001). Improvements in these predictions are found when chemical parameters, such as lignin and nitrogen, are also considered (Meentemeyer, 1978b; Melillo et al., 1982), but we defer a discussion of the role of litter chemistry and the release of plant nutrients during decomposition until Chapter 6.

Humus formation and soil organic matter

Plant litter and soil microbes constitute the cellular fraction of soil organic matter. As decomposition proceeds, there is an increasing content of non-cellular organic matter (i.e., humus) which appears to result from microbial activity. The structure of humus is poorly known, but it contains numerous aromatic rings with phenolic (–OH) and organic acid (–COOH) groups (Flaig et al., 1975; Stevenson, 1986). Humic materials offer a major source of cation exchange capacity in many soils (Chapter 4).

Traditional chemical characterizations of humus was based on the solubility its components in alkaline and acid solutions. These are operational definitions that are only useful in the laboratory. The structure of humic substances changes dramatically in response to changes in the pH, ion strength, and various ions in the soil solution (Myneni et al., 1999). In particular, alkaline extractions produce a number of changes in the molecular structure of soil organic matter and are not indicative of humic substances (Kleber and Lehmann, 2019). Alternatively, size or density fractionations can be used to quantify the labile and resistant organic matter. Density fractionations are performed by adding soil samples to solutions of increasing specific gravity and collecting the material that floats to the surface (Spycher et al., 1983). In size fractionation, soils are passed through screens of varying mesh size (Tisdall and Oades, 1982; Elliott, 1986). Most of the turnover of soil organic matter is in the "light" or large fractions that represent fresh plant materials (Tiessen and Stewart, 1983). The "heavy" fraction is composed of polysaccharides (sugars) and humic materials that are complexed with clay minerals to form microaggregates of relatively high specific gravity (Tisdall and Oades, 1982; Tiessen and Stewart, 1988). The labile and resistant pools are often referred to as the active and passive forms of soil organic matter.

Humic substances tend to be richer in nitrogen than the plant detritus from which they are derived (Fine et al., 2018). A recent proposition suggests that humus is composed of diverse small molecular units, especially peptides that aggregate into supramolecular associations with soil minerals (Knicker, 2011; Sutton and Sposito, 2005; Kögel-Knabner and Rumpel, 2018). Recent progress in elucidating the chemical structure of humus has been made using ^{13}C nuclear magnetic resonance (NMR) spectroscopy (Mahieu et al., 1999; Baldock et al., 2004; Kelleher et al., 2006; Feng et al., 2010). Humus substances are not inherently resistant to decomposition, but rather they are likely to be protected from decomposition by forming complexes with soil minerals (Allison, 2006; Kögel-Knabner and Rumpel, 2018; Hemingway et al., 2019).

Among humic materials, organic acids control the downward movement of Fe and Al in soils. Percolating downward from the forest floor and A-horizon, these acids often account for a large fraction of the soil organic matter in the lower soil profile, where they are complexed with clay minerals and calcium (Beyer et al., 1993; Oades, 1988; Kalbitz et al., 2000; Gaiffe and Schmitt, 1980). Noncrystalline forms of Fe- and Al-oxides are particularly effective in preserving organic matter by surface adsorption (Torn et al., 1997; Mikutta et al., 2006; Powers and Veldkamp, 2005; Porras et al., 2017). Doetterl et al. (2015) found that despite the obvious role of climate, soil carbon storage along a latitudinal gradient extending from Chile to Antarctica could only be explained by considering soil mineralogy. Aluminum and iron minerals also play a dominant role in stabilizing soil organic matter on the Tibetan plateau (Fang et al., 2019) and in subtropical forests in China (Yu et al., 2019c).

Humus is very resistant to microbial attack; extracted humic materials from forest soil in Saskatchewan had a measured mean ^{14}C age of 250–940 years (Campbell et al., 1967), and a recent compilation suggests that the mean age of humic materials worldwide is about 3100 years (He et al., 2016). The radiocarbon age of the different size or weight fractions indicates their rate of turnover. Anderson and Paul (1984) reported a ^{14}C age of 1255 years for organic matter in the clay fraction of a soil for which the overall age was 795 years. The resistance of humus substances to decay is most likely due to their preservation in association with soil minerals, rather than due to inherent chemical resistance to decomposition (Allison, 2006).

Under most vegetation, the mass of humus in the soil profile exceeds the combined content of organic matter in the forest floor and aboveground biomass. Globally the pool of organic carbon in world soils amounts to about 1500×10^{15} g to 1-m depth (Schlesinger, 1977; Batjes, 1996; Amundson, 2001). At least a portion of that total, perhaps 5%, is composed of charcoal from past wildfires (Landry and Damon Matthews, 2017). Many tropical soils contain small amounts of soil organic matter dispersed in the lower profile, some at great depth (Harper and Tibbett, 2013; Wade et al., 2019). Table 5.4 provides a global inventory of plant detritus and soil organic matter to 3-m depth, totaling 2344×10^{15} g C (Jobbágy and Jackson, 2000; Jackson et al., 2017). Even that value may underestimate the total mass of organic material in regions of permafrost (Tarnocai et al., 2009; Hugelius et al., 2014). Northern peatlands may contain up to 547×10^{15} g C (Yu et al., 2010b; Bradshaw and Warkentin, 2015), and large

TABLE 5.4   Distribution of soil organic matter by ecosystem types.

Biome	World area (10^6 km^2)	Mean soil profile carbon (kg C/m^2) 0–100 cm	Mean soil profile carbon (kg C/m^2) 0–300 cm	Total soil carbon pool (10^{15} g C) 0–300 cm
Tropical forests				
Deciduous	7.5	15.8	29.1	218
Evergreen	17.0	18.6	27.9	474
Temperate forests				
Deciduous	7	17.4	22.8	160
Evergreen	5	14.5	20.4	102
Boreal forests	12	9.3	12.5	150
Mediterranean shrublands	8.5	8.9	14.6	124
Tropical savannas/grasslands	15	13.2	23.0	345
Temperate grasslands	9	11.7	19.1	172
Deserts	18	6.2	11.5	208[a]
Arctic tundra	8	14.2	18.0	144
Crops	14	11.2	17.7	248
Extreme desert, rock and ice	15.5			
Total	136.5			2344

[a] Excludes soil carbonates, which may contain an additional 930×10^{15} g C (Schlesinger, 1985).
From Jobbágy and Jackson (2000).

accumulations of peat are also found in the Congo and other areas of the tropics (Dargie et al., 2017).

The global estimate of soil organic matter, divided by the estimate of global litterfall, suggests a mean residence time of about 40 years for the total pool of organic carbon in soils, but the mean residence time varies over several orders of magnitude between the surface litter and the various humus fractions (Fig. 5.17). In the temperate zone, the mass of soil organic matter and its mean residence time increase from warm-temperate to boreal forests (Schlesinger, 1977; Garten, 2011; Frank et al., 2012; Wang et al., 2018a). Regional inventories of the distribution and abundance of soil organic carbon (0–100 cm) are available for the United States (46 to 74×10^{15} g C; Guo et al., 2006; Guevara et al., 2020), North America (367×10^{15} gC; Liu et al., 2013), China (84 to 89×10^{15} g C; Li et al., 2007; Tang et al., 2018), India (63×10^{15} g C; Lal, 2004) and Brazil (71×10^{15} g C; Gomes et al., 2019), and other nations as part of recent national accounts of carbon storage in vegetation and soils.

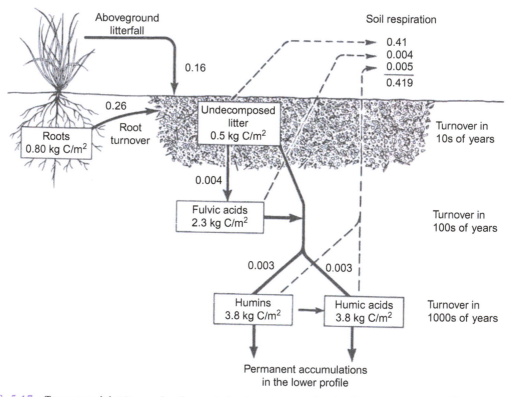

FIG. 5.17   Turnover of detritus and soil organic fractions in a grassland soil, in units of kg C/m^2/yr. Note that mean residence time can be calculated for each fraction from measurements of the quantity in the soil and the annual production or loss (respiration) from that fraction. *From Schlesinger (1977).*

Turnover and soil respiration

The incorporation of nuclear-bomb-derived radiocarbon (^{14}C) into different fractions of soil organic matter shows promise as a means of estimating their turnover (Trumbore, 1993). For grasslands in California, Torn et al. (2013) reported that <10% of the soil organic matter was "fast-cycling," but this fraction was responsible for 90% of the annual soil carbon flux. In British deciduous woodlands, the distribution of radiocarbon is also compatible with two pools of carbon, each with about 3.5 kg/m^2 to 15-cm depth (Tipping et al., 2010). In general the light fraction of soil organic matter with rapid turnover is found in the upper part of the soil profile, whereas the heavy fraction is found in the lower profile, where its turnover time is slow (Schrumpf et al., 2013). Because of different turnover times, there is no universal decomposition constant, k, that can be applied to the entire mass of organic matter in the soil profile (Trumbore, 1997; Gaudinski et al., 2000).

Field measurements of the flux of CO_2 from the soil surface provide an estimate of the total respiration in the soil. Most of the production of CO_2 occurs in the surface litter where decomposition is rapid and a large proportion of the fine root biomass is found (Bowden et al., 1993). Edwards and Sollins (1973) found that only 17% of the annual production of CO_2 in a temperate forest soil was contributed by soil layers below 15 cm. Flux of CO_2 from the deeper soil layers is presumably due to the decomposition of humus substances. Production of CO_2 in the soil leads to the accumulation of CO_2 in the soil pore space, particularly at depth, which drives carbonation weathering in the lower profile (Fig. 4.4). Geologic sources of CO_2 diffusing upward to the soil surface are normally very small (Keller and Bacon, 1998).

Unfortunately, the respiration of living roots makes it impossible to use estimates of CO_2 flux from the soil surface to calculate turnover of the soil organic pool (Fahey et al., 2005). In a compilation of values, Schlesinger (1977) found that CO_2 respiration from soils exceeded the deposition of aboveground litter by a factor of about 2.5 (Fig. 5.18). The additional CO_2 is presumably derived from root and mycorrhizal metabolism and the decomposition of root detritus (Raich and Nadelhoffer, 1989; Subke et al., 2011). In a field experiment using girdling of trees to eliminate the transport of new photosynthate to roots, Högberg et al. (2001) reported a 54% decline in soil respiration; presumably the remaining respiration was due to decomposers in the soil (cf. Andrews et al., 1999; Hanson et al., 2000). Globally, soil respiration is $80-100 \times 10^{15}$ g C/yr, with about half derived from the respiration of live roots and the remainder from decomposition (Raich et al., 2002; Subke et al., 2006; Bond-Lamberty and Thomson, 2010; Hursh et al., 2017; Konings et al., 2018). Soil respiration shows a strong correlation with NPP and detritus inputs in world ecosystems (Raich and Tufekcioglu, 2000; Bond-Lamberty et al., 2004; Hibbard et al., 2005), and has perhaps increased in recent years due to climatic warming (Bond-Lamberty et al., 2018).

The global distribution of soil organic matter shows how moisture and temperature control the balance between primary production and decomposition in surface and lower soil layers (Amundson, 2001). Among forests, accumulations in the forest floor increase from tropical to boreal climates. Net primary productivity shows the opposite trend, so the accumulation of soil organic matter is largely due to differences in decomposition. Thus, soil microbes are more sensitive than vegetation to regional differences in temperature and moisture. Worldwide, the accumulation of organic matter in surface litter seems more related to factors

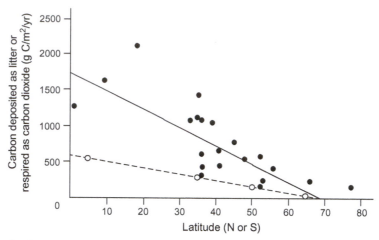

FIG. 5.18 Latitudinal trends for carbon dynamics in forest and woodland soils of the world. The *dashed line* shows the mean annual input of organic carbon to the soil by litterfall. The *solid line* shows the loss of carbon, measured as the flux of CO_2 from the soil surface. The difference between these lines represents the loss of CO_2 from root respiration, including mycorrhizae, and the decomposition of root detritus and exudates. *From Schlesinger (1977).*

controlling decomposition than to the NPP of terrestrial ecosystems (Cebrián and Duarte, 1995; Valentini et al., 2000). Nevertheless, when inputs of plant litter are reduced, soil organic matter declines (Dove et al., 2019).

Parton et al. (1987) developed a model based on the differential turnover of soil organic fractions to predict the accumulation of soil organic matter in grassland ecosystems. Accurate predictions were achieved when temperature, moisture, soil texture, and plant lignin content were included as variables. Despite relatively low NPP, soils of temperate grasslands contain large amounts of soil organic matter (Sanchez et al., 1982b) due to relatively low rates of decomposition and a larger fraction of plant debris that is derived from root turnover (Ma et al., 2019; Xu et al., 2019a). In contrast, tropical grasslands and savannas have relatively small accumulations of surface litter, perhaps due to frequent fire (Kadeba, 1978; Jones, 1973; Pellegrini et al., 2018).

Storage of soil organic matter represents a component of net ecosystem production (NEP) in terrestrial ecosystems. Studies of soil chronosequences show that soil organic matter accumulates rapidly on disturbed sites, but rates decline to 1–12 g $C/m^2/yr$ during long-term soil development (Fig. 5.19; Schlesinger, 1990; Chadwick et al., 1994), with the highest rates under cool, wet conditions. Many wetland soils also show large rates of organic accumulation due to anoxic conditions in their sediments (Chapter 7). The low rate of accumulation of soil organic matter in upland soils speaks strongly for the efficiency of decomposers using aerobic metabolic pathways of degradation (Gale and Gilmour, 1988). With their relatively high nutrient content, humic substances are not inherently resistant to decomposition, but they are stabilized by interactions with soil minerals (Schmidt et al., 2011; Allison, 2006). Globally, the annual net production of humus substances is $<0.4 \times 10^{15}$ g C/yr or only about 0.7% of NPP (Schlesinger, 1990).

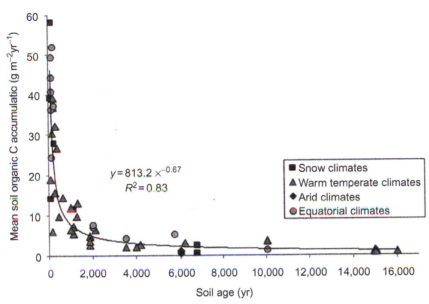

FIG. 5.19 The rate of accumulation of organic matter in soil chronosequences of different age and climate zones, all derived from volcanic materials. *From Zehetner (2010).*

The mass of soil organic matter in most upland ecosystems is likely to have been fairly constant before widespread human disturbance of soils. When soils show a steady state in organic content, the production of humic compounds must be equal to their removal from soils by erosion. The small losses of dissolved organic carbon in runoff are of major importance to the metabolism of stream ecosystems (Chapter 8). Coincidentally, estimates of the global transport of organic carbon in rivers are also about 0.4×10^{15} g C/yr (Schlesinger and Melack, 1981; Meybeck, 1982), suggesting that before human impacts, NEP for the Earth's land surface was essentially zero.

It is important not to forget that large amounts of carbon were stored in vegetation and soils of high northern latitudes during the Holocene, when these areas experienced warmer climates at the end of the last continental glaciation (Harden et al., 1992; Treat et al., 2019; cf. Chen et al., 2006). For areas covered by the last continental glaciation, the total accumulation of soil organic matter represents NEP for the last 10,000 years. The maximum extent of the last glaciations, covering 29.5×10^6 km^2 of the present land area (Flint, 1971), now contains roughly 300×10^{15} g C—that is, more than 10% of the organic carbon contained in all the world's soils (Table 5.4). In these areas, soil organic matter has accumulated at rates of about 1.35 g C/m^2/yr during the Holocene period. The current rate of storage in northern ecosystems (0.015–0.035×10^{15} g C/yr) is too small to be a significant sink for human releases of CO_2 to the atmosphere from fossil fuels, nor is it likely to have increased significantly during the last century (Gorham, 1991; Harden et al., 1992).

Total storage of carbon in soils, $\sim 2000 \times 10^{15}$ g or 166×10^{15} mol, can account for only 0.45% of the O_2 content of the atmosphere, given that the storage of organic carbon and the release of

O_2 occur on a mole-for-mole basis during photosynthesis (Eq. 5.2). Thus, accumulations of atmospheric O_2 cannot be the result of the storage of organic carbon on land. Long-term storage of organic carbon appears to be dominated by accumulations in anoxic marine sediments (Chapter 9).

Soil organic matter and global change

Before widespread human occupation of the landscape, the organic matter in soil probably persisted in equilibrium with inputs of detritus from vegetation. Most cultivation reduces the inputs of fresh plant debris and increases the rate of decomposition of soil organic matter as a result of better soil moisture and aeration conditions. Losses of soil organic matter from many agricultural soils are typically 20–30% within the first few decades of cultivation (e.g., Fig. 5.20; Kopittke et al., 2017). The loss is greatest during the first years of cultivation. Eventually a new, lower level of soil organic matter is achieved that is in equilibrium with the plant production and decomposition in cropland (Jenkinson and Rayner, 1977). Some of the soil organic matter is lost as a result of erosion and buried elsewhere, but most is probably oxidized to CO_2 and released to the atmosphere (Van Oost et al., 2007).

The dynamics of soil organic matter are illustrated by the pattern of loss after land is converted to agriculture. Recall that soil organic matter consists of a labile and a resistant fraction. The rapid, early decline in soil organic matter in agricultural soils is largely the result of losses from the light/labile fraction (Buyanovsky et al., 1994; Cambardella and Elliott, 1994). Continuing losses occur more slowly from the resistant phases (e.g., Fujisaki et al., 2015).

FIG. 5.20 Decline of soil organic matter following the conversion of native soil to agriculture in two grassland soils. *From Schlesinger (1986).*

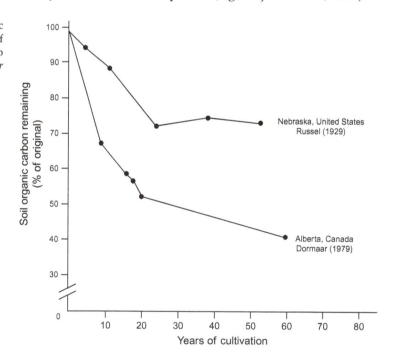

With about 10% of the world's soils under cultivation (refer to Table 5.4), losses of organic matter from agricultural soils have been a major component of the increase in atmospheric CO_2 during the past several centuries (Schlesinger, 1984; Jackson et al., 2017). As much as 133×10^{15} g C is likely to have been lost from soils since the beginning of organized agriculture (Sanderman et al., 2017). The current rate of release from soils, as much as 0.8×10^{15} g C/yr, is largely dependent on the rate at which natural ecosystems, especially in the tropics, are being converted to agriculture (Maia et al., 2010; Don et al., 2011; Assad et al., 2013). Especially large losses of soil carbon are seen when organic soils in wetlands and peatlands are drained (Armentano and Menges, 1986; Hutchinson, 1980; Chapter 7).

Losses of soil organic matter with cultivation contribute to the rise of atmospheric CO_2, but also to the hope that better management of agricultural soils might restore their native soil carbon and act to store CO_2 that would otherwise accumulate in the atmosphere. A reduced frequency of tillage results in the accumulation of carbon in the surface layers of the soil (West and Marland, 2002), but relatively little change in the entire soil profile (Powlson et al., 2014). Successful management of agricultural soils may benefit from the preservation of their organic microaggregate structure by reduced tillage.

Greater returns of crop residues, stimulated by fertilizer and irrigation, can also result in accumulations of soil organic matter. Soils in urban environments—lawns, parks, and golf courses—can also show increases in soil carbon storage with intensive management (Golubiewski, 2006; Pouyat et al., 2009; Raciti et al., 2011). In most circumstances the enhanced soil carbon storage under intensive management is less than the CO_2 emitted during the manufacture or delivery of soil amendments (Schlesinger, 2000; Russell et al., 2005; Khan et al., 2007; Townsend-Small and Czimczik, 2010). When ancillary and offsite emissions are considered, most agricultural lands in the United States show negative NEP (West et al., 2010).

Soil organic matter is lower in ecosystems that experience frequent fires (Pellegrini et al., 2018), but many of these soils contain significant amounts of charcoal (Landry and Damon Matthews, 2017; Jauss et al., 2017; Koele et al., 2017; Mao et al., 2012). Although charcoal is resistant to decay, it is not entirely immune to decomposition (Hammes et al., 2008), and additions of biochar are recommended to increase soil organic matter in some circumstances (Roberts et al., 2010).

Allowing vegetation to regrow on abandoned agricultural lands seems the only certain way to increase soil carbon (Vuichard et al., 2008; McLauchlan et al., 2006). Afforestation of abandoned croplands has resulted in significant increases in soil organic matter in Europe (Bárcena et al., 2014) and globally (Laganiere et al., 2010; Li et al., 2012). Significantly, soil organic matter can accumulate fairly rapidly when agricultural soils are abandoned, with rates averaging 33 g C/m^2/yr in a wide review of published values (Post and Kwon, 2000; Guo and Gifford, 2002; Clark and Johnson, 2011)—much higher than the accumulation under undisturbed vegetation (see Fig. 5.20). In all cases of natural succession, the accumulation of carbon in soil organic matter is dwarfed by the accumulation in regrowing woody vegetation (Richter et al., 2005; Johnson et al., 2003; Hooker and Compton, 2003).

Growth at high CO_2 increases the productivity of vegetation, especially belowground, so it is natural to expect increasing storage of soil carbon as a result. Nevertheless, various high CO_2 experiments using FACE technology in forests report only small differences in soil carbon storage (Hagedorn et al., 2003; Lichter et al., 2008; Hungate et al., 2013). Soil carbon shows no significant increase in Russian grassland soils during the last 100 years, while atmospheric

CO_2 has increased 30% (Torn et al., 2002). Rates of soil respiration increase in forests exposed to high CO_2, reflecting greater rates of decomposition (Bernhardt et al., 2006; Dieleman et al., 2010). Greater inputs of plant residues appear to stimulate the decomposition of existing soil organic matter, in what is known as the "priming effect" (Fontaine et al., 2007; Langley et al., 2009; Phillips et al., 2012; Kuzyakov et al., 2019).

Global patterns in the distribution and abundance of soil organic carbon (refer to Table 5.4) show the greatest accumulations in cold, wet conditions. Future, warmer conditions are likely to increase the decomposition of organic matter in boreal and arctic ecosystems, where much organic matter is held in the permafrost (Zimov et al., 2006). The increasing flux of CO_2 to the atmosphere from large pools of soil carbon at these latitudes could exacerbate global warming (Oechel et al., 2014; Schuur et al., 2009; Dorrepaal et al., 2009; Belshe et al., 2013). In the tundra of Alaska, changes in the depth of the water table—as might be expected by melting permafrost—appear to have a greater effect on soil carbon than changes in soil temperature alone (Huemmrich et al., 2010). Throughout the arctic and boreal regions, the organic carbon in deep soil horizons is relatively labile and subject to rapid decomposition in warmer, drier climates (Waldrop et al., 2010; Dorrepaal et al., 2009; Nowinski et al., 2010; Voigt et al., 2019). Predicted losses of 0.47×10^{15} g C/yr from northern high latitudes during the remainder of this century would add about 5% to the current yearly release from fossil fuel combustion (Zhuang et al., 2006).

Even in the temperate zone, soils appear to have lost significant amounts of organic carbon during hot dry years of 1978–2003 (Bellamy et al., 2005; Prietzel et al., 2016). Several compilations of the literature suggest increasing soil respiration globally in a warmer, future climate will add large amounts of CO_2 to Earth's atmosphere (Bond-Lamberty et al., 2018; Crowther et al., 2016).

Nearly all experiments imposing soil warming in intact ecosystems report an increase in the decomposition of soil organic matter and a greater release of nutrients (Van Cleve et al., 1990; Melillo et al., 2017; Conant et al., 2011; Harte et al., 2006; Teramoto et al., 2018). There is little evidence that soil microbial metabolism acclimates to elevated temperature (Schindlbacher et al., 2015; Carey et al., 2016; Karhu et al., 2014; Walker et al., 2018). Various soil-warming experiments have reported both greater soil respiration and greater plant growth, especially in cold regions (Rustad et al., 2001; Lu et al., 2013; Crowther et al., 2016). Thus, with global warming, some of the carbon lost from soils is likely to be balanced by enhanced growth of vegetation, which may mitigate the total loss of carbon from the site (Melillo et al., 2011; Sistla et al., 2013). Experimental ecosystem warming of a montane meadow in Colorado showed that the greatest changes in NPP appear to derive from an early start to the spring growing season, which subsequently leads to increasing drought later in the year (Saleska et al., 1999). Woody vegetation invaded the meadow, but soil organic matter declined in both control and warmed plots as a result of increasing temperature (Harte et al., 2015). Experimental warming of arctic tundra resulted in a large growth response of evergreen shrubs (Zamin et al., 2014).

In temperate and tropical regions, losses of soil organic matter due to climate change may be partially mediated by soil mineralogy, especially in areas where organic compounds are bound to Fe- and Al minerals (Rasmussen et al., 2006; Powers and Veldkamp, 2005). Nevertheless, losses of organic matter are found throughout the soil profile, from both labile and resistant fractions (Hicks Pries et al., 2017).

Summary

Photosynthesis provides the energy that powers the biochemical reactions of life (Fig. 5.21). That energy is captured from sunlight and stored in carbohydrates (organic matter). Globally, net primary production of about 60×10^{15} g C/yr is available in the terrestrial biosphere. Although that is a large value, NPP typically captures less than 1% of the available sunlight energy. Most of the remaining energy evaporates water and heats the air, resulting in the global circulation of the atmosphere (Chapters 3 and 10). Thus, the terrestrial biosphere is fueled by a relatively inefficient initial process.

During photosynthesis, plants take up moisture from the soil and lose it to the atmosphere in the process of transpiration. Available moisture appears to be a primary factor determining global variation in leaf area and NPP. Among communities with adequate soil moisture, net primary production is determined by the length of the growing season and by mean annual temperature; both are an index of the receipt of solar energy. Soil nutrients appear to be of secondary importance to NPP on land, perhaps because plants have various adaptations for obtaining and recycling nutrients efficiently when they are in short supply (Chapter 6).

Most net primary production is delivered to the soil, where it is decomposed by a variety of organisms. The decomposition process is remarkably efficient, so only small amounts of NPP are added to the long-term storage of soil organic matter or humus each year. Soil organic matter consists of a dynamic pool near the surface in which there is rapid turnover of fresh

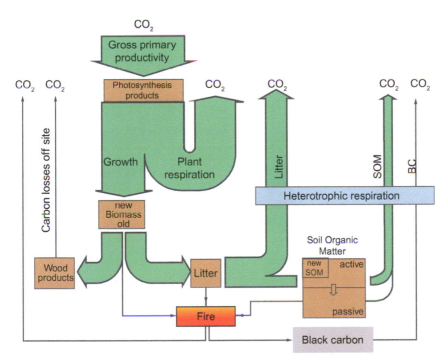

FIG. 5.21   Schematic for tracing the fate of carbon fixed by plants in photosynthesis. *From Schulze et al. (2000).*

plant detritus and little long-term accumulation, and a large refractory pool of humic substances that are dispersed throughout the soil profile. Thus, the turnover time of organic carbon in the soil ranges from about 3 years for the litter to thousands of years for humus. For the United States, the mean residence time of carbon in terrestrial ecosystems is about 46 years (Zhou and Luo, 2008), and the turnover time for carbon in vegetation and soils ranges from 15 to 255 between the tropics and high northern latitudes, respectively (Carvalhais et al., 2014).

Humans have altered the processes of net primary production and decomposition on land, resulting in the transfer of organic carbon to the atmosphere, and perhaps a permanent reduction in the global rate of NPP. This disruption has produced global changes in the biogeochemical cycle of carbon, but little change in the atmospheric concentration of O_2.

Biogeochemical Cycling on Land

Introduction

Living tissue is primarily composed of carbon, hydrogen, and oxygen in the approximate proportion of CH_2O, but more than 25 other elements are necessary for biochemical reactions and for the growth of structural biomass. For example, phosphorus (P) is required for adenosine triphosphate (ATP), the universal molecule for energy transformations in organisms, and calcium (Ca) is a major structural component in both plants and animals. The enzymes and structural proteins found in plants and animals contain about 16% nitrogen (N) by weight. Earlier we saw that the enzyme ribulose bisphosphate carboxylase determines the rate of photosynthesis—carbon (C) uptake—by many plant species (Chapter 5). The link between C and N that begins in cellular biochemistry extends to the global biogeochemical cycles of these elements.

The various elements essential to biochemical structure and function are often found in predictable proportions—stoichiometry—in living tissues (e.g., wood, leaf, bone, and muscle; Reiners, 1986; Sterner and Elser, 2002). For instance, the ratio of C to N in leaf tissue ranges from 25 to 50 (i.e., 1–2% N). At the global level, our estimate of net primary production (NPP) on land, 60×10^{15} gC/yr, implies that at least 1200×10^{12} g of nitrogen must be supplied to plants each year through biogeochemical cycling to achieve the level of NPP that we observe. As we shall see, the availability of some elements, such as N and P, is often limited, and the supply of these elements determines the rate of net primary production in many terrestrial ecosystems (Elser et al., 2007; LeBauer and Treseder, 2008; Xia and Wan, 2008; Fisher et al., 2012). Moreover, the cycling of various plant nutrients is often not independent; plant uptake of P requires enzymes containing N and energy from carbohydrate (CH_2O).

Conversely, for elements that are typically available in greater quantities in soil relative to plant demand (e.g., Ca and S), the rate of net primary production often determines the rate of cycling in the ecosystem and losses to stream waters. In every case, the biosphere exerts a strong control on the geochemical behavior of the major elements of life. Much less biological control is seen in the cycling of elements such as sodium (Na) and chloride (Cl), which are less important constituents of biomass (Gorham et al., 1979).

In earlier chapters, we found that the atmosphere is the dominant source of C, N, and S in terrestrial ecosystems. Except on old, highly weathered soils, rock weathering is the major source for most of the remaining biochemical elements (e.g., Ca, Mg, K, Fe, and P). In any terrestrial ecosystem, the receipt of elements from the atmosphere and the lithosphere represents an input of new quantities of nutrients for plant growth (Cleveland et al., 2013).[a] However, as a result of internal cycling and retention of past inputs, plant growth is not solely dependent on new inputs to the ecosystem. In fact, the annual circulation of important elements such as N within an ecosystem is often $10–20 \times$ greater than the amount received from outside the system (Table 6.1).[b] This large internal, or intrasystem, cycle is achieved by the long-term retention and recycling of elements derived from the atmosphere and the lithosphere. Important biochemical elements are accumulated in terrestrial ecosystems by biotic uptake, whereas nonessential elements pass through these systems under simple geochemical control (Vitousek and Reiners, 1975).

In this chapter, we analyze the cycles of biochemical elements in terrestrial ecosystems. We begin by examining aspects of plant uptake, allocations during growth, and losses due to the death of plants and plant tissues. Then we see how elements such as N, P, and S in dead organic matter are transformed in the soil, leading to their release for plant uptake or for loss from the ecosystem. The yearly uptake, allocation, return, and release of nutrient elements in

---

[a] Unlike the well-known models developed for the Hubbard Brook Ecosystem (e.g., Likens and Bormann, 1995), the nutrient budgets in this book consider rock weathering as an *external* source of nutrients that enter a terrestrial ecosystem each year (Gorham et al., 1979).

[b] Volk (1998) defines the recycling ratio as the ratio of the amount cycling in a system to the amount exiting it, which is roughly 6 for N and 4 for P in terrestrial ecosystems (see Chapter 12).

TABLE 6.1   Percentage of the annual requirement of nutrients for plant growth in the Northern Hardwoods Forest at Hubbard Brook, New Hampshire, which could be supplied by various sources of available nutrients.

Process	N	P	K	Ca	Mg
Growth requirement (kg/ha/yr)	115.4	12.3	66.9	62.2	9.5
Percentage of the requirement that could be supplied by:					
Intersystem inputs					
Atmospheric	18	0	1	4	6
Rock weathering	0	1	11	34	37
Intrasystem transfers					
Reabsorptions	31	28	4	0	2
Detritus turnover (includes return in throughfall and stemflow)	69	67	87	85	87

Note: *Calculated using Eqs. (6.2) and (6.3).*
Reabsorption data are from Ryan and Bormann (1982). Data for N, K, Ca, and Mg are from Likens and Bormann (1995) and for P from Yanai (1992).

an ecosystem constitute the nutrient cycle. Throughout, we stress interactions between C and other biochemical elements and examine how land plants have adapted to the widespread limitations of N and P in terrestrial ecosystems. We deduce changes in the sources of nutrients that determine plant growth during ecosystem development.

Biogeochemical cycling in land plants

Nutrient uptake

It is easy to forget the essential, initial role played by plants in all of biogeochemistry. Plants obtain essential elements from the soil (e.g., N from NO_3^-) and incorporate them into biochemical molecules (e.g., amino acids) (Oaks, 1994). Animals may eat plants, and each other, and synthesize new amino acids, but the building blocks of the amino acids in animal protein are those originally synthesized in plants. Only in isolated instances, for example, in animals at natural salt licks, do we find a direct transfer of elements from inorganic form to animal biochemistry—geophagy (Jones and Hanson, 1985; Panichev et al., 2013). There are few vitamin pills in the natural biosphere!

Soil chemical characteristics, including mineralogy and ion exchange, set the initial constraints on the availability of essential elements for plant uptake. However, when plant nutrient demand is large, plants can release organic compounds that enhance the solubility of elements, such as P, from soil minerals (Chapter 4). Thus, plants can affect the availability of nutrients needed for their own growth and adapt to a wide range of soil fertility (Forde and Lorenzo, 2001). Although foliar uptake is known, the vast majority of plant nutrient uptake passes through roots and their associated fungi, as we shall see in this chapter. A few unusual, insectivorous plants obtain N and P by digesting captured organisms (Adamec, 1997;

Wakefield et al., 2005), including vertebrates (Moldowan et al., 2019). For instance, Dixon et al. (1980) found that 11–17% of the annual uptake of N in *Drosera erythrorhiza* (sundew) can be obtained from captured insects. Ants deliver N to plants that host colonies of ants in small chambers known as domatia (Gegenbauer et al., 2012).

Delivery of ions to plant roots can occur by several pathways (Barber, 1962). The concentration of some elements in the soil solution is such that their passive uptake with water is adequate for plant nutrition (Turner, 1982). In some cases, the delivery is excessive, and the ions must be actively excluded at the root surface. For example, it is not unusual to see accumulations of Ca, as $CaCO_3$, surrounding the roots of desert shrubs growing in calcareous soils (Klappa, 1980; Wullstein and Pratt, 1981) or Fe-oxide plaques coating the roots of wetland plants (Mendelssohn et al., 1995).

Mass flow of water to the root, driven by transpiration from leaf surfaces, aids in the delivery of N, P, and K to the plant (Oyewole et al., 2014; McMurtrie and Näsholm, 2018), but the concentrations of these elements in the soil solution are often so low, that plant uptake must be enhanced by enzymes—transporters—that carry ions through channels in the root membrane using active transport (Hirsch et al., 1998; Williams and Miller, 2001; Nacry et al., 2013; Che et al., 2018). For zinc, the transporter enzymes in the roots are activated when zinc is deficient in the plant shoot (Sinclair et al., 2018). In contrast, NO_3^- uses signal transduction to activate the enzymes that promote its own uptake (Zhang and Forde, 1998; Tischner, 2000). These transporters must be phosphorylated, consuming energy and enhancing root respiration (Sun et al., 2014; Parker and Newstead, 2014). Ion transport accounts for a large portion of the respiration that is associated with roots (Chapter 5) and links the biogeochemical cycles of C (for energy), N (for enzymatic synthesis), and P (for phosphorylation) with the uptake of required plant nutrients.

The transporter systems embedded in root membranes achieve increasing rates of nutrient uptake as a function of increasing concentrations in the soil solution until the activity of the enzyme system is saturated. Chapin and Oechel (1983) found that populations of the arctic sedge, *Carex aquatilis*, from colder habitats had higher rates of uptake than those from warmer habitats, presumably reflecting adaptation to the lower availability of phosphorus in cold soils (Fig. 6.1). For such comparative studies, root physiologists define the specific absorption rate (SAR) as the rate of uptake of a nutrient from the soil per unit of root mass over a specified period of time.

Because more nutrients occur as positively charged ions than as negatively charged ions in the soil solution, one might expect that plant roots would develop a charge imbalance as a result of nutrient uptake. When ions such as K^+ are removed from the soil solution in excess of the uptake of negatively charged ions, the plant releases H^+ to maintain an internal balance of charge (Maathuis and Sanders, 1994). This H^+ may, in turn, replace K^+ on a cation exchange site, driving another K^+ into the soil solution. The high concentration of N in plant tissues causes the form in which N is taken up to dominate this process (Table 6.2). Oaks (1992) has shown how plants that use NH_4^+ as an N source tend to acidify the immediate zone around their roots (Fig. 6.2). The uptake of NO_3^- has the opposite effect as a result of plant releases of HCO_3^- and organic anions to balance the negative charge (Nye, 1981; Hedley et al., 1982a; Schöttelndreier and Falkengren-Grerup, 1999). These changes in the pH of the rhizosphere can affect the solubility and availability of soil P (Fig. 4.10).

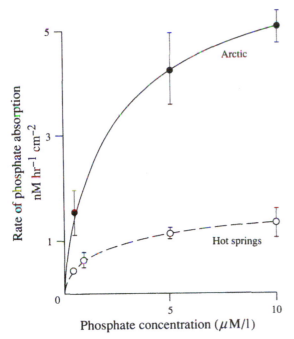

FIG. 6.1 **FIG. 6.1** Rate of phosphate absorption per unit of root surface area in populations of *Carex aquatilis* from cold (arctic) and warm (hot springs) habitats, measured at 5°C. *From Chapin (1974). Used with permission of the Ecological Society of America.*

TABLE 6.2 Chemical composition and ionic balance for perennial ryegrass.

	N	P	S	Cl	K	Na	Mg	Ca
Percent in leaf tissue	4.00	0.40	0.30	0.20	2.50	0.20	0.25	1.00
Equivalent weight (g)	14.00	30.98	16.03	35.46	39.10	22.99	12.16	20.04
mEq present	285.7	12.9	18.7	5.6	63.9	8.8	20.6	49.9
Sum of mEq	±285.7	−37.2			+143.1			
Imbalance in mEq %								

(a) Where ammonium nitrogen is taken up: $285.7 + 143.1 - 37.2 = +391.6$.
(b) Where nitrate nitrogen is taken up: $143.1 - 285.7 - 37.2 = -179.8$.
From Middleton and Smith (1979). Used with permission of Springer.

Normally, the uptake of N and P is so rapid and the concentrations in the soil solution are so low that these elements are effectively absent in the soil solution surrounding roots, and the rate of uptake is determined by diffusion to the root from other areas (Nye, 1977). Phosphate is particularly immobile in most soils, and the rate of diffusion strongly limits P supply to plant roots (Robinson, 1986; Santner et al., 2012). Although adaptations to enhance the efficiency of nutrient uptake are seen in some species (Pennell et al., 1990), the most apparent

FIG. 6.2   The pH of the soil in plants fertilized with nitrate (left) and ammonium (right), shown with a dye that changes color as a function of acidity. *From Oaks (1994).*

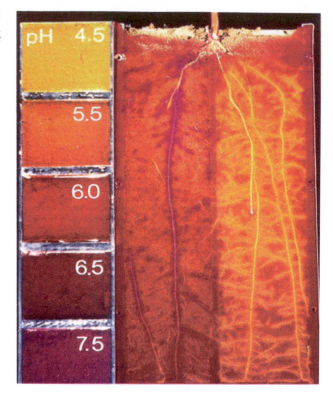

response of plants to low nutrient concentrations is an increase in the root/shoot ratio, which increases the volume of soil exploited and decreases diffusion distances (Aerts and Chapin, 2000; Clarkson and Hanson, 1980; Robinson, 1994; Ma et al., 2018). In many species, the relative growth rate of roots determines the uptake of nitrogen and phosphorus (Newman and Andrews, 1973; Fig. 6.3). Enhanced root growth is found in low phosphorus soils (Bates and Lynch, 1996; Ma et al., 2001), and roots rapidly proliferate in nutrient-rich patches (Jackson et al., 1990; Black et al., 1994; Muller et al., 2017). When forests are fertilized with nitrogen, there is lower overall allocation to root growth and a greater allocation to aboveground net primary productivity (Li et al., 2019c).

Alterations of physical and chemical conditions by soil microbes in close association with roots create the rhizosphere. The rhizosphere, operationally defined as the soil that adheres to roots when they are separated from soil, is often chemically and functionally distinct from the surrounding bulk soil (Jones et al., 2004). Chemical conditions in the rhizosphere are regulated by plant roots to minimize nutrient stress (Castrillo et al., 2017). In Chapter 4, we learned that plants can release organic acids into the soil that enhance the release of P from soil minerals. Plant roots and soil microbes also release enzymes into the soil to extract inorganic phosphorus from organic matter, where it is bound to carbon in ester groups (C—O—P). These extracellular enzymes from roots and the associated microbial community are known as phosphatases, which have different forms in acid and alkaline soils (Malcolm, 1983;

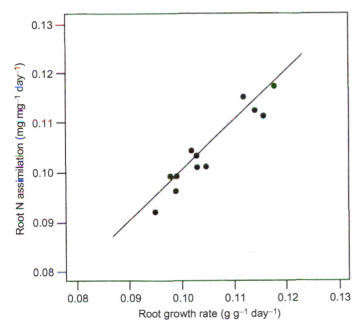

FIG. 6.3 The rate of N uptake in tobacco as a function of the relative growth rate of roots. *From Raper et al. (1978). Used with permission of the University of Chicago Press.*

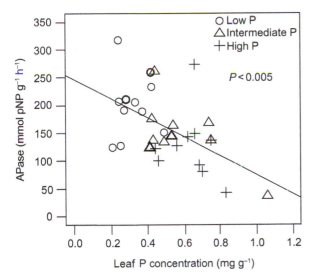

FIG. 6.4 Root phosphatase activity as a function of the phosphorus content of canopy leaves from high and low nutrient sites in tropical Borneo. *From Ushio et al. (2015).*

Dinkelaker and Marschner, 1992; Duff et al., 1994). In many cases, root phosphatase activity is inversely proportional to available soil P (Fox and Comerford, 1992; Treseder and Vitousek, 2001), especially in the low-P soils of the moist tropics (Kitayama, 2013, Fig. 6.4). Phosphatase activity may supply up to 75% of the annual phosphorus demand of some highly P-limited tundra and boreal forest species (Kroehler and Linkins, 1991; Firsching and Claassen, 1996). Increases in soil organic matter stocks during ecosystem development are often correlated with a corresponding increase in phosphatase activity, as seen in the development of *Eucalyptus* plantations after fire (Polglase et al., 1992). Experimental additions of N to soils have been shown to increase phosphatase activity to enhance P uptake for plant growth at the higher levels of N (Marklein and Houlton, 2012; Godin et al., 2015; Deng et al., 2017).

Mycorrhizal fungi

Most plants benefit from symbiotic associations with soil fungi known as mycorrhiza[c] (Allen, 1992). This symbiosis is important for the nutrition of plants and may have even determined the origin of land plants (Simon et al., 1993; Courty et al., 2010). In a mycorrhizal association a fungus colonizes a plant root, either extracellularly as ectomycorrhizal (EM) fungi, or intracellularly as arbuscular mycorrhizal (AM) fungi. EM fungi form a sheath around the active fine roots and extend additional hyphae into the surrounding soil,[d] while AM fungal hyphae actually penetrate root cells. Like roots and soil microbes, EM fungi produce a wide variety of extracellular enzymes, whereas AM fungi are more limited in their enzymatic arsenal.

By virtue of their large surface area, enzymatic capabilities, and efficient absorption capacity, mycorrhizal fungi are able to extend the nutrient acquisition capabilities of plants well beyond their roots. Mycorrhizal fungi transfer soil nutrients to plants in exchange for the supply of organic carbon as exudates and sloughed cells from plant roots. There is some indication that plants with thick roots are more dependent on mycorrhizae than plants with thin roots, which can explore the soil more thoroughly (Ma et al., 2018). Like fine roots, mycorrhizae proliferate in local areas of nutrients (Chen et al., 2018b).

Ectomycorrhizal fungi are directly involved in the breakdown of soil organic materials through the release of extracellular enzymes such as cellulases and phosphatases (Antibus et al., 1981; Dodd et al., 1987; Hodge et al., 2001) and in the weathering of soil minerals through the release of organic acids (Bolan et al., 1984; Illmer et al., 1995; van Breemen et al., 2000; van Scholl et al., 2008; Blum et al., 2002; see also Chapter 4). It is important to remember that some of these reactions are also associated with plant roots; mycorrhizae enhance these mechanisms in the rhizosphere, increasing the overall rate of plant nutrient uptake (Bolan, 1991). In return, mycorrhizal fungi depend on the host plant for supplies of carbohydrate.

Under conditions of nutrient deficiency, plant growth usually slows, whereas photosynthesis continues at relatively high rates (Chapin, 1980) leading to increases in the amount of soluble carbohydrates in plant tissues. Marx et al. (1977) found that high concentrations of carbohydrate in root tissues of loblolly pine (*Pinus taeda*) stimulated mycorrhizal infection (Fig. 6.5). Plants grown at high CO_2 also appear to allocate additional carbohydrate to fine roots to support mycorrhizal development and activity (DeLucia et al., 1997; Pritchard

[c]From Greek *mykēs*, "fungus," and *rhiza*, "root."

[d]Truffles are the fruiting body of certain ectomycorrhizae, which emit a volatile compound that resembles the mating pheromone of pigs, which are often used to locate them (Claus et al., 1981).

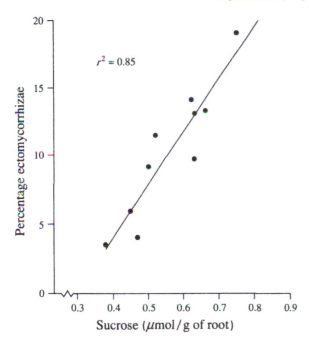

FIG. 6.5   Relationship between the infection of the roots of loblolly pine (*Pinus taeda*) by ecto-mycorrhizal fungi and the sucrose concentration in the root. *From Marx et al. (1977).*

et al., 2008a; Phillips et al., 2011a; Treseder, 2004). Thus, internal plant allocation of carbohydrates to roots may result in increased nutrient uptake by mycorrhizae and an alleviation of nutrient deficiencies (Bücking and Shachar-Hill, 2005; Ryan et al., 2012). Recent work has elucidated the genetic development of the symbiosis and the molecular structure of a transporter protein in mycorrhizae that moves phosphorus into the roots (Harrison and van Buuren, 1995; Bucher, 2007).

The importance of mycorrhizae in infertile sites is well known. Many species of conifers harbor ectomycorrhizae, which perhaps accounts for the success of pines in nutrient-poor soils, especially in boreal forests (Steidinger et al., 2019; Averill et al., 2019). Similarly, mycorrhizal fungi are widespread among the *Eucalyptus* species growing in the low-phosphorus soils of Australia, where the fungal spores are spread by feeding marsupials (Johnson, 1995; Weirich et al., 2018). Berliner et al. (1986) report the complete exclusion of the shrub *Cistus incanus* from basaltic soils in Israel due to a failure of ectomycorrhizal development. The same species grows well on adjacent calcareous soils or on basaltic soils supplied with fertilizer.

Most tropical trees appear to require AM associations for proper growth (Janos, 1980; Steidinger et al., 2019). Mycorrhizal fungi are especially important in the transfer of those soil nutrients, such as P, with low diffusion rates in the soil. Treseder and Vitousek (2001) measured greater mycorrhizal colonization, greater extracellular phosphatase, and greater phosphorus uptake by plants on P-deficient soils in Hawaii.

While a large number of studies document the importance of mycorrhizae in P nutrition (Koide, 1991), absorption of N and other nutrients is also known (Bowen and Smith, 1981;

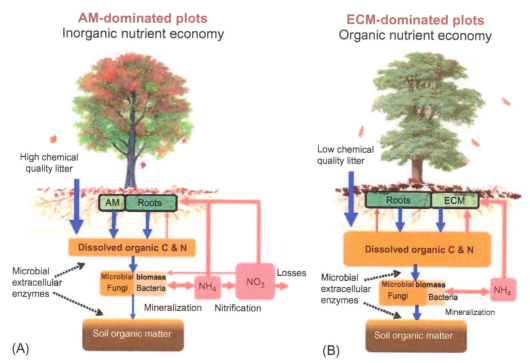

FIG. 6.6 Contrasting properties of forest communities dominated by arbuscular mycorrhizae (AM) versus ectotrophic mycorrhizae (ECM), showing major differences in the forms of nitrogen cycling in these ecosystems. *From Phillips et al. (2013).*

Ames et al., 1983; Govindarajulu et al., 2005). Ectomycorrhizae mobilize N from soil organic matter and mycorrhizae appear responsible for 61–86% of N uptake in Arctic tundra (Hobbie and Hobbie, 2006) and perhaps globally (Shi et al., 2016b). Ectomycorrhizae are particularly effective in mobilizing amino acids from organic matter, whereas AM mycorrhizae mobilize NH_4 and NO_3 for plant uptake (Hodge and Storer, 2015; Fig. 6.6).

Often plants with mycorrhizal fungi show higher levels of various nutrients in foliage, and frequently the enhanced uptake of nutrients results in higher rates of growth (Schultz et al., 1979). Mycorrhizal fungi use a fraction of the carbon fixed by the host plant, representing a diversion of net primary production that might otherwise be allocated to plant growth (Rygiewicz and Andersen, 1994). Overall, the costs of phosphorus acquisition are greater in low-phosphorus soils (Raven et al., 2018). That the cost of symbiotic fungi is significant is underscored by experiments in which the degree of colonization declined and plant growth increased when plants were fertilized (e.g., Blaise and Garbaye, 1983; Treseder, 2004; Teste and Laliberté, 2019; Ven et al., 2019). Across a wide variety of species, the carbon allocation to mycorrhizae appears to lower net primary productivity by about 15% (Hobbie, 2006; see also Table 5.1).

Nitrogen assimilation

Among various habitats, the availability of soil nitrogen as NH_4^+ or NO_3^- differs largely depending on the environmental conditions that affect the conversion of NH_4^+ to NO_3^- in the microbial process known as nitrification (Eqs. 2.16 and 2.17). For example, in waterlogged soils, almost all nitrogen is found as NH_4^+ (Barsdate and Alexander, 1975), whereas in some deserts and forests, nearly all mineralized NH_4 is converted to NO_3^- (Virginia and Jarrell, 1983; Nadelhoffer et al., 1984). Many species show a preference for NO_3^-, although species occurring in sites where nitrification is slow or inhibited often tend to show superior growth with ammonium ((Haynes and Goh, 1978; Adams and Attiwill, 1982; Falkengren-Grerup, 1995; Wang and Macko, 2011).

Derived from the breakdown of proteins, amino acids are found in many soils (Yu et al., 2002; Hofmockel et al., 2010) and used as a source of N by plants in a wide range of habitats, including tundra (Kielland, 1994; Schimel and Chapin, 1996; Nordin et al., 2004), boreal and temperate forest (Nasholm et al., 1998), and desert ecosystems (Jin and Evans, 2010). Direct uptake of amino acids has been demonstrated using isotopically-labeled amino acids (Nasholm et al., 1998) and nanoscale labels known as "quantum dots," which are attached to amino acids (Whiteside et al., 2009). Generally, the uptake of amino acids is greatest when the availability of inorganic N is low (Finzi and Berthrong, 2005). In a British grassland, most species showed preferential uptake of inorganic N over amino acids (Harrison et al., 2007).

Once inside the plant, NO_3^- and NH_4^+ are converted to amino groups ($-NH_2$) that are attached to soluble organic compounds. In many woody species these conversions occur in the roots, and N is transported to the shoot as amides, amino acids, and ureide compounds in the xylem stream (Andrews, 1986; Tischner, 2000). However, in some species, N in the xylem is found as NO_3^-, and the reduction of NO_3^- to $-NH_2$ occurs in leaf tissues (Smirnoff et al., 1984). Eventually, most plant N is incorporated into protein, and the amino acid arginine is used for storage of excess N (Llácer et al., 2008).

The conversion of NO_3^- to $-NH_2$ is a biochemical reduction reaction that requires metabolic energy and is catalyzed by an enzyme, *nitrate reductase*. One might puzzle why most plants do not show a clear preference for NH_4^+, which is assimilated more easily. Several explanations have been offered. Recall that NH_4^+ is bound to soil cation exchange sites, whereas NO_3^- is extremely mobile in most soils. The rate of delivery of NO_3^- to the root by diffusion or mass flow is thus much higher than that of NH_4^+ under otherwise equivalent conditions (Raven et al., 1992). Plants that use NH_4^+ may have to compensate for the differences in diffusion by investing more energy in root growth (Gijsman, 1990; Oaks, 1992; Bloom et al., 1993). Uptake of NO_3^- also avoids the competition that occurs in root enzyme carriers between NH_4^+ and other positively charged nutrient ions. For example, the presence of large amounts of K^+ in the soil solution can reduce the uptake of NH_4^+ (Haynes and Goh, 1978). Finally, relatively low concentrations of NH_4^+ are potentially toxic to plant tissues. These potential disadvantages in the uptake of NH_4^+ may explain why many plants take up NO_3^- when thermodynamic calculations suggest that the metabolic costs of reducing NO_3^- are greater than for plants that assimilate NH_4^+ or amino acids directly (Middleton and Smith, 1979; Gutschick, 1981; Bloom et al., 1992; Zerihun et al., 1998).

It is unclear why so many woody species concentrate nitrate reductase in their roots, when the same reaction can be performed in leaf tissues, where it can be coupled to the photosynthetic reaction and is therefore energetically much less costly (Gutschick, 1981; Andrews, 1986). Addition of NO_3^- to the soil often induces the production of root enzymes for NO_3^- uptake and the synthesis of more nitrate reductase in plant tissues (Lee and Stewart, 1978; Hoff et al., 1992; Oaks, 1994; Tischner, 2000). There is some evidence that the proportion of nitrate reductase in the shoot increases at high levels of available NO_3^- (Andrews, 1986). Both photosynthetic rates and nitrate uptake increase when plants are grown at high CO_2, but the response is not universal (Bassirirad, 2000).

Nitrogen fixation

Several types of bacteria possess the enzyme *nitrogenase*, which converts atmospheric N_2 to NH_3 in local conditions of cellular anoxia (see Eq. 2.9). Some of these exist as free-living (asymbiotic) forms in soils, but others, such as *Rhizobium* and *Frankia*, form symbiotic associations with the roots of higher plants. The symbiotic bacteria reside in root nodules that are easily recognized in the field (Fig. 6.7). Nitrogen fixation is especially well known among species of legumes (Leguminosae) (Bryan et al., 1996), but is also found in other plant families (Santi et al., 2013).

Free-living heterotrophic bacteria that conduct asymbiotic nitrogen fixation are usually found in organic soils or local areas with high levels of organic matter that provide a ready source of energy (Granhall, 1981; Billings et al., 2003). These organic-rich environments also foster the development of cellular anaerobiosis required by the nitrogenase enzyme (Marchal and Vanderleyden, 2000). For instance, nitrogen fixation is frequently observed in rotten logs (Roskoski, 1980; Silvester et al., 1982; Griffiths et al., 1993), where it is probably associated with anaerobic cellulolytic bacteria (Leschine et al., 1988).

The nitrogen that enters terrestrial ecosystems by fixation is a "new" input in the sense that it is derived from outside the boundaries of the ecosystem—that is, from the atmosphere (Cleveland et al., 2011). The reduction of N_2 to NH_3 has large metabolic costs that require the respiration of organic carbon. Nevertheless, Gutschick (1981) suggests that symbiotic fixation of N by higher plants is not greatly less efficient than the uptake of NO_3^- for those species in which the nitrate reductase is located in plant roots. Only a few land plants support symbiotic nitrogen fixation, and it is interesting to speculate why nitrogen fixation is not more widespread, when nitrogen

FIG. 6.7 Clusters of root nodules on a nitrogen-fixing shrub, *Calicotome villosa*, in Israel. *Photo courtesy of Guy Dovrat, Volcani Center, Agricultural Research Organization, Ramat Yishay.*

limitations of net primary production are so frequent (Vitousek and Howarth, 1991; Crews, 1999). Globally, plants "spend" about 2.5% of NPP on nitrogen fixation (Gutschick, 1981).

The energy cost of nitrogen fixation links this biogeochemical process to the availability of organic carbon, provided by net primary production. In plants with symbiotic nitrogen fixation, the rate of N fixation is often directly related to the rate of photosynthesis in the host plant and the efficiency of plant growth (Bormann and Gordon, 1984). N fixation is stimulated in seedlings of various species grown at high CO_2 (Tissue et al., 1997; Millett et al., 2012; Nasto et al., 2019), but it is unclear if this initial effect is persistent in long-term field experiments (Hungate et al., 2014).

N-fixation is also found in a wide variety of local microenvironments where other organisms provide abundant organic matter, such as the root zone of desert grasses (Herman et al., 1993), the hind-gut of termites (Breznak et al., 1973; Yamada et al., 2006; Hongoh et al., 2008), the interior of pineapples (Tapia-Hernández et al., 2000), the carpets of feather-moss in boreal forests (DeLuca et al., 2002), and the fungus gardens of leaf-cutter ants in tropical rainforests (Pinto-Tomás et al., 2009). Nitrogen fixation has also been reported in the foliage (endophytic) of some conifers (Moyes et al., 2016). Studies of nitrogen fixation in these habitats are often aided by the identification of the genes coding for nitrogenase (nifH) using molecular techniques (Widmer et al., 1999; Reed et al., 2010).

In both symbiotic and asymbiotic forms, nitrogen fixation is generally inhibited at high levels of available nitrogen (Dynarski and Houlton, 2018). In many cases, the rate of fixation appears to be controlled by the N:P ratio in the soil (Chapin et al., 1991; Smith et al., 1992b; Fig. 6.8), and added phosphorus stimulates symbiotic N fixation especially in warm temperate and tropical forests (Augusto et al., 2013; Ament et al., 2018). In bacteria, phosphorus

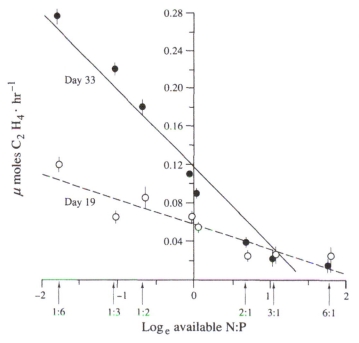

FIG. 6.8 Acetylene reduction as an index of nitrogen fixation in asymbiotic N-fixing bacteria as a function of the ratio of N to P in the soil. *From Eisele et al. (1989). Used with permission of Springer.*

appears to activate the gene for the synthesis of nitrogenase (Stock et al., 1990), illustrating how the linkage between the global cycles of nitrogen and phosphorus has a basis in molecular biology. Requirements for Mo and Fe as structural components of nitrogenase also link nitrogen fixation to the availability of these elements in natural ecosystems (Kim and Rees, 1994; O'Hara et al., 1988).[e] Low availability of Mo may limit asymbiotic N fixation in many forests (Silvester, 1989; Barron et al., 2009), and low P may limit N fixation in tropical forests (Dynarski and Houlton, 2018; Batterman et al., 2013a).

N-fixing species tend to have higher levels of phosphatase activity in their rhizosphere (Houlton et al., 2008), and some plants with symbiotic N-fixing bacteria appear to acidify their rooting zone to make Fe and P more available (Ae et al., 1990; Raven et al., 1990; Gillespie and Pope, 1990). Legumes appear to have several mechanisms to aid the uptake and retention of P in phosphorus-poor soils (He et al., 2011; Nasto et al., 2014). Nitrogen-fixing Red alder (*Alnus rubra*) appear to enhance the weathering and uptake of soil minerals (especially Ca and P) to maintain a favorable stoichiometric balance of nutrients (Perakis and Pett-Ridge, 2019). Available P may limit the extent of N-fixation in newly developing soils after glacial retreat (Darcy et al., 2018).

Rose and Youngberg (1981) provide an insightful experiment on the biogeochemical linkages of C, N, and P in *Ceanothus velutinus* growing in nitrogen-deficient soils with and without mycorrhizae and symbiotic nitrogen-fixing bacteria (Table 6.3). The highest rates of growth

TABLE 6.3   Effects of mycorrhizae and N-fixing nodules on growth and nitrogen fixation in *Ceanothus velutinus* seedlings.

	Control	+ Mycorrhizae	+ Nodules	+ Mycorrhizae and nodules
Mean shoot dry weight (mg)	72.8	84.4	392.9	1028.8
Mean root dry weight (mg)	166.4	183.4	285.0	904.4
Root/shoot	2.29	2.17	0.73	0.88
Nodules per plant	0	0	3	5
Mean nodule weight (mg)	0	0	10.5	44.6
Acetylene reduction (mg/nodule/hr)	0	0	27.85	40.46
Percent mycorrhizal colonization	0	45	0	80
Nutrient concentration (in shoot, %)				
N	0.32	0.30	1.24	1.31
P	0.08	0.07	0.25	0.25
Ca			1.07	1.15

From Rose and Youngberg (1981). Used with permission of NRC Research Press.

[e]In some cyanobacteria, vanadium can substitute for Mo in nitrogenase (Bellenger et al., 2014; Zhang et al., 2016), especially in boreal forest soils where Mo is often limiting (Darnajoux et al., 2019).

were seen when both of these symbiotic associations were present, which also allowed a decrease in the root/shoot ratio. Nitrogen fixation enhanced the uptake of phosphorus by mycorrhizal fungi; presumably mycorrhizal associations aided the supply of P for N-fixation. These results illustrate the strong interactions between these elements in the nutrition of higher plants.

The isotopic ratio of N in plant tissues is expressed as $\delta^{15}N$, using a calculation analogous to what we saw for the isotopes of carbon in Chapter 5 (Robinson, 2001). In the case of nitrogen, the standard is the atmosphere, which contains 99.63% ^{14}N and 0.37% ^{15}N. Nitrogenase shows only a slight discrimination between these isotopes of N, that is, between $^{15}N_2$ and $^{14}N_2$ (Handley and Raven, 1992; Hogberg, 1997), so differences in the isotopic ratio of nitrogen among plant species growing in the same soil can be used to suggest which species may be involved in nitrogen fixation (Virginia and Delwiche, 1982; Yoneyama et al., 1993). Nitrogen-fixing species typically show values of $\delta^{15}N$ that are slightly negative or close to the atmospheric ratio ($\delta^{15}N = 0$), whereas non-fixing species show a wide range of values (usually positive) depending on various N transformations in the soil (Garten and Van Miegroet, 1994; Hogberg, 1997; Fig. 6.9).

Shearer et al. (1983) used the difference in isotopic ratio between *Prosopis* grown in the laboratory without added N (i.e., all nitrogen was derived from fixation) and the same species in the field to estimate that the field plants derived 43–61% of their nitrogen from fixation. Of course, when nitrogen-fixing plants die, their nitrogen content is available for

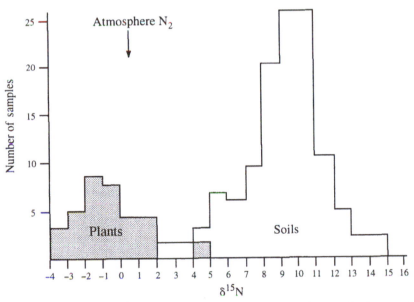

FIG. 6.9   Frequency distribution of $\delta^{15}N$ in the tissues of 34 nitrogen-fixing plants and in the organic matter of 124 soils from throughout the United States. *Plotted using data from Shearer and Kohl (1988, 1989).*

other species in the ecosystem (Huss-Danell, 1986; van Kessel et al., 1994). Lajtha and Schlesinger (1986) found that the desert shrub *Larrea tridentata*, growing adjacent to nitrogen-fixing *Prosopis*, had lower δ^{15}N than when *Larrea* was growing alone. During the development of vegetation following glacial retreat in Alaska, available nitrogen was first derived from N-fixation, and later as the nitrogen stock accumulated, from internal cycling (Malone et al., 2018).

Nitrogenase activity can be measured using the acetylene-reduction technique, which is based on the observation that this enzyme also converts acetylene to ethylene under experimental conditions. Plants or nodules are placed in small chambers or small chambers are placed over field plots, and the conversion of injected acetylene to ethylene over a known time period is measured using gas chromatography. The conversion of acetylene (in moles) is not exactly equivalent to the potential rate of fixation of N_2 because the enzyme has different affinities for these substrates. However, appropriate conversion ratios can be determined using other techniques (Schwintzer and Tjepkema, 1994; Liengen, 1999). For instance, many investigators have added $^{15}N_2$, the heavy stable isotope of N, to closed chambers and used the increase in organic compounds containing ^{15}N in test plants or soil as a measure of nitrogen fixation (Chalk et al., 2017).

Heterotrophic N-fixing bacteria and cyanobacteria (blue-green algae) are widespread, and their nitrogen fixation can be an important source of N for some terrestrial ecosystems (Reed et al., 2011). Exceptionally high rates of fixation have been recorded in cyanobacterial crusts that cover the soil surface in some desert ecosystems (Rychert et al., 1978), but in most cases, the total input from these sources of asymbiotic N fixation is in the range of 1 to 5 kg N/ha/yr (Boring et al., 1988; Cushon and Feller, 1989; Son, 2001; Cleveland et al., 2010). This input is usually less than the annual deposition of nitrogen in wetfall and dryfall from the atmosphere (Schwintzer and Tjepkema, 1994; Staccone et al., 2020).

The importance of symbiotic nitrogen fixation in terrestrial ecosystems varies widely depending on the presence of species that harbor symbiotic bacteria (Reed et al., 2011). Grazed pastures with clover routinely show N fixation at rates of 100–200 kg N/ha/yr (Bolan et al., 2004). In natural ecosystems, some of the greatest rates of N fixation are seen in species that invade after disturbance, where high light levels allow maximum photosynthesis (Vitousek and Howarth, 1991; Batterman et al., 2013b). For example, in the recovery of Douglas fir forests after fire, Youngberg and Wollum (1976) found that the nodulated shrub *Ceanothus velutinus* contributed up to 100 kg N/ha/yr on some sites. Invasion of the exotic, nitrogen-fixing tree *Myrica fay*, provides important inputs of nitrogen (18 kg/ha/yr) on fresh volcanic ashflows in Hawaii (Vitousek et al., 1987). In most cases the importance of plants with symbiotic nitrogen fixation declines with the recovery of mature vegetation, and their occurrence in undisturbed communities is limited (e.g., Taylor et al., 2019). In this regard, the widespread occurrence of leguminous species in tropical forests, versus their paucity at higher latitudes, is deserving of further study (Menge et al., 2017). The sporadic occurrence of symbiotic nitrogen fixation in terrestrial ecosystems makes it difficult to extrapolate from local studies to provide a global estimate of its importance. The global rate of N fixation (asymbiotic + symbiotic) in natural ecosystems may supply $60–100 \times 10^{12}$ g N/yr, or about 10% of the annual plant demand for nitrogen on land (Davies-Barnard and Friedlingstein, 2020; Chapter 12).

Nutrient balance

In addition to an adequate supply of nutrient elements, plant growth is affected by the balance of nutrients in the soil, relative to the essential biochemical proportions—the stoichiometry—in plant tissues. For seedlings of several tree species, Ingestad (1979b) found that a solution containing 100 parts N, 15 parts P, 50 parts K, 5 parts Ca and Mg, and 10 parts S was ideal for maximum growth. In a compilation of data from nearly 10,000 species, leaf N and P contents were related by a 2/3-power-law function, with a mean N/P ratio of 10.9 (by mass; (Reich et al., 2010; cf. Kerkhoff et al., 2005).[f] Despite wide variations in nutrient availability in the environment, most land plants show an N/P ratio of about 14–15 (by mass) in leaf tissues (Gusewell, 2004; McGroddy et al., 2004; Han et al., 2005; Koerselman and Meuleman, 1996), with N deficiency at lower and P deficiency at higher values. This constraint in cellular biochemistry is seen amongst the geographic distribution of plant stoichiometry in the forests of China (Zhang et al., 2018c).

Nutrient allocations and cycling in land vegetation

The annual intrasystem cycle

Plant uptake of nutrients from the soil is allocated to the growth of new tissues. Although short-lived tissues (leaves and fine roots) compose a small fraction of total plant biomass, they receive the largest proportion of the annual nutrient uptake (Pregitzer et al., 2010). Growth of leaves and roots received 87% of the N and 79% of the P allocated to new tissues in a deciduous forest in England (Cole and Rapp, 1981, p. 404). In a perennial grassland dominated by *Bouteloua gracilis* in Colorado, new growth of aboveground tissues accounted for 45% of the annual uptake of N (Clark, 1977).

When leaf buds break and new foliage begins to grow, these new leaf tissues often have high concentrations of N, P, and K. As the foliage matures, these concentrations often decrease (Van den Driessche, 1974). Some of these changes are due to the increasing accumulation of photosynthetic products with time and to leaf thickening during development. Leaf mass per unit area (mg/cm^2) may increase as much as 50% during the growing season and then decline as the leaf senesces (Smith et al., 1981). The initial concentrations of N and P are diluted as the leaf tissues accumulate carbohydrates and cellulose. In contrast, the concentrations of other nutrients (e.g., Ca, Mg, and Fe) often increase with leaf age (Van den Driessche, 1974). Increases in calcium concentration with leaf age result from secondary thickening, including calcium pectate deposition in cell walls, and from increasing storage of calcium oxalate in cell vacuoles.

Although there are variations among species, nutrient concentrations in mature foliage are related to the rate of photosynthesis (Chapter 5) and plant growth (e.g., Tilton, 1978), and analysis of foliage is often a good index of site fertility (Van den Driessche, 1974; Ordonez et al., 2009). Worldwide, higher forest net primary production is associated with higher leaf

[f]See: Tian et al. (2019). A global database of paired leaf nitrogen and phosphorus concentrations of terrestrial plants. Ecology doi: 10.1002/ecy.2812.

N concentrations in temperate forests and higher leaf P concentrations in tropical forests (Šímová et al., 2019). Among tropical forests, concentrations of major nutrients in leaves are significantly higher on more fertile soils (Vitousek and Sanford, 1986). Leaf concentrations of trace metals often reflect the content of the underlying soil, such that leaf tissues are useful for mineral prospecting in some areas (Cannon, 1960; Brooks, 1973). Some species "hyperaccumulate" metals, especially on mine spoils (Reeves et al., 2018).[g]

Yin (1993) found that concentration of N and P in the foliage of deciduous trees varied systematically with higher values among species in colder habitats than in the tropics. The higher leaf nutrient contents in colder climates may allow for higher photosynthetic rates and rapid growth of these species in response to a short growing season (Mooney and Billings, 1961; Körner, 1989; Reich and Oleksyn, 2004).

Unless the supply of a nutrient reaches very low levels, plants usually do not show deficiency symptoms; they simply grow more slowly (Clarkson and Hanson, 1980). Inherent slow growth is a characteristic of plants adapted to infertile habitats, and it often persists in these species even when nutrients are added experimentally (Chapin et al., 1986a). After fertilization with a specific nutrient, the concentrations of other leaf nutrients can show surprising changes. When nitrogen fertilization of N-deficient stands stimulates photosynthesis, the concentrations of other nutrients in foliage may be diluted by new accumulations of carbohydrate (Fowells and Krauss, 1959; Timmer and Stone, 1978; Jarrell and Beverly, 1981). In some cases, uptake of P from the soil may fall behind the rates needed for maximum growth at the newly established levels of N availability. For example, leaf N increased when Miller et al. (1976) fertilized Corsican pine (*Pinus nigra*) with N, but in the same samples, concentrations of P, Ca, and Mg declined. In areas of high deposition of N from the atmosphere, forests may show Mg, Mn or P deficiency (Schulze, 1989; Gonzales and Yanai, 2019). In other cases, improvements in plant nitrogen status enhance the uptake of other elements as well (e.g., Table 6.3). Plant responses to multiple element fertilizations are illustrative of the importance of considering balanced nutrition for maximum plant growth.

Once leaves are fully expanded, changes in the nutrient content per unit of leaf area indicate movements of nutrients between the foliage and the stem. Woodwell (1974) found that oak leaves rapidly accumulated N during the early summer, presumably as a component of photosynthetic enzymes. The leaf content of N, P, and K remained relatively constant at high levels throughout the growing season, but concentrations of all three elements declined rapidly in autumn prior to leaf abscission. Such losses often represent active withdrawal of nutrients from foliage for reuse during the next year. Some trace micronutrients are also withdrawn before leaf-fall (Killingbeck, 1985), but usually reabsorption of foliar Ca and Mg is limited. Fife and Nambiar (1984) observed that reabsorption of N, P, and K was not just related to leaf senescence in *Pinus radiata*; these nutrients could also move from tissues produced early and later during the same growing season. In apple trees, zinc can be remobilized from older tissues to the growing shoot in zinc-deficient conditions (Xie et al., 2019).

Leaf nutrient contents are also affected by rainfall that leaches nutrients from the leaf surface (Tukey, 1970; Parker, 1983). In particular, seasonal changes in the content of K, which is

---

[g] Brazil nuts (*Bertholletia excels*) are known to accumulate selenium, which may provide a protective agency in diet against toxic accumulations of mercury.

highly soluble and especially concentrated in cells near the leaf surface, may represent leaching. The losses of nutrients in leaching often follow the order:

$$K \gg P > N > Ca. \tag{6.1}$$

Leaching rates generally increase as foliage senesces before abscission; thus, care must be taken to recognize changes due to leaching versus changes due to active nutrient resorption by the plant.

Nutrient losses by leaching differ among leaf types. Luxmoore et al. (1981) calculated lower rates of leaching loss from pines than from broad-leaf deciduous species in a forest in Tennessee. Such differences may be the result of variation in leaf nutrient concentration, surface-area-to-volume ratio, surface texture, and leaf age. Among the trees of the humid tropics, the smooth surface of broad sclerophylls may be an adaptive response to reducing leaching by minimizing the length of time that rainwater is in contact with the leaf surface (Dean and Smith, 1978). Species-specific differences in rates of leaching from potential host trees may explain differences in their epiphyte loads (Benzing and Renfrow, 1974; Awasthi et al., 1995), with many epiphytes showing P deficiency (Wanek and Zotz, 2011).

Rainwater that passes through a vegetation canopy is called *throughfall*, which is usually collected in funnels or troughs placed on the ground. Throughfall contains nutrients leached from leaf surfaces and is most important in the cycling of nutrients such as K (Parker, 1983; Schaefer and Reiners, 1989). In forests, rainwater that travels down the surface of stems is called *stemflow* (Levia and Germer, 2015). The concentrations of nutrients in stemflow waters are high, but usually much more water reaches the ground as throughfall. Stemflow is significant to the extent that it returns highly concentrated nutrient solutions to the soil at the base of plants (Gersper and Holowaychuk, 1971).

Leaching varies seasonally depending on forest type and climate. Not surprisingly, in temperate deciduous forests, the greatest losses are during the summer months (Lindberg et al., 1986). Some of the nutrient content in throughfall is derived from aerosols that are deposited on leaf surfaces (Chapter 3). Indeed, Lindberg and Garten et al. (1988) found that about 85% of the SO_4^{2-} collected in throughfall was due to dry deposition on leaf surfaces, and some studies have used the SO_4^{2-} content of throughfall to estimate dry deposition in the canopy (Garten et al., 1988; Ivens et al., 1990). Similarly, atmospheric contaminants like Hg^+ can be deposited on and then rinsed off leaf surfaces (Wright et al., 2016).

For most elements, however, leaching of nutrients from vegetation makes it difficult to use nutrient concentrations in the rainfall collected under a canopy to calculate dry deposition on leaf surfaces (Chapter 3). In some cases, leaves appear to take up nutrients from rainfall, particularly soluble forms of N (Carlisle et al., 1966; Garten and Hanson, 1990; Lovett and Lindberg, 1993). Various reactive nitrogen gases (e.g., NH_3, NO_x, and peroxyacetyl nitrate) are also absorbed at the leaf surface (Gessler et al., 2000; Sparks, 2009). In grassland species, foliar uptake of S gases is more efficient when soil S contents are low (Cliquet and Lemauviel-Lavenant, 2019).

Litterfall

When the biomass of vegetation is not changing, the annual production of new tissues is balanced by the senescence and loss of plant parts (Chapter 5). In the intrasystem cycle,

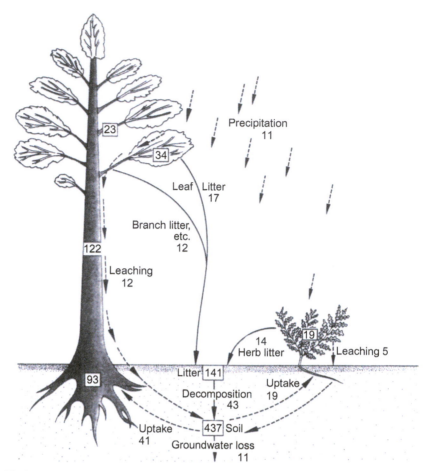

FIG. 6.10 The intrasystem cycle of Ca in a forest ecosystem in Great Britain. Pools are shown in kg/ha and annual flux in kg/ha/yr. *From Whittaker (1975, p. 110). Reprinted by permission of Prentice Hall, Upper Saddle River, New Jersey.*

plant litterfall is the dominant pathway for nutrient return to the soil, especially for N and P (Fig. 6.10). Below ground, root death also makes a major (but much harder to measure) contribution of nutrients to the soil each year (Cox et al., 1978; Vogt et al., 1983).

The nutrient concentrations in litterfall differ from the nutrient concentrations in mature foliage by the reabsorption of constituents during leaf senescence (Killingbeck, 1996). Nutrient reabsorption is also known to occur during the senescence of fine roots (Freschet et al., 2010) and during the aging of sapwood to heartwood in trees (Laclau et al., 2001). Nutrient reabsorption potentially confers a second type of nutrient-use efficiency on vegetation (see Chapter 5 for a discussion of nutrient-use efficiency in photosynthesis). Nutrients that are reabsorbed can be used in net primary production in future years, increasing the carbon fixed per unit of nutrient uptake (Salifu and Timmer, 2003).

Compiling data from a wide range of species, Aerts (1996) found a mean reabsorption of 50% N and 52% P during leaf senescence. A recent report by Vergutz et al. (2012) suggests that N and P reabsorption from leaves frequently may exceed 60%.[h] Somewhat lower values are seen in a California shrubland (Table 6.4), in the Hubbard Brook forest (Table 6.1), and in grassland ecosystems (Woodmansee et al., 1978). Lajtha (1987) found exceptionally high values for P reabsorption (72–86%) in the desert shrub *Larrea tridentata* growing in calcareous soils in which P availability is limited due to the precipitation of calcium phosphate minerals (see Fig. 4.10). DeLucia and Schlesinger (1995) report 94% reabsorption of leaf P in *Cyrilla racemiflora* in a P-limited bog in the southeastern United States.

Plants grown with low nutrient availability or occurring on infertile sites tend to have low nutrient concentrations in mature leaves and litter; they generally reabsorb a smaller *amount* but a larger *proportion* of the nutrient pool in senescent leaves, compared to individuals of the same species under conditions of greater nutrient availability (Chapin, 1988; Killingbeck, 1996; Kobe et al., 2005; Gerdol et al., 2019). In contrast, fertilization tends to reduce nutrient reabsorption by plants (Yuan and Chen, 2015; Brant and Chen, 2015). In most cases, however, species appear to have only a limited ability to adjust the efficiency of reabsorption of leaf nutrients as a function of site fertility (Birk and Vitousek, 1986; Chapin and Moilanen, 1991). For birch (*Betula pendula*), N-reabsorption efficiency has a genetic basis, such that higher leaf reabsorption and slower soil nutrient turnover are found under certain trees (Mikola et al., 2018).

Among trees throughout France, nutrient reabsorption efficiency was inversely related to site fertility (Achat et al., 2018). Differences in nutrient-use efficiency in reabsorption between nutrient-rich and nutrient-poor sites are not as likely to be due to a direct response of plants as to the tendency for species with higher inherent capabilities for nutrient reabsorption to dominate nutrient-poor sites (Pastor et al., 1984; Chapin et al., 1986b; Schlesinger et al., 1989). In an Australian soil chronosequence, reabsorption of P is greatest among species on older, highly weathered soils (Hayes et al., 2014). Among tropical forests, reabsorption of N and P varies as an inverse function of site fertility (Silver, 1994; Kitayama et al., 2000; Tsujii et al., 2017; Fig. 6.11). As a result of mycorrhizal associations and internal conservation of P, it appears that tropical trees are well adapted to P-deficient soils, which are widespread in these regions (Cuevas and Medina, 1986; Cleveland et al., 2011). This results in a worldwide pattern of greater P reabsorption among tropical species and greater N reabsorption among boreal species (Reed et al., 2012; Gill and Finzi, 2016).

Mass balance of the intrasystem cycle

The annual circulation of nutrients in land vegetation, the intrasystem cycle (Fig. 6.10), can be studied using the mass-balance approach. Nutrient requirement is equal to the peak nutrient content in newly produced tissues during the growing season (refer to Table 6.1). Nutrient uptake cannot be measured directly, but uptake must equal the annual storage in

---

[h] Vergutz et al. (2012) indicate an average 70% reabsorption of K from senescing leaves, but the value is likely biased upward by substantial leaching of K during the senescence process. For instance, Likens et al. (1994) report only 10–32% reabsorption of K from deciduous leaves at Hubbard Brook and Gray (1983) finds none in Table 6.4 (see also Ostman and Weaver, 1982).

TABLE 6.4 Nutrient cycling in a 22-year-old stand of the Chaparral Shrub (*Ceanothus megacarpus*) near Santa Barbara, California.

	Biomass	N	P	K	Ca	Mg
Atmospheric input (g/m²/yr)						
Deposition		0.15		0.06	0.19	0.10
N-fixation		0.11				
Total input		0.26		0.06	0.19	0.10
Compartment pools (g/m²)						
Foliage	553	8.20	0.38	2.07	4.50	0.98
Live wood	5929	32.60	2.43	13.93	28.99	3.20
Reproductive tissues	81	0.92	0.08	0.47	0.32	0.06
Total live	6563	41.72	2.89	16.47	33.81	4.24
Dead wood	1142	6.28	0.46	2.68	5.58	0.61
Surface litter	2027	20.5	0.6	4.7	26.1	6.7
Annual flux (g/m²/yr)						
Requirement for production						
Foliage	553	9.35	0.48	2.81	4.89	1.04
New twigs	120	1.18	0.06	0.62	0.71	0.11
Wood increment	302	1.66	0.12	0.71	1.47	0.16
Reproductive tissues	81	0.92	0.08	0.47	0.32	0.07
Total in production	1056	13.11	0.74	4.61	7.39	1.38
Reabsorption before abscission		4.15	0.29	0	0	0
Return to soil						
Litterfall	727	6.65	0.32	2.10	8.01	1.41
Branch mortality	74	0.22	0.01	0.15	0.44	0.02
Throughfall		0.19	0	0.94	0.31	0.09
Stemflow		0.24	0	0.87	0.78	0.25
Total return	801	7.30	0.33	4.06	9.54	1.77
Uptake (= increment+return)		8.96	0.45	4.77	11.01	1.93
Stream-water loss (g/m²/yr)		0.03	0.01	0.06	0.09	0.06
Comparisons of turnover and flux						
Foliage requirement/total requirement (%)		71.3	64.9	61.0	66.2	75.4
Litter fall/total return (%)		91.1	97.0	51.7	84.0	79.7
Uptake/total live pool (%)		21.4	15.6	29.0	32.6	45.5
Return/uptake (%)		81.4	73.3	85.1	86.6	91.7
Reabsorption/requirement (%)		31.7	39.0	0	0	0
Surface litter/litter fall (yr)	2.8	3.1	1.9	1.2	3.3	4.8

Modified from Gray (1983) and Schlesinger et al. (1982).

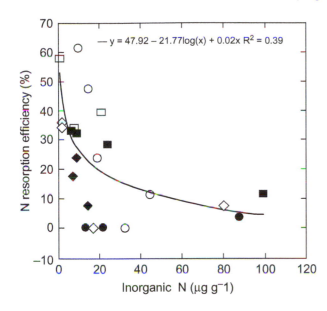

FIG. 6.11   Nitrogen retranslocation efficiency in leaves of six canopy species as a function of extractable nitrogen in the soils of a humid tropical rainforest of Costa Rica (Reed et al., 2012).

perennial tissues, such as wood, plus the replacement of losses in litterfall and leaching (e.g., Soper et al., 2018a); thus, the following equation:

$$Uptake = Retained + Returned. \tag{6.2}$$

Uptake is less than the annual requirement by the amount reabsorbed from leaves and fine roots before abscission; namely:

$$Requirement = Uptake + Retranslocation. \tag{6.3}$$

The requirement is the nutrient flux needed to complete a mass balance; it should not be taken as indicative of biological requirements. In fact, this equation can be solved for nonessential elements such as Na and Si. For a forest in Tennessee, the mass-balance approach was used to show that accumulations of Ca and Mg in vegetation were directly related to decreases in the content of exchangeable Ca and Mg in the soil during 11 years of growth (Johnson et al., 1988). Similarly, during 100 years of forest growth on abandoned agricultural land, Hooker and Compton (2003) showed that the nitrogen accumulation by vegetation was largely supplied from the soil—not new inputs. Mass-balance studies also show that some silicon (Si), which is often used as an index of rock weathering, is retained and recycled by terrestrial vegetation (Markewitz and Richter, 1998; Conley, 2002; Derry et al., 2005; Cornelis et al., 2010; Clymans et al., 2016; Turpault et al., 2018).

The mass-balance approach was used to analyze the internal storage and the annual transfers of nutrients in the aboveground portion of a California shrubland (Table 6.4). These data serve to summarize many aspects of the intrasystem cycle. Note that 71% of the annual requirement of N is allocated to foliage, whereas much less is allocated to stem wood. Nevertheless, total nutrient storage in short-lived tissues is small compared to storage in wood, which has lower nutrient concentrations than leaf tissue but has accumulated during 22 years of growth. For most nutrients in this ecosystem, the storage in wood increases by about 5% each year.

In this community the nutrient flux in stemflow is unusually large but the total annual return in leaching is relatively small, except for K. Despite substantial reabsorption of N and P before leaf abscission, litterfall is the dominant pathway of return of these elements from aboveground vegetation to the soil. Ca is actively exported to the leaves before abscission (i.e., requirement < uptake). In this shrubland, annual uptake is 16–46% of the total storage in vegetation, but 73–92% of the uptake is returned each year. As in most studies, some of these calculations would be revised if belowground transfers were better understood.

There are changes in the nutrient cycles in vegetation, which accompany changes in the allocation of net primary production with stand age. During forest regrowth after disturbance, leaf area develops rapidly and the nutrient movements dependent on leaf area (i.e., litterfall and leaching) are quickly reestablished (Marks and Bormann, 1972; Davidson et al., 2007). Gholz et al. (1985) found that the proportion of the annual requirement met by internal cycling (i.e., nutrient reabsorption from leaves) increased with time during the development of pine forests in Florida. Nutrients are accumulated in plant biomass most rapidly during the early development of forests and more slowly as the aboveground biomass reaches a steady state (Pearson et al., 1987; Reiners, 1992). Percentage turnover in vegetation declines as the mass and nutrient storage in vegetation increase. In mature forests, leaf biomass is <5% of the total, and leaves contain only 5–20% of the total nutrient pool in vegetation (Waring and Schlesinger, 1985).

Vitousek et al. (1988) have compiled data showing the proportions of biomass (i.e., carbon) and major nutrient elements in various types of mature forest (Table 6.5). The nutrient ratios vary over a surprisingly small range, so the global pattern of element stocks in vegetation is similar to that for biomass; that is, tropical > temperate > boreal forests (cf. Zhang et al., 2018c; Table 5.3). It is important to remember that these ratios are calculated for the total plant biomass; the concentrations of nutrients in leaf tissues are higher, and C/N and C/P ratios in leaves are correspondingly smaller. Thus, nutrient ratios for whole-plant biomass increase

TABLE 6.5   Biomass and element accumulations in biomass of mature forests.

Forest biome	Number of stands	Total biomass (t/ha)	Percent of total biomass				Mass ratio		
			Leaf	Branch	Bole	Roots	C/N	C/P	N/P
Northern/subalpine conifer	12	233	4.5	10.2	62.8	22.6	143	1246	8.71
Temperate broadleaf deciduous	13	286	1.1	16.2	63.1	19.5	165	1384	8.40
Giant temperate conifer	5	624	2.5	10.2	66.4	20.8	158	1345	8.53
Temperate broadleaf evergreen	15	315	2.7	14.7	66.2	16.5	159	1383	8.73
Tropical/subtropical closed forest	13	494	1.9	21.8	59.8	16.4	161	1394	8.65
Tropical/subtropical woodland and savanna	13	107	3.6	19.1	60.4	16.9	147	1290	8.80

From Vitousek et al. (1988). Used with permission of Springer.

with time as the vegetation becomes increasingly dominated by structural tissues with lower nutrient content (Reiners, 1992).

Nutrient use efficiency

A mass balance for the intrasystem cycle of vegetation allows us to calculate an integrated measure of nutrient-use efficiency by vegetation—net primary production per unit nutrient uptake (Pastor and Bridgham, 1999). This measure is affected by various factors that we have examined individually, including the rate of photosynthesis per unit leaf nutrient (Chapter 5), uptake per unit of root growth (refer to Fig. 6.3), and leaching and nutrient retranslocations from leaves. As a result of changes in these factors, net primary production per unit of nitrogen or phosphorus taken from the soil increased by factors of 5 and 10, respectively, during the growth of pine forests in central Florida (Gholz et al., 1985).

Among temperate forests, the annual circulation of nutrients in coniferous forests is much lower than the circulation in deciduous forests, largely as a result of lower leaf nutrient concentrations and lower leaf turnover in coniferous forest species (Cole and Rapp, 1981; Aerts, 1996; Neumann et al., 2018). The foliage of some coniferous species persists for 8–10 years. Also, leaching losses are lower in coniferous forests (Parker, 1983), and photosynthesis per unit of leaf nitrogen tends to be greater in coniferous species (Reich et al., 1995). Ectomycorrhizal species, such as pine, tend to have higher nutrient reabsorption from senescent foliage (Zhang et al., 2018a). Together these mechanisms result in greater nutrient-use efficiency in coniferous forests compared to deciduous forests of the world (Table 6.6). Higher nutrient-use efficiency in coniferous species may explain their frequent occurrence on nutrient-poor sites and in boreal climates where soil nutrient turnover is slow. Significantly, larch (*Larix* sp.), one of the few deciduous species in the boreal forest, has exceptionally high fractional reabsorption of foliar nutrients (Carlyle and Malcolm, 1986).

The high nutrient-use efficiency of most conifers may also extend to the occurrence of broad-leaf evergreen vegetation on nutrient-poor soils in other climates (Monk, 1966; Beadle, 1966; Goldberg, 1982; DeLucia and Schlesinger, 1995). Escudero et al. (1992) suggest that leaf longevity was the most important factor increasing nutrient-use efficiency among various trees and shrubs in central Spain (compare Reich et al., 1992), since deciduous and evergreen species have roughly similar amounts of nutrient reabsorption during leaf senescence (del Arco et al., 1991; Aerts, 1996; Eckstein et al., 1999). Changes in the seasonal timing of leaf senescence and litterfall as a result of changes in climate may alter nutrient-use efficiency by vegetation (Estiarte and Peñuelas, 2015).

TABLE 6.6   Net primary production (kg/ha/yr) per unit of nutrient uptake, as an index of nutrient-use efficiency in deciduous and coniferous forests.

Forest type	Production per unit nutrient uptake				
	N	P	K	Ca	Mg
Deciduous	143	1859	216	130	915
Coniferous	194	1519	354	217	1559

From Cole and Rapp (1981).

For biogeochemical cycling in vegetation, we have seen that the leaves and fine roots contain only a small portion of the nutrient content in biomass, but the growth, death, and replacement of these tissues largely determine the annual intrasystem cycle of nutrients. Net primary production is positively correlated to soil N availability in both coniferous and deciduous forests (Zak et al., 1989; Reich et al., 1997), but differences in nutrient-use efficiency tend to weaken the correlation, so that light and moisture are the primary determinants of net primary production on a global basis (Fig. 5.13). When nutrient concentrations in litter are low, as might be expected after reabsorption of nutrients, decomposition is slower (Scott and Binkley, 1997; Lovett et al., 2004; Hobbie, 2015). Thus, intrasystem cycling contains a positive feedback to the extent that an increase in nutrient-use efficiency by vegetation may reduce the future availability of soil nutrients for plant uptake (Shaver and Melillo, 1984; Mikola et al., 2018).

Because of the uptake of nutrients from the soil and the intrasystem cycling of nutrients, terrestrial vegetation leaves a marked imprint on the nutrient distribution in soils (Zinke, 1962; Waring et al., 2015). Some nutrients are actively accumulated from deep in the soil profile and deposited at the surface (Lawrence and Schlesinger, 2001; Jobbágy and Jackson, 2004). This effect is most pronounced for nutrients that are strongly recycled within the plant community, whereas others are more evenly distributed through the soil profile. As a result, a global compilation of 10,000 soil profiles shows the following rank-order of nutrient concentrations with depth (shallow to deep) (Jobbágy and Jackson, 2001):

$$P > K > Ca > Mg > Na = Cl = SO_4. \tag{6.4}$$

Nutrient uplift by plants is mediated by precipitation—if the site is very wet, leaching tends to dominate over plant uplift (Porder and Chadwick, 2009). In deserts and other areas where there is patchy vegetation, plant nutrients are strongly concentrated under shrubs, whereas nonlimiting or nonessential nutrients accumulate in the barren spaces between shrubs (Schlesinger et al., 1996; Gallardo and Parama, 2007). Even in closed-canopy forests individual species leave an imprint on soil chemistry beneath their canopy (Boettcher and Kalisz, 1990; Rodriguez et al., 2011; Keller et al., 2013). And, as discussed in Chapter 4, uptake of Si by forest trees enriches the soil surface with Si that might otherwise be depleted by rock weathering and leaching.

Biogeochemical cycling in the soil

Despite new inputs from the atmosphere and from rock weathering, and adaptations in plants to minimize their loss of nutrients, most of the annual nutrient requirement of land plants is supplied by the uptake of nutrients released from the decomposition of dead materials in the soil (refer to Table 6.1). Decomposition of dead organic matter completes the intrasystem cycle. Decomposition is a general term that refers to the breakdown of organic matter. *Mineralization* is a more specific term[i] that refers to processes that release carbon as CO_2 and nutrients in inorganic form (e.g., N as NH_4^+ and P as PO_4^{3-}).

[i]This use of the term *mineralization* differs from its common usage in the literature of geology, in which mineralization refers to various processes (e.g., precipitation from hydrothermal fluids) that result in the deposition of metals in an ore deposit of economic significance.

Soil microbial biomass and the decomposition process

A variety of soil animals, including earthworms, fragment and mix fresh litterfall (Swift et al., 1979; Wolfe, 2001); however, the main biogeochemical transformations are performed by fungi and bacteria in the soil. Most of the mineralization reactions are the result of the activity of extracellular degradative enzymes, released by soil microbes and mycorrhizae (Burns, 1982; Linkins et al., 1990; Sinsabaugh et al., 2002, 2008). The release of a wide variety of extracellular enzymes, including cellulases and proteases, increases in response to freshly deposited organic matter available for decomposition (Brzostek et al., 2013). Soil microbial biomass and activity vary greatly in space and time, and we are just beginning to sample it at appropriate scales (Pedersen et al., 2015).

Only a small fraction of the measured soil microbial biomass is active at any given time (Blagodatskaya and Kuzyakov, 2013); however, total microbial biomass is often measured as an index of its activity (Booth et al., 2005; Colman and Schimel, 2013). Determination of microbial biomass is often performed by one of several techniques involving fumigation with chloroform. For instance, in a subdivided soil sample, total soluble nitrogen (NH_4^+, NO_3^-, and dissolved organic N) is measured before and after fumigation with chloroform. A higher content in the fumigated sample is assumed to result from the lysis of microbes that were killed by chloroform (Brookes et al., 1985; Joergensen, 1996). Microbial biomass is then calculated by assuming a standard nitrogen content in microbial tissue and a correction factor, Kn, to account for microbial N that is not immediately released by fumigation (Voroney and Paul, 1984; Joergensen and Mueller, 1996). The technique is justified by the observation of relatively constant C/N and C/P ratios in soil microbial biomass from many different environments (Brookes et al., 1984). Soil microbial biomass can also be estimated by extractions of the phospholipid fatty acid (PLFA) content (Zelles, 1999; Leckie et al., 2004; Bailey et al., 2002) or DNA from soils (Fig. 6.12), both of which are proxies for the microbial biomass that is present.

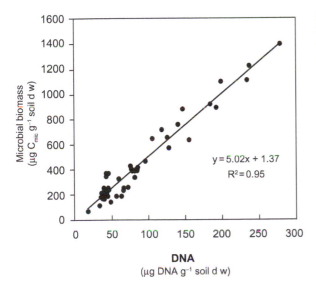

FIG. 6.12   Relationship between the amount of DNA extracted from soil and the microbial biomass contained therein. *From Anderson and Martens (2013).*

Soil microbial activity can be estimated by substrate-induced respiration or SIR, measurements of respiration in response to additions of labile carbon compounds to soil samples (Anderson and Domsch, 1978). The incorporation of isotopically labeled glucose into microbial DNA is also used as a measure of microbial growth (Rousk and Bååth, 2011). Ruess and Seagle (1994) found a direct correlation between soil microbial biomass and soil respiration in the grasslands of the African Serengeti, and soil microbial biomass is directly correlated with nitrogen mineralization in a worldwide sampling of soils (Li et al., 2019e).

Microbial biomass (living bacteria + fungi) typically comprises <3% of the mass of organic carbon found in soils (Wardle, 1992; Zak et al., 1994). High levels of microbial biomass are found in most forest soils and lower levels in deserts (Insam, 1990; Gallardo and Schlesinger, 1992). Fungi dominate over bacteria in most well-drained, upland soils (Anderson and Domsch, 1980). Soil moisture is a major determinant of microbial biomass, whereas temperature controls its activity (Serna-Chavez et al., 2013). More than half of the accumulation of soil organic matter in forests may derive from dead microbial biomass (Liang et al., 2019).

Soil microbes have high nutrient concentrations relative to the organic matter they decompose (Diaz-Ravina et al., 1993; Cleveland and Liptzin, 2007). The C:N:P ratio for soils and soil microbial biomass are 287:17:1 and 42:6:1, respectively (Xu et al., 2013). Globally the stoichiometry of soil microbial biomass is a major determinant of decomposition activity (Zechmeister-Boltenstern et al., 2015; Buchkowski et al., 2015). Calculations of carbon-use efficiency (CUE), defined as the ratio of growth to assimilation by soil microbes, are useful for comparative studies of soil nutrient turnover (Sinsabaugh et al., 2016). The N:P ratio in microbial biomass is generally higher in tropical soils versus those at higher latitudes (Li et al., 2014). Microbial biomass contained 2.5–5.6% of the organic carbon but up to 19.2% of the organic phosphorus in tropical soils of central India (Srivastava and Singh, 1988). During the decomposition of plant material, respiration of soil microbes converts organic carbon to CO_2, while the N and P contents are initially retained in microbial biomass. The accumulation of N, P, and other nutrients in soil microbes is known as *immobilization*.

As a result of immobilization, when the decomposition of fresh litter is observed in litterbags (Chapter 5), the C/N and C/P ratios decline as decomposition proceeds and the remaining materials are progressively dominated by microbial biomass that has colonized and grown on the substrate (Table 6.7; Sinsabaugh et al., 1993; Manzoni et al., 2010; Fahey et al., 2011). Immobilization is most significant for N and P, which are limiting to microbial growth and usually less obvious for Mg and K, which are available in greater quantities (Jorgensen et al., 1980; Staaf and Berg, 1982). Immobilization is least obvious and the release of nitrogen is most rapid in substrates of greatest initial nitrogen content (Parton et al., 2007; Manzoni et al., 2008).

In the process of immobilization, soil microbes not only retain the nutrients released from their substrate but also accumulate nutrients from the soil solution—leading to *net* immobilization (Fig. 6.13; Drury et al., 1991). In cases of low nutrient availability, the total nutrient content of decaying litter appears to increase during the initial phases of decomposition, owing to net nutrient immobilization (Aber and Melillo, 1980; Berg, 1988). In a beech forest near Vienna, Austria, microbial uptake exceeded mineralization during the winter (Kaiser et al., 2011). Microbial uptake of NH_4^+ is rapid, sequestering available NH_4^+ that might otherwise be available for plant uptake or nitrifying bacteria (Jackson et al., 1989; Schimel and Firestone, 1989). Thus, microbes respond to poor substrates by immobilization of nutrients from the soil solution

TABLE 6.7   Ratios of nutrient elements to carbon in the litter of scots pine (*Pinus sylvestris*) at sequential stages of decomposition.

	C/N	C/P	C/K	C/S	C/Ca	C/Mg	C/Mn
Needle litter							
Initial	134	2630	705	1210	79	1350	330
After incubation of:							
1 year	85	1330	735	864	101	1870	576
2 year	66	912	867	ND	107	2360	800
3 year	53	948	1970	ND	132	1710	1110
4 year	46	869	1360	496	104	704	988
5 year	41	656	591	497	231	1600	1120
Fungal biomass							
Scots pine forest	12	64	41	ND	ND	ND	ND

Note: C/N and C/P ratios decline with time, which indicates retention of these nutrients as C is lost, whereas C/Ca and C/K ratios increase, which indicates that these nutrients are lost more rapidly than carbon.
From Staaf and Berg (1982). Used with permission of NRC Research Press.

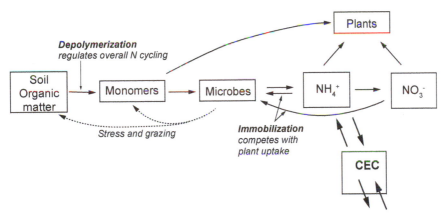

FIG. 6.13   A conceptual model for the soil nitrogen cycle. *Modified from Schimel and Bennett (2004); see also Drury et al. (1991).*

and the release of specific extracellular enzymes to enhance mineralization from their substrates (Mooshammer et al., 2014).

Mineralization slows when fresh residues with high C/N ratio are added to the soil (Turner and Olson, 1976; Gallardo and Merino, 1998; Buchkowski et al., 2015). Many field experiments use this to induce nutrient limitations by adding wood chips or sawdust to soils. Fallen logs have low N contents, and long-term immobilization of N is especially evident during log decay in temperate forests (Lambert et al., 1980; Fahey, 1983), especially if they harbor

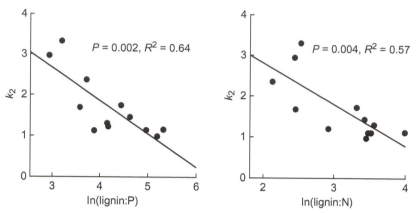

FIG. 6.14 Fractional decomposition of leaf litter from wet tropical forests as a function of the lignin-to-P and lignin-to-N ratio in newly fallen litter. *From Wieder et al. (2009). Used with permission of the Ecological Society of America.*

N-fixing bacteria (Rinne et al., 2017). Vogt et al. (1986) suggest that immobilization of N in soils is greatest in temperate and boreal forests, whereas the immobilization of P is more important in tropical forests (Marklein et al., 2016; Chen et al., 2016b).

Ecologists have long used the C/N ratio of litterfall as an index of its potential rate of decomposition. Lignin/nitrogen (Melillo et al., 1982) and lignin/phosphorus (Wieder et al., 2009) ratios are also good predictors of the decomposition of leaf litter (Fig. 6.14) and fine roots (See et al., 2019). These relationships allow us to predict the rate of decomposition and mineralization over large regions using remote sensing of forest canopy characteristics (Fan et al., 1998a; Ollinger et al., 2002). Recent work suggests that the concentration of Mn in litter may also control the rate of decomposition and accumulation of soil organic matter in boreal forests, due to limited synthesis of Mn-peroxidases by fungi (Keiluweit et al., 2015; Stendahl et al., 2017).

When microbial growth on a substrate slows, there is little further nutrient immobilization. Thus, immobilization of nutrients predominates in the layer of fresh litter on the soil surface, while net mineralization of N, P, and S is usually greatest in the lower forest floor (Federer, 1983). Sollins et al. (1984) found that the "light" fraction of soil organic matter, representing fresh plant residues, had a higher C/N ratio and lower mineralization than the "heavy" fraction, composed of humic substances. Net mineralization of N often begins as substrates approach a C/N ratio near 30:1, but this can vary depending on the substrate and the assimilation efficiency of the decomposer (Manzoni et al., 2010; Ågren et al., 2013).

During the decomposition process, the nutrients released from the original litter are transported to lower soil horizons by organic acids or other dissolved organic compounds (Schoenau and Bettany, 1987; Qualls and Haines, 1991), which add to the nutrient pool in humus. Decaying plant litter appears to adsorb Al and Fe (Rustad, 1994; Laskowski et al., 1995), perhaps in compounds that are precursors to organic acids that carry Al and Fe to the lower soil profile in the process of podzolization (Mimmo et al., 2014; Chapter 4).

Differential losses of nutrients and nutrient immobilizations mean that the loss of mass from litterbags cannot be directly equated with the proportional release of its original nutrient content (Jorgensen et al., 1980; Rustad, 1994). Table 6.8 shows the mean residence time for organic matter and its nutrient content in the surface litter of various ecosystems.

TABLE 6.8 Mean residence time, in years, of organic matter and nutrients in the surface litter of forest and woodland ecosystems.

Region	Mean residence time (year)					
	Organic matter	N	P	K	Ca	Mg
Boreal forest	353	230	324	94	149	455
Temperate forest						
Coniferous	17	17.9	15.3	2.2	5.9	12.9
Deciduous	4	5.5	5.8	1.3	3.0	3.4
Mediterranean	3.8	4.2	3.6	1.4	5.0	2.8
Tropical rainforest	0.4	2.0	1.6	0.7	1.5	1.1

Note: *Values are calculated by dividing the forest floor mass by the mean annual litterfall.*
Sources: *Boreal and temperate values are from Cole and Rapp (1981); tropical values are from Edwards and Grubb (1982), Edwards (1977, 1982); Mediterranean values are from Gray and Schlesinger (1981).*

Some nutrients, such as K, that are easily leached from litter may show mineralization rates in excess of the loss of litter mass. Others, such as N, turn over more slowly due to immobilization in microbial tissues. Pregitzer et al. (2010) used experimental additions of $^{15}NO_3$ to a sugar maple forest in Michigan to measure a mean residence time of 6.5 years for N in the forest floor.

Given the nutrient limitations facing soil microbes, it is perhaps not surprising that their activity increases with nutrient additions (e.g., Cleveland and Townsend, 2006; Allen and Schlesinger, 2004), leading to losses of labile organic matter from the soil profile (Neff et al., 2002b; Mack et al., 2004). Most nutrient-addition experiments have led to shifts in microbial community composition along with decreased microbial biomass and enhanced microbial activity (Wallenstein et al., 2006; Treseder, 2008; Lu et al., 2011; Liu and Greaver, 2010). Surprisingly, nutrient enrichment seems to slow the degradation of recalcitrant organic molecules such as lignin (Wang et al., 2019c).

In Chapter 5 we saw that the pool of soil organic matter greatly exceeds the mass of vegetation in most ecosystems. As a result of its high nutrient content, humus also dominates the storage of biogeochemical elements in most ecosystems. Aboveground biomass contains only 4–8% of the total quantity of N in temperate forests (Cole and Rapp, 1981) and 3–32% in tropical forests (Edwards and Grubb, 1982). Generally, the ratio of C, N, P, and S in humus is close to 140:10:1.3:1.3 (Stevenson, 1986; Cleveland and Liptzin, 2007; Tipping et al., 2016), so the global pool of nitrogen in soil, estimated at $95–140 \times 10^{15}$ g (Post et al., 1985; Batjes, 1996), dwarfs the pool of nitrogen in vegetation, 3.8×10^{15} g.[j] Owing to the stability of humic

[j] Calculated using the global biomass of 615×10^{15} g C (Table 5.3) and a C/N ratio in vegetation of 160 (Table 6.5).

compounds in soil, the large nutrient pool in humus turns over very slowly. Typically soil microbes mineralize only about 1–3% of the pool of nitrogen in the soil each year (Connell et al., 1995).

Simple measurements of extractable nutrients, such as NH_4^+ or PO_4^{3-}, are unlikely to give a good index of nutrient availability in terrestrial ecosystems. These nutrients are subject to active uptake by plant roots, immobilization by soil microbes, and a variety of other processes that rapidly remove available forms from the soil solution. At any moment, the quantity extractable from a soil sample may be only a small fraction of what is made available by mineralization during the course of a growing season (Davidson et al., 1990). Thus, studies of biogeochemical cycling in the soil need to be based on measurements that record the dynamic nature of nutrient turnover.

Nitrogen cycling

The mineralization of N from decomposing materials begins with the release of amino acids and other simple organic-N molecules by the microbes involved in decomposition (Schimel and Bennett, 2004; Geisseler et al., 2010; Fig. 6.13). Some of the amino acids are taken up directly by plants and soil microbes. Using ^{15}N as a tracer, Marumoto et al. (1982) have shown that much of the N mineralized in the soil is released from dead microbes. The presence of soil animals that feed on bacteria and fungi can increase the rates of release of N and P from microbial tissues (Cole et al., 1977a; Anderson et al., 1983). The release, or mineralization, of NH_4^+ from organic forms is known as *ammonification*.

Subsequently, a variety of abiotic and biotic processes may remove NH_4^+ from the soil solution, including uptake by plants, immobilization by microbes, and incorporation in clay minerals (Johnson et al., 2000a). Some of the remaining NH_4^+ may undergo nitrification, in which the oxidation of NH_4^+ to NO_3^- is coupled to the fixation of carbon by chemoautotrophic bacteria, traditionally classified within the genera *Nitrosomonas* and *Nitrobacter* (Meyer, 1994; see Eqs. 2.16 and 2.17).[k] Recent work suggests that nitrification is also performed by prokaryotes in the more primitive group archaea, although it is unclear whether they normally achieve the same level of activity in soils as bacteria (Leininger et al., 2006; Di et al., 2009; Lu et al., 2015b; Lu et al., 2020). In some cases, NH_4^+ is also oxidized by heterotrophic nitrification, producing NO_3^- (Schimel et al., 1984; Brierley and Wood, 2001; Pedersen et al., 1999).

Nitrate may be taken up by plants and microbes or lost from the ecosystem in runoff waters or in emissions of N-containing gases during denitrification. An intermediate product in the nitrification reaction (see Eq. 2.16), NO_2^- appears to bind to soil organic matter by abiotic processes (Fitzhugh et al., 2003; Davidson et al., 2003). Nitrate taken up by soil microbes (immobilization) is reduced to NH_4^+ by nitrate reductase and used in microbial growth (Davidson et al., 1990; DeLuca and Keeney, 1993). This process is known as *assimilatory reduction*. Nitrate can also be utilized by dissimilatory nitrate-reducing bacteria to produce NH_4^+, which can cycle back through these pathways (Fig. 12.1). Dissimilatory nitrate reduction to ammonium

---

[k]CO_2 fixation by soil chemoautotrophs typically accounts for <1% of photosynthesis in temperate forests (Spohn et al., 2019).

(DNRA) is best known from its occurrence in wet tropical soils, where it can exceed denitrification—the conversion of NO_3^- to N_2 (Silver et al., 2001; Rutting et al., 2008; Templer et al., 2008). A variety of factors determine whether DNRA or denitrification predominates in soils (Kraft et al., 2014). DNRA potentially reduces the loss of NO_3^- to stream waters.

At any time, the extractable quantities of NH_4^+ and NO_3^- in the soil represent the net result of all of these processes. A low concentration of NH_4^+ is not necessarily an indication of low mineralization rates; it can also indicate rapid nitrification or plant uptake (Rosswall, 1982; Davidson et al., 1990). A variety of techniques are available to study the transformations of nitrogen in the soil (Binkley and Hart, 1989). Most of these involve the isolation of soil samples in a plastic bag, a tube, or a weeded and trenched plot, which allows soil microbial activity to continue in the absence of plant uptake (Raison et al., 1987; Vitousek et al., 1982). An increase in the quantity of available N in these isolated samples is taken to represent *net* mineralization (i.e., the mineralization in excess of microbial immobilization). Repeated samples taken through an annual cycle allow an estimate of annual net mineralization, which can be correlated with plant uptake and cycling (Pastor et al., 1984).

A more expensive approach involves the use of $^{15}NH_4^+$ to label the initial pool of available NH_4^+ in the soil (Van Cleve and White, 1980; Davidson et al., 1991; Di et al., 2000). After a period of time, the pool is remeasured for the ratio of $^{15}NH_4^+$ to $^{14}NH_4^+$. A decline in proportion of $^{15}NH_4^+$ is assumed to result from the microbial mineralization of $^{14}NH_4^+$ from the pool of N in soil organic matter. This technique gives a measure of total (gross) mineralization under natural field conditions. Using this approach, Davidson et al. (1992) found that net mineralization was only 14% of the total in a coniferous forest; the remainder was immobilized by the microbial community. Known as the isotope-pool dilution technique, this approach has also been applied to determine turnover of amino acids (Wanek et al., 2010) and NO_3^- (Stark and Hart, 1997). For instance, $^{15}NO_3$ is used to label the pool of NO_3^- in the soil, and gross nitrification is measured as the label is diluted by $^{14}NO_3^-$. Net nitrification can also be studied by measuring changes in the concentration of NH_4^+ and NO_3^- after application of compounds that specifically inhibit nitrification, including nitrapyrin (Bundy and Bremner, 1973), acetylene (Berg et al., 1982), or chlorate (Belser and Mays, 1980).

Mineralization and nitrification have been studied in a wide variety of ecosystems. Net mineralization typically ranges from 20 to 120 kg N/ha/yr in forests (Pastor et al., 1984; Fan et al., 1998b; Perakis and Sinkhorn, 2011), 40–90 kg N/ha/yr in grasslands (Hatch et al., 1990), and 10–30 kg N/ha/yr in deserts (Schlesinger et al., 2006). Generally, net mineralization is directly related to the soil microbial biomass (Li et al., 2019e), the total content of organic nitrogen in the soil (e.g., Marion and Black, 1988; Accoe et al., 2004), and the availability of carbon (Booth et al., 2005). Mineralization is also closely linked to temperature (Pierre et al., 2017) and responds to temperature with a Q_{10} averaging 2.21 (Durán et al., 2017; Hu et al., 2016). Vegetation with a high C/N ratio in litterfall often shows low rates of mineralization in the soil (Gosz, 1981; Vitousek et al., 1982). When field plots are fertilized with sugar, net mineralization and nitrification slow because of increased immobilization of NH_4^+ by soil microbes (DeLuca and Keeney, 1993; Zagal and Persson, 1994). Fertilization of a Douglas fir forest with sugar resulted in lower N content in leaves and greater nutrient reabsorption before leaf-fall, showing a direct link between microbial processes in the soil and the nutrient-use efficiency of vegetation (Turner and Olson, 1976).

Although soil microbial populations may adapt to a wide variety of field conditions, nitrification is generally lower at low pH, low O_2, low soil moisture content, and high litter C/N ratios (Wetselaar, 1968; Robertson, 1982; Bramley and White, 1990; Booth et al., 2005). Compared to mineralization, nitrification is more sensitive to low soil water content, so NH_4^+ accumulates in seasonally dry desert soils (Hartley and Schlesinger, 2000) and in experiments imposing drought conditions (Homyak et al., 2017). Generally, however, nitrification rates are high whenever both NH_4^+ and O_2 are readily available (Robertson and Vitousek, 1981; Vitousek and Matson, 1988).

A large amount of effort has been directed toward understanding the control of nitrification following disturbances, such as forest harvest or fire (Likens et al., 1969; Vitousek and Melillo, 1979; Vitousek et al., 1982). When vegetation is removed, soil temperature and moisture content are generally higher, and rapid ammonification increases the availability of NH_4^+. Subsequently, nitrification may be so rapid that uptake by regrowing vegetation and immobilization by soil microbes are insufficient to prevent large losses of NO_3^- in stream water. However, not all disturbed sites show large losses of NO_3^-. In pine forests in the southeastern United States, microbial immobilization in harvest debris accounted for 83% of the uptake of ^{15}N that was applied as an experimental tracer following forest harvest (Vitousek and Matson, 1984). Immobilization of nitrate accounts for a large fraction of the nitrogen turnover in coniferous forests (Stark and Hart, 1997), and microbial immobilization also retards the loss of nitrate following burning of tallgrass prairie (Seastedt and Hayes, 1988).

In general, nitrification and losses of NO_3^- in stream water after disturbance are greatest in forests with high nitrogen availability (Krause, 1982; Vitousek et al., 1982). Rates of nitrification decline during the early recovery of vegetation, and only minor differences are seen between middle- and old-age forests (Robertson and Vitousek, 1981; Christensen and MacAller, 1985; Oda et al., 2018). There is some evidence that nitrification is inhibited by terpenoid, tannin, and pinene compounds released by some types of vegetation (Olson and Reiners, 1983; White, 1988; Uusitalo et al., 2008; Subbarao et al., 2009).

Increases in nitrification following disturbance affect other aspects of ecosystem function. Nitrification generates acidity (Eq. 2.16), so losses of NO_3^- in stream water are often accompanied by increased losses of cations, which are removed from cation exchange sites in favor of H^+ (Likens et al., 1970). Stream-water losses of nearly all biogeochemical elements increased following harvest of the Hubbard Brook forest in New Hampshire. Sulfate was a curious exception (Fig. 6.15). Nodvin et al. (1988) showed that the decline in stream water SO_4^{2-} concentrations after forest harvest was a result of the acidity generated from nitrification, which increased soil anion adsorption capacity (Chapter 4; Mitchell et al., 1989). These observations are a good example of a link between the biogeochemical cycles of N and S in terrestrial ecosystems.

Since the various transformations of N in the soil favor ^{14}N over [15]N (Hogberg, 1997), ^{15}N increases in the undecomposed residues (Nadelhoffer and Fry, 1988), with depth in the soil profile (Koba et al., 1998; Hobbie and Ouimette, 2009; Piccolo et al., 1996), and with time of ecosystem development (Brenner et al., 2001; Billings and Richter, 2006). In contrast, nitrate in runoff waters is depleted in [15]N (Spoelstra et al., 2007). Nearly all soils show $\delta^{15}N > 0$, with the greatest values seen in ecosystems with rapid nitrogen cycling (Templer et al., 2007) and with substantial losses of NO_3^- in runoff and emissions of nitrogen gases from the ecosystem (Amundson et al., 2003; Pardo et al., 2002). Since plants depend on nitrogen mineralization

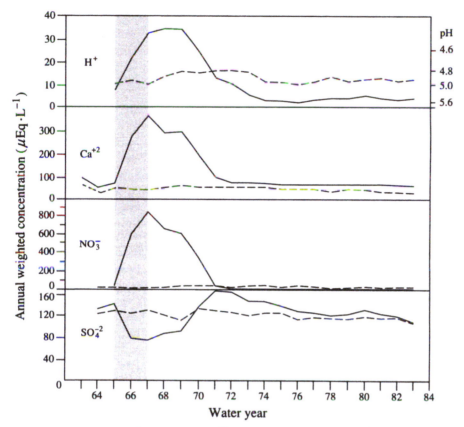

FIG. 6.15 Concentration of H^+, Ca^{2+}, NO_3^-, and SO_4^{2-} in streams draining the Hubbard Brook Experimental Forest for the years 1964–1984. Streams draining undisturbed forest are shown with the dashed line. The solid line depicts the concentrations in a stream draining a single watershed that was disturbed between 1965 and 1967 (shaded). Losses of Ca and NO_3 increased strongly during the period of disturbance, and then recovered to normal values as the vegetation regenerated. The budget for SO_4 shows greater retention during and after the period of disturbance, presumably as a result of increased acidity and anion adsorption in the soil. *Modified from Nodvin et al. (1988).*

for uptake, plant $\delta^{15}N$ is usually lower than that of soil organic matter, but $\delta^{15}N$ in plants increases as a function of soil nitrogen cycling, which progressively enriches ^{15}N remaining in the soil (Garten and Van Miegroet, 1994; Templer et al., 2007).

Emission of nitrogen gases from soils

During transformations of nitrogen in the soil, a variety of nitrogen gases, including NH_3, NO, N_2O, and N_2, are produced as products and byproducts of microbial activity (Fig. 6.16). Several chemical pathways are also a potential source of N_2O (Zhu-Barker et al., 2015). Some of these gases may escape from the ecosystem, contributing to a loss of local soil fertility. More significantly, terrestrial ecosystems are a significant source of these gases in the atmosphere (Chapters 3 and 12).

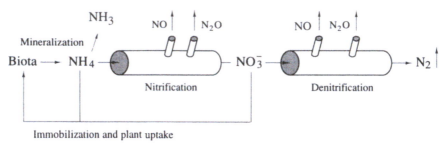

FIG. 6.16   Microbial processes that yield nitrogen gases during nitrification and denitrification in the soil. *Modified from Firestone and Davidson (1989).*

In soils, ammonium may be converted to ammonia gas (NH_3), which is lost to the atmosphere. The reaction:

$$NH_4{}^+ + OH^- \rightarrow NH_3 \uparrow + H_2O \tag{6.5}$$

is favored in dry soils and deserts, where accumulations of $CaCO_3$ maintain alkaline pH. Low cation exchange capacity and low rates of nitrification also maximize the production and loss of NH_3 in these environments (Nelson, 1982; Freney et al., 1983). Small losses of NH_3 have been measured in a variety of natural forest and grassland soils worldwide (Schlesinger and Hartley, 1992; Fig. 6.17). Losses of NH_3 are greatest in fertilized soils (e.g., Griffis et al., 2019) and during the decomposition of urea excreted by wild and domestic animals (Terman, 1979). As much as 3% of the nitrogen in penguin guano can be volatilized as

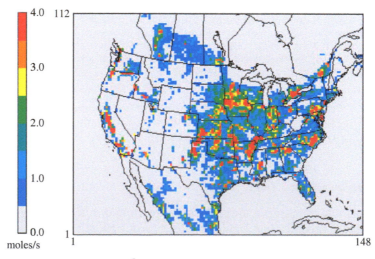

FIG. 6.17   Soil emissions of NH_3 moles/m^2/sec for the continental United States, showing relatively high emissions from various agricultural regions and lower emissions from undisturbed ecosystems. *From Gilliland et al. (2006).*

NH_3 (Riddick et al., 2016). During the loss of NH_3 from soils, isotopic fraction occurs, leaving soils enriched in ^{15}N (Mizutani et al., 1986; Mizutani and Wada, 1988).

Vegetation can also be a source of NH_3 during leaf senescence (Whitehead et al., 1988; Heckathorn and DeLucia, 1995), but some of the NH_3 emitted by soils may be taken up by plants, so in many cases natural ecosystems are only a small net source of NH_3 to the atmosphere (Sutton et al., 1993; Pryor et al., 2001; Wentworth et al., 2014). The loss of ammonia is typically $< 1\,kg/ha/yr$ in soils that are not impacted by fertilizers or domestic animals. The global flux from natural soils is about $2.4 \times 10^{12}\,gN/yr$ (Chapter 12). This flux to the atmosphere is significant inasmuch as NH_3 is the only substance that is a net source of alkalinity in the atmosphere, where it can reduce the acidity of rain (Eqs. 3.5 and 3.40). Extremely high NH_3 volatilization from barns and animal feedlots may result in enhanced atmospheric deposition of NH_4^+ in areas immediately downwind (Draaijers et al., 1989; Aneja et al., 2003). Somewhat counterintuitively, these large inputs of NH_4^+ may acidify soils, as the NH_4^+ is nitrified (Eqs. 2.16 and 2.17) and the nitrate is taken up by vegetation (van Breemen et al., 1982; Verstraten et al., 1990).

Both nitric oxide (NO) and nitrous oxide (N_2O) are generated as microbial byproducts of nitrification, with NO generally being the more abundant (Williams et al., 1992; Medinets et al., 2015). Specifically, NO is released during both steps in the oxidation of NH_4^+ by chemoautotrophic bacteria (Eqs. 2.16 and 2.17; Venterea and Rolston, 2000; Mushinski et al., 2019). Typically, about 1–3% of the nitrogen passing through the nitrification pathway is volatilized as NO each year (Baumgärtner and Conrad, 1992; Hutchinson et al., 1993), and the net flux from soils to the atmosphere is estimated to be about $12 \times 10^{12}\,gN/yr$ globally (Ganzeveld et al., 2002). Losses of NO from arid and semi-arid soils may account for 10% of the global total (Weber et al., 2015). In the southeastern United States, Davidson et al. (1998) estimate that soils supply a significant fraction of the NO_x emitted, with industrial and transportation sources accounting for the rest. (Recall that NO plays a major role in the chemistry of ozone in the troposphere—see Chapter 3.)

The flux of NO from soils is highest under conditions that stimulate nitrification, including fertilization with NH_4^+ (Skiba et al., 1993; Roelle et al., 1999). Soil NO efflux is directly related to nitrification in desert soils (Fig. 6.18). In various ecosystems, the flux appears to increase as a function of soil temperature (Williams et al., 1992; Roelle et al., 1999; Van Dijk and Duyzer, 1999) and immediately following the wetting of dry soils (Homyak et al., 2017). When the atmospheric concentration of nitric oxide is high, some NO is taken up by plants and soils, reducing the net flux to the atmosphere (Ganzeveld et al., 2002). The atmospheric concentration that produces no net uptake or loss is known as the *compensation point*. In most cases, the background concentration in the atmosphere, about 10 ppbv (refer to Table 3.5), is below the compensation point, so terrestrial ecosystems are a net source of NO to the atmosphere (Kaplan et al., 1988; Ludwig et al., 2001). In remote areas, the concentration of NO is greatest in the lower atmosphere, and it declines with altitude (Luke et al., 1992).

Several chemical pathways are a potential source of N_2O in soils (Zhu-Barker et al., 2015).[1] Losses of NO and N_2O increase with factors that increase the rate of nitrification in soils, including the clearing, cultivation, and fertilization of agricultural soils (Conrad et al., 1983;

---

[1]The production of N_2O during nitrification is sometimes called nitrifier denitrification (Wrage-Mönnig et al., 2018; Kool et al., 2011).

FIG. 6.18 Emission of NO as a function of nitrification rates in soils of the Chihuahuan Desert, New Mexico. *From Hartley and Schlesinger (2000).*

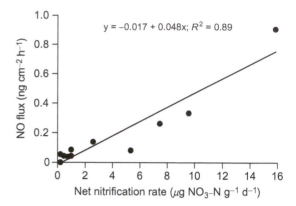

Mosier et al., 1991; Bouwman et al., 2002a; Liu et al., 2017a). Shepherd et al. (1991) report that 11% of fertilizer N was lost as NO and 5% as N_2O in some cultivated fields in Ontario. When tropical forests are cleared, the losses of NO and N_2O from soils increase dramatically (Sanhueza et al., 1994; Keller et al., 1993; Weitz et al., 1998), but older pastures often have lower N_2O emissions than uncut forest (Melillo et al., 2001; Verchot et al., 1999). Thus, fertilized and newly cleared fields may be responsible for the rising concentration of N_2O in Earth's atmosphere (Chapter 12).

Beyond nitrification, nitrate is also converted to NO, N_2O, and N_2 in the process of denitrification (Knowles, 1982; Firestone, 1982; Goregues et al., 2005). This reaction (Eq. 2.19) is performed by soil bacteria that are aerobic heterotrophs in the presence of O_2 but facultative anaerobes at low concentrations of O_2. During anoxia, heterotrophic activity continues, with nitrate serving as the terminal electron acceptor in metabolism. Normally, N_2 is the expected product of denitrification; NO and N_2O are intermediates that can escape the reaction. The structure of the various denitrification enzymes, *nitrite reductases*, contains Fe and Cu (Godden et al., 1991; Tavares et al., 2006; Glass and Orphan, 2012). Because NO_3^- is reduced, but not incorporated into microbial tissue, denitrification is also known as *dissimilatory nitrate reduction*. Bacteria in the genus *Pseudomonas* are the best known denitrifiers, but many others are reported (Knowles, 1982; Tiedje et al., 1989).

For a long time, denitrification was thought to occur only in flooded, anoxic soils (Chapter 7), and its importance in upland ecosystems was overlooked. Indeed, the activity of denitrification enzymes is often greatest at low soil O_2 concentrations (Burgin et al., 2010). Now, soil scientists have shown that oxygen diffusion to the center of soil aggregates is so slow that anoxic microsites are common, even in well-drained soils (Fig. 6.19; Tiedje et al., 1984; Sexstone et al., 1985a; Ebrahimi and Or, 2018). Thus, denitrification is widespread in terrestrial ecosystems, especially those in which organic carbon and nitrate are readily available (Burford and Bremner, 1975; Wolf and Russow, 2000; Qin et al., 2017).

Davidson and Swank (1987) found that additions of NO_3^- stimulated denitrification in the surface litter of forests in western North Carolina, and the addition of organic carbon stimulated denitrification in the mineral soil. Additions of organic carbon stimulated the expression of genes related to denitrification in some cultivated soils (Miller et al., 2012). Rainfall generally increases the rate of denitrification because the diffusion of oxygen is slower in

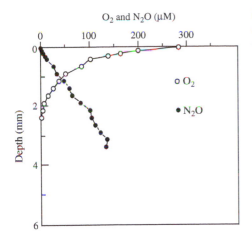

O₂ and N₂O (μM)

FIG. 6.19 Concentration of O_2 and N_2O (μM) determined in a soil aggregate as a function of the depth of penetration of a micro-electrode. *From Hojberg et al. (1994).*

wet soils (Sexstone et al., 1985b; Smith and Tiedje, 1979; Rudaz et al., 1991), and denitrification losses appear to increase with soil N enrichment (Morse et al., 2015).

Typically the production of both NO and N_2O produced by nitrification declines with increasing soil moisture content (or declining O_2), and in anoxic conditions N_2O is entirely due to denitrification (Khalil et al., 2004; Wolf and Russow, 2000; Wrage et al., 2001). In Germany, well-drained soils with near-neutral pH produced NO only from nitrification, whereas in acid, anoxic soils, NO was produced from denitrification (Remde and Conrad, 1991). In studies of a semi-desert ecosystem, Mummey et al. (1994) found that nitrification accounted for 61–98% of the N_2O produced in moist soils, but denitrification was the predominant reaction in saturated conditions. In the wet soils of Amazon rainforests, N_2O appeared to be mostly from denitrification (Livingston et al., 1988; Keller et al., 1988).

The relative importance of NO, N_2O, and N_2 as products of denitrification varies depending on environmental conditions (Firestone and Davidson, 1989; Bonin et al., 1989). Typically, in denitrification, the production of N_2O dwarfs the production of NO, so the proportional and total loss (nitrification + denitrification) of NO from soils declines with increasing moisture content, while the flux of N_2O increases (Fig. 6.20; Drury et al., 1992; Bollmann and Conrad, 1998; Wolf and Russow, 2000). Factors affecting the relative loss of N_2O and N_2 by denitrification are poorly understood, but they include soil pH and the relative abundance of NO_3^- and O_2 as oxidants and organic carbon as a reductant (Firestone et al., 1980; Morley and Baggs, 2010; Burgin and Groffman, 2012; Bakken et al., 2012). Soil pH was a major factor affecting N_2O emission from agricultural soils in China (Wang et al., 2018c). When NO_3^- is abundant relative to the supply of organic carbon, N_2O can be an important product (Firestone and Davidson, 1989; Mathieu et al., 2006; Huang et al., 2004). Reduction to N_2 can lower the net flux of N_2O to the atmosphere, and in some circumstances, N_2O can diffuse into soils, where it is consumed by microbes, producing N_2 (Sanford et al., 2012; Schlesinger, 2013).

The ratio of N_2O to N_2 produced in denitrification varies widely (Weier et al., 1993). On average, N_2O represents about 50% of the total gas loss by denitrification (Schlesinger, 2009; Wang and Yan, 2016). Whereas the ratio for the emission of these gases from upland

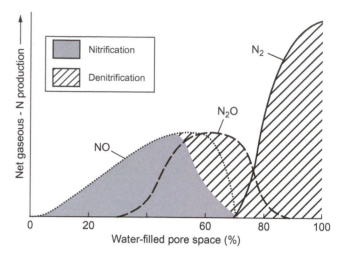

soils appears to be about 1:1, but the proportional emission of N_2O from wetter soils is much lower (Schlesinger, 2009). The total loss of N from soils by denitrification is typically <3 kg/ha/yr in forests and grasslands (Bai et al., 2014; Teh et al., 2014; Morse et al., 2015).

The sum of N_2O losses from nitrification and denitrification in croplands are typically 2–4 kg N/ha/yr (Roelandt et al., 2005; Kim et al., 2013; Shcherbak et al., 2014), but occasional values as high as 13 kg N/ha/yr are reported from some agricultural soils (Barton et al., 1999; Sgouridis and Ullah, 2015). Globally, the total loss of N_2 from soils ($\sim$45 $\times$ 10^{12} g N/yr; Chapter 12) dwarfs the loss of N_2O (<10 $\times$ 10^{12} g N/yr) or NO (12 $\times$ 10^{12} g N/yr). Denitrification is a major process that returns N_2 to the atmosphere, completing the global biogeochemical cycle of nitrogen (Fang et al., 2015; Brookshire et al., 2017; Chapter 12). The flux of N_2O is significant inasmuch as its concentration in the atmosphere is increasing, and N_2O plays an important role as both a greenhouse gas and a catalyst of ozone destruction in the stratosphere (Chapters 3 and 12).

Field measurements of denitrification are problematic, since the production of N_2 is difficult to measure against the background concentration of 78% in Earth's atmosphere. Recently, various workers have measured denitrification using an inert gas, such as helium or argon, to fill collection chambers in the field, allowing the efflux of N_2 from the soil to be observed more easily (Scholefield et al., 1997; Butterbach-Bahl et al., 2002; Burgin et al., 2010).

Many laboratory studies have also measured denitrification using acetylene to block the conversion of the intermediate denitrification product, N_2O, to the final product, N_2 (refer to Fig. 6.16; Yoshinari and Knowles, 1976; Burton and Beauchamp, 1984; Davidson et al., 1986; Tiedje et al., 1989). With the application of acetylene to laboratory soils or field plots, the sole product of denitrification is N_2O, which is easy to measure with gas chromatography against its background concentration of about 320 ppb in the atmosphere. Alternatively, many field workers have measured denitrification by the application of $^{15}NO_3^-$ to field plots and by measurements of the release of ^{15}N gases or the decline in $^{15}NO_3^-$ remaining in the soil (Parkin et al., 1985; Mathieu et al., 2006; Kulkarni et al., 2014).

Field estimates of denitrification are complicated by high spatial and temporal variability. At the local scale, a large portion of the total variability is found at distances of <10 cm, which Parkin (1987) link to the local distribution of soil aggregates that provide anaerobic microsites. In one case, Parkin (1987) found that 85% of the total denitrification in a 15-cm-diameter soil core was located under a 1-cm^2 section of decaying pigweed (*Amaranthus*) leaf! In desert ecosystems, soil nitrogen content and nitrification rates are localized under shrubs, and denitrification is largely confined to those areas (Virginia et al., 1982; Peterjohn and Schlesinger, 1991).

Robertson et al. (1988) documented the pattern of mineralization, nitrification, and denitrification in a field in Michigan. All these processes showed large spatial variation, but the coefficient-of-variation for denitrification, 275%, was the largest measured. Significant correlations were found among these processes. Soil respiration and potential nitrification explained 37% of the variation in denitrification, presumably due to the dependence of denitrification on organic carbon and NO_3^- as substrates.

The high variability in these processes makes it difficult to use measurements from a few sample chambers to calculate a mean or total flux from an ecosystem (Ambus and Christensen, 1994; Mathieu et al., 2006). High rates of denitrification are often confined to particular landscape positions, where conditions are favorable. For example, Peterjohn and Correll (1984) suggested that the runoff of nitrate from agricultural fields was largely denitrified in streamside forests, minimizing the losses in rivers (see also, Pinay et al., 1993; Jordan et al., 1993). In upland forests, losses from denitrification are often highest in local areas with saturated soils (Duncan et al., 2013; Anderson et al., 2015). In calculating regional averages for denitrification, investigators must evaluate the relative contributions from local areas of high and low activity (e.g., Groffman and Tiedje, 1989; Yavitt and Fahey, 1993; Morse et al., 2012). Differences in the microbial communities between natural and disturbed ecosystems are also likely to complicate regional estimates of denitrification (Cavigelli and Robertson, 2000). Rather than trying to aggregate individual field measurements of N_2O flux, Soper et al. (2018b) used remote sensing to map the nitrogen content of the canopy of tropical rainforests, which was highly correlated to the N_2O emissions from their soils.

As in the case of NH_3 volatilization, the losses of N gases as products and byproducts of nitrification and denitrification leave soils enriched in ^{15}N. Denitrifying bacteria fractionate among the isotopes of available nitrogen—that is, between $^{14}NO_3^-$ and $^{15}NO_3^-$ (Handley and Raven, 1992; Robinson, 2001; Snider et al., 2009). Preference for $^{14}NO_3^-$ leads to positive $\delta^{15}N$ in most soils (refer to Fig. 6.9), as $^{14}N_2$ is lost from the soil by denitrification (Shearer and Kohl, 1988; Knöller et al., 2011; Lewicka-Szczebak et al., 2014). Evans and Ehleringer (1993) show a strong inverse relation between $\delta^{15}N$ and the nitrogen content in soils (Fig. 6.21), suggesting that soils with low nitrogen are enriched in ^{15}N as a result of the loss of N gases (Garten, 1993). Strong enrichments in ^{15}N are also seen in saturated soils with low redox potential (Chapter 7) compared to adjacent well-drained soils in the same field (Sutherland et al., 1993). Strong enrichments in tropical soils suggest that gaseous losses may be a major pathway of nitrogen loss (Martinelli et al., 1999; Houlton et al., 2006; Koba et al., 2012).

FIG. 6.21 $\delta^{15}N$ of soil organic matter as a function of the total N content of the soil in juniper woodlands of Utah. *From Evans and Ehleringer (1993).*

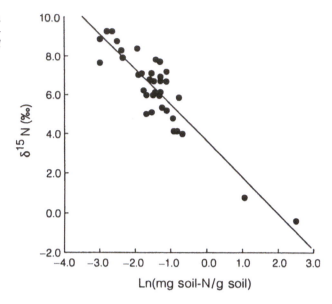

Soil phosphorus cycling

Mineralization of phosphorus from organic compounds is difficult to study because of the variety of forms of soil phosphorus (Weihrauch and Opp, 2018) and the immediate reaction of available phosphorus with various soil minerals (Figs. 4.10 and 6.22). A few workers have examined phosphorus mineralization using the buried-bag approach (e.g., Pastor et al., 1984), but in many cases there is no apparent net mineralization because of the rapid physical adsorption of P onto soil minerals. Thus, most studies of phosphorus cycling have followed the decay of radioactively labeled plant materials (Harrison, 1982) or measured the dilution of radioactive ^{32}P or ^{33}P that is applied to the soil pool as a tracer (Walbridge and Vitousek, 1987; Bünemann, 2015; Helfenstein et al., 2018; Wanek et al., 2019). With the isotope-dilution technique, one must assume that ^{32}P equilibrates with the chemical pools in the soil and that the only dilution of its concentration is by the mineralization of organic phosphorus (Kellogg et al., 2006). Unfortunately, these assumptions are not always valid, making the technique difficult to apply in many instances (Di et al., 1997; Bunemann et al., 2007). To avoid using radioactive materials, some workers have measured dilution of non-radioactive isotope of oxygen that can be used to label the PO_4^{3-} ion in experimental additions to the soil (i.e., $^{18}\delta O$; Weiner et al., 2018).

In the face of difficulty measuring P mineralization directly, many workers have used sequential extractions to quantify phosphorus availability in the soil (Hedley et al., 1982b; Tiessen et al., 1984). Extraction with 0.5 M NaHCO$_3$ is a convenient index of labile inorganic and soluble organic phosphorus in many soils (Olsen et al., 1954; Sharpley et al., 1987). Organic P is often determined as the difference between PO$_4$ in a sample that has been combusted at high temperatures and an untreated sample (Stevenson, 1986), and microbial P by the change in extractable phosphorus after fumigation with chloroform (Brookes et al., 1984). Extraction with NaOH (to raise pH and lower anion adsorption capacity) indicates the amount

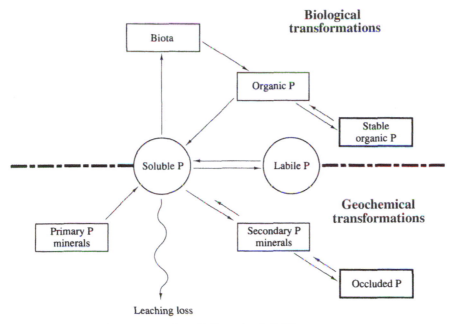

FIG. 6.22 Phosphorus transformations in the soil. *From Smeck (1985).*

of P that is held on Fe and Al minerals, whereas extraction with HCl releases P from many Ca-bound forms, including $CaCO_3$ (Tiessen et al., 1984; Cross and Schlesinger, 1995). Acid-extractable phosphorus also includes P derived from apatite (Chapter 4), including secondary hydroxyapatite ($Ca_5OH(PO_4)_3$) from bones and fluoroapatite ($Ca_5F(PO_4)_3$) in teeth. These biominerals in soils are sometimes used by archeologists to determine the location of past human activity and settlements (Sjöberg, 1976; Vitousek et al., 2004). These various fractions correspond to the generalized changes in phosphorus availability during soil development (Fig. 4.11). Among soil P fractions, the mean residence time of inorganic P in labile forms is minutes to hours, versus days to months for NaOH-extractable forms and years to millennia for acid-extractable P (Helfenstein et al., 2019).

In most ecosystems, much of the phosphorus available for biogeochemical cycling is held in organic forms (Chapin et al., 1978; Yanai, 1992; Gressel et al., 1996), especially inositol phosphates (Turner and Millward, 2002). Phosphorus monoesters (i.e., C—O—P) tend to accumulate with time (Wang et al., 2019). These various organic forms can be isolated and identified using phosphorus-31 nuclear magnetic resonance spectroscopy (Turner et al., 2007; Turner and Engelbrecht, 2011; Huang et al., 2017; Deiss et al., 2018). Earlier we discussed the ability of soil microbes and plant roots, including associated mycorrhizae, to release phosphatase enzymes and organic acids that mineralize P from organic forms (see also Chapter 4).

Much of the P in decomposing materials is found in ester linkages. These groups may be mineralized by the release of extracellular enzymes (e.g., phosphatases) in response to specific microbial demand for P (McGill and Cole, 1981). Release of acid phosphatases by soil microbes is directly related to levels of soil organic matter (Tabatabai and Dick, 1979; Polglase et al., 1992; Godin et al., 2015). During forest development, the phosphorus taken up from labile pools in the soil is replenished by P released from the anion adsorption and non-occluded pools, which presumably equilibrate with the soil solution over longer periods (Richter et al., 2006; Helfenstein et al., 2019).

In most instances, transformations of phosphorus held in organic forms dominate its cycling near the soil surface, whereas physiochemical reactions dominate at depth (Wood et al., 1984; Achat et al., 2013; Bünemann et al., 2016; Wilcke et al., 2019). Walbridge et al. (1991) found that microbial biomass held up to 35% of the organic P in the undecomposed litter in a warm-temperate forest, and Gallardo and Schlesinger (1994) found that additions of inorganic P increased the microbial biomass in the lower horizons of a forest soil in North Carolina, where the mineralogy is dominated by Fe and Al-oxide minerals with strong phosphorus adsorption capacity. Similar results are reported for tropical forests (Cleveland et al., 2002; Liu et al., 2012). Chemical weathering of phosphorus minerals predominates in young soils (Schlesinger et al., 1998), whereas organic cycling dominates in low-phosphorus soils (Lang et al., 2017; Turner et al., 2014) and highly weathered soils of tropical rainforests.

Sulfur cycling

Similar to phosphorus, the cycle of sulfur in the soil is affected by both chemical and biological reactions. Sulfur is derived from atmospheric deposition (Chapter 3) and from the weathering of sulfur-bearing minerals in rocks (Chapter 4), and the proportion from each source varies with location and soil development (Novak et al., 2005; Bern and Townsend, 2008; Mitchell et al., 2011b). The concentration of SO_4^{2-} in the soil solution exists in equilibrium with sulfate adsorbed on soil minerals (Chapter 4). Plant uptake of SO_4^{2-} is followed by assimilatory reduction and incorporation of sulfur into glutathione (Köstner et al., 1998) and the amino acids cysteine and methionine, which are incorporated into protein (Johnson, 1984). The molecular structure of the S-reducing enzyme contains Fe as a cofactor (Crane et al., 1995). A small quantity of sulfur in plants is found in ester-bonded sulfates ($-C-O-SO_4$), and when soil sulfate concentrations are high, plants may accumulate SO_4 in leaf tissues (Turner et al., 1980).

In most soils, the majority of the S is held in a variety of organic forms (Bartel-Ortiz and David, 1988; Mitchell et al., 1992; Schroth et al., 2007). Decomposition of plant tissues is accompanied by microbial immobilization of S (Saggar et al., 1981; Staaf and Berg, 1982; Fitzgerald et al., 1984). Using ^{35}S as a radio-isotopic tracer, Wu et al. (1995) found high rates of microbial immobilization of $^{35}SO_4^{2-}$ when glucose was added to soils (cf. Houle et al., 2001). Downward movement of organic acids appears to transport organic sulfur compounds to the lower soil profile (Schoenau and Bettany, 1987; Kaiser and Guggenberger, 2005), where they are mineralized (Houle et al., 2001; Dail and Fitzgerald, 1999). Typically, mineralization of SO_4^{2-} begins at C/S ratios <200 (Stevenson, 1986). Sulfur in soil organic matter shows higher $\delta^{34}S$ than soil sulfate, suggesting that soil microbes discriminate against the heavy isotope of S in favor of ^{32}S during mineralization (Mayer et al., 1995). Most of the SO_4^{2-} in runoff waters appears to have passed through the organic pool (Likens et al., 2002; Novak et al., 2005; Yi-Balan et al., 2014).

In forest soils, the microbial immobilization of added SO_4^{2-} is greatest in the upper soil profile, whereas anion adsorption of inorganic SO_4^{2-} dominates the B horizons, where sesquioxide minerals are present (Schindler et al., 1986; Randlett et al., 1992; Houle et al., 2001). In most cases, the majority of microbial S is found in carbon-bonded forms (David et al., 1982; Schindler et al., 1986; Dhamala and Mitchell, 1995). Organic S appears to accumulate in areas of high inputs of SO_4^{2-} from acid rain (Likens et al., 2002; Armbruster et al., 2003). At the Coweeta Experimental Forest in North Carolina, a large portion of the immobilization of sulfur by soil microbes has accumulated as ester sulfates (Fitzgerald et al., 1985; Watwood and Fitzgerald, 1988), yielding a significant sink for SO_4^{2-} deposited from the atmosphere (Swank et al., 1984). Despite the predominance of organic forms, the pool of SO_4^{2-} in most soils is not insignificant. In the study of a forest in Tennessee, Johnson et al. (1982) found that the pool of adsorbed SO_4^{2-} was larger than the total pool of S in vegetation by a factor of 15.

To maintain a charge balance, plant uptake and reduction of SO_4^{2-} consumes H^+ from the soil, whereas the mineralization of organic sulfur returns H^+ to the soil solution, producing no net increase in acidity (Binkley and Richter, 1987). In contrast, reduced inorganic sulfur is found in association with some rock minerals (e.g., pyrite), and the oxidative weathering of reduced sulfide minerals accounts for highly acidic solutions draining mine tailings (Eqs. 2.15 and 4.6). This oxidation is performed by chemoautotrophic bacteria, generally in the genus *Thiobacillus*.

Production of reduced sulfur gases such as H_2S, COS (carbonyl sulfide), and $(CH_3)_2$ S (dimethylsulfide) is largely confined to wetland soils, since highly reducing, anaerobic conditions are required (Chapter 7). Globally, upland soils are only a small source of sulfur gases in the atmosphere (Goldan et al., 1987; Staubes et al., 1989; Yi et al., 2010). However, many plants (e.g., garlic) produce a variety of volatile organic sulfur compounds that activate sensory receptors in humans and presumably other herbivores (Bautista et al., 2005).[m] The smell of CS_2 (carbon disulfide) is often found when excavating the roots of the tropical tree *Stryphnodendron excelsum* (Haines et al., 1989), and many plant leaves are known to release sulfur gases during photosynthesis (Winner et al., 1981; Kesselmeier et al., 1993).

Often the total net flux of sulfur gases from an ecosystem (soil+plant) is estimated by examining the vertical profile of gas concentrations in the atmosphere (e.g., Andreae and Andreae, 1988). Hydrogen sulfide appears to dominate the release of sulfur gases from plants (Delmas and Servant, 1983; Andreae et al., 1990; Rennenberg, 1991). Terrestrial ecosystems also appear to be a source of $(CH_3)_2S$ during the day (Andreae et al., 1990; Berresheim and Vulcan, 1992; Jardine et al., 2015b; Whelan and Rhew, 2016), but vegetation is a major sink for COS globally (Chapter 13).

Transformations in fire

During fires, nutrients are lost in gases and particles of smoke (Andreae, 2019), but soil nutrient availability increases with the addition of ash to soil (Raison, 1979; Giardina et al., 2000). Following fire, there is often increased runoff and erosion from bare, ash-covered soils. High rates of nitrification in these nutrient-rich soils can stimulate the loss of NO and N_2O after fire.

---

[m] The tasty compound in garlic is diallyl thiosulfinate or diallyl disulfide.

Increases in the rate of forest burning worldwide have the potential to deplete the nutrient content of soils and add trace gases to the atmosphere (Mahowald et al., 2005a). However, before human intervention, fires were a natural part of the environment in many regions; thus, nutrient losses as a result of fire occurred at infrequent but somewhat regular intervals (Clark, 1990). Using a mass-balance approach we can estimate the length of time it takes to replace the nutrients that are lost in a single fire. For instance, 11–40 kg/ha of N are lost in small ground fires in southeastern pine forests (Richter et al., 1982), equivalent to 3–12 times the annual deposition of N from the atmosphere in this region (Swank and Henderson, 1976). In contrast, periodic large losses of N in fires may dominate the long-term nitrogen budget in semi-arid forests—requiring hundreds of years of new inputs to replace the losses from a single fire (Johnson et al., 1998).

When leaves and twigs are burned under laboratory conditions, up to 90% of their N content can be lost, presumably as N_2 or as one or more forms of nitrogen oxide gases (DeBell and Ralston, 1970; Lobert et al., 1990). Forest fires volatilize nitrogen in proportion to the heat generated and the organic matter consumed (DeBano and Conrad, 1978; Raison et al., 1985; McNaughton et al., 1998); the rate of loss declines dramatically as fires pass from flaming to smoldering phases (Crutzen and Andreae, 1990). Typically N losses in forest fires range from 100 to 600 kg/ha, or 10–40% of the amount in aboveground vegetation and surface litter (Johnson et al., 1998; Dannenmann et al., 2018). Especially large losses are reported from slash fires in the Amazon rainforest (Kauffman et al., 1993).

Studies of the gaseous products of fires are often conducted by flying aircraft through the smoke plume to gather gas samples (e.g., Cofer et al., 1990; Nance et al., 1993; Hurst et al., 1994). Methane and ammonia released during fires can also be monitored by satellites (Whitburn et al., 2015; Ross et al., 2013). The enrichment of CO_2 and CO over the atmospheric background is measured, as well as the ratio of other gases to CO_2 in the smoke (e.g., NH_3/CO_2). Assuming that the carbon in the fuel is all converted to CO_2 and CO, the loss of other fuel constituents as gases and particles can be calculated from estimates of the carbon in the biomass consumed by fire and the ratio of the constituent in question to the total carbon ($CO_2 + CO$) in smoke (Laursen et al., 1992; Delmas et al., 1995; Luo et al., 2015). Thus, global estimates of the volatilization of nitrogen from forest fires can be calculated from global estimates of the amount of carbon lost in forest fires each year (Andreae, 2019; Schultz et al., 2008). N_2 dominates the gaseous loss of nitrogen (Kuhlbusch et al., 1991), constituting a form of "pyrodenitrification" that removes fixed nitrogen from the biosphere (Chapters 3 and 12).

The losses of nitrogen gases in forest fires account for 4% of the N_2O, 15% of the NH_3, and 16% of the NO_x emitted annually to the atmosphere (Chapter 12). Tropospheric circulation can carry the plume of NO_x from fires in the boreal forest of Canada to Europe (Spichtinger et al., 2001). Forest fires are also a major global source of CO (Seiler and Conrad, 1987; refer to Table 11.3) and smaller sources of CH_4 (Delmas et al., 1991; Quay et al., 1991), CH_3Br, CH_3Cl (refer to Table 3.7), and SO_2 in the atmosphere (Sanborn and Ballard, 1991; Crutzen and Andreae, 1990).

Air currents and updrafts during fire carry particles of ash that remove other nutrients from the site. These losses are usually much smaller than gaseous losses (Arianoutsou and Margaris, 1981; Gaudichet et al., 1995). Expressed as a percentage of the amount present in aboveground vegetation and litter before fire, the total loss of plant nutrients in gases and particulates often follows the order $N > K > Mg > Ca > P > 0\%$. Differential rates of loss change the balance of nutrients available in the soil after fire (Raison et al., 1985), and nutrient losses to the atmosphere in fire may enhance the atmospheric deposition of nutrients in adjacent locations (Clayton, 1976; Lewis, 1981). Phosphorus aerosols derived from forest fires in Africa fertilize the Amazon Basin and the southern oceans (Barkley et al., 2019).

Depending on intensity, fire kills aboveground vegetation and transfers varying proportions of its mass and nutrient content to the soil as ash. There are a large number of changes in chemical and biological properties of soil as a result of the addition of ash (Raison, 1979). Cations and P may be readily available in ash, which usually increases soil pH (Butler et al., 2018). Burning increases extractable P, but reduces the levels of organic P and phosphatase activity in soil (DeBano and Klopatek, 1988; Saa et al., 1993; Serrasolsas and Khanna, 1995). Nitrogen in the ash is subject to rapid mineralization and nitrification (Christensen, 1977; Dunn et al., 1979; Matson et al., 1987), so available NH_4^+ and NO_3^- usually increase after fire, even though total soil N may be unchanged (Wan et al., 2001). In a Mediterranean shrubland of Spain, losses after fire of N_2, NO and N_2O amounted to about 15% of the losses of these gases in the fire itself (Dannenmann et al., 2018). Enhanced emissions of NO and accumulations of nitrite after fire stimulate the germination of post-fire species (Keeley and Fotheringham, 1998). The increase in available nutrients as a result of ashfall is usually short-lived, as nutrients are taken up by vegetation or lost to leaching and erosion (Lewis, 1974; Uhl and Jordan, 1984; Goodridge et al., 2018).

Stream-water runoff typically increases following fire because of reduced water losses in transpiration. High nutrient availability in the soil coupled with greater runoff can lead to large hydrologic losses of nutrients from ecosystems following burning. The loss of nutrients in runoff depends on many factors, including the season, rainfall pattern, and the growth of postfire vegetation (Dyrness et al., 1989). Wright (1976) noted significant increases in the loss of K and P from burned forest watersheds in Minnesota. These losses were greatest in the first 2 years after fire; by the third year there was actually less P lost from burned watersheds than from adjacent mature forests, presumably due to uptake by regrowing vegetation (McColl and Grigal, 1975; Saá et al., 1994). Although there are exceptions, the relative increase in the loss of Ca, Mg, Na, and K in runoff waters after fire often exceeds that of N and P (Chorover et al., 1994).

Lake and ocean sediments contain layers of buried ash that indicate the frequency of fires (Mensing et al., 1999; Clark et al., 1996). Layers of ice with buried ash often contain especially high concentrations of NH_4^+, indicating substantial NH_3 volatilization during fires (Legrand et al., 1992). Cores taken from the Greenland ice cap show episodes of increased biomass burning that appear to be related to the European colonization of North America (Whitlow et al., 1994; Savarino and Legrand, 1998). Past changes in climate and human activities have influenced the global annual rates of biomass consumed by fires (Nicewonger et al., 2018; Ward et al., 2018) Forest fires and their biogeochemical consequences are likely to increase further in response to future droughts (Schlesinger et al., 2016).

The role of animals

Discussions of terrestrial biogeochemistry usually center on the role of plants and soil microbes. Having seen that animals harvest about 5% of terrestrial net primary production

(Chapter 5), it is legitimate to ask if they might play a significant role in nutrient cycling. Certainly an impressive nutrient influx is observed in the soils below roosting birds (Gilmore et al., 1984; Mizutani and Wada, 1988; Lindeboom, 1984; Simas et al., 2007; Maron et al., 2006). Soils rich in ammonium and $\delta^{15}N$ are used to track the past location of penguin colonies in Antarctica (Mizutani et al., 1986) and the roosting areas of birds in other regions (Leblans et al., 2014; Ellis et al., 2006). In Yellowstone National Park, elk appear to redistribute plant materials among habitats in the landscape, increasing the nitrogen content and nitrogen mineralization in soils where they congregate (Frank et al., 1994; Frank and Groffman, 1998). Similarly, sea turtles return nutrients to dune habitats, where they lay eggs (Bouchard and Bjorndal, 2000; Hannan et al., 2007), and nutrients are delivered to African rivers by the feeding and wallowing of hippopotamus (Subalusky et al., 2018). It is likely that nutrient redistributions by animals are significantly lower today than before the extirpation of Pleistocene megafauna (Doughty et al., 2013).

Soil $\delta^{15}N$ is higher in grazed soils, indicating greater N losses from the ecosystem (Frank and Evans, 1997). Grazing in the Serengeti of Africa appears to stimulate nutrient cycling and plant productivity, providing better habitat for the animals (McNaughton et al., 1997). In the African savannah, movements of ancient herders can be traced by enrichments in soil nutrients and $\delta^{15}N$ that persist after several thousand years (Marshall et al., 2018).

Various workers have suggested that the grazing of vegetation, especially by insects, stimulates the intrasystem cycle of nutrients and might even be advantageous for terrestrial vegetation (Owen and Wiegert, 1976). Trees that are susceptible to herbivory are often those that are deficient in minerals or otherwise stressed (Waring and Schlesinger, 1985). Periodic herbivory may stimulate nutrient return to the soil via insect frass and alleviate nutrient deficiencies (Mattson and Addy, 1975; Yang, 2004). Risley and Crossley (1988) also noted significant premature leaf-fall in a forest that was subject to insect grazing. These leaves delivered large quantities of nutrients to the soil, since nutrient reabsorption had not yet occurred. In the same forest, Swank et al. (1981) noted an increase in stream-water nitrate when the trees were defoliated by grazing insects. Herbivory makes a significant contribution to nutrient cycling in tropical rainforests (Metcalfe et al., 2014).

An enormous literature exists on the characteristics of plant tissues that are selected for food. Seasonal variations in the plants selected as food by large mammals may help these grazing animals avoid mineral deficiency (McNaughton, 1990; Ben-Shahar and Coe, 1992; Grasman and Hellgren, 1993). Many studies report that herbivory is centered on plants with high nitrogen contents (Mattson, 1980; Griffin et al., 1998), suggesting that animal populations might be limited by N. However, the preference for such tissues may be related more to their high water (Scriber, 1977) and low phenolic contents (Jonasson et al., 1986) than to a specific search for leaves with high amino acid content. Grazing often reduces plant photosynthesis while nutrient uptake continues, resulting in high nutrient contents in the aboveground tissues that remain (McNaughton and Chapin, 1985). Grazing may even enhance nitrogen uptake in some species (Jaramillo and Detling, 1988). Thus, consumers sometimes increase the nutritional quality of the forage available for future consumption, although the quantity of plant defensive compounds may also increase (White, 1984; Seastedt, 1985).

In extreme cases, defoliations may be the dominant form of nutrient turnover in the ecosystem (Hollinger, 1986). Usually, however, the role of grazing animals in terrestrial ecosystems is rather minor (Gosz et al., 1978; Pletscher et al., 1989), and certainly of limited benefit to plants (Lamb, 1985). In fact, plants often show large allocations of net primary production to defensive compounds (Coley et al., 1985) and higher net primary production when they are relieved of pests (Cates, 1975; Morrow and Lamarche, 1978; Marquis and Whelan, 1994). Accumulations of Si in plant tissues, especially in grasses, may be a defense mechanism against herbivory (Hartley and DeGabriel, 2016).

At higher trophic levels, predators may also affect nutrient cycling in terrestrial ecosystems, inasmuch as they create local patches of nutrient-rich soil with the delivery of dead animals (Schmitz et al., 2010; Carter et al., 2007; Barton et al., 2013). Soil enrichment under decomposing carrion can persist for more than a year (Keenan et al., 2018, 2019). Spawning salmon deliver nitrogen from the marine environment to streamside forests (Helfield and Naiman, 2001; Ben-David et al., 1998; Quinn et al., 2018), especially when bears prey on them (Hilderbrand et al., 1999). When predators—foxes—were introduced to an Aleutian island, they disrupted nesting seabirds, which normally delivered nutrients from the marine environment to the soil (Maron et al., 2006).

Animals play and important role in litter decomposition (Swift et al., 1979; Hole, 1981; Seastedt and Crossley, 1980). Nematodes, earthworms, and termites are particularly widespread and important in the initial breakdown of litter and the turnover of nutrients in the soil. Schaefer and Whitford (1981) found that termites were responsible for the turnover of 8% of litter N in a desert soil (Fig. 6.23). An additional 2% of the pool of nitrogen in surface litter was transported belowground by their burrowing activities. When termites were excluded using an application of pesticides, decomposition slowed and surface litter accumulated. Because soil animals have short lifetimes, their nutrient contents are rapidly decomposed and returned to the intrasystem cycle (Seastedt and Tate, 1981). Larger animals enhance nutrient cycling by disturbing and digging in soils (Schmitz et al., 2018; Mallen-Cooper et al., 2019).

It is interesting to view the biogeochemistry of animals from another perspective: What is the role of biogeochemistry in determining the distribution and abundance of animals? The death of ducks and cattle feeding in areas of high soil selenium (Se) suggests that such interactions might also be of widespread significance (Dhillon et al., 2018).

Plants have no essential role for sodium in their biochemistry, and naturally have low Na contents due to limited uptake and exclusion at the root surface (Smith, 1976). On the other hand, sodium is an important, essential element for all animals. The wide ratio between the Na content of herbivores to that in their foodstuffs suggests that Na might limit mammal populations generally (Prather et al., 2018; Welti et al., 2019). Observations of Na deficiency are supported by the interest that many animals show in natural salt licks (Jones and Hanson, 1985; Freeland et al., 1985; Smedley and Eisner, 1995) and Na-rich plants (Botkin et al., 1973; Rothman et al., 2006). The home range of snowshoe hares in Alaska appears to include at least one area of barren rock, where the hares can obtain soil minerals (Kielland et al., 2019).

Weir (1972) suggested that the distribution of elephants in central Africa was at least partially dependent on sodium in seasonal waterholes, and McNaughton (1988) found that the abundance of ungulates in the Serengeti area was linked to Na, P, and Mg in plant tissues available for grazing. Thus, animal populations may be affected by the availability of Na in natural ecosystems. Aumann (1965) found high rodent populations in areas of Na-rich soils, and speculated that the increased abundance of rodents in the eastern United States during the 1930s might have been due to a large deposition of Na-rich soil dust that was derived

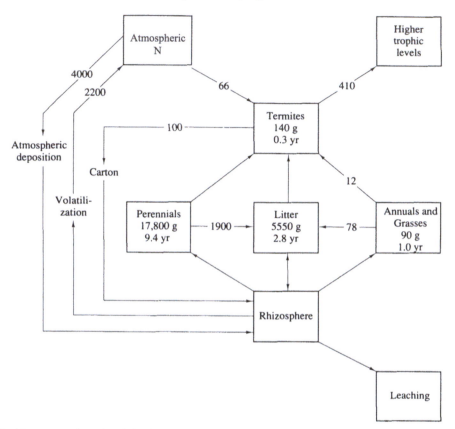

FIG. 6.23   Nitrogen cycle in the Chihuahuan Desert of New Mexico, showing the role of termites in nitrogen transformations. Flux of nitrogen is shown along arrows in g N/ha/yr; nitrogen pools are shown in boxes with turnover
time in years. *From Schaefer and Whitford (1981). Used with permission of Springer.*

from the prairies during the "Dust Bowl." Such a case would link the abundance of animals to
the biogeochemistry of soils and to soil erosion by wind in a distant region. Similarly, Kaspari
et al. (2014) suggest that the decomposition of litter by termites in tropical forests is mediated
by the availability of sodium, derived from atmospheric deposition, which decreases inland
from coastal areas.

Calculating landscape mass balance

Elements are retained in terrestrial ecosystems when they play a functional role in biochemistry or are incorporated into organic matter. In most instances, the pool of nutrients held
in the soil and vegetation is many times larger than the annual receipt of nutrients from the
atmosphere and rock weathering (e.g., Table 6.4). Often phosphorus has the longest mean

residence time among the nutrient elements in terrestrial ecosystems (Spohn and Sierra, 2018). In the Hubbard Brook Experimental Forest in New Hampshire, turnover times (mass/input) range from 21 years for Mg to >100 years for P in the vegetation and forest floor (Likens and Bormann, 1995; Yanai, 1992). In contrast, for a nonessential element, sodium (Na), the turnover time is rapid (1.2 years), because Na is not retained by biota or incorporated into humus.

Because chlorine is highly soluble, not strongly involved in soil chemical reactions (Chapter 4), and only a trace element in plant nutrition (White and Broadley, 2001), chloride (Cl^-) has traditionally been used as a tracer of hydrologic flux through ecosystems (Juang and Johnson, 1967). However, several studies (Oberg et al., 2005; Bastviken et al., 2007; Leri and Myneni, 2010, Table 4.5) show that some Cl is incorporated and retained in soil organic matter, partially compromising its use as a conservative tracer of geochemical processes, especially when the overall Cl flux is small (Svensson et al., 2012). Similarly, silicon (Si) is not an essential element for plants, yet a substantial amount of Si cycles as biogenic Si in terrestrial ecosystems (Sommer et al., 2013; Turpault et al., 2018), especially agricultural systems (Carey and Fulweiler, 2012). Some nonessential—even toxic—elements such as lead (Pb) that bind to organic matter may also accumulate in soils (Smith and Siccama, 1981; Friedland and Johnson, 1985; Kaste et al., 2005; Anastopoulos et al., 2019). Even though Pb is not involved in biochemistry, its retention in the ecosystem is the result of the presence of biotic processes. Thus, studies of the movement of Si, Cl, Pb, mercury (Hg), and other non-essential elements in the Earth's terrestrial ecosystems all fall into the realm of biogeochemistry.

Annual mineralization, plant uptake, and litterfall result in a large internal cycle of elements in most ecosystems. In areas not affected by air pollution, annual nitrogen inputs are typically 1–5kg/ha/yr, while mineralization of soil nitrogen is 50–100kg/ha/yr (Bowden, 1986). Despite such large movements of available nutrients within the ecosystem, there are usually only small losses of N in streams draining forested landscapes (~3kgN/ha/yr; Lewis, 2002; Alvarez-Cobelas et al., 2008). The minor loss of nitrogen in stream water speaks strongly for the efficiency of biological processes that retain elements essential to biochemistry. Tracing the dual isotopes in NO_3^-, i.e., $\delta^{15}N$ and $\delta^{18}O$, has proven useful in separating the source of nitrate in runoff—atmospheric inputs versus microbial nitrification (Kaneko and Poulson, 2013; Yu et al., 2016). Where plants are present, most nitrogen deposited from the atmosphere is taken up and recycled in the ecosystem (Durka et al., 1994), whereas in deserts, a substantial portion is lost (Michalski et al., 2004b).

Relatively few studies have included measurements of gaseous flux in nutrient budgets (Schlesinger, 2009). Losses of nitrogen in denitrification may explain why the retention of N applied in fertilizer is often somewhat lower than that of other elements (e.g., P and K), which have no gaseous phase (Stone and Kszystyniak, 1977). Mass-balance studies implicate large losses of N by denitrification to balance the ecosystem budget for the Hubbard Brook ecosystem in New England (Yanai et al., 2013; Morse et al., 2015). In many regions, soil nitrogen shows positive values for $\delta^{15}N$, reflecting losses of nitrogen by denitrification (Houlton et al., 2006). Outside of the highly weathered soils of the humid tropics, where net primary production is limited by P (Alvarez-Clare et al., 2013; Fisher et al., 2013), the growth of most vegetation worldwide is controlled by the availability of N (LeBauer and Treseder, 2008; Šímová et al., 2019). Globally, denitrification may explain the tendency for vegetation to be N-limited, despite efficient plant uptake of N from the soil and only minor losses in stream water (Houlton et al., 2006; Chapter 12).

Allan et al. (1993) compared the mass balance of elements in small patches of forest occupying rock outcrops in Ontario. Areas of bare rock showed net losses of various elements, whereas adjacent patches of forest showed accumulations of N, P, and Ca in vegetation. We should not, however, expect that the essential biological nutrients will accumulate indefinitely in all ecosystems. The incorporation of N and P in biomass should be greatest when structural biomass and soil organic matter are accumulating rapidly—that is, in young ecosystems where there is positive net ecosystem production (Chapter 5). Losses of N should be higher in mature, steady-state ecosystems where the total biomass is stable (Vitousek and Reiners, 1975; Davidson et al., 2007). The extent to which N is incorporated into biota may depend on its availability relative to other elements. For instance, lowland tropical rainforests forests, where vegetation growth is generally limited by P, appear to be "leaky" with respect to N relative to temperate forests (Martinelli et al., 1999; Brookshire et al., 2012).

Using the mass-balance approach, where

$$Input - Output = \Delta Storage, \tag{6.6}$$

Vitousek (1977) found greater losses of available N from old-growth forests than from younger sites in New Hampshire. Hedin et al. (1995) confirmed high nutrient losses in old-growth forests of Chile, with forms of dissolved organic nitrogen being an important fraction of the total loss (cf. Taylor et al., 2015). The relatively high losses of N and P from the Caura River in Venezuela (Table 6.9) are consistent with the mature vegetation covering most of its watershed (Lewis, 1986; cf. Davidson et al., 2007).

In seasonal climates, losses of N and K in stream waters are usually minor during the growing season and greater during periods of reduced biological activity in the soil (Likens et al., 1994; Goodale et al., 2015). Intact ecosystems have a large capacity to retain NO_3 as a result of microbial processes in soils (Nakagawa et al., 2013; Sudduth et al., 2013; Sabo et al., 2016). In contrast, often there is little seasonal variation in the loss of Na and Cl, which pass through the system under simple geochemical control (Johnson et al., 1969; Belillas and F. Rodà, 1991). Stream-water losses of nutrients give old-growth ecosystems

TABLE 6.9   Annual chemical budgets for undisturbed forests in various world regions.

Location and reference	Precipitation (cm)	Chemical (kg/ha/yr)			
		Ca	Cl	N	P
British Columbia (Feller and Kimmins, 1979)	240	15.8	2.9	−2.6	0
Oregon (Martin and Harr, 1988)	219	41.2	–	−1.2	0.3
New Hampshire (Likens and Bormann, 1995)	130	11.7	−1.6	−16.7	0
North Carolina (Swank and Douglass, 1977)	185	3.9	1.7	−5.5	−0.1
Venezuela (Lewis et al., 1987; Lewis, 1988)	450	14.2	−1.4	8.5	0.32
Brazil (Lesack and Melack, 1996)	240	−0.52	3.58	−2.4	−0.04

Note: *Total stream-water losses minus atmospheric deposition.*

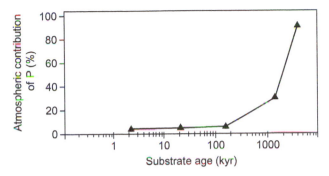

FIG. 6.24   The relative importance of atmospheric inputs of phosphorus to ecosystems on the Hawaiian islands as a function of the age of the landscape. *From Chadwick et al. (1999).*

the appearance of being "leaky," but it is important to recognize that outputs represent the excess of inputs over the seasonal demand for nutrients by vegetation and soil microbes (Gorham et al., 1979).

During long periods of soil development, chemical weathering depletes soils of the essential nutrients traditionally thought to be derived from bedrock, especially Ca and P (Fig. 4.11). Thus, studying a 4-million-year-old sequence of soils in Hawaii, Chadwick et al. (1999) showed that small atmospheric inputs of phosphorus assume great importance to vegetation growing on ancient soils (Fig. 6.24). In this humid, tropical climate, vegetation progressively shifts from N limitation to P limitation with soil age (Vitousek, 2004; Richardson et al., 2004). It is widely believed that forests of the humid tropics are limited by phosphorus, whereas those elsewhere are more likely to be limited by N. Nevertheless, among 48 fertilization experiments, Wright (2019) found that many tropical forests were equally likely to respond to additions of N and P.

Atmospheric inputs of phosphorus from the long-distance transport of desert dust seem essential to the continued productivity of tropical rainforests on highly weathered soils in Hawaii, the Caribean, and the Amazon Basin (Gardner, 1990; Okin et al., 2004; Bristow et al., 2010; Yu et al., 2015; Dessert et al., 2020; Gallardo et al., 2020). Using strontium (Sr) as a tracer, some workers have suggested that the vegetation in many regions is dependent on atmospheric inputs of elements traditionally associated with rock weathering (Graustein and Armstrong, 1983; Miller et al., 1993; Kennedy et al., 1998). Similarly, in a forest in Sri Lanka, Mg appears to be derived largely from atmospheric inputs and not from chemical weathering of the underlying bedrock (Schuessler et al., 2018).

Generally, ecosystems shift from N limitation to P-limitation with age and soil development (Huston, 2012; Augusto et al., 2017), and the long-term productivity of terrestrial vegetation may depend on periodic renewal of weatherable minerals (Wardle et al., 2004). In areas of rapid geologic uplift and erosion, weathering is a persistent source of phosphorus and other plant nutrients that are derived from bedrock (Porder et al., 2006; Eger et al., 2018; Chadwick and Asner, 2018). Thus, biogeochemists must include both atmospheric and bedrock sources in ecosystem nutrient budgets, especially when these budgets are used to evaluate the impacts of changing levels of air pollution and atmospheric deposition on forest growth (e.g., Drouet et al., 2005; Mitchell et al., 2011a).

Among elements in short supply to biota, nitrogen is unique in that it is largely derived from the atmosphere (Table 4.5).[n] Net primary production appears to show a correlation to N inputs in precipitation in temperate forests (Cole and Rapp, 1981) and grasslands (Stevens et al., 2015). Comparing forests from Oregon, Tennessee, and North Carolina, Henderson et al. (1978) noted strong N retention in each, despite a tenfold difference in N input from the atmosphere. The data suggest that plant growth is limited by N in each region. In contrast, losses of Ca were always a large percentage of the amount cycling in these forests. Especially on limestone soils, ample supplies of Ca were derived from rock weathering and Ca was not in short supply. Thus, abundant (e.g., Ca) and nonessential (e.g., Na) elements are most useful in estimating the rate of rock weathering (Chapter 4), whereas biogeochemistry controls the loss of elements that are essential to life. Desertification derived from increasing aridity exacerbates the losses of C and N from desert soils, uncoupling their stoichiometric linkage to P as organic matter is lost (Delgado-Baquerizo et al., 2013).

While most studies of ecosystem mass balance have considered watersheds, Baker et al. (2001) developed a nitrogen budget for the metropolitan ecosystem of Phoenix, Arizona, in which anthropogenic inputs (from food and pet food, combustion, and fertilizer) and gaseous outputs (NO_x and N_2) composed the largest movements of nitrogen (Table 6.10). Metson et al. (2012b) formulated a similar assessment of the phosphorus budget for Phoenix. A nitrogen budget for the city of Paris (Fig. 6.25) shows the low efficiency by which nitrogen applied in farmland fertilizers is incorporated into the food eaten by Parisians. The ecosystem concept can even be applied to individual households to produce biogeochemical budgets for their inputs and outputs (Fissore et al., 2011). At the other extreme, Cui et al. (2013) compiled an input-output budget for nitrogen in all of mainland China. They found that chemical fertilizers, N fixation, and precipitation dominated the sources of nitrogen in this region, while denitrification, ammonia volatilization, and hydrologic export dominated the losses. Similar regional budgets for nitrogen have been developed for African tropical forests (Bauters et al., 2019) and the grassland steppe of China (Giese et al., 2013).

Many of the transformations in biochemistry involve oxidation and reduction reactions that generate or consume acidity (H^+). For instance, H^+ is produced during nitrification and consumed in the plant uptake and reduction of NO_3^-. Binkley and Richter (1987) review these processes and show how ecosystem budgets for H^+ may be useful as an index of net change in ecosystem function, particularly as soils acidify during ecosystem development (Chapter 4). H^+-ion budgets are also useful as an index of human impact, especially from acid rain and excess nitrogen deposition (Driscoll and Likens, 1982). For example, a net increase in acidity is expected when excess NH_4^+ deposition is subject to nitrification, with the subsequent loss of NO_3^- in stream water (van Breemen et al., 1982). H^+ budgets are analogous to measurements of human body temperature. When we see a change, we suspect that the ecosystem is stressed, but we must look carefully within the system for the actual diagnosis.

---

[n]Sedimentary and metasedimentary rocks usually contain a small amount of nitrogen, which can be released on weathering (Holloway and Dahlgren, 2002). Nitrogen derived from bedrock can make a significant contribution to the nitrogen budget of some forests (Morford et al., 2011) and potentially to the global sources of nitrogen for land plants (Chapter 12).

TABLE 6.10   Nitrogen budget for the Phoenix
metropolitan area.

Inputs	N flux (G g/y)
Surface water	1.2
Wet deposition	3.0
Human food	9.9
N-containing chemicals	5.8
Food for dairy cows	0.8
Pet food	2.7
Commercial fertilizer	24.3
Biological fixation	
Alfalfa	7.5
Desert plants	7.1
Fixation by combustion	36.3
Total inputs of fixed N	**98.4**
Outputs	
Surface water	2.6
Cows for slaughter	0.1
Milk	2.4
Atmospheric NO_x	17.1
Atmospheric NH_3	3.8
N_2O from denitrification	4.6
N_2 from denitrification	46.9
Total outputs	**77.5**
Accumulation (inputs-outputs)	20.9
Net subsurface storage	8.3
Landfills	8.6
Increase in human biomass	0.2

Modified from Baker et al. (2001).

Ecosystem models and remote sensing

Various models have been developed to show links between plant and soil processes in terrestrial biogeochemistry. Walker and Adams (1958) suggested that the level of available phosphorus during soil development was the primary determinant of terrestrial net primary production, since nitrogen-fixing bacteria depend on a supply of organic carbon and available

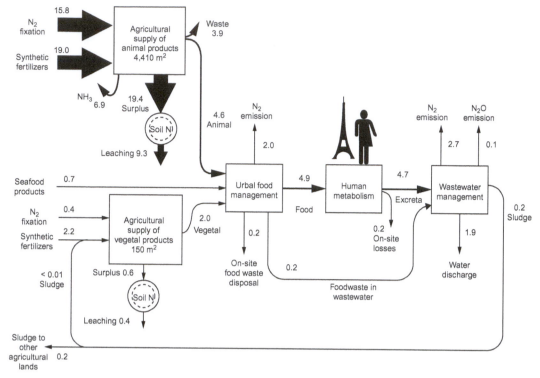

FIG. 6.25   The flow of nitrogen from farmlands to and through the city of Paris, France, in units of kilograms per person per year. *From Esculier et al. (2019).*

phosphorus. They used the level of organic carbon in the soil as an index of terrestrial productivity and suggested that organic carbon peaks midway during soil development and then declines as an increasing fraction of the phosphorus is rendered unavailable by precipitation with secondary minerals (Fig. 4.11). The model is consistent with observations of the increasing limitation of NPP by phosphorus during soil development in many areas, especially the humid tropics (Chadwick et al., 1999; Richardson et al., 2004).

Numerous workers have examined the Walker and Adams (1958) hypothesis in various ecosystems. Tiessen et al. (1984) found that available phosphorus explained 24% of the variability of organic carbon in a collection of 168 soils from eight different soil orders. Roberts et al. (1985) found a similar relationship between bicarbonate-extractable P and organic carbon in several grassland soils of Saskatchewan. Raghubanshi (1992) found that phosphorus was well correlated to soil organic matter, soil nitrogen, and nitrogen mineralization rates in dry tropical forests of India. Thus, available phosphorus explains some, but not all, of the variation in soil organic carbon, which is ultimately derived from the production of vegetation. The linkage of phosphorus and carbon is likely to be strongest during early soil development, when both organic phosphorus and carbon are accumulating. The importance of organic phosphorus increases during soil development, and through the release of phosphatase enzymes, vegetation interacts with this soil pool to control the mineralization of P.

About 30 years ago, Parton et al. (1988) began to develop a model linking the cycling of C, N, P, and S in grassland ecosystems. The model, known as CENTURY, has been widely used as a tool to predict long-term changes in various ecosystems. The flow of carbon is shown in Fig. 6.26. The nitrogen cycling submodel has a similar structure, since the model assumes that most nitrogen is bonded directly to carbon in amino groups (McGill and Cole, 1981).

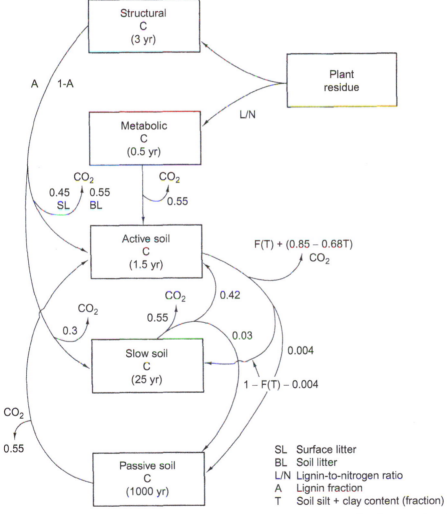

FIG. 6.26 Flow diagram for carbon in the CENTURY model. The proportion of carbon moving along each flowpath is shown as a fraction, and turnover times for reservoirs are shown in parentheses. *From Parton et al. (1988). Used with permission of Springer.*

Phosphorus availability is controlled by a modification of a model first presented by Cole et al. (1977b), which includes C/P control over mineralization of organic pools and geochemical control over the availability of inorganic forms as in Fig. 6.22 (Parton et al., 1988). More recently, the DAYCENT model was developed to predict ecosystem flux for shorter periods, such as the loss of N_2O from fertilized soils (Parton et al., 2001).

In CENTURY, lignin controls decomposition rates, and nitrogen is mineralized from soil pools when critical C/N ratios are achieved during the respiration of carbon. However, unlike nitrogen, C/P ratios in plant tissues and soil organic matter are allowed to vary widely as a function of P availability. When CENTURY is used to predict patterns of primary production and nutrient mineralizations during 10,000 years of soil development, net primary production and accumulations of soil organic matter are strongly linked to P availability during the first 800 years, after which increases in plant production are related to increases in soil N mineralization (Fig. 6.27). Organic P increases throughout the 10,000-year sequence. In simulations of the response of native soils to cultivation, the model predicted a correlated decline in the native levels of organic carbon and nitrogen in the soil, but a relatively small decline in P. Validation of the model is seen in the data of Tiessen et al. (1982), who found declines of 51% for C and 44% for N, but only 30% for P in a silt loam soil cultivated for 90 years in Saskatchewan.

Linkages among components of the intrasystem cycle suggest that an integrative index of terrestrial biogeochemistry might be derived from the measure of a single component, such as the chemical characteristics of the leaf canopy (Matson et al., 1994). Wessman et al. (1988) analyzed the spectral reflectance of leaf tissues in the laboratory as a first step toward developing an index of forest canopies by remote sensing. Their data show a strong correlation between nitrogen and lignin measured by infrared reflectance and by traditional laboratory analyses of leaf nutrient content. The spectral reflectance characteristics of many species have now been related to properties of their foliage (Serbin et al., 2014), and several studies have used satellite measurements of reflectance to characterize canopy properties (Martin and Aber, 1997; Asner and Vitousek, 2005; Kokaly et al., 2009; Ollinger, 2011).

In the White Mountains of New Hampshire, forest productivity appears related to canopy nitrogen content, as measured by remote sensing (Fig. 6.28; Ollinger and Smith, 2005; Ollinger et al., 2008). Variations in canopy nitrogen are also related to evapotranspiration and water-use efficiency (Guerrieri et al., 2016). Recognizing that decomposition is frequently controlled by the lignin, nitrogen content, and C/N ratio of litter (refer to Fig. 6.14), remote sensing of canopy characteristics has potential for comparative regional studies of nutrient cycling (Myrold et al., 1989; Ollinger et al., 2002) and for a mechanistic basis in models such as CENTURY. Canopy lignin, measured by aircraft remote sensing, was highly correlated to soil nitrogen mineralization in Wisconsin forests (Wessman et al., 1988; Pastor et al., 1984). Osborne et al. (2017) used measures of canopy nitrogen from remote sensing to differentiate patterns of nutrient cycling under different species and different landscape positions in the Osa Peninsula of Costa Rica. Canopy nitrogen is correlated to emissions of N_2O in tropical rainforests (Soper et al., 2018b). These studies reinforce our appreciation of the linkage between vegetation and soil characteristics.

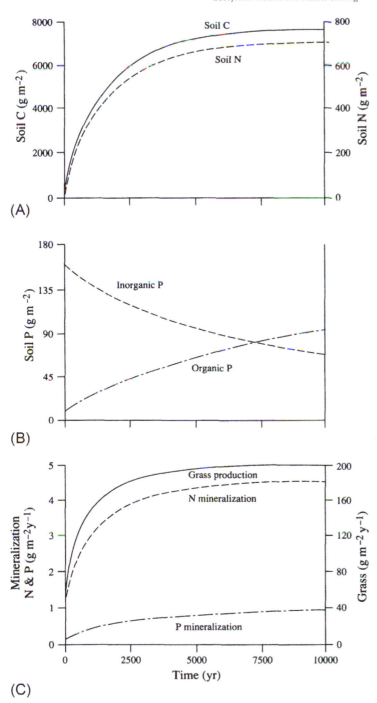

FIG. 6.27 Simulated changes in soil C, N, and P during 10,000 years of soil development in a grassland, using the CENTURY model. *Source: Parton et al. (1988). Used with permission of Springer.*

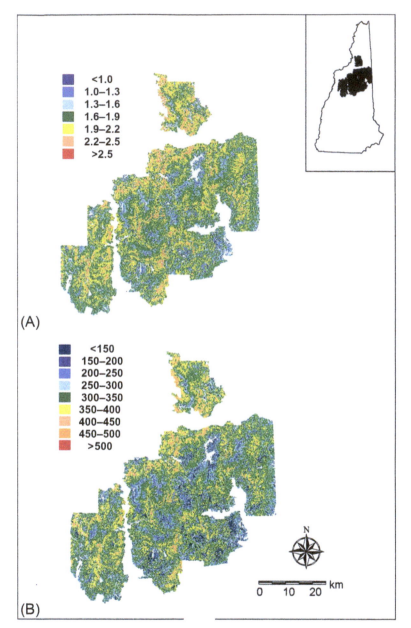

FIG. 6.28 Spatial variation of canopy nitrogen and forest wood production in central New Hampshire, as estimated from the AVIRIS satellite. (A) Whole-canopy nitrogen concentration (%). (B) Aboveground wood production $(g/M^2/yr)$. *From Smith et al. (2002). Used with permission of the Ecological Society of America.*

Human impacts on terrestrial biogeochemistry

Acid rain

Forest growth has declined in many areas that are downwind of air pollution (Savva and Berninger, 2010). In addition to the direct effects of ozone and other gaseous pollutants on their growth, plants in these areas are subject to "acid rain"—perhaps better named *acid deposition*, since some of the acidity is delivered as dryfall (Chapter 3). While all rain is naturally somewhat acidic (Chapter 13), human activities can produce rain of exceptional low pH as a result of NO_3^- and SO_4^{2-} that are derived from the dissolution of gaseous pollutants in raindrops (Eqs. 3.26–3.29). The chemical inputs in acid rain affect several aspects of soil chemistry and plant nutrition, leading to changes in plant growth rates.

Inputs of H^+ in acid rain increase the rate of weathering of soil minerals, the release of cations from cation exchange sites, and the movement of Al^{3+} into the soil solution (Chapter 4). At the Rothamsted Experimental Station in the United Kingdom, soil pH declined from 6.2 to 3.8 between 1883 and 1991, in association with acid rainfall (Blake et al., 1999). Similar declines in soil pH were observed under forests in Russia (Lapenis et al., 2004) and China (Yang et al., 2015). Depending on the underlying parent rocks, the forest floor and soil exchange capacity may be substantially depleted of Ca^{2+} in areas of acid deposition (Miller et al., 1993; Wright et al., 1994; Likens et al., 1996; Johnson et al., 2008a). Forests in the Adirondack Mountains of New York lost 64% of their soil Ca between 1930 and 2006 (Bedison and Johnson, 2010). Graveland et al. (1994) suggest that the effects of acid rain are also seen at higher trophic levels; in the Netherlands, birds showed poor reproduction in forests subject to acid rain as a result of a decline in the abundance of snails, which are the main source of Ca for eggshell development. Similar effects are potentially related to the decline of the Wood Thrush and Ovenbird in eastern North America (Hames et al., 2002; Pabian and Brittingham, 2007).

Field experiments simulating acid rain show the depletion of cations on soil cation exchange sites and mobilization of Al^{+3} (Fernandez et al., 2003). Between 1984 and 2001 the loss of Ca from soils in the northeastern United States was closely balanced with an increase in Al on the cation exchange sites (Warby et al., 2009). High concentrations of Al^{3+} may reduce the plant uptake of Ca^{2+} and other cations (Godbold et al., 1988; Bondietti et al., 1989), and in the northeastern United States forest growth appears to decline as a result of a decreased Ca/Al ratio in the soil solution (Shortle and Smith, 1988; Cronan and Grigal, 1995). Bernier and Brazeau (1988a,b) link dieback of sugar maple to deficiencies of K on areas of low-K rocks and to deficiencies of Mg on low-Mg granites in southeastern Quebec. Magnesium deficiencies are also seen in the forests of central Europe, where forest decline is linked to an imbalance in the supply of Mg and N to plants (Oren et al., 1988; Berger and Glatzel, 1994). In a spruce forest of New England, Bullen and Bailey (2005) document decreasing Ca, Sr, and other cations and increasing Al in tree rings during the past century of acid inputs. Losses of Al from the mineral soil horizons are also associated with the mobilization of Al-bound phosphorus (SanClements et al., 2010).

With an abatement of air pollution, the acidity of rainfall has declined in many areas of Europe and the eastern United States (Fig. 3.15), commencing a slow recovery of soils and vegetation, especially in sites where Ca has been strongly depleted from the soil

(Palmer et al., 2004). Soil pH increased in England and Wales between 1978 and 2003, presumably as a result of lower emissions of sulfur dioxide and lower levels of acid rain (Kirk et al., 2010). Atmospheric inputs of Ca in dryfall can assume special significance in the rejuvenation of soil cation exchange capacity (Drouet et al., 2005). When researchers added 1.2 tons Ca per hectare as the mineral wollastonite ($CaSiO_3$) to the forest at Hubbard Brook, New Hampshire, the addition restored soil Ca (Cho et al., 2010), ameliorated the effects of acidity on the growth of sugar maple (Juice et al., 2006), and increased soil nitrogen turnover (Rosi-Marshall et al., 2016; Marinos et al., 2018).

It is important to note that soil acidity can derive from a number of causes. During forest growth, the accumulation of cations in biomass leads to greater soil acidity (Berthrong et al., 2009). For a U.S. forest in South Carolina, Markewitz et al. (1998) attribute 62% of soil acidity to plant uptake and 38% to atmospheric deposition. Permanent increases in soil acidity will result if the new inputs of cations during a forest growth cycle are less than the removals in harvest. Similarly, the use of NH_4^+ fertilizers, which generate H^+ during nitrification, appears responsible for the acidification of agricultural soils in China (Guo et al., 2010).

Nitrogen saturation

Currently, the deposition of available nitrogen from the atmosphere in the northcentral United States and western Europe ($\sim 8\,kg\,N/ha/yr$) is about four times greater than that recorded under pristine conditions.[o] Elevated nitrogen deposition is also reported in China (Tian et al., 2018). The excess nitrogen derives from combustion of fossil fuels (releasing NO_x) and agricultural activities (releasing NH_3) upwind (Fang et al., 2011; Li et al., 2016c). Many workers have speculated that excess nitrogen deposition may act as fertilizer—stimulating the growth of trees. Nitrogen deposition also enhances net primary production in grasslands (Stevens et al., 2015). In forests, the added nitrogen could lead to significant enhanced growth and carbon storage, enhancing the removal of CO_2 from the atmosphere (Thomas et al., 2010; Magnani et al., 2007). In contrast, some field observations have shown relatively small effects of nitrogen deposition on forest net primary productivity (Lovett et al., 2013). Surprisingly, even though most forests are nitrogen-limited, plant uptake of exogenous nitrogen is usually only 10–30% of that applied (Schlesinger, 2009; Pregitzer et al., 2010; Templer et al., 2012).

Fertilizer experiments show that some of the added nitrogen, especially NH_4^+, accumulates in soil organic matter (Nave et al., 2009; Gardner and Drinkwater, 2009; Liu et al., 2017b), enhancing carbon storage in soils (Nadelhoffer et al., 2004; Pregitzer et al., 2008; Hyvonen et al., 2008; Liu and Greaver, 2010). Excess nitrogen deposition from the atmosphere retards the decomposition of organic matter, which increases in the soil (Lovett et al., 2013; Maaroufi et al., 2015; Frey et al., 2014; Tian et al., 2019). This effect stems from a reduction in the abundance of lignolytic fungi (Entwistle et al., 2018; Xia et al., 2017; Wang et al., 2016; Zak et al., 2019). The overall effect of N deposition is to enhance carbon sequestration in terrestrial ecosystems, largely due to greater accumulations of soil organic matter (Chapter 11).

In some areas of high nitrogen deposition, particularly at high elevations, forest decline is observed as the ecosystem becomes saturated with nitrogen (Aber et al., 1998, 2003;

[o]See, for example, http://nadp.slh.wisc.edu/maplib/pdf/2017/N_dep_2017.pdf.

McNulty et al., 2005; Lovett and Goodale, 2011). In these sites, nitrification rates increase dramatically, yielding higher losses of NO_3^- in stream waters (Peterjohn et al., 1996; Corre et al., 2003; Lu et al., 2011; Liu et al., 2017b) and greater emissions of N_2O to the atmosphere (Peterjohn et al., 1998; Venterea et al., 2003; Zhu et al., 2013). Losses of nitrogen to stream waters are normally dominated by dissolved organic nitrogen (DON), but in areas of excessive nitrogen deposition, NO_3^- becomes increasingly dominant (Perakis and Hedin, 2002; Lovett et al., 2000; Lutz et al., 2011). Forests receiving high nitrogen deposition show higher $\delta^{15}N$ in canopy foliage, indicative of a high rate of nitrification and high NO_3^- losses in streams (Pardo et al., 2007).

Along three gradients of increasing air pollution in southern California, Zinke (1980) showed that N content in the foliage of Douglas fir increased from 1% to more than 2%, while P content decreased abruptly, changing the ratios of N to P from about 7 in relatively pristine areas to 20–30 in polluted areas. Such an imbalance in leaf N/P ratios is also seen in *Pinus sylvestrris* across much of Europe in areas of excessive inputs of NH_4^+ from the atmosphere (Sardans and Peñuelas, 2015). Historical collections show increasing nitrogen concentrations in some plants during the past century (Peñuelas and Filella, 2001), perhaps indicating a shift of the terrestrial biosphere away from N deficiency (Elser et al., 2007). Nevertheless, in areas of high N deposition, forests show only scattered evidence of P deficiency (Finzi, 2009; Goswami et al., 2018; Gonzales and Yanai, 2019).

The symptoms of nitrogen saturation vary as a function of underlying site fertility, species composition, and other factors. Lovett and Goodale (2011) stress the importance of the rate of N input to the rate of N uptake by plants and soil microbes in controlling the appearance of nitrogen saturation and enhanced N loss. Low fertility sites may show only small changes in nitrification because plants take up the excess N deposition from the atmosphere (Fenn et al., 1998; Lovett et al., 2000). Without specific field experiments, it is often difficult to separate the effects of acid rain from those of excess nitrogen, since a large fraction of the nitrogen deposited from the atmosphere arrives as nitric acid, and inputs of NH_4^+ generate acidity if they are nitrified (Stevens et al., 2011). Nitrogen additions stimulated the loss of soil Ca in forests in Oregon (Hynicka et al., 2016). It is likely that the response of tropical forests to excessive nitrogen deposition will differ from observations in temperate regions, as a result of soil P limitation in tropical soils (Corre et al., 2010; Cusack et al., 2011; Hietz et al., 2011).

Nitrogen saturation is reversible. With experimental reductions of nitrogen inputs in areas of high deposition, the rates of nitrification and the loss of NO_3^- to stream waters decline (Quist et al., 1999; Corre and Lamersdorf, 2004; Lutz et al., 2012b).

Rising CO_2 and global warming

Rising concentrations of CO_2 in Earth's atmosphere appear to stimulate the growth and carbon storage of land plants by enhancing plant photosynthesis (Chapter 5). Early greenhouse studies suggested that this response might be short-lived because of soil nutrient limitations (Thomas et al., 1994). Several workers postulated a progressive nutrient limitation of field plants grown at high CO_2 (Luo et al., 2004); however, some experiments that exposed intact forests to high CO_2 show a positive response to CO_2 that lasts up to a decade (Finzi et al., 2006). Some of the greater nutrient demand by faster growing plants is met by greater nutrient-use efficiency in photosynthesis (Springer et al., 2005), greater nutrient reabsorption before leaf abscission (Finzi et al., 2002; Norby et al., 2010), and greater allocation of carbon to root exudates that stimulate the decomposition of soil organic matter and nutrient mineralization (Drake et al., 2011; Phillips et al., 2011b).

Plants also respond to elevated CO_2 with greater root growth and mycorrhizae, which appear to explore the soil nutrient pool more fully (Norby and Iversen, 2006; Pritchard et al., 2008b; Finzi et al., 2007). Pines with ectomycorrhizal fungi show a positive response to CO_2-fertilization in N-poor soils (Terrer et al., 2016). The duration of the positive growth response of plants to high CO_2 in field experiments is surprising; eventually, stoichiometric constraints, such as the C/N ratio in plant biomass, will limit the amount of carbon that can be sequestered in woody biomass and soils in the absence of exogenous inputs of N (Johnson, 2006; van Groenigen et al., 2006; Wieder et al., 2015). Indeed, nitrogen appears to constrain the long-term growth response of the deciduous forest at the FACE experiment at Oak Ridge, Tennessee (Norby et al., 2010; Garten Jr et al., 2011).

Soil-warming experiments, designed to simulate ongoing climate change, typically show an increase in soil nitrogen mineralization (Van Cleve et al., 1990; Rustad et al., 2001; Shaw and Harte, 2001; Melillo et al., 2002; Bai et al., 2013). The change in soil microbial activity mobilizes nitrogen for plant uptake, potentially enhancing plant growth and carbon uptake. In wet tundra, soil warming stimulated the rate of decomposition, but caused only a small increase in plant growth in field experiments (Johnson et al., 2000b; Mack et al., 2004; Shaver et al., 2006). In contrast, in temperate forest ecosystems, soil warming stimulates nitrogen mineralization, plant carbon uptake, and net carbon sequestration in aboveground tissues (Melillo et al., 2011). In some areas, where the loss of the insulating effect of a winter snow pack results in frozen soils, nitrogen mineralization rates are likely to decline under warmer, future climatic conditions (Groffman et al., 2009).

Summary

Interactions between plants, animals, and soil microbes link the internal biogeochemistry of terrestrial ecosystems. Plants adapted to low nutrient availability have low nutrient contents and higher nutrient reabsorption before leaf-fall, yielding higher nutrient-use efficiency (Fig. 6.29). In some cases these characteristics can be induced by experimental treatments that reduce nutrient availability. For instance, when Douglas fir were fertilized with sugar, which increased the C/N ratio of the soil and the immobilization of N by microbes, reabsorption of foliar N increased, implying greater nutrient-use efficiency by the trees (Turner and Olson, 1976). Internal cycling by the vegetation may partially alleviate nutrient deficiencies, but decomposition of nutrient-poor litterfall is slow, further exacerbating the availability of nutrients in the soil (Hobbie, 1992; Lovett et al., 2004). Thus, nutrient-poor sites are likely to be occupied by vegetation that is specially adapted for long-term persistence under such conditions (Chapin et al., 1986b). In turn, the vegetation leaves its imprint on microbial activity and soil properties (Lovett et al., 2004; Reich et al., 2005).

Biogeochemistry controls the distribution and characteristics of vegetation at varying scales. Whereas nitrogen limits the net primary productivity of vegetation in most regions,

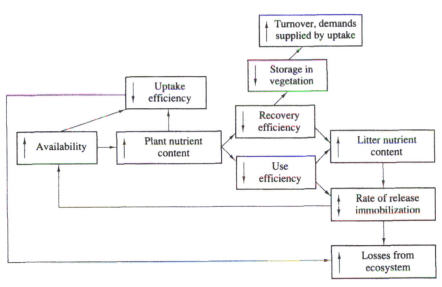

FIG. 6.29   Changes in internal nutrient cycling that are expected with changes in nutrient availability. *From Shaver and Melillo (1984). Used with permission of the Ecological Society of America.*

the humid tropics are often limited by available phosphorus, and regional vegetation shows a variety of adaptations to conserve and recycle P. Continental distributions of vegetation, such as the widespread dominance of conifers in the boreal regions, are likely to be related to the higher nutrient-use efficiency of evergreen vegetation under conditions of limited nutrient turnover in the soil. The effect of soil properties on the distribution of vegetation is also seen in the occurrence of evergreen vegetation on nutrient-poor, hydrothermally altered soils in arid and semiarid climates (Fig. 6.30). Fine-scale spatial heterogeneity of soil properties, as

FIG. 6.30 Occurrence of *Pinus ponderosa and Pinus jeffreyi* on acid, nutrient-poor hydrothermally altered andesites in the Great Basin Desert of Nevada, with *Artemisia tridentata* occurring on adjacent desert soils, of higher pH and phosphorus availability. *Sources: Schlesinger et al. (1989) and Gallardo and Schlesinger (1996).*

recorded by Robertson et al. (1988) for a field in Michigan, determines the diversity of land plant communities (Tilman, 1985), and several studies show the importance of local soil conditions to the distribution and abundance of forest and grassland herbs (Snaydon, 1962; Pigott and Taylor, 1964; Lechowicz and Bell, 1991; John et al., 2007). Additions of fertilizer tend to reduce the species diversity of plant communities (Huenneke et al., 1990; Stevens et al., 2006; Cleland and Stanley Harpole, 2010); thus, perturbations of biogeochemical cycling have a direct impact on the preservation of biodiversity.

Introduction

Aquatic and terrestrial ecosystems differ in the relative importance of hydrology as a critical factor in their biogeochemistry. In terrestrial ecosystems water may directly limit autotrophic or heterotrophic activity, especially during droughts. In contrast, all wetlands have water at or near the surface for at least some portion of the year; as a result they have hydric soils.[a] Hydrophytic (water-loving) plants capable of living in saturated soils dominate wetland vegetation.

[a] Soils that are formed under conditions of periodic or continuous saturation sufficient to develop anoxic conditions in the upper horizons.

By definition water is seldom in short supply in wetlands, but hydrology plays a fundamental role in these ecosystems. First, because oxygen diffuses 10^4 more slowly in water than it does in air, water indirectly limits biogeochemical activity in wetlands by constraining oxygen supply. As a result there are many places in aquatic ecosystems where oxygen consumption exceeds rates of delivery, so anoxia is the typical condition of most sediments in wetlands, lakes, and streams. The plants and microbes controlling the biogeochemistry of aquatic ecosystems must cope with limited oxygen supplies. Second, because of their low topographic positions where surface waters collect or groundwaters emerge, aquatic ecosystems receive substantial inputs of organic materials and minerals from the surrounding terrestrial catchments.

The importance of these subsidies from terrestrial ecosystems depends to a great extent on the ratio of shoreline to the volume of the ecosystem. In many aquatic ecosystems these *allochthonous* (i.e., externally derived) inputs of energy and elements can exceed *autochthonous* (in situ) inputs by photosynthesis, such that many (perhaps most) aquatic ecosystems are net heterotrophic, or reliant on surrounding terrestrial ecosystems to provide the majority of their annual supply of organic matter and essential elements.

For many aquatic ecosystems, ecosystem boundaries can be difficult to assess on the ground, and all estimates of wetland area are fraught with uncertainty. Even delineating wetland boundaries in the field can be challenging, requiring the identification of hydric soils or hydrophytic plants in the absence of detailed water elevation data. Small wonder then that attempts to map the global wetland area using remote sensing have considerable uncertainty. A recent synthesis of the literature reported estimates that range from 3.6 to $17.3 \times 10^6 \, km^2$, which represent anywhere from 2.4% to 11.6% of the land surface area (Davidson et al., 2018). Such mapping efforts are complicated by the fact that traditional remote sensing efforts are limited to detecting open water surfaces (Pekel et al., 2016) and cannot detect inundated wetlands under thick vegetation (i.e., forested wetlands or swamps), and non-inundated wetlands (Döll et al., 2019). The most widely used estimate is that of Lehner and Doll (2004), who mapped a global wetland area of 9 ± 1 million km^2 (6.2–7.6% of global land surface excluding Antarctica and Greenland). We will use that estimate here, while acknowledging that more sophisticated multispectral remote sensing approaches are likely to expand this estimate in the near future (Mahdavi et al., 2018). The best current global estimates of lakes, rivers, and wetlands are shown in Fig. 7.1 and Table 7.1 (Döll et al., 2019).

Despite their relatively small proportional area, wetlands play an important role in global carbon cycling. By some accounts, wetlands have the highest average productivity of any ecosystem type ($1300 \, g \, C \, m^{-2} \, yr^{-1}$) (Houghton and Skole, 1990). Wetlands contribute 7–15% of global terrestrial productivity, and collectively store more than half of all the soil carbon on Earth (peatlands alone are estimated to store >50% of the world's soil carbon) (Gorham, 1991; Eswaran et al., 1993; Roulet, 2000; Tarnocai et al., 2009), with (Yu, 2012) estimating that northern peatlands store $500 \pm 100 \, Gt$ of C. Wetlands are important sources of dissolved organic matter (and organic nutrients) to downstream and coastal ecosystems (Schiff et al., 1998; Pellerin et al., 2004; Harrison et al., 2005). The importance of wetlands as a global soil carbon sink is considerably offset by their production of the methane for which wetlands are the dominant natural source, contributing more than 35% of total global emissions (refer to Table 11.2; Saunois et al., 2016; Bousquet et al., 2006; Bloom et al., 2010; Ringeval et al., 2010).

Wetlands support ideal conditions for the removal of reactive nitrogen by denitrification (Jordan et al., 2011) and for the sequestration of nitrogen and phosphorus in sediments

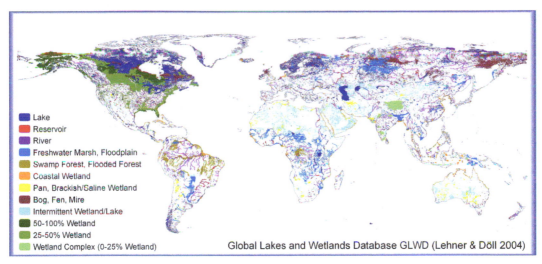

FIG. 7.1   The distribution of global wetlands. *From Lehner and Doll (2004).*

TABLE 7.1   Estimated global spatial extent of inland waters.

Class	Global area	
	$10^3 \, km^2$	%
1. Lake	2428	1.8
2. Reservoir	251	0.2
3. River	360	0.3
4. Freshwater marsh, floodplain	2529	1.9
5. Swamp forest, flooded forest	1165	0.9
6. Coastal wetland	660	0.5
7. Pan, brackish/saline wetland	435	0.3
8. Bog, fen, mire	708	0.5
9. Intermittent wetland/lake	690	0.5
10. Wetland complexes		
50–100% wetland	882–1764	0.7–1.3
35–50% wetland	790–1580	0.6–1.2
0–25% wetland	0–228	0–0.2
Total lakes and reservoirs (1–3)	2679	2.0
Total wetlands (4–10)	8219–10,119	6.2–7.6

Data from Lehner and Doll (2004).
In these analyses, they assumed a total global land surface area (excluding Antarctica and glaciated Greenland) of 133 million km².

(Reddy et al., 1999). Despite the high potential for denitrification in wetlands, there are no good estimates of their contribution to global N_2O emission, and it has generally been assumed that N_2O production is much lower in wetlands than in upland soils (Bridgham et al., 2006; Schlesinger, 2009; Mitsch and Gosselink, 2007).

The heightened capacity for denitrification, CH_4 production, and soil carbon storage in wetlands are all a result of a lack of oxygen in wetland sediments. While aerobic oxidation ($CH_2O + O_2 \rightarrow CO_2 + H_2O$) dominates decomposition in most terrestrial ecosystems, microbes in flooded soils must use a variety of anaerobic metabolic pathways to obtain energy from organic matter. Microbial consumption of oxygen in wet soils and sediments frequently exceeds O_2 supply through diffusion. Without oxygen, microbes cannot use oxidative phosphorylation to decompose organic polymers to CO_2, and instead must rely on the alternate electron acceptors NO_3^-, Fe^{3+}, Mn^{4+}, and SO_4^{2-} or, in the most highly reducing environments, fermentation products such as acetate or CO_2 itself. Many of these metabolic pathways evolved prior to the oxygenation of the Earth (Chapter 2), and continue to dominate the biogeochemistry of anoxic wetland sediments. These pathways yield less energy than aerobic respiration and the supply of alternate electron acceptors is often limiting, leading to far less efficient decomposition in wetlands and ultimately to large stores of organic matter. Over geologic time, the organic detritus accumulated in wetlands of the Carboniferous was buried and lithified to become modern coal deposits (Cross and Phillips, 1990; McCabe, 2009).

Types of wetlands

The great variety of wetland types are classified using hydrologic and physical properties that determine how wetlands influence the form, timing, and magnitude of chemical exports from catchments (Brinson, 1993). Among the most important factors are water residence time, the degree of hydrologic connectivity between the wetland and regional rivers or groundwater, and the frequency, intensity, and duration of inundation. Wetlands may also be characterized by their soils and vegetation, which both respond to and exert control over wetland hydrology.

Wetland hydrology

Water may enter a wetland by precipitation, tributary inflows, near-surface seepage, and exchange with deeper groundwater; water leaves wetlands through groundwater recharge, surface outflows, and evapotranspiration (Fig. 7.2).

$$Wetland\ volume\ (V) = Inputs\ (P_n + S_i + G_i) - Outputs\ (ET - G_o - S_o). \tag{7.1}$$

FIG. 7.2   A wetland water budget. P_n represents precipitation inputs; ET represents evapotranspiration losses. S denotes surface water; G denotes groundwater. Subscript i indicates inputs; o indicates outputs.

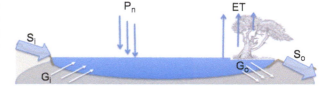

The residence time of water within a wetland is calculated as the wetland volume (V) divided by the total inputs or outputs (MRT = V/inputs or V/outputs).[b] Wetlands in which precipitation and evapotranspiration are the only modes of water exchange usually have long residence times for water and are sometimes referred to as closed systems because the internal turnover vastly exceeds the exchange of elements across ecosystem boundaries. In contrast, wetlands where runoff and outflow dominate the water budget are known as open systems, since water residence times are short and the flux of materials through the system may approach or exceed nutrient turnover within the ecosystem. The residence time of water ultimately limits the capacity of biota to change the composition of waters passing through a wetland.

The degree of hydrologic connectivity between a wetland and its catchment affects not only the source but the composition of biogeochemically important elements in the wetland, as well as the degree to which the wetland can affect biogeochemical patterns at larger, catchment scales. Comparisons of northern peatlands and riverine wetlands are instructive. The extensive peatlands at high latitudes tend to be hydrologically isolated wetlands, known as *ombrotrophic*[c] bogs, which receive all or most of their water from precipitation (Gorham, 1957). In contrast, the wetlands that border many rivers experience episodic flooding and inundation accompanied by high rates of sediment deposition and erosion, and thus active biogeochemical exchange with adjacent surface waters.

Just as the amount and timing of annual precipitation is a key determinant of vegetation composition and productivity in terrestrial ecosystems, the annual hydroperiod of wetlands is a defining characteristic of their biological and biogeochemical properties (Brinson, 1993). A wetland's hydroperiod describes the depth, duration, and frequency of inundation during a typical year (Fig. 7.3). Some wetlands are permanently flooded, while others never have standing surface water. The periodicity and predictability of flooding in wetlands varies, with some wetlands flooded seasonally while others are only inundated after heavy rainfall. Wetlands subject to lunar tides will have predictable daily flooding, whereas wind tides lead to dynamic and unpredictable flooding in wetlands on low-lying coastal plains. The duration and intensity of flooding drives temporal variation in soil oxygen availability in wetland sediments, which places strong constraints on wetland vegetation. Flood pulses can also provide critical subsidies of material from upland and upstream ecosystems that fuel wetland productivity and food webs (Brinson et al., 1981; Megonigal et al., 1997; Junk, 1999).

Salinity merits special consideration. Wetlands that are hydrologically connected to the oceans or to saline lakes have intermittently or permanently high concentrations of salts. In salt marshes salinity may change from freshwater to full-strength seawater over the course

[b] See Eq. (3.3).

[c] Ombrotrophic, literally meaning "to feed on rain," is a term often used to describe vegetation dependent on atmospheric nutrients.

I. Processes and reactions

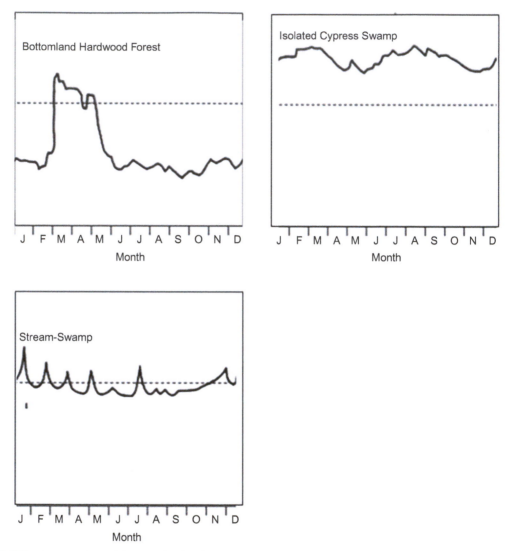

FIG. 7.3 A comparison of three wetland hydroperiods. Note that the duration and frequency of surface water vary greatly between wetland types. The Bottomland Hardwood Forest is seasonally flooded, the hydrologically isolated cypress swamp is permanently flooded. In contrast, wetlands adjacent to rivers such as the stream-swamp shown here may undergo periodic inundation in conjunction with river floods throughout the year. Tidal marshes may be flooded and drained on a daily basis, while some rich fens and bogs never have standing surface water. *From Brinson (1993). Used with permission of Springer.*

of a single tidal cycle, whereas more inland coastal wetlands may experience saltwater intrusion only during rare droughts or extreme wind tides associated with hurricanes. Only a few species of herbaceous vegetation have successfully adapted to life in full-strength seawater, since the high ionic strength of saltwater makes it difficult for plants to maintain osmotic balance.

Wetland soils

Generally, hydric soils can be classified into three categories:

1. Soils permanently inundated with water above the soil surface,
2. Saturated soils with the water table at or just below the soil surface,
3. Soils where the water table depth is always below the surface.

In saturated wetland soils, oxygen typically does not diffuse more than a few millimeters below the water table and reduced compounds and trace gases (N_2O, H_2S, CH_4) produced from anaerobic metabolic pathways may accumulate at high concentrations. Iron is a convenient indicator of anoxic conditions in soils because oxidized iron is easily recognized by its red color, whereas reduced iron is grayish (Megonigal et al., 1993). Soil layers with reduced iron are called gley (Fig. 7.4). In saturated wetland soils, the soil volume is generally 50% solids and 50% water, while in upland soils as much as 25% of the soil volume can consist of air-filled pore space. In upland soils only the interior of soil aggregates is typically anoxic (Chapter 6) and gases diffuse readily between the soil and the atmosphere. Wetland soils can be converted to upland soils through drainage. The global loss of wetlands has largely resulted from efforts to drain wetlands so that formerly saturated sediments can support agriculture.

Wetland vegetation

With saturated sediments but sufficiently shallow surface water to allow vascular plants to dominate, wetlands occupy a special place along the terrestrial-to-aquatic continuum. The

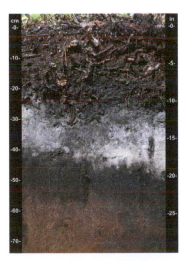

FIG. 7.4   A hydric soil profile, with a thick dark layer of organic soil overlying a gray mineral soil characteristic of reduced iron. The traditional soil horizons for this Spodosol are indicated on the right. The organic (O) horizon overlies a mineral (A) horizon enriched in humic materials. The zone of eluviation (E) is characterized by a loss of silicate clays, iron, or aluminum and overlies the B horizon, or zone of illuviation. *Source: Image from NRCS 2010 Field Indicators of Hydric Soils in the United States; see https://www.nrcs.usda.gov/Internet/FSE_DOCUMENTS/nrcs142p2_053171.pdf.*

dominant autotrophs in wetlands have similar light and nutrient requirements as the plants of upland systems (Chapter 6) but must overcome the additional constraint of rooting in waterlogged soils with low O_2 concentration. Wetlands may be dominated by autotrophs ranging from sphagnum moss, sedges, reeds, and grasses to shrubs or trees—with the dominant vegetation both a function of and a control on wetland water balance and hydroperiod. Vascular plants growing in wetlands must cope with periodic or permanent saturation of their root tissues, a physiological challenge that prevents many plants from growing successfully in wetlands (Bailey-Serres and Voesenek, 2008). A lack of oxygen in saturated soils directly interferes with root metabolism, creating root oxygen deficiency (Keeley, 1979; Gibbs and Greenway, 2003). In addition, anoxic sediments support microbial production of some metabolic products that are toxic to plants (e.g., H_2S from sulfate reduction; Eq. 2.12) (Lamers et al., 1998; Wang and Chapman, 1999).

Wetland plants have developed a variety of morphological and physiological traits that allow them to persist in saturated sediments. Some wetland plants have the capacity to use anaerobic fermentation in their roots during periods of low or no oxygen (Keeley, 1979; Gibbs and Greenway, 2003); however, metabolism of organic compounds via fermentation is far less efficient than aerobic respiration. Many wetland plants have airspaces within their cortex (*aerenchyma*) that facilitate gas exchange between the atmosphere and the sediments surrounding their roots (Brix et al., 1992; Jackson and Armstrong, 1999; Fig. 7.5). For many wetland plants these tissues allow passive gas exchange, but some species have specialized mechanisms to promote active ventilation (Dacey, 1981). Still other wetland plants, such as the bald cypress and coastal mangrove trees, have specialized aerial rooting structures (*pneumatophores*) that appear to facilitate gas exchange (Kurz and Demaree, 1934; Scholander et al., 1955).

Facilitated gas exchange between the atmosphere and the rhizosphere, in addition to alleviating oxygen stress for plants, can also increase the oxygen content of wetland soils around plant roots (Wolf et al., 2007; Schmidt et al., 2010b), allowing aerobic metabolism by soil microbes in flooded soils. For many wetland plants the extent of aerynchymous tissues depends on the intensity or duration of inundation, suggesting that there is some physiological cost associated with these specialized tissues (Justin and Armstrong, 1987). These adaptations allow wetland plants to persist in suboxic or anoxic conditions and to alter soil oxygen

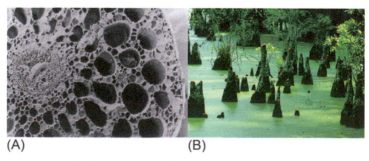

(A)   (B)

FIG. 7.5   (A) Electron micrograph of a cross-sectioned stem of the aquatic macrophyte *Potamageton*; (B) photograph of Cypress pneumatophores. *(A) From Jackson and Armstrong (1999). (B) From Wikipedia Commons.*

availability. Some plants are only found in wetland ecosystems (obligate wetland plants) whereas others are capable of growing across a broader range of hydrologic conditions (facultative wetland plants) and are merely more common in wetland ecosystems.

Productivity in wetland ecosystems

Emergent plants dominate the vegetation of most wetlands and their net primary production is usually estimated using the harvest or eddy-covariance approaches outlined in Chapter 5. Net primary productivity varies widely across wetland ecosystems depending on nutrient supply (Brinson et al., 1981; Brown, 1981). Unlike terrestrial ecosystems, where variation in vegetation type and stature is largely predictable from climate, the differences in wetland productivity are more strongly influenced by soil type and hydrology (Brinson, 1993). Variation in wetland hydroperiod has important consequences for productivity, because autotrophic respiration is less efficient in saturated soils (as discussed in the previous section) and because a high proportion of nutrients are sequestered in undecomposed soil organic matter, leaving low concentrations (and slow turnover) of available nutrients in the soil. Areas that are less frequently flooded tend to have higher productivity, since periodic soil drying allows for more rapid nutrient mineralization by aerobic microbes (Fig. 7.6).

In contrast to upland terrestrial ecosystems, where numerous experiments have documented nutrient limitation of primary productivity, there have been far fewer experimental manipulations of nutrient supply in wetland ecosystems (Bedford et al., 1999). Venterink et al. (2001) reported that among 50 fertilization experiments in wetlands nearly half found significant N limitation of plant biomass, eight experiments reported P limitation, and 13 reported colimitation by N and either P or K. Among wetlands located in regions of differing atmospheric N deposition across Europe and Canada, N inputs appear to be correlated with increasing vascular plant biomass and reduced biomass of low stature

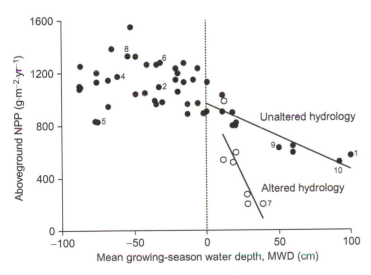

FIG. 7.6 Water depth was negatively correlated with aboveground NPP for southern coastal wetlands. The effect of inundation was more pronounced when levees were built to maintain permanent flooding (plots shown in open circles). *From Megonigal et al. (1997). Used with permission of Springer.*

mosses (Berendse et al., 2001; Turunen et al., 2004; Limpens et al., 2008). The consequences for ecosystem carbon storage can be hard to predict. For instance, in a long-term (5-year) fertilization experiment in the Mer Bleue peatland in Ottawa, Canada, Bubier et al. (2007) measured lower net ecosystem production (NEP) in fertilized plots. They found that nutrient stimulation of vascular plant growth was accompanied by declines in the abundance of mosses, particularly *Sphagnum,* leading to a decline in organic matter accumulation. The loss of *Sphagnum* mosses due to N deposition, fertilization, or drainage-induced increases in N turnover could lead to substantial reductions in peat accumulation because replacement species typically produce higher-quality litter and have higher rates of evapotranspiration than *Sphagnum* (van Breemen, 1995).

In closed wetland systems such as the extensive boreal peatlands at high latitudes in the Northern Hemisphere, nitrogen and phosphorus are typically both in short supply for plant growth and decomposition (Chapin et al., 1978; Damman, 1988). In the tundra of Alaska, Chapin et al. (1978) found that the soil organic matter contained 64% of the total phosphorus in the ecosystem and had a mean residence time of 220 years, while available phosphorus in soil solution comprised 0.3% of the total phosphorus and had a residence time of 10 h. Low temperatures and high water tables together limit nutrient mineralization in the tundra (Marion and Black, 1987), and as a result of slow decomposition, many boreal bogs show a net accumulation of nitrogen and phosphorus in peat (Hemond, 1983; Urban et al., 1989b; Damman, 1988). In a fertilization experiment, Shaver and Chapin (1986) found that the response of *Eriophorum vaginatum* in tussock tundra was greater for N than for P. Rates of nitrogen fixation within boreal wetlands can be very high (Barsdate and Alexander, 1975; Waughman and Bellamy, 1980; Schwintzer, 1983). A variety of arctic plants are capable of assimilating low-molecular-weight organic nitrogen molecules (e.g., Chapin et al., 1993; Nasholm et al., 1998), which suggests that in isolated arctic wetlands, nitrogen limitation is frequently severe.

Experimental determination of nutrient limitation of primary productivity in wetlands is more difficult because hydrologic losses complicate fertilization experiments. Wetlands receiving surface runoff can have high inputs of phosphorus and other elements derived from rock weathering (Mitsch et al., 1979; Waughman and Bellamy, 1980; Craft, 1996). In these ecosystems phosphorus and sulfur are retained on iron and aluminum minerals that are constituents of soil organic matter (Richardson, 1985; Mowbray and Schlesinger, 1988). With greater surface and groundwater inputs, net primary production is more likely to be limited by N than P (e.g., Tilton, 1978) because large amounts of nitrogen can be lost through denitrification, while P tends to accumulate in soil organic material. Many wetlands receive high inorganic N loading from fertilizer, sewage-derived runoff, or N deposition, which can lead to substantial changes in plant composition (Bedford et al., 1999). When the primary form of N pollution is NO_3^-, the high capacity for denitrification in wetland sediments may reduce the impacts of nitrogen eutrophication for vegetation (e.g., Johnson et al., 2016).

Net primary production (NPP) is highest in wetlands receiving nutrient enrichment or with high nutrient turnover. The degree, duration, and periodicity of flooding affect wetland productivity more than rainfall or temperature. Drainage can promote enhanced productivity by increasing nutrient mineralization. Tree growth and nitrogen content increase when northern wetlands are drained (Fig. 7.7; Lieffers and Macdonald, 1990; Westman and Laiho, 2003; Choi et al., 2007; Turetsky et al., 2011). Flooding can enhance wetland productivity when it brings subsidies of nutrients from the contributing catchment, but flooding can also stress

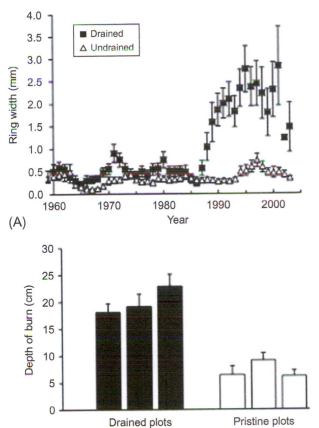

(A)

(B)

FIG. 7.7 Drainage of a forested boreal fen in western Canada in 1986 doubled the rate of peat C accumulation through increases in tree biomass and detritus (indicated here by tree ring growth) (A), but also made the drained fen more susceptible to catastrophic losses of carbon in fire (B). In 2001, a wildfire burned ~450 years of accumulated peat in the drained portion while removing only ~58 years of accumulated peat in the undrained portions of the fen. *Modified from Turetsky et al. (2011).*

wetland plants by suppressing organic matter mineralization and promoting the production of H_2S. This subsidy-stress relationship (sensu Odum et al., 1979) precludes a general relationship between water availability and NPP in wetland ecosystems. In a survey of temperate forested wetlands, Megonigal et al. (1997) found that intermittently flooded wetlands had higher litterfall and NPP than permanently flooded wetlands (Fig. 7.6) and suggested that intermittent flooding allows soils to dry, which increases decomposition and promotes nutrient mineralization. In contrast, other studies have suggested that inundation by flowing water can deliver nutrients from upland areas to wetland forests (Conner and Day, 1976; Conner et al., 2011). Several studies in forested floodplain wetlands found the highest litterfall in the wettest sites but little clear evidence that plant growth was affected by flooding regime (Clawson et al., 2001; Conner et al., 2011). The discrepancy in findings may be due to the type of inundation, with stagnant inundation suppressing nutrient mineralization and reducing productivity (e.g., Schlesinger, 1978; Megonigal et al., 1997) and flowing water providing nutrient subsidies (e.g., Conner and Day, 1976; Clawson et al., 2001).

Organic matter storage in wetlands

Decomposition is impeded in flooded and saturated soils, and primary production often exceeds decomposition, leading to a net accumulation of soil organic matter. As a result, over decadal to millennial timescales many wetlands have accumulated large standing stocks of soil organic matter (Table 7.2). While rates of soil carbon accumulation are reported by biome for terrestrial ecosystems, the variation in organic matter storage in wetland sediments is more closely tied to their hydrology than to their latitude and we simply report the full range of estimates. If the plant remains are still recognizable, these organic materials are called peat. As decomposition removes carbon and the relative mineral fraction increases, the peat forms a darker muck without recognizable plant tissues (see Fig. 7.4). The rate of peat accumulation is determined by the rates of decomposition in both the oxic upper level and the lower level of the deposit. Through time, older layers of organic material are buried and compacted beneath the weight of newly deposited plant detritus (Fig. 7.8). The transport of solutes and diffusion of gases are impeded with increasing depth as a result of water saturation and compaction of organic matter. It is useful to differentiate between the biogeochemically active *acrotelm*, the surface layer of peat that experiences fluctuations between oxic and anoxic conditions, and the *catotelm*, or underlying layers that are permanently saturated.

Peatland ecosystems can be perceived as a special category of wetland wherein plants build the terrain through the deposition of litter into saturated soils. Accretion occurs in these low-energy environments through biogenic processes rather than by sediment deposition (Gosselink and Turner, 1978; Brinson et al., 1981). Clymo (1984) proposed a model for peat accumulation which predicts that peatlands will eventually attain a steady state when the input of detritus from primary production at the surface is balanced by the loss of organic matter by decomposition throughout the peat profile. The saturated soils of tundra and boreal forest region contain about 50% of the total storage of organic matter in soils of the world (Tarnocai et al., 2009; Frolking et al., 2011). Many of these ecosystems have accumulated soil carbon since the retreat of the last continental glaciers (Harden et al., 1992; Roulet et al., 2007; Yu, 2012).

The unique aspect of wetland ecosystems is the dominance and diversity of anaerobic metabolic pathways employed by microbes for metabolism in the absence of oxygen. The drainage of wetland soils (through natural droughts or anthropogenic drainage) leads to rapid oxidation of their large stocks of organic matter by aerobic microbes (Armentano and Menges, 1986; Turner, 2004; Reddy and Graetz, 1988). The resulting decrease in soil elevation, or subsidence, has been notably documented for inland wetlands in England, Germany, and the Florida Everglades, where posts have been anchored in a stable subsurface layer and changing surface elevations are recorded relative to the immobile post. Soil elevation has declined more than 4m in the past 130 and 150 years in German and English sites, respectively (Heathwaite et al., 1990), and more than 3m since 1924 at an Everglades site (Stephens and Stewart, 1976). The rapid oxidation of soil organic matter in drained wetlands provides clear evidence that much of the organic material stored in wetland sediments is not inherently recalcitrant. Instead, wetlands accumulate large standing stocks of organic matter because decomposition in flooded soils is constrained by a lack of oxygen (Fig. 7.9).

To understand how flooding constrains decomposition, we must compare the mechanisms and energy yield derived from anaerobic respiration and aerobic respiration. Both decomposition pathways are initiated by the cleavage of simple organic molecules, such as acetate,

TABLE 7.2   Carbon accumulation in wetland sediments, a compilation of reported rates.

Location	Wetland/vegetation type	Accumulation interval yrs	Accumulation rate $g\,Cm^{-2}\,yr^{-1}$	Reference
Peatlands			12 – 25	Malmer (1975)
Global Wetlands			20–140	Mitra et al. (2005)
North America	Peatlands		29	Gorham (1991)
Boreal wetlands			**8–80**	
Alaska and Canada	Peatlands		8–61	Ovenden (1990)
Alaska	*Picea and Sphagnum*	4790	11–61	Billings (1987)
Russia	*Mires, bogs and fens*	3000–7000	12–80	Botch et al. (1995)
Manitoba	*Picea and Sphagnum*	2960–7939	13–26	Reader and Stewart (1972)
Western Canada	*Sphagnum* bogs	9000	13.6–34.9	Kuhry and Vitt (1996)
Sweden	Bogs		20–30	Armentano and Menges (1986)
Alaska	*Eriphorum vaginatum*	7000	27	Viereck (1966)
Ontario	*Sphagnum* bogs	5300	30–32	Belyea and Warner (1996)
Russia	*Siberian mires*	8000–10,000	12.1–23.7	Turunen et al. (2001)
Finland	*mires*		18.5	Turunen et al. (2002)
Canada	*Mer Bleue ombrotrophic bog*	2700	10–25	Roulet et al. (2007)
Canada	*23 ombrotrophic bogs*	150	73±17	Moore et al. (2005)
Finland	*795 bogs and fens*	5000	21	Clymo et al. (1998)
Sweden	*Store Mosse mire*	5000	14–72	Belyea and Malmer (2004)
Canada	*Continental western Canadian peatlands*	current	19.4	Vitt et al. (2000)
Temperate wetlands			**17–317**	
Georgia	Floodplain cypress gum forests	100	107	Craft and Casey (2000)
Georgia	Depressional wetlands	100	70	Craft and Casey (2000)
Wisconsin	*Sphagnum*	8260	17–38	Kratz and DeWitt (1986)
Massachussets (USA)	Thoreau's bog		90	Hemond (1980)

Continued

TABLE 7.2 Carbon accumulation in wetland sediments, a compilation of reported rates—cont'd

Location	Wetland/vegetation type	Accumulation interval yrs	Accumulation rate $g\,Cm^{-2}\,yr^{-1}$	Reference
North America	Protected Prairie Potholes		83	Euliss et al. (2006)
Ohio (USA)	Created marshes, OH		180–190	Anderson and Mitsch (2006)
North America	Restored Prairie Potholes		305	Euliss et al. (2006)
Ohio (USA)	Depressional wetlands	42	317 ± 93	Bernal and Mitsch (2012)
Ohio (USA)	Riverine, flow-through	42	140 ± 16	Bernal and Mitsch (2012)
Eastern U.S.	Circumneutral freshwater peatlands	30	49 ± 11	Craft et al. (2008)
Eastern U.S.	Acidic freshwater peatlands	30	88 ± 20	Craft et al. (2008)
Subtropical wetlands			**70 to 387**	
Lousiana	Salt marsh		200–300	Hatton et al. (1983)
Florida	*Cladium* swamp	25–30	70–105	Craft and Richardson (1993)
Florida Everglades	*Cladium* sp.		86–140	Reddy et al. (1993)
Florida Everglades	*Typha* sp.		163–387	Reddy et al. (1993)
Tropical wetlands			**39–480**	
Amazon	Lowland peatlands	1700–2850	39–85	Lahteenoja et al. (2009)
Kenya	Papyrus wetlands		160	Jones and Humphries (2002)
Uganda	Papyrus wetlands		480	Saunders et al. (2007)
Costa Rica	humid tropical wetland	42	255	Mitsch et al. (2010)
Mexico	Mangroves		100	Twilley et al. (1992)
Range of reported values			**8–480**	

from large complex organic polymers by extracellular enzymes. Aerobic respiration can completely degrade the resulting organic molecules to CO_2 using glycolysis followed by the Kreb's cycle (Fig. 7.9). When oxygen is available, a single molecule of glucose yields 2 mol of ATP from glycolysis and a further 36 mol of ATP through the Kreb's cycle (Madigan and Martinko, 2006). Without oxygen, this reaction sequence stops at pyruvate,

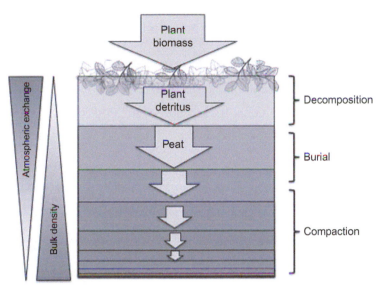

FIG. 7.8 A model of peat accumulation and compaction over time. Fresh litter is deposited in the surface layers, where decomposition rates are highest due to oxygen diffusion and the supply of alternate electron acceptors. Organic matter that escapes decomposition is buried beneath new litter inputs and over time becomes compacted through the accumulation of overlying material. Models of peat accumulation predict that eventually peatlands reach a steady state where new biomass inputs are balanced by carbon losses through decomposition. *Adapted from Clymo (1984).*

and further degradation requires fermentative metabolism, which has a low energy yield (Fig. 7.9). Whereas aerobic respiration results in the complete degradation of molecules to CO_2, fermentation results in the accumulation of a variety of organic acids and alcohols. The resulting fermentation products are subsequently further degraded to CO_2 by bacteria using NO_3^-, Mn^{4+}, Fe^{3+}, or SO_4^{2-} as alternative electron acceptors in place of O_2, or they may undergo additional fermentation steps to produce CH_4. These alternative respiratory pathways have lower energy yields, and thus support a smaller microbial biomass that in turn produces lower concentrations of extracellular enzymes (McLatchey and Reddy, 1998).

There are two mechanistic explanations for the inefficient decomposition typical of wetlands. Until recently, decomposition was primarily assumed to be limited by the supply of oxygen and alternative electron acceptors necessary for the terminal steps in organic matter decomposition. Recent work has suggested additional enzyme-mediated constraints at earlier stages of the decomposition pathway (Limpens et al., 2008). The activity of phenol oxidase, a critical extracellular enzyme involved in the degradation of lignin and phenolics (Freeman et al., 2004b),[d] is substantially reduced under low-oxygen conditions, leading to an accumulation of phenolic compounds in wetland sediments (McLatchey and Reddy, 1998; Freeman et al.,

---

[d] Phenolics are a class of chemical compound consisting of a hydroxyl group bonded directly to an aromatic hydrocarbon. In wetlands, soluble organic acids make up a large fraction of phenolics.

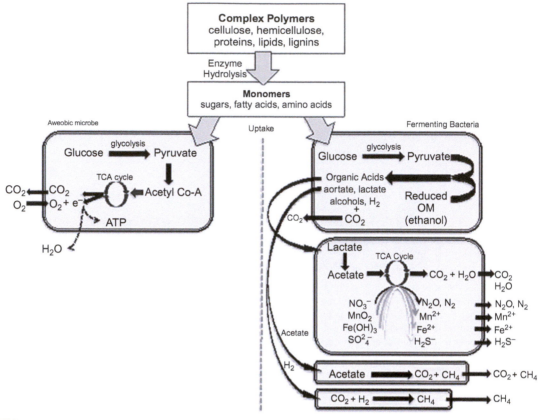

FIG. 7.9 Contrasting the single aerobic respiration pathway with the multistage pathway involved in decomposition in the absence of oxygen. *Figure drawn with inspiration from Megonigal et al. (2003a, b) and Reddy and DeLaune (2008).*

2001b). High concentrations of phenolic compounds can then further inhibit organic matter decomposition (Freeman et al., 2001b; Yu, 2012; Appel, 1993; Wang et al., 2015).

When alternate electron acceptors are abundant, the rate of soil organic matter decomposition will be limited by the pace of enzymatic hydrolysis or fermentation (Megonigal et al., 2003a, b). In contrast, when alternate electron acceptors are in short supply, fermentation products may accumulate. Decomposition of soil organic matter in wetlands can be enhanced either by lowering the water table (allowing oxygen to penetrate to deeper soil layers) or by increasing the supply of alternate electron acceptors. Nitrogen deposition, amendments with oxidized Fe, and enhanced SO_4 availability resulting from acid rain or saltwater intrusion significantly enhance decomposition rates (Van Bodegom et al., 2005; Bragazza et al., 2006; Gauci and Chapman, 2006; Weston et al., 2006). Decomposition in wetland sediments is typically highest at the wetland surface, where recently synthesized, more labile organic material comes into contact with the greatest potential supply of electron acceptors.

Although conditions at the oxic interface promote more efficient decomposition, there is new evidence that the formation of iron hydroxides at these oxic interfaces can lead to precipitation of significant quantities of dissolved organic matter. During the experimental aeration of two boreal fens, Riedel et al. (2013) found that 90% of the dissolved iron (Fe(II)) and 27% of the dissolved organic carbon (DOC) were removed from solution by the coprecipitation of aromatic DOC molecules with iron hydroxides. This "iron gate" may place an important constraint on the loss of dissolved organic matter from wetlands during drying periods, yet this mechanism of retention may be reversed following reflooding as iron reduction leads to the release of Fe-protected DOC (Huang and Hall, 2017). If phenolics are coprecipitated with iron hydroxides at oxic interfaces, iron oxidation may actually stimulate decomposition in organic rich wetlands by removing their inhibitory effect on phenol oxidase activity (see above paragraph) (Wang et al., 2019a).

Microbial metabolism in saturated sediments

In a closed aqueous system containing a large supply of organic material together with appreciable concentrations of oxidants (O_2, NO_3^-, Mn^{4+}, Fe^{3+}, and SO_4^{2-}), we can easily predict the order in which the oxidants will be utilized in metabolism (Table 7.3). The exergonic (energy-yielding) oxidation of organic matter (A) would first be paired with (B) oxygen respiration, then (C) NO_3^- respiration, (D) Mn^{4+}, (E) Fe^{3+}, and (F) SO_4^{2-}-respiration would follow in sequence (Table 7.3). If organic matter remains after all of these oxidants are depleted, we might subsequently measure an accumulation of CH_4 in the closed vessel. This predictable sequence of biologically-mediated chemical reactions occurs because there is a tendency for the microbes performing the highest energy yielding metabolic pathways to outcompete microbes reliant on lower energy yielding processes for the limited supplies of fermentation products (Fig. 7.9) (Stumm and Morgan, 1996). The same reaction sequence observed in a closed vessel can also be observed in wetland ecosystems examined through time following flooding or with depth in the soil profile (Fig. 7.10). Under extremely reducing conditions, phosphine gas (PH_3) can be produced from phosphate (PO_4^{3-}) acting as a terminal electron acceptor in highly reducing environments (Bartlett, 1986).[e]

The most common reduction and oxidation half reactions are shown in Table 7.3 together with the standard electrical potential of each reaction. Standard electrical potentials are expressed per mol of electrons transferred; thus, each reaction has been written to transfer one mol of electrons. Where $E° > 0$, the reaction will proceed spontaneously as written. Where

---

[e]The flickering light reported at night over bogs, swamps and cemeteries and known in folklore as the "Will-O'-the Wisp," may derive from the spontaneous combustion of phosphine gas as it enters the Earth's oxygen-rich atmosphere.

7. Wetland Ecosystems

TABLE 7.3   Common reduction and oxidation half reactions are shown here together with the standard electrical potential of each reaction.

Part A

Reduction	$E°$ (V)	Oxidation	$E°$(V)
(A) $1/4O_2(g) + H^+ + e^- = 1/2\,H_2O$	+0.813	(L) $1/4CH_2O + 1/4H_2O = 1/4CO_2 + H^+ + e^-$	−0.485
(B) $1/5NO_3^- + 6/5H^+ + e^- = 1/10N_2 + 3/5H_2O$	+0.749	(M) $1/2CH_4 + 1/2H_2O = 1/2CH_3OH + H^+ + e^-$	+0.170
(C) $1/2MnO_2(s) + 1/2HCO_3^- + 3/2H^+ + e^- = 1/2MnCO_3 + H_2O$	+0.526	(N) $1/8HS^- + 1/2H_2O = 1/8SO_4^{2-} + 9/8H^+ + e^-$	−0.222
(D) $1/8NO_3^- + 5/4H^+ + e^- = 1/8NH_4^+ + 3/8H_2O$	+0.363	(O) $FeCO_3(s) + 2H_2O = FeOOH(s) + HCO_3^- + 2H^+ + e^-$	−0.047
(E) $FeOOH(s) + HCO_3^- + 2H^+ + e^- = FeCO_3(s) + 2H_2O$	−0.047	(P) $1/8NH_4^+ + 3/8H_2O = 1/8NO_3^- + 5/4H^+ + e^-$	+0.364
(F) $1/2CH_2O + H^+ + e^- = 1/2CH_3OH$	−0.178	(Q) $1/2MnCO_3(s) + H_2O = 1/2MnO_2(s) + 1/2HCO_3^- + 3/2H^+ + e^-$	+0.527
(G) $1/8SO_4^{2-} + 9/8H^+ + e^- = 1/8HS^- + 1/2H_2O$	−0.222		
(H) $1/8CO_2 + H^+ + e^- = 1/8CH_4 + 1/4H_2O$	−0.244		
(I) $1/6N_2 + 4/3H^+ + e^- = 1/3NH_4$	−0.277		

Part B

Examples	Combinations	$\Delta G°$ (W) pH = 7 (kJ eq^{-1})
Aerobic respiration	A+L	−125
Denitrification	B+L	−119
Nitrate reduction to ammonium	D+L	−82
Fermentation	F+L	−27
Sulfate reduction	G+L	−25
Methane fermentation	H+L	−23
Methane oxidation	A+M	−62
Sulfide oxidation	A+N	−100
Nitrification	A+P	−43
Ferrous oxidation	A+O	−88
Mn(II) oxidation	A+Q	−30

Modified from Stumm and Morgan (1996).

Standard electrical potentials are expressed per mol of electrons transferred thus each reaction has been written to transfer one mol of electrons. Where $E° > 0$ the reaction will proceed spontaneously as written. Where $E° < 0$ the reaction will proceed in the opposite direction. The greater the difference in $E°$ between two half reactions, the greater the resulting free energy yield from their combination will be. In Table B the standard free energies of common redox couplets are shown. These are calculated from the $E°$ values in Table A using the formula $\Delta G° = -nF\Delta E$ (Eq. 7.1). Here {CH_2O} represents an "average" organic substance. The actual free energy yield of different organic substances may differ for that given for CH_2O. This difference may be very large, particularly for anoxic processes involving substrates whose carbon has a very different oxidation state than that assumed for CH_2O.

$E° < 0$ the reaction will proceed in the opposite direction. The greater the difference in $E°$ between two half reactions, the greater the free energy yield from their combination. In Part B of Table 7.3, the standard free energies of common redox couplets are shown. These are calculated from the $E°$ values in Part A using Eq. (7.2). In the table, CH_2O represents an "average" organic substance. The actual free energy yield of different organic substances may differ from that given for CH_2O. This difference may be very large, particularly for anoxic processes involving carbon substrates with very different oxidation states than that assumed for CH_2O.

Decomposition in wetlands is dominated by anaerobic metabolic pathways that yield a variety of reaction products in addition to the CO_2 and H_2O generated by aerobic oxidation. These pathways are responsible for the production of N_2, N_2O, and CH_4, the abundance of reduced H_2, Fe^{2+}, and H_2S, and the production of pyrite (FeS_2) in wetland soils. In any wetland the relative importance of these metabolic pathways to overall ecosystem carbon and nutrient cycling depends on the availability of the various electron acceptors. Since fermentation is slow and fermentation products are scarce, the metabolic pathways that maximize energy gain are highly favored. Successful metabolic strategies (and thus successful microbes) are those that garner the greatest energy given available substrates. The "redox ladder," the predictable sequence of reactions following flooding or with depth, thus arises from competitive interactions between microbes (Postma and Jakobsen, 1996; Stumm and Morgan, 1996). It seems at first paradoxical that in these carbon-rich systems we see such fierce competition for carbon substrates. This paradox can be explained if we consider that the rate of decomposition is determined by fermentation while the order and relative dominance of terminal electron acceptor processes is predictable from their energy yield (Postma and Jakobsen, 1996; Megonigal et al., 2003a, b).

To understand and predict which microbial metabolisms will dominate at any given time or place in wetland sediments, we must understand how the possible reactions vary in the amount of energy generated (free energy yield). Thermodynamics allows us to predict the dominant metabolisms because particular microbial species, with different metabolic strategies, become competitively superior under different chemical conditions.

Free energy yield

To calculate the energy yield from the oxidation of organic matter paired with the reduction of any electron acceptor, we calculate the standard Gibb's free energy yield ($\Delta G°$) for a redox couplet as:

$$\Delta G° = -nF\Delta E, \tag{7.2}$$

where n is the number of electrons, F is Faraday's constant ($23.061\,kcal\,V^{-1}$), and ΔE is the difference in electrical potential (V) between the oxidation and reduction reactions (Table 7.3). Reactions with a negative ΔG are energy yielding (exergonic), while reactions with a positive ΔG require an input of energy (endergonic).

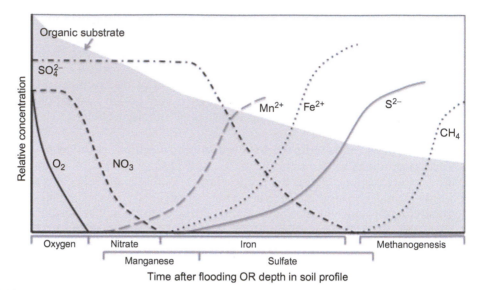

FIG. 7.10 The concentrations of reactants and products of terminal decomposition pathways are shown for a wetland sediment over time following flooding. Rotating the figure 90° to the right shows the pattern of substrate concentrations (and the order of metabolic pathways) with depth in a soil profile.

For aerobic respiration of a standard organic molecule (CH_2O), we can calculate the standard free energy yield $\Delta G°$. The standard free energy assumes that all substrates are available in abundance and that the reaction occurs at a standard temperature of 25 °C. The free energy yield of each reaction is higher for carbon compounds that have more reduced chemical bonds and lower for more oxidized organic molecules. Here we will use a generic carbon molecule with 1/6 the free energy of glucose, which has six carbon atoms.

The oxidation 1/2 reaction:

$$[CH_2O] + H_2O \rightarrow CO_2(g) + 4H^+ + 4e^- \quad E° = -0.485V \tag{7.3}$$

And the reduction 1/2 reaction:

$$O_2(g) + 4H^+(W) + 4e^- \rightarrow 2H_2O \quad E° = +0.813V \tag{7.4}$$

are coupled for a joint reaction:

$$[CH_2O] + O_2 \rightarrow CO_2 + H_2O \quad \Delta E° = +1.30V, \tag{7.5}$$

$$\Delta G° = -nF\Delta E$$
$$= -(4)(23.061\,kcal)(+1.30V)$$
$$= -119.9\,kcal\,per\,mol\,CH_2O$$

or,

$$= -29.9 \, \text{kcal per e}^{-1}$$

Since $1 \, \text{kcal} = 4.184 \, \text{kJ} = -502 \, \text{kJ mol}^{-1} \, CH_2O$ or $-125 \, \text{kJ per e}^{-1}$.

Note that the reduction step requires energy ($+E°$) while the oxidation step yields energy ($-E°$). We can express the energy yields for each equation per mole of carbon substrate or per mole of electrons transferred. The net energy yield of the paired reaction is ~125 kJ for every mole of electrons transferred. Expressing energy yield per mole of electrons is useful for comparing processes that oxidize inorganic energy sources (e.g., sulfide, Fe^{2+}, or Mn^{2+}) with those that oxidize organic matter.

If we pair the same organic matter oxidation reaction with the reduction of NO_3^- as an alternative electron acceptor, we find a lower energy yield.

The oxidation 1/2 reaction:

$$[CH_2O] + H_2O \rightarrow CO_2(g) + 4H^+ + 4e^- \quad E° = -0.485V \tag{7.6}$$

and the reduction 1/2 reaction:

$$0.8NO_3^- + 4.8H^+ + 4e^- \rightarrow 0.4N_2 + 2.4H_2O \quad E° = +0.749 \tag{7.7}$$

Yield the joint reaction:

$$[CH_2O] + 0.8NO_3^- + 0.8H^+ \rightarrow CO_2 + 0.4N_2 + 1.4H_2O \quad \Delta E° = +1.23V \tag{7.8}$$

And since

$$\begin{aligned}
\Delta G° &= -nF\Delta E \\
&= -(4)(23.061 \, \text{kcal V}^{-1})(1.23 \, \text{V}) \\
&= -113 \, \text{kcal per mol } CH_2O \\
\text{or} &= -474 \, \text{kJ mol}^{-1} \, CH_2O \\
\text{or} &= -28.5 \, \text{kcal per e}^{-1} \\
\text{or} &= -119 \, \text{kJ per e}^{-1}
\end{aligned}$$

In comparing the $\Delta G°$ for these two reactions, we see that in denitrification, nitrate respiration releases 95% of the energy contained in the same organic molecule (CH_2O) relative to aerobic respiration. Because of this difference in efficiency, whenever O_2 is available, heterotrophs utilizing aerobic respiration should outcompete denitrifiers for organic substrates. The "actual free energy yields" (ΔG) for these reactions, which take into account the concentrations of all reactants, indicate that under the conditions found in most oxic soils, aerobic respiration has a much higher ΔG than denitrification because oxygen is far more available than nitrate. In contrast, in wet soils where oxygen concentrations are low and nitrate concentrations are high (as in wet agricultural fields or wetlands receiving nitrogen-rich runoff), the two pathways may have nearly equivalent ΔG.

To calculate the actual free energy yield (ΔG) of a reaction we use the equation:

$$\Delta G = \Delta G° + RT \ln Q, \tag{7.9}$$

where R is the universal gas constant ($1.987 \times 10^{-3} \, kcal \, K^{-1} \, mol^{-1}$), T is temperature in °K, and Q represents the reaction quotient, or the concentration of reaction products relative to the concentration of reactants. For a generic reaction $^{a}Ox_1 + ^{b}Red_2 \rightarrow {}^{c}Red_1 + ^{d}Ox_2$, the reaction quotient would be calculated as:

$$Q = [Red_1]^c [Ox_2]^d / [Ox_1]^a [Red_2]^b. \tag{7.10}$$

Actual free energies thus modify the prediction of energy yield by taking into account the relative abundance of reactants and products in the environment. This is a critical adjustment because the assumption of standard activities of all reactants inherent to the standard free energy calculations is rarely met in natural ecosystems. In most salt marsh ecosystems, for example, a dominant pathway for organic matter decomposition is sulfate reduction (Howes et al., 1984; Howarth and Teal, 1979), a pathway that is not energetically favored according to standard free energy predictions (Table 7.4). Sulfate reduction becomes important in anoxic sediments where sulfate is abundant.

In a comparison of a freshwater and a brackish wetland in coastal Maryland, Neubauer et al. (2005) measured high rates of Fe^{3+} reduction in the early summer, giving way later in the season to methanogenesis in the freshwater wetland (low SO_4^{2-}) and sulfate reduction in the brackish wetland receiving marine-derived SO_4^{2-} (Fig. 7.11). We can use the actual free energy calculations to understand the environmental conditions under which we would predict these shifts in the dominant metabolic pathways. Terminal electron acceptors that are extremely abundant will be likely to dominate decomposition pathways.

While the energy yield from sulfate reduction and methane fermentation is very low, the reduced products of sulfate reduction (HS^-) and methanogenesis (CH_4) are themselves energy-rich substrates for other organisms. In the presence of oxygen, sulfide oxidation generates $100 \, kJ \, eq^{-1}$ of free energy, which, when combined with the free energy yield of sulfate reduction ($25 \, kJ \, eq^{-1}$), achieves the same total free energy yield as aerobic respiration of the same

TABLE 7.4   Examples of chemical reactions and associated free energy yield (per mol of organic matter or sulfur compound) and catalyzing organisms performing fermentation.

Reaction		$-\Delta G^O$ (kJ mol^{-1})	Organisms catalyzing these reactions
(1) 3(CH$_2$O)	$\rightarrow CO_2 + C_2H_6O$	23.4	E.g., Yeasts, *Sarcina ventriculi*, *Zymonas*, *Leuconostoc* sp., *Clostridia*, *Thermoanaerobium brockii*, etc.
(2) n(CH$_2$O)	$\rightarrow m CO_2$ and/or fatty acids and/or alcohols and/or H$_2$	5–60	E.g., Yeasts, *Clostridia*, *Enterobacteria*, *Lactobacilli*, *Streptococci*, *Propionibacteria*, etc.

From Zehnder and Stumm (1988).

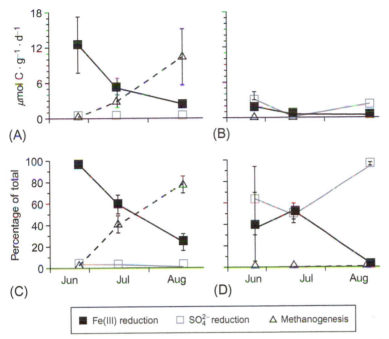

FIG. 7.11 Seasonal changes in the rates and relative importance of Fe(III) reduction, SO_4^{2-} reduction, and methanogenesis during summer 2002 in (A) and (C) Jug Bay, a freshwater wetland, and (B) and (D) Jack Bay, a brackish wetland on the coastal plain of Maryland. *From Neubauer et al. (2005). Used with permission of the Ecological Society of America.*

original carbon molecules (Table 7.3, Part B). This explains why very little sulfide gas or methane escapes from wetlands relative to the amounts produced in wetland sediments. Energy is extracted from these highly reduced gases as they pass upward through sediments with O_2.

Most wetlands undergo both flooding and drying cycles, and many wetlands are shallow enough to support rooted vegetation that can transport gases between the atmosphere and sediments. Because of this variation in water level and in plant-facilitated gas exchange, oxygen depletion within wetlands is far from uniform in either time or space. As a result, wetland sediments are characterized by complex temporal and spatial gradients in the dominant metabolic pathways and rates of organic matter oxidation. Oxygen concentrations are typically high in shallow surface waters or drained surface sediments that are in direct contact with the atmosphere, but O_2 levels decline rapidly with depth in organic-rich sediments. The rhizosphere (root-associated sediments) of aerenchymous plants can remain well oxygenated and can support higher rates of mineralization at depth that may help alleviate nutrient limitation of wetland plant growth (Weiss et al., 2005; Laanbroek, 2010; Schmidt et al., 2010b; Fig. 7.12). Greater root biomass can thus be indirectly associated with higher rates of organic matter mineralization due to rhizosphere oxidation (Wolf et al., 2007).

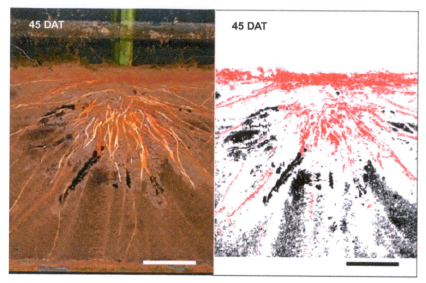

FIG. 7.12 Schmidt et al. (2010b) grew rice seedlings (*Oryza sativa*) in experimental containers (rhizotrons) with a clear plastic side. On the *left* is a photograph of the rhizosphere of a single rice seedling grown in a paddy soil for 45 days. On the *right* the same image is highlighted to show areas that are oxic in red and anoxic in black. *From Schmidt et al. (2010b). Used with permission of Springer.*

Redox potential in wetlands

Just as pH expresses the concentration of H⁺ in solution, redox potential is used to express the tendency of an environmental sample (usually measured in situ) to either receive or supply electrons. We can measure the development of anoxic conditions in sediments by measuring redox potential (pe). Oxic environments have high redox potential because they have a high capacity to attract electrons (oxygen is the most powerful electron acceptor), while anoxic environments have a low redox potential (reducing conditions) because of an abundance of reduced compounds already replete with electrons. When a metal probe is inserted into a soil or sediment, the metal surface will begin to exchange electrons with its surroundings, and the net direction of the exchange will depend on both the reactivity of the metal and the relative availability of electrons in the environment. To measure the direction and strength of this electron exchange, the metal probe is connected to a reference electrode, with a voltmeter placed in the circuit. The redox potential is then measured as the voltage necessary to stop the flow of electrons to the metal electrode.

In a laboratory setting, the redox potential of chemical mixtures is determined by connecting the redox probe to a standard hydrogen electrode. The relative abundance of electrons in the solution will alter the equilibrium constant for the exchange of electrons within the electrode, where electrons are shuttled between sulfuric acid and a hydrogen gas atmosphere:

$$4H^+ + 4e^- \longleftrightarrow 2H_2\,(g). \tag{7.11}$$

In the field it is not easy to maintain standard hydrogen electrodes, so investigators typically use either a Ag/AgCl electrode or a calomel reference electrode that has been calibrated against a hydrogen electrode (Fiedler, 2004; Rabenhorst et al., 2009). The Ag/AgCl electrode consists of a silver wire surrounded by AgCl salt that is contained within a concentrated KCl solution. The solid Ag exchanges electrons with the AgCl solution:

$$Ag \longleftrightarrow Ag^+ + e^-. \tag{7.12}$$

When the reference electrode is connected to the platinum probe and inserted into an oxic soil, oxygen will consume electrons along the platinum probe:

$$O_2 + 4e^- + 4H^+ \rightarrow 2H_2O. \tag{7.13}$$

The reaction within the Ag/AgCl electrode (Eq. 7.12) will go to the right (Ag is oxidized) and the voltmeter will record a positive flow of electrons, $E_h = +400$ to $+700\,mV$ (Fig. 7.13A). If instead the platinum probe is inserted into a highly reduced sediment, electrons will flow toward the reference electrode. The reaction (Eq. 7.12) will proceed to the left (Ag is reduced), and the voltmeter will record a negative flow of electrons (Fig. 7.13B). Charge balance is maintained within the reference electrode by the diffusion of ions through a porous ceramic or membrane tip. Potassium ions (K^+) will be released through the ceramic tip when the redox potential is positive, and Cl^- ions will be released to the soil when the redox potential is negative (Fig. 7.13).

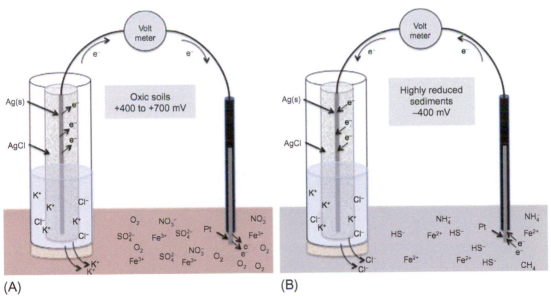

FIG. 7.13 Schematic of an Ag/AgCl reference electrode and platinum electrode inserted into (A) an oxic soil and (B) a highly reduced sediment, demonstrating how the direction of electron flow depends on the availability of electron acceptors at the platinum electrode. Depending on the direction of e^- flow, either K^+ or Cl^- will be released through the ceramic or membrane tip of the reference electrode into the soil to maintain charge balance within the electrode.

The oxidizing or reducing potential of the soil is thus estimated relative to the reference electrode. Redox potential must be corrected if one uses Ag/AgCl rather than a standard hydrogen electrode using a correction factor (~200 mV for an Ag/AgCl electrode and ~250 mV for a calomel electrode). Typically, the redox potential in soils with any oxygen present varies only between +400 and +700 mV. As oxygen is depleted, other constituents, such as Fe^{3+} may accept electrons, but a lower voltage will be recorded since Fe^{3+} has a reduced capacity for attracting electrons relative to O_2.

$$Fe^{2+} + 3H_2O \leftrightarrow Fe(OH)_3 + 3H^+ + e^- \quad \text{and} \quad E_h = +100 \, to - 100 \, mV \tag{7.14}$$

In aqueous environments, redox potential can range from –400 mV to +800 mV (Fig. 7.14). A negative sign means that the reducing environment has excess electrons, a situation in which excess electrons will form H_2 gas via Eq. 7.11.

The pH of the environment will affect the redox potential established by oxygen and other alternate electron acceptors (Fig. 7.14). Aerobic oxidation, which generates H^+ ions (shown in Eq. 7.13), is more likely to proceed under neutral or alkaline conditions while anaerobic

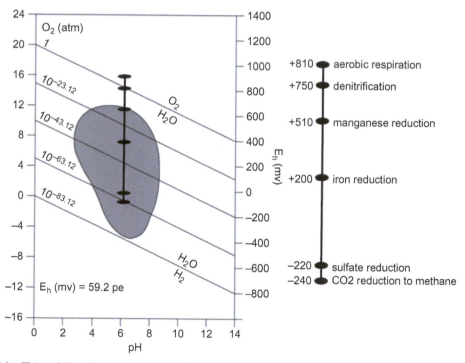

FIG. 7.14 This stability diagram shows the relationship between pH and redox potential. Diagonal lines are the redox potential at different oxygen partial pressures. The *shaded zone* in the center of the figure shows the range of redox potentials that are found in natural aqueous environments. The black bar in the center of the diagram is expanded on the right of the graph to indicate the redox potential (in mV) at which individual microbial metabolic pathways are optimized at neutral pH. *Modified from Lindsay (1979), which was based on the original compilation by Baas Becking et al. (1960). Used with permission of The University of Chicago Press.*

pathways that consume H^+ ions (shown in Eq. 7.14) are chemically more favorable in acidic environments (e.g., Weier and Gilliam, 1986). Because of the pH sensitivity of redox reactions, it is often useful to express the redox potential of a reaction in units of pe, a constant that is derived from the equilibrium constant of the oxidation–reduction reaction (K), which incorporates information on pH. For any reaction,

$$Oxidized\ Species + e^- + H^+ \longleftrightarrow Reduced\ Species, \tag{7.15}$$

and the equilibrium constant can be determined by

$$\log K = \log[reduced] - \log[oxidized] - \log[e^-] - \log[H^+]. \tag{7.16}$$

If we assume that the concentrations of oxidized and reduced species are equal, then

$$pe + pH = \log K. \tag{7.17}$$

Here pe is the negative logarithm of the electron activity $(-\log[e^-])$, and it expresses the energy of electrons in the system (Bartlett, 1986). Because the sum of pe and pH is constant, if one goes up, the other must decline. When a given reaction occurs at a lower pH, it will occur at a higher redox potential, expressed as pe. Measurements of redox potential that are expressed as voltage can be converted to pe following

$$pe = E_h/(RT/F)2.3, \tag{7.18}$$

where R is the universal gas constant $(1.987\,cal\,mol^{-1}\,K^{-1})$, F is Faraday's constant $(23.06\,kcal\,V^{-1}\,mol^{-1})$, T is temperature in Kelvin, and 2.3 is a constant to convert natural to base-10 logarithms.

Environmental chemists use stability diagrams to predict the likely oxidation state of any element in natural environments at each combination of E_h-pH or pe-pH (e.g., Fig. 7.15). Two lines bound all such diagrams. At any redox potential above the upper line, even water would be oxidized (Eq. 7.13, reverse)—a condition not normally found on the surface of the Earth.[f] Similarly any condition below the lower line would allow the reduction of water, again a condition rarely seen on Earth. These boundary redox conditions vary predictably with pH, with E_h declining by 59 mV with each unit of pH increase, reflecting that oxidation proceeds at a lower redox potential under more alkaline conditions.

In most cases organic matter contributes a large amount of "reducing power" that lowers the redox potential in flooded soils and sediments (Bartlett, 1986). High concentrations of Fe^{2+} will be found in flooded low-redox potential environments where impeded decomposition leaves undecomposed organic matter in the soil and humic substances impart acidity to the soil solution. Where organic matter is sparse, iron may persist in its oxidized form (Fe^{3+}) even when the soils are flooded (e.g., Couto et al., 1985). Aeration and liming are each used as techniques for ameliorating acid mine drainage because iron tends to precipitate in oxidized forms at high redox potential or high pH.

Redox potential measurements do not actually measure the number of electrons in the environment, but instead provide a standard method for comparing their relative availability. Because individual anaerobic metabolic pathways vary predictably in their energy yield, each

---

[f] The water-splitting reaction of photosynthesis (Eq. 5.1) is one such instance.

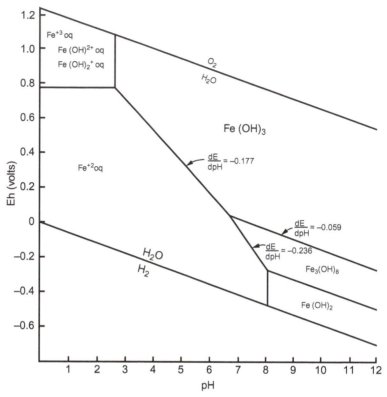

FIG. 7.15 This stability diagram shows the expected form of iron in natural environments of varying pH and E_h. In interpreting such diagrams it is important to remember that E_h and pH are properties of the environment determined by the total suite of chemical species present. Thus, E_h predicts the forms of iron that will be present under a set of chemical conditions using a set of simplifying assumptions. Because iron is interacting with a variety of chemicals in solution, these predictions are not always accurate due to competing reactions. Important to biogeochemistry, E_h can be used to predict what modes of microbial activity are possible in a given environment, and E_h-pH predictions can be useful against which to compare field patterns. *From Ponnamperuma et al. (1967).*

tends to dominate in a fairly narrow range of redox potentials so that a measure of field redox potential can provide a good prediction of the likely metabolic pathways at the point of measurement. The measure of a field redox potential is not equivalent to the measurement of the redox potential of a chemical reaction in laboratory solutions where only one equilibrium reaction is possible. Reflecting the mix of reactions in the soil or sediment, field redox potential provides a way of comparing the degree to which electron acceptor abundance and oxidizing efficiency vary in space and time.

Soils and sediments that resist change in their redox potential are said to be highly poised. Conceptually, poise is to redox as buffering capacity is to pH (Bartlett, 1986). As long as soils are exposed to the atmosphere, they will appear to be highly poised, since O_2 will maintain a high redox potential under nearly all conditions. Soils with high concentrations of Mn^{4+} and

Fe^{3+} are less likely to produce substantial amounts of H_2S or CH_4 during short-term flooding events because it is unlikely that microbes will be able to deplete Mn^{4+} and Fe^{3+} as electron acceptors and turn to the less efficient oxidizing constituents (SO_4^{2-}, CO_2) (Lovley and Phillips, 1988a; Achtnich et al., 1995; Maynard et al., 2011). Such soils are also poised but at a lower redox potential than oxic soils. Because of the large differences in free energy yield from the various anaerobic metabolic pathways, we can predict what metabolic products are likely to accumulate from a measure of redox potential (see Fig. 7.14).

Anaerobic metabolic pathways

Prior to the oxidation of the Earth's atmosphere about 2.5 bya, anaerobic metabolism dominated the biosphere (Chapter 2). In the anoxic sediments of wetlands, lakes, rivers, and oceans, these metabolic pathways continue to dominate biogeochemical cycling today. Table 7.3 listed the more important overall reactions involved in anaerobic oxidation of organic matter in wetland sediments and allows calculations of the resulting free energy yield from each reaction. These reactions have different biogeochemical consequences.

Fermentation

Fermentation of organic matter (Table 7.4) is a required precursor or accompaniment for each of the subsequent anaerobic metabolic pathways in Table 7.3 (Fig. 7.9). Fermenting bacteria in wetlands are obligate anaerobes that use a variety of organic substrates, including alcohols, sugars, and organic and amino acids, and convert them into CO_2 and various reduced fermentation products (predominantly organic acids and alcohols, molecular hydrogen, and CO_2) (Zehnder and Stumm, 1988; Megonigal et al., 2003a, b). Fermentation occurs inside microbial cells and does not require an external supply of electron acceptors. During fermentation ATP is produced, but fermentation pathways have very low energy yield. The low-molecular-weight organic acids, alcohols, and molecular hydrogen that are produced during fermentation ultimately determine the rate of other metabolic reactions.

Although often ignored as an important pathway for the production of CO_2, a number of recent studies have suggested that fermentation (together with reduction of humic acids; Lovley et al., 1996) can account for a significant fraction of anaerobic carbon mineralization in wetland sediments (Keller and Bridgham, 2007). In addition to producing CO_2 or facilitating organic matter mineralization to CO_2 or CH_4, fermentation products can also accumulate as dissolved organic compounds (DOC) that are susceptible to leaching and hydrologic export. In any case, it is important to remember that the products of fermentation are the reactants for all the anaerobic metabolisms that follow.

Dissimilatory nitrate reduction

After O_2 is depleted by aerobic respiration, nitrate becomes the best alternative electron acceptor, and in anoxic sediments nitrate will be rapidly consumed whenever there are suitable organic substrates or reduced chemical compounds from fermentation are available for metabolism (Table 7.5). The term denitrification is typically used to refer to the process by which bacteria convert nitrate to gaseous N_2O or N_2 during the oxidation of organic matter

TABLE 7.5 The chemical reactions and associated free energy yield (per mol of organic matter or sulfur compound) for dissimilatory nitrate reduction pathways.

Reaction		$-\Delta G^O$ (kJ mol^{-1})	Organisms catalyzing these reactions
(1) $2NO_3^- + (CH_2O)$	$\rightarrow 2NO_2^- + CO_2 + H_2O$	82.2	E.g., Members of the genus *Enterobacter, E. coli* and many others
(2) $4/5NO_3^- + (CH_2O) + 4/5H^+$	$\rightarrow 2/5N_2 + CO_2 + 7/5H_2O$	112	E.g., Members of the genus *Pseudomonas, Bacillus lichenformis, Paracoccus denitrificans*, etc.
(3) $\frac{1}{2}NO_3^- + (CH_2O) + H^+$	$\rightarrow 1/2NH_4^+ + CO_2 + 1/2H_2O$	74	Members of the genus *Clostridium*
(4) $6/5NO_3^- + S^0 + 2/5H_2O$	$\rightarrow 3/5N_2 + SO_4^{2-} + 4/5H^+$	91.3	Member of the genus *Thiobacillus*
(5) $8/5NO_3^- + HS^- + 3/5H^+$	$\rightarrow 4/5N_2 + SO_4^{2-} + 4/5H_2O$	93	*Thoiosphaera pantotropha* and members of the genus *Thiobacillus*

From Zehnder and Stumm (1988).

(see Eqs. 1 and 2, Table 7.5). These reactions are dissimilatory since the nitrogen used for denitrification is not incorporated into the biomass of denitrifying microbes. Aerobic heterotrophs and denitrifiers often coexist in upland soils; indeed, many heterotrophic microbes are facultative denitrifiers that switch between aerobic respiration and denitrification depending on the supply of O_2 versus NO_3^- (Carter et al., 1995; Chapter 6). The standard free energy yield of the reaction is only marginally lower than that of aerobic respiration (refer to Table 7.3). Standard free energy calculations assume unlimited supplies of reactants; however, the NO_3^- in flooded soils is never as abundant as the O_2 in oxic soils. Thus even highly NO_3^--enriched wetland sediments are likely to have lower soil respiration rates once oxygen is depleted. In many flooded soils denitrification is limited by the availability of NO_3^-, a problem exacerbated by the fact that nitrification (Eqs. 2.17, 2.18) cannot proceed without oxygen.

Although often discussed as a single process, denitrification occurs in multiple steps whereby NO_3^- is converted sequentially to NO_2^-, NO, N_2O, and N_2 (Fig. 6.16). Particularly in nitrogen-enriched agricultural fields, N_2O can be a significant fraction of the total gaseous nitrogen produced (Stehfest and Bouwman, 2006) and dissolved nitrite (NO_3^-) can become a more important dissolved export than NH_4^+(Stanley and Maxted, 2008). It appears that the ratio of N_2O:N_2 production from denitrification in wetlands is typically lower than in upland soils (Schlesinger, 2009; Beaulieu et al., 2011), but there are concerns that continued nitrogen loading to wetlands from nitrogen deposition or fertilizer runoff could enhance N_2O yields since incomplete denitrification (with N_2O as a terminal product) becomes increasingly energetically favorable when the supply of NO_3^- is high (Verhoeven et al., 2006). In permanently flooded wetlands this outcome is unlikely, since nitrification is inhibited and NO_3^- is in high demand as an alternative electron acceptor, but N_2O may be a more important product of both nitrification and denitrification in intermittently flooded wetlands (Morse et al., 2012; Ardón et al., 2018; Helton et al., 2019).

An alternative metabolic process to denitrification is dissimilatory nitrogen reduction to ammonium (usually referred to by the acronym DNRA, Chapter 6), in which NO_3^- is converted to NH_4^+ through fermentation by obligate anaerobes (Zehnder and Stumm, 1988; Megonigal et al., 2003a, b; Eq. 3 in Table 7.5). DNRA appears to be a dominant process in some wet soils and wetland sediments (Silver et al., 2001; Scott et al., 2008; Dong et al., 2011), particularly in anoxic habitats where nitrate availability is very low and labile carbon supplies are high. Under these conditions, selection may favor microbes that retain fixed N over denitrifiers that further deplete limited N supplies (Tiedje, 1988; Burgin and Hamilton, 2007).

Anaerobic oxidation of NH_4^+, or anammox, is another dissimilatory pathway, wherein NH_4^+ is oxidized by reaction with NO_2^- under anaerobic conditions to produce N_2. This process was only recently identified in wastewater treatment systems, but appears to be an important pathway for N_2 production in some coastal and marine sediments (Dalsgaard and Thamdrup, 2002; Zehr and Ward, 2002). Anammox appears to be competitively advantageous when carbon is highly limiting, but thus far there has been little research on its importance in freshwater ecosystems.

Nitrate may also be used in the oxidation of reduced sulfur, iron, or manganese compounds in anoxic sediments. Indeed, anaerobic sulfide oxidation using NO_3^- as an electron acceptor is an energetically favorable process accomplished by chemolithotrophic sulfur bacteria, which may occur in preference to denitrification or DNRA in situations where reduced sulfur compounds are abundant (refer to Table 7.5, Zehnder and Stumm, 1988). Anaerobic oxidation of Fe^{2+} using NO_3^- by microbes is known to occur (Fig. 1.6), but the ecosystem-level importance of this process is not currently well understood (Clement et al., 2005; Burgin and Hamilton, 2007). Anaerobic Fe and Mn oxidation might both be expected to be important consumption pathways for NO_3^- when organic substrates are limiting and the concentrations of reduced Fe or Mn compounds are high.

Burgin and Hamilton (2007) proposed a useful conceptual model summarizing current understanding of the many alternative pathways for NO_3^- reduction under anoxic conditions

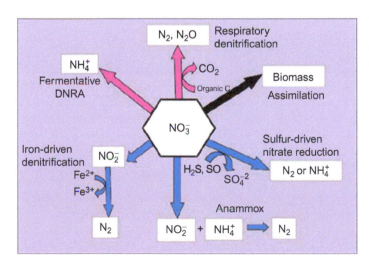

FIG. 7.16 A conceptual diagram of important nitrate removal pathways in the absence of oxygen. *Blue arrows* denote autotrophic pathways and *dark pink* arrows denote heterotrophic pathways. In addition to using nitrate to acquire energy through dissimilatory reactions, all microbes require N assimilation into biomass (*black arrow*). *From Burgin and Hamilton (2007). Used with permission of Ecological Society of America.*

(Fig. 7.16). The relative importance of these processes is a function of the actual free energy yield of competing reactions resulting from the relative concentration of chemical substrates (e.g., heterotrophic pathways will likely dominate when suitable organic molecules are available).

Iron and manganese reduction

Manganese reduction, although thermodynamically favorable in many anoxic environments, is only locally important because Mn^{4+} is rarely found at high concentrations. The product of the reaction, soluble Mn^{2+}, is toxic to many plants and can affect productivity or species composition. In contrast, Fe reduction is a dominant metabolic pathway in many wetlands (e.g., Fig. 7.11). In many cases there appears to be some overlap between the zone of denitrification and the zone of Mn reduction in sediments (e.g., Klinkhammer, 1980; Kerner, 1993; Fig. 7.10), and most of the microbes in this zone are facultative anaerobes that can tolerate periods of oxic conditions (Chapter 6). In contrast, there is little overlap between the zone of Mn reduction and that of Fe reduction because soil bacteria show an enzymatic preference for Mn^{4+}, and Fe^{3+} reduction will not begin until Mn^{4+} is completely depleted (Lovley and Phillips, 1988a). Below the zone of Mn^{4+} reduction, most redox reactions are performed by obligate anaerobes. Our earlier emphasis on the redox state of iron (refer to Fig. 7.15) reflects the widespread use of Fe as an index of the transition from mildly oxidizing to strongly reducing conditions.

Certain types of bacteria (e.g., *Shewanella putrefaciens*) can couple the reduction of Mn and Fe directly to the oxidation of simple organic substances (Caccavo et al., 1992; Lovley et al., 1993; Lovley and Phillips, 1988a), but usually these reactions are catalyzed by a suite of coexisting bacteria—with some species using fermentation to obtain metabolic energy (Eqs. 1 and 2 in Table 7.6), while others oxidize hydrogen, using Mn^{4+} and Fe^{3+} as electron acceptors (Eqs. 2 and 4 in Table 7.6; (Weber et al., 2006). Below the depth of iron reduction, the redox potential progressively drops and sulfate reduction and then methanogenesis become the dominant terminal decomposition pathways.

TABLE 7.6   The chemical reactions and associated free energy yield (per mol of organic matter or hydrogen compound) for Fe and Mn reduction pathways.

Reaction		$-\Delta G^O$ $(kJ\,mol^{-1})$	Organisms catalyzing these reactions
$2MnO_2 + (CH_2O) + 2H^+$	$\rightarrow MnCO_3 + Mn^{2+} + 2H_2O$	94.5	Members of the genus *Bacillus*, *Micrococcus*, and *Pseudomonas*
$2Mn^{3+} + H_2$	$\rightarrow 2Mn^{2+} + 2H^+$	285.3	Members of the genus *Shewanella*
$4FeOOH + (CH_2O) + 6H^+$	$\rightarrow FeCO_3 + 3Fe^{2+} + 6H_2O$	24.3	Members of the genus *Bacillus*
$2Fe^{3+} + H_2$	$\rightarrow 2Fe^{2+} + 2H^+$	148.5	Members of the genus *Pseudomonas* and *Shewanella*

From Zehnder and Stumm (1988) and Lovley (1991).

Sulfate reduction

In Chapter 6 we examined the reduction of sulfate that accompanied the uptake, or assimilation, of sulfur by soil microbes and plants. In contrast, dissimilatory sulfate reduction in anaerobic soils is analogous to denitrification in which SO_4^{2-} acts as an electron acceptor during the oxidation of organic matter by bacteria (Table 7.7). This metabolic pathway evolved at least 2 bya (Chapter 2). Sulfate-reducing bacteria produce a variety of sulfur gases including hydrogen sulfide, H_2S; dimethylsulfide, $(CH_3)_2 S$); and carbonyl sulfide, COS (Conrad, 1996). Before widespread human air pollution the release of sulfur gases from wetlands was the dominant source of sulfur in the atmosphere (Chapter 13). In brackish or saline waters, sulfate is typically the dominant electron acceptor and sulfate reduction can be an important fate for organic matter (e.g., Howarth and Teal, 1979; Neubauer et al., 2005).

Although the production of reduced-sulfur gases in wetlands may be high, the escape of H_2S from wetland soils is often much less than the rate of sulfate reduction at depth as a result of reactions between H_2S and other soil constituents (e.g., NO_3, Table 7.5). Hydrogen sulfide can react abiotically with Fe^{2+} to precipitate FeS, which gives the characteristic black color to anaerobic soils. FeS may be subsequently converted to pyrite in the reaction

$$FeS + H_2S \rightarrow FeS_2 + 2H^+ + 2e^-. \tag{7.19}$$

This process is accompanied by the generation of CH_4 by co-occurring microbial populations engaging in CO_2 reduction (Eq. 2.8; Thiel et al., 2019). When H_2S diffuses upward through the zone of oxidized Fe^{3+}, pyrite (FeS_2) is precipitated following

$$2Fe(OH)_3 + 2H_2S + 2H^+ \rightarrow FeS_2 + 6H_2O + Fe^{2+}. \tag{7.20}$$

Thus, not all the reduced iron in wetland soils is formed directly by iron-reducing bacteria. In some cases the indirect pathways (Eqs. 7.19, 7.20) may account for most of the total (Canfield, 1989a; Jacobson, 1994).

TABLE 7.7 The chemical reactions and associated free energy yield (per mol of organic matter or H_2 compound) for sulfate reduction pathways.

Reaction		$-\Delta G^O$ (kJ mol^{-1})	Organisms catalyzing these reactions
$1/2SO_4^{2-} + (CH_2O) + 1/2H^+$	$\rightarrow 1/2HS^- + CO_2 + H_2O$	18.0	*Desulfobacter* sp., *Desulfovibrio* sp., *Desulfonema* sp., etc.
$S^0 + (CH_2O) + H_2O$	$\rightarrow HS^- + H^+$	12.0	*Desulfomonas acetoxidans*, *Campylobacter* sp., *Thermoproteus tenax*, *Pyrobaculum islandicum*
$S^0 + H_2$	$HS^- + H^+$	14.0	*Thermoproteus* sp., *Thermodiscus* sp., *Pyrodictum* sp., various bacteria

From Zehnder and Stumm (1988).

A low iron content limits the accumulation of iron sulfides in many wetland sediments (Rabenhorst and Haering, 1989; Berner, 1984; Giblin, 1988). During periods of low water, specialized bacteria may reoxidize the iron sulfides (Ghiorse, 1984), releasing SO_4^{2-} that can diffuse back down to the zone of sulfate-reducing bacteria. Thus, high rates of sulfate reduction may be maintained in soils and sediments that have relatively low SO_4^{2-} concentrations, owing to the recycling of sulfur between oxidized and reduced forms (Marnette et al., 1992; Urban et al., 1989b; Wieder et al., 1990).

Hydrogen sulfide also reacts with organic matter to form carbon-bonded sulfur that accumulates in peat (Casagrande et al., 1979; Anderson and Schiff, 1987). In many areas, the majority of the sulfur in wetland soils is carbon-bonded and only small amounts are found in reduced inorganic forms—that is, H_2S, FeS, and FeS_2 (Spratt and Morgan, 1990; Wieder and Lang, 1988). Carbon-bonded forms—from the original plant debris, from the reaction of H_2S with organic matter, and from the direct immobilization of SO_4 by soil microbes—are relatively stable (Rudd et al., 1986; Wieder and Lang, 1988). Carbon-bonded sulfur accounts for a large fraction of the sulfur in many coals (Casagrande and Siefert, 1977; Altschuler et al., 1983) and thus for the sulfuric acid content of air pollution from power plants. Organic sediments and coals containing carbon-bonded sulfur that is the result of dissimilatory sulfate reduction show negative value for $\delta^{34}S$ as a result of bacterial discrimination against the rare heavy isotope $^{34}SO_4^{2-}$ in favor of $^{32}SO_4^{2-}$ during sulfate reduction (Chambers and Trudinger, 1979; Hackley and Anderson, 1986).

These various sinks for reduced sulfur initially led many researchers to believe that sulfate reduction was not a particularly important pathway for organic matter decomposition in wetland ecosystems since the emission of reduced sulfur gases from wetlands was so low. Experimental studies using radiolabeled $^{34}SO_4^{2-}$ have demonstrated that, where reduced iron is available, a substantial fraction of the reduced sulfur products of sulfate reduction are sequestered in the sediments as $Fe^{34}S_2$ (Howarth and Teal, 1979; Howes et al., 1984; Cornwell et al., 2013). Indeed iron sequestration of reduced S in soils can buffer ecosystems against sulfidation caused by saltwater intrusion (Schoepfer et al., 2014).

Because H_2S can react with various soil constituents and is oxidized by sulfur bacteria as it passes through the overlying sediments and water (Eq. 2.13), many biogeochemists once believed that various organic sulfur gases might be the dominant forms escaping from wetland soils. Most studies, however, have found that H_2S accounts for most of the emission from wetland soils (Adams et al., 1981; Kelly and Smith, 1990). Castro and Dierberg (1987) reported a flux of H_2S containing 1–110 mg S m^{-2} yr^{-1} from various wetlands in Florida. Nriagu et al. (1987) reported a total flux of sulfur gases ranging from 25 to 184 mg m^{-2} yr^{-1} from swamps in Ontario and found that the sulfate in rainfall in the surrounding region had a lower $\delta^{34}S$ value during the summer than during the winter. Presumably a portion of the SO_4^{2-} content in rain is derived from the oxidation of sulfur gases released to the atmosphere by sulfate reduction in local wetlands.

Methanogenesis

Methanogenesis degrades carbon when all alternative electron acceptors have been exhausted. Despite its extremely low energy yield (refer to Table 7.3), methanogenesis is a dominant pathway for organic matter decomposition in many wetlands due a limited supply of oxidants at depth. Methane can be produced from two different pathways in flooded sediments, both of which are accomplished by methanogens, a diverse group of strict anaerobes in the archaea (Zehnder and Stumm, 1988). When fermentation of organic matter produces organic acids at concentrations in excess of the availability of alternative electron donors (NO_3^-, Fe^{3+}, SO_4^{2-}), then methanogens can split acetate to produce methane in a process called acetoclastic methanogenesis or acetate fermentation (Eq. 2.7 and Table 7.8; Reaction 1). The energy yield of acetoclastic methanogenesis is very low compared to other anaerobic metabolic pathways (Table 7.4), and the product of this reaction produces a $\delta^{13}C$ of $-50‰$ to $-65‰$ in CH_4 (Woltemate et al., 1984; Whiticar et al., 1986; Cicerone and Oremland, 1988). Acetoclastic methanogenesis is performed by only two genera of methanogens: the *Methanosarcina* and *Methanosaeta* (Megonigal et al., 2003a, b).

When acetate is unavailable, a wider variety of methanogens engage in CO_2 reduction (Eq. 2.8 and Table 7.8, Reaction 2), in which the hydrogen from fermentation serves as a source of electrons and energy while the CO_2 serves as the source of C and as an electron acceptor. Methanogenesis by CO_2 reduction accounts for the limited release of H_2 gas from wetland soils (Conrad, 1996). This methane is even more highly depleted in ^{13}C than that produced from acetoclastic methanogenesis, with $\delta^{13}C$ of $-60‰$ to $-100‰$ (Whiticar et al., 1986). The isotope ratios in methane are useful to ascertain the source of rising concentrations of methane in Earth's atmosphere (Chapter 11).

Methanogenesis is often limited by the supply of fermentation products (H_2 or acetate), and can be stimulated through experimental additions of either organic matter or hydrogen (Coles and Yavitt, 2002). Thus, methanogenesis generally declines with depth below the oxic-anoxic interface in wetland sediments (Megonigal and Schlesinger, 1997), and with greater depth, methane production is increasingly from CO_2 reduction (Hornibrook et al., 1997).

Methanogenic bacteria can use only certain organic substrates for acetate splitting and in many cases there is evidence that sulfate-reducing bacteria are more effective competitors for the same compounds (Kristjansson et al., 1982; Schönheit et al., 1982). Sulfate-reducing bacteria also use H_2 as a source of electrons, and they are more efficient in the uptake of H_2 than methanogens engaging in CO_2 reduction (Kristjansson et al., 1982; Achtnich et al., 1995). Thus in most environments there is little to no overlap between the zone of methanogenesis and the

TABLE 7.8 The chemical reactions and associated free energy yield (per mol of organic matter or H_2) for the methane producing pathways of acetate splitting and CO_2 reduction (or hydrogen fermentation).

Reaction		$-\Delta G^O$ (kJ mol^{-1})	Organisms catalyzing these reactions
Acetate splitting			
(1) CH_3COOH	$\rightarrow CH_4 + CO_2$	28	Some methanogens (*M. barkeri*, *M. mazei*, *M. söhngenii*
CO_2 reduction			
(2) $CO_2 + 4H_2$	$\rightarrow CH_4 + H_2O$	17.4	Most methane bacteria

From Zehnder and Stumm (1988).

zone of sulfate reduction in sediments (Kuivila et al., 1989; Lovley and Phillips, 1987). Methanogenesis in marine sediments is inhibited by the high concentrations of SO_4 in seawater, and where methanogenesis occurs in marine environments CO_2 reduction is much more important than acetate splitting because acetate is entirely depleted within the zone of sulfate-reducing bacteria (Chapter 9). Because sulfate reduction provides a higher energy alternative, sulfate inputs to freshwater wetlands via acid deposition appear to suppress CH_4 flux (Gauci et al., 2004; Gauci and Chapman, 2006; Dise and Verry, 2001).

Methane flux varies widely across wetland ecosystem types, making global extrapolations challenging. Current estimates suggest that approximately 3% of the net ecosystem production in wetlands is released to the atmosphere as CH_4 (an estimated 217×10^{12} g CH_4 annually) (Bridgham et al., 2013; Dlugokencky et al., 2011 refer to Table 11.2). Much of the interannual variation in global CH_4 fluxes can be explained by annual variation in wetland emissions (Bousquet et al., 2006). The flux of methane from wetland ecosystems results from the net effects of methanogenesis and methanotrophy. Methanogenesis is limited by the supply of labile organic matter (Bridgham and Richardson, 1992; Cicerone et al., 1992; Valentine et al., 1994; Denier van der Gon and Neue, 1995), and CH_4 flux shows a direct correlation to net ecosystem production across a variety of wetland ecosystems (Whiting and Chanton, 2001; Updegraff et al., 2001; Vann and Megonigal, 2003; Fig. 7.17).

The positive association between plant productivity and methane flux may be due at least in part to the facilitated gas exchange provided by many wetland plants (Sebacher et al., 1985;

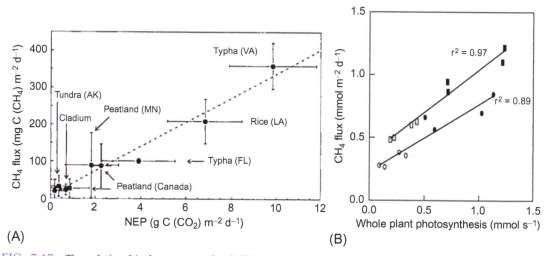

(A)   (B)

FIG. 7.17   The relationship between wetland CH_4 emissions and various measures of primary productivity. (A) Emissions versus NEP in North American ecosystems ranging from the subtropics to the subarctic; here the slope is 0.033 g methane C/g CO_2. (B) Emissions versus whole-plant net photosynthesis in marsh microcosms planted with the emergent macrophyte *Orantium aquaticum* that were exposed to elevated and ambient concentrations of atmospheric CO_2. (A) From Whiting and Chanton (1993); (B) from Vann and Megonigal (2003). Used with permission of Nature Publishing Group and Springer.

Chanton and Dacey, 1991; Yavitt and Knapp, 1995; Carmichael et al., 2014). Many wetland plants, including rice, have hollow stems composed of aerenchymous tissue, which allows O_2 to reach the roots. These hollow stems inadvertently act as a conduit for CH_4 transport to the surface (Kludze et al., 1993). When oxic soil layers overlay zones of highly reduced anoxic sediments, much of the CH_4 that diffuses to the sediment surface will be oxidized. Higher methane fluxes typically occur when soils are completely saturated. Under these conditions obligate anaerobic methanogens can operate in shallower sediments, where high-quality organic matter is concentrated and there is a greater chance for CH_4 escape to the atmosphere (Sebacher et al., 1986; Moore and Knowles, 1989; Shannon and White, 1994; von Fischer et al., 2010). In flooded soils CH_4 flux tends to increase with soil temperature (Roulet et al., 1992; Bartlett and Harriss, 1993).

Aerobic oxidation of CH_4

Methanotrophs tend to outcompete nitrifiers for O_2 when CH_4 is abundant since more energy can be released from oxidizing methane than from oxidizing NH_4. Thus much of the methane produced at depth is oxidized to CO_2 before it is released from wetland sediments (Fig. 7.18). When CH_4 concentrations are very high at depth—high enough to exceed the hydrostatic pressure of the overlying water—then CH_4-rich gas bubbles can escape to the surface in the process of *ebullition* (Fig. 7.18). Effectively bypassing the methanotrophs, ebullition can account for a large fraction of the methane flux to the atmosphere (Neue et al., 1997; Goodrich et al., 2011; Baird et al., 2004; Comas and Wright, 2012). Bubbles escaping by ebullition may be nearly pure CH_4, whereas bubbles emerging from vegetation are often diluted with N_2 from the atmosphere (Chanton and Dacey, 1991).

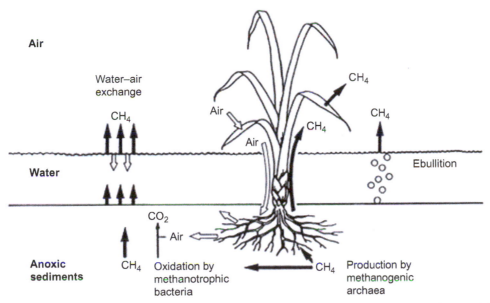

FIG. 7.18 Processes of methane production, oxidation, and escape from wetland soils. *From Schutz et al. (1991).*

The effect of plant-facilitated gas exchange on methane transport and oxidation is complex because, by oxidizing their rhizosphere, aerenchymous plants create a larger volume of oxidized soil where methanotrophs can oxidize CH_4 at the same time that aerenchymous tissues can facilitate CH_4 escape through direct transport (Laanbroek, 2010).

Anaerobic oxidation of CH_4

In marine sediments anaerobic methane oxidation (AMO) by sulfate-reducing bacteria can be a major sink for methane as it passes upward through the overlying sediments (Reeburgh, 1983; Henrichs and Reeburgh, 1987; Blair and Aller, 1995):

$$CH_4 + SO_4^{2-} \rightarrow HS^- + HCO_3^- + H_2O, \tag{7.21}$$

and it has been suggested that AMO may be an important mechanism regulating methane emissions in freshwater wetlands as well (Smemo and Yavitt, 2011). Due to both anaerobic and aerobic methane oxidation, the net ecosystem fluxes of CH_4 from wetlands are often far lower than gross rates of methanogenesis. Reeburgh et al. (1993) have estimated that the global rate of methane production in wetlands is about 20% larger than the net release of methane from wetland soils. To understand the relative importance of methane consumption, investigators note that methane-oxidizing bacteria alter the $\delta^{13}C$ of CH_4 escaping to the atmosphere, and comparisons of the isotopic ratio of CH_4 in sediments and surface collections can indicate the importance of oxidation (Happell et al., 1993). A study in the Florida Everglades, for example, found that more than 90% of methane production is consumed by methanotrophs before it diffuses to the atmosphere (King et al., 1990).

The flux of methane from wetlands shows great spatial variability as a result of differences in soil properties, topography, and vegetation (Bridgham et al., 2013), making global extrapolations difficult. Methane fluxes may increase with increasing temperatures if wetland sediments remain saturated (Christensen et al., 2003). Wetlands may shift from being net CH_4 sources during wet seasons to net CH_4 sinks as temperatures rise and water tables are lowered (Harriss et al., 1982). In an early comprehensive synthesis of wetland methane efflux studies, Bartlett and Harriss (1993) reported that tropical wetlands were responsible for >60% of total global wetland CH_4 emissions while northern wetlands (north of 45°N) were responsible for nearly one-third of global emissions from wetlands (Table 7.9).

Although the range of measured rates of CH_4 emissions overlaps considerably across latitudes, the longer growing seasons and greater spatial extent of tropical wetlands explain their greater contribution to global atmospheric CH_4 (Fig. 7.19; Bartlett and Harriss, 1993). Net regional losses of methane from wetlands are partially balanced by the consumption of atmospheric methane in adjacent upland soils, where it is consumed by methane-oxidizing bacteria (Whalen et al., 1990; Le Mer and Roger, 2001; see also Chapter 11). In wet tropical forests, contributions of methane emissions to global climate change can exceed the benefits of carbon uptake (Dalmagro et al., 2019). Efforts over the last two decades to constrain the global estimates of CH_4 emissions from natural wetlands have suggested that wetlands may produce $217 \times 10^{12} g\ CH_4\ yr^{-1}$, or approximately one third of annual CH_4 emissions (Bridgham et al., 2013; refer to Table 11.2).

TABLE 7.9 Compiled estimates of global emissions of CH_4 from natural wetlands.

Authors	Method	Global emissions from wetlands (Tg CH_4 yr^{-1})	Global natural emissions (Tg CH_4 yr^{-1})	% Natural sources
Matthews and Fung (1987)	Upscaling from field estimates	110	–	–
Aselmann and Crutzen (1989)	Upscaling from field estimates	40–160	–	–
Hein et al. (1997)	Global inverse modeling	231	–	–
Houweling et al. (2000)	Global inverse modeling	163	222	73%
Wuebbles and Hayhoe (2002)		100	145	69%
Wang et al. (2004)	Global inverse modeling	176	200	88%
Fletcher et al. (2004)	Global inverse modeling	231	260	89%
Chen and Prinn (2006)	Global inverse modeling	145	168	86%
Our synthesis——Table 11.2	Data synthesis	217	347	63%

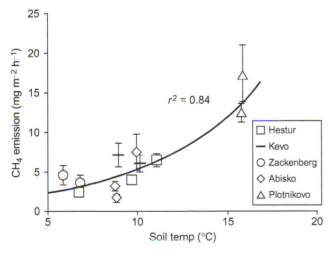

FIG. 7.19 The relationship between mean seasonal soil temperature (at 5-cm depth) and mean seasonal CH_4 flux (measured at least 8 times throughout the growing season at each site) at the measurement site during all years. *From Christensen et al. (2003). Used with permission of American Geophysical Union.*

The redox ladder

The large differences in free energy yield between the various decomposition pathways in wetlands can be used to predict the dominant metabolic processes under many environmental conditions. When NO_3^- or oxidized Fe or Mn is available in anoxic sediments, sulfate reduction is generally suppressed. Similarly, methanogenesis is suppressed by the provision of SO_4. Indeed, it has been suggested that sulfate delivered by acid rain could be suppressing current global CH_4 emissions by as much as 8% (Vile et al., 2003; Gauci et al., 2004). Thus the redox ladder is a useful tool for predicting the dominant metabolic pathways in wetlands.

That said, most of our theoretical understanding of anaerobic metabolism is based on using the measured free energy yields from pure culture laboratory studies. The expansion of our ability to study microbiomes in the environment is shifting our understanding of anaerobic metabolism considerably. Though redox potential may predict the winning competitive strategy, it turns out that not all microbes are in direct competition, and indeed syntrophic[g] associations regularly "break the rules" of the free energy based redox ladder. A variety of recent studies have demonstrated that consortia of co-occurring microbial species can collectively perform chemical reactions that would not be predicted from studies of individual species (Lovley and Phillips, 1988a; Boetius et al., 2000; Raghoebarsing et al., 2006). Microbial consortia have been described in which sulfate-reducing or nitrate-reducing microbes grow in intimate association with methanogens, allowing them to use methane as a carbon source to fuel sulfate and nitrate reduction (Conrad, 1996; Boetius et al., 2000; Raghoebarsing et al., 2006). These syntrophic interactions result in anaerobic methane oxidation, and the role of methanogens in fueling these reactions would not be apparent from traditional measures of substrate accumulation or loss at the ecosystem scale.

Wetlands and water quality

Thus far we have primarily focused on the retention of C in the organic matter in wetland sediments and the gaseous loss of C as CO_2 or CH_4. Hydrologic losses of dissolved organic carbon (DOC) from wetlands represent a major export term and an important source of organic carbon and nutrients for many rivers and coastal estuaries. Water in contact with peat leaches DOC from the peat matrix (Dalva and Moore, 1991), and concentrations of DOC in peatland soils and overlying surface waters typically range from 20 to 100 mg C/L, with the vast majority of the DOC characterized as organic acids (Thurman, 1985). Higher water tables tend to increase the rate of DOC production and increase rates of DOC loss for wetlands that are hydrologically connected to downstream ecosystems or groundwaters (Blodau et al., 2004). Much of the variation in DOC fluxes between rivers can be explained by differences in the wetland area of their watersheds (Dillon and Molot, 1997; Gergel et al., 1999; Pellerin et al., 2004; Johnston et al., 2008; Raymond and Saiers, 2010).

[g]Syntrophy, (derived from the Greek syn meaning together and trophe meaning nourishment) is used to describe the phenomenon in which one species lives off the products of another species.

Without oxygen, decomposition is slow in saturated sediments, maintaining nutrient limitation of biomass growth and leading to the gradual buildup of soil organic matter in wetland soils. Wetlands can thus sequester large quantities of nutrients and trace elements delivered from their catchments, incorporating these elements into plant biomass and eventually into soil organic matter. In addition to retaining elements in tissues and soils, wetland sediments provide ideal conditions for denitrifers and possess a great capacity to remove excess NO_3 and convert inorganic phosphorus and other trace elements into organic molecules (e.g.,Johnston et al., 1990; Emmett et al., 1994; Zedler and Kercher, 2005; Fergus et al., 2011). In addition to sequestering trace elements, microbes in wetland sediments are responsible for the methylation of a wide variety of metals, some of which are toxic to biota and more rapidly assimilated in the methyl form (e.g., CH_3Hg; Chapter 13).

Wetlands and global change

The areal extent of wetlands has declined through human history as wetlands have been drained, filled, or cultivated for agricultural or urban development. Although it is impossible to determine the global area of wetlands that have been lost through direct human intervention, recent estimates suggest that 50% of all terrestrial wetlands have been destroyed by human activities in the United States, with higher rates for the developed regions of Canada, Europe, Australia, and Asia (Mitsch and Gosselink, 2007).

Sea level rise and saltwater intrusion

In some regions freshwater wetlands may become brackish or saline as a result of saltwater intrusion into coastal wetlands or through drought-induced evaporative concentration of salts (Herbert et al., 2015; Tully et al., 2019). The resulting changes in ionic strength drive "internal eutrophication," as PO_4^{3-} formerly bound to anion adsorption sites is displaced by Cl^- and SO_4^{2-} and reduced S compounds bind with Fe, reducing the efficiency of Fe-PO_4 binding (Caraco et al., 1989; Lamers et al., 1998; Beltman et al., 2000; Ardón et al., 2017). Base cations in marine salts may displace NH_4^+ on cation exchange sites (Chapter 4), releasing large quantities of nitrogen from salinizing wetlands (Ardón et al., 2013). The provision of the important electron acceptor SO_4^{2-}, a dominant constituent of seawater, enhances organic matter decomposition (through sulfate reduction), increasing concentrations of HS^-, and potentially suppressing methanogenesis and denitrification (Lamers et al., 1998; Weston et al., 2006; Sutton-Grier et al., 2011).

Climate change

Climate change is likely to accelerate the rate of wetland loss, as wetlands are particularly vulnerable to altered patterns of precipitation and evaporation on the world's continents (Chapter 10). It is difficult to anticipate the net effect of wetland loss on the global C cycle (Avis et al., 2011), but it is clear the extent of wetland inundation is a primary driver of interannual variation in wetland methane emissions at all latitudes (Ringeval et al., 2010). Less frequent inundation is likely to foster oxidation of soil organic matter, but the loss of organic matter may be coupled with enhanced NPP (Megonigal et al., 1997; Choi et al., 2007) and reduced CH_4 emissions (Bousquet et al., 2006; Ringeval et al., 2010).

Predicting the effect of rising temperatures on wetland biogeochemistry is difficult because it requires reconciling sometimes opposite direct effects and complex positive and negative feedbacks. For instance, higher temperatures will lead to a lower capacity of water to hold oxygen, perhaps further reducing the zones of aerobic respiration in many wetlands and slowing organic matter decomposition. At the same time, both methanogenesis and CO_2 production may increase in wetland sediments at higher temperatures (Avery et al., 2002). A synthesis of experimental warming studies in wetlands found no consistent effect of higher temperatures on fluxes of CO_2, CH_4, and N_2O, with responses ranging in both sign and magnitude for all three greenhouse gases (Dijkstra et al., 2012).

The effect of warmer temperatures on plant evapotranspiration and the resulting changes in wetland hydroperiods will likely have the greatest impact on wetland processes. Of particular concern is the potential for a loss of high-latitude wetlands due to permafrost thaw (Limpens et al., 2008; Schuur et al., 2008; Shaver et al., 2011; Koven et al., 2011; McGuire et al., 2010). More than 50% of all wetlands are located at high latitudes (refer to Fig. 7.1) and recent estimates suggest that as much as 1300 ± 200 Pg of soil C is held within the northern permafrost region, of which about 60% occurs in perennially frozen soils (Hugelius et al., 2014). The hydrology of many boreal wetlands is constrained by permafrost barriers to drainage; thus, permafrost melting may lead to wetland drainage, with the potential for positive feedbacks to climate change through the decomposition of previously saturated organic matter (Avis et al., 2011) and the hydrologic export of large amounts of DOC (Guo et al., 2007; Olefeldt and Roulet, 2014).

The abundance of stored soil organic matter in peatland ecosystems results from inefficient decomposition over long periods of time rather than high rates of organic matter production. Peat decomposition will increase as water tables drop (Yu et al., 2003; Fenner and Freeman, 2011); thus, drier arctic climates will lead to lower water table elevations and increased oxidation of organic matter. Changes in the flux of methane to the atmosphere may occur if global warming causes changes in the saturation of peatlands, particularly at northern latitudes. If these soils become warm and dry, the flux of methane may be lower while the flux of CO_2 to the atmosphere may increase (Freeman et al., 1993; Moore and Roulet, 1993; Funk et al., 1994; Whalen, 2005). Drainage of wetlands (either intentionally or as a result of a drier climate) can make wetlands more susceptible to wildfire and fires that burn more deeply into peat or spread over a greater spatial extent (Grosse et al., 2011). Within the last two decades peatland fires in Alaska (Mack et al., 2011) and Canada (Turetsky et al., 2011) each removed centuries of accumulated peat. In a comparison of burned and unburned peatlands in northern Canada, Gibson et al. (2019) found that deep soil respiration was four times higher in burned sites even 16 years after the wildfire. By increasing soil temperatures and the active layer depth, the long-term impacts of fire on SOM loss can extend for many years after the disturbance itself.

Elevated CO_2

Investigations of the effects of elevated CO_2 on wetlands provide little evidence for increasing plant biomass, and raise concerns that increased CO_2 may stimulate enhanced rates of soil organic matter oxidation. Several experimental CO_2 enrichments in peatlands have shown that higher atmospheric CO_2 increases vascular plant biomass, but the effect is considerably offset by accompanying losses of peat-building bryophytes (especially *Sphagnum*) (Berendse et al., 2001). In long-term CO_2 enrichment experiments in Virginia, USA coastal marshes, the response of vascular plant biomass to CO_2 was constrained by low nutrient availability, and the low nutrient turnover in many wetlands may substantially limit their ability to sequester additional C in biomass in response to elevated CO_2 (Langley and Megonigal, 2010; Pastore et al., 2016). Both field and laboratory evidence suggests that elevated CO_2 leads to enhanced organic matter exudation by wetland plants that can exacerbate wetland DOC losses (Freeman et al., 2001b), promote organic matter oxidation (Wolf et al., 2007), and increase CH_4 production (Megonigal and Schlesinger, 1997).

Because CH_4 production has been found to be highly correlated with NEP in a number of studies (refer to Fig. 7.17), if rising atmospheric CO_2 stimulates the growth of wetland plants, it may simultaneously provide a greater supply of organic substances for methane-producing bacteria (Fenner et al., 2007; Vann and Megonigal, 2003; Megonigal and Schlesinger, 1997). A meta-analysis of CO_2-enrichment experiments from 16 studies of natural wetlands and 21 studies in rice paddies showed that increased CO_2 stimulated CH_4 emissions in wetlands by 13.2% and in rice paddies by 43.4% (van Groenigen et al., 2011), suggesting that enhanced production of CH_4 may offset C sequestration in wetlands exposed to elevated CO_2.

The flux of methane from natural wetlands accounts for $\sim 217 \times 10^{12}$ g of CH_4 annually, which is a large portion of the total global flux to the atmosphere (Chapter 11). With atmospheric CH_4 concentrations increasing at about 1% yr^{-1}, various workers have asked whether changes in ecosystem processes within wetlands or changes in the spatial extent of wetlands might be responsible. Certainly, global methanogenesis has increased with the increasing cultivation of rice, which now accounts for $\sim 10\%$ of the global production of methane from wetlands (Chapter 11).

Summary

Wetlands occupy a small (and shrinking) proportion of the continental land surface, but play a critical role in global biogeochemistry by storing as much as 50% of soil organic matter, producing more than 20% of CH_4, and by substantially reducing the inorganic nutrient export to river networks and the sea. The unique role of wetlands in local to global biogeochemical cycling results from the tendency for wetlands to have waterlogged, anoxic soils for some or all of each year. The limited availability of oxygen to wetland heterotrophs impedes decomposition, slows nutrient turnover, and allows a variety of alternative metabolic pathways to dominate ecosystem C cycling and to link the cycling of C with that of nitrate, manganese, iron, sulfate, and hydrogen through microbial energetics.

Inland Waters

Introduction

Freshwater ecosystems (lakes and rivers) hold less than 0.02% of all water on Earth and occupy about 6% of the land surface area (Fig. 10.1; Wetzel, 2001; Lehner and Doll, 2004). This small footprint of surface waters on the land belies their importance to human civilization and to global biogeochemistry. Because surface waters are constantly replenished and easily accessed, rivers and lakes provide the water supply for the vast majority of humans on Earth. In addition to providing the water necessary to grow most crops, inland waters support freshwater fisheries that currently provide ~1/3 of global annual fish

Biogeochemistry: An Analysis of Global Change, Fourth Edition
https://doi.org/10.1016/B978-0-12-814608-8.00008-6

293

catch.[a] At the same time, surface waters are highly managed for recreation, disposal of wastes, transport of cargo, and generation of electricity, with important consequences for their hydrology, chemistry, biodiversity and biogeochemical cycles.

In Chapter 4 we discussed the role of rivers in moving chemical elements from the continents to the oceans. Over geologic time the movement of water has weathered primary minerals, producing sedimentary rocks and salty seas (Eq. 4.1). For most elements, their transfer from land to sea is considerably delayed by their interactions with aquatic organisms and by their long residence times in the sediments of lakes and floodplains. In this chapter, we will describe the biogeochemistry of aquatic ecosystems—focusing particularly on the capacity of aquatic ecosystems to alter the form, timing, and magnitude of downstream transport of C, N, and P from terrestrial ecosystems to the oceans.

Much of this chapter is devoted to the distinguishing features of lakes, rivers and estuaries. At this point in the text we have covered all of the key metabolic pathways that occur in the biosphere, and we merely have to describe how their distribution, timing and magnitudes differ between ecosystems with different attributes. Throughout this chapter it is important to keep in mind that although traditional biogeochemical studies have treated these as discrete systems, in reality streams, ponds, lakes, rivers and estuaries exist along a continuum of freshwater habitats that stretch from the smallest headwaters to their ultimate disposition in oceans or terminal lakes. Over sufficiently long periods of time, even water bodies that have no apparent connection to rivers are connected via groundwater or occasional floodwaters. Examining satellite imagery of any landscape demonstrates the network structure of rivers and the frequent intersections between lotic (flowing water) and lentic (standing water) habitats (Fig. 8.1).

The structure of aquatic ecosystems

In our discussion of terrestrial ecosystems and wetlands (Chapters 5–7) we regularly differentiated between aboveground and belowground biogeochemical pools and fluxes. In aquatic ecosystems we make a similar distinction between benthic ("bottom") and pelagic ("open water") compartments. Just as in terrestrial ecosystems surface topography can drive variation in light and water availability, in aquatic ecosystems bathymetry[b] determines the extent of benthic habitat that is exposed to sufficient light to support photosynthesis. The benthos of any aquatic ecosystem is always physically and often chemically distinct from the overlying water column, being a solid surface where materials are deposited from the overlying water column or eroded from upslope. In shallow freshwater systems or along the shores of larger water bodies, rooted macrophytes may occupy both zones, but in deeper aquatic systems there is significant specialization. Life in the pelagic zone requires adaptation or effort to remain in suspension, but has the advantage of higher light and oxygen

[a] From FAO's 2010 State of World Fisheries and Aquaculture report, available from http://www.fao.org/docrep/013/i1820e/i1820e.pdf.

[b] The topography of the bottom of an aquatic ecosystem, as determined by depth measurements from the surface.

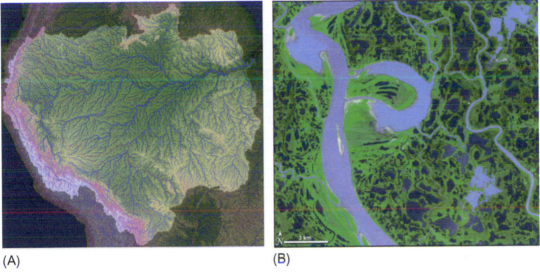

(A)   (B)

FIG. 8.1   Two views of river networks: (A) the Amazon River Basin, which is currently estimated to be 6800 km long and drains a landscape that ranges in elevation from 4500 m (in *white*) to sea level (in *dark green*); (B) a segment of Canada's Mackenzie River Delta from August 4, 2005, when lakes throughout the Mackenzie's floodplain had thawed. *Both images from NASA's Earth Observatory, (A) Created using Shuttle Radar Topography Mission (SRTM) data together with river data developed by the World Wildlife Fund's HydroSHEDS program (Lehner et al. (2008)). (B) From NASA's Terra Satellite Advanced Spaceborne Thermal Emission and Reflection Radiometer (ASTER).*

availability. Life in the benthos must regularly contend with low oxygen and limited light, but benthic organisms can more easily resist being moved with the flow.

In terrestrial ecosystems, much of the physical structure within the ecosystem is created by vegetation, with plants investing significant amounts of carbon in structural tissues that enable them to outcompete their neighbors for limiting supplies of light, nutrients or water. In contrast, most of the physical structure within aquatic ecosystems is abiotic. Except in the case of macrophytes or moss which provide a biomass canopy in which other organisms may live, it is sediment, rocks and coarse woody debris that provide surfaces and structure. For the majority of aquatic ecosystems, algae are the dominant autotrophs. These single celled or filamentous colonial organisms take many growth forms; they may float or attach to surfaces, but they are quite small in size relative to the majority of their consumers. Rather than investing in structural tissues, many aquatic organisms secrete exopolysaccharides that create biofilms on solid surfaces[c] or allow collections of buoyant organisms and particles to aggregate in the water column.

[c]It is these biofilms that make rocks 'slippery' in so many rivers and lakes.

The special properties of water

The unique physical and chemical properties of water control the biogeochemistry of aquatic ecosystems. In most aquatic ecosystems water is the medium where aquatic organisms live rather than a constraint on their rates of activity, and it is only the extremes of low and high water supply (droughts and floods) that water constrains biological activity via physical stress. Among the most important distinctions between terrestrial and aquatic ecosystems are the attenuation of incident light by water, the potential for density stratification of water masses, and the slow diffusion of gases through water relative to air (as discussed in Chapter 7).

Aquatic habitats can be further described based on their light regimes. Much of the light that reaches the surface of a river or lake is scattered or absorbed, so that the amount of solar energy delivered to a freshwater ecosystem diminishes with depth. The euphotic zone extends from the water surface to the depth at which underwater light is <1% of full sunlight. Photosynthesis only occurs where sufficient light can penetrate. In shallow, clear water systems where light penetrates all the way to the sediments, the entire system is in the euphotic zone. In deep lakes, turbid rivers and blackwater systems, productivity will be confined to the small proportion of the system with sufficient light to support photosynthesis.

Because liquid water becomes less dense as it warms, solar energy on the water surface can generate density gradients that separate warmer surface waters from cooler deep-water habitats. In deepwater lakes, solar energy warms the surface waters (the epilimnion), reducing its density and causing these warmer surface waters to float above cold, dense deep waters (the hypolimnion). While the surface waters are mixed by the wind and readily exchange gases with the overlying atmosphere, the deeper waters of stratified systems have low rates of gas exchange. As a result, bottom waters can become oxygen-depleted whenever the rate of respiration exceeds the delivery of O_2 by diffusion. Even in aquatic ecosystems with well-oxygenated water columns, the slow rate of O_2 exchange and high O_2 demand in their sediments results in the dominance of anaerobic metabolic pathways in the benthos (Chapter 7).

In aquatic ecosystems, the majority of inorganic carbon is in the form of bicarbonate ions rather than gaseous CO_2. Carbon dioxide dissolved in water is partitioned between dissolved CO_2, bicarbonate and carbonate ions. Collectively these three forms are referred to as dissolved inorganic carbon, or DIC. The relative proportion of forms depends upon pH:

$$CO_2 + H_2O <=> H^+ + HCO_3^- <=> 2H^+ + CO_3^{2-} \tag{8.1}$$

At pHs below 4.3 most carbon dioxide is found as a dissolved gas, between 4.3 and 8.3 the dominant form is bicarbonate (HCO_3^-), while at pHs above 8.3 carbonate (CO_3^-) dominates. The majority of freshwater ecosystems have a pH between 5 and 8, thus bicarbonate dominates the inorganic C content of most freshwaters. To photosynthesize, aquatic autotrophs must first convert bicarbonate ions to CO_2 using the enzyme carbonic anhydrase, which contains Zn (Morel et al., 1994).

Inland waters and their watersheds

Water and solute supply

Aquatic ecosystems exist because precipitation exceeds evapotranspiration across most of the land surface and excess water flows downslope. This excess water may leach vertically through soils into groundwater, move laterally down slopes through subsurface flow paths, or flow over the soil surface when rainfall exceeds the rate of soil infiltration. Thus much of the water found within an aquatic ecosystem has arrived after passing through soil and mixing with groundwater in the surrounding watershed. Water entering a receiving water body is referred to as either groundwater flow, subsurface flow, or overland flow according to these differences in routing. Permanent surface waters are found wherever the land surface is below the level of local or regional groundwater (Chapter 7). When water flow from the surrounding landscape is sufficient to erode sediments and materials from the land, channels are formed (Montgomery and Dietrich, 1988).

The route and the rate at which water moves along these various flow paths will determine its chemical properties. Differences in the chemistry of waters can be used to understand the relative importance of different flow paths. For example the groundwater that flows into Walker Branch, a small stream in eastern Tennessee, has high concentrations of Ca^{2+} because of its long residence time in carbonate bedrock. In contrast, storm waters that pass through surface soils (i.e., the vadose zone) pick up high concentrations of SO_4 that have been deposited by acid deposition over the last several decades (Mulholland, 1993). These chemical signatures make it possible to determine the relative contribution of each flow path by measuring the ratio of Ca^{2+} to SO_4^{2-} in stream water (Fig. 8.2). Similar comparisons can distinguish the dominant sources of water in many aquatic ecosystems. For instance, in a survey of freshwaters across northern Wisconsin, Lottig et al. (2011) found that streams tended to have Ca^{2+} concentrations ~4 times higher than lakes, suggesting that stream flow in that region was primarily derived from groundwater with long mineral-water contact times,

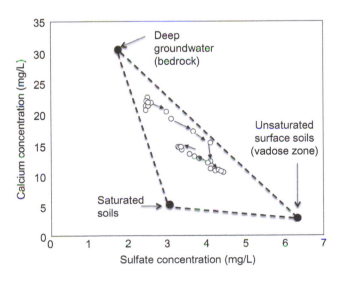

FIG. 8.2 Using a three component model of hydrologic flowpaths in Walker Branch Watershed, changes in stream-water calcium and sulfate concentrations over the course of a March 1991 storm (*open symbols*) are shown relative to the average [Ca^{2+}] and [SO_4^{2-}] of the three contributing flowpaths. Note that as the storm progresses (progression of time indicated by *arrows*), stream water quickly becomes dominated by shallow vadose zone flowpaths and later by saturated soil flowpaths. *Modified from Mulholland (1993).*

whereas lakes were dominated by solutes that predominate in precipitation and shallow sub-surface flowpaths that have had less contact with the underlying bedrock.

Since surface waters integrate the chemistry of their watersheds, they are often used to understand the biogeochemistry of terrestrial ecosystems (Chapters 4 and 6, Likens and Bormann, 1974a; Bormann and Likens, 1969). For example, seasonal variations in stream NO_3 concentrations, with very low concentrations during the growing season, are commonly used as to suggest that N uptake by vegetation controls watershed export (e.g., Bear Brook, NH, Bernhardt et al., 2005) (Fig. 8.3). This effect of terrestrial vegetation on stream nutrient concentrations is only observed in watersheds where water residence times are relatively short (<1 year; Lutz et al., 2012b). In watersheds with deep soils and long groundwater residence times, nitrogen uptake by terrestrial vegetation may be insufficient to deplete N concentrations in groundwater, and the N concentrations in stream water may be relatively constant. In the Walker Branch watershed in Tennessee, the lowest annual NO_3^- concentrations are observed during spring algal blooms and following autumn litter fall when rates of biological activity in the stream are highest (Mulholland, 2004; Roberts et al., 2007).

Water budgets

At any given time, the volume of water (V) in a river or lake reflects the balance between inputs in precipitation and surface inflows, exports as evaporation or surface flows and the net subsurface exchange with soil waters and groundwater (as in Eq. 7.1). The mean residence time (MRT, Eq. 3.3) of water within an aquatic ecosystem ultimately constrains the extent to which organisms can affect the form, magnitude and timing of element exports. Aquatic ecosystem inflows and precipitation are relatively easy (though time consuming) inputs to measure; however, an accurate determination of net groundwater flows is often difficult. The *benthos* (or bottom) of many lakes and rivers are sites of active exchange between surface water and groundwater, and net flux estimates may considerably underestimate the gross

FIG. 8.3   A comparison of average monthly stream NO_3 concentrations for Walker Branch. Tennessee (*black circles*) in the Oak Ridge National Lab and for Bear Brook, New Hampshire (*gray circles*) in the Hubbard Brook Experimental Forest. *Drawn with data from Mulholland (2004) and Bernhardt et al. (2005).*

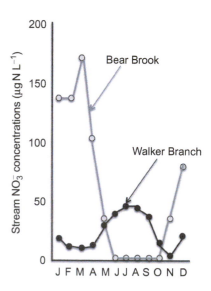

exchange of water across the sediment water interface (Covino, 2017; Poole et al., 2008). Net groundwater inputs are typically calculated as the difference in flow between two points along a river channel or between the inflows and outflows of a lake or reservoir.

Ion chemistry

The composition of freshwater ecosystems is typically dominated by four major cations (Ca^{2+}, Mg^{2+}, Na^+ and K^+) and four major anions (HCO_3^-, CO_3^-, SO_4^{2-}, and Cl^-) with dissolved forms of N, P, Fe and other trace elements at lower concentrations (Livingstone, 1963; Meybeck, 1979, 2003; Table 4.7). The concentration of these ions in surface waters can vary considerably as a function of catchment geology, the chemistry of precipitation, and the extent of evaporative concentration (Fig. 4.19). Atmospheric deposition of marine salts contributes substantial amounts of certain cations (Na^+, Mg^{2+}) and anions (Cl^-, SO_4^{2-}) to catchments that are near the sea. In areas downwind of coal-fired power plants and cities, a significant fraction of the acid volatiles emitted as atmospheric pollutants (SO_4^{2-} and NO_3^-) (Chapter 3) are delivered to surface waters.

The total mass of dissolved ions in water is known as its *ionic strength*. Ionic strength is typically reported in units of milliequivalents (mEq) of charge[d] (Chapter 4) and is often measured in the field as electrical conductivity.[e] The ionic composition of surface waters determines their alkalinity, which is roughly equivalent to the balance of cations and anions in water, viz.:

$$\text{Alkalinity} = \left[2Ca^{2+} + 2Mg^{2+} + Na^+ + K^+ + NH_4^+\right] - \left[2SO_4^{2-} + NO_3^- + Cl^-\right]. \quad (8.3)$$

This measure of alkalinity is roughly equivalent to:

$$\text{Alkalinity} = \left[2CO_3^{2-} + HCO_3^- + OH^-\right] - H^+. \quad (8.4)$$

Generally, alkalinity is measured in milliequivalents per liter by titration of a water sample to a pH of 4.3. Alkalinity is high in systems where H^+ inputs are low or when there are high rates of H^+ removal from the water column via base cation exchange (Chapter 4), H^+ protonation of organic acids (Hedin et al., 1990), or the release of the acid cation Al^{3+} through dissolution or exchange processes (Eqs. 4.10–4.12). Thus, the titration of a water sample to a pH of 4.3 is often said to represent its *acid-neutralizing capacity* (ANC), a term that embraces the total suite of inorganic and organic constituents that allow a surface water to resist acidification (sensu Schindler, 1988). In practice, alkalinity and ANC are often used interchangeably.

About 2/3 of the HCO_3^- in rivers is derived from the atmosphere and the remainder is produced from carbonate weathering reactions (Chapter 4). Carbon dioxide can enter aquatic ecosystems: (1) through gas exchange with the overlying atmosphere; (2) as the product of respiration within the aquatic ecosystem; or (3) as the products of organic decomposition and root respiration delivered via subsurface flow from the land (Meybeck, 1987;

---

[d] An equivalent (Eq) equals the weight of an element divided by its molecular weight and multiplied by its valence. For example 1 mg of Ca $=0.05$ mEq.

[e] The capacity of water to conduct electricity increases with ionic strength, typically reported in $\mu S\ cm^{-1}$.

Raymond and Cole, 2003; Stets et al., 2017). Because the majority of terrestrial ecosystems export both soil CO_2 and organic matter to water bodies, river and lake waters are typically supersaturated with CO_2 and are a net source of CO_2 to the atmosphere (Table 8.1) (Cole et al., 1994; Mayorga et al., 2005; Raymond et al., 2013). This displaced soil respiration represents a substantial fraction of CO_2 outgassing from many freshwater ecosystems (Stets et al., 2017; Raymond et al., 2013).

In most freshwaters, the cations delivered from the drainage basin are the dominant source of ANC, because the runoff of cations is usually balanced by HCO_3^- (Stoddard et al., 1999). Land use change can contribute additional alkalinity through enhanced erosion of base cations from soils and enhanced weathering associated with cultivation or mining (Raymond and Hamilton, 2018; Ross et al., 2018; Renberg et al., 1993; Raymond and Cole, 2003; Edmondson and Lehman, 1981).

The easiest way to understand the acid-neutralizing capacity of a freshwater system is to examine its charge balance. Negatively and positively charged ions in solution must balance, so if more acid anions (SO_4^{2-} and NO_3^-) are added to a watershed by acid rain or acid mine drainage, this charge must be balanced either by protons (H^+) or by cations (Na^+, K^+, Ca^{2+}, Mg^{2+}, Al^{3+}) that are released through dissolution or ion exchange processes. In decades of study at the Hubbard Brook Experimental Forest, inputs of the acid anions SO_4^{2-} and

TABLE 8.1   Estimate of CO_2 outgassing from inland waters.

Zone-class	Area of inland waters (1000s km^2) Min-max	pCO_2 (ppm) Median	Gas exchange velocity (k_{600} cm h^{-1}) Median	Areal outgassing (g C m^{-2} yr^{-1}) Median	Zonal outgassing (Pg C yr^{-1}) Median
Tropical (0°–25°)					
Lakes and reservoirs	1840–1840	1900	4.0	240	0.45
Rivers (>60–100 m wide)	146–146	3600	12.3	1600	0.23
Streams (<60–100 m wide)	60–60	4300	17.2	2720	0.16
Wetlands	3080–6170	2900	2.4	240	1.12
Temperate (25°–50°)					
Lakes and reservoirs	880–1050	900	4.0	80	0.08
Rivers (> 60–100 m wide)	70–84	3200	6.0	720	0.05
Streams (< 60–100 m wide)	29–34	3500	20.2	2630	0.08
Wetlands	880–3530	2500	2.4	210	0.47
Boreal and Arctic (50°–90°)					
Lakes and reservoirs	80–1650	1100	4.0	130	0.11
Rivers (> 60–100 m wide)	7–131	1300	6.0	260	0.02
Streams (< 60–100 m wide)	3–54	1300	13.1	560	0.02
Wetlands	280–5520	2000	2.4	170	0.49

TABLE 8.1 Estimate of CO_2 outgassing from inland waters—cont'd

Zone-class	Area of inland waters (1000s km²) Min-max	pCO₂ (ppm) Median	Gas exchange velocity (k_{600} cm h⁻¹) Median	Areal outgassing (g C m⁻² yr⁻¹) Median	Zonal outgassing (Pg C yr⁻¹) Median
Global		Global land area			
Lakes and reservoirs	2800–4540	2.1–3.4%			0.64
Rivers (> 60–100 m wide)	220–360	0.2–0.3%			0.30
Streams (< 60–100 m wide)	90–150	0.1–0.1%			0.26
Wetlands	4240–15,220	3.2–11.4%			2.08
All inland waters	7350–20,260	5.5–15.2%			3.28

Source: Aufdenkampe et al. (2011). Used with permission of the Ecological Society of America.

NO_3^- in precipitation have been strongly correlated with the base cation concentrations of receiving streams over several decades of study (Likens and Buso, 2012; Fig. 8.4). It is through proton displacement of base cations that acid rain or pyrite oxidation in mine spoils can raise the pH of receiving surface waters for watersheds dominated by carbonate rock (Kilham, 1982; Lajewski et al., 2003; Ross et al., 2018; Raymond and Oh, 2009).

Surface waters draining watersheds where precipitation passes through soils of low cation-exchange capacity and limited carbonate mineral material are more likely to become

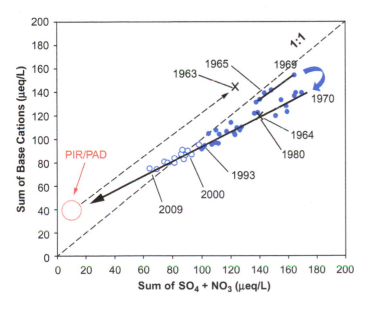

FIG. 8.4 Relation between the average annual, volume-weighted concentrations of the base cations (Ca, Mg, Na) and acid anions (SO_4 and NO_3) in streamwater from the Hubbard Brook Experimental Forest over the period 1963–2009. Units are in microequivalents of charge. The input of acidity and the loss of cations increased until the implementation of the Clean Air Act in 1970, following which both declined in parallel. The projected PIR (pre-industrial revolution) PAD (pre-acid deposition) range of values is shown in the *red* circle. *From Likens and Buso (2012).*

FIG. 8.5 The nutrient content of surface waters reflect nutrient loading to their catchments. The extractable soil phosphorus in agricultural watersheds is a good predictor of the concentrations of dissolved P in receiving streams. *Source: Sharpley et al. (1996). Used with permission of the Soil and Water Conservation Society.*

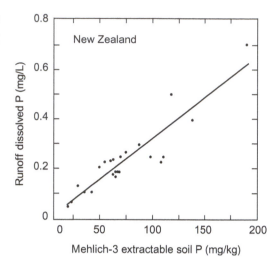

acidified in response to acid rain. Similarly freshwater systems with inputs that are dominated by precipitation or short flow paths with minimal water residence times also tend to have low ANC (Lottig et al., 2011). In either case, the receiving freshwaters are more susceptible to acidification. Freshwater bodies themselves have some buffering capacity, and alkalinity is increased by processes that consume SO_4^{2-} or NO_3^- from the water column, including sulfate reduction, sulfate-adsorption on minerals, and denitrification (Cook and Schindler, 1983; Schindler, 1986; Kelly et al., 1987). In contrast, the production of organic carbon by photosynthesis and the deposition of calcite by phytoplankton reduce alkalinity by consuming HCO_3^-, converting it into biomass or sedimentary carbonates known as marl.

Just as water in excess of evapotranspiration results in surface runoff, nutrients supplied to the land surface in excess of biological demand and soil sorption capacity result in increasing nutrient loading to receiving surface waters. Concentrations of N and P thus rise predictably with nutrient loading in the watershed (Figs. 8.5 and 8.6; Vollenweider, 1976; Sharpley et al., 1996; Boyer et al., 2002). The positive correlation between watershed nutrient loading and nutrient concentrations in receiving waters is enhanced by watershed alterations (e.g., tile drains, storm water pipes, soil compaction, and pavement) that reduce the residence time of water in terrestrial soils and increase the proportion of rainfall that is readily transmitted to surface waters (McCrackin et al., 2014; Green et al., 2004; Bouwman et al., 2005).

Carbon cycling in freshwater ecosystems

A distinguishing characteristic of carbon cycling within freshwater ecosystems is the importance of organic carbon inputs from their contributing watersheds in supporting aquatic food webs and fueling nutrient cycling (Fig. 8.7). These allochthonous[f] inputs supplement

[f]Not formed where it is found.

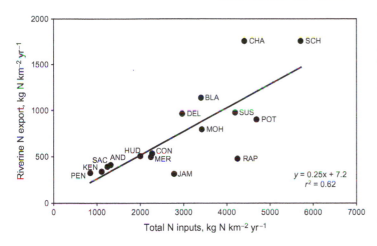

FIG. 8.6 In an analysis of 16 large rivers in the northeastern U.S. nitrogen exports in streamflow were strongly related to the total new inputs of nitrogen to each catchment measured From north to south, the catchments are: Penobscot (PEN), Kennebec (KEN), Androscoggin (AND), Saco (SAC), Merrimack (MER), Charles (CHA), Blackstone (BLA), Connecticut (CON), Hudson (HUD), Mohawk (MOH), Delaware (DEL), Schuylkill (SCH), Susquehanna (SUS), Potomac (POT), Rappahannock (RAP), and James (JAM). *From Boyer et al. (2002). Used with permission of the Ecological Society of America.*

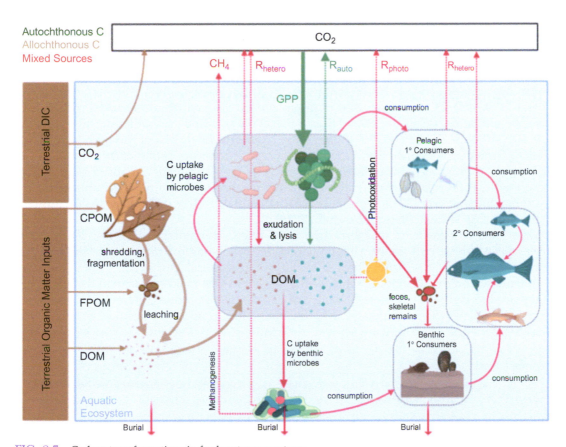

FIG. 8.7 Carbon transformations in freshwater ecosystems.

I. Processes and reactions

and in many cases exceed the organic matter generated from autochthonous[g] primary productivity by freshwater algae, bryophytes and macrophytes.

Allochthonous carbon inputs

Organic C that is not stored or respired on land is exported as dissolved organic molecules (collectively DOM[h]), organic compounds attached to eroded soil minerals, or plant litter. Dissolved organic carbon (DOC) compounds dominate this flux while particulate carbon inputs to freshwaters typically represent only a small fraction (~10%) of total allochthonous carbon exports from terrestrial watersheds (Schlesinger and Melack, 1981). Generally, particulate organic matter (POM) is subdivided into coarse (CPOM) >1 mm and fine (FPOM) <1 mm particulate organic matter (Hope et al., 1994). POM can represent important seasonal or pulsed inputs to aquatic habitats with a high edge-to-volume ratio, such as small streams, braided rivers, and ponds (Hotchkiss et al., 2015a; Fisher and Likens, 1972; Wallace et al., 1997; Meyer et al., 1998).

Riverine DOM is composed of a diverse and complex array of organic molecules, much of which is derived from the surrounding terrestrial watershed (Singer et al., 2012; Aufdenkampe et al., 2007; Hope et al., 1994; Findlay and Sinsabaugh, 2003). In general, the amount of DOM carried by rivers increases with river flow (Fig. 8.8; Raymond et al., 2016; Schlesinger and Melack, 1981) and it carries C:N ratios that resemble catchment soils (Aitkenhead and McDowell, 2000). While this DOM represents a small loss of C and organic nutrients from terrestrial ecosystems (Chapter 5), it can be a primary energy source for receiving aquatic ecosystems (Hotchkiss et al., 2015a; Wetzel, 1992). The majority of freshwater DOM (50–75%) is comprised of organic acids derived from terrestrial soils and upstream wetlands (Wetzel, 1992; Hope et al., 1994) while a small but high turnover fraction of labile C is produced in situ (Hotchkiss and Hall, 2015b; Raymond and Bauer, 2001b; Creed et al., 2018). The autochthonous production of labile carbon by algae and its supply to microbes through exudation or lysis may stimulate enhanced microbial decomposition of terrestrially derived organic matter through priming[i] mechanisms in some systems (Bianchi et al., 2015; Ward et al., 2016) though this does not appear to be a universal pattern across studies (Bengtsson et al., 2018).

It is generally assumed that terrestrial DOC (the C contained in DOM) is leached from upland ecosystems only because it is nonlabile, yet DOC is rapidly metabolized by freshwater biota (Hotchkiss et al., 2015a; Creed et al., 2015; Kaushal and Lewis, 2005; Lutz et al., 2011; Guillemette and del Giorgio, 2011) and little terrestrial OM makes its way to the open ocean (Hedges and Keil, 1997; Meyers-Schulte and Hedges, 1986). Indeed, the net CO_2 losses from some rivers is equal to or greater than the estimated inputs (Raymond et al., 1997, 2000; Mayorga et al., 2005). There are several explanations for this seeming paradox. First, recall

---

[g]Formed within the system.

[h]Dissolved organic matter (DOM) is often analyzed for its carbon content and referred to as dissolved organic carbon (DOC). Similarly POM may be expressed in terms of its carbon content (POC).

[i]Priming was originally defined for soil organic matter (SOM) degradation as "strong short-term changes in the turnover of SOM caused by comparatively moderate treatments of soil" (sensu Kuzyakov et al., 2006). See page 180.

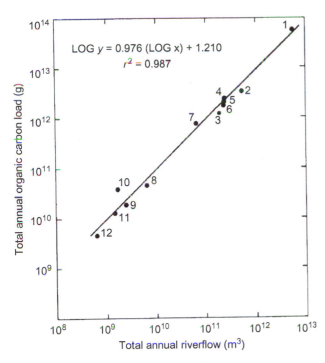

$$\text{LOG } y = 0.976 \text{ (LOG } x) + 1.210$$
$$r^2 = 0.987$$

Total annual organic carbon load (g)

Total annual riverflow (m³)

Rivers 1–7 are among the 50 largest.
1 = Amazon; 2 = Mississippi; 3 = St. Lawrence;
4 = MacKenzie; 5 = Danube; 6 = Volga; 7 = Rhine

FIG. 8.8 Total annual load of organic carbon shown as a logarithmic function of total annual riverflow for major rivers of the world. *From Schlesinger and Melack (1981) with a revision of the data for the St. Lawrence derived from Pocklington and Tan (1987). Used with permission of the Ecological Society of America.*

that the primary constraint on the decomposition of organic matter in soils is mineral protection (Chapter 4) and in wetlands is the absence of electron acceptors and the slow speed of fermentation (Chapter 7). The DOM that is eroded or leached from upland soils where it is associated with mineral surfaces may undergo dissolution reactions in freshwaters. The DOM that cannot be metabolized by microbes in anoxic sediments may be leached or flushed into rivers or lakes where high oxygen is available, allowing for aerobic respiration. Second, most freshwater ecosystems differ from terrestrial soils and wetland sediments in the availability of sunlight. Complex organic molecules within DOM can be degraded by ultraviolet light into a wide variety of photoproducts that are less refractory (Cory et al., 2014; Wetzel, 1992; Moran and Zepp, 1997; Bertilsson et al., 1999). Exposing field DOM samples to sunlight can substantially decrease their resistance to decomposition or result in complete photomineralization to CO_2. For example, by simply exposing lake water DOM to natural sunlight prior to incubation, Lindell et al. (1996) were able to increase bacterial growth in incubations by 83–175% above assays without a prior sunlight exposure. In shallow arctic rivers in midsummer, Cory et al. (2014) found that photomineralization and photooxidation were responsible for 70–95% of all instream transformation of organic matter.

In many aquatic ecosystems, the annual input organic C from land (allochthonous carbon) exceeds C fixation in situ by aquatic algae and plants. In these "donor-controlled" aquatic systems, the annual rates of heterotrophic respiration, export and storage of carbon may each be significantly larger than GPP, a condition that is rarely observed in terrestrial ecosystems (Solomon et al., 2013; Savoy et al., 2019; Battin et al., 2009; Delgiorgio and Peters, 1994). In addition to providing organic matter to aquatic heterotrophs, some of the terrestrial DOC delivered to surface waters absorbs light. This 'colored DOC' reduces the depth of light penetration and can substantially limit NPP in freshwater ecosystems (Karlsson et al., 2009, 2015; Solomon et al., 2015; Carpenter et al., 1998). Increases in terrestrial DOC loading can drive aquatic ecosystems towards greater heterotrophy through both of these mechanisms.

Dissolved organic matter is not only an important source of energy, but is also the dominant form of nutrient input in most freshwaters. Dissolved organic nitrogen (DON) dominates annual nitrogen losses from many unpolluted terrestrial ecosystems (Meybeck, 1982; Perakis and Hedin, 2002; Scott et al., 2007) while dissolved organic phosphorus comprises 25–50% of the total P pool in most freshwater ecosystems (Wetzel, 2001). The fraction of dissolved organic nutrients that are readily bioavailable is highly variable across systems, but both DOC:DOP and DOC:DON ratios tend to increase in aquatic ecosystems where productivity is strongly nutrient limited (Cotner and Wetzel, 1992; Thompson and Cotner, 2018; Stepanauskas et al., 1999; Wiegner et al., 2006). In polluted watersheds, excess inorganic N loading tends to be associated with corresponding increases in DON, either via increased DOM production or reduced demand for organic nutrients (Pellerin et al., 2006).

Primary production in freshwater ecosystems

Inverted biomass pyramids and secondary production

On an annual basis, most freshwater ecosystems respire more CO_2 than is fixed by photosynthesis and are classified as net heterotrophic (NEP < 1). Freshwater ecosystems may be classified along a gradient of heterotrophy by the ratio of production-to-respiration (P:R ratio) (Tanentzap et al., 2017; Webster et al., 2012).[j] Generally, we expect the P:R ratio to increase as the size of lakes or rivers increases (Vannote et al., 1980), because the receipt of solar energy in an aquatic ecosystem is a function of its surface area and because the relative importance of terrestrial inputs declines with decreasing edge-to-volume ratios (Fig. 8.9) (Solomon et al., 2013; Finlay, 2011; Alin and Johnson, 2007).

In contrast to terrestrial food webs where the majority of grazers are smaller than the plants they are consuming, the dominant autotrophs in inland waters are algae, which are consumed by much larger zooplankton, grazing insects, and fish. Unlike woody plants, algae do not invest heavily in structural tissues and often during herbivory the entire autotroph is consumed. As a result, the biomass of autotrophs in aquatic ecosystems tends to turn over rapidly and to be lower than the biomass of higher trophic levels.[k] As a result standing stocks of algal biomass are often not a good index of GPP. Thus, the carbon budgets for aquatic ecosystems often focus on an estimate of *secondary production*[l] rather than NPP.

---

[j]A widely used abbreviation for the ratio of GPP:ER.

[k]This is referred to as an inverted biomass pyramid (sensu Elton, 1927).

[l]The rate of heterotrophic biomass production within an ecosystem (Benke and Huryn, 2006).

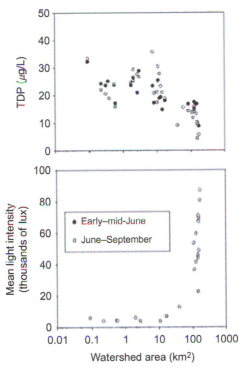

FIG. 8.9   As the contributing watershed area increases, rivers grow wider and deeper. In a survey of streams in the Coast Range of northern California, Finlay (2011) found that phosphorus availability declines and light availability increases with watershed (~stream) size. *Used with permission of the Ecological Society of America.*

Estimates of secondary production typically consider only a limited subset of aquatic heterotrophs (e.g., bacterial, insect or fish productivity). Stable isotope ratios (i.e., $\delta^{13}C$; see Chapter 5) are often used to estimate the relative contribution of autochthonous algae ($\delta^{13}C \sim$ atmospheric CO_2) compared to allochthonous inputs from land vegetation which is depleted in ^{13}C relative to atmospheric CO_2 (Finlay, 2001; Cole et al., 2002; McCutchan and Lewis, 2002; Pace et al., 2004). Rivers and lakes with low $P < R$ ratios can still produce large numbers of insects and fish from the processing of terrestrial organic matter (Webster and Meyer, 1997). In aquatic ecosystems with higher P:R ratios, the $\delta^{13}C$ of herbivores is similar to atmospheric CO_2 (indicative of autochthonous sources). However, allochthonous sources of C can remain an important energy source for the aquatic food webs even in well lit rivers and lakes (Tanentzap et al., 2017; Wilkinson et al., 2013; Pace et al., 2004).

Lakes

A lake is a permanent body of water surrounded by land. Lakes vary in size from small, shallow ponds to Lake Baikal in southeastern Siberia, which is the oldest (25 million years) and deepest (1700 m) lake in the world. The most recent satellite imagery based global inventory estimates that Earth has ~117 million lakes of >0.002 km^2 in size, that collectively comprise 5×10^6 km^2 (or 3.7%) of the planet's nonglaciated land area (Verpoorter et al., 2014). The vast majority of natural lakes are located in the northern temperate zone in formerly glaciated landscapes (Verpoorter et al., 2014; Lehner and Doll, 2004). This areal estimate is surprisingly dynamic due to both the creation of new impoundments and reservoirs and the reductions in water inputs to many lakes due to extractive water use in their watersheds (Downing et al., 2006; Pekel et al., 2016; Busker et al., 2019). Recent high resolution satellite imagery analysis estimated that between 1984 and 2015, new permanent bodies of water were formed over an area of 184,000 km^2, while over that same time period permanent surface water disappeared from an area of almost 90,000 km^2 (Pekel et al., 2016). These shifts in the total lake area, its global distribution and the differences in biogeochemical cycling within natural and manmade lakes complicate efforts to calculate the influence of lakes on global element cycles (Lauerwald et al., 2019; Prairie et al., 2018).

Lake water budgets and mixing

Lakes receive water from precipitation, surface inflows, and subsurface exchange and lose water through evaporation, surface outflows, and export to groundwater. The mean residence time (MRT) of water in a lake varies as a function of both lake volume and watershed area. River and groundwater inflows increase with increasing catchment size; thus, lakes that are small relative to their watersheds have shorter residence times. Small lake basins that are set within large river networks may have substantial river inputs relative to their volume. These "open systems" effectively function as slow-moving pools within a network of rivers, and the movement of elements through the lake may be much higher than the turnover of elements within the lake itself. Most reservoirs, typically built on large rivers, are small relative to their watersheds and thus tend to have short MRTs. In contrast, some terminal lakes, such as Utah's Great Salt Lake or Siberia's Aral Sea have nearly closed water budgets, in which inputs from rivers and precipitation are balanced by evaporative losses. In these closed systems, elements without a gaseous phase accumulate over time, so most terminal lakes are saline.

Because less dense warm waters float above colder bottom waters, density stratification can occur in water bodies of sufficient depth. In lakes, the warmer (low density) surface layer is referred to as the epilimnion, which floats atop the cooler (high density) hypolimnion. The zone of rapid change in temperature in between these two water masses is known as the thermocline or metalimnion (Fig. 8.10). The persistence of the density-driven separation of surface and bottom waters depends upon the extent of mixing by floods and winds, the bathymetery[m] of the water body, and variation in external climate drivers. Many lakes are seasonally stratified, and shallow lakes can have periods of stratification punctuated by storm-driven mixing events. The depth of the epilimnion depends upon air temperature, lake surface area (which determines the total capacity to absorb heat), and the extent of wind mixing (which acts to mix surface and deep waters and reduce the density gradient) (Gorham and Boyce, 1989).

Very deep tropical lakes may be permanently stratified. Lake Nyos in Cameroon is a permanently stratified lake in a volcanic crater that receives high inputs of carbon dioxide from deep geothermal seeps. In 1986, the concentrations of CO_2 at depth became so high (up to 5 L of CO_2 per liter of hypolimnetic water) that the lake explosively outgassed CO_2 as if the lake

---

[m]See footnote b, this chapter.

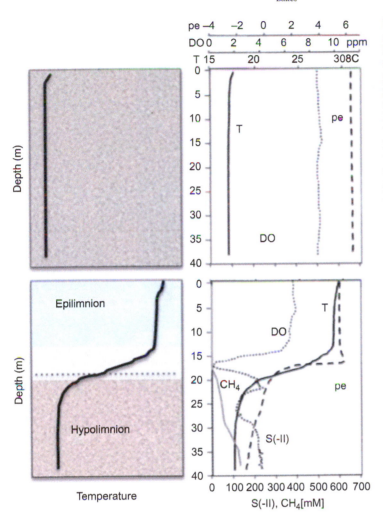

FIG. 8.10 In the upper panels are shown a hypothetical and an actual lake temperature profile during winter (data from January). The lower panels show profiles during the period of summer stratification (data from July). The *dashed line* in the lower left panel indicates the lake thermocline. Depth profiles for temperature (T), dissolved oxygen (DO), redox potential (pe), total sulfide (S(-II)) and methane (CH$_4$) measured in the water column of Lake Kinneret in the Afro-Syrian rift valley during 1999. *From Eckert and Conrad (2007). Used with permission of Springer.*

were a carbonated soda (Kling et al., 1994, 2005). The hypolimentic and epilimnetic waters mixed and a gas cloud of nearly pure CO$_2$ was released from the lake surface. This gas was denser than air, and flowed down the slopes of the ancient volcano, killing 1700 people and many livestock (Kling et al., 1994, 2005).

Trophic status of lakes

Lakes are often classified by their "trophic status" which can be determined via measures of productivity or nutrient load (Table 8.2). Oligotrophic, or low productivity, lakes are classified as having rates of primary productivity (GPP) of <300 mg C m^{-2} day^{-1} (Likens, 1975). Oligotrophic lakes are often of relatively recent geologic origin (e.g., postglacial) with deep and cold hypolimnetic waters. Large lakes within small catchments are often oligotrophic, as nutrient inputs are dominated by precipitation and water residence times are longer when the lake area-to-catchment area ratio is high (Dingman and Johnson, 1971). In contrast,

TABLE 8.2 Lake classification by trophic status.

Trophic type	Mean gross primary productivity (mg C m^{-2} d^{-1})	Phytoplankton biomass (mg C m^{-3})	Chlorophyll a (mg m^{-3})	Light extinction coefficient (h m^{-1})	Total organic carbon (mg L^{-1})	Total P (mg L^{-1})	Total N (mg L^{-1})
Ultra-oligotrophic	<50	<50	0.01–0.05	0.03–0.08		<1–5	<1–250
Oligotrophic	50–300	20–100	0.3–3	0.05–1	< 1–3		
Oligomesotrophic						5–10	250–600
Mesotrophic	250–1000	100–300	2–15	0.1–2.0	< 1–5		
Mesoeutrophic						10–30	500–1100
Eutrophic	>1000	>300	10–500	0.5–4.0	5–30		
Hypertrophic						30–>5000	500–>15,000
Dystrophic	<50–500	<50–200	0.1–10	1.0–4.0	3–30	<1–10	<1–500

Modified from Wetzel (2001, Table 15.13, p. 389).

eutrophic lakes receive high nutrient inputs from the surrounding watershed. Typically eutrophic lakes have lower N:P ratios as a result of higher P inputs in runoff, with the most eutrophic lakes having a TN:TP ratio < 10. These nutrient-rich lakes are often shallow, with warm, highly productive waters.

Highly colored lakes are sometimes classified separately as dystrophic lakes, since as DOC increases, light penetration is reduced (Thienemann, 1921) and the production of algae in such lakes is likely to be limited by light rather than by nutrient inputs (Karlsson et al., 2009, 2015). Recent studies have shown that rising concentrations of DOC in lakes can interact with rising global temperatures to increase the duration and severity of thermal stratification (Solomon et al., 2015; Heiskanen et al., 2015).

Carbon cycling in lakes

In the earliest efforts to understand carbon cycling in lakes, investigators attempted to balance a carbon budget for lakes, quantifying all inputs and outputs of organic matter during the course of a year (e.g. Richey et al., 1978). The idealized organic carbon budget for a lake can be expressed as:

$$\Delta \text{Storage} = [\text{Inputs}] - [\text{Outputs}] \tag{8.5}$$

$$\Delta S = [P_W + P_B + A_I] - [R_W + R_B + B + H_O]. \tag{8.6}$$

where ΔS = change in C storage within the lake, P_W = water column photosynthesis, P_B = benthic photosynthesis, A_I = allochthonous input of organic carbon, R_W = water column respiration, R_B = benthic respiration, B = permanent burial in sediments, H_O = hydrologic loss of organic carbon in outflows.

Studies of the production and fate of organic carbon are useful in understanding the overall biogeochemistry of lakes. Rich and Wetzel (1978) present a carbon budget for Lawrence Lake, a small shallow lake in Michigan in which rooted aquatic plants (macrophytes) contribute ~51.3% of annual autochthonous primary production while phytoplankton, the free-floating algae that normally contribute most of the net production in large lakes, contribute only 25.4% (Table 8.3). In contrast, Jordan and Likens (1975) found that phytoplankton accounted

TABLE 8.3 Organic matter budgets for Lawrence Lake in Michigan and Mirror Lake in New Hampshire.

Inputs	Lawrence Lake, MI		Mirror Lake, NH	
	g C m^{-2} yr^{-1}	Inputs	g C m^{-2} yr^{-1}	Inputs
Net primary production (NPP)	191.4	88%	87.5	83%
POC				
Phytoplankton	43.3	20%	78.5	74%
Epiphytic algae	37.9	18%	2.2	2%
Epipelic algae	2	1%	–	–
Macrophytes	87.9	41%	2.8	3%
Bacterial CO_2 fixation	–	–	4	4%
DOC released by macrophytes				
Littoral	5.5	3%	–	–
Pelagic	14.7	7%	–	–
Imports	25.1	12%	17.93	17%
POC	4.1	2%	6.63	6%
DOC	21	10%	113	11%
Total available organic inputs	216.5		105.43	
Outputs and storage	g C m^{-2} yr^{-1}	% of Total	g C m^{-2} yr^{-1}	% of Total
Respiration	159.7	74%	87.53	83%
Benthic	117.5	55%	43.13	41%
Water column	42.2	20%	44.4	42%
C storage	16.8	8%	7.6	7%
Sedimentation	16.8	8%	7.6	7%
Exports	38.6	18%	10.2	10%
POC	2.8	1%	1.05	1%
DOC	35.8	17%	9.15	9%
Total removal of carbon	215.1		105.33	

Sources: Rich and Wetzel (1978) and Jordan and Likens (1975).

I. Processes and reactions

for ~90% of annual NPP in Mirror Lake, a deep oligotrophic lake in New Hampshire. In Lawrence Lake, with its abundant macrophytes, NPP exceeds total ecosystem respiration (R), while in Mirror Lake NPP and R are equivalent. The more productive shallow lake has higher rates of carbon burial in sediments and exports more DOC in outflow than seen in Mirror Lake.

Primary production in lakes

Light attenuates rapidly through the water column, and the depth at which light levels are insufficient to support photosynthesis in excess of respiration is known as the compensation depth. During periods of stratification, phytoplankton are confined to the epilimnion. If the thermocline is above the compensation depth, phytoplankton will have sufficient light to meet their respiratory demands and productivity is likely to be limited by factors other than light. In contrast, if air temperatures are cold or winds mix the water column (e.g., Fig. 8.10A), phytoplankton can be carried out of the surface waters (or photic zone) to below the compensation depth.

Nutrients often limit the productivity of surface waters because the nutrients that are incorporated into biomass sink out of the well lit epilimnion in the form of dead organisms and fecal pellets. During prolonged periods of stratification surface waters are increasingly depleted in nutrients. Organic matter that sinks from the epilimnion is decomposed by heterotrophs in the hypolimnion, enriching bottom waters in inorganic nutrients and resulting in a depletion of oxygen and lower redox potential in lake sediments (Fig. 8.10). Nutrients typically accumulate in the hypolimnion as sinking organic matter is mineralized by heterotrophs which are dependent on the supply of fixed carbon from the lake surface. Periods of mixing may bring these hypolimnetic waters back into contact with sunlight where their nutrient content can stimulate phytoplankton growth. The spring algal blooms characteristic of many temperate lakes occur when nutrients are mixed throughout the volume of the lake, just before warmer summer temperatures constrain phytoplankton to the photic zone.

Measuring primary productivity

Rooted and floating aquatic plants in shallow lakes or along the margins of deep lakes contribute to lake productivity (Table 8.3), but their importance diminishes with lake size and depth. In deep lakes, the dominant primary producers are phytoplankton. Historically, methods for assessing net primary production in lakes ignored benthic productivity and used bottle assays to estimate NPP. Two different bottle assay approaches are still widely used to determine NPP in lakes (Wetzel and Likens, 2000). In the first method, lake water is placed in gas-tight glass bottles that are either clear or opaque to sunlight. Bottles are incubated (often in the lake itself) and changes in oxygen concentration over time are measured in both the light and dark bottles. In the light bottles, photosynthesis and respiration co-occur. An increase in O_2 concentration over the course of the incubation is taken as the equivalent of net primary production—i.e., photosynthesis in excess of respiration by the plankton. Over the same period, the reduction of O_2 in the dark bottle is taken as a measure of phytoplankton respiration. By summing the O_2 consumption of the dark bottles and the O_2 production in the light bottles, researchers can estimate gross primary production (GPP).[n]

$$NPP = ([O_2]_{t2} - [O_2]_{t1})_{LIGHT} \tag{8.7}$$

$$GPP = ([O_2]_{t2} - [O_2]_{t1})_{LIGHT} - ([O_2]_{t2} - [O_2]_{t1})_{DARK} \tag{8.8}$$

Although widely used and inexpensive, these bottle assays suffer from a series of experimental artifacts (reviewed by Peterson, 1980). The sensitivity of most oxygen measurements is relatively low, so incubations must be long in order to ensure measurable changes. Confining a small water sample in a bottle for long periods can exacerbate nutrient or CO_2 limitations that may not occur in the well-mixed surface waters of a large lake. The method also makes a simplifying assumption that O_2 consumption in the dark bottle is only by phytoplankton, ignoring that planktonic bacteria and zooplankton are also responsible for some fraction of respiration.

A refinement of the classic light vs. dark bottle approach uses ^{14}C-labeled DIC that is added to light bottles to measure the incorporation of ^{14}C into biomass over the course of short-term incubations (Wetzel and Likens, 2000). [14]C can be measured with high precision, so these assays can be conducted quickly and thus avoid many of the bottle artifacts. Because the pH of most surface waters is in the range of 4.3–8.3, these assays often use radiolabelled bicarbonate, frequently as $NaH^{14}CO_3$. After a short incubation period, the sample is filtered, and the accumulation of ^{14}C in the solids collected on the filter is determined with a scintillation counter, and assumed to represent NPP.

A major shortcoming of the ^{14}C method is that any DIC fixed and subsequently released from an algal cell as DOC (e.g., as exudates or cell contents are released during zooplankton feeding) is not captured on the filter and thus not counted as part of carbon fixation. Filters may also not capture very small picoplankton that can be an important component of productivity in some surface waters. When the two methods are compared side by side, the oxygen method typically gives higher values for net primary production than does the ^{14}C method. The ^{14}C method also provides no estimate of respiration, so it is not possible to use this method to estimate gross primary productivity. Each bottle method provides an instantaneous assessment of net primary production, but in order to determine annual production, or to make comparisons of productivity across lakes, it is necessary to repeat the assays many times over the course of a year.

With the development of field-deployable oxygen sensors, limnologists increasingly use daily variation in dissolved oxygen concentrations to calculate net ecosystem production (NEP) in lakes (Staehr et al., 2010b; Winslow et al., 2016; Solomon et al., 2013). In concept, such measurements are similar to estimates of NEP derived from eddy-flux tower measurements of terrestrial ecosystems (Chapter 5). To estimate NEP in lakes, an oxygen sensor is suspended from a floating buoy, usually in the center of a lake. Changes in oxygen concentration between measurements (often 5–30 min apart) are used to calculate instantaneous oxygen production (by photosynthesis) or consumption (by respiration) (Fig. 8.11). At each time

[n]In aquatic systems, the low standing stocks and rapid turnover of autotrophs has led researchers to focus on measuring the fixation of C (GPP) rather than the accumulation of autotrophic biomass (NPP) which is the more common measure of productivity in terrestrial ecosystems (Chapter 5).

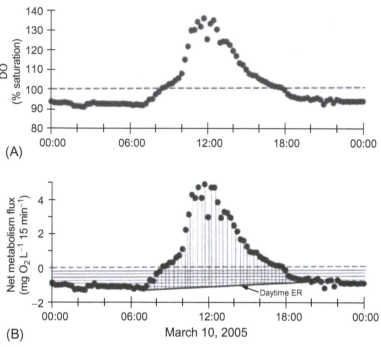

(A)

(B) March 10, 2005

FIG. 8.11 GPP and ER are typically derived from diel oxygen concentrations like the one shown here for a small stream in Tennessee. Diel profiles of (A) percent saturation of DO and (B) net metabolism flux showing the area representing gross primary production (GPP, *vertical lines*) and ecosystem respiration (ER, *horizontal lines*). *Dashed lines* indicate 100% saturation in (A) and a net metabolism flux of zero in (B). The *solid line* in (B) indicates the inter-polated values of daytime ER. *Data are from March 2005 in Walker Branch, Tennessee. Source: From Roberts et al. (2007). Used with permission of Springer.*

step, changes in oxygen concentration must be corrected for the physical processes of oxygen exchange with the overlying atmosphere, by simultaneous measurements of an inert tracer gas, such as SF_6. The sum of oxygen produced over the course of daylight hours is taken as an estimate of NEP.

$$NEP_{\Delta t} = \Delta O_2 - D/Z_{mix}, \tag{8.9}$$

where ΔO_2 (mmol O_2 m^{-3} time interval^{-1}) is the change in oxygen concentration over the time interval, D is the physical exchange with the atmosphere over the time interval, and Z_{mix} is the mixing depth of the lake during the interval.

Using this method, nighttime oxygen consumption is used to estimate ecosystem respiration:

$$ER = \text{average nighttime } R \times 24\,h. \tag{8.10}$$

The estimate of ER (always a negative number) is added to daytime estimates of oxygen production to calculate a cumulative daily estimate of GPP:

$$GPP = NEP - ER. \qquad (8.11)$$

This technique makes the simplifying assumption that daytime respiration rates are similar to nighttime rates and is thus likely to provide a conservative estimate of GPP (Staehr et al., 2010b). A major constraint on sensor-derived estimates of NPP in lakes is the difficulty of determining the volume of water from which the oxygen signal is derived, and thus the appropriate scaling term for extrapolation. The difficulty in attributing metabolism to the appropriate lake volume is akin to the challenge of attributing eddy-flux measurements to the appropriate area of terrestrial vegetation (Chapter 5).

Nutrient limitation of lake NPP

As for terrestrial systems, growing season length is an important determinant of NPP in lakes. While peak productivity in arctic lakes may reach that of temperate and tropical lakes, arctic lakes have a far shorter growing season. In compilations of annual NPP estimates for large, low-nutrient lakes, Wetzel (2001) and Alin and Johnson (2007) each found the lowest annual NPP in arctic, Antarctic, and alpine lakes and the highest in tropical lakes (Fig. 8.12). Lake size is also an important determinant of productivity, because lake area is positively correlated with wind exposure and mixing depth and negatively

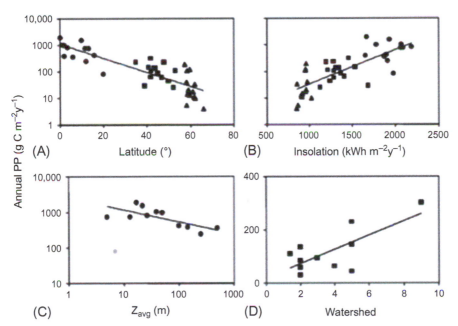

FIG. 8.12   Global-scale relationships between annual primary production (PP) and the environmental variables: (A) latitude, (B) incident solar radiation, (C) depth, (D) watershed to lake area ratio. *From Alin and Johnson (2007). Used with permission of the American Geophysical Union.*

correlated with nutrient input per water volume (Brylinski and Mann, 1973; Duarte and Kalff, 1989; Fee et al., 1994). The drainage ratio of a lake (the ratio of catchment area-to-lake area) is positively related to the external inputs of nutrients and CO_2 (Gergel et al., 1999) and thus to increasing primary productivity (Alin and Johnson, 2007). Put simply, the greater the edge-to-volume ratio, the larger the influence of terrestrial vegetation and soils on the biogeochemistry of a lake and thus the greater the supply of nutrients from the landscape. Larger lakes, with longer residence times and low edge-to-volume ratios, are more dependent upon internal nutrient cycling to sustain productivity.

Aquatic algae have body mass N:P ratios of ~7.2 on average (although this can vary between 3 and 20; Klausmeier et al., 2004),[o] but runoff sources can vary more widely. Runoff from unfertilized forests and fields typically has N: P mass ratios of 20–200 while many pollutant sources are enriched in both N and P—with raw sewage, urban storm waters and feedlot runoff each having N:P ratios between 1 and 10 (Downing and Mccauley, 1992). Both the absolute amount of N and P and their relative proportions are important drivers of freshwater biogeochemistry.

Lakes receiving higher supplies of nitrogen and phosphorus (from urban or agricultural pollution) support greater phytoplankton NPP than do lakes with low nutrient loading (Wetzel, 2001). In a series of influential studies in the 1970s, researchers consistently demonstrated that the variation in NPP or algal biomass across northern temperate lakes was closely associated with the concentration or annual loading rates of phosphorus in the lake (Schindler, 1974; Vollenweider, 1976; Smith, 1983; Correll, 1998; Fig. 8.13). Phosphorus concentrations in the epilimnion are directly related to the total chlorophyll content of the water column, which is directly correlated to net primary productivity (Schindler, 1978). In 1974, David Schindler published the results of a whole-lake experiment using a small hourglass-shaped lake (#226) in the

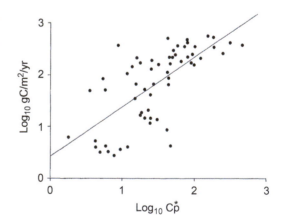

FIG. 8.13   The relationship between net primary production and the phosphorus concentration of lakes of the world is fit by the line log [P]=0.83 log NPP+0.56 ($r=0.69$). Schindler excluded lakes with N:P ratios in inputs of <5:1 from this analysis. *Adapted with permission from Schindler (1978).*

[o]Convert N:P mass ratios to N:P molar ratios by multiplying by 2.21. The mass ratio of 7.2 is equivalent to a molar ratio of 16, and the range of molar N:P ratios in phytoplankton reported by Klausmeier et al. (2004) is 6.6–44.2.

Experimental Lakes Area of Canada (Schindler, 1974). The two basins of this lake were separated with a plastic curtain. Both sides of the lake received sucrose and nitrogen additions, but only one basin received phosphate addition. A large algal bloom erupted in the basin receiving C, N, and P, while the water remained clear and algal biomass low in the basin receiving only C and N (Schindler, 1974). The aerial photo of this experimental effect (Fig. 1.5) is undoubtedly one of the most influential environmental science photographs ever taken. This experimental demonstration convinced regulators to remove phosphates from detergents, dramatically reducing municipal P loading to surface waters throughout North America.

Despite the limited availability of phosphorus in surface waters, we might expect that, as for land vegetation, processes such as denitrification might limit the nitrogen supply in lakes. In a series of studies conducted over the last two decades, nitrogen limitation and colimitation by N and P of lake phytoplankton have been reported frequently (reviewed by Paerl et al., 2016; Conley et al., 2009b; Elser et al., 2007). Authors of several of these studies have suggested that the preponderance of evidence for P limitation of lake productivity from research throughout the 1970s may be a consequence of making such measurements in areas of North America and Canada that had experienced decades of excess atmospheric nitrogen deposition (Chapter 6) that over time had increased the N:P ratio of nutrient inputs to lakes (Elser et al., 2009). Other authors argue that sewage and fertilizer inputs to lakes are typically phosphorus rich and thus may lower the N:P ratio of inputs and enhance internal stores of phosphorus, thereby increasing the potential for N limitation (Downing and Mccauley, 1992).

While phytoplankton productivity in lakes may be limited by N, P, or a combination, there are several fundamental reasons to expect increases in phosphorus loading to play a disproportionate role in what is known as cultural eutrophication (Schindler, 1977; Schindler et al., 2008). When phytoplankton grow with limited supplies of nitrogen, the prevalence of nitrogen-fixing algae, primarily cyanobacteria, typically increases, adding nitrogen through fixation and raising the N:P ratio (cf. Fig. 6.8). In a literature synthesis, Howarth et al. (1988) found that significant nitrogen fixation by lake phytoplankton occurred only when lake N:P ratios were below 16. A subsequent analysis by Smith (1990) suggested that total P loading, rather than N:P ratios was the best predictor of N fixation rates across lakes globally. When phosphorus is added as a pollutant to lakes with low N:P, the algal community shifts to species of blue-green algae and primary productivity increases, with nitrogen inputs through fixation tending to maintain a phosphorus shortage for the growth and photosynthesis of phytoplankton (Smith, 1982).

Håkanson et al. (2007) suggest that for N:P ratios below 15, the N:P ratio is a good predictor of cyanobacterial biomass but that in lakes with TN:TP ratios >15, cyanobacterial biomass can be predicted from total phosphorus concentrations or loading alone (Fig. 8.14). With high P loading, nitrogen fixation can supply up to 82% of the nitrogen input to the phytoplankton community (Howarth et al., 1988). Thus lake phytoplankton have a mechanism for acquiring additional inputs of nitrogen, but there is no equivalent biogeochemical process that can increase the supply of phosphorus in lakes when it is in short supply (Schindler, 1977).

When the input of phosphorus to a lake ceases, blue-green algae typically decrease in importance and algal productivity declines (Edmondson and Lehman, 1981; Schindler et al.,

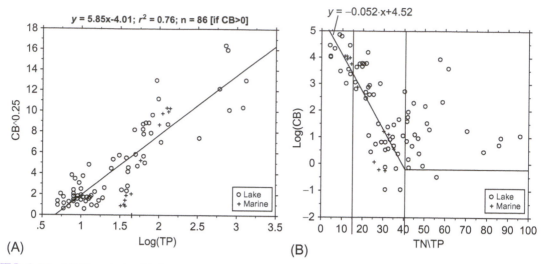

FIG. 8.14 (A) The relationship between cyanobacterial biomass (CB) and total phosphorus concentrations across 86 lakes and coastal estuaries for which CB > 0. (B) The relationship between log CB and the TN:TP molar ratio. The regression equation given is for systems with TN:TP ratios <15. *From Håkanson et al. (2007).*

2008). A particularly influential demonstration of the potential for eutrophic lakes to recover following P reductions was recorded in Lake Washington (Edmondson and Lehman, 1981). Until 1967 the city of Seattle disposed of sewage directly into Lake Washington, whereas after that year sewage was diverted to the ocean. The resulting decline in noxious algal blooms in the lake provided strong evidence for P limitation of lake productivity. The time required for the recovery from eutrophication depends on the extent to which the legacy of P inputs are mineralized and released from lake sediments (Genkai-Kato and Carpenter, 2005; Mehner et al., 2008; Zhao et al., 2020). Historic nitrogen loading is more quickly ameliorated because significant quantities of nitrogen may be lost by denitrification (McCrackin and Elser, 2012). Because of this difference in the persistence of N and P pollution in lakes, Schindler et al. (2008) argued that efforts to reduce N loading are likely to be less effective at reversing cultural eutrophication than reducing P inputs. In contrast, preventing future eutrophication in unpolluted lakes will require reductions in nitrogen as well as phosphorus loading (Conley et al., 2009b; Finlay et al., 2013; Paerl et al., 2016). A further argument for mitigating both nutrients simultaneously is that, in addition to stimulating algal productivity, nitrogen pollution can increase lake N_2O emissions (Kortelainen et al., 2019; McCrackin and Elser, 2010) and there is some evidence that higher N:P ratios enhance the growth of toxic algae and the production of cyanotoxins that lead to harmful algal blooms (Orihel et al., 2012; Paerl et al., 2018, 2016).

Micronutrient limitation

The potential for lake phytoplankton to respond to increasing N and P loading may be constrained by micronutrient availability. When phosphorus is added to nutrient-poor lakes, the growth of diatoms, single-celled algae with silicate cell walls, may reduce the concentrations of silica to low levels, so that diatoms are competitively replaced by non-siliceous green algae or cyanobacteria (e.g., Schelske and Stoermer, 1971; Tilman et al., 1986). Similarly, enhanced loading of phosphorus may lead to Fe limitation in clear oligotrophic lakes (e.g., Sterner et al., 2004). All phytoplankton require Fe for photosynthesis (see Chapter 2) so Fe limitation may reduce whole ecosystem NPP. Because cyanobacteria typically have higher Fe requirements than eukaryotic algae (Morton and Lee, 1974; Brand, 1991), and the nitrogenase enzyme for N_2 fixation requires Fe and Mo (Murphy et al., 1976; Glass et al., 2009), increased competition for Fe or Mo could also lead to a decline in cyanobacteria or rates of N fixation (Goldman, 1960; Glass et al., 2012). Changes in micronutrient loading or nutrient ratios may lead to more subtle changes in algal community structure. Titman (1976) showed that slight differences in the ratio of silica to phosphorus altered the outcome of competition between two dominant species of diatoms (*Asterionella* and *Cyclotella*). Other studies have shown that N fixation rates by algae can be suppressed by the addition of trace micronutrients, such as B (Rao1981), Fe (Vrede and Tranvik, 2006) or Cu (Horne and Goldman, 1974).

Light limitation of NPP

Much of the research on nutrient limitation has been conducted in clear lakes without high concentrations of dissolved organic carbon. In lakes stained with DOC, light is the primary limitation on productivity. Across a series of lakes in northern Sweden that ranged in DOC concentrations from ~10 to 100 mg L, Ask et al. (2012) found that higher DOC was negatively correlated with GPP and positively correlated with ecosystem respiration (ER) (Fig. 8.15). In these lakes, light availability was also a good predictor of secondary production (Karlsson et al., 2009). For dystrophic lakes (see Table 8.3), nutrient inputs are unlikely

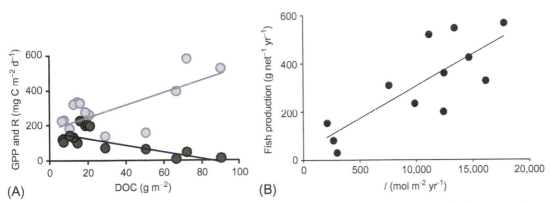

FIG. 8.15 Light limitation of primary and secondary production in Swedish lakes, (A) whole-lake gross primary production (GPP, *dark gray circles*) and respiration (R, *light gray circles*) for 15 lakes in northern Sweden. (B) Fish production as a function of the annual light climate (I, representing the mean PAR in the whole-lake volume during the ice-free period) for 12 lakes in northern Sweden ($r_2 = 0.63$, $P = .002$). (A) From Ask et al. (2012), used with permission of American Geophysical Union; (B) from Karlsson et al. (2009), used with permission of Nature Publishing Group.

to stimulate algal blooms, but may instead fuel secondary production and enhance the rates of DOC decomposition and CO_2 (Sadro et al., 2014; Kelly et al., 2014; Pace et al., 2007; Tanentzap et al., 2017).

Herbivore control of GPP

In lakes, much of the algal biomass produced by phytoplankton is consumed by zooplankton, so changes in the food web can affect whole ecosystem GPP. More productive lakes typically produce higher fish biomass per unit volume (e.g., Melack, 1976; Karlsson et al., 2009). This trophic linkage works in more than one direction. Nutrient and light availability may increase the productivity of higher trophic levels ("bottom-up" control), whereas an abundance of predators can influence the intensity of herbivory and alter GPP ("top down" control; Carpenter et al., 1985). For example, in a series of whole-lake experiments, Carpenter et al. (2001) showed that adding pike (*Esox lucius*), a predatory fish, reduced lake GPP by a factor of three compared to that in a nearby lake that lacked predatory fishes. Piscivores consume fish that normally eat zooplankton, allowing abundant zooplankton to substantially reduce standing stocks of phytoplankton. This effect has been termed the "trophic cascade". In an analysis of 54 experimental manipulations of top predators in lakes and ponds, Brett and Goldman (1996) found that nearly all such experimental manipulations of piscivores led to increases in phytoplankton. Not all lake manipulations of top predators, however, have led to changes in NPP (e.g., Elser et al., 1998; MacKay and Elser, 1998) and food web interactions are less likely to affect ecosystem productivity in oligotrophic lakes where algae are extremely nutrient limited or in highly eutrophic lakes where algal growth rates are high and algal communities become dominated by unpalatable species (Kitchell and Carpenter, 1993). In highly eutrophic lakes, algal blooms cause bottom water hypoxia and the production of algal toxins that may further constrain the growth of grazers (Heisler et al., 2008; Brooks et al., 2016; Mallin et al., 2006).

The fate of organic carbon in lakes

Organic carbon delivered to lakes from the surrounding catchment or derived from photosynthesis in the lake may be incorporated into consumer biomass, respired, sequestered in lake sediments, or transported downstream (Cole and Caraco, 2001). Because the sediments of lake bottoms are frequently anoxic, organic material that reaches the lake bed will decompose slowly (Chapter 7), and lakes typically accumulate sediment carbon over time. Einsele et al. (2001) estimated that throughout the Holocene the amount of carbon stored in lake sediments globally has been ~820 Pg, with small lakes (<500 km^2) containing about 70% of this total. In a survey of 228 European lakes, Kastowski and Hinderer (2011) measured highly variable rates of sediment organic carbon accumulation (0.1–57.8 g C m^{-2} yr^{-1}) with an average of 5.6 g C m^{-2} yr^{-1}. The world's 250 largest lakes are estimated to accumulate an estimated 7.5 Tg C annually[P] in their sediments (Alin and Johnson, 2007).

[P] This represents 3% of the total production of organic matter in these systems—compare to a burial rate in the world's oceans that is <1% of NPP (see Chapter 9).

Over geologic time lake basins fill with sediments and organic material. This can happen very quickly in reservoirs with large inputs of river sediment and organic materials (Downing et al., 2006). Cole et al. (2007) compiled annual estimates of C storage in lakes, which showed a global total ranging from 0.03 to 0.07 Pg C yr^{-1} (Table 8.4). While this global rate is miniscule in comparison with the annual C storage in terrestrial vegetation and soils (about 2.7 Pg C yr^{-1}), it is roughly equivalent to the annual organic C storage in marine sediments ($\sim$0.12 Pg yr^{-1}; Sarmiento and Sundquist, 1992) but occurs within an area equivalent to only 2% of the ocean area. When lake carbon accumulation is corrected for area, long-term carbon burial rates range from 4.5 to 14 g C m^{-2} yr^{-1} (Dean and Gorham, 1998; Stallard, 1998) which is higher than the long-term soil C accumulation rates estimated for most terrestrial soils (Schlesinger, 1990; Fig. 5.19). An estimated $\sim$20% of modern oil and natural gas currently under production is derived from ancient large-lake basins (Bohacs et al., 2000).

Carbon burial in lake sediments is predicted to increase with eutrophication as a result high autochthonous production coupled with more constant and severe oxygen depletion in sediments (Alin and Johnson, 2007; Hutchinson, 1938; Downing et al., 2008). Examining several lakes, Hutchinson (1938) suggested that the rate of depletion of O$_2$ in the hypolimnion during seasonal stratification was related to the productivity of the overlying waters. Highly productive waters should contribute large quantities of organic carbon for respiration in the hypolimnion, which is seasonally isolated from sources of oxygen. The areal hypolimnetic oxygen deficit (abbreviated as AHOD) is a useful concept, but attempts to predict AHOD as a function of nutrient loading have been problematic. Cornett and Rigler (1979) concluded that "a simple proportionality between biomass in the epilimnion and area hypolimnetic oxygen deficit (AHOD) does not appear to exist." They suggested instead that the greatest O$_2$ consumption occurred in deep lakes with higher water temperatures and a thick hypolimnion (Cornett and Rigler, 1979, 1980). It is logical that warmer water temperatures support higher rates of bacterial respiration in the hypolimnion. The relationship between AHOD and

TABLE 8.4   Carbon burial rates in lakes and impoundments.

Environment	Mean or median OC burial (g m^{-2} yr^{-1})	Range
Eutrophic impoundments	2122	148–17,392
Impoundments (Asia)	980	20–3300
Impoundments (Central Europe)	465	14–1700
Impoundments (United States)	350	52–2000
Impoundments (Africa)	260	
Small mesoeutrophic lakes	94	11–198
Small oligotrophic lakes	27	3–128
Large mesoeutrophic lakes	18	10–30
Large oligotrophic lakes	6	2–9

Sources: Downing et al, (2008) and Mulholland and Elwood (1982).

hypolimnion thickness, however, was unexpected, because it suggests that the greatest deficits are found in deep lakes with large hypolimnetic volume. Their findings, while not without criticism (Chang and Moll, 1980), suggest that the consumption of oxygen in the hypolimnion may be largely the result of respiration within the water column, which is greatest in deep lakes where the transit time for sinking detritus is long (cf. Cole and Pace, 1995).

Some of the controversy over whether nutrients, temperature, or lake depth drive hypolimnetic anoxia and carbon burial may arise because authors are attempting to find a single relationship that fits lakes of varying nutrient status. In a synthesis of carbon-burial rates Downing et al. (2008) found extremely low carbon-burial rates in large nutrient-poor lakes, but very high rates of carbon burial in midsize eutrophic impoundments despite the fact that these impoundments are typically shallow (Table 8.4).

Undoubtedly global carbon storage in lake sediments is rising because of the rapid increase in constructed reservoirs. Rates of carbon burial in reservoirs are high (estimated at $\sim$400 g C m^{-2} yr^{-1}) (Mulholland and Elwood, 1982; Dean and Gorham, 1998; Downing et al., 2008) and because of the rapid proliferation of reservoirs, they now bury more organic carbon annually than all natural lake basins combined. Based on the data in Table 8.4, the global rate of burial in reservoirs is between 0.16 and 0.2 Tg C annually. These estimates are likely conservative as they are based on outdated (and low) measures of reservoir area. Extrapolating the areal rate of C accumulation in reservoirs to the estimated 1,500,000 km^2 of global reservoirs (St. Louis et al., 2000) indicates that global C storage in reservoirs could exceed 0.6 Pg C yr^{-1} (Cole et al., 2007).

Carbon export from lakes

Carbon can be exported from lakes through the degassing of CO_2 or CH_4 or through the downstream export of particulate and dissolved organic matter. Respiration exceeds GPP in most *oligotrophic* lakes (Sand-Jensen and Staehr, 2009; Staehr et al., 2010b; Ask et al., 2012; Jansson et al., 2012; Solomon et al., 2013). In many lakes, the degree of net heterotrophy is positively correlated with CO_2 supersaturation and lake DOC concentrations (Roehm et al., 2009; Ask et al., 2012). Synthesis efforts have suggested that lakes outgas $\sim$0.64 Pg C yr^{-1} as CO_2 (Table 8.1; Cole et al., 2007; Aufdenkampe et al., 2011). This estimate is constantly being refined as a result of more intensive and widespread field measurements of lake CO_2 outgassing and increasingly precise estimates of lake area (Verpoorter et al., 2014). Recent estimates suggest that the CO_2 flux from lakes and reservoirs is closer to 0.3–0.5 Pg C yr^{-1} (Raymond et al., 2013; DelSontro et al., 2018).

In low sulfate, oligotrophic lakes, methanogenesis is often the dominant form of anaerobic metabolism, but rates of methanogenesis tend to be highly variable in both space and time (Rudd and Hamilton, 1978; Kuivila et al., 1989; Zimov et al., 1997). As long as the hypolimnion remains oxygenated, most of the methane produced in hypolimnetic sediments is oxidized by methanotrophs during upward diffusion; however, in shallow sediments a much higher proportion of gross methanogenesis can escape to the atmosphere. Gaining an accurate estimate of CH_4 emissions from lakes is particularly problematic, since a dominant pathway for CH_4 releases from lakes is through ebullition, the episodic movement of bubbles from the sediments to the surface (Bastviken et al., 2004; Chapter 7). In a survey of Siberian thaw lakes, Walter et al. (2006) estimated that ebullition accounted for 95% of CH_4 emissions.

Bastviken et al. (2011) estimated the total global CH_4 emission from natural lakes at $\sim$54 Tg of CH_4 yr^{-1}, and Deemer et al. (2016) estimated an additional 13.3 Tg CH_4 yr^{-1} derived from global reservoirs. Although these CH_4 emissions represent a relatively small flux in most lake carbon budgets, the warming potential of CH_4 is $\sim$25 to 35 times higher than CO_2, so that CH_4 production in lakes represents an important link between lakes and climate change (DelSontro et al., 2018; Bastviken et al., 2011; Deemer et al., 2016). Since low oxygen, high chlorophyll a and high temperatures are all positively correlated with lake methane emissions, it is anticipated that methane emission from lakes and impoundments is likely increasing over time (Prairie et al., 2018; Tranvik et al., 2009).

There is growing evidence that the construction of reservoirs is enhancing CH_4 fluxes by increasing lake area and providing ideal conditions for methanogenesis. Vegetation and organic soils are often flooded when impoundments are created, providing a large source of organic matter to fuel methanogenesis immediately after flooding (Prairie et al., 2018; Kelly et al., 1997; St. Louis et al., 2000; Sobek et al., 2012; Teodoru et al., 2012). In some tropical hydroelectric reservoirs, large amounts of CH_4 are released as deep, methane-enriched waters are passed through turbines. For some reservoirs the resulting greenhouse gas emissions can be greater than from the power plants that these hydropower dams were intended to replace (Fearnside, 1995; Abril et al., 2005). Methane emissions from natural lakes and reservoirs are estimated at 54Tg CH_4 yr^{-1}, a substantial fraction of total global methane emissions from natural sources (Table 11.2).

Nutrient cycling in lakes

Nutrient budgets in lakes are constructed by assessing the inputs of nutrients in precipitation, runoff, and N-fixation and the losses of nutrients from lakes due to sedimentation, outflow, and the release of reduced gases. In many cases human inputs now dominate the nutrient budgets of lakes (e.g., Edmondson and Lehman, 1981). For elements without a substantial gaseous phase (e.g. Fe, Si, P), patterns of element concentration in lake sediments can be used to reconstruct historical loading and retention (Dillon and Evans, 1993; Rippey and Anderson, 1996). Lake sediments retain a record of changes in exports from terrestrial catchments and can be used to understand how land use change or climate change influence both regional and internal nutrient cycling (Davis et al., 1985).

Most lakes show a net retention of N, P and Si (Table 7.4; Cross and Rigler, 1983; Muller et al., 2005; Harrison et al., 2008), although lakes with short MRTs tend to have low net storage of N and P during periods of high flow (Windolf et al., 1996). Iron derived from terrestrial runoff may also be sequestered in lake sediments (Dillon and Evans, 2001). Net retention of Ca is seen in lakes where mollusk shells are accumulating in the sediments (Brown et al., 1992) and in some highly productive, alkaline lakes (pH $\sim$9) where calcite ($CaCO_3$) may precipitate directly as *marl* (Rosen et al., 1995; Hamilton et al., 2009):

$$Ca^{2+} + 2HCO_3^- \rightarrow CaCO_3 \downarrow + H_2O + CO_2. \qquad (8.12)$$

These lakes will show a net retention of Ca and a relatively short mean residence time for Ca in the water column (Canfield et al., 1984). Calcite deposition is inhibited in lakes with high allochthonous DOC (Reynolds, 1978; Hoch et al., 2000; Lin et al., 2005) and enhanced by high rates of photosynthesis that consume CO_2 (Hartley et al., 1995; Couradeau et al., 2012). In a phosphorus-enrichment experiment in a Michigan Lake, Hamilton et al. (2009) found that calcite deposition was enhanced by P fertilization and that the resulting biogenic calcite-phosphorus sedimentation was a substantial sink for the additional phosphorus. By this mechanism calcite precipitation has the potential to ameliorate eutrophication caused by nutrient loading (Koschel et al., 1983; Robertson et al., 2007).

In general, nutrient retention in lakes increases with nutrient loading and water residence time (Seitzinger et al., 2006), and declines with lake depth. In deeper lakes, it takes longer for organic matter produced in the epilimnion to fall through the water column to the sediments. Falling organic matter or carbonates can be decomposed or mineralized during transport through the hypolimnion, reducing the absolute amount of material that reaches the sediments. It is only upon sedimentation that elements have the potential to be permanently buried or, in the case of N, denitrified in anoxic sediments (Mendonça et al., 2017; Carignan and Lean, 1991; Dillon and Molot, 2005; Maavara et al., 2015).

Nitrogen

Nitrogen fixation rates in lakes range from 0.1 to >90 kg N ha^{-1} yr^{-1} (Howarth et al., 1988), roughly spanning the range of nitrogen fixation reported for terrestrial ecosystems (Chapter 6). Lakes with high rates of nitrogen fixation can show large apparent accumulations of N (Horne and Galat, 1985). Because N fixation is competitively advantageous when N concentrations are low, rates of N fixation within individual lakes typically respond to changes in nitrogen availability. In many seasonally stratified lakes, a succession from eukaryotic algae to N-fixing cyanobacteria is observed as epilimnetic nitrogen concentrations decline over the period of stratification (Sterner, 1989), and N-fixation rates are reduced when the supply of nitrogen is high (Doyle and Fisher, 1994).

Fewer studies have assessed denitrification and other processes of gaseous loss in lakes. Denitrification can be studied by the application of acetylene-block techniques, and ^{15}N isotopic labels as discussed in Chapter 6 (Seitzinger et al., 1993). Denitrification in individual lakes ranges from 1.8 to 383 kg ha^{-1} yr^{-1} from a literature synthesis by Piña-Ochoa and Álvarez-Cobelas (2006). The total loss of nitrogen by denitrification exceeds the input of nitrogen by fixation in almost all lakes where both processes have been measured simultaneously (Seitzinger, 1988). In a synthesis of more than 100 studies, Harrison et al. (2008) found that some lakes remove nearly 100% of N inputs. Higher rates of N removal were found for lakes or reservoirs with high N loading, high catchment area-to-lake area ratios, and high settling velocities for N. Based on the statistical relationships in this dataset, they estimated that lakes and reservoirs collectively return 19.7 Tg N yr^{-1} to the atmosphere by denitrification, with small lakes (<50 km^2) responsible for nearly 50% of the global total (Harrison et al., 2008). Often denitrification appears limited by the production and availability of NO_3^- in the sediments (Rysgaard et al., 1994), and the proportion of N removed increases with water residence time (Fig. 8.16; Yoh et al., 1983, 1988; Mengis et al., 1997; Seitzinger et al., 2006). Because atmospheric N_2O concentrations are so low, most lakes are often supersaturated in N_2O (Whitfield et al., 2011); however, the loss of N_2 from lakes greatly exceeds the loss of N_2O (Seitzinger, 1988; Beaulieu et al., 2011). As NO_3 loading to lakes increases (through atmospheric deposition or polluted surface runoff) higher N_2O production is expected

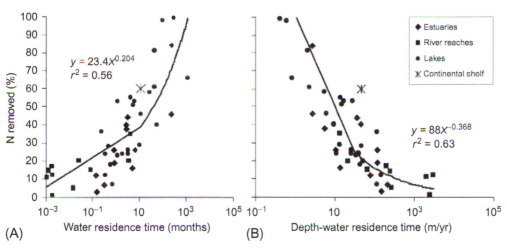

FIG. 8.16   Relationship between the percentage of N removed (via burial or denitrification) and (A) water residence time (months) or (B) depth/water residence time (m/yr) for lakes, river reaches, estuaries, and continental shelves. *From Seitzinger et al. (2006). Used with permission of the Ecological Society of America.*

(McCrackin and Elser, 2011), but current models estimate this flux at 0.05 Tg N annually, less than 0.5% of N gas exports (Lauerwald et al., 2019).

Phosphorus

Because P is weathered slowly from rock and effectively bound in soils or assimilated by terrestrial vegetation, under most natural conditions phosphorus runoff to lakes is relatively small (Ahl, 1988; Reynolds and Davies, 2001; Chapter 4). Much of the phosphorus entering lakes is carried with soil minerals, which are rapidly deposited in the sediments (Froelich, 1988; Dillon and Evans, 1993). The small amount of inorganic phosphorus entering lakes undergoes rapid precipitation with Fe, Ca, or Mn minerals that are insoluble in well oxygenated waters, with the form depending on pH (Fig. 4.10; Mortimer, 1941; Blomqvist et al., 2004; Hamilton et al., 2009).

Analyses of lake water typically show that a large proportion of the phosphorus is contained in the plankton biomass and only a small portion is found in available forms (Lean, 1973; Schindler, 1977; Lewis and Wurtsbaugh, 2008). Uptake of phosphorus by phytoplankton is an active process that shows a curvilinear relationship to increasing P concentration (Jansson, 1993). Continued net primary production by phytoplankton depends on the rapid cycling of phosphorus between dissolved (e.g., HPO_4^{3-}) and organic forms in the epilimnion (Fee et al., 1994).

Studies of phosphorus cycling have shown that the turnover of phosphorus in the epilimnion is dominated by bacterial decomposition of organic material. Phytoplantkon and bacteria excrete extracellular phosphatases to aid in the mineralization of P (Stewart and Wetzel, 1982; Wetzel, 1992), and planktonic bacteria may immobilize phosphorus when the C/P ratio of their substrate is high (Vadstein et al., 1993). Globally, the molar N:P ratio of freshwater

phytoplankton ranges from 6 to 44 (Klausmeier et al., 2004) and net phosphorus mineralization begins at N:P < 16 (Tezuka, 1990). Immobilization of N is less common, because the C:N
ratio of phytoplankton (8–20) is similar to that of bacterial biomass (Tezuka, 1990; Downing
and Mccauley, 1992; Elser et al., 2000b). Nutrient turnover in lakes is enhanced by the activities of grazing zooplankton (Porter, 1976; Lehman, 1980; Elser and Hassett, 1994) and fish
(Vanni et al., 2013; Vanni, 2002). Grazing zooplankton vary in N:P ratios, with the common
cladoceran *Daphnia* having a low N:P ratio (~14:1) relative to most copepods (~30–50:1;
Sterner et al., 1992); thus, changes in the identity of dominant grazers can alter the ratio as
well as the rate of N and P turnover (Elser et al., 2000a).

During a period of stratification, the phosphorus pool in the surface waters is progressively
depleted as phytoplankton and other organisms die and sink to the hypolimnion (Levine
et al., 1986; Rippey and McSorley, 2009). Baines and Pace (1994) found that 10–50% of the
NPP was exported to the hypolimnion of 12 lakes of the eastern United States, with a tendency
for a greater fractional export in lakes of lower productivity (Fig. 8.17). Higher rates of sinking
particles are correlated with higher rates of bacterial respiration in the hypolimnion (Cole and
Pace, 1995). When fecal pellets and dead organisms sink through the thermocline, phosphorus mineralization continues in the lower water column and sediments (Gachter et al., 1988;
Lehman, 1988; Carignan and Lean, 1991). Anoxic hypolimnetic waters often show high concentrations of P, which is returned to the surface during periods of seasonal mixing.

As long as the hypolimnetic waters contain oxygen, a layer of Fe-oxide minerals at the
sediment-water interface traps phosphorus that diffuses upward from bacterial decomposition in the sediments or from the dissolution of Fe-P minerals at lower redox potential at
depth. However, when hypolimnetic waters become anoxic, this "iron trap" for phosphorus
is lost and oxidized Fe minerals are reduced, releasing P to the overlying waters (Mortimer,
1941; Caraco et al., 1990; Golterman, 2001; Blomqvist et al., 2004). Interactions between elements may be important in determining the release of P from sediments. In most freshwaters
the concentration of SO_4^{2-} is low, and P is strongly adsorbed by Fe minerals in the sediment.

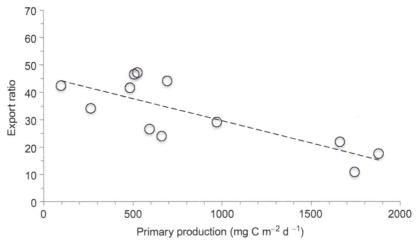

FIG. 8.17   The percentage of planktonic productivity that sinks to the hypolimnion in lakes as a function of their net
primary production. *Modified from Baines and Pace (1994). Used with permission of NRC Research Press.*

Increasing concentrations of SO_4^{2-} in lakes due to acid rain or acid mine drainage may act through the anion exchange reactions (Eq. 4.12) to drive P into solution (Caraco et al., 1989; Wang and Chapman, 1999).

In many cases the dissolution of Fe minerals is incomplete, and there is no regeneration of phosphorus from sediments (Davison et al., 1982; Levine et al., 1986; Caraco et al., 1990; Davison, 1993; Golterman, 1995). Sedimentary accumulations of undecomposed organic matter and Fe minerals contain P that is effectively lost from the ecosystem (Cross and Rigler, 1983).

Sulfur

In the sediments and suboxic bottom waters of lakes, as in the saturated sediments of wetlands, sulfur can play an important role in both carbon and nitrogen cycling. Sulfate concentrations in lakes are generally low, but where hypolimnetic oxygen concentrations are reduced, sulfate reduction can be an important component of lake carbon cycling, particularly in eutrophic lakes (Holmer and Storkholm, 2001; Fig. 8.18).

From a review of the literature, Holmer and Storkholm (2001) concluded that sulfate reduction can account for a significant fraction (12–81%) of the total anaerobic carbon mineralization in lake sediments. Sulfate reduction is enhanced in eutrophic lakes where more organic carbon is supplied to the sediments and where shallow sediments are more likely to be anoxic. When reduced forms of sulfur are cycled through reoxidation pathways, high rates of SO_4^{2-} reduction can occur in lake sediments, despite low concentrations of SO_4^{2-} in lake water (Holmer and Storkholm, 2001). Urban et al. (1994) found that rates of sulfide oxidation in the surface sediments of an oligotrophic lake in Wisconsin were nearly as rapid as sulfate reduction at depth, indicating rapid fluxes of S despite the small pool sizes of S in the sediments.

Recent work has shown that anaerobic sulfide oxidation during dissimilatory nitrate reduction or denitrification can be an important sink for NO_3^- in lake sediments, with the sulfur bacteria (*Thiobacillus denitrificans* or *Thioploca*; Fig. 1.6) gaining energy by oxidizing reduced

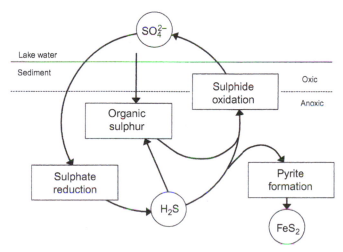

FIG. 8.18 A simplified lake sediment sulfur cycle. *From Holmer and Storkholm (2001).*

sulfide to SO_4^{2-} and reducing NO_3^- either to N_2 or NH_4^+ (Chapter 7; Burgin and Hamilton, 2008; Laverman et al., 2012). Although volatile losses of sulfur occur (Brinkman and Santos, 1974), most H_2S appears to be oxidized as it passes through the upper sediments (Dornblaser et al., 1994) or the water column (Mazumder and Dickman, 1989), so little escapes to the atmosphere (Nriagu et al., 1987; see Fig. 8.10).

Streams and rivers

Unlike the situation in lakes, the flow of water in rivers maintains a constant supply of nutrients, water turbulence constantly mixes particulates into the water column, and frequent scouring limits the capacity for permanent burial of elements in sediments (Fig. 8.19). In addition, the boundaries of rivers, both laterally and longitudinally are highly dynamic, with rivers expanding to encompass their floodplains and ephemeral headwaters during periods of high runoff and contracting to a limited portion of their channels during periods of drought.

As we saw for lakes, the organisms in rivers can alter the magnitude, timing, and form of chemical delivery to downstream waters (Meyer et al., 1988). Although some think of streams as "the gutters down which flow the ruins of continents" (Leopold et al., 1964), stream channels are physically and biologically complex and thus trap, transform, delay, and attenuate the flow of water, chemicals, and sediment pulses received from upslope and upstream (Bencala and Walters, 1983). A significant fraction of the energy supporting the community of organisms in streams is derived from terrestrial materials, and their aggregate decomposition and consumption of terrestrial organic matter significantly alters the timing and quantity of chemical exports (Wallace and Webster, 1996; Wallace et al., 1997).

Materials accumulate in river floodplains as channels migrate across them (Wohl et al., 2012), but river channels themselves are not aggrading systems and are less retentive of

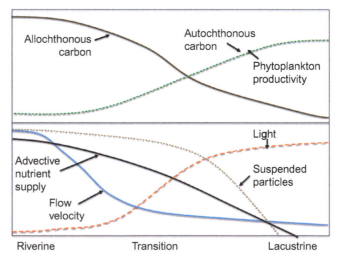

FIG. 8.19 Commonly observed shifts in flow, light, nutrients, and sources of organic matter in the transition between rivers and lakes.

chemicals and solutes than wetlands or lakes (Wagener et al., 1998; Essington and Carpenter, 2000; Grimm et al., 2003). Because of the comparatively high velocities of flow in channels (i.e., limited residence time), chemicals and solutes that reach streams are routed much more rapidly downstream than would occur via subsurface flow. Once introduced to streams, the primary fate for elements that have no gaseous form is downstream transport towards the sea. This transport may take anywhere from days to millennia depending upon the amount of time elements spend buried in the sediments of floodplains, lakes, reservoirs or coastal zones, or incorporated into the tissues of aquatic organisms. Some small fraction may also be transported against gravity in the bodies of aquatic or semiaquatic organisms that either disperse into terrestrial ecosystems or are captured by terrestrial predators (Helfield and Naiman, 2001; Baxter et al., 2005; Sabo and Power, 2002; Richmond et al., 2018).

For elements with a gaseous form at ambient conditions, substantial conversion and thus permanent loss to the atmosphere can occur in rivers. In particular, denitrification can convert ~16 to 50% of dissolved nitrate (NO_3^-) inputs to N_2 (Galloway et al., 2004; Seitzinger, 1988; Mulholland et al., 2008) and metabolic respiration can convert >50% of organic carbon inputs to CO_2 (Raymond et al., 2013; Battin et al., 2009).

Riverbeds provide ideal conditions for a variety of metabolic processes. Because oxygenated waters are in close proximity to anoxic sediments, rivers typically have disproportionately high rates of nutrient transformations and decomposition compared to adjacent surrounding soils (Lohse et al., 2009; Stelzer et al., 2011). Anoxic habitats within streams and their associated riparian zones are often the primary locations where denitrification and methanogenesis occur within landscapes (Bernhardt et al., 2017a; Vidon and Hill, 2004; Burgin and Groffman, 2012).

River water budgets and mixing

Water may enter a segment of a river channel as flow delivered from upstream, as direct precipitation received on the surface, or as lateral inflows that are delivered over land, either through shallow subsurface pathways or via exchange with deep groundwater. In small headwaters streams, lateral inputs can dominate the water balance, but as streams grow in size, inflow from upstream segments quickly becomes the dominant water source. Water may leave the channel through downstream flow, through evaporation, or through net losses to regional groundwater, known as transmission loss. In some arid ecosystems the entire flow of large rivers is ultimately lost to groundwater. Even when there is no net loss to groundwater, the exchange of water between the channel and groundwater can be quite large (Covino and McGlynn, 2007; Poole et al., 2008), and during periods of high flow, rivers may overtop their banks and export water to their floodplains where it is stored or evapotranspired. Under most conditions, however, the dominant route of water export from a river segment is downstream flow.

Long-term observations of streams show that flow is affected by topography, vegetation, and soil characteristics, as well as the pattern and intensity of rainfall (Ward, 1967; Bosch and Hewlett, 1982; McGuire et al., 2005). Stormflows tend to increase when vegetation is removed because of reduced terrestrial ET and because bare soil allows a greater proportion of precipitation to leave via overland flow (Bosch and Hewlett, 1982; Schlesinger et al., 2000b; Likens, 2013). The temporal pattern of flow in a stream (or its *hydrograph*) provides information about

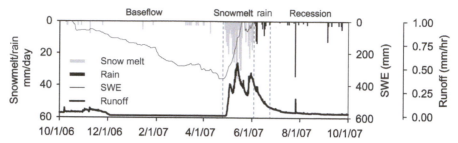

FIG. 8.20 The annual hydrograph for a snowmelt-dominated river in Montana is shown in *black*. Stream flow reaches its peak during the late spring snowmelt period, but also floods following an intense summer rainstorm. The mass of snowpack is recorded as snow water equivalent (SWE) and is indicated by the *gray line*. Rain is shown with *black bars* and snowmelt is shown in *gray bars*. *From Jencso et al. (2009). Used with permission of the American Geophysical Union.*

how rapidly rainwater or snowmelt is transported to channels, how frequently channels are flooded (Fig. 8.20), and the relative proportion of water contributed from various flow paths (Fig. 8.2; Bonell, 1993; Sidle et al., 2000). The relative importance of short flowpaths ($\sim$quickflow = overland flow + shallow soil flowpaths) versus baseflow (provided by groundwater and flow from permanently saturated soil) can vary greatly as a function of climate, and as a function of soil depth and topography in the watershed (McGuire et al., 2005; Lutz et al., 2012b). Streams dominated by groundwater have consistent baseflows and tend to have stable channels with obvious headwaters. In contrast, in streams with flows derived predominantly from precipitation, such as those in arid or urban watersheds, the flow is dynamic and inconsistent (Stanley et al., 1997; Doyle and Bernhardt, 2011), such that the extent of headwater streams and the expanse of the channel network vary dramatically over time (Fig. 8.21).

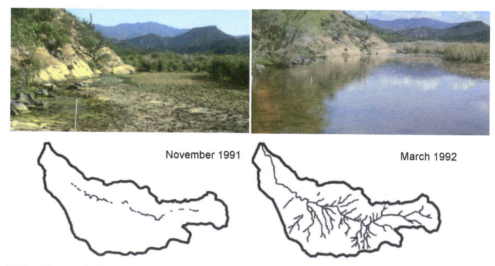

FIG. 8.21 Changes in the longitudinal and lateral surface water area of Sycamore Creek, a desert stream in Arizona. *Maps from Stanley et al. (1997). Used with permission of the University of California Press. Photos provided by Emily Stanley.*

Carbon cycling in rivers

Allochthonus inputs

While some fraction of the organic carbon in rivers is derived from internal productivity, particulate and dissolved C (POC and DOC) derived from the terrestrial landscape is an important input to all river C budgets (Fisher and Likens, 1973; Webster and Meyer, 1997; Mayorga et al., 2005). The leaves, needles, twigs, branches and trunks of vegetation that fall into streams support diverse aquatic food webs as these materials are shredded and decomposed during downstream transport (Webster et al., 1999). Particulate matter dominates the carbon budget of many small streams, but along rivers, the ratio of particulate to dissolved allochthonous carbon inputs generally decreases as larger rivers have a smaller proportion of terrestrial "edge" from which to acquire particulate material. CPOM delivered to tributaries is degraded during downstream transport (Vannote et al., 1980; Webster and Meyer, 1997; Webster et al., 1999); thus, dissolved organic carbon (DOC) becomes an increasingly important fraction of allochthonous carbon in larger rivers. The contents of DOC include soluble carbohydrates and amino acids, which are leached from decomposing leaves and plant roots (Suberkropp et al., 1976), as well as organic acids leached from soils (McDowell and Likens, 1988; Qualls and Haines, 1992; Chapter 5).

Using ^{14}C dating, Raymond and Bauer (2001b) report DOC ages in the Susquehanna, the Rappahannock and the Hudson River of 688, 736, and 1384 years bp.[q] Although the age and the composition of this terrestrial DOC would suggest that it is highly recalcitrant, aquatic microbes assimilate and respire most DOC during river transport (Hotchkiss et al., 2015a; Wallace et al., 1999; Richey et al., 2002; Mayorga et al., 2005; Battin et al., 2009) and very little terrestrial DOC makes its way to the open ocean (Bauer et al., 2013; Bianchi, 2011; Hedges and Keil, 1997). The most reasonable explanation for this seeming paradox is that small amounts of ancient, potentially petrogenic DOM are contained within riverine DOM pools and have a disproportionate effect on ^{14}C based dating of riverine DOC. Most of the DOC in the Amazon River is of recent origin (Mayorga et al., 2005), and DOC that was recalcitrant in soils or anoxic sediments may become labile in well lit and well oxygenated rivers.

Autochthonus inputs—Primary productivity in rivers

Net primary production in streams and rivers is typically estimated using one of two approaches, respirometer chambers or in situ changes in dissolved oxygen concentrations (Bott, 2006). Respirometer chambers are analogous to the light/dark bottle methods described for lakes, and involve isolating stream sediments and water in closed containers and measuring changes in the dissolved oxygen concentration in the overlying water over time. Although chamber estimates are useful for comparative studies and experimental manipulations, NPP estimates derived from chambers are particularly difficult to extrapolate to river ecosystems. First, enclosing stream sediments in a closed vessel reduces or eliminates

---

[q]Note that this ^{14}C dating provides a single age for the bulk pool of DOC which will include DOC molecules that were fixed in photosynthesis seconds, hours, days, seasons, years and millennia before the sample was collected.

flow, nutrient supply, and the gas exchange conditions of natural streams (Bott, 2006). Second, because river sediments are typically very heterogeneous, scaling to the whole ecosystem requires extensive sampling of all benthic habitat types (Hondzo et al., 2013). Finally, chambers typically do not include subsurface sediments, so they tend to considerably underestimate rates of ecosystem respiration by ignoring oxygen consumption in the hyporheic zone[r] (Fellows et al., 2001).

In contrast, open-channel techniques involve measuring daily fluctuations of stream water oxygen, or less commonly CO_2 concentrations, and linking these changes to the processes of production, respiration and exchange with groundwater or the atmosphere (Odum, 1956) (Fig. 8.11). For any time step:

$$\Delta O_2 = GPP - R \pm E, \tag{8.13}$$

where E is atmospheric exchange as estimated using a gas tracer (Wanninkhof et al., 1990; Hall and Ulseth, 2020). These data are analyzed as described for lakes. The chief difference is that, in rivers, turbulence is a more important driver of gas diffusion than is wind, so gas tracer-derived estimates of diffusion must be made at the same flows for which oxygen changes are measured. In general, chamber methods indicate that primary production often exceeds respiration in well lit streams (Minshall et al., 1983; Bott et al., 1985), whereas open-channel methods are more likely to find net heterotrophy (Finlay, 2011; Hoellein et al., 2013; Hall and Tank, 2003; Bernot et al., 2010). A compilation of whole ecosystem measures of primary productivity and ecosystem respiration from flowing waters finds that the majority of both small streams and large rivers are net heterotophic (Table 8.5, Battin et al., 2009) and that smaller streams tend to have higher rates of ecosystem respiration than large rivers.

TABLE 8.5 A Compilation of Literature Estimates of GPP, R, and NEP for Streams, Rivers, and Estuaries from Whole-Ecosystem Metabolism Estimates.

Ecosystem	GPP (g C $m^{-2}d^{-1}$)	R (g C $m^{-2}d^{-1}$)	NEP (g C $m^{-2}d^{-1}$)	Global R (PgC yr^{-1})	Global net heterotrophy (PgC yr^{-1})
Streams (n = 62)	0.73±0.14 (0.02–5.62)	1.93±0.19 (0.29–8.16)	−1.20±0.15 (−5.86–2.51)	0.19	0.12
River (n = 37)	0.91±0.10 (0.06–2.28)	1.53±0.15 (0.20–3.54)	−0.66±0.11 (−2.06–1.60)	0.16	0.07
Estuaries (n = 31)	3.14±0.41 (0.72–10.4)	3.51±0.32 (0.83–7.58)	−0.39±0.21 (−2.98–2.86)	1.20	0.13

Note: Given is the mean standard error and the minimum and maximum in brackets.
Adapted from Battin et al. (2009).

[r] The hyporheic zone is the volume of subsurface sediments where groundwaters and surface waters are actively exchanged. Hyporheic zone volume varies widely across streams, there is no hyporheic zone when bedrock underlies the channel but the hyporheos may extend 10s to 100s of meters in coarse alluvial channels.

Limits to autochthonous production in flowing waters

In small forested streams light is often the primary factor limiting GPP (Hill, 1996; Hall and Tank, 2003; Roberts et al., 2007). In 2 years of continuous monitoring of GPP in the Walker Branch River in Tennessee, Roberts et al. (2007) showed high day-to-day and seasonal variability in GPP that was constrained by seasonal forest canopy cover (Fig. 8.22). In this stream, and probably most small steams draining deciduous forests, the highest rates of stream metabolism are recorded outside of the forest growing season (Roberts et al., 2007; Savoy et al., 2019).

In streams, nutrient limitation of benthic algae is often assessed using nutrient diffusing substrates (NDS), or agar-filled containers that slowly release a nutrient solution through a porous membrane. The biomass of algae growing on NDS containing a single or multiple nutrients can be compared to the algae on substrates containing nutrient-free agar (Pringle, 1987). In a comprehensive synthesis of 237 NDS experiments, Francoeur (2001) found that 43% reported no algal response to either N or P additions, while co-limitation by both N and P (23%) was more common than limitation by either N (17%) or P (18%) alone. Results may vary seasonally according to light availability. For example Bernhardt and Likens (2004) measured significant nutrient limitation of benthic algae for 10 small forested streams in New England in the spring before canopy leaf-out, with no nutrient enrichment effects observed during summer or fall.

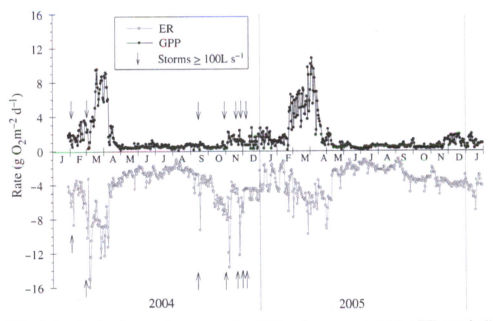

FIG. 8.22 Daily estimates of gross primary production (GPP) and ecosystem respiration (ER) rates for Walker Branch, Tennessee, derived from open channel oxygen measurements over 2 years. *Arrows* indicate when large storms occurred. *From Roberts et al. (2007). Used with permission of Springer.*

Compared to the pattern in lakes, algal biomass and nutrient concentrations in streams and rivers are seldom correlated (Biggs, 1996; Francoeur, 2001). In well-lit streams, factors other than nutrient supply are likely to constrain the response of autotrophs to nutrient loading. Many rivers experience floods, which either scour or bury riverbeds with transported sediments. Thus the algae and macrophytes in many rivers are nearly always at some stage of recovery from the most recent flood (Grimm and Fisher, 1989; Death and Winterbourn, 1995). Some of the most productive river systems on earth are fed by springs and dominated by lush growths of aquatic macrophytes that never experience scouring flows (Odum, 1957; Heffernan and Cohen, 2010).

Biomass estimates or measures of chlorophyll are often misleading indicators of river productivity because flow disturbance and variable light regimes cause substantial temporal variation in NPP (Grimm and Fisher, 1989; Hill, 1996; Bernhardt et al., 2018) and because grazing invertebrates are capable of consuming a large fraction of NPP (Wallace and Webster, 1996; Taylor et al., 2006). Algae and macrophyte tissues are more nutrient rich and palatable than terrestrial organic matter, so autotrophic production often contributes disproportionately to secondary production in river food webs, even in cases where rates of NPP are quite low (Hotchkiss and Hall, 2015; McCutchan and Lewis, 2002). Grazing by invertebrates and fish can constrain responses to nutrient enrichment (Fig. 8.23; Rosemond and Elwood, 1993; Taylor et al., 2006).

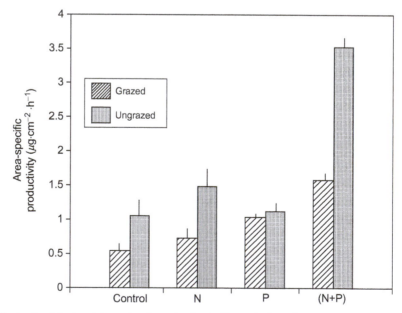

FIG. 8.23   Effects of nutrient enrichment and grazing by snails on algal productivity in a series of streamside channels (measured by ^{14}C incorporation) in Walker Branch, Tennessee. *From Rosemond et al. (1993). Used with permission of the Ecological Society of America.*

Carbon budgets for rivers

Carbon budgets are often used to understand the relative importance of autochthonous and allochthonous carbon inputs to river systems. For a small headwater stream known as Bear Brook within the Hubbard Brook Experimental Forest, Fisher and Likens (1972) found that stream bryophytes were the only autotrophs, and that they contributed less than ~0.1% of the annual C inputs (Table 8.6). In 1 year of monitoring surface and lateral carbon inputs

TABLE 8.6 Annual carbon budget for Bear Brook, a small New Hampshire stream.

Item	Kg—whole stream[a]	Kcal/m^2	Percentage
Inputs			
Litterfall			
Leaf	1990	1370	22.7
Branch	740	520	8.6
Miscellaneous	530	370	6.1
Wind transport			
Autumn	422	290	4.8
Spring	125	90	1.5
Throughfall	43	31	0.5
Fluvial transport			
CPOM	640	430	7.1
FPOM	155	128	2.1
DOM, surface	1580	1300	21.5
DOM, subsurface	1800	1500	24.8
Moss production	13	10	0.2
Input total	8051	6039	99.9
Outputs			
Fluvial transport			
CPOM	1370	930	15.0
FPOM	330	274	5.0
DOM	3380	2800	46.0
Respiration			
Macroconsumers	13	9	0.2
Microconsumers	2930	2026	34.0
Output total	8020	6039	100.2

[a] Budget in kg doesnot balance because of different caloric equivalents of budgetary components.
Source: Adapted from Fisher and Likens (1973).

and outputs, Fisher and Likens (1973) estimated that ~3260 kg of C was delivered to the stream as leaves and branches, of which ~2930 kg were respired by stream microbes. The energetics of this stream was nearly completely dominated by allochthonous inputs with a P:R ratio < 0.01.

In a compilation of organic matter budgets for 35 streams, Webster and Meyer et al. (1997) documented a wide range in P:R ratios, from 0 (as in Bear Brook) to 1.7. Generally the P:R ratio increased with stream size, however desert streams were far more productive, and the single blackwater river was far less productive than predicted for their size (Fig. 8.24; Webster and Meyer, 1997). Desert streams, which drain a landscape of low stature vegetation, are not light-limited (Jones et al., 1997), whereas the blackwater Ogeechee River, as for dystrophic lakes, had high concentrations of allochthonous DOC that attenuated light availability (Meyer et al., 1997).

Much of the coarse particulate material provided to rivers from terrestrial vegetation is respired by microbes (Fisher and Likens, 1973; Battin et al., 2009; Hotchkiss et al., 2015a) and serves as a primary resource for aquatic food webs. Terrestrial carbon thus contributes substantially to secondary production of aquatic insects. Wallace et al. (1997) excluded terrestrial litter from a small mountain stream in North Carolina for four years and observed substantial declines in the biomass production of aquatic insects (Fig. 8.25).

In narrow, forested streams allochthonus inputs are the dominant energy source and R greatly exceeds NPP. Authochthonous production by benthic algae (periphyton) and aquatic macrophytes (vascular hydrophytes) becomes increasingly important in wider channels (Vannote et al., 1980) and in streams with limited or seasonal canopy cover (Savoy et al., 2019; Koenig et al., 2019). This frequently observed pattern, led Vannote et al. (1980) to predict that productivity increases with stream size, and the ratio of POC to DOC will decline downstream. They proposed that in very large rivers, allochthonous inputs and heterotrophy

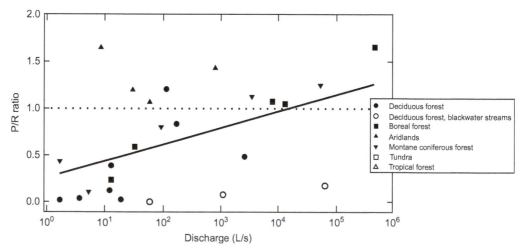

FIG. 8.24   P:R ratios from a synthesis of 26 organic matter budgets reported in the literature. *From Webster and Meyer (1997). Used with permission of the North American Benthological Society.*

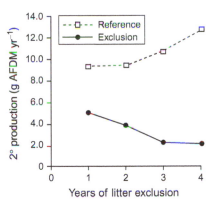

FIG. 8.25 Effects of a multiyear leaf litter exclusion experiment on the secondary production (~total biomass of stream macroinvertebrates) of a small forested stream (Coweeta, North Carolina). *From Wallace et al. (1997). Used with permission of the American Association for the Advancement of Science.*

will again dominate the carbon budget, because large rivers are often too deep to support benthic autotrophs and too turbulent or turbid to support significant phytoplankton productivity.

Nutrient spiraling in rivers

An effective theory for nutrient cycling in stream ecosystems is the concept of nutrient spiraling, which recognizes that nutrients are constantly carried downstream (Fig. 8.26; Webster and Patten, 1979; Newbold et al., 1981; Newbold, 1992; Webster and Ehrman, 1996). During downstream transport, dissolved ions are accumulated by bacteria and other stream organisms and converted to organic forms. When these organisms die, they are degraded to inorganic forms that are returned to the water, only to be taken up again by

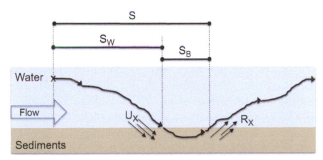

FIG. 8.26 A nutrient spiral is a nutrient cycle that is displaced by advective flow. In this diagram the black line represents the path of an average nutrient molecule during downstream transport as it is moved downstream prior to uptake into sediments (U) and then transported yet further downstream prior to remineralization (R) and release to the water column. The spiraling length (S) is composed of the uptake length (S_W), which is the transport distance prior to removal from the water column, and the remineralization length (S_B), which is the downstream distance transported within benthic sediments or biota. *Modified from Newbold (1992).*

I. Processes and reactions

organisms that are involved in the further degradation of organic materials. The cycle between inorganic and organic forms may be completed many times while a nutrient atom moves downstream to the ocean.

Biogeochemists often compare the downstream transport of nutrients and conservative solutes[s] to understand the role of biota in determining element fate and transport. Ecosystem processes within streams will often lead to more rapid attenuation of nutrient pulses than can be explained by dilution alone (Alexander et al., 2000; Seitzinger et al., 2002; Bernhardt et al., 2003; Green et al., 2004). With the injection of conservative ions (e.g., Cl, Br) into a river, their concentration will decline with distance downstream of the injection point as a function of dilution (Stream Solute Workshop, 1990) or permanent loss to groundwater (Covino and McGlynn, 2007; Poole et al., 2008). If the decline in concentration of an injected nutrient is more rapid than measured for the conservative tracer, this indicates biological or chemical uptake in addition to dilution and mass loss. Solute concentrations (A_x) are corrected for dilution by dividing them by the concentration of the tracer (Tr) measured at the same location:

$$A_x = [N_x]/[Tr_x]. \tag{8.14}$$

The uptake length of a solute (S_w) represents the average downstream distance that a dissolved nutrient molecule travels before it enters the particulate phase:

$$S_w = 1/k_A, \tag{8.15}$$

where K_A is the per meter uptake rate of solute A. The nutrient spiral (between dissolved and particulate phases) is equivalent to a nutrient cycle (between organic and inorganic forms) for solutes that are not strongly adsorbed onto mineral particles, but conflates both physical sorption and biological assimilation for ions such as NH_4^+ or PO_4^{3-} (Chapter 4). Elements that are in greater biological demand will have shorter spiraling lengths. The spiraling length can be used to estimate whole ecosystem uptake rates (U) by:

$$U = [Q \times C_{bkgrnd}]/[S_w \times w], \tag{8.16}$$

where Q is stream flow in $m^3 \, min^{-1}$, C_{bkgrnd} is the concentration of the solute of interest prior to the measurement, and w is the average width (meters) of the study reach. Spiraling length estimates are made by enriching nutrient concentrations above background or by injecting isotopic tracers that do not alter nutrient concentrations (^{15}N and ^{32}P) together with a conservative tracer.

Nutrient uptake rates decline strongly with increasing flow velocities, and increase with increasing contact of water and sediments, indicating that hydrology ultimately constrains biological assimilation (Peterson et al., 2001; Hall et al., 2002; Webster et al., 2003; Wollheim et al., 2006). To allow comparisons of uptake efficiency across streams with very different flows, biogeochemists sometimes calculate uptake velocities, or mass-transfer coefficients (V_f):

$$V_f = [Q/w]/S_w, \tag{8.17}$$

---

[s] These are ions such as Cl and Br which are not normally absorbed by stream sediments and not appreciably accumulated by biota.

which allows comparisons of uptake efficiencies between rivers with significant differences in flow.

Modifications of the spiraling technique (Covino et al., 2010, 2018) have made it possible to calculate from a single addition both the ambient uptake rate and the maximum potential uptake rate of nutrients in a stream. Simultaneous injection of several nutrients can show the relative strength of nutrient limitations (Bernhardt and McDowell, 2008; Lutz et al., 2012a).

Nitrogen cycling

Ecosystem nutrient demand can also be assessed using mass-balance approaches, but studies of annual nutrient mass balances are very time consuming and rare in the literature (e.g. Table 8.7; Meyer et al., 1981; Triska et al., 1984). In an exhaustive annual budget for nitrogen in a small stream draining a temperate rainforest in Oregon, Triska et al. (1984) showed that dissolved organic nitrogen (DON) was the largest input and export term. Assuming steady state, the ~100-m stream segment he studied lost nearly one third of the organic nitrogen it received by converting it to nitrogen gas or secondary production, and transformed nearly 2/3 of coarse terrestrial organic matter inputs (leaves, needles, wood) into CO_2 or fine particles (Triska et al., 1984).

Measurements and models of stream nitrogen cycling using spiraling approaches generally conclude that biological uptake within stream ecosystem significantly reduces the downstream flux of inorganic nitrogen (e.g., Alexander et al., 2000; Peterson et al., 2001; Mulholland et al., 2008; Wollheim et al., 2008). Across sites, much of the variation in nitrogen-uptake

TABLE 8.7 Yearly fluxes of organic carbon, nitrogen, and phosphorus in Bear Brook.

	Organic carbon (g/m^2)	Nitrogen (g/m^2)	Phosphorus (g/m^2)	Atomic ratio C:N:P
Inputs				
Total dissolved	200	56	0.39	1700:320:1
Total fine particulate	12	0.27	0.55	54:1:1
Total coarse particulate	340	8.2	0.7	1300:26:1
Total gaseous	1	< 0.1	0	
Total inputs	620	64	1.6	990:89:1
Outputs				
Total dissolved	260	57	0.29	2300:440:1
Total fine particulate	25	0.43	1.1	59:0.9:1
Total coarse particulate	100	1.8	0.38	720:10:1
Total gaseous	230	?	0	
Total outputs	620	59	1.8	890:72:1

Source: Meyer et al. (1981). Used with permission of VG Wort, Germany.

lengths can be explained by differences in stream flow and depth (Peterson et al., 2001). Nutrients travel longer distances in streams with faster flow and reduced water-sediment contact.

In a cross-biome survey of $^{15}NH_4$ uptake in streams of North America, Peterson et al. (2001) found that ammonium was typically removed from the water column within 10 to 100 m, while nitrate tended to travel much longer distances in the same streams. Rapid uptake of dissolved NH_4^+ by stream sediments is due both to preferential uptake of NH_4^+ by phytoplankton and bacteria (Chapter 6) and cation exchange on sediments (Chapter 4) (Peterson et al., 2001). Much of the NH_4^+ that enters streams is nitrified with the proportion returning to the water column increasing with stream nitrate concentrations (Peterson et al., 2001; Bernhardt et al., 2002). Nitrification facilitates the downstream transport of nitrogen, so that NO_3^- loading may exacerbate downstream nitrogen pollution (Koenig et al., 2017; Bernhardt et al., 2002). Nitrate spiraling is controlled almost entirely by biological processes because NO_3^- is not strongly adsorbed on sediments. Generally, nitrate uptake tends to increase with nitrate loading, although this capacity can be easily saturated by high loading rates (Mulholland et al., 2008).

Variations in carbon supply and carbon processing (metabolism) are good correlates of nitrogen uptake both within and among streams (Baker et al., 1999; Bernhardt and Likens, 2002; Roberts and Mulholland, 2007). Tight linkages between carbon and nutrient processing are expected given stoichiometric and thermodynamic constraints on organisms. Experimental addition of labile carbon to stream sediment samples decreases nitrification (Strauss and Lamberti, 2000, 2002; Strauss et al., 2002) and stimulates N assimilation and denitrification (Hedin et al., 1998; Baker et al., 1999; Sobczak et al., 2003; Crenshaw et al., 2002). During a two-month DOC enrichment to a stream in the Hubbard Brook Experimental Forest, Bernhardt and Likens et al. (2002) were able to reduce inorganic nitrogen export from an entire watershed to below analytical detection by stimulating microbial N demand.

Since streams are not aggrading systems, a large portion of the net loss of N from streams must be attributed to denitrification. In a series of $^{15}NO_3$ addition experiments in 72 North American streams, Mulholland et al. (2008) found that NO_3^- uptake rates varied widely (from 0.01 to 1000 mg N m^{-2} h^{-1}). In these experiments, up to 100% of the $^{15}NO_3$ was converted, within 24 h, to $^{15}N_2$ (Mulholland et al., 2008, 2009), although the median rate was 16%. The N_2O yield from this series of experiments was very low, with only 0.04 to 5.6% of the total gaseous ^{15}N produced as N_2O (Beaulieu et al., 2011). Although the total N_2O produced increased at high denitrification rates, Beaulieu et al. (2011) did not observe an increase in the proportional yield of N_2O with N loading. These observations suggest that streams provide ideal conditions for complete denitrification and do not contribute large quantities to the global budget for N_2O despite their effective removal of inorganic N.

Sunlight can have a major influence on nutrient cycling in streams. Roberts et al. (2007) compared upstream and downstream nitrogen fluxes in a small stream in Tennessee, finding that variation in GPP could explain nearly 80% of instantaneous NO_3^- flux (Fig. 8.22). Since GPP in this stream is well correlated to light, they were able to show that solar energy directly and rapidly drives nitrogen-uptake capacity. Heffernan and Cohen (2010) provide a more extreme example of the tight linkages between light availability, stream GPP, and NO_3^- uptake

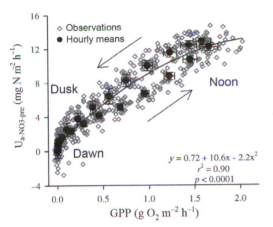

FIG. 8.27 Nitrate uptake rate (U_{NO3}) varies as a function of GPP over the course of a single day in the macrophyte-dominated Ichetucknee River. Florida. The *small circles* represent individual data points, the *large circles* represent hourly averages. The *arrows* show the hysteresis of the relationship from predawn on day 1 to predawn on day 2. *From Bernhardt et al. (2018). Used with permission of the Association for the Sciences of Limnology and Oceanography. All Rights Reserved.*

rates. Using oxygen and nitrate sensors in a macrophyte-dominated spring in south Florida, GPP and NO_3^- uptake could be almost perfectly predicted by light availability (Fig. 8.27).

Hydrology constrains nitrogen assimilation and denitrification. Benthic algae and microbes are less able to assimilate N from the water column at high flow rates. The capacity of river ecosystems to retain versus transport additional N depends not only on the biological capacity (set by either light or carbon availability) but also on the timing of N delivery. Nitrogen that enters streams in storm flows will travel much farther downstream than nitrogen molecules that seep into the channel during baseflows (Shields et al., 2008). Nitrate removal rates are much higher in river segments with long mean residence times (Seitzinger, 1988; McCrackin et al., 2014).

Phosphorus cycling

Globally rivers transport ∼21×10^{12} g of phosphorus to the oceans each year, with nearly all of this phosphorus in particulate form (Meybeck, 1982, 1993; Ittekkot and Zhang, 1989). Only about 10% of the particulate phosphorus is biologically available; the rest is strongly bound to soil minerals (Meyer and Likens, 1979; Ramirez and Rose, 1992). For rivers without significant wastewater or fertilizer inputs, very little inorganic phosphorus is found in the water column. Phosphorus mineralized from organic matter is rapidly adsorbed or assimilated, keeping P out of solution (Meyer, 1979, 1980; Meyer and Likens, 1979).

Phosphorus spiraling is more difficult to interpret than nitrogen spiraling, because a large proportion of inorganic phosphorus may be removed from the water through physical adsorption rather than biological assimilation (Demars, 2008; Stutter et al., 2010). Phosphorus is less likely to be limiting to productivity and respiration in rivers because P that is deposited in river sediments remains available to most river biota. The abundance of physical barriers (debris dams and slow-moving pools) in stream channels explained >90% of the variation in phosphorus uptake rates across streams of the Hubbard Brook valley in New Hampshire (Warren et al., 2007) consistent with previous observations that sediment adsorption and trapping dominates P uptake in these streams (Meyer and Likens, 1979). Changes in biological demand can affect P uptake rates. Grazing invertebrates reduced P uptake rates by suppressing algal productivity in laboratory streams (Mulholland, 1993) and autumn litter fall increased P immobilization by microbes on coarse particulate organic matter in the Walker Branch stream in Tennessee (Mulholland et al., 1985).

Estuaries: Where the turf meets the surf

An estuarine ecosystem consists of the river channel, to the maximum upstream extent of tidal influence, and the adjacent ocean waters, to the maximum seaward extent that they are affected by the addition of freshwater. The estuary also includes any salt marshes that may develop along the shoreline. Estuaries are zones of mixing; within an estuary there is a strong gradient in salinity from land to sea. Estuaries are among the most challenging environments on Earth to study biogeochemistry, because in addition to the underlying salinity gradient, turbulent mixing of fresh and salt water generates abrupt changes in temperature, salinity, pH, redox, and element concentrations—all with implications for biogeochemical cycling.

When large rivers reach the sea, their velocity slows, reducing their ability to carry sediment. The load of suspended materials carried by coastal rivers is deposited in the river channel and on the continental shelf. Rivers carrying large sediment loads, such as the Mississippi, may form obvious deltas of deposited sediments, which may support broad, flat areas of salt marsh vegetation (Fig. 8.28).

Estuarine water budgets and mixing

The mixing of freshwater from rivers and salt water from the sea occurs in the central channel of an estuary. If the estuary is well mixed, the transition from freshwater to seawater is

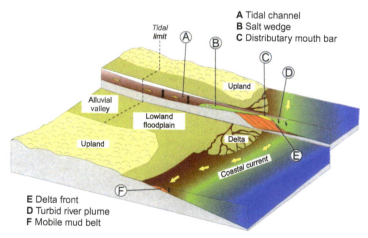

FIG. 8.28   Generic diagram of a river estuary. The estuary boundaries are defined as the upper limit of the tidal influence within the river inflow to the coastal boundary of freshwater influence. *From Bianchi and Allison (2009). Used with permission of the National Academy of Sciences.*

gradual and progressive as one moves downstream. In many cases, inflowing freshwater may extend over a "wedge" of denser saltwater, creating a sharp vertical gradient in salinity throughout much of the estuary (Fig. 8.29). In either case, this transition zone is an arena of rapid biogeochemical transformations and high productivity (Burton, 1988; Dagg et al., 2004).

Seawater has high pH (~8.3), redox potential (>+200mV), and ionic strength relative to freshwater (Fig. 4.19, Table 9.1). The mixing of freshwater and seawater causes a rapid precipitation of the dissolved humic compounds carried by rivers. The cations in seawater replace H^+ on the exchange sites of the humic materials (Chapter 4), causing these materials to flocculate[t] and sink to estuarine sediments (Sholkovitz, 1976; Boyle et al., 1977). Although organic acids make up only a small fraction of total riverine DOC, this flocculation is also responsible for the "salting out" of hydrocarbons and organometallic complexes which are precipitated in the estuary or within a short distance of the mouth of the river (Boyle et al., 1974; Sholkovitz, 1976; Jickells, 1998; Turner and Millward, 2002; Blair and Aller, 2012). The flocculation of dissolved organic compounds and the deposition of larger plant debris account for a major portion of the organic carbon in estuarine sediments (Hedges and Keil, 1997; Blair and Aller, 2012), and there is little evidence that organic matter from land contributes much to marine sediments beyond the continental shelf (Hedges and Parker, 1976; Prahl et al., 1994; Hedges and Keil, 1997). As a result of the removal of terrestrial organic matter, the majority of the organic carbon in estuarine waters is contributed by net primary production in the estuary and its salt marshes (Fox, 1983; Nixon et al., 1996).

Thermal stratification in estuaries is reinforced by salinity differences, which separate more dilute, well-lit estuarine surface waters from darker and more saline bottom waters. Just as in lakes, this density gradient can lead to well oxygenated and productive surface waters overlying deep waters where oxygen consumption can exceed diffusion. Together with enhanced nutrient loading, this density separation contributes to the widespread occurrence of coastal hypoxia (Rabalais et al., 2002; Diaz and Rosenberg, 2008; Conley et al., 2009a).

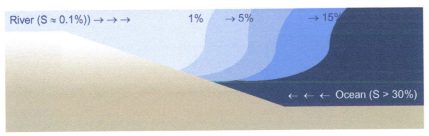

FIG. 8.29   A diagram of a generic salinity gradient within a coastal estuary. River waters interacting with ocean waters lead to a gradient from fresh to full-strength ocean waters. The greater density of saltwater often leads to a salt wedge underlying a plume of less saline surface waters.

[t]See footnote # e, Chapter 4.

Carbon cycling in estuaries

The upper reaches of estuaries receive large subsidies of organic matter from inflowing rivers. Despite having highly productive vegetation along their margins (mangroves, salt marshes) these systems tend to be net heterotrophic, because much of their metabolism is sustained by the allochthonous carbon inputs from rivers, groundwaters, and, for many estuaries, urban wastewaters (Odum and Hoskin, 1958; Kemp et al., 1997; Wang and Cai, 2004). As a result, most estuaries are supersaturated in CO_2. In a comprehensive review, Chen and Borges (2009) found only one reported instance where an estuary was a net sink for CO_2 (air-water flux (FCO_2[u])) of $3.9\,mol\,C\,m^{-2}\,yr^1$ (from Kone and Borges, 2008), while the rest of the estuaries had FCO_2 values ranged from $-3.6\,mol\,C\,m^{-2}\,yr^1$ in Finland's Bothnian Bay (Algesten et al., 2004) to $-76\,mol\,C\,m^{-2}\,yr^1$ in Portugal's Douro estuary (Frankignoulle et al., 1998). Chen and Borges (2009) estimate that globally estuaries emit $\sim0.50\,PgC\,yr^{-1}$ to the atmosphere as CO_2 (Fig. 8.30).

Mass balance calculations and pCO_2 measurements indicate that the continental shelves are sinks of atmospheric CO_2 (Borges, 2005; Chen and Borges, 2009). Annual FCO_2 estimates on continental shelves average $\sim1.1\,mol\,CO_2\,m^{-2}\,yr^{-1}$, which, scaled to a global continental shelf area of $26 \times 10^6\,km^2$, yields an annual CO_2 uptake of about $0.35\,PgC\,yr^{-1}$ (Chen and Borges, 2009). Because of their much greater spatial extent, coastal shelves nearly offset the atmospheric CO_2 source of inland estuaries and their fringing marshes and mangroves (Table 8.8).

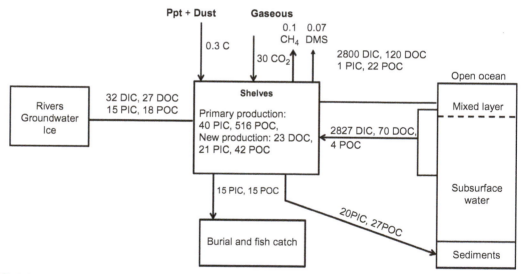

FIG. 8.30 Mass balance of carbon in continental shelves (flows are in $10^{12}\,mol\,C\,yr^{-1}$). *From Chen and Borges (2009).*

[u]By convention, a positive fCO_2 indicates the ecosystem is a net sink for atmospheric CO_2 while a negative fCO_2 value indicates that the system is a net source of CO_2 to the atmosphere.

TABLE 8.8 Air-water CO_2 flux in open oceanic waters and major coastal ecosystems (including inner estuariesand salt marshes).

	Surface ($10^6\,km^2$)	fCO$_2$ Air-water CO$_2$ flux (mol C m^{-2} yr^{-1})	Air-water CO$_2$ flux (Pg C yr^{-1})
60°–90°			
Open oceanic waters	30.77	−0.75	−0.28
Inner estuaries	0.4	46	0.22
Open shelf	6.79	−1.88	−0.15
Subtotal	37.96	−0.46	−0.21
30°–60°			
Open oceanic waters	122.44	−1.4	−2.05
Inner estuaries	0.29	46	0.16
Non-estuarine salt marshes	0.14	23.45	0.04
Coastal upwelling systems	0.24	1.09	0.003
Open shelf	14.47	−1.74	−0.3
Subtotal	137.58	−1.3	−2.15
30°N–30°S			
Open oceanic waters	182.77	0.35	0.77
Inner estuaries	0.25	16.83	0.05
Coastal upwelling systems	1.25	1.09	0.02
Coral reefs	0.62	1.52	0.01
Mangroves	0.2	18.66	0.04
Open shelf	1.35	1.74	0.03
Subtotal	186.44	0.41	0.92
Coastal ocean	26	0.381	0.12
Open ocean	336	−0.388	−1.56
Global ocean	362	−0.331	−1.44

Compiled by Borges (2005, Table 3). Used with permission of Springer.

Primary production in estuaries

Many estuaries show a peak in net primary productivity at intermediate salinities, reflecting the zone of maximum nutrient availability and phytoplankton abundance where fresh and salt waters coalesce (Anderson, 1986; Lohrenz et al., 1999; Dagg et al., 2004). In other

cases, mixing hides any obvious relationship between net primary production and conservative properties, such as salinity, in the estuary (Powell et al., 1989).

Phytoplankton productivity and organic matter derived from the surrounding salt
marshes fuel the high productivity of fish and shellfish in estuarine waters. For many years
the large production of fish and shellfish in estuaries was attributed to an abundance of organic carbon flushing from salt marshes to the open water. Indeed, the losses of organic carbon from salt marshes are usually $>100\,g\,Cm^{-2}\,yr^{-1}$, compared to losses of $1-5\,g\,Cm^{-2}\,yr^{-1}$
from uplands (Nixon, 1980; Schlesinger and Melack, 1981). Haines (1977), however,
suggested that this paradigm was questionable, because the isotopic ratio of carbon in estuarine animals did not match that of *Spartina*, the dominant salt marsh grass. Using the natural
abundance of stable isotopes of both sulfur and carbon, Peterson et al. (1985, 1986) showed
that the organic carbon in primary consumers within the Great Sippewissett Marsh is about
equally derived from *Spartina* and from phytoplankton production in the open water
(Fig. 8.31). The shellfish, crabs, fish and shrimp at the base of the marsh food web show

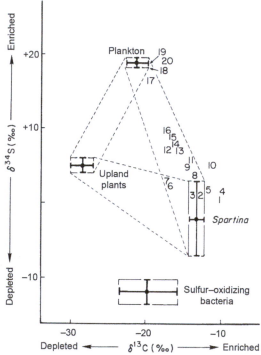

FIG. 8.31   The isotope ratio for C and S in consumers is shown in relation to their ratios in upland plants, plankton,
the salt marsh grass *Spartina*, and sulfur-oxidizing bacteria for Cape Cod's Great Sippewissett Salt Marsh in Massachusetts. The isotope ratios in sulfur-oxidizing bacteria are very different from those of consumers, indicating that
sulfur oxidizers are not a major source of carbon for higher trophic levels in the estuary. Similarly, the C isotope ratio
for terrestrial plants is considerably more depleted than in consumer biomass, suggesting that allochthonous C is less
important than autochthonous C in this marsh. Consumers include shellfish, snails, shrimp, crabs, and fish. The
values for each consumer represent pooled samples of 10 to 200 individuals except in the case of the flounder (9)
and swordfish (19). *From Peterson et al. (1986). Used with permission of the Ecological Society of America.*

isotopic ratios for C and S that are midway between these sources. Similar results were found in the Sapelo Island marsh (Peterson and Howarth, 1987). In contrast, carbon from upland, terrestrial vegetation and carbon fixed by sulfur-oxidizing bacteria in salt marsh soils appears to play a minor role in supporting the abundant marine life of estuaries.

Nutrient cycling in estuaries

Nitrogen

A great deal of effort has been directed towards understanding the nitrogen budget of estuaries. Nitrogen has been implicated as the nutrient responsible for eutrophication in the Chesapeake Bay (Cooper and Brush, 1991; Bronk et al., 1998; Boesch et al., 2001), the Gulf of Mexico (Turner and Rabalais, 1994; Rabalais et al., 2002), Narangassett Bay (Nixon et al., 1995; Howarth and Marino, 2006), the Baltic Sea (Conley et al., 2007, 2009a) and many estuaries in the developing world (Fig. 8.32).

Nitrogen fixation is often limited in estuarine systems because of the low availability of molybdenum (a critical cofactor in the nitrogenase enzyme) as well as sulfate interference with Mo uptake (Howarth and Cole, 1985; Cole et al., 1993; Marino et al., 2003). The higher turbulence of estuarine waters (Howarth et al., 1995b; Paerl, 1996) and the abundance of zooplankton grazers (Marino et al., 2006) also constrain the growth of filamentous cyanobacteria that are the common N fixers.

Most river waters do not contain large concentrations of available nitrogen (NO_3 and NH_4), and these forms are removed when the waters pass over coastal salt marshes. Indeed, the filtering action of land and marsh vegetation is so effective that the direct inputs of nitrogen in rain can make a substantial contribution to the nitrogen budget of the central waters of unpolluted estuaries (Correll and Ford, 1982). However, as is the case for terrestrial ecosystems (Chapter 6), most of the nitrogen that supports estuarine productivity is not derived from new inputs but from mineralization and recycling of organic nitrogen within the estuary and its sediments (Stanley and Hobbie, 1981).

At the pH and redox potential of seawater, nitrification occurs rapidly in estuarine waters (Billen, 1975; Capone et al., 1990), and in the upper layers of sediment (Admiraal and Botermans, 1989). Denitrification in the lower, anaerobic layers of sediment is primarily

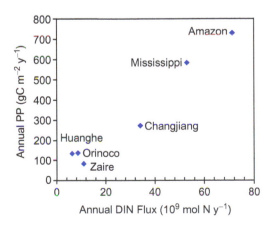

FIG. 8.32   Relationship between primary production in coastal shelf waters impacted by major rivers and riverine DIN flux. *Adapted from Dagg et al. (2004).*

supported by nitrate diffusing down from the upper sediment (Seitzinger, 1988; Kemp et al., 1990), although nitrate in the water column may also diffuse into the sediments, where it is reduced (Simon, 1988; Law et al., 1991). In Narragansett Bay, Rhode Island, Seitzinger et al. (1980, 1984) found that denitrification removed about 50% of the available NO_3 entering in rivers and about 35% of that derived from mineralization within the estuary. The major product of denitrification was N_2.

In Chesapeake Bay, denitrification leaves the nitrate in the lower water column enriched in $\delta^{15}N$ (Horrigan et al., 1990). When the nitrification rate in the sediments is low, available NO_3 may limit the rate of denitrification, and more NH_4^+ remains to support the growth of phytoplankton in the estuary (Kemp et al., 1990). Storms and tidal currents stir up the sediments in an estuary, releasing large quantities of NH_4^+ to the water column (Simon, 1989). In oligotrophic estuaries where little carbon accumulates in the sediments, denitrification may be far less important. Fulweiler et al. (2007) found that during periods of low productivity (and low organic matter deposition) the sediments of Narangasset Bay became N sources (through N fixation) rather than N sinks (through denitrification) to the water column.

Phosphorus

Most river waters are supersaturated with dissolved CO_2, which is derived from the degradation of organic materials during downstream transport. High concentrations of dissolved CO_2 and organic acids cause river waters to be slightly acid. Under these conditions, most phosphate ions are bound to Fe-hydroxide minerals and are transported in the load of suspended sediment (Fig. 4.4, Table 4.8; Eyre, 1994). Upon mixing with the higher pH of seawater, phosphorus can desorb from these minerals and contribute to dissolved phosphorus in the estuary (Lebo, 1991; Conley et al., 1995; Lin et al., 2012). Seitzinger (1991) found that an increase in pH in the Potomac River estuary caused a release of P from sediments, stimulating a bloom of N-fixing blue-green algae, a scenario that is analogous to the shifts in species dominance that are seen in P-polluted lakes. Under the anoxic conditions common to a growing number of coastal estuaries, Fe bound phosphorus can be released at high rates (Ghaisas et al., 2019), creating conditions for a positive feedback to coastal eutrophication through release of legacy P.

The abundance of SO_4 found in saline waters limits Fe-P binding in estuarine sediments, such that organic P, once mineralized, is likely to remain in the water column (Fig. 8.33; Caraco et al., 1990; Blomqvist et al., 2004; Jordan et al., 2008). The "iron trap" that so effectively sequesters PO_4^{3-} in freshwater sediments is ineffective in estuaries where Fe is rapidly bound to FeS_2 (Fig. 8.35; Blomqvist et al., 2004). De Jonge and Villerius (1989) additionally suggest that the phosphorus bound to carbonate particles delivered to estuaries from the open ocean is released as seawater mixes with freshwater and the carbonates dissolve under the acidic conditions of the upper estuary. For both reasons, phosphorus is often more available in the waters of estuaries than in either freshwater or seawater. Thus, N, rather than P, is likely to be the most limiting nutrient in estuaries though reductions in boost nutrients are required to mitigate coastal hypoxia (Paerl et al., 2016; Conley et al., 2009b).

Anaerobic metabolism in estuarine sediments

Estuarine sediments show high rates of sulfate reduction (Chapter 7), since they are rich in organic matter, flushed with high concentrations of SO_4^{2-} from seawater, and are frequently anoxic. Although the exact magnitude of sulfate reduction is the subject of some controversy

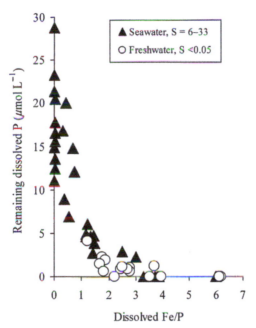

FIG. 8.33 In experimental oxygenation of anoxic waters collected from the brackish northwestern Baltic Sea and adjoining freshwater Lake Malaren of Sweden, the amount of P remaining in solution following oxygenation was dependent on the dissolved molar Fe:P ratio, irrespective of salinity. Very little P remained in solution at Fe:P ratios >2, while below this ratio, dissolved phosphate removal was related to the proportion of Fe(II) present. *From Blomqvist et al. (2004). Used with permission of the Association for the Sciences of Limnology and Oceanography. All Rights Reserved.*

(Howes et al., 1984), various investigators have suggested that more than half of the CO_2 released during decomposition of organic matter in salt marshes and coastal marine sediments is associated with sulfate reduction (Jorgensen, 1982; Howarth, 1984; Henrichs and Reeburgh, 1987; Skyring, 1987; King, 1988) (see Chapter 7). Very little of the sulfide produced escapes to the atmosphere; most is often reoxidized in the upper sediment or the water column. The importance of sulfate reduction depends upon the concentration, and the rate of oxidation of iron in estuarine sediments. When Fe^{3+} is abundant or when Fe^{3+} is continuously resupplied in oxidized rhizospheres or animal burrows, Fe reduction is often the most prevalent form of anaerobic metabolism (Gribsholt et al., 2003; Hyun et al., 2009; Attri et al., 2011; Kostka et al., 2012).

At a series of sites along the York River in the Chesapeake Bay estuary, Bartlett et al. (1987) found a gradient of decreasing methanogenesis with increasing salinity, as the SO_4^{2-} from seawater progressively inhibits methanogenesis (Fig. 8.34; see also Kelley et al., 1990). Howes et al. (1984a, b) found that only about 0.3% of total carbon input to the sediments of Sippewissett marsh in Massachusetts was lost through methanogenesis. Slightly higher rates have been reported for the Sapelo Island estuary in Georgia (King and Wiebe, 1978), but globally, the methane emission from saltwater marshes contributes little to the flux of CH_4 to the atmosphere (Chapter 11). Some salt marsh soils also appear to be a small source of carbonyl sulfide (Aneja et al., 1979), methylchloride (Chapter 3) or phosphine (PH_3) gases to the atmosphere (Hou et al., 2011).

FIG. 8.34 Annual methane flux as a function of average soil salinity across three southeastern United States salt marshes. *From Bartlett et al. (1987). Used with permission of Springer.*

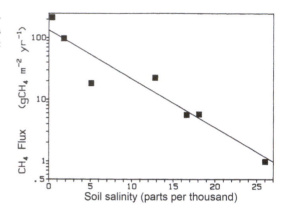

Human impacts on inland waters

Through the creation of infrastructure, such as levees and dams, and alterations of the land surface, including drainage of wetlands, storm sewers, and pavements, humans have fundamentally changed the routing and the timing of the hydrologic connections between terrestrial and aquatic ecosystems.

Water infrastructure

Intensive land use by humans, whether for agriculture or for settlement, tends to dramatically reduce water residence time in upland soils while increasing water residence time in impoundments (e.g., Changnon and Demissie, 1996; Walsh et al., 2005). Collectively, water control structures have tripled the average residence time of water in the world's rivers (Vorosmarty et al., 1997). Coinciding with this hydrologic rerouting, human activities have greatly increased nutrient and sediment loading to rivers and greatly reduced sediment export to coastal seas. There are no large rivers remaining in the world that are not directly impacted by human infrastructure and human wastes.

Humans now appropriate 20% of the global river volume (Jaramillo and Destouni, 2015). In many arid regions this proportion is much higher; in the American Southwest humans appropriate 76% of annual river flows (Sabo et al., 2010). Where annual water extraction exceeds annual runoff, groundwaters are depleted, small streams dry up more frequently, river flows decline, and lakes are dewatered (Chapter 10). Dramatic effects of human water extraction have diminished once large iconic water bodies (Fig. 8.35). Both the Aral Sea in central Asia and Lake Chad in northern Africa are examples of lakes that have shrunk to <20% of their former surface area as a result of water extracted for irrigation. The lack of freshwater flows has also led to the salinization of lakes through evaporative concentration. The same process is occurring globally in arid regions; as water extraction increases, catchment evapotranspiration and the water inputs to rivers and lakes decline. In the southwestern United States, satellite remote sensing calculated that Lake Mead, Lake Powell and the Great Salt Lake lost 11, 6 and 16 km^3 respectively of average volume between 1984 and 2000 and 2000–15 (Busker et al., 2019).

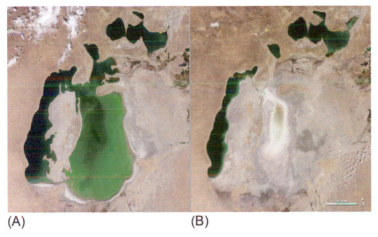

(A)   (B)

FIG. 8.35   NASA satellite imagery shows the boundaries of the Aral Sea in 2000 (A) and in 2009 (B). Once the fourth largest lake in the world, the Aral Sea today is an example of a terminal lake that is both shrinking and becoming saltier as irrigation in the lake basin reduces annual river inflows below annual evaporative losses.

For thousands of years humans have constructed dams to support irrigation, to grow flooded crops, and to ensure long-term water supply. Both the number and size of dams have increased markedly since 1950, with more than 55,000 large dams (>15 m height) in operation worldwide (Berga et al., 2006; Lehner et al., 2011; Pekel et al., 2016). Collectively the river segments converted to reservoirs by these dams are estimated to store 7000–8300 km^3 of water (Vörösmarty et al., 2003; Chao et al., 2008), a volume equivalent to 10% of the water stored in all natural freshwater lakes on Earth (Gleick, 2000). Estimates of the current area of impounded surface waters in manmade lakes range from 0.3 to 1.5×10^6 km^2 (Downing et al., 2006; Barros et al., 2011; Lehner and Doll, 2004; St. Louis et al., 2000) (Table 8.9).

Both the total amount of reservoir water storage, and the net effect of their expansion on global C cycling remains uncertain, but qualitatively it is clear that reservoirs are strong sources of both CO_2 and CH_4 to the atmosphere. Since reservoirs often replace terrestrial ecosystems that were likely carbon sinks, their creation represents a significant shift in local carbon budgets (Prairie et al., 2018). New reservoirs drown terrestrial vegetation and can have high rates of CO_2 and CH_4 emissions, as large quantities of flooded organic matter are decomposed (St. Louis et al., 2000; Downing et al., 2008). Though the construction of new hydropower dams is often advertised as a green source of renewable energy, there are some cases of new reservoirs associated with hydropower dams releasing substantial quantities of CO_2 and CH_4 that considerably offset their supply of renewable energy relative to traditional fossil fuels (Prairie et al., 2018). In contrast, reservoir construction appears to confer significant benefits with respect to the amelioration of nitrogen pollution, since reservoirs have high rates of N removal and provide conditions conducive to denitrification—anoxic, carbon-rich sediments receiving high concentrations of nitrogen in surface runoff (Harrison et al., 2005). Reservoirs also trap substantial amounts of particulate phosphorus in their sediments. A recent modeling effort estimates that reservoirs currently trap ~12% of riverine P loads (Maavara et al., 2015).

TABLE 8.9   Estimated total global area of small and large water impoundments.

A_{min} (km^2)	A_{max} (km^2)	Number of impoundments	Average impoundment area (km^2)	Total impoundment area (km^2)	d_L (impoundments per 10^6 km^2)
0.01	0.1	444,800	0.027	12,040	2965
0.1	1	60,740	0.271	16,430	405
1	10	8295	2.71	22,440	55.3
10	100	1133	27.1	30,640	7.55
100	1000	157	271	41,850	1.05
1000	10,000	21	2706	57,140	0.14
10,000	100,000	3	27,060	78,030	0.02
All impoundments		515,149	0.502	258,570	

Source: Downing et al. (2008). Used with permission of the American Geophysical Union.

Reservoirs intercept more than 40% of global river discharge (Vörösmarty et al., 2003), and more than 50% of large river systems are affected by dams (Nilsson et al., 2005; Lehner et al., 2011). Dams lead to dramatic alterations in the timing and magnitude of river flows. In many cases dam construction leads to reduced flows over the entire year (e.g., Figure 8.36), leading to less frequent and extensive hydrologic exchange between rivers and their surrounding floodplains. In the case of hydropower dams, dams may be operated to release peak flows several times per day during the hottest days of the summer in a practice known as "hydropeaking"—with frequent, scouring flows reducing biological activity downstream.

Sediment delivery to many river deltas has been dramatically reduced as a result of flow regulation of rivers (Syvitski et al., 2005; Day et al., 2007; Syvitski and Saito, 2007). Many of the major river deltas on earth are now sinking at rates many times faster than global sea level is rising (Syvitski et al., 2009). Sinking deltas and rising seas are a bad combination because

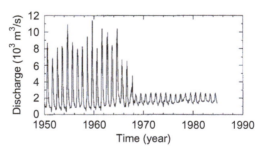

FIG. 8.36   The effect of dam construction on river flows for the Nile River before and after construction of Aswan High Dam Discharge is recorded just below the dam. The stabilization of flow is apparent, and it is not difficult to identify the time at which the dam was constructed and the Lake Nasser reservoir filled. The post-impoundment Nile shows reduced overall discharge, substantially truncated peak flows, higher low flows, and a many-month shift in the timing of the natural hydrograph. *From Vörösmarty et al. (2003). Used with permission of the American Geophysical Union.*

storm surges can inundate increasingly large areas of low deltas. Within estuaries, fringing salt marsh vegetation exists in a dynamic equilibrium between the rate of sediment accumulation and the rate of coastal subsidence or change in sea level (Kirwan and Murray, 2007; Langley et al., 2009; Kirwan and Blum, 2011). As deposits accumulate, the rate of erosion and the oxidation of organic materials increase, slowing the rate of further accumulation. Conversely, as sea level rises, deposits are inundated more frequently, leading to greater rates of sediment deposition and peat accumulation. Along the Gulf Coast of the United States, the rate of sedimentation has not kept pace with coastal subsidence, and substantial areas of marshland have been lost (DeLaune et al., 1983; Baumann et al., 1984). This loss of protective fringing wetlands is considered to be a critical factor in the extensive flooding caused by Hurricane Katrina in 2005 (Tornqvist et al., 2008). Current models suggest that from 5% to 20% of coastal wetlands will be lost by 2080 as a result of coastal subsidence and sea level rise (Schuerch et al., 2018; Nicholls, 2004).

Reservoirs, because they trap and retain a large portion of the sediments transported by their contributing rivers, are very effective at retaining mineral elements. The construction of dams can significantly reduce the transport of Fe, P, and Si in rivers. When the High Aswan Dam on the Nile river was built, N and P exports to the Nile River estuary dropped precipitously (Table 8.10), and fish and shrimp abundance declined by 80% (Nixon, 2003). The fisheries began to recover ~15 years later as the cities of Cairo and Alexandria grew and released larger fluxes of sewage N and P into the Nile (Oczkowski et al., 2009) (Table 8.10).

Productivity in reservoirs can sequester and trap silica in their sediments, reducing Si inputs to coastal waters (Teodoru and Wehrli, 2005; Humborg et al., 2006). After construction of the dams on the Nile, diatoms were far less dominant, probably because urban wastewaters failed to replace the Si carried in rivers. In many coastal waters, reductions in Si coupled with urban nutrient inputs are leading to increases in N:Si or P:Si element ratios that may favor the growth of nuisance algae over siliceous diatoms (Howarth et al., 2011).

Eutrophication

Humans are causing rapid "cultural eutrophication" of many inland and coastal waters by increasing the amount of nitrogen and phosphorus in the biosphere (Vitousek et al., 1997; Galloway et al., 2008; Childers et al., 2011; Schipanski and Bennett, 2012). While better sewage treatment and a ban on detergent phosphates has reduced P loading to many freshwaters, it is unclear if cultural eutrophication can be completely reversed when pollution controls are implemented. Some lakes show rapid declines in algal productivity following reductions in nutrient loading (Edmondson and Lehman, 1981; Jeppesen et al., 2005; Kronvang et al., 2005), but other systems show only limited response (Jeppesen et al., 2005; Kemp et al., 2009).

Many culturally eutrophied lakes contain large quantities of "legacy P" in their sediments, and the extent to which this P is susceptible to mineralization and mixing into the epilimnion appears to be a major constraint on reversing eutrophication (Goyette et al., 2018; Martin et al., 2011). In addition, sulfate loading to lakes from acid rain or salt water intrusion can counteract reductions in P loading if SO_4^{2-} displaces PO_4^{3-} from the sedimentary minerals (Caraco et al., 1989; Smolders et al., 2006). Under reducing conditions, most Fe is found in sulfide minerals, leaving little oxidized Fe available to bind P in lake sediments; thus, the "iron trap" for phosphorus becomes much less effective (Blomqvist et al., 2004).

TABLE 8.10 Potential release of P and N by the urban populations of Greater Cairo and Alexandria and the total urban population of Egypt compared with the estimated flux of nutrients from the Nile.

	$10^3 t yr^{-1}$	
	P	N
The Nile		
Pre-Aswan High Dam		
Dissolved	3.2	6.7
On sediments	4–8	?
Total	7–11	6.7
Post-High Dam		
Dissolved	0.03	0.2
On sediment	0	0
Total	0.03	0.2
Human waste		
Total generated in Cairo and Alexandria		
1965	4.4	21
1985	8.9	55
1995	12.6	87
Potential N and P in wastewater discharge		
Cairo and Alexandria[a]		
1965	1.1	5
1985	3.6	22
1995	9.5	65
Potential N and P in wasterwater discharge		
Total urban population[b]		
1965	2.4	12
1985	6.7	41
1995	15.8	108

[a] *Assuming that the population connected to the sewers was 25% in 1965, 40% in 1985, and 75% in 1995. The 1965 estimate is very uncertain.*
[b] *Extrapolated from Cairo and Alexandria assuming that it accounted for 45% of the total urban population in 1965, 54% in 1985, and 65% in 1995 (see text).*
From Nixon (2003). Used with permission of Springer.

Because primary productivity in streams and rivers is less likely to be nutrient limited (Dodds et al., 2002), eutrophication of rivers is less studied than in lake or coastal ecosystems (Hilton et al., 2006). The supply of anthropogenic nutrients to rivers can lead to algal blooms where there is sufficient light and limited flow disturbance (Peterson and Melillo, 1985; Hilton et al., 2006). In shaded streams, nutrient loading may instead speed the decomposition of terrestrial organic matter (Rosemond et al., 2015; Benstead et al., 2009; Woodward et al., 2012) and enhance C consumption by invertebrate consumers (Cross et al., 2007). Where allochthonous DOC concentrations are very high, nutrient loading is likely to directly stimulate heterotrophic activity which may exacerbate and expand problems of river hypoxia (Mallin et al., 2006; Blaszczak et al., 2019).

The management of polluted estuaries has been the subject of much controversy. Increases in nutrient loading in rivers (Green et al., 2004; Boyer et al., 2006; Alexander et al., 2008; Howarth et al., 2012) are typically linked to issues of coastal eutrophication. Some workers argue that an improvement in estuarine conditions will be directly related to efforts to reduce nutrients in inflowing waters (Boesch, 2002; Howarth and Marino, 2006; Smith and Schindler, 2009). Others suggest that the retention of prior inputs and the recirculation of nitrogen within estuaries will hamper immediate improvements in water quality (Kunishi, 1988; Van Cappellen and Ingall, 1994). Some have argued that the prevalence of nitrogen limitation in estuaries with long histories of human impacts occurs simply because of historic phosphorus loading.

The controversy over which element is most limiting, and thus what sorts of nutrient controls should be implemented, is similar to the ongoing debates about the relative importance of regulating N vs. P inputs to lakes (Schindler et al., 2008; Conley et al., 2009b; Lewis et al., 2011). In fact, both elements can enhance eutrophication and exacerbate the duration and extent of anoxia (Paerl et al., 2016). Anoxic conditions in turn, can enhance rates of phosphorus regeneration from sediments and provide a positive feedback to eutrophication (Van Cappellen and Ingall, 1994; Vahtera et al., 2007). Legacy phosphorus, because it cannot be removed through a process analogous to denitrification, will likely have longer-term impacts than nitrogen loading; however, the impacts of enhanced N loads may have more immediate effects on phytoplankton growth (Conley et al., 2009b) and may contribute to freshwater ecosystem releases of N_2O (Seitzinger et al., 1980; Beaulieu et al., 2007, 2011). Regardless of what nutrient is targeted, efforts to reduce nutrient loads to freshwaters and coastal waters will be hampered by legacy fertilizers that have accumulated in regional groundwater and deep soils from decades of intensive agriculture. These nutrient legacies are likely to result in substantial time lags between efforts to reduce fertilizer inputs and the abatement of cultural eutrophication of our inland waters and coastal estuaries (Van Meter et al., 2018).

Global climate change

The extensive direct impacts of human activities on inland waters makes it difficult to detect and predict the effects of climate change on freshwaters (Vorosmarty et al., 2000; Barnett et al., 2008; Milly et al., 2008; Arrigoni et al., 2010; Wang and Hejazi, 2011). In the rare instance where concentrations of dissolved CO_2 are low and nutrients are abundant, rising atmospheric CO_2 may increase DIC and stimulate enhanced productivity in freshwaters (Schippers et al., 2004). The majority of lakes, rivers and estuaries, however, are already supersaturated with CO_2, and in these ecosystems rising atmospheric CO_2 is unlikely to fundamentally alter aquatic biogeochemistry. In contrast, the rising temperatures of aquatic ecosystems caused by rising atmospheric CO_2 is likely to have much greater impacts on the water and nutrient budgets of freshwaters.

A warmer climate is predicted to generate a more rapid hydrologic cycle with higher rainfall across much of the planet. At the same time greater evapotranspiration will lead to declining soil moisture and surface runoff (Rodell et al., 2018, see Chapter 10), making it difficult to predict the consequences of climate change for the flow of water from land into inland waters (Lehner et al., 2019). Climate models have suggested that climate warming will lead to a 10–40% increase in surface runoff by midcentury (Milly et al., 2005) and that in many regions a significant proportion of this increase will occur during extreme seasonal precipitation events (Milly et al., 2002; Palmer and Räisänen, 2002). Thus, somewhat paradoxically, climate models have predicted a rising frequency of both droughts and floods, though to date, there is limited evidence for this as a global trend (Greve et al., 2014). Instead, the effects of climate change on runoff are regionally heterogeneous and often difficult to detect against the strong impact of flow regulation and extractive water uses (Vorosmarty et al., 2000; Arheimer et al., 2017; Liu et al., 2019a). In a data-based global modeling exercise, Greve et al. (2014) estimate that only 10.8% of the global land area shows a robust "dry gets drier, wet gets wetter" pattern, while 9.5% of the global area follows opposite trends, with dry areas becoming wetter and wet areas becoming drier (Fig. 8.37).

Across the conterminous United States, an increase in precipitation throughout the twentieth century and in stream flow since at least 1940 has been observed (Karl and Knight, 1998; Lins and Slack, 1999; McCabe and Wolock, 2002; Groisman et al., 2004; Krakauer and Fung, 2008), although these patterns are primarily seen in data from the eastern United States, with evidence of declines or no change in the Pacific Northwest (Luce and Holden, 2009). These trends cannot be attributed solely to climate change. In many regions irrigation,

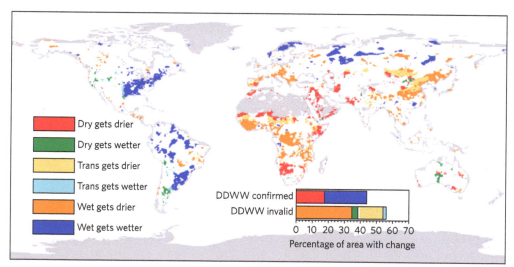

FIG. 8.37   Significant trends in drying (ΔE (evaporation) > ΔP (precipitation)) or wetting (ΔE < ΔP) trends between 1948 and 2005 are shown in color on the global map. Dark red denotes arid regions that have become drier and dark blue denotes wet regions that have become wetter. Humid areas getting drier (orange) are widely found. *From Greve et al. (2014).*

damming and urbanization are having much greater effects on runoff patterns than climate change (Arrigoni et al., 2010; Schilling et al., 2010). Generally, stream flows have increased in watersheds in proportion to their population density and the percent of land converted to urban areas or cropland and declined in proportion to reservoir volume and the area of irrigated land (Wang and Hejazi, 2011). Despite these differences in water management across watersheds, new analyses find coherent regional shifts in streamflow across both unmanaged and flow managed watersheds (Fig. 8.38; Ficklin et al., 2018), indicating rising streamflow in the northern Great Plains, Appalachia and New England and declining streamflow in most other basins.

More frequent storm events are likely to contribute larger pollutant loads to rivers during peak flows when river biota has a limited capacity to assimilate excess nutrients (Kaushal et al., 2008). Increasing storm pulses of nutrients are thus likely to exacerbate problems of freshwater and coastal eutrophication (e.g., Paerl et al., 2001). Hotter temperatures combined with higher nutrient loading will likely expand and extend the already widespread problems of anoxia in freshwater and estuarine ecosystems.

Rising air temperatures are leading to later freeze dates and earlier thaw dates for ice- covered lakes and rivers, earlier dates for snowmelt flows in rivers, and increasing contributions of glacial and permafrost meltwater to rivers (Magnuson et al., 1997; Peterson et al., 2002; Barnett et al., 2005). For many northern temperate lakes, the period of ice cover has declined over the last 50 years (Fig. 8.39; Magnuson et al., 2000; Benson et al., 2012). The longer period without ice and rising air temperatures are lengthening the period of thermal stratification in northern temperate lakes, a change that is expected to lead to increases in lake productivity (Carpenter et al., 1992). For the many arid land rivers that are fed primarily by snowmelt, earlier and smaller snowmelt may lead to substantial declines in annual flows and the spatial extent of river networks.

Changes in DOC loading to freshwaters may also affect their response to climate change or other anthropogenic impacts (Freeman et al., 2004a). For example, increasing DOC concentrations in rivers throughout eastern North American and western Europe could potentially

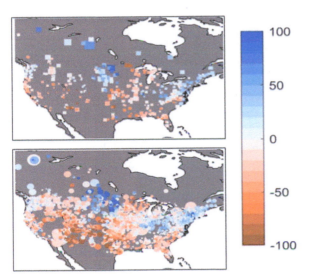

FIG. 8.38 Percent change in median streamflows between 1981 and 2015 for 3, 119 stream gauges throughout Canada and the United States. The upper panel shows change for 570 watersheds for which streamflow reflects prevailing meteorological conditions with minimal land use change (Natural Flows) while the lower panel shows the change in the remaining watersheds with hydrologic management and significant land use change. Similar regional trends are observed with both datasets. *From Ficklin et al. (2018).*

FIG. 8.39 Graphic of the ice cover duration and extremes for the winters of 1855–56 through 2004–05 for Lake Mendota, Wisconsin. Over the 150-year record the six most extreme short and long ice duration winters (25-year events) are marked as *closed* or *open circles. From Benson et al. (2012). Used with permission of Springer.*

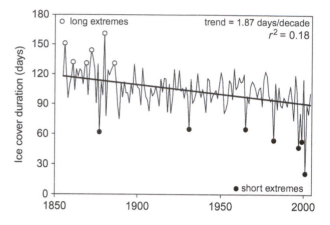

constrain productivity responses to increased growing season length, while exacerbating thermal pollution (Creed et al., 2018; Evans et al., 2005; Stanley et al., 2012).

Globally, freshwater ecosystems are becoming less fresh due to sea level rise, irrigation and road salting. Marine salts are encroaching into coastal rivers and lakes as a result of sea level rise, declining coastal sediment accumulation, and drought- induced saltwater intrusion (see Chapter 7). Further inland, the evaporative concentration of salts in irrigation water can increase the salinity of inflows during periods of drought. Finally, the widespread use of road salts is salinizing many inland waters. This 'freshwater salinization syndrome' has important consequences for freshwater biota, many of which are unable to cope with the osmotic stress of high salinity waters (Kaushal et al., 2018).

Summary

The biogeochemical cycles of freshwater ecosystems are embedded within the biogeochemistry of their surrounding terrestrial ecosystems. The rate of water delivery and the chemical properties of freshwaters are largely determined by soil properties, vegetation, and hydrology of the contributing watershed. Most inland waters are heterotrophic, with an excess of respiration over net primary production supported by subsides of terrestrial organic matter. During transport through freshwaters, nutrients are removed from the water and sequestered in organic and inorganic forms in sediments or, in the case of N and C, lost as gaseous products.

Because most inland waters are hydrologically connected from the smallest headwater streams, through rivers and lakes, and ultimately to estuaries or terminal lakes, their global importance must be considered collectively. Although inland waters occupy only a small portion of the terrestrial land surface and a small fraction of the total liquid water volume on Earth, the relatively high rates of carbon and nutrient transformations in freshwater ecosystems makes them more important in global nutrient cycles than surface area alone would suggest. Collectively the biota of inland waters respires ~40% and stores ~20% of the 2.7 Pg of allochthonous carbon and denitrifies or stores ~73% of the 118 Tg of nitrogen received each year from terrestrial ecosystems (Fig. 8.40). The construction of reservoirs has likely enhanced

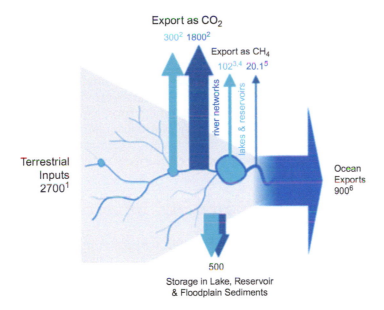

Export as CO$_2$

300[2] 1800[2]

Export as CH$_4$

102[3,4] 20.1[5]

river networks

lakes & reservoirs

Terrestrial
Inputs
2700[1]

Ocean
Exports
900[6]

500
Storage in Lake, Reservoir
& Floodplain Sediments

Riparian
Denitrification

5.9[1]

0.9[1]

River
Denitrification   Lake &
Reservoir
35[3]   Denitrification

0.07[4]   19.7[5]

0.06[6]

N Exports from Land
to Surface Waters
71[1]–118[2]

Ocean
Exports
43[7]

??
Storage in Lake, Reservoir
& Floodplain Sediments

FIG. 8.40   The cumulative effect of inland waters on global C and N cycling (in Tg yr^{-1}). Note that rivers and lakes deliver more C and N to the atmosphere than to the ocean, indicating that biological processing of these elements within freshwaters is as important as their physical transport.

I. Processes and reactions

the storage and removal of C, N, and P (St. Louis et al., 2000; Downing et al., 2008; Harrison et al., 2008; Heathcote and Downing, 2012; Maavara et al., 2015; Maranger et al., 2018).

Humans have a dramatic impact on inland waters throughout the world, regulating the flow of water and altering the load of dissolved and suspended materials. The mixing of freshwater and seawater occurs in estuaries, located at the mouth of major rivers. In response to changes in pH, redox potential, and salinity, river waters provide estuaries with a rich solution of available N and P, and high rates of net primary production fuel a productive coastal marine ecosystem. Despite a temporary storage of nutrients in salt marshes and estuarine sediments, river waters are always a net source of nutrients to their estuary and the coastal ocean. As we shall see, rivers are an important source of nutrients in the global budgets of biogeochemical elements in the ocean.

9

The Oceans

Biogeochemistry: An Analysis of Global Change, Fourth Edition
https://doi.org/10.1016/B978-0-12-814608-8.00009-8

Introduction

Most of Earth's water resides in the sea. In this chapter we examine the biogeochemistry of seawater and the contributions that oceans make to global biogeochemical cycles. We will begin with a brief overview of the circulation of the oceans and the mass balance of the major elements that contribute to the salinity of seawater. We will examine the biogeochemical cycles of essential elements in the surface ocean and the processes that lead to the exchange of gaseous components between the oceans and the atmosphere. The biotic processes fueled by photosynthesis strongly affect the chemistry of many elements in seawater. Net primary productivity in the oceans is related to the availability of essential nutrient elements, particularly nitrogen and phosphorus. Finally, we will consider the vast volume of the oceans below the well lit surface waters, examining the exchange of solutes and particles between the surface and deep oceans, and the fate of materials that sink to the ocean sediments.

The ocean covers 71% ($361 \times 10^6 \, \text{km}^2$) of Earth's surface. Estimates of the ocean's volume are constantly being refined as we gain higher resolution bathymetric maps of the seafloor. Earlier estimates tended to overestimate the mean ocean depth due to undersampling of seamounts and ocean ridges (Charette and Smith, 2010). The most up to date bathymetric maps of the world's oceans suggest a mean depth of 3897 m (Weatherall et al., 2015). At this depth, the oceans contain an estimated $\sim 1.4 \times 10^9 \, \text{km}^3$ of Earth's inventory of water. Over much of the open ocean, evaporation and precipitation represent the only water fluxes. These fluxes are not balanced and there is considerable uncertainty in their estimates (Durack, 2015). Across the ocean's surface, each year an estimated 129 ± 10 cm evaporates ($\sim 4.1 \times 10^5 \, \text{km}^3 \, \text{yr}^{-1}$), while only $118 \pm 10 \, \text{cm} \, \text{yr}^{-1}$ ($\sim 3.7 \times 10^5 \, \text{km}^3 \, \text{yr}^{-1}$) returns to the ocean surface in precipitation (Trenberth et al., 2007; Durack, 2015; Yu et al., 2017).[a] The ocean-to-continent transfer of water vapor contributes to precipitation on land that is balanced by the annual inputs of river water and groundwater into coastal oceans ($\sim 0.4 \times 10^5 \, \text{km}^3 \, \text{yr}^{-1}$) (Chapter 8).

Precipitation, river water and coastal groundwater enter the surface of the ocean. The ocean's surface waters are warmed by the sun, mixed by winds, and isolated through density stratification from the much larger volume of the deep ocean. This thin layer of warm, less dense surface water is typically between 75 and 200 m deep with an average temperature of 18°C. The warm surface layer floats atop the colder and denser waters of the deep ocean that contain about 95% of the ocean's volume at a mean temperature of 3°C with little variation. The strongest stratification and highest temperatures occur in tropical seas, where the surface layer can reach maximum temperatures of ~ 30°C (Fig. 9.1).

---

[a] Large fluxes of seawater are often expressed in Svedrups. A Svedrup, abbreviated Sv, is equivalent to $10^6 \, \text{m}^3 \, \text{s}^{-1}$ or $3.2 \times 10^{13} \, \text{m}^3 \, \text{yr}^{-1}$.

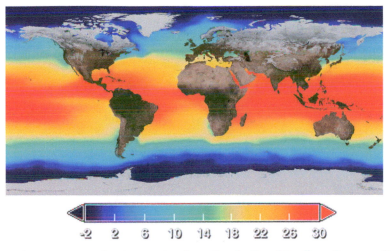

FIG. 9.1   A map of average sea surface temperature in centigrade. *From https://svs.gsfc.nasa.gov/3652.*

Ocean circulation

Surface ocean circulation

At the surface, seawater is relatively well mixed by the wind (Thorpe, 1985; Archer, 1995). The frictional drag of prevailing winds over the ocean's surface (Chapter 3) lead to the formation of surface currents (Fig. 9.2).[b] These surface currents have been mapped at high resolution using the TOPEX/Poseidon satellite, which monitors small changes in the relative height of sea level (Ducet et al., 2000).[c] Global wind patterns drive large gyres, circular rotating currents, within all ocean basins. Circulating gyres are impacted by the Earth's rotation such that they move anticyclonically (clockwise in the northern hemisphere) around low-pressure centers (where surface waters have diverged). In the northern hemisphere, the currents are deflected to the right by the Coriolis force, which is observed when fluid motions occur in a rotating frame (see Fig. 3.3). Thus, the Gulf Stream crosses the North Atlantic and delivers warm waters to northern Europe. In the southern hemisphere, the gyres move in a counter-clockwise direction and are deflected to the left. Winds cause surface waters to converge in some areas and diverge in others. The horizontal pressure gradients that result from these areas of convergence and divergence drive motions below the wind-driven flow at the surface.

The global circulation of the oceans transfers heat from the tropics to the polar regions of the Earth (Oort et al., 1994). About 10–20% of the net excess of solar energy received in the

[b] View an animation of the planet's ocean currents developed by NASA's Estimating the Circulation and Climate of the Ocean (ECCO) project at https://www.nasa.gov/topics/earth/features/perpetual-ocean.html.

[c] Learn more about satellite-derived estimates of sea levels at https://climatedataguide.ucar.edu/climate-data/global-mean-sea-level-topex-jason-altimetry.

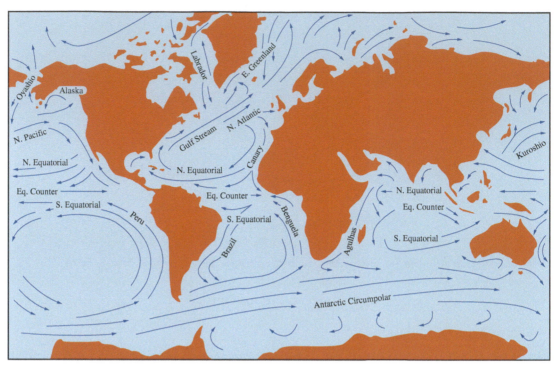

FIG. 9.2 Major currents in the surface waters of the world's oceans. *From Knauss (1978). Used with permission of Dr. John Knauss.*

tropics is transferred to the poles by ocean circulation; the remainder is transferred through the atmosphere (Trenberth and Caron, 2001). With the loss of heat at polar latitudes, their surface waters cool, until their density exceeds that of the underlying water. The resulting convective mixing permits exchange between the surface ocean and the deep waters. In contrast, deep convective mixing driven by winds is rarely observed (Gascard et al., 2002; Wadhams et al., 2002).

Our understanding of the circulation of the oceans has advanced with the deployment of sensors that are designed to follow currents at particular depths. Several thousand floats compose the ARGO network, which transmit data to a satellite for subsequent relay to ground stations.[d] Among these, RAFOS floats are subsurface floats that remain at depth, recording data for location, temperature, and pressure, which are transmitted to a satellite when the float rises to the sea surface after its mission is complete, typically after 2 years of deployment (Fig. 9.3). These floats have documented the interior (mid-depth) circulation in the Atlantic, which returns polar waters to lower latitudes (Bower et al., 2009; Lozier et al., 2013, 2017).

---

[d] ARGO derives from an ocean explorer in Greek mythology. RAFOS is SOFAR spelled backwards, reflecting that these floats receive signals from devices moored on the seafloor to record their position, whereas SOFAR (Sound Fixing and Ranging) devices emit signals that are received by moored devices.

FIG. 9.3 Deployment of a RAFOS float in the North Atlantic Ocean. *Photograph courtesy of Susan Lozier, Duke University.*

Deep ocean

The zone of rapid increase in density between the surface and deep ocean is known as the *pycnocline*. Although organisms, organic matter, solutes and gases may swim, fall or diffuse across the pycnocline, the direct exchange of water masses between the surface and the deep ocean is limited to areas of the ocean where this density gradient breaks down. The most important areas for the transfer of surface waters into the deep ocean occur near the poles. These "deep-water formation" zones occur because as water cools to near freezing its density exceeds that of deep waters. The resulting loss of buoyancy of surface waters at polar latitudes is what drives the overturning circulation of the world's oceans (Broecker and Others, 1991; Fig. 9.4).

The extent of deep water formation at the poles varies seasonally. During the winter, the density of some polar waters increases when fresh water is "frozen out" of seawater and added to floating sea ice, leaving behind waters of greater salinity that sink to the deep ocean.[e] In contrast, during the summer, the polar oceans have lower surface salinity as less saline waters are delivered from the seasonal melting of ice caps (Peterson et al., 2006). Because the downwelling of cold polar waters is driven by both temperature and salinity, it is known as *thermohaline circulation*. In the Atlantic Ocean, the formation of deep water is known as the Atlantic Meridional Overturning Circulation (AMOC).

The penetration of cold waters to the deep ocean at both poles, due to downwelling, forms a "conveyor belt" that transfers seawaters between the major ocean basins via the Antarctic circumpolar current (ACC), which carries $>4.1 \times 10^6 \, km^3$ at any given time (Cunningham et al., 2003; Barker and Thomas, 2004; Firing et al., 2011). The winds that drive this circumpolar current also serve to isolate cold air masses over the continent of Antarctica.[f] The ACC appears to have developed during the Cenozoic era and its onset has been considered to be responsible for the glaciation of Antarctica (Barker and Thomas, 2004). This deep ocean circulation allows for complete ocean mixing, or overturn. For example, the North Atlantic deep water (NADW), which forms near Greenland, transports dense, saline water out of the deep Atlantic and into the Southern Ocean, where the ACC carries it around the tip of Africa and into the Indian and Pacific Oceans (Dickson and Brown, 1994; Lozier, 2012) Fig. 9.4).

In the Atlantic Ocean, downward mixing of ^3H_2O and $^{14}CO_2$ produced from the testing of atomic bombs (Fig. 9.5) and downward mixing of anthropogenic chemicals of recent origin (e.g., see Krysell and Wallace, 1988; Tanhua et al., 2009) allow us to trace the entry of surface waters to the deep sea and the movement of deep water towards the equator. The downward transport in the North Atlantic is estimated at $4.8 \times 10^5 \, km^3 \, yr^{-1}$, roughly 10 times greater than the total annual river flow to the oceans (Fig. 9.6) (Dickson and Brown, 1994; Ganachaud, 2003; Luo and Ku, 2003). About $6.7 \times 10^5 \, km^3 \, yr^{-1}$ sink in the Southern Ocean, including $1.3–1.6 \times 10^5 \, km^3 \, yr^{-1}$ in the Weddell Sea (Hogg et al., 1982; Schmitz, 1995). To maintain a mass-balance in the overall volume of the deep sea, the formation of deep waters in polar regions must be associated with equivalent upwelling of deep waters into the surface ocean in other regions. The most important zones of upwelling are in the circumpolar Southern Ocean around 65°S latitude (Kuhlbrodt et al., 2007; Marshall and Speer, 2012) and in the Eastern Equatorial Pacific (Toggweiler and Samuels, 1993; Zhang et al., 2017b).

Mean residence times

The patterns of ocean circulation have important implications for biogeochemistry. The overall mean residence time for seawater is ~35,000 years with respect to river flow (i.e., total ocean volume/global annual river flow). Since most river inputs mix only with the smaller volume of the surface ocean, the mean residence time in the surface ocean is

[e]Sea ice salt content is generally <20% of seawater (Cox and Weeks, 1974)

[f]Recall from Chapter 3 that this isolation of cold air masses and polar ice clouds over Antarctica is what allows the ozone hole to form over the continent.

FIG. 9.4 Major ocean circulations patterns depicted to show zones of deep water formation and surface-depp ocean exchange near the poles and the major zones of upwelling in the Southern ocean. Colors represent temperature, with coldest waters in blue and warmest waters in orange. *This is Fig 4.2 from the IPCC AR3 report. https://www.ipcc. ch/report/ar3/syr/question-1-9/fig4-2-5/.*

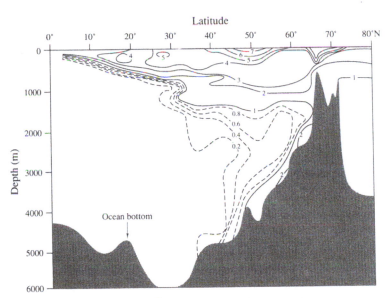

FIG. 9.5 Penetration of bomb-derived tritium (^{3}H$_2$O) into the North Atlantic Ocean. Data are expressed as the ratio of ^{3}H/H $\times$ 10^{-18} for samples collected in 1972. *From Ostlund (1983).*

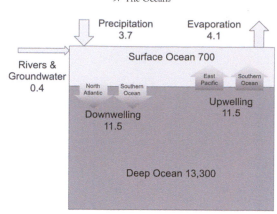

FIG. 9.6   A water budget for the world's oceans. To facilitate comparisons the volume of the surface and deep ocean are shown in units of 10^5km^3 and fluxes are shown in units of $10^5 \text{km}^3 \text{yr}^{-1}$.

~1750 years with respect to riverflow (i.e., surface ocean volume/global annual river flow). If we account for the addition of rain waters and upwelling waters to the surface ocean, the actual turnover time of the surface waters is even faster. For example, the mean residence time of surface waters in the North Pacific Ocean is about 9–15 years (Michel and Suess, 1975).

Because the volume of water entering the deep ocean at the poles is much greater than the total annual river flow to the sea, the mean residence time of water in the deep ocean is much less than 35,000 years. Estimates of the mean age of bottom waters using ^{14}C dating of total dissolved CO_2 range from 275 years for the Atlantic Ocean to 510 years for the Pacific (Stuiver et al., 1983), and the overall renewal of the oceans' bottom waters is normally assumed to occur in 500–1000 years. Thus, the deep waters maintain a historical record of the conditions of the surface ocean several centuries ago.

Climate oscillations in ocean circulation patterns

Changes in ocean currents, particularly the formation of deep waters, may be associated with changes in global climate. At the end of the last glacial epoch, the concentration of atmospheric CO_2 rose from 200 ppm to about 280 ppm (Fig. 1.4). An increase in the rate of upwelling of deep waters in the Southern Ocean may have been associated with the release of CO_2 to the atmosphere (Burke and Robinson, 2012). Some researchers see evidence that the formation of polar deep waters has slowed in recent years, perhaps indicative of global warming and a greater density stratification of the surface waters in the North Atlantic (Cunningham and Marsh, 2010). Climate change is also likely to affect the pattern of surface currents. During the last glacial epoch, the Gulf Stream, which carries about 30 Sv, appears to have shifted southward, producing a humid climate in southern Europe (Keffer et al., 1988). During the Little Ice Age, 200–600 years ago, a weakened flow of the Gulf Stream appears to have been related to cold conditions in Europe (Lund et al., 2006).

One of the best known variations in current occurs in the central Pacific Ocean. Under normal conditions, the trade winds drive warm surface waters to the western Pacific, allowing cold subsurface waters to upwell along the coast of Peru. Periodically, the surface transport breaks down in an event known as the El Niño-Southern Oscillation (ENSO). During El Niño years, the warm surface waters remain along the coast of Peru, reducing the upwelling of nutrient-rich water (Fig. 9.7).[g] Phytoplankton growth is limited and the normally productive anchovy fishery collapses (Glynn, 1988). Associated with these occasional warm surface waters in the eastern Pacific are changes in regional climate, for example, exceptionally warm winters and greater rainfall in western North America (Molles and Dahm, 1990; Swetnam and Betancourt, 1990; Redmond and Koch, 1991). At the same time the absence of warm surface waters in the western Pacific reduces the intensity of the monsoon rainfalls in Southeast Asia and India.

Working with atmospheric scientists, oceanographers now recognize that El Niño events are part of a cycle that yields opposite but equally extreme conditions during non-El Niño years, which are known as La Niña (Philander, 1989). Although the switch from El Niño to La Niña is poorly understood, it is likely that the conditions at the beginning of each phase

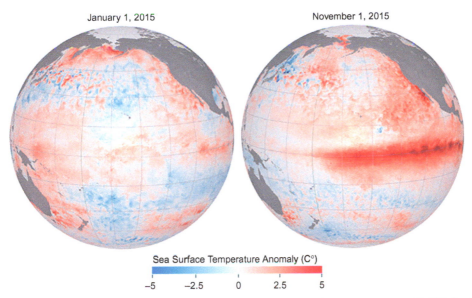

FIG. 9.7 El Niño is associated with above-average equatorial sea surface temperatures. El Niño's signature warmth is apparent in the November 2015 map. *NASA Earth Observatory maps by Joshua Stevens, using data from Coral Reef Watch available at https://earthobservatory.nasa.gov/features/ElNino.*

[g] A thorough description of El Nino and accompanying data visualizations is available at https://earthobservatory.nasa.gov/features/ElNino.

reinforce its development, with the cycle averaging between 3 and 5 years between El Niño events, which are recognized in sedimentary records extending to 5000 years ago (Rodbell et al., 1999).

A similar but less powerful cyclic pattern of ocean circulation is seen in the Atlantic Ocean (van Loon and Rogers, 1978; Hurrell, 1995). The North Atlantic Oscillation (NAO) describes year to year variation in the atmospheric pressure between a permanent zone of low pressure normally located over Iceland and a zone of high pressure near the Azores. This pressure differential causes variation in the strength and direction of the prevailing winds over the North Atlantic, called westerlies or anti-trade winds, that blow east across the ocean. When the pressure differential is higher than average (a positive NAO)—the westerlies are strengthened and more heat and precipitation moves across the North Atlantic towards Northern Europe. In years with a negative NAO, limited differences in atmospheric pressure direct this heat and precipitation towards southern Europe. Variation in the direction and amplitude of the NAO has important consequences for climate, streamflows and alpine glaciers in Europe (Osborn, 2006).[h]

Recent climate reconstructions have determined that the sea surface temperature of the North Atlantic has undergone distinct oscillations during the last 8000 years, with shifts between warm and cool phases occurring with a periodicity of ∼55–70 years (Schlesinger and Ramankutty, 1994). This pattern, called the Atlantic Meridional Oscillation (AMO) also affects regional and global climate. The warm phase of the AMO appears to be associated with periods of drought in North America and northern South America (Knudsen et al., 2011).

Oscillations in ocean circulation have important consequences for the amount of atmospheric heat that is absorbed by the ocean. In El Nino years, less heat is absorbed by the Pacific Ocean leading to positive anomalies in the long term global warming trend. Thus, El Niño-La Niña cycles add variation to the global temperature record, complicating efforts to perceive atmospheric warming caused by the greenhouse effect. Similarly, when the AMO shifted into its warm phase in the 1990s, it may have accentuated global warming patterns during this period. As we will discuss below, these major climate oscillations can also weaken or strengthen the oceanic uptake of CO_2 in any given year due to differences in the rate at which water masses are transferred between the surface and the deep ocean (Landschützer et al., 2016; DeVries et al., 2017).

The composition of seawater

Major ions

Seawater has relative uniform global composition with an average salinity of 35‰ or 35 g of salt for every kg of seawater (Millero et al., 2008). This gives seawater an average density of 1.025 g mL^{-1}. Although seawater varies slightly in salinity throughout the world (Fig. 9.8), the major ions are conservative in the sense that they maintain the same concentrations relative to one another in most ocean waters. For example, recent changes in the salinity of seawater, due

[h] A recent study found that the best vintages of Spanish cava were produced in negative NAO years (Real and Báez, 2013).

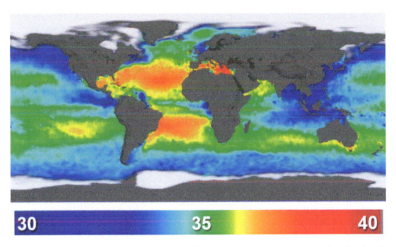

FIG. 9.8   Flat map projection showing average Sea Surface Salinity measurements taken by the Aquarius spacecraft from September 2011 through September 2014. Salinity is expressed in units of parts per thousand (o/00) or g kg^{-1}. *From https://svs.gsfc.nasa.gov/4233.*

to freshening of the North Atlantic by Greenland ice melt (Boyer et al., 2005; Curry and Mauritzen, 2005), do not change the relationship between the concentration of various ions and Cl– they are all diluted in the same proportion. Thus, a good estimate of total salinity can be calculated from the concentration of a single ion. Often chloride is used, and the relationship is:

$$\text{Salinity} = (1.81) \times \text{Cl}^- \qquad (9.1)$$

Both values are typically reported in parts per thousand (‰).

Table 9.1 shows the mean ratio between chloride and other major ions in seawater of average salinity (35‰). In addition to containing a much higher concentration of ions, seawater diverges widely in its concentration of various elements relative to the river waters from which they are derived (Chapter 4). These differences reflect not only evaporative concentration of elements in seawater, but differential mechanisms of loss. As we found for gases in the atmosphere (Chapter 3), constituents that are least reactive accumulate to the highest concentrations and have the longest mean residence time (MRT). Whitfield and Turner (1979) showed an indirect correlation between the mean residence time of elements in seawater and their tendency to incorporate into one or more sedimentary forms (Fig. 9.9). Sodium and chloride dominate the mixture of seasalts, not because they have the highest rates of delivery, but because there are fewer mechanisms for their removal. Elements that are more reactive have shorter MRTs, as a result of biological demand or their potential for adsorption onto mineral or organic particles. For example, Ca^{2+} is the dominant cation in river water (Table 9.1), but its concentration in seawater ranks behind Na^+ and Mg^{2+}. This difference arises because calcium is incorporated into the carbonate skeletons of many marine organisms, and is thus under much higher biological demand than either Na^+ or Mg^{2+}. Despite these differences in biological demand, the mean residence times for all of the major ions are much longer than the MRT for water in the oceans, allowing plenty of time for mixing

TABLE 9.1 Major ion composition of seawater, showing relationships to total chloride and mean residence times for the elements with respect to river inputs (Holland, 1978; Meybeck, 1979; Millero et al., 2008; Lee et al., 2010; Schlesinger and Vengosh, 2016).

Constituent	Concentration in seawater[a] ($g\,kg^{-1}$)	Chlorinity ratio[a,b] ($g\,kg^{-1}$)	Concentration in river water[c] ($mg\,kg^{-1}$)	Mean residence time[c,d] ($10^6\,yr^{-1}$)
Water	964.83496	49.85529		0.034
Chloride	19.35271	1.00000	5.75	120
Sodium	10.78145	0.55710	5.15	75
Sulfate	2.71235	0.14015	8.25	12
Magnesium	1.28372	0.06633	3.35	14
Calcium	0.41208	0.02129	13.4	1.1
Potassium	0.3991	0.02062	1.3	11
Bicarbonate	0.10481	0.00542	52	0.1
Bromide	0.06728	0.00348	0.02	100
Boron	0.00467	0.00024	0.01	1.5
Strontium	0.00795	0.00041	0.03	12
Fluoride	0.0013	0.00007	0.1	0.50

[a] Source: Millero et al. (2008).
[b] Source for B: Lee et al. (2010).
[c] Source: Meybeck (1979) and Holland (1978).
[d] Source for B, Schlesinger and Vengosh (2016) and F from Schlesinger et al. (in prep.).

and maintaining the predictable relationship between each major ion and the concentration of Cl^- ions across all oceans.

Variations in salinity across the global oceans (Fig. 9.8) arise from an imbalance between inputs of freshwater relative to evaporative losses. Local variation in sea surface salinity is caused by vertical advection, river inflows (Gordon, 2016) and submarine groundwater inputs (Moore, 2010). In the Atlantic Ocean, evaporation exceeds the sum of river flow and precipitation, yielding higher seawater salinity than in the Pacific. The Atlantic receives a net inflow of less saline waters from the Pacific to restore the water balance. These salinity differences across the world's oceans are becoming more extreme in response to changing climate and an intensifying hydrologic cycle (see Chapter 10). The net result is that low salinity areas of the ocean (e.g., the tropical and high-latitude precipitation-dominant regions) are getting fresher while high salinity areas (e.g., the evaporation-dominant subtropical gyres) are getting saltier (Durack and Wijffels, 2010). These shifts are increasing the salinity contrast between the Atlantic and Pacific oceans with important implications for ocean circulation (Gordon, 2016).

Although each of the major ions is well mixed throughout the world's oceans, Table 9.1 shows that the time required for river fluxes to supply the elemental mass in the ocean,

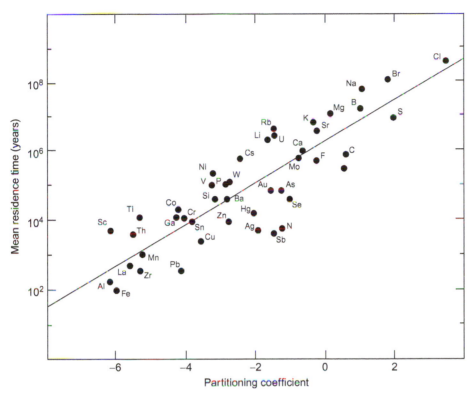

FIG. 9.9 Mean residence time of elements in seawater as a function of their concentration in seawater divided by their mean concentration in the Earth's crust—with high values of the index indicating elements that are very soluble. *From Whitfield and Turner (1979). Reprinted with permission from* Nature, *copyright 1979 Macmillan Magazines Limited.*

the mean residence time, varies from 120 million years for Cl to 1.1 million years for Ca. Even for Cl the MRT is much shorter than the age of the oceans indicating that Cl has not simply accumulated in seawater through Earth's history.

Through geologic time, changes in the composition of seawater have occurred when inputs and outputs of individual constituents were not in balance. For example, the Ca content of seawater has declined during the past 28 million years, not only as a result of changes in the relative importance of inputs from rock weathering but also due to rising rates of losses to carbonate sediments (De La Rocha and DePaolo, 2000; Griffith et al., 2008). The salinity of the ocean has varied considerably over our planetary history, declining during periods when large amounts of salts and brines were sequestered on the continents, and increasing when these salts were returned via rivers to the sea (Knauth, 1998; Knauth and Knauth, 2011). Salinity also shifts between glacial and interglacial periods, as the influx of major ions derived from continental weathering tends to peak at the onset of interglacial periods when large volumes of weathered substrate are exposed to the hydrologic cycle by glacial retreat (Vance et al., 2009; Frings et al., 2016).

Over long periods, the inputs of ions derived from continental weathering are balanced by processes that remove ions from the oceans. Some of these processes are nonfractionating such as the transfer of seaspray aerosols to the land surface (Chapter 3) or the creation of evaporites (Chapter 4). Other removal processes, such as the sorption to sediment particles, depend upon ionic charge. Biological processes can be highly selective, such as the accumulation of silica into the frustules of diatoms or strontium into the skeletons of radiolarians.

Earlier, we saw that wind blowing across the ocean surface produces sea spray and marine aerosols that contain the elements of seawater (Chapter 3). Thus, a significant portion of the river transport of Cl from land is derived directly from the sea (Fig. 3.17). The atmospheric transport of these "cyclic salts" removes ions from the sea roughly in proportion to their concentration in seawater (Table 4.9).

During some periods of the Earth's history, vast deposits of minerals have formed when seawater evaporated from shallow, closed basins. Today, the extensive salt flats, or *sabkhas*, in the Persian Gulf region are the best examples. Huge deposits of salt laid down 400,000,000 years ago are now mined beneath Lake Erie, near Cleveland, Ohio. Although the area of such evaporites is limited, the formation of evaporite minerals has been an important mechanism for the removal of Na, Cl and SO_4 from the oceans in the geologic past (Holland, 1974; Hardie, 1991). Weathering of evaporite minerals is an important source of ions in river water, exacerbated in modern times by rock salt and potash mining, which greatly increase the loading of ancient marine salts onto the land surface and into inland waters (Kaushal et al., 2018).

Other physical and geochemical mechanisms of loss occur in ocean sediments. Sediments are porous and the pores contain seawater. Burial of ocean sediments and their pore waters is a significant loss term for Na^+ and Cl^-, which are the most concentrated ions in seawater. Ions are also removed from the oceans when the clays in the suspended sediments of rivers undergo ion exchange with seawater. In rivers, most of the cation exchange sites (Chapter 4) are occupied by Ca^{2+}. When these clays are delivered to the sea, Ca^{2+} is released and replaced by other cations, especially Na^+ (Sayles and Mangelsdorf, 1977; James and Palmer, 2000). As a result, most deep sea clays contain much higher concentrations of Na^+, K^+, and Mg^{2+} than are found in the suspended matter of river water (Martin and Meybeck, 1979; Viers et al., 2009). These clays eventually settle to the ocean floor, causing a net loss of these cations from ocean waters.

Biological processes are also involved in the burial of elements in sediments. As we will discuss in more detail below, the deposition of $CaCO_3$ by organisms is the major process removing Ca^{2+} from seawater. Sulfate reduction results in substantial uptake of SO_4^{2-} in sediments, where it is reduced and binds with reduced Fe that is deposited as pyrite (see Chapters 7 and 8). In anoxic bottom waters and sediments, sulfate reduction can account for ~50% of sedimentary organic matter respiration (Kasten and Jørgensen, 2000; Jørgensen and Kasten, 2006).

So far, the processes that we have discussed for the removal of elements from seawater cannot explain the removal of much of the annual river flow of Mg^{2+} and K^+ to the sea. Marine geochemists have postulated several reactions of "reverse weathering," whereby silicate minerals are reconstituted (authigenic) in ocean sediments, removing Mg^{2+} and other cations from the ocean (Mackenzie and Kump, 1995; Michalopoulos and Aller, 1995; Misra and

Froelich, 2012). In this form of diagenesis, clays are formed by the association of biogenic opal, metal hydroxides, and dissolved cations via:

$$SiO_2 + Al(OH)_4^- + (K^+, Mg^{2+}, Li^+, etc.) + HCO_3 \rightarrow clay\ minerals + H_2O + CO_2. \qquad (9.2)$$

Reverse weathering is a major sink for Li from seawater (Misra and Froelich, 2012), and the formation of authigenic clay minerals is apparently a small sink for Mg and K (Kastner, 1974; Berner, 2004). Michalopoulos and Aller (1995) found that alumniosilicate minerals were reconstituted in laboratory incubations of marine sediments from the Amazon River, suggesting that this mechanism could sequester as much as 10% of the annual flux of K to the sea (cf. (Hover et al., 2002).

Inputs from hydrothermal vents and seafloor volcanoes

Continental runoff is not the only source of elements to the sea. About 80% of the volcanic activity on Earth occurs on the seafloor of the deep ocean, primarily along plate-spreading ocean ridges. These hydrothermal emissions represent an important flux of heat and chemicals from the mantle to the ocean. In the late 1970s (Corliss et al., 1979), examined the emissions from hydrothermal (volcanic) vents in the sea. One of the best-known hydro-thermal systems is found at a depth of 2500 m near the Galapagos Islands in the eastern Pacific Ocean. Hot fluids emanating from these vents are substantially depleted in Mg and SO_4 and enriched in Ca, Li, Rb, Si, and other elements compared to seawater (Elderfield and Schultz, 1996; de Villiers and Nelson, 1999) (Fig. 9.10). Globally the annual sink of dissolved Mg in hydrothermal vents, where it is incorporated into Mg-rich silicate rocks, exceeds the delivery of Mg to the oceans in river water. The flux of Ca to the oceans in rivers, $480 \times 10^{12}\ g\,yr^{-1}$, is incremented by an additional flux of up to $170 \times 10^{12}\ g\,yr^{-1}$ from hydrothermal vents (Edmond et al., 1979). Changes in the Mg/Ca ratio in seawater through geologic time are a good index of the relative importance of hydrothermal activities (Horita et al., 2002; Coggon et al., 2010). In addition to sustained venting, deep sea hydrothermal eruptions appear to be important and episodic sources of heat and chemicals to the world's oceans (Baker et al., 2019).

In sum, it appears that most Na and Cl ions are removed from the sea in pore water burial, sea spray, and evaporites. Magnesium is largely removed in hydrothermal ex-change, and calcium and sulfate by the deposition of biogenic sediments. The mass balance of potassium is not well understood, but K^+ appears to be removed by exchange with clay minerals, leading to the formation of illite, and by some reactions with basaltic sediments (Gieskes and Lawrence, 1981). Over long periods of time, ocean sediments are subducted to the Earth's mantle, where they are converted into primary silicate minerals, with volatile components being released in volcanic gases (H_2O, CO_2, Cl_2, SO_2, etc.; Fig. 1.3). The entire oceanic crust appears to circulate through this pathway in less than 300 million years (Müller et al., 2008) transferring sedimentary deposits to the mantle (Plank and Langmuir, 1998).

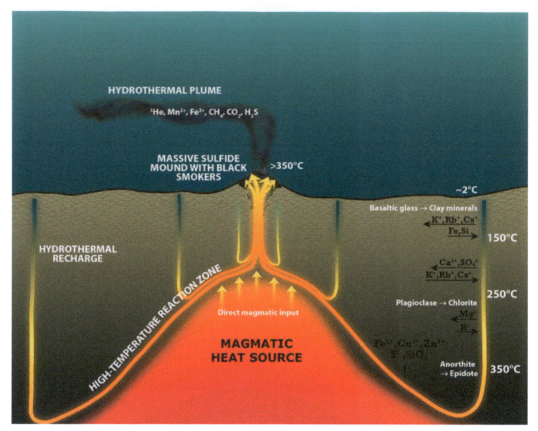

FIG. 9.10   Schematic of a hydrothermal system at a fast-spreading mid-ocean ridge showing fluid temperature gradients and fluid-rock chemical exchange. *From Jamieson et al. (2016).*

Nutrients

In contrast to the major ions, the concentrations of macronutrients (N, P) and micronutrients (Fe, Mo, Cu, Zn) in seawater are variable in both space and time. As we discussed in Chapter 8, the nutrient supply along coastal margins can be quite high, and increasing nutrient loads over the last century are implicated in the eutrophication of many coastal estuaries (Diaz and Rosenberg, 2008). In the open (pelagic) ocean, the supply of most nutrients is quite low, depending on atmospheric inputs or, in the case of N, biological fixation. Compounding these low inputs, nutrients are lost via the sinking of organic matter out of the photic zone (as discussed for lakes in Chapter 8). Although much of the nutrient content of this sinking organic matter is mineralized during downward transport to ocean sediments, vertical stratification prevents these nutrients from returning to the photic zone except in areas of upwelling. As a result, most nutrients share a similar depth profile throughout the world's oceans, with concentrations depleted throughout the photic zone, and increasing with depth (Fig. 9.11). As deep-water parcels move along the global conveyor

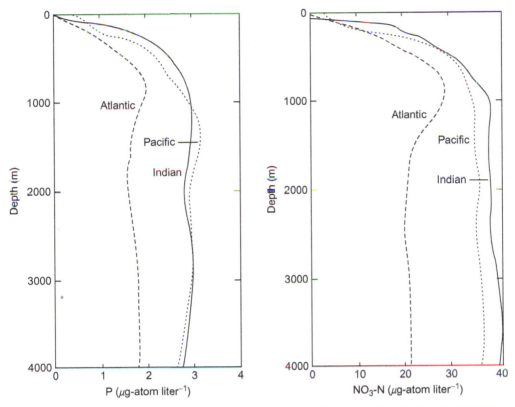

FIG. 9.11   Vertical distribution of phosphate and nitrate in the world's oceans. *From Svedrup et al. (1942).*

belt, they accumulate the nutrients mineralized from organic matter generated in the overlying surface waters. These accumulated nutrients are then returned to the surface ocean in areas of upwelling (Fig. 9.12).

Vertical mixing includes upwelling, upward convection, and diffusion from the deep ocean. Upwelling accounts for about half of the global upward flux, and it is centered in coastal areas where the resulting nutrient-rich waters yield high productivity. Away from areas of upwelling, diffusion and convection dominate the upward flux (Table 9.2), but diffusion rates are low (Ledwell et al., 1993), so the total supply of nutrients is limited in most of the open ocean (Lewis et al., 1986; Martin and Gordon, 1988). Diffusion appears globally significant only as a result of the large area of pelagic ocean compared to the small area of upwellings.

Dissolved gases

All of the gases in Earth's atmosphere are also found as dissolved gases in seawater. The capacity of any volume of water to hold gas declines with rising temperatures and rising

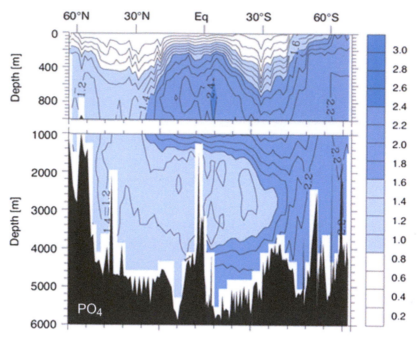

FIG. 9.12 Phosphorus in the Atlantic Ocean, showing the increase in its concentration in deep waters as they travel from north to south. Note that in this figure deepwater ages as it moves from downwelling areas of the North Atlantic towards the circumpolar current around Antarctica. *Source: From Sarmiento and Gruber (2006). Used with permission of Princeton University Press.*

TABLE 9.2 Sources of Fe, PO_4, and NO_3 in surface waters of the North Pacific Ocean.

Source	Fe	PO_4	NO_3
Concentration at 150 m ($mmol\,m^{-3}$)	0.075	330	4300
Upwelling ($mmol\,m^{-2}\,day^{-1}$)	0.00090	4.0	52
Net upward diffusion ($mmol\,m^{-2}\,day^{-1}$)	0.0034	30	400
Atmospheric flux ($mmol\,m^{-2}\,day^{-1}$)	0.16	0.102	26
Total fluxes ($mmol\,m^{-2}\,day^{-1}$)	0.164	34	480
Percent from advective input	0.5	12	11
Percent from diffusive input	2	88	83
Percent from atmospheric input	98	0	5

From Martin and Gordon (1988).

salinity and increases with depth as a function of the high pressure of overlying water.[i] Because atmospheric gases vary widely in solubility, reactivity, and importance to organisms, the composition of gases in the ocean is quite distinct from the composition of the atmosphere, even at the air-sea interface.

At the ocean's surface, the rate of gas exchange[j] (F) between the air and the surface ocean can be calculated from the equation:

$$F_{GAS} = k_w \cdot S_{GAS} \cdot (1 - f_{ice}) \cdot (pGAS^{wat} - pGAS^{atm}) \tag{9.3}$$

In this equation k is gas transfer velocity, which is a function of wind speed and turbulence at the ocean surface. The solubility (S) of each gas differs, but S for every gas declines with rising water temperature. Because exchange between the air and the sea is eliminated when sea ice covers the surface ($1 - f_{ice}$ is the ice free zone of the world's oceans), areas with more ice have less air-sea exchange. Finally differences in the partial pressure of each gas between the ocean surface and the overlying atmosphere drive exchange. For nonreactive noble gases, this partial pressure differential is close to equilibrium, with variation due solely to the mixing of water parcels of differing history.

Biological consumption and production of gases determines their concentration in surface waters and alters the difference in partial pressure between air and water. As is true in the atmosphere (Chapter 3), the dominant gases in ocean water are N_2, O_2, Ar and CO_2, but their relative concentrations in seawater are quite different as a function of solubility. Concentrations of N_2 and Ar are relatively constant across the surface ocean. In consequence of its high content in the atmosphere and low rates of biological production and consumption in seawater, N_2 concentrations exhibit little variation globally (ocean N_2 ranges from 8.4–14.5 mL L^{-1}). In contrast, surface ocean oxygen concentrations range from essentially zero (anoxic) in polluted coastal oceans to values that can exceed 8.5 mL L^{-1} in cold polar waters. Much of the surface ocean is supersaturated in O_2 ($pO_{2wat} > pO_{2atm}$) (Locarnini et al., 2013) and releases O_2 to the atmosphere. This efflux from the ocean's surface is balanced by very high rates of O_2 influx in more limited portions of the ocean. Global ocean dissolved oxygen concentrations have declined in recent decades as a result of rising ocean temperatures and coastal eutrophication (Keeling et al., 2010; Breitburg et al., 2018).

Carbon dioxide, which makes up only 3% of the mass of the atmosphere, is the dominant gas in the surface ocean (with concentrations sometimes above 50 mL L^{-1}) (Fig. 9.13). Across the global ocean pCO$_2$ ranges from 100 to 1000 µatm,[k] with most of the surface ocean undersaturated relative to CO_2 in the atmosphere (Takahashi et al., 2006) allowing for a net influx of CO_2 into the ocean. The solubility of CO_2 in seawater depends on temperature; CO_2 is about twice as soluble at 0 °C as it is at 20 °C (Broecker, 1974). The temperature of the upper 1 mm of the ocean's surface, the "skin" temperature, is thus critical to determining the atmosphere-to-ocean flux. Average global ocean pCO$_2$ concentrations are increasing in conjunction with the

[i] Recall the high CO_2 content in the bottom of Lake Nyos from Chapter 7.

[j] Gas exchange between a natural water body and the atmosphere is more complicated than the simple predictions of solubility from Henry's law (Eq. 2.6).

[k] In comparison to an atmospheric pCO$_2$ of 411 µatm recorded at Mauna Loa Observatory in Hawaii on the date this section was written (August 2019).

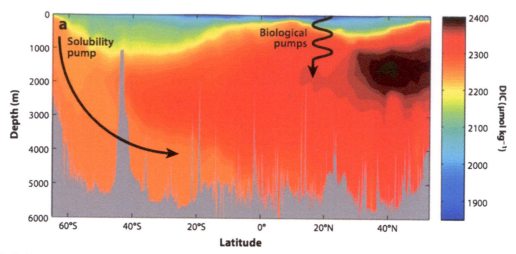

FIG. 9.13 The distribution of DIC across a south-to-north transect of the Pacific ocean. Dissolved CO_2 enters in downwelling waters around the Antarctic ice sheet. As it moves northward and "ages" this deepwater accumulates additional DIC from the dissolution and mineralization of sinking organic particles—what is known as the biological pump. Gray shading shows the bathymetry along the transect. *From Hamme et al. (2019).*

rising concentration of CO_2 in the atmosphere (Peng et al., 1998; Sabine et al., 2004; Takahashi et al., 2006). Recall from Chapter 7 that CO_2 forms carbonic acid (H_2CO_3) when it dissolves in water, and carbonic acid can then further dissociate through the loss of H^+ to form bicarbonate (HCO_3^-) and carbonate (CO_3^{2-}) ions in the chemical equilibrium:

$$CO_{2(atmos)} \leftrightarrow CO_{2(aq)} + H_2O \leftrightarrow H_2CO_3 \leftrightarrow H^+ + HCO_3 \leftrightarrow 2H^+ + CO_3{}^{2-}. \tag{9.4}$$

In average surface seawater (pH of ~8.1) all three forms of DIC are in chemical equilibrium with 90% of the DIC as HCO_3^-, 9% as CO_3^{2-} and only 1% as CO_2. The marine carbonate buffer system allows the ocean to absorb far more atmospheric CO_2 that would occur based on solubility alone.

The carbonate buffering system has several important implications for ocean carbon cycling. First, most of the CO_2 absorbed at the ocean's surface is converted to forms of DIC that are not subject to gas exchange with the atmosphere. The carbonate buffering system thus greatly increases the residence time of CO_2 in seawater.[1] Second, the addition of CO_2 to the world's oceans increases the concentration of H^+ which leads to the conversion of CO_3^{2-} to HCO_3^- and reduces the acid buffering potential of the ocean (Doney et al., 2009). Long-term records of atmospheric and seawater CO_2 from the North Pacific document increases in CO_2 in the atmosphere and in seawater that are accompanied by declines in ocean pH (Fig. 9.14).

While the surface of the ocean is in gaseous equilibrium with the atmosphere, the ventilation of the deep ocean is confined primarily to downwelling areas where the buoyancy difference between surface and deep waters is overcome (Fig. 9.4). The cold, high pressure waters of the deep ocean can hold far higher gas concentrations than the surface ocean,

---

[1]In the surface waters of the Atlantic Ocean the mean residence time of CO_2 is about 6 years (Stuiver, 1980).

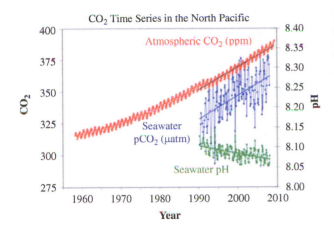

FIG. 9.14 The long-term increase in atmospheric CO_2 at the Mauna Loa observatory is contrasted with long-term trends in seawater CO_2 and seawater pH at nearby Station Aloha. *Figure modified from original Dore et al. (2009).*

and water residence times in the deep ocean are on the order of centuries. When cold waters form in equilibrium with today's atmospheric CO_2 concentrations, they carry more CO_2 than when they formed in equilibrium with an atmosphere of 280 ppm CO_2 (the atmospheric CO_2 concentration of 300–500 years ago). Brewer et al. (1989) report that North Atlantic deep water now carries a *net* flux of 0.26×10^{15} g C yr southward, due to the global rise in atmospheric CO_2 during this century. This transfer of atmospheric CO_2 to the deep ocean is often referred to as the solubility pump (Volk and Hoffert, 1985).

One of the critical zones for contact between the deep ocean and the atmosphere is the Southern Ocean around Antarctica. Rae et al. (2018) used a 40,000-year record of boron isotopes in deep corals to reconstruct the history of CO_2 accumulation and release from the southern deep ocean. Their data are consistent with the idea that sea ice acts as a "lid"—allowing for a buildup of deep sea CO_2 during glacial periods and reducing the deep ocean CO_2 sink during interglacial periods (Skinner et al., 2017). This mechanism may account for oscillations in atmospheric CO_2 in mid glacial periods, and raises concerns about how rising global temperatures and shrinking sea ice volumes are likely to alter global carbon cycling.

The degree of ventilation of the deep ocean varies over much shorter time scales as well. The upwelling of cold, deep ocean waters during the La Niña years leads to lower atmospheric temperatures, whereas the release of CO_2 from cold, upwelling waters is lower during years of El Niño (Bacastow, 1976, 1981; Inoue and Sugimura, 1992). During the 1991–92 El Niño, the ocean released 0.3×10^{15} g C as CO_2 to the atmosphere, compared to its normal efflux of 1.0×10^{15} g C (Murray et al., 1994, 1995; Feely et al., 1999; Chavez et al., 1999), and the rate of CO_2 increase in the atmosphere slowed for several years (Keeling et al., 1995).

Biogenic carbonates

A large number of marine organisms precipitate carbonate in their skeletal and protective tissues by the reaction:

$$Ca^{2+} + 2HCO_3^- \rightarrow CaCO_3 \downarrow + H_2O + CO_2. \tag{9.5}$$

During the formation of each mole of $CaCO_3$, alkalinity is reduced by two moles and total DIC decreases by one mole, leading to a shift in the carbonate system towards higher CO_2 (Buitenhuis et al., 2001; Koeve, 2002). Whether these skeletal remains are preserved in shallow-water calcareous sediments or sink to the deep ocean, each molecule of $CaCO_3$ carries the equivalent of one CO_2 and leaves behind the equivalent of one CO_2 in the surface ocean.[m]

In addition, the surface waters in many areas of the oceans are variably undersaturated in CO_2 as a result of photosynthesis. Sinking organic materials remove carbon from the surface ocean, and it is replaced by the dissolution of new inputs of atmosphere CO_2. Taylor et al. (1992) found that during a 46-day period there was a net downward transport of carbon in the northeast Atlantic Ocean due to the sinking of live ($2\,g\,C/m^2$) and dead ($17\,g\,C/m^2$) cells and the downward mixing of living cells by turbulence ($3\,g\,C/m^2$). Thus, biotic processes act to convert inorganic carbon (CO_2) in the surface waters to organic carbon that is delivered to the deep waters of the ocean.

Thus, the ocean sink for atmospheric CO_2 consists of the solubility pump along with three biological processes that remove carbon dioxide from the atmosphere. Carbon can be lost to the deep ocean in sinking carbonate skeletal debris (the "hard tissue" pump). Carbon fixed into organic matter in the surface waters can be lost as sinking dead tissues (the "soft tissue" pump). Finally, organic carbon released as DOC via lysis, excretion, exudation or leaching can be transported to the deep sea in downwelling zones (Fig. 9.15). These *biogeochemical* processes are superimposed on a much larger, background flux of CO_2 that enters the oceans by dissolution of CO_2 in cold downwelling waters. The biotic pump responds to changes in the activity of the biosphere; whereas net downwelling of dissolved CO_2 responds only to changes in the concentration of CO_2 in the atmosphere and the circulation of deep ocean waters.

As atmospheric carbon dioxide increases, we would expect an increased dissolution of CO_2 in the oceans, following Henry's Law. However, the surface ocean provides only a limited volume for CO_2 uptake, and the atmosphere is not in immediate contact with the much larger volume of the deep ocean. In the absence of large changes in NPP, it is the rate of formation of bottom waters in polar regions that limits the rate at which the oceans can take up CO_2. Reductions in the upwelling and CO_2 degassing of deep waters in the Southern Ocean may have caused the low concentrations of atmospheric CO_2 during the last glacial epoch (Kumar et al., 1995; Francois et al., 1997; Kohfeld et al., 2005; Sigman et al., 2010). Conversely, greater upwelling may have led to rising CO_2 and warmer temperatures at the end of the last glacial period (Burke and Robinson, 2012).

Biogeochemistry of the surface ocean

Now that we have considered the circulation, stratification and composition of the world's oceans, we can begin to explore the details of the ocean's biogeochemical cycles. Here we will discuss the biogeochemistry of the surface ocean, the deep ocean, and ocean sediments in

---

[m] Note that the precipitation of carbonate by phytoplankton supplies some of the CO_2 needed for photosynthesis, reducing the net uptake of CO_2 from the atmosphere (Robertson et al., 1994).

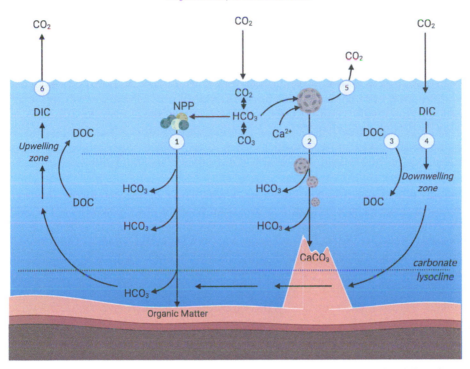

FIG. 9.15   The modes by which atmospheric CO_2 is transported to the deep ocean, include: (1) the soft tissue pump: the formation and sinking of organic matter (POC); (2) the hard tissue pump: the sinking of carbonate skeletons; (3) the downwelling of DOC, and (4) the solubility pump: the transfer of DIC to the deep ocean. These sinks of C out of the surface ocean are offset by: (5) the calcium carbonate counterpump, which liberates CO_2 during the formation of carbonate skeletons and (6) the degassing of DIC rich upwelling waters. *Drawn by ESB.*

series while recognizing the important exchange of materials between each layer. Most of the ocean's volume receives no sunlight. It is only in the well lit waters of the surface ocean and shallow coastal sediments that photosynthesis can occur. Elsewhere in the ocean, organisms must rely on the supply of organic matter settling out of the photic zone[n] or from the reduced chemical substances (H_2, HS^-) that are delivered from hydrothermal vents and volcanic ridges (Jannasch, 1989; Petersen et al., 2011; Worman et al., 2016). The basic energetic and nutrient constraints thus differ considerably between the photic surface zone, the aphotic ocean depths, and the ocean floor.

Net primary production

Compared to the vegetation that dominates the biomass of terrestrial ecosystems, marine autotrophs are dominated by single-celled algae with short life spans and limited investments in structural tissues. Much of the autotrophic production of the ocean is rapidly consumed,

[n]These falling particles can range in size from individual diatom frustules to whale carcasses; a large proportion are small particles, which aggregate to form "marine snow."

thus rates of NPP are very high relative to the standing biomass of phytoplankton in the world's oceans.

The earliest estimates of marine productivity were derived using modifications of the oxygen-bottle and ^{14}C techniques, as outlined for lake waters in Chapter 8. In addition to the artifacts common to all bottle assays, early efforts to measure marine productivity were confounded by trace element contamination of samples[o] and the very small size and high exudation rates characteristic of many marine algae. In the waters of the eastern tropical Pacific Ocean, Li et al. (1983) found that 25% to 90% of the photosynthetic biomass passed through a 1-μm filter, and thus was not captured as fixed C in traditional ^{14}C-labeling experiments. These picoplankton are more important components of the phytoplankton in warm, nutrient-poor waters than in the sub-polar oceans (Agawin et al., 2000), but are common throughout the world's oceans. Stockner and Antia (1986) suggest that picoplankton may regularly account for up to 50% of ocean production. Marine phytoplankton also release large amounts of dissolved organic carbon to seawater as exudates (Baines and Pace, 1991), and these compounds—technically a component of NPP—also pass through the filtration procedures of the ^{14}C method. Further complicating traditional GPP estimates, mixotrophic bacteria, which metabolize organic carbon when it is available and produce it by anoxygenic photosynthesis when it is not, may constitutes 2–5% of total production in some waters. This productivity would not be captured by the O_2-bottle method (Kolber et al., 2000).

Modern open water estimates of NPP measure net increases in O_2 (Craig and Hayward, 1987; Najjar and Keeling, 1997; Emerson and Stump, 2010) or decreases in HCO_3^- (Lee, 2001) in the upper water column to calculate rates of photosynthesis as the divergence from the gas or bicarbonate concentrations predicted by physical processes alone. Supersaturation of O_2 or depletion of HCO_3^- can be used to estimate NPP in the water column (using mathematical models that are conceptually similar to eddy flux measurements in terrestrial ecosystems (Chapter 5) or diel oxygen time series in freshwaters (Chapter 7)). As with bottle methods, a supersaturation of O_2 or depletion of HCO_3^- are best taken as an index of net community production (NCP) in the water column.[p] The precision of such techniques can be further improved by simultaneous measurement of dissolved O_2 and $\delta^{18}O$ of H_2O (Luz and Barkan, 2009). Oxygen concentrations are affected by the rates of both photosynthesis and respiration, whereas $\delta^{18}O$ is solely affected by photosynthesis. Differential changes in these parameters can be used to estimate NPP in ocean waters (Juranek and Quay, 2010). In side-by-side comparisons, this $\delta^{18}O$ method can produce values more than twice as high as the ^{14}C method for measuring marine production (Quay et al., 2010).

Scaling estimates from either bottle- or open-water estimates to regional or global estimates is difficult. Fortunately remote sensing offers the potential to link measured rates of productivity and chlorophyll with remotely sensed measures of chlorophyll in the world's oceans. Where ocean waters contain little phytoplankton, there is limited absorption of incident

---

[o] For many years the iron hulls of ships and the copper tubing used to collect seawater artificially elevated the concentration of these trace elements in samples and delayed our appreciation of the potential for trace element limitation of marine primary production.

[p] NCP is analogous to net ecosystem production (NEP) on land (Chapter 5), except that NCP does not include deep-water and sediment respiration, which is supported by surface photosynthesis in the oceans.

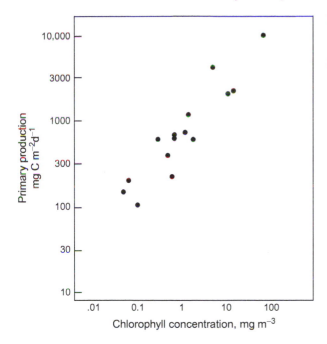

FIG. 9.16 Net primary productivity as a function of surface chlorophyll in waters of coastal California. *From Eppley et al. (1985). Reprinted by permission of Oxford University Press.*

radiation by chlorophyll, and the reflected radiation is blue. Where chlorophyll and other pigments are abundant, the reflectance contains a greater proportion of green wavelengths (Prézelin and Boczar, 1986). The reflected light is indicative of algal biomass in the upper 20–30% of the euphotic zone, where most NPP occurs (Balch et al., 1992). The reflectance data can be used to calculate the concentration of chlorophyll or organic carbon in the water column and hence production (Fig. 9.16, Platt and Sathyendranath, 1988; Falkowski, 2005). With the deployment of the MODIS satellite (Chapter 5), multispectral images of the ocean are now available for worldwide estimates of marine NPP.

Estimating global ocean productivity

Early efforts to estimate ocean productivity from bottle-assay experiments suggested that marine NPP was 23 to 27×10^{15} g C yr^{-1} (Berger, 1989). With the benefit of higher precision empirical estimates of NPP and higher frequency and resolution of satellite imagery, modern estimates of marine NPP are roughly $2\times$ larger and equivalent to NPP on land (Table 9.3) (Field et al., 1998). Behrenfeld and Falkowski (1997) used satellite measurements of pigment concentration in surface waters to estimate marine NPP at 43.5×10^{15} gC yr^{-1}. More recently, (Friend et al., 2009) estimated marine NPP at 52.5×10^{15} gC yr^{-1} (Fig. 9.17), with a spatial pattern similar to that for dissolved O_2 in seawater (Falkowski et al., 2011). Net community production in the surface waters, yielding organic matter that can sink into the deep ocean, is about 10–20% of total NPP (Laws et al., 2000; Lee, 2001; Falkowski, 2005; Quay et al., 2010), with higher values in cold polar waters where bacterial respiration is lower.

TABLE 9.3   Estimates of total marine primary productivity and the proportion that is new production.

Province	% of ocean	Area $(10^{12}\,m^2)$	Mean production $(g\,C\,m^{-2}\,yr^{-1})$	Total global production $(10^{15}\,g\,C\,yr^{-1})$	New production[a] $(g\,C\,m^{-2}\,yr^{-1})$	Global new production $(10^{15}\,g\,C\,yr^{-1})$
Open ocean	90	326	130	42	18	5.9
Coastal zone	9.9	36	250	9.0	42	1.5
Upwelling area	0.1	0.36	420	0.15	85	0.03
Total		362		51		7.4

[a] New productivity definedas C flux at 100 m.
From Knauer (1993). Used with permission of Springer-Verlag.

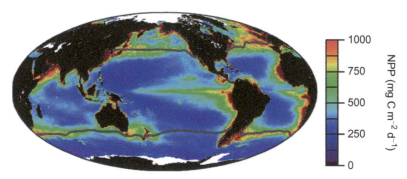

FIG. 9.17   Global map of marine NPP. *From Behrenfeld et al. (2006).*

The highest rates of NPP are measured in coastal regions, where nutrient-rich estuarine waters mix with seawater, and in regions of upwelling, where nutrient-rich deep waters are mixed into the photic zone. In contrast, among open-water environments, even the most productive areas show NPP of $<300\,g\,C\,m^{-2}\,yr^{-1}$, roughly similar to the NPP of an arid woodland (Table 5.3). Much of the open ocean has rates of NPP more akin to a desert. Nevertheless, as a result of their large area, the open oceans account for about 80% of total marine NPP, with continental shelf areas accounting for the remainder (Table 9.3). NPP in beds of intertidal rockweed, *Ascophyllum nodosum*, in Cobscook Bay of eastern Maine is nearly $900\,g\,C\,m^{-2}\,yr^{-1}$, similar to that of a temperate forest (Table 5.3; Vadas et al., 2004). Although massive beds of kelp are found along some coasts, such as the *Macrocystis* kelps of southern California, seaweed accounts for only about 0.1% of marine production globally (Smith, 1981; Walsh, 1984).

Nutrient limitation of NPP

Net primary productivity in the sea is limited by a scarcity of nutrients. Production is highest in regions of high nutrient availability—the continental shelf and regions of upwelling (Fig. 9.17)—and lower in the open ocean, where the concentrations of available N, P, Fe and Si are normally very low. A glance at a map of ocean NPP (Fig. 9.17) reveals that the highest rates of NPP are observed along the coasts of continents, particularly in areas where major rivers enter the ocean. In addition, the band of high chlorophyll stretching across the equatorial Pacific reflects an important zone of deep-water upwelling. The challenge for autotrophs in the open ocean is the limited supply of elements essential for building photosynthetic machinery and the high potential for the loss of nutrients through particle sinking to the deep ocean. Nutrients are continuously removed from the surface water by the downward sinking of dead organisms and fecal pellets. Working off the California coast, Shanks and Trent (1979) found that 4–22% of the nitrogen contained in particles (PON) was removed from the surface waters each day. The mean residence time of N, P, and Si in the surface ocean is much less than the mean residence time of the surface waters, and there are wide differences in the concentration of these elements between the surface and the deep ocean (Fig. 9.11). N, P, and Si are non-conservative elements in seawater; their behavior is strongly controlled by the presence of life.

Away from the coast, primary production in the open ocean is heavily reliant on local mineralization and the influx of nutrient-rich deep water masses. Because biological nutrient demand in the surface ocean is high relative to supply, we see depth profiles for most biologically important elements that are strongly depleted in the photic zone with higher concentrations at depth (Fig. 9.11). Nutrient ratios relative to the stoichiometry of phytoplankton can be used to infer which nutrients are limited within any study area.

Macronutrient Limitation of NPP: In 1958, Albert Redfield noted that wherever they were collected, marine phytoplankton contained N and P in a fairly constant molar ratio to the content of carbon, 106C:16N:1P (Redfield, 1963),[q] as a result of the incorporation of these elements during photosynthesis and growth:

$$106CO_2 + 16NO_3^- + HPO_4^{2-} + 122H_2O + 18H^+ \rightarrow (CH_2O)_{106}(NH_3)_{16}(H_3PO_4) + 138O_2. \quad (9.6)$$

The molar N:P ratio of 16 may reflect fundamental relationships between the requirements for protein and RNA synthesis common to all plants (Loladze and Elser, 2011). Despite differences in nutrient concentration among the major oceans, upwelling waters contain available C, N, and P (i.e., HCO_3^-, NO_3^-, and HPO_4^{2-}) in the approximate ratio of 800C:16N:1P. Thus, even in the face of the high productivity found in upwelling waters, only about 10% of the HCO_3^- can be consumed by photosynthesis before the N and P are exhausted. Significantly, (Redfield, 1958) noted that the biota determined the relative concentrations of N and P in the deep sea, and that the biotic demand for N and P was closely matched to the availability of these elements in upwelling waters (Holland, 1978).

It is important to remember that the Redfield ratio is an average value. The nutrient concentrations in individual plankton species may differ from the Redfield ratio depending on

---

[q]This molar ratio is equivalent to a mass ratio of 40C:7N:1P. Thus, the N:P mass ratio in marine phytoplankton is lower than in terrestrial plants (viz. ~15; Chapter 6).

season and environmental conditions (Klausmeier et al., 2004; Weber and Deutsch, 2010). Nevertheless, as an average value, the Redfield ratio allows us to compare the importance of riverflow, upward transport, and internal recycling for their contributions to the annual net primary production of the surface ocean. To sustain a global marine NPP of 50×10^{15} g C yr^{-1} (Table 9.3), phytoplankton must take up about 8.8×10^{15} g N and 1.2×10^{15} g P each year (Table 9.4). Rivers supply about 0.05×10^{15} g N yr^{-1} and 0.002×10^{15} g yr^{-1} of reactive P to the oceans (Chapters 8 and 12). However, the total nutrient supply from rivers, atmospheric inputs, and vertical movements (upwelling + diffusion + eddy convection) provides only a small fraction (11% for N and 9% for P) of the total nutrient requirement in the surface ocean, so nutrient recycling in the surface waters must supply the rest. Rapid turnover of nutrients is consistent with the rapid turnover of 80 to 90% of the organic carbon in the surface ocean.

Nitrogen: Just as it does on land, nitrogen appears to limit marine primary producers over much of the surface ocean (Elser et al., 2007). In most areas of the ocean, nitrate is not measurable in surface waters, and phytoplankton grow rapidly in response to nanomolar additions of nitrogen to seawater (Glover et al., 1988). In the oceans, nitrogen fixation, the process that alleviates N limitation for most freshwater ecosystems, is often highly constrained by the limited supply of micronutrients, particularly Fe and Mo, that are necessary for building nitrogenase enzymes. In coastal oceans with high rates of N loading, the dissimilatory processes return N to the atmosphere at rates that often match or exceed N fixation (Paerl, 2018).

In the face of nutrient-limited growth and efficient nutrient uptake, phytoplankton maintain very low concentrations of N in surface waters (Fig. 9.11). McCarthy and Goldman (1979) showed that much of the nutrient cycling in the surface waters may occur in a small zone, perhaps in a nanoliter (10^{-9} L) of seawater, which surrounds a dying phytoplankton cell. Growing phytoplankton in the immediate vicinity are able to assimilate the nitrogen as soon as it is released. Often it is difficult to study nutrient cycling on such a small scale, so various workers have applied isotopic tracers (e.g., $^{15}NH_4$ and $^{15}NO_3$) to measure nutrient uptake by phytoplankton and bacteria (Glibert et al., 1982; Goldman and Glibert, 1982; Dickson and Wheeler, 1995). Leakage of dissolved organic nitrogen compounds (DON) from

TABLE 9.4 Calculation of the sources of nutrients that would sustain a global net primary productivity of 50×10^{15} gC yr^{-1} in the surface waters of the oceans.

Flux	Carbon (10^{12} g)	Nitrogen (10^{12} g)	Phosphorus (10^{12} g)
New primary production[a]	50,000	8800	1200
Amounts supplied			
By rivers[b]		50	2
By atmospheric deposition[c]		67	1
By N fixation[d]		150	—
By upwelling		700	100
Recycling (by difference)		7800	1100

[a] *Assuming a Redfield atom ratio of 106:16:1.*
[b] *N from Galloway et al. (2004); P from Meybeck (1982).*
[c] *Duce et al. (2008).*
[d] *Deutsch et al. (2007).*

phytoplankton may also account for a significant amount of the bacterial uptake and turnover of nitrogen in the surface waters (Kirchman et al., 1994; Bronk et al., 1994; Kroer et al., 1994). During decomposition of organic particles in the surface ocean, nitrogen is mineralized more rapidly than carbon, so that surviving particles carry C/N ratios that are somewhat greater than the Redfield ratio (Sambrotto et al., 1993) and that increase with depth (Honjo et al., 1982; Takahashi et al., 1985; Anderson and Sarmiento, 1994; Alldredge, 1998; Schneider et al., 2003).

Nutrient demand by phytoplankton is so great that it has been traditional to assume that little of the NH_4 released by mineralization remains for nitrification in the surface waters, and NH_4 dominates phytoplankton uptake of recycled N (Dugdale and Goering, 1967; Harrison et al., 1992, 1996; Yool et al., 2007). In contrast, most of the nitrogen mineralized in the deep ocean is converted to NO_3—some by nitrifiers recently recognized as archaea (Könneke et al., 2005; Francis et al., 2005).[r] Nitrate also dominates the nitrogen supply in rivers, so oceanographers can use the fraction of NPP that derives from the uptake of NH_4 versus that derived from NO_3 to estimate the recycling of nutrients that sustains NPP in the surface waters. For example, Jenkins (1988) estimated that the upward flux of NO_3 from the deep ocean near Bermuda would support an NPP of about $36\,g\ C\ m^{-2}\ yr^{-1}$—about 38% of observed NPP (Michaels et al., 1994). The remaining production must depend on NH_4 supplied by recycling in the surface waters.

The fraction of NPP that is sustained by nutrients delivered from atmospheric inputs (including N fixation), rivers, and upwelling is known as *new production* (Table 9.3). Rivers dominate the sources of N in many coastal waters, but in the pelagic oceans, the relative role of nutrient recycling in the surface waters and nutrient delivery by upwelling depends strongly on location. Outside of the major areas of thermohaline upwelling (Fig. 9.4), nutrients are delivered to the surface waters by convection in eddies (McGillicuddy et al., 1998; Oschlies and Garcon, 1998; Siegel et al., 1999; Johnson et al., 2010) and other processes of large-scale mixing (Uz et al., 2001) and advection (Palter et al., 2005). The vertical migration of diatoms is known to transport nitrate to the surface (Villareal et al., 1999), while the vertical movements of large animals can transport nutrients from great depths; for example, some whales feed at depth and defecate in surface waters (Roman and McCarthy, 2010).

Globally, new production is about 10–20% of total NPP, but the fraction, *f*n, is greatest in areas of cold, upwelling waters (Sathyendranath et al., 1991). Low, steady-state nutrient concentrations in the surface waters of the oceans—near the Redfield N/P ratio of 16—indicate that the sources of nutrients that sustain new production globally are about equal to the annual losses of nutrients in organic debris that sink through the thermocline to the deep ocean—that is, export production (f_e) (Eppley and Peterson, 1979). However, it is now recognized that the traditional separation of new and recycled production based solely on the form of nitrogen taken up is complicated by significant nitrogen fixation in the marine environment, the presence of nitrifying bacteria and archaea that produce NO_3 in the surface waters (Yool et al., 2007; Martens-Habbena et al., 2009) and atmospheric deposition of NH_4 and NO_3 (Duce et al., 2008; Kim et al., 2014b).

Anthropogenic N deposition to the world's oceans is typically quite low relative to the continents, but there are periods during the summer months when parts of the North Atlantic

---

[r]Nitrification in the deep sea is a form of chemoautotrophy (Eqs. 2.17 and 2.18) in which carbon is fixed in the dark. Chemoautotrophic nitrification and sulfide oxidation in the oceans are estimated to account for the addition of $0.77\,PgC/yr$ or 1.5% to marine NPP (Middelburg, 2011).

receive NO_3^- deposition that rivals the highest deposition measured on land (St-Laurent et al., 2017). These high N deposition events lead to substantially enhanced surface ocean productivity (8% annual new production increase) (St-Laurent et al., 2017). Similarly, rising rates of anthropogenic N deposition in areas of the western Pacific represent a substantial increase above otherwise low background N inputs to the open ocean (Kim et al., 2014b). Even on a remote atoll in the western Pacific, shifts in the ^{15}N isotopic signature of corals suggest that anthropogenic N deposition currently makes up 15–25% of total annual N inputs and is increasing (Ren et al., 2017a). In contrast, coral records from Bermuda show little evidence of directional change in N loading throughout the last century despite being downwind of major anthropogenic N sources from North America (Wang et al., 2018b).

Enhanced N loading to the oceans may be increasing marine NPP, but, somewhat counterintuitively, these N inputs do not appear to be reducing the prevalence of marine N limitation. Paerl (2018) has proposed that estuarine and coastal waters are becoming chronically "nitrogen hungry" as they receive increasing doses of both anthropogenic nitrogen (a large fraction of it as nitrate) and organic matter—reactants that stimulate N removal processes (denitrification and anammox).

Phosphorus limitation: Given its relative global rarity and its key role in limiting freshwater NPP, it is curious that phosphorus rarely appears to be the primary limiting nutrient in the oceans (Howarth and Marino, 2006). One key difference in P cycling in marine surface waters results from the high concentrations of sulfate. In freshwater systems, coprecipitation of phosphate ions with Fe(III) oxyhydroxides (Masion et al., 1997a, 1997b) is an important abiotic sink for P. This "iron trap" effectively maintains P limitation for many freshwaters (Blomqvist et al., 2004). Riverine inputs of P to the ocean are delivered as FeOOH-bound P (Caraco et al., 1990). When these oxidized minerals are deposited into anoxic sediments, iron reduction results in the release of Fe(II) and PO_4^{3-} ions. In marine sediments, where high rates of sulfate reduction in sediments result in the production of sulfides, the primary fate for reduced Fe is the formation of iron sulfides (FeS_2), reducing the availability of iron and the effectiveness of the iron sink. In estuaries, this transformation of Fe results in declines in soluble Fe concentrations and lower soluble N:P ratios with increasing salinity (Fig. 9.18, Jordan et al., 2008). The freshwater "iron trap" for P is effectively replaced with a highly efficient sulfur trap for Fe, reducing the abiotic sink for P and maintaining higher P availability in saltwater (Blomqvist et al., 2004; Jordan et al., 2008).

While the shifting efficiency of the iron trap helps explain why P rarely limits NPP in the coastal ocean, different mechanisms must explain the rarity of P limitation in pelagic regions, where inorganic P availability (DIP) is very low. In much of the open ocean, dissolved organic phosphorus (DOP) is the dominant pool of phosphorus, often an order of magnitude higher in concentration than DIP (Wu et al., 2000; Browning et al., 2017). Dissolved organic phosphorus is a complex mixture of organic molecules, from very labile phospholipids to highly refractory organic acids that are derived from lysis, excretion and cellular rupture. Because labile constituents are rapidly assimilated by biota, refractory DOP dominates the pool. To access the P contained in these molecules, algae and microbes produce alkaline phosphatase enzymes to liberate PO_4^{3-} ions (Browning et al., 2017). Two dominant bacterial alkaline phosphatase enzymes produced in the oceans, PhoX and PhoZ (Sebastian and Ammerman, 2009), and both require Fe as cofactors (Yong et al., 2014; Rodriguez et al., 2014). Thus, access to P may be limited by the very low availability of iron in the open ocean. While phosphorus

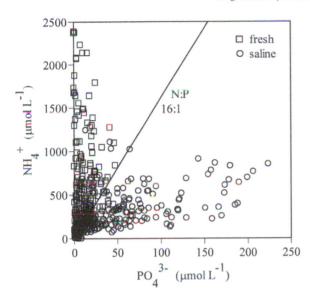

FIG. 9.18 Pore-water NH_4^+ versus PO_4^{3-} concentrations (mmol L^{-1}) at freshwater and more saline sites along the Patuxent River estuary (a tributary of the Chesapeake bay). The line shows the 16: 1 atomic N: P ratio. *From Jordan et al. (2008).*

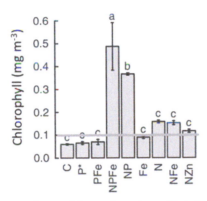

FIG. 9.19 Chlorophyll response to nutrient enrichment in waters collected from the subtropical Atlantic. Shown are means ± standard errors. The gray horizontal line is the mean initial time point measurement. C indicates control assays (no nutrients added) while all other treatments include N, P, Fe, and Zn added alone or in the combinations listed. For this location (off the North African coast) there is strong evidence for N, P and Fe colimitation, as none of these elements added alone resulted in enhanced algal growth. *From Browning et al. (2017).*

is in limited supply over much of the ocean, its availability is generally higher in proportion to demand than is true for either N or Fe (Browning et al., 2017, 2017a). Because of this, P enrichment in marine systems rarely results in a NPP response unless it is added in conjunction with N (Fig. 9.19).

Silica limitation of diatoms: The availability of dissolved Si (DSi) controls the rate of primary production by diatoms, algae that are encased in a silicate frustule, which may account for up to 40% of ocean NPP. Diatom productivity can be constrained by Si supply and the frustules themselves serve as ballast that increases the rate of sinking of many other elements, including carbon (Smetacek, 1998; Jin et al., 2006; Tréguer et al., 2017). This is particularly important in coastal areas where high Si loading from rivers fuels diatom production. There is evidence that Si is becoming an increasingly limiting nutrient in coastal areas where anthropogenic enrichment of N and P alleviates macronutrient limitation (Officer and Ryther, 1980; Conley and Malone, 1992; Rabalais et al., 1996; Bristow et al., 2017).

Iron limitation: Iron is a required element in the ferrodoxin proteins that transfer electrons in photosynthesis (Arnon, 1965). Iron is required for both proteins that make up the nitrogenase enzyme complex (Carpenter and Capone, 2008), and Fe is necessary to construct the alkaline phosphatases (Browning et al., 2017). While Fe is essential, the absolute requirement for Fe is low in comparison with N and P. Compare Redfield's molar C:N:P of 106:16:1 to the C:Fe molar ratio observed for marine autotrophs that ranges up to 100,000:1 in Fe-deficient systems (Anderson and Morel, 1982; Morel et al., 1991). These low requirements are satisfied in most terrestrial and freshwater environments where crustal minerals and soils contain large quantities of Fe. The potential for Fe to limit these processes in the oceans arises from its limited supply and high rates of removal from the photic zone and sequestration within marine sediments. In response to low concentrations of Fe, some bacterioplankton release organic compounds, known as siderophores, that chelate dissolved Fe and enhance its assimilation from seawater (Wilhelm et al., 1996; Butler, 1998; Mendez et al., 2010).

It wasn't until the mid 1980s that analytical techniques improved sufficiently for biological oceanographers to measure the low Fe concentrations are in much of the world's oceans, finding many instances in which the relative supply of N:P:Fe was consistent with Fe limitation. In the central Pacific Ocean, Martin and Gordon (1988) found that internal sources of Fe could sustain only a small percentage of the observed NPP. They suggested that as much as 98% of the new production in this area is supported by Fe derived from dust deposited from the atmosphere (Table 9.2). Most of the dust deposited in this region is probably transported from the deserts of central China (Duce and Tindale, 1991; Uematsu et al., 2003). Growth of phytoplankton appears to be limited by iron, so small quantities of NO_3 and PO_4 remain in surface seawater even during periods of peak production. These waters became known as the HNLC—high nutrient, low chlorophyll—zones of the ocean.

These observations and early experimental evidence of Fe limitation of NPP in bottle assays led (Martin, 1990a) to propose "the iron hypothesis", which suggested that not only was Fe a critical limiting element for ocean NPP, but also variation in Fe inputs could explain much of the variation in atmospheric CO_2 associated with glacial interglacial cycles (Fig. 9.20).

During the last glacial epoch, arid environments were more widespread and more continental shelf area was exposed because of lower sea level. It is likely that there was greater wind erosion and atmospheric transport of Fe (Jickells et al., 2005; Wolff et al., 2006; Pollard et al., 2009). This added Fe deposition may have enhanced marine NPP (Kumar et al., 1995; Martínez-Garcia et al., 2011) and export production of carbon to the deep sea. Bubbles of gas trapped in Antarctic ice 18,000 years ago show lower levels of atmospheric CO_2 (Fig. 9.20). Export production is correlated to natural variations in dust deposition in the North Pacific (Bishop et al., 2002) and Southern Oceans (Thöle et al., 2019), whereas the upwelling of Fe-rich waters controls NPP in other areas (Coale et al., 1996a; Blain et al., 2007; Tagliabue et al., 2017).

Over a period of about 10 years, a number of iron fertilization experiments were conducted in the Pacific (e.g., the IronEX experiment) and the Southern Ocean (SOFeX) to test the "iron

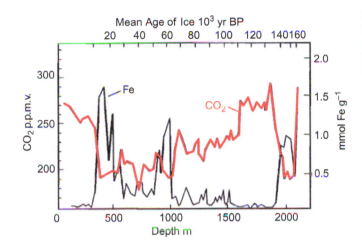

FIG. 9.20 In this figure from the original Iron Hypothesis paper, (Martin, 1990a) contrasted variation in CO_2 and Fe from the Vostok ice core over the last 160,000 years.

hypothesis" (Boyd et al., 2007). In every case, the rate of net primary production by phytoplankton increased significantly—sometimes by a factor of $10\times$—when Fe was added to the ocean surface (Martin et al., 1994; Coale et al., 1996b, 2004; Boyd et al., 2000), which lowered the concentration of CO_2 dissolved in the surface waters (Cooper et al., 1996; Watson et al., 2000). Export production was calculated from the C/Fe ratio in POC, an extension of the Redfield ratio. For Fe-enrichment experiments in the Southern Ocean, this ratio was 3000, but increases in the export of POC to the deep ocean were modest (Buesseler, 2004).

Fe in desert dust is normally found as Fe^{3+}, which is much less soluble in seawater than the organic complexes that normally dominate the pool of available Fe (Rue and Bruland, 1997). Most Fe in the oceans is found in particulate matter (Johnson et al., 1997), which sinks quickly (Croot et al., 2004). In the oceanographic experiments, Fe added to the surface waters rapidly disappeared from the upper water column, so the bloom of phytoplankton was short-lived (Boyd et al., 2007). In experiments of longer duration, the development of an active community of zooplankton might have regenerated Fe in surface waters through their grazing activities (Reinfelder and Fisher, 1991; Hutchins et al., 1993). Heterotrophic bacteria accumulate Fe (Tortell et al., 1996), and are sometimes fed on by phytoplankton (Maranger et al., 1998).

These experimental results have led some to suggest iron fertilization of the oceans as a "geoengineering" strategy to mitigate climate change (Buesseler and Boyd, 2003; Lampitt et al., 2008), but many caution that iron fertilization is probably not a cure for rising anthropogenic atmospheric CO_2 (Zeebe and Archer, 2005; Aumont and Bopp, 2006). In some iron-fertilization experiments, the flux of N_2O from the ocean surface increased significantly, potentially negating the uptake of CO_2 (Law and Ling, 2001). Marine biologists warn of the disruptive effects of iron fertilization on the marine biosphere (Chisholm et al., 2001), by leading to other nutrient deficiencies. Iron fertilization effects are likely to be short-lived, and the use of fossil fuels to mine, refine, and distribute Fe to the oceans may release more CO_2 than can be balanced by enhanced uptake in seawater.

Trace element colimitation: The combination of low inputs, high biological demand, and high rates of loss through particle sinking creates conditions for micronutrient limitation throughout much of the ocean. In some regions high nitrogen and phosphorus concentrations are found in surface waters but micronutrients constrain NPP, producing high nutrient-low-chlorophyll zones (HNLC). Numerous large-scale enrichment experiments have shown that NPP in the HNLC zones of the equatorial Pacific ocean is Fe limited (Kolber et al., 1994; Martin et al., 1994; Behrenfeld et al., 1996). Similarly, Middag et al. (2019) provide strong evidence that Zn is a primary limiting nutrient in Nordic and Antarctic origin waters. Zinc is an essential trace metal in the carbonic anhydrase enzyme that converts HCO_3^- to CO_2 for photosynthesis. Zn is also required in some phosphatase enzymes that are used to mobilize P from organic forms (Morel et al., 1994; Shaked et al., 2006). Recently, Kellogg et al. (2020) suggest that Co may substitute for Zn in some biochemical processes.

As a result, colimitation by several nutrients is very common in the world's oceans, with three types of colimitation commonly observed (Arrigo, 2005; Saito et al., 2008). In many regions of the ocean, increased productivity in response to nutrient amendment only occurs when two nutrients are added together (Fig. 9.19). Such nutrient colimitation (Type I) is very commonly observed for N and P, but other combinations are also known. In the South Atlantic gyre, phytoplankton production increased only when N and Fe were added together (Browning et al., 2017a). Biochemical substitution colimitation (Type II) is sometimes found when two or more elements can substitute in the same enzyme complex (e.g. Zn and Cd can substitute for one another in carbonic anhydrase). Finally, phytoplankton can be limited either by a macronutrient (N or P), or by the micronutrients required to build enzymes necessary for the acquisition of N and P (Type III). The addition of Fe can stimulate the production of nitrogenase enzymes for N fixation (Mills et al., 2004; Moore et al., 2009) and the addition of Zn or Fe can increase the production of phosphatase enzymes for P acquisition (Shaked et al., 2006; Browning et al., 2017). Under these conditions, phytoplankton productivity will increase following the addition of either the micronutrient necessary to synthesize acquisition enzymes or the macronutrient that those enzymes are designed to make more available.

Fate of marine NPP

There are three primary fates for marine NPP (Fig. 9.21). Algal biomass may be consumed by grazers and contribute to secondary production (as defined in Chapter 8) in the world's oceans. Algae may release organic compounds through exudation or cell lysis, fueling microbial food webs in the surface ocean. Algal cells or cell remains may sink below the thermocline where they will be consumed by heterotrophs. The mineralization of fixed organic matter in the surface ocean releases nutrients to fuel NPP, while incorporation of algal biomass into long lived organisms or loss below the thermocline removes nutrients from surface waters.

Consumption: Most marine NPP is consumed by zooplankton or bacterioplankton in surface waters. Bacterioplankton respire dissolved organic carbon and use extracellular enzymes to break down particulate organic carbon (POC) and colloids produced by phytoplankton (Druffel et al., 1992). Cho and Azam (1988) concluded that bacteria were more important than zooplankton in the consumption of POC in the North Pacific Ocean. Reviewing a large number of studies from marine and freshwater systems, Cole et al. (1988) found that net bacterial growth (production) is about twice that of zooplankton and accounted for the disappearance of 30% of NPP from the water column (cf. Ducklow and Carlson, 1992; del Giorgio and Cole, 1998). In some areas, gross consumption by bacteria may reach 70% of NPP, especially when NPP is low and water temperature is high (Biddanda et al., 1994). This rapid return of fixed C to respired C via bacteria has been called the "microbial loop" (Azam et al., 1983; Fenchel, 2008).

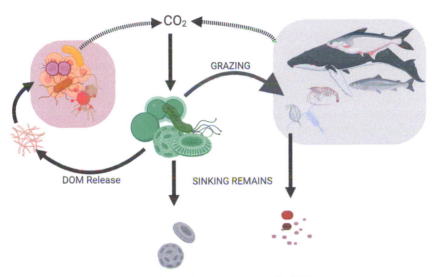

FIG. 9.21   Fates for algal productivity in the surface ocean. *Drawn by ESB.*

Whereas zooplankton represent the first step in a trophic chain that eventually leads to large animals such as fish, bacteria are consumed by a large population of bacteriovores that mineralize nutrients and release CO_2 to the surface waters (Fuhrman and McManus, 1984). Thus, when bacteria are abundant, a large fraction of the carbon fixed by NPP in the sea is not passed to higher trophic levels (Ducklow et al., 1986). In areas where bacterial growth is inhibited, such as in cold waters, more NPP is available to pass to higher trophic levels, including commercial fisheries (Pomeroy and Deibel, 1986; Laws et al., 2000; Rivkin and Legendre, 2001). Many of the world's most productive fisheries are found in cold, polar waters.

Fisheries production is directly linked to the primary production in the sea (Iverson, 1990; Ware and Thomson, 2005). Already humans extract a large harvest of fish and shellfish from the oceans, equivalent to the consumption of 8% of marine net primary productivity at the base of the food web (Pauly and Christensen, 1995). Twentieth century declines in the populations of many important commercial fishes make it clear that high catch rates are unsustainable for future generations without significant regulatory and conservation efforts (Worm et al., 2009; Galbraith et al., 2017). If warmer surface waters accompany climate change, we can expect that more of oceanic NPP will be consumed by microbes and ultimately less energy will support the commercial harvest of fish, meanwhile polar oceans may become more productive (Sumaila et al., 2011; Barange et al., 2014).

Exudation, lysis and leaching of DOC: A large amount of dissolved organic matter is found in seawater (Martin and Fitzwater, 1992). Based on its ^{14}C age and molecular properties, only a small fraction appears to derive from humic substances that have been delivered to the oceans by rivers (Opsahl and Benner, 1997, Raymond and Bauer, 2001b, Hansell et al., 2004). Most marine DOC is derived from marine photosynthesis. It is well known that phytoplankton and bacteria leak dissolved organic compounds, and globally about 17% of net community production may leak from phytoplankton cells as dissolved organic carbon (DOC;

(Hansell and Carlson, 1998). DOC is also released upon the death and lysis of phytoplankton cells (Agusti et al., 1998). Most of this DOC is labile and rapidly decomposed in the surface ocean, where it supports bacterioplankton—free-floating bacteria (Kirchman et al., 1991; Druffel et al., 1992). However, a fraction of the DOC is relatively refractory, apparently composed of nitrogen compounds resynthesized by bacteria (Barber, 1968; McCarthy et al., 1998; Ogawa et al., 2001; Aluwihare et al., 2005; Jiao et al., 2010). This DOC is entrained in downwelling waters and delivered to the deep sea (Carlson et al., 1994; Aluwihare et al., 1997; Loh et al., 2004). The mean residence time for DOC in the oceans, up to 6000 years, is longer than the time for deep-water renewal, implying that some DOC has made more than one cycle through the deep sea (Williams and Druffel, 1987; Bauer et al., 1992; Shen and Benner, 2018).

Production and dissolution of biogenic carbonates

As we described above, many marine organisms (e.g., bivalves, corals, brachiopods, coccolithophores) build calcium carbonate skeletons (Eq. 9.5), ultimately fixing more than 1.6 Pg of C into biogenic carbonates (Berelson et al., 2007; Hopkins and Balch, 2018). These organisms use enzymes to precipitate calcium carbonate minerals from seawater which are then cemented together with proteins to form shells, casings and tests. While corals, oysters and mussels may be the most well known and charismatic producers of biogenic carbonates, it is coccolithophores—plankton that produce an outer covering of calcite plates—that contribute to the largest geological sink for C (Monteiro et al., 2016). Marine organisms may construct their carbonate skeletons with one of two polymorphic forms of calcium carbonate, aragonite or calcite (Falini et al., 1996; Berner, 2004). Aragonite is less stable, with a structure that forms needlelike crystals, whereas calcite has a trigonal structure and forms crystalline blocks. Pteropods, the dominant zooplankton in much of the southern ocean, and corals, a critical foundational species in coastal oceans, each form aragonite skeletons, while coccolithophores form calcite plates. Many bivalves make use of both forms, lining a hard calcite outer shell with interior layers of aragonite.[s] Because aragonite is less resistant to mineralization and more sensitive to increasing ocean acidity than calcite, these differences in skeletal composition are critically important to predicting the response of foundational marine species to ocean acidification (Orr et al., 2005; Bednaršek et al., 2012).

Both the rate of biogenic carbonate formation (Eq. 9.5) and its rate of dissolution via:

$$CaCO_3 + H_2CO_3 \rightarrow Ca^{2+} + 2HCO_3^{-} \tag{9.7}$$

are affected by ocean acidity, leading to concerns that ocean acidification is reducing both the production and the preservation of biogenic carbonates in sediments (Hofmann and Schellnhuber, 2010; Sulpis et al., 2018) and making it more difficult for organisms with aragonite or calcite skeletal structures to compete with more acid-tolerant organisms for limiting resources. Already, large-scale observations show rising dissolved CO_2 and lower pH in the surface of Pacific Ocean, amounting to a drop of 0.06 in pH during the past 15 years

---

[s] The mother of pearl on the interior of oyster and mussel shells, and indeed pearls themselves, are constructed of aragonite.

(Takahashi et al., 2006; Dore et al., 2009; Byrne et al., 2010); Fig. 9.14). Losses of corals are likely to be the most immediate consequence (Kleypas, 1999; Hoegh-Guldberg et al., 2007; Kleypas and Yates, 2009), but higher seawater acidity could affect the ability of a wide variety of plankton to build carbonate skeletons (Riebesell et al., 2000; Orr et al., 2005). Differential responses among species can be expected (Iglesias-Rodriguez et al., 2008), disrupting the current food web of marine ecosystems. Of course, the drop in seawater pH is buffered by the dissolution of carbonates (Eq. 9.7), but unregulated CO_2 emissions could lead to a decline of seawater pH of 0.7 units in the next several centuries—a greater change than any observed during the past 300,000,000 years (Zeebe et al., 2008; Hönisch et al., 2012).

Deep ocean biogeochemistry

Theorganic carbon and skeletal remains that are lost from the surface ocean become critical fodder for the organisms living in the deep sea. Here we explore the fate of sinking particles and the accumulation of the products of decomposition that dominate the deep sea biogeochemistry. The average depth of the ocean is about 3.9 km, but the Challenger Deep, which is located at the southern end of the Mariana Trench in the western Pacific Ocean, extends to 11 km below the water surface. Very little light enters the "twilight zone" which begins about 200 m below the surface, and the ocean volume below 1000-m depths is completely devoid of light. In addition to the lack of light, the immense weight of overlying water generates exceptionally high water pressures that range from 40 to 110 times that of the overlying atmosphere. This pressure differential, and the evolutionary adaptations required for organisms to survive under such pressures reduces the movement of organisms between the deep and surface ocean.

Sinking organic matter, the soft tissue pump and marine snow

There is general agreement among oceanographers that about 80–90% of the NPP is degraded to inorganic compounds (CO_2, NO_3^-, PO_4^{3-}, etc.) within the surface ocean, while the remainder sinks to the deep ocean. These sinking materials consist of particulate organic carbon (POC), including dead phytoplankton, fecal pellets, and organic aggregates, known as marine snow (Alldredge and Gotschalk, 1990). The fraction of net primary production that sinks out of the surface waters is referred to as export production—often called fe or the f-ratio. At steady-state, we expect that the amount of sinking NPP must be equivalent to the fraction of NPP that is sustained by nutrients delivered from atmospheric inputs (including N-fixation), rivers and upwelling–the fraction is known as *new production*, abbreviated fn, with global estimates of 10–20% of total NPP (Table 9.3). This fraction varies spatially across the world's oceans, with the highest fn measurements in areas of cold, upwelling waters (Sathyendranath et al., 1991). Low, steady-state nutrient concentrations in the surface waters of the oceans indicate that the sources of nutrients that sustain new production globally are about equal to the annual losses of nutrients in organic debris that sink through the thermocline to the deep ocean—i.e., export production (f_e) (Eppley and Peterson, 1979; Li and Cassar, 2017). Higher rates of sinking organic particles, fe, would remove unreasonably large quantities of nutrients from the surface ocean and constrain NPP well below current estimates (Broecker, 1974; Eppley and Peterson, 1979). If the current, higher estimates of marine NPP are correct, then at least 7.4×10^{15} g C yr^{-1} (i.e., global $f_e = 0.15$) sinks to the deep waters of the ocean (Knauer, 1993; Falkowski, 2005). This export production includes both particulate organic carbon (5×10^{15} g C yr^{-1}; (Henson et al., 2011) and dissolved organic carbon (Hopkinson and Vallino, 2005).

Most of the particulate organic matter that enters the deep ocean is mineralized en route to the ocean floor. The downward flux of organic matter varies seasonally depending on productivity in the surface water (Deuser et al., 1981; Asper et al., 1992; Sayles et al., 1994; Legendre, 1998). With a mean sinking rate of about 350 m day^{-1}, the average particle spends about 10 days in transit to the bottom (Honjo et al., 1982). Bacterial respiration accounts for the consumption of O_2 and the production of CO_2 in the deep water. Fragmentation of sinking particles is important to the efficiency of decomposition and the ultimate delivery of organic carbon to the sediments (Briggs et al., 2020). Globally, Seiter et al. (2005) suggest that only 0.5×10^{15} g C yr^{-1} pass 1000-m depth. From a compilation of data from sediment cores taken throughout the oceans, Berner (1982) estimated that the rate of incorporation of organic carbon in sediments is 0.157×10^{15} g C yr^{-1}. These values suggest that about 98% of the sinking organic materials are degraded in the deep sea (cf. Martin et al., 1991).

Nutrients contained in sinking organic matter are regenerated in the deep ocean, where the concentrations are much higher than in surface waters. Recalling that the age of deep water in the Pacific Ocean is older than that in the Atlantic, we note that nutrient concentrations are higher in deep Pacific Ocean (Fig. 9.11), because its waters have had a longer time to receive sinking debris that are remineralized at depth. Similarly, in the Atlantic Ocean, nutrient concentrations increase progressively as North Atlantic Deep Water "ages" during its journey southward (Fig. 9.12) Nutrients are also remineralized from DOC that mixes into the deep sea (Hopkinson and Vallino, 2005). These nutrients are returned to the surface waters in the upwelling zones of the global thermohaline circulation (Fig. 9.4; Sarmiento et al., 2004).

Recognizing that the downward flux of biogenic particles carries $CaCO_3$ as well as organic carbon, Broecker (1974) recalculated Redfield's ratios to include $CaCO_3$. His modified Redfield ratio in sinking particles is 120C:15N:1P:40Ca. The ratio in upwelling waters is 800C:15N:1P:3200Ca. Based on these quantities, net production in the surface water could remove all the N and P but only 1.25% of the Ca in upwelling waters. Although biogenic $CaCO_3$ is the main sink for Ca in the ocean, biogenic carbonates represent only a tiny sink for Ca in surface waters relative to the total amount of dissolved Ca available. Thus, calcium is a well-mixed and conservative element in seawater (Table 9.1).

Carbonate dissolution in the deep ocean

Estimates of the global rate of biogenic carbonate export out of the photic zone range from 0.6 to 1.8×10^{15} g C yr^{-1} (Berelson et al., 2007; Hopkins and Balch, 2018). The likelihood of this material reaching to the ocean sediments depends upon the size of the particle and the depth of the water column it must travel through. The waters of the deep ocean are supersaturated with CO_2 with respect to the atmosphere as a result of their long isolation from the surface and the progressive accumulation of respiratory CO_2. Carbon dioxide is also more soluble at the low temperatures and high pressures that are found in deep ocean water.[t] The accumulation of CO_2 simultaneously makes the deep waters undersaturated with respect to $CaCO_3$, as a result of the formation of carbonic acid (Eq. 9.4). As the skeletal remains of carbonate-producing organisms sink through the deep ocean, they dissolve (Eq. 9.7). Their dissolution increases the alkalinity of the deep ocean through the production of HCO_3^-. Small particles may dissolve totally during transit to the bottom, while large particles may survive the journey, with subsequent dissolution occurring as part of sediment diagenesis.

The depth at which the dissolution of $CaCO_3$ begins in the water column is called the carbonate lysocline, which is an index of the carbonate saturation depth (CSD), the depth at which seawater is undersaturated with respect to $CaCO_3$. This depth is roughly 3000 m in the southern Pacific and 4000–4500 m in the Atlantic Ocean (Berger et al., 1976; Biscaye et al., 1976), but is rising as a result of increasing ocean CO_2 concentrations (Sulpis et al., 2018; Fig. 9.22). Slightly deeper, the carbonate compensation depth (CCD) is the depth where the downward flux of carbonate balances the rate of dissolution, so there are no carbonate sediments (Kennett, 1982). The tendency for a shallower saturation depth and CCD in the Pacific is the result of the older age of Pacific deep water, which allows a greater accumulation of

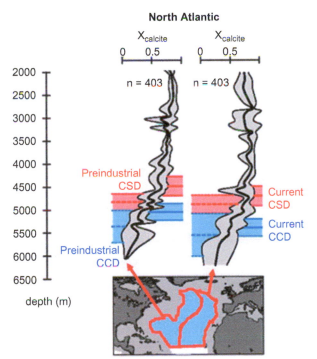

FIG. 9.22 Shifts in the depth of the calcite saturation depth (CSD) and the calcite compensation depth (CCD) in the North Atlantic between preindustrial and modern times are estimated from sediment calcite profiles and calcite marker horizons from Atlantic Ocean sediments. n represents the number of samples from each basin on which estimates are based. *Modified from Sulpis et al. (2018).*

[t]You can easily observe the importance of temperature and pressure in determining gas solubility by opening a warm bottle of soda and observing the effervescence of CO_2 from the resulting reduction in air pressure.

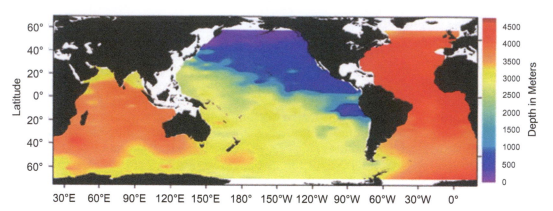

FIG. 9.23 Calcite saturation depth in the world's oceans. *Source: Feely et al. (2004). Used with permission of the American Association for the Advancement of Science.*

respiratory CO_2 (Li et al., 1969) (Fig. 9.23). Dissolution of sinking $CaCO_3$ means that calcareous sediments are found only in shallow ocean basins, and no carbonate sediments are found over much of the pelagic area where the ocean is greater than 4500 m deep. Of the estimated 10×10^{15} g yr^{-1} of biogenic $CaCO_3$ that is produced in surface waters, only about 0.8×10^{15} g $CaCO_3$ are preserved in deep-ocean sediments (Feely et al., 2004; Berelson et al., 2007). Added to the estimated carbonate preservation in shallow-water sediments (2.2×10^{15} g $CaCO_3$; (Milliman, 1993), this estimate of total carbonate deposition consumes more than the estimated flux of Ca to the oceans, suggesting that the Ca budget of the oceans is not currently in steady-state. Today, the preservation ratio of organic carbon to carbonate-carbon in ocean sediments is about 0.30 by weight, close to the ratio in the Earth's sedimentary inventory (Table 2.3).

Many studies of carbonate dissolution have employed sediment traps that are anchored at varying depths to capture sinking particles. In most areas, biogenic particles constitute most of the material caught in sediment traps, and most of the $CaCO_3$ is found in the form of calcite. The downward movement of aragonite has been long overlooked because it is more easily dissolved than calcite and often disappears from sediment traps that are deployed for long periods. The carbonate lysocline for aragonite is typically found at 500- to 1000-m depth (Milliman et al., 1999; Feely et al., 2004). As much as 12% of the movement of biogenic carbonate to the deep ocean may occur as aragonite (Berner and Honjo, 1981; Betzer et al., 1984).

Biogeochemistry in ocean sediments

As a result of decomposition and dissolution, very little of the organic carbon and biogenic carbonate produced in the surface ocean is deposited on the ocean floor. Degradation of organic carbon continues in marine sediments, and the ultimate rate of burial of organic carbon in the ocean is about 0.12×10^{15} g C yr^{-1} (Berner, 1982; Seiter et al., 2005)—less than 1% of marine NPP. The small fraction of organic matter that persists during transit to the deep sea is refractory, but susceptible to decomposition by heterotrophic microbes in the subseafloor.

FIG. 9.24   (A) The remains of a 35 ton gray whale carcass photographed in 2004. The carcass settled on the seafloor at 1674 m depth in 1998. (B) Giant tubeworms, *Riftia pachyptila*, from hydrothermal vents in the East Pacific at 2500 m depth. (C) Deep-sea corals, Paramuricea sp. form a reef at 1000 m depth in the Gulf of Mexico. *(A) Image from NOAA. (B) Image courtesy of Monika Bright, University of Vienna, Austria. (C) Image courtesy of Craig McClain, LUMCON.*

Despite the low inputs, ocean sediments support fascinating communities of organisms. Slow growing deep sea corals harbor high densities of organisms (Fig. 9.24). Large carcasses that fall to the seafloor can create ephemeral carbon-rich habitats that attract dense assemblages of scavengers. Many of the materials that are deposited on the sediment surface are mineralized or released into the overlying water column while the remainder are gradually incorporated into sedimentary rocks, accumulating large reservoirs of carbon and nutrients over geologic time. (Chapters 2 and 4). In this section we will discuss the fate of carbon and nutrients that are deposited onto the ocean floor.

The fate of organic matter on the seafloor

Significant rates of decomposition are found at the surface of ocean sediments (Emerson et al., 1985; Cole et al., 1987; Bender et al., 1989; Smith, 1992), where the rate of decay is determined by the length of exposure of the organic matter to oxygen (Fig. 9.25; Gélinas et al.,

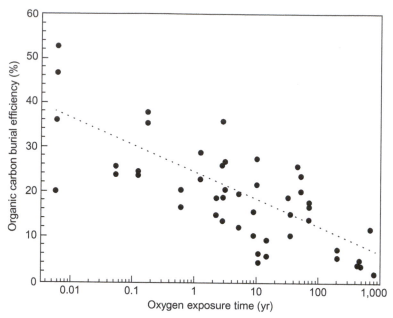

FIG. 9.25   Organic carbon burial efficiency versus the time of its exposure to O_2 in sediments of the eastern North Pacific Ocean. *From Hartnett et al. (1998).*

2001; Arnarson and Keil, 2007). Where burrowing organisms stir or *bioturbate* sediments, O_2 may penetrate to considerable depth (e.g., (Ziebis et al., 1996; Lohrer et al., 2004), stimulating degradation of buried organic matter (Hulthe et al., 1998; Middelburg, 2017, 2019). Only a small fraction of benthic carbon that deposits on the seafloor escapes mineralization. This fraction is calculated as the burial efficiency (BE):

$$BE = \frac{F_B}{F_C} = \frac{F_B}{(F_B + R)} \tag{9.8}$$

where F_C is the incoming carbon flux, F_B is the rate of carbon burial and R is the total mineralization rate. Burial efficiencies range from a fraction of a percent in deep-sea sediments up to tens of a percent in rapidly accumulating coastal sediments (Canfield, 1994; Aller, 2014). R is far greater than F_B in most sediments, so rates of respiration are often used as a reliable proxy for estimating the rate of organic carbon delivery to sediments (F_C).

Since heterotrophs are degrading organic matter continuously during its descent through the water column (a residence time of weeks to months), what lands on the ocean floor is both reduced in mass and in quality compared to what falls out of the photic zone. The long residence time for organic materials in surficial ocean sediments (a century or more) more than compensates for the reduction in quality, so decomposition consumes a large fraction of even the most recalcitrant organic materials. The amount of time that sedimentary organic matter is exposed to oxygen (its residence time in surface sediments) predicts much of the variation in burial efficiency (Fig. 9.25).

Across the vast majority of the seafloor, microbes dominate the biota. Bacteria are found at 500-m depth in pelagic sediments (Parkes et al., 1994; D'Hondt et al., 2004; Schippers et al., 2005). Viable bacteria collected from 1626 m below the seafloor near Newfoundland, Canada, where ambient temperatures range from 60°C to 100°C, currently represent the maximum known extent of the biosphere into the Earth's crust (Roussel et al., 2008). Given the age of some deep sediments (e.g., 111,000,000 years; Roussel et al., 2008), it is not surprising that all but the most refractory organic compounds have disappeared. Recent efforts to collect and analyze microbial DNA and the isotopic signatures of C and S reveal the presence of viable methane and sulfur cycling microbes in 3.5-million-year old subseafloor basalts (Lever et al., 2013). In these deep sediments, geogenic methane and hydrogen may fuel chemoautotrophic activity as well as decomposition of ancient organic matter. Genomic analyses of the "deep biosphere" document the presence of all three domains of life, with active bacteria, fungi and archaea, and suggest that sulfate reduction, methanogenesis and anaerobic methane oxidation are primary forms of metabolism in the deep biosphere (Orsi et al., 2013). Although their activity is slow (D'Hondt et al., 2002), the vast quantity of deep ocean sediments is estimated to contain 2.9×10^{29} microbial cells, an estimated 4.1×10^{15} g of C and 0.6% of Earth's total living biomass—equivalent to our estimates of soil microbial biomass (Kallmeyer et al., 2012). Note that this living biomass represents only a small fraction of the total organic carbon contained in sedimentary environments (Hartgers et al., 1994; Table 2.3).

Once organic matter is buried on the ocean floor, oxygen depletion below the shallowest surficial sediments reduces rates of decay. Within a few centimeters of the surface of sediments, NO_3^- and Mn^{4+} are also exhausted by the anaerobic oxidation of organic matter. Supplementing the supply of NO_3^- that diffuses into the sediment from the overlying waters, some motile microbes appear to transport NO_3^- into the sediment, where it can be denitrified (Prokopenko et al., 2011). Mat-forming bacteria in the genus *Thioploca* provide the spatial connection within their own cells, by capturing NO_3^- from the overlying water column and transferring it intracellularly to the sediments, where they oxidize sulfides produced by sulfate-reduction at depth (Fossing et al., 1995). Recently, it has been discovered that microbes can promote the extracellular transfer of electrons across local redox gradients in marine sediments by constructing electrically conductive pilli, or nanowires, or by facilitating the precipitation of extracellular metallic matrices (Meysman et al., 2015; Shi et al., 2016a).

Sulfate, as the most abundant alternate terminal electron acceptor (TEA) in seawater, becomes the dominant TEA below the zones of Mn-reduction and Fe-reduction (Fig. 9.26). Sulfate reduction is estimated to be responsible for oxidizing 12–29% of the organic C flux to the seafloor (Bowles et al., 2014). Assuming that 12% of the 0.2×10^{15} g organic matter that reaches the ocean floor[u] is remineralized through sulfate reduction, this should result in the conversion of an estimated 24×10^{12} g of SO_4^{2-} to sulfides (Bowles et al., 2014). Strong isotopic enrichment in the $\delta^{18}O_{SO4}$ relative to $\delta^{34}S_{SO4}$ in deep sea sediments suggests that as much as 99% of the sulfite produced within microbial cells through sulfate reduction is reoxidized to sulfate helping maintain microbial activity in areas where the supply of new terminal electron acceptors is very low (Antler et al., 2013; Findlay et al., 2020). In the deep ocean sulfate

[u]This estimate comes from (Friedlingstein et al., 2019).

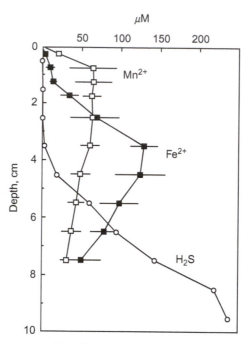

FIG. 9.26   Pore water distribution of Mn^{2+}, Fe^{2+}, and H_2S in coastal sediments of Denmark, showing the approximate depth of Mn-reduction, Fe-reduction, and SO_4-reduction, respectively. *From Thamdrup et al. (1994).*

penetrates deeply into sediments because of the low supply of reduced organic matter. In contrast, sulfates are rapidly depleted in ocean sediments with high organic inputs.

There is a critical transition in all marine sediments between the zone of sulfate reducing vs methanogenic sediments, referred to as the sulfate-methane transition zone, or SMTZ. Current estimates suggest that there may be as much as $10^8 km^3$ of ocean sediments below the SMTZ (defined as <0.1 mM sulfate) where microbes are supported entirely by fermentation and methanogenesis. In the Earth's deepest sediments, the majority of methane is produced from CO_2 reduction, because normally acetate is depleted before SO_4 is fully removed from the sediment (Sansone and Martens, 1981; Crill and Martens, 1986; Whiticar et al., 1986). However, acetate can also be supplied by autotrophic acetogenesis from the reduction of CO_2 (Heuer et al., 2009).

Globally, 9–14% of sedimentary organic carbon may be oxidized through anaerobic respiration, especially sulfate reduction (Lein, 1984; Henrichs and Reeburgh, 1987). The importance of sulfate reduction is much greater in organic-rich, near-shore sediments than in sediments of the open ocean (Skyring, 1987; Canfield, 1989a). Near-shore environments are characterized by high rates of NPP and a large delivery of organic particles to the sediment surface. Sulfate-reduction generally increases with the overall rate of sedimentation, which is also greatest near the continents (Canfield, 1989a, b, 1993). Anoxic conditions develop rapidly as organic matter is buried in these sediments. In a marine basin off the coast of North Carolina (USA), Martens and Val Klump (1984) found that $149 \, mol \, C \, m^{-2} \, yr^{-1}$ were deposited, of which 35.6 mol were respired annually. The respiratory pathways included 27% in aerobic respiration, 57% in sulfate reduction leading to CO_2, and 16% in methanogenesis. Near-shore environments promote the hydrogenation of sedimentary organic carbon, often by H_2S from sulfate-reducing bacteria (Hebting et al., 2006). These reduced organic residues are more resistant to decay and likely precursors to the formation of fossil petroleum (Gélinas et al., 2001).

In contrast, pelagic areas have lower NPP, lower downward flux of organic particles, and lower overall rates of sedimentation. In pelagic sediments of the Pacific Ocean, net carbon burial was 0.005 mol C m^{-2} yr^{-1} (D'Hondt et al., 2004). The sediments in these areas are generally oxic (Murray and Grundmanis, 1980; Murray and Kuivila, 1990), so aerobic respiration exceeds sulfate reduction by a large factor (Canfield, 1989a, b). Little organic matter remains to support sulfate reduction at depth (Berner, 1984).

In some sediments, methane produced in cold, high pressure conditions crystallizes with water to form methane hydrates or *clathrates*, which are unstable and volatilize CH_4 when brought to the surface of the Earth (Zhang et al., 2011a, b; Fig. 9.27). There is great interest in clathrates as a commercial source of natural gas as well as concern that a catastrophic degassing of clathrates in response to global warming might release vast quantities of methane to the atmosphere, exacerbating further warming (Archer et al., 2009). However, evidence for large releases of methane from clathrates during climate warming at the end of the last glacial epoch is equivocal (Kennett et al., 2000; Sowers, 2006; Petrenko et al., 2009).

When methane produced at depth diffuses upward through the sediment, it is subject to anaerobic oxidation by methanotrophs (AOM) that use SO_4^{-2}, Mn^{4+} and Fe^{3+} as alternative electron acceptors in the absence of O_2 (Reeburgh, 2007; Beal et al., 2009). Some anaerobic methanotrophs are archaea that appear to coexist in consortia with sulfate-reducing bacteria (Hinrichs et al., 1999; Boetius et al., 2000; Michaelis et al., 2002). When sediments are low in organic matter, the rate of sulfate-reduction may be determined solely by the upward flux of

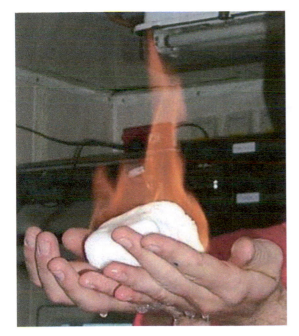

FIG. 9.27 Methane volatilized from a frozen clathrate can be burned at the Earth's surface. *Photo by Gary Klinkhammer, courtesy of NASA.*

methane that provides an organic substrate for metabolism (Hensen et al., 2003; Sivan et al., 2007). Methane-consuming archaea are known to fix nitrogen to support their growth in deep sediments, where nitrate has been depleted by denitrification (Dekas et al., 2009).

Methane released from ocean sediments, natural seeps and hydrothermal vents is easily oxidized by microbes before it reaches the surface (Iversen, 1996). A large amount of natural gas associated with the blow out of the Deep-Water Horizon oil well in the Gulf of Mexico was apparently oxidized before it reached the surface (Kessler et al., 2011).

Sediment diagenesis

The rate of burial of organic carbon depends strongly on the sedimentation rate (Fig. 9.28; Muller and Suess, 1979 ; Betts and Holland, 1991). Greater preservation of organic matter in near-shore environments is likely to be due to the greater NPP in these regions (Bertrand and Lallier-Vergès, 1993), rapid burial (Henrichs and Reeburgh, 1987; Canfield, 1991), and some-what less efficient decomposition under anoxic conditions (Canfield, 1994; Kristensen et al.,

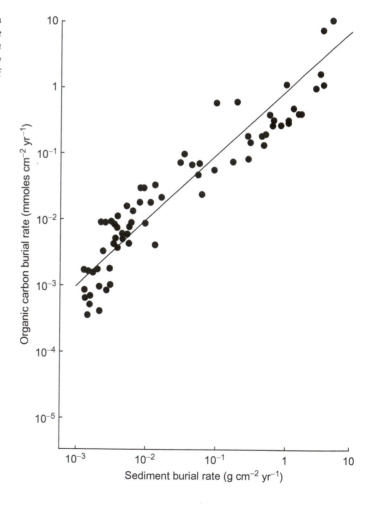

FIG. 9.28 Burial of organic carbon in marine sediments as a function of the overall rate of sedimentation. *From Berner and Canfield (1989). Reprinted by permission of* American Journal of Science.

1995). As seen in soils (Chapter 5), the long-term persistence of organic matter in marine sediments is also enhanced by association with mineral surfaces. Adsorption of organic matter to clays and iron minerals tends to retard its degradation and lead to preservation (Keil et al., 1994; Mayer, 1994; Kennedy et al., 2002a, 2002b; Lalonde et al., 2012; Blattmann et al., 2019).

Change in the chemical composition of sediments after deposition is known as *diagenesis*. Many forms of diagenesis are the result of microbial activities that proceed following the order of redox reactions outlined above and in Chapter 7 (Thomson et al., 1993; D'Hondt et al., 2004). Manganese, Mn^{3+}, accumulates in Mn oxides in the suboxic zone, where it can act as an electron donor or an electron acceptor, depending upon upward or downward shifts in redox potential (Anschutz et al., 2005; Trouwborst et al., 2006). Organic marine sediments undergo substantial diagenesis as a result of sulfate reduction (Froelich et al., 1979; Berner, 1984). In marine environments, sulfate reduction leads to the release of reduced sulfur compounds (e.g., H_2S) and to the deposition of pyrite in sediments (Eqs. 7.19, 7.20). The rate of pyrite formation is often limited by the amount of available iron (Boudreau and Westrich, 1984; Morse et al., 1992), so only a small fraction of the sulfide is retained as pyrite and the remainder escapes to the upper layers of sediment where it is reoxidized (Jorgensen, 1977; Thamdrup et al., 1994). Related to *Thioploca*, the bacteria *Beggiatoa* oxidize upward diffusing hydrogen sulfide at the sediment surface.

Among near-shore and pelagic habitats, there is a strong positive correlation between the content of organic carbon and pyrite sulfur in sediments (Fig. 9.29), but it is important to remember that the deposition of pyrite occurs at the expense of organic carbon (Eq. 9.9), viz.

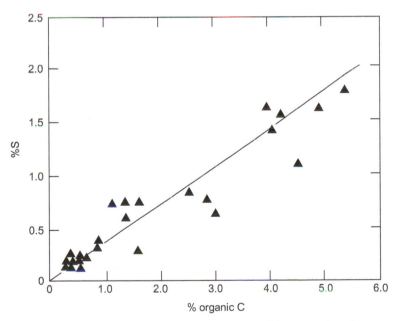

FIG. 9.29 Pyrite sulfur content in marine sediments as a function of their organic carbon content. *From Berner (1984).*

$$8FeO + 16CH_2O + 16SO_4^{2-} \rightarrow 16O_2 + 8FeS_2 \downarrow + 16HCO_3^- + 8H_2O. \quad (9.9)$$

Thus, the net ecosystem production of marine environments is represented by the *total* of sedimentary organic carbon + sedimentary pyrite—with the latter resulting from the transformation of organic carbon to reduced sulfur (Table 7.7).

Permanent burial of reduced compounds (organic carbon and pyrite) accounts for the accumulation of O_2 in Earth's atmosphere (Chapter 3). The molar ratio is 1.0 for organic carbon, but the burial of 1 mol of reduced sulfur accounts for nearly 2.0 mol of O_2 (Raiswell and Berner, 1986; Berner and Canfield, 1989). The weight ratio of C/S in most marine shales is about 2.8—equivalent to a molar ratio of 7.5 (Raiswell and Berner, 1986). Thus, through geologic time the deposition of reduced sulfur in pyrite may account for about 20% of the O_2 in the atmosphere. During periods of rapid continental uplift, erosion, and sedimentation, large amounts of organic substances and pyrite were buried and the oxygen content of the atmosphere increased (Des Marais et al., 1992). Rising atmospheric O_2 increases aerobic decomposition in marine sediments, consuming O_2 and limiting the further growth of O_2 in the atmosphere (Walker, 1980).

During sediment diagenesis, organic- and Fe-bound P are converted to phosphorites (authigenic apatite), which ultimately dominate the P storage in sediments (Ruttenberg, 1993; Filippelli and Delaney, 1996; Rasmussen, 1996). Phosphorite is formed when the PO_4^{3-} produced from the mineralization of organic P combines with Ca and F to form fluorapatite, in a mineral known as francolite (Ruttenberg and Berner, 1993; Krajewski et al., 1994; Anderson et al., 2001). Organic P disappears from sediments in parallel with an increase in francolite formed within them (Filippelli and Delaney, 1996; Delaney, 1998). In coastal California sediments, (Kim et al., 1999) found that francolite sequestered about 30% of the P mineralized from organic compounds or adsorbed to Fe-oxides. The F is supplied by inward diffusion from seawater (Froelich et al., 1983; Schuffert et al., 1994). In some areas of the ocean, phosphorite nodules up to several centimeters in diameter accumulate on the sea floor. These nodules are an enigma; they remain on the surface of the sediment despite growing at rates slower than the rate of sediment accumulation (Burnett et al., 1982).

Geochemists have long puzzled that dolomite—$(Ca, Mg)CO_3$—does not appear to be deposited abundantly in the modern oceans, despite the large concentration of Mg in seawater and the occurrence of massive dolomites in the geologic record. There are few organisms that precipitate Mg calcites in their skeletal carbonates, but thermodynamic considerations would predict that calcite should be converted to dolomite in marine sediments (e.g., Malone et al., 1994). Baker and Kastner (1981) show that the formation of dolomite is inhibited by SO_4^{2-}, but dolomite can form in organic-rich marine sediments in which HCO_3^- is enriched and SO_4^{2-} is depleted by sulfate reduction (Eq. 9.9; Baker and Burns, 1985). Dolomite is precipitated in laboratory cultures of the sulfate-reducing bacterium *Desulfovibrio* (Vasconcelos et al., 1995). Thus, the precipitation of dolomite is directly linked to biogeochemical processes in marine sediments. Although dolomite has been a significant sink for marine Mg in the geologic past, its contribution to the removal of Mg from modern seawater is likely to be minor.

Recent budgets estimate that 9.55×10^{12} mol of dissolved Si enters the ocean each year, with 64% delivered from rivers, and 25% delivered through the dissolution of eolian and riverborne sediments (Frings et al., 2016). The majority of the Si supplied by the weathering of the continents is taken up by diatoms to form biogenic opal, and much of this is recycled in the water column (Tréguer and De La Rocha, 2013). When diatoms and other siliceous organisms sink to the ocean floor, they are initially deposited as biogenic opal. After burial, biogenic opal may undergo dissolution, partitioning into metal oxides, or diagenetic alteration to form a range of alternate states including silicate clays via reverse weathering (Aller, 2014). Biogenic opal and authigenic silicate clays are particularly important in coastal areas where high Si loading from rivers fuels diatom production. Globally, 4.5–4.9 Tg yr^{-1} Si are deposited in nearshore siliceous sediments (Rahman et al., 2017).

Hydrothermal vent communities

In 1977, the first hydrothermal vent communities were discovered at 2500 m depth in the east Pacific Ocean. Hydrothermal vents are formed where cracks in the ocean crust allows seawater to percolate into the subsurface. Cold seawater in contact with hot rocks causes reactions that create plumes of chemically enriched, highly reduced hot fluids. Hot water is less dense than the cold waters of the deep ocean, so these buoyant hydrothermal fluids rise from the sea floor and rapidly mix with cold seawater (Fig. 9.10). These plumes create strong redox interfaces where reduced materials derived from the mantle are exposed to oxygen rich ocean waters, supporting a suite of chemosynthetic microbes that rely on the oxidation of molecular hydrogen, sulfide, methane and reduced iron emitted from the vents (Dick, 2019). These microbes in turn fuel a complex food web of tube worms, mollusks, and a wide variety of specialized microbes, many of which are recognized as new species (Corliss et al., 1979; Grassle, 1985; Levesque et al., 2005); Fig. 9.24B). While chemosynthesis at hydrothermal vents may have been operating since the earliest origins of life (Chapter 2), the amount of energy derived in an oxygenated ocean is far higher than what was available in the primitive ocean. In the modern ocean, sulfur bacteria at hydrothermal vents are able to use the energy gained from oxidizing hydrogen sulfide (H_2S) to produce carbohydrate by the reaction (Jannasch and Wirsen, 1979; Jannasch and Mottl, 1985):

$$O_2 + 4H_2S + CO_2 \rightarrow CH_2O + 4S \downarrow + 3H_2O. \tag{9.10}$$

Note that this reaction is ultimately dependent upon the supply of O_2, which links chemosynthesis in the deep sea to photosynthesis in the surface ocean. Other bacteria at hydrothermal vents employ chemosynthetic reactions based on molecular hydrogen and methane that are emitted in conjunction with H_2S (Jannasch and Mottl, 1985; Petersen et al., 2011; Dick, 2019).

Remarkable mutualisms between chemosynthetic microbes and animals are responsible for much of the physical structure that characterizes hydrothermal vents in the modern ocean. Symbiotic bacteria in the tube worm *Riftia* deposit elemental sulfur, leading to the rapid growth of tubular columns of sulfur up to 1.5 m long (Cavanaugh et al., 1981; Lutz et al., 1994). Filter-feeding clams up to 30 cm in diameter occur in dense mats near the vents. These communities are dynamic, and a particular vent may be active for only about 10 years. Because they are below the carbonate compensation depth, the clam shells slowly dissolve when the vent activity ceases (Grassle, 1985). The offspring of these organisms must continually disperse to colonize new vent systems.

Various metallic elements and silicon are soluble in the hot, low-redox conditions of hydrothermal vents. Upon mixing with seawater, the precipitation of metallic sulfides may remove as much as $100 \times 10^{12}\,\mathrm{g\,S\,yr^{-1}}$ from the ocean (Edmond et al., 1979; Jannasch, 1989), although a lower value ($27 \times 10^{12}\,\mathrm{g\,S\,yr^{-1}}$) is used in the sulfur cycle of Fig. 9.36 (Elderfield and Schultz, 1996). A larger proportion of Mn and Fe are also deposited as insoluble oxides (MnO_2, FeO) and nodules on the sea-floor, but new research suggests that a significant fraction of the Fe emitted from hydrothermal events binds with coemitted ligands and contributes to regional dissolved Fe inputs to the ocean (Tagliabue et al., 2010). Iron and manganese oxides act to scavenge vanadium (V) and other elements from seawater and may remove 25% of the annual riverine input of V to the oceans each year (Trefry and Metz, 1989; Schlesinger et al., 2017).

The sedimentary record of biogeochemistry

Marine sediments contain a record of the conditions of the oceans through geologic time (Kastner, 1999). Sediments and sedimentary rocks rich in $CaCO_3$ (calcareous ooze) show the past location of shallow, productive seas, where foraminifera and coccolithopores were abundant. Sediments deposited in the deep sea are dominated by silicate clay minerals, with high concentrations of Fe and Mn (red clays). Opal indicates the past environment of diatoms, whereas sediments with abundant organic carbon are associated with near-shore areas, where burial of organic materials is rapid. Changes in the species composition of preserved organisms have also been used to infer patterns of ocean climate, circulation, and productivity during the geologic past (Weyl, 1978; Corliss et al., 1986). For instance, the ratio of germanium (Ge) to silicon (Si) in diatomaceous sediments has been used to infer variations in the past rates of continental weathering (Froelich et al., 1992).

Calcareous sediments contain a record of paleotemperature. When the continental ice caps grew during glacial periods, the water they contained was depleted in $H_2^{18}O$, relative to ocean water, because $H_2^{16}O$ evaporates more readily from seawater and subsequently contributes more to continental rainfall and snowfall. When large quantities of water were lost from the ocean and stored in ice, the waters that remained in the ocean were enriched in $H_2^{18}O$ compared to today. Carbonates precipitate in an equilibrium reaction with seawater (Eqs. 9.4, 9.5), so an analysis of changes in the ^{18}O content of sedimentary carbonates is an indication of past changes in ocean volume and temperature (Fig. 9.30).

The history of the Sr content of seawater is also of particular interest to geochemists, because its isotopic ratio changes as a result of changes in the rate of rock weathering on land (Dia et al., 1992). Most strontium is ultimately removed from the oceans by coprecipitation with $CaCO_3$ (Kinsman, 1969; Pingitore and Eastman, 1986). During periods of extensive weathering, the ^{87}Sr content of seawater increases as a result of high content of that isotope in continental rocks. Thus, changes in the ^{87}Sr content of marine carbonate rocks offer an index of the relative rate of rock weathering over long periods (Richter et al., 1992). For calcium, changes in the rate of rock weathering versus the rate of carbonate sedimentation are reflected in the $\delta^{44}Ca$ ratio of calcium in carbonate sediments (De La Rocha and DePaolo, 2000; Griffith et al., 2008). High rates of weathering are implicated for the Miocene, when atmospheric CO_2 levels were higher than today.

Carbonates are a small sink (20%) of boron in the oceans (Park and Schlesinger, 2002), and the isotopic ratio of boron in carbonate varies as a function of seawater pH. The ratio

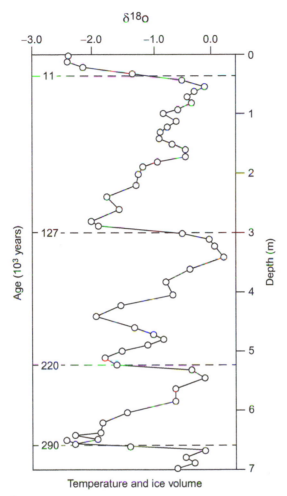

FIG. 9.30 Changes in the $\delta^{18}O$ in sedimentary carbonates of the Caribbean Sea during 300,000 years. Enrichment of $\delta^{18}O$ during the last glacial epoch (20,000 years ago) is associated with lower sea levels and a greater proportion of in seawater. *From Broecker (1973).*

measured in sedimentary foraminifera of the Miocene (21 million years ago) indicates that seawater pH was lower (7.4) than that of today (8.2), consistent with suggestions of higher atmospheric CO_2 during that period (Spivack et al., 1993; Pearson and Palmer, 2000). Similarly, the boron isotope ratios of sedimentary carbonate indicate a higher seawater pH during the last glacial, when atmospheric CO_2 was low (Sanyal et al., 1995). As in all studies of sediments, the time resolution of the method is constrained by the mean residence time of the element in seawater, which for boron is >1,000,000 years (Park and Schlesinger, 2002).

Sedimentary deposits of ^{13}C in organic matter and in $CaCO_3$ contain a record of the biotic productivity of Earth. Recall that photosynthesis discriminates against $^{13}CO_2$ relative to

$^{12}CO_2$, slightly enriching plant materials in ^{12}C compared to the atmosphere (Chapter 5). When large amounts of organic matter are stored on land and in ocean sediments, $^{13}CO_2$ accumulates in the atmosphere and the ocean (i.e., $^{13}HCO_3$). (Arthur et al., 1988) suggest that the relatively high ^{13}C content of marine carbonates during the late Cretaceous reflects a greater storage of organic carbon from photosynthesis. Similar changes are seen in the ^{13}C of coal age (Permian) brachiopods (Brand, 1989). When the storage of organic carbon is greater, there is the potential for an increase in atmospheric O_2, as postulated for the Permian (Berner and Canfield, 1989).

Marine element cycles

The marine C cycle

The largest carbon reservoir at the Earth's surface is the dissolved inorganic carbon in the deep ocean. This pool is constantly replenished with the delivery of DIC in areas of deep-water formation and through the mineralization of DOC and particulate OM that sink out of the surface ocean (Fig. 9.31). Deep oceans return CO_2 to the surface ocean in areas of upwelling. Over the last decade the ocean solubility pump is currently estimated to remove 2.5 ± 0.5 Pg C yr^{-1} of the annual fossil fuel emissions of atmospheric CO_2 (Friedlingstein

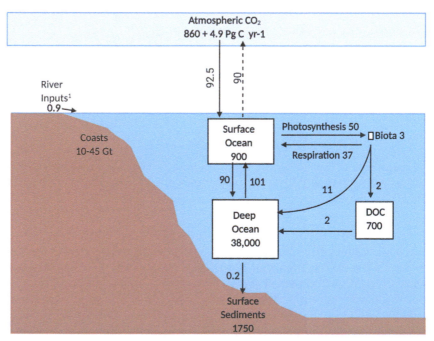

FIG. 9.31   The ocean C cycle. Pools are in Pg C and Fluxes are in Pg C yr^{-1}. Estimates in this diagram are derived from the 2013 IPCC report (Ciais et al., 2013) updated with more recent estimates from the 2019 Global Carbon Budget (Friedlingstein et al., 2019).

et al., 2019), a significant increase from an estimated annual sink of 1.0 ± 0.5 Pg C yr^{-1} in the 1960s (Le Quéré et al., 2018). There is considerable interannual variation in the net flux of CO_2 between the ocean and atmosphere, with the highest annual ocean uptake during large El Niño events (e.g. 1997–98) (Rödenbeck et al., 2018).

The current consensus estimate for the total inventory of anthropogenic CO_2 emissions that are now stored in the ocean is 155 ± 30 Pg C (Khatiwala et al., 2012), or roughly a quarter of the total CO_2 produced by human activities since the preindustrial era (Ciais et al., 2013). Ocean NPP is a large flux, with 50 Pg C annually across the global ocean, but there is no evidence that the rates of ocean NPP are changing in response to rising atmospheric CO_2. Most global models currently do not include the effects of anthropogenic changes in nutrient supply. Anthropogenic N deposition to the world's oceans could have stimulated increases in the ocean soft tissue sink for C by 0.15 to 0.3 Pg yr^{-1} over the industrial period (Duce et al., 2008; Jickells et al., 2017).

Carbon fixation rates and carbon stocks are highest in estuarine, coastal, and continental shelf waters (Chapter 8). These areas occupy only about 8% of the ocean's surface, but they account for about 18% of ocean productivity (Table 9.3), and 83% of the carbon that is buried in sediments. Globally-averaged models (e.g., Figs. 9.31) mask the comparative importance of these regions to the overall biogeochemical cycles of the sea. For example, a significant amount of organic carbon may be transported from the continental shelf to the deep sea (Walsh, 1991; Wollast, 1993).

The marine N cycle

Inputs: Biological N fixation is the dominant input of N to the oceans. Current estimates suggest N-fixation may account for about 164×10^{12} g N yr^{-1} added to the sea—roughly $10 \times$ higher than estimates only a couple of decades ago (Großkopf et al., 2012; Jickells et al., 2017; Tang et al., 2019); Fig. 9.32), although there remains considerable uncertainty (Großkopf et al., 2012; Jickells et al., 2017; Tang et al., 2019). As in terrestrial plants, nitrogen fixation in the phytoplankton is associated with δ^{15}N of $\sim$0‰ in plankton biomass and in organic debris in sediments (Karl et al., 2002). This isotopic signature can be used to estimate nitrogen fixation in the water column (Mahaffey et al., 2003). In many areas primary productivity in the sea appears limited by available N, but the amount of N is largely determined by the availability of Fe for the synthesis and activity of nitrogenase (Falkowski, 1997; Wu et al., 2000; Moore and Doney, 2007; Moore et al., 2009). Through glacial-interglacial cycles, the rate of nitrogen fixation in the oceans appears to relate to the distribution of desert dust in the atmosphere; high marine NPP during the last glacial may have derived from a greater deposition of Fe in the oceans from desert dust (Martin, 1990b; Jickells et al., 2005). The deposition of desert dust in the oceans links marine NPP to soil biogeochemistry of distant terrestrial ecosystems.

Riverine inputs of reactive N have increased throughout the industrial era, from a background of $\sim$27 $\times$ 10^{12} g N yr^{-1} to modern inputs of $\sim$43 $\times$ 10^{12} N yr^{-1} (Galloway et al., 2004; Seitzinger et al., 2010) (Fig. 9.32). Dissolved organic forms of nitrogen likely dominated this flux in the pre-industrial era, but inorganic forms are now estimated to account for $\sim$75% of river N fluxes to the modern ocean (Seitzinger et al., 2010; Tian et al., 2020). Current models

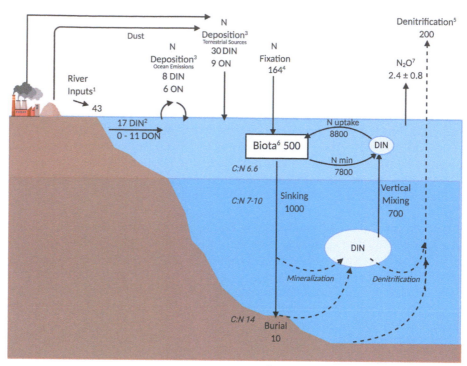

FIG. 9.32 Nitrogen budget for the world's oceans, showing major fluxes in units of 10^{12} g N yr^{-1}. (1) Riverine inputs of Dissoved inorganic N (DIN) and dissolved organic N (DON) are from Seitzinger et al. (2005, 2010). (2) DIN transport from continental shelf to open ocean from Sharples et al. (2017). (3) Rates of atmospheric deposition are derived from the TM4-ECPL model (Tsigaridis et al., 2014; Daskalakis et al., 2015). (4 and 5) Rates of biological N fixation and denitrification are estimated from the PlankTOM model reported in Jickells et al. (2017). The modeled N fixation rate is consistent with measurement based estimates of 177 Tg N yr^{-1} from Großkopf and Laroche (2012). (6) Marine biota N pool from Galloway et al. (2004). (7) Nitrous oxide flux is derived from Buitenhuis et al. (2018) and Ji et al. (2018).

estimate that roughly 25% of these riverine N inputs are consumed on continental shelves, primarily via denitrification in the organic-rich sediments formed by coastal phytoplankton (Seitzinger, 1988; Seitzinger et al., 2005; Duce et al., 2008) with the remaining 75% of riverine N delivered to the open ocean (Sharples et al., 2017). As discussed above, humans have also increased the transport of reactive nitrogen from land to sea, largely as air pollutants (Duce et al., 2008; Jickells et al., 2017). In some coastal regions downwind of pollution sources, an increased deposition of reactive nitrogen compounds from air pollution may cause higher marine NPP (Fanning, 1989; Paerl, 1995; Kim et al., 2011). In combination, elevated riverine and atmospheric N loading have resulted in large areas of the ocean having elevated nitrogen concentrations and N:P ratios (Fig. 9.33) (Deutsch and Weber, 2012).

Gaseous Losses of Nitrogen from the Sea: Denitrification is an important loss term for nitrogen, and may occur in zones of low O_2 concentration. Direct measurements in the eastern Pacific Ocean, suggested that denitrification could result in the loss of $50–60 \times 10^{12}$ g N from the ocean water column each year (Codispoti and Christensen, 1985; Deutsch et al., 2001). Other regions

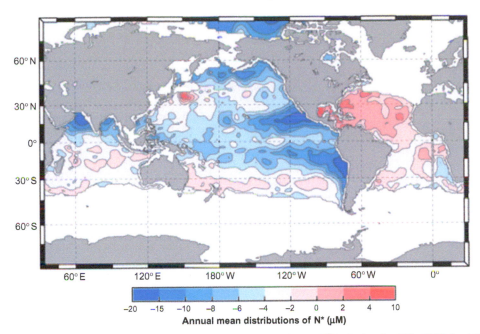

FIG. 9.33   An interpolated map of excess nitrate relative to phosphorus, N^* (calculated as $N^* = [NO_3^-] - 16[PO_4^{3-}]$).
From Deutsch and Weber (2012).

with suboxic conditions in the water column also provide local areas for marine denitrification. Often these are regions of high NPP, with oxygen depletion in the mid-water column due to the decomposition of sinking organic debris. The anoxic microzones created by flocculations of organic matter allow significant rates of denitrification in the oceans, despite the high redox potential of seawater (Alldredge and Cohen, 1987). Rates of water column denitrification are likely increasing in concert with enhanced N loading, as increased marine productivity leads to expansion of low oxygen zones that promote denitrification, providing a negative feedback mechanism for maintaining low N availability in the surface ocean (Landolfi et al., 2015; Somes et al., 2016; Yang et al., 2017).

As we saw in terrestrial ecosystems (Chapter 6), $^{14}NO_3$ is used preferentially (over $^{15}NO_3$) as a substrate in the production of N_2 during denitrification. Denitrification results in a high content of ^{15}N in the residual nitrate in seawater (Liu and Kaplan, 1989; Sigman et al., 2000; Voss et al., 2001). Denitrification is estimated from measures of $\delta^{15}N$ in the residual nitrate pool, as well as excess concentrations of dissolved N_2 gas in seawater (Chang et al., 2010). The pool of nitrate left in the ocean, which is taken up by phytoplankton, leaves an enriched ^{15}N signature in organic sediments. Recall we used this signature to ascertain the origin of denitrification in the geologic history of the Earth (Chapter 2).

Substantial denitrification is observed in ocean sediments, where it is performed by bacteria and a few specialized benthic eukaryotes (Brandes and Devol, 2002; Piña-Ochoa and Álvarez-Cobelas, 2006; Wang et al., 2019b). Christensen et al. (1987) estimated that over 50×10^{12} g N yr^{-1} may be lost from the sea by sedimentary denitrification in coastal regions.

Devol (1991) found that nitrification occurring within the sediments supplied most of the nitrate for denitrification on the continental shelf of the western United States. Denitrification in sediments leaves the NO_3^- pool in the pore space enriched in ^{15}N (Lehmann et al., 2007). Including sediments, the overall rate of denitrification in the oceans is estimated between 150 and $300 \times 10^{12}\, g\, N\, yr^{-1}$ (Codispoti et al., 2001; Brandes and Devol, 2002; DeVries et al., 2012; Jickells et al., 2017; Wang et al., 2018b; Paerl, 2018), which is somewhat in excess of the current estimate of N inputs to the oceans. Changes in the balance of nitrogen fixation and denitrification have controlled the nitrogen content of the oceans through geologic time (Ganeshram et al., 1995; Ren et al., 2009).

Some additional N_2 is produced by annamox, the anaerobic oxidation of ammonium, using nitrite as an alternative electron acceptor in place of oxygen (Mulder et al., 1995; Strous et al., 1999; Schmidt et al., 2002; Kuypers et al., 2003). While significant in some high productivity waters (Dalsgaard et al., 2003; Kuypers et al., 2005), the flux of N_2 from anammox is lower than that from denitrification in suboxic waters of the Arabian Sea where both have been measured together (Ward et al., 2009; Bulow et al., 2010). In coastal sediments of the eastern U.S. and Sweden, (Engström et al., 2005) found that 7 to 79% of the N_2 flux from sediments was due to anammox.

Most of the gaseous nitrogen lost from marine environments by denitrification is N_2; however, seawater is supersaturated with N_2O in many regions (e.g., Walter et al., 2004b), and the oceans supply about 30% of the natural source of N_2O to the atmosphere annually (Voss et al., 2013) (Table 12.5). This N_2O may is derived from two sources, the oxidation of ammonia by archaea and other nitrifying species in the water column, as well as by denitrifiers in low O_2 areas of the oceans (Koeve and Kähler, 2010; Santoro et al., 2011; Bianchi et al., 2012; Zamora and Oschlies, 2014).

Finally, a small amount of ammonia is lost from the surface of the sea, where NH_4^+ is deprotonated to form gaseous NH_3 in the slightly alkaline conditions of seawater (Eq. 6.4; Quinn et al., 1988; Jickells et al., 2003). This source of ammonia contributes about 5% to the annual global flux of NH_3 to the atmosphere (Chapter 12).

A global model for the N cycle of the oceans (Fig. 9.32) offers a deceptive level of tidiness to our understanding of marine biogeochemistry, and the reader should realize that many fluxes, for example, nitrogen fixation, denitrification, and sedimentary preservation, are not known to better than a factor of two. Nevertheless, the model shows that most NPP is supported by nutrient recycling in the surface waters, and only small quantities of nutrients are lost to the deep ocean. Assuming that the total N pool in marine biota is $\sim 500 \times 10^{12}\, g\, N$ (Galloway et al., 2004), the mean residence time of the available nitrogen (inorganic and organic) in the surface ocean is ~ 125 days, whereas the mean residence time of organic N is about 20 days. Thus, each atom of N cycles through the biota many times. In the absence of upwelling, the biotic pump would remove the pool of nutrients in the surface water in less than a year. Upon sinking and mineralization in the deep ocean, N enters pools with a mean residence time of about 500 years—largely controlled by the circulation of water through the deep ocean. The nitrogen cycle is dynamic; the overall mean residence time for N in the oceans is estimated to be ~ 2000 years, so the nitrogen cycle is responsive to global changes over relatively short periods (Brandes and Devol, 2002).

The marine phosphorus cycle

Phosphorus is nearly undetectable in most surface seawaters. Upwelling waters deliver N and P in amounts close to the Redfield ratio of 16, so it is reasonable to surmise that N and P would be depleted in tandem by the growth of phytoplankton. As seen in Chapter 8, only a small portion of the total phosphorus transport in rivers (21×10^{12} g P yr^{-1}) is carried in dissolved forms; the remainder is adsorbed to Fe and Al oxide minerals that are carried as suspended particles. Some of the adsorbed P is released upon the mixing of freshwater and seawater (Chase and Sayles, 1980; Caraco et al., 1990), but most is probably buried with the deposition of terrigenous sediments on the continental shelf (Filippelli, 1997). The total flux of "bioreactive" P to the oceans is about 2.0×10^{12} g yr^{-1} (Ramirez and Rose, 1992; Delaney, 1998), giving an atom ratio of ~55 for N/P (reactive) in global riverflow (Fig. 9.34).

Deposition of P on the ocean surface from the dust of deserts may play a special role in stimulating new production in areas of the open ocean that are distant from rivers and zones of upwelling (Wu et al., 2000; Mills et al., 2004). However, as seen for N, recycling of P in the surface waters accounts for the vast majority of the P uptake by phytoplankton (Fig. 9.34). Most of the phosphorus pool in plankton is remineralized within a few days (Benitez-Nelson

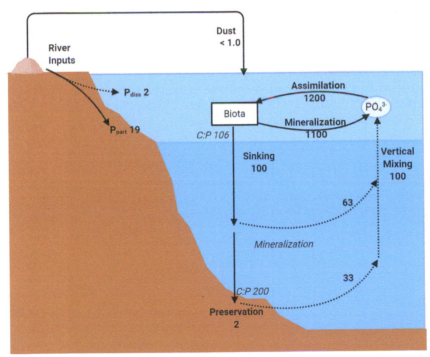

FIG. 9.34 A phosphorus budget for the world's oceans, with important fluxes shown in units of 10^{12} g P yr^{-1}. From an original conception by Wollast (1981), but with newer data added for dust inputs (Graham and Duce, 1979), riverflow (Meybeck, 1982), sedimentary preservation (Wallmann, 2010), and nutrient regeneration in surface waters (see Table 9.4). The global values have been rounded.

and Buesseler, 1999). Phosphorus appears to be mineralized selectively from the phosphonate component of DOC (Clark et al., 1999). Each year a small amount of organic debris, with C/P ratios somewhat greater than the Redfield ratio, sinks through the thermocline to the deep ocean (Honjo et al., 1982). An average of 500 years later, mineralized P (i.e., HPO_4^{2-}) returns to the surface waters in upwelling.

The C/P ratio in organic matter that is buried in marine sediments is about 200 (Mach et al., 1987; Ingall and Van Cappellen, 1990; Ramirez and Rose, 1992), suggesting that P is mineralized more rapidly than C during the downward transport and sedimentary diagenesis of organic matter in the sea (Froelich et al., 1979; Honjo et al., 1982; Loh and Bauer, 2000). Phosphorus release and C/P ratios are greatest in anoxic sediments (Ingall et al., 1993; Ingall and Jahnke, 1997; Fig. 9.35). Anoxic environments have lower concentrations of oxidized Fe minerals that can adsorb P as it is mineralized from organic matter (Krom and Berner, 1981; Sundby et al., 1992; Berner and Rao, 1994; Blomqvist et al., 2004). Both dissolved and inorganic P are adsorbed to iron oxide minerals, especially less crystalline forms (Ruttenberg and Sulak, 2011; cf. Chapter 4). (See Fig. 9.35.)

In contrast to N, there are no significant gaseous losses of P from the sea. At steady state, the inputs to the sea in river water must be balanced by the burial of phosphorus in ocean sediments. Most of the P carried in suspended sediments is probably deposited near the coast of continents, where P burial parallels overall rates of sedimentation (Filippelli, 1997). Burial of biogenic P compounds in sediments of the open ocean is estimated between about 2.0×10^{12} g P yr^{-1}—similar to the delivery of bioreactive P in rivers (Howarth et al., 1995a; Delaney, 1998; Wallmann, 2010). Burial occurs with the deposition of organic matter or $CaCO_3$ (Froelich et al., 1982), with at least a portion of the burial being in biogenic polyphosphates (Diaz et al., 2008).

The mean residence time for reactive P in the oceans, relative to the input in rivers or the loss to sediments, is >25,000 years (Ruttenberg, 1993; Filippelli and Delaney, 1996; Delaney, 1998). Thus, each atom of P that enters the sea may complete 50 cycles between the surface and

FIG. 9.35 Flux of phosphorus from sediments to the water column as a function of the decomposition of organic carbon in areas of high and low O_2 in the overlying waters. *From Ingall and Jahnke (1997).*

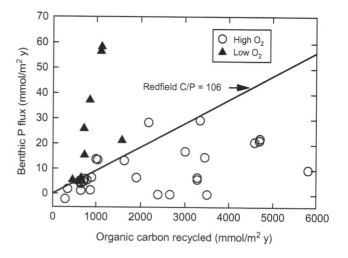

the deep ocean before it is lost to sediments. The major sinks include the formation of authigenic phosphorites (francolite) and uptake at hydrothermal vents (Elderfield and Schultz, 1996; Wheat et al., 2003). All forms of buried phosphorus complete a global biogeochemical cycle when geologic processes lift sedimentary rocks above sea level and weathering begins again. Thus, relative to N, the global cycle of P turns very slowly (Chapter 12).

The marine sulfur cycle

Sulfur is the second most abundant anion in the oceans, where it is found overwhelmingly as SO_4^{2-} (Table 9.1). Most of the atmospheric deposition is derived from seasalt aerosols that are released from and then quickly redeposited on the ocean's surface (i.e., cyclic salt). Rivers and atmospheric deposition derived from land are the major new sources of SO_4 in the sea (Fig. 9.36). Metallic sulfides precipitated at hydrothermal vents and biogenic pyrite forming in sediments are the major marine sinks. Although S is a necessary nutrient incorporated into proteins by assimilatory reduction of SO_4 from seawater by marine phytoplankton and bacteria (Giordano et al., 2005), marine sulfate concentrations are very high relative to

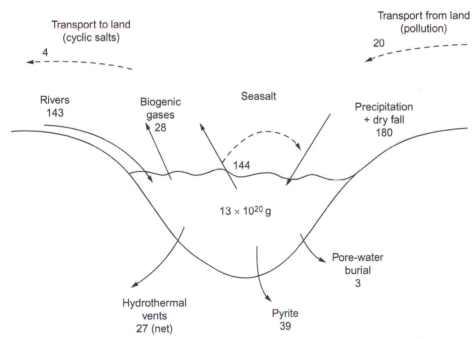

FIG. 9.36 Sulfur budget for the world's oceans, showing important fluxes in units of 10^{12} g S yr^{-1} (See also Fig. 13.1.). *Riverflux from Meybeck (1979); gaseous output from Lana et al. (2011); hydrothermal flux from Elderfield and Schultz (1996); and pyrite deposition from Berner (1982).*

biological demand. Sulfate thus shows a highly conservative behavior in seawater, with a mean residence time of about 10 million years relative to inputs from rivers.

Although it represents a small flux in the S cycle, the biological production of dimethylsulfide [(CH₃)₂S] aerosols by ocean phytoplankton may play an important role in regulating the Earth's climate (as will be discussed in Chapter 13). Trace quantities of this gas contribute to the "odor of the sea" in coastal regions (Ishida, 1968). DMS was first proposed as a major gaseous output of the sea by (Lovelock et al., 1972), but it wasn't until 1977 that researchers were able to measure DMS as an atmospheric constituent along the eastern coast of the United States (Maroulis and Bandy, 1977). DMS is now widely recognized as a trace constituent in seawater and in the marine atmosphere (Lana et al., 2011; Fig. 9.37). Current estimates indicate a global DMS flux of 19.6 Tg S y⁻¹ to the atmosphere (Land et al., 2014), making this the largest natural emission of a sulfur gas (Kjellstrom et al., 1999).

Dimethylsulfide (DMS) is produced during the decomposition of dimethylsulfoniopropionate (DMSP) from dying phytoplankton cells (Kiene, 1990). The reaction is mediated by the enzyme DMSP-lyase and grazing by zooplankton seems to be important to the release of DMS to seawater (Dacey and Wakeham, 1986; Wolfe et al., 1997). In the atmosphere, DMS is

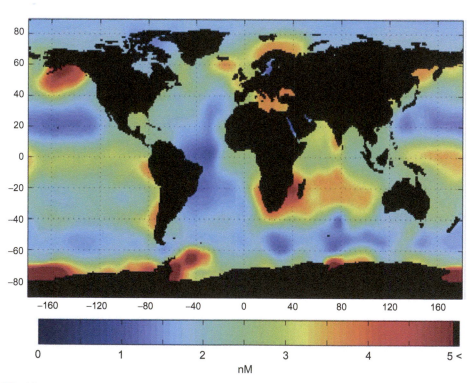

FIG. 9.37 Mean annual dimethylsulfide concentration in the surface ocean (nM), showing zones of high concentrations in the high-latitude oceans. *Redrawn from Lana et al. (2011). Used with permission of the American Geophysical Union. All rights reserved.*

rapidly oxidized by OH radicals, forming SO_2 and then sulfate aerosols that are deposited in precipitation (Shon et al., 2001; Faloona et al., 2009; Chapter 3). Nearly 80% of the non-seasalt sulfate in the atmosphere over the North Pacific Ocean appears to be derived from DMS, with the soil dust and pollution contributing the rest (Savoie and Prospero, 1989). Marine DMS is estimated to contribute up to 10% of the atmospheric sulfur over industrial Europe (Tarrasón et al., 1995).

It has been proposed that marine DMS production provides an important climate regulating feedback mechanism because DMS oxidation products in the atmosphere are effective cloud condensing nuclei (Charlson et al., 1987; Bates and Quinn, 1997; Liss and Lovelock, 2008). By this theory, marine algal DMS production could lead to cooling by increasing cloud cover (Chapter 3). Thus far evidence for a strong climate feedback by this mechanism is weak (Bates and Quinn, 1997; Liss and Lovelock, 2008), however climate models effectively demonstrate that large declines in DMS production would exacerbate global warming trends (Land et al., 2014).

Trace element cycles

The marine Si cycle: Diatoms compose a large proportion of the marine phytoplankton, and they require silicon (Si) as a constituent of their cell walls, where it is deposited as opal (as discussed above). As a result of biotic uptake, the concentration of dissolved Si in the surface waters is very low, usually $<2\,\mu M$, with the highest concentrations in the Southern and the North Pacific Oceans (Ragueneau et al., 2000). Globally, the annual uptake of Si by diatoms is more than $6000 \times 10^{12}\,g$ (Nelson et al., 1995; Tréguer and De La Rocha, 2013). Upon the death of diatoms, a large fraction of the opal dissolves, and the Si is recycled in the surface waters, where bacterioplankton mediate the dissolution process (Bidle and Azam, 1999). Silicate concentrations generally increase with depth, but the dissolution of opal is dependent on temperature, so the rate of dissolution in the deep ocean is relatively slow (Honjo et al., 1982; Van Cappellen et al., 2002; Bidle et al., 2002). The average Si concentration in deep waters is about $70\,\mu M$. In sinking particles and in ocean sediments, Si/C ratios increase with depth, suggesting that C is mineralized more readily than Si (Nelson et al., 2002). Globally, the burial efficiency for opaline Si is about 3% of production—significantly greater than for organic carbon (<1%) (DeMaster, 2002). Silicon loss is retarded when dissolved silicon complexes with Al in the sediments (Dixit et al., 2001).

A mass-balance model for Si in the oceans shows that rivers ($156 \times 10^{12}\,g\,yr^{-1}$), dust ($14 \times 10^{12}\,g\,yr^{-1}$), and hydrothermal vents ($17 \times 10^{12}\,g\,yr^{-1}$) are the main sources, and sedimentation of biogenic opal is the only significant sink (DeMaster, 2002; Tréguer and De La Rocha, 2013). The mean residence time for Si in the oceans is about 15,000 years, which is consistent with its non-conservative behavior in seawater. Most of the Si input is delivered by tropical rivers, as a result of high rates of rock weathering in tropical climates (Chapter 4). In contrast, sedimentation in the cold waters of the Antarctic Ocean accounts for 70% of the global sink (Ragueneau et al., 2000; DeMaster, 2002), largely as a result of massive seasonal diatom blooms in the Southern Ocean near Antarctica (Nelson et al., 2002). About 10% of the sink is found in coastal regions, where the growth of diatoms in nutrient-enriched waters may be limited by silicon (Justic et al., 1995). Increasing deposition of desert dusts to the ocean

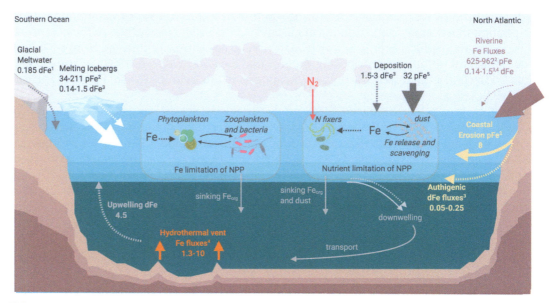

FIG. 9.38 A conceptual model of the ocean Fe cycle (modified from (Tagliabue et al., 2017) including estimates of the dominant Fe inputs to the ocean in Tg Fe yr^{-1}. Note that inputs are dominated by particulate Fe (pFe) while inputs of more biologically accessible dissolved Fe (dFe) are small. Estimates of inputs are from (1) Li et al. (2019d), (2) Poulton and Raiswell (2002), (3) Raiswell and Canfield (2012), (4) Elderfield and Schultz (1996), and (5) Duce and Tindale (1991). Physical transport is indicated with gray arrows and biological transformations are shown with black arrows. In much of the southern ocean Fe supplies are sufficiently low to directly limit NPP while in the North Atlantic higher inputs of dust and riverine Fe are more likely to secondarily limit NPP and alter N cycling by constraining rates of N fixation.

surface, potentially stimulating diatom productivity and export production with a higher organic carbon content relative to $CaCO_3$, may explain periods with lower CO_2 in Earth's atmosphere (Harrison, 2000).

The marine iron cycle: Using the Fe:C ratio of sinking organic matter produced in a series of iron fertilization experiments, we can assume that phytoplankton require $\sim 1.3 \times 10^{-4}$ mols Fe for every mole of C (Buesseler and Boyd, 2003). Under these assumptions 30 Tg of Fe would be required to supply the Fe requirement associated with 50 Pg of C fixation by marine phytoplankton annually.[v] The inputs of total Fe to the ocean are estimated at 700–1200 Tg yr^{-1} (Fig. 9.38), greatly exceeding this estimate; nevertheless, NPP, N fixation, and carbon export are constrained by the availability of Fe in some 40% of the surface ocean (reviewed by (Tagliabue et al., 2017). This seeming paradox arises because much of the Fe that is delivered to the world's oceans is either not in available forms or is rapidly removed from the water column through abiotic reactions that result in iron being sequestered in particles which sink

---

[v] 4×10^{15} mols of carbon are fixed annually by marine phytoplankton. With cell ratios of 1.3×10^{-4} mols of Fe per mole of C, this equates to an annual requirement of 5.4×10^{11} mols of Fe (or $\sim$ 30 Tg Fe) for all marine phytoplankton.

out of the surface ocean. Inputs of readily bioavailable dissolved iron (dFe) to the surface ocean amount to <10Tg Fe annually (Fig. 9.38).

The major flux of Fe into the surface ocean is via surface runoff in rivers, but much of this delivery is particulate Fe that is buried in estuarine and coastal sediments. Very little of the Fe delivered to the ocean is in soluble forms, with estimates suggesting that about 2% of aerosol deposition is in soluble forms, supplying 1.5–3 Tg Fe yr^{-1} (Fung et al., 2000; Raiswell and Canfield, 2012), a flux equivalent to 30–70% of the soluble iron flux associated with deep water upwelling (Archer and Johnson, 2000; Fung et al., 2000; Moore et al., 2001). In the well oxygenated waters of the surface ocean, oxidized and highly insoluble Fe(III) dominates over the soluble, reduced form Fe(II) by a factor of $\sim10^{10}$ (Waite, 2001), with most of the Fe(III) within the water column is found as $Fe(OH)_2^+$ and $Fe(OH)_4^-$ or bound to organic ligands (Johnson et al., 1997; Boyd and Ellwood, 2010). The high biological demand for Fe, coupled with the strong tendency for Fe to bind to mineral surfaces, removes any available Fe from the water column and shunts Fe to sediments.

There is still much to learn about the marine Fe cycle, and our current understanding is constrained by inconsistencies in the methodology of sampling and analysis. Through the coordinated efforts of the GEOTRACES programme[w] and detailed regional measurements, researchers have begun to understand that there are multiple sources of iron to the ocean that are critically important. While early work emphasized the importance of eolian dust inputs (Martin, 1990a; Jickells et al., 2005; Duce et al., 2008), more comprehensive sampling has shown the importance of Fe release from coastal sediments and from hydrothermal vents as dominant controls on Fe cycling in some regions (Tagliabue et al., 2017) (Fig. 9.38).

Both biological and physical "scavenging" mechanisms trap much of the riverbourne Fe in estuaries and the sediments of coastal shelves. These Fe-rich sediments can become a source of Fe to offshore waters in situations where high organic matter inputs result in favorable conditions for Fe reduction and release Elrod et al. (2004). Since the rate of Fe reduction and release to overlying waters is linked to the oxidation of organic matter, (Elrod et al., 2004) estimated that as much as 5.0 Tg Fe could be released from organic matter on continental shelves annually, an estimate that is equivalent to the global estimates of Fe derived from aerosol inputs. In settings where this subsurface supply of iron is released or mixed into well lit surface waters, these sediment fluxes can be key sources of bioavailable Fe (Lam and Bishop, 2008). The naturally occurring algal blooms around islands in the otherwise low productivity south Pacific are now understood to be caused by Fe release from shelf sediments and vertical mixing into well lit surface waters (Blain et al., 2007; Pollard et al., 2009). These blooms in turn lead to high rates of carbon export and enhanced organic matter oxidation in the underlying shelf sediments, sustaining high rates of Fe supply.

Although hydrothermal vents are known to release significant volumes of metal rich fluids, until recently most researchers assumed that most of these metals were rapidly incorporated into sulfide or oxide minerals and deposited to the seafloor, with little escaping to fuel productivity in the surface ocean. A reassessment of the biogeochemistry of hydrothermal vent plumes indicates that a significant fraction of the Fe released from hydrothermal activity is bound to organic ligands and remains soluble (Bennett et al., 2008; Toner et al.,

[w] Access these datasets at http://www.geotraces.org/.

2009; Sander and Koschinsky, 2011). Modeling efforts with sparse data suggest that modern Fe concentration patterns in the Southern Ocean are consistent with a hydrothermal source (Tagliabue et al., 2010).

Other trace elements: Similar to the use of Si by diatoms, marine protists known as acantharians require strontium (Sr). These organisms precipitate celestite ($SrSO_4$) as a skeletal component. As a result of the uptake of Sr in surface waters and the sinking of acantharians to the deep sea, the Sr/Cl ratio in seawater varies from about $392 \mu g\, g^{-1}$ in surface waters to >$405 \mu g\, g^{-1}$ with depth—relatively conservative behavior (Bernstein et al., 1987). The biotic demand for Sr leaves only a slight imprint on the large pool of Sr in the oceans; its overall mean residence time is about 12,000,000 years (Table 9.1).

The concentrations of many essential elements, such as Si, Fe, Zn, Cu, Co, and Ni, are depleted in the surface waters and increase with depth (Bruland et al., 1991; Donat and Bruland, 1995; Shelley et al., 2012). A few nonessential elements, such as Ti and Ba, which adsorb to sinking particles, also show this nonconservative behavior in seawater. Cherry et al. (1978) showed that the mean residence time for 14 trace elements in seawater was inversely related to their concentration in sinking fecal pellets (Fig. 9.39). Some of these elements are remineralized by grazing zooplankton (Reinfelder and Fisher, 1991) or by the degradation of POC by bacteria in the deep ocean, while the fate for many trace constituents is downward transport in organic particles and burial in the sediments of the deep sea (Turekian, 1977; Lal, 1977; Li, 1981). Elements that are not required by biota or that are available in great excess of biological demand remain as the major constituents of seawater (refer to Table 9.1).

As discussed earlier, zinc (Zn) is an essential component of carbonic anhydrase—the enzyme that allows phytoplankton to convert bicarbonate in seawater to CO_2 for photosynthesis (Morel et al., 1994). Low concentrations of Zn in surface waters can limit the growth of phytoplankton in marine environments (Brand et al., 1983; Sunda and Huntsman, 1992). Zn is also an essential cofactor for alkaline phosphatase, which allows phytoplankton to extract P from dissolved organic forms (DOP) in low-phosphorus waters (Shaked et al., 2006). Like Fe, the concentrations of Zn increase with depth in the deep sea (Bruland, 1989). Among samples of surface and deep waters, the concentrations of Fe and Zn are often well correlated to those of N, P, and Si, suggesting that biological processes control the distribution of all these elements in seawater. For example, Zn is correlated to Si in the northeast Pacific (Bruland et al., 1978).

Uptake of trace metals also occurs for some nonessential, toxic metals, such as cadmium (Cd), which accumulates in phytoplankton. Cadmium appears to substitute for zinc in biochemical molecules, allowing diatoms to maintain growth in zinc-deficient seawater (Price and Morel, 1990; Lane et al., 2005; Park et al., 2008; Xu et al., 2008a, 2008b). Cadmium is well correlated with available P in waters of the Pacific Ocean (Fig. 9.40; Boyle et al., 1976; Abe, 2002), and the concentration of Cd in marine sediments is sometimes used as an index of the availability of P in seawater in the geologic past (Hester and Boyle, 1982; Elderfield and Rickaby, 2000). When marine phosphate rock is used as a fertilizer, cadmium is often an undesirable trace contaminant (Smil, 2000; Longanathan et al., 2003).

When nonessential elements (e.g., Al, Ti, Ba, and Cd) and essential elements (e.g., Si and P) show similar variations in concentration with depth, it is tempting to suggest that both are affected by biotic processes, but the correlation does not indicate whether the association is active or passive. Organisms actively accumulate essential micronutrients by enzymatic uptake, whereas other elements may show passive accumulations, as a result of

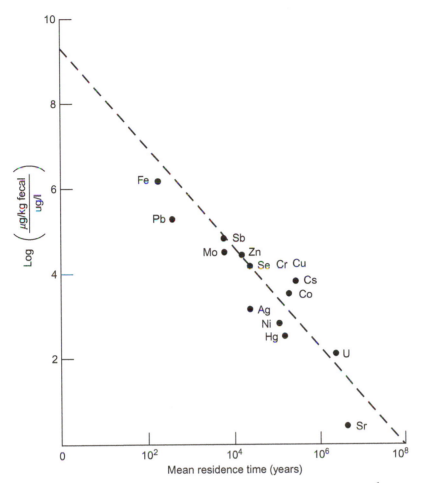

FIG. 9.39 The ratio between the concentration of an element in sinking fecal pellets ($\mu g\,kg^{-1}$) and its concentration in seawater ($\mu g\,l^{-1}$), plotted as a function of its mean residence time in the ocean. *From Cherry et al. (1978). Reprinted with permission from* Nature, *copyright 1978 Macmillan Magazines Limited.*

coprecipitation or adsorption on dead, sinking particles. For instance, titanium (Ti), which is not essential for biochemistry, shows nonconservative behavior in seawater, with concentrations ranging from $10\,\mu M$ at the surface to $>200\,\mu M$ at depth (Orians et al., 1990). The mean residence times for Ti, Ga (gallium), and Al (aluminum) in seawater range from 70 to 150 years (Orians et al., 1990).

Widespread observations of nonconservative behavior of barium (Ba) in seawater do not appear to result from direct biotic uptake (Sternberg et al., 2005). $BaSO_4$ precipitates on dead, sinking phytoplankton, especially diatoms and acantharians, as a result of the high concentrations of SO_4 that surround these organisms during decomposition (Bishop, 1988; Bernstein and Byrne, 2004). The precipitation of barite ($BaSO_4$) is an indication of the productivity of oceans in the past (Paytan et al., 1996).

FIG. 9.40 **FIG. 9.40** Depth distribution of nitrate, phosphate, and cadmium in the coastal waters of California. *From Bruland et al. (1978b).*

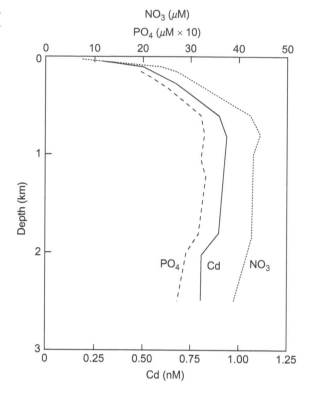

In the Mediterranean Sea, Al (aluminum) shows a concentration minimum at a depth of 60m, where Si and NO_3 are also depleted. (Mackenzie et al., 1978) suggested that this distribution is the result of biotic activity, and active uptake has been confirmed in laboratory studies (Moran and Moore, 1988). Other workers have found that organic particles carry Al to the deep ocean, but that the association is passive (Hydes, 1979; Deuser et al., 1983). High Al in surface waters is due to atmospheric inputs of dust (Orians and Bruland, 1986; Measures and Vink, 2000; Kramer et al., 2004). Aluminum declines in concentration with depth as a result of scavenging by organic particles and sedimentation of mineral particles.

Manganese (Mn), an essential element for photosynthesis (Chapter 5), is found at higher concentrations in the surface waters (0.1 μg L) than in the deep waters (0.02 μg L) of the ocean. Calculating an Mn budget for the oceans, Bender et al. (1977) attribute the high surface concentrations to the input of dust to the ocean surface (Guieu et al., 1994; Shiller, 1997; Mendez et al., 2010). Manganese appears less limiting than Fe and Zn for the growth of marine phytoplankton in surface waters (Brand et al., 1983). As in the case of Al, the deposition of Mn in dust in surface waters must exceed the rate of biotic uptake, downward transport, and remineralization of Mn in the deep sea.

The Mn budget of the ocean has long puzzled oceanographers, who recognized that the Mn concentration in ocean sediments greatly exceeds that found in the average continental rock (Broecker, 1974; Martin and Meybeck, 1979a, b). Other sources of Mn are found in riverflow and in releases from hydrothermal vents (Edmond et al., 1979). Various deep-sea bacteria appear to concentrate Mn by oxidizing Mn^{2+} in seawater to Mn^{4+} that is deposited in sediment (Krumbein, 1971; Ehrlich, 1975, 1982). The most impressive sedimentary accumulations are seen in Mn nodules that range in diameter from 1 to 15 cm and cover large portions of the seafloor (Broecker, 1974; McKelvey, 1980). The rate of growth of Mn nodules, about 1 to 300 mm per million years (Odada, 1992), is slower than the mean rate of sediment accumulation (1000 mm per million years; (Sadler, 1981), yet they remain on the surface of the seafloor. Various hypotheses invoking sediment stirring by biota have been suggested to explain the enigma, but none is proven. In addition to a high concentration of Mn (15–25%), these nodules also contain high concentrations of Fe, Ni, Cu, and Co and are a potential economic mineral resource.

The oceans and global change

The ocean contains nearly 12 times as much carbon as the atmosphere, terrestrial biosphere, and soils combined (Ciais et al., 2013), has taken up 93% of the excess heat generated by global warming increase since 1971 (Rhein et al., 2013), and is responsible for 20–30% of the uptake of anthropogenic derived CO_2 over the industrial era (Ciais et al., 2013). The enormous capacity of the world's oceans to absorb excess heat and CO_2 is dependent upon the connections between the deep ocean and the atmosphere. A warmer ocean not only absorbs less CO_2, but the resulting intensification of stratification in a warmer ocean will likely reduce the extent of deepwater mixing (Riebesell et al., 2009). It is possible that global warming will enhance the density gradient between the surface and deep ocean, thereby reducing the formation of North Atlantic Deep Water (Cunningham and Marsh, 2010; Smeed et al., 2014). Changes in stratospheric ozone over Antarctica appear to be altering westerly winds in the southern hemisphere (Waugh et al., 2013), potentially weakening the CO_2 sink strength of the downwelling in the Southern Ocean (Le Quéré et al., 2007; DeVries, 2014; Gruber et al., 2019b). Any reduction in the magnitude of downwelling will reduce the capacity for the ocean to continue to serve as a critical planetary sink for rising CO_2 and heat.

As the ocean warms, the melting of sea ice in the polar oceans is altering their climate, salinity and rates of atmospheric gas exchange. When white ice melts into a blue ocean, the albedo of the polar landscape is reduced, amplifying radiative forcing. The observed loss of sea ice in the Arctic 1979–2011 led to a reduction in Arctic planetary albedo from 0.52 to 0.48 that increased radiative forcing by $6.4 \pm 0.9\,W\,m^{-2}$ over this 32-year record (Pistone et al., 2014).[x] Sea ice formation is an important mechanism that facilitates isopyncnal mixing at the poles by freezing out water and leaving behind more concentrated marine salts. Declines in sea ice extent may act in concert with melting land-based ice to freshen the surface waters of the North Atlantic (Dickson et al., 2002; Curry and Mauritzen, 2005; Morison et al., 2012). The combined trends of a warming, freshening Arctic are both likely to intensify water column stratification and slow the rate of deep water formation (Broecker et al., 1985; Broecker,

[x] Adding complexity to predictions, cloud cover increases with future sea ice melting could offset the decline in surface albedo.

1997; Alley, 2007). Reductions in the rate of the Atlantic Meridional Overturning Circulation (AMOC) are expected to reduce the ocean C sink, creating a positive feedback to rising CO_2. Finally the loss of sea ice from the southern ocean may increase the rate of deep ocean CO_2 venting to the atmosphere by increasing the extent of direct air-sea gas exchange over upwelling waters (Waugh et al., 2013).

Warm water is less dense than cold water, and thus as the ocean absorbs excess heat, it increases in volume. This thermal expansion acts in concert with the melting of ice on land to cause sea level rise. Global mean sea level has risen by 16–21 cm since 1900 and is expected to continue to rise by 30–130[y] cm by 2100 (Sweet et al., 2017b). While the mechanisms apply to the entire ocean, the actual rates of sea level rise along continental margins varies widely due to local geomorphology and differences in the rates of uplift and subsidence of continental plates. In addition to the gradual encroachment of the sea onto land, sea level rise is associated with an increasing frequency of coastal flooding and a shift in the upstream extent of tides and salinity into coastal ecosystems (Sweet et al., 2017b; Tully et al., 2019).

Beyond the physical changes in ocean circulation, atmospheric exchange and the shift from frozen to liquid water, there are many recent changes in ocean biogeochemisty that are also of concern. Chief among these is that rising CO_2 concentrations in the world's oceans are leading to ocean acidification, with important energetic consequences for the many pelagic and benthic organisms that build carbonate and aragonite skeletons, and for the collective efficiency of these organisms to sequester fixed carbon in shallow marine sediments via the hard tissue pump (Doney et al., 2009; Riebesell et al., 2009). The problem is not just a reduction in new carbonate inputs to the seafloor, but also a substantial increase in the rates of sediment carbonate dissolution. Sulpis et al. (2018) find evidence of rising $CaCO_3$ dissolution rates in the North Atlantic deep ocean sediments that are consistent with the increasingly acidic waters that are highly enriched in anthropogenic CO_2. The geologic record contains evidence of prior deep sea $CaCO_3$ dissolution events in response to natural acidification, for example, at the Paleocene-Eocene Thermal Maximum 54 million years ago (Zachos et al., 2005; Cui et al., 2011). The increasing dissolution of benthic carbonates will have a buffering effect on deep ocean acidity, but one that comes at a substantial cost to the many benthic organisms that build their skeletons from carbonate and aragonite.

Nutrient pollution of the coastal oceans by fertilizer and wastewater enriched river discharges and enhanced nitrogen deposition to the open ocean are both leading to zones of high productivity, often accompanied by low oxygen conditions (Doney, 2010; Howarth et al., 2012). Coastal zone hypoxia is becoming more common and more extensive (Diaz, 2016). These eutrophic and low oxygen coastal areas can effectively trap and sequester large amounts of organic matter in sediments (Bauer et al., 2013) and more efficiently denitrify excess nitrogen (Fulweiler et al., 2007). This capacity however comes with considerable declines in the fisheries, biodiversity and aesthetic and recreational value of coastal ecosystems.

---

[y] This range represents the difference between the highly certain increase of at least 30 cm to the far less certain upper bound estimates derived from some forecast models.

Summary

Biogeochemistry in the oceans offers striking contrasts to that on land. The environment on land is spatially heterogeneous; within short distances there are great variations in soil characteristics, including redox potential and nutrient turnover. In contrast, the sea is relatively well mixed. Large, long-lived plants dominate the primary production on land, versus small, ephemeral phytoplankton in the sea. A fraction of the organic matter in the sea escapes decomposition and accumulates in sediments, whereas soils contain little permanent storage of organic matter. Terrestrial plants are rooted in the soil, which harbors most of the nutrient recycling by bacteria and fungi. The soil is sometimes dry, which limits NPP in many areas. In contrast, marine phytoplankton are never limited by water.

Through their buffering of atmospheric composition and temperature, the oceans exert enormous control over the climate of Earth. At a pH of 8.1 and a redox potential of +200mV, seawater sets the conditions for biogeochemistry on the 71% of the Earth's surface that is covered by seawater. Most of the major ions in the oceans have long mean residence times and their concentration in seawater has been relatively constant for at least the past 1 million years or more. All of this reinforces the traditional, and unfortunate, view that the ocean is a body that offers nearly infinite dilution potential for the effluents of modern society. As we find high concentrations of mercury and other toxins in pelagic fish and birds, we realize that this is no longer true (Monteiro and Furness, 1997; Vo et al., 2011).

Looking at the sedimentary record, we see that the ocean has been subject to large changes in volume, nutrients, and productivity, due to changes in global climate. Already, we have strong reason to suspect that the productivity of coastal waters is affected by human inputs of N and P (e.g., Beman et al., 2005). Changes in the temperature and productivity of the central ocean basins may well indicate that global changes are affecting the oceans as a whole (Behrenfeld et al., 2006; Polovina et al., 2008). Several studies report a decline in the oxygen content of the global ocean waters (Whitney et al., 2007; Helm et al., 2011). With warming climate, the overturning circulation of the oceans will decline, leading to lower NPP in the surface waters (Schmittner, 2005; Lozier et al., 2011). The oceans of the future are likely to be warmer, more acidic, and less productive than those of today—just at a time when the growing human population expects greater productivity from them.

Global cycles

The Global Water Cycle

Introduction

The annual circulation of water is the largest movement of a chemical substance at the surface of the Earth. Through evaporation and precipitation, water transfers much of the heat energy received by the Earth from the tropics to the poles, just as a steam-heating system transfers heat from the furnace to the rooms of a home. Movements of water vapor through the atmosphere determine the distribution of rainfall on Earth, and the annual availability of water on land is the single most important factor that determines the growth of plants (Jung et al., 2017; Humphrey et al., 2018). Where precipitation exceeds evapotranspiration on land, there is runoff. Runoff carries the products of mechanical and chemical weathering to the sea (Chapter 4).

In this chapter, we examine a general outline of the global hydrologic cycle and then look briefly at some indications of past changes in the hydrologic cycle and global water balance. Finally, we look somewhat speculatively at changes in the water cycle that may accompany global climate change and other human impacts in the future. These changes will have direct effects on global patterns of plant growth, the height of sea level, and the movement of materials in biogeochemical cycles.

Too often, we forget that adequate freshwater is the most essential resource for human survival. Changes in the water cycle have strong implications for the future of agricultural productivity and for the social and economic well-being of human society (Vörösmarty et al., 2000). Widespread drought seems associated with the collapse of the early Mesopotamian civilization in the Middle East around 2200 BC (Weiss et al., 1993) and the disappearance of the Mayan civilization in Mexico around 900 AD(Peterson and Haug, 2005; Medina-Elizalde and Rohling, 2012; Evans et al., 2018). In other areas, archeologists find the historical ruins of elaborate infrastructure, such as aqueducts, built to supply the human demands for water (e.g., Bono and Boni, 1996; Sandor et al., 2007). Despite its critical role, water is frequently wasted. The average daily consumption of drinking water in the United States is about 560 L (Chini and Stillwell, 2018), compared to about 178 L for Chinese citizens. If we include the water used to provide the food and products we use, water use in the U.S. increases to about 17,000 L/person/day (Chini et al., 2017). About 7200 L of water are needed to grow and deliver a one pound steak of beef to your table.

The global water cycle

The quantities of water in the global hydrologic cycle are so large that it is traditional to describe the pools and transfers in units of km^3 (Fig. 10.1). Each km^3 contains 10^{12} L and weighs 10^{15} g. The flux of water in the water cycle may also be expressed in units of average depth. For example, if all the rainfall on land were spread evenly over the surface, each weather station would record a depth of about 700 mm/yr. Units of depth can also be used to express runoff and evaporation (e.g., Fig. 4.17). For instance, annual evaporation from the oceans removes the equivalent of 1000 mm of water each year from the surface area of the sea.

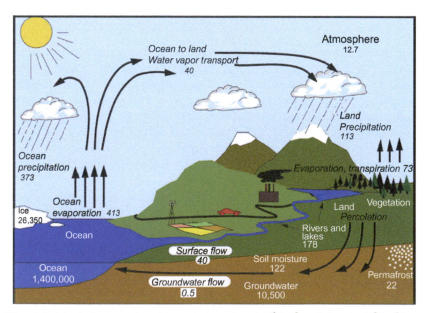

FIG. 10.1   The global hydrologic cycle, with pools in units of 10^3 km^3 and flux in 10^3 km^3/yr. *Modified from Trenberth et al. (2007). Used with permission of the American Meteorological Society.*

Not surprisingly, the oceans are the dominant pool in the global water cycle (Fig. 10.1). Seawater composes more than 97% of all the water at the surface of the Earth, about 1.4×10^9 km^3 (Bodnar et al., 2013). The mean depth of the oceans is 3900 m (Chapter 9). Water held in polar ice caps and in continental glaciers is the next largest contributor to the global pool. Only 42×10^6 km^3 (3%) of Earth's water is freshwater, and about 70% of that is frozen in Antarctica (Parkinson, 2006). Less than 0.01% of the Earth's water is freshwater that is accessible in rivers and lakes, which occupy less than 3% of the land's surface (Fig. 7.1). The mean residence time of freshwater in lakes and rivers, totaling about $\sim$200,000 km^3, is about 1.7 years (Bodnar et al., 2013). In many areas the amount of accessible surface water has declined in recent years, as measured by the GRACE satellite (Rodell et al., 2018).[a]

Soils contain 121,800 km^3 of water, of which about 58,100 km^3 is within the rooting zone of plants (Webb et al., 1993). In the eastern United States, fluctuations in soil water are the dominant contributor to fluctuations in total water storage on the landscape, as measured by GRACE (Cao et al., 2019). The large pool of freshwater below the unsaturated or vadose zone is known as groundwater, which contributes to the baseflow of rivers (Chapter 8). Global estimates of the volume of groundwater are poorly constrained—4,200,000 to 23,400,000 km^3—but, except for the ingenuity of humans and a few deep-rooted plants (e.g., Dawson and Ehleringer, 1991), groundwater is largely inaccessible to the biosphere. Plant roots can reach shallow groundwater on less than 20% of the land surface (Fan et al., 2013). We use a value of 10,500,000 km^3 as a median value for the total pool of groundwater (Bodnar et al., 2013), of which only a small fraction is recharged within a few decades (Gleeson et al., 2016; Evaristo et al., 2015; Jasechko, 2019).

The pool of water in the atmosphere is tiny, equivalent to about 25 mm of rainfall at any given time. Nevertheless, enormous quantities of water move through the atmosphere each year. Precipitation delivers about 500,000 km^3 to the surface of the Earth each year, giving a mean residence time for a water molecule in the atmosphere of about 9 days or less (Eq. 3.4).

Evaporation removes about 413,000 km^3/yr of water from the world's oceans (cf. Syed et al., 2010). Thus, the mean residence time of ocean water with respect to losses to the atmosphere is about 3100 years. Only about 373,000 km^3/yr of this water returns to the oceans in rainfall; the rest contributes to precipitation on land, which totals 113,000 km^3/yr. Plant transpiration and surface evaporation return $\sim$73,000 km^3/yr from soils to the atmosphere. On average plant transpiration (T) returns 39% of land precipitation to the atmosphere, but the relative importance of transpiration varies regionally (Table 10.1). Across world ecosystems, the ratio of T to ET is unrelated to precipitation (Schlesinger and Jasechko, 2014; Fatichi and Pappas, 2017), but the fraction T/ET increases as a function of leaf area (Wang et al., 2014; Wei et al., 2017). Transpiration is directly related to photosynthesis (Scott and Biederman, 2017) and to the uptake of carbonyl sulfide by plants, which can be used to measure photosynthesis (Wehr et al., 2017; see also Chapter 5). Globally plants appear to be responsible for about 60% of total terrestrial evapotranspiration (ET) to the atmosphere (Schlesinger and Jasechko, 2014; Good et al., 2015; Wei et al., 2017). Not surprisingly, ET is greatest in tropical

[a] The GRACE (Gravity Recovery and Climate Experiment) satellite measures changes in the Earth's gravitational attraction that stem from changes in the mass of water beneath its view. These changes have been used to estimate the water content of surface reservoirs, groundwater, and ice packs (Syed et al., 2008). See Famiglietti and Rodell (2013) and https://www.nasa.gov/mission_pages/Grace/index.html.

TABLE 10.1   A comparison of transpiration (T) and evaporation (E) among different world ecosystems (Schlesinger and Jasechko, 2014).

Ecoregion	T/ET percent average ± 1 s.d.	Land area (%)	Precipitation (mm/yr)	Percent of terrestrial precipitation (%)	ET[a] (mm/yr)	Percent of terrestrial ET (%)
Tropical rainforest	70 ± 14 (n = 8)	16	1830	35	1076 (927)	33.1 (28.5)
Tropical grassland	62 ± 19 (n = 5)	12	950	14	583 (726)	13.9 (17.3)
Temperate deciduous forests	67 ± 14 (n = 9)	9	850	10	549 (506)	10.1 (9.3)
Boreal forest	65 ± 18 (n = 5)	14	500	8	356 (315)	9.5 (8.4)
Temperate grassland	57 ± 19 (n = 8)	8	470	5	332 (406)	5.4 (6.6)
Desert	54 ± 18 (n = 14)	18	180	4	209 (186)	7.3 (6.5)
Temperate coniferous forest	55 ± 15 (n = 13)	4	880	4	458 (404)	3.4 (3.0)
Steppe	48 ± 12 (n = 3)	4	440	2	467 (343)	3.4 (2.5)
Mediterranean shrubland	47 ± 10 (n = 4)	2	480	1	302 (393)	1 (1.3)

[a] *Evapotranspiration (ET) data from MODIS (Mu et al., 2011) and FAO (in parentheses, from www.fao.org/geonnetwork/).*

forests and lowest in arid lands (Table 10.1). With respect to precipitation inputs or evapotranspiration losses, the mean residence time of soil water is about 1 year.

When precipitation is below average, droughts develop relatively quickly, impacting a wide variety of ecosystem processes, especially plant growth. Even in moist climates, only 21% of the moisture from a rainfall event remains in the surface soil after 3 days (Kim and Lakshmi, 2019). During a 2005 drought, the rainforests of the Amazon Basin lost $>1.2 \times 10^{15}$ g C to the atmosphere (Phillips et al., 2009; cf. Tian et al., 1998; Potter et al., 2011). Owing to the excess of precipitation over evapotranspiration on land, about 40,000 km³/yr becomes runoff, derived from surface flow and groundwater. The runoff in rivers is supplemented by the subterranean flow of groundwater to the sea in coastal areas, which is potentially large but difficult to estimate (Moore, 2010; Sawyer et al., 2016). One study estimates this flow as 489 km³/yr or 1.3% of riverflow (Zhou et al., 2019); another, suggesting 120,000 km³/yr globally (Kwon et al., 2014), is almost certainly much too large.

These global average values obscure large regional differences in the water cycle. Evaporation from the oceans is not uniform, but ranges from 4 mm/day in tropical latitudes to <1 mm/day at the poles (Mitchell, 1983). Although much precipitation falls at tropical latitudes, an excess of evaporation over precipitation from the tropical oceans provides a net regional flux of water vapor to the atmosphere. Net evaporative loss accounts for the high salinity in tropical oceans (Fig. 9.8), and the movement of water vapor in the atmosphere carries latent heat to polar regions (Trenberth and Caron, 2001).

On land the relative balance of precipitation and evaporation differs strongly between regions. In tropical rainforests, precipitation may greatly exceed evapotranspiration. Shuttleworth (1988) calculates that ~50% of the rainfall becomes runoff in the Amazon rainforests (Table 10.1). In desert regions, precipitation and evapotranspiration are essentially equal, so there is no runoff and only limited recharge of groundwater (Scanlon et al., 2006). As a global average, rivers carry about one-third of the precipitation from land to the sea (cf. Alton et al., 2009). About 11% of precipitation becomes groundwater ($12{,}666 \, km^3/yr$; Doll and Fiedler, 2008; Zektser and Loaiciga, 1993). Precipitation is the major determinant of groundwater recharge (Kim and Jackson, 2012). The mean residence time of groundwater is 690 (Bodnar et al., 2013) to 1000 years (Slutsky and Yen, 1997), but there is much regional variation among aquifers (Befus et al., 2017). A small amount of groundwater turns over rapidly; groundwater older than 12,000 years dominates the global aquifer storage (Jasechko, 2019).

The concept of *potential evapotranspiration* (PET) developed by hydrologists expresses the maximum evapotranspiration that would be expected to occur under the climatic conditions of a particular site, assuming that water is always present in the soil and plant cover is 100%. Potential evapotranspiration is greater than the evaporation from an open pond, as a result of the plant uptake of water from the deep soil and a leaf area index >1.0 in many plant communities (Chapter 5). In tropical rainforests, PET and actual evapotranspiration (AET) are about equal (Vörösmarty et al., 1989). In deserts, PET greatly exceeds actual AET, owing to long periods when the soils are dry. In southern New Mexico, precipitation averages about 210 mm/year, but the receipt of solar energy could potentially evaporate over 2000 mm/yr from the soil (Phillips et al., 1988).

With higher values in warm, wet conditions, actual evapotranspiration (Fig. 10.2) is often useful as a predictor of primary production (Fig. 5.8), decomposition (Fig. 5.16), and microbial activity (Fig. 4.3). Changes in climate that affect rainfall and AET have a dramatic effect on the biosphere. Annual variability in AET is greatest in ecosystems with low AET, causing large year-to-year variations in net primary production related to annual precipitation in deserts (Frank and Inouye, 1994; Prince et al., 1998; Ahlström et al., 2015). Actual evapotranspiration is more constant in tundra and boreal forest ecosystems, where wet soils do not constrain the supply of water to plants. The net primary productivity of land plants (60×10^{15} g C/yr) and the plant transpiration of water from land (~60% of 73×10^{18} g/yr) indicate that the global average water-use efficiency of vegetation is about 2.0 mmol of CO_2 fixed per mole of water lost (Eq. 5.3)—somewhat higher than the range measured by physiologists studying individual leaves (Chapter 5). There is evidence that plant water-use efficiency has increased globally as a result of rising concentrations of atmospheric CO_2 and decreased transpiration (Keenan et al., 2013; Keeling et al., 2017). This may not translate into greater runoff if vegetation responds with an increasing leaf area index (Tor-ngern et al., 2015).

The sources of water contributing to precipitation also differ greatly among different regions of the Earth. Nearly all the rainfall over the oceans is derived from the oceans. On land, much of the rainfall in maritime and monsoonal climates is also derived from evaporation from the sea. Estimates of the percentage of rainfall derived from evapotranspiration on land

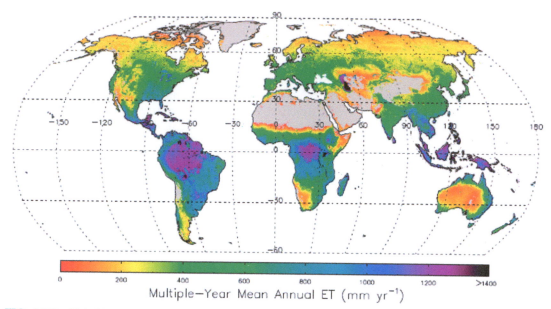

FIG. 10.2 Global loss of water from the land surface by evapotranspiration. *From Zhang et al. (2010). Used with permission of the American Geophysical Union.*

are poorly constrained and regionally variable, with global estimates varying by 10–60% (Trenberth, 1998; van der Ent et al., 2010).[b] In the boreal forest, local evaporation is an important determinant of local precipitation (Wei and Dirmeyer, 2019). In the Amazon Basin, 25–50% of the rainfall may be derived from evapotranspiration within the basin, with the rest derived from long-distance transport through the atmosphere from other regions (Salati and Vose, 1984; Eltahir and Bras, 1994). Evapotranspiration in Amazon forests is maximized by deep-rooted plants (Nepstad et al., 1994), and the regional importance of evapotranspiration in the Amazon basin speaks strongly for the long-term implications of forest destruction in that region (Spracklen et al., 2012). Using a general circulation model of the Earth's climate, Lean and Warrilow (1989) show that a replacement of the Amazon rainforest by a savanna would decrease regional evaporation and precipitation and increase surface temperatures (compare Shukla et al., 1990). Such a transition might represent a threshold effect, beyond which the ecosystem could not return to its prior state (Scheffer et al., 2009).

Regional cooling from plant transpiration is well known from comparisons of urban and rural areas (Juang et al., 2007). Often we refer to the "heat-island effect" for urban areas, which are warmer than the surrounding countryside, due to a reduction in transpiration. In semiarid regions, precipitation may decline as a result of the removal of vegetation, leading to soil warming (Balling, 1989; Kurc and Small, 2004; He et al., 2010) and increasing desertification (Chahine,

[b]Note the scale-dependence of such a statistic. At a local scale, nearly all the precipitation will be derived outside of the local area, whereas at larger regional scales an increased percentage will be derived within the area of interest (Trenberth, 1998).

1995; Koster et al., 2004). Thus, the transpiration of land plants is an important factor determining the movement of water in the hydrologic cycle and Earth's climate.

Estimates of global riverflow range from 33,500 to 47,000 km^3/yr (Lvovitch, 1973; Speidel and Agnew, 1982). A recent satellite-derived global estimate indicates 36,055 km^3 of annual riverflow to the sea between 1994 and 2006 (Syed et al., 2010). For global models, many workers assume a value of about 40,000 km^3/yr (Fig. 10.1). The distribution of flow among rivers is highly skewed. The 50 largest rivers carry about 43% of total riverflow; the Amazon alone carries ~20%. Reasonable estimates of the global transport of organic carbon, inorganic nutrients, and suspended sediments from land to sea can be based on data from a few large rivers. As a result of the positions of the continents and their surface topography, there are large regional differences in the delivery of runoff to the sea (Fig. 10.3). The average runoff from North America is about 320 mm/yr, whereas the average runoff from Australia, which has a large area of internal drainage and deserts, is only 40 mm/yr (Tamrazyan, 1989). Thus, the delivery of dissolved and suspended sediment to the oceans varies greatly between rivers draining different continents (Table 4.8).

Globally, only 37% of rivers longer than 1000 km remain free-flowing to the ocean (Grill et al., 2019). In the Northern Hemisphere, 77% of the runoff is carried in rivers in which the flow is now regulated by dams and other human structures (Dynesius and Nilsson, 1994; Nilsson et al., 2005), which strongly affect the sediment transport to the sea. The mean transit time of water to the sea has increased by an average of 60 days in rivers with significant impoundments and reservoirs (Vörösmarty and Sahagian 2000). Jaramillo and Destouni (2015) calculate that humans now use ~20% of the volume of rivers globally (~10,688 km^2), converting a large portion of it to water vapor as a result of irrigated agriculture (Rost et al., 2008; Jaramillo and Destouni, 2015). In some regions, such as the southwestern United States, human appropriation of surface runoff is as high as 76% (Sabo et al., 2010). Globally, the consumption of water for

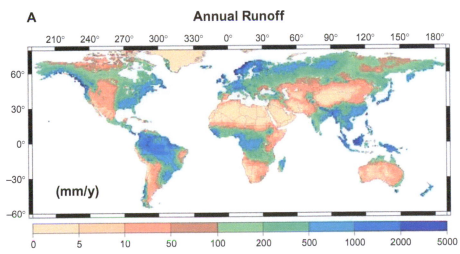

FIG. 10.3 Annual runoff to the oceans, in mm/yr. *From Oki and Kanae (2006). Used with permission of the American Association for the Advancement of Science.*

farmland irrigation is estimated to amount to $874\,km^3/yr$ (Chen et al., 2013). In the United States, as much as 10% of the freshwater extracted from rivers is evaporated in the cooling towers of power plants (Dieter et al., 2018).

Globally, the net human depletion of groundwater, $113–283\,km^3/yr$ (Wada et al., 2010; Konikow, 2011; Döll et al., 2014), is less than 2% of groundwater recharge, but it is centered in arid and semiarid regions, which show large groundwater depletion (Rodell et al., 2009; Gleeson et al., 2012). Much of this groundwater is used for irrigation (Siebert et al., 2010; Wada et al., 2012). In the central valley of California, groundwater withdrawal of about $3\,km^3/yr$ lowered the water table by ~20.3 mm/yr between 2003 and 2010 (Fig. 10.4). Similar rates of loss are reported from the GRACE satellite for the High Plains of Texas (Breña-Naranjo et al., 2014). Worldwide, depletion of groundwater is likely to reduce streamwater flow (de Graaf et al., 2019).

The mean residence time of the oceans with respect to riverflow is about 34,000 years. The mean residence times differ among ocean basins. The mean residence time for water in the Pacific Ocean is 57,500 years—significantly longer than that for the Atlantic (18,000 years; Speidel and Agnew, 1982, p. 30). This is consistent with the greater accumulation of nutrients in deep Pacific waters and a shallower carbonate compensation depth in the Pacific Ocean (Chapter 9). Despite the enormous riverflow of the Amazon, continental runoff to the Atlantic Ocean is less than the loss of water through evaporation. Thus, the Atlantic Ocean has a net water deficit, which accounts for its greater salinity (refer to Fig. 9.8). Conversely, the Pacific

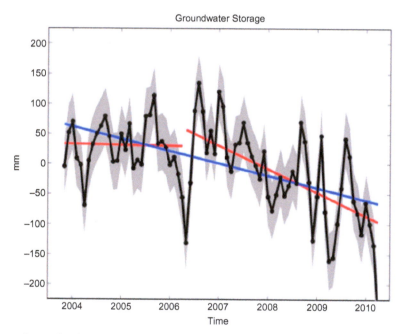

FIG. 10.4   Groundwater levels in the Central Valley of California, showing changes in mm from a 2004 baseline value. *From Famiglietti et al. (2011). Used with permission of the American Geophysical Union.*

Ocean receives a greater proportion of the total freshwater returning to the sea each year. Ocean currents carry water from the Pacific and Indian oceans to the Atlantic Ocean to restore the balance (refer to Fig. 9.3).

Models of the hydrologic cycle

A variety of models have been developed to predict the movement of water in hydrologic cycles from regional to global scales. Watershed models follow the fate of water received in precipitation and calculate runoff after subtraction of losses due to evaporation and plant uptake (Waring et al., 1981; Moorhead et al., 1989; Ostendorf and Reynolds, 1993). In these models, the soil is considered as a collection of small boxes, in which the annual input and output of water must be equal. Water entering the soil in excess of its water-holding capacity is routed to the next lower soil layer or to the next downslope soil unit via subsurface flow (Chapter 8). Models of water movement in the soil can be coupled to models of soil chemistry to predict the loss of elements in runoff (e.g., Nielsen et al., 1986; Knight et al., 1985; Furrer et al., 1990).

A major challenge in building these models is the calculation of plant uptake and transpiration loss. This flux is usually computed using a formulation of the basic diffusion law, in which the loss of water is determined by the gradient, or vapor pressure deficit, between plant leaves and the atmosphere. The loss is mediated by a resistance term, which includes stomatal conductance and wind speed (Chapter 5). In a model of forest hydrology, Running et al. (1989) assume that canopy conductance decreases to zero when air temperatures fall below $0\,°C$ or soil water potential declines below $-1.6\,MPa$. Their model appears to give an accurate regional prediction of evapotranspiration and primary productivity for a variety of forest types in western Montana.

Similar approaches can be used to formulate models for the flow of water over large regions. Where urban areas are involved, these models must account for inter-basin transfers to supply human demands via pipelines and aqueducts. For example, Good et al. (2014) used the isotopic composition to show a non-local origin of drinking water in 31% of 614 cities of the western United States.

Larger-scale models have been developed to assess the contribution of continental land areas to the global hydrologic cycle. For example, Vörösmarty et al. (1989) divided South America into 5700 boxes, each $1/2° \times 1/2°$ in size. Large-scale maps of each nation were used to characterize the vegetation and soils in each box, and data from local weather stations were used to characterize the climate. A model (Fig. 10.5) was used to calculate the water balance in each unit. During periods of rainfall, soil moisture storage is allowed to increase up to a maximum water-holding capacity determined by soil texture. During dry periods, water is lost to evapotranspiration, with the rate becoming a declining fraction of PET as the soil dries.

This type of model can be coupled to other models, including general circulation models of the Earth's climate (Chapter 3), to predict global biogeochemical phenomena. For example, a monthly prediction of soil moisture content for the South American continent can be used with known relationships between soil moisture and denitrification (Fig. 6.20) to predict the loss of N_2O and the total loss of gaseous nitrogen from soils to the atmosphere (Potter et al., 1996). The excess water in the water-balance model is routed to stream channels, where

FIG. 10.5   Components of a model for the hydrologic cycle of South America. *From Vörösmarty et al. (1989). Used with permission of the American Geophysical Union.*

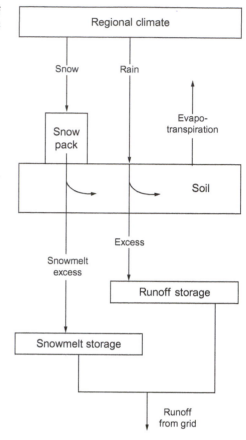

it can be used to predict the flow of the major rivers draining the continent (Russell and Miller, 1990; Milly et al., 2005). Changes in land use and the destruction of vegetation are easily added to these models, allowing a prediction of future changes in continental-scale hydrology and biogeochemistry.

The history of the water cycle

As we learned in Chapter 2, water was delivered to the primitive Earth during its accretion from planetesimals, meteors, and comets. The accretion of the planets was largely complete by 4.5 billion years ago (bya). Then water was released from the Earth's mantle in volcanic eruptions (i.e., degassing), which continue to deliver water vapor to the atmosphere today—equivalent to $\sim2.5\,km^3/yr$ (Wallmann, 2001). As long as the Earth's temperature was greater than the boiling point of water, water vapor remained in the atmosphere. When the Earth cooled, nearly all the water condensed to form the oceans. Even then, a small amount of water vapor and CO_2 remained in the Earth's atmosphere—enough to maintain the temperature of the Earth above freezing via the greenhouse effect (Chapter 3). Today, water vapor and clouds account for 75% of the greenhouse effect on Earth (Lacis et al., 2010; Schmidt et al., 2010a). Without this greenhouse effect the surface of the Earth might be coated with a thick layer of ice, and biogeochemistry would be much less interesting.

There is good evidence of liquid oceans on Earth as early as 3.8 bya, and it is likely that the volume of water in the hydrologic cycle has not changed appreciably since that time. The total inventory of volatiles at the surface of the Earth (refer to Table 2.3) indicates that about 155×10^{22} g of water have been degassed from its crust. The difference between this value and the total of the pools of water shown in Fig. 10.1 is largely contained in sedimentary rocks. Weathering of continental rocks by water, suggestive of the modern hydrologic cycle, may have begun about 2.5 billion years ago (Bindeman et al., 2018).

Each year about 0.1–0.2 km^3 of water is carried to the Earth's lower mantle by subducted sediments (van Keken et al., 2011; Kendrick et al., 2017). This is less than the return of water vapor to the atmosphere in volcanic emissions because much of the water in subducted marine sediments is degassed before it reaches the lower mantle (Dixon et al., 2002; Green et al., 2010; van Keken et al., 2011). Nevertheless, a large volume of water, perhaps equivalent to the current volume of the oceans, is found in Earth's mantle (Marty, 2012; Bodnar et al., 2013; Nakagawa and Spiegelman, 2017).

Owing to the low content of water vapor in the stratosphere, the Earth appears to have lost only a small amount of H_2O by photolysis, perhaps less than 35% of its total degassed water through Earth's history (Yung et al., 1989; Yung and DeMore, 1999, p. 346; Pope et al., 2012). Much larger quantities appear to have been lost from Venus, where all water remains as vapor and exposed to ultraviolet light (Chapter 2). The loss of water from Venus is consistent with the high D/H ratio of its current inventory, reflecting the slower loss of its heavier isotope (Chapter 2). The accumulation of O_2 in the atmosphere and in oxidized minerals of the Earth's crust suggests that about 2% of Earth's water has been consumed by net photosynthesis through geologic time (refer to Table 2.3).

There is substantial evidence that the Earth may have undergone several periods of frozen conditions in the Precambrian, and perhaps even more recently, yielding a "snowball" Earth, on which ice covered most of the planet (Hoffman et al., 1998; Kirschvink et al., 2000). Such large changes in global temperature lead to changes in sea level. The geologic record shows large changes in ocean volume during the 16 continental glaciations that occurred during the Pleistocene Epoch, beginning about 2 million years ago (Bintanja et al., 2005; Dutton and Lambeck, 2012). During the most recent glaciation, which reached a peak 22,000 years ago, an increment of $52,500 \times 10^3$ km^3 of seawater was sequestered in polar and other glacial ice (Lambeck et al., 2002; Yokoyama et al., 2000). This represents nearly 4% of the ocean volume, and it lowered the sea level about 120–130 m from that of present day (Fairbanks, 1989; Siddall et al., 2003). As we saw in Chapter 9, the Pleistocene glaciations are recorded in carbonate sediments. During periods of glaciation, the ocean was relatively rich in $H_2^{18}O$, which evaporates more slowly than $H_2^{16}O$. Calcium carbonate precipitated in these oceans shows higher values of $\delta^{18}O$, which can be used as an index of paleotemperature (refer to Fig. 9.32).

Although many causes have been suggested, most climate scientists now believe that ice ages are related to small variations in the Earth's orbit around the Sun (Hays et al., 1976; Harrington, 1987). These variations lead to differences in the receipt of solar energy, particularly in polar regions, and large changes in the Earth's hydrologic cycle. Once polar ice begins to accumulate, the cooling accelerates because snow has a high reflectivity or albedo to incoming solar radiation. Low concentrations of atmospheric CO_2 (refer to Fig. 1.4) and high concentrations of sulfate aerosols (Legrand et al., 1991) and atmospheric dust (Lambert et al., 2008) during the last ice age were probably an effect, rather than the cause, of global cooling; however, these changes in the atmosphere may have reinforced the rate and onset of the cooling trend (Harvey, 1988; Shakun et al., 2012). At the end of the last glacial epoch, the global temperature and the concentration of CO_2 in Earth's atmosphere rose in concert (Parrenin et al., 2013), and periods of high CO_2 are correlated with high sea level over the past 20 million years (Tripati et al., 2009).

Earth appears to enter each ice age relatively slowly ($\sim$50,000 years), whereas the warming to interglacial conditions occurs over a relatively short period ($<$1000 years; Fig. 1.4). Some periods of climate change have been remarkably rapid; records of paleoclimate show periods when the mean annual temperature in Greenland rose as much as 9 °C over a couple of decades (Taylor, 1999; Severinghaus and Brook, 1999). At the present time, the Earth is unusually warm; we are about halfway through an interglacial period, which should end about 12,000 AD.

Continental glaciations represent a major disruption—a loss of steady-state conditions—in Earth's water cycle. Global cooling yields lower rates of evaporation, reducing the circulation of moisture through the atmosphere and reducing precipitation. One model of global climate suggests that during the last glacial epoch, total precipitation was 14% lower than that of today (Gates, 1976). Throughout most of the world, the area of deserts expanded, and total net primary productivity and plant biomass on land were much lower (Chapter 5). Greater wind erosion of desert soils contributed to the accumulation of dust in ocean sediments, polar ice caps, and loess deposits (Yung et al., 1996; Lambert et al., 2008), and provided a source of additional nutrient inputs to the oceans (Jickells et al., 2005). The southwestern United States appears to have been an exception. Over most of this desert area, the climate of 18,000 years ago was wetter than today (Van Devender and Spaulding, 1979; Wells, 1983; Marion et al., 1985).

Worsley and Davies (1979) show that deep-sea sedimentation rates throughout geologic time have been greatest during periods of relatively low sea level, when a greater area of continents is displayed. During the last glacial epoch, changes in the rate of global riverflow produced changes in the delivery of dissolved and suspended matter to the sea. Broecker (1982) suggests that erosion of exposed continental shelf sediments during the glacial sea-level minimum may have led to a greater nutrient content of seawater and higher marine net primary productivity in glacial times. These observations are consistent with the current imbalance in the ocean nitrogen cycle—the oceans may still be responding to large nitrogen inputs during the last glacial period (McElroy, 1983; Figure 9.34).

The water cycle and climate change

Our ability to monitor ongoing changes in the global water cycle has improved dramatically with the deployment of the GRACE satellite system (Syed et al., 2008). For example, GRACE measurements show increasing water storage in the Amazon Basin during the seasonal period of heavy rains (Fig. 10.6), and GRACE allows the calculation of evapotranspiration by subtracting changes in soil moisture from precipitation. The GRACE satellite has also allowed

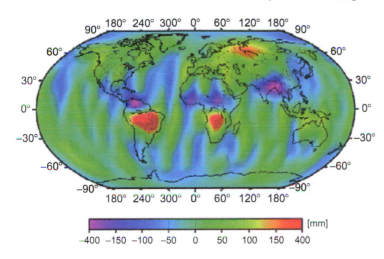

FIG. 10.6 Changes in water content of the Earth's land surface, as measured by the GRACE satellite, April to August 2003. Note the large increase in water storage during the wet season in the Amazon. *From Schmidt et al. (2006).*

measurements of groundwater depletion in California and India, due to irrigation (Rodell et al., 2009; and Fig. 10.4). Separately, changes in sea level and ocean currents are accurately monitored by the TOPEX/Poseidon satellite (Chapter 9).

Rise in sea level

Deduced from a variety of methods, the average rise in sea level has been about 1 to 2 mm/yr for the last 100 years (Church and White, 2011; Merrifield et al., 2009; Kemp et al., 2011; Fig. 10.7), but recent rates are now ≥3 mm/yr and accelerating (Jevrejeva et al., 2014; Hay et al., 2015; Dieng et al., 2017; Nerem et al., 2018). Recent measurements by the TOPEX/Poseidon satellite suggest that relative sea level rose at a rate of 3.3 mm/yr for 1993–2009 (Cazenave and Llovel, 2010), which is consistent with tide-gauge measurements during the same interval (Merrifield et al., 2009; Church and White, 2011).

Throughout geologic history, changes in relative sea level have accompanied periods of tectonic activity that increase (or decrease) the volume of submarine mountains. The recent, rapid changes in sea level are largely derived from changes in the height of coastal land, changes in the temperature of seawater, and changes in the volume of ice held on land. It is widely believed that the current global warming will cause a melting of the ice packs on Greenland and Antarctica, leading to an acceleration in the rise of sea level and flooding of coastal areas during the next century.

Observations of sea-level rise are complicated by the continuing isostatic adjustments[c] of continental elevations in response to the melting of ice from the last continental glaciation (Sella et al., 2007). Areas of subsidence show greater rates of sea-level rise, varying from 2

[c] Isostatic adjustments result from the depression of the Earth's crust due to the weight of the glacier, and a rebound of its elevation as the ice melts. Thus, much of the coast of the northeastern United States, which was glaciated 18,000 years ago, is now rising. Conversely, the eastern coast of the U.S. south of the terminal glacial boundary (moraine), which was squeezed upward by the weight of the glacier to the north, is now falling, adding to the manifestation of sea-level rise (Piecuch et al., 2018).

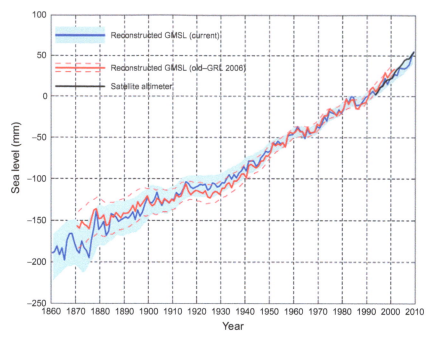

FIG. 10.7   Global average sea level from 1860 to 2009. *From Church and White (2011). Used with permission of Springer.*

to 7 mm/yr along the eastern coast of the United States (Piecuch et al., 2018). In addition, over much of the oceans, sea-surface temperatures have also risen over the last 100 years (Levitus et al., 2001, 2005; Barnett et al., 2005), so some of the rise in sea level—perhaps 0.6 mm/yr— must be attributed to the thermal expansion of water at warmer temperatures (Miller and Douglas, 2004; Antonov et al., 2005; Chen et al., 2013).

Some rise in sea level may also stem from human activities, including the extraction of groundwater, which is then delivered to the sea by rivers (Sahagian et al., 1994; Konikow, 2011; Pokhrel et al., 2012). On the other hand, reservoirs and irrigation are thought to have reduced global riverflow by about 2% (930 km^3/yr; Biemans et al., 2011; Haddeland et al., 2006) and reduced the rise in sea level by 0.55–0.71 mm/yr (Chao et al., 2008; Reager et al., 2016). Impoundments now contain 8070 km^3 of water, and massive new dams are planned or under construction in many areas of Asia and South America (Lehner et al., 2011). In recent years, the net storage of water on land may have increased by 3200 km^3 (Reager et al., 2016) although there is much uncertainty in such estimates (Kim et al., 2019).

Some rise in sea level is likely due to the melting of mountain glaciers throughout the world—an indication of a global warming trend (Gardner et al., 2013; Parkes and Marzeion, 2018). The collective losses from mountain glaciers worldwide may exceed 335 km^3/yr (Zemp et al., 2019) and account for 25–30% of the total observed rise in sea level. The rest of the current rise in sea level is thought to be due to melting of the Greenland and Antarctic ice packs. Loss of the ice caps on Greenland and Antarctica is associated with dramatic rise in sea level during past glacial cycles (Raymo and Mitrovica, 2012; Deschamps et al., 2012), and perhaps as much as 50 cm of the sea-level rise expected by 2050 (Golledge et al., 2019).

GRACE satellite measurements show that the ice pack on Greenland is in decline (van den Broeke et al., 2009; Shepherd et al., 2012), and the melting on Greenland is consistent with observations of lower salinity in North Atlantic waters (Curry and Mauritzen, 2005; Dickson et al., 2002). The ice pack on Greenland shows the greatest decline at its margins and may actually be accumulating some mass in northern and interior regions (Pritchard et al., 2009; Kjær et al., 2012). The recent net loss of ice on Greenland is estimated at $222 \times$ km^3/yr (IMBIE, 2020) to $286 \times$ km^3/yr (Mouginot et al., 2019).

Overall, the massive ice cap on Antarctica shows smaller changes than in Greenland, but it also appears to be decreasing in volume (Shepherd et al., 2012), especially in ice shelves that surround the continent (Rignot et al., 2013). Rapid melting on the Antarctic Peninsula may be balanced by slight accumulations of snowfall in East Antarctica (Rignot et al., 2008; Shepherd et al., 2019; Martín-Español et al., 2016; IMBIE, 2018). Although the record is not long, measurements from GRACE show an accelerating loss of ice mass from Antarctica, now approaching 252 km^3/yr (Rignot et al., 2008). Declining salinity in the Ross Sea is consistent with these trends (Jacobs et al., 2002). Melting in Antarctica is thought to be responsible for 0.36 mm/yr of sea-level rise during the past decade (Rignot et al., 2008).

The measured losses of ice volume from Greenland, Antarctica, and various mountain glaciers throughout the world yield a global rise in sea level of about 1.5–1.8 mm/yr (Meier et al., 2007; Jacob et al., 2012; Golledge et al., 2019), about half of the observed rate. Presumably the thermal expansion of warmer seawater explains the difference. By 2100, losses from these ice packs may yield a total sea level rise >1 m (DeConto and Pollard, 2016), yielding widespread flooding of coastal environments and most of the world's major cities. Overall the Greenland ice pack contains the equivalent of 7 m of sea level rise, whereas Antarctica contains about 65 m. Thus, the expected melting during this century will release only a small fraction of the water in frozen ice packs.

Sea ice

Just as the volume of a glass of water is not affected by ice cubes that may melt within it, sea level is not affected by changes in the area or volume of ice, known as *sea ice*, which floats on the ocean surface. Nevertheless, trends in sea ice are a useful index of Earth's climate.

Dramatic losses of sea ice are seen in the Arctic, which may have ice-free summers within a few decades (Serreze et al., 2007; Comiso et al., 2008; Parkinson and Cavalieri, 2008; Fig. 10.8).[d] This decline in sea ice is unprecedented during the past 1450 years (Kinnard et al., 2011). During the past few decades, the loss of sea ice has been highly correlated with anthropogenic emissions of CO_2—namely, each year, there is 3 m^2 less ice at the September minimum per ton of CO_2 emitted from fossil fuel combustion (Notz and Stroeve, 2016). The remaining ice has also lost thickness (Kwok and Rothrock, 2009; Laxon et al., 2013). Loss of the natural reflectivity or albedo of snow cover and sea ice in the Arctic is estimated to have contributed

---

[d]The dynamics of arctic sea ice can be followed at https://climate.nasa.gov/climate_resources/155/video-annual-arctic-sea-ice-minimum-1979-2018-with-area-graph/.

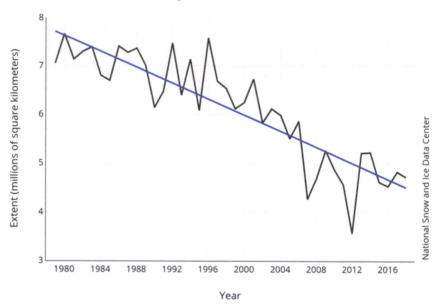

FIG. 10.8 September sea ice extent, 1979 to 2010 in the Arctic. *From National Snow and Ice Data Center.*

$0.1–0.45\,\mathrm{W\,m}^{-2}$ to global warming in the interval from 1979 to 2008—reinforcing the ongoing climate change due to the accumulation of greenhouse gases in the atmosphere (Hudson, 2011; Flanner et al., 2011).[e]

Historical records of whaling near Antarctica show an increasing frequency of kills at extreme southern latitudes, consistent with a long-term decline in the extent of sea ice in the Southern Ocean (de la Mare, 1997; Cotte and Guinet, 2007). Until recently satellite measurements of the Southern Ocean had shown relatively little change in the area of sea ice surrounding Antarctica (Zwally et al., 2002; Cavalieri and Parkinson, 2008), before beginning a precipitous decline in 2014 (Parkinson, 2019). Future losses of sea ice are of particular concern to biogeochemistry in the Southern Ocean, where penguins and other species depend on algae that inhabit the underside of the ice pack (McClintock et al., 2008; Kohlbach et al., 2018).

Terrestrial water balance

In response to future global warming, most climate models predict a more humid world, in which the movements of water in the hydrologic cycle through precipitation, evaporation, and runoff are enhanced (Loaiciga et al., 1996; Huntington, 2006). Increased cloudiness may moderate the degree of warming, but a new steady state in Earth's temperature will

[e]Compare to the $2.3\ \mathrm{W\ m}^{-2}$ estimated to derive from the current accumulation of greenhouse gases from human activities (Chapter 3).

be found at a higher value than that of today (Chapter 3). Not all areas of the land will be affected equally. Most of the anticipated temperature change is confined to high latitudes. Permafrost is melting and arctic rivers now carry a significantly enhanced flow of water compared to the early 1960s (Peterson et al., 2006; McClelland et al., 2006).

Over much of the world, the historical record of precipitation is scanty. There is no discernable trend in global annual precipitation from 1850 to the present (van Wijngaarden and Syed, 2015), but increased precipitation is reported from some shorter periods of record (Bradley et al., 1987; Wentz et al., 2007). The changes are small and areas with greater precipitation partially compensate for those reporting lesser amounts (Milly et al., 2005; Smith et al., 2006). In many areas, precipitation seems to be becoming more variable; that is, floods and droughts are more frequent—consistent with the predictions of several general circulation models of future climate (Min et al., 2011).

We must hope that global estimates of precipitation will improve dramatically with further application of satellite remote sensing (Petty, 1995), especially since the launch of the Global Precipitation Measurement (GPM) satellite in 2013.[f] Because water vapor absorbs microwave energy, the relative transmission of microwave radiation through the atmosphere is related to water vapor content and rainfall, and satellite remote sensing of the microwave emission from Earth can measure the rainfall and soil moisture over large areas (e.g., Weng et al., 1994). Various applications of radar are also promising for global estimates of precipitation (Tang et al., 2017).

Greater precipitation should lead to greater runoff from land (Miller and Russell, 1992). Probst and Tardy (1987) found a 3% increase in stream flow in major world rivers over the last 65 years (cf. Labat et al., 2004; McCabe and Wolock, 2002; Syed et al., 2010). This increased stream flow may be an indication of global climate change, but it may also relate to the human destruction of vegetation leading to greater runoff (DeWalle et al., 2000; Brown et al., 2005; Rost et al., 2008; Peel et al., 2010). We might also speculate that greater stream flow is expected due to greater water-use efficiency by vegetation growing in a high-CO_2 atmosphere (Gedney et al., 2006; Betts et al., 2007; Chapter 5). In cold temperate regions, wintertime snowpack will be smaller and the spring runoff earlier in a warmer climate (Barnett et al., 2005; Burns et al., 2007).

The historical patterns of runoff for each continent and for the world as a whole show a cyclic pattern (Probst and Tardy, 1987). The cycles for the continents are not synchronous, so the trends in the global record are "damped," relative to those on each continent. Runoff is generally predicted to increase in models of future climate (e.g., Milly et al., 2005; Arnell and Gosling, 2013), but there differences among major regions of the world, with some showing increases and some showing extensive decreases (Fig. 10.9). Impoundments of water in reservoirs may reduce regional impacts on the availability of water to humans. Depletion of groundwater, which is a source of the base-flow of rivers, is likely to cause reductions in riverflow worldwide by 2050 (de Graaf et al., 2019).

One might expect that a warmer land surface would increase rates of evaporation and a warmer atmosphere might contain more water vapor. Water-balance studies of the terrestrial watersheds show an increase in evapotranspiration during the past 50 years (Walter

[f] http://pmm.nasa.gov/GPM/.

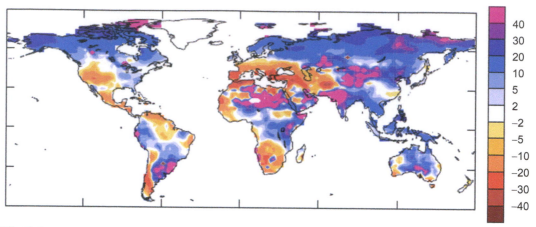

FIG. 10.9   Projected percent changes in runoff from the Earth's land surface for 2041–2060, compared to the mean for 1900–1970. *From Milly et al. (2005). Used with permission of the American Association for the Advancement of Science.*

et al., 2004a), correlated with an increased occurrence of drought (Dai et al., 2004), especially in the southwestern United States (Andreadis and Lettenmaier, 2006; Seager et al., 2007). Analyses of long-term records show increasing evaporation in recent years (Szilagyi et al., 2001; Golubev et al., 2001; Brutsaert, 2006). Global evapotranspiration has increased during most of the past 25 years, except during periods of drought associated with El Niño events (Jung et al., 2010). There are also indications of increases in humidity (Willett et al., 2007), particularly in regions where irrigated agriculture is widespread (Sorooshian et al., 2011).

In future climates, large areas of the central United States and Asia may experience a reduction in soil moisture, leading to more arid conditions. Due to the thermal buffering capacity of water, the oceans may warm more slowly than the land surface. Because most precipitation is generated from the oceans, land areas may experience severe drought during the transient period of global warming (Rind et al., 1990; Dirmeyer and Shukla, 1996). Such changes in precipitation and temperature will lead to large-scale adjustments in the distribution of vegetation and global net primary production (Emanuel et al., 1985; Neilson and Marks, 1994; Smith et al., 1992b).

In sum, the recent increases in precipitation, evaporation, and stream flow are consistent with predicted changes in the water cycle with global warming, but such observations must be evaluated in the context of long-term cycles in climate that have occurred through geologic time.

Summary

Through evaporation and precipitation, the hydrologic cycle transfers water and heat throughout the global system. Receipt of water in precipitation is one of the primary factors controlling net primary production on land. Changes in the hydrologic cycle through

C

Water Scarcity Index

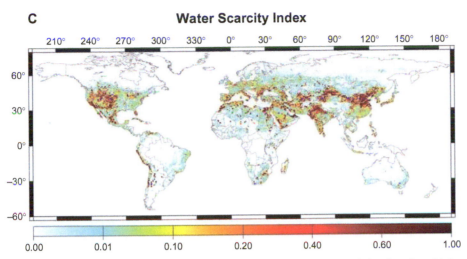

FIG. 10.10   A water scarcity index for the Earth's land surface. Water scarcity is defined as the withdrawal from surface water divided by recycling, adjusted for regional supplements from desalinization. *From Oki and Kanae (2006).*

geologic time are associated with changes in global temperature. All evidence suggests that movements of water in the hydrologic cycle were slower in glacial time, higher in inter-glacial periods, and expected to increase with global warming. Movements of water on the surface of the Earth affect the rate of rock weathering and other biogeochemical phenomena.

Management of water resources will be critical to sustain the world's growing population. Data suggest that there is substantial room for improved management options. People use water very differently in different nations, with per capita consumption ranging from $700\,m^3/yr$ in China to $2480\,m^3/yr$ in the United States (Hoekstra and Chapagain, 2007). Decreases in the per capita availability of water are likely to accompany changes in climate and human population during the next several decades, leaving substantial areas of the world with an increasing scarcity of water (Vörösmarty et al., 2000; Fig. 10.10).

The Global Carbon and Oxygen Cycles

Introduction

The carbon cycle is of central importance to biogeochemistry. Life is composed primarily of carbon, so estimates of the global production and destruction of organic carbon give us an overall index of the health of the biosphere—both past and present. Photosynthetic organisms capture sunlight energy in organic compounds that fuel the biosphere and account for the presence of molecular O_2 in our atmosphere. Thus, the carbon and oxygen cycles on Earth are inextricably linked, and the presence of O_2 in Earth's atmosphere sets the redox potential for organic metabolism in most habitats. Through oxidation and reduction, organisms transform the other important elements of life (e.g., N, P, and S) in reactions that capitalize on the presence of organic carbon and oxygen on Earth. When we understand the carbon cycle, we can make accurate first approximations of the movement of elements in other global cycles, recognizing the predictable stoichiometry of the chemical elements in organic matter. Finally, there is good evidence that through the burning of fossil fuels and other activities, humans have altered the global cycle of carbon, causing the atmospheric concentration of CO_2 to rise to levels that have never been experienced during the evolutionary history of most species that now occupy our planet, including us (Beerling and Royer, 2011).

In this chapter, we consider a simple model for the carbon cycle on the Earth and for assessing human impacts on that cycle. We then consider the magnitude of past fluctuations in the carbon cycle to gain some perspective on the current human impact. We look briefly at the budget of methane (CH_4) in the atmosphere. Because increasing concentrations of carbon dioxide and methane are associated with global warming through the greenhouse effect (Fig. 3.2), the global carbon cycle is directly linked to considerations of global climate change and to international efforts to combat global warming. Finally, we examine the linkage of the carbon and oxygen cycles on Earth as a means of "cross-checking" our estimated budgets for these elements at the global level.

The modern carbon cycle

The Earth contains about 32×10^{23} g of carbon (Marty, 2012), mostly held in the lower mantle and core (Fischer et al., 2020). About 7×10^{22} g C is found in the upper mantle (Zhang and Zindler, 1993), and a similar amount is contained in the atmosphere, oceans, land plants, and sedimentary rocks at Earth's surface (Table 2.3). Carbon in sedimentary rocks is found in organic compounds (1.56×10^{22} g C; Des Marais et al., 1992) and carbonates (5.4–6.5×10^{22} g C; Li, 1972; Lecuyer et al., 2000). Conventional fossil fuels are estimated to contain only 4×10^{18} g C (Sundquist and Visser, 2005). Live biomass contains a relatively small pool of carbon—about 713×10^{15} g C, distributed between the land and ocean (Kallmeyer et al., 2012; Bar-On et al., 2018).

The sum of the active pools of carbon at the Earth's surface is about 40×10^{18} g C (Fig. 11.1). Dissolved inorganic carbon in the ocean is the largest near-surface pool, which has an

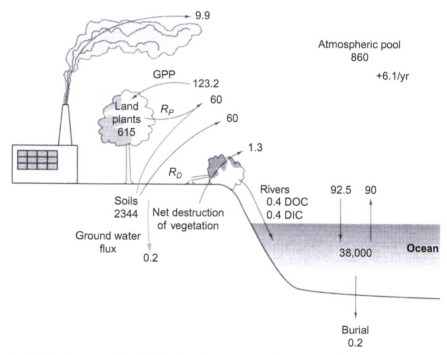

FIG. 11.1　The global carbon cycle for 2016. All pools are expressed in units of 10^{15} g C and all annual fluxes in units of 10^{15} g C/yr. Values are derived in the text.

enormous capacity to buffer changes in the atmosphere via Henry's Law (Eq. 2.7). At equilibrium, the sea contains about 50 times as much carbon as the atmosphere (Raven and Falkowski, 1999). On land, the largest pool of carbon is contained in soils (Table 5.4), followed by vegetation (Table 5.3). Surprisingly, the atmosphere contains more carbon (829×10^{15} g) than all of the Earth's living vegetation (615×10^{15} gC).

The largest fluxes of the global carbon cycle are those that link atmospheric carbon dioxide to land vegetation and the sea (Fig. 11.1). Net primary production (NPP) on land is estimated at 60×10^{15} gC/yr (Chapter 5), consistent with recent estimates of gross primary production (GPP) from the MODIS satellite and the assumption that NPP/GPP ~ 0.5 (Beer et al., 2010; Ryu et al., 2011; Jung et al., 2011). Half of GPP is respired (R_p), and about half of the plant respiration is belowground, where it contributes to soil respiration of $80–100 \times 10^{15}$ gC/yr (Raich et al., 2002; Konings et al., 2018; Warner et al., 2019). Konings et al. (2018) suggest that 43.6×10^{15} gC/yr is derived from soil heterotrophs and $2–4 \times 10^{15}$ gC/yr may derive from biomass burning (Randerson et al., 2012; Andreae, 2019). The mean residence time of carbon in terrestrial vegetation and soil organic matter, about 23 years, varies between 15 and 255 years as a function of latitude, with an inverse relationship to temperature and surprisingly rapid turnover in semi-arid regions (Carvalhais et al., 2014; Wang et al., 2018a).

Considering land vegetation alone, we find that each molecule of CO_2 in the atmosphere has the potential to be captured in gross primary production in about 6 years. The annual exchange of CO_2 with the oceans is somewhat greater, so the overall mean residence time of CO_2 in the atmosphere is about 3 years (cf. Fig. 3.5).[a] Because this mean residence time is similar to the mixing time of the atmosphere, CO_2 shows only small regional and seasonal variations that are superimposed on its global average concentration (Fig. 3.6).

The oscillations in the atmospheric content of CO_2 are the result of the seasonal uptake of CO_2 by photosynthesis in each hemisphere and seasonal differences in the use of fossil fuels and in the exchange of CO_2 with the oceans. The role of photosynthesis is indicated inasmuch as the oscillation in CO_2 is mirrored by slight oscillations in atmospheric O_2, which has a much longer mean residence time in the atmosphere and a much larger pool size (Fig. 1.2). Globally, about two-thirds of the terrestrial vegetation occurs in regions with seasonal periods of growth, and the remainder occurs in the moist tropics, where growth occurs throughout the year (Box, 1988). The seasonal effect of photosynthesis on atmospheric CO_2 is most pronounced in the Northern Hemisphere (Fig. 3.6; Hammerling et al., 2012), which contains most of the world's land area. At high northern latitudes, vegetation accounts for about half of the seasonal variation in atmospheric CO_2 (D'Arrigo et al., 1987). In the Southern Hemisphere, smaller fluctuations in atmospheric CO_2 are seasonally reversed relative to the Northern Hemisphere, and they appear to be dominated by exchange with ocean waters (Keeling et al., 1984). El Nino events also appear to affect the CO_2 uptake by terrestrial vegetation (Zhang et al., 2019).

---

[a] This calculation is based on the traditional definition of mean residence time for a steady-state pool (i.e., mass/input) see footnote b, Chapter 3. A recent report of the Intergovernmental Panel on Climate Change (Houghton et al., 1995; Table 3, p. 25) indicates a mean residence time of 50–200 years for atmospheric CO_2. This is actually the time it would take for the current human perturbation of the atmosphere to disappear into other pools at the Earth's surface (e.g., the ocean) if the use of fossil fuels were to cease. Thus, it would take several centuries to return steady-state conditions to the carbon cycle on Earth.

The oscillation CO_2 at Mauna Loa, Hawaii, located at 19° N latitude, is about 6 ppm/yr (Fig. 1.2), equivalent to a transfer of about 13×10^{15} g C/yr to and from the atmosphere. This value is less than the annual net primary productivity of land plants (Table 5.3) because of the asynchrony of terrestrial photosynthesis and respiration throughout the globe and the buffering of atmospheric CO_2 concentrations by exchange with the oceans.

The release of CO_2 in fossil fuels, currently nearly 10×10^{15} g C/yr, is one of the best-known values in the global carbon cycle.[b] If all of this CO_2 accumulated in the atmosphere, the annual increment would be >1.0%/yr. In fact, the atmospheric increase is about 0.4%/yr (2 ppm) because only about 45% of the fossil fuel release remains in the atmosphere (Sabine et al., 2004; Raupach et al., 2014). This is known as the "airborne fraction," which appears to have declined slightly in recent years (Ballantyne et al., 2015; Keenan et al., 2016). Where is the rest?

Oceanographers believe that about 32% ($\sim 2.8 \times 10^{15}$ g C/yr) of the CO_2 released from fossil fuels enters the oceans each year (Khatiwala et al., 2013; Kouketsu and Murata, 2014; Gruber et al., 2019a), with considerable year-to-year variability (Landschützer et al., 2014; Goto et al., 2017; DeVries et al., 2019). This estimate is derived from repeat measurements of dissolved CO_2 in ocean water (e.g., Takahashi et al., 2006; Carter et al., 2019) and measurements of the simultaneous uptake of trace contaminants, such as chloroflurocarbons, in seawater as a result of human activities (e.g., McNeil et al., 2003; Sweeney et al., 2007; Quay et al., 2003). Thus, Fig. 11.1 shows an annual uptake by the oceans (92.5×10^{15} g/yr) that is slightly greater than the return of CO_2 to the atmosphere (90×10^{15} g C/yr). Following Henry's Law (Eq. 2.7), the excess CO_2 dissolves in seawater, where it causes ocean acidification and dissolution of marine carbonates (Eqs. 9.4, 9.5). Most of the uptake occurs in the large areas of downwelling water in the North Atlantic and Southern oceans (Fig. 11.2) and in continental shelf areas throughout the world (Thomas et al., 2004b; Gruber et al., 2019b). We can expect the uptake of CO_2 by the oceans to increase as the concentration of CO_2 in the atmosphere rises (Gruber et al., 2019a). Owing to low levels of nutrients in seawater, changes in marine NPP are believed to be relatively unimportant to the current oceanic uptake of anthropogenic CO_2, which is largely determined by the dissolution of CO_2 in the surface waters (Shaffer, 1993).

Remembering that the exchange of CO_2 between the atmosphere and the oceans takes place only in the surface waters (Chapter 9), we can calculate the mean residence time of CO_2 in the surface ocean—about 10 years—by dividing the pool of carbon in surface waters (921×10^{15} g C) by the rate of influx (92.6×10^{15} g C/yr). A similar mixing time is calculated from the distribution of ^{14}C in the surface ocean (Chapter 9). Turnover of carbon in the entire ocean is much slower, about 350 years—consistent with the age of deep ocean waters. Thus, the uptake of CO_2 by the oceans is constrained by mixing of surface and deep waters—not by the rate of dissolution of CO_2 across the surface (Chapter 9). It is the rate of release of CO_2 from fossil fuels relative to the rate that CO_2 can mix into the deep ocean that accounts for the current increase of CO_2 in the atmosphere. If the release of CO_2 from fossil fuels were curtailed, nearly all the CO_2 that has accumulated in the atmosphere would eventually dissolve in the oceans and the global carbon cycle would return to a steady state in a couple of hundred years (Laurmann, 1979).

[b] CO_2 that is released during the production of cement, when $CaCO_3$ is roasted to produce CaO, is often included in the emissions from fossil fuels. With time, the $Ca(OH)_2$ in cement absorbs CO_2 from the atmosphere, partially balancing the emissions in its manufacture (Xi et al., 2016).

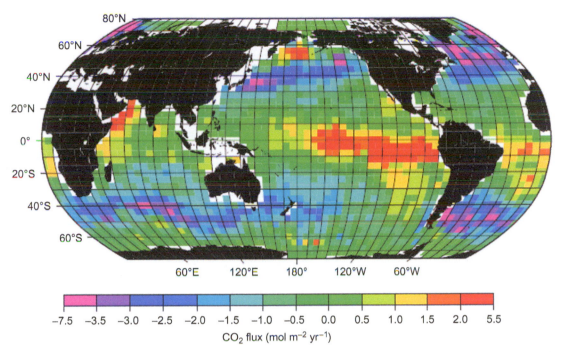

FIG. 11.2 Estimates of the flux of CO_2 between the atmosphere and the ocean's surface for 1995. *From Denman et al. (2007). Used with permission of Cambridge University Press.*

Taken alone, the atmospheric increase and oceanic uptake of CO_2 account for almost 80–90% of the annual emissions from fossil fuels (Sabine et al., 2004). Considering the errors associated with these global estimates, it would seem that we have a fairly tidy picture of the global carbon cycle. However, many terrestrial ecologists believe that there have also been substantial releases of CO_2 from terrestrial vegetation and soils, caused by the destruction of forest vegetation in favor of agriculture, especially in the tropics (Chapter 5). If their calculations are accurate, then the atmospheric budget is unbalanced, and a large amount of carbon dioxide that ought to be in the atmosphere is "missing" (Table 11.1).

The net release of CO_2 from vegetation is difficult to estimate globally (Chapter 5). At any time, some land is being cleared, while agriculture is abandoned in other areas that are allowed to regrow (Chapter 5). For example, the Amazon Basin appears to be a net source of CO_2 to the atmosphere, comprised of losses of CO_2 to the atmosphere by deforestation (0.86×10^{15} gC/yr, Baccini et al., 2017), offset by an uptake of 0.44×10^{15} gC/yr in forest regrowth. Estimates of carbon flux from the Amazonian rainforest are complicated by significant year-to-year variation, and the region is a net source of CO_2 during droughts (Tian et al., 1998; Gatti et al., 2014). Similar observations are reported for rainforests in the Congo (Zhou et al., 2014).

TABLE 11.1   Estimated global budget for anthropogenic co_2 in Earth's atmosphere.

	Fossil fuel combustion and cement production	Biomass destruction	=	Atmospheric increase	Ocean uptake	Terrestrial uptake (inferred)	References
1990s	6.4+	1.6	=	3.2+	2.2+	2.6	IPCC, 2007
2000s	7.8	1.1		4.0	1.6	3.3	IPCC, 2013
2016	9.9	1.3		6.1	2.6	2.7	Le Quéré et al., 2018

Note: *All data in 10^{15} g C/yr.*

Historical changes in the ^{13}C and ^{14}C isotopic ratios in atmospheric CO_2 show unequivocal evidence of a net release of carbon from the biosphere by forest cutting and land clearing in the late 1800s (Wilson, 1978).[c] Both isotopes are diluted by burning of fossil fuels, whereas only ^{13}C is reduced by burning vegetation. Until about 1960, the release from land clearing may have exceeded the release from fossil fuel combustion (Houghton et al., 1983). Cumulative emissions from vegetation and soils to the atmosphere by human activities may have exceeded 400×10^{15} gC since the beginning of civilization (Kaplan et al., 2011; Erb et al., 2018), about half from biomass (Li et al., 2017) and half from soils (Sanderman et al., 2017).

For recent decades, global estimates of the net CO_2 released from deforestation in the tropics range from 0.9 to 1.5×10^{15} gC (Pan et al., 2011; Harris et al., 2012; Houghton and Nassikas, 2017), but some estimates of the release of carbon from tropical deforestation are lower—0.81×10^{15} gC/yr (Harris et al., 2012; Achard et al., 2014). These estimates are fraught with difficulty stemming from irregularities in reported deforestation rates and in the occurrence of drought. Global estimates of changes in the carbon held in vegetation and soils will improve with the application of remote sensing by satellites, which show a significant flux of carbon from selective logging, not normally considered in estimates of deforestation (Asner et al., 2010; Federici et al., 2015). Old-growth tropical forests also appear to provide net carbon storage from the atmosphere (Luyssaert et al., 2008; Lewis et al., 2009; Davidson et al., 2012). Any net release of carbon from land complicates our ability to balance a carbon dioxide budget for the atmosphere (Table 11.1).

The release of carbon from tropical deforestation is partially balanced by the carbon captured by the regrowth of forests in temperate and boreal regions, estimated from 0.65 to 2.1×10^{15} gC (Goodale et al., 2002; Myneni et al., 2001; Pan et al., 2011; Houghton et al., 2018). Forests in the United States—mostly in New England—now appear to accumulate $0.1–0.2 \times 10^{15}$ gC/yr (Birdsey et al., 1993; Turner et al., 1995; Woodbury et al., 2007; Zhang et al., 2012a). Similar amounts of forest regrowth are estimated for Europe (0.165×10^{15} gC/yr; Peters et al. 2010), China ($<0.26 \times 10^{15}$ gC/yr; Piao et al., 2009), and Russia ($<0.13 \times 10^{15}$ gC/yr; Beer et al., 2006). The scrub vegetation that dominates arid and semiarid lands globally may account for a large portion of the year-to-year variability in carbon uptake by the terrestrial biosphere (Ahlström et al., 2015; Biederman et al., 2017). Overall, the global net sink for carbon in

[c]The changes in $^{14}CO_2$ in the atmosphere were only meaningful until the early 1960s, when the atmospheric signal was overwhelmed by the release of radiocarbon from atmospheric testing of nuclear weapons.

vegetation may be about 1.1×10^{15} gC/yr (Pan et al., 2011)—a release of 4.2×10^{15} gC/yr balanced by an uptake of 3.1×10^{15} gC/yr (Houghton et al., 2018). At 0.6×10^{15} gC/yr, the Northern hemisphere appears to dominate the net sink for carbon in vegetation (Ciais et al., 2019).

The successful launch of the OCO-2 (Orbiting Carbon Observatory) satellite in July 2014, provides a means to monitor the emissions of CO_2 from specific regions and nations, as a means of validating the emissions-reduction commitments specified by international agreements to mitigate climate change (Hakkarainen et al., 2016).[d] The satellite measures the amount of sunlight reflected off CO_2 molecules in the atmosphere above a 1.3×2.25-km^2 "footprint" on the Earth's surface. Recently, this satellite was used to detect emissions of 42×10^9 g CO_2 ($=11 \times 10^9$ gC) per day from Mt. Yasur, an island volcano in the remote Pacific, and the satellite appears capable of monitoring the emissions from individual coal-fired power plants (Nassar et al., 2017). OCO-2 also monitors chlorophyll fluorescence, the emission of red light by photosynthesizing plants, which is well correlated to photosynthesis (Frankenberg and Berry, 2018). This offers a promising way to constrain estimates of the global gross primary productivity of vegetation (Sun et al., 2017; Li et al., 2018b), recently suggested to be 167×10^{15} gC/yr (Norton et al., 2019). Estimates using near-infrared reflectance of vegetation suggest 131 to 163×10^{15} gC/yr (Badgley et al., 2019).

What might stimulate net primary production?

We might reconcile the carbon dioxide budget of the atmosphere if we find evidence that the pool of carbon in land vegetation and soils has increased as a result of a global stimulation of plant growth by higher concentrations of atmospheric CO_2 (Chapter 5). Despite widespread forest destruction, enhanced uptake of CO_2 in areas of undisturbed vegetation could act as a sink for atmospheric CO_2 and add to the pool of carbon on land. The overall stimulation of terrestrial photosynthesis by human activities is informally known as the "beta" factor in models of the global carbon cycle. Beta is usually defined as the change in NPP that would derive from a doubling of atmospheric CO_2 concentration. In controlled experiments with tree seedlings, the beta factor usually lies in a range of 32–41% as a result of CO_2 fertilization (Poorter, 1993; Curtis and Wang, 1998; Wang et al., 2012). Free-Air CO_2 Enrichment (FACE) experiments (Chapter 5) show an average 18% stimulation of net primary production in forests grown at 1.5 times current levels of CO_2 (Norby et al., 2005). Vegetation also responds to rising atmospheric CO_2 concentrations by increasing water-use efficiency, which results in higher (less negative) values for $\delta^{13}C$ in wood and lower $\delta^{13}C$ in atmospheric CO_2 (Keeling et al., 2017, Chapter 5).

Most of the net carbon uptake of forests grown at high CO_2 is stored in woody biomass (McCarthy et al., 2010); changes in soils are less dramatic (Lichter et al., 2008). Through a reallocation of photosynthate to roots, some trees appear to avoid the nutrient deficiencies that might be expected to develop as a result of faster growth at high CO_2 (Drake et al., 2011). However, soil nitrogen appears to constrain the long-term growth response at the FACE experiment at Oak Ridge, Tennessee (Norby et al., 2010; Garten Jr et al., 2011), and

[d]OCO-3 was launched and deployed on the International Space Station on May 10, 2019.

nutrient deficiencies are likely to limit enhanced carbon storage by vegetation globally (van Groenigen et al., 2006; Wieder et al., 2015).

The historical record of CO_2 in the atmosphere offers several indirect approaches for estimating changes in global net primary production and the potential for a significant, positive beta factor. For example, in the record of atmospheric CO_2 in the Northern Hemisphere, the seasonal decline each summer is largely due to photosynthesis, while the seasonal upswing in the autumn derives from decomposition. An increasing *amplitude* of the CO_2 oscillation, after the removal of fossil fuel and El Niño effects, implies a greater activity of the terrestrial biosphere outside the tropics, where growth and decomposition occur year-round. Such a trend is evident in an analysis of the Mauna Loa record of CO_2 (Fig. 11.3), in which the amplitude has increased by about 0.54%/yr from 1958 until the mid 1990s (Bacastow et al., 1985; Keeling, 1993). The increase of amplitude at high northern latitudes has been about 50% since 1960, perhaps as a result of rising CO_2 and climatic warming in this region over the same period (Graven et al., 2013; Forkel et al., 2016; Piao et al., 2018). Although an increasing annual oscillation in the concentration of atmospheric CO_2 suggests that biospheric processes have been stimulated, we should not necessarily assume that a greater amount of carbon is being stored on land. Greater rates of decomposition may simply balance increased rates of photosynthesis (Houghton, 1987; Keeling et al., 1996; Piao et al., 2008).

An independent analysis based on trends in atmospheric carbonyl sulfide (COS) suggests that the global uptake of CO_2 by vegetation (i.e., GPP) has increased in recent decades. Namely, vegetation is a major sink for COS (Table 3.8), which enters plants through stomates and is destroyed biochemically in their cells (Spielmann et al., 2019). The uptake of COS should be related to GPP as a function of the ratio of COS to CO_2 in Earth's atmosphere

FIG. 11.3 Increasing amplitude of the seasonal oscillations in atmospheric CO_2 at Point Barrow, Alaska, and Mauna Loa, Hawaii. *From Graven et al. (2013).*

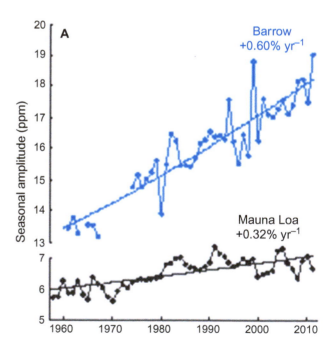

(Asaf et al., 2013; Whelan et al., 2018). The recent decline in COS in the atmosphere suggests that GPP may have increased as much as 31% (Campbell et al., 2017). This suggestion is certainly worthy of further investigation as an independent measure of changes in gross primary production globally. These various lines of evidence suggest that the net uptake by vegetation has decreased the airborne fraction of CO_2 that remains in the atmosphere in recent years (Ballantyne et al., 2015; Keenan et al., 2016).

Despite substantial theoretical and indirect evidence that it *should* occur, the direct evidence that an enhanced growth of land plants (e.g., changes in tree-ring thickness) currently acts as a large sink for atmospheric CO_2 is equivocal (Chapter 5). Our knowledge of the overall response of the terrestrial biosphere to future conditions is limited to a few studies that have examined changes in both CO_2 and temperature in field experiments. Whereas plant uptake of CO_2 might increase as a result of longer growing seasons, one might expect substantial carbon losses from warmer soils (Lu et al., 2013; Crowther et al., 2016). For example, tundra ecosystems appear to be net sources of CO_2 to the atmosphere under the current warmer climate and rising CO_2 that they experience (Oechel et al., 2014; Belshe et al., 2013). Nevertheless, in a 20-year soil-warming experiment, Sistla et al. (2013) report no significant changes in carbon storage in Arctic tundra. It is possible that the warmer soil temperatures enhanced decomposition, improving the supply of nutrients for plant growth (Van Cleve et al., 1990). Because the C/N ratio of soil organic matter (12–15) is lower than the C/N ratio of plant tissues (about 160; Table 6.5), a small amount of additional nitrogen mineralization in soils could yield a large enhancement of NPP and carbon sequestration in plants (Rastetter et al., 1992; McGuire et al., 1992). In a soil-warming experiment at the Harvard Forest in Massachusetts. Melillo et al. (2011) found that the mobilization of nitrogen from decomposing soil organic matter stimulated carbon uptake by the trees, so that changes in net carbon storage in the entire ecosystem were relatively minor.

Many areas of the world receive an excess atmospheric deposition of nitrogen derived from anthropogenic emissions of NO_x and NH_3 (Chapters 3 and 12). In some areas the N input is so extreme that symptoms of forest decline are observed (Chapter 6), but in other areas the added nitrogen has the potential to act as a fertilizer, stimulating plant growth (Townsend et al., 1996). If the global human production of fixed N ($\sim100 \times 10^{12}$ g/yr; Chapter 12) were all stored in the woody tissues of plants with a C/N ratio of ~160 (Table 6.5), then as much as 16×10^{15} gC/yr might be stored in terrestrial ecosystems. Such a large storage is unlikely because not all nitrogen falls on forests and some nitrogen is removed from the land by runoff and denitrification (Schlesinger, 2009). In some areas, the fate of the excess nitrogen deposited from the atmosphere is unclear—some may accumulate in soil organic matter, leading to greater carbon storage (Hyvonen et al., 2008; Nave et al., 2009; Templer et al., 2015). In forests, the change in carbon storage (kg) per unit nitrogen deposition (kg) is about 40 ± 20 (Hogberg, 2012)—about half in woody biomass (Schulte-Uebbing and de Vries, 2018; Goodale, 2017). Using a slightly higher value, Thomas et al. (2010) indicate that forests may accumulate 0.31×10^{15} gC/yr as a result of inadvertent nitrogen fertilization worldwide.

Over longer periods of time, changes in the distribution of vegetation as a result of global climate change could also affect the concentration of atmospheric CO_2 (Chapter 5). Coupled to models of climate change, most models of the global carbon cycle suggest an increase in the carbon content of vegetation and soils when vegetation is in equilibrium with a warmer and wetter world of the future (Smith et al., 1992b). If the adjustment of vegetation to climate occurs over 100 years, these models suggest that the net uptake by the terrestrial biosphere could be as high as 1.8×10^{15} gC/yr—mostly in vegetation. However, other models suggest that changes in vegetation and soils during the transition in climate may yield the opposite effect. Smith and Shugart (1993) estimate large losses of carbon from vegetation during the transient period of drought that is likely to accompany global warming over the next century.

In sum, whole-ecosystem response will be determined by various factors—CO_2, nutrient availability, and global patterns of temperature and rainfall, which are all affected by human activities. The increase in CO_2 in Earth's atmosphere appears particularly sensitive to changes in soil moisture (Humphrey et al., 2018; Green et al., 2019). Although it seems unlikely that enhanced growth by terrestrial vegetation will ultimately stem the rise of CO_2 that is derived from fossil fuels (e.g., Idso and Kimball, 1993), the response of the terrestrial biosphere could have a dramatic impact on the future composition of the atmosphere.

In our view of the global carbon cycle, it is important to recognize that the annual movements of carbon, rather than the amount stored in various reservoirs, are most important. Desert soil carbonates contain more carbon (930×10^{15} g) than land vegetation, but the exchange between desert soils and the atmosphere is tiny (0.023×10^{15} gC/yr), yielding a turnover time of 85,000 years in that pool (Schlesinger, 1985). In the global carbon cycle, a flux that has not changed in recent times, no matter how large, is not likely to affect the concentration of atmospheric CO_2 (Houghton et al., 1983). For example, the release of CO_2 in forest fires is of no consequence to changes in atmospheric CO_2 unless the frequency or area of forest fires has changed in recent times (Auclair and Carter, 1993; Kasischke et al., 1995; Walker et al., 2019). The carbon flux in rivers or sinking pteropods cannot serve as a net sink for anthropogenic CO_2 in the ocean, unless the flux in these pathways is greater as a result of human activities. Similarly, the storage of carbon in peatland soils is not a sink for fossil fuel CO_2, unless the rate of storage in these areas has increased significantly during the Industrial Revolution. These peatlands were accumulating carbon throughout the Holocene when atmospheric CO_2 has been relatively constant (Harden et al., 1992; Treat et al., 2019).[e] On the other hand, landfills (0.12×10^{15} gC/yr; Barlaz, 1998), wooden structures (Churkina et al., 2010) and other wood products (0.10×10^{15} gC/yr; Johnston and Radeloff, 2019) are sinks for CO_2 released from fossil fuel combustion during the past century. The urge to create new sinks for CO_2 has motivated interest in afforestation worldwide (Griscom et al., 2017).

Relatively small changes in large pools of carbon can have a dramatic impact on the carbon dioxide content of the atmosphere, especially if they are not balanced by simultaneous changes in other components of the carbon cycle. A 1% increase in the rate of decomposition on land, as a result of global warming, could release nearly 30 to 203×10^{15} gC/yr to the atmosphere by mid-century (Crowther et al., 2016). Recent losses of carbon from soils throughout England are attributed to a warming climate (Bellamy et al., 2005). Losses of carbon from the pool of soil organic matter frozen in permafrost have the potential to release large

---

[e]Many who deny the potential impact of humans on climate argue that the flux from fossil fuels is so small compared to CO_2 from natural sources that it couldn't possibly have much effect on the concentration of atmospheric CO_2. This argument is easily dismissed, since the flux to and from the land and sea must have been balanced when CO_2 concentrations were relatively constant before the Industrial Revolution. The flux from fossil fuels is new to the carbon cycle and not balanced by other large human-induced sinks for carbon.

amounts of CO_2 to the atmosphere (Dorrepaal et al., 2009; Schuur et al., 2009, 2015; Schaefer et al., 2011), perhaps as much as 0.47×10^{15} gC/yr (Zhuang et al., 2006). Bond-Lamberty et al. (2018) suggest that soil heterotrophic respiration has increased globally over recent decades due to global warming.

Similarly, a 1%/yr increment to the biomass of carbon on land, as a result of a greater storage of NPP, could balance the CO_2 budget in the atmosphere and stem the rise of CO_2 (Table 11.1). We can speculate that this increment should be first realized in vegetation, which has a faster turnover time than soils; only a small percentage of NPP that enters the soil survives to become a component of soil organic matter, which has very slow turnover (Schlesinger, 1990; He et al., 2016; Chapter 5).

'At the surface of the Earth, the largest global pool of organic carbon is found in sedimentary rocks, including fossil fuels (Table 2.3). This organic matter was derived from net primary production in the oceans and by particulate organic carbon (POC; 0.2×10^{15} gC/yr) and charcoal (0.026×10^{15} gC/yr) carried by rivers to the sea (Galy et al., 2015; Jaffé et al., 2013). Storage of organic carbon in sedimentary deposits accounts for the accumulation of O_2 in the atmosphere through geologic time (Chapter 2). In the absence of human perturbations, the exchange between the fossil pool and the atmosphere could be ignored in global models. Only a small amount of buried sedimentary organic matter is exposed to uplift, erosion, and oxidation each year—0.043 to 0.097×10^{15} gC/yr (Di-Giovanni et al., 2002; Copard et al., 2007). In extracting fossil fuels from the Earth's crust, humans affect the global system by creating a large biogeochemical flux, $\sim 10 \times 10^{15}$ gC/yr, where essentially none existed before. The rise in atmospheric CO_2 is closely correlated with the rise in human population (Hofmann et al., 2009), economic activity (Gozgor et al., 2019), and the "carbon-intensity" of the modern economy (Canadell et al., 2007). Carbon dioxide emission from fossil fuel combustion dwarfs the CO_2 exhaled by the human population worldwide, 0.6×10^{15} gC/yr (Prairie and Duarte, 2006; West et al., 2009).

Temporal perspectives of the carbon cycle

Studies of the biogeochemistry of carbon on Earth must begin with a consideration of the origin of carbon as an element and with theories that explain its differential abundance on the planets of our solar system (Chapter 2). During the early development of Earth, the carbon cycle was decidedly non-steady-state: The carbon content of the planet grew with the receipt of planetesimals and meteorites—especially carbonaceous chondrites—and the atmospheric content increased as volcanoes released CO_2 trapped in Earth's mantle. The oldest geologic sediments suggest that atmospheric CO_2 may have been as high as 3% on primitive Earth, contributing to a greenhouse effect during a time of low solar output (Walker, 1985; Rye et al., 1995). Today small amounts of water vapor and CO_2 in our atmosphere maintain the surface temperature of the Earth above freezing—obviously an essential condition for the persistence of the biosphere (Ramanathan, 1988; Lacis et al., 2010; Schmidt et al., 2010a).

As discussed in Chapter 4, CO_2 in the atmosphere interacts with the crust of the Earth, causing rock weathering (Eq. 4.3). Carbon dioxide is removed from the atmosphere and transferred via rivers to the oceans, where it is eventually deposited on the seafloor in carbonate rocks, returning to the Earth's crust (Fig. 1.3). Along with carbonates, undecomposed organic materials also accumulate in marine sediments from photosynthesis on land and in the sea. As early as 1918, Arrhenius (1918, p. 177) speculated that the consumption of CO_2 by rock weathering might eventually cool the planet by causing a loss of its natural "greenhouse effect":

> As the crust grew thicker, the supply of this gas [CO_2] diminished and was further used up in the process of disintegration [weathering]. As a consequence the temperature slowly decreased, although decided fluctuations occurred with the changing volcanic activity during different periods. Supply and consumption of carbon dioxide fairly balanced as disintegration ran parallel with the proportion of this gas in the air.

Geologic epochs of enhanced rock weathering are associated with lower concentrations of atmospheric CO_2 (Misra and Froelich, 2012; Macdonald et al., 2019). Fortunately, CO_2 is returned to the atmosphere as a result of tectonic activity. In the complete geochemical cycle of carbon (Fig. 1.3), subduction of the oceanic crust carries carbon deposited on the seafloor to the interior of the Earth, where CO_2 and other volatile elements are once again released by hydrothermal and volcanic emissions. Plank and Manning (2019) estimate that up to 0.082×10^{15} gC/yr are subducted worldwide. Some of the subducted oceanic crust may mix to 1000-km depth in the mantle (Walter et al., 2011; Drewitt et al., 2019), and its carbon returns to the atmosphere in volcanoes at mid-ocean ridges and volcanic islands. Presumably the rest of the carbon is returned to the atmosphere from shallower depths by the emissions of arc volcanoes.

Today, the Earth's mantle appears to release about 0.079×10^{15} gC to the atmosphere each year in volcanic eruptions (Kerrick, 2001; Dasgupta and Hirschmann, 2010; Plank and Manning, 2019). Total volcanic emanation of volatile carbon from the Earth's surface, including both degassing of the mantle and of recycled sedimentary materials, is estimated at about 0.05 to 0.10×10^{15} gC/yr (Morner and Etiope, 2002; Fischer and Aiuppa, 2020)[f]—similar to estimates of carbon accumulation in ocean sediments. The isotopic composition ($\delta^{13}C$) of volcanic emissions is similar to that of sedimentary rocks, especially limestone, implying a common origin (Mason et al., 2017). At the current estimated rates of subduction and volcanic emissions, the entire mass of carbon in the upper mantle, 7×10^{22} g, would recycle in less than a billion years (Dasgupta and Hirschmann, 2010); thus, it is likely that much of the carbon in the mantle has spent some time in the biosphere in the geologic past. If this cycle were not complete, rock weathering would deplete CO_2 in the atmosphere and oceans in <500,000 years, and all carbon would be stored in sedimentary rocks (Moon et al., 2014; Colbourn et al., 2015).

On Earth, this geochemical cycle has helped to maintain the concentration of atmospheric CO_2 below 1% (10,000 ppm) for the last 100 million years (Berner and Lasaga, 1989). On Mars, where this cycle has slowed or stopped, the atmosphere contains a small amount of CO_2, and the planet is very cold (Chapter 2). On Venus, which is too hot for CO_2 to react with crustal minerals, the atmosphere contains a large amount of CO_2, reinforcing its greenhouse effect

---

[f]Those who deny the reality of human-induced climate change often love to cite volcanic emissions as the source of rising CO_2 in Earth's atmosphere. This argument is easily dismissed, based on the $\delta^{13}C$ in the CO_2 of volcanoes (Fischer and Lopez, 2016), measurements of their overall flux to the atmosphere compared to that from fossil fuels, and no indication that the record of atmospheric CO_2 shows any response to recent large volcanic eruptions.

(Walker, 1977; Nozette and Lewis, 1982). During periods of extensive volcanism, the atmospheric concentration of CO_2 on Earth may have been greater than today, leading to warmer climates (Owen and Rea, 1985; Gutjahr et al., 2017)[g]; however, a continuous geologic record of liquid oceans on Earth indicates that CO_2 and other greenhouse gases have always remained at levels that produce relatively moderate surface temperatures. Deposition of carbon in ocean sediments by various biotic activities is an example of how life may help promote long-term stability in Earth's climate, favorable to the persistence of life.

Despite their long-term significance in buffering atmospheric CO_2, the annual transfers of carbon in the geochemical cycle are relatively small. The massive quantities of CO_2 that are now tied up in the carbonate minerals of the Earth's crust are the result of a slow accumulation of these materials over long periods of Earth's history. Today rivers carry about 500×10^{12} g/yr of Ca^{2+} (Milliman, 1993) and 0.40×10^{15} g/yr of carbon as HCO_3^- to the sea (Sarmiento and Sundquist, 1992). About 65% of the HCO_3^- is derived from the atmosphere and the rest from carbonate minerals (Gaillardet et al., 1999; Suchet et al., 2003; Hartmann et al., 2009). For seawater to maintain fairly constant concentrations of calcium, an equivalent amount of Ca must be deposited as $CaCO_3$ in ocean sediments, carrying 0.15×10^{15} gC/yr to the oceanic crust. Dividing the mass of carbonate rocks by their annual rate of formation, we find that each atom of carbon sequestered in marine carbonate spends more than 400 million years in that reservoir.

With the appearance of life, a *bio*geochemical cycle was added on top of the underlying geochemical cycle of carbon on Earth. Models of the modern biogeochemical cycle of carbon focus on the large annual transfer of CO_2 from the atmosphere to plants as a result of photosynthesis and the large return of CO_2 to the atmosphere as a result of decomposition (Fig. 11.1). Today, the fluxes of carbon in the biogeochemical cycle of carbon, mostly expressed in units of 10^{15}–10^{17} gC/yr, dwarf the fluxes of the underlying geochemical cycle of carbon, where the movements are typically 10^{13}–10^{14} g C/yr.

During Earth's history, at times when the production of organic carbon by photosynthesis has exceeded its decomposition, organic carbon has accumulated in geologic sediments. The earliest organic carbon is present in rocks from 3.8 bya, with the pool increasing to 1.56×10^{22} g by about 540 mya (Des Marais et al., 1992). During that interval, about 20% of all the carbon buried in marine sediments was organic—similar to the ratio found in modern marine sediments (Li, 1972; Holser et al., 1988; Dobrovolsky, 1994, p. 163). During the Carboniferous Period (300 mya), large deposits of organic carbon were also stored in freshwater environments, leading to modern economic deposits of coal.[h] During the Tertiary Epoch, the precursors to modern deposits of petroleum were added to marine sediments.

Net storage of organic carbon in sediments has varied between about 0.04 and 0.07×10^{15} gC/yr during the last 300 million years (Fig. 11.4; Berner and Raiswell, 1983). A greater rate for the present day is likely due to increased soil erosion by human activities (Regnier et al., 2013). Today, about 10–20% of the global organic burial occurs in the Bengal Fan as a result of rapid burial by Himalayan sediment (Galy et al., 2007; France-Lanord and Derry, 1997).

[g]Interestingly, an emission of only about 0.9×10^{15} gC/yr during the Paleocene-Eocene epochs led to a 5–8°C increase in global temperature (Bowen et al., 2015). Today's fossil fuel emissions are ten times greater.

[h]It is possible that the end of the Carboniferous coincided with the evolution of "white-rot" fungi, which have the capacity to decompose lignin in woody tissues (Floudas et al., 2012).

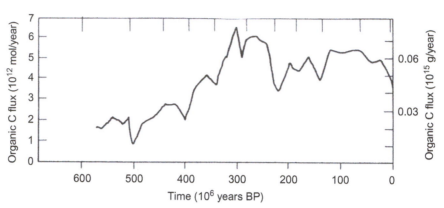

FIG. 11.4　Burial of organic carbon on Earth during the past 600 million years. *From Olson et al. (1985).*

Life also stimulated some of the reactions in the underlying geochemical cycle of carbon. Various marine organisms enhance the deposition of calcareous sediments, which now cover more than half of the ocean's seafloor (Kennett, 1982). Land plants, by maintaining high concentrations of CO_2 in the soil pore space, raise the rate of carbonation weathering, speeding the reaction of CO_2 with the Earth's crust (Moulton et al., 2000; Chapter 4). Land plants and soil microbes also excrete a variety of organic compounds, byproducts of photosynthesis, which enhance rock weathering. About 0.2×10^{15} gC/yr is transported to groundwater as HCO_3^- derived from rock weathering (Kessler and Harvey, 2001). Various models developed and summarized by Robert Berner of Yale University suggest that the atmospheric concentration of CO_2 declined precipitously as land plants gained dominance about 350 mya (Berner, 1992; Berner and Kothavala, 2001; Royer et al., 2001; Rothman, 2002). The record of CO_2 in Earth's atmosphere suggests that concentrations have remained between 150 and 500 ppm for the past 5 million years (Stap et al., 2016) and below 1500 ppm for the past 50 million years (Pagani et al., 2005; Zachos et al., 2008; Franks et al., 2014; Beerling and Royer, 2011). Calcium levels in seawater were higher, atmospheric CO_2 levels were lower, and the Earth's climate was colder during the Miocene, 13,000,000 years ago (Griffith et al., 2008). Indeed, atmospheric CO_2 concentrations and global temperature are well correlated, especially for the last 20 million years (Came et al., 2007; Tripati et al., 2009).

Collections of gas trapped in ice cores from the Antarctic provide a historical record of atmospheric CO_2 for the last 800,000 years (Fig. 1.4). The cyclic fluctuations of CO_2 in Earth's atmosphere imply non-steady-state conditions of the carbon cycle, with a periodicity of 100,000 years. Concentrations varied between 190 and 280 ppm, with the lowest values found in layers of ice that were deposited during glacial periods.[i] CO_2 concentrations and global

---

[i]The consistent minimum at ~200 ppm and maximum at ~280 ppm during glacial-interglacial intervals (Fig. 1.4) is striking. Holland (1965) suggests that the minimum may be related to the onset of gypsum precipitation in ocean sediments, which raises the pH of seawater and leads to the precipitation of calcite (Lindsay, 1979, p. 49). Galbraith and Eggleston (2017) suggest that photosynthesis is markedly lower at CO_2 levels below 190 ppm, potentially limiting the further storage of organic carbon in sediments.

temperatures have been well correlated for the past 160,000 years (Cuffey and Vimeux, 2001). The lowest concentrations of CO_2 were likely to invoke significant physiological effects on land plants, reducing their photosynthesis (Gerhart and Ward, 2010). Although the exact magnitude is controversial, the mass of carbon stored in vegetation and soils was also lower during the last glacial, as a result of the advance of continental ice sheets and widespread desertification of land habitats (Adams et al., 1990; Servant, 1994; Bird et al., 1994; Beerling, 1999). Thus, glacial conditions must have produced changes in the oceans that allowed a large uptake of CO_2. An enhanced transfer of carbon to the deep ocean by the biological pump (Chapter 9) could account for this carbon storage and thus low concentrations of CO_2 in the atmosphere (Rae et al., 2018; Anderson et al., 2019). This carbon dioxide was returned to the atmosphere at the end of the glacial epoch (Yu et al., 2010b; Burke and Robinson, 2012; Schmitt et al., 2012).

A decrease in the amount of carbon stored as carbonates could also lead to a greater retention of CO_2 in the oceans, following Eqs. (9.4) and (9.5). Most of the $CaCO_3$ dissolution in marine sediments is driven by CO_2 released during the decomposition of organic matter (Berelson et al., 1990), so Archer and Maier-Reimer (1994) suggest that an increase in the ratio of organic carbon to carbonate in sinking particles could lead to a greater dissolution of carbonate in marine sediments. More than one paper has suggested adding carbonates to the oceans in an effort to enhance CO_2 uptake through their dissolution (Renforth and Henderson, 2017).

Whatever the mechanism for enhanced carbon storage in the glacial oceans, at the end of the last glacial, 17,000 years ago, atmospheric CO_2 rose to about 280 ppm, where it remained with minor variations until the beginning of the Industrial Revolution (Fig. 11.5;

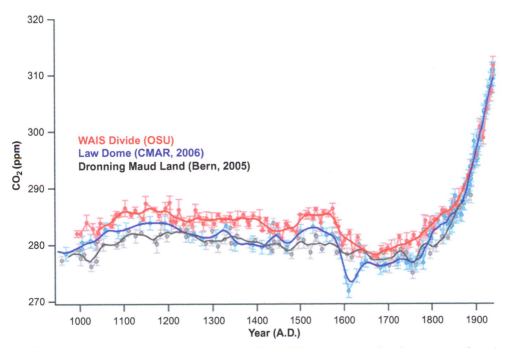

FIG. 11.5 Concentrations of atmospheric CO_2 measured in bubbles of gas trapped in three ice cores from Antarctica. *From Ahn et al. (2012).*

Indermuhle et al., 1999; Meure et al., 2006; Ahn et al., 2012). The rise in CO_2 at the end of the last glacial epoch preceded the rise in temperature and may have been a cause of the rapid post-glacial warming (Shakun et al., 2012; Parrenin et al., 2013; Gadens-Marcon et al., 2014). The increase in concentration from 280 ppm to today's value more than 400 ppm represents a global change of $\sim$50% in less than 200 years! Although the current level of CO_2 is not unprecedented in the geologic record, our concern is the speed at which a basic characteristic of the planet has changed to levels not previously experienced during human history or during the evolution of current ecosystems. Since global temperature and atmospheric CO_2 are related, we seem destined for significant global warming during the near future (Fig. 1.4).

These perspectives on the global carbon cycle extend from processes that occur on a time scale of 10^9 years to those that occur annually. The global carbon cycle is composed of large, rapid transfers in the *bio*geochemical cycle superimposed on the underlying, small, slow transfers of the geologic cycle. Buffering of atmospheric CO_2 over geologic time involves small net changes in carbon storage that occur relatively slowly. Thus, an increase in the natural rate of rock weathering (which consumes $\sim 0.25 \times 10^{15}$ g C/yr) as a result of high CO_2 and rising global temperature (Chapter 4) is not likely to be an effective buffer to the rapid release of CO_2 from fossil fuels ($\sim 10 \times 10^{15}$ g C/yr), and enhanced terrestrial weathering is not likely to be an effective "geo-engineering" solution to the problem of climate change (Schlesinger and Amundson, 2019). In contrast, the current net exchange of CO_2 between the atmosphere and the biosphere is about 150×10^{15} g C/yr, so the biosphere is more likely to buffer the rise of CO_2 due to human activities. The current increase in atmospheric CO_2 results from our ability to change the flux of CO_2 to the atmosphere by an amount that is significant relative to the biogeochemical reactions that buffer the system over long periods of time.

Atmospheric methane

At first glance, the annual flux of methane (CH_4) would seem to be a minor component of the global carbon cycle. All sources of methane in the atmosphere are in the range of 10^{12}–10^{14} g C/yr, which is several orders of magnitude lower than the movements of CO_2 shown in Fig. 11.1. Globally, the atmospheric methane concentration is about 1.83 ppm, versus 400 ppm for CO_2 (Table 3.1). However, over the next century each molecule of methane added to the atmosphere has the potential to contribute 25–35$\times$ or more to the human-induced greenhouse effect compared to each molecule of CO_2 (Lashof and Ahuja, 1990; Shindell et al., 2009; Etminan et al., 2017). Substantial progress in mitigating global warming could accompany a reduction in methane emissions (Montzka et al., 2011a; Shindell et al., 2012).

The ice-core record of CH_4 extending to 800,000 years ago shows repeated cycles, with low values during glacial periods and higher values during interglacial periods (Fig. 11.6). These records show that CH_4 concentrations were about 0.4 ppm during the last glacial period (20,000 years ago), increasing abruptly to the preindustrial value of 0.7 ppm as the glaciers melted (Chappellaz et al., 1990; Loulergue et al., 2008). The increase during deglaciation seems to have occurred while many northern wetlands were still covered with ice, suggesting that changes in tropical wetlands may have caused the initial methane increase, which reinforced the global warming during deglaciation (Chappellaz et al., 1993; Petrenko et al., 2017). With the

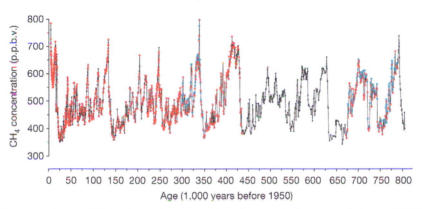

FIG. 11.6   Concentrations of CH$_4$ in air that is extracted from ice cores in Antarctica dating to 800,000 years before present. *From Loulergue et al. (2008). Compare to Fig. 1.4.*

progression of deglaciation, high-latitude, northern wetlands may have also contributed to the accumulation of methane in the atmosphere (Zimov et al., 1997; Walter et al., 2007; Smith et al., 2004), although the amounts may have been small (Dyonisius et al., 2020).

Concentrations of atmospheric methane showed minor variation during the Holocene ($\pm15\%$; Blunier et al., 1995; Sapart et al., 2012; Mitchell et al., 2013), but beginning about 200 years ago the concentration began to increase rapidly at an average rate of about 1%/yr (Fig. 11.7), which was much faster than the rate of CO$_2$ increase over the same interval

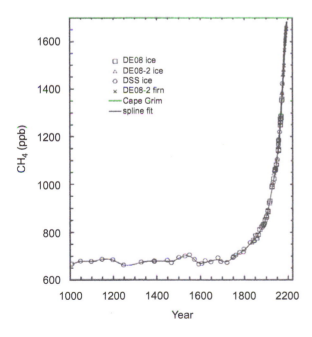

FIG. 11.7   Concentrations of CH$_4$ in air extracted from Antarctic ice cores. *From Etheridge et al. (1998). Used with permission of the American Geophysical Union.*

(Fig. 11.5). The methane concentration in Earth's atmosphere has more than doubled since the beginning of the Industrial Revolution.

Strangely, the rise in methane concentration slowed in the mid-1990s, but the upward trend resumed in 2007 (Dlugokencky et al., 2011; Terao et al., 2011). Whether this was due to changes in sources or changes in the hydroxyl-radical sink (Rigby et al., 2017, see Chapter 3) is hotly debated among biogeochemists, but substantial evidence suggests increased emissions from tropical wetlands (Nisbet et al., 2016; Schaefer et al., 2016).

The cause of the increase in methane in Earth's atmosphere is not obvious, because a wide variety of sources contribute to the total annual production of about 680×10^{12} g/yr (Table 11.2). The sum of anthropogenic sources is about twice the sum of natural sources, so it is perhaps surprising that the annual increase of methane in the atmosphere is not larger. The estimate of total flux is fairly well constrained because it yields a mean residence time for atmospheric CH_4 of about 9 years, which is consistent with independent calculations based on methane consumption (Khalil and Rasmussen, 1990; Prinn et al., 1995; Dentener et al., 2003) and with the spatial variation in CH_4 concentration in the atmosphere (Fig. 3.5).

TABLE 11.2   Estimated sources and sinks of methane in the atmosphere.

	Flux estimates (10^{12} g CH_4/yr) IPCC	Alternatives/ updates	References
Natural sources			
Wetlands	217	164–245	IPCC, 2013; Bridgham et al., 2013, Zhu et al., 2015
Tropics		46	Bloom et al., 2010, [42.7 for the Amazon Basin (Pangala et al., 2017)]
Northern latitude		20	Christensen et al., 1996
Lakes and rivers	40		IPCC, 2013
Rivers and streams		27	Stanley et al., 2016
Natural lakes		54.3	Bastviken et al., 2011
Upland vegetation		10 (estimate)	Megonigal and Guenther, 2008; Kirschbaum et al., 2006
Natural fires	3		IPCC, 2013
Wild animals	15		IPCC, 2013
		11	Smith et al., 2016
Termites	11		IPCC, 2013
		19	Sanderson, 1996
Geologic seepage including hydrates	61		IPCC, 2013
Marine sediments		10	Reeburgh, 2007
Terrestrial		33	Etiope et al., 2008
Total natural sources	331		

TABLE 11.2 Estimated sources and sinks of methane in the atmosphere—Cont'd

	Flux estimates (10^{12} g CH_4/yr) IPCC	Alternatives/ updates	References
Anthropogenic sources			
Fossil fuel related	96		IPCC, 2013
Coal mines		30	Prather et al., 1995
Coal combustion		15	Prather et al., 1995
Oil and gas		72	Neef et al., 2010
Waste and waste management	75		IPCC, 2013
Landfills		18	Bogner and Matthews, 2003
Animal waste		25	Prather et al., 1995
Sewage treatment		25	Prather et al., 1995
Ruminants	89		IPCC, 2013
		97	Dangal et al., 2017
		119	Wolf et al., 2017 (includes emissions from animal wastes, above)
Reservoirs[a]		13	Deemer et al., 2016
Biomass burning	35	50	Andreae, 2019
Rice cultivation	36		IPCC, 2013
Total anthropogenic sources	331		
Total sources	**678**		
Sinks			
Reaction with OH radicals	528		IPCC, 2013
Reaction with Cl	25		IPCC, 2013
Removal in the stratosphere	51		IPCC, 2013
Removal by soils	28		IPCC, 2013
Total sinks	**632**		
Atmospheric increase (2007)	23		Dlugokencky et al., 2009

[a] This was not estimated by IPCC.

For each flux, mean values are taken from Kirschke et al. (2013), subtended by alternative estimates by others. All data in 10^{12} g CH_4/yr.

The concentration of CH_4 is slightly higher in the Northern Hemisphere, suggesting that it is the location of major emissions (Fig. 3.4).

Like CO_2, the concentration of methane in the atmosphere oscillates, showing a minimum concentration in midsummer in the Northern Hemisphere (Steele et al., 1987; Khalil et al., 1993a; Dlugokencky et al., 1994). While methane emissions from wetlands are greatest during warm periods, the summer is also the time of the most rapid destruction of atmospheric methane by OH radicals (Khalil et al., 1993b).

Methanogenesis in wetland habitats is widely acknowledged as the dominant natural source of atmospheric methane (Chapter 7), and the rate of methane release would undoubtedly be much higher if it were not for high rates of methanotrophy by bacteria in wetland soils (Megonigal and Schlesinger, 2002). Matthews and Fung (1987) estimated that 110×10^{12} g/yr stem from anaerobic decomposition in natural wetlands globally, but more recent estimates are twice as high (Table 11.2; Zhang et al., 2017a; Zhu et al., 2015). The rate of production is higher in tropical wetlands than in boreal wetlands (Schütz et al., 1991; Bartlett and Harriss, 1993; Cao et al., 1996), reflecting the positive relationship between temperature, net ecosystem production, and the rate of methanogenesis in many wetland ecosystems (Chapter 7). Because tropical wetlands cover a large area of the world, they dominate the methane flux from wetlands globally (Aselmann and Crutzen, 1989; Fung et al., 1991; Bartlett and Harriss, 1993). Eddy-covariance methods (Chapter 5) have been used to monitor methane release from seasonally flooded tropical forests (Pantanal) in the Amazon Basin (Dalmagro et al., 2019) and in tundra ecosystems in Alaska, where permafrost is melting (Taylor et al., 2018).

A large portion of the current increase in atmospheric methane may derive from an increase in the worldwide cultivation of rice (Sass and Fisher, 1997). Because most rice paddies are found in warm climates, they often yield a large CH_4 flux, which is enhanced by the upward transport of CH_4 through the hollow stems of rice (Chapter 7). Matthews et al. (1991) provide maps of the global distribution of CH_4 production from rice cultivation, which has likely increased by more than 1%/yr during the past several decades (Anastasi et al., 1992). Better management of rice cultivation may be responsible for a reduced flux of CH_4 to the atmosphere in recent years (Kai et al., 2011; Zhang et al., 2016a).

Many grazing animals and termites maintain a population of anaerobic microbes that conduct fermentation at low redox potentials in their digestive tract. Digestion in these animals provides the functional equivalent of a mobile wetland soil! As much as 5% of the dietary intake by ruminants can appear as CH_4 (Charmley et al., 2016). Thus, the belches[j] of grazing animals make a significant contribution to the global sources of methane (Table 11.2). In the early 1980s, about 78×10^{12} g/yr of CH_4 were derived from domestic and wild animals, while human flatulence contributed $<1 \times 10^{12}$ g/yr (Crutzen et al., 1986; Lerner et al., 1988; Polag and Keppler, 2019). A recent global estimate of CH_4 emissions from grazing animals (119×10^{12} g/yr, Wolf et al., 2017) greatly exceeds the methane emissions calculated for the wild herbivores that humans have extirpated (Smith et al., 2016). Increasing demand for meat products is likely to increase methane emissions from animals worldwide (Anastasi and Simpson, 1993; Dangal et al., 2017). Some termites and other insects also make a small but

---

[j]Although many believe that farts are the major pathway for methane release from cattle, about 90% of the methane is released from digestive burps from the rumen (Kebreab et al., 2006). Wilkinson et al. (2012) suggest that similar methane emissions from dinosaurs may have warmed Earth in the Mesozoic.

significant contribution to atmospheric methane as a result of anaerobic decomposition in their hindgut (Khalil et al., 1990; Brauman et al., 1992; Hackstein and Stumm, 1994). It is not likely, however, that the flux of CH_4 from termites has increased significantly in recent years.

Some upland trees are small sources of methane (Zeikus and Ward, 1974; Wang et al., 2016), including production in heartwood (Wang et al., 2016) and in the wet interior of hollow trees (Covey et al., 2012). These local methane sources are normally masked by the otherwise dominant process of methane destruction in upland soils (Pitz and Megonigal, 2017; Covey and Megonigal, 2019). However, do Carmo et al. (2006) reported anomalously high methane concentrations above the canopy of tropical rainforest in Brazil, also seen in satellite observations of atmospheric methane over tropical rainforests (Frankenberg et al., 2005; Beck et al., 2012). Keppler et al. (2006) suggested a large global flux of methane ($62-236 \times 10^{12}$ g/yr) from upland vegetation by an unknown aerobic biochemical pathway. Several biochemical mechanisms have been proposed to explain aerobic methanogenesis (Fraser et al., 2015; Lenhart et al., 2015; Liu et al., 2015), but the global importance of this methane source has now been tempered by the failure of laboratory measurements to confirm the observation and by more cautious extrapolations to a global estimate (Kirschbaum et al., 2006; Megonigal and Guenther, 2008; Dueck et al., 2007; Nisbet et al., 2009). A large flux of methane from upland vegetation cannot be accommodated in the current budget for atmospheric methane (refer to Table 11.2) without yielding a mean residence time that is incompatible with atmospheric measurements and conflicting with the observed midsummer minimum for methane concentrations in the atmosphere. The observations of methane flux from tropical rainforests are likely due to the extensive areas of flooded soils within them (Pangala et al., 2017). Trees in wetland soils may provide a conduit for methane from methanogenesis from soils to the atmosphere. In assessments of contributions to global climate change, the methane from tropical forests may roughly balance their CO_2 uptake (Dalmagro et al., 2019).

Forest fires produce methane as a product of incomplete combustion. We know little about the annual area of burning in the preindustrial world (Marlon et al., 2013), but it is likely that the current release of CH_4 from forest fires has increased as a result of high, recent rates of biomass burning in the tropics (Andreae, 1991). Kaufman et al. (1990) used remote sensing of fires in Brazil to calculate a loss of 7×10^{12} g CH_4/yr in that region in 1987, and (Delmas et al., 1991) found a flux of 9.2×10^{12} g CH_4 from burning of African savannas. CH_4 typically accounts for 1% of the total carbon lost by fire (Levine et al., 1993), so the estimate of methane flux from forest fires (50×10^{12} g CH_4/yr; Table 11.2 is compatible with a range of global estimates of the carbon released in forest fires (2.5×10^{15} gC/yr; Chapter 5).

Humans contribute directly to atmospheric methane during the production and use of fossil fuels and due to the disposal of wastes. The flux of methane from landfills increases linearly with the amount of material buried, which presumably decomposes under anoxic conditions (Thorneloe et al., 1993). The global flux from landfills is estimated at 16 to 20 Tg CH_4/yr (Bogner and Matthews, 2003).

Natural seepage of CH_4 from the Earth and inadvertent releases of fossil CH_4 during the mining and use of coal and natural gas must account for about 20–30% of the total annual flux of CH_4 to the atmosphere, based on the ^{14}C age of atmospheric methane (Ehhalt, 1974; Wahlen et al., 1989; Quay et al., 1999; Etiope et al., 2008). As methane concentrations have risen during the past decade, increasing attention has focused on methane emissions from the oil and gas industry, where fugitive emissions of methane can amount to about 2% of natural gas production (Schwietzke et al., 2014; Marchese et al., 2015; Ren et al., 2019; Alvarez et al., 2018). This is hotly debated. One side suggests that this source of methane has been stable or declining in recent years, as reflected by decreasing concentrations of ethane, which is emitted with methane during natural gas production (Aydin et al., 2011; Simpson et al., 2012). Others believe that oil and gas production, especially increased production by hydraulic fracking, is responsible for the recent rise in methane and ethane concentrations in the atmosphere (Rice et al., 2016; Helmig et al., 2016). The isotopic composition of atmospheric methane tells a different story. Until 2006, the $\delta^{13}C$ of atmospheric methane had increased from $-50‰$ to $-47‰$, consistent with emissions from fossil fuels (Craig et al., 1988; Quay, 1988), but in recent years, the $\delta^{13}C$ of atmospheric methane has decreased from $-47.2‰$ to $-47.4‰$, which is compatible with an increasing flux from wetlands and agriculture (Schaefer et al., 2016; Nisbet et al., 2019). Reconciliation of the changing sources of methane in recent years is much in need of better data (Turner et al., 2019).

Our understanding of the sources of methane in the atmosphere will improve with continued development of methods to measure CH_4 using satellite technology (Frankenberg et al., 2005; Jacob et al., 2016). In one comparison, annual CH_4 emissions for the U.S. were estimated at 30×10^{12} g/yr by satellite vs. $25–28 \times 10^{12}$ g/yr in independent ground-based inventories. The largest source in the U.S. is seen at the Four Corners region—0.59×10^{12} g/yr (Kort et al., 2014). Methane emissions from many urban areas are dominated by leakage from the natural gas distribution systems (Plant et al., 2019; He et al., 2019). In California, methane emissions are dominated by landfills (Duren et al., 2019).

The major sink for atmospheric methane is reaction with hydroxyl radicals in the atmosphere (Chapter 3). Each year about 520×10^{12} g is removed from the troposphere by this process (Neef et al., 2010). The rise in atmospheric methane is paralleled by a rise in formaldehyde—a methane oxidation product (Eq. 3.21)—in polar ice cores (Staffelbach et al., 1991). Some workers have suggested that the rise in atmospheric methane during the Industrial Revolution was derived from a reduction in the sink strength offered by hydroxyl radicals, which react more rapidly with CO, which is also increasing in the atmosphere (Khalil and Rasmussen, 1985). A rise in OH concentrations in the atmosphere may have slowed the growth rate of atmospheric CH_4 between 1999 and 2006 (McNorton et al., 2016; Rigby et al., 2017). A decline in the reactions that destroy CH_4 in the atmosphere is inconsistent with indirect observations that the concentration of hydroxyl radicals has not decreased, and may even have increased, in the atmosphere in recent years (Dentener et al., 2003; Prinn et al., 2005; Chapter 3). As a result of its mean atmospheric lifetime of 9 years, about 50×10^{12} g of the tropospheric CH_4 mixes into the stratosphere, where it is destroyed by reaction with OH, producing CO_2 and water vapor (Thomas et al., 1989; Oltmans and Hofmann, 1995).

A small amount of methane diffuses from the atmosphere into upland soils, where it is oxidized by methanotrophic bacteria (King, 1992; Covey and Megonigal, 2019). In Mojave Desert soils, where the supply of labile organic matter is limited, soil bacteria consume an average of $0.66\,mg\,CH_4\,m^{-2}\,day^{-1}$, with the greatest rates observed after rainstorms (Striegl et al., 1992). Consumption of CH_4 in temperate and tropical forest soils typically ranges from 1.0 to $5.0\,mg\,CH_4\,m^{-2}\,day^{-1}$ (Crill, 1991; Adamsen and King, 1993; Ishizuka et al., 2000; Smith et al., 2000; Price et al., 2004), with lower values after rainstorms, which tend to retard the diffusion of O_2 and CH_4 into clay-rich soils (Koschorreck and Conrad, 1993;

Castro et al., 1994; Castro et al., 1995; Ni and Groffman, 2018). Methanotrophic bacteria remain active at extremely low CH_4 concentrations (Conrad, 1994), so the global significance of soil methanotrophy appears limited by the rate of diffusion of methane into the soil (Born et al., 1990; King and Adamsen, 1992).

Some of the methanotrophic activity in soils derives from the activities of nitrifying bacteria, which can use CH_4 as an alternative substrate to NH_4^+ (Jones and Morita, 1983; Hyman and Wood, 1983; Bédard and Knowles, 1989). Steudler et al. (1989) suggested that the consumption of CH_4 by nitrifying bacteria may be lower in forests that currently receive a large atmospheric deposition of NH_4^+, because the NH_4^+/CH_4 ratio in soils has greatly increased in these regions. Thus, the impact of nitrogen deposition on methane uptake depends on the amount of nitrogen received, with the threshold being about $60\,kg\,N/ha/yr$ (Du et al., 2019). A number of workers reported reduced methane uptake when forest and grassland soils were fertilized with nitrogen, with the threshold being about $100\,kg\,N/ha/yr$ (Aronson and Helliker, 2010; Kim et al., 2012). Methane uptake by soils is also lower after land clearing, which stimulates nitrification (Hütsch et al., 1994; Keller and Reiners, 1994). With fertilization or land clearing, ammonium oxidation produces small amounts of nitrite (NO_2), which may cause a persistent inhibition of methanotrophic bacteria in soils (King and Schnell, 1994; Schnell and King, 1994). Atmospheric deposition of nitrate is also known to reduce CH_4 uptake by forest soils (Steudler and Peterson, 1985; Mochizuki et al., 2012).

Over large regions, the sink for methane in upland soils consumes only a small fraction of the production of methane in adjacent, wet lowland soils (e.g., Whalen et al., 1991; Delmas et al., 1992; Yavitt and Fahey, 1993; Ullah and Moore, 2011; Yu et al., 2019a). The global estimate of the sink for atmospheric methane in soils is about 20 to 30×10^{12} g/yr (Curry, 2007; Dutaur and Verchot, 2007). Given this relatively small value, it is unlikely that changes in this process by human activities can account for the current increase in atmospheric CH_4 globally (e.g., Willison et al., 1995).

Future changes in the global methane budget as a result of rising CO_2 and global temperatures are difficult to predict. Warmer conditions may shift the balance between aerobic and anaerobic decomposition in wetlands and increase the ratio of CO_2 to CH_4 emitted from these ecosystems (Whalen and Reeburgh, 1990; Funk et al., 1994; Moore and Dalva, 1993). On the other hand, methanogenic bacteria show a greater positive response to temperature than methane-oxidizing bacteria, suggesting that the flux of methane from wetland soils could increase with global warming (King and Adamsen, 1992; Dunfield et al., 1993; Megonigal and Schlesinger, 2002; Gill et al., 2017). Changes in microbial community structure in response to changing climate may mediate the production of CO_2 and CH_4 (McCalley et al., 2014). An increasing flux of CH_4 may also accompany a CO_2-induced stimulation of wetland plants, which leak excess carbohydrate from their roots fueling methanogenesis in wetland soils (Dacey et al., 1994; Hutchin et al., 1995; Megonigal and Schlesinger, 1997; van Groenigen et al., 2011). In contrast, methane consumption in upland soils appears to decline when vegetation is grown at high CO_2 (Phillips et al. 2001a; McLain et al., 2002; Dijkstra et al., 2013).

Catastrophic release of methane from marine sediments, where it is held as methane hydrate (clathrate; Chapter 9) could also yield a large increase in atmospheric methane and greenhouse warming in the future (Macdonald, 1990). Estimates of the global pool of CH_4 in clathrates center on 500×10^{15} g CH_4-C (Burwicz et al., 2011; Piñero et al., 2013)—more than $1000 \times$ current annual emissions (refer to Table 11.2). The geologic record indicates release of methane from hydrates in the past (Jahren et al., 2001; Katz et al., 1999), although evidence for such degassing at the end of the last glacial epoch is equivocal (Kennett et al., 2000; Sowers, 2006; Petrenko et al., 2009). Given methane's potential as a greenhouse gas and indications that increasing concentrations of CH_4 may have preceded the global warming 10,000 years ago, a better understanding of the global methane budget is paramount if biogeochemists are to contribute to the development of effective international policy to combat global warming (Nisbet and Ingham, 1995).

Carbon monoxide

Carbon monoxide has a low concentration (45–250 ppb) and a short lifetime (2 months) in the atmosphere (Table 3.5). The short lifetime is consistent with wide regional and seasonal variations in its concentration (Fig. 3.5); the concentration of CO in the Northern Hemisphere is typically three times larger than that in the Southern Hemisphere (Dianov-Klokov et al., 1989; Novelli et al., 2003). The budget for CO is dominated by anthropogenic sources (Table 11.3), especially fossil fuel combustion and biomass burning, which are concentrated in the Northern Hemisphere. Year-to-year variations in the occurrence of forest fires account for much of the variation in CO concentrations in downwind regions (Wotawa et al., 2001; Novelli et al., 2003; Vasileva et al., 2011). Variations in the concentration of CO and its isotopic composition (i.e., $\delta^{13}C$) in the Antarctic icepack have been used to trace biomass burning in the Southern Hemisphere for the past 650 years (Wang et al., 2010b).

TABLE 11.3   Budget for major sources and sinks for CO in the atmosphere.

Sources	Flux
Fossil fuel combustion	400
Biofuel combustion	160
Biomass burning	460[a]
Oxidation of methane	820
Oxidation of other volatile carbon compounds	521
Total	**2361**
Sinks	
Uptake by soils (Sanhueza et al., 1998)	115–230
Oxidation by OH reactions (Prather et al., 1995)	1400–2600
Stratospheric destruction	100
Total	**1615–3030**

[a] *Kaiser et al. (2012), Jain (2007), Mieville et al. (2010) and Andreae (2019) give alternative estimates ranging from 350 to 820 × 10^{12} g CO/yr.*
Note: *All units are 10^{12} g CO/yr; from Duncan et al. (2007) unless otherwise noted.*

Until recently, the concentration of CO was increasing at a rate of >1%/yr (Khalil and Rasmussen, 1988; Dianov-Klokov et al., 1989), presumably as a result of increasing emissions from fossil fuel combustion and an increasing production of CO as a methane-oxidation product (Chapter 3). Surprisingly, CO concentrations began to decline slightly during the early 1990s (Novelli et al., 1994; Khalil and Rasmussen, 1994; Zheng et al., 2019), reflecting a greater emphasis on the control of air pollution in the United States and Europe (Novelli et al., 2003; Hudman et al., 2008). The decline in CO may also be related to the slower growth rate of atmospheric CH_4, inasmuch as methane oxidation to CO accounts for 28–35% of the inputs of CO to the troposphere (Granier et al., 2000; Table 11.3).

A small amount of CO is taken up by vegetation and soils, but the dominant sink for CO is oxidation by hydroxyl radicals in the atmosphere (Eqs. 3.30–3.35). Because it is oxidized to CO_2 so rapidly, carbon monoxide is normally included as a component of the CO_2 flux in most accounts of the global carbon cycle (e.g., Fig. 11.1). Actually, the direct release of CO may account for about 5% of the total carbon emitted during fossil fuel combustion and perhaps as much as 15% of the carbon released during biomass burning (Andreae, 1991).

Carbon monoxide shows limited absorption of infrared radiation. Its main effect on the greenhouse warming of Earth is probably indirect—by slowing the destruction of methane in the atmosphere (Lashof and Ahuja, 1990). More important, carbon monoxide plays a major role in atmospheric chemistry by controlling levels of tropospheric ozone (Chapter 3). High concentrations of atmospheric ozone over the tropical regions of South America and Africa appear to be related to the production of CO by forest burning (Fig. 3.9), followed by the reaction of CO with OH radicals to produce ozone. About 2% of net primary production on land may be lost as CO or as volatile hydrocarbons that are oxidized to CO in the atmosphere (Chapter 5).

Synthesis: Linking the carbon and oxygen cycles

Even on a lifeless Earth, photolysis of water vapor would produce small amounts of O_2 in Earth's atmosphere, as it has on planets such as Mars (Chapter 2). Early in its history, Earth may have lost up to 35% of the water degassed from the mantle by photolysis (Chapter 10). The process is now limited by cold temperatures and the ozone layer in the stratosphere, which minimize the amount of water vapor exposed to ultraviolet light. During the geologic history of Earth, significant amounts of atmospheric O_2 appeared only about 2.4 billion years ago, well after the advent of autotrophic photosynthesis (Lyons et al., 2014; Canfield, 2014). O_2 began to accumulate to its present level when the annual production exceeded the reaction of O_2 with reduced minerals in the Earth's crust. The early, low levels of O_2 in Earth's atmosphere are likely to have delayed the rise of multicellular animals (Reinhard et al., 2016; Planavsky et al., 2014b).

The current atmospheric pool of O_2 is only a small fraction of the total O_2 produced over geologic time; most of the rest has been consumed in the oxidation of Fe and S (refer to Fig. 2.8). The total net production of O_2 over geologic time is balanced stoichiometrically by the storage of reduced organic carbon (1.56×10^{22} g) and sedimentary pyrite (4.97×10^{21} gS; refer to Table 13.1) in the Earth's crust. Today, the content of oxygen in Earth's atmosphere is determined by the balance between the burial of organic carbon in sediments and the subsequent weathering of ancient sedimentary rocks, which have been uplifted to form the continents (Petsch et al., 2001). Oxygen is likely to have accumulated most rapidly during periods when large amounts of organic sediments were buried (Des Marais et al., 1992; France-Lanord and Derry, 1997). The burial of organic carbon in marine sediments may have been enhanced by the simultaneous deposition of clay minerals, which increased in abundance in response to the colonization of the land surface by plants about 500 million years ago (Kennedy et al., 2006).

Today, the uplift and weathering of sedimentary rocks expose about 0.1–0.2×10^{15} g/yr of organic carbon to oxidation (Di-Giovanni et al., 2002; Fig. 11.8). Much of the organic matter in exposed sedimentary rocks is labile (Schillawski and Petsch, 2008; Galy et al., 2008; Hemingway et al., 2018). For instance, the biomass of microbes degrading organic matter in 365-million-year-old shales exposed to weathering in Kentucky is almost entirely derived from the shale (Petsch et al., 2001).

We have little evidence of historical variations in atmospheric O_2, but geochemical models suggest that the concentrations may have ranged from 15% to 35% during the past 500 million years (Berner and Canfield, 1989; Berner, 2001). The highest values would be expected in the Carboniferous and Permian periods, when large amounts of organic matter were buried in sediments (Berner et al., 2000; Fig. 11.4).

High concentrations of O_2 would have dramatic implications for the physiology, morphology, and evolution of most organisms (Graham et al., 1995). Giant insects are known to have been common during this period of high O_2 concentrations (Harrison et al., 2010). Fortunately, the pool of atmospheric O_2 is well buffered over geologic time because increases in O_2 expand the area and depth of aerobic respiration in marine sediments, leading to a greater consumption of O_2 and lower storage of organic carbon (Chapters 7 and 9). High levels of oxygen also reduce the rate of photosynthesis by causing increasing amounts of photorespiration (Tolbert et al., 1995; Chapter 5). The geologic record of sedimentary charcoal suggests that higher levels of O_2 may increase the occurrence of forest fires (Scott and Glasspool, 2006) and suppress the release of P by rock weathering (Lenton, 2001). Finally, high dissolved O_2 in seawater would increase the adsorption of P to iron minerals in marine sediments, subsequently lowering nutrient availability and NPP in the sea (Van Cappellen and Ingall, 1996). These interactions between the carbon, oxygen, and phosphorus cycles buffer the concentration of O_2 in Earth's atmosphere. Unlike geologic uplift and weathering, these processes are directly responsive to changes in the Earth's oxygen concentration.

Like the carbon cycle, the modern oxygen cycle is composed of a set of large, annual fluxes superimposed on the smaller, slow fluxes of the geologic cycle (Walker, 1980). The current atmospheric pool of O_2 is maintained in a dynamic equilibrium between the production of O_2 by photosynthesis and its consumption in respiration, including fires (Fig. 11.8). For the terrestrial biosphere, the oxidative ratio of CO_2 sequestered to O_2 produced is about 1.04 (Worrall et al., 2013).[k] The oxygen content of Earth's atmosphere has declined very slightly since the last glacial epoch, perhaps as a result of lower rates of erosion and carbon burial during the Holocene (Stolper et al., 2016). Against the average background concentration of 20.946% in today's atmosphere, the annual fluctuation of O_2 in the atmosphere due to photosynthesis and respiration is about $\pm 0.0020\%$ (Keeling and Shertz, 1992). The mean

[k]See Footnote #q, Chapter 5.

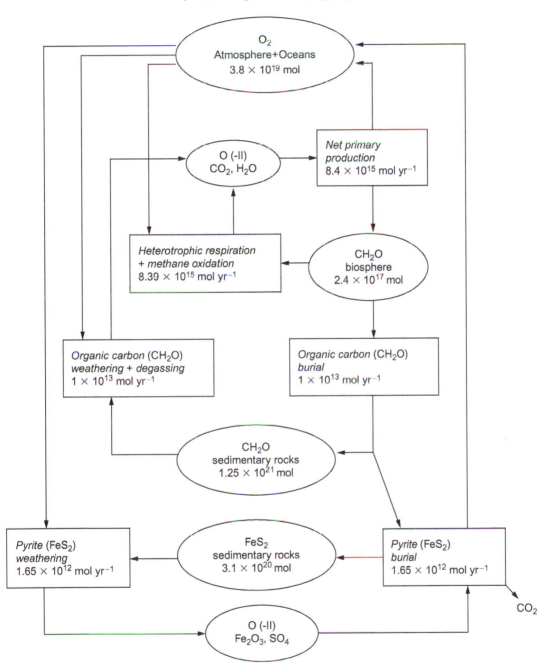

FIG. 11.8 Linkage of the global carbon and oxygen cycles. Ovals contain estimates of the reservoirs of O_2 or the equivalent amount of reduced molecules that could be oxidized by O_2. Boxes indicate fluxes of O_2 or reduced molecules in moles/year. *Modified from Lenton (2001).*

residence time of O_2 in the atmosphere is about 4000 years—significantly shorter than what would be predicted merely by the reaction of O_2 with the Earth's crust, about 100,000,000 years (Fig. 11.8).

Measured changes in the concentration of atmospheric O_2 provide a check on current estimates of CO_2 uptake by the terrestrial and marine biosphere (Bender et al., 1998). Photosynthesis on land produces O_2 and consumes CO_2 on an equal molar basis; the reverse is true when biomass and fossil fuels are burned. In contrast, uptake of CO_2 by the oceans, largely driven by Henry's Law, is not accompanied by a significant flux of O_2 to the atmosphere; the O_2 from marine photosynthesis remains dissolved in seawater and is consumed in the oceans during the degradation of organic matter. Thus, with fossil fuel combustion, the molar content of CO_2 should increase in Earth's atmosphere less rapidly than the molar decrease of O_2; the difference will reflect the dissolution of CO_2 in the oceans.

This method, developed by Keeling and Shertz (1992), in conjunction with precise measurements of changes in O_2 in the atmosphere, confirms a sink for about 1.7×10^{15} gC in the world's oceans and thus, an apparent net sink of 1.0×10^{15} gC in the terrestrial biosphere for the 1990s (compare to Bender et al., 2005; Table 11.1). This net sink for carbon in vegetation is relatively recent; the same methods show relatively little carbon accumulation on land during the 1980s (Langenfelds et al., 1999; Battle et al., 2000; Bopp et al., 2002). The sink for CO_2 in the oceans, which increases ocean acidification, is projected to exceed what is observed in sedimentary records for the past 300,000,000 years (Hönisch et al., 2012).

An examination of the isotopic composition of atmospheric O_2 (i.e., $\delta^{18}O$) also allows us to constrain the atmospheric CO_2 budget within certain limits. Photosynthesis does not discriminate among the oxygen isotopes of water—the O_2 released has an isotopic composition that is identical to that in the seawater or the soil water in which the plant is growing. Respiration discriminates among oxygen isotopes, consuming $^{16}O_2$ in preference to $^{18}O_2$; as a result $^{18}O_2$ is enriched in the atmosphere—known as the Dole effect (Luz and Barkan, 2011). The $\delta^{18}O$ of atmospheric O_2 (+ 23.5‰) suggests that gross primary production must be $>170 \times 10^{15}$ gC/yr on land and about 140×10^{15} gC/yr in the oceans (Bender et al., 1994).[1] Assuming that net primary production is one-half of gross primary production, in both cases, these values would imply that NPP is somewhat higher than what we have estimated independently for land (Table 5.3; Beer et al., 2010) and marine (Table 9.2) habitats globally. Nevertheless, these values offer an upper limit for NPP, which helps constrain our estimates for the global carbon cycle (Fig. 11.1).

The oxygen cycle is directly linked to other biogeochemical cycles. For example, assuming that NO_3 accounts for about half the plant uptake of N on land (1200×10^{12} g) and about 10% of the N cycle in the oceans (8000×10^{12} g; Fig. 9.21), then about 2% of the annual production of O_2 by photosynthesis is used to oxidize NH_4 in the nitrification reactions (compare to Ciais et al., 2007).

The formation and oxidation of sedimentary pyrite, through sulfate reduction, also affects the concentration of O_2 in the atmosphere. For every mole of pyrite-S oxidized, nearly 2 mol of

---

[1] This calculation is based on the amount of GPP, evolving O_2 identical to that in soil or seawater (i.e., $\delta^{18}O = 0$), that is necessary to maintain a constant level of $\delta^{18}O$ in the atmosphere in the face of preferential respiratory consumption of the light isotope.

O_2 are consumed from the atmosphere (Eq. 9.2). Currently, the annual burial of pyrite in marine sediments accounts for about 20% of the oxygen in our atmosphere (Chapter 9).

Methanogenesis in freshwater sediments returns CH_4 to the atmosphere, where it is oxidized (Henrichs and Reeburgh, 1987). Methane oxidation in the atmosphere accounts for about 1% of the total consumption of atmospheric O_2 each year. In the absence of methanogenesis, the burial of organic carbon in freshwater sediments might be greater and the atmospheric content of O_2 might be slightly higher. Thus, methanogenesis acts as a negative feedback in the regulation of atmospheric O_2 (Watson et al., 1978; Kump and Garrels, 1986). It is perhaps entertaining to speculate whether the carbon cycle on Earth drives the oxygen cycle, or vice versa. Over geologic time, the answer is obvious: The conditions on our neighboring planets provide ample evidence that O_2 is derived from life. Now, however, the carbon and oxygen cycles are inextricably linked, and the metabolism of eukaryotic organisms, including humans, depends on the flow of electrons from reduced organic molecules to oxygen.

Summary

Humans harvest about 20% of the annual production of organic carbon on land (i.e., NPP; Imhoff et al., 2004b; Haberl et al., 2007). In many areas we have destroyed land vegetation, while in other areas we have planted productive crops and forests and perhaps stimulated NPP by raising levels of atmospheric CO_2 and nitrogen deposition. At the moment, it appears that humans have created a net sink for carbon in the terrestrial biosphere, which mitigates some of the anticipated rise in atmospheric CO_2 from fossil fuel combustion (Table 11.1). With the use of fossil fuels, humans have, of course, supplemented the energy available to power modern society, including vast supplements to the agricultural systems that feed us. Dukes (2003) calculates that each year we burn the equivalent of the organic matter stored during 400 years of primary production in the geologic past. Accelerating use of fossil fuels is destined to lead to large changes in the Earth's conditions, which have otherwise been relatively stable during the 8000-year history of organized human society.

The Global Cycles of Nitrogen, Phosphorus and Potassium

Introduction

Nitrogen and phosphorus control many aspects of ecosystem function and global biogeochemistry. Nitrogen often limits the rate of net primary production on land and in the sea (Chapters 6 and 9). In living tissues, nitrogen is an integral part of enzymes, which mediate the biochemical reactions in which carbon is reduced (e.g., photosynthesis) or oxidized (e.g., respiration). Nearly all of the nitrogen in biomass is first assimilated by the attachment of an amine group ($-NH_2$) to the 5-carbon sugar, oxoglutarate, linking the C and N cycles at the level of cellular biochemistry (Williams, 1996, p. 158). Phosphorus is an essential component of DNA, ATP, and the phospholipid molecules of cell membranes. The molar ratio of N to P in phytoplankton, about 16, finds its basis in the ratio of protein to RNA in protoplasm (Loladze and Elser, 2011). Somewhat overlooked, potassium is gaining considerable attention for its role as a limiting nutrient, particularly in tropical soils. Changes in the availability of N, P and K and their relative abundance are likely to have controlled the size and activity of the biosphere through geologic time.

A large number of biochemical transformations of nitrogen are possible, since nitrogen is found at valence states ranging from -3 (in NH_3) to $+5$ (in NO_3^-). A variety of microbes capitalize on the potential for transformations of N among these states and use the energy released by the changes in redox potential to maintain their life processes (Rosswall, 1982). Collectively, these microbial reactions drive the cycle of nitrogen (Fig. 12.1). In contrast, whether it occurs in soils or in biochemistry, phosphorus is almost always found in combination with oxygen (i.e., as PO_4^{3-}). Most metabolic activity is associated with the synthesis or destruction of high-energy bonds between a phosphate ion and various organic molecules, but in nearly all cases the phosphorus atom remains at a valence of $+5$ in these reactions. Potassium is nearly always found in ionic form, K^+, and is involved in osmotic balance and electrochemical reactions in biochemistry.

The most abundant form of nitrogen at the surface of the Earth, N_2, is the least reactive species. Nitrogen fixation converts atmospheric N_2 to one of the forms of reactive ("fixed"

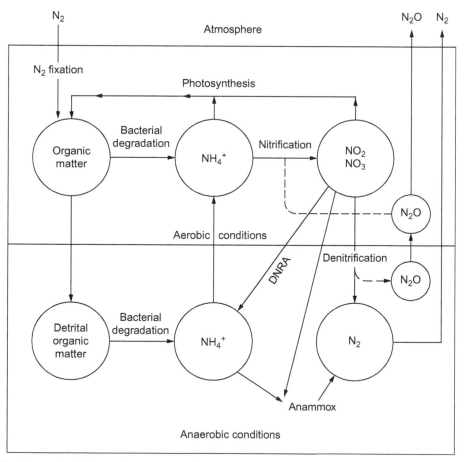

FIG. 12.1   Some microbial transformations in the nitrogen cycle. *Modified from Wollast (1981).*

or "odd," Chapter 3) nitrogen that can be used by biota. Nitrogen-fixing species are most abundant in nitrogen-poor habitats, where their activity increases the availability of nitrogen for the biosphere (Eq. 2.9). At the same time, denitrifying bacteria return N_2 to the atmosphere (Eq. 2.19), lowering the overall stock of nitrogen available for life on Earth.

Rocks of the continental crust hold the reservoir of phosphorus and potassium that becomes available to the biosphere through rock weathering. Land plants can increase the rate of rock weathering in P-deficient habitats (Chapter 4), but in nearly all cases, the phosphorus content of rocks is relatively low. Subsequent reactions between dissolved P and other minerals reduce the availability of P in soil solutions or seawater (Fig. 4.10). Thus, in most habitats—both on land and in the sea—the availability of P is controlled by the degradation of organic forms of P (e.g., Fig. 6.22). This *bio*geochemical cycle temporarily retains and recycles some P from the unrelenting flow of P from weathered rock to ocean sediments. The global P cycle is complete only when sedimentary rocks are lifted above sea level and the weathering begins again.

Because supplies of N, P and K often define soil fertility, humans have added enormous quantities of these elements to soils to enhance crop production. The production of fertilizer has more than doubled the supply of N and P on the land surface, altering biogeochemical cycling and leading to inadvertent enrichments of ecosystems downwind or downstream of the point of application.

In this chapter, we will examine our current understanding of the global cycles of N, P and K. We will attempt to balance N and P budgets for the world's land area and for the sea. For N, the balance between N-fixation and denitrification through geologic time determines the nitrogen available to biota and the global nitrogen cycle. One of the byproducts of nitrification and denitrification is N_2O (nitrous oxide), which is both a greenhouse gas and a cause of ozone destruction in the stratosphere (Chapter 3). We will present a budget for N_2O based on our current understanding of the sources of this gas in the atmosphere.

The global nitrogen cycle

Land

Fig. 12.2 presents the global nitrogen cycle, showing the linkage between the atmosphere, the land, and the oceans.[a] The atmosphere contains the largest pool (3.9×10^{21} gN; Table 3.1). Relatively small amounts of N are found in terrestrial plant biomass (3.8×10^{15} g)[b] and in soil organic matter (95–140×10^{15} g to 1-m depth; (Post et al., 1985; Batjes, 1996). About 3–4% of the soil nitrogen is held in microbial biomass (Xu et al., 2013). The mean C/N ratios for terrestrial

---

[a] For the purposes of biogeochemistry, we ignore the huge quantity, perhaps equivalent of 7 atm, of N that is held in the Earth's mantle (Johnson and Goldblatt, 2015), either as N_2 or NH_4 depending upon temperature, pressure and pH (Li and Keppler, 2014, Mikhail et al., 2017). There is some exchange between surface pools and the mantle due to subduction and degassing, but the *bio*geochemical cycle of N largely plays out at the Earth's surface.

[b] This value is derived from an estimate of 600×10^{15} g C in terrestrial biomass (Table 5.3) and a C/N ratio of 160 in forest biomass (Table 6.4). Given the convergence of estimates for the carbon pool in biomass, higher estimates for the N pool in vegetation would require a lower estimate of the C/N ratio in biomass, which seems unlikely.

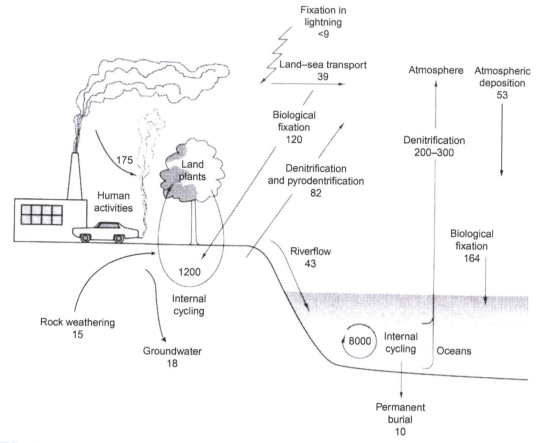

FIG. 12.2   The global nitrogen cycle. Each flux is shown in units of 10^{12} g N/yr. Values as derived in the text.

biomass and soil organic matter are about 160 and 12, respectively. At any time, the pool of inorganic nitrogen, NH_4^+ and NO_3^-, in soils is very small. The uptake of N by organisms is so rapid that little nitrogen remains in inorganic forms, despite the large annual flux through this pool (Chapter 6). Since most N in soils is held in organic forms, soil organic matter is a good predictor of total nitrogen content in most circumstances (Glendining et al., 2011).[c]

The nitrogen in the atmosphere is not available to most organisms, because the great strength of the triple bond in N_2 makes this molecule practically inert.[d] All nitrogen that is available to biota was originally derived from nitrogen fixation—either in lightning or in a few specialized species of microbes, which convert N_2 to forms of reactive nitrogen (Chapter 6). The rate of nitrogen fixation by lightning, which produces momentary conditions

---

[c]Typically the C/N ratio in desert soils is very low (Post et al. 1985). In deserts, where soil biotic activity is limited, a large amount of nitrate may accumulate in the soil profile below the rooting zone of plants, perhaps amounting to $3–15 \times 10^{15}$ gN globally (Walvoord et al., 2003, Nettleton and Peterson, 2011).

[d]The mean bond energy in N_2 is 226 kcal/mol, versus N—H (93), N—C (70) or N—O (48) (Davies 1972).

of high pressure and temperature allowing N_2 and O_2 to form NO_x, is relatively small. A recent, global estimate is 9×10^{12} g N/yr (Nault et al., 2017), and the estimated total annual deposition of oxidized N (NO_y) from the preindustrial atmosphere precludes an estimate higher than about 12×10^{12} g N/yr (Galloway et al., 2004). Assuming that lightning is distributed uniformly over land and sea, a liberal estimate for the deposition of N fixed by lightning over land would be about 3×10^{12} g N/yr. The present-day deposition of oxidized nitrogen on land is about 25×10^{12} g/yr, owing to the additional NO_x that is emitted from soils, biomass burning, and human activities (Table 12.1).

Biological nitrogen fixation is performed by several species of microbes, which are either free-living in lake waters, soils, and sediments or found in symbiotic association with the roots of plants (Chapter 6). Prior to widespread human activities, total biological nitrogen fixation on land is likely to have been 60 to 195×10^{12} g N/yr (Cleveland et al., 1999; Vitousek et al., 2013; Davies-Barnard and Friedlingstein, 2020). The current rate of biological nitrogen fixation is estimated at 120 to 180×10^{12} g N/yr, for the sum of natural and agricultural ecosystems (Burns and Hardy, 1975; Wang and Houlton, 2009). As natural lands have given way to cultivation, the nitrogen fixation in agricultural systems has replaced (and even exceeded) the nitrogen-fixation that was lost from natural ecosystems (Galloway et al., 2004). Estimates of N fixation in agriculture are as high as $50-70 \times 10^{12}$ g N/yr, or nearly half of the land total (Herridge et al., 2008).

These estimates of N-fixation on land are equivalent to about 10 kg N/yr for each hectare of the Earth's land surface. Most studies of nitrogen fixation in free-living soil bacteria report values ranging from 1 to 5 kg/ha/yr (Chapter 6). A value of 3 kg N/ha/yr suggests that asymbiotic fixation contributes about one-third of the global total. The remainder is assumed

TABLE 12.1 A global budget for atmospheric NO_x (Values are Tg N (10^{12} g N)/yr as NO.

Fossil fuel combustion	25	Galloway et al. (2004)
Net emissions from soils	12	Ganzeveld et al. (2002) (Gross flux $\sim$21 Tg N/yr; Davidson and Kingerlee, 1997)
Biomass burning	9	Andreae (2019) and Kaiser et al. (2012) (cf. 9.8 Tg N/yr; Mieville et al., 2010)
Lightning	9	Nault et al. (2017)
NH_3 oxidation	1	cf. Table 12.2 (Warneck, 2000)
Aircraft	0.4	Prather et al. (1995)
Transport from the stratosphere	0.6	(for total NO_y; Prather et al., 1995)
Total sources	57.0	(cf. 37 Tg N/yr from satellite measurements Martin et al., 2003; 46 Tg N/yr, Galloway et al., 2004; and 48.4 Tg N/yr, IPCC, 2013)
Deposition on land	24.8	Galloway et al. (2004)
Deposition on the ocean surface	23.0	Duce et al. (2008) and Dentener et al. (2006)
Total sinks	47.8	

The authors acknowledge that presentations of global budgets for reactive gases such as NO, NH3 and SO2 are not entirely realistic as the emissions and depositions occur regionally. Nevertheless, global compilations allow easy comparisons of the magnitude of various sources.

to come from symbiotic association of bacteria with higher plants. This flux is not distributed uniformly among natural ecosystems; the greatest values are often found in tropical forests and in areas of disturbed or successional vegetation (Vitousek and Howarth, 1991). In any case, biotic N-fixation dwarfs abiotic fixation by lightning as the source of fixed N. The evolution of life and of nitrogen-fixation on Earth greatly speeded the movement of nitrogen in a biogeochemical cycle. Taking all forms of N-fixation as the only source, the mean residence time of nitrogen in the terrestrial biosphere is about 700 years (i.e., pool/input). This calculation is not altered significantly by considering the small amount of fixed nitrogen that is contained in sedimentary and metasedimentary rocks, which may contribute more than $11\text{--}18 \times 10^{12}$ g N/yr to the terrestrial biosphere by chemical weathering (Holloway and Dahlgren, 2002; Houlton et al., 2018).

Assuming that the estimate of terrestrial net primary production, 60×10^{15} g C/yr, is roughly correct and that the mean C/N ratio of net primary production is about 50, the nitrogen requirement of land plants is about 1200×10^{12} g/yr (Chapter 6).[e] Thus, nitrogen fixation supplies only about 9–15% of the nitrogen that is assimilated by land plants each year (Shi et al., 2016b). The remaining nitrogen must be derived from internal recycling and the decomposition of dead materials in the soil (Chapter 6). When the turnover in the soil is calculated with respect to the input of dead plant materials, the mean residence time of nitrogen in soil organic matter is >100 years. Thus, the mean residence time of N exceeds that of C in both land vegetation and soils (10 and 40 years for C, respectively; Chapter 5).

Humans have a dramatic impact on the global N cycle. In addition to planting N-fixing species for crops, humans produce nitrogen fertilizers through the Haber process, viz.,

$$3CH_4 + 6H_2O \rightarrow 3CO_2 + 12H_2 \tag{12.1}$$

$$4N_2 + 12H_2 \rightarrow 8NH_3, \tag{12.2}$$

in which natural gas is burned to produce hydrogen, which is combined with N_2 to form ammonia under conditions of high temperature and pressure (Smil, 2001). The industrial production of reactive N by these reactions supplies $\sim 140 \times 10^{12}$ g N/yr for agricultural and chemical uses[f]—roughly matching the natural rate of nitrogen fixation on land.

A substantial fraction, perhaps half, of the annual application of nitrogen fertilizer to agricultural lands is lost to the atmosphere and to runoff waters (Erisman et al., 2007). Some NH_3 is volatilized directly to the atmosphere from cultivated soils (Fig. 6.17), while some N is lost indirectly as NH_3 volatilized from the excrement of domestic livestock that are fed forage crops (Table 12.2). The loss of NH_3 from agricultural lands to the atmosphere carries fixed N to adjacent natural ecosystems where it is deposited and enters biogeochemical cycles (Draaijers et al., 1989; Hesterberg et al., 1996). Deposition of reduced forms of nitrogen

---

[e]Most primary production consists of short-lived tissues with a C/N ratio that is much lower than that of wood (160), which composes most of the terrestrial biomass. Mineralization of $\sim 1000 \times 10^{12}$ g N/yr is consistent with experimental observations suggesting 1–3% turnover of nitrogen in soils annually (Chapter 6). Raven et al. (1993) give an alternative estimate of 2338×10^{12} g/yr for the nitrogen uptake by land plants. Another recent global model for the N cycle assumes a C/N ratio of about 10 and uptake of 6207×10^{12} g N/yr by terrestrial NPP, which seems unlikely (Lin et al., 2000).

[f]http://minerals.usgs.gov/minerals/pubs/commodity/nitrogen/mcs-2017-nitro.pdf

TABLE 12.2 A global budget for atmospheric ammonia.

Domestic animals	18.5	Bouwman et al. (2002a,b)
Wild animals	0.1	
Sea surface	2.5	Paulot et al. (2015)
Undisturbed soils	2.4	
Agricultural crops and soils	16.7	Xu et al. (2019b)
Biomass burning	8.0	Kaiser et al. (2012) and Andreae (2019)
Human excrement	2.6	
Combustion and industry	3.5	Behera et al. (2013)
Automobiles	0.46	Behera et al. (2013)
Total sources	54.5	(cf. 58.2 TgN/yr; Galloway et al., 2004; 50.7 TgN/yr, IPCC, 2013)
Deposition on land	38.4	
Deposition on the ocean surface	24.0	Duce et al. (2008) and Dentener et al. (2006)
Reaction with OH radical	1.0	Schlesinger and Hartley (1992)
Total sinks	63.7	

All values are Tg N (10^{12} g N)/yr as NH_3 or NH_4^+ (in deposition).
Unless noted otherwise, sources are derived from Bouwman et al. (1997) and sinks from Galloway et al. (2004).

(i.e., NH_4^+) from the atmosphere, presumably from agricultural sources, has increased significantly in recent years (Li et al., 2016a; Ackerman et al., 2019). Fertilized agricultural soils are also a source of emissions of NO_x from nitrification that is enhanced in fertilized soils (Chapter 6).

Fossil fuel combustion produces about 25×10^{12} g of fixed N (viz., NO_x) annually (Galloway et al., 2004). Some of this is derived from the organic nitrogen contained in fuels (Bowman, 1991), but is best regarded as a source of new, fixed N for the biosphere because in the absence of human activities, this N would remain inaccessible in the Earth's crust. NO_x also forms directly from N_2 and O_2 during the combustion of fossil fuels, especially in automobiles and coal-fired power plants (Bowman, 1992; Davidson et al., 1998; Kim et al., 2006). Attribution of NO_x emissions to power plants has been controversial, but aided by lower airborne concentrations of NO_x that are observed during unusual large-scale power blackouts (Marufu et al., 2004). Owing to its short residence time (Table 3.5), most of the NO_x from fossil fuel combustion is deposited by precipitation over land, where it enters biogeochemical cycles (Fig. 12.3). A small portion of NO_x undergoes long-distance transport in the troposphere, accounting for excess nitrogen deposition in the oceans (Duce et al., 2008; Kim et al., 2011) and Greenland snow (Fig. 12.4).

Anthropogenic emissions of nitrogen to the atmosphere, largely from agricultural soils, date to about 1900 as shown by declining $\delta^{15}N$ in the sedimentary record of ice cores and lakes (Hastings et al., 2009; Holtgrieve et al., 2011; Felix and Elliott, 2013). In some areas, the atmospheric deposition of N may fertilize plant growth (Chapter 6), leading to an increment in the

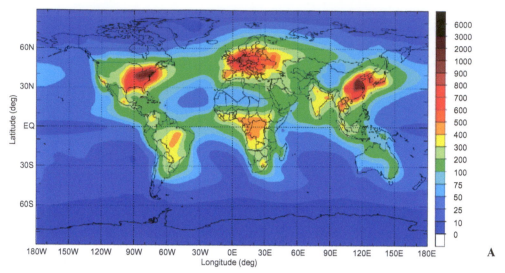

FIG. 12.3 Deposition of NO_y on the Earth's surface. All values are $mgN/m^2/yr$. *From Dentener et al. (2006).*

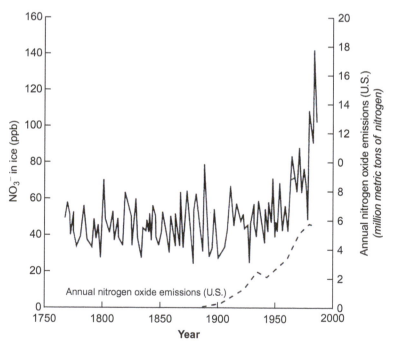

FIG. 12.4 The 200-year record of nitrate in layers of the Greenland ice pack and the annual production of nitric oxides by fossil fuel combustion in the United States. *Modified from Mayewski et al. (1990).*

sink for carbon in vegetation (Chapter 11). However, some high-elevation forests downwind of major population centers now receive enormous nitrogen inputs, both NH_4^+ and NO_3^-, which may be related to forest decline (Chapter 6). Various workers have tried to define the "critical load"—the level of nitrogen deposition that can be expected to cause changes in ecosystem properties—e.g., increased nitrification and leaching. In many ecosystems, noticeable effects begin at nitrogen depositions of 1 kg/ha/yr (Pardo et al., 2011; Liu et al., 2011; Clark et al., 2018). Globally, nitrogen deposition from the atmosphere appears to have increased about 8% between 1984 and 2016 (Ackerman et al., 2019), but as a result of air-pollution control legislation, deposition of nitrogen from the atmosphere is now declining in North America (Zbieranowski and Aherne, 2011; Butler et al., 2011). Satellite remote sensing may be particularly useful to measure future trends and sources of NH_3 and NO_x emissions globally (Schneider and van der A., 2012; Van Damme et al., 2018).

In total, about 300×10^{12} g of newly fixed N is delivered from the atmosphere to the Earth's land surface each year—28% from natural and 72% from human-derived sources (Table 12.3, Battye et al., 2017). Some of this nitrogen remains sequestered in the terrestrial biosphere, primary in forests and in agricultural soils, which often store about 30–40% of the applied nitrogen (Schlesinger, 2009; Gardner and Drinkwater, 2009; Sebilo et al., 2013). During the past century, nitrogen has also accumulated in the vadose zone of soils (Ascott et al., 2017). In the absence of processes removing nitrogen, a very large pool of nitrogen would be found on land in a relatively short time. About 23% of the nitrogen deposited on the land surface is lost to runoff (Howarth, 1998; van Breemen et al., 2002), especially during periods of peak flow (Chapter 8). Runoff of nitrogen into Lake Michigan is well correlated to fertilizer and land-use changes in its basin (Han and Allan, 2012). Globally, rivers carry $40–60 \times 10^{12}$ gN/yr from land to the sea (Boyer et al., 2006; Van Drecht et al., 2003). Despite increased N retention in reservoirs (Beusen et al., 2016) and its removal in sewage treatment (Hale et al., 2015), more than half of the present-day riverine transport of N in rivers is of human origin (Schlesinger, 2009). The fluvial transport of nitrogen is dominated by sewage pollution in Europe, by fertilizer in the U.S., and by a combination of both sources in China (Liu et al., 2019b).

Human additions of fixed nitrogen to the terrestrial biosphere have also resulted in marked increases in the nitrogen content of groundwaters, especially in many agricultural areas (Spalding and Exner, 1993; Ju et al., 2006; Rupert, 2008; Scanlon et al., 2010; Exner et al., 2014; Zhou et al., 2016). For example, the loss of nitrate to groundwater was the largest single fate of nitrogen added to the fields of a dairy farm in Ontario (Barry et al., 1993). Nitrogen and other biochemical elements are also enriched in the drainage from cemeteries (Zychowski, 2012). The global transport of N to groundwaters may approach 18×10^{12} gN/yr—calculated from an estimate of the annual flux of groundwater ($12,666 \, km^3$/yr; (Doll and Fiedler, 2008) and the median concentration of 1.9 mgN/L in groundwaters of the U.S. (Nolan et al., 2002; Schlesinger, 2009). In some instances, denitrification in the saturated zone can partially mitigate NO_3^- in polluted groundwaters (Korom, 1992; Hinkle and Tesoriero, 2014; Jahangir et al., 2013).

Despite these large transports, riverflow and groundwater cannot account for all of the nitrogen that is lost from land. The remaining nitrogen is assumed to be lost by denitrification and other gaseous pathways in terrestrial soils (Chapter 6), in wetlands (Chapter 7), and during forest fires (Chapter 6). Denitrification likely represents the largest loss of N from many

TABLE 12.3   Mass-balance for nitrogen on the Earth's land surface.

	Pre-industrial	Human-derived	Total
Inputs			
Biological N-fixation	60[a]	60[b]	120
Lightning	3[c]	0	3
Rock weathering	15[d]	0	15
Industrial N-fixation	0	150[e]	150
Fossil fuel combustion	0	25	25
Total	78	235	313
Fates			
Biospheric increment	0	9	9
Soil accumulation	0	48	48
Riverflow[f]	27	16	43
Groundwater	0	18	18
Denitrification	28[g]	17	45
Pyrodenitrification	13[h]	24	37
Atmospheric land-sea transport[i]	10	29	39
Total	78	157	239

[a] *Vitousek et al. (2013).*
[b] *Herridge et al. (2008); value is net from human activities.*
[c] *Nault et al. (2017), scaled to land surface.*
[d] *Houlton et al. (2018) give a range of 11–18 TgN/yr.*
[e] *https://minerals.usgs.gov/minerals/pubs/commodity/nitrogen/mcs-2018-nitro.pdf.*
[f] *Seitzinger et al. (2010).*
[g] *Houlton and Bai (2009).*
[h] *By difference, to balance pre-industrial inputs.*
[i] *Jickells et al. (2017).*
All values are in Tg N/yr (=10^{12} gN/yr).
From Schlesinger (2009), with updated values as footnoted.

forest ecosystems (Houlton et al., 2006; Fang et al., 2015; Brookshire et al., 2017; Yu et al., 2019b). Seitzinger et al. (2006) has compiled rates of denitrification, showing that as much as 124×10^{12} gN/yr might occur in soils and 110×10^{12} g/yr in freshwater environments (Table 12.4). If N-fixation and denitrification were once in balance, then a terrestrial denitrification rate of ~60×10^{12} gN/yr was most likely in the preindustrial world (i.e., fixation minus riverflow). Most of the loss occurs as N_2, but the small fraction that is lost as N_2O during nitrification and denitrification (Chapter 6) contributes significantly to the global budget of this gas. Indeed, the current rise in atmospheric N_2O can be used to estimate the overall increase in global denitrification as a result of human activities (Schlesinger, 2009). If we assume that the ($N_2 + N_2O$)/N_2O ratio for denitrification is about 4.0 and that the recent increase of

TABLE 12.4   A global estimate of denitrification of nitrogen on or applied to land.

System	Denitrification (Tg N/yr)
Terrestrial	
Soils	124 (65–175)
Freshwater	
Groundwater	44 (>0–138)
Lakes and reservoirs	31 (19–43)
Rivers	35 (20–35)
Subtotal	110 (39–216)
Marine	
Estuaries	8 (3–10)
Continental shelves	46 (>0–70)
Oxygen minimum zones	25 (>0–30)
Subtotal	79 (3–145)

From Seitzinger et al. (2006).

N_2O in the atmosphere (nearly 4×10^{12} gN/yr) all derives from increased denitrification, then it is possible that the overall loss of N_2 from denitrification has increased by as much as 17×10^{12} gN/yr, helping to balance the present-day N budget on land (Fig. 12.2).[g] Denitrification leaves soils enriched in $\delta^{15}N$ globally (Amundson et al., 2003; Houlton and Bai, 2009); especially high rates of denitrification and high soil $\delta^{15}N$ are reported in the tropics (Houlton et al., 2006).

Nitrogen in biomass is volatilized as NH_3, NO_x, and N_2 during fires—the latter constituting a form of *pyrodenitrification* (Chapter 6). About 30% of the nitrogen in fuel is converted to N_2, so globally, biomass burning may return about 37×10^{12} gN/yr to the atmosphere as N_2 (Kuhlbusch et al., 1991). Uncertainties in the historical rate of biomass burning (Mouillot et al., 2006 vs. Ward et al., 2018) make it difficult to estimate the change in pyrodenitrification since the pre-industrial period. The budget presented in Table 12.3 assumes a doubling of the nitrogen losses by pyrodenitrification globally in recent years.

In balancing the terrestrial N cycle, we concentrate on processes that affect the net production or loss of fixed nitrogen (Table 12.3). We do not include processes that recycle N that was fixed at an earlier time. Thus, NH_3 volatilization from biomass burning (Table 12.2) and the natural emission of NO_x from soils (Table 12.1) can be ignored to the extent that these forms

---

[g] The ratio ranges from 2.0 in upland soils to >12 in wetlands (Schlesinger, 2009), so the total increase in denitrification caused by human inputs may range from 8 to 68×10^{12} gN/yr.

are redeposited on land in precipitation. Ammonia and NO_x have relatively short atmospheric lifetimes, so they are usually deposited in precipitation and dryfall near their point of origin (Chapter 3). Indeed, some have suggested that nitrogen "hop-scotches" across the landscape, where losses from one area result in increased deposition and local cycling in other areas. The cycle is complete only when nitrogen is returned to the atmosphere as N_2. By focusing on the new sources of available nitrogen, we find that nearly all of the human perturbation of the global nitrogen cycle has occurred in the past 150 years (Battye et al., 2017).

Sea

The world's oceans receive about 43×10^{12} gN/yr in dissolved forms from rivers (Chapter 8), about 164×10^{12} gN/yr via biological N-fixation,[h] and about 53×10^{12} gN in atmospheric deposition (Duce et al., 2008; Jickells et al., 2017). Some of the precipitation flux is NH_4^+ that is derived from NH_3 volatilized from the sea (Quinn et al., 1988), but about 80% of the atmospheric deposition of N in the oceans derives from human activities on land (Duce et al., 2008). A significant fraction of the deposition of atmospheric nitrogen on the oceans is found in various types of organic compounds (Cornell et al., 2003), perhaps amounting to 16×10^{12} g/yr (Kanakidou et al., 2012). Through their various activities, humans have created a large flux of nitrogen through the atmosphere, from land to sea.

As we have shown for terrestrial ecosystems, most of the net primary production in the sea is supported by nitrogen recycling in the water column (Table 9.3). N inputs from the atmosphere have the greatest influence in the open oceans, where the pool of inorganic nitrogen is very small. The riverflux of N assumes its greatest importance in coastal seas and estuaries, but some is transported to the open ocean (Sharples et al., 2017). Runoff of excess nitrogen has increased the productivity of coastal ecosystems, leading to eutrophication and hypoxia (Michael Beman et al., 2005; Goolsby et al., 2001; Kim et al., 2011).

The deep ocean contains a large pool of inorganic nitrogen (720×10^{15} gN)[i] derived from the decomposition of sinking organic debris and mineralization from dissolved organic compounds. Permanent burial of organic nitrogen in sediments is small, so most of the nitrogen input to the oceans must be returned to the atmosphere as N_2 by denitrification and the anammox reaction (Figs. 9.21 and 12.2). Important areas of denitrification are found in the anaerobic deep waters of the eastern tropical Pacific Ocean, the Benguela upwelling, and the Arabian Sea (Chapter 9). Seitzinger et al. (2006) estimate denitrification of $\sim 50 \times 10^{12}$ gN/yr in near-shore waters. Globally, coastal and marine denitrification may account for the return of 134–230×10^{12} gN/yr to the atmosphere as N_2 (DeVries et al., 2012; Eugster and Gruber, 2012), substantially lower estimates than a few years ago (Codispoti, 2007). We balance the marine nitrogen cycle using a range of 200–300×10^{12} g N/yr in Fig. 12.2.

Although these estimates are subject to large uncertainty, the model of Fig. 12.2 indicates a small net loss of nitrogen from the oceans. The overall gaseous losses of nitrogen from the

---

[h] As seen in Chapter 9, the overall rate of N-fixation in the oceans is poorly constrained, potentially ranging from 68 to 177 TgN/yr (Tang et al., 2019; Großkopf et al. 2012). We use an estimate of 164 TgN/yr in Figure 12.2

[i] Volume of the deep oceans ($0.95 \times 1.335 \times 10^{24}$ g; Fig. 10.4) multiplied by the NO_3^- in the deep ocean (40 μmol NO_3/kg; Figs. 9.18 and 9.27) multiplied by 14 g/mol.

ocean exceed the inputs from rivers and the atmosphere, so that the oceans may be declining in nitrogen content (McElroy, 1983; Codispoti, 2007; Ren et al., 2017b; Wang et al., 2019f). Various workers have suggested that, in the absence of denitrification, higher concentrations of NO_3 would be found in the ocean and lower concentrations of N_2 in the atmosphere. The balance of nitrogen fixation and denitrification has probably controlled marine NPP through past glacial cycles (Ganeshram et al., 1995). Globally, suboxic waters, in which available nitrogen is depleted by denitrification, are also sites of N-fixation, providing long-term self-regulation to the nitrogen cycle of the oceans (Deutsch et al., 2007).

Temporal variations in the global nitrogen cycle

The earliest atmosphere on Earth is thought to have been dominated by nitrogen, since N is abundant in volcanic emissions and only sparingly soluble in seawater (Chapter 2). The early rate of degassing of N_2 from the Earth's mantle must have been much greater than today. The present-day flux of N_2 from the mantle is about 78×10^9 g N/yr (Sano et al., 2001) to 123×10^9 g N/yr (Tajika, 1998), which could not result in the observed accumulations of nitrogen at the Earth's surface (Table 2.3), even if it occurred over 4.5 billion years of Earth's history.[j]

Before the origin of life, nitrogen was fixed by lightning and in the shock waves of meteors, which create local conditions of high temperature and pressure in the atmosphere (Mancinelli and McKay, 1988). The rate of N-fixation was very low, perhaps about 6% of the present-day rate, because abiotic N-fixation in an atmosphere dominated by N_2 and CO_2 is much slower than in an atmosphere of N_2 and O_2 (Kasting and Walker, 1981). The best estimates of abiotic N-fixation suggest that it had a limited effect on the content of atmospheric nitrogen, but it provided a small but important supply of fixed nitrogen, largely NO_3^-, to the waters of the primitive Earth (Kasting and Walker, 1981; Mancinelli and McKay, 1988). Similar abiotic N-fixation on Mars is postulated to result in accumulations of nitrate or cyanide (HCN) on its surface (Segura and Navarro-Gonzalez, 2005; Manning et al., 2008). Nitrate also accumulates in extremely arid soils on Earth, although here it is likely to be derived from distant biogenic sources (Michalski et al., 2004a). The limited supply of fixed nitrogen in the primitive oceans on Earth is likely to have led to the early evolution of N-fixation in marine biota, perhaps 3.2 billion years ago (Stüeken et al., 2015; Chapter 2).

With respect to N-fixation by lightning, the mean residence time of N_2 in the atmosphere is about 500 million years. The mean residence time of atmospheric nitrogen decreases to about 9,000,000 years when biological nitrogen fixation is included. This is much shorter than the history of life on Earth, and it speaks strongly for the importance of denitrification in returning N_2 to the atmosphere over geologic time. Denitrification closes the global biogeochemical cycle of nitrogen, but it also means that nitrogen remains in short supply for the biosphere. In the absence of denitrification, most nitrogen on Earth would be found as NO_3^- in seawater, and the oceans would be quite acidic (Sillen, 1966).

---

[j]Some nitrogen degassed to the Earth's surface is carried back to the mantle by subduction of sediments (760×10^9 g N/yr; Busigny et al., 2003), with the contemporary net flux to the mantle perhaps amounting to 330×10^9 g/yr to 960×10^9 g N/yr (Goldblatt et al., 2009; Busigny et al., 2011). Thus the pool of nitrogen at the Earth's surface is decreasingly slightly each year by net entrainment in the mantle.

It is likely that denitrification appeared later than the other major metabolic pathways. Denitrifying bacteria are facultative anaerobes, switching from simple heterotrophic respiration to NO_3 respiration under suboxic conditions (Broda, 1975; Betlach, 1982). Denitrification enzymes are somewhat tolerant of low concentrations of O_2, allowing denitrifying bacteria to persist in environments with fluctuations in redox potential (Bonin et al., 1989; McKenney et al., 1994; Carter et al., 1995).

Because NO_3^- is very soluble, there is little reliable record of changes in the content of NO_3^- in seawater through geologic time. Only changes in the deposition of organic nitrogen are recorded in sediments. Altabet and Curry (1989) show that the $^{15}N/^{14}N$ record in sedimentary foraminifera is useful in reconstructing the past record of ocean N chemistry. The isotope ratio in sedimentary organic matter increases when high rates of denitrification remove NO_3^- from the oceans, leaving the residual pool of nitrate enriched in ^{15}N (Altabet et al., 1995; Ganeshram et al., 1995; Kast et al., 2019). Today, the $\delta^{15}N$ of NO_3^- in seawater is $\sim$+5‰. The geologic record offers some insight regarding the origin of denitrification. Sedimentary rocks with enrichments of $\delta^{15}N$, indicating denitrification, are first found in rocks of 2.3–2.7 billion years ago—well after the origin of oxygenic photosynthesis (Beaumont and Robert, 1999; Godfrey and Falkowski, 2009; Thomazo et al., 2011; Zerkle et al., 2017).

Requiring oxygen as a reactant, nitrification clearly arose after photosynthesis and the development of an O_2-rich atmosphere (Eqs. 2.16–2.17). Some of the earliest nitrifying organisms may have been archaea, which are found in many soils (Leininger et al., 2006). Today, the rate of denitrification is controlled by the rate of nitrification, which supplies NO_3^- as a substrate (Eq. 2.19; Fig. 6.16). Thus, the major microbial reactions in the nitrogen cycle (Fig. 12.1) are all likely to have been in place at least 2 billion years ago.

Assuming a steady state in the ocean nitrogen cycle, the mean residence time for an atom of fixed N in the sea is >4000 years. During this time, this atom will make several trips through the deep ocean, each lasting 200–500 years (Chapter 9). Because the turnover of N is much longer than the mixing time for ocean water, NO_3 shows a relatively uniform distribution in deep ocean water. The nitrogen budget in the oceans appears not in steady-state; current estimates of the rate of denitrification slightly exceed known inputs (cf. Fig. 12.2). McElroy (1983) suggests that the oceans received a large input of nitrogen during the continental glaciation 20,000 years ago, and they have been recovering from this input ever since (Christensen et al., 1987). This suggestion is consistent with sedimentary evidence of greater net primary production in the oceans during the last ice age (Broecker, 1982), and with the low ratio of $^{15}N/^{14}N$ in sedimentary organic matter of glacial age implying low rates of denitrification (Ganeshram et al., 1995; Altabet et al., 2002; Gruber and Galloway, 2008).

McElroy's paper should remind us to question the assumption of steady-state conditions whenever we construct global models of Earth's biogeochemistry, such as Fig. 12.2. The Earth has experienced large fluctuations in its biogeochemical function through geologic time. Changes in the global distribution and circulation of nitrogen may have accompanied climatic changes—just as the rise in CO_2 concentrations at the end of the last glacial indicates a period of non-steady-state conditions in the global carbon cycle (Straub et al., 2013).

At present, human activities have certainly disrupted the potential for steady-state conditions in the nitrogen cycle on land (Galloway et al., 2004; Liu et al., 2010). The production of N-fertilizers and the cultivation of leguminous crops, especially soybean, have increased dramatically since World War II, allowing higher crop yields to feed the world's growing human

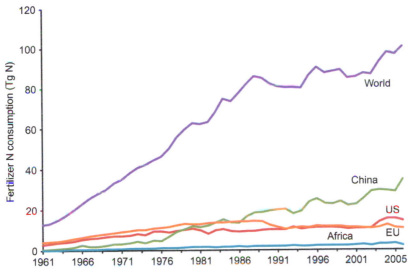

FIG. 12.5   The production history of nitrogen fertilizer. *From Robertson and Vitousek (2009).*

population (Fig. 12.5). Anthropogenic nitrogen now accounts for about 80% of the available nitrogen in China (Cui et al., 2013). Globally the human distribution of nitrogen fertilizers has roughly doubled the annual inputs of available nitrogen to the land surface. As a result of these activities, each year about 24×10^{12} gN moves in the international trade of grains and food, including livestock—a flux that rivals those of the natural biogeochemical cycle (Lassaletta et al., 2014).

What is most surprising is how little of this added nitrogen, about 10%, ends up in the food we eat (Galloway and Cowling, 2002; Esculier et al., 2019). Nitrogen-use efficiency in agriculture (the fraction of N input that is harvested as product) averages about 42% globally (Zhang et al., 2015a; Conant et al., 2013; Liu et al., 2011). The rest of the added nitrogen enters its global biogeochemical cycle via losses to runoff, to groundwater, and to the atmosphere, from which it is deposited in areas downwind (Ti et al., 2018). Archival collections show decreasing δ^{15}N in plant tissues, ice cores and lake sediments, consistent with greater anthropogenic nitrogen deposition in recent years (Peñuelas and Filella, 2001; Hastings et al., 2009; Holtgrieve et al., 2011).

Changes in the global nitrogen cycle have important implications for human health (Townsend et al., 2003), carbon storage in the biosphere (Townsend et al., 1996), and the persistence of species diversity in nature (Bobbink et al., 2010; Stevens et al., 2010; Midolo et al., 2018). In many regions the fertilizer application of N are greater than removals in harvested crops (Vitousek et al., 2009). In China, use of nitrogen fertilizer accounts for widespread acidification of agricultural soils, via the nitrification reactions (Eq. 2.16; Guo et al., 2010). In the U.S., about half of the value of fertilizer-enhanced crop production in the Midwest is consumed by higher costs for health care in regions downwind, due to secondary aerosols derived from ammonia emitted to the atmosphere (Paulot et al., 2015). Enrichments of nitrogen in terrestrial ecosystems, stimulating the rates of nitrification and denitrification, are likely to account for the rapid rise in the concentration of N_2O in Earth's atmosphere. Controls on N_2O emissions may be one of the most effective ways to reduce the threat of global climate change (Montzka et al., 2011a).

Nitrous oxide

Nitrous oxide, N_2O, has a mean concentration of 320 ppb in Earth's atmosphere, which indicates a global pool of 2.5×10^{15} g N_2O or 1.6×10^{15} gN (Table 3.1). The concentration of N_2O is increasing at a rate of 0.3%/yr (IPCC, 2007). Each molecule of N_2O has the potential to contribute about 300 times more to the greenhouse effect relative to each molecule of CO_2, so the current increase in the atmosphere has the potential to impact global climate over the next century (Shindell et al., 2009). Also, as a result of reductions in the emissions of chlorofluorocarbons, nitrous oxide is now the dominant reactant that causes ozone depletion in Earth's stratosphere (Ravishankara et al., 2009; Eq. 3.47–3.52).

Nitrous oxide shows minor seasonal variation in the atmosphere, with greater amplitude at high northern latitudes (±1.15 ppb) than at the South Pole (±0.29 ppb; (Jiang et al., 2007), cf. Ishijima et al., 2009). The mean residence time for N_2O in the atmosphere is about 120 years (Prather et al., 2015), consistent with observations of a relatively uniform (320 ± 1 ppb) concentration of atmospheric N_2O around the world (Ishijima et al., 2009; see Fig. 3.5). The only significant sink for N_2O—stratospheric destruction (Eqs. 3.47 to 3.49)—consumes about 12.2×10^{12} gN as N_2O per year (Minschwaner et al., 1993). A few soils also appear to consume N_2O, but the global sink in soils is probably very small (Schlesinger, 2013). Unfortunately, estimates of sources—particularly sources that have changed greatly in recent years—are poorly constrained (Table 12.5).

The oceans are a large natural source of N_2O to the atmosphere as a result of nitrification in the deep sea (Cohen and Gordon, 1979; Oudot et al., 1990). Incubation experiments and the isotopic content of N_2O from the marine environment are consistent with its source from nitrifying archaea in seawater (Santoro et al., 2011; Ji and Ward, 2017). Some of this N_2O may subsequently be *consumed* by denitrification, converting N_2O to N_2 as it passes upward through zones of low O_2 (Cohen and Gordon, 1978; Kim and Craig, 1990; Frame et al., 2014). However, denitrification also produces a small amount of N_2O (Chapter 6), and denitrification is the dominant process contributing to N_2O production in the anoxic waters of the Eastern Tropical Pacific (Ji et al., 2015).

In many areas, surface waters are supersaturated in N_2O with respect to the atmosphere (Walter et al., 2004b). Specifically, the waters of the northwest Indian Ocean, a local zone of upwelling, may account for 20% of the total flux of N_2O from the oceans to the atmosphere (Law and Owens, 1990). Based on the belief that the N_2O supersaturation of seawater was worldwide, calculated emissions from the ocean dominated the earliest global estimates of N_2O sources (Liss and Slater, 1974; Hahn, 1974). When more extensive sampling showed that the areas of supersaturation were regional, these workers substantially lowered their estimate of N_2O production in marine ecosystems. The most extensive surveys of ocean waters suggest a flux of about 2 to 6×10^{12} gN/yr, emitted as N_2O to the atmosphere (Bianchi et al., 2012; Battaglia and Joos, 2018; Buitenhuis et al., 2018). A large portion of this may derive from coastal

TABLE 12.5 A global budget for nitrous oxide (N_2O) in the atmosphere.

Sources		
Natural		
Soils	6.6	Bouwman et al. (1995)[a,b,c]
Ocean Surface	2.5	Buitenhuis et al. (2018)
Total natural	9.1	
Anthropogenic		
Agricultural soils	3.75	Aneja et al. (2019)
Cattle and feed lots	2.8	Davidson (2009)
Biomass burning	0.9	Kaiser et al. (2012)
		Andreae (2019)
Industry	0.66	Davidson (2009)[d]
Transportation	0.14	Wallington and Wiesen (2014)
Human sewage	0.2	Mosier et al. (1998)
Total anthropogenic	8.45	
Total sources	17.6	(cf. 17.9 TgN/yr; IPCC, 2013
Sinks		
Stratospheric destruction	12.3	Prather et al. (1995)
Uptake by soils	0.3	Schlesinger (2013)
Atmospheric increase	3.6	IPCC (2013)
Total Identified Sinks	16.2	

[a] *The flux from undisturbed soils includes the flux from lakes (0.2 Tg N/yr; Lauerwald et al., 2019) and rivers (0.7 Tg N/yr; Beaulieu et al., 2011).*
[b] *Alternative estimates for the flux of N_2O from undisturbed soils include 3.4 Tg N/yr (Zhuang et al., 2012), 6.1 Tg N/yr (Potter et al., 1996), 6.3 Tg N/yr (Tian et al., 2018), 8.2–9.5 Tg N/yr (Xu-Ri et al., 2012); Tian et al. (2018) give an estimate of 10 Tg N/yr as the total current flux from natural and agricultural soils.*
[c] *Davidson's 2009 value of 0.8 less global transportation now separately estimated by Wallington and Wiesen (2014).*
[d] *The sum of emission from agriculture and domestic animals given here, 6.55 Tg N/yr is in close agreement with the value of 5.0 Tg N/yr estimated by Syakila and Kroeze (2011). These estimates of N_2O flux from agricultural activities include emissions of N_2O from downstream ecosystems and groundwaters impacted by agricultural inputs in these regions.*
All values are Tg N/yr (10^{12} g/yr) nitrogen, as N_2O.

waters, where increasing N_2O flux may derive from seawater enriched with NO_3^- from terrestrial runoff (Bange et al., 1996; Nevison et al., 2004; Naqvi et al., 2000; Frame et al., 2014).

Soil emissions from nitrification and denitrification (Chapter 6) are now thought to compose the largest global source of N_2O (Table 12.5). Particularly large emissions of N_2O are found from tropical soils (Bouwman et al., 1993; Kort et al., 2011; van Lent et al., 2015). Conversion of tropical forests to cultivated lands and pasture results in greater N_2O emissions (Matson and Vitousek, 1990; Keller and Reiners, 1994), and the flux of N_2O increases when agricultural lands and forests are fertilized or manured (Shcherbak et al., 2014; Della Chiesa et al., 2019), particularly on acidic soils (Wang et al., 2018c). Typically about 1% of the application of nitrogen fertilizer is lost to the atmosphere as N_2O (Bouwman et al., 2002a; Lesschen et al., 2011), with even higher percentages in areas of intensive fertilizer use (Nevison et al., 2018). N_2O emissions have increased exponentially with increasing fertilizer applications during the past century (Gao et al., 2011). Presumably the increased flux of N_2O from disturbed and fertilized soils stems from higher rates of nitrification and greater availability of NO_3^- to denitrifying bacteria (Chapter 6). Globally the flux of N_2O from agricultural soils is about $2.2–3.7 \times 10^{12}$ gN/yr (Bouwman et al., 2002a; Davidson, 2009; Tian et al., 2018), and the total production of N_2O from all soils is $\sim 10 \times 10^{12}$ gN/yr (Table 12.5). In most ecosystems, soil CO_2 and N_2O efflux are correlated, and a N_2O flux as high as 13.3×10^{12} gN/yr from soils is possible given current estimates of CO_2 efflux (Xu et al., 2008a).

Downward leaching of fertilizer nitrate also has the potential to stimulate denitrification in groundwaters. Ronen et al. (1988) suggest that groundwater may be an important source of N_2O to the atmosphere—up to 1×10^{12} gN/yr, but most recent assessments suggest significantly lower values (Bottcher et al., 2011; Keuskamp et al., 2012). Excess nitrogen in surface runoff also causes a small flux of N_2O from streams and rivers (Beaulieu et al., 2011; Hu et al., 2016; Grant et al., 2018; Audet et al., 2019), as does the disposal of human sewage (Kaplan et al., 1978; McElroy and Wang, 2005).

Relatively small emissions of N_2O result from the combustion of fossil fuels or biomass (Table 12.5), but the industrial production of nylon and other chemicals results in significant emissions of N_2O to the atmosphere (Thiemens and Trogler, 1991). Total anthropogenic sources of N_2O are more than enough to explain its rate of increase in the atmosphere, but the total compilation of sources—both natural and anthropogenic—is slightly more than the known sinks, including the rate of N_2O accumulation in the atmosphere (Table 12.5).

Some constraints on the global budget for N_2O in Earth's atmosphere are set by studies of $\delta^{15}N$ and $\delta^{18}O$ in the gas. The flux of N_2O from soils is depleted in $\delta^{15}N$ and $\delta^{18}O$, since denitrifiers discriminate against the heavy isotopes in the pool of available NO_3^- (Chapter 6). Conversely, the "backflux" of N_2O from the stratosphere is *enriched* in both $\delta^{15}N$ and $\delta^{18}O$, since the photochemical destruction of N_2O in the stratosphere also discriminates against the heavy isotopes (Morgan et al., 2004; Bernath et al., 2017). Estimates of changes in the sources of N_2O in the troposphere are constrained by its isotopic content, which is determined by N_2O from the stratosphere mixing with the weighted average of the isotopic content in various sources of N_2O at the Earth's surface (Kim and Craig, 1993). In the past few decades, both $\delta^{15}N$ and $\delta^{18}O$ in atmospheric N_2O have declined—consistent with an increasing flux from soils to the atmosphere (Ishijima et al., 2007; Rockmann and Levin, 2005; Park et al., 2012).[k]

---

[k]N_2O is a linear molecule (NNO). Analyses of the isotopic composition of the nitrogen atoms at the central (alpha) and end (beta) positions (isotopomers) show promise to distinguish among various sources of N_2O, including nitrification and denitrification contributing to N_2O flux (Yoshida and Toyoda, 2000; Sutka et al., 2006; Toyoda et al., 2011). The Site Preference (SP) is defined as the nitrogen isotope ratio of the central nitrogen atom minus that of the terminal nitrogen atom. However, interpretations of various isotope ratios in N_2O (viz. $\delta^{18}O$, $\delta^{15}N$ and SP) are problematic because the net flux of N_2O measured in the field is composed of production by nitrifiers and denitrifiers and by consumption by denitrifiers (e.g., Conen and Neftel, 2007; Koba et al., 2009; Lewicka-Szczebak et al., 2014; Well et al., 2012).

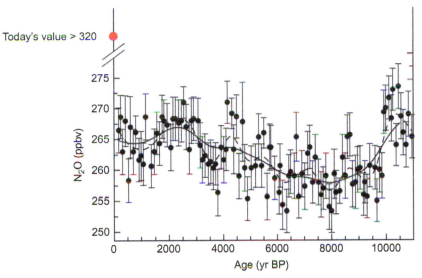

FIG. 12.6 Nitrous oxide measurements from ice-core samples in Antarctica. *From Flückiger et al. (2002).*

Cores extracted from the Antarctic ice cap show that the concentration of N_2O was much lower (180 ppb) during the last glacial period (Leuenberger and Siegenthaler, 1992; Sowers et al., 2003; Schilt et al., 2010). At the end of the Pleistocene, concentrations rose to ~265 ppb and remained fairly constant until the Industrial Revolution, when they increased to the present value of about 320 ppb (Fig. 12.6; Flückiger et al., 2002; Sowers et al., 2003; Schilt et al., 2014). Anticipating the future, field experiments show that rising CO_2 in Earth's atmosphere, excess N deposition in precipitation, and additions of Fe to ocean waters are all likely to increase the flux of N_2O to Earth's atmosphere and exacerbate global warming (Kammann et al., 2008; Liu and Greaver, 2009; Law and Ling, 2001; van Groenigen et al., 2011; Kim et al., 2012). Indeed, increasing N_2O emissions from soils may negate some of the benefits seen in the use of fertilizer to enhance crop growth for biofuels and soil carbon sequestration (Adler et al., 2007; Melillo et al., 2009).

The global phosphorus cycle

The global biogeochemical cycle of P has no significant gaseous component (Fig. 12.7). The redox potential of most soils is too high to allow for the production of phosphine gas (PH_3), except under very specialized, local conditions (Bartlett, 1986; Chapter 7). Phosphine emissions are reported for sewage treatment ponds in Hungary (Dévai et al., 1988), marshes in Louisiana and Florida (Devai and Delaune, 1995), and a lake in China (Geng et al., 2005). Phosphine is also found in the marine atmosphere over the Atlantic and Pacific Oceans, where it may be derived from the impact of lightning on P-containing soil dusts (Glindemann et al., 2003; Zhu et al.,

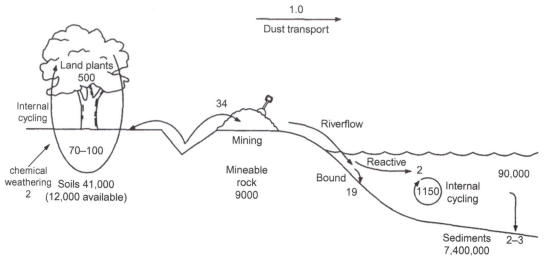

FIG. 12.7 The global phosphorus cycle. Each flux is shown in units of 10^{12} g P/yr. Values for P production and reserves are taken from the U.S. Geological Survey. Estimate for sediments is from Van Cappellen and Ingall (1996), and estimates for other pools and flux as derived in the text.

2007). Overall, global flux of P in phosphine is probably $<0.04 \times 10^{12}$ g P/yr (Gassmann and Glindemann, 1993).

The flux of P through the atmosphere in soil dust and seaspray (1×10^{12} g P/yr; Graham and Duce, 1979; Mahowald et al., 2008; Myriokefalitakis et al., 2016) is also much smaller than other transfers in the global P cycle, which are largely derived from chemical weathering in soils. A small amount of P in the atmosphere is derived from plant pollen (Bigio and Angert, 2018). Atmospheric deposition of P is known to make a critical contribution to the supply of available P when it is deposited in some tropical forests on highly weathered soils (Swap et al., 1992; Chadwick et al., 1999; Okin et al., 2004) and in the open ocean (Talbot et al., 1986; Bristow et al., 2010). Newman (1995) reports P deposition from the atmosphere (0.07–1.7 kg/ha/yr) that rivals P derived from rock weathering (0.05–1.0 kg/ha/yr) in various terrestrial ecosystems. The Bodélé depression in Chad is thought to be the source of 0.12×10^{12} g P/yr in soil dusts that blow westward over the Atlantic Ocean and into the Amazon basin (Bristow et al., 2010; Ben-Ami et al., 2010). A significant fraction of the phosphorus in this dust is biogenic apatite, derived from fish bones (Hudson-Edwards et al., 2014).

Nearly all the phosphorus in terrestrial ecosystems is originally derived from the weathering of calcium phosphate minerals, especially apatite ($Ca_5(PO_4)_3OH$; Eq. 4.7). The phosphorus content of most rocks is not large, and in most soils only a small fraction of the total P is available to biota (Chapter 4). Land vegetation appears to contain about 0.5×10^{15} g P (Smil, 2000), while soils, to 50-cm depth, hold about 41×10^{15} g P, of which only 12.2×10^{15} g P is in labile or organic forms (Yang et al., 2013b; Wang et al., 2010a,b). Estimated plant uptake on land, $70–100 \times 10^{12}$ g P/yr (Smil, 2000), implies a turnover of 0.5%/yr in soils. Root exudates and mycorrhizae may increase the rate of rock weathering on land (Chapter 4), but there is no process, equivalent to N-fixation, that can produce dramatic increases in phosphorus availability for plants in P-deficient habitats. Thus, on both land and at sea, biota persists as a result of a well-developed recycling of phosphorus in organic forms (Figs. 6.22 and 9.36).

The main flux of P in the global cycle is carried by rivers, which transport about 21×10^{12} g P/yr to the sea (Meybeck, 1982; Smil, 2000)—about twice as much as 300 years ago (Wallmann, 2010). Only about 10% of the flux delivered to the oceans is potentially available to marine biota; the remainder is strongly bound to soil particles that are rapidly sedimented on the continental shelf (Chapter 9). The solubility product of apatite is only about 10^{-58} (Lindsay and Vlek, 1977). At a seawater pH of 8.3, the phosphorus concentration in equilibrium with apatite is about 1.3×10^{-7} M ($\sim 4 \mu g$ P/L; Atlas and Pytkowicz, 1977; cf. Fig. 4.10). In seawater, organic and colloidal forms maintain the concentration of P in excess of that in equilibrium with respect to apatite; the average content of P in deep ocean water is about 3×10^{-6} M ($\sim 93 \mu g$/L; cf. Figs. 9.18 and 9.27). The concentration of PO_4^{3-} in the surface oceans is low, but the large volume of the deep sea accounts for a substantial pool of P (Fig. 12.7). The overall mean residence time for reactive P in the sea is about 25,000 years (Chapter 9).

The turnover of P through the organic pools in the surface ocean occurs in a few days. Nearly 90% of the phosphorus taken up by marine biota is regenerated in the surface ocean, and most of the rest is mineralized in the deep sea (Fig. 9.36). Eventually, however, phosphorus is deposited in ocean sediments, which contain the largest phosphorus pool near the surface of the Earth (Van Cappellen and Ingall, 1996). About $2–3 \times 10^{12}$ g P/yr are added to sediments of the open ocean—roughly equivalent to the delivery of reactive P to the oceans by rivers (Wallmann, 2010; Baturin, 2007; Fig. 9.22). On a time scale of hundreds of millions of years, these sediments are uplifted and subject to rock weathering, completing the global cycle. Today, most of the phosphorus in rivers is derived from the weathering of sedimentary rocks, and it represents P that has made at least one complete journey through the global cycle (Griffith et al., 1977).

It is likely that PO_4 has always been the dominant form of P available to biota. Samples of the geologic rock record suggest that limited availability of P in seawater until about 800 million years ago (Bjerrum and Canfield, 2002; Reinhard et al., 2017). Griffith et al. (1977) calculates that it took about 3 billion years for the weathering of igneous rocks to saturate seawater with PO_4, allowing the development of phosphorite skeletons in organisms (Cook and Shergold, 1984) and the precipitation of authigenic apatites in sediments (Chapter 9). Today, these minerals are carried by subduction to the mantle, completing the global phosphorus cycle upon tectonic uplift that returns rocks to the surface (Guidry et al., 2000; Buendia et al., 2010). The present-day burial of phosphorus appears similar to the rate during much of Earth's history (Filippelli and Delaney, 1992), but periods of massive uplift and erosion may have fueled high net primary productivity in the oceans (Filippelli and Delaney, 1994; Filippelli, 2008).

Students of Earth's earliest life have speculated on mechanisms by which PO_4 could be polymerized and condensed, so it could be incorporated into biochemical molecules. It is possible that lightning, volcanic eruptions, and other local high-energy environments may have been involved in the production of phosphite and polyphosphates (Yamagata et al., 1991; Pasek and Block, 2009).

Van Cappellen and Ingall (1996) suggest a negative-feedback mechanism by which changes in the concentration of O_2 in Earth's atmosphere determine the availability of P in the deep sea, where the net mineralization of P is controlled by its adsorption on Fe minerals in oxic sediments. This cycle stabilizes the level of O_2 in Earth's atmosphere through geologic time. For instance, if the concentration of O_2 in the atmosphere were to fall to low levels, available P would become more plentiful in the waters upwelling from the deep sea, marine NPP would increase, and more O_2 would be released to the atmosphere from the oceans. An analogous model centers on the terrestrial biosphere, in which the frequency of fire modulates changes in phosphorus availability, primary productivity, and changes in the levels of O_2 in the atmosphere (Kump, 1988; Lenton, 2001).

In many areas, humans have enhanced the availability of P by mining phosphate rocks—34×10^{12} g P/yr,[1] which can be used as fertilizer. Huge increases in phosphate production have occurred in China during the past decade. Most of the economic deposits of phosphate are found in sedimentary rocks of marine origin, so the mining activity directly enhances the turnover of the global P cycle. By mining phosphate rock, humans impact the global P cycle at a rate that rivals our impact on the global N cycle. The ratio of industrial nitrogen fixation to the mining of phosphate rock is about 9.1 (molar) is somewhat less than the ratio in the uptake of these elements by land plants (31.3) but similar to the Redfield ratio for the uptake by marine phytoplankton (16). The largest deposits of phosphate rock are found in Morocco (Cooper et al., 2011). In the United States, deposits of phosphate rock are found in Florida and North Carolina. Current global reserves are estimated at about 9×10^{15} g P.[1] At current rates of use, the global estimate of phosphorus reserves will last about 300 years (Cooper et al., 2011; Koppelaar and Weikard, 2013). Much of the phosphorus mobilized by mining is transported internationally in fertilizer and food products grown from fertilizer, amounting to as much as 3.0 to 5.2×10^{12} g P/yr (Nesme et al., 2018; Yang et al., 2019). As we saw for N, this transfer of P across the oceans is a new flux of P in its global cycle, wrought by humans.

In many areas, the flux of P in rivers is significantly higher than it was in prehistoric times as a result of erosion, pollution, and fertilizer runoff (Bennett et al., 2001; Yuan et al., 2011; Morée et al., 2013; Ostrofsky et al., 2018). Indeed, the extensive use of phosphate rock, a non-renewable resource, may lead to future phosphorus shortages for agriculture, which accounts for the majority of its use. Expansion of cultivation onto P-fixing soils in the tropics demands better nutrient management in agriculture (Roy et al., 2016). Various suggestions to improve the phosphorus-use efficiency in agriculture focus on greater phosphorus-use efficiency and recovery from livestock (Suh and Yee, 2011; Wang et al., 2011; MacDonald et al., 2012; Metson et al., 2012a). Other proposals recommend a greater effort to recycle P from effluent sewage waters to farmland (Zhou et al., 2017; Venkatesan et al., 2016; Tonini et al., 2019). Losses of P to lakes and rivers are greatest in developing economies (Fink et al., 2018), and in many regions a substantial amount of the riverine P flux is captured in the sediments behind dams (Maavara et al., 2015), indicating substantial opportunity to capture and reuse phosphorus as fertilizer.

---

[1]http://minerals.usgs.gov/minerals/pubs/commodity/phosphate_rock/. All data derived from this source are calculated assuming phosphate rock is 30% P_2O_5, and converted to elemental P content (Steve Jasinki, USGS, personal communication)

Potassium: An overlooked limiting element?

Every year nearly 32.2×10^{12} g of potassium (K) are mined from the Earth's crust, where K is largely found in evaporate minerals, especially sylvinite.[m] An essential nutrient for plants and animals, nearly all of this K is applied to agricultural crops worldwide. Potassium is involved in osmotic and ionic balance in both plants and animals. Movements of potassium are the major determinant of stomatal function in plants. In humans potassium controls the electrolyte balance of blood, and a cellular sodium/potassium pump controls the contraction of muscles. Sterner and Elser (2002) calculate a C/K ratio of 484 in the human body, and Reiners (1986) indicates a ratio of 46 for herbaceous plants. Not overlooked by agronomists, K has garnered relatively little interest among biogeochemists (Zörb et al., 2014; Sardans and Peñuelas, 2015). Together with N and P, potassium should be expected to exert a major control on terrestrial plant productivity.

Potassium is contained in a variety of primary minerals, especially orthoclase feldspar, biotite and muscovite, from which it is released by chemical weathering. Some plants actively access K from primary minerals via the release of organic acids (Boyle and Voigt, 1973). As it is lost from terrestrial ecosystems in runoff, K is eventually delivered to the sea, where it accumulates in evaporate deposits and secondary clay minerals, including illite (Siever, 1974; Michalopoulos and Aller, 1995; Berner and Berner, 2012, p. 375). The content of δ^{41}K in rivers is indicative of the rate of silicate weathering in their watershed (Li et al., 2019b).

Rock weathering dominates the exogenous sources of K in most terrestrial ecosystems. At the Hubbard Brook Experimental Forest in New Hampshire (USA), chemical weathering supplies 11% of the annual plant uptake of K, versus 1% from atmospheric deposition. The remaining K in vegetation is derived from the turnover of plant materials accumulated in prior years (Table 6.1). K held on soil cation exchange sites is likely the dominant source of K for plant nutrition (Chapter 4).

Potassium presents a conundrum to biogeochemists. K is cycled wastefully at the plant level, but it appears to be conserved in the nutrient budgets of entire ecosystems. For instance K is easily leached from plant foliage and often is the most concentrated element in throughfall precipitation collected beneath the forest canopy (Tukey, 1970; Parker, 1983). Particles containing K are lost from plant leaves, forming organic aerosols (Crozat, 1979; Pohlker et al., 2012). With autumnal senescence of foliage, K is reabsorbed for reuse, but calculations of reabsorption efficiency are complicated by leaching losses during the senescence process. In the Hubbard Brook Experimental Forest, resorption accounted for 10 to 32% of the loss of K from senescing leaves, depending upon rainfall during the autumn (Likens et al., 1994). In California chaparral, about 12% of the K accumulation in vegetation is derived from resorption (Gray, 1982). Thus, the internal cycle of K in plants is not as tightly controlled as that for N and P (cf. Vergutz et al., 2012).

K is normally found at its highest concentration in the upper layers of the soil (Jobbágy and Jackson, 2001). Even in hyper-arid regions, the highest concentrations of K are often at the soil surface (e.g., Schlesinger, 1985), where K is held in illite of pedogenic origin (Singer, 1989). In areas of sparse vegetation, K is concentrated under plant canopies, where it may be 1.5 times

[m]https://minerals.usgs.gov/minerals/pubs/commodity/potash/mcs-2016-potas.pdf

more concentrated under plants than in the barren soils between them (Schlesinger et al., 1996). These patterns reflect the recycling of K by plants (Barre et al., 2009). Using rubidium as a tracer for K, Stone and Kszystyniak (1977) document the retention of Rb in forest ecosystems via a one-time fertilization experiment (112 kg/ha) in which 40% of applied Rb was found within the system after 23 years.

Potassium appears to be a limiting nutrient for plant growth in some lowland tropical forests (Wright et al., 2011; Santiago et al., 2012) and in the boreal region (Ouimet and Moore, 2015). In a temperate deciduous forest, the seasonal retention of K results in the lowest concentrations of K in runoff waters collected during the growing season despite the lowest volume of flow at that season (Likens et al., 1994; Likens, 2013; Tripler et al., 2006). The loss of K in global riverflow ($\sim$52 $\times$ 10^{12} g/yr, Berner and Berner, 2012) is roughly 1% of the annual plant demand for K to support global net primary productivity on land.

The global net primary production of agricultural crops is estimated at 5.25 $\times$ 10^{15} g C/yr (Wolf et al., 2015), which is equivalent to 0.43 $\times$ 10^{15} mol of carbon fixed by photosynthesis. Assuming a C/K ratio of 46 in herbaceous plants (Reiners, 1986), this crop growth would demand 9.5 $\times$ 10^{12} mol of K/yr or 371 $\times$ 10^{12} gK—more than 10 times larger than the amount of K mined each year (32.2 $\times$ 10^{12} gK). The difference is presumably made up by the decomposition of crop residues left in fields and mineral weathering of K-bearing minerals in agricultural soils.

We estimate that the annual global rate of nitrogen made available by nitrogen-fixing crops and N-fertilizers and applied to crops, 200 $\times$ 10^{12} gN ($\sim$14.3 $\times$ 10^{12} molN) should be accompanied by 0.62 $\times$ 10^{12} mol of P and 3.11 $\times$ 10^{12} mol of K, to match the stoichiometry of plant demand by crops (Reiners, 1986). This is equivalent to 19 $\times$ 10^{12} g P and 121 $\times$ 10^{12} gK. For P, this is somewhat less than the annual global mining of P from phosphate rock (34 $\times$ 10^{12} g P), but for K the amount is much more than K supplied by potash (32.2 $\times$ 10^{12} gK above). Again, the implication is that rock weathering supplies a considerable amount of the K demand by crop plants or, alternatively, that K is sorely under-applied to most fertilized crops. In the future, an increasing demand for K fertilizer may accompany the expansion of agriculture into highly weathered tropical soils, where K appears to control the distribution and productivity of natural vegetation (Yavitt et al., 2011; Lloyd et al., 2015).

Linking global biogeochemical cycles

The cycles of important biogeochemical elements are linked at many levels. Stock et al. (1990) describe how P is used to activate a transcriptional protein, stimulating nitrogen fixation in bacteria when nitrogen is in short supply. In this case, an understanding of the interaction between these elements is gained through the study of molecular biology. In Chapter 5, we saw that the photosynthetic rate of land plants is related to the N and P content of their leaves, linking the net production of organic carbon to the availability of these elements in plant cells. In marine ecosystems, net primary productivity is often calculated from the Redfield ratio of C:N:P in phytoplankton biomass (Chapter 9). The molar N/P ratio of 16 may reflect fundamental relationships between the requirements for protein and RNA synthesis common to all plants (Loladze and Elser, 2011). Whatever our viewpoint—from molecules to whole ecosystems—the movements of N, P, and C are strongly linked in biogeochemistry (Reiners, 1986).

Nitrogen fixation by free-living bacteria appears inversely related to the N/P ratio in soil (Fig. 6.8), and the rate of accumulation of N is greatest in soils with high P content (Walker and Adams, 1958). Similarly, N/P ratios <29 appear to stimulate N-fixation in freshwater ecosystems (Chapter 7). One might speculate that the high demand for P by N-fixing organisms links the global cycles of N and P, with P being the ultimate limit on nitrogen availability and net primary production. Indeed, in many soils the accumulation of organic carbon is correlated to available P (Chapter 6). N-fixation in ocean waters is stimulated by P and Fe deposition from soil dusts (Falkowski et al., 1998). Despite these theoretical arguments for a phosphorus limitation of the biosphere through geologic time, net primary production in most terrestrial and marine ecosystems usually shows an immediate response to additions of N (Fig. 12.8). Denitrification appears to maintain small supplies of N in most ecosystems, and co-limitation by N and P is commonplace (Harpole et al., 2011).

In the face of rising concentrations of atmospheric CO_2 and increasing dispersal of reactive N, the biosphere is showing unprecedented increases in the C/P and C/N ratio in various ecosystems (Peñuelas et al., 2020). It is urgent that biogeochemists offer improved predictions of the response of the biosphere to ongoing changes in the availability of N, P and K, which may allow a sustained increase in plant growth in response to rising CO_2 in Earth's atmosphere. Theoretical considerations based on stoichiometry indicate that the sink for CO_2 in land plants

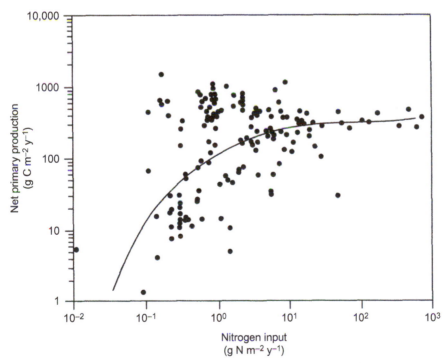

FIG. 12.8   Net primary production versus nitrogen inputs to terrestrial, aquatic, and marine ecosystems. Net primary production increases in direct response to added nitrogen up to inputs of about 10 g N/m/yr (100 kg/ha). Inputs in excess of that level are rarely found in natural ecosystems, but are seen in polluted environments and agricultural soils. *Modified from Levin (1989).*

will be limited (van Groenigen et al., 2006; Reay et al., 2008), but changes in soil nutrient turnover and plant uptake may provide enhanced nutrient supplies (Finzi et al., 2007; Drake et al., 2011; Dieleman et al., 2010). Similarly, changes in the supply of N and P to the surface oceans of the world, as determined by changes in circulation and nutrient turnover, may determine the long-term sink for carbon in the sea. Such changes in biogeochemistry are likely to have affected the concentration of CO_2 in the atmosphere and Earth's climate during glacial-interglacial cycles, implying that the marine biosphere could respond rapidly to future climate change (Falkowski et al., 1998; Altabet et al., 2002; Gruber and Galloway, 2008).

Summary

For both N and P, a small biogeochemical cycle with relatively rapid turnover is coupled to a large global pool with relatively slow turnover. For N, the major pool is found in the atmosphere. For P, the large pool is found in unweathered rock and soil.

The biogeochemical cycle of N begins with the fixation of atmospheric nitrogen, which transfers a small amount of inert N_2 to the biosphere. This transfer is balanced by denitrification, which returns N_2 to the atmosphere. The balance of these processes maintains a steady-state concentration of N_2 in the atmosphere with a turnover time of $\sim 10^7$ years. In the absence of denitrification, most of the N inventory on the Earth would eventually be sequestered in the ocean and in organic sediments. Denitrification closes the global nitrogen cycle, and it causes nitrogen to cycle more rapidly than phosphorus, which has no gaseous phase. The mean residence time of phosphorus in sedimentary rocks is measured in 10^8 yr, and the phosphorus cycle is complete only as a result of tectonic movements of the Earth's crust.

Once within the biosphere, the movements of N and P are more rapid than in their global cycles, showing turnover times ranging from hours (for soluble P in seawater) to hundreds of years (for N in biomass). In response to nutrient limitations, biotic recycling in terrestrial and marine habitats allows much greater rates of net primary production than rates of N-fixation and rock weathering alone would otherwise support (Tables 6.1 and 9.3). The high efficiency of nutrient recycling may explain why, in the face of widespread nitrogen limitation, only about 2.5% of global net primary production is diverted to nitrogen fixation (Gutschick, 1981).

Human perturbations of the global N, P and K cycles are widespread and dramatic. Through the production of fertilizers, humans have doubled the rate at which nitrogen enters the biogeochemical cycle on land. It is unclear how rapidly denitrification will respond to this global increase in nitrogen availability, but the rising concentrations of atmospheric N_2O are perhaps one indication of an ongoing biotic response (Vitousek, 1994). Increasing nitrogen availability has led to the local extinction of species from polluted ecosystems and shifted the limitation of net primary production in some systems from N to P (e.g., Mohren et al., 1986; Elser et al., 2007; Peñuelas et al., 2012). Losses of N and P to surface waters have increased the ratio of N-to-P (in dissolved inorganic forms) from 18 to 27 during the past century (Vilmin et al., 2018). Increasing transport of both N and P in rivers has shifted many estuarine and coastal ecosystems to a condition of Si deficiency (Justic et al., 1995). All these changes indicate the effect of a single species—the human—in upsetting previous steady-state conditions in global nutrient cycling (Peñuelas et al., 2020).

The Global Cycles of Sulfur and Mercury

Introduction

The original pool of sulfur on Earth was held in igneous rocks, largely as pyrite (FeS_2). Degassing of the mantle and, later, weathering of the crust under an atmosphere containing O_2 transferred a large amount of S to the oceans, where it is now found as SO_4^{2-} (Table 2.3). When SO_4^{2-} is assimilated by organisms, it is reduced and converted into organic sulfur, which is an essential component of protein. However, the live biosphere contains relatively little sulfur. Today, the major global pools of S at the Earth's surface are found in biogenic pyrite, seawater, and evaporites derived from ocean water (Table 13.1).

Like sulfur, the majority of the Earth's mercury (Hg) is found in the crust, and it is released to the atmosphere and oceans by volcanic eruptions, rock weathering, and human activities. Human and microbial activities yield a global biogeochemical cycle for this element, which has long been recognized as toxic to life.

As in the case of nitrogen, microbial transformations between valence states drive the global S and Hg cycles. Sulfur is found in valence states ranging from +6 in SO_4^{2-} to −2 in sulfides. Under anoxic conditions, SO_4 is a substrate for sulfate reduction, which may lead

TABLE 13.1   Active reservoirs of sulfur near the
surface of the earth.

Reservoir	10^{18} g S
Atmosphere	0.0000028
Seawater—SO_4^{2-}	1280
Dissolved organic sulfur	0.0067
Sedimentary rocks	
Evaporites	2470
Shales	4970
Land plants	0.0085
Soil organic matter	0.0155
Total	8720

From Holser et al. (1989), Dobrovolsky (1994), and Ksionzek et al. (2016),
for DOS in seawater.

to the release of reduced gases to the atmosphere and to the deposition of biogenic pyrite in sediments (Chapters 7 and 9). Anoxic environments can also support sulfur-based photosynthesis, which is likely to have been one of the first forms of photosynthesis on Earth (Eq. 2.11). In the presence of oxygen, reduced sulfur compounds are oxidized by microbes. In some cases, the oxidation of S is coupled to the reduction of CO_2 in the reactions of S-based chemosynthesis (Eq. 4.6).

Mercury is transformed among valence states that include dissolved ions (Hg^{2+}), elemental mercury (Hg^0), and an especially toxic form known as methyl-mercury (CH_3Hg). Hg^0 and CH_3Hg can exist as vapor, allowing transport of Hg in the atmosphere. Sulfate-reducing bacteria are implicated in the production of methyl-mercury in sediments, so the global cycles of S and Hg are linked by metabolism.

Understanding the biogeochemistry of S and Hg has enormous economic significance. Many metals are mined from sulfide minerals in hydrothermal ore deposits (Meyer, 1985). Microbial reactions involving sulfur bacteria are used to concentrate metals from water (e.g., Zn, Labrenz et al., 2000; Cu, Sillitoe et al., 1996; and Au, Lengke and Southam, 2006) and are used to remove metals from relatively low-grade ore (Lundgren and Silver, 1980). Sulfur is an important constituent of both coal and oil. The organic S in coal is oxidized to sulfuric acid when coal is exposed to air. SO_2 is emitted to the atmosphere when coal and oil are burned. A large amount of SO_2 is also emitted during the smelting of copper ores (Cullis and Hirschler, 1980; Oppenheimer et al., 1985).

An understanding of the relative importance of natural sulfur compounds in the atmosphere compared to anthropogenic SO_2 is essential to evaluate the causes of acid rain and the impact of acid rain on natural ecosystems. Similarly, understanding the natural and anthropogenic sources of mercury in the atmosphere allows rational policy decisions regarding the regulation of mercury emissions from power plants and other sources (see Table 1.1).

In this chapter, we compare the global cycles of S and Hg, since both are characterized by microbial reduction reactions that produce volatile forms. Assembling quantitative global cycles for these elements allows us to put the human emissions of S and Hg into a larger context. As we did for carbon (Chapter 11), nitrogen, and phosphorus (Chapter 12), we will attempt to establish a budget for S and Hg on land and in the atmosphere. Then we will couple these to marine budgets to form an overall picture of the global cycles for S and Hg. The biogeochemical cycle of S has varied through Earth's history as a result of the appearance of new metabolic pathways and changes in their importance. We will review the history of the S cycle as it is told by sedimentary rocks.

The global sulfur cycle

No sulfur gas is a long-lived or major constituent in the atmosphere. Thus, all attempts to model the global S cycle must explain the fate of the large annual input of sulfur compounds to the atmosphere. The short mean residence time, about 5 days, for atmospheric sulfur compounds, as a result of their oxidation to SO_4, allows us to express all the fluxes in the global budget in units of 10^{12} g of S, regardless of the original form of emission. Despite the small atmospheric content of S compounds (totaling $\sim 4 \times 10^{12}$ g S at any moment; Rodhe, 1999), the annual flux of S compounds through the atmosphere (about 300×10^{12} g S/yr) rivals the movements of N in the global nitrogen cycle (compare Fig. 13.1 to Fig. 12.2).[a]

Land

Eriksson (1960) examined the potential origins of SO_4 in Swedish rainfall and hence, indirectly, sources of SO_4 in the atmosphere. He reasoned that if all the Cl^- in rainfall is derived from the ocean, then seaspray should also carry SO_4 roughly in proportion to the ratio of SO_4^{2-} to Cl^- in seawater. His calculation suggested that about 4×10^{12} g S/yr deposited on land must be derived from the sea. At about the same time, Junge (1960) was evaluating the SO_4 content of rainfall, and he estimated that about 73×10^{12} g S/yr was deposited on land globally. Clearly, there were other sources of SO_4 in the atmosphere and in rainfall. Junge's maps showed that SO_4 was abundant in the rainfall of industrial regions and in areas downwind of deserts (Fig. 3.14A). Desert soils are a source of gypsum ($CaSO_4 \cdot 2H_2O$) in atmospheric dust (Reheis and Kihl, 1995), and the burning of fossil fuels in industrial regions contributes SO_2 to air pollution (Langner et al., 1992; Spiro et al., 1992). By the mid 1970s, "acid rain" was clearly linked to coal-fired power plants, which release SO_2 (Likens and Bormann 1974). In the intervening years, new sources of S in the atmosphere have been recognized, and global flux estimates have been revised repeatedly. Our understanding of the

[a] Formulating a global picture for S in the atmosphere, while useful for comparative and budgetary purposes, is not entirely realistic. As pointed out by the late Ralph Cicerone, all S compounds have such short atmospheric mean residence times that regional to hemispheric differences in the emission, deposition and concentration of sulfur compounds in the atmosphere are expected, and there is no uniform mixing of S compounds into a global circulation.

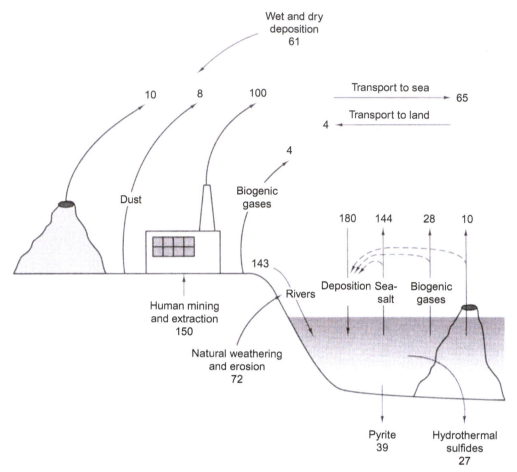

FIG 13.1 The global S cycle with annual flux shown in units of 10^{12} g S/yr. The derivation of most values is described in the text, with the marine values taken from Fig. 9.38.

global S cycle is much improved, but many of the estimates illustrated in Fig. 13.1 remain subject to considerable uncertainty.

Episodic events, including volcanic eruptions and dust storms, contribute to the global biogeochemical cycle of S. Sulfur emissions from volcanoes are especially difficult (and dangerous!) to measure. SO_2 dominates the volcanic release, but significant H_2S is reported in some eruptions (Aiuppa et al., 2005; Clarisse et al., 2011); both are oxidized to SO_4 in the atmosphere (Eqs. 3.27–3.28). Legrand and Delmas (1987) used the deposition of SO_4 in the Antarctic ice pack to estimate the contribution of volcanoes to the global S cycle during the last 220 years. The Tambora eruption of 1815 was the largest, releasing 50×10^{12} g S to the atmosphere. Typically, major eruptions, such as that of Mt. Pinatubo (15 June 1991), release $5–10 \times 10^{12}$ g S each (Bluth et al., 1993).

When the volcanic emissions are averaged over many years, the annual global flux is about 7.5 to 10.5 × 10^{12} g S/yr (Halmer et al., 2002). Remote sensing from satellites during the past decade suggests that emissions have averaged about 23 x 10^{12} g S/yr (Carn et al., 2017). About 70% of the sulfur gases leak passively from volcanoes and the remainder is derived from periodic, explosive events (Bluth et al., 1993; Allard et al., 1994). Huge eruptions in the Late Cretaceous (66 mya) may have released more than 1000 Tg S to the atmosphere (Self et al., 2008). Following a major volcanic eruption, the Earth's climate is cooled for several years, as a result of SO$_4$ aerosols in the stratosphere (Sigl et al., 2015; Chapter 3).

The movement of S in soil dust is also episodic and poorly understood. Many of the large particles are deposited locally, while smaller particles may undergo long-range transport in the atmosphere (Chapter 3). Savoie et al. (1987) found that dust from the deserts of the Middle East contributes SO$_4$ to the waters of the northwest Indian Ocean. Ivanov (1983) suggests a global flux of 8 × 10^{12} g S/yr owing to dust transport in the troposphere—about 10% of the fossil fuel release.

The total flux of biogenic sulfur gases from land is likely <4 × 10^{12} g S/yr (Watts, 2000). The dominant sulfur gas emitted from freshwater wetlands and anoxic soils is H$_2$S, with dimethylsulfide and carbonyl sulfide (COS) playing lesser roles (Chapter 7). The Amazon basin is an apparent source of dimethylsulfide to its atmosphere (Jardine et al. 2015). Emissions from other ecosystems are poorly understood and deserving of further study (Chapter 6). Forest fires emit an additional 2 × 10^{12} g S/yr (Andreae and Merlet, 2001; Kaiser et al., 2012).

Compared to these various natural sources, it seems certain that direct emissions from human industrial activities are the largest sources of S gases in the atmosphere. Ice cores from Greenland show a large increase in the deposition of SO$_4$ from the atmosphere at the beginning of the Industrial Revolution (Herron et al., 1977; Mayewski et al., 1986, 1990; Fischer et al., 1998). In Europe and the United States, these emissions have been sharply curtailed by the control of air pollution in recent years (Likens et al., 2005; Stern, 2006; Velders et al., 2011), and the SO$_4$ content in rainfall in these areas has fallen in parallel (Fig. 13.2). Estimates

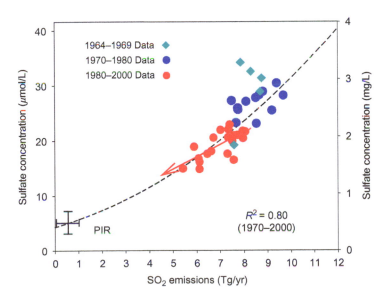

FIG. 13.2 Sulfate concentration in wetfall precipitation at New Hampshire's Hubbard Brook Experimental Forest, as a function of SO$_2$ emissions in the estimated 24-hr source area. This shows the decline in both parameters as a result of the implementation of the Clean Air Act. (PRI shows the levels of the pre-industrial revolution.) *From Likens et al. (2005). Used with permission of RSC Publishing, copyright Royal Society of Chemistry.*

of the recent global flux from anthropogenic activities range from 50 to 100×10^{12} g S/yr, with large, additions from China and India, which have major expansion of industry and nascent concerns for air pollution (Lee et al., 2011; Klimont et al., 2013).

Owing to the reactivity of S gases in the atmosphere, most of the anthropogenic emission of SO_2 is deposited locally in precipitation and dryfall. Total deposition of S on land may be as high as 120×10^{12} g S/yr (Andreae and Jaeschke, 1992), but a value of $<100 \times 10^{12}$ g S/yr would balance the global S cycle shown in Fig. 13.1. Deposition in dryfall and the direct absorption of SO_2 by plants are poorly understood, so this global estimate is subject to revision. The estimate of atmospheric deposition on land accounts for a large fraction of the total emissions from land. The remainder undergoes long-distance transport in the atmosphere and accounts for a net transfer of S from land to sea (Whelpdale and Galloway, 1994).

A small fraction of the natural river load of SO_4 is derived from rainfall, which includes cyclic salts that are carried through the atmosphere from the ocean (4×10^{12} g S/yr in Fig. 13.1). Weathering of pyrite and gypsum also contributes to the SO_4 content of river water, totaling about 90×10^{12} g S/yr (Burke et al., 2018). The remainder of the global flux of S in rivers is derived from human activities. Brimblecombe (2003) calculated that at least 60% of the SO_4 content of modern rivers is derived from air pollution, mining, erosion, and other human activities. The current river transport of sulfate, estimated at $143–225 \times 10^{12}$ g S/yr, is significantly higher than in preindustrial conditions.

Marine environments

The marine portion of the global S cycle is largely taken from Fig. 9.38. The oceans receive SO_4 from rivers and deposit sulfides in hydothermal vents (27×10^{12} g S/yr; Elderfield and Schultz, 1996) and biogenic pyrite in sediments (39×10^{12} g S/yr, Berner, 1982). The latter constitutes about 40% of the total removal of S from the oceans (cf. Tostevin et al., 2014), but only accounts for about 10% of the gross rate of sulfate-reduction in ocean sediments (Bowles et al., 2014). The ocean is a large source of seasalt aerosols that contain SO_4, but most of these are redeposited in the ocean in precipitation and dryfall.

Dimethylsulfide—$(CH_3)_2$ S, or DMS—is the major biogenic gas emitted from the sea, with recent global estimates ranging from 19.6 to 28 Tg S/yr (Lana et al., 2011; Land et al., 2014; Chapter 9). Thus, DMS from the oceans is the largest natural source of sulfur gases in the atmosphere. The mean residence time of DMS in the atmosphere is <2 days (Table 3.5), as a result of its oxidation to SO_2. Much of the SO_2 is further oxidized to SO_4, which is redeposited in the oceans (Faloona et al., 2009). SO_4 aerosols in the marine atmosphere or deposited from it that are not derived from seawater are known as non-seasalt sulfate (nss-sulfate), and are largely indicative of emissions of DMS and volcanoes (Xie et al., 2002). Only about 6×10^{12} g S are emitted as SO_2 by international shipping (Smith et al., 2011), and a very small additional amount by aircraft (Kjellstrom et al., 1999). A significant portion of the SO_4 in the Antarctic icepack and exposed soils is nss-sulfate, likely from DMS (Legrand et al., 1991; Bao et al., 2000).

In addition to helping balance the marine sulfur budget, dimethylsulfide attains global significance for its potential effects on climate (Shaw, 1983). Charlson et al. (1987) postulated that the oxidation of DMS to sulfate aerosols would increase the scattering of solar radiation and the abundance of cloud condensation nuclei in the troposphere, leading to greater cloudiness (Bates et al., 1987). Low clouds over the sea reflect incoming sunlight, increasing global albedo and leading to global cooling. Note that this effect of sulfate aerosols in the troposphere is analogous to the effect of sulfate aerosols from carbonyl sulfide (COS) and from large volcanic eruptions that reach the stratosphere (Chapter 3).

The production of DMS in the surface ocean is often related to the growth of certain marine phytoplankton (Andreae et al., 1995; Steinke et al., 2002). DMS flux is directly correlated to solar irradiance (Vallina and Simó, 2007; Gali et al., 2011), so the flux of DMS from the sea is greater in summer than in winter, as a result of greater sea surface temperature (Prospero et al., 1991; Tarrasón et al., 1995). The concentration of DMS in seawater is well correlated to that in the air over the North Pacific Ocean (Watanabe et al., 1995). Cloud condensation nuclei also appear to be well correlated to the atmospheric burden of DMS in nonpolluted areas (Ayers and Gras, 1991; Putaud et al., 1993; Andreae et al., 1995).

The hypothesis that DMS might provide a biotic regulation on global temperature is intriguing, for it may be responsible for the moderation of global climate throughout geologic time. If higher marine NPP is associated with warmer sea surface temperatures, then a greater flux of DMS would have the potential to act as a negative feedback on global warming through the greenhouse effect. Quinn and Bates (2011) question the impact of SO_4^{2-} aerosols derived from DMS on climate versus the production of a variety of other aerosols in the marine atmosphere. Given the strong arguments pointing to global warming by increased atmospheric CO_2, the potential negative feedbacks of DMS remain a subject of intense scientific scrutiny and debate.

Examining the deposition record in the Antarctic ice pack, Iizuka et al. (2012) report a strong correlation between nss-SO_4 and cooler global conditions during the past 300,000 years. The ice-core record of methanesulphonate (MSA)—a DMS degradation product—suggests higher concentrations during the last glacial epoch than today (Legrand et al., 1991; but see Castebrunet et al., 2006). Certainly the ocean's temperature was lower during the last glacial, but if, for other reasons (e.g., a greater deposition of iron-rich dust), marine NPP productivity was higher during the last glacial, an increased flux of DMS may have reinforced the global cooling of Earth's climate. Indeed, Turner et al. (1996) report more than a threefold increase in DMS flux to the atmosphere during an iron-enrichment experiment in the Pacific Ocean. If natural emissions of DMS increase in the future, it is possible that they will act to dampen the greenhouse effect during the next century.

Schwartz (1988) argued that anthropogenic emissions of SO_2 should have the same effect on global climate as natural emissions of DMS, because SO_2 is also oxidized to produce condensation nuclei in the atmosphere. Because SO_2 has a short atmospheric lifetime (Chapter 3), its effect is regional and centered on areas of industry (Falkowski et al., 1992; Langner et al., 1992). While the Earth's albedo does not seem to show large trends globally (Chapter 3), it is noteworthy that regional trends in albedo show recent "dimming" over China (He et al., 2018). Using a general circulation model for global climate, Wigley (1989) suggested that climatic cooling by SO_2, largely from coal-fired power plants, may have offset some of the temperature change expected from the greenhouse effect, especially before the institution of air pollution controls in the United States and Europe. Reductions in man-made sulfur emissions might exacerbate global warming because the cooling effect of sulfur aerosols will be removed (Andreae et al., 2005; Samset et al., 2018).

Interactions of the sulfur and oxygen cycles

Two forms of pyrite occur in the Earth's crust and sediments. Igneous pyrite is formed by abiotic processes, accounting for crystals of pyrite in granite and hydrothermal deposits. Biogenic pyrite is formed by the action of sulfate-reducing bacteria, which produce H_2S that reacts with Fe to form FeS_2 (Eq. 7.19 and 7.20). Most of the biogenic pyrite occurs in near-shore organic rich marine sediments (Chapter 9). A smaller amount forms in freshwater wetlands, which normally have lower Fe contents. Biogenic pyrite can also form from the activities of sulfate-reducing bacteria in groundwater environments, which receive dissolved organic carbon percolating downward from soils (Drake et al., 2013).

Permanent burial of reduced compounds (organic carbon and biogenic pyrite) accounts for the accumulation of O_2 in Earth's atmosphere (Chapter 3). The molar ratio is 1.0 for organic carbon, but the burial of 1 mole of reduced sulfur accounts for nearly 2.0 moles of O_2 (Raiswell and Berner, 1986; Berner and Canfield, 1989). The weight ratio of C/S in most marine shales is about 2.8—equivalent to a molar ratio of 7.5 (Raiswell and Berner, 1986). Thus, through geologic time the deposition of reduced sulfur in pyrite may account for about 20% of the O_2 in today's atmosphere. During periods of rapid continental uplift, erosion, and sedimentation, large amounts of organic substances and pyrite were buried and the oxygen content of the atmosphere increased (Des Marais et al., 1992).

Both forms of pyrite are oxidized by chemoautotrophic bacteria when they are exposed to oxygen, accounting for the production of acidic waters draining areas of coal mining. It is worth asking how much of Earth's oxygen is consumed each year by aerobic rock weathering of pyrite (Eq. 2.15 and 4.6). Burke et al. (2018) suggest that about 30% of the load of SO_4 in rivers is derived from the weathering of pyrite—about 42×10^{12} g S/yr or $\sim 1.3 \times 10^{12}$ moles of S (see Table 4.9). The oxidation of the reduced S in pyrite would thus consume about 2.6×10^{12} moles of O_2 each year (see Fig. 11.9). This is only a tiny fraction of the O_2 content of Earth's atmosphere, where O_2 has accumulated at slow rates by the deposition of organic carbon and pyrite through eons. Nevertheless, these interactions of the S and O cycles foster some long-term stability to atmospheric O_2. Falling atmospheric O_2 would increase anoxic conditions in marine sediments, fostering the storage of organic carbon, the formation of biogenic pyrite, and the accumulation of O_2 in the atmosphere (Walker, 1980).

Temporal perspectives on the global sulfur cycle

During the accretion of the primordial Earth, sulfur was among the gases that were released by degassing of the Earth's mantle to form the secondary atmosphere (Chapter 2). Even today, volcanic emissions contain appreciable concentrations of SO_2 and H_2S (Table 2.2). When the oceans condensed on Earth, the atmosphere was essentially swept clear of these S gases, owing to their high solubility in water (Eq. 2.6). On Venus, where there is no ocean, crustal degassing has resulted in a large concentration of SO_2 in the atmosphere (Oyama et al., 1979).

The dominant form of S in the earliest seawater is likely to have been SO_4^{2-}; high concentrations of Fe^{2+} in the primitive ocean would have precipitated any sulfides, which are insoluble under anoxic conditions (Walker and Brimblecombe, 1985). Nevertheless, the

SO_4 content of the earliest oceans was low (Habicht et al., 2002; Crowe et al., 2014) and increased in concert with the rise of O_2 in Earth's atmosphere (Canfield et al., 2000). The total inventory of S compounds on the surface of the Earth (nearly 10^{22} g S) represents the total crustal outgassing of S through geologic time (Table 2.3).

The ratio of ^{32}S to ^{34}S in the total S inventory on Earth is thought to be similar to the ratio of 22.22 measured in the Canyon Diablo Troilite (CDT), a meteorite collected in Arizona. The sulfur isotope ratio in this rock is accepted as an international standard and assigned a value of 0.00. In other samples, deviations from this ratio are expressed as $\delta^{34}S$, with the units of parts per thousand parts (‰)—a convention that we also used for the isotopes of carbon (Chapter 5) and nitrogen (Chapter 6). Presumably the $\delta^{34}S$ isotope ratio in the earliest oceans was 0.00, because there is no reason to expect any discrimination between the isotopes of S during crustal degassing. When evaporite minerals precipitate from seawater, there is little differentiation among the isotopes of sulfur, so geologic deposits of gypsum ($CaSO_4 \cdot 2H_2O$) and barite ($BaSO_4$) carry a record of the isotopic composition of S in seawater. In the earliest sedimentary rocks, dating to 3.8 bya, $\delta^{34}S$ is close to 0.00 (Schidlowski, 1983).

Dissimilatory sulfate reduction by bacteria strongly differentiates among the isotopes of sulfur, as a result of a more rapid enzymatic reaction with $^{32}SO_4$. By itself, sulfate metabolism results in an isotopic depletion of -18‰ to -25‰ $\delta^{34}S$ in H_2S and sedimentary sulfides relative to the ratio in the source reservoir (Canfield and Teske, 1996; Bradley et al., 2016). Stronger fractionations, sometimes as much as -66‰ (Sim et al., 2011) can result from repeated cycles of oxidation and reduction (Canfield and Teske, 1996) and the reduction of thiosulfate ($S_2O_3^{2-}$) and sulfite ($S_2O_3^{2-}$) in sediments (Canfield and Thamdrup, 1994; Habicht et al., 1998).

The evolution of sulfate reduction dates to 2.4–2.7 bya, based on the first occurrence of sedimentary rocks with depletion of ^{34}S (Cameron, 1982; Schidlowski, 1983; Habicht and Canfield, 1996; Fig. 13.3).[b] This also marks the time of the first substantial SO_4 concentrations in the primitive oceans. Today, the average $\delta^{34}S$ in sedimentary sulfides in all of the Earth's crust is about -10‰ to -12‰ (Holser and Kaplan, 1966; Migdisov et al., 1983).

Fig. 13.4 shows a three-box model for the S cycle, in which marine SO_4 and sedimentary sulfides are connected through microbial oxidation and reduction reactions, which discriminate between the sulfur isotopes (Garrels and Lerman, 1981). During periods of Earth's history when large amounts of sedimentary pyrite were formed as a result of sulfate reduction, seawater SO_4 became enriched in residual $^{34}SO_4$. Meanwhile, deposition or dissolution of evaporate deposits has no effect on the isotopic composition of seawater, but can strongly affect the concentration of SO_4 in seawater and thus the rate of sulfate-reduction and pyrite deposition (Wortmann and Paytan, 2012).

Currently, about 50% of the pool of S near the surface of the Earth is found in reduced form (Li, 1972; Holser et al., 1989), and the $\delta^{34}S$ of seawater is +21‰ (Kaplan, 1975; Rees et al., 1978; Tostevin et al., 2014). Because there is little differentiation among isotopes of S during the precipitation of evaporites, the sedimentary record of evaporites indicates changes in the $\delta^{34}S$ of seawater over geologic time. Changes in the isotopic composition of seawater indicate changes in the relative size of the reservoir of sedimentary pyrite, which occur as a result of changes in the net global balance between sulfate reduction and the oxidation of sedimentary sulfides

---

[b]Scattered evidence is found for S-reduction as early as 3.4 bya (Ohmoto et al., 1993; Shen et al., 2001).

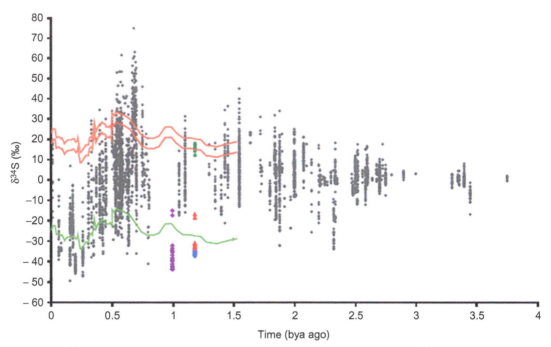

FIG. 13.3 δ^{34}S in sedimentary pyrites through geologic time. The orange line traces the δ^{34}S in seawater for the past 1.5 billion years and the green line shows the expected offset by bacterial sulfate reduction, which appears in the sedimentary record about 2.5 billion years ago. *Modified from Parnell et al. (2010).*

($\sim$100 $\times$ 10^{12} g S/yr). By contrast, the annual uptake of S by marine phytoplankton and release through their decomposition ($\sim$1390 $\times$ 10^{12} g S/yr; Dobrovolsky, 1994, p. 183) have little effect on the isotopic composition of the major reservoirs of the global S cycle.

During the last 600,000,000 years, seawater SO$_4$ has varied between +10‰ and +30‰ in δ^{34}S (Fig. 13.5), with an average value close to that of today. Seawater sulfate shows a marked positive excursion (+32‰) in δ^{34}S during the Cambrian (550 mya), when the deposition of pyrite must have been greatly in excess of the oxidation of sulfide minerals exposed on land. Seawater sulfate was less concentrated in ^{34}S, that is, δ^{34}S of +10‰, during the Carboniferous and Permian, when a large proportion of the Earth's net primary production occurred in freshwater swamps, where SO$_4$, sulfate reduction, and pyrite deposition are less important (Berner, 1984). Presumably the concentration of SO$_4$ in seawater was also greater during that interval, because the global rate of pyrite formation was depressed. The record of δ^{34}S in marine barite fluctuates in an inverse correlation with records of marine productivity during the past 130 million years (Paytan et al., 2004).

Although the sulfur cycle has shown shifts between net sulfur oxidation and net sulfur reduction in the geologic past, the current human impact is probably unprecedented in the geologic record. As we found for the carbon cycle, the present-day cycle of S is not in steady state. Human activities have added a large flux of gaseous sulfur to the atmosphere, some of which is transported globally. Humans are mining coal and extracting petroleum from the Earth's

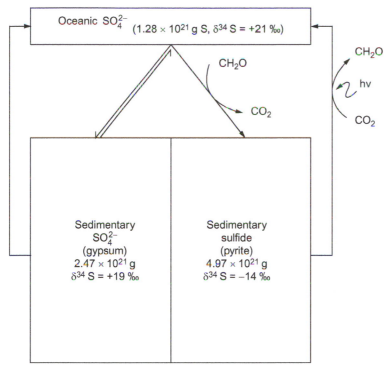

FIG. 13.4   A model for the global sulfur cycle, showing the linkage and partitioning of S between oxidized and reduced pools near the surface of the Earth. Transfers of S from seawater to pyrite involve a major fractionation between ^{34}S and ^{32}S isotopes, whereas exchange between seawater SO_4 and sedimentary SO_4 (largely gypsum) involves only minor fractionation. The sum of all pools, nearly 10^{22} g S, represents the total outgassing of S from the mantle (compare to Table 2.3). About 15% now resides in the ocean. Estimates of the pool of S in sedimentary sulfides show a wide range of values. The value here, from Holser et al. (1989), is close to that estimated from the pool of sedimentary organic carbon (1.56×10^{22} g; Des Marais et al., 1992) divided by the mean C/S ratio in marine sediments (2.8; Raiswell and Berner, 1986). Isotope ratios in seawater and gypsum are taken from Holser et al. (1989). The isotope rate of S in sedimentary sulfides is derived by mass balance to yield $\delta^{34}S$ of +4.2 in the global inventory.

crust at a rate that mobilizes 150×10^{12} g S/yr, more than double the rate of 100 years ago (Brimblecombe et al., 1989). The net effect of these processes is to increase the pool of oxidized sulfur (SO_4) in the global cycle, at the expense of the storage of reduced sulfur in the Earth's crust. The current estimates of inputs to the ocean are slightly in excess of the estimate of total sinks, implying that the oceans are increasing in SO_4 by over 10^{13} g S/yr. Such an increase will be difficult to document because the content in the oceans is 1.28×10^{21} g S. As calculated in Chapter 9, the mean residence time for SO_4 in seawater is about 10,000,000 years with respect to the current inputs from rivers.

Various workers have attempted to use measurements of $\delta^{34}S$ to deduce the origin of the SO_4 in rainfall and the extent of human impact on the movement of S in the atmosphere (Grey and Jensen, 1972; Nriagu et al., 1991; Mast et al., 2001; Puig et al., 2008). Unfortunately, the potential sources of SO_4 show a wide range of values for $\delta^{34}S$, making the identification of specific sources difficult (Nielsen, 1974). For example, the sulfur in coal may be depleted

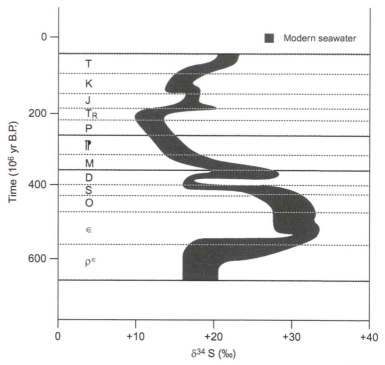

FIG. 13.5   Variations in the isotopic composition of seawater SO₄ through geologic time. *From Kaplan (1975). Used with permission from the Royal Society of London.*

in $\delta^{34}S$ if it is found as pyrite or enriched in $\delta^{34}S$ if it is derived from the sulfur that was originally assimilated by the plants forming coal (Hackley and Anderson, 1986). Thus, coals show a wide range in $\delta^{34}S$. Similarly, petroleum shows a range of $-10.0‰$ to $+25‰$ in ^{34}S (Krouse and McCready, 1979). Desert dusts containing SO₄ show a wide range in $\delta^{34}S$, averaging $+5.8‰$ (Bao and Reheis, 2003). In the eastern United States, $\delta^{34}S$ of SO_4^{2-} in rainfall varies seasonally between $+6.4‰$ in winter and $+2.9‰$ in summer, consistent with any of these sources or a combination of them (Nriagu and Coker, 1978). The lower values of summer are thought to reflect the influence of biogenic sulfur derived from sulfate reduction in wetlands (Nriagu et al., 1987).

When SO_2 is emitted as an air pollutant, it forms sulfuric acid through heterogeneous reactions with water in the atmosphere (Eqs. 3.26–3.29). As a strong acid that is completely dissociated in water, H_2SO_4 suppresses the disassociation of natural, weak acids in rainfall. For example, in the absence of strong acids, the dissolution of CO_2 in water will form a weak solution of carbonic acid, H_2CO_3, and rainfall pH will be about 5.6:

$$CO_2 + H_2O \rightarrow H^+ + HCO_3^-. \tag{13.1}$$

In the presence of strong acids that lower the pH below 4.3, this reaction moves to the left, and carbonic acid makes no contribution to free acidity. In many industrialized areas, free acidity in precipitation is almost wholly determined by the concentration of the strong acid

TABLE 13.2   Sources of acidity in acid rainfall collected in Ithaca, New York, on July 11, 1975 (ambient pH 3.84).

Component	Concentration in precipitation (mg/L)	Contribution to	
		Free acidity at pH 3.84 (μeq/L)	Total acidity in a titration to pH 9.0 (μeq/L)
H_2CO_3	0.62	0	20
Clay	5	0	5
NH_4^+	0.53	0	29
Dissolved Al	0.050	0	5
Dissolved Fe	0.040	0	2
Dissolved Mn	0.005	0	0.1
Total organic acids	0.43	2	5.7
HNO_3	2.80	40	40
H_2SO_4	5.60	102	103
Total		144	210

anions, SO_4^{2-} and NO_3^- (Table 13.2). Rock weathering that was primarily driven by carbonation weathering in the preindustrial age is now driven by anthropogenic H^+ (Johnson et al., 1972).

It is interesting to estimate the global sources of acidity in the atmosphere. In this analysis, we are only interested in reactions that are net sources of H^+, so we can ignore the movements of soil dusts and seaspray because the strong-acid anions they contain, NO_3^- and SO_4^{2-}, are largely balanced by cations (especially Ca and Na) that are emitted at the same time. If the pH of all rainfall on Earth was 5.6 as a result of an equilibrium with atmospheric CO_2, the total deposition of H^+ ions would be 1.24×10^{12} moles/year. The production of NO by lightning produces additional acidity because NO dissolves in rainwater, forming HNO_3. Globally, N fixation by lightning contributes 0.64×10^{12} moles of H^+/yr, and other natural sources of NO_x (soils and forest fires) contribute 1.54×10^{12} moles of H^+/yr. In the years following massive eruptions, SO_2 from volcanoes is distributed globally and dominates the atmospheric deposition of S (Mayewski et al., 1990; Langway et al., 1995). On average, however, volcanic emanations of SO_2 contribute $\sim 1.24 \times 10^{12}$ moles H^+/yr, and the oxidation of biogenic S gases produces 2.0×10^{12} moles H^+/yr. Thus, the total, natural production of H^+ in the atmosphere is normally about 6.66×10^{12} moles/yr. In contrast, the anthropogenic production of NO_x and SO_2 results in about 8.06×10^{12} moles of H^+/year—more than all natural sources of acidity combined.

The only net source of alkalinity in the atmosphere comes from the reaction of NH_3 with the strong acids H_2SO_4 and HNO_3 to form aerosols, $(NH_4)_2SO_4$ and NH_4NO_3 (Eq. 3.5). However, the "natural" global emission of NH_3, about 13×10^{12} g N/yr (Table 12.2), reduces the "natural" production of H$^+$ by only about 0.91×10^{12} moles/year, or 14% (cf. Savoie et al., 1993). Thus, even though the current acidity of the atmosphere is much higher as a result of human activities, the atmosphere has always acted as an acidic medium with respect to the Earth's crust throughout geologic time. Anthropogenic emissions of NH$_3$ neutralize 3.0×10^{12} moles H$^+$/yr, or 37% of H$^+$ derived from the anthropogenic emissions of the acid-forming gases, NO$_x$ and SO$_2$.

The global mercury cycle

Mercury is found as a trace metal in igneous rocks and is locally concentrated in economic deposits of a red mineral known as cinnabar (HgS) that is often associated with hydrothermally-altered rock. Volcanoes are a large natural source of mercury emissions to Earth's atmosphere, in the form of elemental mercury, Hg^0—a volatile metal (Engle et al., 2006; Pyle and Mather, 2003; Bagnato et al., 2011); Fig. 13.6). Hg^0 is oxidized in the atmosphere to Hg^{2+}, which is the dominant form found in wet and dry deposition over most of the Earth's surface (Lin and Pehkonen, 1999). The lifetime of Hg in the atmosphere ranges from about one year for Hg^0 to a few weeks for Hg^{2+}, so mercury shows regional patterns of deposition that reflect local emissions from natural sources and power plants (Engle et al., 2010; Prestbo and Gay, 2009; Selin and Jacob, 2008). A variety of stable isotopes of Hg (e.g., ^{202}Hg, ^{201}Hg, and ^{198}Hg), which are differentially fractionated by physical and biological

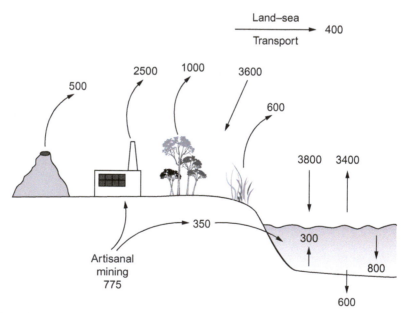

FIG. 13.6 The global mercury cycle of the modern world. All values are 10^6 g Hg/yr. *From Outridge et al. (2018), except for artisanal mining (Streets et al. (2019)).*

processes, show promise in elucidating the factors affecting mercury cycling in the environment (Blum, 2012; Kritee et al., 2013; Sun et al., 2019).

Rock weathering releases a small amount of mercury to runoff waters as Hg^{2+} (Fig. 13.6). Humans have enhanced the flux of Hg from the Earth's crust by direct mining of mercury and other ores and by the combustion of coal (Pirrone et al., 2010; Gratz and Keeler, 2011). As it circulates at the Earth's surface, mercury is rapidly converted among Hg^0, Hg^{2+}, and particulate forms. When mercury is deposited in forests, some is reduced to Hg^0 and revolatilized to the atmosphere (Graydon et al., 2012), while the remainder accumulates in soil organic matter (Demers et al., 2007; Gabriel et al., 2012) or binds to dissolved organic compounds that are transported in stream water (Dittman et al., 2010; Stoken et al., 2016; Lavoie et al., 2019; Fig. 13.7). In some lakes, mercury in the surface water appears to originate from atmospheric deposition, whereas mercury in the sediments derives from the fluvial transport of particles from the surrounding watershed (Chen et al., 2016a, b).

The concentration of Hg in seawater has been enhanced by a factor of 3–4 as a result of human contamination of the environment (Lamborg et al., 2014; Zhang et al., 2014). Some of the mercury deposited in the oceans is removed from the surface waters by adsorption to particles (Archer and Blum, 2018), and uptake by diatoms (Zaferani et al., 2018). Some Hg^{2+} that is deposited at sea is photoreduced to elemental mercury (Hg^0) that volatilizes from the ocean's surface, as a function of wind speed and Henry's Law for the dissolution of soluble gases in water (Andersson et al., 2008; Mason and Fitzgerald, 1993; Kuss et al., 2011). Mercury tends to accumulate in polar ecosystems (Johnson et al., 2008a, b, c; Schuster et al., 2018), where large amounts are deposited from the atmosphere as Hg^0 (Obrist et al., 2017). The global cycle of mercury may reflect distillation of Hg^0 in warmer regions and deposition of Hg^0 in cooler regions, where revolatilization is limited. The large quantities of Hg in polar regions may be subject to mobilization by runoff and volatilization as permafrost melts under a warmer climate (Olson et al., 2018; Pérez-Rodríguez et al., 2019).

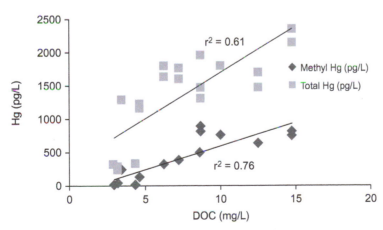

FIG. 13.7   Concentration of total and methylmercury in stream waters draining into Lake Sunapee, New Hampshire as a function of the concentration of dissolved organic carbon. *From Kathleen Weathers et al., unpublished.*

Mercury has a long history of economic uses that led to the early human exploitation of ore deposits (Pacyna et al., 2006; Pirrone et al., 2010; Selin, 2009). Before its toxic properties were well known, mercury was widely used as a fungicide and as a preservative for leather, paint, and other products subject to microbial degradation.[c] Mercury has long been used to extract trace quantities of gold and other metals from raw ores,[d] and emissions from such mining activities (artisanal and small scale gold mining (ASGM)) are now the largest sources of mercury pollution today, constituting about 37% of anthropogenic emissions to the atmosphere (UNEP, 2019), and as much as 70% of the mercury in mine-drainage waters (Goix et al., 2019). Mercury accumulates in organic sediments and coal, and is released to the atmosphere when these are burned (Billings and Matson, 1972; Lee et al., 2006). Emissions of Hg rose to high values in the late 1800s, associated with the gold rush in the western United States, and again after World War II, associated with the proliferation of coal-fired power plants (Streets et al., 2011). Lake sediments in New England and the Rocky Mountains show higher, recent levels of mercury accumulation than several centuries ago (Kamman and Engstrom, 2002; Mast et al., 2010). Analyses of Hg in an ice core in Peru (Beal et al., 2014) and a peat core from central China (Li et al., 2016a, b, c) show increasing environmental contamination of Hg began more than 1000 years ago (Fig. 13.8). In several peat bogs, modern rates of mercury accumulation are more than 15 times greater than those of the preindustrial period (Martinez-Cortizas et al., 1999; Roos-Barraclough et al., 2002; Shotyk et al., 2003). Accumulations are often greater in colder periods, when low temperatures reduce the revolatilization of Hg from sediments (Martinez-Cortizas et al., 1999).

An especially toxic form of mercury, methylmercury (CH_3Hg^+),[e] is produced by various sulfate-reducing bacteria in anoxic sediments (Compeau and Bartha, 1985; King et al., 2001; Gilmour et al., 1992, 2011) and in seawater (Munson et al., 2018). The methylation process appears to have a common genetic pathway across various taxonomic groups (Parks et al., 2013; Bravo and Cosio, 2019). The methylation process proceeds most rapidly with dissolved mercury (Hg^{2+}) and mercury associated with nanoparticles (Zhang et al., 2012b). Mercury accumulates in fishes as they age (Bache et al., 1971; Barber et al., 1972) and at progressively higher levels of food chains, so that large predatory fish often have the highest levels (Cabana and Rasmussen, 1994; Rumbold et al., 2018). Higher concentrations of methylmercury are found in modern seabirds than in museum specimens of the same species collected a century ago (Monteiro and Furness, 1997; Vo et al., 2011; Bond et al., 2015). Population declines of some birds may be linked to methylmercury concentrations in their tissues (Driscoll et al., 2007).

Methylmercury is found in fishes in nearly all waters of the United States (Scudder et al., 2009). When toxic levels of mercury were found in many freshwater and saltwater fishes in

---

[c]See Act V, Scene 1 in Shakespeare's *Hamlet*, where the gravediggers comment that a hatmaker's corpse will decompose slowly, presumably because its content of mercury will impede microbial activity.

[d]The use of liquid mercury to extract gold and silver from raw ore stems from its tendency to bind to these metals, so they can be decanted from the remaining spoil. Mercury is then boiled off the supernatant to the atmosphere, leaving the desired metals behind. This is known as artisanal mining, which is in widespread use in the undeveloped world (Pfeiffer et al., 1991; Artaxo et al., 2000).

[e]Often abbreviated MeHg.

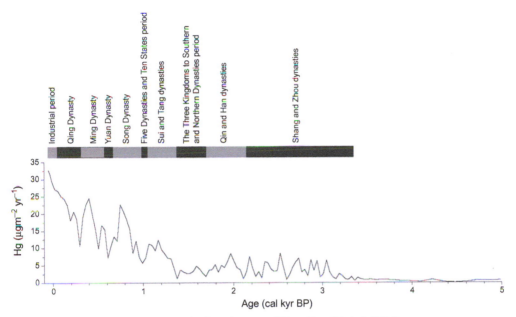

FIG. 13.8 Accumulation of mercury in the Dajiuhu mire, China. *From Li et al. (2016).*

the 1970s, environmental regulations were implemented to reduce airborne emissions, especially from coal-fired power plants. As a result, mercury emissions have decreased in some regions (Streets et al., 2019), which is expected to result in declining deposition of mercury from the atmosphere (Butler et al., 2008) and declining levels in fish (Harris et al., 2007) and birds (Braune et al., 2016). Unfortunately, the rate of decline in mercury in the environment is slowed by virtue of the legacy of past contamination, which can be volatilized from soils and waters and recirculate in nature (Amos et al., 2013). Methylmercury is degraded by sunlight to Hg^0, which is revolatilized from the surface of lakes and ocean waters (Mason and Fitzgerald, 1993; Sellers et al., 1996; Black et al., 2012). Hg^0 is also produced from microbial transformations of methylmercury in seawater (Lee and Fisher, 2019).

Enormous controversy has accompanied the establishment of policies to regulate mercury emissions to the atmosphere.[f] Biogeochemists have helped to elucidate how much of the deposition in various areas is due to local, regional, or global sources (Sigler and Lee, 2006; Selin et al., 2008; Weiss-Penzias et al., 2011; Gratz and Keeler, 2011). Similarly, biogeochemistry shows how methylmercury forms during sulfate reduction in sediments and explains why the fish in some lakes have relatively low concentrations while those in nearby lakes, often those with organic sediments, may have higher concentrations (Chen et al., 2005; Ward et al., 2010). Sulfate-reducing bacteria and SO_4-based acid rain are implicated in the

---

[f]92 nations have pledged to protect human health and the environment from anthropogenic emissions of mercury by signing the Minamata Convention on Mercury, named for one of the earliest recognized industrial contaminations of mercury in Japan.

methylation process (Gilmour and Henry, 1991). The mercury content of river waters is controlled by chemical reactions in the water and by the types of environments that are a source of drainage (Burns et al., 2012). For Hg in the marine environment, we still have a poor understanding of the processes that produce methylmercury in seawater and contamination of marine fishes (Malcolm et al., 2010).

Summary

The major pool of S in the global cycle is found in the crustal minerals gypsum and pyrite. Additional S is found dissolved in ocean water. Thus, with respect to pools, the global S cycle resembles the global cycle of phosphorus (Chapter 12). In contrast, the largest pool of the global N cycle is found in the atmosphere.

In other respects, however, there are strong similarities between the global cycles of N, S, and Hg. In all cases, the major annual movement of the element is through the atmosphere, and under natural conditions a large portion of the movement is through the production of reduced gases by biological activity. These gases return N, S, and Hg to the atmosphere, providing a closed global cycle with a relatively rapid turnover. In contrast, the ultimate fate for P is incorporation into ocean sediments, where the cycle of P is complete only as a result of long-term sedimentary uplift.

Biogeochemistry exerts a major influence on the global S cycle. The largest pool of S near the surface of the Earth is found in sedimentary pyrite, as a result of sulfate reduction. The sedimentary record shows that the relative extent of sulfate reduction has varied through geologic time. In the absence of sulfate-reducing bacteria, the concentration of SO_4 in seawater would be higher and O_2 in the atmosphere would be lower than present-day values (Eq. 9.9).

Current human perturbations of the sulfur and mercury cycles are extreme—roughly doubling the annual mobilization of these elements from the crust of the Earth. As a result of fossil fuel combustion, areas that are downwind of industrial regions now receive massive amounts of acidic deposition and mercury from the atmosphere. The excess acidity is likely to lead to changes in rock weathering (Chapter 4), forest growth (Chapter 6), and ocean productivity (Chapter 9). The deposition of mercury in aquatic ecosystems often leads to health advisories for fish consumption.

Coda

Save for a few incoming meteors from space and volcanic emanations from the deep Earth, the surface of our planet, the arena of life, is a closed chemical system. The surface environment, including the atmosphere and the oceans, is a thin peel, roughly 50 km in thickness at the outside of the Earth's 6371-km radius. Certainly, the characteristics of Earth's surface have changed dramatically through geologic time, especially as Earth cooled and violent degassing of the mantle diminished. The most profound change appears due to life itself—the oxygenation of Earth's atmosphere about 2.5 billion years ago.

We study biogeochemistry with the recognition that many of the characteristics of the Earth's surface are determined by life. Earth is very different from its nearest neighbors, Mars and Venus, where pure geochemistry prevails. On Earth, rock weathering and the fluvial transport of materials to the sea are driven by plant root and microbial activity in soils. The composition of seawater is determined by biotic processes that remove materials from the surface waters and deposit them in marine sediments. Many of the trace gases in the atmosphere are derived from the biosphere; they have short residence times in the atmosphere and must be restored to it by biotic activity. The constancy of Earth's atmospheric composition for the past 10,000 years is rather surprising, reflecting a close balance between biotic activities that produce and consume its gaseous constituents.

Today, when we study biogeochemistry, we see the pervasive influence of the human species on our planet's chemistry. With our use of fossil fuels, we extract carbon from the Earth's crust at a rate that is more than $36\times$ greater than the natural exposure of organic carbon in rocks at the Earth's surface; the mobilization of some chemical elements by mining is several-fold greater than their natural release by rock weathering (Table 14.1; Sen and Peucker-Ehrenbrink, 2012). We consume 20% of the world's freshwater and much of what we leave behind is of diminished quality for the support of life. The atmosphere's composition is changing rapidly as our expanding population taps into the Earth's resources in search of a better life for all peoples (Table 1.1; Hofmann et al., 2009). These changes in the mobilization of chemical elements worldwide are seen in the stratigraphic record, allowing earth scientists to recognize the Anthropocene as a new geologic epoch of human domination of planet Earth (Williams et al., 2016).

A few decades ago, we missed the opportunity to institute worldwide programs for family planning that might have stabilized the Earth's human population at levels that would minimize our lasting impact on the chemical conditions of our planet. Our similar efforts today

Biogeochemistry: An Analysis of Global Change, Fourth Edition
https://doi.org/10.1016/B978-0-12-814608-8.00014-1

TABLE 14.1   Estimates of the global flux in the biogeochemical cycles of certain elements, illustrating the human impact.

Element	Juvenile flux[a] (1)	Chemical weathering (2)	Natural cycle[b] (3)	Biospheric recycling ratio[c] 3/(1+2)	Human mobilization[d] (4)	Human enhancement 4/(1+2)	Reference
B	0.02	0.19	6.5	31	2.3	11	Schlesinger and Vengosh (2016)
C	30	250	110,000	393	10,000	36	Chapter 11
N	9	15[e]	9200	383	235	9.8	Chapter 12
P	~0	2	1250	625	34	17	Chapter 12
S	20	72	450	4.9	171	1.9	Chapter 13
Cl	2	260	120	0.46	170	0.65	Fig. 3.17
Ca	120	500	2300	3.7	65	0.10	Milliman et al. (1999) and Caro et al. (2010)
Hg	0.0005	0.0002	0.003	4.3	0.0076	10.9	Table 1.1; Chapter 13

[a] Degassing from the Earth's crust and mantle; sum of volcanic emissions to the atmosphere (subaerial) and net hydrothermal flux to the sea (Elderfield and Schultz, 1996) and for N, fixation by lightning (Chapter 12).
[b] Annual biogeochemical cycle to (uptake) and loss (decomposition) from the Earth's biota on land and in the oceans, in the absence of humans.
[c] Following Volk (1998).
[d] Direct and indirect mobilization by extraction and mining from the Earth's crust or (for N) industrial fixation (From U.S. Geological Survey, http://minerals.usgs.gov/minerals/pubs/commodity).
[e] Houlton et al. (2018)
Note: All data 10^{12} g/yr.

may trim only a billion or so from the population that is expected in 2050—9.3 billion neighbors for each of us (United Nations, 2010). With the increasing number of global citizens now expected, we must strive for a planet that will be habitable by many people for the foreseeable future. With the changes in Earth's chemistry that are now in progress, we are not on a sustainable course.

Preservation of natural habitat is the foundation for the preservation of biodiversity, and night-time satellite photographs of the Earth's surface show that we have already left little of nature that is not fragmented, traversed, or converted to human use (Fig. 14.1; Ellis et al., 2010; Hannah et al., 1995; Watts et al., 2007). As humans capture an increasing fraction of the planet's productive capacity (NPP), there will be a diminishing proportion left for the other species to persist with us (Haberl et al., 2007; Butchart et al., 2010).

And those species matter! Numerous studies show that sustainable levels of ecosystem function, on land and in the sea, depend on the rich diversity of life on Earth (Naeem et al., 1994; Worm et al., 2006; Liang et al., 2016; Huang et al., 2018). With fewer species, ecosystems are less productive, more vulnerable to disturbance, and less likely to recover. Table 14.2 shows how a relatively stable total ecosystem NPP in arctic tundra derives from

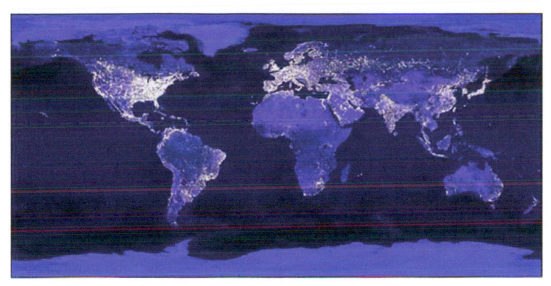

FIG. 14.1   Night-time view of the world from satellite-derived measures of brightness. *From http://eoimages.gsfc. nasa.gov/images/imagerecords/55000/55167/earth_lights.gif. See Román et al. (2018).*

TABLE 14.2   Annual variation of NPP (% of 5-year mean) of major species and of total aboveground NPP in an Alaskan Tundra ecosystem.

| | Production (% of average) | | | | | |
	1968	1969	1970	1978	1981	Coefficient of Variation (%)
Eriophorum	77	58	148	101	116	35
Betula	30	52	55	248	121	88
Ledum	106	138	62	103	91	27
Vaccinium	135	172	96	28	71	56
Total production	93	110	106	84	107	11

From Chapin et al. (1996). Used with permission from Cambridge University Press.

the compensation of individual species to varying conditions in individual years. If any are lost, the remaining productivity is lower and more variable from year to year. Often, increasing variability in ecosystem properties is the vanguard of impending collapse, analogous to similar observations in economics and medicine (Scheffer et al., 2009; Carpenter et al., 2011). When losses of species pass a certain threshold, the ecosystem undergoes a "state-change," beyond which a return to prior conditions will not yield a return to prior productivity. The stable conditions of our planet's chemistry are vulnerable to progressive biotic impoverishment. Biodiversity is the basis of stability in our life support system; maintaining biodiversity is more than a pastime for nature lovers.

If there are lessons in biogeochemistry that should guide the human practice of life, we should look for ways to transition to a zero-emission society, in which every product extracted from the Earth's surface is used and reused, without effluents delivered to the common pool of air and freshwater. Soil management is paramount, especially for elements such as phosphorus, for which there are limited supplies and no substitutes. We should look to ways to grow our food, without exogenous inputs, that will maximize yields while allowing us to maintain as much natural land as possible.

At root, almost every global environmental problem is the direct result of the continuing increase in the number of humans on Earth. As our population rises, the material expectations of individuals will raise the cumulative impact on the planet's resources (Krausmann et al., 2009). The paradigm of economic growth, which has so long dominated economic theory, is obsolete. Rising standards of living often reduce human fertility rates, but not fast enough to keep total resource consumption from increasing (Moses and Brown, 2003). While the biosphere will no doubt survive our species, without our stewardship, life will become increasingly unpleasant for most citizens, who will live on a hot, dirty, thirsty, and crowded planet.

In every population of organisms studied to date, exponential population growth is followed by dramatic collapse (e.g., Klein, 1968). In the history of our own species, wars and plagues have set us back temporarily along our inexorable trajectory toward the current world population peak (Turchin, 2009; Zhang et al., 2011a; Bevan et al., 2017). It is almost inevitable that water scarcity will grow worse as population increases in arid regions, and that food security will become harder to achieve in many regions when the climate grows hotter and rainfall becomes less predictable (Lobell et al., 2011). For those living in affluent countries, this will be unpleasant, while for those living in regions already suffering from food or water scarcity, this decline will be disastrous (Miranda et al., 2011).

Making meaningful changes will be inconvenient, but it is not too late to build a future where we waste much less energy, fertilizer, food, and water and in which we build innovative tools to better capture, store, and use the unlimited energy of the Sun to fuel our daily lives. For this to make a difference, we need to embrace a new culture of personal and corporate responsibility that values minimizing environmental impacts and maximizing long-term resource sustainability, over short-term economic gains and growth driven by human numbers. The only other viable alternative to slow our impact on the biosphere would be to reverse our own trajectory of population growth. If we do not intentionally choose to use less, it is nearly certain that future droughts, famines, disease, and war will choose an alternative future for us.

References

Abbott, D., Sparks, D., Herzberg, C., Mooney, W., Nikishin, A., Zhang, Y.S., 2000. Quantifying precambrian crustal extraction: the root is the answer. Tectonophysics 322, 163–190.

Abe, K., 2002. Preformed Cd and PO_4 and the relationship between the two elements in the Northwestern Pacific and the Okhotsk Sea. Mar. Chem. 79, 27–36.

Aber, J.D., Melillo, J.M., 1980. Litter decomposition—measuring relative contributions of organic matter and nitrogen to forest soils. Can. J. Bot. 58, 416–421.

Aber, J., McDowell, W., Nadelhoffer, K., Magill, A., Berntson, G., Kamakea, M., McNulty, S., Currie, W., Rustad, L., Fernandez, I., 1998. Nitrogen saturation in temperate forest ecosystems—hypotheses revisited. Bioscience 48, 921–934.

Aber, J.D., Goodale, C.L., Ollinger, S.V., Smith, M.L., Magill, A.H., Martin, M.E., Hallett, R.A., Stoddard, J.L., 2003. Is nitrogen deposition altering the nitrogen status of northeastern forests? Bioscience 53, 375–389.

Abramov, O., Mojzsis, S.J., 2009. Microbial habitability of the hadean earth during the late heavy bombardment. Nature 459 (7245), 419–422.

Abril, G., Guérin, F., Richard, S., Delmas, R., Galy-Lacaux, C., Gosse, P., Tremblay, A., Varfalvy, L., Dos Santos, M.A., Matvienko, B., 2005. Carbon dioxide and methane emissions and the carbon budget of a 10-year old tropical reservoir (Petit Saut, French Guiana). Glob. Biogeochem. Cycles 19(4).

Accoe, F., Boeckx, P., Busschaert, J., Hofman, G., Van Cleemput, O., 2004. Gross N transformation rates and net N mineralisation rates related to the C and N contents of soil organic matter fractions in grassland soils of different age. Soil Biol. Biochem. 36, 2075–2087.

Achard, F., Eva, H.D., Stibig, H.-J., Mayaux, P., Gallego, J., Richards, T., Malingreau, J.-P., 2002. Determination of deforestation rates of the world's humid tropical forests. Science 297 (5583), 999–1002.

Achard, F., Beuchle, R., Mayaux, P., Stibig, H.-J., Bodart, C., Brink, A., Carboni, S., et al., 2014. Determination of tropical deforestation rates and related carbon losses from 1990 to 2010. Glob. Chang. Biol. 20 (8), 2540–2554.

Achat, D.L., Bakker, M.R., Augusto, L., Morel, C., 2013. Contributions of microbial and physical–chemical processes to phosphorus availability in podzols and arenosols under a temperate forest. Geoderma 211–212 (December), 18–27.

Achat, D.L., Pousse, N., Nicolas, M., Augusto, L., 2018. Nutrient remobilization in tree foliage as affected by soil nutrients and leaf life span. Ecol. Monogr. 88 (3), 408–428.

Achtnich, C., Bak, F., Conrad, R., 1995. Competition for electron donors among nitrate reducers, ferric iron redures, sulfate reducers, and methanogens in anoxic paddy soil. Biol. Fertil. Soils 19, 65–72.

Ackerman, S.A., Chung, H., 1992. Radiative effects of airborne dust on regional energy budgets at the top of the atmosphere. J. Appl. Meteorol. 31, 223–233.

Ackerman, D., Millet, D.B., Chen, X., 2019. Global estimates of inorganic nitrogen deposition across four decades. Glob. Biogeochem. Cycles 33 (1), 100–107.

Adair, E.C., Parton, W.J., Del Grosso, S.J., Silver, W.L., Harmon, M.E., Hall, S.A., Burke, I.C., Hart, S.C., 2008. Simple three-pool model accurately describes patterns of long-term litter decomposition in diverse climates. Glob. Chang. Biol. 14, 2636–2660.

Adamec, L., 1997. Mineral nutrition of carnivorous plants: a review. Bot. Rev. 63, 273–299.

Adams, M.A., Attiwill, P.M., 1982. Nitrate reductase activity and growth response of forest species to ammonium and nitrate sources of nitrogen. Plant Soil 66, 373–381.

Adams, W.A., Raza, M.A., Evans, L.J., 1980. Relationships between net redistribution of Al and Fe and extractable levels in podzolic soils derived from lower paleozoic sedimentary rocks. J. Soil Sci. 31, 533–545.

Adams, D.F., Farwell, S.O., Robinson, E., Pack, M.R., Bamesberger, W.L., 1981. Biogenic sulfur source strengths. Environ. Sci. Technol. 15 (12), 1493–1498.

Adams, J.M., Faure, H., Fauredenard, L., McGlade, J.M., Woodward, F.I., 1990. Increases in terrestrial carbon storage from the last glacial maximum to the present. Nature 348, 711–714.

Adamsen, A.P.S., King, G.M., 1993. Methane consumption in temperate and sub-arctic forest soils—rates, vertical zonation, and responses to water and nitrogen. Appl. Environ. Microbiol. 59, 485–490.

Adler, P.R., Del Grosso, S.J., Parton, W.J., 2007. Life-cycle assessment of net greenhouse-gas flux for bioenergy cropping systems. Ecol. Appl. 17 (3), 675–691.

Admiraal, W.I.M., Botermans, Y.J.H., 1989. Comparison of nitrification rates in three branches of the lower river rhine. Biogeochemistry 8, 135–151.

Ae, N., Arihara, J., Okada, K., Yoshihara, T., Johansen, C., 1990. Phosphorus uptake by pigeon pea and its role in cropping systems of the Indian subcontinent. Science 248 (4954), 477–480.

Aerts, R., 1996. Nutrient resorption from senescing leaves of perennials: Are there general patterns? J. Ecol. 84, 597–608.

Aerts, R., Chapin, F.S., 2000. The mineral nutrition of wild plants revisited: a re-evaluation of processes and patterns. Adv. Ecol. Res. 30, 1–67.

Agawin, N.S.R., Duarte, C.M., Agusti, S., 2000. Nutrient and temperature control of the contribution of picoplankton to phytoplankton biomass and production. Limnol. Oceanogr. 45, 591–600.

Agbenin, J.O., 2003. Extractable iron and aluminum effects on phosphate sorption in a savanna alfisol. Soil Sci. Soc. Am. J. 67, 589–595.

Ågren, G.I., Hyvönen, R., Berglund, S.L., Hobbie, S.E., 2013. Estimating the critical N:C from litter decomposition data and its relation to soil organic matter stoichiometry. Soil Biol. Biochem. 67 (December), 312–318.

Aguilar, R., Heil, R.D., 1988. Soil organic carbon, nitrogen, and phosphorus quantities in northern great plains rangeland. Soil Sci. Soc. Am. J. 52, 1076–1081.

Agusti, S., Satta, M.P., Mura, M.P., Benavent, E., 1998. Dissolved esterase activity as a tracer of phytoplankton lysis: evidence of high phytoplankton lysis rates in the northwestern mediterranean. Limnol. Oceanogr. 43, 1836–1849.

Ahl, T., 1988. Background yield of phosphorus from drainage area and atmosphere—an empirical approach. Hydrobiologia 170, 35–44.

Ahlström, A., Raupach, M.R., Schurgers, G., Smith, B., Arneth, A., Jung, M., Reichstein, M., et al., 2015. The dominant role of semi-arid ecosystems in the trend and variability of the land CO_2 sink. Science 348 (6237), 895–899.

Ahn, J., Brook, E.J., Mitchell, L., Rosen, J., McConnell, J.R., Taylor, K., Etheridge, D., Rubino, M., 2012. Atmospheric CO_2 over the last 1000 years: a high-resolution record from the west antarctic ice sheet (WAIS) divide ice core. Glob. Biogeochem. Cycles 26.

Aikawa, M., Kajino, M., Hiraki, T., Mukai, H., 2014. The contribution of site to washout and rainout: precipitation chemistry based on sample analysis from 0.5 Mm precipitation increments and numerical simulation. Atmos. Environ. 95 (October), 165–174.

Aitkenhead, J.A., McDowell, W.H., 2000. Soil C:N ratio as a predictor of annual riverine DOC flux at local and global scales. Glob. Biogeochem. Cycles 14 (1), 127–138.

Aiuppa, A., Inguaggiato, S., McGonigle, A.J.S., O'Dwyer, M., Oppenheimer, C., Padgett, M.J., Rouwet, D., Valenza, M., 2005. H_2S fluxes from Mt Etna, Stromboli, and Vulcano (Italy) and implications for the sulfur budget at volcanoes. Geochim. Cosmochim. Acta 69, 1861–1871.

Albarède, F., 2009. Volatile accretion history of the terrestrial planets and dynamic implications. Nature 461 (7268), 1227–1233.

Alexander, E.B., 1988. Rates of soil formation—implications for soil loss tolerance. Soil Sci. 145, 37–45.

Alexander, R.B., Smith, R.A., Schwarz, G.E., 2000. Effect of stream channel size on the delivery of nitrogen to the Gulf of Mexico. Nature 403 (6771), 758–761.

Alexander, C.O., Manhes, G., Gopel, C., 2001. The early evolution of the inner solar systems: a meteoritic perspective. Science 293, 64–68.

Alexander, R.B., Smith, R.A., Schwarz, G.E., Boyer, E.W., Nolan, J.V., Brakebill, J.W., 2008. Differences in phosphorus and nitrogen delivery to the Gulf of Mexico from the Mississippi River basin. Environ. Sci. Technol. 42 (3), 822–830.

Alexandre, A., Meunier, J.D., Colin, F., Koud, J.M., 1997. Plant impact on the biogeochemical cycle of silicon and related weathering processes. Geochim. Cosmochim. Acta 61, 677–682.

Algesten, G., Wikner, J., Sobek, S., Tranvik, L.J., Jansson, M., 2004. Seasonal variation of CO_2 saturation in the Gulf of Bothnia: indications of marine net heterotrophy. Glob. Biogeochem. Cycles 18.

Alin, S.R., Johnson, T.C., 2007. Carbon cycling in large lakes of the world: a synthesis of production, burial, and lake-atmosphere exchange estimates. Glob. Biogeochem. Cycles 21, GB3002.

Allan, C.J., Roulet, N.T., 1994. Solid phase controls of dissolved aluminum within upland precambrian shield catchments. Biogeochemistry 26, 85–114.

Allan, C.J., Roulet, N.T., Hill, A.R., 1993. The biogeochemistry of pristine, headwater precambrian shield watersheds—an analysis of material transport within a heterogeneous landscape. Biogeochemistry 22, 37–79.

Allan, M., Le Roux, G., De Vleeschouwer, F., Bindler, R., Blaauw, M., Piotrowska, N., Sikorski, J., Fagel, N., 2013. High-resolution reconstruction of atmospheric deposition of trace metals and metalloids since AD 1400 recorded by ombrotrophic peat cores in Hautes-Fagnes, Belgium. Environ. Pollut. 178 (July), 381–394.

Allard, P., Carbonnelle, J., Metrich, N., Loyer, H., Zettwoog, P., 1994. Sulfur output and magma degassing budget of stromboli volcano. Nature 368, 326–330.

Alldredge, A., 1998. The carbon, nitrogen and mass content of marine snow as a function of aggregate size. Deep-Sea Res. Part I—Oceanogr. Res. Pap. 45, 529–541.

Alldredge, A.L., Cohen, Y., 1987. Can microscale chemical patches persist in the sea? Microelectrode study of marine snow, fecal pellets. Science 235 (4789), 689–691.

Alldredge, A.L., Gotschalk, C.C., 1990. The relative contribution of marine snow of different origins to biological processes in coastal waters. Cont. Shelf Res. 10, 41–58.

Allègre, C.J., Manhes, G., Gopel, C., 1995. The age of the earth. Geochim. Cosmochim. Acta 59(1445–1456).

Allen, M.F., 1992. Mycorrhizas: Plant-Fungus Relationships. Chapman and Hall, London, United Kingdom.

Allen, A.S., Schlesinger, W.H., 2004. Nutrient limitations to soil microbial biomass and activity in loblolly pine forests. Soil Biol. Biochem. 36, 581–589.

Aller, R., 2014. Sedimentary diagenesis, depositional environments, and benthic fluxes. In: Holland, H.D., Turekian, K.K. (Eds.), Treatise on Geochemistry. In: vol. 8. Elsevier, Amsterdam, pp. 293–334.

Alley, R.B., 2007. Wally was right: Predictive ability of the North Atlantic "conveyor belt" hypothesis for abrupt climate change. Annu. Rev. Earth Planet. Sci. 35, 241–272.

Allison, S.D., 2006. Brown ground: a soil carbon analogue for the green world hypothesis? Am. Nat. 167 (5), 619–627.

Allwood, A.C., Rosing, M.T., Flannery, D.T., Hurowitz, J.A., Heirwegh, C.M., 2018. Reassessing evidence of life in 3,700-million-year-old rocks of Greenland. Nature 563, 241–244.

Altabet, M.A., Curry, W.B., 1989. Testing models of past ocean chemistry using foraminifera $^{15}N/^{14}N$. Glob. Biogeochem. Cycles 3, 107–119.

Altabet, M.A., Francois, R., Murray, D.W., Prell, W.L., 1995. Climate-related variations in denitrification in the arabian sea from sediment $^{15}N/^{14}N$ ratios. Nature 373, 506–509.

Altabet, M.A., Higginson, M.J., Murray, D.W., 2002. The effect of millennial-scale changes in arabian sea denitrification on atmospheric CO_2. Nature 415 (6868), 159–162.

Alton, P., Fisher, R., Los, S., Williams, M., 2009. Simulations of global evapotranspiration using semi-empirical and mechanistic schemes of plant hydrology. Glob. Biogeochem. Cycles 23.

Altschuler, Z.S., Schnepfe, M.M., Silber, C.C., Simon, F.O., 1983. Sulfur diagenesis in everglades peat and origin of pyrite in coal. Science 221, 221–227.

Aluwihare, L.I., Repeta, D.J., Chen, R.F., 1997. A major biopolymeric component to dissolved organic carbon in surface sea water. Nature 387, 166–169.

Aluwihare, L.I., Repeta, D.J., Pantoja, S., Johnson, C.G., 2005. Two chemically distinct pools of organic nitrogen accumulate in the ocean. Science 308 (5724), 1007–1010.

Alvarez, E., Martinez, A., Calvo, R., 1992. Geochemical aspects of aluminum in forest soils in galicia (NW Spain). Biogeochemistry 16, 167–180.

Alvarez, R.A., Zavala-Araiza, D., Lyon, D.R., Allen, D.T., Barkley, Z.R., Brandt, A.R., Davis, K.J., et al., 2018. Assessment of methane emissions from the U.S. oil and gas supply chain. Science 361 (6398), 186–188.

Alvarez-Clare, S., Mack, M.C., Brooks, M., 2013. A direct test of nitrogen and phosphorus limitation to net primary productivity in a lowland tropical wet forest. Ecology 94, 1540–1551.

Alvarez-Cobelas, M., Angeler, D.G., Sánchez-Carrillo, S., 2008. Export of nitrogen from catchments: a worldwide analysis. Environ. Pollut. 156 (2), 261–269.

Ambrose, S.H., Sikes, N.E., 1991. Soil carbon isotope evidence for holocene habitat change in the Kenya Rift Valley. Science 253 (5026), 1402–1405.

Ambus, P., Christensen, S., 1994. Measurement of N_2O emission from a fertilized grassland: An analysis of spatial variability. J. Geophys. Res.-Atmos. 99, 16549–16555.

Amend, J.P., Shock, E.L., 1998. Energetics of amino acid synthesis in hydrothermal ecosystems. Science 281 (5383), 1659–1662.

Ament, M.R., Tierney, J.A., Hedin, L.O., Hobbie, E.A., Wurzburger, N., 2018. Phosphorus and species regulate N_2 fixation by herbaceous legumes in longleaf pine savannas. Oecologia 187 (1), 281–290.

Ames, R.N., Reid, C.P.P., Porter, L.K., Cambardella, C., 1983. Hyphal uptake and transport of nitrogen from two N-15 labeled sources by glomus mosseae, a vesicular-arbuscular mycorrhizal fungus. New Phytol. 95, 381–396.

Amos, H.M., Jacob, D.J., Streets, D.G., Sunderland, E.M., 2013. Legacy impacts of all-time anthropogenic emissions on the global mercury cycle: Global impacts of legacy mercury. Glob. Biogeochem. Cycles 27 (2), 410–421.

Amthor, J.S., 1984. The role of maintenance respiration in plant growth. Plant Cell Environ. 7, 561–569.

Amthor, J.S., 1989. Respiration and Crop Productivity. Springer.

Amthor, J.S., 2010. From sunlight to phytomass: on the potential efficiency of converting solar radiation to phyto-energy. New Phytol. 188 (4), 939–959.

Amundson, R., 2001. The carbon budget in soils. Annu. Rev. Earth Planet. Sci. 29, 535–562.

Amundson, R.G., Davidson, E.A., 1990. Carbon dioxide and nitrogenous gases in the soil atmosphere. J. Geochem. Explor. 38, 13–41.

Amundson, R., Jenny, H., 1991. The place of humans in the state factor theory of ecosystems and their soils. Soil Sci. 151, 99–109.

Amundson, R.G., Chadwick, O.A., Sowers, J.M., 1989. A comparison of soil climate and biological activity along an elevation gradient in the eastern Mojave Desert. Oecologia 80 (3), 395–400.

Amundson, R., Austin, A.T., Schuur, E.A.G., Yoo, K., Matzek, V., Kendall, C., Uebersax, A., Brenner, D., Baisden, W.T., 2003. Global patterns of the isotopic composition of soil and plant nitrogen. Glob. Biogeochem. Cycles 17.

Amundson, R., Heimsath, A., Owen, J., Yoo, K., Dietrich, W.E., 2015. Hillslope soils and vegetation. Geomorphology 234 (April), 122–132.

Anastasi, C., Simpson, V.J., 1993. Future methane emissions from animals. J. Geophys. Res. 98, 7181–7186.

Anastasi, C., Dowding, M., Simpson, V.J., 1992. Future CH_4 emissions from rice production. J. Geophys. Res. 97, 7521–7525.

Anastopoulos, I., Massas, I., Pogka, E.-E., Chatzipavlidis, I., Ehaliotis, C., 2019. Organic materials may greatly enhance Ni and Pb progressive immobilization into the oxidisable soil fraction, acting as providers of sorption sites and microbial substrates. Geoderma 353 (July), 482–492.

Anbeek, C., 1993. The effect of natural weathering on dissolution rates. Geochim. Cosmochim. Acta 57, 4963–4975.

Andela, N., Morton, D.C., Giglio, L., Chen, Y., van der Werf, G.R., Kasibhatla, P.S., DeFries, R.S., et al., 2017. A human-driven decline in global burned area. Science 356 (6345), 1356–1362.

Anderegg, W.R.L., Konings, A.G., Trugman, A.T., Yu, K., Bowling, D.R., Gabbitas, R., Karp, D.S., et al., 2018. Hydraulic diversity of forests regulates ecosystem resilience during drought. Nature.

Anders, E., 1989. Pre-biotic organic matter from coments and asteroids. Nature 342, 255–257.

Anders, E., Grevesse, N., 1989. Abundances of the elements—meteoritic and solar. Geochim. Cosmochim. Acta 53, 197–214.

Anders, E., Owen, T., 1977. Mars and earth—origin and abundance of volatiles. Science 198, 453–465.

Anderson, G.F., 1986. Silica, diatoms and a fresh water Productivity maximum in Atlantic coastal plain estuaries, Chesapeake Bay. Estuar. Coast. Shelf Sci. 22 (2), 183–197.

Anderson, J.P.E., Domsch, K.H., 1978. A physiological method for the quantitative measurement of microbial biomass in soils. Soil Biol. Biochem. 10 (3), 215–221.

Anderson, J.P.E., Domsch, K.H., 1980. Quantities of plant nutrients in the microbial biomass of selected soils. Soil Sci. 130, 211–216.

Anderson, T.-H., Martens, R., 2013. DNA determinations during growth of soil microbial biomasses. Soil Biol. Biochem. 57 (February), 487–495.

Anderson, C.J., Mitsch, W.J., 2006. Sediment, carbon, and nutrient accumulation at two 10-year-old created riverine marshes. Wetlands 26, 779–792.

Anderson, M.A., Morel, F.M.M., 1982. The influence of aqueous iron chemistry on the uptake of iron by the coastal diatom *Thalassiosira weissflogii*. Limnol. Oceanogr.

Anderson, D.W., Paul, E.A., 1984. Organo-mineral complexes and their study by radiocarbon dating. Soil Sci. Soc. Am. J. 48, 298–301.

Anderson, L.A., Sarmiento, J.L., 1994. Redfield ratios of remineralization determined by nutrient data analysis. Glob. Biogeochem. Cycles 8, 65–80.

Anderson, R.F., Schiff, S.L., 1987. Alkalinity generation and the fate of sulfur in lake sediments. Can. J. Fish. Aquat. Sci. 44, 188–193.

Anderson, J.M., Ineson, P., Huish, S.A., 1983. Nitrogen and cation mobilization by soil fauna feeding on leaf litter and soil organic matter from deciduous woodlands. Soil Biol. Biochem. 15, 463–467.

Anderson, L.D., Delaney, M.L., Faul, K.L., 2001. Carbon to phosphorus ratios in sediments: implications for nutrient cycling. Glob. Biogeochem. Cycles 15, 65–79.

Anderson, S.P., Dietrich, W.E., Brimhall, G.H., 2002. Weathering profiles, mass-balance analysis, and rates of solute loss: linkages between weathering and erosion in a small, steep catchment. Geol. Soc. Am. Bull. 114, 1143–1158.

Anderson, T.R., Groffman, P.M., Todd Walter, M., 2015. Using a soil topographic index to distribute denitrification fluxes across a Northeastern headwater catchment. J. Hydrol. 522 (March), 123–134.

Anderson, R.F., Sachs, J.P., Fleisher, M.Q., Allen, K.A., Yu, J., Koutavas, A., Jaccard, S.L., 2019. Deep-sea oxygen depletion and ocean carbon sequestration during the last ice age. Glob. Biogeochem. Cycles 33 (3), 301–317.

Anderson-Teixeira, K.J., Wang, M.M.H., McGarvey, J.C., Herrmann, V., Tepley, A.J., Bond-Lamberty, B., LeBauer, D.S., 2018. ForC: a global database of forest carbon stocks and fluxes. Ecology 99 (6), 1507.

Andersson, M.E., Gårdfeldt, K., Wängberg, I., Strömberg, D., 2008. Determination of Henry's Law constant for elemental mercury. Chemosphere 73 (4), 587–592.

Andreadis, K.M., Lettenmaier, D.P., 2006. Trends in 20th century drought over the continental United States. Geophys. Res. Lett. 33.

Andreae, M.O., 1991. Biomass burning: its history, use, and distribution and its impact on environmental quality and global climate. In: Levine, J.S. (Ed.), Global Biomass Burning: Atmospheric, Climatic, and Biospheric Implications. MIT Press, Cambridge, MA, pp. 3–21.

Andreae, M.O., 2019. Emission of trace gases and aerosols from biomass burning—an updated assessment. Atmos. Chem. Phys. 19 (13), 8523–8546.

Andreae, M.O., Andreae, T.W., 1988. The cycle of biogenic sulfur compounds over the Amazon Basin. 1. Dry season. J. Geophys. Res. D: Atmos. 93, 1487–1497.

Andreae, M.O., Jaeschke, W.A., 1992. Exchange of sulphur between biosphere and atmosphere over temperate and tropical regions. In: Howarth, R.W., Stewart, J.W.B., Ivanov, M.V. (Eds.), Sulphur Cycling on the Continents: Wetlands, Terrestrial Ecosystems and Associated Water Bodies. Wiley, pp. 27–61.

Andreae, M.O., Merlet, P., 2001. Emission of trace gases and aerosols from biomass burning. Glob. Biogeochem. Cycles 15, 955–966.

Andreae, M.O., Rosenfeld, D., 2008. Aerosol-cloud-precipitation interactions: Part 1. The nature and sources of cloud-active aerosols. Earth Sci. Rev. 89, 13–41.

Andreae, M.O., Charlson, R.J., Bruynseels, F., Storms, H., Grieken, R.V.A.N., Maenhaut, W., 1986. Internal mixture of sea salt, silicates, and excess sulfate in marine aerosols. Science 232 (4758), 1620–1623.

Andreae, M.O., Berresheim, H., Bingemer, H., Jacob, D.J., Lewis, B.L., Li, S.M., Talbot, R.W., 1990. The atmospheric sulfur cycle over the Amazon Basin. 2. Wet season. J. Geophys. Res. D: Atmos. 95, 16813–16824.

Andreae, M.O., Anderson, B.E., Blake, D.R., Bradshaw, J.D., Collins, J.E., Gregory, G.L., Sachse, G.W., Shipham, M.C., 1994. Influence of plumes from biomass burning on atmospheric chemistry over the equatorial and tropical South Atlantic during CITE-3. J. Geophys. Res. 99, 12793–12808.

Andreae, M.O., Elbert, W., Demora, S.J., 1995. Biogenic sulfur emissions and aerosols over the tropical South Atlantic: 3. Atmospheric dimethylsulfide, aerosols and cloud condensation nuclei. J. Geophys. Res. 100, 11335–11356.

Andreae, M.O., Jones, C.D., Cox, P.M., 2005. Strong present-day aerosol cooling implies a hot future. Nature 435 (7046), 1187–1190.

Andreu-Hayles, L., Planells, O., Gutierrez, E., Muntan, E., Helle, G., Anchukaitis, K.J., Schleser, G.H., 2011. Long tree-ring chronologies reveal 20th century increases in water-use efficiency but no enhancement of tree growth at five Iberian pine forests. Glob. Chang. Biol. 17, 2095–2112.

Andrews, M., 1986. The partitioning of nitrate assimilation betweeen root and shoot of higher plants. Plant Cell Environ. 9, 511–519.

Andrews, J.A., Schlesinger, W.H., 2001. Soil CO_2 dynamics, acidification, and chemical weathering in a temperate forest with experimental CO_2 enrichment. Glob. Biogeochem. Cycles 15, 149–162.

Andrews, J.A., Harrison, K.G., Matamala, R., Schlesinger, W.H., 1999. Separation of root respiration from total soil respiration using carbon-13 labeling during free-air carbon dioxide enrichment (FACE). Soil Sci. Soc. Am. J. 63, 1429–1435.

Aneja, V.P., Overton, J.H., Cupitt, L.T., Durham, J.L., Wilson, W.E., 1979. Carbon disulfide and carbonyl sulfide from biogenic sources and their contributions to the global sulfur cycle. Nature 282, 493–496.

Aneja, V.P., Kim, D.S., Das, M., Hartsell, B.E., 1996. Measurements and analysis of reactive nitrogen species in the rural troposphere of Southeast United States: southern oxidant study site SONIA. Atmos. Environ. 30, 649–659.

Aneja, V.P., Nelson, D.R., Roelle, P.A., Walker, J.T., Battye, W., 2003. Agricultural ammonia emissions and ammonium concentrations associated with aerosols and precipitation in the Southeast United States. J. Geophys. Res. 108.

Aneja, V., Schlesinger, W., Li, Q., Nahas, A., Battye, W., 2019. Characterization of atmospheric nitrous oxide emissions from global agricultural soils. SN Appl. Sci. 1 (November), 1662.

Anschutz, P., Dedieu, K., Desmazes, F., Chaillou, G., 2005. Speciation, oxidation state, and reactivity of particulate manganese in marine sediments. Chem. Geol. 218, 265–279.

Antibus, R.K., Croxdale, J.G., Miller, O.K., Linkins, A.E., 1981. Ectomycorrhizal fungi of *Salix rotundifolia*. 3. Resynthezied mycorrhizal complexes and their surface phosphatase activities. Can. J. Bot. 59, 2458–2465.

Antler, G., Turchyn, A.V., Rennie, V., Herut, B., Sivan, O., 2013. Coupled sulfur and oxygen isotope insight into bacterial sulfate reduction in the natural environment. Geochim. Cosmochim. Acta 118 (October), 98–117.

Antonov, J.I., Levitus, S., Boyer, T.P., 2005. Thermosteric sea level rise. Geophys. Res. Lett. 32, 1955–2003.

Antweiler, R.C., Drever, J.I., 1983. The weathering of a late tertiary volcanic ash: importance of organic solutes. Geochim. Cosmochim. Acta 47, 623–629.

Appel, H.M., 1993. Phenolics in ecological interactions: the importance of oxidation. J. Chem. Ecol. 19 (7), 1521–1552.

April, R., Keller, D., 1990. Mineralogy of the rhizosphere in forest soils of the Eastern United States—mineralogic studies of the rhizosphere. Biogeochemistry 9, 1–18.

April, R., Newton, R., Coles, L.T., 1986. Chemical weathering in two Adirondack watersheds: past and present-day rates. Geol. Soc. Am. Bull. 97, 1232–1238.

Aragão, L.E.O.C., Anderson, L.O., Fonseca, M.G., Rosan, T.M., Vedovato, L.B., Wagner, F.H., Silva, C.V.J., et al., 2018. 21st century drought-related fires counteract the decline of Amazon deforestation carbon emissions. Nat. Commun. 9 (1), 536.

Arai, Y., Sparks, D.L., 2007. Phosphate reaction dynamics in soils and soil minerals: a multiscale approach. Adv. Agron. 94, 135–179.

Archer, D., 1995. Upper ocean physics as relevant to ecosystem dynamics: a tutorial. Ecol. Appl. 5, 724–739.

Archer, D.E., Blum, J.D., 2018. A model of mercury cycling and isotopic fractionation in the ocean. Biogeosciences 15 (20), 6297–6313.

Archer, D.E., Johnson, K., 2000. A model of the iron cycle in the ocean. Glob. Biogeochem. Cycles 14 (1), 269–279.

Archer, D., Maier-Reimer, E., 1994. Effect of deep sea sedimentary calcite preservation on atmospheric CO_2 concentration. Nature 367, 260–263.

Archer, D., Buffett, B., Brovkin, V., 2009. Ocean methane hydrates as a slow tipping point in the global carbon cycle. Proc. Natl. Acad. Sci. U. S. A. 106 (49), 20596–20601.

Ardón, M., Morse, J.L., Colman, B.P., Bernhardt, E.S., 2013. Drought-induced saltwater incursion leads to increased wetland nitrogen export. Glob. Chang. Biol. 19 (10), 2976–2985.

Ardón, M., Helton, A.M., Scheuerell, M.D., Bernhardt, E.S., 2017. Fertilizer legacies meet saltwater incursion: challenges and constraints for coastal plain wetland restoration. Elem. Sci. Anth. 5.

Ardón, M., Helton, A.M., Bernhardt, E.S., 2018. Salinity effects on greenhouse gas emissions from wetland soils are contingent upon hydrologic setting: a microcosm experiment. Biogeochemistry 140 (2), 217–232.

Arheimer, B., Donnelly, C., Lindström, G., 2017. Regulation of snow-fed rivers affects flow regimes more than climate change. Nat. Commun. 8 (1), 62.

Arianoutsou, M., Margaris, N.S., 1981. Fire-induced nutrient losses in a phryganic (east mediterranean) ecosystem. Int. J. Biometeorol. 25, 341–347.

Arkley, R.J., 1963. Calculation of carbonate and water movement in soil from climatic data. Soil Sci. 92, 239–248.

Armbruster, M., Abiy, M., Feger, K.H., 2003. The biogeochemistry of two forested catchments in the black forest and the eastern ore mountains (Germany). Biogeochemistry 65, 341–368.

Armentano, T.V., Menges, E.S., 1986. Patterns of change in the carbon balance of organic-soil wetlands of the temeprate zone. J. Ecol. 74, 755–774.

Armstrong, K., Frost, D.J., McCammon, C.A., Rubie, D.C., Ballaran, T.B., 2019. Deep Magma ocean formation set the oxidation state of Earth's mantle. Science 365 (6456), 903–906.

Arnarson, T.S., Keil, R.G., 2007. Changes in organic matter-mineral interactions for marine sediments with varying oxygen exposure times. Geochim. Cosmochim. Acta 71, 3545–3556.

Arnell, N.W., Gosling, S.N., 2013. The impacts of climate change on river flow regimes at the global scale. J. Hydrol. 486 (April), 351–364.

Arnold, F., Bührke, T., 1983. New H_2SO_4 and HSO_3 vapour measurements in the stratosphere—evidence for a volcanic influence. Nature 301 (5898), 293–295.

Arnon, D.I., 1965. Ferredoxin and photosynthesis. Science 149 (3691), 1460–1470.

Aronson, E.L., Helliker, B.R., 2010. Methane flux in non-wetland soils in response to nitrogen addition: a meta-analysis. Ecology 91 (11), 3242–3251.

Arrhenius, S., 1918. The Destinies of the Stars. G.P. Putnam's Sons, New York.

Arrigo, K.R., 2005. Marine microorganisms and global nutrient cycles. Nature 437 (7057), 349–355.

Arrigo, K.R., Lubin, D., van Dijken, G.L., Holm-Hansen, O., Morrow, E., 2003. Impact of a deep ozone hole on southern ocean primary production. J. Geophys. Res. 108.

Arrigoni, A.S., Greenwood, M.C., Moore, J.N., 2010. Relative impact of anthropogenic modifications versus climate change on the natural flow regimes of rivers in the northern rocky mountains, United States. Water Resour. Res. 46.

Art, H.W., Bormann, F.H., Voigt, G.K., Woodwell, G.M., 1974. Barrier island forest ecosystems: the role of meteorological inputs. Science 184, 60–62.

Artaxo, P., de Campos, R.C., Fernandes, E.T., Martins, J.V., Xiao, Z.F., Lindqvist, O., Fernandez-Jimenez, M.T., Maenhaut, W., 2000. Large scale mercury and trace element measurements in the Amazon Basin. Atmos. Environ. 34, 4085–4096.

Arthur, M.A., Dean, W.E., Pratt, L.M., 1988. Geochemical and climatic effects of increased marine organic carbon burial at the Cenomanian-Turonian boundary. Nature 335, 714–717.

Asaf, D., Rotenberg, E., Tatarinov, F., Dicken, U., Montzka, S.A., Yakir, D., 2013. Ecosystem photosynthesis inferred from measurements of carbonyl sulphide flux. Nat. Geosci. 6 (3), 186–190.

Ascott, M.J., Gooddy, D.C., Wang, L., Stuart, M.E., Lewis, M.A., Ward, R.S., Binley, A.M., 2017. Global patterns of nitrate storage in the Vadose zone. Nat. Commun. 8 (1), 1416.

Aselmann, I., Crutzen, P.J., 1989. Global distribution of natural fresh water wetlands and rice paddies, their net primary productivity, seasonality and possible methane emissions. J. Atmos. Chem. 8, 307–358.

Ask, J., Karlsson, J., Jansson, M., 2012. Net ecosystem production in clear-water and brown-water lakes. Glob. Biogeochem. Cycles 26, 1–7.

Asman, W.A.H., van Jaarsveld, H.A., 1992. A variable-resolution transport model applied for $NH\chi$ in Europe. Atmos. Environ. A, Gen. Top. 26 (3), 445–464.

Asner, G.P., Vitousek, P.M., 2005. Remote analysis of biological invasion and biogeochemical change. Proc. Natl. Acad. Sci. U. S. A. 102 (12), 4383–4386.

Asner, G.P., Powell, G.V.N., Mascaro, J., Knapp, D.E., Clark, J.K., Jacobson, J., Kennedy-Bowdoin, T., et al., 2010. High-resolution forest carbon stocks and emissions in the Amazon. Proc. Natl. Acad. Sci. U. S. A. 107 (38), 16738–16742.

Asper, V.L., Deuser, W.G., Knauer, G.A., Lohrenz, S.E., 1992. Rapid coupling of sinking particle fluxesbetween surface and deep ocean waters. Nature 357, 670–672.

Asplund, M., Nicolas, G., Jacques Sauval, A., Scott, P., 2009. The chemical composition of the sun. Annu. Rev. Astron. Astrophys. 47 (1), 481–522.

Assad, E.D., Pinto, H.S., Martins, S.C., Groppo, J.D., Salgado, P.R., Evangelista, B., Vasconcellos, E., et al., 2013. Changes in soil carbon stocks in Brazil due to land use: paired site comparisons and a regional pasture soil survey. Biogeosciences 10 (10), 6141–6160.

Atkin, O.K., Bloomfield, K.J., Reich, P.B., Tjoelker, M.G., Asner, G.P., Bonal, D., Bönisch, G., et al., 2015. Global variability in leaf respiration in relation to climate, plant functional types and leaf traits. New Phytol. 206 (2), 614–636.

Atkinson, R., 2000. Atmospheric chemistry of VOCs and NOx. Atmos. Environ. 34, 2063–2101.

Atkinson, R., Arey, J., 2003. Gas-phase tropospheric chemistry of biogenic volatile organic compounds: a review. Atmos. Environ. 37, S197–S219.

Atkinson, L.J., Campbell, C.D., Zaragoza-Castells, J., Hurry, V., Atkin, O.K., 2010. Impact of growth temperature on scaling relationships linking photosynthetic metabolism to leaf functional traits. Funct. Ecol. 24, 1181–1191.

Atlas, E., Pytkowicz, R.M., 1977. Solubility behavior of apatites in seawater. Limnol. Oceanogr. 22, 290–300.

Atreya, S.K., Trainer, M.G., Franz, H.B., Wong, M.H., Manning, H.L.K., Malespin, C.A., Mahaffy, P.R., et al., 2013. Primordial argon isotope fractionation in the atmosphere of mars measured by the SAM instrument on curiosity and implications for atmospheric loss. Geophys. Res. Lett. 40 (21), 5605–5609.

Attri, K., Kerkar, S., LokaBharathi, P.A., 2011. Ambient iron concentration regulates the sulfate reducing activity in the mangrove swamps of Diwar, Goa, India. Estuar. Coast. Shelf Sci. 95, 156–164.

Auclair, A.N.D., Carter, T.B., 1993. Forest wildfires as a recent source of CO_2 at northern latitudes. Can. J. For. Res. 23, 1528–1536.

Audet, J., Bastviken, D., Bundschuh, M., Buffam, I., Alexander, F., Klemedtsson, L., Laudon, H., et al., 2019. Forest streams are important sources for nitrous oxide emissions. Glob. Change Biol. 26, 626–641.

Aufdenkampe, A.K., Mayorga, E., Hedges, J.I., Llerena, C., Quay, P.D., Gudeman, J., Krusche, A.V., Richey, J.E., 2007. Organic matter in the Peruvian headwaters of the Amazon: compositional evolution from the Andes to the lowland Amazon mainstem. Org. Geochem. 38 (3), 337–364.

Aufdenkampe, A.K., Mayorga, E., Raymond, P.A., Melack, J.M., Doney, S.C., Alin, S.R., Aalto, R.E., Yoo, K., 2011. Riverine coupling of biogeochemical cycles between land, oceans, and atmosphere. Front. Ecol. Environ. 9, 53–60.

Augusto, L., Delerue, F., Gallet-Budynek, A., Achat, D.L., 2013. Global assessment of limitation to symbiotic nitrogen fixation by phosphorus availability in terrestrial ecosystems using a meta-analysis approach: P availability and symbiotic N_2 fixation. Glob. Biogeochem. Cycles 27 (3), 804–815.

Augusto, L., Achat, D.L., Jonard, M., Vidal, D., Ringeval, B., 2017. Soil parent material—a major driver of plant nutrient limitations in terrestrial ecosystems. Glob. Chang. Biol. 23 (9), 3808–3824.

Aumann, G.D., 1965. Microtine abundance and soil sodium levels. J. Mammal. 46, 594–604.

Aumont, O., Bopp, L., 2006. Globalizing results from ocean in situ iron fertilization studies. Glob. Biogeochem. Cycles 20.

Austin, A.T., Vivanco, L., 2006. Plant litter decomposition in a semi-arid ecosystem controlled by photodegradation. Nature 442 (7102), 555–558.

Autry, A., Fitzgerald, J.W., 1993. Saturation potentials for sulfate adsorption by field-moist forest soils. Soil Biol. Biochem. 25, 833–838.

Averill, C., Bhatnagar, J.M., Dietze, M.C., Pearse, W.D., Kivlin, S.N., 2019. Global imprint of mycorrhizal fungi on whole-plant nutrient economics. Proc. Natl. Acad. Sci. U. S. A. 116 (46), 23163–23168.

Avery, G.B., Shannon, R.D., White, J.R., Christopher, S., Alperin, M.J., Brooks Avery Jr., G., Jeffrey, R., Martens, C.S., 2002. Controls on methane production in a tidal freshwater estuary and a Peatland: methane production via acetate fermentation and CO_2 reduction. Biogeochemistry 62 (1), 19–37.

Avila, A., Alarcon, M., Queralt, I., 1998. The chemical composition of dust transported in red rains—its contribution to the biogeochemical cycle of a Holm oak forest in Catalonia (Spain). Atmos. Environ. 32, 179–191.

Avis, C.A., Weaver, A.J., Meissner, K.J., 2011. Reduction in areal extent of high-latitude Wetlands in response to permafrost thaw. Nat. Geosci. 4 (7), 444–448.

Avitabile, V., Herold, M., Heuvelink, G.B.M., Lewis, S.L., Phillips, O.L., Asner, G.P., Armston, J., et al., 2016. An integrated pan-tropical biomass map using multiple reference datasets. Glob. Chang. Biol. 22 (4), 1406–1420.

Avnimelech, Y., McHenry, J.R., 1984. Enrichment of transported sediments with organic carbon, nutrients, and clay. Soil Sci. Soc. Am. J. 48, 259–266.

Awasthi, O.P., Sharma, E., Palni, L.M.S., 1995. Stemflow—a source of nutrients in some naturally growting epiphytic orchids of the Sikkim Himalaya. Ann. Bot. 75, 5–11.

Aydin, M., De Bruyn, W.J., Saltzman, E.S., 2002. Preindustrial atmospheric carbonyl sulfide (OCS) from an Antarctic Ice Core. Geophys. Res. Lett. 29.

Aydin, M., Verhulst, K.R., Saltzman, E.S., Battle, M.O., Montzka, S.A., Blake, D.R., Tang, Q., Prather, M.J., 2011. Recent decreases in fossil-fuel emissions of ethane and methane derived from firn air. Nature 476 (7359), 198–201.

Ayers, G.P., Gras, J.L., 1991. Seasonal relationship between cloud condensation nuclei and aerosol methanesulfonate in marine air. Nature 353, 834–835.

Ayers, G.P., Penkett, S.A., Gillett, R.W., Bandy, B., Galbally, I.E., Meyer, C.P., Elsworth, C.M., Bentley, S.T., Forgan, B.W., 1992. Evidence for photochemical control of ozone concentrations in unpolluted marine air. Nature 360, 446–449.

Azam, F., Fenchel, T., Field, J.G., Gray, J.S., Meyer-Reil, L.A., Thingstad, F., 1983. The ecological role of water-column microbes in the sea. Mar. Ecol. Prog. Ser.

Baas Becking, L.G.M., Kaplan, I.R., Moore, D., 1960. Limits of the natural environment in terms of pH and oxidation-reduction potentials. J. Geol. 68, 243–284.

Babst, F., Bouriaud, O., Papale, D., Gielen, B., Janssens, I.A., Nikinmaa, E., Ibrom, A., et al., 2014. Above-ground woody carbon sequestration measured from tree rings is coherent with net ecosystem productivity at five eddy-covariance sites. New Phytol. 201, 1289–1303.

Bacastow, R.B., 1976. Modulation of atmospheric carbon dioxide by the southern oscillation. Nature 261 (5556), 116–118.

Bacastow, R.B., 1981. Atmospheric CO_2 and the southern oscillation: effects associated with recent El Nino events. In: Proceedings of WMO/ICSU/UNEP Scientific Conference on Analysis and Interpretation of Atmospheric CO_2 Data, pp. 109–112.

Bacastow, R.B., Keeling, C.D., Whorf, T.P., 1985. Seasonal amplitude increase in atmospheric CO_2 concentration at Mauna Loa, Hawaii, 1959–1982. J. Geophys. Res. 90, 10529–10540.

Baccini, A., Asner, G.P., 2013. Improving pantropical forest carbon maps with airborne LiDAR sampling. Carbon Manag. 4 (6), 591–600.

Baccini, A., Walker, W., Carvalho, L., Farina, M., Sulla-Menashe, D., Houghton, R.A., 2017. Tropical forests are a net carbon source based on aboveground measurements of gain and loss. Science 358 (6360), 230–234.

Bache, C.W., 1984. The role of calcium in buffering soils. Plant Cell Environ. 7, 391–395.

Bache, C.A., Gutenmann, W.H., Lisk, D.J., 1971. Residues of total mercury and methyl-mercuric salts in lake trout as a function of age. Science 172, 951–952.

Bachmann, P.A., Luisi, P.L., Lang, J., 1992. Autocatalytic self-replicating micelles as models for prebiotic structures. Nature 357, 57–59.

Bader, M.K.-F., Leuzinger, S., Keel, S.G., Siegwolf, R.T.W., Hagedorn, F., Schleppi, P., Körner, C., 2013. Central European hardwood trees in a high-CO_2 future: synthesis of an 8-year forest canopy CO_2 enrichment project. J. Lee (Ed.), J. Ecol. 101 (6), 1509–1519.

Badgley, G., Anderegg, L.D.L., Berry, J.A., Field, C.B., 2019. Terrestrial gross primary production: using NIRV to scale from site to globe. Glob. Chang. Biol. 25 (11), 3731–3740.

Baethgen, W.E., Alley, M.M., 1987. Nonexchangeable ammonium nitrogen contribution to plant available nitrogen. Soil Sci. Soc. Am. J. 51, 110–115.

Bagnato, E., Aiuppa, A., Parello, F., Allard, P., Shinohara, H., Liuzzo, M., Giudice, G., 2011. New clues on the contribution of Earth's volcanism to the global mercury cycle. Bull. Volcanol. 73, 497–510.

Bahlmann, E., Keppler, F., Wittmer, J., Greule, M., Schöler, H.F., Seifert, R., Zetzsch, C., 2019. Evidence for a major missing source in the global chloromethane budget from stable carbon isotopes. Atmos. Chem. Phys. 19 (3), 1703–1719.

Bai, E., Li, S., Xu, W., Li, W., Dai, W., Jiang, P., 2013. A meta-analysis of experimental warming effects on terrestrial nitrogen pools and dynamics. New Phytol. 199 (2), 431–440.

Bai, Z., Yang, G., Chen, H., Zhu, Q., Chen, D., Li, Y., Xu, W., Wu, Z., Zhou, G., Peng, C., 2014. Nitrous oxide fluxes from three forest types of the tropical mountain rainforests on Hainan Island, China. Atmos. Environ. 92 (August), 469–477.

Bailey, J., Chrysostomou, A., Hough, J.H., Gledhill, T.M., McCall, A., Clark, S., Menard, F., Tamura, M., 1998. Circular polarization in star-formation regions: implications for biomolecular homochirality. Science 281 (5377), 672–674.

Bailey, V.L., Peacock, A.D., Smith, J.L., Bolton, H., 2002. Relationships between soil microbial biomass determined by chloroform fumigation-extraction, substrate-induced respiration, and phospholipid fatty acid analysis. Soil Biol. Biochem. 34, 1385–1389.

Bailey-Serres, J., Voesenek, L.A.C.J., 2008. Flooding stress: acclimations and genetic diversity. Annu. Rev. Plant Biol. 59 (January), 313–339.

Baines, S.B., Pace, M.L., 1991. The production of dissolved organic matter by phytoplankton and its importance to bacteria—patterns across marine and freshwater systems. Limnol. Oceanogr. 36, 1078–1090.

Baines, S.B., Pace, M.L., 1994. Relationships between suspended particulate matter and sinking flux along a trophic gradient and implications for the fate of planktonic primary production. Can. J. Fish. Aquat. Sci. 51, 25–36.

Baird, A.J., Beckwith, C.W., Waldron, S., Waddington, J.M., 2004. Ebullition of methane-containing gas bubbles from near-surface sphagnum peat. Geophys. Res. Lett. 31 (21).

Baker, V.R., 1977. Stream-channel response to floods, with examples from central Texas. Geol. Soc. Am. Bull. 88, 1057–1071.

Baker, P.A., Burns, S.J., 1985. Occurrence and formation of dolomite in organic-rich continental margin sediments. Bull. Am. Assoc. Pet. Geol. 69, 1917–1930.

Baker, P.A., Kastner, M., 1981. Constraints on the formation of sedimentary dolomite. Science 213 (4504), 214–216.

Baker, M.A., Dahm, C.N., Valett, H.M., 1999. Acetate retention and metabolism in the hyporheic zone of a mountain stream. Limnol. Oceanogr. 44, 1530–1539.

Baker, L.A., Hope, D., Xu, Y., Edmonds, J., Lauver, L., 2001. Nitrogen balance for the central Arizona-phoenix (CAP) ecosystem. Ecosystems 4, 582–602.

Baker, E.T., Walker, S.L., Chadwick Jr., W.W., Butterfield, D.A., Buck, N.J., Resing, J.A., 2019. Posteruption enhancement of hydrothermal activity: a 33-year, multieruption time series at axial seamount (Juan de Fuca Ridge). Geochem. Geophys. Geosyst. 20 (2), 814–828.

Bakken, L.R., Bergaust, L., Liu, B., Frostegård, A., 2012. Regulation of denitrification at the cellular level: a clue to the understanding of N_2O emissions from soils. Philos. Trans. R. Soc. Lon. B Biol. Sci. 367 (1593), 1226–1234.

Balch, W., Evans, R., Brown, J., Feldman, G., McClain, C., Esaias, W., 1992. The remote sensing of ocean primary productivity: use of a new data compilation to test satellite algorithms. J. Geophys. Res. Lect. Notes Coast. Estuar. Stud. 97 (C2), 2279.

Baldocchi, D.D., 2003. Assessing the eddy covariance technique for evaluating carbon dioxide exchange rates of ecosystems: past, present and future. Glob. Chang. Biol. 9, 479–492.

Baldock, J.A., Masiello, C.A., Gelinas, Y., Hedges, J.I., 2004. Cycling and composition of organic matter in terrestrial and marine ecosystems. Mar. Chem. 92, 39–64.

Ballantyne, A.P., Andres, R., Houghton, R., Stocker, B.D., Wanninkhof, R., Anderegg, W., Cooper, L.A., et al., 2015. Audit of the global carbon budget: estimate errors and their impact on uptake uncertainty. Biogeosciences 12 (8), 2565–2584.

Balling, R.C., 1989. The impact of summer rainfall on the temperature gradient along the United States-Mexico Border. J. Appl. Meteorol. 28, 304–308.

Balmford, A., Amano, T., Bartlett, H., Chadwick, D., Collins, A., Edwards, D., Field, R., et al., 2018. The environmental costs and benefits of high-yield farming. Nat. Sustain. 1 (9), 477–485.

Banfield, J.F., Barker, W.W., Welch, S.A., Taunton, A., 1999. Biological impact on mineral dissolution: application of the Lichen model to understanding mineral weathering in the rhizosphere. Proc. Natl. Acad. Sci. U. S. A. 96 (7), 3404–3411.

Bange, H.W., Rapsomanikis, S., Andreae, M.O., 1996. Nitrous oxide in coastal waters. Glob. Biogeochem. Cycles 10, 197–207.

Banin, A., Navrot, J., 1975. Origin of life—clues from relations between chemical compositions of living organisms and natural environments. Science 189, 550–551.

Bao, H.M., Reheis, M.C., 2003. Multiple oxygen and sulfur isotopic analyses on water-soluble sulfate in bulk atmospheric deposition from the Southwestern United States. J. Geophys. Res. D: Atmos. 108.

Bao, H., Campbell, D.A., Bockheim, J.G., Thiemens, M.H., 2000. Origins of sulphate in antarctic dry-valley soils as deduced from anomalous 17O compositions. Nature 407 (6803), 499–502.

Barange, M., Merino, G., Blanchard, J.L., Scholtens, J., Harle, J., Allison, E.H., Allen, J.I., Holt, J., Jennings, S., 2014. Impacts of climate change on marine ecosystem production in societies dependent on fisheries. Nat. Clim. Chang. 4 (February), 211.

Barber, S.A., 1962. A diffusion and mass-flow concept of soil nutrient availability. Soil Sci. 93, 39–43.

Barber, R.T., 1968. Dissolved organic carbon from deep water resists microbial oxidation. Nature 220, 274–275.

Barber, R.T., Vijayakumar, A., Cross, F.A., 1972. Mercury concentrations in recent and ninety-year-old benthopelagic fish. Science 178 (4061), 636–639.

Barber, V.A., Juday, G.P., Finney, B.P., 2000. Reduced growth of alaskan white spruce in the twentieth century from temperature-induced drought stress. Nature 405 (6787), 668–673.

Barboni, M., Boehnke, P., Keller, B., Kohl, I.E., Schoene, B., Young, E.D., McKeegan, K.D., 2017. Early formation of the moon 4.51 billion years ago. Sci. Adv. 3 (1), e1602365.

Bárcena, T.G., Kiær, L.P., Vesterdal, L., Stefánsdóttir, H.M., Gundersen, P., Sigurdsson, B.D., 2014. Soil carbon stock change following afforestation in northern Europe: a meta-analysis. Glob. Chang. Biol. 20 (8), 2393–2405.

Barford, C.C., Wofsy, S.C., Goulden, M.L., Munger, J.W., Pyle, E.H., Urbanski, S.P., Hutyra, L., Saleska, S.R., Fitzjarrald, D., Moore, K., 2001. Factors controlling long- and short-term sequestration of atmospheric CO_2 in a mid-latitude forest. Science 294 (5547), 1688–1691.

Barge, L.M., Flores, E., Baum, M.M., VanderVelde, D.G., Russell, M.J., 2019. Redox and pH gradients drive amino acid synthesis in iron oxyhydroxide mineral systems. Proc. Natl. Acad. Sci. U. S. A. 116 (11), 4828–4833.

Barker, P.F., Thomas, E., 2004. Origin, signature and palaeoclimatic influence of the Antarctic circumpolar current. Earth Sci. Rev. 66 (1), 143–162.

Barkley, A.E., Prospero, J.M., Mahowald, N., Hamilton, D.S., Popendorf, K.J., Oehlert, A.M., Pourmand, A., et al., 2019. African biomass burning is a substantial source of phosphorus deposition to the Amazon, tropical Atlantic Ocean, and Southern Ocean. Proc. Natl. Acad. Sci. U. S. A. 116 (33), 16216–16221.

Barlaz, M.A., 1998. Carbon storage during biodegradation of municipal solid waste components in laboratory-scale landfills. Glob. Biogeochem. Cycles 12, 373–380.

Barnett, T.P., Pierce, D.W., Achutarao, K.M., Gleckler, P.J., Santer, B.D., Gregory, J.M., Washington, W.M., 2005. Penetration of human-induced warming into the world's oceans. Science 309 (5732), 284–287.

Barnett, T.P., Pierce, D.W., Hidalgo, H.G., Bonfils, C., Santer, B.D., Das, T., Bala, G., et al., 2008. Human-induced changes in the hydrology of the Western United States. Science 319 (5866), 1080–1083.

Bar-On, Y.M., Phillips, R., Milo, R., 2018. The biomass distribution on earth. Proc. Natl. Acad. Sci. U. S. A. 115 (25), 6506–6511.

Barre, P., Berger, G., Velde, B., 2009. How element translocation by plants may stabilize illitic clays in the surface of temperate soils. Geoderma 151, 22–30.

Barrett, L.R., Schaetzl, R.J., 1992. An examination of podzolization near lake Michigan using chronofunctions. Can. J. Soil Sci. 72, 527–541.

Barron, A.R., Wurzburger, N., Bellenger, J.P., Wright, S.J., Kraepiel, A.M.L., Hedin, L.O., 2009. Molybdenum limitation of asymbiotic nitrogen fixation in tropical forest soils. Nat. Geosci. 2, 42–45.

Barros, N., Cole, J.J., Tranvik, L.J., Prairie, Y.T., Bastviken, D., Huszar, V.L.M., del Giorgio, P., Roland, F., 2011. Carbon emission from hydroelectric reservoirs linked to reservoir age and latitude. Nat. Geosci. 4 (9), 593–596.

Barrow, N.J., 2017. The effects of pH on phosphate uptake from the soil. Plant Soil 410 (1–2), 401–410.

Barry, D.A.J., Goorahoo, D., Goss, M.J., 1993. Estimation of nitrate concentrations in groundwater using a whole-farm nitrogen budget. J. Environ. Qual. 22, 767–775.

Barsdate, R.J., Alexander, V., 1975. Nitrogen balance of arctic tundra—pathways, rates, and environmental implications. J. Environ. Qual. 4 (1), 111–117.

Bartel-Ortiz, L.M., David, M.B., 1988. Sulfur constituents and transfomations in upland and floodplain forest soils. Can. J. For. Res. 18, 1106–1112.

Bartlett, R.J., 1986. Soil redox behavior. In: Sparks, D.L. (Ed.), Soil Physical Chemistry. CRC Press, Boca Raton, FL, pp. 179–207.

Bartlett, K.B., Harriss, R.C., 1993. Review and assessment of methane emissions from wetlands. Chemosphere 26 (1–4), 261–320.

Bartlett, K.B., Bartlett, D.S., Harriss, R.C., Sebacher, D.I., 1987. Methane emissions along a salt marsh salinity gradient. Biogeochemistry 4, 183–202.

Barton, L., McLay, C.D.A., Schipper, L.A., Smith, C.T., 1999. Annual denitrification rates in agricultural and forest soils: a review. Aust. J. Soil Res. 37, 1073–1093.

Barton, P.S., Cunningham, S.A., Lindenmayer, D.B., Manning, A.D., 2013. The role of carrion in maintaining biodiversity and ecological processes in terrestrial ecosystems. Oecologia 171 (4), 761–772.

Bassirirad, H., 2000. Kinetics of nutrient uptake by roots: responses to global change. New Phytol. 147, 155–169.

Bastviken, D., Cole, J., Pace, M., Tranvik, L., 2004. Methane emissions from lakes: dependence of lake characteristics, two regional assessments, and a global estimate. Glob. Biogeochem. Cycles 18.

Bastviken, D., Thomsen, F., Svensson, T., Karlsson, S., Sanden, P., Shaw, G., Matucha, M., Oberg, G., 2007. Chloride retention in forest soil by microbial uptake and by natural chlorination of organic matter. Geochim. Cosmochim. Acta 71, 3182–3192.

Bastviken, D., Tranvik, L.J., Downing, J.A., Crill, P.M., Enrich-Prast, A., 2011. Freshwater methane emissions offset the continental carbon sink. Science 331 (6013), 50.

Bates, T.R., Lynch, J.P., 1996. Stimulation of root hair elongation in *Arabidopsis thaliana* by low phosphorus availability. Plant Cell Environ. 19, 529–538.

Bates, T.S., Quinn, P.K., 1997. Dimethylsulfide (DMS) in the equatorial Pacific Ocean (1982 to 1996): evidence of a climate feedback? Geophys. Res. Lett. 24 (8), 861–864.

Bates, T.S., Charlson, R.J., Gammon, R.H., 1987. Evidence for the climatic role of marine biogenic sulfur. Nature 329, 319–321.

Batjes, N.H., 1996. Total carbon and nitrogen in the soils of the world. Eur. J. Soil Sci. 47, 151–163.

Bator, A., Collett, J.L., 1997. Cloud chemistry varies with drop size. J. Geophys. Res. D: Atmos. 102, 28071–28078.

Battaglia, G., Joos, F., 2018. Marine N₂O emissions from nitrification and denitrification constrained by modern observations and projected in multimillennial global warming simulations. Glob. Biogeochem. Cycles 32 (1), 92–121.

Batterman, S.A., Hedin, L.O., van Breugel, M., Ransijn, J., Craven, D.J., Hall, J.S., 2013a. Key role of symbiotic dinitrogen fixation in tropical forest secondary succession. Nature 502 (7470), 224–227.

Batterman, S.A., Wurzburger, N., Hedin, L.O., 2013b. Nitrogen and phosphorus interact to control tropical symbiotic N₂ fixation: a test in *Inga punctata*. Amy Austin J. Ecol. 101 (6), 1400–1408.

Battin, T.J., Luyssaert, S., Kaplan, L.A., Aufdenkampe, A.K., Richter, A., Tranvik, L.J., 2009. The boundless carbon cycle. Nat. Geosci. 2 (9), 598–600.

Battipaglia, G., Saurer, M., Cherubini, P., Calfapietra, C., McCarthy, H.R., Norby, R.J., Francesca Cotrufo, M., 2013. Elevated CO₂ increases tree-level intrinsic water use efficiency: insights from carbon and oxygen isotope analyses in tree rings across three forest FACE sites. New Phytol. 197 (2), 544–554.

Battle, M., Bender, M.L., Tans, P.P., White, J.W.C., Ellis, J.T., Conway, T., Francey, R.J., 2000. Global carbon sinks and their variability inferred from atmospheric O₂ and Delta¹³C. Science 287, 2467–2470.

Battle, M.O., William Munger, J., Conley, M., Sofen, E., Perry, R., Hart, R., Davis, Z., et al., 2019. Atmospheric measurements of the terrestrial O₂:CO₂ exchange ratio of a midlatitude forest. Atmos. Chem. Phys. 19 (13), 8687–8701.

Battye, W., Aneja, V.P., Schlesinger, W.H., 2017. Is nitrogen the next carbon? Earth's Future 5 (9), 894–904.

Baturin, G.N., 2007. Issue of the relationship between primary productivity of organic carbon in ocean and phosphate accumulation (Holocene-Late Jurassic). Lithol. Miner. Resour. 42, 318–348.

Bauer, J.E., Williams, P.M., Druffel, E.R.M., 1992. ¹⁴C activity of dissolved organic carbon fractions in the North-Central Pacific and Sargasso Sea. Nature 357 (6380), 667–670.

Bauer, J.E., Cai, W.-J., Raymond, P.A., Bianchi, T.S., Hopkinson, C.S., Regnier, P.A.G., 2013. The changing carbon cycle of the coastal ocean. Nature 504 (7478), 61–70.

Baumann, R.H., Day Jr., J.W., Miller, C.A., 1984. Mississippi deltaic wetland survival: sedimentation versus coastal submergence. Science 224 (4653), 1093–1095.

Baumgärtner, M., Conrad, R., 1992. Effects of soil variables and season on the production and consumption of nitric oxide in oxic soils. Biol. Fertil. Soils 14 (3), 166–174.

Bauters, M., Verbeeck, H., Rütting, T., Barthel, M., Mujinya, B.B., Bamba, F., Bodé, S., et al., 2019. Contrasting nitrogen fluxes in African tropical forests of the Congo Basin. Ecol. Monogr. 89(1), e01342.

Bautista, D.M., Movahed, P., Hinman, A., Axelsson, H.E., Sterner, O., Högestätt, E.D., Julius, D., Jordt, S.-E., Zygmunt, P.M., 2005. Pungent products from garlic activate the sensory ion channel TRPA1. Proc. Natl. Acad. Sci. U. S. A. 102 (34), 12248–12252.

Bauwens, M., Compernolle, S., Stavrakou, T., Muller, J.-F., van Gent, J., Eskes, H., Levelt, P.F., van der, A.R., Veefkind, J.P., Vlietinck, J., Yu, H., Zehner, C., 2020. Impact of coronavirus outbreak on NO2 pollution assessed using TROPOMI and OMI observations. Geophys. Res. Lett. https://doi.org/10.1029/2020GL087978.

Baxter, C.V., Fausch, K.D., Carl Saunders, W., 2005. Tangled webs: reciprocal flows of invertebrate prey link streams and Riparian zones. Freshw. Biol. 50 (2), 201–220.

Bazilevskaya, E., Archibald, D.D., Martínez, C.E., 2018. Mineral colloids mediate organic carbon accumulation in a temperate forest spodosol: depth-wise changes in pore water chemistry. Biogeochemistry 141 (1), 75–94.

Bazzaz, F.A., 1990. The response of natural ecosystems to the rising global CO₂ levels. Annu. Rev. Ecol. Syst. 21, 167–196.

Beadle, N.C.W., 1966. Soil phosphate and its role in molding segments of the Australian flora and vegetation, with special reference to xeromorphy and sclerophylly. Ecology 47, 992–1007.

Beal, E.J., House, C.H., Orphan, V.J., 2009. Manganese- and iron-dependent marine methane oxidation. Science 325 (5937), 184–187.

Beal, S.A., Kelly, M.A., Stroup, J.S., Jackson, B.P., Lowell, T.V., Tapia, P.M., 2014. Natural and anthropogenic variations in atmospheric mercury deposition during the Holocene near Quelccaya Ice Cap, Peru. Glob. Biogeochemical Cycles 28 (4), 437–450.

Beaulieu, J.J., Arango, C.P., Hamilton, S.K., Tank, J.L., 2007. The production and emission of nitrous oxide from headwater streams in the midwestern United States. Glob. Chang. Biol. 14 (4), 878–894.

Beaulieu, J.J., Tank, J.L., Hamilton, S.K., Wollheim, W.M., Hall, R.O., Mulholland, P.J., Peterson, B.J., et al., 2011. Nitrous oxide emission from denitrification in stream and river networks. Proc. Natl. Acad. Sci. U. S. A. 108 (1), 214–219.

Beaumont, V., Robert, F., 1999. Nitrogen isotope ratios of Kerogens in Precambrian Cherts: a record of the evolution of atmosphere chemistry? Precambrian Res. 96, 63–82.

Beck, V., Chen, H., Gerbig, C., Bergamaschi, P., Bruhwiler, L., Houweling, S., Röckmann, T., et al., 2012. Methane airborne measurements and comparison to global models during Barca: methane in the Amazon during Barca. J. Geophys. Res. Geophys. Monogr. Ser. 117(D15).

Beckwith, S.V., Sargent, A.I., 1996. Circumstellar disks and the search for neighbouring planetary systems. Nature 383 (6596), 139–144.

Bédard, C., Knowles, R., 1989. Physiology, biochemistry, and specific inhibitors of CH_4, NH_4^+, and CO oxidation by methanotrophs and nitrifiers. Microbiol. Rev. 53 (1), 68–84.

Bedford, B.L., Walbridge, M.R., Aldous, A., 1999. Patterns in nutrient availability and plant diversity of temperate North American wetlands. Ecology 80 (7), 2151–2169.

Bedison, J.E., Johnson, A.H., 2010. Seventy four years of calcium loss from forest soils of the Adirondack Mountains, New York. Soil Sci. Soc. Am. J. 74, 2187–2195.

Bednaršek, N., Tarling, G.A., Bakker, D.C.E., Fielding, S., Jones, E.M., Venables, H.J., Ward, P., et al., 2012. Extensive dissolution of live pteropods in the Southern Ocean. Nat. Geosci. 5 (November), 881.

Beer, C., Lucht, W., Schmullius, C., Shvidenko, A., 2006. Small net carbon dioxide uptake by Russian forests during 1981–1999. Geophys. Res. Lett. 33.

Beer, C., Reichstein, M., Tomelleri, E., Ciais, P., Jung, M., Carvalhais, N., Rödenbeck, C., et al., 2010. Terrestrial gross carbon dioxide uptake: global distribution and covariation with climate. Science 329 (5993), 834–838.

Beerling, D.J., 1999. New estimates of carbon transfer to terrestrial ecosystems between the last glacial maximum and the holocene. Terra Nova 11, 162–167.

Beerling, D.J., Chaloner, W.G., 1993. Evolutionary responses of stomatal density to global CO_2 change. Biol. J. Linn. Soc. 48, 343–353.

Beerling, D.J., Royer, D.L., 2011. Convergent cenozoic CO_2 history. Nat. Geosci. 4 (7), 418–420.

Befus, K.M., Jasechko, S., Luijendijk, E., Gleeson, T., Bayani Cardenas, M., 2017. The rapid yet uneven turnover of earth's groundwater. Geophys. Res. Lett. 44 (11), 5511–5520.

Behera, S.N., Sharma, M., 2011. Degradation of SO_2, NO_2 and NH_3 leading to formation of secondary inorganic aerosols: an environmental chamber study. Atmos. Environ. 45, 4015–4024.

Behera, S.N., Sharma, M., Aneja, V.P., Balasubramanian, R., 2013. Ammonia in the atmosphere: a review on emission sources, atmospheric chemistry and deposition on terrestrial bodies. Environ. Sci. Pollut. Res. Int. 20 (11), 8092–8131.

Behrenfeld, M.J., Falkowski, P.G., 1997. Photosynthetic rates derived from satellite-based chlorophyll concentration. Limnol. Oceanogr. 42, 1–20.

Behrenfeld, M.J., Bale, A.J., Kolber, Z.S., Aiken, J., Falkowski, P.G., 1996. Confirmation of iron limitation of phytoplankton photosynthesis in the equatorial Pacific Ocean. Nature 383 (6600), 508–511.

Behrenfeld, M.J., O'Malley, R.T., Siegel, D.A., McClain, C.R., Sarmiento, J.L., Feldman, G.C., Milligan, A.J., Falkowski, P.G., Letelier, R.M., Boss, E.S., 2006. Climate-driven trends in contemporary ocean productivity. Nature 444 (7120), 752–755.

Beilke, S., Lamb, D., 1974. On the absorption of SO_2 in ocean water. Tellus 26, 268–271.

Bekker, A., Holland, H.D., Wang, P.-L., Rumble 3rd, D., Stein, H.J., Hannah, J.L., Coetzee, L.L., Beukes, N.J., 2004. Dating the rise of atmospheric oxygen. Nature 427 (6970), 117–120.

Belcher, C.M., McElwain, J.C., 2008. Limits for combustion in low O_2 redefine paleoatmospheric predictions for the mesozoic. Science 321 (5893), 1197–1200.

Belillas, C.M., F. Rodà, 1991. Nutrient budgets in a dry heathland watershed in Northeastern Spain. Biogeochemistry 13, 137–157.

Bell, R.A., 1993. Cryptoendolithic algae of hot semiarid lands and deserts. J. Phycol. 29, 133–139.

Bell, J.F., 2012. The search for habitable worlds: planetary exploration in the 21st century. Daedalus 141 (3), 8–22.

Bell, D.R., Rossman, G.R., 1992. Water in earth's mantle: the role of nominally anhydrous minerals. Science 255 (5050), 1391–1397.

Bell, N., Hsu, L., Jacob, D.J., Schultz, M.G., Blake, D.R., Butler, J.H., King, D.B., Lobert, J.M., Maier-Reimer, E., 2002. Methyl iodide: atmospheric budget and use as a tracer of marine convection in global models. J. Geophys. Res. D: Atmos. 107.

Bellamy, P.H., Loveland, P.J., Ian Bradley, R., Murray Lark, R., Kirk, G.J.D., 2005. Carbon losses from all soils across England and Wales 1978–2003. Nature 437 (7056), 245–248.

Bellenger, J.P., Xu, Y., Zhang, X., Morel, F.M.M., Kraepiel, A.M.L., 2014. Possible contribution of alternative nitrogenases to nitrogen fixation by asymbiotic N_2-fixing bacteria in soils. Soil Biol. Biochem. 69 (February), 413–420.

Bellini, G., Sumner, M.E., Radcliffe, D.E., Qafoku, N.P., 1996. Anion transport through columns of highly weathered acid soil: adsorption and retardation. Soil Sci. Soc. Am. J. 60, 132–137.

Bellouin, N., Boucher, O., Haywood, J., Shekar Reddy, M., 2005. Global estimate of aerosol direct radiative forcing from satellite measurements. Nature 438 (7071), 1138–1141.

Belser, L.W., Mays, E.L., 1980. Specific inhibition of nitrite oxidation by chlorate and its use in assessing nitrification in soils and sediments. Appl. Environ. Microbiol. 39 (3), 505–510.

Belshe, E.F., Schuur, E.A.G., Bolker, B.M., 2013. Tundra ecosystems observed to be CO_2 sources due to differential amplification of the carbon cycle. Ecol. Lett. 16 (10), 1307–1315.

Beltman, B., Rouwenhorst, T.G., Van Kerkhoven, M.B., Van der Krift, T., 2000. Internal eutrophication in peat soils through competition between chloride and sulphate with phosphate for binding sites. Biogeochemistry 50, 183–194.

Belyea, L.R., Malmer, N., 2004. Carbon sequestration in Peatland: patterns and mechanisms of response to climate change. Glob. Chang. Biol. 10, 1043–1052.

Belyea, L.R., Warner, B.G., 1996. Temporal scale and the accumulation of peat in a *Sphagnum* bog. Can. J. Bot. 74, 366–377.

Beman, J.M., Arrigo, K.R., Matson, P.A., 2005. Agricultural runoff fuels large phytoplankton blooms in vulnerable areas of the ocean. Nature 434 (7030), 211–214.

Ben-Ami, Y., Koren, I., Rudich, Y., Artaxo, P., Martin, S.T., Andreae, M.O., 2010. Transport of north African dust from the Bodele depression to the Amazon Basin: A case study. Atmos. Chem. Phys. 10, 7533–7544.

Bencala, K.E., Walters, R.A., 1983. Simulation of solute transport in a mountain pool and riffle stream—a transient storage model. Water Resour. Res. 19 (3), 718–724.

Ben-David, M., Hanley, T.A., Schell, D.M., 1998. Fertilization of terrestrial vegetation by spawning Pacific Salmon: the role of flooding and predator activity. Oikos 83, 47–55.

Bender, M.L., Klinkhammer, G.P., Spencer, D.W., 1977. Manganese in seawater and marine manganese balance. Deep-Sea Res. 24, 799–812.

Bender, M., Jahnke, R., Weiss, R., Martin, W., Heggie, D.T., Orchardo, J., Sowers, T., 1989. Organic carbon oxidation and benthic nitrogen and silica dynamics in San Clemente Basin, a continental borderland site. Geochim. Cosmochim. Acta 53, 685–697.

Bender, M., Sowers, T., Labeyrie, L., 1994. The dole effect and its variations during the last 130,000 years as measured in the vostock ice core. Glob. Biogeochem. Cycles 8, 363–376.

Bender, M.L., Battle, M., Keeling, R.F., 1998. The O_2 balance of the atmosphere: a tool for studying the fate of fossil-fuel CO_2. Annu. Rev. Energy Environ. 23, 207–223.

Bender, M.L., Ho, D.T., Hendricks, M.B., Mika, R., Battle, M.O., Tans, P.P., Conway, T.J., Sturtevant, B., Cassar, N., 2005. Atmospheric O_2/N_2 changes, 1993–2002: implications for the partitioning of fossil fuel CO_2 sequestration. Glob. Biogeochem. Cycles 19.

Bengtsson, M.M., Attermeyer, K., Catalán, N., 2018. Interactive effects on organic matter processing from soils to the ocean: are priming effects relevant in aquatic ecosystems? Hydrobiologia 822 (1), 1–17.

Benitez-Nelson, C.R., Buesseler, K.O., 1999. Variability of inorganic and organic phosphorus turnover rates in the coastal ocean. Nature 398, 502–505.

Benke, A.C., Huryn, A.D., 2006. Secondary production of macroinvertebrates. In: Hauer, F.R., Lamberti, G.A. (Eds.), Methods in Stream Ecology. Elsevier, pp. 691–710.

Bennett, P.C., Melcer, M.E., Siegel, D.I., Hassett, J.P., 1988. The dissolution of quartz in dilute aqueous solutions of organic acids at 25°C. Geochim. Cosmochim. Acta 52, 1521–1530.

Bennett, E.M., Carpenter, S.R., Caraco, N.F., 2001. Human impact on erodable phosphorus and eutrophication: a global perspective. Bioscience 51, 227–234.

Bennett, S.A., Achterberg, E.P., Connelly, D.P., Statham, P.J., Fones, G.R., German, C.R., 2008. The distribution and stabilisation of dissolved Fe in deep-sea hydrothermal plumes. Earth Planet. Sci. Lett. 270 (3), 157–167.

Benoit, G., Rozan, T.F., 1999. The influence of size distribution on the particle concentration effect and trace metal partitioning in rivers. Geochim. Cosmochim. Acta 63, 113–127.

Ben-Shahar, R., Coe, M.J., 1992. The relationships between soil factors, grass nutrients and the foraging behavior of wildebeest and zebra. Oecologia 90, 422–428.

Benson, B.J., Magnuson, J.J., Jensen, O.P., Card, V.M., Hodgkins, G., Korhonen, J., Livingstone, D.M., Stewart, K.M., Weyhenmeyer, G.A., Granin, N.G., 2012. Extreme events, trends, and variability in northern hemisphere lake-ice phenology (1855–2005). Clim. Chang. 112, 299–323.

Benstead, J.P., Rosemond, A.D., Cross, W.F., Bruce Wallace, J., Eggert, S.L., Suberkropp, K., Gulis, V., Greenwood, J.L., Tant, C.J., 2009. Nutrient enrichment alters storage and fluxes of detritus in a headwater stream ecosystem. Ecology 90 (9), 2556–2566.

Benzing, D.H., Renfrow, A., 1974. Mineral nutrition of Bromeliaceae. Bot. Gaz. 135, 281–288.

Berelson, W.M., Hammond, D.E., Cutter, G.A., 1990. In situ measurements of calcium carbonate dissolution rates in deep-sea sediments. Geochim. Cosmochim. Acta 54, 3013–3020.

Berelson, W.M., Balch, W.M., Najjar, R., Feely, R.A., Sabine, C., Lee, K., 2007. Relating estimates of $CaCO_3$ production, export, and dissolution in the water column to measurements of $CaCO_3$ rain into sediment traps and dissolution on the sea floor: a revised global carbonate budget. Glob. Biogeochem. Cycles.

Berendse, F., Van Breemen, N., Rydin, H., Buttler, A., Heijmans, M., Hoosbeek, M.R., Lee, J.A., et al., 2001. Raised atmospheric CO_2 levels and increased N deposition cause shifts in plant species composition and production in *Sphagnum* bogs. Glob. Chang. Biol. 7 (5), 591–598.

Berg, B., 1988. Dynamics of nitrogen (^{15}N) in decomposing scots pine (*Pinus sylvestris*) needle litter. Long-term decompositon in a scots pine forest. VI. Can. J. Bot. 66, 1539–1546.

Berg, P., Klemedtsson, L., Rosswall, T., 1982. Inhibitory effect of low partial pressures of acetylene on nitrification. Soil Biol. Biochem. 14, 301–303.

Berg, B., Berg, M.P., Bottner, P., Box, E., Breymeyer, A., Deanta, R.C., Couteaux, M., et al., 1993. Litter mass loss rates in pine forests of Europe and eastern United States: some relationships with climate and litter quality. Biogeochemistry 20, 127–159.

Berg, B., Albrektson, A., Berg, M.P., Cortina, J., Johansson, M.B., Gallardo, A., Madeira, M., et al., 1999. Amounts of litter fall in some pine forests in a European transect, in particular scots pine. Ann. For. Sci. 56, 625–639.

Berga, L., Buil, J.M., Bofill, E., DeCea, J.C., Garcia Perez, J.A., Manueco, G., Polimon, J., Soriano, A., Yaque, J., 2006. Dams and Reservoirs, Societies and Environment in the 21st Century. Taylor and Francis Group.

Berge, E., Jakobsen, H.A., 1998. A regional scale multilayer model for the calculation of long-term transport and deposition of air pollution in Europe. Tellus Ser. B Chem. Phys. Meteorol. 50 (3), 205–223.

Berger, W.H., 1989. Global maps of ocean productivity. In: Berger, W.H., Smetacek, V.S., Wefer, G. (Eds.), Productivity of the Ocean: Present and Past. Wiley, pp. 429–455.

Berger, T.W., Glatzel, G., 1994. Deposition of atmospheric constituents and its impact on nutrient budgets of oak forests (*Quercus petraea* and *Quercus robur*) in lower Austria. For. Ecol. Manag. 70, 183–193.

Berger, W.H., Adelseck, C.G., Mayer, L.A., 1976. Distribution of carbonate in surface sediments of pacific ocean. J. Geophys. Res. Oceans Atmos. 81, 2617–2627.

Bergeron, O., Strachan, I.B., 2011. CO_2 sources and sinks in urban and suburban areas of a northern mid-latitude city. Atmos. Environ. 45, 1564–1573.

Berggren, D., Mulder, J., 1995. The role of organic matter in controlling aluminum solubility in acidic mineral soil horizons. Geochim. Cosmochim. Acta 59, 4167–4180.

Bergstrom, A.-K., Jansson, M., 2006. Atmospheric nitrogen deposition has caused nitrogen enrichment and eutrophication of lakes in the northern hemisphere. Glob. Chang. Biol. 12 (4), 635–643.

Berkner, L.V., Marshall, L.C., 1965. On the origin and rise of oxygen concentration in the Earth's atmosphere. J. Atmos. Sci. 22, 225–261.

Berliner, R., Jacoby, B., Zamski, E., 1986. Absence of *Cistus incanus* from basaltic soils in Israel—effect of mycorrhizae. Ecology 67, 1283–1288.

Bern, C.R., Townsend, A.R., 2008. Accumulation of atmospheric sulfur in some Costa Rican soils. J. Geophys. Res. 113.

Bernal, B., Mitsch, W.J., 2012. Comparing carbon sequestration in temperate freshwater wetland communities. Glob. Chang. Biol. 18, 1636–1647.

Bernath, P.F., Yousefi, M., Buzan, E., Boone, C.D., 2017. A near-global atmospheric distribution of N_2O isotopologues: N_2O isotopologues. Geophys. Res. Lett. 44 (20), 10,735–10,743.

Berner, R.A., 1982. Burial of organic carbon and pyrite sulfur in the modern ocean: its geochemical and environmental significance. Am. J. Sci. 282, 451–473.

Berner, R.A., 1984. Sedimentary pyrite formation: an update. Geochim. Cosmochim. Acta 48, 605–615.

Berner, R.A., 1992. Weathering, plants, and the long-term carbon cycle. Geochim. Cosmochim. Acta 56, 3225–3231.

Berner, R.A., 1997. The rise of plants and their effect on weathering and atmospheric CO_2. Science 276, 544–546.

Berner, R.A., 2001. Modeling atmospheric O_2 over phanerozoic time. Geochim. Cosmochim. Acta 65, 685–694.

Berner, R.A., 2004. A model for calcium, magnesium and sulfate in seawater over phanerozoic time. Am. J. Sci. 304 (5), 438–453.

Berner, E.K., Berner, R.A., 2012. Global Environment: Water, Air, and Geochemical Cycles. Princeton University Press.

Berner, R.A., Canfield, D.E., 1989. A new model for atmospheric oxygen over phanerozoic time. Am. J. Sci. 289 (4), 333–361.

Berner, R.A., Honjo, S., 1981. Pelagic sedimentation of aragonite: its geochemical significance. Science 211 (4485), 940–942.

Berner, R.A., Kothavala, Z., 2001. GEOCARB III: a revised model of atmospheric CO_2 over phanerozoic time. Am. J. Sci. 301, 182–204.

Berner, R.A., Lasaga, A.C., 1989. Modeling the geochemical carbon cycle. Sci. Am. 260, 74–81.

Berner, R.A., Raiswell, R., 1983. Burial of organic carbon and pyrite sulfur in sediments over phanerozoic time: a new theory. Geochim. Cosmochim. Acta 47, 855–862.

Berner, R.A., Rao, J.L., 1994. Phosphorus in sediments of the Amazon river and estuary: impilcations for the global flux of phosphorus to the sea. Geochim. Cosmochim. Acta 58, 2333–2339.

Berner, R.A., Rao, J.-L., 1997. Alkalinity buildup during silicate weathering under a snow cover. Aquat. Geochem. 2, 301–312.

Berner, R.A., Petsch, S.T., Lake, J.A., Beerling, D.J., Popp, B.N., Lane, R.S., Laws, E.A., et al., 2000. Isotope fractionation and atmospheric oxygen: implications for phanerozoic O(2) evolution. Science 287 (5458), 1630–1633.

Bernhardt, E.S., Likens, G.E., 2002. Dissolved organic carbon enrichment alters nitrogen dynamics in a forest stream. Ecology 83, 1689–1700.

Bernhardt, E.S., Likens, G.E., 2004. Controls on periphyton biomass in heterotrophic streams. Freshw. Biol. 49, 14–27.

Bernhardt, E.S., McDowell, W.H., 2008. Twenty years apart: comparisons of DOM uptake during leaf leachate releases to Hubbard Brook valley streams in 1979 versus 2000. J. Geophys. Res. Biogeosci. 113, G03032.

Bernhardt, E.S., Hall, R.O., Likens, G.E., 2002. Whole-system estimates of nitrification and nitrate uptake in streams of the Hubbard Brook experimental forest. Ecosystems 5, 419–430.

Bernhardt, E.S., Likens, G.E., Buso, D.C., Driscoll, C.T., 2003. In-stream uptake dampens effect of major forest disturbance on watershed nitrogen export. Proc. Natl. Acad. Sci. 100, 10304–10308.

Bernhardt, E.S., Likens, G.E., Hall, R.O., Buso, D.C., Fisher, S.G., Burton, T.M., Meyer, J.L., et al., 2005. Can't see the forest for the stream?—in-stream processing and terrestrial nitrogen exports. Bioscience 55 (3), 219–230.

Bernhardt, E.S., Barber, J.J., Pippen, J.S., Taneva, L., Andrews, J.A., Schlesinger, W.H., 2006. Long-term effects of free air CO_2 enrichment (FACE) on soil respiration. Biogeochemistry 77 (1), 91–116.

Bernhardt, E.S., Blaszczak, J.R., Ficken, C.D., Fork, M.L., Kaiser, K.E., Seybold, E.C., 2017a. Control points in ecosystems: moving beyond the hot spot hot moment concept. Ecosystems 20 (4), 665–682.

Bernhardt, E.S., Rosi, E.J., Gessner, M.O., 2017b. Synthetic chemicals as agents of global change. Front. Ecol. Environ. 15 (2), 84–90.

Bernhardt, E.S., Heffernan, J.B., Grimm, N.B., Stanley, E.H., Harvey, J.W., Arroita, M., Appling, A.P., et al., 2018. The metabolic regimes of flowing waters. Limnol. Oceanogr. 63 (S1), S99–118.

Bernier, B., Brazeau, M., 1988a. Foliar nutrient status in relation to sugar maple dieback and decline in the Quebec Appalachians. Can. J. For. Res. 18, 754–761.

Bernier, B., Brazeau, M., 1988b. Magnesium-deficiency symptoms associated with sugar maple dieback in a lower laurentians site in southeastern Quebec. Can. J. For. Res. 18, 1265–1269.

Bernot, M.J., Sobota, D.J., Hall Jr., R.O., Mulholland, P.J., Dodds, W.K., Webster, J.R., Tank, J.L., et al., 2010. Inter-regional comparison of land-use effects on stream metabolism. Freshw. Biol. 55 (9), 1874–1890.

Bernstein, R.E., Byrne, R.H., 2004. Acantharians and marine barite. Mar. Chem. 86, 45–50.

Bernstein, R.E., Betzer, P.R., Feely, R.A., Byrne, R.H., Lamb, M.F., Michaels, A.F., 1987. Acantharian fluxes and strontium to chlorinity ratios in the North Pacific Ocean. Science 237 (4821), 1490–1494.

Berresheim, H., Vulcan, V.D., 1992. Vertical distributions of COS,CS2, DMS and other sulfur compounds in a loblolly pine forest. Atmos. Environ. A—Gen. Top. 26, 2031–2036.

Berry, J., Adam, W., Elliott Campbell, J., Baker, I., Blake, N., Don, B., Scott Denning, A., et al., 2013. A coupled model of the global cycles of carbonyl sulfide and CO_2: a possible new window on the carbon cycle: carbonyl sulfide as a global carbon cycle tracer. J. Geophys. Res. Biogeosci. 118 (2), 842–852.

Berthrong, S.T., Jobbágy, E.G., Jackson, R.B., 2009. A global meta-analysis of soil exchangeable cations, pH, carbon, and nitrogen with afforestation. Ecol. Appl. 19 (8), 2228–2241.

Bertilsson, S., Stepanauskas, R., Cuadros-Hansson, R., Graneli, W., Wikner, J., Tranvik, L., 1999. Photochemically induced changes in bioavailable carbon and nitrogen pools in a boreal watershed. Aquat. Microb. Ecol. Int. J. 19 (1), 47–56.

Bertine, K.K., Goldberg, E.D., 1971. Fossil fuel combustion and the major sedimentary cycle. Science 173 (3993), 233–235.

Bertram, T.H., Perring, A.E., Wooldridge, P.J., Crounse, J.D., Kwan, A.J., Wennberg, P.O., Scheuer, E., et al., 2007. Direct measurements of the convective recycling of the upper troposphere. Science 315 (5813), 816–820.

Bertrand, P., Lallier-Vergès, E., 1993. Past sedimentary organic matter accumulation and degradation controlled by productivity. Nature 364 (6440), 786–788.

Betlach, M.R., 1982. Evolution of bacterial denitrification and denitrifier diversity. Antonie Van Leeuwenhoek 48 (6), 585–607.

Betts, J.N., Holland, H.D., 1991. The oxygen content of ocean bottom waters, the burial efficiency of organic carbon, and the regulation of atmospheric oxygen. Glob. Planet. Chang. 97, 5–18.

Betts, R.A., Boucher, O., Collins, M., Cox, P.M., Falloon, P.D., Gedney, N., Hemming, D.L., et al., 2007. Projected increase in continental runoff due to plant responses to increasing carbon dioxide. Nature 448 (7157), 1037–1041.

Betzer, P.R., Byrne, R.H., Acker, J.G., Lewis, C.S., Jolley, R.R., Feely, R.A., 1984. The oceanic carbonate system: a reassessment of biogenic control. Science 226, 1074–1077.

Beusen, A.H.W., Bouwman, A.F., Van Beek, L.P.H., Mogollón, J.M., Middelburg, J.J., 2016. Global riverine N and P transport to ocean increased during the 20th century despite increased retention along the aquatic continuum. Biogeosciences 13 (8), 2441–2451.

Bevan, A., Colledge, S., Fuller, D., Fyfe, R., Shennan, S., Stevens, C., 2017. Holocene fluctuations in human population demonstrate repeated links to food production and climate. Proc. Natl. Acad. Sci. U. S. A. 114 (49), E10524–E10531.

Beyer, L., Schulten, H.R., Fruend, R., Irmler, U., 1993. Formation and properties of organic matter in a forest soil, as revealed by its biological activity, wet chemical analysis, CPMAS C-13 NMR spectroscopy and pyrolysis field ionization mass spectrometry. Soil Biol. Biochem. 25, 587–596.

Beyersdorf, A.J., Blake, D.R., Swanson, A., Meinardi, S., Rowland, F.S., Davis, D., 2010. Abundances and variability of tropospheric volatile organic compounds at the south pole and other Antarctic locations. Atmos. Environ. 44, 4565–4574.

Bhatti, J.S., Comerford, N.B., Johnston, C.T., 1998. Influence of oxalate and soil organic matter on sorption and desorption of phosphate onto a spodic horizon. Soil Sci. Soc. Am. J. 62, 1089–1095.

Bianchi, T.S., 2011. The role of terrestrially derived organic carbon in the coastal ocean: a changing paradigm and the priming effect. Proc. Natl. Acad. Sci. U. S. A. 108 (49), 19473–19481.

Bianchi, T.S., Allison, M.A., 2009. Large-river delta-front estuaries as natural 'recorders' of global environmental change. Proc. Natl Acad. Sci. U S A 106 (20), 8085–8092.

Bianchi, D., Dunne, J.P., Sarmiento, J.L., Galbraith, E.D., 2012. Data-based estimates of Suboxia, denitrification, and N_2O production in the ocean and their sensitivities to dissolved O_2. Glob. Biogeochem. Cycles 26 (2).

Bianchi, T.S., Thornton, D.C.O., Yvon-Lewis, S.A., King, G.M., Eglinton, T.I., Shields, M.R., Ward, N.D., Curtis, J., 2015. Positive priming of terrestrially derived dissolved organic matter in a freshwater microcosm system. Geophys. Res. Lett. 42, 5460–5467.

Biddanda, B., Opsahl, S., Benner, R., 1994. Plankton respiration and carbon flux through bacterioplankton on the louisiana shelf. Limnol. Oceanogr. 39, 1259–1275.

Bidle, K.D., Azam, F., 1999. Accelerated dissolution of diatom silica by marine bacterial assemblages. Nature 397, 508–512.

Bidle, K.D., Manganelli, M., Azam, F., 2002. Regulation of oceanic silicon and carbon preservation by temperature control on bacteria. Science 298 (5600), 1980–1984.

Biederman, J.A., Scott, R.L., Bell, T.W., Bowling, D.R., Dore, S., Garatuza-Payan, J., Kolb, T.E., et al., 2017. CO_2 exchange and evapotranspiration across dryland ecosystems of Southwestern North America. Glob. Chang. Biol. 23 (10), 4204–4221.

Biemans, H., Haddeland, I., Kabat, P., Ludwig, F., Hutjes, R.W.A., Heinke, J., von Bloh, W., Gerten, D., 2011. Impact of reservoirs on river discharge and irrigation water supply during the 20th century. Water Resour. Res. 47.

Biggs, B.J.F., 1996. Patterns of benthic algae in streams. In: Stevenson, R.J., Bothwell, M.L., Lowe, R.L. (Eds.), Algal Ecology. Academic Press, pp. 31–56.

Bigio, L., Angert, A., 2018. Isotopic signature of atmospheric phosphate in airborne tree pollen. Atmos. Environ. 194 (December), 1–6.

Billen, G., 1975. Nitrification in Scheldt estuary—(Belgium and Netherlands). Estuar. Coast. Mar. Sci. 3 (1), 79–89.

Billings, W.D., 1987. Carbon balance of Alaskan Tundra and Taiga Ecosystems: past, present and future. Quat. Sci. Rev. 6, 165–177.

Billings, C.E., Matson, W.R., 1972. Mercury emissions from coal combustion. Science 176 (4040), 1232–1233.

Billings, S.A., Richter, D.D., 2006. Changes in stable isotopic signatures of soil nitrogen and carbon during 40 years of forest development. Oecologia 148 (2), 325–333.

Billings, S.A., Schaeffer, S.M., Evans, R.D., 2003. Nitrogen fixation by biological soil crusts and heterotrophic bacteria in an intact Mojave Desert ecosystem with elevated CO_2 and added soil carbon. Soil Biol. Biochem. 35, 643–649.

Bindeman, I.N., Zakharov, D.O., Palandri, J., Greber, N.D., Dauphas, N., Retallack, G.J., Hofmann, A., Lackey, J.S., Bekker, A., 2018. Rapid emergence of subaerial landmasses and onset of a modern hydrologic cycle 2.5 billion years ago. Nature 557 (7706), 545–548.

Binford, M.W., Gholz, H.L., Starr, G., Martin, T.A., 2006. Regional carbon dynamics in the Southeastern US Coastal plain: balancing land cover type, timber harvesting, fire, and environmental variation. J. Geophys. Res. 111.

Binkley, D., 1986. Forest Nutrition Management. Wiley.

Binkley, D., Hart, S.C., 1989. The components of nitrogen availability assessments in forest soils. Adv. Soil Sci. 10, 57–112.

Binkley, D., Richter, D., 1987. Nutrient cycles and H+ budgets of forest ecosystems. Adv. Ecol. Res. 16, 1–51.

Bintanja, R., van de Wal, R.S.W., Oerlemans, J., 2005. Modelled atmospheric temperatures and global sea levels over the past million years. Nature 437 (7055), 125–128.

Bird, M.I., Lloyd, J., Farquhar, G.D., 1994. Terrestrial carbon storage at the LGM. Nature 371, 566.

Birdsey, R.A., Plantinga, A.J., Heath, L.S., 1993. Past and prospective carbon storage in United States' forests. For. Ecol. Manag. 58, 33–40.

Birk, E.M., Vitousek, P.M., 1986. Nitrogen availability and nitrogen-use efficiency in loblolly pine stands. Ecology 67, 69–79.

Birkeland, P.W., 1984. Soils and Geomorphology, second ed. Oxford.

Biscaye, P.E., Kolla, V., Turekian, K.K., 1976. Distribution of calcium carbonate in surface sediments of Atlantic ocean. J. Geophys. Res. Oceans Atmos. 81, 2595–2603.

Bishop, J.K.B., 1988. The Barite-Opal-organic carbon association in oceanic particulate matter. Nature 332, 341–343.

Bishop, J.K.B., Davis, R.E., Sherman, J.T., 2002. Robotic observations of dust storm enhancement of carbon biomass in the North Pacific. Science 298 (5594), 817–821.

Bjerrum, C.J., Canfield, D.E., 2002. Ocean productivity before about 1.9 Gyr ago limited by phosphorus adsorption onto iron oxides. Nature 417 (6885), 159–162.

Black, R.A., Richards, J.H., Manwaring, J.H., 1994. Nutrient uptake from enriched soil microsites by three Great Basin perennials. Ecology 75, 110–122.

Black, F.J., Poulin, B.A., Flegal, A.R., 2012. Factors controlling the abiotic photo-degradatioon of monomethylmercury in surface waters. Geochim. Cosmochim. Acta 84, 492–507.

Blackard, J.A., Finco, M.V., Helmer, E.H., Holden, G.R., Hoppus, M.L., Jacobs, D.M., Lister, A.J., et al., 2008. Mapping US forest biomass using nationwide forest inventory data and moderate resolution information. Remote Sens. Environ. 112, 1658–1677.

Blagodatskaya, E., Kuzyakov, Y., 2013. Active microorganisms in soil: critical review of estimation criteria and approaches. Soil Biol. Biochem. 67 (December), 192–211.

Blain, S., Quéguiner, B., Armand, L., Belviso, S., Bombled, B., Bopp, L., Bowie, A., et al., 2007. Effect of natural iron fertilization on carbon sequestration in the Southern Ocean. Nature 446 (7139), 1070–1074.

Blair, N.E., Aller, R.C., 1995. Anaerobic methane oxidation on the Amazon shelf. Geochim. Cosmochim. Acta 59 (18), 3707–3715.

Blair, N.E., Aller, R.C., 2012. The fate of terrestrial organic carbon in the marine environment. Annu. Rev. Mar. Sci. 4, 401–423.

Blaise, T., Garbaye, J., 1983. Effects of mineral fertilization on the mycorrhization of roots in a beech forest. Acta Oecol.-Oecol. Plantar. 4, 165–169.

Blake, L., Goulding, K.W.T., Mott, C.J.B., Johnston, A.E., 1999. Changes in soil chemistry accompanying acidification over more than 100 years under woodland and grass at Rothamsted Experimental Station, UK. Eur. J. Soil Sci. 50, 401–412.

Blake, R.E., Chang, S.J., Lepland, A., 2010. Phosphate oxygen isotopic evidence for a temperate and biologically active Archaean Ocean. Nature 464 (7291), 1029–1032.

Blaszczak, J.R., Delesantro, J.M., Urban, D.L., Doyle, M.W., Bernhardt, E.S., 2019. Scoured or suffocated: urban stream ecosystems oscillate between hydrologic and dissolved oxygen extremes. Limnol. Oceanogr. 64 (3), 877–894.

Blatt, H., Jones, R.L., 1975. Proportions of exposed igneous, metamorphic, and sedimentary rocks. Geol. Soc. Am. Bull. 86, 1085–1088.

Blättler, C.L., Claire, M.W., Prave, A.R., Kirsimäe, K., Higgins, J.A., Medvedev, P.V., Romashkin, A.E., et al., 2018. Two-billion-year-old evaporites capture Earth's great oxidation. Science 360, 320–323.

Blattmann, T.M., Liu, Z., Zhang, Y., Zhao, Y., Haghipour, N., Montluçon, D.B., Plötze, M., Eglinton, T.I., 2019. Mineralogical control on the fate of continentally derived organic matter in the ocean. Science 366 (6466), 742–745.

Blodau, C., Basiliko, N., Moore, T.R., 2004. Carbon turnover in peatland mesocosms exposed to different water table levels. Biogeochemistry 67 (3), 331–351.

Blomqvist, S., Gunnars, A., Elmgren, R., 2004. Why the limiting nutrient differs between temperate coastal seas and freshwater lakes: a matter of salt. Limnol. Oceanogr. 49 (6), 2236–2241.

Blonquist Jr., J.M., Montzka, S.A., Munger, J.W., Yakir, D., Desai, A.R., Dragoni, D., Griffis, T.J., Monson, R.K., Scott, R.L., Bowling, D.R., 2011. The potential of carbonyl sulfide as a proxy for gross primary production at flux tower sites. J. Geophys. Res. 116.

Bloom, A.J., Sukrapanna, S.S., Warner, R.L., 1992. Root respiration associated with ammonium and nitrate absorption and assimilation by barley. Plant Physiol. 99 (4), 1294–1301.

Bloom, A.J., Jackson, L.E., Smart, D.R., 1993. Root growth as a function of ammonium and nitrate in the root zone. Plant Cell Environ. 16, 199–206.

Bloom, A.A., Palmer, P.I., Fraser, A., Reay, D.S., Frankenberg, C., 2010. Large-scale controls of methanogenesis inferred from methane and gravity space-borne data. Science 327, 322–325.

Bloomfield, C., 1972. The oxidation of iron sulphides in soils in relation to the formation of acid sulphate soils, and ofochre deposits in field drains. J. Soil Sci. 23, 1–16.

Blum, J.D., 2012. Applications of stable mercury isotopes to biogeochemistry. In: Baskaran, M. (Ed.), Handbook of Environmental Isotope Geochemistry. In: vol. I. Springer Berlin Heidelberg, Berlin, Heidelberg, pp. 229–245.

Blum, J.D., Klaue, A., Nezat, C.A., Driscoll, C.T., Johnson, C.E., Siccama, T.G., Eagar, C., Fahey, T.J., Likens, G.E., 2002. Mycorrhizal weathering of apatite as an important calcium source in base-poor forest ecosystems. Nature 417 (6890), 729–731.

Blunier, T., Chappellaz, J., Schwander, J., Stauffer, B., Raynaud, D., 1995. Variations in atmospheric methane concentration during the Holocene epoch. Nature 374, 46–49.

Bluth, G.J.S., Kump, L.R., 1994. Lithologic and climatologic controls of river chemistry. Geochim. Cosmochim. Acta 58, 2341–2359.

Bluth, G.J.S., Schnetzler, C.C., Krueger, A.J., Walter, L.S., 1993. The contribution of explosive volcanism to global atmospheric sulphur dioxide concentrations. Nature 366, 327–329.

Bobbink, R., Hicks, K., Galloway, J., Spranger, T., Alkemade, R., Ashmore, M., Bustamante, M., et al., 2010. Global assessment of nitrogen deposition effects on terrestrial plant diversity: a synthesis. Ecol. Appl. 20 (1), 30–59.

Bockheim, J.G., 1980. Solution and use of chrono functions in studying soil development. Geoderma 24, 71–85.

Bodnar, R.J., Azbej, T., Becker, S.P., Cannatelli, C., Fall, A., Severs, M.J., 2013. Whole earth geohydrologic cycle, from the clouds to the core: the distribution of water in the dynamic earth system. Geol. Soc. Am. Spec. Pap. 500, 431–461.

Boesch, D.F., 2002. Challenges and opportunities for science in reducing nutrient over-enrichment of coastal ecosystems. Estuaries 25, 886–900.

Boesch, D.F., Brinsfield, R.B., Magnien, R.E., 2001. Chesapeake bay eutrophication: scientific understanding, ecosystem restoration, and challenges for agriculture. J. Environ. Qual. 30 (2), 303–320.

Boetius, A., Ravenschlag, K., Schubert, C.J., Rickert, D., Widdel, F., Gieseke, A., Amann, R., Jørgensen, B.B., Witte, U., Pfannkuche, O., 2000. A marine microbial consortium apparently mediating anaerobic oxidation of methane. Nature 407 (6804), 623–626.

Boettcher, S.E., Kalisz, P.J., 1990. Single-tree influence on soil properties in the mountains of eastern Kentucky. Ecology 71, 1365–1372.

Bogner, J., Matthews, E., 2003. Global methane emissions from landfills: new methodology and annual estimates 1980–1996. Glob. Biogeochem. Cycles 17.

Bohacs, K.M., Carroll, A.R., Neal, J.E., Mankiewicz, P.J., Others, 2000. Lake-basin type, source potential, and hydrocarbon character: an integrated sequence-stratigraphic-geochemical framework. In: Lake Basins through Space and Time. vol. 46. AAPG Studies in Geology, pp. 3–34.

Bolan, N.S., 1991. A critical review on the role of mycorrhizal fungi in the uptake of phosphorus by plants. Plant Soil 134, 189–207.

Bolan, N.S., Robson, A.D., Barrow, N.J., Aylmore, L.A.G., 1984. Specific activity of phosphorus in mycorrhizal and non-mycorrhizal plants in relation to the availability of phosphorus to plants. Soil Biol. Biochem. 16, 299–304.

Bolan, N.S., Saggar, S., Luo, J.F., Bhandral, R., Singh, J., 2004. Gaseous emissions of nitrogen from grazed pastures: processes, measurements and modelling, environmental implications, and mitigation. Adv. Agron. 84, 37–120.

Bollmann, A., Conrad, R., 1998. Influence of O_2 availability on NO and N_2O release by nitrification and denitrification in soils. Glob. Chang. Biol. 4, 387–396.

Bolton, E.W., Berner, R.A., Petsch, S.T., 2006. The weathering of sedimentary organic matter as a control on atmospheric O_2: II. Theoretical modeling. Am. J. Sci. 306 (8), 575–615.

Bonanno, A., Schlattl, H., Paterno, L., 2002. The age of the sun and the relativistic corrections in the EOS. Astron. Astrophys. Suppl. Ser. 390, 1115–1118.

Bond, A.L., Hobson, K.A., Branfireun, B.A., 2015. Rapidly increasing methyl mercury in endangered ivory gull (*Pagophila eburnea*) feathers over a 130 year record. Proc. R. Soc. B Biol. Sci. 282 (1805), 20150032.

Bondarenko, N.V., Head, J.W., Ivanov, M.A., 2010. Present-day volcanism on venus: evidence from microwave radiometry. Geophys. Res. Lett. 37.

Bondietti, E.A., Baes, C.F., McLaughlin, S.B., 1989. Radial trends in cation ratios in tree rings as indicators of the impact of atmospheric deposition on forests. Can. J. For. Res. 19, 586–594.

Bond-Lamberty, B., Thomson, A., 2010. Temperature-associated increases in the global soil respiration record. Nature 464 (7288), 579–582.

Bond-Lamberty, B., Wang, C.K., Gower, S.T., 2004. A global relationship between the heterotrophic and autotrophic components of soil respiration? Glob. Chang. Biol. 10, 1756–1766.

Bond-Lamberty, B., Peckham, S.D., Ahl, D.E., Gower, S.T., 2007. Fire as the dominant driver of central Canadian boreal forest carbon balance. Nature 450 (7166), 89–92.

Bond-Lamberty, B., Bailey, V.L., Chen, M., Gough, C.M., Vargas, R., 2018. Globally rising soil heterotrophic respiration over recent decades. Nature 560 (7716), 80–83.

Bonell, M., 1993. Progress in the understanding of runoff generation dynamics in forests. J. Hydrol. 150, 217–275.

Bonin, P., Gilewicz, M., Bertrand, J.C., 1989. Effects of oxygen on each step of denitrification on *Pseudomonas nautica*. Can. J. Microbiol. 35, 1061–1064.

Bonner, M.T.L., Schmidt, S., Shoo, L.P., 2013. A meta-analytical global comparison of aboveground biomass accumulation between tropical secondary forests and monoculture plantations. For. Ecol. Manag. 291 (March), 73–86.

Bono, P., Boni, C., 1996. Water supply of Rome in antiquity and today. Environ. Geol. 27, 126–134.

Booth, M.S., Stark, J.M., Rastetter, E., 2005. Controls on nitrogen cycling in terrestrial ecosystems: a synthetic analysis of literature data. Ecol. Monogr. 75, 139–157.

Bopp, L., Le Quere, C., Heimann, M., Manning, A.C., Monfray, P., 2002. Climate-induced oceanic oxygen fluxes: implications for the contemporary carbon budget. Glob. Biogeochem. Cycles 16.

Borges, A.V., 2005. Do we have enough pieces of the jigsaw to integrate CO_2 fluxes in the Coastal Ocean? Estuaries 28, 3–27.

Boring, L.R., Swank, W.T., Waide, J.B., Henderson, G.S., 1988. Sources, fates, and impacts of nitrogen inputs to terrestrial ecosystems: review and synthesis. Biogeochemistry 6, 119–159.

Bormann, B.T., Gordon, J.C., 1984. Stand density effects in young red alder plantations: productivity, photosynthate partitioning, and nitrogen fixation. Ecology 65, 394–402.

Bormann, F.H., Likens, G.E., 1969. The watershed-ecosystem concept and studies of nutrient cycles. In: Van Dyne, G. (Ed.), The Ecosystem Concept in Natural Resource Management. Academic Press, New York and London.

Bormann, F.H., Likens, G.E., Siccama, T.G., Pierce, R.S., Eaton, J.S., 1974. The export of nutrients and recovery of stable conditions following deforestation at Hubbard Brook. Ecol. Monogr. 44, 255–277.

Born, M., Dorr, H., Levin, I., 1990. Methane consumption in aerated soils of the temperate zone. Tellus Ser. B Chem. Phys. Meteorol. 42, 2–8.

Bosch, J.M., Hewlett, J.D., 1982. A review of catchment experiments to determine the effect of vegetation changes on water yield and evapotranspiration. J. Hydrol. 55, 3–23.

Boss, A.P., 1988. High temperatures in the early solar nebula. Science 241 (4865), 565–567.

Botch, M.S., Kobak, K.I., Vinson, T.S., Kolchugina, T.P., 1995. Carbon pools and accumulation in peatlands of the former Soviet Union. Glob. Biogeochem. Cycles 9, 37–46.

Botkin, D.B., Malone, C.R., 1968. Efficiency of net primary production based on light intercepted during the growing season. Ecology 49, 438–444.

Botkin, D.B., Simpson, L.G., 1990. Biomass of the North American Boreal Forest: a step toward accurate global measures. Biogeochemistry 9, 161–174.

Botkin, D.B., Jordan, P.A., Dominski, A.S., Lowendorf, H.S., Hutchinson, G.E., 1973. Sodium dynamics in a northern ecosystem. Proc. Natl. Acad. Sci. U. S. A. 70 (10), 2745–2748.

Botkin, D.B., Simpson, L.G., Nisbet, R.A., 1993. Biomass and carbon storage of the North American deciduous forest. Biogeochemistry 20, 1–17.

Bott, T.L., 2006. Primary productivity and community respiration. In: Hauer, F.R., Lamberti, G.A. (Eds.), Methods in Stream Ecology. Academic Press, San Diego, CA, pp. 663–689.

Bott, T.L., Brock, J.T., Dunn, C.S., Naimann, R.J., Ovink, R.W., Peterson, R.C., 1985. Benthic community metabolism in four temperate stream systems: an inter-biome comparison and evaluation of the river continuum concept. Hydrobiologia 123, 3–45.

Bottcher, J., Weymann, D., Well, R., von der Heide, C., Schwen, A., Flessa, H., Duijnisveld, W.H.M., 2011. Emission of groundwater-derived nitrous oxide into the atmosphere: model simulations based on a ^{15}N-field experiment. Eur. J. Soil Sci. 62, 216–225.

Bouchard, S.S., Bjorndal, K.A., 2000. Sea turtles as biological transporters of nutrients and energy from marine to terrestrial ecosystems. Ecology 81, 2305–2313.

Boudreau, B.P., Westrich, J.T., 1984. The dependence of bacterial sulfate reduction on sulfate concentration in marine sediments. Geochim. Cosmochim. Acta 48, 2503–2516.

Boudreau, J., Nelson, R.F., Margolis, H.A., Beaudoin, A., Guindon, L., Kimes, D.S., 2008. Regional aboveground forest biomass using airborne and spaceborne LiDAR in Quebec. Remote Sens. Environ. 112, 3876–3890.

Bousquet, P., Ciais, P., Miller, J.B., Dlugokencky, E.J., Hauglustaine, D.A., Prigent, C., Van der Werf, G.R., et al., 2006. Contribution of anthropogenic and natural sources to atmospheric methane variability. Nature 443 (7110), 439–443.

Boutron, C.F., Gorlach, U., Candelone, J.P., Bolshov, M.A., Delmas, R.J., 1991. Decrease in anthropogenic lead, cadmium and zinc in Greenland snows since the late 1960s. Nature 353, 153–156.

Boutron, C.F., Candelone, J.P., Hong, S.M., 1994. Past and recent changes in the large-scale tropospheric cycles of lead and other heavy metals as documented in Antarctic and Greenland Snow and Ice—a review. Geochim. Cosmochim. Acta 58, 3217–3225.

Bouvier, A., Wadhwa, M., 2010. The age of the solar system redefined by the oldest Pb-Pb age of a meteoritic inclusion. Nat. Geosci. 3, 637–641.

Bouwman, A.F., Fung, I., Matthews, E., John, J., 1993. Global analysis of the potential for N$_2$O production in natural soils. Glob. Biogeochem. Cycles 7, 557–597.

Bouwman, A.F., Vanderhoek, K.W., Olivier, J.G.J., 1995. Uncertainties in the global source distribution of nitrous oxide. J. Geophys. Res. D: Atmos. 100, 2785–2800.

Bouwman, A.F., Lee, D.S., Asman, W.A.H., Dentener, F.J., VanderHoek, K.W., Olivier, J.G.J., 1997. A global high-resolution emission inventory for ammonia. Glob. Biogeochem. Cycles 11, 561–587.

Bouwman, A.F., Boumans, L.J.M., Batjes, N.H., 2002a. Emissions of N$_2$O and NO from fertilized fields: summary of available measurement data. Glob. Biogeochem. Cycles 16.

Bouwman, A.F., Boumans, L.J.M., Batjes, N.H., 2002b. Estimation of global NH$_3$ volatilization loss from synthetic fertilizers and animal manure applied to arable lands and grasslands. Glob. Biogeochem. Cycles. 16.

Bouwman, A.F., Van Drecht, G., Knoop, J.M., Beusen, A.H.W., Meinardi, C.R., 2005. Exploring changes in river nitrogen export to the world's oceans. Glob. Biogeochem. Cycles 19 (1).

Bouwman, A.F., Beusen, A.H.W., Griffioen, J., Van Groenigen, J.W., Hefting, M.M., Oenema, O., Van Puijenbroek, P.J.T.M., Seitzinger, S., Slomp, C.P., Stehfest, E., 2013. Global trends and uncertainties in terrestrial denitrification and N$_2$O emissions. Philos. Trans. R. Soc. Lond. Ser. B Biol. Sci. 368 (1621), 20130112.

Bowden, W.B., 1986. Gaseous nitrogen emissions from undisturbed terrestrial ecosystems: an assessment of their impacts on local and global nitrogen budgets. Biogeochemistry 2, 249–279.

Bowden, R.D., Nadelhoffer, K.J., Boone, R.D., Melillo, J.M., Garrison, J.B., 1993. Contributions of aboveground litter, belowground litter, and root respiration to total soil respiration in a temperate mixed hardwood forest. Can. J. For. Res. 23, 1402–1407.

Bowen, G.D., Smith, S.E., 1981. The effects of mycorrhizas on nitrogen uptake by plants. In: Clark, F.E., Rosswall, T. (Eds.), Terrestrial Nitrogen Cycles. Swedish Natural Science Research Council, Stockholm, pp. 237–247.

Bowen, G.J., Maibauer, B.J., Kraus, M.J., Röhl, U., Westerhold, T., Steimke, A., Gingerich, P.D., Wing, S.L., Clyde, W.C., 2015. Two massive, rapid releases of carbon during the onset of the Palaeocene–Eocene thermal maximum. Nat. Geosci. 8 (1), 44–47.

Bower, A.S., Susan Lozier, M., Gary, S.F., Böning, C.W., 2009. Interior pathways of the North Atlantic meridional overturning circulation. Nature 459 (7244), 243–247.

Bowles, M.W., Mogollón, J.M., Kasten, S., Zabel, M., Hinrichs, K.-U., 2014. Global rates of marine sulfate reduction and implications for sub-sea-floor metabolic activities. Science 344 (6186), 889–891.

Bowman, C.T., 1991. The chemistry of gaseous pollutant formation and destruction. In: Bartok, W., Sarofim, A.F. (Eds.), Fossil Fuel Combustion: A Source Book. Wiley, pp. 215–260.

Bowman, C.T., 1992. Control of combustion-generated nitrogen oxide emissions. Technology driven by regulation. In: Twenty-Fourth Symposium on Combustion. Symposium on Combustion. vol. 24, pp. 859–878.

Bowman, D.M.J.S., Balch, J.K., Artaxo, P., Bond, W.J., Carlson, J.M., Cochrane, M.A., D'Antonio, C.M., et al., 2009. Fire in the Earth system. Science 324 (5926), 481–484.

Bowring, S.A., Housh, T., 1995. The Earth's early evolution. Science 269 (5230), 1535–1540.

Box, E.O., 1988. Estimating the seasonal carbon source-sink geography of a natural, steady-state terrestrial biosphere. J. Appl. Meteorol. 27, 1109–1124.

Box, E.O., Holben, B.N., Kalb, V., 1989. Accuracy of the AVHRR vegetation index as a predictor of biomass, primary productivity and net CO_2 flux. Veg. Hist. Archaeobotany 80, 71–89.

Boyd, P.W., Ellwood, M.J., 2010. The biogeochemical cycle of iron in the ocean. Nat. Geosci. 3 (10), 675–682.

Boyd, P.W., Watson, A.J., Law, C.S., Abraham, E.R., Trull, T., Murdoch, R., Bakker, D.C., et al., 2000. A meso-scale phytoplankton bloom in the polar southern ocean stimulated by iron fertilization. Nature 407 (6805), 695–702.

Boyd, P.W., Jickells, T., Law, C.S., Blain, S., Boyle, E.A., Buesseler, K.O., Coale, K.H., et al., 2007. Mesoscale iron en-richment experiments 1993–2005: synthesis and future directions. Science 315 (5812), 612–617.

Boyd, E.S., Anbar, A.D., Miller, S., Hamilton, T.L., Lavin, M., Peters, J.W., 2011. A late methanogen origin for molybdenum-dependent nitrogenase. Geobiology 9 (3), 221–232.

Boyer, E.W., Goodale, C.L., Jaworsk, N.A., Howarth, R.W., 2002. Anthropogenic nitrogen sources and relationships to riverine nitrogen export in the Northeastern United States. Biogeochemistry 57, 137–169.

Boyer, T.P., Levitus, S., Antonov, J.I., Locarnini, R.A., Garcia, H.E., 2005. Linear trends in salinity for the world ocean, 1955–1998. Geophys. Res. Lett. 32.

Boyer, E.W., Howarth, R.W., Galloway, J.N., Dentener, F.J., Green, P.A., Vorosmarty, C.J., 2006. Riverine nitrogen export from the continents to the coasts. Glob. Biogeochem. Cycles 20.

Boyle, J.R., Voigt, G.K., 1973. Biological weathering of silicate minerals: implications for tree nutrition and soil genesis. Plant Soil 38, 191–201.

Boyle, E., Collier, R., Dengler, A.T., Edmond, J.M., Ng, A.C., Stallard, R.F., 1974. Chemical mass-balance in estuaries. Geochim. Cosmochim. Acta 38 (11), 1719–1728.

Boyle, E.A., Sclater, F., Edmond, J.M., 1976. Marine geochemistry of cadmium. Nature 263, 42–44.

Boyle, E.A., Edmond, J.M., Sholkovitz, E.R., 1977. Mechanism of iron removal in estuaries. Geochim. Cosmochim. Acta 41, 1313–1324.

Bradford, J.B., Lauenroth, W.K., Burke, I.C., 2005. The impact of cropping on primary production in the US great plains. Ecology 86, 1863–1872.

Bradley, R.S., Diaz, H.F., Eischeid, J.K., Jones, P.D., Kelly, P.M., Goodess, C.M., 1987. Precipitation fluctuations over northern hemisphere land areas since the mid-19th century. Science 237 (4811), 171–175.

Bradley, A.S., Leavitt, W.D., Schmidt, M., Knoll, A.H., Girguis, P.R., Johnston, D.T., 2016. Patterns of sulfur isotope fractionation during microbial sulfate reduction. Geobiology 14 (1), 91–101.

Bradshaw, C.J.A., Warkentin, I.G., 2015. Global estimates of boreal forest carbon stocks and flux. Glob. Planet. Chang. 128 (May), 24–30.

Bragazza, L., Freeman, C., Jones, T., Rydin, H., Limpens, J., Fenner, N., Ellis, T., et al., 2006. Atmospheric nitrogen deposition promotes carbon loss from peat bogs. Proc. Natl. Acad. Sci. U. S. A. 103 (51), 19386–19389.

Braissant, O., Cailleau, G., Aragno, M., Verrecchia, E.P., 2004. Biologically induced mineralization in the tree milicia excelsa (Moraceae): its causes and consequences to the environment. Geobiology 2, 59–66.

Bramley, R.G.V., White, R.E., 1990. The variability of nitrifying activity in field soils. Plant Soil 126, 203–208.

Brand, U., 1989. Biogeochemistry of late paleozoic North American brachiopods and secular variation of seawater composition. Biogeochemistry 7, 159–193.

Brand, L.E., 1991. Minimum iron requirements of marine phytoplankton and the implications for the biogeochemical control of new production. Limnol. Oceanogr. 36, 1756–1771.

Brand, L.E., Sunda, W.G., Guillard, R.R.L., 1983. Limitation of marine phytoplankton reproductive rates by zinc, man-ganese, and iron. Limnol. Oceanogr. 28, 1182–1198.

Brandes, J.A., Devol, A.H., 2002. A global marine-fixed nitrogen isotopic budget: implications for holocene nitrogen cycling. Glob. Biogeochem. Cycles 16 (4), 67.

Brant, A.N., Chen, H.Y.H., 2015. Patterns and mechanisms of nutrient resorption in plants. Crit. Rev. Plant Sci. 34 (5), 471–486.

Brasseur, G., Hitchman, M.H., 1988. Stratospheric response to trace gas perturbations: changes in ozone and temperature distributions. Science 240 (4852), 634–637.

Brauman, A., Kane, M.D., Labat, M., Breznak, J.A., 1992. Genesis of acetate and methane by gut bacteria of nutritionally diverse termites. Science 257 (5075), 1384–1387.

Braune, B.M., Gaston, A.J., Mallory, M.L., 2016. Temporal trends of mercury in eggs of five sympatrically breeding seabird species in the Canadian Arctic. Environ. Pollut. 214 (July), 124–131.

Bravard, S., Righi, D., 1989. Geochemical differences in an oxisol-spodosol toposequence of Amazonia, Brazil. Geoderma 44, 29–42.

Bravo, A., Cosio, C., 2019. Biotic formation of methylmercury: a bio-physico-chemical conundrum. Limnol. Oceanogr. 65, 1010–1027.

Breitburg, D., Levin, L.A., Oschlies, A., Grégoire, M., Chavez, F.P., Conley, D.J., Garçon, V., et al., 2018. Declining oxygen in the global ocean and coastal waters. Science 359 (6371).

Breña-Naranjo, J.A., Kendall, A.D., Hyndman, D.W., 2014. Improved methods for satellite-based groundwater storage estimates: a decade of monitoring the high plains aquifer from space and ground observations: improving groundwater storage estimates. Geophys. Res. Lett. 41 (17), 6167–6173.

Brenner, D.L., Amundson, R., Baisden, W.T., Kendall, C., Harden, J., 2001. Soil N and N-15 variation with time in a California annual grassland ecosystem. Geochim. Cosmochim. Acta 65, 4171–4186.

Brett, M.T., Goldman, C.R., 1996. A meta-analysis of the freshwater trophic cascade. Proc. Natl. Acad. Sci. U. S. A. 93 (15), 7723–7726.

Brewer, P.G., Goyet, C., Dyrssen, D., 1989. Carbon dioxide transport by ocean currents at 25° N latitude in the Atlantic Ocean. Science 246, 477–479.

Breznak, J.A., Brill, W.J., Mertins, J.W., Coppel, H.C., 1973. Nitrogen fixation in termites. Nature 244, 577–579.

Bridgham, S.D., Richardson, C.J., 1992. Mechanisms controlling soil respiration (CO_2 and CH_4) in southern peatlands. Soil Biol. Biochem. 24 (11), 1089–1099.

Bridgham, S.D., Megonigal, J.P., Keller, J.K., Bliss, N.B., Trettin, C., Megonigal, J.P., 2006. The carbon balance of North American wetlands. Wetlands 26 (4), 889.

Bridgham, S.D., Cadillo-Quiroz, H., Keller, J.K., Zhuang, Q., 2013. Methane emissions from wetlands: biogeochemical, microbial, and modeling perspectives from local to global scales. Glob. Chang. Biol. 19 (5), 1325–1346.

Brierley, E.D.R., Wood, M., 2001. Heterotrophic nitrification in an acid forest soil: isolation and characterisation of a nitrifying bacterium. Soil Biol. Biochem. 33, 1403–1409.

Briffa, K.R., Jones, P.D., Schweingruber, F.H., Osborn, T.J., 1998. Influence of volcanic eruptions on northern hemisphere summer temperature over the past 600 years. Nature 393, 450–455.

Briggs, N., Dall'Olmo, G., Claustre, H., 2020. Major role of particle fragmentation in regulating biological sequestration of CO_2 by the oceans. Science 367, 791–793.

Brimblecombe, P., 2003. The global sulfur cycle. In: Holland, H.D., Turekian, K.K. (Eds.), Treatise on Geochemistry. In: vol. 10. Pergamon, Oxford, pp. 645–682.

Brimblecombe, P., Dawson, G.A., 1984. Wet removal of highly soluble gases. J. Atmos. Chem. 2, 95–107.

Brimblecombe, P., Hammer, C., Rodhe, H., Ryaboshapko, A., Boutron, C.G., 1989. Human influence on the sulphur cycle. In: Brimblecombe, P., Lein, A.Y. (Eds.), Evolution of the Global Biogeochemical Sulphur Cycle. Wiley, pp. 77–121.

Brimhall, G.H., Chadwick, O.A., Lewis, C.J., Compston, W., Williams, I.S., Danti, K.J., Dietrich, W.E., Power, M.E., Hendricks, D., Bratt, J., 1991. Deformational mass transport and invasive process in soil evolution. Science 255, 695–702.

Brinkman, W.l., Santos, U.D.M., 1974. Emission of biogenic hydrogen sulfide from Amazonian floodplain lakes. Tellus 26, 261–267.

Brinson, M.M., 1993. Changes in the functioning of wetlands along environmental gradients. Wetlands 13 (2), 65–74.

Brinson, M.M., Lugo, A.E., Place, J., 1981. Primary productivity, decomposition and consumer activity in freshwater wetlands. Annu. Rev. Ecol. Syst. 12, 123–161.

Bristow, C.S., Hudson-Edwards, K.A., Chappell, A., 2010. Fertilizing the Amazon and equatorial Atlantic with West African dust. Geophys. Res. Lett. 37.

Bristow, L.A., Mohr, W., Ahmerkamp, S., Kuypers, M.M.M., 2017. Nutrients that limit growth in the ocean. Curr. Biol. 27 (11), R474–R478.

Brix, H., Sorrell, B.K., Orr, P.T., Orr, T., 1992. Internal pressurization and convective gas flow in some emergent freshwater macrophytes. Limnology 37 (7), 1420–1433.

Broda, E., 1975. The history of inorganic nitrogen in the biosphere. J. Mol. Evol. 7 (1), 87–100.

Broecker, W.S., 1973. Factors controlling CO_2 content in the oceans and atmosphere. In: Woodwell, G.M., Pecan, E.V. (Eds.), Carbon and the Biosphere. National Technical Information Service, Washington, DC, pp. 32–50.

Broecker, W.S., 1974. Chemical Oceanography. Hartcourt Brace Jovanovich.

Broecker, W.S., 1982. Ocean chemistry during glacial time. Geochim. Cosmochim. Acta 46, 1689–1705.

Broecker, W., 1997. Thermohaline circulation, the Achilles heel of our climate system: Will man-made CO_2 upset the current balance? Science 278 (5343), 1582–1588.

Broecker, W.S., Others, 1991. The great ocean conveyor. Oceanography 4 (2), 79–89.

Broecker, W.S., Peteet, D.M., Rind, D., 1985. Does the ocean–atmosphere system have more than one stable mode of operation? Nature 315 (6014), 21–26.

Bronk, D.A., Glibert, P.M., Ward, B.B., 1994. Nitrogen uptake, dissolved organic nitrogen release, and new production. Science 265 (5180), 1843–1846.

Bronk, D.A., Glibert, P.M., Malone, T.C., Banahan, S., Sahlsten, E., 1998. Inorganic and organic nitrogen cycling in chesapeake bay: autotrophic versus heterotrophic processes and relationships to carbon flux. Aquat. Microb. Ecol. Int. J. 15, 177–189.

Brook, G.A., Folkoff, M.E., Box, E.O., 1983. A world model of soil carbon dioxide. Earth Surf. Process. Landf. 8, 79–88.

Brookes, P.C., Powlson, D.S., Jenkinson, D.S., 1984. Phosphorus in the soil microbial biomass. Soil Biol. Biochem. 16, 169–175.

Brookes, P.C., Landman, A., Pruden, G., Jenkinson, D.S., 1985. Chloroform fumigation and the release of soil nitrogen: a rapid direct extraction method to measure microbial biomass nitrogen in soil. Soil Biol. Biochem. 17, 837–842.

Brooks, R.R., 1973. Biogeochemical parameters and their significance for mineral exploration. J. Appl. Ecol. 10, 825–836.

Brooks, B.W., Lazorchak, J.M., Howard, M.D.A., Johnson, M.-V.V., Morton, S.L., Perkins, D.A.K., Reavie, E.D., Scott, G.I., Smith, S.A., Steevens, J.A., 2016. Are harmful algal blooms becoming the greatest inland water quality threat to public health and aquatic ecosystems? Environ. Toxicol. Chem./SETAC 35 (1), 6–13.

Brookshire, E.N.J., Gerber, S., Menge, D.N.L., Hedin, L.O., 2012. Large losses of inorganic nitrogen from tropical rainforests suggest a lack of nitrogen limitation. Ecol. Lett. 15 (1), 9–16.

Brookshire, E.N.J., Gerber, S., Greene, W., Jones, R.T., Thomas, S.A., 2017. Global bounds on nitrogen gas emissions from humid tropical forests. Geophys. Res. Lett. 44 (5), 2502–2510.

Brosens, L., Campforts, B., Robinet, J., Vanacker, V., Opfergelt, S., Ameijeiras-Marino, Y., Minella, J.P.G., Covers, G., 2020. Slope gradient controls soil thickness and chemical weathering in subtropical Brazil: understanding rates and timescales of regional soilscape evolution through a combination of field data and modeling. J. Geophys. Res. Earth Surf. https://doi.org/10.1029/2019JF005321.

Brown, S., 1981. A comparison of the structure, primary productivity, and transpiration of cypress ecosystems in Florida. Ecol. Monogr. 51 (4), 403–427.

Brown, B.E., Fassbender, J.L., Winkler, R., 1992. Carbonate production and sediment transport in a marl lake of Southeastern Wisconsin. Limnol. Oceanogr. 37, 184–191.

Brown, A.E., Zhang, L., McMahon, T.A., Western, A.W., Vertessy, R.A., 2005. A review of paired catchment studies for determining changes in water yield resulting from alterations in vegetation. J. Hydrol. 310, 28–61.

Brown, R.H., Clark, R.N., Buratti, B.J., Cruikshank, D.P., Barnes, J.W., Mastrapa, R.M.E., Bauer, J., et al., 2006a. Composition and physical properties of Enceladus' surface. Science 311 (5766), 1425–1428.

Brown, S.S., Ryerson, T.B., Wollny, A.G., Brock, C.A., Peltier, R., Sullivan, A.P., Weber, R.J., et al., 2006b. Variability in nocturnal nitrogen oxide processing and its role in regional air quality. Science 311 (5757), 67–70.

Browning, T.J., Achterberg, E.P., Rapp, I., Engel, A., Bertrand, E.M., Tagliabue, A., Moore, C.M., 2017. Nutrient co-limitation at the boundary of an oceanic gyre. Nature 551 (7679), 242–246.

Browning, T.J., Achterberg, E.P., Yong, J.C., Rapp, I., Utermann, C., Engel, A., Moore, C.M., 2017a. Iron limitation of microbial phosphorus acquisition in the tropical North Atlantic. Nat. Commun. 8. https://doi.org/10.1038/ncomms15465.

Brownlee, D.E., 1992. The origin and evolution of the Earth. In: Butcher, S.S., Charlson, R.J., Orians, G.H., Wolfe, G.V. (Eds.), Global Biogeochemical Cycles. Academic Press, pp. 9–20.

Brown-Steiner, B., Hess, P., 2011. Asian influence on surface ozone in the United States: a comparison of chemistry, seasonality, and transport mechanisms. J. Geophys. Res. 116.

Bruland, K.W., 1989. Complexation of zinc by natural organic ligands in the central North Pacific. Limnol. Oceanogr. 34, 269–285.

Bruland, K.W., Knauer, G.A., Martin, J.H., 1978a. Cadmium in northeast pacific waters. Limnol. Oceanogr. 23, 618–625.

Bruland, K.W., Knauer, G.A., Martin, J.H., 1978b. Zinc in northeast pacific water. Nature 271, 741–743.

Bruland, K.W., Donat, J.R., Hutchins, D.A., 1991. Interactive influences of bioactive trace metals on biological production in oceanic waters. Limnol. Oceanogr. 36, 1555–1577.

Brunner, D., Staehelin, J., Jeker, D., 1998. Large-scale nitrogen oxide plumes in the tropopause region and implications for ozone. Science 282 (5392), 1305–1309.

Brutsaert, W., 2006. Indications of increasing land surface evaporation during the second half of the 20th century. Geophys. Res. Lett. 33.

Bryan, J.A., Berlyn, G.P., Gordon, J.C., 1996. Toward a new concept of the evolution of symbiotic nitrogen fixation in the Leguminosae. Plant Soil 186, 151–159.

Brylinski, M., Mann, K.H., 1973. Analysis of factors governing productivity in lakes and reservoirs. Limnol. Oceanogr. 18, 1–14.

Brzostek, E.R., Greco, A., Drake, J.E., Finzi, A.C., 2013. Root carbon inputs to the rhizosphere stimulate extracellular enzyme activity and increase nitrogen availability in temperate forest soils. Biogeochemistry 115 (1–3), 65–76.

Bu, H., Yuan, P., Liu, H., Dong, L., Qin, Z., Zhong, X., Song, H., Li, Y., 2019. Formation of macromolecules with peptide bonds via the thermal evolution of amino acids in the presence of montmorillonite: insight into prebiotic geochemistry on the early earth. Chem. Geol. 510 (April), 72–83.

Bubier, J.L., Moore, T.R., Bledzki, L.A., 2007. Effects of nutrient addition on vegetation and carbon cycling in an ombrotrophic bog. Glob. Chang. Biol. 13 (6), 1168–1186.

Bucher, M., 2007. Functional biology of plant phosphate uptake at root and mycorrhiza interfaces. New Phytol. 173 (1), 11–26.

Buchkowski, R.W., Schmitz, O.J., Bradford, M.A., 2015. Microbial stoichiometry overrides biomass as a regulator of soil carbon and nitrogen cycling. Ecology 96 (4), 1139–1149.

Bücking, H., Shachar-Hill, Y., 2005. Phosphate uptake, transport and transfer by the arbuscular mycorrhizal fungus glomus intraradices is stimulated by increased carbohydrate availability. New Phytol. 165 (3), 899–911.

Buendia, C., Kleidon, A., Porporato, A., 2010. The role of tectonic uplift, climate, and vegetation in the long term terrestrial phosphorous cycle. Biogeosciences 7, 2025–2038.

Buesseler, K.O., 2004. The effects of iron fertilization on carbon sequestration in the southern ocean. Science 304, 414–417.

Buesseler, K.O., Boyd, P.W., 2003. Will ocean fertilization work? Science 300 (5616), 67–68.

Buitenhuis, E.T., van der Wal, P., de Baar, H.J.W., 2001. Blooms of *Emiliania huxleyi* are sinks of atmospheric carbon dioxide: a field and mesocosm study derived simulation. Glob. Biogeochem. Cycles 15 (3), 577–587.

Buitenhuis, E.T., Suntharalingam, P., Le Quéré, C., 2018. Constraints on global oceanic emissions of N_2O from observations and models. Biogeosciences 15 (7), 2161–2175.

Bullen, T.D., Bailey, S.W., 2005. Identifying calcium sources at an acid deposition-impacted spruce forest: a strontium isotope, alkaline earth element multi-tracer approach. Biogeochemistry 74, 63–99.

Bulow, S.E., Rich, J.J., Naik, H.S., Pratihary, A.K., Ward, B.B., 2010. Denitrification exceeds anammox as a nitrogen loss pathway in the Arabian sea oxygen minimum zone. Deep-Sea Res. Part I, Oceanogr. Res. Pap. 57, 384–393.

Bultel, B., Viennet, J.-c., Poulet, F., Carter, J., Werner, S.C., 2019. Detection of carbonates in martian weathering profiles. J. Geophys. Res. Planets 124 (4), 989–1007.

Bundy, L.G., Bremner, J.M., 1973. Inhibition of nitrification in soils. Proc. Soil Sci. Soc. Am. 37, 396–398.

Bünemann, E.K., 2015. Assessment of gross and net mineralization rates of soil organic phosphorus—a review. Soil Biol. Biochem. 89 (October), 82–98.

Bunemann, E.K., Marschner, P., McNeill, A.M., McLaughlin, M.J., 2007. Measuring rates of gross and net mineralisation of organic phosphorus in soils. Soil Biol. Biochem. 39, 900–913.

Bünemann, E.K., Augstburger, S., Frossard, E., 2016. Dominance of either physicochemical or biological phosphorus cycling processes in temperate Forest soils of contrasting phosphate availability. Soil Biol. Biochem. 101 (October), 85–95.

Burbidge, E.M., Burbidge, G.E., Fowler, W.A., Hoyle, F., 1957. Synthesis of the elements in stars. Rev. Mod. Phys. 29, 547–650.

Burford, J.R., Bremner, J.M., 1975. Relationships between denitrification capacities of soils and total, water-soluble and readily decomposable soil organic matter. Soil Biol. Biochem. 7, 389–394.

Burgin, A.J., Groffman, P.M., 2012. Soil O_2 controls denitrification rates and N_2O yield in a riparian wetland. J. Geophys. Res. Biogeosci. 117 (G1).

Burgin, A.J., Hamilton, S.K., 2007. Have we overemphasized the role of denitrification in aquatic ecosystems? A review of nitrate removal pathways. Front. Ecol. Environ. 5 (2), 89–96.

Burgin, A.J., Hamilton, S.K., 2008. NO_3-driven SO_4^{2-} production in freshwater ecosystems: implications for N and S cycling. Ecosystems 11, 908–922.

Burgin, A.J., Groffman, P.M., Lewis, D.N., 2010. Factors regulating denitrification in a riparian wetland. Soil Sci. Soc. Am. J. 74, 1826–1833.

Burke, A., Robinson, L.F., 2012. The southern ocean's role in carbon exchange during the last deglaciation. Science 335 (6068), 557–561.

Burke, A., Present, T.M., Paris, G., Rae, E.C.M., Sandilands, B.H., Gaillardet, J., Peucker-Ehrenbrink, B., et al., 2018. Sulfur isotopes in rivers: insights into global weathering budgets, pyrite oxidation, and the modern sulfur cycle. Earth Planet. Sci. Lett. 496 (August), 168–177.

Burnard, P., Graham, D., Turner, G., 1997. Vesicle-specific Noble gas analyses of "popping Rock": implications for primordial noble gases in earth. Science 276, 568–571.

Burnett, W.C., Beers, M.J., Roe, K.K., 1982. Growth rates of phosphate nodules from the continental margin off Peru. Science 215 (4540), 1616–1618.

Burns, R.G., 1982. Enzyme activity in soil: location and a possible role in microbial ecology. Soil Biol. Biochem. 14, 423–427.

Burns, R.C., Hardy, R.W.F., 1975. Nitrogen Fixation in Bacteria and Higher Plants. Springer-Verlag, New York.

Burns, D.A., Klaus, J., McHale, M.R., 2007. Recent climate trends and implications for water resources in the Catskill Mountain Region, New York. J. Hydrol. 336, 155–170.

Burns, D.A., Riva-Murray, K., Bradley, P.M., Aiken, G.R., Brigham, M.E., 2012. Landscape controls on total and methyl Hg in the Upper Hudson River Basin, New York. J. Geophys. Res. Biogeosci. 117.

Burrows, A., 2000. Supernova explosions in the universe. Nature 403 (6771), 727–733.

Burton, J.D., 1988. Riverborne materials and the continent-ocean interface. In: Lerman, A., Meybeck, M. (Eds.), Physical and Chemical Weathering in Geochemical Cycles. Kluwer Academic Publishers, pp. 299–321.

Burton, D.L., Beauchamp, E.C., 1984. Field techniques using the acetylene blockage of nitrous oxide reduction to measure denitrification. Can. J. Soil Sci. 64, 555–562.

Burwicz, E.B., Ruepke, L.H., Wallmann, K., 2011. Estimation of the global amount of submarine gas hydrates formed via microbial methane formation based on numerical reaction-transport modeling and a novel parameterization of holocene sedimentation. Geochim. Cosmochim. Acta 75, 4562–4576.

Busemann, H., Young, A.F., O'd Alexander, C.M., Hoppe, P., Mukhopadhyay, S., Nittler, L.R., 2006. Interstellar chemistry recorded in organic matter from primitive meteorites. Science 312 (5774), 727–730.

Busigny, V., Cartigny, P., Philippot, P., Ader, M., Javoy, M., 2003. Massive recycling of nitrogen and other fluid-mobile elements (K, Rb, Cs, H) in a cold slab environment: evidence from HP to UHP oceanic metasediments of the schistes lustres nappe (Western Alps, Europe). Earth Planet. Sci. Lett. 215, 27–42.

Busigny, V., Cartigny, P., Philippot, P., 2011. Nitrogen isotopes in ophiolitic metagabbros: a re-evaluation of modern nitrogen fluxes in subduction zones and implication for the early earth atmosphere. Geochim. Cosmochim. Acta 75, 7502–7521.

Busigny, V., Lebeau, O., Ader, M., Krapež, B., Bekker, A., 2013. Nitrogen cycle in the late *Archean ferruginous* ocean. Chem. Geol. 362 (December), 115–130.

Busker, T., de Roo, A., Gelati, E., Schwatke, C., Adamovic, M., Bisselink, B., Pekel, J.-F., Cottam, A., 2019. A global lake and reservoir volume analysis using a surface water dataset and satellite altimetry. Hydrol. Earth Syst. Sci. 23 (2), 669–690.

Butchart, S.H.M., Walpole, M., Collen, B., van Strien, A., Scharlemann, J.P.W., Almond, R.E.A., Baillie, J.E.M., et al., 2010. Global biodiversity: indicators of recent declines. Science 328 (5982), 1164–1168.

Butler, A., 1998. Acquisition and utilization of transition metal ions by marine organisms. Science 281 (5374), 207–210.

Butler, J.H., Battle, M., Bender, M.L., Montzka, S.A., Clarke, A.D., Saltzman, E.S., Sucher, C.M., Severinghaus, J.P., Elkins, J.W., 1999. A record of atmospheric halocarbons during the twentieth century from polar firn air. Nature 399, 749–755.

Butler, T.J., Likens, G.E., Vermeylen, F.M., Stunder, B.J.B., 2003. The relation between NOx emissions and precipitation NO_3^- in the eastern USA. Atmos. Environ. 37, 2093–2104.

Butler, T.J., Likens, G.E., Vermeylen, F.M., Stunder, B.J.B., 2005. The impact of changing nitrogen oxide emissions on wet and dry nitrogen deposition in the northeastern USA. Atmos. Environ. 39, 4851–4862.

Butler, T.J., Cohen, M.D., Vermeylen, F.M., Likens, G.E., Schmeltz, D., Artz, R.S., 2008. Regional precipitation mercury trends in the eastern USA, 1998–2005: declines in the northeast and midwest, no trend in the southeast. Atmos. Environ. 42, 1582–1592.

Butler, T.J., Vermeylen, F.M., Rury, M., Likens, G.E., Lee, B., Bowker, G.E., McCluney, L., 2011. Response of ozone and nitrate to stationary source NOx emission reductions in the eastern USA. Atmos. Environ. 45 (5), 1084–1094.

Butler, O.M., Elser, J.J., Lewis, T., Mackey, B., Chen, C., 2018. The phosphorus-rich signature of fire in the soil-plant system: a global meta-analysis. Ecol. Lett. 21 (3), 335–344.

Butterbach-Bahl, K., Willibald, G., Papen, H., 2002. Soil core method for direct simultaneous determination of N_2 and N_2O emissions from forest soils. Plant Soil 240, 105–116.

Buyanovsky, G.A., Wagner, G.H., 1983. Annual cycles of carbon dioxide levels in soil air. Soil Sci. Soc. Am. J. 47, 1139–1145.

Buyanovsky, G.A., Aslam, M., Wagner, G.H., 1994. Carbon turnover in soil physical fractions. Soil Sci. Soc. Am. J. 58, 1167–1173.

Byrne, R.H., Mecking, S., Feely, R.A., Liu, X.W., 2010. Direct observations of basin-wide acidification of the North Pacific Ocean. Geophys. Res. Lett. 37.

Cabana, G., Rasmussen, J.B., 1994. Modeling food chain structure and contaminant bioaccumulation using stable nitrogen isotopes. Nature 372, 255–257.

Caccavo, F., Blakemore, R.P., Lovley, D.R., 1992. A hydrogen-oxidizing, Fe(III)-reducing microorganism from the Great Bay Estuary, New Hampshire. Appl. Environ. Microbiol. 58 (10), 3211–3216.

Cachier, H., Ducret, J., 1991. Influence of biomass burning on equatorial African rains. Nature 352, 228–230.

Cachier, H., Bremond, M.P., Buatmenard, P., 1989. Carbonaceous aerosols from different tropical biomass burning sources. Nature 340, 371–373.

Cadoux, A., Scaillet, B., Bekki, S., Oppenheimer, C., Druitt, T.H., 2015. Stratospheric ozone destruction by the bronze-age Minoan eruption (Santorini Volcano, Greece). Sci. Rep. 5 (1), 12243.

Cahoon, D.R., Stocks, B.J., Levine, J.S., Cofer, W.R., Oneill, K.P., 1992. Seasonal distribution of African savanna fires. Nature 359, 812–815.

Cai, T., Flanagan, L.B., Syed, K.H., 2010. Warmer and drier conditions stimulate respiration more than photosynthesis in a boreal peatland ecosystem: analysis of automatic chambers and eddy covariance measurements. Plant Cell Environ. 33 (3), 394–407.

Cailleau, G., Braissant, O., Verrecchia, E.P., 2004. Biomineralization in plants as a long-term carbon sink. Die Naturwissenschaften 91 (4), 191–194.

Cairns, M.A., Brown, S., Helmer, E.H., Baumgardner, G.A., 1997. Root biomass allocation in the world's upland forests. Oecologia 111 (1), 1–11.

Cairns-Smith, A.G., 1985. The first organisms. Sci. Am. 252 (6), 90–100.

Caldeira, K., Kasting, J.F., 1992. The life span of the biosphere revisited. Nature 360 (6406), 721–723.

Caldwell, M.M., Flint, S.D., 1994. Stratospheric ozone reduction, solar UvB radiation and terrestrial ecosystems. Clim. Chang. 28, 375–394.

Callaway, R.M., Delucia, E.H., Schlesinger, W.H., 1994. Biomass allocation of montane and desert ponderosa pine: an analog for response to climate change. Ecology 75, 1474–1481.

Cama, J., Ganor, J., 2006. The effects of organic acids on the dissolution of silicate minerals: a case study of oxalate catalysis of kaolinite dissolution. Geochim. Cosmochim. Acta 70, 2191–2209.

Cambardella, C.A., Elliott, E.T., 1994. Carbon and nitrogen dynamics of soil organic matter fractions from cultivated grassland soils. Soil Sci. Soc. Am. J. 58, 123–130.

Came, R.E., Eiler, J.M., Veizer, J., Azmy, K., Brand, U., Weidman, C.R., 2007. Coupling of surface temperatures and atmospheric CO_2 concentrations during the Palaeozoic Era. Nature 449 (7159), 198–201.

Cameron, E.M., 1982. Sulfate and sulfate reduction in early Precambrian oceans. Nature 296, 145–148.

Campbell, C.A., Paul, E.A., Rennie, D.A., McCallum, K.J., 1967. Factors affecting the accuracy of the carbon-dating method in soil humus studies. Soil Sci. 104, 81–85.

Campbell, J.E., Carmichael, G.R., Chai, T., Mena-Carrasco, M., Tang, Y., Blake, D.R., Blake, N.J., et al., 2008. Photosynthetic control of atmospheric carbonyl sulfide during the growing season. Science 322 (5904), 1085–1088.

Campbell, J.E., Whelan, M.E., Seibt, U., Smith, S.J., Berry, J.A., Hilton, T.W., 2015. Atmospheric carbonyl sulfide sources from anthropogenic activity: implications for carbon cycle constraints: atmospheric OCS sources. Geophys. Res. Lett. 42 (8), 3004–3010.

Campbell, J.E., Berry, J.A., Seibt, U., Smith, S.J., Montzka, S.A., Launois, T., Belviso, S., Bopp, L., Laine, M., 2017. Large historical growth in global terrestrial gross primary production. Nature 544 (7648), 84–87.

Canadell, J.G., Le Quéré, C., Raupach, M.R., Field, C.B., Buitenhuis, E.T., Ciais, P., Conway, T.J., Gillett, N.P., Houghton, R.A., Marland, G., 2007. Contributions to accelerating atmospheric CO_2 growth from economic activity, carbon intensity, and efficiency of natural sinks. Proc. Natl. Acad. Sci. U. S. A. 104 (47), 18866–18870.

Canfield, D.E., 1989a. Reactive iron in maine sediments. Geochim. Cosmochim. Acta 53, 619–632.

Canfield, D.E., 1989b. Sulfate reduction and oxic respiration in marine sediments: implications for organic carbon preservation in euxinic environments. Deep-Sea Res. Part A, Oceanogr. Res. Pap. 36 (1), 121–138.

Canfield, D.E., 1991. Sulfate reduction in deep-sea sediments. Am. J. Sci. 291 (2), 177–188.

Canfield, D.E., Mackenzie, F.T., Wollast, R., Chou, L., 1993. Organic matter oxidation in marine sediments. In: Interactions of C, N, P and S in Biogeochemical Cycles and Global Change. Springer, pp. 333–363.

Canfield, D.E., 1994. Factors influencing organic carbon preservation in marine sediments. Chem. Geol. 114, 315–329.

Canfield, D.E., 1997. The geochemistry of river particulates from the continental USA: major elements. Geochim. Cosmochim. Acta 61 (16), 3349–3365.

Canfield, D.E., 1998. A new model for Proterozoic Ocean chemistry. Nature 396, 450–453.

Canfield, D.E., 2014. Oxygen: A Four Billion Year History. Princeton University Press.

Canfield, D.E., Teske, A., 1996. Late Proterozoic rise in atmospheric oxygen concentration inferred from phylogenetic and sulphur-isotope studies. Nature 382 (6587), 127–132.

Canfield, D.E., Thamdrup, B., 1994. The production of 34S-depleted sulfide during bacterial disproportionation of elemental sulfur. Science 266 (December), 1973–1975.

Canfield, D.E., Green, W.J., Gardner, T.J., Ferdelman, T., 1984. Elemental residence times in Acton Lake, Ohio. Arch. Hydrobiol. 100, 501–519.

Canfield, D.E., Habicht, K.S., Thamdrup, B., 2000. The Archean sulfur cycle and the early history of atmospheric oxygen. Science 288 (5466), 658–661.

Canfield, D.E., Rosing, M.T., Bjerrum, C., 2006. Early anaerobic metabolisms. Philos. Trans. R. Soc. Lond. Series B Biol. Sci. 361 (1474), 1819–1834 (discussion 1835–1836).

Cannon, H.L., 1960. Botanical prospecting for ore deposits. Science 132 (3427), 591–598.

Cao, M.K., Marshall, S., Gregson, K., 1996. Global carbon exchange and methane emissions from natural wetlands: application of a process-based model. J. Geophys. Res. 101, 14399–14414.

Cao, Q., Clark, E.A., Mao, Y., Lettenmaier, D.P., 2019. Trends and interannual variability in terrestrial water storage over the eastern United States, 2003–2016. Water Resour. Res. 55 (3), 1928–1950.

Capo, R.C., Chadwick, O.A., 1999. Sources of strontium and calcium in desert soil and calcrete. Earth Planet. Sci. Lett. 170, 61–72.

Capone, D.G., Horrigan, S.G., Dunham, S.E., Fowler, J., 1990. Direct determination of nitrification in marine waters by using the short-lived radioisotope of nitrogen, 13N. Appl. Environ. Microbiol. 56, 1182–1184.

Caraco, N.F., Cole, J.J., Likens, G.E., 1989. Evidence for sulfate-controlled phosphorus release from sediments of aquatic ecosystems. Nature 341, 316–318.

Caraco, N., Cole, J., Likens, G., 1990. A comparison of phosphorus immobilization in sediments of freshwater and coastal marine systems. Biogeochemistry 9 (3), 277–290.

Carey, J.C., Fulweiler, R.W., 2012. Human activities directly alter watershed dissolved silica fluxes. Biogeochemistry 111 (1/3), 125–138.

Carey, J.C., Tang, J., Templer, P.H., Kroeger, K.D., Crowther, T.W., Burton, A.J., Dukes, J.S., et al., 2016. Temperature response of soil respiration largely unaltered with experimental warming. Proc. Natl. Acad. Sci. U. S. A. 113 (48), 13797–13802.

Carignan, R., Lean, D.R.S., 1991. Regeneration of dissolved substances in a seasonally anoxic lake—the relative importance of processes occurring in the water column and in the sediments. Limnol. Oceanogr. 36 (4), 683–707.

Carlisle, A., Brown, A.H.F., White, E.J., 1966. The organic matter and nutrient elements in the precipitation beneath a sessile oak (Quercus Patraea) canopy. J. Ecol. 54, 87–98.

Carlson, C.A., Ducklow, H.W., Michaels, A.F., 1994. Annual flux of dissolved organic carbon from the euphotic zone in the Northwestern Sargasso Sea. Nature 371, 405–408.

Carlyle, J.C., Malcolm, D.C., 1986. Larch litter and nitrogen availability in mixed larch-spruce stands. 1. Nutrient withdrawal, redistribution, and leaching loss from larch foliage at senescence. Can. J. For. Res. 16, 321–326.

Carmichael, M.J., Bernhardt, E.S., Bräuer, S.L., Smith, W.K., 2014. The role of vegetation in methane flux to the atmosphere: should vegetation be included as a distinct category in the global methane budget? Biogeochemistry 119 (1–3).

Carn, S.A., Fioletov, V.E., McLinden, C.A., Li, C., Krotkov, N.A., 2017. A decade of global volcanic SO$_2$ emissions measured from space. Sci. Rep. 7 (March), 44095.

Caro, G., Papanastassiou, D.A., Wasserburg, G.J., 2010. ^{40}K/^{40}Ca isotopic constraints on the oceanic calcium cycle. Earth Planet. Sci. Lett. 296, 124–132.

Carpenter, E.J., Capone, D.G., 2008. (Chapter 4). Nitrogen fixation in the marine environment. In: Capone, D.G., Bronk, D.A., Mulholland, M.R., Carpenter, E.J. (Eds.), Nitrogen in the Marine Environment, second ed. Academic Press, San Diego, pp. 141–198.

Carpenter, S.R., Kitchell, J.F., Hodgson, J.R., 1985. Cascading trophic interactions and lake productivity. Bioscience 35, 634–639.

Carpenter, S.R., Fisher, S.G., Grimm, N.B., Kitchell, J.F., 1992. Global change and freshwater ecosystems. Annu. Rev. Ecol. Syst. 23, 119–139.

Carpenter, S.R., Cole, J.J., Kitchell, J.F., 1998. Impact of dissolved organic carbon, phosphorus, and grazing on phytoplankton biomass and production in experimental lakes. Limnol. Ocanogr. 43, 73–80.

Carpenter, S.R., Cole, J.J., Hodgson, J.R., Kitchell, J.F., Pace, M.L., Bade, D., Cottingham, K.L., Essington, T.E., Houser, J.N., Schindler, D.E., 2001. Trophic cascades, nutrients, and lake productivity: whole-lake experiments. Ecol. Monogr. 71, 163–186.

Carpenter, S.R., Cole, J.J., Pace, M.L., Batt, R., Brock, W.A., Cline, T., Coloso, J., et al., 2011. Early warnings of regime shifts: a whole-ecosystem experiment. Science 332 (6033), 1079–1082.

Carr, M.H., 1987. Water on mars. Nature 326, 30–35.

Carroll, M.A., Heidt, L.E., Cicerone, R.J., Prinn, R.G., 1986. OCS, H$_2$S, and CS$_2$ fluxes from a saltwater marsh. J. Atmos. Chem. 4, 375–395.

Carter, J.P., Hsiao, Y.H., Spiro, S., Richardson, D.J., 1995. Soil and sediment bacteria capable of aerobic nitrate respiration. Appl. Environ. Microbiol. 61 (8), 2852–2858.

Carter, D.O., Yellowlees, D., Tibbett, M., 2007. Cadaver decomposition in terrestrial ecosystems. Die Naturwissenschaften 94 (1), 12–24.

Carter, B.R., Feely, R.A., Wanninkhof, R., Kouketsu, S., Sonnerup, R.E., Pardo, P.C., Sabine, C.L., et al., 2019. Pacific anthropogenic carbon between 1991 and 2017. Glob. Biogeochem. Cycles 8 (May), 65.

Carvalhais, N., Forkel, M., Khomik, M., Bellarby, J., Jung, M., Migliavacca, M., Mingquan, M., et al., 2014. Global covariation of carbon turnover times with climate in terrestrial ecosystems. Nature 514 (7521), 213–217.

Casagrande, D., Siefert, K., 1977. Origins of sulfur in coal—importance of ester sulfate content of peat. Science 195 (4279), 675–676.

Casagrande, D.J., Idowu, G., Friedman, A., Rickert, P., Siefert, K., Schlenz, D., 1979. H$_2$S incorporation in coal precursors—origins of organic sulfur in coal. Nature 282 (5739), 599–600.

Casey, W.H., Westrich, H.R., Banfield, J.F., Ferruzzi, G., Arnold, G.W., 1993. Leaching and reconstruction at the surfaces of dissolving chain-silicate minerals. Nature 366, 253–256.

Castebrunet, H., Genthon, C., Martinerie, P., 2006. Sulfur cycle at last glacial maximum: model results versus Antarctic ice core data. Geophys. Res. Lett. 33.

Castelle, A.J., Galloway, J.N., 1990. Carbon dioxide dynamics in acid forest soils in Shenandoah National Park, Virginia. Soil Sci. Soc. Am. J. 54, 252–257.

Castresana, J., Saraste, M., 1995. Evolution of energetic metabolism: the respiration-early hypothesis. Trends Biochem. Sci. 20 (11), 443–448.

Castrillo, G., Teixeira, P.J.P.L., Paredes, S.H., Law, T.F., de Lorenzo, L., Feltcher, M.E., Finkel, O.M., et al., 2017. Root microbiota drive direct integration of phosphate stress and immunity. Nature 543 (7646), 513–518.

Castro, M.S., Dierberg, F.E.E., 1987. Biogenic hydrogen sulfide emissions from selected Florida wetlands. Water Air Soil Pollut. 33 (1–2), 1–13.

Castro, M.S., Melillo, J.M., Steudler, P.A., Chapman, J.W., 1994. Soil moisture as a predictor of methane uptake by temperate forest soils. Can. J. For. Res. 24, 1805–1810.

Castro, M.S., Steudler, P.A., Melillo, J.M., Aber, J.D., Bowden, R.D., 1995. Factors controlling atmospheric methane consumption by temperate forest soils. Glob. Biogeochem. Cycles 9, 1–10.

Cates, R.G., 1975. Interface between slugs and wild ginger: some evolutionary aspects. Ecology 56, 391–400.

Catling, D.C., Zahnle, K.J., McKay, C., 2001. Biogenic methane, hydrogen escape, and the irreversible oxidation of early earth. Science 293 (5531), 839–843.

Cavalieri, D.J., Parkinson, C.L., 2008. Antarctic sea ice variability and trends. J. Geophys. Res. 113, 1979–2006.

Cavanaugh, C.M., Gardiner, S.L., Jones, M.L., Jannasch, H.W., Waterbury, J.B., 1981. Prokaryotic cells in the hydrothermal vent tube worm *Riftia pachyptila* Jones: possible chemoautotrophic symbionts. Science 213 (4505), 340–342.

Cavigelli, M.A., Robertson, G.P., 2000. The functional significance of denitrifier community composition in a terrestrial ecosystem. Ecology 81, 1402–1414.

Cayrel, R., Hill, V., Beers, T.C., Barbuy, B., Spite, M., Spite, F., Plez, B., et al., 2001. Measurement of stellar age from uranium decay. Nature 409 (6821), 691–692.

Cazenave, A., Llovel, W., 2010. Contemporary sea level rise. Annu. Rev. Mar. Sci. 2, 145–173.

Cebrián, J., Duarte, C.M., 1995. Plant growth-rate dependence of detrital carbon storage in ecosystems. Science 268 (5217), 1606–1608.

Cebrian, J., Lartigue, J., 2004. Patterns of herbivory and decomposition in aquatic and terrestrial ecosystems. Ecol. Monogr. 74, 237–259.

Cernusak, L.A., Ubierna, N., Winter, K., Holtum, J.A.M., Marshall, J.D., Farquhar, G.D., 2013. Environmental and physiological determinants of carbon isotope discrimination in terrestrial plants. New Phytol. 200 (4), 950–965.

Ceulemans, R., Mousseau, M., 1994. Effects of elevated atmospheric CO_2 on woody plants. New Phytol. 127, 425–446.

Chacon, N., Silver, W.L., Dubinsky, E.A., Cusack, D.F., 2006. Iron reduction and soil phosphorus solubilization in humid tropical forests soils: the roles of labile carbon pools and an electron shuttle compound. Biogeochemistry 78, 67–84.

Chadwick, K.D., Asner, G.P., 2018. Landscape evolution and nutrient rejuvenation reflected in Amazon forest canopy chemistry. Ecol. Lett. 21 (7), 978–988.

Chadwick, O.A., Kelly, E.F., Merritts, D.M., Amundson, R.G., 1994. Carbon dioxide consumption during soil development. Biogeochemistry 24, 115–127.

Chadwick, O.A., Derry, L.A., Vitousek, P.M., Huebert, B.J., Hedin, L.O., 1999. Changing sources of nutrients during four million years of ecosystem development. Nature 397, 491–497.

Chahine, M.T., 1995. Observation of local cloud and moisture feedbacks over high ocean and desert surface temperatures. J. Geophys. Res. D: Atmos. 100, 8919–8927.

Chalk, P.M., He, J.-Z., Peoples, M.B., Chen, D., 2017. 15N2 as a tracer of biological N_2 fixation: a 75-year retrospective. Soil Biol. Biochem. 106 (March), 36–50.

Chambers, L.A., Trudinger, P.A., 1979. Thiosulfate formation and associated isotope effects during sulfite reduction by *Clostridium pasteurianum*. Can. J. Microbiol. 25 (6), 719–721.

Chameides, W.L., Davis, D.D., 1982. Chemistry in the troposphere. Chem. Eng. News 60, 38–52.

Chameides, W.L., Lindsay, R.W., Richardson, J., Kiang, C.S., 1988. The role of biogenic hydrocarbons in urban photochemical smog: Atlanta as a case study. Science 241 (4872), 1473–1475.

Chameides, W.L., Fehsenfeld, F., Rodgers, M.O., Cardelino, C., Martinez, J., Parrish, D., Lonneman, W., et al., 1992. Ozone precursor relationships in the ambient atmosphere. J. Geophys. Res. 97, 6037–6055.

Chameides, W.L., Kasibhatla, P.S., Yienger, J., Levy 2nd., H., 1994. Growth of continental-scale metro-agro-plexes, regional ozone pollution, and world food production. Science 264 (5155), 74–77.

Chan, M.K., Kim, J.S., Rees, D.C., 1993. The nitrogenase FeMo-cofactor and P-cluster pair: 2.2-angstrom resolution structures. Science 260, 792–794.

Chandler, A.S., Choularton, T.W., Dollard, G.J., Eggleton, A.E.J., Gay, M.J., Hill, T.A., Jones, B.M.R., Tyler, B.J., Bandy, B.J., Penkett, S.A., 1988. Measurements of H_2O_2 and SO_2 in clouds and estimates of their reaction rate. Nature 336, 562–565.

Chang, W.Y., Moll, R.A., 1980. Prediction of hypolimnetic oxygen deficits: problems of interpretation. Science 209 (4457), 721–722.

Chang, S., Des Marais, D., Mack, R., Miller, S.L., Strathearn, G.E., 1983. Prebiotic organic synthesis and the origin of life. In: Schopf, J.W. (Ed.), Earth's Earliest Biosphere. Princeton University Press, pp. 53–92.

Chang, B.X., Devol, A.H., Emerson, S.R., 2010. Denitrification and the nitrogen gas excess in the eastern tropical South Pacific oxygen deficient zone. Deep-Sea Res. Part I—Oceanogr. Res. Pap. 57, 1092–1101.

Changnon, S.A., Demissie, M., 1996. Detection of changes in streamflow and floods resulting from climate fluctuations and land use-drainage changes. Clim. Chang. 32, 411–421.

Chanton, J.P., Dacey, J.W.H., 1991. Effects of vegetation on methane flux reservoirs and carbon isotopic composition. In: Sharkey, T.D., Holland, E.A., Mooney, H.A. (Eds.), Trace Gas Emissions by Plants. Academic Press, San Diego, pp. 65–92.

Chao, B.F., Wu, Y.H., Li, Y.S., 2008. Impact of artificial reservoir water impoundment on global sea level. Science 320 (5873), 212–214.

Chapelle, F.H., O'Neill, K., Bradley, P.M., Methé, B.A., Ciufo, S.A., Knobel, L.L., Lovley, D.R., 2002. A hydrogen-based subsurface microbial community dominated by methanogens. Nature 415 (6869), 312–315.

Chapin, F.S., 1974. Morphological and physiological mechanisms of temperature compensation in phosphate absorption along a latitudinal gradient. Ecology 55, 1180–1198.

Chapin, F.S., 1980. The mineral nutrition of wild plants. Annu. Rev. Ecol. Syst. 11, 233–260.

Chapin, F.S., 1988. Ecological aspects of plant mineral nutrition. Adv. Mineral Nutr. 3, 161–191.

Chapin, F.S., Moilanen, L., 1991. Nutritional controls over nitrogen and phosphorus resorption from Alaskan Birch leaves. Ecology 72, 709–715.

Chapin, F.S., Oechel, W.C., 1983. Photosynthesis, respiration, and phosphate absorption by *Carex aquatilis* ecotypes along latitudinal and local environmental gradients. Ecology 64, 743–751.

Chapin, F.S., Barsdate, R.J., Barel, D., 1978. Phosphorus cycling in Alaskan Coastal Tundra: a hypothesis for the regulation of nutrient cycling. Oikos 31, 189–199.

Chapin, F.S., Shaver, G.R., Kedrowski, R.A., 1986a. Environmental controls over carbon, nitrogen and phosphorus fractions in *Eriophorum vaginatum* in Alaskan tussock tundra. J. Ecol. 74, 167–195.

Chapin, F.S., Vitousek, P.M., Van Cleve, K., 1986b. The nature of nutrient limitation in plant communities. Am. Nat. 127, 48–58.

Chapin, D.M., Bliss, L.C., Bledsoe, L.J., 1991. Environmental regulation of nitrogen fixation in a high arctic lowland ecosystem. Can. J. Bot. 69, 2744–2755.

Chapin, F.S., Moilanen, L., Kielland, K., 1993. Preferential use of organic nitrogen for growth by a non-mycorrhizal Arctic sedge. Nature 361 (6408), 150–153.

Chapin, F.S., Reynolds, H.L., D'Antonio, C.M., Eckhart, V.M., 1996. The functional role of species in terrestrial ecosystems. In: Walker, B., Steffen, W. (Eds.), Global Change and Terrestrial Ecosystems. Cambridge University Press, pp. 403–428.

Chapin, F.S., Woodwell, G.M., Randerson, J.T., Rastetter, E.B., Lovett, G.M., Baldocchi, D.D., Clark, D.A., et al., 2006. Reconciling carbon-cycle concepts, terminology, and methods. Ecosystems 9, 1041–1050.

Chapman, D.J., Schopf, J.W., 1983. Biological and biochemical effects of the development of an aerobic environment. In: Schopf, J.W. (Ed.), Earth's Earliest Biosphere. Princeton University Press, pp. 302–320.

Chappellaz, J., Barnola, J.M., Raynaud, D., Korotkevich, Y.S., Lorius, C., 1990. Ice-core record of atmospheric methane over the past 160,000 years. Nature 345, 127–131.

Chappellaz, J., Blunier, T., Raynaud, D., Barnola, J.M., Schwander, J., Stauffer, B., 1993. Synchronous changes in atmospheric CH_4 and greenland climate between 40-Kyr and 8-Kyr BP. Nature 366, 443–445.

Charette, M.A., Smith, W.H.F., 2010. The volume of Earth's ocean. Oceanography 23 (2), 112–114.

Charlou, J.L., Donval, J.P., Fouquet, Y., Jean-Baptiste, P., Holm, N., 2002. Geochemistry of high H_2 and CH_4 vent fluids issuing from ultramafic rocks at the rainbow hydrothermal field (36°14′N, MAR). Chem. Geol. 191 (4), 345–359.

Charlson, R.J., Lovelock, J.E., Andreae, M.O., Warren, S.G., 1987. Oceanic phytoplankton, atmospheric sulphur, cloud albedo and climate. Nature 326 (6114), 655–661.

Charmley, E., Williams, S.R.O., Moate, P.J., Hegarty, R.S., Herd, R.M., Oddy, V.H., Reyenga, P., Staunton, K.M., Anderson, A., Hannah, M.C., 2016. A universal equation to predict methane production of forage-fed cattle in Australia. Anim. Prod. Sci. 56 (3), 169–180.

Chase, E.M., Sayles, F.L., 1980. Phosphorus in suspended sediments of the Amazon river. Estuar. Coast. Mar. Sci. 11, 383–391.

Chavez, F.P., Strutton, P.G., Friederich, G.E., Feely, R.A., Feldman, G.C., Foley, D.G., McPhaden, M.J., 1999. Biological and chemical response of the equatorial Pacific Ocean to the 1997-98 El Niño. Science 286 (5447), 2126–2131.

Che, J., Yamaji, N., Ma, J.F., 2018. Efficient and flexible uptake system for mineral elements in plants. New Phytol. 219 (2), 513–517.

Chen, C.T.A., Borges, A.V., 2009. Reconciling opposing views on carbon cycling in the coastal ocean: continental shelves as sinks and near-shore ecosystems as sources of atmospheric CO_2. Deep-Sea Res. Part II, Top. Stud. Oceanogr. 56, 578–590.

Chen, Y.H., Prinn, R.G., 2006. Estimation of atmospheric methane emissions between 1996 and 2001 using a three-dimensional global chemical transport model. J. Geophys. Res. D: Atmos. 111.

Chen, C.Y., Stemberger, R.S., Kamman, N.C., Mayes, B.M., Folt, C.L., 2005. Patterns of Hg bioaccumulation and transfer in aquatic food webs across multi-lake studies in the Northeast US. Ecotoxicology 14 (1–2), 135–147.

Chen, J.M., Chen, B.Z., Higuchi, K., Liu, J., Chan, D., Worthy, D., Tans, P., Black, A., 2006. Boreal ecosystems sequestered more carbon in warmer years. Geophys. Res. Lett. 33.

Chen, J.L., Wilson, C.R., Tapley, B.D., 2013. Contribution of ice sheet and mountain glacier melt to recent sea level rise. Nat. Geosci. 6 (7), 549–552.

Chen, J., Hintelmann, H., Wang, Z., Feng, X., Cai, H., Wang, Z., Yuan, S., Wang, Z., 2016a. Isotopic evidence for distinct sources of mercury in lake waters and sediments. Chem. Geol. 426 (May), 33–44.

Chen, Y., Sayer, E.J., Li, Z., Mo, Q., Li, Y., Ding, Y., Wang, J., Lu, X., Tang, J., Wang, F., 2016b. Nutrient limitation of woody debris decomposition in a tropical forest: contrasting effects of N and P addition. Larjavaara, M. (Ed.), Funct. Ecol. 30 (2), 295–304.

Chen, S., Jiang, N., Huang, J., Xu, X., Zhang, H., Zang, Z., Huang, K., et al., 2018a. Quantifying contributions of natural and anthropogenic dust emission from different climatic regions. Atmos. Environ. 191 (October), 94–104.

Chen, W., Koide, R.T., Eissenstat, D.M., 2018b. Nutrient foraging by mycorrhizas: from species functional traits to ecosystem processes. Funct. Ecol. 32 (4), 858–869.

Cheng, D.-L., Niklas, K.J., 2007. Above- and below-ground biomass relationships across 1534 forested communities. Ann. Bot. 99 (1), 95–102.

Cherry, R.D., Higgo, J.J.W., Fowler, S.W., 1978. Zooplankton fecal pellets and element residence times in the ocean. Nature 274, 246–248.

Chesworth, W., Macias-Vasquez, F., 1985. pE, pH, and podzolization. Am. J. Sci. 285, 128–146.

Chesworth, W., Dejou, J., Larroque, P., 1981. The weathering of basalt and relative mobilities of the major elements at Belbex, France. Geochim. Cosmochim. Acta 45, 1235–1243.

Chevalier, R.A., Sarazin, C.L., 1987. Hot gas in the universe. Am. Sci. 75, 609–618.

Childers, D.L., Corman, J., Edwards, M., Elser, J.J., 2011. Sustainability challenges of phosphorus and food: solutions from closing the human phosphorus cycle. Bioscience 61, 117–124.

Chin, M., Davis, D.D., 1993. Global sources and sinks of OCS and CS_2 and their distributions. Glob. Biogeochem. Cycles 7, 321–337.

Chin, M., Davis, D.D., 1995. A reanalysis of carbonyl sulfide as a source of stratospheric background sulfur aerosol. J. Geophys. Res. 100, 8993–9005.

Chini, C.M., Stillwell, A.S., 2018. The state of U.S. urban water: data and the energy-water nexus. Water Resour. Res. 54 (3), 1796–1811.

Chini, C.M., Konar, M., Stillwell, A.S., 2017. Direct and indirect urban water footprints of the United States. Water Resour. Res. 53 (1), 316–327.

Chipperfield, M.P., Bekki, S., Dhomse, S., Harris, N.R.P., Hassler, B., Hossaini, R., Steinbrecht, W., Thiéblemont, R., Weber, M., 2017. Detecting recovery of the stratospheric ozone layer. Nature 549 (7671), 211–218.

Chisholm, S.W., Falkowski, P.G., Cullen, J.J., 2001. Oceans. Dis-crediting ocean fertilization. Science 294 (5541), 309–310.

Chivian, D., Brodie, E.L., Alm, E.J., Culley, D.E., Dehal, P.S., DeSantis, T.Z., Gihring, T.M., et al., 2008. Environmental genomics reveals a single-species ecosystem deep within Earth. Science 322 (5899), 275–278.

Cho, B.C., Azam, F., 1988. Major role of bacteria in biogeochemical fluxes in the ocean's interior. Nature 332, 441–443.

Cho, Y., Driscoll, C.T., Johnson, C.E., Siccama, T.G., 2010. Chemical changes in soil and soil solution after calcium silicate addition to a Northern Hardwood forest. Biogeochemistry 100, 3–20.

Choi, W.-J., Chang, S.X., Bhatti, J.S., 2007. Drainage affects tree growth and C and N dynamics in a Minerotrophic Peatland. Ecology 88 (2), 443–453.

Chorover, J., Sposito, G., 1995. Surface-charge characteristics of kaolinitic tropical soils. Geochim. Cosmochim. Acta 59, 875–884.

Chorover, J., Vitousek, P.M., Everson, D.A., Espesperanza, A.M., Turner, D., 1994. Solution chemistry profiles of mixed-conifer forests before and after fire. Biogeochemistry 26, 115–144.

Chorover, J., Amistadi, M.K., Chadwick, O.A., 2004. Surface charge evolution of mineral-organic complexes during pedogenesis in Hawaiian Basalt. Geochim. Cosmochim. Acta 68, 4859–4876.

Christ, M.J., Driscoll, C.T., Likens, G.E., 1999. Watershed- and plot-scale tests of the mobile anion concept. Biogeochemistry 47, 335–353.

Christensen, N.L., 1977. Fire and soil-plant nutrient relations in a pine-wiregrass Savanna on the coastal plain of North Carolina. Oecologia 31 (1), 27–44.

Christensen, N.L., MacAller, T., 1985. Soil mineral nitrogen transformations during succession in the piedmont of North Carolina. Soil Biol. Biochem. 17, 675–681.

Christensen, J.P., Murray, J.W., Devol, A.H., Codispoti, L.A., 1987. Denitrification in continental shelf sediments has major impact on the oceanic nitrogen budget. Glob. Biogeochem. Cycles 1, 97–116.

Christensen, T.R., Prentice, I.C., Kaplan, J., Haxeltine, A., Sitch, S., 1996. Methane flux from Northern Wetlands and Tundra—an ecosystem source modelling approach. Tellus Ser. B Chem. Phys. Meteorol. 48, 652–661.

Christensen, T.R., Ekberg, A., Strom, L., Mastepanov, M., Panikov, N., Oquist, M., Svensson, B., Nykanen, H., Martikainen, P.J., Oskarsson, H., 2003. Factors controlling large scale variations in methane emissions from wetlands. Geophys. Res. Lett. 30 (7), 10–13.

Church, J.A., White, N.J., 2011. Sea-level rise from the late 19th to the early 21st century. Surv. Geophys. 32, 585–602.

Churkina, G., Brown, D.G., Keoleian, G., 2010. Carbon stored in human settlements: the conterminous United States. Glob. Chang. Biol. 16 (1), 135–143.

Chyba, C.F., 1990a. Extraterrestrial amino acids and terrestrial life. Nature 348, 113–114.

Chyba, C.F., 1990b. Impact delivery and Erosion of planetary oceans in the early inner solar system. Nature 343, 129–133.

Chyba, C., Sagan, C., 1992. Endogenous production, exogenous delivery and impact-shock synthesis of organic molecules: an inventory for the origins of life. Nature 355 (January), 125–132.

Ciais, P., Reichstein, M., Viovy, N., Granier, A., Ogée, J., Allard, V., Aubinet, M., et al., 2005. Europe-wide reduction in primary productivity caused by the heat and drought in 2003. Nature 437 (7058), 529–533.

Ciais, P., Manning, A.C., Reichstein, M., Zaehle, S., Bopp, L., 2007. Nitrification amplifies the decreasing trends of atmospheric oxygen and implies a larger land carbon uptake. Glob. Biogeochem. Cycles 21.

Ciais, P., Sabine, C., Bala, G., Bopp, L., Brovkin, V., Canadell, J., Chhabra, A., et al., 2013. Carbon and Other Biogeochemical Cycles. Climate Change 2013: The Physical Science Basis. Contribution of Working Group I to the Fifth Assessment Report of the Intergovernmental Panel on Climate Change. Cambridge University Press, Cambridge, United Kingdom/New York, NY, pp. 465–570.

Ciais, P., Tan, J., Wang, X., Roedenbeck, C., Chevallier, F., Piao, S.-L., Moriarty, R., et al., 2019. Five decades of northern land carbon uptake revealed by the interhemispheric CO_2 gradient. Nature 568 (7751), 221–225.

Cicerone, R.J., 1987. Changes in stratospheric ozone. Science 237 (4810), 35–42.

Cicerone, R.J., Oremland, R.S., 1988. Biogeochemical aspects of atmospheric methane. Glob. Biogeochem. Cycles 2 (4), 299–327.

Cicerone, R.J., Delwiche, C.C., Tyler, S.C., Zimmerman, P.R., 1992. Methane emissions from California rice paddies with varied treatments. Glob. Biogeochem. Cycles 6 (3), 233–248.

Clarisse, L., Clerbaux, C., Dentener, F., Hurtmans, D., Coheur, P.F., 2009. Global ammonia distribution derived from infrared satellite observations. Nat. Geosci. 2, 479–483.

Clarisse, L., Coheur, P.-F., Chefdeville, S., Lacour, J.-L., Hurtmans, D., Clerbaux, C., 2011. Infrared satellite observations of hydrogen sulfide in the volcanic plume of the August 2008 Kasatochi Eruption. Geophys. Res. Lett. 38.

Clark, F.E., 1977. Internal cycling of nitrogen in shortgrass Prairie. Ecology 58 (6), 1322–1333.

Clark, J.S., 1990. Fire and climate change during the last 750 Yr in northwestern Minnesota. Ecol. Monogr. 60, 135–159.

Clark, J.D., Johnson, A.H., 2011. Carbon and nitrogen accumulation in post-agricultural forest soils of western new England. Soil Sci. Soc. Am. J. 75, 1530–1542.

Clark, J.S., Stocks, B.J., Richard, P.J.H., 1996. Climate implications of biomass burning since the 19th century in eastern North America. Glob. Chang. Biol. 2, 433–442.

Clark, K.L., Nadkarni, N.M., Schaefer, D., Gholz, H.L., 1998. Cloud water and precipitation chemistry in a tropical montane forest, Monteverde, Costa Rica. Atmos. Environ. 32, 1595–1603.

Clark, L.L., Ingall, E.D., Benner, R., 1999. Marine organic phosphorus cycling: novel insights from nuclear magnetic resonance. Am. J. Sci. 299, 724–737.

Clark, D.A., Brown, S., Kicklighter, D.W., Chambers, J.Q., Thomlinson, J.R., Ni, J., 2001. Measuring net primary production in forests: concepts and field methods. Ecol. Appl. 11, 356–370.

Clark, K.L., Gholz, H.L., Castro, M.S., 2004. Carbon dynamics along a chronosequence of slash pine plantations in North Florida. Ecol. Appl. 14, 1154–1171.

Clark, M.L., Roberts, D.A., Ewel, J.J., Clark, D.B., 2011. Estimation of tropical rain forest aboveground biomass with small-footprint lidar and hyperspectral sensors. Remote Sens. Environ. 115, 2931–2942.

Clark, C.M., Phelan, J., Doraiswamy, P., Buckley, J., Cajka, J.C., Dennis, R.L., Lynch, J., Nolte, C.G., Spero, T.L., 2018. Atmospheric deposition and exceedances of critical loads from 1800–2025 for the conterminous United States. Ecol. Appl. 28 (4), 978–1002.

Clarkson, D.T., Hanson, J.B., 1980. The mineral nutrition of higher plants. Annu. Rev. Plant Physiol. Plant Mol. Biol. 31, 239–298.

Claus, R., Hoppen, H.O., Karg, H., 1981. The secret of truffles: a steroidal pheromone? Experientia 37, 1178–1179.

Clawson, R.G., Robin, G., Graeme Lockaby, B.G., Rummer, B., 2001. Changes in production and nutrient cycling across a wetness gradient within a floodplain forest. Ecosystems 4 (2), 126–138.

Clayton, J.L., 1976. Nutrient gains to adjacent ecosystems during a forest fire: an evaluation. For. Sci. 22, 162–166.

Cleland, E.E., Stanley Harpole, W., 2010. Nitrogen enrichment and plant communities. Ann. N. Y. Acad. Sci. 1195 (May), 46–61.

564   References

Clement, J., Shrestha, J., Ehrenfeld, J., Jaffe, P., 2005. Ammonium oxidation coupled to dissimilatory reduction of iron under anaerobic conditions in wetland soils. Soil Biol. Biochem. 37 (12), 2323–2328.

Cleveland, C.C., Liptzin, D., 2007. C:N:P stoichiometry in soil: is there a "Redfield Ratio" for the microbial biomass? Biogeochemistry 85, 235–252.

Cleveland, C.C., Townsend, A.R., 2006. Nutrient additions to a tropical rain forest drive substantial soil carbon dioxide losses to the atmosphere. Proc. Natl. Acad. Sci. U. S. A. 103 (27), 10316–10321.

Cleveland, C.C., Townsend, A.R., Schimel, D.S., Fisher, H., Howarth, R.W., Hedin, L.O., Perakis, S.S., et al., 1999. Global patterns of terrestrial biological nitrogen (N_2) fixation in natural ecosystems. Glob. Biogeochem. Cycles 13, 623–645.

Cleveland, C.C., Townsend, A.R., Schmidt, S.K., 2002. Phosphorus limitation of microbial processes in moist tropical forests: evidence from short-term laboratory incubations and field studies. Ecosystems 5, 680–691.

Cleveland, C.C., Houlton, B.Z., Neill, C., Reed, S.C., Townsend, A.R., Wang, Y.P., 2010. Using indirect methods to constrain symbiotic nitrogen fixation rates: a case study from an Amazonian rain forest. Biogeochemistry 99, 1–13.

Cleveland, C.C., Townsend, A.R., Taylor, P., Alvarez-Clare, S., Bustamante, M.M.C., Chuyong, G., Dobrowski, S.Z., et al., 2011. Relationships among net primary productivity, nutrients and climate in tropical rain forest: a pan-tropical analysis. Ecol. Lett. 14 (9), 939–947.

Cleveland, C.C., Houlton, B.Z., Kolby Smith, W., Marklein, A.R., Reed, S.C., Parton, W., Del Grosso, S.J., Running, S.W., 2013. Patterns of new versus recycled primary production in the terrestrial biosphere. Proc. Natl. Acad. Sci. U. S. A. 110 (31), 12733–12737.

Cliquet, J.-B., Lemauviel-Lavenant, S., 2019. Grassland species are more efficient in acquisition of S from the atmosphere when pedospheric S availability decreases. Plant Soil 435, 69–80.

Cloud, P., 1973. Paleoecological significance of the banded iron formation. Geology 68, 1135–1145.

Clow, D.W., Mast, M.A., 1999. Long-term trends in streamwater and precipitation chemistry at five headwater basins in the northeastern United States. Water Resour. Res. 35, 541–554.

Clymans, W., Conley, D.J., Battles, J.J., Frings, P.J., Koppers, M.M., Likens, G.E., Johnson, C.E., 2016. Silica uptake and release in live and decaying biomass in a northern hardwood forest. Ecology 97 (11), 3044–3057.

Clymo, R.S., 1984. The limits to peat bog growth. Philos. Trans. R. Soc. Lond. Series B Biol. Sci. 303, 605.

Clymo, R.S., Turunen, J., Tolonen, K., 1998. Carbon accumulation in peatland. Oikos 81, 368–388.

Coale, K.H., Johnson, K.S., Fitzwater, S.E., Gordon, R.M., Tanner, S., Chavez, F.P., Ferioli, L., et al., 1996a. A massive phytoplankton bloom induced by an ecosystem-scale iron fertilization experiment in the equatorial pacific ocean. Nature 383 (6600), 495–501.

Coale, K.H., Fitzwater, S.E., Michael Gordon, R., Johnson, K.S., Barber, R.T., 1996b. Control of community growth and export production by upwelled iron in the equatorial Pacific Ocean. Nature 379 (6566), 621–624.

Coale, K.H., Johnson, K.S., Chavez, F.P., Buesseler, K.O., Barber, R.T., Brzezinski, M.A., Cochlan, W.P., et al., 2004. Southern ocean iron enrichment experiment: carbon cycling in high- and low-Si waters. Science 304 (5669), 408–414.

Codispoti, L.A., 2007. An oceanic fixed nitrogen sink exceeding 400 Tg N a^{-1} vs the concept of homeostasis in the fixed-nitrogen inventory. Biogeosciences 4, 233–253.

Codispoti, L.A., Christensen, J.P., 1985. Nitrification, denitrification and nitrous oxide cycling in the eastern tropical South Pacific Ocean. Mar. Chem. 16, 277–300.

Codispoti, L.A., Brandes, J.A., Christensen, J.P., Devol, A.H., Naqvi, S.W.A., Paerl, H.W., Yoshinari, T., 2001. The oceanic fixed nitrogen and nitrous oxide budgets: moving targets as we enter the anthropocene? Sci. Mar. 65 (S2), 85–105.

Cody, G.D., Boctor, N.Z., Filley, T.R., Hazen, R.M., Scott, J.H., Sharma, A., Yoder Jr., H.S., 2000. Primordial carbonylated iron-sulfur compounds and the synthesis of pyruvate. Science 289 (5483), 1337–1340.

Cofer, W.R., Levine, J.S., Winstead, E.L., Stocks, B.J., 1990. Gaseous emissions from Canadian boreal forest fires. Atmos. Environ. 24, 1653–1659.

Cogbill, C.V., Likens, G.E., 1974. Acid precipitation in the northeastern United States. Water Resour. Res. 10, 1133–1137.

Coggon, R.M., Teagle, D.A.H., Smith-Duque, C.E., Alt, J.C., Cooper, M.J., 2010. Reconstructing past seawater Mg/Ca and Sr/Ca from mid-ocean ridge flank calcium carbonate veins. Science 327 (5969), 1114–1117.

Cohen, Y., Gordon, L.I., 1978. Nitrous oxide in the oxygen minimum of eastern tropical North Pacific—evidence for its consumption during denitrification and possible mechanisms for its production. Deep-Sea Res. 25, 509–524.

Cohen, Y., Gordon, L.I., 1979. Nitrous oxide production in the ocean. J. Geophys. Res. Oceans Atmos. 84, 347–353.

Cohen, B.A., Swindle, T.D., Kring, D.A., 2000. Support for the lunar cataclysm hypothesis from lunar meteorite impact melt ages. Science 290 (5497), 1754–1756.

Colbourn, G., Ridgwell, A., Lenton, T.M., 2015. The time scale of the silicate weathering negative feedback on atmospheric CO_2: silicate weathering feedback time scale. Glob. Biogeochem. Cycles 29 (5), 583–596.

Cole, J.J., Caraco, N.F., 2001. Carbon in catchments: connecting terrestrial carbon losses with aquatic metabolism. Mar. Freshw. Res. 52, 101–110.

Cole, C.V., Olsen, S.R., 1959. Phosphorus solubility in calcareous soils. I. Dicalcium phosphate activities in equilibrium solutions. Proc. Soil Sci. Soc. Am. 23, 116–118.

Cole, J.J., Pace, M.L., 1995. Bacterial secondary production in oxic and anoxic freshwaters. Limnol. Oceanogr. 40, 1019–1027.

Cole, D.W., Rapp, M., 1981. Element cycling in forest ecosystems. In: Reichle, D.E. (Ed.), Dynamics Properties of Forest Ecosystems. Cambridge University Press, pp. 341–409.

Cole, C.V., Elliott, E.T., Hunt, H.W., Coleman, D.C., 1977a. Trophic interactions in soils as they affect energy and nutrient dynamics. V. Phosphorus transformations. Microb. Ecol. 4 (4), 381–387.

Cole, C.V., Innis, G.S., Stewart, J.W.B., 1977b. Simulation of phosphorus cycling in semiarid grasslands. Ecology 58, 1–15.

Cole, J.J., Honjo, S., Erez, J., 1987. Benthic decomposition of organic matter at a deep-water site in the Panama Basin. Nature 327, 703–704.

Cole, J.J., Findlay, S., Pace, M.L., 1988. Bacterial production in fresh and saltwater ecosystems: a cross-system overview. Mar. Ecol. Prog. Ser. 43, 1–10.

Cole, J.J., Lane, J.M., Marino, R., Howarth, R.W., 1993. Molybdenum assimilation by Cyanobacteria and Phytoplankton in freshwater and saltwater. Limnol. Oceanogr. 38, 25–35.

Cole, J.J., Caraco, N.F., Kling, G.W., Kratz, T.K., 1994. Carbon dioxide supersaturation in the surface waters of lakes. Science 265 (5178), 1568–1570.

Cole, J.J., Carpenter, S.R., Kitchell, J.F., Pace, M.L., 2002. Pathways of organic carbon utilization in small lakes: results from a whole-lake ^{13}C addition and coupled model. Limnol. Oceanogr. 47, 1664–1675.

Cole, J.J., Prairie, Y.T., Caraco, N.F., McDowell, W.H., Tranvik, L.J., Striegl, R.G., Duarte, C.M., et al., 2007. Plumbing the global carbon cycle: integrating inland waters into the terrestrial carbon budget. Ecosystems 10, 171–184.

Coles, J.R.P., Yavitt, J.B., 2002. Control of methane metabolism in a forested Northern Wetland, New York state, by aeration, substrates, and peat size fractions. Geomicrobiol J. 19 (3), 293–315.

Coley, P.D., Bryant, J.P., Chapin 3rd., F.S., 1985. Resource availability and plant antiherbivore defense. Science 230 (4728), 895–899.

Collalti, A., Prentice, I.C., 2019. Is NPP proportional to GPP? Waring's hypothesis 20 years on. Tree Physiol. 39 (8), 1473–1483.

Colman, B.P., Schimel, J.P., 2013. Drivers of microbial respiration and net N mineralization at the continental scale. Soil Biol. Biochem. 60 (May), 65–76.

Colman, J.J., Blake, D.R., Rowland, F.S., 1998. Atmospheric residence time of CH_3Br estimated from the junge spatial variability relation. Science 281 (5375), 392–396.

Colwyn, D., Nathan, S., Maynard, J., Gaines, R., Hofmann, A., Wang (王相力), X., Gueguen, B., Asael, D., Reinhard, C., Planavsky, N., 2019. A paleosol record of the evolution of Cr redox cycling and evidence for an increase in atmospheric oxygen during the neoproterozoic. Geobiology. 17(August).

Comas, X., Wright, W., 2012. Heterogeneity of biogenic gas ebullition in subtropical peat soils is revealed using time-lapse cameras. Water Resour. Res. 48(4).

Comiso, J.C., Parkinson, C.L., Gersten, R., Stock, L., 2008. Accelerated decline in the arctic sea ice cover. Geophys. Res. Lett. 35.

Compeau, G.C., Bartha, R., 1985. Sulfate-reducing bacteria: principal methylators of mercury in anoxic estuarine sediment. Appl. Environ. Microbiol. 50 (2), 498–502.

Conant, R.T., Ryan, M.G., Agren, G.I., Birge, H.E., Davidson, E.A., Eliasson, P.E., Evans, S.E., et al., 2011. Temperature and soil organic matter decomposition rates: synthesis of current knowledge and a way forward. Glob. Chang. Biol. 17, 3392–3404.

Conant, R.T., Berdanier, A.B., Grace, P.R., 2013. Patterns and trends in nitrogen use and nitrogen recovery efficiency in world agriculture: global n recovery efficiencies. Glob. Biogeochem. Cycles 27 (2), 558–566.

Conen, F., Neftel, A., 2007. Do increasingly depleted $\delta_{15}N$ values of atmospheric N_2O indicate a decline in soil N_2O reduction? Biogeochemistry 82 (3), 321–326.

Conley, D.J., 2002. Terrestrial ecosystems and the global biogeochemical silica cycle. Glob. Biogeochem. Cycles 16.

Conley, D.J., Malone, T.C., 1992. Annual cycle of dissolved silicate in Chesapeake bay: implications for the production and fate of phytoplankton biomass. Mar. Ecol. Prog. Ser. 81 (2), 121–128.

Conley, D.J., Smith, W.M., Cornwell, J.C., Fisher, T.R., 1995. Transformation of particle-bound phosphorus at the land-sea interface. Estuar. Coast. Shelf Sci. 40, 161–176.

Conley, D.J., Carstensen, J., Aertebjerg, G., Christensen, P.B., Dalsgaard, T., Hansen, J.L.S., Josefson, A.B., 2007. Long-term changes and impacts of hypoxia in danish coastal waters. Ecol. Appl. 17, S165–S184.

Conley, D.J., Björck, S., Bonsdorff, E., Carstensen, J., Destouni, G., Gustafsson, B.G., Hietanen, S., et al., 2009a. Hypoxia-related processes in the Baltic Sea. Environ. Sci. Technol. 43 (10), 3412–3420.

Conley, D.J., Paerl, H.W., Howarth, R.W., Boesch, D.F., Seitzinger, S.P., Havens, K.E., Lancelot, C., Likens, G.E., 2009b. Controlling eutrophication: nitrogen and phosphorus. Science 323 (5917), 1014–1015.

Connell, M.J., Raison, R.J., Khanna, P.K., 1995. Nitrogen mineralization in relation to site history and soil properties for a range of Australian forest soils. Biol. Fertil. Soils 20, 213–220.

Conner, W.H., Day Jr., J.W., 1976. Productivity and composition of a baldcypress-water tupelo site and a bottomland hardwood site in a Louisiana swamp. Am. J. Bot. 63 (10), 1354–1364.

Conner, W.H., Song, B., Williams, T.M., Vernon, J.T., 2011. Long-term tree productivity of a South Carolina coastal plain forest across a hydrology gradient. J. Plant Ecol. 4 (1–2), 67–76.

Conrad, R., 1994. Compensation concentration as critical variable for regulating the flux of trace gases between soil and atmosphere. Biogeochemistry 27, 155–170.

Conrad, R., 1996. Soil microorganisms as controllers of atmospheric trace gases (H_2, CO, CH_4, OCS, N_2O, and NO). Microbiol. Rev. 60 (4), 609–640.

Conrad, R., Seiler, W., Bunse, G., 1983. Factors influencing the loss of fertilizer nitrogen into the atmosphere as N_2O. J. Geophys. Res. D: Atmos. 88, 6709–6718.

Conway, T.J., Tans, P., Waterman, L.S., Thoning, K.W., Masarie, K.A., Gammon, R.H., 1988. Atmospheric carbon dioxide measurements in the remote global atmosphere, 1981–1984. Tellus Ser. B Chem. Phys. Meteorol. 40, 81–115.

Cook, R.B., Schindler, D.W., 1983. The biogeochemistry of sulfur in an experimentally acidified lake. Ecol. Bull. 35, 115–127.

Cook, P.J., Shergold, J.H., 1984. Phosphorus, phosphorites and skeletal evolution at the Precambrian-Cambrian boundary. Nature 308, 231–236.

Cook, E.R., Woodhouse, C.A., Mark Eakin, C., Meko, D.M., Stahle, D.W., 2004. Long-term aridity changes in the Western United States. Science 306 (5698), 1015–1018.

Cooper, S.R., Brush, G.S., 1991. Long-term history of Chesapeake bay anoxia. Science 254 (5034), 992–996.

Cooper, D.J., Watson, A.J., Nightingale, P.D., 1996. Large decrease in ocean-surface CO_2 fugacity in response to in situ iron fertilization. Nature 383, 511–513.

Cooper, G., Kimmich, N., Belisle, W., Sarinana, J., Brabham, K., Garrel, L., 2001. Carbonaceous meteorites as a source of sugar-related organic compounds for the early earth. Nature 414 (6866), 879–883.

Cooper, O.R., Parrish, D.D., Stohl, A., Trainer, M., Nédélec, P., Thouret, V., Cammas, J.P., et al., 2010. Increasing springtime ozone mixing ratios in the free troposphere over Western North America. Nature 463 (7279), 344–348.

Cooper, J., Lombardi, R., Boardman, D., Carliell-Marquet, C., 2011. The future distribution and production of global phosphate rock reserves. Resour. Conserv. Recycl. 57, 78–86.

Copard, Y., Amiotte-Suchet, P., Di-Giovanni, C., 2007. Storage and release of fossil organic carbon related to weathering of sedimentary rocks. Earth Planet. Sci. Lett. 258, 345–357.

Čorić, I., Mercado, B.Q., Bill, E., Vinyard, D.J., Holland, P.L., 2015. Binding of dinitrogen to an iron-sulfur-carbon site. Nature 526 (7571), 96–99.

Corliss, J.B., Dymond, J., Gordon, L.I., Edmond, J.M., Herzen, R.P.V., Ballard, R.D., Green, K., et al., 1979. Submarine thermal springs on the Galapagos rift. Science 203, 1073–1083.

Corliss, B.H., Martinson, D.G., Keffer, T., 1986. Late quaternary deep-ocean circulation. Geol. Soc. Am. Bull. 97, 1106–1121.

Cornelis, J.T., Delvaux, B., Cardinal, D., Andre, L., Ranger, J., Opfergelt, S., 2010. Tracing mechanisms controlling the release of dissolved silicon in forest soil solutions using Si isotopes and Ge/Si ratios. Geochim. Cosmochim. Acta 74, 3913–3924.

Cornell, S.E., Jickells, T.D., Cape, J.N., Rowland, A.P., Duce, R.A., 2003. Organic nitrogen deposition on land and coastal environments: a review of methods and data. Atmos. Environ. 37, 2173–2191.

Cornett, R.J., Rigler, F.H., 1979. Hypolinimetic oxygen deficits: their prediction and interpretation. Science 205 (4406), 580–581.

Cornett, R.J., Rigler, F.H., 1980. The areal hypolimnetic oxygen deficit: an empirical test of the model. Limnol. Oceanogr. 25, 672–679.

Cornwell, J.C., DeLaune, R.D., Reddy, K.R., Richardson, C.J., Megonigal, J.P., 2013. Measurement of sulfate reduction in wetland soils. In: DeLaune, R.D., Reddy, K.R., Richardson, C.J., Megonigal, J.P. (Eds.), Methods in Biogeochemistry of Wetlands, vol. 10. Soil Science Society of America, pp. 765–773.

Corre, M.D., Lamersdorf, N.P., 2004. Reversal of nitrogen saturation after long-term deposition reduction: impact on soil nitrogen cycling. Ecology 85, 3090–3104.

Corre, M.D., Beese, F.O., Brumme, R., 2003. Soil nitrogen cycle in high nitrogen deposition forest: changes under nitrogen saturation and liming. Ecol. Appl. 13, 287–298.

Corre, M.D., Veldkamp, E., Arnold, J., Joseph Wright, S., 2010. Impact of elevated N input on soil N cycling and losses in old-growth lowland and montane forests in Panama. Ecology 91 (6), 1715–1729.

Correll, D.L., 1998. The role of phosphorus in the eutrophication of receiving waters: a review. J. Environ. Qual. 27, 261–266.

Correll, D.L., Ford, D., 1982. Comparison of precipitation and land runoff as sources of estuarine nitrogen. Estuar. Coast. Shelf Sci. 15 (1), 45–56.

Correll, D.L., Clark, C.O., Goldberg, B., Goodrich, V.R., Hayes, D.R., Klein, W.H., Schecher, W.D., 1992. Spectral ultraviolet-B radiation fluxes at the Earth's surface—long-term variations at 39°N, 77°W. J. Geophys. Res. 97, 7579–7591.

Cory, R.M., Ward, C.P., Crump, B.C., Kling, G.W., 2014. Sunlight controls water column processing of carbon in Arctic fresh waters. Science 345 (6199), 925–928.

Cosby, B.J., Hornberger, G.M., Wright, R.F., Galloway, J.N., 1986. Modeling the effects of acid deposition: control of long-term Sulfate dynamics by soil sulfate adsorption. Water Resour. Res. 22, 1283–1291.

Cotner Jr., J.B., Wetzel, R.G., 1992. Uptake of dissolved inorganic and organic bphosphorus compounds by phytoplankton and bacterioplankton. Limnol. Oceanogr. 37 (2), 232–243.

Cotte, C., Guinet, C., 2007. Historical whaling records reveal major regional retreat of Antarctic Sea ice. Deep-Sea Res. Part I—Oceanogr. Res. Pap. 54, 243–252.

Cotton, J.M., Louise Jeffery, M., Sheldon, N.D., 2013. Climate controls on soil respired CO_2 in the United States: implications for 21st century chemical weathering rates in temperate and arid ecosystems. Chem. Geol. 358 (November), 37–45.

Couradeau, E., Benzerara, K., Gérard, E., Moreira, D., Bernard, S., Brown Jr., G.E., López-García, P., 2012. An early-branching microbialite Cyanobacterium forms intracellular carbonates. Science 336 (6080), 459–462.

Courchesne, F., Hendershot, W.H., 1989. Sulfate retention in some podzolic soils of the southern Laurentians, Quebec. Can. J. Soil Sci. 69, 337–350.

Courty, P.E., Buee, M., Diedhiou, A.G., Frey-Klett, P., Le Tacon, F., Rineau, F., Turpault, M.P., Uroz, S., Garbaye, J., 2010. The role of ectomycorrhizal communities in forest ecosystem processes: new perspectives and emerging concepts. Soil Biol. Biochem. 42, 679–698.

Couto, W., Sanzonowicz, C., Barcellos, A.D.E.O., 1985. Factors affecting oxidation-reduction processes in an oxisol with a seasonal water table. Soil Sci. Soc. Am. J. 49, 1245–1248.

Covey, K.R., Megonigal, J.P., 2019. Methane production and emissions in trees and forests. New Phytol. 222 (1), 35–51.

Covey, K.R., Wood, S.A., Warren, R.J., Lee, X., Bradford, M.A., 2012. Elevated methane concentrations in trees of an upland forest. Geophys. Res. Lett. 39.

Covino, T., 2017. Hydrologic connectivity as a framework for understanding biogeochemical flux through watersheds and along fluvial networks. Geomorphology 277 (SI), 133–144.

Covino, T.P., McGlynn, B.L., 2007. Stream gains and losses across a mountain-to-valley transition: impacts on watershed hydrology and stream water chemistry. Water Resour. Res. 43.

Covino, T.P., McGlynn, B.L., McNamara, R.A., 2010. Tracer additions for spiraling curve characterization (TASCC): quantifying stream nutrient uptake kinetics from ambient to saturation. Limnol. Oceanogr. 8, 484–498.

Covino, T.P., Bernhardt, E.S., Heffernan, J.B., 2018. Measuring and interpreting relationships between nutrient supply, demand, and limitation. Freshw. Sci. 37 (3), 448–455.

Cowan, J.J., Sneden, C., 2006. Heavy element synthesis in the oldest stars and the early universe. Nature 440 (7088), 1151–1156.

Cox, G.F.N., Weeks, W.F., 1974. Salinity variations in sea ice. J. Glaciol. 13 (67), 109–120.

Cox, T.L., Harris, W.F., Ausmus, B.S., Edwards, N.T., 1978. Role of roots in biogeochemical cycles in an eastern deciduous forest. Pedobiologia 18, 264–271.

Cox, N., Retegan, M., Neese, F., Pantazis, D.A., Boussac, A., Lubitz, W., 2014. Photosynthesis. electronic structure of the oxygen-evolving complex in photosystem II prior to O-O bond formation. Science 345 (6198), 804–808.

Craft, C.B., 1996. Dynamics of nitrogen and phosphorus retention during wetland ecosystem succession. Wetl. Ecol. Manag. 4 (3), 177–187.

Craft, C.B., Casey, W.P., 2000. Sediment and nutrient accumulation in floodplain and depressional freshwater Wetlands of Georgia, USA. Wetlands 20, 323–332.

Craft, C.B., Richardson, C.J., 1993. Peat accretion and N, P, and organic C accumulation in nutrient-enriched and unenriched everglades peatlands. Ecol. Appl. 3 (3), 446–458.

Craft, C., Washburn, C., Parker, A., 2008. Latitudinal trends in organic carbon accumulation in temperate freshwater peatlands. In: Vymagal, J. (Ed.), Wastewater Treatment, Plant Dynamics and Management in Constructed and Natural Wetlands. Springer, pp. 23–31.

Craig, H., Hayward, T., 1987. Oxygen supersaturation in the ocean: biological versus physical contributions. Science 235 (4785), 199–202.

Craig, H., Chou, C.C., Welhan, J.A., Stevens, C.M., Engelkemeir, A., 1988. The isotopic composition of methane in polar ice cores. Science 242 (4885), 1535–1539.

Crane, B.R., Siegel, L.M., Getzoff, E.D., 1995. Sulfite reductase structure at 1.6 A: evolution and catalysis for reduction of inorganic anions. Science 270 (5233), 59–67.

Crawford, B., Grimmond, C.S.B., Christen, A., 2011. Five years of carbon dioxide fluxes measurements in a highly vegetated suburban area. Atmos. Environ. 45, 896–905.

Creed, I.F., McKnight, D.M., Pellerin, B.A., Green, M.B., Bergamaschi, B.A., Aiken, G.R., Burns, D.A., et al., 2015. The river as a chemostat: fresh perspectives on dissolved organic matter flowing down the river continuum. Can. J. Fish. Aquat. Sci. 72 (8), 1272–1285.

Creed, I.F., Bergström, A.K., Trick, C.G., Grimm, N.B., Hessen, D.O., Karlsson, J., Kidd, K.A., et al., 2018. Global change-driven effects on dissolved organic matter composition: implications for food webs of Northern Lakes. Glob. Chang. Biol. 24 (8), 3692–3714.

Crenshaw, C.L., Valett, H.M., Webster, J.R., 2002. Effects of augmentation of coarse particulate organic matter on metabolism and nutrient retention in hyporheic sediments. Freshw. Biol. 47, 1820–1831.

Crews, T.E., 1999. The presence of nitrogen-fixing legumes in terrestrial communities: evolutionary vs ecological considerations. Biogeochemistry 46, 233–246.

Crews, T.E., Kitayama, K., Fownes, J.H., Riley, R.H., Herbert, D.A., Muellerdombois, D., Vitousek, P.M., 1995. Changes in soil phosphorus fractions and ecosystem dynamics across a long chronosequence in Hawaii. Ecology 76, 1407–1424.

Crill, P.M., 1991. Seasonal patterns of methane uptake and carbon dioxide release by a temperate woodland soil. Glob. Biogeochem. Cycles 5, 319–334.

Crill, P.M., Martens, C.S., 1986. Methane production from bicarbonate and acetate in an anoxic marine sediment. Geochim. Cosmochim. Acta 50, 2089–2097.

Croft, H., Chen, J.M., Luo, X., Bartlett, P., Chen, B., Staebler, R.M., 2017. Leaf chlorophyll content as a proxy for leaf photosynthetic capacity. Glob. Chang. Biol. 23 (9), 3513–3524.

Cromack, K., Sollins, P., Graustein, W.C., Speidel, K., Todd, A.W., Spycher, G., Li, C.Y., Todd, R.L., 1979. Calcium-oxalate accumulation and soil weathering in mats of the hypogeous fungus *Hysterangium crassum*. Soil Biol. Biochem. 11, 463–468.

Cronan, C.S., 1980. Solution chemistry of a new hampshire subalpine ecosystem: a biogeochemical analysis. Oikos 34, 272–281.

Cronan, C.S., Aiken, G.R., 1985. Chemistry and transport of soluble humic substances in forested watersheds of the Adirondack Park, New York. Geochim. Cosmochim. Acta 49, 1697–1705.

Cronan, C.S., Grigal, D.F., 1995. Use of calcium-aluminum ratios as indicators of stress in forest ecosystems. J. Environ. Qual. 24, 209–226.

Cronan, C.S., Driscoll, C.T., Newton, R.M., Kelly, J.M., Schofield, C.L., Bartlett, R.J., April, R., 1990. A comparative analysis of aluminum biogeochemistry in a northeastern and a southeastern forested watershed. Water Resour. Res. 26, 1413–1430.

Croot, P.L., Streu, P., Baker, A.R., 2004. Short residence time for iron in surface seawater impacted by atmospheric dry deposition from Saharan dust events. Geophys. Res. Lett. 31.

Cross, A.T., Phillips, T.L., 1990. Coal-forming plants through time in North America. Int. J. Coal Geol. 16 (1–3), 1–46.

Cross, P.M., Rigler, F.H., 1983. Phosphorus and iron retention in sediments measured by mass budget calculations and directly. Can. J. Fish. Aquat. Sci. 40, 1589–1597.

Cross, A.F., Schlesinger, W.H., 1995. A literature review and evaluation of the hedley fractionation: applications to the biogeochemical cycle of soil phosphorus in natural ecosystems. Geoderma 64, 197–214.

Cross, W.F., Bruce Wallace, J., Rosemond, A.D., 2007. Nutrient enrichment reduces constraints on material flows in a detritus-based food web. Ecology 88 (10), 2563–2575.

Crowe, S.A., Døssing, L.N., Beukes, N.J., Bau, M., Kruger, S.J., Frei, R., Canfield, D.E., 2013. Atmospheric oxygenation three billion years ago. Nature 501 (7468), 535–538.

Crowe, S.A., Paris, G., Katsev, S., Jones, C., Kim, S.-T., Zerkle, A.L., Nomosatryo, S., et al., 2014. Sulfate was a trace constituent of Archean seawater. Science 346 (6210), 735–739.

Crowley, T.J., 2000. Causes of climate change over the past 1000 years. Science 289 (5477), 270–277.

Crowther, T.W., Todd-Brown, K.E.O., Rowe, C.W., Wieder, W.R., Carey, J.C., Machmuller, M.B., Snoek, B.L., et al., 2016. Quantifying global soil carbon losses in response to warming. Nature 540 (7631), 104–108.

Crozat, G., 1979. Emission of potassium aerosols in tropical forest. Tellus 31, 52–57.

Crutzen, P.J., 1983. Atmospheric interactions: homogeneous gas reactions of C, N, and S containing compounds. In: Bolin, B., Cook, R.B. (Eds.), The Major Biogeochemical Cycles and Their Interactions. Wiley, pp. 67–114.

Crutzen, P.J., 1988. Variability atmospheric chemical systems. In: Rosswall, R.G.W.T., Risser, P.G. (Eds.), Scales and Global Change. Wiley, pp. 81–108.

Crutzen, P.J., 2006. Albedo enhancement by stratospheric sulfur injections: a contribution to resolve a policy dilemma? Clim. Chang. 77, 211–219.

Crutzen, P.J., Andreae, M.O., 1990. Biomass burning in the tropics: impact on atmospheric chemistry and biogeochemical cycles. Science 250 (4988), 1669–1678.

Crutzen, P.J., Zimmermann, P.H., 1991. The changing photochemistry of the troposphere. Tellus A Dyn. Meteorol. Oceanogr. 43, 136–151.

Crutzen, P.J., Delany, A.C., Greenberg, J., Haagenson, P., Heidt, L., Lueb, R., Pollock, W., Seiler, W., Wartburg, A., Zimmerman, P., 1985. Tropospheric chemical composition measurements in Brazil during the dry season. J. Atmos. Chem. 2, 233–256.

Crutzen, P.J., Aselmann, I., Seiler, W., 1986. Methane production by domestic animals, wild ruminants, and other herbivorous fauna, and humans. Tellus Ser. B Chem. Phys. Meteorol. 38, 271–284.

Crutzen, P.J., Lawrence, M.G., Poschl, U., 1999. On the background photochemistry of tropospheric ozone. Tellus A Dyn. Meteorol. Oceanogr. 51, 123–146.

Cuevas, E., Medina, E., 1986. Nutrient dynamics within amazonian forest ecosystems: I. Nutrient flux in fine litter fall and efficiency of nutrient utilization. Oecologia 68 (3), 466–472.

Cuevas, E., Medina, E., 1988. Nutrient dynamics within Amazonian forests: II. Fine root growth, nutrient availability and leaf litter decomposition. Oecologia 76 (2), 222–235.

Cuffey, K.M., Vimeux, F., 2001. Covariation of carbon dioxide and temperature from the Vostok ice core after deuterium-excess correction. Nature 412 (6846), 523–527.

Cui, Y., Kump, L.R., Ridgwell, A.J., Charles, A.J., Junium, C.K., Diefendorf, A.F., Freeman, K.H., Urban, N.M., Harding, I.C., 2011. Slow release of fossil carbon during the palaeocene—eocene thermal maximum. Nat. Geosci. 4, 481–485.

Cui, S., Shi, Y., Groffman, P.M., Schlesinger, W.H., Zhu, Y.-G., 2013. Centennial-scale analysis of the creation and fate of reactive nitrogen in China (1910–2010). Proc. Natl. Acad. Sci. U. S. A. 110 (6), 2052–2057.

Cullis, C.F., Hirschler, M.M., 1980. Atmospheric sulphur: natural and man-made sources. Atmos. Environ. 14, 1263–1278.

Cunningham, S.A., Marsh, R., 2010. Observing and modeling changes in the Atlantic MOC. Wiley Interdiscip. Rev. Clim. Chang. 1 (2), 180–191.

Cunningham, S.A., Alderson, S.G., King, B.A., Brandon, M.A., 2003. Transport and variability of the Antarctic circumpolar current in drake passage. J. Geophys. Res. C: Oceans 108 (C5).

Curry, C.L., 2007. Modeling the soil consumption of atmospheric methane at the global scale. Glob. Biogeochem. Cycles 21.

Curry, R., Mauritzen, C., 2005. Dilution of the northern North Atlantic Ocean in recent decades. Science 308 (5729), 1772–1774.

Curtis, P.S., Wang, X., 1998. A meta-analysis of elevated CO_2 effects on woody plant mass, form, and physiology. Oecologia 113 (3), 299–313.

Cusack, D.F., Silver, W.L., Torn, M.S., McDowell, W.H., 2011. Effects of nitrogen additions on above and belowground carbon dynamics in two tropical forests. Biogeochemistry 104, 203–225.

Cushon, G.H., Feller, M.C., 1989. Asymbiotic nitrogen fixation and denitrification in a mature forest in coastal British Columbia. Can. J. For. Res. 19, 1194–1200.

Cyr, H., Pace, M.L., 1993. Magnitude and patterns of herbivory in aquatic and terrestrial ecosystems. Nature 361, 148–150.

Czaja, A.D., Johnson, C.M., Beard, B.L., Roden, E.E., Li, W., Moorbath, S., 2013. Biological Fe oxidation controlled deposition of banded iron formation in the Ca. 3770 Ma Isua supracrustal belt (West Greenland). Earth Planet. Sci. Lett. 363 (February), 192–203.

Cziczo, D.J., Thomson, D.S., Murphy, D.M., 2001. Ablation, flux, and atmospheric implications of meteors inferred from stratospheric aerosol. Science 291 (5509), 1772–1775.

D'Arrigo, R., Jacoby, G.C., Fung, I.Y., 1987. Boreal forests and atmosphere-biosphere exchange of carbon dioxide. Nature 329, 321–323.

D'Hondt, S., Rutherford, S., Spivack, A.J., 2002. Metabolic activity of subsurface life in deep-sea sediments. Science 295 (5562), 2067–2070.

D'Hondt, S., Jørgensen, B.B., Jay Miller, D., Batzke, A., Blake, R., Cragg, B.A., Cypionka, H., et al., 2004. Distributions of microbial activities in deep subseafloor sediments. Science 306 (5705), 2216–2221.

da Silva, J.J.R., Williams, R.J.P., 2001. The Biological Chemistry of the Elements: The Inorganic Chemistry of Life. OUP Oxford.

Dacey, J.W.H., 1981. Pressurized ventilation in the yellow waterlily. Ecology 62 (5), 1137–1147.

Dacey, J.W., Wakeham, S.G., 1986. Oceanic dimethylsulfide: production during zooplankton grazing on phytoplankton. Science 233 (4770), 1314–1316.

Dacey, J.W.H., Drake, B.G., Klug, M.J., 1994. Stimulation of methane emission by carbon dioxide enrichment of marsh vegetation. Nature 370, 47–49.

Dagg, M., Benner, R., Lohrenz, S., Lawrence, D., 2004. Transformation of dissolved and particulate materials on continental shelves influenced by large rivers: plume processes. Cont. Shelf Res. 24, 833–858.

Dahlgren, R.A., Walker, W.J., 1993. Aluminum release rates from selected spodosol Bs horizons: effect of pH and solid-phase aluminum pools. Geochim. Cosmochim. Acta 57, 57–66.

Dahlgren, R.A., Ugolini, F.C., Casey, W.H., 1999. Field weathering rates of Mt St. Helens tephra. Geochim. Cosmochim. Acta 63, 587–598.

Dai, A., Trenberth, K.E., Qian, T.T., 2004. A global dataset of palmer drought severity index for 1870–2002: relationship with soil moisture and effects of surface warming. J. Hydrometeorol. 5, 1117–1130.

Dail, D.B., Fitzgerald, J.W., 1999. S cycling in soil and stream sediment: influence of season and in situ concentrations of carbon, nitrogen and sulfur. Soil Biol. Biochem. 31, 1395–1404.

Daims, H., Lebedeva, E.V., Pjevac, P., Han, P., Herbold, C., Albertsen, M., Jehmlich, N., et al., 2015. Complete nitrification by nitrospira bacteria. Nature 528 (7583), 504–509.

Dalmagro, H.J., Zanella de Arruda, P.H., Vourlitis, G.L., Lathuillière, M.J., Nogueira, J.d.S., Couto, E.G., Johnson, M.S., 2019. Radiative forcing of methane fluxes offsets net carbon dioxide uptake for a tropical flooded forest. Glob. Chang. Biol. 25 (6), 1967–1981.

Dalponte, M., Jucker, T., Liu, S., Frizzera, L., Gianelle, D., 2019. Characterizing forest carbon dynamics using multi-temporal Lidar data. Remote Sens. Environ. 224 (April), 412–420.

Dalsgaard, T., Thamdrup, B., 2002. Production of N_2 through anaerobic ammonium oxidation coupled to nitrate reduction in marine sediments. Appl. Environ. Microbiol. 68 (3), 1312–1318.

Dalsgaard, T., Canfield, D.E., Petersen, J., Thamdrup, B., Acuña-González, J., 2003. N_2 production by the anammox reaction in the anoxic water column of Golfo Dulce, Costa Rica. Nature 422 (6932), 606–608.

Dalva, M., Moore, T.R., 1991. Sources and sinks of dissolved organic carbon in a forested swamp catchment. Biogeochemistry 15 (1), 1–19.

Damman, A.W.H., 1988. Regulation of nitrogen removal and retention in sphagnum bogs and other peatlands. Oikos 51 (3), 291–305.

Dangal, S.R.S., Tian, H., Zhang, B., Pan, S., Lu, C., Yang, J., 2017. Methane emission from global livestock sector during 1890–2014: magnitude, trends and spatiotemporal patterns. Glob. Chang. Biol. 23 (10), 4147–4161.

Daniels, W.L., Zelazny, L.W., Everett, C.J., 1987. Virgin hardwood foerst soils of the southern Appalachian Mountains. II. Weathering, mineralogy, and chemical properties. Soil Sci. Soc. Am. J. 51, 730–738.

Dannenmann, M., Díaz-Pinés, E., Kitzler, B., Karhu, K., Tejedor, J., Ambus, P., Parra, A., et al., 2018. Postfire nitrogen balance of mediterranean Shrublands: direct combustion losses versus gaseous and leaching losses from the postfire soil mineral nitrogen flush. Glob. Chang. Biol. 24 (10), 4505–4520.

Darcy, J.L., Schmidt, S.K., Knelman, J.E., Cleveland, C.C., Castle, S.C., Nemergut, D.R., 2018. Phosphorus, not nitrogen, limits plants and microbial primary producers following glacial retreat. Sci. Adv. 4 (5), eaaq0942.

Dargie, G.C., Lewis, S.L., Lawson, I.T., Mitchard, E.T.A., Page, S.E., Bocko, Y.E., Ifo, S.A., 2017. Age, extent and carbon storage of the Central Congo Basin Peatland complex. Nature 542 (7639), 86–90.

Darling, M.S., 1976. Interpretation of global differences in plant calorific values: the significance of desert and arid woodland vegetation. Oecologia 23 (2), 127–139.

Darnajoux, R., Magain, N., Renaudin, M., Lutzoni, F., Bellenger, J.-P., Zhang, X., 2019. Molybdenum threshold for ecosystem scale alternative vanadium nitrogenase activity in boreal forests. Proc. Natl. Acad. Sci. 116 (49), 24682–24688.

Dasgupta, R., Hirschmann, M.M., 2010. The deep carbon cycle and melting in Earth's interior. Earth Planet. Sci. Lett. 298, 1–13.

Daskalakis, N., Myriokefalitakis, S., Kanakidou, M., 2015. Sensitivity of tropospheric loads and lifetimes of short lived pollutants to fire emissions. Atmos. Chem. Phys. 15, 3543–3563.

Dauphas, N., 2005. The U/Th production ratio and the age of the milky way from meteorites and galactic halo stars. Nature 435 (7046), 1203–1205.

Dauphas, N., 2017. The isotopic nature of the Earth's accreting material through time. Nature 541 (7638), 521–524.

David, M.B., Mitchell, M.J., Nakas, J.P., 1982. Organic and inorganic sulfur constitutents of a forest soil and their relationship to microbial activity. Soil Sci. Soc. Am. J. 46, 847–852.

David, M.B., Reuss, J.O., Walthall, P.M., 1988. Use of a chemical-equilibrium model to understand soil chemical processes that influence soil solution and surface water alkalinity. Water Air Soil Pollut. 38, 71–83.

David, M.B., Vance, G.F., Fasth, W.J., 1991. Forest soil response to acid and salt additions of sulfate. II. Aluminum and base cations. Soil Sci. 151, 208–219.

Davidson, E.A., 2009. The contribution of manure and fertilizer nitrogen to atmospheric nitrous oxide since 1860. Nat. Geosci. 2, 659–662.

Davidson, E.A., Kingerlee, W., 1997. A global inventory of nitric oxide emissions from soils. Nutr. Cycl. Agroecosyst. 48, 37–50.

Davidson, E.A., Swank, W.T., 1987. Factors limiting denitrification in soils from mature and disturbed southeastern hardwood forests. For. Sci. 33, 135–144.

Davidson, E.A., Swank, W.T., Perry, T.O., 1986. Distinguishing between nitrification and denitrification as sources of gaseous nitrogen production in soil. Appl. Environ. Microbiol. 52 (6), 1280–1286.

Davidson, E.A., Stark, J.M., Firestone, M.K., 1990. Microbial production and consumption of nitrate in an annual grassland. Ecology 71, 1968–1975.

Davidson, E.A., Hart, S.C., Shanks, C.A., Firestone, M.K., 1991. Measuring gross nitrogen mineralization, immobilization, and nitrification by ^{15}N isotopic pool dilution in intact soil cores. J. Soil Sci. 42, 335–349.

Davidson, E.A., Hart, S.C., Firestone, M.K., 1992. Internal cycling of nitrate in soils of a mature coniferous forest. Ecology 73, 1148–1156.

Davidson, E.A., Potter, C.S., Schlesinger, P., Klooster, S.A., 1998. Model estimates of regional nitric oxide emissions from soils of the Southeastern United States. Ecol. Appl. 8, 748–759.

Davidson, E.A., Keller, M., Erickson, H.E., Verchot, L.V., Veldkamp, E., 2000. Testing a conceptual model of soil emissions of nitrous and nitric oxides. Bioscience 50, 667–680.

Davidson, E.A., Chorover, J., Dail, D.B., 2003. A mechanism of abiotic immobilization of nitrate in forest ecosystems: the ferrous wheel hypothesis. Glob. Chang. Biol. 9, 228–236.

Davidson, E.A., Reis de Carvalho, C.J., Figueira, A.M., Ishida, F.Y., Ometto, J.P.H.B., Nardoto, G.B., Sabá, R.T., et al., 2007. Recuperation of nitrogen cycling in Amazonian forests following agricultural abandonment. Nature 447 (7147), 995–998.

Davidson, E.A., de Araújo, A.C., Artaxo, P., Balch, J.K., Foster Brown, I., Bustamante, M.M.C., Coe, M.T., et al., 2012. The Amazon basin in transition. Nature 481 (7381), 321–328.

Davidson, N.C., Fluet-Chouinard, E., Finlayson, C.M., 2018. Global extent and distribution of wetlands: trends and issues. Mar. Freshw. Res. 69 (4), 620–627.

Davies, W.G., 1972. Introduction to Chemical Thermodynamics. W.B. Saunders, Philadelphia.

Davies-Barnard, T., Friedlingstein, P., 2020. The global distribution of biological nitrogen fixation in terrestrial natural ecosystems. Glob. Biogeochem. Cycles. https://doi.org/10.1029/2019GB006387.

Davis, R.B., Anderson, D.S., Berge, F., 1985. Paleolimnological evidence that lake acidification is accompanied by loss of organic matter. Nature 316 (6027), 436–438.

Davison, W., 1993. Iron and manganese in lakes. Earth Sci. Rev. 34, 119–163.

Davison, W., Woof, C., Rigg, E., 1982. The dynamics of iron and manganese in a seasonally anoxic lake—direct measurement of fluxes using sediment traps. Limnol. Oceanogr. 27 (6), 987–1003.

Dawson, T.E., Ehleringer, J.R., 1991. Streamside trees that do not use stream water. Nature 350, 335–337.

Day Jr., J.W., Boesch, D.F., Clairain, E.J., Paul Kemp, G., Laska, S.B., Mitsch, W.J., Orth, K., et al., 2007. Restoration of the Mississippi delta: lessons from Hurricanes Katrina and Rita. Science 315 (5819), 1679–1684.

Day, T.A., Neale, P.J., 2002. Effects of UV-B radiation on terrestrial and aquatic primary producers. Annu. Rev. Ecol. Syst. 33, 371–396.

de Bergh, C., 1993. The D/H ratio and the evolution of water in the terrestrial planets. Orig. Life Evol. Biosph. 23, 11–21.

de Bergh, C., Bézard, B., Owen, T., Crisp, D., Maillard, J.-P., Lutz, B.L., 1991. Deuterium on Venus: observations from Earth. Science 251 (February), 547–549.

de Duve, C., 1995. The beginnings of life on earth. Am. Sci. 83, 428–437.

de Graaf, I.E.M., Gleeson, T., van Beek, R.L.P.H., Sutanudjaja, E.H., Bierkens, M.F.P., 2019. Environmental flow limits to global groundwater pumping. Nature 574 (7776), 90–94.

De Jonge, V.N., Villerius, L.A., 1989. Possible role of carbonate dissolution in estuarine phosphate dynamics. Limnol. Oceanogr. 34, 332–340.

De Kimpe, C.R., Martel, Y.A., 1976. Effects of vegetation on distribution of carbon, iron, and aluminum in B horizons of Horthern Appalachian Spodosols. Soil Sci. Soc. Am. J. 40, 77–80.

de la Mare, W.K., 1997. Abrupt mid-twentieth-century decline in Antarctic sea-ice extent from whaling records. Nature 389, 57–60.

De La Rocha, C.L., DePaolo, D.J., 2000. Isotopic evidence for variations in the marine calcium cycle over the Cenozoic. Science 289, 1176–1178.

De Sanctis, M.C., Ammannito, E., McSween, H.Y., Raponi, A., Marchi, S., Capaccioni, F., Capria, M.T., et al., 2017. Localized aliphatic organic material on the surface of ceres. Science 355 (6326), 719–722.

de Villiers, S., Nelson, B.K., 1999. Detection of low-temperature hydrothermal fluxes by seawater Mg and Ca anomalies. Science 285 (5428), 721–723.

Dean, W.E., Gorham, E., 1998. Magnitude and significance of carbon burial in lakes, reservoirs, and peatlands. Geology 26, 535–538.

Dean, J.M., Smith, A.P., 1978. Behavioral and morphological adaptations of a tropical plant to high rainfall. Biotropica 10, 152–154.

Death, R.G., Winterbourn, M.J., 1995. Diversity patterns in stream benthic invertebrate communities: the influence of habitat stability. Ecology 76, 1446–1460.

DeBano, L.F., Conrad, C.E., 1978. Effect of fire on nutrients in a chaparral ecosystem. Ecology 59, 489–497.

DeBano, L.F., Klopatek, J.M., 1988. Phosphorus dynamics of Pinyon-Juniper soils following simulated burning. Soil Sci. Soc. Am. J. 52, 271–277.

DeBell, D.S., Ralston, C.W., 1970. Release of nitrogen by burning light forest fuels. Proc. Soil Sci. Soc. Am. 34, 936–938.

DeConto, R.M., Pollard, D., 2016. Contribution of Antarctica to past and future sea-level rise. Nature 531 (7596), 591–597.

Deemer, B.R., Harrison, J.A., Li, S., Beaulieu, J.J., DelSontro, T., Barros, N., Bezerra-Neto, J.F., Powers, S.M., dos Santos, M.A., Arie Vonk, J., 2016. Greenhouse gas emissions from reservoir water surfaces: a new global synthesis. Bioscience 66 (11), 949–964.

Deevey Jr., E.S., 1970. Mineral cycles. Sci. Am. 223 (3), 149–158.

DeFries, R.S., Houghton, R.A., Hansen, M.C., Field, C.B., Skole, D., Townshend, J., 2002. Carbon emissions from tropical deforestation and regrowth based on satellite observations for the 1980s and 1990s. Proc. Natl. Acad. Sci. U. S. A. 99 (22), 14256–14261.

Deirmendjian, L., Loustau, D., Augusto, L., Lafont, S., Chipeaux, C., Poirier, D., Abril, G., 2018. Hydrological and ecological controls on dissolved carbon concentrations in groundwater and carbon export to surface waters in a temperate pine forest watershed. Biogeosciences 15, 669–691.

Deiss, L., de Moraes, A., Maire, V., 2018. Environmental drivers of soil phosphorus composition in natural ecosystems. Biogeosciences 15 (14), 4575–4592.

Dekas, A.E., Poretsky, R.S., Orphan, V.J., 2009. Deep-Sea Archaea fix and share nitrogen in methane-consuming microbial consortia. Science 326 (5951), 422–426.

del Arco, J.M., Escudero, A., Garrido, M.V., 1991. Effects of site characteristics on nitrogen retranslocation from senescing leaves. Ecology 72, 701–708.

del Giorgio, P.A., Cole, J.J., 1998. Bacterial growth efficiency in natural aquatic systems. Annu. Rev. Ecol. Syst. 29, 503–541.

Del Grosso, S., Parton, W., Stohlgren, T., Zheng, D., Bachelet, D., Prince, S., Hibbard, K., Olson, R., 2008. Global potential net primary production predicted from vegetation class, precipitation, and temperature. Ecology 89 (8), 2117–2126.

Delaney, M.L., 1998. Phosphorus accumulation in marine sediments and the oceanic phosphorus cycle. Glob. Biogeochem. Cycles 12, 563–572.

DeLaune, R.D., Baumann, R.H., Gosselink, J.G., 1983. Relationships among vertical accretion, coastal submergence, and erosion in a Louisiana Gulf Coast Marsh. J. Sediment. Petrol. 53, 147–157.

Delcourt, H.R., Harris, W.F., 1980. Carbon budget of the southeastern U.S. biota: analysis of historical change in trend from source to sink. Science 210 (4467), 321–323.

Delgado-Baquerizo, M., Maestre, F.T., Gallardo, A., Bowker, M.A., Wallenstein, M.D., Quero, J.L., Ochoa, V., et al., 2013. Decoupling of soil nutrient cycles as a function of aridity in global drylands. Nature 502 (7473), 672–676.

Delgiorgio, P.A., Peters, R.H., 1994. Patterns in planktonic P-R ratios in lakes—influence of lake trophy and dissolved organic carbon. Limnol. Oceanogr. 39 (4), 772–787.

Della Chiesa, T., Piñeiro, G., Yahdjian, L., 2019. Gross, background, and net anthropogenic soil nitrous oxide emissions from soybean, corn, and wheat croplands. J. Environ. Qual. 48 (1), 16–23.

Delmas, R., Servant, J., 1983. Atmospheric balance of sulfur above an equatorial forest. Tellus Ser. B Chem. Phys. Meteorol. 35, 110–120.

Delmas, R.A., Marenco, A., Tathy, J.P., Cros, B., Baudet, J.G.R., 1991. Sources and sinks of methane in the African Savanna: CH_4 emissions from biomass burning. J. Geophys. Res.-Atmos. 96, 7287–7299.

Delmas, R.A., Tathy, J.P., Cros, B., 1992. Atmospheric methane budget in Africa. J. Atmos. Chem. 14, 395–409.

Delmas, R., Lacaux, J.P., Menaut, J.C., Abbadie, L., Leroux, X., Helas, G., Lobert, J., 1995. Nitrogen compound emission from biomass burning in tropical African Savanna FOS/DECAFE-1991 experiment (Lamto, Ivory Coast). J. Atmos. Chem. 22, 175–193.

DelSontro, T., Beaulieu, J.J., Downing, J.A., 2018. Greenhouse gas emissions from lakes and impoundments: upscaling in the face of global change. Limnol. Oceanogr. Lett. 3 (3), 64–75.

DeLuca, T.H., Keeney, D.R., 1993. Glucose-induced nitrate assimilation in prairie and cultivated soils. Biogeochemistry 21, 167–176.

DeLuca, T.H., Zackrisson, O., Nilsson, M.-C., Sellstedt, A., 2002. Quantifying nitrogen-fixation in feather moss carpets of boreal forests. Nature 419 (6910), 917–920.

DeLucia, E.H., Schlesinger, W.H., 1991. Resource-use efficiency and drought tolerance in adjacent Great Basin and Sierran plants. Ecology 72, 51–58.

DeLucia, E.H., Schlesinger, W.H., 1995. Photosynthetic rates and nutrient-use efficiency among evergreen and deciduous shrubs in Okefenokee swamp. Int. J. Plant Sci. 156, 19–28.

DeLucia, E.H., Callaway, R.M., Thomas, E.M., Schlesinger, W.H., 1997. Mechanisms of phosphorus acquisition for Ponderosa pine seedlings under high CO_2 and temperature. Ann. Bot. 79, 111–120.

DeLucia, E.H., Drake, J.E., Thomas, R.B., Gonzalez-Meler, M., 2007. Forest carbon use efficiency: is respiration a constant fraction of gross primary production? Glob. Chang. Biol. 13, 1157–1167.

DeLucia, E.H., Schlesinger, W.H., Billings, W.D., 1988. Water relations and the maintenance of Sierran conifers on hydrothermally-altered rock. Ecology 69, 303–311.

Demars, B.O.L., 2008. Whole-stream phosphorus cycling: testing methods to assess the effect of saturation of sorption capacity on nutrient uptake length measurements. Water Res. 42 (10–11), 2507–2516.

DeMaster, D.J., 2002. The accumulation and cycling of biogenic silica in the Southern Ocean: revisiting the marine silica budget. Deep-Sea Res. Part I—Top. Stud. Oceanogr. 49, 3155–3167.

Demers, J.D., Driscoll, C.T., Fahey, T.J., Yavitti, J.B., 2007. Mercury cycling in litter and soil in different forest types in the Adirondack Region, New York, USA. Ecol. Appl. 17 (5), 1341–1351.

Deng, Q., Hui, D., Dennis, S., Chandra Reddy, K., 2017. Responses of terrestrial ecosystem phosphorus cycling to nitrogen addition: a meta-analysis: DENG et al. Global Ecol. Biogeogr. J. Macroecol. 26 (6), 713–728.

Denier van der Gon, H.A.C., Neue, H.U., 1995. Influence of organic matter incorporation on the methane emission from a wetland rice field. Glob. Biogeochem. Cycles 9 (1), 11–22.

Denman, K.L., Brasseur, G., Chidthaisong, A., Ciais, P., Cox, P.M., Dickinson, R.E., Hauglustaine, D., et al., 2007. Couplings between changes in the climate system and biogeochemistry. In: Qin, D., Solomon, S., Manning, M., Chen, Z., Marquis, M., Averyt, K.B., Tignor, M., Miller, H.L. (Eds.), Climate Change 2007: The Physical Science Basis. Contribution of Working Group I to the Fourth Assessment Report of the Intergovernmental Panel on Climate Change. Cambridge University Press, pp. 499–587.

Dentener, F., Peters, W., Krol, M., van Weele, M., Bergamaschi, P., Lelieveld, J., 2003. Interannual variability and trend of CH_2 lifetime as a measure for OH changes in the 1979–1993 time period. J. Geophys. Res.-Atmos. 108.

Dentener, F., Drevet, J., Lamarque, J.F., Bey, I., Eickhout, B., Fiore, A.M., Hauglustaine, D., et al., 2006. Nitrogen and sulfur deposition on regional and global scales: a multimodel evaluation. Glob. Biogeochem. Cycles 20.

Dere, A.L., White, T.S., April, R.H., Reynolds, B., Miller, T.E., Knapp, E.P., McKay, L.D., Brantley, S.L., 2013. Climate dependence of feldspar weathering in shale soils along a latitudinal gradient. Geochim. Cosmochim. Acta 122 (December), 101–126.

Derry, L.A., Kurtz, A.C., Ziegler, K., Chadwick, O.A., 2005. Biological control of terrestrial silica cycling and export fluxes to watersheds. Nature 433 (7027), 728–731.

Des Marais, D.J., Strauss, H., Summons, R.E., Hayes, J.M., 1992. Carbon isotope evidence for the stepwise oxidation of the proterozoic environment. Nature 359 (6396), 605–609.

Deschamps, P., Durand, N., Bard, E., Hamelin, B., Camoin, G., Thomas, A.L., Henderson, G., Okuno, J., Yokoyama, Y., 2012. Ice-sheet collapse and sea-level rise at the bolling warming 14,600 years ago. Nature 483, 559–564.

Després, V., Huffman, J.A., Burrows, S.M., Hoose, C., Safatov, A., Buryak, G., Fröhlich-Nowoisky, J., et al., 2012. Primary biological aerosol particles in the atmosphere: a review. Tellus Ser. B Chem. Phys. Meteorol. 64 (1), 15598.

Dessert, C., Clergue, C., Rousteau, A., Crispi, O., Benedetti, M.F., 2020. Atmospheric contribution to cations cycling in highly weathered catchment, Guadeloupe (Lesser Antilles). Chem. Geol. 531. https://doi.org/10.1016/j.chemgeo.2019.119354v.

Dethier, D.P., Jones, S.B., Feist, T.P., Ricker, J.E., 1988. Relations among sulfate, aluminum, iron, dissolved organic carbon and pH in upland forest soils of northwestern Massachusetts. Soil Sci. Soc. Am. J. 52, 506–512.

Deuser, W.G., Ross, E.H., Anderson, R.F., 1981. Seasonality in the supply of sediment to the deep Sargasso Sea and implications for the rapid transfer of matter to the deep ocean. Deep-Sea Res. Part A—Oceanogr. Res. Pap. 28, 495–505.

Deuser, W.G., Brewer, P.G., Jickells, T.D., Commeau, R.F., 1983. Biological control of the removal of abiogenic particles from the surface ocean. Science 219 (4583), 388–391.

Deutsch, C., Weber, T., 2012. Nutrient ratios as a tracer and driver of ocean biogeochemistry. Annu. Rev. Mar. Sci. 4, 113–141.

Deutsch, C., Gruber, N., Key, R.M., Sarmiento, J.L., Ganachaud, A., 2001. Denitrification and N_2 fixation in the Pacific Ocean. Glob. Biogeochem. Cycles 15, 483–506.

Deutsch, C., Sarmiento, J.L., Sigman, D.M., Gruber, N., Dunne, J.P., 2007. Spatial coupling of nitrogen inputs and losses in the ocean. Nature 445 (7124), 163–167.

Devai, I., Delaune, R.D., 1995. Evidence for phosphine production and emission from Louisiana and Florida Marsh soils. Org. Geochem. 23 (3), 277–279.

Dévai, I., Felföldy, L., Wittner, I., Plósz, S., 1988. Detection of phosphine: new aspects of the phosphorus cycle in the hydrosphere. Nature 333 (6171), 343–345.

Deventer, M.J., Jiao, Y., Knox, S.H., Anderson, F., Ferner, M.C., Lewis, J.A., Rhew, R.C., 2018. Ecosystem-scale measurements of methyl halide fluxes from a brackish tidal Marsh invaded with perennial pepperweed (*Lepidium latifolium*). J. Geophys. Res. Biogeosci. 123 (7), 2104–2120.

Devol, A.H., 1991. Direct measurement of nitrogen gas fluxes from continental shelf sediments. Nature 349, 319–321.

DeVries, T., 2014. The oceanic anthropogenic CO_2 sink: storage, air-sea fluxes, and transports over the industrial era. Glob. Biogeochem. Cycles 28 (7), 631–647.

DeVries, T., Deutsch, C., Primeau, F., Chang, B., Devol, A., 2012. Global rates of water-column denitrification derived from nitrogen gas measurements. Nat. Geosci. 5 (8), 547–550.

DeVries, T., Holzer, M., Primeau, F., 2017. Recent increase in oceanic carbon uptake driven by weaker upper-ocean overturning. Nature 542 (7640), 215–218.

DeVries, T., Le Quéré, C., Andrews, O., Berthet, S., Hauck, J., Ilyina, T., Landschützer, P., et al., 2019. Decadal trends in the ocean carbon sink. Proc. Natl. Acad. Sci. U. S. A. 116 (24), 11646–11651.

DeWalle, D.R., Swistock, B.R., Johnson, T.E., McGuire, K.J., 2000. Potential effects of climate change and urbanization on mean annual streamflow in the United States. Water Resour. Res. 36, 2655–2664.

Dey, S., Di Girolamo, L., 2011. A decade of change in aerosol properties over the Indian subcontinent. Geophys. Res. Lett. 38.

Dhamala, B.R., Mitchell, M.J., 1995. Sulfur speciation, vertical distribution, and seasonal variation in a northern hardwood forest soil, USA. Can. J. For. Res. 25, 234–243.

Dhillon, K.S., Dhillon, S.K., et al., 2018. Genesis of seleniferous soils and associated animal and human health problems. Adv. Agron. 154, 1–80.

Dhingra, R.D., Barnes, J.W., Brown, R.H., Burrati, B.J., Sotin, C., Nicholson, P.D., Baines, K.H., et al., 2019. Observational evidence for summer rainfall at Titan's north pole. Geophys. Res. Lett. 46 (3), 1205–1212.

Di, H.J., Condron, L.M., Frossard, E., 1997. Isotope techniques to study phosphorus cycling in agricultural and forest soils: a review. Biol. Fertil. Soils 24, 1–12.

Di, H.J., Cameron, K.C., McLaren, R.G., 2000. Isotopic dilution methods to determine the gross transformation rates of nitrogen, phosphorus, and sulfur in soil: a review of the theory, methodologies, and limitations. Aust. J. Soil Res. 38, 213–230.

Di, H.J., Cameron, K.C., Shen, J.P., Winefield, C.S., O'Callaghan, M., Bowatte, S., He, J.Z., 2009. Nitrification driven by bacteria and not archaea in nitrogen-rich grassland soils. Nat. Geosci. 2, 621–624.

Dia, A.N., Cohen, A.S., O'Nions, R.K., Shackleton, N.J., 1992. Seawater Sr isotope variation over the past 300 Kyr and influence of global climate cycles. Nature 356, 786–788.

Dianov-Klokov, V.I., Yurganov, L.N., Grechko, E.I., Dzhola, A.V., 1989. Spectroscopic measurements of atmospheric carbon monoxide and methane. 1: Latitudinal distribution. J. Atmos. Chem. 8, 139–151.

Diaz, R.J., 2016. Anoxia, hypoxia, and dead zones. In: Kennish, M.J. (Ed.), Encyclopedia of Estuaries. Springer, The Netherlands, pp. 19–29.

Diaz, R.J., Rosenberg, R., 2008. Spreading dead zones and consequences for marine ecosystems. Science 321 (5891), 926–929.

Diaz, J., Ingall, E., Benitez-Nelson, C., Paterson, D., de Jonge, M.D., McNulty, I., Brandes, J.A., 2008. Marine polyphosphate: a key player in geologic phosphorus sequestration. Science 320 (5876), 652–655.

Diaz-Ravina, M., Acea, M.J., Carballas, T., 1993. Microbial biomass and its contribution to nutrient concentrations in forest soils. Soil Biol. Biochem. 25, 25–31.

Dick, G.J., 2019. The microbiomes of deep-sea hydrothermal vents: distributed globally, shaped locally. Nat. Rev. Microbiol. 17 (5), 271–283.

Dickerson, R.E., 1978. Chemical evolution and the origin of life. Sci. Am. 239 (3), 70–86.

Dickson, R.R., Brown, J., 1994. The production of north Atlantic deep water: sources, rates, and pathways. J. Geophys. Res. NATO ASI Ser. 99 (C6), 12319.

Dickson, M.L., Wheeler, P.A., 1995. Nitrate uptake rates in a coastal upwelling regime: a comparison of pN-specific, absolute, and Chl A-specific rates. Limnol. Oceanogr. 40, 533–543.

Dickson, B., Yashayaev, I., Meincke, J., Turrell, B., Dye, S., Holfort, J., 2002. Rapid freshening of the deep North Atlantic Ocean over the past four decades. Nature 416 (6883), 832–837.

DiDonato, N., Chen, H., Waggoner, D., Hatcher, P.G., 2016. Potential origin and formation for molecular components of humic acids in soils. Geochim. Cosmochim. Acta 178 (April), 210–222.

Dieleman, W.I.J., Luyssaert, S., Rey, A., de Angelis, P., Barton, C.V.M., Broadmeadow, M.S.J., Broadmeadow, S.B., et al., 2010. Soil [N] modulates soil C cycling in CO_2-fumigated tree stands: a meta-analysis. Plant Cell Environ. 33 (12), 2001–2011.

Dieng, H.B., Cazenave, A., Meyssignac, B., Ablain, M., 2017. New estimate of the current rate of sea level rise from a sea level budget approach. Geophys. Res. Lett. 44 (8), 3744–3751.

Dieter, C.A., Maupin, M.A., Caldwell, R.R., Harris, M.A., Ivahnenko, T.I., Lovelace, J.K., Barber, N.L., Linsey, K.S., 2018. Circular. Circular, Reston, VA.

Di-Giovanni, C., Disnar, J.R., Macaire, J.J., 2002. Estimation of the annual yield of organic carbon released from carbonates and shales by chemical weathering. Glob. Planet. Chang. 32, 195–210.

Dijkstra, F.A., Prior, S.A., Brett Runion, G., Allen Torbert, H., Tian, H., Lu, C., Venterea, R.T., 2012. Effects of elevated carbon dioxide and increased temperature on methane and nitrous oxide fluxes: evidence from field experiments. Front. Ecol. Environ. 10 (10), 520–527.

Dijkstra, F.A., Morgan, J.A., Follett, R.F., Lecain, D.R., 2013. Climate change reduces the net sink of CH_4 and N_2O in a semiarid grassland. Glob. Chang. Biol. 19 (6), 1816–1826.

Dillaway, D.N., Kruger, E.L., 2010. Thermal acclimation of photosynthesis: a comparison of boreal and temperate tree species along a latitudinal transect. Plant Cell Environ. 33 (6), 888–899.

Dillon, P.J., Evans, H.E., 1993. A comparison of phosphorus retention in lakes determined from mass-balance and sediment core calculations. Water Res. 27, 659–668.

Dillon, P.J., Evans, H.E., 2001. Comparison of iron accumulation in lakes using sediment core and mass balance calculations. Sci. Total Environ. 266 (1–3), 211–219.

Dillon, P.J., Molot, L.A., 1997. Effect of landscape form on export of dissolved organic carbon, iron, and phosphorus from forested stream catchments. Water Resour. Res.

Dillon, P.J., Molot, L.A., 2005. Long-term trends in catchment export and lake retention of dissolved organic carbon, dissolved organic nitrogen, total iron, and total phosphorus: the dorset, ontario, study, 1978–1998. J. Geophys. Res. Biogeosci. 110(G1).

Dingman, S.L., Johnson, A.H., 1971. Pollution potential of some New Hampshire Lakes. Water Resour. Res. 7, 1208–1215.

Dinkelaker, B., Marschner, H., 1992. In vivo demonstration of acid-phosphatase activity in the rhizosphere of soil-grown plants. Plant Soil 144, 199–205.

Dirmeyer, P.A., Shukla, J., 1996. The effect on regional and global climate of expansion of the world's deserts. Q. J. R. Meteorol. Soc. 122, 451–482.

Dise, N.B., Verry, E.S., 2001. Suppression of peatland methane emission by cumulative sulfate deposition in simulated acid rain. Biogeochemistry 53, 143–160.

Dittman, J.A., Shanley, J.B., Driscoll, C.T., Aiken, G.R., Chalmers, A.T., Towse, J.E., Selvendiran, P., 2010. Mercury dynamics in relation to dissolved organic carbon concentration and quality during high flow events in three northeastern US streams. Water Resour. Res. 46.

Dixit, S., Van Cappellen, P., van Bennekom, A.J., 2001. Processes controlling solubility of biogenic silica and pore water build-up of silicic acid in marine sediments. Mar. Chem. 73, 333–352.

Dixon, K.W., Pate, J.S., Bailey, W.J., 1980. Nitrogen nutrition of the tuberous Sundew Drosera Erythrorhiza Lindl with special reference to catch of arthropod fauna by its glandular leaves. Aust. J. Bot. 28, 283–297.

Dixon, J.E., Leist, L., Langmuir, C., Schilling, J.-G., 2002. Recycled dehydrated lithosphere observed in plume-influenced mid-ocean-ridge basalt. Nature 420 (6914), 385–389.

Dixon, J.L., Hartshom, A.S., Heimsath, A.M., DiBiase, R.A., Whipple, K.X., 2012. Chemical weathering response to tectonic forcing: a soils perspective from the San Gabriel Mountains, California. Earth Planet. Sci. Lett. 323 (324), 40–49.

Dlugokencky, E.J., Steele, L.P., Lang, P.M., Masarie, K.A., 1994. The growth rate and distribution of atmospheric methane. J. Geophys. Res. 99, 17021–17043.

Dlugokencky, E.J., Bruhwiler, L., White, J.W.C., Emmons, L.K., Novelli, P.C., Montzka, S.A., Masarie, K.A., Lang, P.M., Crotwell, A.M., Miller, J.B., Gatti, L.V., 2009. Observational constraints on recent increases in the atmospheric CH4 burden. Geophys. Res. Lett. 36.

Dlugokencky, E.J., Nisbet, E.G., Fisher, R., Lowry, D., 2011. Global atmospheric methane: budget, changes and dangers. Philos. Trans. A Math. Phys. Eng. Sci, 369 (1943), 2058–2072.

do Carmo, J.B., Keller, M., Dias, J.D., de Camargo, P.B., Crill, P., 2006. A source of methane from upland forests in the Brazilian Amazon. Geophys. Res. Lett. 33.

Dobrovolsky, V.V., 1994. Biogeochemistry of the World's Land. CRC Press.

Dodd, J.C., Burton, C.C., Burns, R.G., Jeffries, P., 1987. Phosphatase activity associated with the roots and the rhizosphere of plants infected with vesicular-arbuscular mycorrhizal fungi. New Phytol. 107, 163–172.

Dodd, M.S., Papineau, D., Grenne, T., Slack, J.F., Rittner, M., Pirajno, F., O'Neil, J., Little, C.T.S., 2017. Evidence for early life in earth's oldest hydrothermal vent precipitates. Nature 543 (7643), 60–64.

Dodds, W.K., Smith, V.H., Lohman, K., 2002. Nitrogen and phosphorus relationships to benthic algal biomass in temperate streams. Can. J. Fish. Aquat. Sci. 59, 865–874.

Doetterl, S., Stevens, A., Six, J., Merckx, R., Van Oost, K., Pinto, M.C., Casanova-Katny, A., et al., 2015. Soil carbon storage controlled by interactions between geochemistry and climate. Nat. Geosci. 8 (10), 780–783.

Doll, P., Fiedler, K., 2008. Global-scale modeling of groundwater recharge. Hydrol. Earth Syst. Sci. 12, 863–885.

Döll, P., Schmied, H.M., Schuh, C., Portmann, F.T., Eicker, A., 2014. Global-scale assessment of groundwater depletion and related groundwater abstractions: combining hydrological modeling with information from well observations and GRACE satellites. Water Resour. Res. 50 (7), 5698–5720.

Döll, P., Trautmann, T., Göllner, M., 2019. A global-scale analysis of water storage dynamics of inland wetlands: quantifying the impacts of human water use and man-made reservoirs as well as the unavoidable and avoidable impacts of climate change. Ecohydrology 13, https://doi.org/10.1002/eco.2175.

Don, A., Schumacher, J., Freibauer, A., 2011. Impact of tropical land-use change on soil organic carbon stocks: a meta-analysis. Glob. Chang. Biol. 17, 1658–1670.

Donahue, T.M., Hoffman, J.H., Hodges Jr., R.R., Watson, A.J., 1982. Venus was wet: a measurement of the ratio of deuterium to hydrogen. Science 216 (4546), 630–633.

Donarummo, J., Ram, M., Stoermer, E.F., 2003. Possible deposit of soil dust from the 1930's US dust bowl identified in Greenland ice. Geophys. Res. Lett. 30.

Donat, J.R., Bruland, K.W., 1995. Trace elements in the oceans. In: Salbu, B., Steinnes, E. (Eds.), Trace Elements in Natural Waters. CRC Press, pp. 247–281.

Doney, S.C., 2010. The growing human footprint on coastal and open-ocean biogeochemistry. Science 328 (5985), 1512–1516.

Doney, S.C., Fabry, V.J., Feely, R.A., Kleypas, J.A., 2009. Ocean acidification: the other CO_2 problem. Annu. Rev. Mar. Sci. 1, 169–192.

Dong, L.F., Sobey, M.N., Smith, C.J., Rusmana, I., Phillips, W., Andrew, S., Mark Osborn, A., Nedwell, D.B., 2011. Dissimilatory reduction of nitrate to ammonium, not denitrification or anammox, dominates benthic nitrate reduction in tropical estuaries. Limnol. Oceanogr. 56 (1), 279–291.

Doolittle, R.F., Feng, D.F., Tsang, S., Cho, G., Little, E., 1996. Determining divergence times of the major kingdoms of living organisms with a protein clock. Science 271 (5248), 470–477.

Dore, J.E., Lukas, R., Sadler, D.W., Church, M.J., Karl, D.M., 2009. Physical and biogeochemical modulation of ocean acidification in the Central North Pacific. Proc. Natl. Acad. Sci. U. S. A. 106 (30), 12235–12240.

Dorn, H.-P., Callies, J., Platt, U., Ehhalt, D.H., 1988. Measurement of tropospheric OH concentration by laser long-path absorption spectroscopy. Tellus Ser. B Chem. Phys. Meteorol. 40, 437–445.

Dornblaser, M., Giblin, A.E., Fry, B., Peterson, B.J., 1994. Effects of sulfate concentration in the overlying water on sulfate reduction and sulfur storage in lake sediments. Biogeochemistry 24, 129–144.

Dorrepaal, E., Toet, S., van Logtestijn, R.S.P., Swart, E., van de Weg, M.J., Callaghan, T.V., Aerts, R., 2009. Carbon respiration from subsurface peat accelerated by climate warming in the subarctic. Nature 460, 616–619.

Doughty, C.E., Wolf, A., Malhi, Y., 2013. The legacy of the *Pleistocene megafauna* extinctions on nutrient availability in Amazonia. Nat. Geosci. 6 (9), 761–764.

Dove, N.C., Stark, J.M., Newman, G.S., Hart, S.C., 2019. Carbon control on terrestrial ecosystem function across contrasting site productivities: the carbon connection revisited. Ecology. 100 (7), e02695.

Downing, J.A., Mccauley, E., 1992. The nitrogen: phosphorus relationship in lakes. Limonol. Oceanogr. 37 (5), 936–945.

Downing, J.a., Prairie, Y.T., Cole, J.J., Duarte, C.M., Tranvik, L.J., Striegl, R.G., McDowell, W.H., Kortelainen, P., Caraco, N.F., Melack, J.M., 2006. The global abundance and size distribution of lakes, ponds, and impoundments. Limnol. Oceanogr. 51 (5), 2388–2397.

Downing, J.A., Cole, J.J., Middelburg, J.J., Striegl, R.G., Duarte, C.M., Kortelainen, P., Prairie, Y.T., Laube, K.A., 2008. Sediment organic carbon burial in agriculturally eutrophic impoundments over the last century. Glob. Biogeochem. Cycles 22.

Doyle, M.W., Bernhardt, E.S., 2011. What is a stream? Environ. Sci. Technol. 45 (2), 354–359.

Doyle, R.D., Fisher, T.R., 1994. Nitrogen fixation by periphyton and plankton on the Amazon floodplain at Lake Calado. Biogeochemistry 26, 41–66.

Draaijers, G.P., Ivens, W.P., Bos, M.M., Bleuten, W., 1989. The contribution of ammonia emissions from agriculture to the deposition of acidifying and eutrophying compounds onto forests. Environ. Pollut. 60 (1–2), 55–66.

Drake, J.B., Knox, R.G., Dubayah, R.O., Clark, D.B., Condit, R., Blair, J.B., Hofton, M., 2003. Aboveground biomass estimation in closed canopy neotropical forests using Lidar remote sensing: factors affecting the generality of relationships. Global Ecol. Biogeogr. J. Macroecol. 12, 147–159.

Drake, J.E., Gallet-Budynek, A., Hofmockel, K.S., Bernhardt, E.S., Billings, S.A., Jackson, R.B., Johnsen, K.S., et al., 2011. Increases in the flux of carbon belowground stimulate nitrogen uptake and sustain the long-term enhancement of forest productivity under elevated CO_2. Ecol. Lett. 14, 349–357.

Drake, H., Åström, M.E., Tullborg, E.-L., Whitehouse, M., Fallick, A.E., 2013. Variability of sulphur isotope ratios in pyrite and dissolved sulphate in granitoid fractures down to 1km depth—evidence for widespread activity of sulphur reducing bacteria. Geochim. Cosmochim. Acta 102 (February), 143–161.

Dregne, H., 1976. Soils of Arid Regions, second ed. Elsevier.

Drever, J.I., 1994. The effect of land plants on weathering rates of silicate minerals. Geochim. Cosmochim. Acta 58, 2325–2332.

Drever, J.I., Smith, C.L., 1978. Cyclic wetting and drying of the soil zone as an influence on chemistry of ground water in arid terrains. Am. J. Sci. 278, 1448–1454.

Drewitt, J.W.E., Walter, M.J., Zhang, H., McMahon, S.C., Edwards, D., Heinen, B.J., Lord, O.T., Anzellini, S., Kleppe, A.K., 2019. The fate of carbonate in oceanic crust subducted into Earth's lower mantle. Earth Planet. Sci. Lett. 511 (April), 213–222.

Driscoll, C.T., Likens, G.E., 1982. Hydrogen ion budget of an aggrading forested ecosystem. Tellus 34, 283–292.

Driscoll, C.T., Vanbreemen, N., Mulder, J., 1985. Aluminum chemistry in a forested spodosol. Soil Sci. Soc. Am. J. 49, 437–444.

Driscoll, C.T., Han, Y.-J., Chen, C.Y., Evers, D.C., Lambert, K.F., Holsen, T.M., Kamman, N.C., Munson, R.K., 2007. Mercury contamination in forest and freshwater ecosystems in the northeastern United States. Bioscience 57 (1), 17–28.

Drouet, T., Herbauts, J., Gruber, W., Demaiffe, D., 2005. Strontium isotope composition as a tracer of calcium sources in two forest ecosystems in Belgium. Geoderma 126, 203–223.

Drout, M.R., Piro, A.L., Shappee, B.J., Kilpatrick, C.D., Simon, J.D., Contreras, C., Coulter, D.A., et al., 2017. Light curves of the neutron star merger GW170817/SSS17a: implications for R-process nucleosynthesis. Science 358 (6370), 1570–1574.

Druffel, E.R.M., Williams, P.M., Bauer, J.E., Ertel, J.R., 1992. Cycling of dissolved and particulate organic matter in the open ocean. J. Geophys. Res. 97, 15639–15659.

Drury, C.F., Voroney, R.P., Beauchamp, E.G., 1991. Availability of NH4þ-N to microorganisms and the soil internal N cycle. Soil Biol. Biochem. 23, 165–169.

Drury, C.F., McKenney, D.J., Findlay, W.I., 1992. Nitric oxide and nitrous oxide production from soil: water and oxygen effects. Soil Sci. Soc. Am. J. 56, 766–770.

Du, Q., Zhang, C., Yujing, M., Cheng, Y., Zhang, Y., Liu, C., Song, M., et al., 2016. An important missing source of atmospheric carbonyl sulfide: domestic coal combustion: COS from domestic coal combustion. Geophys. Res. Lett. 43 (16), 8720–8727.

Du, E., Xia, N., de Vries, W., 2019, February. Effects of nitrogen deposition on growing-season soil methane sink across global forest biomes. Biogeosci. Dis, 1–16.

Duarte, C.M., Kalff, J., 1989. The influence of catchment geology and lake depth on phytoplankton biomass. Arch. Hydrobiol. 115, 27–40.

Dubayah, R.O., Sheldon, S.L., Clark, D.B., Hofton, M.A., Blair, J.B., Hurtt, G.C., Chazdon, R.L., 2010. Estimation of tropical forest height and biomass dynamics using lidar remote sensing at La Selva, Costa Rica. J. Geophys. Res. Biogeosci. 115.

Dubinsky, E.A., Silver, W.L., Firestone, M.K., 2010. Tropical forest soil microbial communities couple iron and carbon biogeochemistry. Ecology 91 (9), 2604–2612.

Duce, R.A., Tindale, N.W., 1991. Atmospheric transport of iron and its deposition in the ocean. Limnol. Oceanogr. 36, 1715–1726.

Duce, R.A., Unni, C.K., Ray, B.J., Prospero, J.M., Merrill, J.T., 1980. Long-range atmospheric transport of soil dust from Asia to the tropical north Pacific: temporal variability. Science 209 (4464), 1522–1524.

Duce, R.A., Liss, P.S., Merrill, J.T., Atlas, E.L., Buat-Menard, P., Hicks, B.B., Miller, J.M., et al., 1991. The atmospheric input of trace species to the world ocean. Glob. Biogeochem. Cycles 5, 193–259.

Duce, R.A., LaRoche, J., Altieri, K., Arrigo, K.R., Baker, A.R., Capone, D.G., Cornell, S., et al., 2008. Impacts of atmospheric anthropogenic nitrogen on the open ocean. Science 320 (5878), 893–897.

Ducet, N., Le Traon, P.Y., Reverdin, G., 2000. Global high-resolution mapping of ocean circulation from TOPEX/Poseidon and ERS-1 and -2. J. Geophys. Res. 105, 19477–19498.

Ducklow, H.W., Carlson, C.A., 1992. Oceanic bacterial production. Adv. Microb. Ecol. 12, 113–181.

Ducklow, H.W., Purdie, D.A., Williams, P.J., Davies, J.M., 1986. Bacterioplankton: a sink for carbon in a coastal marine plankton community. Science 232 (4752), 865–867.

Dueck, T.A., de Visser, R., Poorter, H., Persijn, S., Gorissen, A., de Visser, W., Schapendonk, A., et al., 2007. No evidence for substantial aerobic methane emission by terrestrial plants: a 13C-labelling approach. New Phytol. 175 (1), 29–35.

Duff, S.M.G., Sarath, G., Plaxton, W.C., 1994. The role of acid phosphatases in plant phosphorus metabolism. Physiol. Plant. 90, 791–800.

Dugdale, R.C., Goering, J.J., 1967. Uptake of new and regenerated forms of nitrogen in primary production. Limnol. Oceanogr. 12, 196–206.

Dukes, J.S., 2003. Burning buried sunshine: human consumption of ancient solar energy. Clim. Chang. 61, 31–44.

Duncan, B.N., Logan, J.A., Bey, I., Megretskaia, I.A., Yantosca, R.M., Novelli, P.C., Jones, N.B., Rinsland, C.P., 2007. Global budget of CO, 1988–1997: source estimates and validation with a global model. J. Geophys. Res. 112.

Duncan, J.M., Groffman, P.M., Band, L.E., 2013. Towards closing the watershed nitrogen budget: spatial and temporal scaling of denitrification: scaling denitrification. J. Geophys. Res. Biogeosci. 118 (3), 1105–1119.

Duncan, B.N., Lamsal, L.N., Thompson, A.M., Yoshida, Y., Lu, Z., Streets, D.G., Hurwitz, M.M., Pickering, K.E., 2016. A space-based, high-resolution view of notable changes in urban NO_x pollution around the world (2005–2014): notable changes in urban NO_x pollution. J. Geophys. Res. D: Atmos. 121 (2), 976–996.

Dunfield, P., Knowles, R., Dumont, R., Moore, T.R., 1993. Methane production and consumption in temperate and sub-Arctic peat soils: response to temperature and pH. Soil Biol. Biochem. 25, 321–326.

Dunn, P.H., DeBano, L.F., Eberlein, G.E., 1979. Effects of burning on chaparral soils. II. Soil microbes and nitrogen mineralization. Soil Sci. Soc. Am. J. 43, 509–514.

Dunne, T., Leopold, L.B., 1978. Water in Environmental Planning. Freeman, San Francisco.

Durack, P., 2015. Ocean salinity and the global water cycle. Oceanography.

Durack, P.J., Wijffels, S.E., 2010. Fifty-year trends in global ocean salinities and their relationship to broad-scale warming. J. Clim. 23 (16), 4342–4362.

Durán, J., Morse, J.L., Rodríguez, A., Campbell, J.L., Christenson, L.M., Driscoll, C.T., Fahey, T.J., et al., 2017. Differential sensitivity to climate change of C and N cycling processes across soil horizons in a northern hardwood forest. Soil Biol. Biochem. 107 (April), 77–84.

Duren, R.M., Thorpe, A.K., Foster, K.T., Rafiq, T., Hopkins, F.M., Yadav, V., Bue, B.D., et al., 2019. California's methane super-emitters. Nature 575 (7781), 180–184.

Durka, W., Schulze, E.D., Gebauer, G., Voerkelius, S., 1994. Effects of forest decline on uptake and leaching of deposited nitrate determined from 15N and 18O measurements. Nature 372, 765–767.

Durr, H.H., Meybeck, M., Durr, S.H., 2005. Lithologic composition of the Earth's continental surfaces derived from a new digital map emphasizing riverine material transfer. Glob. Biogeochem. Cycles 19.

Dusenge, M.E., Duarte, A.G., Way, D.A., 2019. Plant carbon metabolism and climate change: elevated CO_2 and temperature impacts on photosynthesis, photorespiration and respiration. New Phytol. 221 (1), 32–49.

Dutaur, L., Verchot, L.V., 2007. A global inventory of the soil CH_4 sink. Glob. Biogeochem. Cycles 21.

Dutton, A., Lambeck, K., 2012. Ice volume and sea level during the last interglacial. Science 337 (6091), 216–219.

Dynarski, K.A., Houlton, B.Z., 2018. Nutrient limitation of terrestrial free-living nitrogen fixation. New Phytol. 217 (3), 1050–1061.

Dynesius, M., Nilsson, C., 1994. Fragmentation and flow regulation of river systems in the northern third of the world. Science 266 (5186), 753–762.

Dyonisius, M.N., Petrenko, V.V., Smith, A.M., Hua, Q., Yang, B., Schmitt, J., Beck, J., Seth, B., Bock, M., Hmiel, B., Vimont, I., Menking, J.A., Shackleton, S.A., Baggenstos, D., Auska, T.K., Rhodes, R.H., Sperlich, P., Beaudette, R., Harth, C., Kalk, M., Brook, E.J., Fischer, H., Severinghaus, J.P., Weiss, R.F., 2020. Old carbon reservoirs were not important in the deglacial methane budget. Science 367, 907–910.

Dyrness, C.T., Vancleve, K., Levison, J.D., 1989. The effect of wildfire on soil chemistry in four forest types in interior Alaska. Can. J. For. Res. 19, 1389–1396.

Ebrahimi, A., Or, D., 2018. Dynamics of soil biogeochemical gas emissions shaped by remolded aggregate sizes and carbon configurations under hydration cycles. Glob. Chang. Biol. 24 (1), e378–e392.

Eckert, W., Conrad, R., 2007. Sulfide and methane evolution in the hypolimnion of a subtropical lake: a three-year study. Biogeochemistry 82 (1), 67–76.

Eckstein, R.L., Karlsson, P.S., Weih, M., 1999. Leaf life span and nutrient resorption as determinants of plant nutrient conservation in temperate-Arctic regions. New Phytol. 143, 177–189.

Edmond, J.M., Measures, C., McDuff, R.E., Chan, L.H., Collier, R., Grant, B., Gordon, L.I., Corliss, J.B., 1979. Ridge crest hydrothermal activity and the balances of the major and minor elements in the ocean: the Galapagos data. Earth Planet. Sci. Lett. 46 (1), 1–18.

Edmondson, W.T., Lehman, J.T., 1981. The effect of changes in the nutrient income on the condition of Lake Washington. Limnol. Oceanogr. 26, 1–29.

Edwards, N.T., 1975. Effects of temperature and moisture on carbon dioxide evolution in a mixed deciduous forest floor. Soil Sci. Soc. Am. J. 39, 361–365.

Edwards, P.J., 1977. Studies of mineral cycling in a montane rain forest in new guinea. II. Production and disappearance of litter. J. Ecol. 65, 971–992.

Edwards, P.J., 1982. Studies of mineral cycling in a montane rain forest in new guinea. X. Rates of cycling in throughfall and litter fall. J. Ecol. 70, 807–827.

Edwards, P.J., Grubb, P.J., 1982. Studies of mineral cycling in a montane rain forest in new guinea. IV. Soil characteristics and the division of mineral elements between the vegetation and soil. J. Ecol. 70, 649–666.

Edwards, N.T., Harris, W.F., 1977. Carbon cycling in a mixed deciduous forest floor. Ecology 58, 431–437.

Edwards, N.T., Sollins, P., 1973. Continuous measurement of carbon dioxide evolution from partitioned forest floor components. Ecology 34, 406–412.

Eger, A., Yoo, K., Almond, P.C., Boitt, G., Larsen, I.J., Condron, L.M., Wang, X., Mudd, S.M., 2018. Does soil erosion rejuvenate the soil phosphorus inventory? Geoderma 332 (December), 45–59.

Eghbal, M.K., Southard, R.J., Whittig, L.D., 1989. Dynamics of evaporite distribution in soils on a Fan-Playa transect in the Carrizo Plain, California. Soil Sci. Soc. Am. J. 53, 898–903.

Ehhalt, D.H., 1974. The atmospheric cycle of methane. Tellus 26, 58–70.

Ehleringer, J.R., Monson, R.K., 1993. Evolutionary and ecological apsects of photosynthetic pathway variation. Annu. Rev. Ecol. Syst. 24, 411–439.

Ehlmann, B.L., Mustard, J.F., Murchie, S.L., Poulet, F., Bishop, J.L., Brown, A.J., Calvin, W.M., et al., 2008. Orbital identification of carbonate-bearing rocks on mars. Science 322 (5909), 1828–1832.

Ehrlich, H.L., 1975. Formation of ores in the sedimentary environment of the deep sea with microbial participation: the case for ferromanganese concretions. Soil Sci. 119, 36–41.

Ehrlich, H.L., 1982. Enhanced removal of Mn^{2+} from seawater by marine sediments and clay minerals in the presence of bacteria. Can. J. Microbiol. 28, 1389–1395.

Eigenbrode, J.L., Summons, R.E., Steele, A., Freissinet, C., Millan, M., Navarro-González, R., Sutter, B., et al., 2018. Organic matter preserved in 3-billion-year-old mudstones at Gale Crater, Mars. Science 360 (6393), 1096–1101.

Einsele, G., Yan, J., Hinderer, M., 2001. Atmospheric carbon burial in modern lake basins and its significance for the global carbon budget. Glob. Planet. Chang. 30 (3), 167–195.

Eisele, K.A., Schimel, D.S., Kapustka, L.A., Parton, W.J., 1989. Effects of available P-ratio and N-P ratio on non-symbiotic dinitrogen fixation in tallgrass praire soils. Oecologia 79, 471–474.

Eissenstat, D.M., Yanai, R.D., 1997. The ecology of root lifespan. Adv. Ecol. Res. 27, 1–60.

El Albani, A., Bengtson, S., Canfield, D.E., Bekker, A., Macchiarelli, R., Mazurier, A., Hammarlund, E.U., et al., 2010. Large colonial organisms with coordinated growth in oxygenated environments 2.1 Gyr ago. Nature 466 (7302), 100–104.

Elbert, W., Hoffmann, M.R., Kramer, M., Schmitt, G., Andreae, M.O., 2000. Control of solute concentrations in cloud and fog water by liquid water content. Atmos. Environ. 34, 1109–1122.

Elderfield, H., Rickaby, R.E., 2000. Oceanic Cd/P ratio and nutrient utilization in the glacial southern ocean. Nature 405 (6784), 305–310.

Elderfield, H., Schultz, A., 1996. Mid-ocean ridge hydrothermal fluxes and the chemical composition of the ocean. Annu. Rev. Earth Planet. Sci. 24 (1), 191–224.

Elkins, J.W., Thompson, T.M., Swanson, T.H., Butler, J.H., Hall, B.D., Cummings, S.O., Fisher, D.A., Raffo, A.G., 1993. Decrease in the growth rates of atmospheric chlorofluorocarbon-11 and chlorofluorocarbon-12. Nature 364, 780–783.

Elliott, E.T., 1986. Aggregate structure and carbon, nitrogen, and phosphorus in native and cultivated soils. Soil Sci. Soc. Am. J. 50, 627–633.

Ellis, J.C., Fariña, J.M., Witman, J.D., 2006. Nutrient transfer from sea to land: the case of gulls and cormorants in the Gulf of Maine. J. Anim. Ecol. 75 (2), 565–574.

Ellis, E.C., Goldewijk, K.K., Siebert, S., Lightman, D., Ramankutty, N., 2010. Anthropogenic transformation of the biomes, 1700 to 2000. Global Ecol. Biogeogr. J. Macroecol. 19, 589–606.

Ellsworth, D.S., Crous, K.Y., Lambers, H., Cooke, J., 2015. Phosphorus recycling in photorespiration maintains high photosynthetic capacity in woody species. Plant Cell Environ. 38 (6), 1142–1156.

Elrod, V.A., Berelson, W.M., Coale, K.H., Johnson, K.S., 2004. The flux of iron from continental shelf sediments: a missing source for global budgets. Geophys. Res. Lett. 31(12).

Elser, J.J., Hassett, R.P., 1994. A stoichiometric analysis of the zooplankton-phytoplankton interaction in marine and freshwater ecosystems. Nature 370, 211–213.

Elser, J.J., Chrzanowski, T.H., Sterner, R.W., Mills, K.H., 1998. Stoichiometric constraints on food-web dynamics: a whole-lake experiment on the Canadian shield. Ecosystems 1, 120–136.

Elser, J.J., Fagan, W.F., Denno, R.F., Dobberfuhl, D.R., Folarin, A., Huberty, A., Interlandi, S., et al., 2000a. Nutritional constraints in terrestrial and freshwater food webs. Nature 408 (6812), 578–580.

Elser, J.J., Sterner, R.W., Gorokhova, E., Fagan, W.F., Markow, T.A., Cotner, J.B., Harrison, J.F., Hobbie, S.E., Odell, G.M., Weider, L.J., 2000b. Biological stoichiometry from genes to ecosystems. Ecol. Lett. 3, 540–550.

Elser, J.J., Bracken, M.E.S., Cleland, E.E., Gruner, D.S., Harpole, W.S., Hillebrand, H., Ngai, J.T., et al., 2007. Global analysis of nitrogen and phosphorus limitation of primary producers in freshwater, marine and terrestrial ecosystems. Ecol. Lett. 10 (12), 1135–1142.

Elser, J.J., Andersen, T., Baron, J.S., Bergström, A.-K., Jansson, M., Kyle, M., Nydick, K.R., Steger, L., Hessen, D.O., 2009. Shifts in lake N:P stoichiometry and nutrient limitation driven by atmospheric nitrogen deposition. Science 326 (5954), 835–837.

Elser, J.J., Fagan, W.F., Kerkhoff, A.J., Swenson, N.G., Enquist, B.J., 2010. Biological stoichiometry of plant production: metabolism, scaling and ecological response to global change. New Phytol. 186 (3), 593–608.

Elsgaard, L., Isaksen, M.F., Jorgensen, B.B., Alayse, A.M., Jannasch, H.W., 1994. Microbial sulfate reduction in deep sea sediments at the guaymas basin hydrothermal vent area: influence of temperature and substrates. Geochim. Cosmochim. Acta 58, 3335–3343.

Eltahir, E.A.B., Bras, R.L., 1994. Precipitation recycling in the Amazon basin. Q. J. R. Meteorol. Soc. 120, 861–880.

Elton, C.S., 1927. Animal Ecology, by Charles Elton; with an Introduction by Julian S. Huxley.

Emanuel, W.R., Fung, I.Y.-S., Killough, G.G., Moore, B., Peng, T.-H., 1985. Modeling the global carbon cycle and changes in the atmospheric carbon dioxide. In: Trabalka, J.R. (Ed.), Atmospheric Carbon Dioxide and the Global Carbon Cycle. U.S. Department of Energy, Washington, DC, pp. 141–173.

Emerson, S., Stump, C., 2010. Net biological oxygen production in the ocean-II: Remote in situ measurements of O_2 and N_2 in subarctic pacific surface waters. Deep-Sea Res. Part I—Oceanogr. Res. Pap. 57, 1255–1265.

Emerson, S., Fischer, K., Reimers, C., Heggie, D., 1985. Organic carbon dynamics and preservation in deep sea sediments. Deep-Sea Res. Part A—Oceanogr. Res. Pap. 32, 1–21.

Emmett, B.A., Hudson, J.A., Coward, P.A., Reynolds, B., 1994. The impact of a Riparian wetland on streamwater quality in a recently afforested upland catchment. J. Hydrol. 162, 337–353.

Engel, M.H., Macko, S.A., 1997. Isotopic evidence for extraterrestrial non-racemic amino acids in the murchison meteorite. Nature 389 (6648), 265–268.

Engel, M.H., Macko, S.A., 2001. The stereochemistry of amino acids in the murchison meteorite. Precambrian Res. 106, 35–45.

Engel, M.H., Macko, S.A., Silfer, J.A., 1990. Carbon isotope composition of individual amino acids in the murchison meteorite. Nature 348 (6296), 47–49.

Engle, M.A., Gustin, M.S., Goff, F., Counce, D.A., Janik, C.J., Bergfeld, D., Rytuba, J.J., 2006. Atmospheric mercury emissions from substrates and fumaroles associated with three hydrothermal systems in the Western United States. J. Geophys. Res. 111.

Engle, M.A., Tate, M.T., Krabbenhoft, D.P., Schauer, J.J., Kolker, A., Shanley, J.B., Bothner, M.H., 2010. Comparison of atmospheric mercury speciation and deposition at nine sites across central and eastern North America. J. Geophys. Res. D: Atmos. 115.

Engström, P., Dalsgaard, T., Hulth, S., Aller, R.C., 2005. Anaerobic ammonium oxidation by nitrite (anammox): implications for N_2 production in coastal marine sediments. Geochim. Cosmochim. Acta 69 (8), 2057–2065.

Enquist, B.J., Niklas, K.J., 2002. Global allocation rules for patterns of biomass partitioning in seed plants. Science 295 (5559), 1517–1520.

Enquist, B.J., Kerkhoff, A.J., Stark, S.C., Swenson, N.G., McCarthy, M.C., Price, C.A., 2007. A general integrative model for scaling plant growth, carbon flux, and functional trait spectra. Nature 449 (7159), 218–222.

Entwistle, E.M., Romanowicz, K.J., Argiroff, W.A., Freedman, Z.B., Jeffrey Morris, J., Zak, D.R., 2018. Anthropogenic N deposition alters the composition of expressed class II fungal peroxidases. Appl. Environ. Microbiol. 84. https://doi.org/10.1128/AEM.02816-17.

Eppley, R.W., Peterson, B.J., 1979. Particulate organic matter flux and planktonic new production in the deep ocean. Nature 282, 677–680.

Eppley, R.W., Stewart, E., Abbott, M.R., Heyman, U., 1985. Estimating ocean primary production from satellite chlorophyll. Introduction to regional differences and statistics for the southern California bight. J. Plankton Res. 7, 57–70.

Epstein, H.E., Burke, I.C., Lauenroth, W.K., 2002. Regional patterns of decomposition and primary production rates in the US great plains. Ecology 83, 320–327.

Erb, T.J., Zarzycki, J., 2018. A short history of RubisCO: the rise and fall (?) of nature's predominant CO_2 fixing enzyme. Curr. Opin. Biotechnol. 49 (February), 100–107.

Erb, T.J., Kiefer, P., Hattendorf, B., Günther, D., Vorholt, J.A., 2012. GFAJ-1 is an arsenate-resistant, phosphate-dependent organism. Science 337 (6093), 467–470.

Erb, K.-H., Kastner, T., Plutzar, C., Bais, A.L.S., Carvalhais, N., Fetzel, T., Gingrich, S., et al., 2018. Unexpectedly large impact of forest management and grazing on global vegetation biomass. Nature 553 (7686), 73–76.

Eriksson, E., 1960. The yearly circulation of chloride and sulfur in nature; meteorological, geochemical and pedological implications. Part II. Tellus 12, 63–109.

Erisman, J.W., Bleeker, A., Galloway, J., Sutton, M.S., 2007. Reduced nitrogen in ecology and the environment. Environ. Pollut. 150 (1), 140–149.

Escudero, A., Del Arco, J.M., Sanz, I.C., Ayala, J., 1992. Effects of leaf longevity and retranslocation efficiency on the retention time of nutrients in the leaf biomass of different woody species. Oecologia 90 (1), 80–87.

Esculier, F., Le Noë, J., Barles, S., Billen, G., Créno, B., Garnier, J., Lesavre, J., Petit, L., Tabuchi, J.-P., 2019. The biogeochemical imprint of human metabolism in Paris megacity: a regionalized analysis of a water-agro-food system. J. Hydrol. 573 (June), 1028–1045.

Essington, T.E., Carpenter, S.R., 2000. Nutrient cycling in lakes and streams: insights from a comparative analysis. Ecosystems 3, 131–143.

Estiarte, M., Peñuelas, J., 2015. Alteration of the phenology of leaf senescence and Fall in Winter deciduous species by climate change: effects on nutrient proficiency. Glob. Chang. Biol. 21 (3), 1005–1017.

Eswaran, H., Berg, E., Reich, P., Vandenberg, E., 1993. Organic-carbon in soils of the world. Soil Sci. Soc. Am. J. 57 (1), 192–194.

Etheridge, D.M., Steele, L.P., Francey, R.J., Langenfelds, R.L., 1998. Atmospheric methane between 1000 AD and present: evidence of anthropogenic emissions and climatic variability. J. Geophys. Res. 103, 15979–15993.

Etiope, G., Lassey, K.R., Klusman, R.W., Boschi, E., 2008. Reappraisal of the fossil methane budget and related emission from geologic sources. Geophys. Res. Lett. 35.

Etminan, M., Myhre, G., Highwood, E.J., Shine, K.P., 2017. Radiative forcing of carbon dioxide, methane, and nitrous oxide: a significant revision of the methane radiative forcing. Geophys. Res. Lett. 12, 614–623.

Eugster, O., Gruber, N., 2012. A probabilistic estimate of global marine N-fixation and denitrification: probabilistic estimate of N source/sink. Glob. Biogeochem. Cycles 26 (4), 4013.

Euliss, N.H., Gleason, R.A., Olness, A., McDougal, R.L., Murkin, H.R., Robarts, R.D., Bourbonniere, R.A., Warner, B.G., 2006. North American Prairie wetlands are important nonforested land-based carbon storage sites. Sci. Total Environ. 361, 179–188.

Evans, J.R., 1989. Photosynthesis and nitrogen relationships in leaves of C_3 plants. Oecologia 78 (1), 9–19.

Evans, R.D., Ehleringer, J.R., 1993. A break in the nitrogen cycle in Aridlands? Evidence from deltaN-15 of soils. Oecologia 94, 314–317.

Evans, C.D., Monteith, D.T., Cooper, D.M., 2005. Long-term increases in surface water dissolved organic carbon: observations, possible causes and environmental impacts. Environ. Pollut. 137 (1), 55–71.

Evans, N.P., Bauska, T.K., Gázquez-Sánchez, F., Brenner, M., Curtis, J.H., Hodell, D.A., 2018. Quantification of drought during the collapse of the classic maya civilization. Science 361 (6401), 498–501.

Evaristo, J., Jasechko, S., McDonnell, J.J., 2015. Global separation of plant transpiration from groundwater and streamflow. Nature 525 (7567), 91–94.

Ewing, S.A., Sutter, B., Owen, J., Nishiizumi, K., Sharp, W., Cliff, S.S., Perry, K., Dietrich, W., McKay, C.P., Amundson, R., 2006. A threshold in soil formation at Earth's arid-hyperarid transition. Geochim. Cosmochim. Acta 70, 5293–5322.

Exner, M.E., Hirsh, A.J., Spalding, R.F., 2014. Nebraska's groundwater legacy: nitrate contamination beneath irrigated cropland. Water Resour. Res. 50 (5), 4474–4489.

Eyre, B., 1994. Nutrient biogeochemistry in the tropical moresby river estuary system north Queensland, Australia. Estuar. Coast. Shelf Sci. 39, 15–31.

Fahey, T.J., 1983. Nutrient dynamics of above-ground detritus in lodgepole pine (*Pinus contorta* ssp. *latifolia*) ecosystems, southeastern Wyoming. Ecol. Monogr. 53, 51–72.

Fahey, T.J., Hughes, J.W., 1994. Fine root dynamics in a northern hardwood forest ecosystem, Hubbard Brook experimental forest, NH. J. Ecol. 82, 533–548.

Fahey, T.J., Tierney, G.L., Fitzhugh, R.D., Wilson, G.F., Siccama, T.G., 2005. Soil respiration and soil carbon balance in a northern hardwood forest ecosystem. Can. J. For. Res. 35, 244–253.

Fahey, T.J., Yavitt, J.B., Sherman, R.E., Groffman, P.M., Fisk, M.C., Maerz, J.C., 2011. Transport of carbon and nitrogen between litter and soil organic matter in a northern hardwood forest. Ecosystems 14, 326–340.

Fairbanks, R.G., 1989. A 17,000-year glacio-eustatic sealevel record: influence of glacial melting rates on the younger dryas event and deep ocean circulation. Nature 342, 637–642.

Falini, G., Albeck, S., Weiner, S., Addadi, L., 1996. Control of aragonite or calcite polymorphism by Mollusk shell macromolecules. Science 271 (5245), 67–69.

Falkengren-Grerup, U., 1995. Interspecies differences in the preference of ammonium and nitrate in vascular plants. Oecologia 102 (3), 305–311.

Falkowski, P.G., 1997. Evolution of the nitrogen cycle and its influence on the biological sequestration of CO_2 in the ocean. Nature 387, 272–275.

Falkowski, P.G., 2005. Biogeochemistry of primary production in the sea. In: Schlesinger, W.H. (Ed.), Biogeochemistry. Elsevier, Amsterdam, pp. 185–213.

Falkowski, P.G., Kim, Y., Kolber, Z., Wilson, C., Wirick, C., Cess, R., 1992. Natural versus anthropogenic factors affecting low-level cloud albedo over the north atlantic. Science 256 (5061), 1311–1313.

Falkowski, P.G., Barber, R.T., V. Smetacek V., 1998. Biogeochemical controls and feedbacks on ocean primary production. Science 281 (5374), 200–207.

Falkowski, P.G., Fenchel, T., Delong, E.F., 2008. The microbial engines that drive Earth's biogeochemical cycles. Science 320 (5879), 1034–1039.

Falkowski, P.G., Algeo, T., Codispoti, L., 2011. Ocean Deoxygenation: Past, Present, and Future. Eos, Transactions, American Geophysical Union.

Faloona, I., Conley, S.A., Blomquist, B., Clarke, A.D., Kapustin, V., Howell, S., Lenschow, D.H., Bandy, A.R., 2009. Sulfur dioxide in the tropical marine boundary layer: dry deposition and heterogeneous oxidation observed during the pacific atmospheric sulfur experiment. J. Atmos. Chem. 63, 13–32.

Famiglietti, J.S., Rodell, M., 2013. Environmental science. Water in the balance. Science 340 (6138), 1300–1301.

Famiglietti, J.S., Lo, M., Ho, S.L., Bethune, J., Anderson, K.J., Syed, T.H., Swenson, S.C., de Linage, C.R., Rodell, M., 2011. Satellites measure recent rates of groundwater depletion in California's Central Valley. Geophys. Res. Lett. 38.

Fan, S., Gloor, M., Mahlman, J., Pacala, S., Sarmiento, J., Takahashi, T., Tans, P., 1998a. A large terrestrial carbon sink in North America implied by atmospheric and oceanic carbon dioxide data and models. Science 282 (5388), 442–446.

Fan, W.H., Randolph, J.C., Ehman, J.L., 1998b. Regional estimation of nitrogen mineralization in forest ecosystems using geographic information systems. Ecol. Appl. 8, 734–747.

Fan, Y., Li, H., Miguez-Macho, G., 2013. Global patterns of groundwater table depth. Science 339 (6122), 940–943.

Fanale, F.P., 1971. A case for catastrophic early degassing of the earth. Chem. Geol. 8, 79–105.

Fang, Y.T., Yoh, M., Koba, K., Zhu, W.X., Takebayashi, Y., Xiao, Y.H., Lei, C.Y., Mo, J.M., Zhang, W., Lu, X.K., 2011. Nitrogen deposition and forest nitrogen cycling along an urban-rural transect in Southern China. Glob. Chang. Biol. 17, 872–885.

Fang, Y., Koba, K., Makabe, A., Takahashi, C., Zhu, W., Hayashi, T., Hokari, A.A., et al., 2015. Microbial denitrification dominates nitrate losses from forest ecosystems. Proc. Natl. Acad. Sci. U. S. A. 112 (5), 1470–1474.

Fang, K., Qin, S., Chen, L., Zhang, Q., Yang, Y., 2019. Al/Fe mineral controls on soil organic carbon stock across Tibetan alpine grasslands. J. Geophys. Res. Biogeosci. 124 (2), 247–259.

Fanning, K.A., 1989. Influence of atmospheric pollution on nutrient limitation in the ocean. Nature 339, 460–463.

Farley, K.A., Neroda, E., 1998. Noble gases in the Earth's mantle. Annu. Rev. Earth Planet. Sci. 26, 189–218.

Farman, J.C., Gardiner, B.G., Shanklin, J.D., 1985. Large losses of total ozone in antarctica reveal seasonal ClO_x/NO_x interaction. Nature 315, 207–210.

Farquhar, G.D., Hubick, K.T., Condon, A.G., Richards, R.A., 1989. Carbon isotope fractionation and plant water-use efficiency. In: Rundel, P.W., Ehleringer, J.R., Nagy, K.A. (Eds.), Stable Isotopes in Ecological Research. Springer, pp. 21–40.

Farquhar, J., Bao, H., Thiemens, M., 2000. Atmospheric influence of Earth's earliest sulfur cycle. Science 289 (5480), 756–759.

Farquhar, J., Zerkle, A.L., Bekker, A., 2011. Geological constraints on the origin of oxygenic photosynthesis. Photosynth. Res. 107 (1), 11–36.

Farrar, J.F., 1985. The respiratory source of CO_2. Plant Cell Environ. 8, 427–438.

Fassbinder, J.W., Stanjek, H., Vali, H., 1990. Occurrence of magnetic bacteria in soil. Nature 343 (6254), 161–163.

Fatichi, S., Pappas, C., 2017. Constrained variability of modeled T:ET ratio across biomes: transpiration:evapotranspiration ratio. Geophys. Res. Lett. 44 (13), 6795–6803.

Fearnside, P.M., 1995. Hydroelectric dams in the Brazilian Amazon as sources of greenhouse gases. Environ. Conserv. 22, 7–19.

Federer, C.A., 1983. Nitrogen mineralization and nitrification: depth variation in four new England forest soils. Soil Sci. Soc. Am. J. 47, 1008–1014.

Federer, C.A., Hornbeck, J.W., 1985. The buffer capacity of forest soils in new England. Water Air Soil Pollut. 26, 163–173.

Federici, S., Tubiello, F.N., Salvatore, M., Jacobs, H., Schmidhuber, J., 2015. New estimates of CO_2 forest emissions and removals: 1990–2015. For. Ecol. Manag. 352 (September), 89–98.

Fee, E.J., Hecky, R.E., Regehr, G.W., Hendzel, L.L., Wilkinson, P., 1994. Effects of lake size on nutrient availability in the mixed layer during summer stratification. Can. J. Fish. Aquat. Sci. 51, 2756–2768.

Feely, R.A., Lamb, M.F., Greeley, D.J., Wanninkhof, R., 1999. Comparison of the carbon system parameters at the global CO_2 survey crossover locations in the North and South Pacific Ocean, 1990–1996. ORNL/CDIAC-115. Carbon Dioxide Information Analysis Center, Oak Ridge National Laboratory, U.S. Department of Energy. Oak Ridge, TN, 74 pp.

Feely, R.A., Sabine, C.L., Lee, K., Berelson, W., Kleypas, J., Fabry, V.J., Millero, F.J., 2004. Impact of anthropogenic CO_2 on the $CaCO_3$ system in the oceans. Science 305 (5682), 362–366.

Felix, J.D., Elliott, E.M., 2013. The agricultural history of human-nitrogen interactions as recorded in ice core $\delta_{15}N$-NO_3^-: ice core records agricultural history. Geophys. Res. Lett. 40 (8), 1642–1646.

Feller, M.C., Kimmins, J.P., 1979. Chemical Characteristics of Small Streams near Haney in Southwestern British Columbia. Water Resour. Res. 15, 247–258.

Fellows, C.S., Valett, M.H., Dahm, C.N., 2001. Wholestream metabolism in two montane streams: contribution of the hyporheic zone. Limnol. Oceanogr. 46 (3), 523–531.

Fenchel, T., 2008. The microbial loop—25 years later. J. Exp. Mar. Biol. Ecol. 366 (1–2), 99–103.

Feng, X., 1999. Trends in intrinsic water-use efficiency of natural trees for the past 100–200 years: a response to atmospheric CO_2 concentration. Geochim. Cosmochim. Acta 63, 1891–1903.

Feng, X., Xu, Y., Jaffé, R., Schlesinger, W.H., Simpson, M.J., 2010. Turnover rates of hydrolysable aliphatic lipids in duke forest soils determined by compound specific ^{13}C isotopic analysis. Org. Geochem. 41 (6), 573–579.

Feng, M., Sexton, J.O., Huang, C., Anand, A., Channan, S., Song, X.-P., Song, D.-X., Kim, D.-H., Noojipady, P., Townshend, J.R., 2016. Earth science data records of global forest cover and change: assessment of accuracy in 1990, 2000, and 2005 epochs. Remote Sens. Environ. 184 (October), 73–85.

Fenn, M.E., Poth, M.A., Aber, J.D., Baron, J.S., Bormann, B.T., Johnson, D.W., Lemly, A.D., McNulty, S.G., Ryan, D.E., Stottlemyer, R., 1998. Nitrogen excess in North American ecosystems: predisposing factors, ecosystem responses, and management strategies. Ecol. Appl. 8, 706–733.

Fenner, N., Freeman, C., 2011. Drought-induced carbon loss in peatlands. Nat. Geosci. 4 (12), 895–900.

Fenner, N., Ostle, N.J., McNamara, N., Sparks, T., Harmens, H., Reynolds, B., Freeman, C., 2007. Elevated CO_2 effects on peatland plant community carbon dynamics and DOC production. Ecosystems 10 (4), 635–647.

Fergus, C.E., Soranno, P.A., Cheruvelil, K.S., Bremigan, M.T., 2011. Multiscale landscape and wetland drivers of lake total phosphorus and water color. Limnol. Oceanogr. 56, 2127–2146.

Fernandes, A.M., da Conceição, F.T., Junior, E.P.S., de Souza Sardinha, D., Mortatti, J., 2016. Chemical weathering rates and atmospheric/soil CO_2 consumption of igneous and metamorphic rocks under tropical climate in southeastern Brazil. Chem. Geol. 443 (December), 54–66.

Fernandez, I.J., Rustad, L.E., Norton, S.A., Kahl, J.S., Cosby, B.J., 2003. Experimental acidification causes soil base-cation depletion at the bear brook watershed in maine. Soil Sci. Soc. Am. J. 67, 1909–1919.

Ferris, J.P., Hill Jr., A.R., Liu, R., Orgel, L.E., 1996. Synthesis of long prebiotic oligomers on mineral surfaces. Nature 381 (6577), 59–61.

Ferro-Vázquez, C., Nóvoa-Muñoz, J.C., Costa-Casais, M., Klaminder, J., Martínez-Cortizas, A., 2014. Metal and organic matter immobilization in temperate podzols: a high resolution study. Geoderma 217–218 (April), 225–234.

Ficklin, D.L., Abatzoglou, J.T., Robeson, S.M., Null, S.E., Knouft, J.H., 2018. Natural and managed watersheds show similar responses to recent climate change. Proc. Natl. Acad. Sci. U. S. A. 115 (34), 8553–8557.

Fiedler, S.S., 2004. Water and redox conditions in wetland soils—their influence on pedogenic oxides and morphology. Soil Sci. Soc. Am. J. 335, 326–335.

Field, C.B., Mooney, H.A., 1986. The photosynthesis-nitrogen relationship in wild plants. In: Givnish, T.J. (Ed.), On the Economy of Plant Form and Function. Cambridge University Press, Cambridge, pp. 25–55.

Field, C., Merino, J., Mooney, H.A., 1983. Compromises between water-use efficiency and nitrogen-use efficiency in five species of California evergreens. Oecologia 60 (3), 384–389.

Field, C.B., Randerson, J.T., Malmstrom, C.M., 1995. Global net primary production: combining ecology and remote sensing. Remote Sens. Environ. 51, 74–88.

Field, C.B., Behrenfeld, M.J., Randerson, J.T., Falkowski, P., 1998. Primary production of the biosphere: integrating terrestrial and oceanic components. Science 281 (5374), 237–240.

Field, J.P., Belnap, J., Breshears, D.D., Neff, J.C., Okin, G.S., Whicker, J.J., Painter, T.H., Ravi, S., Reheis, M.C., Reynolds, R.L., 2010. The ecology of dust. Front. Ecol. Environ. 8 (8), 423–430.

Fierer, N., Bradford, M.A., Jackson, R.B., 2007. Toward an ecological classification of soil bacteria. Ecology 88 (6), 1354–1364.

Fife, D.N., Nambiar, E.K.S., 1984. Movement of nutrients in Radiata pine needles in relation to the growth of shoots. Ann. Bot. 54, 303–314.

Filippelli, G.M., 1997. Controls on phosphorus concentration and accumulation in oceanic sediments. Mar. Geol. 139, 231–240.

Filippelli, G.M., 2008. The global phosphorus cycle: past, present, and future. Elements 4, 89–95.

Filippelli, G.M., Delaney, M.L., 1992. Similar phosphorus fluxes in ancient phosphorite deposits and a modern phosphogenic environment. Geology 20, 709–712.

Filippelli, G.M., Delaney, M.L., 1994. The oceanic phosphorus cycle and continental weathering during the neogene. Paleoceanography 9, 643–652.

Filippelli, G.M., Delaney, M.L., 1996. Phosphorus geochemistry of equatorial pacific sediments. Geochim. Cosmochim. Acta 60, 1479–1495.

Filippelli, G.M., Souch, C., 1999. Effects of climate and landscape development on the terrestrial phosphorus cycle. Geology 27, 171–174.

Fimmen, R.L., Richter, D.D., Vasudevan, D., Williams, M.A., West, L.T., 2008. Rhizogenic Fe-C redox cycling: a hypothetical biogeochemical mechanism that drives crustal weathering in upland soils. Biogeochemistry 87, 127–141.

Findlay, S.E.G., Sinsabaugh, R.L., 2003. Aquatic ecosystems interactivity of dissolved organic matter. In: Thorp, J.H. (Ed.), Aquatic Ecology Series. Academic Press, San Diego, CA.

Findlay, A.J., Pellerin, A., Laufer, K., Jorgensen, B.B., 2020. Quantification of sulphide oxidation rates in marine sediment. Geochim. Cosmochim. Acta 280, 441–452.

Fine, A.K., Schmidt, M.P., Martínez, C.E., 2018. Nitrogen-rich compounds constitute an increasing proportion of organic matter with depth in Oi-Oe-Oa-A horizons of temperate forests. Geoderma 323 (August), 1–12.

Fink, G., Alcamo, J., Flörke, M., Reder, K., 2018. Phosphorus loadings to the world's largest lakes: sources and trends. Glob. Biogeochem. Cycles 32 (4), 617–634.

Finlay, J.C., 2001. Stable-carbon-isotope ratios of river biota: implications for energy flow in lotic food webs. Ecology 82, 1052–1064.

Finlay, J.C., 2011. Stream size and human influences on ecosystem production in river networks. Ecosphere 2 (8), art87.

Finlay, J.C., Small, G.E., Sterner, R.W., 2013. Human influences on nitrogen. Science 342 (October), 247–250.

Finzi, A.C., 2009. Decades of atmospheric deposition have not resulted in widespread phosphorus limitation or saturation of tree demand for nitrogen in southern New England. Biogeochemistry 92, 217–229.

Finzi, A.C., Berthrong, S.T., 2005. The uptake of amino acids by microbes and trees in three cold-temperate forests. Ecology 86, 3345–3353.

Finzi, A.C., DeLucia, E.H., Hamilton, J.G., Richter, D.D., Schlesinger, W.H., 2002. The nitrogen budget of a pine forest under free air CO_2 enrichment. Oecologia 132 (4), 567–578.

Finzi, A.C., Moore, D.J.P., DeLucia, E.H., Lichter, J., Hofmockel, K.S., Jackson, R.B., Kim, H.-S., et al., 2006. Progressive nitrogen limitation of ecosystem processes under elevated CO_2 in a warm-temperate forest. Ecology 87 (1), 15–25.

Finzi, A.C., Norby, R.J., Calfapietra, C., Gallet-Budynek, A., Gielen, B., Holmes, W.E., Hoosbeek, M.R., et al., 2007. Increases in nitrogen uptake rather than nitrogen-use efficiency support higher rates of temperate forest productivity under elevated CO_2. Proc. Natl. Acad. Sci. U. S. A. 104 (35), 14014–14019.

Firestone, M.K., 1982. Biological denitrification. In: Stevenson, F.J. (Ed.), Nitrogen in Agricultural Soils. American Society of Agronomy, Madison, Wisconsin, pp. 289–326.

Firestone, M.K., Davidson, E.A., 1989. Microbiological basis of NO and N_2O production and consumption in soil. In: Andreae, M.O., Schimel, D.S. (Eds.), Exchange of Trace Gases Between Terrestrial Ecosystems and the Atmosphere. Wiley, pp. 7–21.

Firestone, M.K., Firestone, R.B., Tiedje, J.M., 1980. Nitrous oxide from soil denitrification: factors controlling its biological production. Science 208 (4445), 749–751.

Firing, Y.L., Chereskin, T.K., Mazloff, M.R., 2011. Vertical structure and transport of the antarctic circumpolar current in drake passage from direct velocity observations. J. Geophys. Res. 116 (C8), 170.

Firsching, B.M., Claassen, N., 1996. Root phosphatase activity and soil organic phosphorus utilization by Norway Spruce *Picea abies* (L.) Karst. Soil Biol. Biochem. 28, 1417–1424.

Fischer, T.P., Aiuppa, A., 2020. Volcanoes and deep carbon: global CO_2 emissions from subaerial volcanism: recent progress and future challenges. Geochem. Geophys. Geosyst. https://doi.org/10.1029/GC0086890.

Fischer, T.P., Lopez, T.M., 2016. First airborne samples of a volcanic plume for $\delta_{13}C$ of CO_2 determinations. Geophys. Res. Lett. 43 (7), 3272–3279.

Fischer, H., Wagenbach, D., Kipfstuhl, J., 1998. Sulfate and nitrate firn concentrations on the Greenland ice sheet: 2. Temporal anthropogenic deposition changes. J. Geophys. Res.-Atmos. 103, 21935–21942.

Fischer, R.A., Cottrell, E., Hauri, E., Lee, K.K.M., Le Voyer, M., 2020. The carbon content of Earth and its core. Proc. Natl Acad. Sci. 117, 8743–8749.

Fisher, R.F., 1972. Spodosol development and nutrient distribution under Hydnaceae fungal mats. Proc. Soil Sci. Soc. Am. 36, 492–495.

Fisher, R.F., 1977. Nitrogen and phosphorus mobilization by fairy ring fungus, *Marasmium oreades* (Bolt) Fr. Soil Biol. Biochem. 9, 239–241.

Fisher, S.G., Likens, G.E., 1972. Stream ecosystem: organic energy budget. Bioscience 22, 33–35.

Fisher, S.G., Likens, G.E., 1973. Energy flow in bear brook, new hampshire: an integrative approach to stream ecosystem metabolism. Ecol. Monogr. 43 (4), 421–439.

Fisher, S.G., Minckley, W.L., 1978. Chemical characteristics of a desert stream in flash flood. J. Arid Environ. 1, 25–33.

Fisher, J.B., Badgley, G., Blyth, E., 2012. Global nutrient limitation in terrestrial vegetation: global nutrient limitation in vegetation. Glob. Biogeochem. Cycles 26 (3), 1.

Fisher, J.B., Malhi, Y., Torres, I.C., Metcalfe, D.B., van de Weg, M.J., Meir, P., Silva-Espejo, J.E., Huasco, W.H., 2013. Nutrient limitation in rainforests and cloud forests along a 3,000-M elevation gradient in the Peruvian Andes. Oecologia 172 (3), 889–902.

Fishman, J., Wozniak, A.E., Creilson, J.K., 2003. Global distribution of tropospheric ozone from satellite measurements using the empirically corrected tropospheric ozone residual technique: identification of the regional aspects of air pollution. Atmos. Chem. Phys. 3, 893–907.

Fisk, M.R., Giovannoni, S.J., Thorseth, I.H., 1998. Alteration of oceanic volcanic glass: textural evidence of microbial activity. Science 281 (5379), 978–980.

Fissore, C., Baker, L.A., Hobbie, S.E., King, J.Y., McFadden, J.P., Nelson, K.C., Jakobsdottir, I., 2011. Carbon, nitrogen, and phosphorus fluxes in household ecosystems in the Minneapolis-Saint Paul, Minnesota, urban region. Ecological Applications: A Publication of the Ecological Society of America 21 (3), 619–639.

Fitzgerald, J.W., Andrew, T.L., Swank, W.T., 1984. Availability of carbon-bonded sulfur for mineralization in forest soils. Can. J. For. Res. 14, 839–843.

Fitzgerald, J.W., Strickland, T.C., Ash, J.T., 1985. Isolation and partial characterization of forest floor and soil organic sulfur. Biogeochemistry 1, 155–167.

Fitzhugh, R.D., Lovett, G.M., Venterea, R.T., 2003. Biotic and abiotic immobilization of ammonium, nitrite, and nitrate in soils developed under different tree species in the Catskill Mountains, New York, USA. Glob. Chang. Biol. 9, 1591–1601.

Flaig, W., Beutelspacher, H., Rietz, E., 1975. Chemical composition and physical properties of humic substances. In: Gieseking, J.E. (Ed.), Soil Components: Volume 1. Organic Components. Springer, pp. 1–211.

Flanagan, L.B., Adkinson, A.C., 2011. Interacting controls on productivity in a Northern Great Plains Grassland and implications for response to ENSO events: controls on grassland productivity. Glob. Chang. Biol. 17 (11), 3293–3311.

Flanner, M.G., Shell, K.M., Barlage, M., Perovich, D.K., Tschudi, M.A., 2011. Radiative forcing and albedo feedback from the northern hemisphere cryosphere between 1979 and 2008. Nat. Geosci. 4, 151–155.

Fletcher, S.E.M., Tans, P.P., Bruhwiler, L.M., Miller, J.B., Heimann, M., 2004. CH4 sources estimated from atmospheric observations of CH_4 and its $^{12}C/^{12}C$-isotopic ratios: 1. Inverse modeling of source processes. Glob. Biogeochem. Cycles 18(4).

Flint, R.F., 1971. Glacial and Quaternary Geology. Wiley.

Floudas, D., Binder, M., Riley, R., Barry, K., Blanchette, R.A., Henrissat, B., Martínez, A.T., et al., 2012. The paleozoic origin of enzymatic lignin decomposition reconstructed from 31 fungal genomes. Science 336 (6089), 1715–1719.

Flückiger, J., Monnin, E., Stauffer, B., Schwander, J., Stocker, T.F., Chappellaz, J., Raynaud, D., Barnola, J.-M., 2002. High-resolution holocene N_2O ice core record and its relationship with CH_4 and CO_2. Glob. Biogeochem. Cycles 16 (1), 10–11.

Fontaine, S., Barot, S., Barré, P., Bdioui, N., Mary, B., Rumpel, C., 2007. Stability of organic carbon in deep soil layers controlled by fresh carbon supply. Nature 450 (7167), 277–280.

Ford, B., Val Martin, M., Zelasky, S.E., Fischer, E.V., Anenberg, S.C., Heald, C.L., Pierce, J.R., 2018. Future fire impacts on smoke concentrations, visibility, and health in the contiguous United States. GeoHealth 2 (8), 229–247.

Forde, B., Lorenzo, H., 2001. The nutritional control of root development. Plant Soil 232, 51–68.

Forkel, M., Carvalhais, N., Rödenbeck, C., Keeling, R., Heimann, M., Thonicke, K., Zaehle, S., Reichstein, M., 2016. Enhanced seasonal CO_2 exchange caused by amplified plant productivity in northern ecosystems. Science 351 (6274), 696–699.

Forzieri, G., Alkama, R., Miralles, D.G., Cescatti, A., 2017. Satellites reveal contrasting responses of regional climate to the widespread greening of Earth. Science 356 (6343), 1180–1184.

Fossing, H., Gallardo, V.A., Jørgensen, B.B., Hüttel, M., Nielsen, L.P., Schulz, H., Canfield, D.E., et al., 1995. Concentration and transport of nitrate by the mat-forming sulphur bacterium Thioploca. Nature.

Foukal, P., Fröhlich, C., Spruit, H., Wigley, T.M.L., 2006. Variations in solar luminosity and their effect on the Earth's climate. Nature 443 (7108), 161–166.

Fowells, H.A., Krauss, R.W., 1959. The inorganic nutrition of loblolly pine and Virginia pine with special reference to nitrogen and phosphorus. For. Sci. 5, 95–111.

Fowler, W.A., 1984. The quest for the origin of the elements. Science 226, 922–935.

Fox, L.E., 1983. The removal of dissolved humic acid during estuarine mixing. Estuar. Coast. Shelf Sci. 16, 431–440.

Fox, T.R., Comerford, N.B., 1992. Influence of oxalate loading on phosphorus and aluminum solubility in spodosols. Soil Sci. Soc. Am. J. 56, 290–294.

Fox, G.E., Stackebrandt, E., Hespell, R.B., Gibson, J., Maniloff, J., Dyer, T.A., Wolfe, R.S., et al., 1980. The phylogeny of prokaryotes. Science 209 (4455), 457–463.

Frame, C.H., Deal, E., Nevison, C.D., Casciotti, K.L., 2014. N_2O production in the eastern south atlantic: analysis of N_2O stable isotopic and concentration data: N_2O in the South Atlantic. Glob. Biogeochem. Cycles 28 (11), 1262–1278.

France-Lanord, C., Derry, L.A., 1997. Organic carbon burial forcing of the carbon cycle from Himalayan erosion. Nature 390, 65–67.

Francis, C.A., Roberts, K.J., Beman, J.M., Santoro, A.E., Oakley, B.B., 2005. Ubiquity and diversity of ammonia-oxidizing Archaea in water columns and sediments of the ocean. Proc. Natl. Acad. Sci. U. S. A. 102 (41), 14683–14688.

Francoeur, S.N., 2001. Meta-analysis of lotic nutrient amendment experiments: detecting and quantifying subtle responses. J. N. Am. Benthol. Soc. 20, 358–368.

Francois, R., Altabet, M.A., Yu, E.F., Sigman, D.M., Bacon, M.P., Frank, M., Bohrmann, G., Bareille, G., Labeyrie, L.D., 1997. Contribution of southern ocean surface-water stratification to low atmospheric CO_2 concentrations during the last glacial period. Nature 389, 929–935.

Frank, D.A., Evans, R.D., 1997. Effects of native grazers on grassland N cycling in Yellowstone National Park. Ecology 78, 2238–2248.

Frank, D.A., Groffman, P.M., 1998. Ungulate vs. landscape control of soil C and N processes in grasslands of Yellowstone National Park. Ecology 79, 2229–2241.

Frank, D.A., Inouye, R.S., 1994. Temporal variation in actual evapotranspiration of terrestrial ecosystems: patterns and ecological implications. J. Biogeogr. 21, 401–411.

Frank, D.A., Inouye, R.S., Huntly, N., Minshall, G.W., Anderson, J.E., 1994. The biogeochemistry of a north-temperate grassland with native ungulates: nitrogen dynamics in Yellowstone National Park. Biogeochemistry 26, 163–188.

Frank, D.A., Pontes, A.W., McFarlane, K.J., 2012. Controls on soil organic carbon stocks and turnover among North American ecosystems. Ecosystems 15, 604–615.

Frankenberg, C., Berry, J., 2018. Solar Induced Chlorophyll Fluorescence: Origins, Relation to Photosynthesis and Retrieval. https://doi.org/10.1016/B978-0-12-409548-9.10632-3.

Frankenberg, C., Meirink, J.F., van Weele, M., Platt, U., Wagner, T., 2005. Assessing methane emissions from global space-borne observations. Science 308 (5724), 1010–1014.

Frankenberg, C., Aben, I., Bergamaschi, P., Dlugokencky, E.J., van Hees, R., Houweling, S., van der Meer, P., Snel, R., Tol, P., 2011. Global column-averaged methane mixing ratios from 2003 to 2009 as derived from SCIAMACHY: trends and variability. J. Geophys. Res. 116(D4), D02304.

Frankignoulle, M., Abril, G., Borges, A., Bourge, I., Canon, C., Delille, B., Libert, E., Theate, J.M., 1998. Carbon dioxide emission from European estuaries. Science 282 (5388), 434–436.

Franks, P.J., Royer, D.L., Beerling, D.J., Van de Water, P.K., Cantrill, D.J., Barbour, M.M., Berry, J.A., 2014. New constraints on atmospheric CO_2 concentration for the Phanerozoic: Franks et al.: new constraints on Phanerozoic CO_2. Geophys. Res. Lett. 41 (13), 4685–4694.

Frantseva, K., Mueller, M., ten Kate, I.L., van der Tak, F.F.S., Greenstreet, S., 2018. Delivery of organics to mars through asteroid and comet impacts. Icarus 309 (July), 125–133.

Fraser, W.T., Blei, E., Fry, S.C., Newman, M.F., Reay, D.S., Smith, K.A., McLeod, A.R., 2015. Emission of methane, carbon monoxide, carbon dioxide and short-chain hydrocarbons from vegetation foliage under ultraviolet irradiation. Plant Cell Environ. 38 (5), 980–989.

Freeland, W.J., Calcott, P.H., Geiss, D.P., 1985. Allelochemicals, minerals and herbivore population size. Biochem. Syst. Ecol. 13, 195–206.

Freeman, C., Hawkins, J., Lock, M.A., Reynolds, B., 1993. A laboratory perfusion system for the study of biogeochemical response of wetlands to climate change. In: Gopal, B., Hillbricht-Ilowska, A., Wetzel, R.G. (Eds.), Wetlands and Ecotones: Studies on Land-Water Interactions. National Institute of Ecology, New Delhi, pp. 75–83.

Freeman, C., Evans, C.D., Monteith, D.T., Reynolds, B., Fenner, N., 2001a. Export of organic carbon from peat soils. Nature 412 (6849), 785.

Freeman, C., Ostle, N., Kang, H., 2001b. An enzymic 'latch' on a global carbon store. Nature 409 (6817), 149.

Freeman, C., Fenner, N., Ostle, N.J., Kang, H., Dowrick, D.J., Reynolds, B., Lock, M.A., Sleep, D., Hughes, S., Hudson, J., 2004a. Export of dissolved organic carbon from peatlands under elevated carbon dioxide levels. Nature 430 (6996), 195–198.

Freeman, C., Ostle, N.J., Fenner, N., Kang, H., 2004b. A regulatory role for phenol oxidase during decomposition in peatlands. Soil Biol. Biochem. 36 (10), 1663–1667.

Freissinet, C., Glavin, D.P., Mahaffy, P.R., Miller, K.E., Eigenbrode, J.L., Summons, R.E., Brunner, A.E., et al., 2015. Organic molecules in the Sheepbed Mudstone, Gale Crater, Mars. J. Geophys. Res. Planets 120 (3), 495–514.

Freney, J.R., Simpson, J.R., Denmead, O.T., 1983. Volatilization of ammonia. In: Freney, J.R., Simpson, J.R. (Eds.), Gaseous Loss of Nitrogen from Plant-Soil Systems. Martinus Nijhoff, pp. 1–32.

Freschet, G.T., Cornelissen, J.H.C., van Logtestijn, R.S.P., Aerts, R., 2010. Substantial nutrient resorption from leaves, stems and roots in a subarctic flora: what is the link with other resource economics traits? New Phytol. 186 (4), 879–889.

Frey, S.D., Ollinger, S., Nadelhoffer, K., Bowden, R., Brzostek, E., Burton, A., Caldwell, B.A., et al., 2014. Chronic nitrogen additions suppress decomposition and sequester soil carbon in temperate forests. Biogeochemistry 121 (2), 305–316.

Fridovich, I., 1975. Superoxide dismutases. Annu. Rev. Biochem. 44, 147–159.

Friedland, A.J., Johnson, A.H., 1985. Lead distribution and fluxes in a high-elevation forest in Northern Vermont. J. Environ. Qual. 14, 332–336.

Friedlander, G., Kennedy, J.W., Miller, J.M., 1964. Nuclear and Radiochemistry. John Wiley and Sons.

Friedlingstein, P., Jones, M.W., O'Sullivan, M., 2019. Global carbon budget 2019. Earth Syst. Sci. 11, 1783–1838.

Friedmann, E.I., 1982. Endolithic microorganisms in the Antarctic Cold Desert. Science 215 (4536), 1045–1053.

Friend, A.D., Geider, R.J., Behrenfeld, M.J., Still, C.J., 2009. Photosynthesis in global-scale models. In: Nedbal, L., Laisk, A., Govindjee (Eds.), Photosynthesis in Silico: Understanding Complexity from Molecules to Ecosystems. Springer, pp. 465–497.

Frings, P.J., Clymans, W., Fontorbe, G., De La Rocha, C.L., Conley, D.J., 2016. The continental Si cycle and its impact on the ocean Si isotope budget. Chem. Geol. 425 (May), 12–36.

Froelich, P.N., 1988. Kinetic control of dissolved phosphate in natural rivers and estuaries: a primer on the phosphate buffer mechanism. Limnol. Oceanogr. 33, 649–668.

Froelich, P.N., Klinkhammer, G.P., Bender, M.L., Luedtke, N.A., Heath, G.R., Cullen, D., Dauphin, P., Hammond, D., Hartman, B., Maynard, V., 1979. Early oxidation of organic matter in pelagic sediments of the eastern equatorial Atlantic: Suboxic Diagenesis. Geochim. Cosmochim. Acta 43, 1075–1090.

Froelich, P.N., Bender, M.L., Luedtke, N.A., Heath, G.R., Devries, T., 1982. The marine phosphorus cycle. Am. J. Sci. 282, 474–511.

Froelich, P.N., Kim, K.H., Jahnke, R., Burnett, W.C., Soutar, A., Deakin, M., 1983. Pore-water fluoride in Peru continental margin sediments: uptake from seawater. Geochim. Cosmochim. Acta 47, 1605–1612.

Froelich, P.N., Blanc, V., Mortlock, R.A., Chillrud, S.N., Dunstan, W., Udomkit, A., Peng, T.-H., 1992. River fluxes of dissolved silica to the ocean were higher during glacials: Ge/Si in diatoms, rivers, and oceans. Paleoceanography 7, 739–767.

Frolking, S., Talbot, J., Jones, M.C., Treat, C.C., Boone Kauffman, J., Tuittila, E.-S., Roulet, N., 2011. Peatlands in the earth's 21st century climate system. Environ. Res. 396, 371–396.

Fruchter, J.S., Robertson, D.E., Evans, J.C., Olsen, K.B., Lepel, E.A., Laul, J.C., Abel, K.H., et al., 1980. Mount St. Helens Ash from the 18 May 1980 Eruption: chemical, physical, mineralogical, and biological properties. Science 209 (4461), 1116–1125.

Fueglistaler, S., Wernli, H., Peter, T., 2004. Tropical troposphere-to-stratosphere transport inferred from trajectory calculations. J. Geophys. Res. 109.

Fuhrman, J.A., McManus, G.B., 1984. Do bacteria-sized marine eukaryotes consume significant bacterial production? Science 224 (4654), 1257–1260.

Fujisaki, K., Perrin, A.-S., Desjardins, T., Bernoux, M., Balbino, L.C., Brossard, M., 2015. From forest to cropland and pasture systems: a critical review of soil organic carbon stocks changes in Amazonia. Glob. Chang. Biol. 21 (7), 2773–2786.

Fulweiler, R.W., Nixon, S.W., Buckley, B.A., Granger, S.L., 2007. Reversal of the net dinitrogen gas flux in coastal marine sediments. Nature 448 (7150), 180–182.

Fung, I., John, J., Lerner, J., Matthews, E., Prather, M., Steele, L.P., Fraser, P.J., 1991. Three-dimensional model synthesis of the global methane cycle. J. Geophys. Res. 96, 13033–13065.

Fung, I.Y., Meyn, S.K., Tegen, I., Doney, S.C., John, J.G., Bishop, J.K.B., 2000. Iron supply and demand in the upper ocean. Glob. Biogeochem. Cycles 14 (1), 281–295.

Funk, D.W., Pullman, E.R., Peterson, K.M., Crill, P.M., Billings, W.D., 1994. Influence of water table on carbon dioxide, carbon monoxide and methane fluxes from Taiga Bog microcosms. Glob. Biogeochem. Cycles 8 (3), 271–278.

Füri, E., Barry, P.H., Taylor, L.A., Marty, B., 2015. Indigenous nitrogen in the moon: constraints from coupled nitrogen–noble gas analyses of Mare Basalts. Earth Planet. Sci. Lett. 431 (December), 195–205.

Furrer, G., Westall, J., Sollins, P., 1989. The study of soil chemistry through quasi-steady-state models. 1. Mathematical definition of the model. Geochim. Cosmochim. Acta 53, 595–601.

Furrer, G., Sollins, P., Westall, J.C., 1990. The study of soil chemistry through quasi-steady-state models. 2. Acidity of the soil solution. Geochim. Cosmochim. Acta 54, 2363–2374.

Fuss, C.B., Driscoll, C.T., Johnson, C.E., Petras, R.J., Fahey, T.J., 2011. Dynamics of oxidized and reduced Iron in a northern hardwood forest. Biogeochemistry 104, 103–119.

Gabet, E.J., Mudd, S.M., 2009. A theoretical model coupling chemical weathering rates with denudation rates. Geology 37, 151–154.

Gabriel, M., Kolka, R., Wickman, T., Woodruff, L., Nater, E., 2012. Latent effect of soil organic matter oxidation on mercury cycling within a southern boreal ecosystem. J. Environ. Qual. 41 (2), 495–505.

Gachter, R., Meyer, J.S., Mares, A., 1988. Contribution of bacteria to release and fixation of phosphorus in lake sediments. Limnol. Oceanogr. 33, 1542–1558.

Gadens-Marcon, G.T., Guerra-Sommer, M., Mendonça-Filho, J.G., de Oliveira Mendonça, J., de Araújo Carvalho, M., Hartmann, L.A., 2014. Holocene environmental climatic changes based on Palynofacies and organic geochemical analyses from an inland pond at altitude in southern Brazil. Am. J. Clim. Chang. 3 (01), 95.

Gaidos, E.J., Nealson, K.H., Kirschvink, J.L., 1999. Life in ice-covered oceans. Science 284 (5420), 1631–1633.

Gaiffe, M., Schmitt, A., 1980. Sols et Vegetation à L'étage Montagnard Dans Les Forêts Du Jura Central. Science Du Sol 4, 265–296.

Gaillardet, J., Dupre, B., Louvat, P., Allegre, C.J., 1999. Global silicate weathering and CO_2 consumption rates deduced from the chemistry of large rivers. Chem. Geol. 159, 3–30.

Galbraith, E.D., Eggleston, S., 2017. A lower limit to atmospheric CO_2 concentrations over the past 800,000 years. Nat. Geosci. 10 (4), 295–298.

Galbraith, E.D., Carozza, D.A., Bianchi, D., 2017. A coupled human-earth model perspective on long-term trends in the global marine fishery. Nat. Commun. 8 (March), 14884.

Gale, P.M., Gilmour, J.T., 1988. Net mineralization of carbon and nitrogen under aerobic and anaerobic conditions. Soil Sci. Soc. Am. J. 52, 1006–1010.

Gali, M., Salo, V., Almeda, R., Calbet, A., Simo, R., 2011. Stimulation of gross dimethylsulfide (DMS) production by solar radiation. Geophys. Res. Lett. 38.

Gallardo, A., Merino, J., 1998. Soil nitrogen dynamics in response to carbon increase in a mediterranean shrubland of SW Spain. Soil Biol. Biochem. 30, 1349–1358.

Gallardo, A., Parama, R., 2007. Spatial variability of soil elements in two plant communities of NW Spain. Geoderma 139, 199–208.

Gallardo, A., Schlesinger, W.H., 1992. Carbon and nitrogen limitations of soil microbial biomass in desert ecosystems. Biogeochemistry 18, 1–17.

Gallardo, A., Schlesinger, W.H., 1994. Factors limiting microbial biomass in the mineral soil and forest floor of a warm-temperate forest. Soil Biol. Biochem. 26, 1409–1415.

Gallardo, A., Schlesinger, W.H., 1996. Exclusion of *Artemisia tridentata* Nutt. from hydrothermally altered rock by low phosphorus availability. Madrono 43, 292–298.

Gallardo, A., Fernandez-Palacious, J.M., Bermudez, A., de Nascimento, L., Duran, J., Garcia-Velazquez, L., Mendez, J., Rodriguez, A., 2020. The pedogenic Walker and Syers model under high atmospheric P deposition rates. Biogeochemistry 148, 237–253.

Gallo, M.E., Porras-Alfaro, A., Odenbach, K.J., Sinsabaugh, R.L., 2009. Photoacceleration of plant litter decomposition in an arid environment. Soil Biol. Biochem. 41, 1433–1441.

Galloway, J.N., Cowling, E.B., 2002. Reactive nitrogen and the world: 200 years of change. Ambio 31 (2), 64–71.

Galloway, J.N., Likens, G.E., 1979. Atmospheric enhancement of metal deposition in Adirondack lake sediments. Limnol. Oceanogr. 24, 427–433.

Galloway, J.N., Whelpdale, D.M., 1987. WATOX-86 overview and Western North Atlantic Ocean S and N atmospheric budgets. Glob. Biogeochem. Cycles 1, 261–281.

Galloway, J.N., Likens, G.E., Edgerton, E.S., 1976. Acid precipitation in the Northeastern United States: pH and acidity. Science 194 (4266), 722–724.

Galloway, J.N., Dentener, F.J., Capone, D.G., Boyer, E.W., Howarth, R.W., Seitzinger, S.P., Asner, G.P., et al., 2004. Nitrogen cycles: past, present, and future. Biogeochemistry 70 (2), 153–226.

Galloway, J.N., Townsend, A.R., Erisman, J.W., Bekunda, M., Cai, Z., Freney, J.R., Martinelli, L.A., Seitzinger, S.P., Sutton, M.A., 2008. Transformation of the nitrogen cycle: recent trends, questions, and potential solutions. Science 320 (5878), 889–892.

Galy, V., France-Lanord, C., Beyssac, O., Faure, P., Kudrass, H., Palhol, F., 2007. Efficient organic carbon burial in the Bengal Fan sustained by the Himalayan erosional system. Nature 450 (7168), 407–410.

Galy, V., Beyssac, O., France-Lanord, C., Eglinton, T., 2008. Recycling of graphite during Himalayan erosion: a geological stabilization of carbon in the crust. Science 322 (5903), 943–945.

Galy, V., Peucker-Ehrenbrink, B., Eglinton, T., 2015. Global carbon export from the terrestrial biosphere controlled by erosion. Nature 521 (7551), 204–207.

Ganachaud, A., 2003. Large-scale mass transports, water mass formation, and diffusivities estimated from world ocean circulation experiment (WOCE) hydrographic data. J. Geophys. Res. C: Oceans 108 (C7).

Ganeshram, R.S., Pedersen, T.F., Calvert, S.E., Murray, J.W., 1995. Large changes in oceanic nutrient inventories from glacial to interglacial periods. Nature 376, 755–758.

Ganzeveld, L.N., Lelieveld, J., Dentener, F.J., Krol, M.C., Bouwman, A.J., Roelofs, G.J., 2002. Global soil-biogenic NOx emissions and the role of canopy processes. J. Geophys. Res.-Atmos. 107.

Gao, B., Ju, X.T., Zhang, Q., Christie, P., Zhang, F.S., 2011. New estimates of direct N_2O emissions from chinese croplands from 1980 to 2007 using localized emission factors. Biogeosciences 8, 3011–3024.

Gardner, L.R., 1990. The role of rock weathering in the phosphorus budget of terrestrial watersheds. Biogeochemistry 11, 97–110.

Gardner, J.B., Drinkwater, L.E., 2009. The fate of nitrogen in grain cropping systems: a meta-analysis of ^{15}N field experiments. Ecol. Appl. 19 (8), 2167–2184.

Gardner, L.R., Kheoruenromne, I., Chen, H.S., 1978. Isovolumetric geochemical investigation of a buried granite saprolite near Columbia, SC, USA. Geochim. Cosmochim. Acta 42, 417–424.

Gardner, A.S., Geir, M., Graham Cogley, J., Wouters, B., Arendt, A.A., Wahr, J., Berthier, E., et al., 2013. A reconciled estimate of glacier contributions to sea level rise: 2003 to 2009. Science 340 (6134), 852–857.

Garrels, R.M., Lerman, A., 1981. Phanerozoic cycles of sedimentary carbon and sulfur. Proc. Natl. Acad. Sci. U. S. A. 78 (8), 4652–4656.

Garrels, R.M., MacKenzie, F.T., 1971. Evolution of Sedimentary Rocks. W.W. Norton.

Garten, C.T., 1993. Variation in foliar ^{15}N-abundance and the availability of soil nitrogen on walker branch watershed. Ecology 74, 2098–2113.

Garten, C.T., 2011. Comparison of forest soil carbon dynamics at five sites along a latitudinal gradient. Geoderma 167, 30–40.

Garten, C.T., Hanson, P.J., 1990. Foliar retention of ^{15}N-nitrate and ^{15}N-ammonium by red maple (*Acer rubrum*) and white oak (*Quercus alba*) leaves from simulated rain. Environ. Exp. Bot. 30, 333–342.

Garten Jr., C.T., Iversen, C.M., Norby, R.J., 2011. Litterfall ^{15}N abundance indicates declining soil nitrogen availability in a free-air CO_2 enrichment experiment. Ecology 92 (1), 133–139.

Garten, C.T., Van Miegroet, H., 1994. Relationships between soil nitrogen dynamics and natural ^{15}N abundance in plant foliage from Great Smoky Mountains National Park. Can. J. For. Res. 24, 1636–1645.

Garten, C.T., Bondietti, E.A., Lomax, R.D., 1988. Contribution of foliar leaching and dry deposition to sulfate in net throughfall below deciduous trees. Atmos. Environ. 22, 1425–1432.

Gascard, J.C., Watson, A.J., Messias, M.J., Olsson, K.A., Johannessen, T., Simonsen, K., 2002. Long-lived vortices as a mode of deep ventilation in the Greenland Sea. Nature 416, 525–527.

Gassmann, G., Glindemann, D., 1993. Phosphane (PH_3) in the biosphere. Angew. Chem. Int. Ed. 32, 761–763.

Gates, W.L., 1976. Modeling the ice-age climate. Science 191 (4232), 1138–1144.

Gatti, L.V., Gloor, M., Miller, J.B., Doughty, C.E., Malhi, Y., Domingues, L.G., Basso, L.S., et al., 2014. Drought sensitivity of Amazonian carbon balance revealed by atmospheric measurements. Nature 506 (7486), 76–80.

Gatz, D.F., Dingle, A.N., 1971. Trace substances in rain water: concentration variations during convective rains, and their interpretation. Tellus 23, 14–27.

Gauci, V., Chapman, S.J., 2006. Simultaneous inhibition of CH_4 efflux and stimulation of sulphate reduction in peat subject to simulated acid rain. Soil Biol. Biochem. 38 (12), 3506–3510.

Gauci, V., Matthews, E., Dise, N., Walter, B., Koch, D., Granberg, G., Vile, M., 2004. Sulfur pollution suppression of the wetland methane source in the 20th and 21st centuries. Proc. Natl. Acad. Sci. U. S. A. 101 (34), 12583–12587.

Gaudichet, A., Echalar, F., Chatenet, B., Quisefit, J.P., Malingre, G., Cachier, H., Buatmenard, P., Artaxo, P., Maenhaut, W., 1995. Trace elements in tropical African Savanna biomass burning aerosols. J. Atmos. Chem. 22, 19–39.

Gaudinski, J.B., Trumbore, S.E., Davidson, E.A., Zheng, S.H., 2000. Soil carbon cycling in a temperate forest: radiocarbon-based estimates of residence times, sequestration rates and partitioning of fluxes. Biogeochemistry 51, 33–69.

Gaudinski, J., Trumbore, S., Davidson, E., Cook, A., Markewitz, D., Richter, D., 2001. The age of fine-root carbon in three forests of the eastern United States measured by radiocarbon. Oecologia 129 (3), 420–429.

Gaudinski, J.B., Torn, M.S., Riley, W.J., Dawson, T.E., Joslin, J.D., Majdi, H., 2010. Measuring and modeling the spectrum of fine-root turnover times in three forests using isotopes, minirhizotrons, and the radix model. Glob. Biogeochem. Cycles 24.

Gedalof, Z., Berg, A.A., 2010. Tree-ring evidence for limited direct CO_2 fertilization of forests over the 20th century. Glob. Biogeochem. Cycles 24.

Gedney, N., Cox, P.M., Betts, R.A., Boucher, O., Huntingford, C., Stott, P.A., 2006. Detection of a direct carbon dioxide effect in continental river runoff records. Nature 439 (7078), 835–838.

Gegenbauer, C., Mayer, V.E., Zotz, G., Richter, A., 2012. Uptake of ant-derived nitrogen in the myrmecophytic orchid *Caularthron bilamellatum*. Ann. Bot. 110 (4), 757–766.

Geisseler, D., Horwath, W.R., Joergensen, R.G., Ludwig, B., 2010. Pathways of nitrogen utilization by soil microorganisms—a review. Soil Biol. Biochem. 42, 2058–2067.

Gélinas, Y., Baldock, J.A., Hedges, J.I., 2001. Organic carbon composition of marine sediments: effect of oxygen exposure on oil generation potential. Science 294 (5540), 145–148.

Geng, C.M., Mu, Y.J., 2006. Carbonyl sulfide and dimethyl sulfide exchange between trees and the atmosphere. Atmos. Environ. 40, 1373–1383.

Geng, J.J., Niu, X.J., Jin, X.C., Wang, X.R., Gu, X.H., Edwards, M., Glindemann, D., 2005. Simultaneous monitoring of phosphine and of phosphorus species in Taihu lake sediments and phosphine emission from lake sediments. Biogeochemistry 76, 283–298.

Geng, L., Murray, L.T., Mickley, L.J., Lin, P., Qiang, F., Schauer, A.J., Alexander, B., 2017. Isotopic evidence of multiple controls on atmospheric oxidants over climate transitions. Nature 546 (7656), 133–136.

Genkai-Kato, M., Carpenter, S.R., 2005. Eutrophication due to phosphorus recycling in relation to lake morphometry, temperature, and macrophytes. Ecology 86 (1), 210–219.

Georgii, H.-W., Wötzel, D., 1970. On the relation between drop size and concentration of trace elements in rainwater. J. Geophys. Res. 75 (9), 1727–1731.

Gerard, F., Mayer, K.U., Hodson, M.J., Ranger, J., 2008. Modelling the biogeochemical cycle of silicon in soils: application to a temperate forest ecosystem. Geochim. Cosmochim. Acta 72, 741–758.

Gerdol, R., Iacumin, P., Brancaleoni, L., 2019. Differential effects of soil chemistry on the foliar resorption of nitrogen and phosphorus across altitudinal gradients. Wang, F. (Ed.), Funct. Ecol. 33 (7), 1351–1361.

Gergel, S.E., Turner, M.G., Kratz, T.K., 1999. Dissolved organic carbon as an indicator of the scale of watershed influence on lakes and rivers. Ecol. Appl. 9 (4), 1377–1390.

Gerhart, L.M., Ward, J.K., 2010. Plant responses to low [CO_2] of the past. New Phytol. 188, 674–695.

Geron, C., Harley, P., Guenther, A., 2001. Isoprene emission capacity for US tree species. Atmos. Environ. 35 (19), 3341–3352.

Geron, C., Guenther, A., Greenberg, J., Karl, T., Rasmussen, R., 2006a. Biogenic volatile organic compound emissions from desert vegetation of the southwestern US. Atmos. Environ. 40, 1645–1660.

Geron, C., Owen, S., Guenther, A., Greenberg, J., Rasmussen, R., Bai, J.H., Li, Q.-J., Baker, B., 2006b. Volatile organic compounds from vegetation in southern Yunnan Province, China: emission rates and some potential regional implications. Atmos. Environ. 40 (10), 1759–1773.

Gersper, P.L., Holowaychuk, N., 1971. Some effects of stem flow from forest canopy trees on chemical properties of soils. Ecology 52 (4), 691–702.

Gessler, A., Rienks, M., Rennenberg, H., 2000. NH_3 and NO_2 fluxes between beech trees and the atmosphere—correlation with climatic and physiological parameters. New Phytol. 147, 539–560.

Ghaisas, N.A., Maiti, K., White, J.R., 2019. Coupled iron and phosphorus release from seasonally hypoxic Louisiana shelf sediment. Estuar. Coast. Shelf Sci. 219, 81–89.

Ghiorse, W.C., 1984. Biology of iron- and manganese-depositing bacteria. Annu. Rev. Microbiol. 38, 515–550.

Gholz, H.L., 1982. Environmental limits on aboveground net primary production, leaf-area, and biomass in vegetation zones of the Pacific northwest. Ecology 63, 469–481.

Gholz, H.L., Fisher, R.F., Pritchett, W.L., 1985. Nutrient dynamics in slash pine plantation ecosystems. Ecology 66, 647–659.

Gholz, H.L., Wedin, D.A., Smitherman, S.M., Harmon, M.E., Parton, W.J., 2000. Long-term dynamics of pine and hardwood litter in contrasting environments: toward a global model of decomposition. Glob. Chang. Biol. 6, 751–765.

Giardina, C.P., Sanford, R.L., Dockersmith, I.C., 2000. Changes in soil phosphorus and nitrogen during slash-and-burn clearing of a dry tropical forest. Soil Sci. Soc. Am. J. 64, 399–405.

Gibard, C., Bhowmik, S., Karki, M., Kim, E.-K., Krishnamurthy, R., 2018. Phosphorylation, oligomerization and self-assembly in water under potential prebiotic conditions. Nat. Chem. 10, 212–217.

Gibbs, R.J., 1970. Mechanisms controlling world water chemistry. Science 170 (3962), 1088–1090.

Gibbs, J., Greenway, H., 2003. (Review). Mechanisms of anoxia tolerance in plants. II. Energy requirements for maintenance and energy distribution to essential processes. Funct. Plant Biol. 30 (1), 1–47.

Giblin, A.E.E., 1988. Pyrite formation during early Diagenesis. Geomicrobiol J. 6 (2), 77–97.

Gibson, D.G., Glass, J.I., Lartigue, C., Noskov, V.N., Chuang, R.Y., Algire, M.A., Benders, G.A., et al., 2010. Creation of a bacterial cell controlled by achemically synthesized genome. Science 329, 52–56.

Gibson, C.M., Estop-Aragonés, C., Flannigan, M., Thompson, D.K., Olefeldt, D., 2019. Increased deep soil respiration detected despite reduced overall respiration in permafrost peat plateaus following wildfire. Environ. Res. Lett. 14 (12), 125001.

Giese, M., Brueck, H., Gao, Y.Z., Lin, S., Steffens, M., Kögel-Knabner, I., Glindemann, T., et al., 2013. N balance and cycling of inner Mongolia typical steppe: a comprehensive case study of grazing effects. Ecol. Monogr. 83 (2), 195–219.

Gieskes, J.M., Lawrence, J.R., 1981. Alteration of volcanic matter in deep sea sediments: evidence from the chemical composition of interstitial waters from deep sea drilling cores. Geochim. Cosmochim. Acta 45, 1687–1703.

Gijsman, A.J., 1990. Nitrogen nutrition of douglas fir (*Pseudotsuga menziesii*) on strongly acid sandy soil. I. Growth, nutrient uptake and ionic balance. Plant Soil 126, 53–61.

Gile, L.H., Peterson, F.F., Grossman, R.B., 1966. Morphological and genetic sequences of carbonate accumulation in desert soils. Soil Sci. 101, 347–360.

Gill, A.L., Finzi, A.C., 2016. Belowground carbon flux links biogeochemical cycles and resource-use efficiency at the global scale. Ecol. Lett. 19 (12), 1419–1428.

Gill, R.A., Jackson, R.B., 2000. Global patterns of root turnover for terrestrial ecosystems. New Phytol. 147, 13–31.

Gill, A.L., Giasson, M.-A., Yu, R., Finzi, A.C., 2017. Deep peat warming increases surface methane and carbon dioxide emissions in a black spruce-dominated ombrotrophic bog. Glob. Chang. Biol. 23 (12), 5398–5411.

Gillespie, A.R., Pope, P.E., 1990. Rhizosphere acidification increases phosphorus recovery of black locust. I. Induced acidification and soil response. Soil Sci. Soc. Am. J. 54, 533–537.

Gillette, D.A., Stensland, G.J., Williams, A.L., Barnard, W., Gatz, D., Sinclair, P.C., Johnson, T.C., 1992. Emissions of alkaline elements calcium, magnesium, potassium, and sodium from open sources in the contiguous United States. Glob. Biogeochem. Cycles 6, 437–457.

Gilliland, A.B., Appel, K.W., Pinder, R.W., Dennis, R.L., 2006. Seasonal NH_3 emissions for the continental United States: inverse model estimation and evaluation. Atmos. Environ. 40, 4986–4998.

Gillmann, C., Lognonne, P., Moreira, M., 2011. Volatiles in the atmosphere of mars: the effects of volcanism and escape constrained by isotopic data. Earth Planet. Sci. Lett. 303, 299–309.

Gillon, M., Triaud, A.H.M.J., Demory, B.-O., Jehin, E., Agol, E., Deck, K.M., Lederer, S.M., et al., 2017. Seven temperate terrestrial planets around the nearby ultracool dwarf star TRAPPIST-1. Nature 542 (7642), 456–460.

Gilmore, A.R., Gertner, G.Z., Rolfe, G.L., 1984. Soil chemical changes associated with roosting birds. Soil Sci. 138, 158–163.

Gilmour, C.C., Henry, E.A., 1991. Mercury methylation in aquatic systems affected by acid deposition. Environ. Pollut. 71 (2–4), 131–169.

Gilmour, C.C., Henry, E.A., Mitchell, R., 1992. Sulfate stimulation of mercury methylation in fresh-water sediments. Environ. Sci. Technol. 26, 2281–2287.

Gilmour, C.C., Elias, D.A., Kucken, A.M., Brown, S.D., Palumbo, A.V., Schadt, C.W., Wall, J.D., 2011. Sulfate-reducing bacterium desulfovibrio desulfuricans ND132 as a model for understanding bacterial mercury methylation. Appl. Environ. Microbiol. 77 (12), 3938–3951.

Giordano, M., Norici, A., Hell, R., 2005. Sulfur and phytoplankton: acquisition, metabolism and impact on the environment. New Phytol. 166 (2), 371–382.

Girardin, M.P., Bouriaud, O., Hogg, E.H., Kurz, W., Zimmermann, N.E., Metsaranta, J.M., de Jong, R., et al., 2016. No growth stimulation of Canada's boreal forest under half-century of combined warming and CO_2 fertilization. Proc. Natl. Acad. Sci. U. S. A. 113 (52), E8406–E8414.

Gislason, S.R., Oelkers, E.H., Eiriksdottir, E.S., Kardjilov, M.I., Gisladottir, G., Sigfusson, B., Snorrason, A., et al., 2009. Direct evidence of the feedback between climate and weathering. Earth Planet. Sci. Lett. 277, 213–222.

Glass, S.J., Matteson, M.J., 1973. Ion enrichment in aerosols dispersed from bursting bubbles in aqueous salt solutions. Tellus 25, 272–280.

Glass, J.B., Orphan, V.J., 2012. Trace metal requirements for microbial enzymes involved in the production and consumption of methane and nitrous oxide. Front. Microbiol. 3 (February), 61.

Glass, J.B., Wolfe-Simon, F., Anbar, A.D., 2009. Coevolution of metal availability and nitrogen assimilation in cyanobacteria and algae. Geobiology 7 (2), 100–123.

Glass, J.B., Axler, R.P., Chandra, S., Goldman, C.R., 2012. Molybdenum limitation of microbial nitrogen assimilation in aquatic ecosystems and pure cultures. Front. Microbiol. 3 (September), 331.

Gleeson, T., Wada, Y., Bierkens, M.F.P., van Beek, L.P.H., 2012. Water balance of global aquifers revealed by groundwater footprint. Nature 488 (7410), 197–200.

Gleeson, T., Befus, K.M., Jasechko, S., Luijendijk, E., Bayani Cardenas, M., 2016. The global volume and distribution of modern groundwater. Nat. Geosci. 9 (2), 161–167.

Gleick, P.H., 2000. The changing water paradigm—a look at twenty-first century water resources development. Water Int. 25, 127–138.

Glendining, M.J., Dailey, A.G., Powlson, D.S., Richter, G.M., Catt, J.A., Whitmore, A.P., 2011. Pedotransfer functions for estimating total soil nitrogen up to the global scale. Eur. J. Soil Sci. 62, 13–22.

Glibert, P.M., Lipschultz, F., McCarthy, J.J., Altabet, M.A., 1982. Isotope-dilution models of uptake and remineralization of ammonium by marine plankton. Limnol. Oceanogr. 27, 639–650.

Glindemann, D., Edwards, M., Kuschk, P., 2003. Phosphine gas in the upper troposphere. Atmos. Environ. 37 (18), 2429–2433.

Glover, H.E., Prézelin, B.B., Campbell, L., Wyman, M., Garside, C., 1988. A nitrate-dependent *Synechococcus* bloom in surface Sargasso Sea water. Nature 331 (6152), 161–163.

Glynn, P.W., 1988. El Nino—southern oscillation 1982–1983: nearshore population, community, and ecosystem responses. Annu. Rev. Ecol. Syst. 19 (1), 309–346.

Godbold, D.L., Fritz, E., Hüttermann, A., 1988. Aluminum toxicity and forest decline. Proc. Natl. Acad. Sci. U. S. A. 85 (11), 3888–3892.

Godden, J.W., Turley, S., Teller, D.C., Adman, E.T., Liu, M.Y., Payne, W.J., LeGall, J., 1991. The 2.3 Angstrom X-ray structure of nitrite reductase from *Achromobacter cycloclastes*. Science 253 (5018), 438–442.

Godfrey, L.V., Falkowski, P.G., 2009. The cycling and redox state of nitrogen in the Archaean ocean. Nat. Geosci. 2, 725–729.

Godin, A.M., Lidher, K.K., Whiteside, M.D., Jones, M.D., 2015. Control of soil phosphatase activities at millimeter scales in a mixed paper birch—douglas-fir forest: the importance of carbon and nitrogen. Soil Biol. Biochem. 80 (January), 62–69.

Goix, S., Maurice, L., Laffont, L., Rinaldo, R., Lagane, C., Chmeleff, J., Menges, J., Heimbürger, L.-E., Maury-Brachet, R., Sonke, J.E., 2019. Quantifying the impacts of artisanal gold mining on a tropical river system using mercury isotopes. Chemosphere 219 (March), 684–694.

Gold, D.A., Caron, A., Fournier, G.P., Summons, R.E., 2017. Paleoproterozoic sterol biosynthesis and the rise of oxygen. Nature 543 (7645), 420–423.

Goldan, P.D., Kuster, W.C., Albritton, D.L., Fehsenfeld, F.C., 1987. The measurement of natural sulfur emissions from soils and vegetation: three sites in the eastern United States revisited. J. Atmos. Chem. 5, 439–467.

Goldan, P.D., Fall, R., Kuster, W.C., Fehsenfeld, F.C., 1988. Uptake of COS by growing vegetation: a major tropospheric sink. J. Geophys. Res.-Atmos. 93, 14186–14192.

Goldberg, D.E., 1982. The distribution of evergreen and deciduous trees relative to soil type: an example from the Sierra-Madre, Mexico, and a general model. Ecology 63, 942–951.

Goldblatt, C., Claire, M.W., Lenton, T.M., Matthews, A.J., Watson, A.J., Zahnle, K.J., 2009. Nitrogen-enhanced greenhouse warming on early earth. Nat. Geosci. 2, 891–896.

Goldewijk, K.K., Beusen, A., Janssen, P., 2010. Long-term dynamic modeling of global population and built-up area in a spatially explicit way: HYDE 3.1. The Holocene 20, 565–573.

Goldich, S.S., 1938. A study in rock weathering. J. Geol. 46, 17–58.

Goldman, C.R., 1960. Molybdenum as a factor limiting primary productivity in Castle Lake, California. Science 132 (3433), 1016–1017.

Goldman, J.C., Glibert, P.M., 1982. Comparative rapid ammonium uptake by four species of marine phytoplankton. Limnol. Oceanogr. 27, 814–827.

Golledge, N.R., Keller, E.D., Gomez, N., Naughten, K.A., Bernales, J., Trusel, L.D., Edwards, T.L., 2019. Global environmental consequences of twenty-first-century ice-sheet melt. Nature 566 (7742), 65–72.

Golley, F.B., 1972. Energy flux in ecosystems. In: Wiens, J.A. (Ed.), Ecosystem Structure and Function. Oregon State University Press, pp. 69–90.

Golterman, H.L., 1995. The role of the ironhydroxide-phosphate-sulfide system in the phosphate exchange between sediments and overlying water. Hydrobiologia 297, 43–54.

Golterman, H.L., 2001. Phosphate release from anoxic sediments or 'what did Mortimer really write?'. Hydrobiologia 450 (1–3), 99–106.

Golubev, V.S., Lawrimore, J.H., Groisman, P.Y., Speranskaya, N.A., Zhuravin, S.A., Menne, M.J., Peterson, T.C., Malone, R.W., 2001. Evaporation changes over the contiguous United States and the former USSR: a reassessment. Geophys. Res. Lett. 28, 2665–2668.

Golubiewski, N.E., 2006. Urbanization increases grassland carbon pools: effects of landscaping in Colorado's front range. Ecol. Appl. 16 (2), 555–571.

Gomes, L.C., Faria, R.M., de Souza, E., Veloso, G.V., Schaefer, C.E.G.R., Filho, E.I.F., 2019. Modelling and mapping soil organic carbon stocks in Brazil. Geoderma 340 (April), 337–350.

Gonzales, K., Yanai, R., 2019. Nitrogen-phosphorous interactions in young northern hardwoods indicate P limitation: foliar concentrations and resorption in a factorial N by P addition experiment. Oecologia 189 (3), 829–840.

Good, S.P., Kennedy, C.D., Stalker, J.C., Chesson, L.A., Valenzuela, L.O., Beasley, M.M., Ehleringer, J.R., Bowen, G.J., 2014. Patterns of local and nonlocal water resource use across the Western U.S. determined via stable isotope intercomparisons. Water Resour. Res. 50 (10), 8034–8049.

Good, S.P., Noone, D., Bowen, G., 2015. Water resources. Hydrologic connectivity constrains partitioning of global terrestrial water fluxes. Science 349 (6244), 175–177.

Goodale, C.L., 2017. Multiyear fate of a ^{15}N tracer in a mixed deciduous forest: retention, redistribution, and differences by mycorrhizal association. Glob. Chang. Biol. 23 (2), 867–880.

Goodale, C.L., Apps, M.J., Birdsey, R.A., Field, C.B., Heath, L.S., Houghton, R.A., Jenkins, J.C., et al., 2002. Forest carbon sinks in the northern hemisphere. Ecol. Appl. 12, 891–899.

Goodale, C.L., Fredriksen, G., Weiss, M.S., McCalley, K., Sparks, J.P., Thomas, S.A., 2015. Soil processes drive seasonal variation in retention of ^{15}N tracers in a deciduous forest catchment. Ecology 96 (10), 2653–2668.

Goode, P.R., Qiu, J., Yurchyshyn, V., Hickey, J., Chu, M.C., Kolbe, E., Brown, C.T., Koonin, S.E., 2001. Earthshine observations of the Earth's reflectance. Geophys. Res. Lett. 28, 1671–1674.

Goodrich, J.P., Varner, R.K., Frolking, S., Duncan, B.N., Crill, P.M., 2011. High-frequency measurements of methane ebullition over a growing season at a temperate peatland site. Geophys. Res. Lett. 38.

Goodridge, B.M., Hanan, E.J., Aguilera, R., Wetherley, E.B., Chen, Y.-J., D'Antonio, C.M., Melack, J.M., 2018. Retention of nitrogen following wildfire in a Chaparral ecosystem. Ecosystems 21 (8), 1608–1622.

Goolsby, D.A., Battaglin, W.A., Aulenbach, B.T., Hooper, R.P., 2001. Nitrogen input to the Gulf of Mexico. J. Environ. Qual. 30 (2), 329–336.

Gordon, A.L., 2016. The marine hydrological cycle: the ocean's floods and droughts. Geophys. Res. Lett. 43 (14), 7649–7652.

Goregues, C.M., Michotey, V.D., Bonin, P.C., 2005. Molecular, biochemical, and physiological approaches for understanding the ecology of denitrification. Microb. Ecol. 49 (2), 198–208.

Gorham, E., 1957. The development of peat lands. Q. Rev. Biol. 32, 145–166.

Gorham, E., 1961. Factors influencing supply of major ions to inland waters, with special reference to the atmosphere. Geol. Soc. Am. Bull. 72, 795–840.

Gorham, E., 1991. Northern peatlands: role in the carbon cycle and probably responses to climatic warming. Ecol. Appl. 1 (2), 182–195.

Gorham, E., Boyce, F.M., 1989. Influence of lake surface area and depth upon thermal stratification and the depth of the summer thermocline. J. Great Lakes Res. 15, 233–245.

Gorham, E., Vitousek, P.M., Reiners, W.A., 1979. Regulation of chemical budgets over the course of terrestrial ecosystem succession. Annu. Rev. Ecol. Syst. 10, 53–84.

Gorham, E., Martin, F.B., Litzau, J.T., 1984. Acid rain: ionic correlations in the eastern United States, 1980–1981. Science 225 (4660), 407–409.

Gosselink, J.G., Turner, R.E., 1978. The role of hydrology in freshwater wetland ecosystems. In: Good, R.E., Whigham, D.F., Simpson, R.L. (Eds.), Freshwater Wetlands: Ecological Processes and Management Potential. Academic Press, New York, pp. 63–78.

Goswami, S., Fisk, M.C., Vadeboncoeur, M.A., Garrison-Johnston, M., Yanai, R.D., Fahey, T.J., 2018. Phosphorus limitation of aboveground production in northern hardwood forests. Ecology.

Gosz, J.R., 1981. Nitrogen cycling in coniferous ecosystems. In: Clark, F.E., Rosswall, T. (Eds.), Terrestrial Nitrogen Cycles. Swedish Natural Science Research Council, Stockholm, Sweden, pp. 405–426.

Gosz, J.R., Holmes, R.T., Likens, G.E., Bormann, F.H., 1978. Flow of energy in a forest ecosystem. Sci. Am. 238, 92–102.

Goto, D., Morimoto, S., Ishidoya, S., Aoki, S., Nakazawa, T., 2017. Terrestrial biospheric and oceanic CO_2 uptakes estimated from long-term measurements of atmospheric CO_2 mole fraction, $\delta_{13}C$, and $\delta(O_2/N_2)$ at Ny-Ålesund, Svalbard: recent global CO_2 budget. J. Geophys. Res. Biogeosci. 122 (5), 1192–1202.

Gough, C.M., Vogel, C.S., Schmid, H.P., Su, H.B., Curtis, P.S., 2008. Multi-year convergence of biometric and meteorological estimates of forest carbon storage. Agric. For. Meteorol. 148, 158–170.

Govindarajulu, M., Pfeffer, P.E., Jin, H., Abubaker, J., Douds, D.D., Allen, J.W., Bücking, H., Lammers, P.J., Shachar-Hill, Y., 2005. Nitrogen transfer in the arbuscular mycorrhizal symbiosis. Nature 435 (7043), 819–823.

Gower, S.T., Vogt, K.A., Grier, C.C., 1992. Carbon dynamics of rocky mountain douglas fir: influence of water and nutrient availability. Ecol. Monogr. 62, 43–65.

Gower, S.T., Pongracic, S., Landsberg, J.J., 1996. A global trend in belowground carbon allocation: can we use the relationship at smaller scales? Ecology 77, 1750–1755.

Goyette, J.O., Bennett, E.M., Maranger, R., 2018. Low buffering capacity and slow recovery of anthropogenic phosphorus pollution in watersheds. Nat. Geosci. 11 (12), 921–925.

Gozgor, G., Tiwari, A.K., Khraief, N., Shahbaz, M., 2019. Dependence structure between business cycles and CO_2 emissions in the U.S.: evidence from the time-varying Markov-switching copula models. Energy 188. https://doi.org/10.1016/j.energy.2019.115995.

Graedel, T.E., Crutzen, P.J., 1993. Atmospheric Change. W.H. Freeman.

Graedel, T.E., Keene, W.C., 1995. Tropospheric budget of reactive chlorine. Glob. Biogeochem. Cycles 9, 47–77.

Graham, W.F., Duce, R.A., 1979. Atmospheric pathways of the phosphorus cycle. Geochim. Cosmochim. Acta 43, 1195–1208.

Graham, J.B., Dudley, R., Aguilar, N.M., Gans, C., 1995. Implications of the late paleozoic oxygen pulse for physiology and evolution. Nature 375, 117–120.

Grandstaff, D.E., 1986. The dissolution rate of forsteritic Olivine from Hawaiian Beach sand. In: Colman, S.M., Dethier, D.P. (Eds.), Rates of Chemical Weathering of Rocks and Minerals. Academic/Elsevier, pp. 41–59.

Granhall, U., 1981. Biological nitrogen fixation in relation to environmental factors and functioning of natural ecosystems. In: Clark, F.E., Rosswall, T. (Eds.), Terrestrial Nitrogen Cycles. Swedish Natural Science Research Council, Stockholm, Sweden, pp. 131–144.

Granier, C., Petron, G., Muller, J.F., Brasseur, G., 2000. The impact of natural and anthropogenic hydrocarbons on the tropospheric budget of carbon monoxide. Atmos. Environ. 34, 5255–5270.

Grant, S.B., Azizian, M., Cook, P., Boano, F., Rippy, M.A., 2018. Factoring stream turbulence into global assessments of nitrogen pollution. Science 359 (6381), 1266–1269.

Grasman, B.T., Hellgren, E.C., 1993. Phosphorus nutrition in white-tailed deer: nutrient balance, physiological responses, and Antler growth. Ecology 74, 2279–2296.

Grassle, J.F., 1985. Hydrothermal vent animals: distribution and biology. Science 229 (4715), 713–717.

Gratz, L.E., Keeler, G.J., 2011. Sources of mercury in precipitation to Underhill, VT. Atmos. Environ. 45, 5440–5449.

Graustein, W.C., Armstrong, R.L., 1983. The use of $^{87}Sr/^{86}Sr$ ratios to measure atmospheric transport into forested watersheds. Science 219, 289–292.

Graustein, W.C., Cromack Jr., K., Sollins, P., 1977. Calcium oxalate: occurrence in soils and effect on nutrient and geochemical cycles. Science 198 (4323), 1252–1254.

Graveland, J., Vanderwal, R., Vanbalen, J.H., Vannoordwijk, A.J., 1994. Poor reproduction in forest Passerines from decline of snail abundance on acidified soils. Nature 368, 446–448.

Graven, H.D., Keeling, R.F., Piper, S.C., Patra, P.K., Stephens, B.B., Wofsy, S.C., Welp, L.R., et al., 2013. Enhanced seasonal exchange of CO_2 by northern ecosystems since 1960. Science 341 (6150), 1085–1089.

Gray, J.T., 1982. Community structure and productivity in *Ceanothus* chaparral and coastal sage scrub of Southern California. Ecol. Monogr. 52, 415–435.

Gray, J.T., 1983. Nutrient use by evergreen and deciduous shrubs in Southern California. I. Community nutrient cycling and nutrient-use efficiency. J. Ecol. 71, 21–41.

Gray, J.T., Schlesinger, W.H., 1981. Nutrient cycling in mediterranean type ecosystems. In: Miller, P.C. (Ed.), Resource Use by Chaparral and Matorral. Springer, pp. 259–285.

Graydon, J.A., St Louis, V.L., Lindberg, S.E., Sandilands, K.A., Rudd, J.W.M., Kelly, C.A., Harris, R., et al., 2012. The role of terrestrial vegetation in atmospheric hg deposition: pools and fluxes of spike and ambient Hg from the metaalicus experiment. Glob. Biogeochem. Cycles 26.

Greber, N.D., Dauphas, N., Bekker, A., Ptáček, M.P., Bindeman, I.N., Hofmann, A., 2017. Titanium isotopic evidence for felsic crust and plate tectonics 3.5 billion years ago. Science 357 (6357), 1271–1274.

Greeley, R., Schneid, B.D., 1991. Magma generation on mars: amounts, rates, and comparisons with earth, moon, and venus. Science 254 (5034), 996–998.

Green 2nd, H.W., Chen, W.-P., Brudzinski, M.R., 2010. Seismic evidence of negligible water carried below 400-km depth in subducting lithosphere. Nature 467 (7317), 828–831.

Green, P.A., Vorosmarty, C.J., Meybeck, M., Galloway, J.N., Peterson, B.J., Boyer, E.W., 2004. Pre-industrial and contemporary fluxes of nitrogen through rivers: a global assessment based on typology. Biogeochemistry 68 (1), 71–105.

Green, J.K., Seneviratne, S.I., Berg, A.M., Findell, K.L., Hagemann, S., Lawrence, D.M., Gentine, P., 2019. Large influence of soil moisture on long-term terrestrial carbon uptake. Nature 565 (7740), 476–479.

Greenland, D.J., 1971. Interactions between humic and fulvic acids and clays. Soil Sci. 111, 34–41.

Greenwood, J.P., Itoh, S., Sakamoto, N., Vicenzi, E.P., Yurimoto, H., 2008. Hydrogen isotope evidence for loss of water from mars through time. Geophys. Res. Lett. 35.

Gregg, J.W., Jones, C.G., Dawson, T.E., 2003. Urbanization effects on tree growth in the vicinity of New York City. Nature 424 (6945), 183–187.

Gregor, B., 1970. Denudation of the continents. Nature 228 (5268), 273–275.

Gressel, N., McColl, J.G., Preston, C.M., Newman, R.H., Powers, R.F., 1996. Linkages between phosphorus transformations and carbon decomposition in a forest soil. Biogeochemistry 33, 97–123.

Greve, P., Orlowsky, B., Mueller, B., Sheffield, J., Reichstein, M., Seneviratne, S.I., 2014. Global assessment of trends in wetting and drying over land. Nat. Geosci. 7 (10), 716–721.

Grey, D.C., Jensen, M.L., 1972. Bacteriogenic sulfur in air pollution. Science 177 (4054), 1099–1100.

Gribsholt, B., Kostka, J.E., Kristensen, E., 2003. Impact of fiddler crabs and plant roots on sediment biogeochemistry in a Georgia Saltmarsh. Mar. Ecol. Prog. Ser. 259, 237–251.

Grier, C.C., Running, S.W., 1977. Leaf area of mature northwestern coniferous forests: relation to site water balance. Ecology 58, 893–899.

Griffin, K.L., Bashkin, M.A., Thomas, R.B., Strain, B.R., 1997. Interactive effects of soil nitrogen and atmospheric carbon dioxide on root/rhizosphere carbon dioxide efflux from loblolly and ponderosa pine seedlings. Plant Soil 190, 11–18.

Griffin, M.P.A., Cole, M.L., Kroeger, K.D., Cebrian, J., 1998. Dependence of herbivory on autotrophic nitrogen content and on net primary production across ecosystems. Biol. Bull. 195 (2), 233–234.

Griffin, D., Zhao, X., McLinden, C.A., Boersma, F., Bourassa, A., Dammers, E., Degenstein, D., et al., 2019. High-resolution mapping of nitrogen dioxide with TROPOMI: first results and validation over the Canadian oil sands. Geophys. Res. Lett. 46 (2), 1049–1060.

Griffis, T., Hu, C., Baker, J., Wood, J., Millet, D., Erickson, M., Yu, Z., Deventer, M., Winker, C., Chen, Z., 2019. Tall tower ammonia observations and emission estimates in the US midwest. J. Geophys. Res. Biogeosci. 124, 3432–3447.

Griffith, E.J., Ponnamperuma, C., Gabel, N.W., 1977. Phosphorus, a key to life on the primitive earth. Origins Life 8 (2), 71–85.

Griffith, E.M., Paytan, A., Caldeira, K., Bullen, T.D., Thomas, E., 2008. A dynamic marine calcium cycle during the past 28 million years. Science 322 (5908), 1671–1674.

Griffiths, R.P., Harmon, M.E., Caldwell, B.A., Carpenter, S.E., 1993. Acetylene reduction in conifer logs during early stages of decomposition. Plant Soil 148, 53–61.

Grill, G., Lehner, B., Thieme, M., Geenen, B., Tickner, D., Antonelli, F., Babu, S., et al., 2019. Mapping the world's free-flowing rivers. Nature 569 (7755), 215–221.

Grimm, N.B., Fisher, S.G., 1989. Stability of periphyton and macroinvertebrates to disturbance by flash floods in a desert stream. J. N. Am. Benthol. Soc. 8 (4), 293–307.

Grimm, N.B., Gergel, S.E., McDowell, W.H., Boyer, E.W., Lisa Dent, C., Groffman, P., Hart, S.C., et al., 2003. Merging aquatic and terrestrial perspectives of nutrient biogeochemistry. Oecologia 137 (4), 485–501.

Griscom, B.W., Adams, J., Ellis, P.W., Houghton, R.A., Lomax, G., Miteva, D.A., Schlesinger, W.H., et al., 2017. Natural climate solutions. Proc. Natl. Acad. Sci. U. S. A. 114 (44), 11645–11650.

Groenendijk, P., van der Sleen, P., Vlam, M., Bunyavejchewin, S., Bongers, F., Zuidema, P.A., 2015. No evidence for consistent long-term growth stimulation of 13 tropical tree species: results from tree-ring analysis. Glob. Chang. Biol. 21 (10), 3762–3776.

Groffman, P.M., Tiedje, J.M., 1989. Denitrification in north temperate forest soils: spatial and temporal patterns at the landscape and seasonal scales. Soil Biol. Biochem. 21, 613–620.

Groffman, P.M., Hardy, J.P., Fisk, M.C., Fahey, T.J., Driscoll, C.T., 2009. Climate variation and soil carbon and nitrogen cycling processes in a northern hardwood forest. Ecosystems 12, 927–943.

Groisman, P.Y., Knight, R.W., Karl, T.R., Easterling, D.R., Sun, B.M., Lawrimore, J.H., 2004. Contemporary changes of the hydrological cycle over the contiguous United States: trends derived from in situ observations. J. Hydrometeorol. 5, 64–85.

Grosse, G., Harden, J., Merritt, T., David McGuire, A., Camill, P., Tarnocai, C., Frolking, S., et al., 2011. Vulnerability of high-latitude soil organic carbon in North America to disturbance. J. Geophys. Res. Biogeosci. 116 (G4).

Großkopf, T., Laroche, J., 2012. Direct and indirect costs of dinitrogen fixation in *Crocosphaera watsonii* WH8501 and possible implications for the nitrogen cycle. Front. Microbiol. 3 (July), 236.

Großkopf, T., Mohr, W., Baustian, T., Schunck, H., Gill, D., Kuypers, M.M.M., Lavik, G., Schmitz, R.A., Wallace, D.W.R., LaRoche, J., 2012. Doubling of marine dinitrogen-fixation rates based on direct measurements. Nature 488 (7411), 361–364.

Grotzinger, J.P., Kasting, J.F., 1993. New constraints on Precambrian ocean composition. J. Geol. 101 (2), 235–243.

Grotzinger, J.P., Sumner, D.Y., Kah, L.C., Stack, K., Gupta, S., Edgar, L., Rubin, D., et al., 2014. A habitable fluvio-lacustrine environment at Yellowknife Bay, Gale Crater, Mars. Science 343 (6169), 1242777.

Grotzinger, J.P., Gupta, S., Malin, M.C., Rubin, D.M., Schieber, J., Siebach, K., Sumner, D.Y., et al., 2015. Deposition, exhumation, and paleoclimate of an ancient lake deposit, Gale Crater, Mars. Science 350 (6257), aac7575.

Gruber, N., Galloway, J.N., 2008. An earth-system perspective of the global nitrogen cycle. Nature 451 (7176), 293–296.

Gruber, N., Clement, D., Carter, B.R., Feely, R.A., van Heuven, S., Hoppema, M., Ishii, M., et al., 2019a. The oceanic sink for anthropogenic CO_2 from 1994 to 2007. Science 363 (6432), 1193–1199.

Gruber, N., Landschützer, P., Lovenduski, N.S., 2019b. The variable southern ocean carbon sink. Ann. Rev. Mar. Sci. 11. https://doi.org/10.1146/annurev-marine-121916-063407.

Gu, B.H., Schmitt, J., Chen, Z., Liang, L.Y., McCarthy, J.F., 1995. Adsorption and desorption of different organic-matter fractions on iron oxide. Geochim. Cosmochim. Acta 59, 219–229.

Guadalix, M.E., Pardo, M.T., 1991. Sulfate sorption by variable-charge soils. J. Soil Sci. 42, 607–614.

Guenther, A., 2002. The contribution of reactive carbon emissions from vegetation to the carbon balance of terrestrial ecosystems. Chemosphere 49 (8), 837–844.

Guenther, A.B., Zimmerman, P.R., Harley, P.C., Monson, R.K., Fall, R., 1993. Isoprene and monoterpene emission rate variability: model evaluations and sensitivity analyses. J. Geophys. Res. Crop Soc. Am. S 98 (D7), 12609.

Guenther, A., Geron, C., Pierce, T., Lamb, B., Harley, P., Fall, R., 2000. Natural emissions of non-methane volatile organic compounds; carbon monoxide, and oxides of nitrogen from North America. Atmos. Environ. 34, 2205–2230.

Guerrieri, R., Lepine, L., Asbjornsen, H., Xiao, J., Ollinger, S.V., 2016. Evapotranspiration and water use efficiency in relation to climate and canopy nitrogen in U.S. forests: ET and WUE scaled with canopy nitrogen. J. Geophys. Res. Biogeosci. 121 (10), 2610–2629.

Guevara, M., Arroyo, C., Brunsell, N., Cruz, C.O., Domke, G., Equihua, J., Etchevers, J., Hayes, D., Hengl, T., Ibelles, A., Johnson, K., 2020. Soil organic carbon across Mexico and the conterminous United States (1991–2010). Global Biogeochem. Cycles. https://doi.org/10.1029/2019GB006219.

Guidry, M.W., MacKenzie, F.T., Arvidson, R.S., 2000. Role of tectonics in phosphorus distribution and cycling. In: Glenn, C.R., Prevot-Lucas, L., Lucas, J. (Eds.), Marine Authigenesis: From Global to Microbial. Society of Sedimentary Geology, Tulsa, Oklahoma, pp. 35–51.

Guieu, C., Duce, R., Arimoto, R., 1994. Dissolved input of manganese to the ocean: aerosol source. J. Geophys. Res.-Atmos. 99, 18789–18800.

Guillemette, F., del Giorgio, P.A., 2011. Reconstructing the various facets of dissolved organic carbon bioavailability in freshwater ecosystems. Limnol. Oceanogr. 56 (2), 734–748.

Gunal, H., Ransom, M.D., 2006. Clay illuviation and calcium carbonate accumulation along a precipitation gradient in kansas. Catena 68, 59–69.

Gunderson, C.A., O'Hara, K.H., Campion, C.M., Walker, A.V., Edwards, N.T., 2010. Thermal plasticity of photosynthesis: the role of acclimation in forest responses to a warming climate. Glob. Chang. Biol. 16, 2272–2286.

Guo, L.B., Gifford, R.M., 2002. Soil carbon stocks and land use change: a meta analysis. Glob. Chang. Biol. 8, 345–360.

Guo, Y.Y., Amundson, R., Gong, P., Yu, Q., 2006. Quantity and spatial variability of soil carbon in the conterminous United States. Soil Sci. Soc. Am. J. 70, 590–600.

Guo, L., Ping, C.-L., Macdonald, R.W., 2007. Mobilization pathways of organic carbon from Permafrost to Arctic rivers in a changing climate. Geophys. Res. Lett. 34(13).

Guo, J.H., Liu, X.J., Zhang, Y., Shen, J.L., Han, W.X., Zhang, W.F., Christie, P., Goulding, K.W.T., Vitousek, P.M., Zhang, F.S., 2010. Significant acidification in major Chinese croplands. Science 327 (5968), 1008–1010.

Gurtz, M.E., Marzolf, G.R., Killingbeck, K.T., Smith, D.L., McArthur, J.V., 1988. Hydrologic and riparian influences on the import and storage of coarse particulate organic matter in a Prairie stream. Can. J. Fish. Aquat. Sci. 45, 655–665.

Gusewell, S., 2004. N:P ratios in terrestrial plants: variation and functional significance. New Phytol. 164, 243–266.

Gutjahr, M., Ridgwell, A., Sexton, P.F., Anagnostou, E., Pearson, P.N., Pälike, H., Norris, R.D., Thomas, E., Foster, G.L., 2017. Very large release of mostly volcanic carbon during the Palaeocene-Eocene thermal maximum. Nature 548 (7669), 573–577.

Gutschick, V.P., 1981. Evolved strategies in nitrogen acquisition by plants. Am. Nat. 118, 607–637.

Haberl, H., Heinz Erb, K., Krausmann, F., Gaube, V., Bondeau, A., Plutzar, C., Gingrich, S., Lucht, W., Fischer-Kowalski, M., 2007. Quantifying and mapping the human appropriation of net primary production in Earth's terrestrial ecosystems. Proc. Natl. Acad. Sci. U. S. A. 104 (31), 12942–12947.

Haberle, R.M., 2013. Estimating the power of Mars' greenhouse effect. Icarus 223 (1), 619–620.

Habicht, K.S., Canfield, D.E., 1996. Sulphur isotope fractionation in modern microbial mats and the evolution of the sulphur cycle. Nature 382, 342–343.

Habicht, K.S., Canfield, D.E., Rethmeier, J., 1998. Sulfur isotope fractionation during bacterial reduction and disproportionation of thiosulfate and sulfite. Geochim. Cosmochim. Acta 62, 2585–2595.

Habicht, K.S., Gade, M., Thamdrup, B., Berg, P., Canfield, D.E., 2002. Calibration of sulfate levels in the Archean ocean. Science 298 (5602), 2372–2374.

Habing, H.J., Dominik, C., de Mulzon, M.J., Kessler, M.F., Laureijs, R.J., Leech, K., Metcalfe, L., Salama, A., Slebenmorgen, R., Trams, N., 1999. Disappearance of stellar debris disks around main-sequencestars after 400 million years. Nature 401, 456–458.

Hackley, K.C., Anderson, T.F., 1986. Sulfur isotopic variations in low sulfur coals from the rocky mountain region. Geochim. Cosmochim. Acta 50 (8), 1703–1713.

Hackstein, J.H., Stumm, C.K., 1994. Methane production in terrestrial arthropods. Proc. Natl. Acad. Sci. U. S. A. 91 (12), 5441–5445.

Haddeland, I., Skaugen, T., Lettenmaier, D.P., 2006. Anthropogenic impacts on continental surface water fluxes. Geophys. Res. Let. 33.

Haff, P.K., 2010. Hillslopes, rivers, plows, and trucks: mass transport on Earth's surface by natural and technological processes. Earth Surf. Process. Landf. 35 (10), 1157–1166.

Hagedorn, F., Spinnler, D., Bundt, M., Blaser, P., Siegwolf, R., 2003. The input and fate of new C in two forest soils under elevated CO_2. Glob. Chang. Biol. 9, 862–872.

Hahn, J., 1974. The North Atlantic ocean as a source of atmospheric N_2O. Tellus 26, 160–168.

Hahn, J., 1980. Organic constituents of natural aerosols. Ann. N. Y. Acad. Sci. 338, 359–376.

Haines, E.B., 1977. Origins of detritus in Georgia salt Marsh estuaries. Oikos 29, 254–260.

Haines, B., Black, M., Bayer, C., 1989. Sulfur emissions from roots of the rain forest tree *Stryphnodendron excelsum*. In: Saltzman, E.S., Cooper, W.J. (Eds.), Biogenic Sulfur in the Environment. American Chemical Society, pp. 58–69.

Håkanson, L., Bryhn, A.C., Hytteborn, J.K., 2007. On the issue of limiting nutrient and predictions of Cyanobacteria in aquatic systems. Sci. Total Environ. 379, 89–108.

Hakkarainen, J., Ialongo, I., Tamminen, J., 2016. Direct space-based observations of anthropogenic CO_2 emission areas from OCO-2: direct observations of anthropogenic CO_2. Geophys. Res. Lett. 43 (21), 11,400–411,406.

Hale, R.L., Grimm, N.B., Vörösmarty, C.J., Fekete, B., 2015. Nitrogen and phosphorus fluxes from watersheds of the northeast U.S. from 1930 to 2000: role of anthropogenic nutrient inputs, infrastructure, and runoff. Glob. Biogeochem. Cycles 29 (3), 341–356.

Halevy, I., Bachan, A., 2017. The geologic history of seawater pH. Science 355 (6329), 1069–1071.

Hall, R.O., Tank, J.L., 2003. Ecosystem metabolism controls nitrogen uptake in streams in grand Teton National Park, Wyoming. Limnol. Oceanogr. 48 (3), 1120–1128.

Hall, R.O., Ulseth, A.J., 2020. Gas exchange in streams and rivers. WIREs Water. 7, e1391.

Hall, D.T., Strobel, D.F., Feldman, P.D., McGrath, M.A., Weaver, H.A., 1995. Detection of an oxygen atmosphere on Jupiter's Moon Europa. Nature 373 (6516), 677–681.

Hall, R.O., Bernhardt, E.S., Likens, G.E., 2002. Relating nutrient uptake with transient storage in forested mountain streams. Limnol. Oceanogr. 47, 255–265.

Halmer, M.M., Schmincke, H.-U., Graf, H.-F., 2002. The annual volcanic gas input into the atmosphere in particular into the stratosphere: a global data set for the past 100 years. J. Volcanol. Geotherm. Res. 115, 511–528.

Ham, J.M., Heilman, J.L., 2003. Experimental test of density and energy-balance corrections on carbon dioxide flux as measured using open-path eddy covariance. Agron. J. 95, 1393–1403.

Hames, R.S., Rosenberg, K.V., Lowe, J.D., Barker, S.E., Dhondt, A.A., 2002. Adverse effects of acid rain on the distribution of the wood thrush *Hylocichla mustelina* in North America. Proc. Natl. Acad. Sci. U. S. A. 99 (17), 11235–11240.

Hamilton, J.G., DeLucia, E.H., George, K., Naidu, S.L., Finzi, A.C., Schlesinger, W.H., 2002. Forest carbon balance under elevated CO_2. Oecologia 131 (2), 250–260.

Hamilton, S.K., Bruesewitz, D.A., Horst, G.P., Weed, D.B., Sarnelle, O., 2009. Biogenic calcite-phosphorus precipitation as a negative feedback to lake eutrophication. Can. J. Fish. Aquat. Sci. 66, 343–350.

Hamme, R.C., Nicholson, D.P., Jenkins, W.J., Emerson, S.R., 2019. Using noble gases to assess the ocean's carbon pumps. Annu. Rev. Mar. Sci. 11 (January), 75–103.

Hammerling, D.M., Michalak, A.M., O'Dell, C., Kawa, S.R., 2012. Global CO_2 distributions over land from the greenhouse gases observing satellite (GOSAT). Geophys. Res. Lett. 39.

Hammes, K., Torn, M.S., Lapenas, A.G., Schmidt, M.W.I., 2008. Centennial black carbon turnover observed in a Russian steppe soil. Biogeosciences 5 (5), 1339–1350.

Han, H., Allan, J.D., 2012. Uneven rise in N inputs to the lake Michigan Basin over the 20th century corresponds to agricultural and societal transitions. Biogeochemistry 109, 175–187.

Han, T.M., Runnegar, B., 1992. Megascopic eukaryotic algae from the 2.1-billion-year-old Negaunee iron-formation, Michigan. Science 257 (5067), 232–235.

Han, W., Fang, J., Guo, D., Zhang, Y., 2005. Leaf nitrogen and phosphorus stoichiometry across 753 terrestrial plant species in China. New Phytol. 168 (2), 377–385.

Hancke, K., Lund-Hansen, L.C., Lamare, M.L., Pedersen, S.H., King, M.D., Andersen, P., Sorrell, B.K., 2018. Extreme low light requirement for algae growth underneath sea ice: a case study from Station Nord, NE Greenland: minimum light requirement for ice algae. J. Geophys. Res. C: Oceans 123 (2), 985–1000.

Hanczyc, M.M., Fujikawa, S.M., Szostak, J.W., 2003. Experimental models of primitive cellular compartments: encapsulation, growth, and division. Science 302 (5645), 618–622.

Handley, L.L., Raven, J.A., 1992. The use of natural abundance of nitrogen isotopes in plant physiology and ecology. Plant Cell Environ. 15, 965–985.

Hanks, T.C., Anderson, D.L., 1969. The early thermal history of the earth. Phys. Earth Planet. Inter. 2, 19–29.

Hannah, L., Carr, J.L., Landerani, A., 1995. Human disturbance and natural habitat: a biome-level analysis of a global data set. Biodivers. Conserv. 4, 128–155.

Hannan, L.B., Roth, J.D., Ehrhart, L.M., Weishampel, J.F., 2007. Dune vegetation fertilization by nesting sea turtles. Ecology 88 (4), 1053–1058.

Hansell, D.A., Carlson, C.A., 1998. Net community production of dissolved organic carbon. Glob. Biogeochem. Cycles.

Hansell, D.A., Kadko, D., Bates, N.R., 2004. Degradation of terrigenous dissolved organic carbon in the Western Arctic Ocean. Science 304 (5672), 858–861.

Hansen, M.C., Potapov, P.V., Moore, R., Hancher, M., Turubanova, S.A., Tyukavina, A., Thau, D., et al., 2013. High-resolution global maps of 21st-century forest cover change. Science 342 (6160), 850–853.

Hanson, P.J., Edwards, N.T., Garten, C.T., Andrews, J.A., 2000. Separating root and soil microbial contributions to soil respiration: a review of methods and observations. Biogeochemistry 48, 115–146.

Happell, J.D., Chanton, J.P., Whiting, G.J., Showers, W.J., 1993. Stable isotopes as tracers of methane dynamics in everglades marshes with and without active populations of methane oxidizing bacteria. J. Geophys. Res.

Hardacre, C.J., Heal, M.R., 2013. Characterization of methyl bromide and methyl chloride fluxes at temperate freshwater wetlands: methyl halide fluxes at wetlands. J. Geophys. Res. D: Atmos. 118 (2), 977–991.

Harden, J.W., 1988. Genetic interpretations of elemental and chemical differences in a soil chronosequence, California. Geoderma 43, 179–193.

Harden, J.W., Mark, R.K., Sundquist, E.T., Stallard, R.F., 1992. Dynamics of soil carbon during deglaciation of the laurentide ice sheet. Science 258 (5090), 1921–1924.

Hardie, L.A., 1991. On the significance of evaporites. Annu. Rev. Earth Planet. Sci. 19 (1), 131–168.

Harmon, M.E., Franklin, J.F., Swanson, F.J., Sollins, P., Gregory, S.V., Lattin, J.D., Anderson, N.H., et al., 1986. Ecology of coarse woody debris in temperate ecosystems. Adv. Ecol. Res. 15, 133–302.

Harmon, M.E., Silver, W.L., Fasth, B., Chen, H., Burke, I.C., Parton, W.J., Hart, S.C., Currie AndLidet, W.S., 2009. Long-term patterns of mass loss during the decomposition of leaf and fine root litter: an intersite comparison. Glob. Chang. Biol. 15, 1320–1338.

Harper, R.J., Tibbett, M., 2013. The hidden organic carbon in deep mineral soils. Plant Soil 368 (1–2), 641–648.

Harpole, W.S., Ngai, J.T., Cleland, E.E., Seabloom, E.W., Borer, E.T., Bracken, M.E.S., Elser, J.J., et al., 2011. Nutrient co-limitation of primary producer communities. Ecol. Lett. 14 (9), 852–862.

Harries, J.E., Brindley, H.E., Sagoo, P.J., Bantges, R.J., 2001. Increases in greenhouse forcing inferred from the outgoing longwave radiation spectra of the earth in 1970 and 1997. Nature 410 (6826), 355–357.

Harrington, J.B., 1987. Climatic change: a review of causes. Can. J. For. Res. 17, 1313–1339.

Harris, W.G., Hollien, K.A., Yuan, T.L., Bates, S.R., Acree, W.A., 1988. Nonexchangeable potassium associated with hydroxy-interlayered vermiculite from coastal plain soils. Soil Sci. Soc. Am. J. 52, 1486–1492.

Harris, R.C., Rudd, J.W.M., Amyot, M., Babiarz, C.L., Beaty, K.G., Blanchfield, P.J., Bodaly, R.A., et al., 2007. Whole-ecosystem study shows rapid fish-mercury response to changes in mercury deposition. Proc. Natl. Acad. Sci. U. S. A. 104 (42), 16586–16591.

Harris, N.L., Brown, S., Hagen, S.C., Saatchi, S.S., Petrova, S., Salas, W., Hansen, M.C., Potapov, P.V., Lotsch, A., 2012. Baseline map of carbon emissions from deforestation in tropical regions. Science 336 (6088), 1573–1576.

Harrison, A.F., 1982. 32P-method to compare rates of mineralization of labile organic phosphorus in woodland soils. Soil Biol. Biochem. 14, 337–341.

Harrison, K.G., 2000. Role of increased marine silica input on paleo-pCO$_2$ levels. Paleoceanography 15, 292–298.

Harrison, M.J., van Buuren, M.L., 1995. A phosphate transporter from the mycorrhizal fungus *Glomus versiforme*. Nature 378 (6557), 626–629.

Harrison, R.B., Johnson, D.W., Todd, D.E., 1989. Sulfate adsorption and desorption reversibility in a variety of forest soils. J. Environ. Qual. 18, 419–426.

Harrison, W.G., Harris, L.R., Karl, D.M., Knauer, G.A., Redalje, D.G., 1992. Nitrogen dynamics at the VERTEX time-series site. Deep-Sea Res. A—Oceanogr. Res. Pap. 39, 1535–1552.

Harrison, W.G., Harris, L.R., Irwin, B.D., 1996. The kinetics of nitrogen utilization in the oceanic mixed layer: nitrate and ammonium interactions at nanomolar concentrations. Limnol. Oceanogr. 41, 16–32.

Harrison, J.A., Caraco, N., Seitzinger, S.P., 2005. Global patterns and sources of dissolved organic matter export to the coastal zone: results from a spatially explicit, global model. Glob. Biogeochem. Cycles 19 (4).

Harrison, K.A., Bol, R., Bardgett, R.D., 2007. Preferences for different nitrogen forms by coexisting plant species and soil microbes. Ecology 88 (4), 989–999.

Harrison, J.A., Maranger, R.J., Alexander, R.B., Giblin, A.E., Jacinthe, P.-A., Mayorga, E., Seitzinger, S.P., Sobota, D.J., Wollheim, W.M., 2008. The regional and global significance of nitrogen removal in lakes and reservoirs. Biogeochemistry 93 (1–2), 143–157.

Harrison, J.F., Kaiser, A., VandenBrooks, J.M., 2010. Atmospheric oxygen level and the evolution of insect body size. Proc. Biol. Sci. 277 (1690), 1937–1946.

Harriss, R.C., Sebacher, D.I., Day, F.P., 1982. Methane flux in the great dismal swamp. Nature 297 (5868), 673–674.

Harte, J., Saleska, S., Shih, T., 2006. Shifts in plant dominance control carbon-cycle responses to experimental warming and widespread drought. Environ. Res. Lett. 1.

Harte, J., Saleska, S.R., Levy, C., 2015. Convergent ecosystem responses to 23-year ambient and manipulated warming link advancing snowmelt and shrub encroachment to transient and long-term climate-soil carbon feedback. Glob. Chang. Biol. 21 (6), 2349–2356.

Hartgers, W.A., Sinninghe Damste, J.S., Requejo, A.G., Allan, J., Hayes, J.M., de Leeuw, J.W., 1994. Evidence for only minor contributions from bacteria to sedimentary organic carbon. Nature 369 (6477), 224–227.

Hartley, S.E., DeGabriel, J.L., 2016. The ecology of herbivore-induced silicon defences in grasses. Cooke, J. (Ed.), Funct. Ecol. 30 (8), 1311–1322.

Hartley, A.E., Schlesinger, W.H., 2000. Environmental controls on nitric oxide emission from Northern Chihuahuan Desert soils. Biogeochemistry 50, 279–300.

Hartley, A.M., House, W.A., Callow, M.E., Leadbeater, B.S.C., 1995. The role of a green alga in the precipitation of calcite and the coprecipitation of phosphate in freshwater. Int. Rev. Gesamten Hydrobiol. 80, 385–401.

Hartmann, J., Moosdorf, N., 2012. The new global lithological map database GLiM: a representation of rock properties at the earth surface: technical brief. Geochem. Geophys. Geosyst. Geol. Surv. Can. Open File 13 (12), 119.

Hartmann, J., Jansen, N., Durr, H.H., Kempe, S., Kohler, P., 2009. Global CO_2-consumption by chemical weathering: what is the contribution of highly active weathering regions? Glob. Planet. Chang. 69, 185–194.

Hartnett, H.E., Keil, R.G., Hedges, J.I., Devol, A.H., 1998. Influence of oxygen exposure time on organic carbon preservation in continental margin sediments. Nature 391, 572–574.

Harvey, L.D.D., 1988. Climatic impact of ice-age aerosols. Nature 334, 333–335.

Hassenkam, T., Andersson, M.P., Dalby, K.N., Mackenzie, D.M.A., Rosing, M.T., 2017. Elements of Eoarchean life trapped in mineral inclusions. Nature 548 (7665), 78–81.

Hastings, M.G., Jarvis, J.C., Steig, E.J., 2009. Anthropogenic impacts on nitrogen isotopes of ice-core nitrate. Science 324 (5932), 1288.

Hatch, D.J., Jarvis, S.C., Philipps, L., 1990. Field measurement of nitrogen mineralization using soil core incubation and acetylene inhibition of nitrification. Plant Soil 124, 97–107.

Hattenschwiler, S., Miglietta, F., Raschi, A., Korner, C., 1997. Thirty years of in situ tree growth under elevated CO_2: a model for future forest responses? Glob. Chang. Biol. 3, 463–471.

Hatton, R.S., Delaune, R.D., Patrick, W.H., 1983. Sedimentation, accretion, and subsidence in marshes of Barataria Basin, Louisiana. Limnol. Oceanogr. 28, 494–502.

Hawkesworth, C.J., Kemp, A.I.S., 2006. Evolution of the continental crust. Nature 443 (7113), 811–817.

Hay, C.C., Morrow, E., Kopp, R.E., Mitrovica, J.X., 2015. Probabilistic reanalysis of twentieth-century sea-level rise. Nature 517 (7535), 481–484.

Hayes, P., Turner, B.L., Lambers, H., Laliberté, E., 2014. Foliar nutrient concentrations and resorption efficiency in plants of contrasting nutrient-acquisition strategies along a 2-million-year dune chronosequence. J. Ecol. 102 (2), 396–410.

Hayne, P.O., Paige, D.A., Heavens, N.G., 2014. The role of snowfall in forming the seasonal ice caps of Mars: models and constraints from the Mars climate sounder. Icarus 231 (March), 122–130.

Haynes, R.J., Goh, K.M., 1978. Ammonium and nitrate nutrition of plants. Biol. Rev. Camb. Philos. Soc. 53, 465–510.

Hays, J.D., Imbrie, J., Shackleton, N.J., 1976. Variations in the Earth's orbit: pacemaker of the ice ages. Science 194 (4270), 1121–1132.

He, Y.F., D'Odorico, P., De Wekker, S.F.J., Fuentes, J.D., Litvak, M., 2010. On the impact of shrub encroachment on microclimate conditions in the Northern Chihuahuan Desert. J. Geophys. Res. 115.

He, H., Bleby, T.M., Veneklaas, E.J., Lambers, H., 2011. Dinitrogen-fixing *Acacia* species from phosphorus-impoverished soils resorb leaf phosphorus efficiently. Plant Cell Environ. 34 (12), 2060–2070.

He, Y., Trumbore, S.E., Torn, M.S., Harden, J.W., Vaughn, L.J.S., Allison, S.D., Randerson, J.T., 2016. Radiocarbon constraints imply reduced carbon uptake by soils during the 21st century. Science 353 (6306), 1419–1424.

He, Y., Wang, K., Zhou, C., Wild, M., 2018. A revisit of global dimming and brightening based on the sunshine duration. Geophys. Res. Lett. 45 (9), 4281–4289.

He, L., Zeng, Z.-c., Pongetti, T.J., Wong, C., Liang, J., Gurney, K.R., Newman, S., et al., 2019. Atmospheric methane emissions correlate with natural gas consumption from residential and commercial sectors in Los Angeles. Geophys. Res. Lett. 46 (14), 8563–8571.

Heathcote, A.J., Downing, J.A., 2012. Impacts of eutrophication on carbon burial in freshwater lakes in an intensively agricultural landscape. Ecosystems 15, 60–70.

Heathwaite, A.L., Eggelsmann, R., Gottlich, K.H., Haule, G., 1990. Ecohydrology, mire drainage and mire conservation. In: Heathwaiter, A.L. (Ed.), Mires: Process, Exploitation and Conservation. John Wiley & Sons, New York, pp. 417–484.

Hebting, Y., Schaeffer, P., Behrens, A., Adam, P., Schmitt, G., Schneckenburger, P., Bernasconi, S.M., Albrecht, P., 2006. Biomarker evidence for a major preservation pathway of sedimentary organic carbon. Science 312 (5780), 1627–1631.

Heckathorn, S.A., DeLucia, E.H., 1995. Ammonia volatilization during drought in perennial C4 grasses of tallgrass prairie. Oecologia 101 (3), 361–365.

Hedges, J.I., Keil, R.G., 1997. What happens to terrestrial organic matter in the ocean? Org. Geochem. 27 (5), 195–212.

Hedges, J.I., Parker, P.L., 1976. Land-derived organic matter in surface sediments from the Gulf of Mexico. Geochim. Cosmochim. Acta 40, 1019–1029.

Hedin, L.O., Likens, G.E., Bormann, F.H., 1987. Decrease in precipitation acidity resulting from decreased SO_4^2 concentration. Nature 325, 244–246.

Hedin, L.O., Likens, G.E., Postek, K.M., Driscoll, C.T., 1990. A field experiment to test whether organic acids buffer acid deposition. Nature 345, 798–800.

Hedin, L.O., Armesto, J.J., Johnson, A.H., 1995. Patterns of nutrient loss from unpolluted old-growth forests: evaluation of biogeochemical theory. Ecology 76, 493–509.

Hedin, L.O., von Fischer, J.C., Ostrom, N.E., Kennedy, B.P., Brown, M.G., Robertson, G.P., 1998. Thermodynamic constraints on the biogeochemical structure and transformation of nitrogen at terrestrial-lotic interfaces. Ecology 79, 684–703.

Hedley, M.J., Nye, P.H., White, R.E., 1982a. Plant-induced changes in the rhizosphere of rape (*Brassica napus* Var-Emerald) seedlings. II. Origin of the pH change. New Phytol. 91, 31–44.

Hedley, M.J., Stewart, J.W.B., Chauhan, B.S., 1982b. Changes in inorganic and organic soil phosphorus fractions induced by cultivation practices and by laboratory incubations. Soil Sci. Soc. Am. J. 46, 970–976.

Heffernan, J.B., Cohen, M.J., 2010. Direct and indirect coupling of primary production and diel nitrate dynamics in a subtropical spring-fed river. Limnol. Oceanogr. 55 (2), 677–688.

Hegglin, M.I., et al., 2015. Twenty Questions and Answers About the Ozone Layer: 2014 Update, Scientific Assessment of Ozone Depletion. World Meteorological Organization, Geneva, Switzerland.

Hein, R., Crutzen, P.J., Heimann, M., 1997. An inverse modeling approach to investigate the global atmospheric methane cycle. Glob. Biogeochem. Cycles 11, 43–76.

Heintzenberg, J., 1989. Fine particles in the global troposphere: a review. Tellus B Chem. Phys. Meteorol. 41, 149–160.

Heiskanen, J.J., Mammarella, I., Ojala, A., Stepanenko, V., Erkkilä, K.-m., Miettinen, H., Sandström, H., et al., 2015. Effects of water clarity on lake stratification and lake-atmosphere heat exchange. J. Geophys. Res. D: Atmos. 120 (15), 7412–7428.

Heisler, J., Glibert, P.M., Burkholder, J.M., Anderson, D.M., Cochlan, W., Dennison, W.C., Dortch, Q., et al., 2008. Eutrophication and harmful algal blooms: a scientific consensus. Harmful Algae 8 (1), 3–13.

Hejcman, M., Karlík, P., Ondráček, J., Klír, T., 2013. Short-term medieval settlement activities irreversibly changed forest soils and vegetation in central Europe. Ecosystems 16 (4), 652–663.

Helfenstein, J., Jegminat, J., McLaren, T.I., Frossard, E., 2018. Soil solution phosphorus turnover: derivation, interpretation, and insights from a global compilation of isotope exchange kinetic studies. Biogeosciences 15 (1), 105–114.

Helfenstein, J., Pistocchi, C., Oberson, A., Tamburini, F., Goll, D.S., Frossard, E., 2019. Estimates of mean residence times of phosphorus in commonly-considered inorganic soil phosphorus pools. Biogeosci. Discuss. 2019, 1–25.

Helfield, J.M., Naiman, R.J., 2001. Effects of Salmon-derived nitrogen on riparian forest growth and implications for stream productivity. Ecology 82 (9), 2403–2409.

Helm, K.P., Bindoff, N.L., Church, J.A., 2011. Observed decreases in oxygen content of the global ocean. Geophys. Res. Lett. 38.

Helmig, D., Daly, R.W., Milford, J., Guenther, A., 2013. Seasonal trends of biogenic terpene emissions. Chemosphere 93 (1), 35–46.

Helmig, D., Rossabi, S., Hueber, J., Tans, P., Montzka, S.A., Masarie, K., Thoning, K., et al., 2016. Reversal of global atmospheric ethane and propane trends largely due to US oil and natural gas production. Nat. Geosci. 9 (7), 490–495.

Helton, A.M., Ardón, M., Bernhardt, E.S., 2015. Thermodynamic constraints on the utility of ecological stoichiometry for explaining global biogeochemical patterns. Ecol. Lett. 18 (10), 1049–1056.

Helton, A.M., Ardón, M., Bernhardt, E.S., 2019. Hydrologic context alters greenhouse gas feedbacks of coastal wetland salinization. Ecosystems 22 (5), 1108–1125.

Hember, R.A., Kurz, W.A., Girardin, M.P., 2019. Tree ring reconstructions of stemwood biomass indicate increases in the growth rate of black spruce trees across boreal forests of Canada. J. Geophys. Res. Biogeosci. 124 (8), 2460–2480.

Hemingway, J.D., Hilton, R.G., Hovius, N., Eglinton, T.I., Haghipour, N., Wacker, L., Chen, M.-C., Galy, V.V., 2018. Microbial oxidation of lithospheric organic carbon in rapidly eroding tropical mountain soils. Science 360 (6385), 209–212.

Hemingway, J.D., Rothman, D.H., Grant, K.E., Rosengard, S.Z., Eglinton, T.I., Derry, L.A., Galy, V.V., 2019. Mineral protection regulates long-term global preservation of natural organic carbon. Nature 570 (7760), 228–231.

Hemond, H.F., 1980. Biogeochemistry of Thoreau's Bog, Concord, Massachusetts. Ecol. Monogr. 50, 507–526.

Hemond, H.F., 1983. The nitrogen budget of Thoreau's Bog. Ecology 64 (1), 99–109.

Henderson, G.S., Swank, W.T., Waide, J.B., Grier, C.C., 1978. Nutrient budgets of Appalachian and Cascade region watersheds: a comparison. For. Sci. 24, 385–397.

Henderson-Sellers, A., McGuffie, K., 1987. A Climate Modeling Primer. Wiley.

Hendren, C.O., Mesnard, X., Dröge, J., Wiesner, M.R., 2011. Estimating production data for five engineered nanomaterials as a basis for exposure assessment. Environ. Sci. Technol. 45 (7), 2562–2569.

Hendrey, G.R., Ellsworth, D.S., Lewin, K.F., Nagy, J., 1999. A free-air enrichment system for exposing tall forest vegetation to elevated atmospheric CO_2. Glob. Chang. Biol. 5, 293–309.

Henrichs, S.M., Reeburgh, W.S., 1987. Anaerobic mineralization of marine sediment organic matter—rates and the role of anaerobic processes in the oceanic carbon economy. Geomicrobiol J. 5 (3–4), 191–237.

Hensen, C., Zabel, M., Pfeifer, K., Schwenk, T., Kasten, S., Riedinger, N., Schulz, H.D., Boettius, A., 2003. Control of sulfate pore-water profiles by sedimentary events and the significance of anaerobic oxidation of methane for the burial of sulfur in marine sediments. Geochim. Cosmochim. Acta 67, 2631–2647.

Henson, S.A., Sanders, R., Madsen, E., Morris, P.J., Le Moigne, F., Quartly, G.D., 2011. A reduced estimate of the strength of the ocean's biological carbon pump. Geophys. Res. Lett. 38.

Herbert, E.R., Boon, P., Burgin, A.J., Neubauer, S.C., Franklin, R.B., Ardón, M., Hopfensperger, K.N., Lamers, L.P.M., Gell, P., 2015. A global perspective on wetland salinization: ecological consequences of a growing threat to freshwater wetlands. Ecosphere 6 (10), art206.

Herd, C.D.K., Blinova, A., Simkus, D.N., Huang, Y., Tarozo, R., O'd, C.M., Alexander, F.G., et al., 2011. Origin and evolution of prebiotic organic matter as inferred from the Tagish Lake meteorite. Science 332 (6035), 1304–1307.

Herman, R.P., Provencio, K.R., Torrez, R.J., Seager, G.M., 1993. Effect of water and nitrogen additions on free-living nitrogen fixer populations in desert grass root zones. Appl. Environ. Microbiol. 59 (9), 3021–3026.

Herridge, D.F., Peoples, M.B., Boddey, R.M., 2008. Global inputs of biological nitrogen fixation in agricultural systems. Plant Soil 311, 1–18.

Herron, M.M., Langway, C.C., Weiss, H.W., Cragin, J.H., 1977. Atmospheric trace metals and sulfate in the Greenland ice sheet. Geochim. Cosmochim. Acta 41, 915–920.

Heskel, M.A., Atkin, O.K., Turnbull, M.H., Griffin, K.L., 2013. Bringing the kok effect to light: a review on the integration of daytime respiration and net ecosystem exchange. Ecosphere 4 (8), art98.

Hess, S.L., Henry, R.M., Leovy, C.B., Ryan, J.A., Tillman, J.E., Chamberlain, T.E., Cole, H.L., et al., 1976. Preliminary meteorological results on Mars from the viking 1 lander. Science 193 (4255), 788–791.

Hester, K., Boyle, E., 1982. Water chemistry control of cadmium content in recent benthic foraminifera. Nature 298, 260–262.

Hesterberg, R., Blatter, A., Fahrni, M., Rosset, M., Neftel, A., Eugster, W., Wanner, H., 1996. Deposition of nitrogen-containing compounds to an extensively managed Grassland in Central Switzerland. Environ. Pollut. 91 (1), 21–34.

Heuer, V.B., Pohlman, J.W., Torres, M.E., Elvert, M., Hinrichs, K.-U., 2009. The stable carbon isotope biogeochemistry of acetate and other dissolved carbon species in deep subseafloor sediments at the Northern Cascadia Margin. Geochim. Cosmochim. Acta 73 (11), 3323–3336.

Hewitt, C.N., Harrison, R.M., 1985. Tropospheric concentrations of the hydroxyl radical—a review. Atmos. Environ. 19, 545–554.

Hibbard, K.A., Law, B.E., Reichstein, M., Sulzman, J., 2005. An analysis of soil respiration across northern hemisphere temperate ecosystems. Biogeochemistry 73, 29–70.

Hicke, J.A., Meddens, A.J.H., Allen, C.D., Kolden, C.A., 2013. Carbon stocks of trees killed by bark beetles and wildfire in the western United States. Environ. Res. Lett. 8 (3), 035032.

Hicks Pries, C.E., Castanha, C., Porras, R.C., Torn, M.S., 2017. The whole-soil carbon flux in response to warming. Science 355 (6332), 1420–1423.

Hidy, G.M., 1970. Theory of diffusive and impactive scavenging. In: Englemann, R.J., Slinn, W.G.N. (Eds.), Precipitation Scavenging. U.S. Atomic Energy Commission, Division of Technical Information, Oak Ridge, Tennessee, pp. 355–371.

Hietz, P., Turner, B.L., Wanek, W., Richter, A., Nock, C.A., Joseph Wright, S., 2011. Long-term change in the nitrogen cycle of tropical forests. Science 334 (6056), 664–666.

Hilderbrand, G.V., Hanley, T.A., Robbins, C.T., Schwartz, C.C., 1999. Role of brown bears (Ursus arctos) in the flow of marine nitrogen into a terrestrial ecosystem. Oecologia 121 (4), 546–550.

Hill, W.R., 1996. Effects of light. In: Stevenson, Lowe, B. (Eds.), Algal Ecology. Academic Press, San Diego, pp. 121–148.

Hilley, G.E., Porder, S., 2008. A framework for predicting global silicate weathering and CO_2 drawdown rates over geologic time-scales. Proc. Natl. Acad. Sci. U. S. A. 105 (44), 16855–16859.

Hilley, G.E., Chamberlain, C.P., Moon, S., Porder, S., Willett, S.D., 2010. Competition between erosion and reaction kinetics in controlling silicate-weathering rates. Earth Planet. Sci. Lett. 293, 191–199.

Hilton, J., O'Hare, M., Bowes, M.J., Iwan Jones, J., 2006. How green is my river? A new paradigm of eutrophication in rivers. Sci. Total Environ. 365 (1–3), 66–83.

Hingston, F.J., Atkinson, R.J., Posner, A.M., Quirk, J.P., 1967. Specific adsorption of anions. Nature 215, 1459–1461.

Hinkle, S.R., Tesoriero, A.J., 2014. Nitrogen speciation and trends, and prediction of denitrification extent, in shallow US groundwater. J. Hydrol. 509 (February), 343–353.

Hinrichs, K.U., Hayes, J.M., Sylva, S.P., Brewer, P.G., DeLong, E.F., 1999. Methane-consuming Archaebacteria in marine sediments. Nature 398 (6730), 802–805.

Hirose, K., Sinmyo, R., Hernlund, J., 2017. Perovskite in Earth's deep interior. Science 358 (6364), 734–738.

Hirsch, A.I., Munger, J.W., Jacob, D.J., Horowitz, L.W., Goldstein, A.H., 1996. Seasonal variation of the ozone production efficiency per unit NOx at Harvard Forest, Massachusetts. J. Geophys. Res. 101, 12659–12666.

Hirsch, R.E., Lewis, B.D., Spalding, E.P., Sussman, M.R., 1998. A role for the AKT1 potassium channel in plant nutrition. Science 280 (5365), 918–921.

Hobbie, S.E., 1992. Effects of plant species on nutrient cycling. Trends Ecol. Evol. 7 (10), 336–339.

Hobbie, E.A., 2006. Carbon allocation to Ectomycorrhizal fungi correlates with belowground allocation in culture studies. Ecology 87 (3), 563–569.

Hobbie, S.E., 2015. Plant species effects on nutrient cycling: revisiting litter feedbacks. Trends Ecol. Evol. 30 (6), 357–363.

Hobbie, J.E., Hobbie, E.A., 2006. ^{15}N in symbiotic fungi and plants estimates nitrogen and carbon flux rates in Arctic tundra. Ecology 87 (4), 816–822.

Hobbie, E.A., Ouimette, A.P., 2009. Controls of nitrogen isotope patterns in soil profiles. Biogeochemistry 95, 355–371.

Hoch, A.R., Reddy, M.M., Aiken, G.R., 2000. Calcite crystal growth inhibition by humic substances with emphasis on hydrophobic acids from the Florida Everglades. Geochim. Cosmochim. Acta 64, 61–72.

Hocking, W.K., Carey-Smith, T., Tarasick, D.W., Argall, P.S., Strong, K., Rochon, Y., Zawadzki, I., Taylor, P.A., 2007. Detection of stratospheric ozone intrusions by Windprofiler radars. Nature 450 (7167), 281–284.

Hodge, A., Storer, K., 2015. Arbuscular mycorrhiza and nitrogen: implications for individual plants through to ecosystems. Plant Soil 386 (1), 1–19.

Hodge, A., Campbell, C.D., Fitter, A.H., 2001. An arbuscular mycorrhizal fungus accelerates decomposition and acquires nitrogen directly from organic material. Nature 413 (6853), 297–299.

Hodson, A., Heaton, T., Langford, H., Newsham, K., 2010. Chemical weathering and solute export by meltwater in a maritime Antarctic Glacier Basin. Biogeochemistry 98, 9–27.

Hoegh-Guldberg, O., Mumby, P.J., Hooten, A.J., Steneck, R.S., Greenfield, P., Gomez, E., Harvell, C.D., et al., 2007. Coral reefs under rapid climate change and ocean acidification. Science 318 (5857), 1737–1742.

Hoekstra, A.Y., Chapagain, A.K., 2007. Water footprints of nations: Water use by people as a function of their consumption pattern. Water Resour. Manag. 21, 35–48.

Hoellein, T.J., Bruesewitz, D.A., Richardson, D.C., 2013. Revisiting Odum (1956): a synthesis of aquatic ecosystem metabolism. Limnol. Oceanogr. 58 (6), 2089–2100.

Hoff, T., Stummann, B.M., Henningsen, K.W., 1992. Structure, function and regulation of nitrate reductase in higher plants. Physiol. Plant. 84, 616–624.

Hoffman, P.F., Kaufman, A.J., Halverson, G.P., Schrag, D.P., 1998. A neoproterozoic snowball earth. Science 281 (5381), 1342–1346.

Hofmann, D.J., 1990. Increase in the stratospheric background sulfuric acid aerosol mass in the past 10 years. Science 248 (4958), 996–1000.

Hofmann, D.J., Rosen, J.M., 1980. Stratospheric sulfuric acid layer: evidence for an anthropogenic component. Science 208 (4450), 1368–1370.

Hofmann, D.J., Rosen, J.M., 1983. Sulfuric acid droplet formation and growth in the stratosphere after the 1982 eruption of El Chichon. Science 222 (4621), 325–327.

Hofmann, M., Schellnhuber, H.J., 2010. Ocean acidification: a millennial challenge. Energy Environ. Sci.

Hofmann, D.J., Butler, J.H., Tans, P.P., 2009. A new look at atmospheric carbon dioxide. Atmos. Environ. 43, 2084–2086.

Hofmockel, K.S., Fierer, N., Colman, B.P., Jackson, R.B., 2010. Amino acid abundance and proteolytic potential in North American soils. Oecologia 163 (4), 1069–1078.

Hofzumahaus, A., Rohrer, F., Lu, K., Bohn, B., Brauers, T., Chang, C.-C., Fuchs, H., et al., 2009. Amplified trace gas removal in the troposphere. Science 324 (5935), 1702–1704.

Hogberg, P., 1997. [15]N natural abundance in soil-plant systems. New Phytol. 137, 179–203.

Hogberg, P., 2012. What is the quantitative relation between nitrogen deposition and forest carbon sequestration? Glob. Chang. Biol. 18, 1–2.

Högberg, P., Nordgren, A., Buchmann, N., Taylor, A.F., Ekblad, A., Högberg, M.N., Nyberg, G., Ottosson-Löfvenius, M., Read, D.J., 2001. Large-scale forest girdling shows that current photosynthesis drives soil respiration. Nature 411 (6839), 789–792.

Hogg, N., Biscaye, P., Gardner, W., Schmitz Jr., W.J., 1982. On the transport and modification of Antarctic bottom water in the Vema channel. J. Marine Res. 40 (23), 1–263.

Hojberg, O., Revsbech, N.P., Tiedje, J.M., 1994. Denitrification in soil aggregates analyzed with microsensors for nitrous oxide and oxygen. Soil Sci. Soc. Am. J. 58, 1691–1698.

Hole, F.D., 1981. Effects of animals on soil. Geoderma 25, 75–112.

Holland, H.D., 1965. The history of ocean water and its effect on the chemistry of the atmosphere. Proc. Natl. Acad. Sci. 53, 1173–1183.

Holland, H.D., 1974. Marine evaporites and the composition of sea water during the phanerozoic. Stud. Paleo-Oceanogr.

Holland, H.D., 1978. The Chemistry of the Atmosphere and Oceans. Wiley.

Holland, H.D., 1984. The Chemical Evolution of the Atmosphere and Oceans. Princeton University Press.

Holland, H.D., Feakes, C.R., Zbinden, E.A., 1989. The flin flon paleosol and the composition of the atmosphere 1.8 Bybp. Am. J. Sci. 289, 362–389.

Hollinger, D.Y., 1986. Herbivory and the cycling of nitrogen and phosphorus in isolated California oak trees. Oecologia 70 (2), 291–297.

Hollinger, D.Y., Aber, J., Dail, B., Davidson, E.A., Goltz, S.M., Hughes, H., Leclerc, M.Y., et al., 2004. Spatial and temporal variability in forest-atmosphere CO_2 exchange. Glob. Chang. Biol. 10, 1689–1706.

Holloway, J.M., Dahlgren, R.A., 2002. Nitrogen in rock: occurrences and biogeochemical implications. Glob. Biogeochem. Cycles 16.

Holloway, J.M., Dahlgren, R.A., Hansen, B., Casey, W.H., 1998. Contribution of bedrock nitrogen to high nitrate concentrations in stream water. Nature 395, 785–788.

Holmer, M., Storkholm, P., 2001. Sulphate reduction and sulphur cycling in lake sediments: a review. Freshw. Biol. 46, 431–451.

Holser, W.T., Kaplan, I.R., 1966. Isotope geochemistry of sedimentary sulfates. Chem. Geol. 1, 93–135.

Holser, W.T., Schidlowski, M., MacKenzie, F.T., Maynard, J.B., 1988. Geochemical cycles of carbon and sulfur. In: Gregor, C.B., Garrels, R.M., MacKenzie, F.T., Maynard, J.B. (Eds.), Chemical Cycles in the Evolution of the Earth. Wiley, pp. 105–173.

Holser, W.T., Maynard, J.B., Cruikshank, K.M., 1989. Modelling the natural cycle of sulphur through phanerozoic time. In: Brimblecombe, P., Lein, A.Y. (Eds.), Evolution of the Global Biogeochemical Sulphur Cycle. Wiley, pp. 21–56.

Holtgrieve, G.W., Schindler, D.E., Hobbs, W.O., Leavitt, P.R., Ward, E.J., Bunting, L., Chen, G., et al., 2011. A coherent signature of anthropogenic nitrogen deposition to remote watersheds of the northern hemisphere. Science 334 (6062), 1545–1548.

Holton, J.R., Haynes, P.H., McIntyre, M.E., Douglass, A.R., Rood, R.B., Pfister, L., 1995. Stratosphere-troposphere exchange. Rev. Geophys. 33, 403–439.

Homyak, P.M., Allison, S.D., Huxman, T.E., Goulden, M.L., Treseder, K.K., 2017. Effects of drought manipulation on soil nitrogen cycling: a meta-analysis. J. Geophys. Res. Biogeosci. 122 (12), 3260–3272.

Hondzo, M., Voller, V.R., Morris, M., Foufoula-Georgiou, E., Finlay, J., Ganti, V., Power, M.E., 2013. Estimating and scaling stream ecosystem metabolism along channels with heterogeneous substrate. Ecohydrology 6, 679–688.

Honeycutt, C.W., Heil, R.D., Cole, C.V., 1990. Climatic and topographic relations of three great-plains soils. I. Soil morphology. Soil Sci. Soc. Am. J. 54, 469–475.

Hong, J.I., Feng, Q., Rotello, V., Rebek Jr., J., 1992. Competition, cooperation, and mutation: improving a synthetic replicator by light irradiation. Science 255 (5046), 848–850.

Hongoh, Y., Sharma, V.K., Prakash, T., Noda, S., Toh, H., Taylor, T.D., Kudo, T., et al., 2008. Genome of an endosymbiont coupling N$_2$ fixation to cellulolysis within protist cells in termite gut. Science 322 (5904), 1108–1109.

Hönisch, B., Ridgwell, A., Schmidt, D.N., Thomas, E., Gibbs, S.J., Sluijs, A., Zeebe, R., et al., 2012. The geological record of ocean acidification. Science 335 (6072), 1058–1063.

Honjo, S., Manganini, S.J., Cole, J.J., 1982. Sedimentation of biogenic matter in the deep ocean. Deep-Sea Res. A—Oceanogr. Res. Pap. 29, 609–625.

Hooke, R.L., 1999. Spatial distribution of human geomorphic activity in the United States: comparison with rivers. Earth Surf. Process. Landf. 24, 687–692.

Hooke, R.L., 2000. On the history of humans as geomorphic agents. Geology 28, 843–846.

Hooker, T.D., Compton, J.E., 2003. Forest ecosystem carbon and nitrogen accumulation during the first century after agricultural abandonment. Ecol. Appl. 13, 299–313.

Hooper, P.R., Herrick, I.W., Laskowski, E.R., Knowles, C.R., 1980. Composition of the Mount St. Helens Ashfall in the Moscow-Pullman area on 18 May 1980. Science 209 (4461), 1125–1126.

Hope, M.F., Billett, D., Cresser, M.S., 1994. A review of the export of carbon in river water: fluxes and processes. Environ. Pollut. 84, 301–324.

Höpfner, M., Milz, M., Buehler, S., Orphal, J., Stiller, G., 2012. The natural greenhouse effect of atmospheric oxygen (O$_2$) and nitrogen (N$_2$): natural greenhouse effect of O$_2$ and N$_2$. Geophys. Res. Lett. 39 (10).

Hopkins, J., Balch, W.M., 2018. A new approach to estimating coccolithophore calcification rates from space. J. Geophys. Res.: Biogeosci. 123, 1447–1459.

Hopkinson Jr., C.S., Vallino, J.J., 2005. Efficient export of carbon to the deep ocean through dissolved organic matter. Nature 433 (7022), 142–145.

Horita, J., Zimmermann, H., Holland, H.D., 2002. Chemical evolution of seawater during the phanerozoic: implications from the record of marine evaporites. Geochim. Cosmochim. Acta 66 (21), 3733–3756.

Horne, A.J., Galat, D.L., 1985. Nitrogen fixation in an oligotrophic, saline desert lake—Pyramid Lake, Nevada. Limnol. Oceanogr. 30 (6), 1229–1239.

Horne, A.J., Goldman, C.R., 1974. Suppression of nitrogen fixation by blue-green algae in a eutrophic lake with trace additions of copper. Science 183 (4123), 409–411.

Hornibrook, E.R.C., Longstaffe Frederick, J., Fyfe, W.S., 1997. Spatial distribution of microbial methane production pathways in temperate zone wetland soils: stable carbon and hydrogen isotope evidence. Geochim. Cosmochim. Acta 61 (4), 745–753.

Horodyski, R.J., Knauth, L.P., 1994. Life on land in the precambrian. Science 263 (5146), 494–498.

Horrigan, S.G., Montoya, J.P., Nevins, J.L., McCarthy, J.J., 1990. Natural isotopic composition of dissolved inorganic nitrogen in the Chesapeake Bay. Estuar. Coast. Shelf Sci. 30, 393–410.

Hosker, R.P., Lindberg, S.E., 1982. (Review). Atmospheric deposition and plant assimilation of gases and particles. Atmos. Environ. 16, 889–910.

Hotchkiss, E.R., Hall Jr., R.O., Sponseller, R.A., Butman, D., Klaminder, J., Laudon, H., Rosvall, M., Karlsson, J., 2015a. Sources of and processes controlling CO_2 emissions change with the size of streams and rivers. Nat. Geosci. 8 (9), 696.

Hotchkiss, E.R., Hall, R.O., Jr., 2015b. Whole-stream ^{13}C tracer addition reveals distinct fates of newly fixed carbon. Ecology 96 (2), 403–416.

Hou, L., Liu, M., Ding, P., Zhou, J., Yang, Y., Zhao, D., Zheng, Y., 2011. Influences of sediment dessication on phosphorus transformations in an intertidal marsh: formation and release of phosphine. Chemosphere 83 (7), 917–924.

Hough, A.M., Derwent, R.G., 1990. Changes in the global concentration of tropospheric ozone due to human activities. Nature 344, 645–648.

Houghton, J.T., 1986. The Physics of Atmospheres, second ed. Cambridge University Press.

Houghton, R.A., 1987. Biotic changes consistent with the increased seasonal amplitude of atmospheric CO_2 concentrations. J. Geophys. Res. D: Atmos. 92, 4223–4230.

Houghton, R.A., Nassikas, A.A., 2017. Global and regional fluxes of carbon from land use and land cover change 1850–2015: carbon emissions from land use. Glob. Biogeochem. Cycles 31 (3), 456–472.

Houghton, R.A., Skole, D.L., 1990. Carbon. In: Clark, W.C., Turner, B.L., Kates, R.W., Richards, J.F., Matthews, J.T., Meyer, W.B. (Eds.), The Earth as Transformed by Human Action. Cambridge University Press, pp. 393–408.

Houghton, R.A., Hobbie, J.E., Melillo, J.M., Moore, B., Peterson, B.J., Shaver, G.R., Woodwell, G.M., 1983. Changes in the carbon content of terrestrial biota and soils between 1860 and 1980: a net release of CO_2 to the atmosphere. Ecol. Monogr. 53, 235–262.

Houghton, J.T., Jenkins, G.J., Ephramus, J.J.E., 1990. Climate Change: The IPCC Scientific Assessment. Cambridge University Press.

Houghton, J.T., Meira Filho, L.G., Bruce, J., Lee, H., Callander, B.A., Haites, E., Harris, N., Maskell, K., 1995. Climate Change 1994: Radiative Forcing of Climate Change and an Evaluation of the IPCC IS92 Emission Scenarios. Cambridge University Press.

Houghton, R.A., House, J.I., Pongratz, J., van der Werf, G.R., DeFries, R.S., Hansen, M.C., Le Quéré, C., Ramankutty, N., 2012. Carbon emissions from land use and land-cover change. Biogeosciences 9 (12), 5125–5142.

Houghton, R.A., Baccini, A., Walker, W.S., 2018. Where is the residual terrestrial carbon sink? Glob. Chang. Biol. 24 (8), 3277–3279.

Houle, D., Carignan, R., Ouimet, R., 2001. Soil organic Sulfur dynamics in a coniferous forest. Biogeochemistry 53, 105–124.

Houlton, B.Z., Bai, E., 2009. Imprint of denitrifying bacteria on the global terrestrial biosphere. Proc. Natl. Acad. Sci. U. S. A. 106 (51), 21713–21716.

Houlton, B.Z., Sigman, D.M., Hedin, L.O., 2006. Isotopic evidence for large gaseous nitrogen losses from tropical rainforests. Proc. Natl. Acad. Sci. U. S. A. 103 (23), 8745–8750.

Houlton, B.Z., Wang, Y.-P., Vitousek, P.M., Field, C.B., 2008. A unifying framework for dinitrogen fixation in the terrestrial biosphere. Nature 454 (7202), 327–330.

Houlton, B.Z., Morford, S.L., Dahlgren, R.A., 2018. Convergent evidence for widespread rock nitrogen sources in Earth's surface environment. Science 360 (6384), 58–62.

Houweling, S., Dentener, F., Lelieveld, J., Walter, B., Dlugokencky, E., 2000. The modeling of tropospheric methane: how well can point measurements be reproduced by a global model? J. Geophys. Res. 105, 8981–9002.

Hover, V.C., Walter, L.M., Peacor, D.R., 2002. K uptake by modern estuarine sediments during early marine diagenesis, Mississippi Delta Plain, Louisiana, USA. J. Sediment. Res. 72 (6), 775–792.

Howard, J.A., Mitchell, C.W., 1985. Phytogeomorphology. Wiley.

Howarth, R.W., 1984. The ecological significance of sulfur in the energy dynamics of salt Marsh and coastal marine sediments. Biogeochemistry 1, 5–27.

Howarth, R.W., 1998. An assessment of human influences on fluxes of nitrogen from the terrestrial landscape to the estuaries and continental shelves of the North Atlantic ocean. Nutr. Cycl. Agroecosyst. 52, 213–223.

Howarth, R.W., Cole, J.J., 1985. Molybdenum availability, nitrogen limitation, and phytoplankton growth in natural waters. Science 229 (4714), 653–655.

Howarth, R.W., Marino, R., 2006. Nitrogen as the limiting nutrient for eutrophication in coastal marine ecosystems: evolving views over three decades. Limnol. Oceanogr. 51, 364–376.

Howarth, R.W., Teal, J.M., 1979. Sulfate reduction in a New England salt marsh 1. Limnol. Oceanogr. 24 (6), 999–1013.

Howarth, R.W., Marino, R., Cole, J.J., 1988. Nitrogen fixation in freshwater, estuarine, and marine ecosystems. 2. Biogeochemical controls. Limnol. Oceanogr. 33 (4), 688–701.

Howarth, R.W., Jensen, H.S., Marino, R., Postma, H., 1995a. Transport to and processing of P in nearshore and oceanic waters. In: Tiessen, H. (Ed.), Phosphorus in the Global Environment. Wiley, pp. 323–345.

Howarth, R.W., Swaney, D., Marino, R., Butler, T., Chu, C.R., 1995b. Turbulence does not prevent nitrogen fixation by plankton in estuaries and coastal seas—reply. Limnol. Oceanogr. 40 (3), 639–643.

Howarth, R., Chan, F., Conley, D.J., Garnier, J., Doney, S.C., Marino, R., Billen, G., 2011. Coupled biogeochemical cycles: eutrophication and hypoxia in temperate estuaries and coastal marine ecosystems. Front. Ecol. Environ. 9, 18–26.

Howarth, R., Swaney, D., Billen, G., Garnier, J., Hong, B., Humborg, C., Johnes, P., Mörth, C.-M., Marino, R., 2012. Nitrogen fluxes from the landscape are controlled by net anthropogenic nitrogen inputs and by climate. Front. Ecol. Environ. 10 (1), 37–43.

Howell, D.G., Murray, R.W., 1986. A budget for continental growth and denudation. Science 233 (4762), 446–449.

Howes, B.L., Dacey, J.W.H., King, G.M., 1984. Carbon flow through oxygen and sulfate reduction pathways in salt marsh sediments. Limnol. Oceanogr. 29 (5), 1037–1051.

Hren, M.T., Tice, M.M., Chamberlain, C.P., 2009. Oxygen and hydrogen isotope evidence for a temperate climate 3.42 billion years ago. Nature 462 (7270), 205–208.

Hsiang, S., Kopp, R., Jina, A., Rising, J., Delgado, M., Shashank, M., Rasmussen, D.J., et al., 2017. Estimating economic damage from climate change in the United States. Science 356 (6345), 1362–1369.

Hu, L., Yvon-Lewis, S.A., Liu, Y., Salisbury, J.E., O'Hern, J.E., 2010. Coastal emissions of methyl bromide and methyl chloride along the Eastern Gulf of Mexico and the East Coast of the United States. Glob. Biogeochem. Cycles 24.

Hu, L., Yvon-Lewis, S., Liu, Y., Bianchi, T.S., 2012. The ocean in near equilibrium with atmospheric methyl bromide: oceanic saturation state of CH_3 Br. Glob. Biogeochem. Cycles Geophys. Monogr. Ser. 26 (3), 165.

Hu, L., Yvon-Lewis, S.A., Butler, J.H., Lobert, J.M., King, D.B., 2013. An improved oceanic budget for methyl chloride: oceanic budget for MeCl. J. Geophys. Res. C: Oceans 118 (2), 715–725.

Hu, M., Chen, D., Dahlgren, R.A., 2016. Modeling nitrous oxide emission from rivers: a global assessment. Glob. Chang. Biol. 22 (11), 3566–3582.

Hu, L., Jacob, D.J., Liu, X., Zhang, Y., Lin, Z., Kim, P.S., Sulprizio, M.P., Yantosca, R.M., 2017. Global budget of tropospheric ozone: evaluating recent model advances with satellite (OMI), aircraft (IAGOS), and ozonesonde observations. Atmos. Environ. 167 (October), 323–334.

Huang, P.M., 1988. Ionic factors affecting aluminum transformations and the impact on soil and environmental sciences. Adv. Soil Sci. 8, 1–78.

Huang, W., Hall, S.J., 2017. Elevated moisture stimulates carbon loss from mineral soils by releasing protected organic matter. Nat. Commun. 8 (1), 1774.

Huang, Y., Zou, J., Zheng, X., Wang, Y., Xu, X., 2004. Nitrous oxide emissions as influenced by amendment of plant residues with different C:N ratios. Soil Biol. Biochem. 36, 973–981.

Huang, X., Li, M., Green, D.C., Williams, D.S., Patil, A.J., Mann, S., 2013. Interfacial assembly of protein-polymer nano-conjugates into stimulus-responsive biomimetic protocells. Nat. Commun. 4 (1), 2239.

Huang, Y., Shen, H., Chen, Y., Zhong, Q., Chen, H., Wang, R., Shen, G., Liu, J., Li, B., Tao, S., 2015. Global organic carbon emissions from primary sources from 1960 to 2009. Atmos. Environ. 122 (December), 505–512.

Huang, L.-M., Jia, X.-X., Zhang, G.-L., Shao, M.-A., 2017. Soil organic phosphorus transformation during ecosystem development: a review. Plant Soil 417 (1–2), 17–42.

Huang, Y., Chen, Y., Castro-Izaguirre, N., Baruffol, M., Brezzi, M., Lang, A., Li, Y., et al., 2018. Impacts of species richness on productivity in a large-scale subtropical forest experiment. Science 362 (6410), 80–83.

Huber, C., Wächtershäuser, G., 1998. Peptides by activation of amino acids with CO on (Ni, Fe)S surfaces: implications for the origin of life. Science 281, 670–672.

Huber, C., Wächtershäuser, G., 2006. Alpha-hydroxy and alpha-amino acids under possible hadean, volcanic origin-of-life conditions. Science 314 (5799), 630–632.

Huber, R., Kurr, M., Jannasch, H.W., Stetter, K.O., 1989. A novel group of abyssal methanogenic archaebacteria (*Methanopyrus*) growing at 110°C. Nature 342, 833–834.

Hudak, A.T., Strand, E.K., Vierling, L.A., Byrne, J.C., Eitel, J.U.H., Martinuzzi, S., Falkowski, M.J., 2012. Quantifying aboveground forest carbon pools and fluxes from repeat LiDAR surveys. Remote Sens. Environ. 123 (August), 25–40.

Hudman, R.C., Murray, L.T., Jacob, D.J., Millet, D.B., Turquety, S., Wu, S., Blake, D.R., Goldstein, A.H., Holloway, J., Sachse, G.W., 2008. Biogenic versus anthropogenic sources of CO in the United States. Geophys. Res. Lett. 35.

Hudson, B.D., 1995. Reassessment of the Polynov ion mobility series. Soil Sci. Soc. Am. J. 59, 1101–1103.

Hudson, S.R., 2011. Estimating the global radiative impact of the sea ice–albedo feedback in the Arctic. J. Geophys. Res. 116 (D16), C08004.

Hudson-Edwards, K.A., Bristow, C.S., Cibin, G., Mason, G., Peacock, C.L., 2014. Solid-phase phosphorus speciation in Saharan Bodélé depression dusts and source sediments. Chem. Geol. 384 (September), 16–26.

Hue, N.V., 1991. Effects of organic acid anions on P sorption and phytoavailability in soils with different mineralogies. Soil Sci. 152, 463–471.

Huemmrich, K.F., Kinoshita, G., Gamon, J.A., Houston, S., Kwon, H., Oechel, W.C., 2010. Tundra carbon balance under varying temperature and moisture regimes. J. Geophys. Res.—Biogeosci. 115.

Huenneke, L.F., Hamburg, S.P., Koide, R., Mooney, H.A., Vitousek, P.M., 1990. Effects of soil resources on plant invasion and community structure in Californian serpentine grassland. Ecology 71, 478–491.

Hueso, R., Sánchez-Lavega, A., 2006. Methane storms on Saturn's Moon titan. Nature 442 (7101), 428–431.

Hugelius, G., Strauss, J., Zubrzycki, S., Harden, J.W., Schuur, E.A.G., Ping, C.-L., Schirrmeister, L., et al., 2014. Estimated stocks of circumpolar permafrost carbon with quantified uncertainty ranges and identified data gaps. Biogeosciences 11 (23), 6573–6593.

Hulett Jr., L.D., Weinberger, A.J., Northcutt, K.J., Ferguson, M., 1980. Chemical species in fly ash from coal-burning power plants. Science 210 (4476), 1356–1358.

Humborg, C., Pastuszak, M., Aigars, J., Siegmund, H., Morth, C.M., Ittekkot, V., 2006. Decreased silica land-sea fluxes through damming in the Baltic Sea catchment—significance of particle trapping and hydrological alterations. Biogeochemistry 77, 265–281.

Humphrey, V., Zscheischler, J., Ciais, P., Gudmundsson, L., Sitch, S., Seneviratne, S.I., 2018. Sensitivity of atmospheric CO_2 growth rate to observed changes in terrestrial water storage. Nature 560 (7720), 628–631.

Hungate, B.A., Dijkstra, P., Wu, Z., Duval, B.D., Day, F.P., Johnson, D.W., Megonigal, J.P., Brown, A.L.P., Garland, J.L., 2013. Cumulative response of ecosystem Carbon and nitrogen stocks to chronic CO_2 exposure in a subtropical oak woodland. New Phytol. 200 (3), 753–766.

Hungate, B.A., Duval, B.D., Dijkstra, P., Johnson, D.W., Ketterer, M.E., Stiling, P., Cheng, W., Millman, J., Hartley, A., Stover, D.B., 2014. Nitrogen inputs and losses in response to chronic CO_2 exposure in a subtropical oak woodland. Biogeosciences 11 (12), 3323–3337.

Hunt, A.G., 2017. Spatiotemporal scaling of vegetation growth and soil formation: explicit predictions. Vadose Zone J. 16.

Hunten, D.M., 1993. Atmospheric evolution of the terrestrial planets. Science 259, 915–920.

Huntington, T.G., 2006. Evidence for intensification of the global water cycle: review and synthesis. J. Hydrol. 319, 83–95.

Hupy, J., Schaetzl, R., 2006. Introducing "bombturbation," a singular type of soil disturbance and mixing. Soil Sci. 171 (11), 823–836.

Hurrell, J.W., 1995. Decadal trends in the North Atlantic oscillation: regional temperatures and precipitation. Science 269, 676–679.

Hursh, A., Ballantyne, A., Cooper, L., Maneta, M., Kimball, J., Watts, J., 2017. The sensitivity of soil respiration to soil temperature, moisture, and carbon supply at the global scale. Glob. Chang. Biol. 23 (5), 2090–2103.

Hurst, D.F., Griffith, D.W.T., Cook, G.D., 1994. Trace gas emissions from biomass burning in tropical Australian Savannas. J. Geophys. Res. D: Atmos. 99, 16441–16456.

Huss-Danell, K., 1986. Nitrogen in shoot litter, root litter and root exudates from nitrogen-fixing *Alnus incana*. Plant Soil 91, 43–49.

Huston, M., 1993. Biological diversity, soils, and economics. Science 262 (5140), 1676–1680.

Huston, M.A., 2012. Precipitation, soils, NPP, and biodiversity: resurrection of Albrecht's curve. Ecol. Monogr. 82 (3), 277–296.

Hutchin, P.R., Press, M.C., Lee, J.A., Ashenden, T.W., 1995. Elevated concentrations of CO_2 may double methane emissions from mires. Glob. Chang. Biol. 1, 125–128.

Hutchins, D.A., Ditullio, G.R., Bruland, K.W., 1993. Iron and regenerated production: evidence for biological iron recycling in two marine environments. Limnol. Oceanogr. 38, 1242–1255.

Hutchinson, G.E., 1938. On the relation between oxygen deficit and the productivity and typology of lakes. Int. Revue Des Gesamten Hydrobiol. Hydrogr. 36, 336–355.

Hutchinson, G.E., 1943. The biogeochemistry of aluminum and of certain related elements. Q. Rev. Biol. 18, 331–363.

Hutchinson, J.N., 1980. Record of peat wastage in the east Anglian fenlands at Holme Post, 1848–1978 AD. J. Ecol. 68, 229–249.

Hutchinson, G.L., Guenzi, W.D., Livingston, G.P., 1993. Soil water controls on aerobic soil emission of gaseous nitrogen oxides. Soil Biol. Biochem. 25, 1–9.

Hütsch, B.W., Webster, C.P., Powlson, D.S., 1994. Methane oxidation in soil as affected by land use, soil pH and N fertilization. Soil Biol. Biochem. 26 (12), 1613–1622.

Huxman, T.E., Smith, M.D., Fay, P.A., Knapp, A.K., Rebecca Shaw, M., Loik, M.E., Smith, S.D., et al., 2004. Convergence across biomes to a common rain-use efficiency. Nature 429 (6992), 651–654.

Hydes, D.J., 1979. Aluminum in seawater: control by inorganic processes. Science 205 (4412), 1260–1262.

Hyman, M.R., Wood, P.M., 1983. Methane oxidtaion by *Nitrosomonas europaea*. Biochem. J. 212, 31–37.

Hynicka, J.D., Pett-Ridge, J.C., Perakis, S.S., 2016. Nitrogen enrichment regulates calcium sources in forests. Glob. Chang. Biol. 22 (12), 4067–4079.

Hyun, J.H., Mok, J.S., Cho, H.Y., Kim, S.H., Lee, K.S., Kostka, J.E., 2009. Rapid organic matter mineralization coupled to iron cycling in intertidal mud flats of the Han River estuary, Yellow Sea. Biogeochemistry 92 (3), 231–245.

Hyvonen, R., Persson, T., Andersson, S., Olsson, B., Agren, G.I., Linder, S., 2008. Impact of long-term nitrogen addition on carbon stocks in trees and soils in northern Europe. Biogeochemistry 89, 121–137.

Idso, S.B., Kimball, B.A., 1993. Tree growth in carbon dioxide enriched air and its implications for global carbon cycling and maximum levels of atmospheric CO_2. Glob. Biogeochem. Cycles 7, 537–555.

Iglesias-Rodriguez, M.D., Halloran, P.R., Rickaby, R.E.M., Hall, I.R., Colmenero-Hidalgo, E., Gittins, J.R., Green, D.R.H., et al., 2008. Phytoplankton calcification in a high-CO_2 world. Science 320 (5874), 336–340.

Iizuka, Y., Uemura, R., Motoyama, H., Suzuki, T., Miyake, T., Hirabayashi, M., Hondoh, T., 2012. Sulphate-climate coupling over the past 300,000 years in inland Antarctica. Nature 490 (7418), 81–84.

Illmer, P., Barbato, A., Schinner, F., 1995. Solubilization of hardly soluble $AlPO_4$ with P-solubilizing microorganisms. Soil Biol. Biochem. 27, 265–270.

IMBIE, 2020. Mass balance of the Greenland ice sheet from 1992 to 2018. Nature 579, 233–239.

IMBIE Team, 2018. Mass balance of the Antarctic ice sheet from 1992 to 2017. Nature 558 (7709), 219–222.

Imhoff, M.L., Bounoua, L., DeFries, R., Lawrence, W.T., Stutzer, D., Tucker, C.J., Ricketts, T., 2004a. The consequences of urban land transformation on net primary productivity in the United States. Remote Sens. Environ. 89, 434–443.

Imhoff, M.L., Bounoua, L., Ricketts, T., Loucks, C., Harriss, R., Lawrence, W.T., 2004b. Global patterns in human consumption of net primary production. Nature 429 (6994), 870–873.

Indermuhle, A., Stocker, T.F., Joos, F., Fischer, H., Smith, H.J., Wahlen, M., Deck, B., et al., 1999. Holocene carbon-cycle dynamics based on CO_2 trapped in ice at Taylor Dome, Antarctica. Nature 398, 121–126.

Ingall, E., Jahnke, R., 1997. Influence of watercolumn anoxia on the elemental fractionation of carbon and phosphorus during sediment diagenesis. Mar. Geol. 139, 219–229.

Ingall, E.D., Van Cappellen, P., 1990. Relation between sedimentation rate and burial of organic phosphorus and organic carbon in marine sediments. Geochim. Cosmochim. Acta 54, 373–386.

Ingall, E.D., Bustin, R.M., Van Cappellen, P., 1993. Influence of water column anoxia on the burial and preservation of carbon and phosphorus in marine shales. Geochim. Cosmochim. Acta 57, 303–316.

Ingestad, T., 1979a. Mineral nutrient requirements of *Pinus silvestris* and *Picea abies* seedlings. Physiol. Plant. 45, 373–380.

Ingestad, T., 1979b. Nitrogen stress in birch seedlings. II. N, K, P, Ca and Mg nutrition. Physiol. Plant. 45, 149–157.

Ingham, R.E., Detling, J.K., 1990. Effects of root-feeding nematodes on above-ground net primary production in a North American grassland. Plant Soil 121, 279–281.

Inoue, H.Y., Sugimura, Y., 1992. Variations and distributions of CO_2 in and over the equatorial pacific during the period from the 1986/88 El Nino event to the 1988/89 La Nina event. Tellus B 44, 1–22.

Insam, H., 1990. Are the soil microbial biomass and basal respiration governed by the climatic regime. Soil Biol. Biochem. 22, 525–532.

Inskeep, W.P., Bloom, P.R., 1986. Kinetics of calcite precipitation in the presence of water-soluble organic ligands. Soil Sci. Soc. Am. J. 50, 1167–1172.

Intergovernmental Panel on Climate Change, 2007. Working group 1, science, intergovernmental panel on climate change, intergovernmental panel on climate change. Working group I, and intergovernmental panel on climate change staff. In: Climate Change 2007—The Physical Science Basis: Working Group I Contribution to the Fourth Assessment Report of the IPCC. Cambridge University Press.

IPCC, 2013. Climate Change 2013: The Physical Science Basis. In: Stocker, T.F., Qin, D., Plattner, G.-K., Tignor, M., Allen, S.K., Boschung, J., Nauels, A., Xia, Y., Bex, V., Midgley, P.M. (Eds.), Contribution of Working Group I to the Fifth Assessment Report of the Intergovernmental Panel on Climate Change. Cambridge University Press, Cambridge.

Irvine, W.M., 1998. Extraterrestrial organic matter: a review. Origins Life Evol. Biosph. 28 (4–6), 365–383.

Irwin, J.G., Williams, M.L., 1988. Acid rain: chemistry and transport. Environ. Pollut. 50 (1–2), 29–59.

Isaksen, I.S.A., Hov, O., 1987. Calculation of trends in the tropospheric concentrations of O_3, OH, CO, CH_4, and NO_x. Tellus B Chem. Phys. Meteorol. 39, 271–285.

Ishida, Y., 1968. Physiological Studies on Evolution of Dimethylsulfide. Kyoto University.

Ishijima, K., Sugawara, S., Kawamura, K., Hashida, G., Morimoto, S., Murayama, S., Aoki, S., Nakazawa, T., 2007. Temporal variations of the atmospheric nitrous oxide concentration and its delta N-15 and delta O-18 for the latter half of the 20th century reconstructed from firn-air analyses. J. Geophys. Res. 112.

Ishijima, K., Nakazawa, T., Aoki, S., 2009. Variations of atmospheric nitrous oxide concentration in the northern and western pacific. Tellus B Chem. Phys. Meteorol. 61, 408–415.

Ishizuka, S., Sakata, T., Ishizuka, K., 2000. Methane oxidation in Japanese forest soils. Soil Biol. Biochem. 32, 769–777.

Ito, A., 2011. A historical meta-analysis of global terrestrial net primary productivity: are estimates converging? Glob. Chang. Biol. 17, 3161–3175.

Ittekkot, V., Arain, R., 1986. Nature of particulate organic matter in the river Indus, Pakistan. Geochim. Cosmochim. Acta 50, 1643–1653.

Ittekkot, V., Zhang, S., 1989. Pattern of particulate nitrogen transport in world rivers. Glob. Biogeochem. Cycles 3, 383–392.

Ivanov, M.V., 1983. Major fluxes of the global biogeochemical cycle of sulphur. In: Ivanov, M.V., Freney, J.R. (Eds.), The Global Biogeochemical Sulphur Cycle. Wiley, pp. 449–463.

Ivens, W., Kauppi, P., Alcamo, J., Posch, M., 1990. Sulfur deposition onto European forests: throughfall data and model estimates. Tellus B Chem. Phys. Meteorol. 42, 294–303.

Iversen, N., 1996. Methane oxidation in coastal marine environments. In: Murrell, J.C., Kelly, D.P. (Eds.), Microbiology of Atmospheric Trace Gases. Springer, pp. 51–68.

Iverson, R.L., 1990. Control of marine fish production. Limnol. Oceanogr. 35, 1593–1604.

Jackson, M.B., Armstrong, W., 1999. Formation of Aerenchyma and the processes of plant ventilation in relation to soil flooding and submergence. Plant Biol. 1, 274–287.

Jackson, L.E., Schimel, J.P., Firestone, M.K., 1989. Short-term partitioning of ammonium and nitrate between plants and microbes in an annual grassland. Soil Biol. Biochem. 21, 409–415.

Jackson, R.B., Manwaring, J.H., Caldwell, M.M., 1990. Rapid physiological adjustment of roots to localized Soil enrichment. Nature 344 (6261), 58–60.

Jackson, R.B., Mooney, H.A., Schulze, E.D., 1997. A global budget for fine root biomass, surface area, and nutrient contents. Proc. Natl. Acad. Sci. U. S. A. 94 (14), 7362–7366.

Jackson, R.B., Cook, C.W., Pippen, J.S., Palmer, S.M., 2009. Increased belowground biomass and soil CO_2 fluxes after a decade of carbon dioxide enrichment in a warm-temperate forest. Ecology 90 (12), 3352–3366.

Jackson, R.B., Lajtha, K., Crow, S.E., Hugelius, G., Kramer, M.G., Piñeiro, G., 2017. The ecology of soil carbon: pools, vulnerabilities, and biotic and abiotic controls. Annu. Rev. Ecol. Evol. Syst. 48 (1), 419–445.

Jacob, D.J., Wofsy, S.C., 1990. Budgets of reactive nitrogen, hydrocarbons, and ozone over the Amazon Forest during the wet season. J. Geophys. Res.-Atmos. 95, 16737–16754.

Jacob, D.J., Logan, J.A., Gardner, G.M., Yevich, R.M., Spivakovsky, C.M., Wofsy, S.C., Sillman, S., Prather, M.J., 1993. Factors regulating ozone over the United States and its export to the global atmosphere. J. Geophys. Res.—Atmos. 98, 14817–14826.

Jacob, T., John, W., Tad Pfeffer, W., Swenson, S., 2012. Recent contributions of glaciers and ice caps to sea level rise. Nature 482 (7386), 514–518.

Jacob, D.J., Turner, A.J., Maasakkers, J.D., Sheng, J., Sun, K., Liu, X., Chance, K., Aben, I., McKeever, J., Frankenberg, C., 2016. Satellite observations of atmospheric methane and their value for quantifying methane emissions. Atmos. Chem. Phys. 16 (22), 14371–14396.

Jacobs, S.S., Giulivi, C.F., Mele, P.A., 2002. Freshening of the ross sea during the late 20th century. Science 297 (5580), 386–389.

Jacobson, M.E., 1994. Chemical and biological mobilization of Fe(III) in marsh sediments. Mobilization 25 (1), 41–60.

Jaeglé, L., Shah, V., Thornton, J.A., Lopez-Hilfiker, F.D., Lee, B.H., McDuffie, E.E., Fibiger, D., et al., 2018. Nitrogen oxides emissions, chemistry, deposition, and export over the Northeast United States during the winter aircraft campaign. J. Geophys. Res. D: Atmos. 123 (21), 12368–12393.

Jaffé, R., Ding, Y., Niggemann, J., Vähätalo, A.V., Stubbins, A., Spencer, R.G.M., Campbell, J., Dittmar, T., 2013. Global charcoal mobilization from soils via dissolution and riverine transport to the oceans. Science 340 (6130), 345–347.

Jahangir, M.M.R., Johnston, P., Barrett, M., Khalil, M.I., Groffman, P.M., Boeckx, P., Fenton, O., Murphy, J., Richards, K.G., 2013. Denitrification and indirect N_2O emissions in groundwater: hydrologic and biogeochemical influences. J. Contam. Hydrol. 152 (September), 70–81.

Jahren, A.H., Arens, N.C., Sarmiento, G., Guerrero, J., Amundson, R., 2001. Terrestrial record of methane hydrate dissociation in the early cretaceous. Geology 29, 159–162.

Jain, A.K., 2007. Global estimation of CO emissions using three sets of satellite data for burned area. Atmos. Environ. 41, 6931–6940.

Jain, A.K., Tao, Z.N., Yang, X.J., Gillespie, C., 2006. Estimates of global biomass burning emissions for reactive greenhouse gases (CO, NMHCs, and NO_x) and CO_2. J. Geophys. Res.-Atmos. 111.

Jakosky, B.M., Slipski, M., Benna, M., Mahaffy, P., Elrod, M., Yelle, R., Stone, S., Alsaeed, N., 2017. Mars' atmospheric history derived from upper-atmosphere measurements of $^{38}Ar/^{36}Ar$. Science 355 (6332), 1408–1410.

Jakosky, B.M., Brain, D., Chaffin, M., Curry, S., Deighan, J., Grebowsky, J., Halekas, J., et al., 2018. Loss of the martian atmosphere to space: present-day loss rates determined from MAVEN observations and integrated loss through time. Icarus 315 (November), 146–157.

James, R.H., Palmer, M.R., 2000. Marine geochemical cycles of the alkali elements and boron: the role of sediments. Geochim. Cosmochim. Acta 64 (18), 3111–3122.

James, B.R., Riha, S.J., 1986. pH buffering in forest soil organic horizons: relevance to acid precipitation. J. Environ. Qual. 15, 229–234.

James, P.B., Kieffer, H.H., Paige, D.A., 1992. The seasonal cycle of carbon dioxide on mars. In: Kieffer, H.H., Jakosky, B.M., Snyder, C.W., Matthews, M.A. (Eds.), Mars. University of Arizona Press, pp. 934–968.

Jamieson, J.W., Petersen, S., Bach, W., 2016. Hydrothermalism. In: Harff, J., Meschede, M., Petersen, S., Thiede, J. (Eds.), Encyclopedia of Marine Geosciences. Encyclopedia of Earth Sciences Series. Springer, Dordrecht.

Jannasch, H.W., 1989. Sulphur emission and transformations at deep sea hydrothermal vents. In: Brimblecombe, P., Lein, A.Y. (Eds.), Evolution of the Global Biogeochemical Sulphur Cycle. Wiley, pp. 181–190.

Jannasch, H.W., Mottl, M.J., 1985. Geomicrobiology of deep-sea hydrothermal vents. Science 229 (4715), 717–725.

Jannasch, H.W., Wirsen, C.O., 1979. Chemosynthetic primary production at East Pacific Sea floor spreading centers. Bioscience 29, 592–598.

Janos, D.P., 1980. Vesicular-arbuscular mycorrhizae affect lowland tropical rain forest plant growth. Ecology 61, 151–162.

Jansen, B., Nierop, K.G.J., Verstraten, J.M., 2005. Mechanisms controlling the mobility of dissolved organic matter, aluminium and iron in podzol B horizons. Eur. J. Soil Sci. 56, 537–550.

Janson, R.W., 1993. Monoterpene emissions from scots pine and Norwegian Spruce. J. Geophys. Res. SCOPE Rep. 98 (D2), 2839–2850.

Janssens, I.A., Lankreijer, H., Matteucci, G., Kowalski, A.S., Buchmann, N., Epron, D., Pilegaard, K., et al., 2001. Productivity overshadows temperature in determining soil and ecosystem respiration across European forests. Glob. Chang. Biol. 7, 269–278.

Jansson, M., 1993. Uptake, exchange, and excretion of orthophosphate in phosphate-starved *Scenedesmus quadricauda* and *Pseudomonas* K7. Limnol. Oceanogr. 38, 1162–1178.

Jansson, M., Karlsson, J., Jonsson, A., 2012. Carbon dioxide supersaturation promotes primary production in lakes. Ecol. Lett. 15 (6), 527–532.

Jaramillo, F., Destouni, G., 2015. Local flow regulation and irrigation raise global human water consumption and footprint. Science 350 (6265), 1248–1251.

Jaramillo, V.J., Detling, J.K., 1988. Grazing history, defoliation, and competition: effects on shortgrass production and nitrogen accumulation. Ecology 69, 1599–1608.

Jardine, A.B., Jardine, K.J., Fuentes, J.D., Martin, S.T., Martins, G., Durgante, F., Carneiro, V., Higuchi, N., Manzi, A.O., Chambers, J.Q., 2015a. Highly reactive light-dependent monoterpenes in the Amazon. Geophys. Res. Lett. 42 (5), 1576–1583.

Jardine, K., Yañez-Serrano, A.M., Williams, J., Kunert, N., Jardine, A., Taylor, T., Abrell, L., et al., 2015b. Dimethyl sulfide in the Amazon rain forest: DMS in the Amazon. Glob. Biogeochem. Cycles 29 (1), 19–32.

Jarrell, W.M., Beverly, R.B., 1981. The dilution effect in plant nutrition studies. Adv. Agron. 34, 197–224.

Jasechko, S., 2019. Global isotope hydrogeology-review. Rev. Geophys. 57 (3), 835–965.

Jauss, V., Sullivan, P.J., Sanderman, J., Smith, D.B., Lehmann, J., 2017. Pyrogenic carbon distribution in mineral topsoils of the northeastern United States. Geoderma 296 (June), 69–78.

Javaux, E.J., 2006. Extreme life on earth–past, present and possibly beyond. Res. Microbiol. 157 (1), 37–48.

Javaux, E.J., 2019. Challenges in evidencing the earliest traces of life. Nature 572 (7770), 451–460.

Javoy, M., 1997. The major volatile elements of the earth: their origin, behavior, and fate. Geophys. Res. Lett. 24, 177–180.

Jeandel, C., Oelkers, E.H., 2015. The influence of terrigenous particulate material dissolution on ocean chemistry and global element cycles. Chem. Geol. 395 (February), 50–66.

Jencso, K.G., McGlynn, B.L., Gooseff, M.N., Wondzell, S.M., Bencala, K.E., Marshall, L.A., 2009. Hydrologic connectivity between landscapes and streams: transferring reach-and plot-scale understanding to the catchment scale. Water Resour. Res. 45 (4).

Jenkin, M.E., Clemitshaw, K.C., 2000. Ozone and other secondary photochemical pollutants: chemical processes governing their formation in the planetary boundary layer. Atmos. Environ. 34, 2499–2527.

Jenkins, W.J., 1988. Nitrate flux into the euphotic zone near Bermuda. Nature 331, 521–523.

Jenkins, J.C., Birdsey, R.A., Pan, Y., 2001. Biomass and NPP estimation for the mid-Atlantic region (USA) using plot-level forest inventory data. Ecol. Appl. 11, 1174–1193.

Jenkinson, D.S., Rayner, J.H., 1977. Turnover of soil organic matter in some of the Rothamstead classical experiments. Soil Sci. 123, 298–305.

Jennings, J.N., 1983. Karst landforms. Am. Sci. 71, 578–586.

Jenny, H., 1941. Factors of Soil Formation. Dover Publications.

Jenny, H., 1980. The Soil Resource. Springer.

Jeppesen, E., Sondergaard, M., Jensen, J.P., Havens, K.E., Anneville, O., Carvalho, L., Coveney, M.F., et al., 2005. Lake responses to reduced nutrient loading—an analysis of contemporary long-term data from 35 case studies. Freshw. Biol. 50, 1747–1771.

Jersak, J., Amundson, R., Brimhall, G., 1995. A mass-balance analysis of podzolization: examples from the northeastern United States. Geoderma 66, 15–42.

Jevrejeva, S., Moore, J.C., Grinsted, A., Matthews, A.P., Spada, G., 2014. Trends and acceleration in global and regional sea levels since 1807. Glob. Planet. Chang. 113 (February), 11–22.

Ji, Q., Ward, B.B., 2017. Nitrous oxide production in surface waters of the mid-latitude North Atlantic Ocean: N_2O production in north Atlantic ocean. J. Geophys. Res. C: Oceans 122 (3), 2612–2621.

Ji, Q., Babbin, A.R., Jayakumar, A., Oleynik, S., Ward, B.B., 2015. Nitrous oxide production by nitrification and denitrification in the eastern tropical South Pacific oxygen minimum zone: nitrous oxide production in OMZ. Geophys. Res. Lett. 42 (24), 10,755–10,764.

Ji, Q., Buitenhuis, E., Suntharalingam, P., Sarmiento, J.L., Ward, B.B., 2018. Global nitrous oxide production determined by oxygen sensitivity of nitrification and denitrification. Glob. Biogeochem. Cycles 32 (12), 1790–1802.

Jiang, X., Ku, W.L., Shia, R.L., Li, Q.B., Elkins, J.W., Prinn, R.G., Yung, Y.L., 2007. Seasonal cycle of N_2O: analysis of data. Glob. Biogeochem. Cycles 21.

Jiao, N., Herndl, G.J., Hansell, D.A., Benner, R., Kattner, G., Wilhelm, S.W., Kirchman, D.L., et al., 2010. Microbial production of recalcitrant dissolved organic matter: long-term carbon storage in the global ocean. Nat. Rev. Microbiol. 8 (8), 593–599.

Jickells, T.D., 1998. Nutrient biogeochemistry of the coastal zone. Science 281 (5374), 217–221.

Jickells, T.D., Kelly, S.D., Baker, A.R., Biswas, K., Dennis, P.F., Spokes, L.J., Witt, M., Yeatman, S.G., 2003. Isotopic evidence for a marine ammonia source. Geophys. Res. Lett. 30.

Jickells, T.D., An, Z.S., Andersen, K.K., Baker, A.R., Bergametti, G., Brooks, N., Cao, J.J., et al., 2005. Global iron connections between desert dust, ocean biogeochemistry, and climate. Science 308 (5718), 67–71.

Jickells, T.D., Buitenhuis, E., Altieri, K., Baker, A.R., Capone, D., Duce, R.A., Dentener, F., et al., 2017. A reevaluation of the magnitude and impacts of anthropogenic atmospheric nitrogen inputs on the ocean. Glob. Biogeochem. Cycles 415 (February), 159.

Jimenez, J.L., Canagaratna, M.R., Donahue, N.M., Prevot, A.S.H., Zhang, Q., Kroll, J.H., DeCarlo, P.F., et al., 2009. Evolution of organic aerosols in the atmosphere. Science 326 (5959), 1525–1529.

Jin, V.L., Evans, R.D., 2010. Elevated CO_2 increases plant uptake of organic and inorganic N in the desert shrub Larrea Tridentata. Oecologia 163 (1), 257–266.

Jin, X., Gruber, N., Dunne, J.P., Sarmiento, J.L., Armstrong, R.A., 2006. Diagnosing the contribution of phytoplankton functional groups to the production and export of particulate organic carbon, $CaCO_3$, and opal from global nutrient and alkalinity distributions. Glob. Biogeochem. Cycles. 20. https://doi.org/10.1029/2005GB002532.

Jobbágy, E.G., Jackson, R.B., 2000. The vertical distribution of soil organic carbon and its relation to climate and vegetation. Ecol. Appl. 10 (2), 423–436.

Jobbágy, E.G., Jackson, R.B., 2001. The distribution of soil nutrients with depth: global patterns and the imprint of plants. Biogeochemistry 53 (1), 51–77.

Jobbágy, E.G., Jackson, R.B., 2004. The uplift of soil nutrients by plants: biogeochemical consequences across scales. Ecology 85 (9), 2380–2389.

Jobson, B.T., McKeen, S.A., Parrish, D.D., Fehsenfeld, F.C., Blake, D.R., Goldstein, A.H., Schauffler, S.M., Elkins, J.C., 1999. Trace gas mixing ratio variability versus lifetime in the troposphere and stratosphere: observations. J. Geophys. Res.-Atmos. 104, 16091–16113.

Joergensen, R.G., 1996. Quantification of the microbial biomass by determining ninhydrin-reactive N. Soil Biol. Biochem. 28, 301–306.

Joergensen, R.G., Mueller, T., 1996. The fumigation-extraction method to estimate soil microbial biomass: calibration of the ken value. Soil Biol. Biochem. 28, 33–37.

John, R., Dalling, J.W., Harms, K.E., Yavitt, J.B., Stallard, R.F., Mirabello, M., Hubbell, S.P., et al., 2007. Soil nutrients influence spatial distributions of tropical tree species. Proc. Natl. Acad. Sci. U. S. A. 104 (3), 864–869.

Johnson, D.W., 1984. Sulfur cycling in forests. Biogeochemistry 1, 29–43.

Johnson, C.N., 1995. Interactions between fire, *Mycophagous mammals*, and dispersal of ectromycorrhizal fungi in eucalyptus forests. Oecologia 104 (4), 467–475.

Johnson, D.W., 2006. Progressive N limitation in forests: review and implications for long-term responses to elevated CO_2. Ecology 87 (1), 64–75.

Johnson, D.W., Cole, D.W., 1980. Anion mobility in soils: relevance to nutrient transport from forest ecosystems. Environ. Int. 3, 79–90.

Johnson, B., Goldblatt, C., 2015. The nitrogen budget of the Earth. Earth Sci. Rev. 148, 150–173.

Johnson, J.E., Molnar, P.H., 2019. Widespread and persistent deposition of iron formations for two billion years. Geophys. Res. Lett. 46 (6), 3327–3339.

Johnson, D.W., Todd, D.E., 1983. Relationships among iron, aluminum, carbon, and sulfate in a variety of forest soils. Soil Sci. Soc. Am. J. 47, 792–800.

Johnson, N.M., Likens, G.E., Bormann, F.H., Pierce, R.S., 1968. Rate of chemical weathering of silicate minerals in New Hampshire. Geochim. Cosmochim. Acta 32, 531–545.

Johnson, N.M., Likens, G.E., Bormann, F.H., Fisher, D.W., Pierce, R.S., 1969. A working model for the variation in stream water chemistry at the Hubbard Brook experimental forest, New Hampshire. Water Resour. Res. 5, 1353–1363.

Johnson, N.M., Reynolds, R.C., Likens, G.E., 1972. Atmospheric sulfur: its effect on the chemical weathering of New England. Science 177 (4048), 514–516.

Johnson, D.W., Cole, D.W., Gessel, S.P., Singer, M.J., Minden, R.V., 1977. Carbonic acid leaching in a tropical, temperate, subalpine, and northern forest soil. Arct. Alp. Res. 9, 329–343.

Johnson, D.W., Henderson, G.S., Todd, D.E., 1981. Evidence of modern accumulations of adsorbed sulfate in an east Tennessee forested ultisol. Soil Sci. 132, 422–426.

Johnson, D.W., Henderson, G.S., Huff, D.D., Lindberg, S.E., Richter, D.D., Shriner, D.S., Todd, D.E., Turner, J., 1982. Cycling of organic and inorganic sulfur in a chestnut oak forest. Oecologia 54, 141–148.

Johnson, D.W., Cole, D.W., Vanmiegroet, H., Horng, F.W., 1986. Factors affecting anion movement and retention in four forest soils. Soil Sci. Soc. Am. J. 50, 776–783.

Johnson, D.W., Henderson, G.S., Todd, D.E., 1988. Changes in nutrient distribution in forests and soils of Walker branch watershed, Tennessee, over an 11-year period. Biogeochemistry 5, 275–293.

Johnson, D.B., Ghauri, M.A., Said, M.F., 1992. Isolation and characterization of an acidophilic, heterotrophic bacterium capable of oxidizing ferrous Iron. Appl. Environ. Microbiol. 58 (5), 1423–1428.

Johnson, A.H., Andersen, S.B., Siccama, T.G., 1994. Acid rain and soils of the Adirondacks. I. Changes in pH and available calcium, 1930–1984. Can. Agric. Lib. 24, 39–45.

Johnson, K.S., Michael Gordon, R., Coale, K.H., 1997. What controls dissolved iron concentrations in the world ocean? Mar. Chem. 57 (3), 137–161.

Johnson, D.W., Susfalk, R.B., Dahlgren, R.A., Klopatek, J.M., 1998. Fire is more important than water for nitrogen fluxes in semi-arid forests. Environ. Sci. Pol. 1, 79–86.

Johnson, D.W., Cheng, W., Burke, I.C., 2000a. Biotic and abiotic nitrogen retention in a variety of forest soils. Soil Sci. Soc. Am. J. 64, 1503–1514.

Johnson, L.C., Shaver, G.R., Cades, D.H., Rastetter, E., Nadelhoffer, K., Giblin, A., Laundre, J., Stanley, A., 2000b. Plant carbon-nutrient interactions control CO_2 exchange in Alaskan wet sedge tundra ecosystems. Ecology 81, 453–469.

Johnson, D.W., Todd, D.E., Tolbert, V.R., 2003. Changes in ecosystem carbon and nitrogen in a loblolly pine plantation over the first 18 years. Soil Sci. Soc. Am. J. 67, 1594–1601.

Johnson, A.H., Moyer, A., Bedison, J.E., Richter, S.L., Willig, S.A., 2008a. Seven decades of calcium depletion in organic horizons of Adirondack forest soils. Soil Sci. Soc. Am. J. 72, 1824–1830.

Johnson, A.P., James Cleaves, H., Dworkin, J.P., Glavin, D.P., Lazcano, A., Bada, J.L., 2008b. The Miller volcanic spark discharge experiment. Science 322 (5900), 404.

Johnson, K.P., Blum, J.D., Keeler, G.J., Douglas, T.A., 2008c. Investigation of the deposition and emission of mercury in Arctic snow during an atmospheric mercury depletion event. J. Geophys. Res.-Atmos. 113.

Johnson, K.S., Riser, S.C., Karl, D.M., 2010. Nitrate supply from deep to near-surface waters of the North Pacific subtropical gyre. Nature 465 (7301), 1062–1065.

Johnson, D.W., Walker, R.F., Glass, D.W., Miller, W.W., Murphy, J.D., Stein, C.M., 2012. The effect of rock content on nutrients in a Sierra Nevada forest soil. Geoderma 173 (174), 84–93.

Johnson, D.S., Scott Warren, R., Deegan, L.A., Mozdzer, T.J., 2016. Saltmarsh plant responses to eutrophication. Ecol. Appl. 26 (8), 2647–2659.

Johnston, D.A., 1980. Volcanic contribution of chlorine to the stratosphere: more significant to ozone than previously estimated? Science 209 (4455), 491–493.

Johnston, C.M.T., Radeloff, V.C., 2019. Global mitigation potential of carbon stored in harvested wood products. Proc. Natl. Acad. Sci. U. S. A. 116 (29), 14526–14531.

Johnston, C.A., Detenbeck, N.E., Niemi, G.J., 1990. The cumulative effect of wetlands on stream water quality and quantity—a landscape approach. Biogeochemistry 10 (2), 105–141.

Johnston, C.A., Shmagin, B.A., Frost, P.C., Cherrier, C., Larson, J.H., Lamberti, G.A., Bridgham, S.D., 2008. Wetland types and wetland maps differ in ability to predict dissolved organic carbon concentrations in streams. Sci. Total Environ. 404 (2–3), 326–334.

Jonasson, S., Bryant, J.P., Chapin, F.S., Andersson, M., 1986. Plant phenols and nutrients in relation to variations in climate and rodent grazing. Am. Nat. 128, 394–408.

Jones, M.J., 1973. The organic matter content of the Savanna soils of West Africa. J. Soil Sci. 24, 42–53.

Jones, D.L., 1998. Organic acids in the rhizosphere—a critical review. Plant Soil 205, 25–44.

Jones, R.L., Hanson, H.C., 1985. Mineral Licks, Geophagy, and Biogeochemistry of North American Ungulates. Iowa State University Press.

Jones, M.B., Humphries, S.W., 2002. Impacts of the C-4 sedge *Cyperus papyrus* L. on carbon and water fluxes in an African wetland. Hydrobiologia 488, 107–113.

Jones, R.D., Morita, R.Y., 1983. Methane oxidation by *Nitrosococcus oceanus* and *Nitrosomonas europaea*. Appl. Environ. Microbiol. 45 (2), 401–410.

Jones, J.B., Schade, J.D., Fisher, S.G., Grimm, N.B., 1997. Organic matter dynamics in Sycamore Creek, a desert stream in Arizona. J. N. Am. Benthol. Soc. 16, 78–82.

Jones, D.L., Hodge, A., Kuzyakov, Y., 2004. Plant and mycorrhizal regulation of rhizodeposition. New Phytol. 163 (3), 459–480.

Jordan, C.F., 1971. A world pattern in plant energetics. Am. Sci. 59, 425–433.

Jordan, M., Likens, G.E., 1975. An organic carbon budget for an oligotrophic lake in New Hampshire, U.S.A. Verhandlungen Des Internationalen Verein Limnologie 19, 994–1003.

Jordan, T.E., Correll, D.L., Weller, D.E., 1993. Nutrient interception by a riparian forest receiving inputs from adjacent cropland. J. Environ. Qual. 22, 467–473.

Jordan, T.E., Cornwell, J.C., Boynton, W.R., Anderson, J.T., 2008. Changes in phosphorus biogeochemistry along an estuarine salinity gradient: the iron conveyer belt. Limnol. Oceanogr. 53 (1), 172–184.

Jordan, S.J., Stoffer, J., Nestlerode, J.A., 2011. Wetlands as sinks for reactive nitrogen at continental and global scales: a meta-analysis. Ecosystems 14 (1), 144–155.

Jorgensen, B.B., 1977. Sulfur cycle of a coastal marine sediment (Limfjorden, Denmark). Limnol. Oceanogr. 22, 814–832.

Jorgensen, B.B., 1982. Mineralization of organic matter in the sea bed—the role of sulfate reduction. Nature 296, 643–645.

Jørgensen, B.B., Kasten, S., 2006. Sulfur cycling and methane oxidation. In: Schulz, H.D., Zabel, M. (Eds.), Marine Geochemistry. Springer Berlin Heidelberg, Berlin, Heidelberg, pp. 271–309.

Jorgensen, J.R., Wells, C.G., Metz, L.J., 1980. Nutrient changes in decomposing loblolly pine forest floor. Soil Sci. Soc. Am. J. 44, 1307–1314.

Jorgensen, B.B., Isaksen, M.F., Jannasch, H.W., 1992. Bacterial sulfate reduction above 100°C in deep-sea hydrothermal vent sediments. Science 258, 1756–1757.

Joslin, J.D., Gaudinski, J.B., Torn, M.S., Riley, W.J., Hanson, P.J., 2006. Fine-root turnover patterns and their relationship to root diameter and soil depth in a 14C-labeled hardwood forest. New Phytol. 172 (3), 523–535.

Ju, X.T., Kou, C.L., Zhang, F.S., Christie, P., 2006. Nitrogen balance and groundwater nitrate contamination: comparison among three intensive cropping systems on the North China plain. Environ. Pollut. 143 (1), 117–125.

Juang, F.H.T., Johnson, N.M., 1967. Cycling of chlorine through a forested watershed in New England. J. Geophys. Res. 65, 227–237.

Juang, J.Y., Katul, G.G., Siqueira, M.B.S., Stoy, P.C., Palmroth, S., McCarthy, H.R., Kim, H.S., Oren, R., 2006. Modeling nighttime ecosystem respiration from measured CO_2 concentration and air temperature profiles using inverse methods. J. Geophys. Res.—Atmos. 111.

Juang, J.Y., Katul, G., Siqueira, M., Stoy, P., Novick, K., 2007. Separating the effects of Albedo from ecophysiological changes on surface temperature along a successional chronosequence in the southeastern United States. Geophys. Res. Lett. 34.

Judson, O.P., 2017. The energy expansions of evolution. Nat. Ecol. Evol. 1 (6), 138.

Juice, S.M., Fahey, T.J., Siccama, T.G., Driscoll, C.T., Denny, E.G., Eagar, C., Cleavitt, N.L., Minocha, R., Richardson, A.D., 2006. Response of sugar maple to calcium addition to northern hardwood forest. Ecology 87 (5), 1267–1280.

Jung, M., Reichstein, M., Ciais, P., Seneviratne, S.I., Sheffield, J., Goulden, M.L., Bonan, G., et al., 2010. Recent decline in the global land evapotranspiration trend due to limited moisture supply. Nature 467 (7318), 951–954.

Jung, M., Reichstein, M., Margolis, H.A., Cescatti, A., Richardson, A.D., Arain, M.A., Arneth, A., et al., 2011. Global patterns of land-atmosphere fluxes of carbon dioxide, latent heat, and sensible heat derived from eddy covariance, satellite, and meteorological observations. J. Geophys. Res.—Biogeosci. 116.

Jung, M., Reichstein, M., Schwalm, C.R., Huntingford, C., Sitch, S., Ahlström, A., Arneth, A., et al., 2017. Compensatory water effects link yearly global land CO_2 sink changes to temperature. Nature 541 (7638), 516–520.

Junge, C.E., 1960. Sulfur in the atmosphere. J. Geophys. Res. 65, 227–237.

Junge, C.E., 1974. Residence time and variability of tropospheric trace gases. Tellus 26, 477–488.

Junge, C.E., Werby, R.T., 1958. The concentration of chloride, sodium, potassium, calcium, and sulfate in rain water over the United States. J. Meteorol. 15, 417–425.

Junk, W.J., 1999. The flood pulse concept of large rivers: learning from the tropics. Archiv. Hydrobiol. 3 (suppl. 115), 261–280.

Juranek, L.W., Quay, P.D., 2010. Basin-wide photosynthetic production rates in the subtropical and tropical pacific ocean determined from dissolved oxygen isotope ratio measurements. Glob. Biogeochem. Cycles 24.

Jurinak, J.J., Dudley, L.M., Allen, M.F., Knight, W.G., 1986. The role of calcium oxalate in the availability of phosphorus in soils of semiarid regions: a thermodynamic study. Soil Sci. 142, 255–261.

Justic, D., Rabalais, N.N., Turner, R.E., Dortch, Q., 1995. Changes in nutrient structure of river-dominated coastal waters: stoichiometric nutrient balance and its consequences. Estuar. Coast. Shelf Sci. 40, 339–356.

Justin, S.H.F.W., Armstrong, W., 1987. The anatomical characteristics of roots and plant response to soil flooding. New Phytol. 106 (3), 465–495.

Kadeba, O., 1978. Organic matter status of some savanna soils of northern Nigeria. Soil Sci. 125, 122–127.

Kah, L.C., Lyons, T.W., Frank, T.D., 2004. Low marine sulphate and protracted oxygenation of the proterozoic biosphere. Nature 431 (7010), 834–838.

Kahn, R., 1985. The evolution of CO_2 on mars. Icarus 62, 175–190.

Kai, F.M., Tyler, S.C., Randerson, J.T., Blake, D.R., 2011. Reduced methane growth rate explained by decreased northern hemisphere microbial sources. Nature 476 (7359), 194–197.

Kaiser, K., Guggenberger, G., 2005. Dissolved organic sulphur in soil water under *Pinus sylvestris* L. and *Fagus sylvatica* L. stands in northeastern Bavaria, Germany—variations with seasons and soil depth. Biogeochemistry 72, 337–364.

Kaiser, K., Zech, W., 1998. Soil dissolved organic matter sorption as influenced by organic and sesquioxide coatings and sorbed sulfate. Soil Sci. Soc. Am. J. 62, 129–136.

Kaiser, C., Fuchslueger, L., Koranda, M., Gorfer, M., Stange, C.F., Kitzler, B., Rasche, F., et al., 2011. Plants control the seasonal dynamics of microbial N cycling in a beech forest soil by belowground C allocation. Ecology 92 (5), 1036–1051.

Kaiser, J.W., Heil, A., Andreae, M.O., Benedetti, A., Chubarova, N., Jones, L., Morcrette, J.-J., et al., 2012. Biomass burning emissions estimated with a global fire assimilation system based on observed fire radiative power. Biogeosciences 9, 527–554.

Kalbitz, K., Solinger, S., Park, J.H., Michalzik, B., Matzner, E., 2000. Controls on the dynamics of dissolved organic matter in soils: a review. Soil Sci. 165, 277–304.

Kallmeyer, J., Pockalny, R., Adhikari, R.R., Smith, D.C., D'Hondt, S., 2012. Global distribution of microbial abundance and biomass in subseafloor sediment. Proc. Natl. Acad. Sci. U. S. A. 109 (40), 16213–16216.

Kamman, N.C., Engstrom, D.R., 2002. Historical and present fluxes of mercury to vermont and New Hampshire lakes inferred from Pb-210 dated sediment cores. Atmos. Environ. 36, 1599–1609.

Kammann, C., Muller, C., Grunhage, L., Jager, H.J., 2008. Elevated CO_2 stimulates N_2O emissions in permanent grassland. Soil Biol. Biochem. 40, 2194–2205.

Kanakidou, M., Duce, R.A., Prospero, J.M., Baker, A.R., Benitez-Nelson, C., Dentener, F.J., Hunter, K.A., et al., 2012. Atmospheric fluxes of organic N and P to the global ocean: atmospheric organic N and P to the ocean. Glob. Biogeochem. Cycles Air Pollut. Stud. 26 (3), 4478.

Kaneko, M., Poulson, S.R., 2013. The rate of oxygen isotope exchange between nitrate and water. Geochim. Cosmochim. Acta 118 (October), 148–156.

Kang, D.W., Aneja, V.P., Mathur, R., Ray, J.D., 2003. Nonmethane hydrocarbons and ozone in three rural Southeast United States National Parks: a model sensitivity analysis and comparison to measurements. J. Geophys. Res. Atmos. 108.

Kaplan, I.R., 1975. Stable isotopes as a guide to biogeochemical processes. Proc. R. Soc. Lond. Ser. B Biol. Sci. 189, 183–211.

Kaplan, W.A., Elkins, J.W., Kolb, C.E., McElroy, M.B., Wofsy, S.C., Duran, A.P., 1978. Nitrous oxide in freshwater systems: an estimate for yield of atmospheric N_2O associated with disposal of human waste. Pure Appl. Geophys. 116, 423–438.

Kaplan, W.A., Wofsy, S.C., Keller, M., Dacosta, J.M., 1988. Emission of NO and deposition of O_3 in a tropical forest system. J. Geophys. Res.—Atmos. 93, 1389–1395.

Kaplan, J.O., Krumhardt, K.M., Ellis, E.C., Ruddiman, W.F., Lemmen, C., Goldewijk, K.K., 2011. Holocene carbon emissions as a result of anthropogenic land cover change. Holocene 21 (5), 775–791.

Kappler, A., Pasquero, C., Konhauser, K.O., Newman, D.K., 2005. Deposition of banded iron formations by anoxygenic phototrophic Fe(II)-oxidizing bacteria. Geology 33, 865–868.

Karberg, N.J., Pregitzer, K.S., King, J.S., Friend, A.L., Wood, J.R., 2005. Soil carbon dioxide partial pressure and dissolved inorganic carbonate chemistry under elevated carbon dioxide and ozone. Oecologia 142 (2), 296–306.

Karhu, K., Auffret, M.D., Dungait, J.A.J., Hopkins, D.W., Prosser, J.I., Singh, B.K., Subke, J.-A., et al., 2014. Temperature sensitivity of soil respiration rates enhanced by microbial community response. Nature 513 (7516), 81–84.

Karl, T.R., Knight, R.W., 1998. Secular trends of precipitation amount, frequency, and intensity in the United States. Bull. Am. Meteorol. Soc. 79, 231–241.

Karl, D., Michaels, A., Bergman, B., Capone, D., Carpenter, E., Letelier, R., Lipschultz, F., Paerl, H., Sigman, D., Stal, L., 2002. Dinitrogen fixation in the world's oceans. Biogeochemistry 57 (58), 47–98.

Karlsson, J., Byström, P., Ask, J., Ask, P., Persson, L., Jansson, M., 2009. Light limitation of nutrient-poor lake ecosystems. Nature 460 (7254), 506–509.

Karlsson, J., Bergström, A.-K., Byström, P., Gudasz, C., Rodríguez, P., Hein, C., 2015. Terrestrial organic matter input suppresses biomass production in lake ecosystems. Ecology 96 (11), 2870–2876.

Karltun, E., Gustafsson, J.P., 1993. Interference by organic complexation of Fe and Al on the SO_4^{2-} adsorption in spodic B horizons in Sweden. J. Soil Sci. 44, 625–632.

Kashefi, K., Lovley, D.R., 2003. Extending the upper temperature limit for life. Science 301 (5635), 934.

Kasischke, E.S., Bourgeauchavez, L.L., Christensen, N.L., Haney, E., 1994. Observations on the sensitivity of ERS-1 SAR image intensity to changes in aboveground biomass in young loblolly pine forests. Int. J. Remote Sens. 15, 3–16.

Kasischke, E.S., French, N.H.F., Bourgeauchavez, L.L., Christensen, N.L., 1995. Estimating release of carbon from the 1990 and 1991 forest fires in Alaska. J. Geophys. Res.-Atmos. 100, 2941–2951.

Kaspari, M., Powers, J.S., 2016. Biogeochemistry and geographical ecology: embracing all twenty-five elements required to build organisms. Am. Natur. 188 (Suppl. 1 (S1)), S62–S73.

Kaspari, M., Clay, N.A., Donoso, D.A., Yanoviak, S.P., 2014. Sodium fertilization increases termites and enhances decomposition in an Amazonian forest. Ecology 95 (4), 795–800.

Kass, D.M., Yung, Y.L., 1995. Loss of atmosphere from Mars due to solar wind-induced sputtering. Science 268 (5211), 697–699.

Kast, E.R., Stolper, D.A., Auderset, A., Higgins, J.A., Ren, H., Wang, X.T., Martínez-García, A., Haug, G.H., Sigman, D.M., 2019. Nitrogen isotope evidence for expanded ocean suboxia in the early cenozoic. Science 364 (6438), 386–389.

Kaste, J.M., Friedland, A.J., Miller, E.K., 2005. Potentially mobile lead fractions in montane organic-rich soil horizons. Water Air Soil Pollut. 167, 139–154.

Kasten, S., Jørgensen, B.B., 2000. Sulfate reduction in marine sediments. In: Schulz, H.D., Zabel, M. (Eds.), Marine Geochemistry. Springer Berlin Heidelberg, Berlin, Heidelberg, pp. 263–281.

Kasting, J.F., Walker, J.C.G., 1981. Limits on oxygen concentration in the prebiological atmosphere and the rate of abiotic fixation of nitrogen. J. Geophys. Res.—Oceans Atmos. 86, 1147–1158.

Kasting, J.F., Toon, O.B., Pollack, J.B., 1988. How climate evolved on the terrestrial planets. Sci. Am. 256 (2), 90–97.

Kastner, M., 1974. The contribution of authigenic feldspars to the geochemical balance of alkali metals. Geochim. Cosmochim. Acta.

Kastner, M., 1999. Oceanic minerals: their origin, nature of their environment, and significance. Proc. Natl. Acad. Sci. U. S. A. 96 (7), 3380–3387.

Kastowski, M., Hinderer, M., 2011. Long-term carbon burial in European lakes: analysis and estimate. Glob. Biogeochem. Cycles.

Katterer, T., Reichstein, M., Andren, O., Lomander, A., 1998. Temperature dependence of organic matter decomposition: a critical review using literature data analyzed with different models. Biol. Fertil. Soils 27, 258–262.

Katz, M.E., Pak, D.K., Dickens, G.R., Miller, K.G., 1999. The source and fate of massive carbon input during the latest paleocene thermal maximum. Science 286 (5444), 1531–1533.

Kauffman, J.B., Sanford, R.L., Cummings, D.L., Salcedo, I.H., Sampaio, E., 1993. Biomass and nutrient dynamics associated with slash fires in neotropical dry forests. Ecology 74, 140–151.

Kaufman, Y.J., Tucker, C.J., Fung, I., 1990. Remote sensing of biomass burning in the tropics. J. Geophys. Res.-Atmos. 95, 9927–9939.

Kaufman, Y.J., Tanré, D., Boucher, O., 2002. A satellite view of aerosols in the climate system. Nature 419 (6903), 215–223.

Kaul, M., Mohren, G.M.J., Dadhwal, V.K., 2011. Phytomass carbon pool of trees and forests in India. Clim. Chang. 108, 243–259.

Kaushal, S.S., Lewis, W.M., 2005. Fate and transport of organic nitrogen in minimally disturbed montane streams of Colorado, USA. Biogeochemistry 74 (3), 303–321.

Kaushal, S.S., Groffman, P.M., Likens, G.E., Belt, K.T., Stack, W.P., Kelly, V.R., Band, L.E., Fisher, G.T., 2005. Increased salinization of fresh water in the northeastern United States. Proc. Natl. Acad. Sci. U. S. A. 102 (38), 13517–13520.

Kaushal, S.S., Groffman, P.M., Band, L.E., Shields, C.A., Morgan, R.P., Palmer, M.A., Belt, K.T., Swan, C.M., Findlay, S.E.G., Fisher, G.T., 2008. Interaction between urbanization and climate variability amplifies watershed nitrate export in Maryland. Environ. Sci. Technol. 42 (16), 5872–5878.

Kaushal, S.S., Likens, G.E., Pace, M.L., Utz, R.M., Haq, S., Gorman, J., Grese, M., 2018. Freshwater salinization syndrome on a continental scale. Proc. Natl. Acad. Sci. U. S. A. 115 (4), E574–E583.

Kavouras, I.G., Mihalopoulos, N., Stephanou, E.G., 1998. Formation of atmospheric particles from organic acids produced by forests. Nature 395, 683–686.

Kebreab, E., Clark, K., Wagner-Riddle, C., France, J., 2006. Methane and nitrous oxide emissions from Canadian animal agriculture: a review. Can. J. Anim. Sci. 86 (2), 135–157.

Keeley, J.E., 1979. Population differentiation along a flood frequency gradient: physiological adaptations to flooding in Nyssa Sylvatica. Ecol. Monogr.

Keeley, J.E., Fotheringham, C.J., 1998. Mechanism of smoke-induced seed germination in a post-fire chaparral annual. J. Ecol. 86 (1), 27–36.

Keeling, C.D., 1993. Global observations of atmospheric CO_2. In: Heimann, M. (Ed.), The Global Carbon Cycle. Springer-Verlag, pp. 1–29.

Keeling, R.F., Shertz, S.R., 1992. Seasonal and interannual variations in atmospheric oxygen and implications for the global carbon cycle. Nature 358, 723–727.

Keeling, C.D., Carter, A.F., Mook, W.G., 1984. Seasonal, latitudinal, and secular variations in the abundance and isotopic ratios of atmospheric CO_2: 2. Results from oceanographic cruises in the tropical Pacific Ocean. J. Geophys. Res. D: Atmos. 89, 4615–4628.

Keeling, C.D., Whorf, T.P., Wahlen, M., van der Plichtt, J., 1995. Interannual extremes in the rate of rise of atmospheric carbon dioxide since 1980. Nature 375 (6533), 666–670.

Keeling, C.D., Chin, J.F.S., Whorf, T.P., 1996. Increased activity of northern vegetation inferred from atmospheric CO_2 measurements. Nature 382, 146–149.

Keeling, R.F., Körtzinger, A., Gruber, N., 2010. Ocean deoxygenation in a warming world. Annu. Rev. Mar. Sci. 2 (1), 199–229.

Keeling, R.F., Graven, H.D., Welp, L.R., Resplandy, L., Bi, J., Piper, S.C., Sun, Y., Bollenbacher, A., Meijer, H.A.J., 2017. Atmospheric evidence for a global secular increase in carbon isotopic discrimination of land photosynthesis. Proc. Natl. Acad. Sci. U. S. A. 114 (39), 10361–10366.

Keenan, T.F., Hollinger, D.Y., Bohrer, G., Danilo, D., William Munger, J., Schmid, H.P., Richardson, A.D., 2013. Increase in forest water-use efficiency as atmospheric carbon dioxide concentrations rise. Nature 499 (7458), 324–327.

Keenan, T.F., Colin Prentice, I., Canadell, J.G., Williams, C.A., Wang, H., Raupach, M., James Collatz, G., 2016. Recent pause in the growth rate of atmospheric CO_2 due to enhanced terrestrial carbon uptake. Nat. Commun. 7 (1), 13428.

Keenan, S.W., Schaeffer, S.M., Jin, V.L., DeBruyn, J.M., 2018. Mortality hotspots: nitrogen cycling in forest soils during vertebrate decomposition. Soil Biol. Biochem. 121 (June), 165–176.

Keenan, S.W., Schaeffer, S.M., DeBruyn, J.M., 2019. Spatial changes in soil stable isotopic composition in response to carrion decomposition. Biogeosciences 16 (19), 3929–3939.

Keffer, T., Martinson, D.G., Corliss, B.H., 1988. The position of the Gulf stream during quaternary glaciations. Science 241 (4864), 440–442.

Keil, R.G., Montlucon, D.B., Prahl, F.G., Hedges, J.I., 1994. Sorptive preservation of labile organic matter in marine sediments. Nature 370, 549–552.

Keiluweit, M., Nico, P., Harmon, M.E., Mao, J., Pett-Ridge, J., Kleber, M., 2015. Long-term litter decomposition controlled by manganese redox cycling. Proc. Natl. Acad. Sci. U. S. A. 112 (38), E5253–E5260.

Keith, D.W., Weisenstein, D.K., Dykema, J.A., Keutsch, F.N., 2016. Stratospheric solar geoengineering without ozone loss. Proc. Natl. Acad. Sci. U. S. A. 113 (52), 14910–14914.

Kelleher, B.P., Simpson, M.J., Simpson, A.J., 2006. Assessing the fate and transformation of plant residues in the terrestrial environment using HR-MAS NMR spectroscopy. Geochim. Cosmochim. Acta 70 (16), 4080–4094.

Keller, C.K., Bacon, D.H., 1998. Soil respiration and georespiration distinguished by transport analyses of vadose CO_2, (CO_2)-^{13}C, and (CO_2)-^{14}C. Glob. Biogeochem. Cycles 12, 361–372.

Keller, J.K., Bridgham, S.D., 2007. Pathways of anaerobic carbon cycling across an ombrotrophic-minerotrophic peatland gradient. Limnol. Oceanogr. 52 (1), 96–107.

Keller, M., Reiners, W.A., 1994. Soil-atmosphere exchange of nitrous oxide, nitric oxide, and methane under secondary succession of pasture to forest in the Atlantic lowlands of Costa Rica. Glob. Biogeochem. Cycles 8, 399–409.

Keller, C.K., Wood, B.D., 1993. Possibility of chemical weathering before the advent of vascular land plants. Nature 364, 223–225.

Keller, M., W. A. Kaplan, S. C. Wofsy, and J. M. Dacosta. 1988. Emissions of N_2O from tropical forest soils: response to fertilization with NH_4^+, NO_3^-, and PO_4^{3-}. J. Geophys. Res. D: Atmos. 93, 1600–1604.

Keller, M., Veldkamp, E., Weitz, A.M., Reiners, W.A., 1993. Effect of pasture age on Soil trace gas emissions from a deforested area of Costa Rica. Nature 365, 244–246.

Keller, A.B., Reed, S.C., Townsend, A.R., Cleveland, C.C., 2013. Effects of canopy tree species on belowground biogeochemistry in a lowland wet tropical forest. Soil Biol. Biochem. 58 (March), 61–69.

Kelley, C.A., Martens, C.S., Chanton, J.P., 1990. Variations in sedimentary carbon remineralization rates in the White Oak River estuary, North Carolina. Limnol. Oceanogr. 35, 372–383.

Kelley, D.S., Baross, J.A., Delaney, J.R., 2002. Volcanoes, fluids, and life at mid-ocean ridge spreading centers. Annu. Rev. Earth Planet. Sci. 30, 385–491.

Kelley, D.S., Karson, J.A., Früh-Green, G.L., Yoerger, D.R., Shank, T.M., Butterfield, D.A., Hayes, J.M., et al., 2005. A serpentinite-hosted ecosystem: the lost city hydrothermal field. Science 307 (5714), 1428–1434.

Kellogg, W.W., 1992. Aerosols and global warming. Science 256 (5057), 598.

Kellogg, L.E., Bridgham, S.D., Lopez-Hernandez, D., 2006. A comparison of four methods of measuring gross phosphorus mineralization. Soil Sci. Soc. Am. J. 70, 1349–1358.

Kellogg, M.M., Mclvin, M.R., Vedamati, J., Twining, B.S., Moffett, J.W., Marchetti, A., Moran, D.M., Saito, M.A., 2020. Efficient zinc/cobalt inter-replacement in northeast Pacific diatoms and relationship to high surface dissolved Co:Zn ratios. Limnol. Oceanogr. https://doi.org/10.1002/lno.11471.

Kelly, D.P., Smith, N.A., 1990. Organic sulfur compounds in the environment. Adv. Microb. Ecol. 11, 345–385.

Kelly, C.A., Rudd, J.W.M., Hesslein, R.H., Schindler, D.W., Dillon, P.J., Driscoll, C.T., Gherini, S.A., Hecky, R.E., 1987. Prediction of biological acid neutralization in acid-sensitive lakes. Biogeochemistry 3, 129–140.

Kelly, C.A., Rudd, J.W.M., Bodaly, R.A., Roulet, N.P., StLouis, V.L., Heyes, A., Moore, T.R., et al., 1997. Increases in fluxes of greenhouse gases and methyl mercury following flooding of an experimental reservoir. Environ. Sci. Technol. 31, 1334–1344.

Kelly, E.F., Chadwick, O.A., Hilinski, T.E., 1998. The effect of plants on mineral weathering. Biogeochemistry 42, 21–53.

Kelly, V.R., Lovett, G.M., Weathers, K.C., Likens, G.E., 2002. Trends in atmospheric concentration and deposition compared to regional and local pollutant emissions at a rural site in southeastern New York. Atmos. Environ. 36, 1569–1575.

Kelly, P.T., Solomon, C.T., Weidel, B.C., Jones, S.E., 2014. Terrestrial carbon is a resource, but not a subsidy, for lake zooplankton. Ecology 95 (5), 1236–1242.

Kemp, W.M., Sampou, P., Caffrey, J., Mayer, M., Henriksen, K., Boynton, W.R., 1990. Ammonium recycling versus denitrification in Chesapeake Bay sediments. Limnol. Oceanogr. 35, 1545–1563.

Kemp, W.M., Smith, E.M., MarvinDiPasquale, M., Boynton, W.R., 1997. Organic carbon balance and net ecosystem metabolism in Chesapeake Bay. Mar. Ecol.-Prog. Ser. 150, 229–248.

Kemp, W.M., Testa, J.M., Conley, D.J., Gilbert, D., Hagy, J.D., 2009. Temporal responses of coastal hypoxia to nutrient loading and physical controls. Biogeosciences 6, 2985–3008.

Kemp, A.C., Horton, B.P., Donnelly, J.P., Mann, M.E., Vermeer, M., Rahmstorf, S., 2011. Climate related sea-level variations over the past two millennia. Proc. Natl. Acad. Sci. U. S. A. 108 (27), 11017–11022.

Kempf, K., Allwine, E., Westberg, H., Claiborn, C., Lamb, B., 1996. Hydrocarbon emissions from spruce species using environmental chamber and branch enclosure methods. Atmos. Environ. 30 (9), 1381–1389.

Kendrick, M.A., Hémond, C., Kamenetsky, V.S., Danyushevsky, L., Devey, C.W., Rodemann, T., Jackson, M.G., Perfit, M.R., 2017. Seawater cycled throughout Earth's mantle in partially serpentinized lithosphere. Nat. Geosci. 10 (3), 222–228.

Kennedy, M.J., Chadwick, O.A., Vitousek, P.M., Derry, L.A., Hendricks, D.M., 1998. Changing sources of base cations during ecosystem development, Hawaiian islands. Geology 26, 1015–1018.

Kennedy, M.J., Hedin, L.O., Derry, L.A., 2002a. Decoupling of unpolluted temperate forests from rock nutrient sources revealed by natural $^{87}Sr/^{86}Sr$ and ^{84}Sr tracer addition. Proc. Natl. Acad. Sci. 99, 9639–9644.

Kennedy, M.J., Pevear, D.R., Hill, R.J., 2002b. Mineral surface control of organic carbon in black shale. Science 295 (5555), 657–660.

Kennedy, M., Droser, M., Mayer, L.M., Pevear, D., Mrofka, D., 2006. Late Precambrian oxygenation; inception of the clay mineral factory. Science 311 (5766), 1446–1449.

Kennett, J., 1982. Marine Geology. Prentice Hall.

Kennett, J.P., Cannariato, K.G., Hendy, I.L., Behl, R.J., 2000. Carbon isotopic evidence for methane hydrate instability during quaternary interstadials. Science 288 (5463), 128–133.

Kenrick, P., Crane, P.R., 1997. The origin and early evolution of plants on land. Nature 389, 33–39.

Keppler, F., Hamilton, J.T.G., Brass, M., Röckmann, T., 2006. Methane emissions from terrestrial plants under aerobic conditions. Nature 439 (7073), 187–191.

Kerkhoff, A.J., Enquist, B.J., Elser, J.J., Fagan, W.F., 2005. Plant allometry, stoichiometry and the temperature-dependence of primary productivity. Glob. Ecol. Biogeogr.: J. Macroecol. 14, 585–598.

Kerner, M.J., 1993. Coupling of microbial fermentation and respiration processes in an intertidal mudflat of the Elbe estuary. Limnol. Oceanogr. 38, 314–330.

Kerr, J.B., McElroy, C.T., 1993. Evidence for large upward trends of ultraviolet-B radiation linked to ozone depletion. Science 262 (5136), 1032–1034.

Kerrick, D.M., 2001. Present and past nonanthropogenic CO_2 degassing from the solid earth. Rev. Geophys. 39, 565–585.

Kerrick, D.M., Connolly, J.A., 2001. Metamorphic devolatilization of subducted marine sediments and the transport of volatiles into the Earth's mantle. Nature 411 (6835), 293–296.

Kesselmeier, J., Merk, L., 1993. Exchange of carbonyl sulfide (COS) between agricultural plants and the atmosphere—studies on the deposition of COS to peas, corn and rapeseed. Biogeochemistry 23, 47–59.

Kesselmeier, J., Meixner, F.X., Hofmann, U., Ajavon, A.L., Leimbach, S., Andreae, M.O., 1993. Reduced sulfur compound exchange between the atmosphere and tropical tree species in southern Cameroon. Biogeochemistry 23, 23–45.

Kesselmeier, J., Ciccioli, P., Kuhn, U., Stefani, P., Biesenthal, T., Rottenberger, S., Wolf, A., et al., 2002. Volatile organic compound emissions in relation to plant carbon fixation and the terrestrial carbon budget. Glob. Biogeochem. Cycles 16.

Kessler, T.J., Harvey, C.F., 2001. The global flux of carbon dioxide into groundwater. Geophys. Res. Lett. 28 (2), 279–282.

Kessler, J.D., Valentine, D.L., Redmond, M.C., Mengran, D., Chan, E.W., Mendes, S.D., Quiroz, E.W., et al., 2011. A persistent oxygen anomaly reveals the fate of spilled methane in the deep Gulf of Mexico. Science 331 (6015), 312–315.

Kettle, A.J., Kuhn, U., von Hobe, M., Kesselmeier, J., Andreae, M.O., 2002. Global budget of atmospheric carbonyl sulfide: temporal and spatial variations of the dominant sources and sinks. J. Geophys. Res. 107.

Keuskamp, J.A., van Drecht, G., Bouwman, A.F., 2012. European-scale modelling of groundwter denitrification and associated N_2O production. Environ. Pollut. 165, 67–78.

Khalil, M.A.K., Rasmussen, R.A., 1985. Causes of increasing atmospheric methane: depletion of hydroxyl radicals and the rise of emissions. Atmos. Environ. 19, 397–407.

Khalil, M.A.K., Rasmussen, R.A., 1988. Carbon monoxide in the Earth's atmosphere: indications of a global increase. Nature 332, 242–245.

Khalil, M.A.K., Rasmussen, R.A., 1990. Constraints on the global sources of methane and an analysis of recent budgets. Tellus B Chem. Phys. Meteorol. 42, 229–236.

Khalil, M.A.K., Rasmussen, R.A., 1994. Global decrease in atmospheric carbon monoxide concentration. Nature 370, 639–641.

Khalil, M.A.K., Rasmussen, R.A., French, J.R.J., Holt, J.A., 1990. The influence of termites on atmospheric trace gases: CH_4, CO_2, $CHCl_3$, N_2O, CO, H_2, and light hydrocarbons. J. Geophys. Res. 95, 3619–3634.

Khalil, M.A.K., Rasmussen, R.A., Gunawardena, R., 1993a. Atmospheric methyl bromide: trends and global mass balance. J. Geophys. Res. 98, 2887–2896.

Khalil, M.A.K., Rasmussen, R.A., Moraes, F., 1993b. Atmospheric methane at Cape Meares: analysis of a high-resoultion data base and its environmental implications. J. Geophys. Res. D: Atmos. 98, 14753–14770.

Khalil, K., Mary, B., Renault, P., 2004. Nitrous oxide production by nitrification and denitrification in soil aggregates as affected by O_2 concentration. Soil Biol. Biochem. 36, 687–699.

Khan, S.A., Mulvaney, R.L., Ellsworth, T.R., Boast, C.W., 2007. The myth of nitrogen fertilization for soil carbon sequestration. J. Environ. Qual. 36 (6), 1821–1832.

Kharol, S.K., Shephard, M.W., McLinden, C.A., Zhang, L., Sioris, C.E., O'Brien, J.M., Vet, R., et al., 2018. Dry deposition of reactive nitrogen from satellite observations of ammonia and nitrogen dioxide over North America. Geophys. Res. Lett. 45 (2), 1157–1166.

Khatiwala, S., Tanhua, T., Mikaloff Fletcher, S., Gerber, M., Doney, S.C., Graven, H.D., Gruber, N., et al., 2012. Global ocean storage of anthropogenic carbon. Biogeosciences 10 (7), 8931–8988.

Kieffer, H.H., 1976. Soil and surface temperatures at the viking landing sites. Science 194 (4271), 1344–1346.

Kiehl, J.T., Briegleb, B.P., 1993. The relative roles of sulfate aerosols and greenhouse gases in climate forcing. Science 260 (5106), 311–314.

Kielland, K., 1994. Amino acid absorption by Arctic plants—implications for plant nutrition and nitrogen cycling. Ecology 75, 2373–2383.

Kielland, K., DiFolco, D., Montgomerie, C., 2019. Dining dangerously: geophagy by snowshoe hares. Ecology 100 (3), e02555.

Kiene, R.P., 1990. Dimethyl sulfide production from dimethylsulfoniopropionate in coastal seawater samples and bacterial cultures. Appl. Environ. Microbiol. 56 (11), 3292–3297.

Kietäväinen, R., Ahonen, L., Niinikoski, P., Nykänen, H., Kukkonen, I.T., 2017. Abiotic and biotic controls on methane formation down to 2.5km depth within the Precambrian Fennoscandian shield. Geochim. Cosmochim. Acta 202, 124–145.

Kilham, P., 1982. Acid precipitation: its role in the alkalization of a lake in Michigan 1. Limnol. Oceanogr. 27 (5), 856–867.

Killingbeck, K.T., 1985. Autumnal resorption and accretion of trace metals in gallery forest trees. Ecology 66, 283–286.

Killingbeck, K.T., 1996. Nutrients in senesced leaves: keys to the search for potential resorption and resorption proficiency. Ecology 77, 1716–1727.

Kim, K.R., Craig, H., 1990. Two-isotope characterization of N_2O in the Pacific Ocean and constraints on its origin in deep water. Nature 347, 58–61.

Kim, K.R., Craig, H., 1993. ^{15}N and ^{18}O characteristics of nitrous oxide: a global perspective. Science 262, 1855–1857.

Kim, J.H., Jackson, R.B., 2012. A global analysis of groundwater recharge for vegetation, climate, and soils. Vadose Zone J. 11.

Kim, H., Lakshmi, V., 2019. Global dynamics of stored precipitation water in the topsoil layer from satellite and reanalysis data. Water Resour. Res. 55 (4), 3328–3346.

Kim, D., Ramanathan, V., 2012. Improved estimates and understanding of global albedo and atmospheric solar absorption. Geophys. Res. Lett. 39 (24), 38.

Kim, J., Rees, D.C., 1992. Structural models for the metal centers in the nitrogenase molybdenum-iron protein. Science 257 (5077), 1677–1682.

Kim, J., Rees, D.C., 1994. Nitrogenase and biological nitrogen fixation. Biochemistry 33 (2), 389–397.

Kim, D., Schuffert, J.D., Kastner, M., 1999. Francolite Authigenesis in California continental slope sediments and its implications for the marine P cycle. Geochim. Cosmochim. Acta 63, 3477–3485.

Kim, S.W., Heckel, A., McKeen, S.A., Frost, G.J., Hsie, E.Y., Trainer, M.K., Richter, A., Burrows, J.P., Peckham, S.E., Grell, G.A., 2006. Satellite-observed US power plant NO_x emission reductions and their impact on air quality. Geophys. Res. Lett. 33.

Kim, T.-W., Lee, K., Najjar, R.G., Jeong, H.-D., Jeong, H.J., 2011. Increasing N abundance in the northwestern Pacific Ocean due to atmospheric nitrogen deposition. Science 334 (6055), 505–509.

Kim, Y.S., Imori, M., Watanabe, M., Hatano, R., Yi, M.J., Koike, T., 2012. Simulated nitrogen inputs influence methane and nitrous oxide fluxes from a young larch plantation in northern Japan. Atmos. Environ. 46, 36–44.

Kim, D.-G., Giltrap, D., Hernandez-Ramirez, G., 2013. Background nitrous oxide emissions in agricultural and natural lands: a meta-analysis. Plant Soil 373 (1), 17–30.

Kim, I.-N., Lee, K., Gruber, N., Karl, D.M., Bullister, J.L., Yang, S., Kim, T.-W., 2014a. Chemical oceanography. Increasing anthropogenic nitrogen in the North Pacific Ocean. Science 346 (6213), 1102–1106.

Kim, I.-N., Lee, K., Gruber, N., Karl, D.M., Bullister, J.L., Yang, S., Kim, T.-W., 2014b. Increasing anthropogenic nitrogen in the North Pacific Ocean. Science 346 (6213), 1102–1106.

Kim, H., Stinchcomb, G., Brantley, S.L., 2017. Feedbacks among O_2 and CO_2 in deep soil gas, oxidation of ferrous minerals, and fractures: a hypothesis for steady-state regolith thickness. Earth Planet. Sci. Lett. 460 (February), 29–40.

Kim, J.-S., Seo, K.-W., Jeon, T., Chen, J., Wilson, C., 2019. Missing hydrological contribution to sea level rise. Geophys. Res. Lett. 46 (November).

Kindler, R., Siemens, J., Kaiser, K., Walmsley, D.C., Bernhofer, C., Buchmann, N., Cellier, P., et al., 2011. Dissolved carbon leaching from soil is a crucial component of the net ecosystem carbon balance. Glob. Chang. Biol. 17, 1167–1185.

King, G.M., 1988. Patterns of sulfate reduction and the sulfur cycle in a South Carolina salt marsh. Limnol. Oceanogr. 33, 376–390.

King, G.M., 1992. Ecological aspects of methane oxidation, a key determinant of global methane dynamics. Adv. Microb. Ecol. 12, 431–468.

King, G.M., Adamsen, A.P., 1992. Effects of temperature on methane consumption in a forest soil and in pure cultures of the methanotroph *Methylomonas rubra*. Appl. Environ. Microbiol. 58 (9), 2758–2763.

King, G.M., Schnell, S., 1994. Effect of increasing atmospheric methane concentration on ammonium inhibition of soil methane consumption. Nature 370, 282–284.

King, G.M., Wiebe, W.J., 1978. Methane release from soils of a Georgia salt marsh. Geochim. Cosmochim. Acta 42, 343–348.

King, G.M., Roslev, P., Skovgaard, H., 1990. Distribution and rate of methane oxidation in sediments of the Florida Everglades. Appl. Environ. Microbiol. 56 (9), 2902–2911.

King, J.K., Kostka, J.E., Frischer, M.E., Saunders, F.M., Jahnke, R.A., 2001. A quantitative relationship that demonstrates mercury methylation rates in marine sediments are based on the community composition and activity of sulfate-reducing bacteria. Environ. Sci. Technol. 35 (12), 2491–2496.

King, J.S., Kubiske, M.E., Pregitzer, K.S., Hendrey, G.R., McDonald, E.P., Giardina, C.P., Quinn, V.S., Karnosky, D.F., 2005. Tropospheric O_3 compromises net primary production in young stands of trembling Aspen, paper birch and sugar maple in response to elevated atmospheric CO_2. New Phytol. 168, 623–635.

King, A.W., Andres, R.J., Davis, K.J., Hafer, M., Hayes, D.J., Huntzinger, D.N., de Jong, B., et al., 2015. North America's net terrestrial CO_2 exchange with the atmosphere 1990–2009. Biogeosciences 12 (2), 399–414.

Kinnard, C., Zdanowicz, C.M., Fisher, D.A., Isaksson, E., de Vernal, A., Thompson, L.G., 2011. Reconstructed changes in Arctic Sea ice over the past 1,450 years. Nature 479 (7374), 509–512.

Kinsman, D.J.J., 1969. Interpretation of Sr_2^b concentrations in carbonate minerals and rocks. J. Sediment. Petrol. 39, 486–508.

Kira, T., Shidei, T., 1967. Primary production and turnover of organic matter in different forest ecosystems of the Western Pacific. Jpn. J. Ecol. 17, 70–87.

Kirchman, D.L., Suzuki, Y., Garside, C., Ducklow, H.W., 1991. High turnover rates of dissolved organic carbon during a spring phytoplankton bloom. Nature 352, 612–614.

Kirchman, D.L., Ducklow, H.W., McCarthy, J.J., Garside, C., 1994. Biomass and nitrogen uptake by heterotrophic bacteria during the spring phytoplankton bloom in the North Atlantic Ocean. Deep-Sea Res. Part I—Oceanogr. Res. Pap. 41, 879–895.

Kirk, G.J.D., Bellamy, P.H., Lark, R.M., 2010. Changes in soil pH across England and Wales in response to decreased acid deposition. Glob. Chang. Biol. 16, 3111–3119.

Kirkby, J., Curtius, J., Almeida, J., Dunne, E., Duplissy, J., Ehrhart, S., Franchin, A., et al., 2011. Role of sulphuric acid, ammonia and galactic cosmic rays in atmospheric aerosol nucleation. Nature 476 (7361), 429–433.

Kirschbaum, M.U.F., 1995. The temperature dependence of soil organic matter decomposition, and the effect of global warming on soil organic-C storage. Soil Biol. Biochem. 27, 753–760.

Kirschbaum, M.U.F., Bruhn, D., Etheridge, D.M., Evans, J.R., Farquhar, G.D., Gifford, R.M., Paul, K.I., Winters, A.J., 2006. A comment on the quantitative significance of aerobic methane release by plants. Funct. Plant Biol. 33, 521–530.

Kirschke, S., Bousquet, P., Ciais, P., Saunois, M., Canadell, J.G., Dlugokencky, E.J., Bergamaschi, P., et al., 2013. Three decades of global methane sources and sinks. Nat. Geosci. 6 (10), 813–823.

Kirschvink, J.L., Gaidos, E.J., Bertani, L.E., Beukes, N.J., Gutzmer, J., Maepa, L.N., Steinberger, R.E., 2000. Paleoproterozoic snowball earth: extreme climatic and geochemical global change and its biological consequences. Proc. Natl. Acad. Sci. U. S. A. 97 (4), 1400–1405.

Kirwan, M.L., Blum, L.K., 2011. Enhanced decomposition offsets enhanced productivity and soil carbon accumulation in coastal wetlands responding to climate change. Biogeosciences 8, 987–993.

Kirwan, M.L., Murray, A.B., 2007. A coupled geomorphic and ecological model of tidal marsh evolution. Proc. Natl. Acad. Sci. U. S. A. 104 (15), 6118–6122.

Kitayama, K., 2013. The activities of soil and root acid phosphatase in the nine tropical rain forests that differ in phosphorus availability on Mount Kinabalu, Borneo. Plant Soil 367 (1), 215–224.

Kitayama, K., Majalap-Lee, N., Aiba, S., 2000. Soil phosphorus fractionation and phosphorus-use efficiencies of tropical rainforests along altitudinal gradients of Mount Kinabalu, Borneo. Oecologia 123 (3), 342–349.

Kitchell, J.F., Carpenter, S.R., 1993. Cascading trophic interactions. In: Carpenter, S.R., Kitchell, J.F. (Eds.), The Trophic Cascade in Lakes. Cambridge University Press, pp. 1–14.

Kjær, K.H., Khan, S.A., Korsgaard, N.J., Wahr, J., Bamber, J.L., Hurkmans, R., van den Broeke, M., et al., 2012. Aerial photographs reveal late–20th-century dynamic ice loss in northwestern Greenland. Science 337 (6094), 569–573.

Kjellstrom, E., Feichter, J., Sausen, R., Hein, R., 1999. The contribution of aircraft emissions to the atmospheric sulfur budget. Atmos. Environ. 33, 3455–3465.

Klappa, C.F., 1980. Rhizoliths in terrestrial carbonates: classification, recognition, genesis and significance. Sedimentology 27, 613–629.

Klausmeier, C.A., Litchman, E., Daufresne, T., Levin, S.A., 2004. Optimal nitrogen-to-phosphorus stoichiometry of phytoplankton. Nature 429 (6988), 171–174.

Kleber, M., Lehmann, J., 2019. Humic substances extracted by alkali are invalid proxies for the dynamics and functions of organic matter in terrestrial and aquatic ecosystems. J. Environ. Qual. 48 (2), 207–216.

Klein, D.R., 1968. The introduction, increase, and crash of reindeer on St. Matthew island. J. Wildl. Manag. 32, 350–367.

Klein, T., Siegwolf, R.T.W., Körner, C., 2016. Belowground carbon trade among tall trees in a temperate forest. Science 352 (6283), 342–344.

Kleinman, L., Lee, Y.N., Springston, S.R., Nunnermacker, L., Zhou, X.L., Brown, R., Hallock, K., et al., 1994. Ozone formation at a rural site in the southeastern United States. J. Geophys. Res. 99, 3469–3482.

Kleypas, J.A., 1999. Geochemical consequences of increased atmospheric carbon dioxide on coral reefs. Science 284, 118–120.

Kleypas, J.A., Yates, K.K., 2009. Coral reefs and ocean acidification. Oceanography 22 (4), 108–117.

Klimont, Z., Smith, S.J., Cofala, J., 2013. The last decade of global anthropogenic sulfur dioxide: 2000–2011 emissions. Environ. Res. Lett. 8 (1), 014003.

Kling, G.W., Evans, W.C., Tuttle, M.L., Tanylleke, G., 1994. Degassing of Lake Nyos. Nature 368, 405–406.

Kling, G.W., Evans, W.C., Tanyileke, G., Kusakabe, M., Ohba, T., Yoshida, Y., Hell, J.V., 2005. Degassing Lakes Nyos and Monoun: defusing certain disaster. Proc. Natl. Acad. Sci. U. S. A. 102 (40), 14185–14190.

Klinkhammer, G.P., 1980. Early diagenesis in sediments from the eastern equatorial Pacific II. Pore water metal results. Earth Planet. Sci. Lett. 49, 81–101.

Kludze, H.K., Delaune, R.D., Patrick, W.H., 1993. Aerenchyma formation and methane and oxygen exchange in rice. Soil Sci. Soc. Am. J. 57 (2), 386–391.

Kminek, G., Bada, J.L., 2006. The effect of ionizing radiation on the preservation of amino acids on Mars. Earth Planet. Sci. Lett. 245, 1–5.

Knapp, A.K., Smith, M.D., 2001. Variation among biomes in temporal dynamics of aboveground primary production. Science 291 (5503), 481–484.

Knauer, G.A., 1993. Productivity and new production of the oceanic systems. In: Mackenzie, F.T., Wollast, R., Chou, L. (Eds.), Interactions of C, N, P and S Biogeochemical Cycles and Global Change. Springer, pp. 211–231.

Knauss, J.A., 1978. Introduction to Physical Oceanography. Prentice-Hall.

Knauth, L.P., 1998. Salinity history of the Earth's early ocean [letter]. Nature 395 (6702), 554–555.

Knauth, L.P., Kennedy, M.J., 2009. The late Precambrian greening of the earth. Nature 460 (7256), 728–732.

Knauth, L.P., 2011. Salinity history of the Earth's ocean. In: Reitner, J., Thiel, V. (Eds.), Encyclopedia of Geobiology. Encyclopedia of Earth Sciences Series. Springer, Dordrecht.

Knicker, H., 2011. Soil organic N—an under-rated player for C sequestration in soils? Soil Biol. Biochem. 43 (6), 1118–1129.

Knight, D.H., Fahey, T.J., Running, S.W., 1985. Water and nutrient outflow from contrasting Lodgepole pine forests in Wyoming. Ecol. Monogr. 55, 29–48.

Knoll, A.H., 1992. The early evolution of eukaryotes: a geological perspective. Science 256 (5057), 622–627.

Knoll, A.H., 2015. Life on a Young Planet: The First Three Billion Years of Evolution on Earth—Updated Edition. Princeton University Press.

Knoll, M.A., James, W.C., 1987. Effect of the advent and diversification of vascular land plants on mineral weathering through geological time. Geology 15, 1099–1102.

Knöller, K., Vogt, C., Haupt, M., Feisthauer, S., Richnow, H.-H., 2011. Experimental investigation of nitrogen and oxygen isotope fractionation in nitrate and nitrite during denitrification. Biogeochemistry 103 (1–3), 371–384.

Knowles, R., 1982. Denitrification. Microbiol. Rev. 46 (1), 43–70.

Knudsen, M.F., Seidenkrantz, M.-S., Jacobsen, B.H., Kuijpers, A., 2011. Tracking the Atlantic multidecadal oscillation through the last 8,000 years. Nat. Commun. 2 (February), 178.

Koba, K., Tokuchi, N., Yoshioka, T., Hobbie, E.A., Iwatsubo, G., 1998. Natural abundance of nitrogen-15 in a forest soil. Soil Sci. Soc. Am. J. 62, 778–781.

Koba, K., Osaka, K., Tobari, Y., Toyoda, S., Ohte, N., Katsuyama, M., Suzuki, N., et al., 2009. Biogeochemistry of nitrous oxide in groundwater in a forested ecosystem elucidated by nitrous oxide isotopomer measurements. Geochim. Cosmochim. Acta 73, 3115–3133.

Koba, K., Fang, Y., Mo, J., Zhang, W., Lu, X., Liu, L., Zhang, T., et al., 2012. The ^{15}N natural abundance of the N lost from an N-saturated subtropical forest in Southern China. J. Geophys. Res. 117.

Kobe, R.K., Lepczyk, C.A., Iyer, M., 2005. Resorption efficiency decreases with increasing green leaf nutrients in a global data set. Ecology 86, 2780–2792.

Kodama, H., Schnitzer, M., 1977. Effect of fulvic acid on crystallization of Fe(III) oxides. Geoderma 19, 279–291.

Kodama, H., Schnitzer, M., 1980. Effect of fulvic acid on the crystallization of aluminum hydroxides. Geoderma 24, 195–205.

Koele, N., Bird, M., Haig, J., Marimon-Junior, B.H., Marimon, B.S., Phillips, O.L., de Oliveira, E.A., Quesada, C.A., Feldpausch, T.R., 2017. Amazon basin forest pyrogenic carbon stocks: first estimate of deep storage. Geoderma 306 (November), 237–243.

Koenig, L.E., Song, C., Wollheim, W.M., Rüegg, J., McDowell, W.H., 2017. Nitrification increases nitrogen export from a Tropical River network. Freshw. Sci. 36 (4), 698–712.

Koenig, L.E., Helton, A.M., Savoy, P., Bertuzzo, E., Heffernan, J.B., Hall Jr., R.O., Bernhardt, E.S., 2019. Emergent productivity regimes of river networks. Limnol. Oceanogr. 4 (5), 173–181.

Koerselman, W., Meuleman, A.F.M., 1996. The vegetation N:P ratio: a new tool to detect the nature of nutrient limitation. J. Appl. Ecol. 33, 1441–1450.

Koeve, W., 2002. Upper ocean carbon fluxes in the Atlantic ocean: the importance of the POC:PIC ratio. Glob. Biogeochem. Cycles. 16. https://doi.org/10.1029/2001GB001836.

Koeve, W., Kähler, P., 2010. Heterotrophic denitrification vs. autotrophic anammox—quantifying collateral effects on the oceanic carbon cycle. Biogeosciences 7, 2327–2337.

Kögel-Knabner, I., Rumpel, C., 2018. Advances in molecular approaches for understanding soil organic matter composition, origin, and turnover: a historical overview (chapter one). In: Sparks, D.L. (Ed.), Advances in Agronomy. 149. Academic Press, pp. 1–48.

Kohfeld, K.E., Le Quéré, C., Harrison, S.P., Anderson, R.F., 2005. Role of marine biology in glacial-interglacial CO_2 cycles. Science 308 (5718), 74–78.

Kohlbach, D., Graeve, M., Lange, B.A., David, C., Schaafsma, F.L., van Franeker, J.A., Vortkamp, M., Brandt, A., Flores, H., 2018. Dependency of Antarctic zooplankton species on ice algae-produced carbon suggests a sea ice-driven pelagic ecosystem during winter. Glob. Chang. Biol. 24 (10), 4667–4681.

Kohler, P., Fischer, H., 2004. Simulating changes in the terrestrial biosphere during the last glacial/interglacial transition. Glob. Planet. Chang. 43, 33–55.

Kohler, I.H., Poulton, P.R., Auerswald, K., Schnyder, H., 2010. Intrinsic water-use efficiency of temperate seminatural grassland has increased since 1857: an analysis of carbon isotope discrimination of herbage from the park grass experiment. Glob. Chang. Biol. 16, 1531–1541.

Koide, R.T., 1991. Nutrient supply, nutrient demand and plant response to mycorrhizal infection. New Phytol. 117, 365–386.

Kokaly, R.F., Asner, G.P., Ollinger, S.V., Martin, M.E., Wessman, C.A., 2009. Characterizing canopy biochemistry from imaging spectroscopy and its application to ecosystem studies. Remote Sens. Environ. 113, S78–S91.

Kolari, P., Pumpanen, J., Rannik, U., Ilvesniemi, H., Hari, P., Berninger, F., 2004. Carbon balance of different aged scots pine forests in southern Finland. Glob. Chang. Biol. 10, 1106–1119.

Kolber, Z.S., Barber, R.T., Coale, K.H., Fitzwateri, S.E., Greene, R.M., Johnson, K.S., Lindley, S., Falkowski, P.G., 1994. Iron limitation of phytoplankton photosynthesis in the equatorial Pacific Ocean. Nature 371 (6493), 145–149.

Kolber, Z.S., Van Dover, C.L., Niederman, R.A., Falkowski, P.G., 2000. Bacterial photosynthesis in surface waters of the open ocean. Nature 407 (6801), 177–179.

Kone, Y.J.M., Borges, A.V., 2008. Dissolved inorganic carbon dynamics in the waters surrounding forested mangroves of the Ca Mau Province (Vietnam). Estuar. Coast. Shelf Sci. 77, 409–421.

Konhauser, K.O., Pecoits, E., Lalonde, S.V., Papineau, D., Nisbet, E.G., Barley, M.E., Arndt, N.T., Zahnle, K., Kamber, B.S., 2009. Oceanic nickel depletion and a methanogen famine before the great oxidation event. Nature 458 (7239), 750–753.

Konhauser, K.O., Lalonde, S.V., Planavsky, N.J., Pecoits, E., Lyons, T.W., Mojzsis, S.J., Rouxel, O.J., et al., 2011. Aerobic bacterial pyrite oxidation and acid rock drainage during the great oxidation event. Nature 478 (7369), 369–373.

Konikow, L.F., 2011. Contribution of global groundwater depletion since 1900 to sea-level rise. Geophys. Res. Lett. 38.

Konings, A.G., Bloom, A.A., Liu, J., Parazoo, N.C., Schimel, D.S., Bowman, K.W., 2019. Global, satellite-driven estimates of heterotrophic respiration. Biogeosciences 16, 2269–2284.

Könneke, M., Bernhard, A.E., de la Torre, J.R., Walker, C.B., Waterbury, J.B., Stahl, D.A., 2005. Isolation of an autotrophic ammonia-oxidizing marine archaeon. Nature 437 (7058), 543–546.

Koo, B.-C., Lee, Y.-H., Moon, D.-S., Yoon, S.-C., Raymond, J.C., 2013. Phosphorus in the young supernova remnant Cassiopeia A. Science 342 (6164), 1346–1348.

Kool, D.M., Dolfing, J., Wrage, N., Van Groenigen, J.W., 2011. Nitrifier denitrification as a distinct and significant source of nitrous oxide from soil. Soil Biol. Biochem. 43, 174–178.

Kopittke, P.M., Dalal, R.C., Finn, D., Menzies, N.W., 2017. Global changes in soil stocks of carbon, nitrogen, phosphorus, and sulphur as influenced by long-term agricultural production. Glob. Chang. Biol. 23 (6), 2509–2519.

Koppelaar, R.H.E.M., Weikard, H.P., 2013. Assessing phosphate rock depletion and phosphorus recycling options. Glob. Environ. Change: Hum. Policy Dimens. 23 (6), 1454–1466.

Korablev, O., Vandaele, A.C., Montmessin, F., Fedorova, A.A., Trokhimovskiy, A., Forget, F., Lefèvre, F., et al., 2019. No detection of methane on Mars from early ExoMars trace gas orbiter observations. Nature 568 (7753), 517–520.

Körner, C., 1989. The nutritional status of plants from high altitudes: a worldwide comparison. Oecologia 81 (3), 379–391.

Korom, S.F., 1992. Natural denitrification in the saturated zone: a review. Water Resour. Res. 28 (6), 1657–1668.

Kort, E.A., Patra, P.K., Ishijima, K., Daube, B.C., Jimenez, R., Elkins, J., Hurst, D., Moore, F.L., Sweeney, C., Wofsy, S.C., 2011. Tropospheric distribution and variability of N_2O: evidence for strong tropical emissions. Geophys. Res. Lett. 38.

Kort, E.A., Frankenberg, C., Miller, C.E., Oda, T., 2012. Space-based observations of megacity carbon dioxide: space-based observations of megacity CO_2. Geophys. Res. Lett. 39(17).

Kort, E.A., Frankenberg, C., Costigan, K.R., Lindenmaier, R., Dubey, M.K., Wunch, D., 2014. Four corners: the largest US methane anomaly viewed from space: four corners: largest US methane anomaly. Geophys. Res. Lett. 41 (19), 6898–6903.

Kortelainen, P., Larmola, T., Rantakari, M., Juutinen, S., Alm, J., Martikainen, P.J., 2019. Lakes as nitrous oxide sources in the boreal landscape. Glob. Change Biol.

Koschel, R., Benndorf, J., Proft, G., Recknagel, F., 1983. Calcite precipitation as a natural control mechanism of eutrophication. Arch. Hydrobiol. 98, 380–408.

Koschorreck, M., Conrad, R., 1993. Oxidation of atmospheric methane in soil: measurements in the field in soil cores and in soil samples. Glob. Biogeochem. Cycles 7, 109–121.

Koster, R.D., Dirmeyer, P.A., Guo, Z., Bonan, G., Chan, E., Peter, C., Gordon, C.T., et al., 2004. Regions of strong coupling between soil moisture and precipitation. Science 305 (5687), 1138–1140.

Kostka, J.E., Roychoudhury, A., Van Cappellen, P., 2012. Rates and controls of anaerobic microbial respiration across spatial and temporal gradients in saltmarsh sediments. Biogeochemistry 60, 49–76.

Köstner, B., Schupp, R., Schulze, E.-D., Rennenberg, H., 1998. Organic and inorganic sulfur transport in the xylem sap and the sulfur budget of Picea Abies trees. Tree Physiol. 18 (1), 1–9.

Kouketsu, S., Murata, A.M., 2014. Detecting decadal scale increases in anthropogenic CO_2 in the ocean. Geophys. Res. Lett.

Koven, C.D., Ringeval, B., Friedlingstein, P., Ciais, P., Cadule, P., Khvorostyanov, D., Krinner, G., Tarnocai, C., 2011. Permafrost carbon-climate feedbacks accelerate global warming. Proc. Natl. Acad. Sci. U. S. A. 15 (August).

Kraft, B., Tegetmeyer, H.E., Sharma, R., Klotz, M.G., Ferdelman, T.G., Hettich, R.L., Geelhoed, J.S., Strous, M., 2014. Nitrogen cycling. The environmental controls that govern the end product of bacterial nitrate respiration. Science 345 (6197), 676–679.

Krajewski, K.P., Van Cappellen, P., Trichet, J., Kuhn, O., Lucas, J., Martinalgarra, A., Prevot, L., et al., 1994. Biological processes and apatite formation in sedimentary environments. Eclogae Geol. Helv. 87, 701–745.

Krakauer, N.Y., Fung, I., 2008. Mapping and attribution of change in streamflow in the coterminous United States. Hydrol. Earth Syst. Sci. 12, 1111–1120.

Kral, T.A., Brink, K.M., Miller, S.L., McKay, C.P., 1998. Hydrogen consumptions by methanogens on the early earth. Origins Life Evol. Biosphere 28, 311–319.

Kramer, P.J., 1982. Water and plant productivity of yield. In: Rechcigl, M. (Ed.), Handbook of Agricultural Productivity. CRC Press, pp. 41–47.

Kramer, J., Laan, P., Sarthou, G., Timmermans, K.R., de Baar, H.J.W., 2004. Distribution of dissolved aluminium in the high atmospheric input region of the subtropical waters of the North Atlantic Ocean. Mar. Chem. 88, 85–101.

Krasnopolsky, V.A., Feldman, P.D., 2001. Detection of molecular hydrogen in the atmosphere of Mars. Science 294 (5548), 1914–1917.

Kratz, T.R., DeWitt, C.B., 1986. Internal factors controlling Peatland-Lake ecosystem development. Ecology 67, 100–107.

Krause, H.H., 1982. Nitrate formation and movement before and after clear cutting of a monitored watershed in Central New Brunswick, Canada. Can. J. Forest Res. 12, 922–930.

Krausmann, F., Gingrich, S., Eisenmenger, N., Erb, K.-H., Haberl, H., Fischer-Kowalski, M., 2009. Growth in global materials use, GDP and population during the 20th century. Ecol. Econ.: J. Int. Soc. Ecol. Econ. 68 (10), 2696–2705.

Kristensen, E., Ahmed, S.I., Devol, A.H., 1995. Aerobic and anaerobic decomposition of organic matter in marine sediment: which is fastest? Limnol. Oceanogr. 40, 1430–1437.

Kristjansson, J.K., Schönheit, P., Thauer, R.K., 1982. Different Ks values for hydrogen of methanogenic bacteria and sulfate reducing bacteria: an explanation for the apparent inhibition of methanogenesis by sulfate. Arch. Microbiol. 131 (3), 278–282.

Kritee, K., Blum, J.D., Reinfelder, J.R., Barkay, T., 2013. Microbial stable isotope fractionation of mercury: a synthesis of present understanding and future directions. Chem. Geol. 336 (January), 13–25.

Kroehler, C.J., Linkins, A.E., 1991. The absorption of inorganic phosphate from ^{32}P-labeled inositol hexaphosphate by *Eriophorum vaginatum*. Oecologia 85 (3), 424–428.

Kroer, N., Jørgensen, N.O., Coffin, R.B., 1994. Utilization of dissolved nitrogen by heterotrophic bacterioplankton: a comparison of three ecosystems. Appl. Environ. Microbiol. 60 (11), 4116–4123.

Krom, M.D., Berner, R.A., 1981. The diagenesis of phosphorus in a nearshore marine sediment. Geochim. Cosmochim. Acta 45, 207–216.

Kronvang, B., Jeppesen, E., Conley, D.J., Sondergaard, M., Larsen, S.E., Ovesen, N.B., Carstensen, J., 2005. Nutrient pressures and ecological responses to nutrient loading reductions in Danish streams, lakes and coastal waters. J. Hydrol. 304, 274–288.

Krouse, H.R., McCready, R.G.L., 1979. Reductive reactions in the sulfur cycle. In: Trudinger, P.A., Swaine, D.J. (Eds.), Biogeochemical Cycling of Mineral-Forming Elements. Elsevier, pp. 315–368.

Krumbein, W.E., 1971. Manganese-oxidizing fungi and bacteria in recent shelf sediments of the Bay of Biscay and the North Sea. Die Naturwissenschaften 58, 56–57.

Krumholz, L.R., McKinley, J.P., Ulrich, F.A., Suflita, J.M., 1997. Confined subsurface microbial communities in cretaceous rock. Nature 386, 64–66.

Krysell, M., Wallace, D.W., 1988. Arctic Ocean ventilation studied with a suite of anthropogenic halocarbon tracers. Science 242 (4879), 746–749.

Ksionzek, K.B., Lechtenfeld, O.J., Leigh McCallister, S., Schmitt-Kopplin, P., Geuer, J.K., Geibert, W., Koch, B.P., 2016. Dissolved organic sulfur in the ocean: biogeochemistry of a Petagram inventory. Science 354 (6311), 456–459.

Kuhlbrandt, W., Wang, D.N., Fujiyoshi, Y., 1994. Atomic model of the plant light-harvesting complex by electron crystallography. Nature 367, 614–621.

Kuhlbrodt, T., Griesel, A., Montoya, M., Levermann, A., Hofmann, M., Rahmstorf, S., 2007. On the driving processes of the Atlantic meridional overturning circulation. Rev. Geophys. 45 (2), 1424.

Kuhlbusch, T.A., Lobert, J.M., Crutzen, P.J., Warneck, P., 1991. Molecular nitrogen emissions from denitrificaiton during biomass burning. Nature 351, 135–137.

Kuhn, U., 2002. Isoprene and monoterpene emissions of Amazônian tree species during the wet season: direct and indirect investigations on controlling environmental functions. J. Geophys. Res. 107 (D20), 1616.

Kuhn, U., Ammann, C., Wolf, A., Meixner, F.X., Andreae, M.O., Kesselmeier, J., 1999. Carbonyl sulfide exchange on an ecosystem scale: soil represents a dominant sink for atmospheric COS. Atmos. Environ. 33, 995–1008.

Kuhry, P., Vitt, D.H., 1996. Fossil carbon/nitrogen ratios as a measure of peat decomposition. Ecology 77, 271–275.

Kuivila, K.M., Murray, J.W., Devol, A.H., Novelli, P.C., 1989. Methane proudction, sulfate reduction and competition for substrates in the sediments of Lake Washington. Geochim. Cosmochim. Acta 53 (2), 409–416.

Kulkarni, S.R., 1997. Brown dwarfs: a possible missing link between stars and planets. Science 276, 1350–1354.

Kulkarni, M.V., Burgin, A.J., Groffman, P.M., Yavitt, J.B., 2014. Direct flux and ^{15}N tracer methods for measuring denitrification in forest soils. Biogeochemistry 117, 359–373.

Kumar, N., Anderson, R.F., Mortlock, R.A., Froeroelich, P.N., Kubik, P., Dittrichhannen, B., Suter, M., 1995. Increased biological productivity and export production in the glacial Southern Ocean. Nature 378, 675–680.

Kumar, P., Robins, A., Vardoulakis, S., Britter, R., 2010. A review of the characteristics of nanoparticles in the urban atmosphere and the prospects for developing regulatory controls. Atmos. Environ. 44, 5035–5052.

Kump, L.R., 1988. Terrestrial feedback in atmospheric oxygen regulation by fire and phosphorus. Nature 335, 152–154.

Kump, L.R., Garrels, R.M., 1986. Modeling atmospheric O_2 in the global sedimentary redox cycle. Am. J. Sci. 286, 337–360.

Kump, L.R., Brantley, S.L., Arthur, M.A., 2000. Chemical, weathering, atmospheric CO_2 and climate. Annu. Rev. Earth Planet. Sci. 28, 611–667.

Kump, L.R., Junium, C., Arthur, M.A., Brasier, A., Fallick, A., Melezhik, V., Lepland, A., Crne, A.E., Luo, G., 2011. Isotopic evidence for massive oxidation of organic matter following the great oxidation event. Science 334 (6063), 1694–1696.

Kunishi, H.M., 1988. Sources of nitrogen and phosphorus in an estuary of the Chesapeake Bay. J. Environ. Qual. 17, 185–188.

Kunz, J., Staudacher, T., Allegre, C.J., 1998. Plutonium-fission xenon found in Earth's mantle. Science 280 (5365), 877–880.

Kurc, S.A., Small, E.E., 2004. Dynamics of evapotranspiration in semiarid grassland and Shrubland ecosystems during the summer monsoon season, Central New Mexico. Water Resour. Res. 40.

Kurz, H., Demaree, D., 1934. Cypress buttresses and knees in relation to water and air. Ecology 15 (1), 36–41.

Kurz, W.A., Dymond, C.C., Stinson, G., Rampley, G.J., Neilson, E.T., Carroll, A.L., Ebata, T., Safranyik, L., 2008. Mountain pine beetle and forest carbon feedback to climate change. Nature 452 (7190), 987–990.

Kuss, J., Zuelicke, C., Pohl, C., Schneider, B., 2011. Atlantic mercury emission determined from continuous analysis of the elemental mercury sea-air concentration difference within transects between 50°N and 50°S. Glob. Biogeochem. Cycles 25.

Kuypers, M.M.M., Olav Sliekers, A., Lavik, G., Schmid, M., Jørgensen, B.B., Gijs Kuenen, J., Sinninghe, J.S., Damsté, M.S., Jetten, M.S.M., 2003. Anaerobic ammonium oxidation by *Anammox* bacteria in the Black Sea. Nature 422 (6932), 608–611.

Kuypers, M.M.M., Lavik, G., Woebken, D., Schmid, M., Fuchs, B.M., Amann, R., Jørgensen, B.B., Jetten, M.S.M., 2005. Massive nitrogen loss from the Benguela upwelling system through anaerobic ammonium oxidation. Proc. Natl. Acad. Sci. U. S. A. 102 (18), 6478–6483.

Kuzyakov, Y., Hill, P.W., Jones, D.L., 2006. Root exudate components change litter decomposition in a simulated rhizosphere depending on temperature. Plant Soil 290 (1–2), 293–305.

Kuzyakov, Y., Horwath, W.R., Dorodnikov, M., Blagodatskaya, E., 2019. Review and synthesis of the effects of elevated atmospheric CO_2 on soil processes: no changes in pools, but increased fluxes and accelerated cycles. Soil Biol. Biochem. 128 (January), 66–78.

Kvenvolden, K., Lawless, J., Pering, K., Peterson, E., Flores, J., Ponnamperuma, C., Kaplan, I.R., Moore, C., 1970. Evidence for extraterrestrial amino-acids and hydrocarbons in the Murchison meteorite. Nature 228 (5275), 923–926.

Kwok, R., Rothrock, D.A., 2009. Decline in Arctic Sea ice thickness from submarine and ICESat records: 1958–2008. Geophys. Res. Lett. 36.

Kwon, E.Y., Kim, G., Primeau, F., Moore, W.S., Cho, H.-m., DeVries, T., Sarmiento, J.L., Charette, M.A., Cho, Y.-k., 2014. Global estimate of submarine groundwater discharge based on an observationally constrained radium isotope model. Geophys. Res. Lett. 41 (23), 8438–8444.

Laanbroek, H.J., 2010. Methane emission from natural wetlands: interplay between emergent macrophytes and soil microbial processes. A mini-review. Ann. Bot. 105 (1), 141–153.

Labat, D., Godderis, Y., Probst, J.L., Guyot, J.L., 2004. Evidence for global runoff increase related to climate warming. Adv. Water Resour. 27, 631–642.

Labrenz, M., Druschel, G.K., Thomsen-Ebert, T., Gilbert, B., Welch, S.A., Kemner, K.M., Logan, G.A., et al., 2000. Formation of sphalerite (ZnS) deposits in natural biofilms of sulfate-reducing bacteria. Science 290 (5497), 1744–1747.

Lacis, A.A., Schmidt, G.A., Rind, D., Ruedy, R.A., 2010. Atmospheric CO_2: principal control knob governing Earth's temperature. Science 330 (6002), 356–359.

Laclau, J.P., Bouillet, J.P., Ranger, J., Joffre, R., Gouma, R., Saya, A., 2001. Dynamics of nutrient translocation in stemwood across an age series of a *Eucalyptus* hybrid. Ann. Bot. 88, 1079–1092.

Lagage, P.O., Pantin, E., 1994. Dust depletion in the inner disk of Beta Pictoris as a possible indicator of planets. Nature 369, 628–630.

Laganiere, J., Angers, D.A., Pare, D., 2010. Carbon accumulation in agricultural soils after afforestation: a meta-analysis. Glob. Chang. Biol. 16 (1), 439–453.

Lagrange, A.-M., Bonnefoy, M., Chauvin, G., Apai, D., Ehrenreich, D., Boccaletti, A., Gratadour, D., et al., 2010. A giant planet imaged in the disk of the young star Beta Pictoris. Science 329 (5987), 57–59.

Lahteenoja, O., Ruokolainen, K., Schulman, L., Oinonen, M., 2009. Amazonian Peatlands: an ignored C sink and potential source. Glob. Chang. Biol. 15, 2311–2320.

Lajewski, C.K., Mullins, H.T., Patterson, W.P., Callinan, C.W., 2003. Historic calcite record from the Finger Lakes, New York: impact of acid rain on a buffered terrain. Geol. Soc. Am. Bull. 115 (3), 373–384.

Lajtha, K., 1987. Nutrient-reabsorption efficiency and the response to phosphorus fertilization in the desert shrub *Larrea tridentata* (DC) Cov. Biogeochemistry 4, 265–276.

Lajtha, K., Bloomer, S.H., 1988. Factors affecting phosphate sorption and phosphate retention in a desert ecosystem. Soil Sci. 146, 160–167.

Lajtha, K., Jones, J., 2013. Trends in cation, nitrogen, sulfate and hydrogen ion concentrations in precipitation in the United States and Europe from 1978 to 2010: a new look at an old problem. Biogeochemistry 116 (1), 303–334.

Lajtha, K., Schlesinger, W.H., 1986. Plant response to variations in nitrogen availability in a desert Shrubland community. Biogeochemistry 2, 29–37.

Lajtha, K., Whitford, W.G., 1989. The effect of water and nitrogen amendments on photosynthesis, leaf demography, and resource-use efficiency in Larrea Tridentata, a desert Evergreen shrub. Oecologia 80 (3), 341–348.

Lal, D., 1977. The oceanic microcosm of particles. Science 198 (4321), 997–1009.

Lal, R., 2004. Soil carbon sequestration in India. Clim. Chang. 65, 277–296.

Lalonde, K., Mucci, A., Ouellet, A., Gélinas, Y., 2012. Preservation of organic matter in sediments promoted by iron. Nature 483 (7388), 198–200.

Lam, P.J., Bishop, J.K.B., 2008. The continental margin is a key source of iron to the HNLC North Pacific Ocean. Geophys. Res. Lett. 35 (7).

Lamb, D., 1985. The influence of insects on nutrient cycling in eucalypt forests: a beneficial role. Aust. J. Ecol. 10, 1–5.

Lambeck, K., Esat, T.M., Potter, E.-K., 2002. Links between climate and sea levels for the past three million years. Nature 419 (6903), 199–206.

Lambert, R.L., Lang, G.E., Reiners, W.A., 1980. Loss of mass and chemical change in decaying boles of a subalpine balsam fir forest. Ecology 61, 1460–1473.

Lambert, F., Delmonte, B., Petit, J.R., Bigler, M., Kaufmann, P.R., Hutterli, M.A., Stocker, T.F., Ruth, U., Steffensen, J.P., Maggi, V., 2008. Dust-climate couplings over the past 800,000 years from the EPICA dome C ice core. Nature 452 (7187), 616–619.

Lamborg, C.H., Hammerschmidt, C.R., Bowman, K.L., Swarr, G.J., Munson, K.M., Ohnemus, D.C., Lam, P.J., Heimbürger, L.-E., Rijkenberg, M.J.A., Saito, M.A., 2014. A global ocean inventory of anthropogenic mercury based on water column measurements. Nature 512 (7512), 65–68.

Lamers, L.P.M.P.M., Tomassen, H.B.M.B.M., Roelofs, J.G.M.G.M., 1998. Sulfate-induced entrophication and phytotoxicity in freshwater wetlands. Environ. Sci. Technol. 32 (2), 199–205.

Lampitt, R.S., Achterberg, E.P., Anderson, T.R., Hughes, J.A., Iglesias-Rodriguez, M.D., Kelly-Gerreyn, B.A., Lucas, M., et al., 2008. Ocean fertilization: a potential means of geoengineering? Philos. Trans. Ser. A Math. Phys. Eng. Sci. 366 (1882), 3919–3945.

Lana, A., Bell, T.G., Simo, R., Vallina, S.M., Ballabrera-Poy, J., Kettle, A.J., Dachs, J., et al., 2011. An updated climatology of surface dimethylsulfide concentrations and emission fluxes in the global ocean. Glob. Biogeochem. Cycles 25.

Land, P.E., Shutler, J.D., Bell, T.G., Yang, M., 2014. Exploiting satellite earth observation to quantify current global oceanic DMS flux and its future climate sensitivity. J. Geophys. Res. C: Oceans 119 (11), 7725–7740.

Landais, A., Lathiere, J., Barkan, E., Luz, B., 2007. Reconsidering the change in global biosphere productivity between the last glacial maximum and present day from the triple oxygen isotopic composition of air trapped in ice cores. Glob. Biogeochem. Cycles 21.

Landolfi, A., Koeve, W., Dietze, H., Kähler, P., Oschlies, A., 2015. A new perspective on environmental controls of marine nitrogen fixation. Geophys. Res. Lett. 42 (11), 4482–4489.

Landry, J.-S., Damon Matthews, H., 2017. The global pyrogenic carbon cycle and its impact on the level of atmospheric CO_2 over past and future centuries. Glob. Chang. Biol. 23 (8), 3205–3218.

Landschützer, P., Gruber, N., Bakker, D.C.E., Schuster, U., 2014. Recent variability of the global ocean carbon sink. Glob. Biogeochem. Cycles 28 (9), 927–949.

Landschützer, P., Gruber, N., Bakker, D.C.E., 2016. Decadal variations and trends of the global ocean carbon sink. Glob. Biogeochem. Cycles.

Lane, T.W., Saito, M.A., George, G.N., Pickering, I.J., Prince, R.C., Morel, F.M.M., 2005. Biochemistry: a cadmium enzyme from a marine diatom. Nature 435 (7038), 42.

Lang, G.E., Forman, R.T., 1978. Detrital dynamics in a mature oak forest: Hutcheson memorial forest, New Jersey. Ecology 59, 580–595.

Lang, S.Q., Butterfield, D.A., Schulte, M., Kelley, D.S., Lilley, M.D., 2010. Elevated concentrations of formate, acetate and dissolved organic carbon found at the lost city hydrothermal field. Geochim. Cosmochim. Acta 74, 941–952.

Lang, F., Krüger, J., Amelung, W., Willbold, S., Frossard, E., Bünemann, E.K., Bauhus, J., et al., 2017. Soil phosphorus supply controls P nutrition strategies of beech forest ecosystems in Central Europe. Biogeochemistry 136 (1), 5–29.

Lange, O.L., Schulze, E.-D., Evenari, M., Kappen, L., Buschbom, U., 1974. The temperature-related photosynthetic capacity of plants under desert conditions. Oecologia 17, 97–110.

Langenfelds, R.L., Francey, R.J., Steele, L.P., Battle, M., Keeling, R.F., Budd, W.F., 1999. Partitioning of the global fossil CO2 sink using a 19-year trend in atmospheric O_2. Geophys. Res. Lett. 26, 1897–1900.

Langford, A.O., Fehsenfeld, F.C., 1992. Natural vegetation as a source or sink for atmospheric ammonia: a case study. Science 255 (5044), 581–583.

Langley, J.A., Megonigal, J.P., 2010. Ecosystem response to elevated CO_2 levels limited by nitrogen-induced plant species shift. Nature 466, 96–99.

Langley, J.A., McKee, K.L., Cahoon, D.R., Cherry, J.A., Megonigal, J.P., 2009. Elevated CO_2 stimulates marsh elevation gain, counterbalancing sea-level rise. Proc. Natl. Acad. Sci. U. S. A. 106 (15), 6182–6186.

Langley-Turnbaugh, S.J., Bockheim, J.G., 1998. Mass balance of soil evolution on late quaternary marine terraces in coastal Oregon. Geoderma 84, 265–288.

Langmann, B., Folch, A., Hensch, M., Matthias, V., 2012. Volcanic ash over Europe during the eruption of Eyjafjallajokull on Iceland, April-May 2010. Atmos. Environ. 48, 1–8.

Langmuir, C.H., Broecker, W., 2012. How to Build a Habitable Planet: The Story of Earth from the Big Bang to Humankind. REV–Revi. Princeton University Press.

Langner, J., Rodhe, H., Crutzen, P.J., Zimmermann, P., 1992. Anthropogenic influence on the distribution of tropospheric sulfate aerosol. Nature 359, 712–716.

Langway, C.C., Osada, K., Clausen, H.B., Hammer, C.U., Shoji, H., Mitani, A., 1994. New chemical stratigraphy over the last millennium for Byrd Station, Antarctica. Tellus B—Chem. Phys. Meteorol. 46, 40–51.

Langway, C.C., Osada, K., Clausen, H.B., Hammer, C.U., Shoji, H., 1995. A 10-century comparison of prominent bipolar volcanic events in ice cores. J. Geophys. Res.-Atmos. 100, 16241–16247.

Laothawornkitkul, J., Taylor, J.E., Paul, N.D., Nicholas Hewitt, C., 2009. Biogenic volatile organic compounds in the earth system. New Phytol. 183 (1), 27–51.

Lapenis, A.G., Lawrence, G.B., Andreev, A.A., Bobrov, A.A., Torn, M.S., Harden, J.W., 2004. Acidification of forest soil in Russia: from 1893 to present. Glob. Biogeochem. Cycles 18 (1).

Laronne, J.B., Reid, I., 1993. Very high rates of bedload sediment transport by ephemeral desert rivers. Nature 366, 148–150.

Lashof, D.A., Ahuja, D.R., 1990. Relative contributions of greenhouse gas emissions to global warming. Nature 344, 529–531.

Laskowski, R., Niklinska, M., Maryanski, M., 1995. The dynamics of chemical elements in forest litter. Ecology 76, 1393–1406.

Lassaletta, L., Billen, G., Grizzetti, B., Garnier, J., Leach, A.M., Galloway, J.N., 2014. Food and feed trade as a driver in the global nitrogen cycle: 50-year trends. Biogeochemistry 118 (1), 225–241.

Latimer, J.S., Charpentier, M.A., 2010. Nitrogen inputs to seventy-four southern New England estuaries: application of a watershed nitrogen loading model. Estuar. Coast. Shelf Sci. 89, 125–136.

Laudelout, H., Robert, M., 1994. Biogeochemistry of calcium in a broad-leaved forest ecosystem. Biogeochemistry 27, 1–21.

Lauenroth, W.K., Whitman, W.C., 1977. Dynamics of dry matter production in a mixed-grass Prairie in Western North Dakota. Oecologia 27 (4), 339–351.

Lauerwald, R., Regnier, P., Figueiredo, V., Enrich-Prast, A., Bastviken, D., Lehner, B., Maavara, T., Raymond, P., 2019. Natural lakes are a minor global source of N_2O to the atmosphere. Glob. Biogeochem. Cycles 9, 465.

Laurmann, J.A., 1979. Market penetration characteristics for energy production and atmospheric carbon dioxide growth. Science 205 (4409), 896–898.

Laursen, K.K., Hobbs, P.V., Radke, L.F., Rasmussen, R.A., 1992. Some trace gas emissions from North American biomass fires with an assessment of regional and global fluxes from biomass burning. J. Geophys. Res.-Atmos. 97, 20687–20701.

Laverman, A.M., Pallud, C., Abell, J., Van Cappellen, P., 2012. Comparative survey of potential nitrate and sulfate reduction rates in aquatic sediments. Geochim. Cosmochim. Acta 77, 474–488.

Lavoie, R.A., Amyot, M., Lapierre, J.-f., 2019. Global meta-analysis on the relationship between mercury and dissolved organic carbon in freshwater environments. J. Geophys. Res. Biogeosci. 124 (6), 1508–1523.

Law, C.S., Ling, R.D., 2001. Nitrous oxide flux and response to increased iron availability in the Antarctic circumpolar current. Deep-Sea Res. Part II—Top. Stud. Oceanogr. 48, 2509–2527.

Law, C.S., Owens, N.J.P., 1990. Significant flux of atmospheric nitrous oxide from the Northwest Indian Ocean. Nature 346, 826–828.

Law, C.S., Rees, A.P., Owens, N.J.P., 1991. Temporal variability of denitrification in estuarine sediments. Estuar. Coast. Shelf Sci. 33, 37–56.

Law, B.E., Williams, M., Anthoni, P.M., Baldocchi, D.D., Unsworth, M.H., 2000. Measuring and modelling seasonal variation of carbon dioxide and water vapour exchange of a *Pinus ponderosa* forest subject to soil water deficit. Glob. Chang. Biol. 6, 613–630.

Law, B.E., Falge, E., Gu, L., Baldocchi, D.D., Bakwin, P., Berbigier, P., Davis, K., et al., 2002. Environmental controls over carbon dioxide and water vapor exchange of terrestrial vegetation. Agric. For. Meteorol. 113, 97–120.

Law, B.E., Sun, O.J., Campbell, J., Van Tuyl, S., Thornton, P.E., 2003. Changes in carbon storage and fluxes in a chronosequence of ponderosa pine. Glob. Chang. Biol. 9, 510–524.

Lawless, J.G., Levi, N., 1979. The role of metal ions in chemical evolution: polymerization of alanine and glycine in a cation-exchanged clay environment. J. Mol. Evol. 13 (4), 281–286.

Lawrence, D., Schlesinger, W.H., 2001. Changes in soil phosphorus during 200 years of shifting cultivation in Indonesia. Ecology 82, 2769–2780.

Lawrence, G.B., Hazlett, P.W., Fernandez, I.J., Ouimet, R., Bailey, S.W., Shortle, W.C., Smith, K.T., Antidormi, M.R., 2015. Declining acidic deposition begins reversal of forest-soil acidification in the northeastern U.S. and Eastern Canada. Environ. Sci. Technol. 49 (22), 13103–13111.

Laws, E.A., Falkowski, P.G., Smith, W.O., Ducklow, H., McCarthy, J.J., 2000. Temperature effects on export production in the open ocean. Glob. Biogeochem. Cycles 14, 1231–1246.

Lawson, D.R., Winchester, J.W., 1979. Standard crustal aerosol as a reference for elemental enrichment factors. Atmos. Environ. 13, 925–930.

Laxon, S.W., Giles, K.A., Ridout, A.L., Wingham, D.J., Willatt, R., Cullen, R., Kwok, R., et al., 2013. CryoSat-2 estimates of Arctic Sea ice thickness and volume: CRYOSAT-2 sea ice thickness and volume. Geophys. Res. Lett. 40 (4), 732–737.

Le Mer, J., Roger, P., 2001. Production, oxidation, emission and consumption of methane by soils: a review. Eur. J. Soil Biol. 37 (1), 25–50.

Le Quéré, C., Rödenbeck, C., Buitenhuis, E.T., Conway, T.J., Langenfelds, R., Gomez, A., Labuschagne, C., et al., 2007. Saturation of the southern ocean CO_2 sink due to recent climate change. Science 316 (5832), 1735–1738.

Le Quéré, C., Andrew, R.M., Friedlingstein, P., Sitch, S., Pongratz, J., Manning, A.C., Korsbakken, J.I., et al., 2018. Global carbon budget 2017. Earth Syst. Sci. Data Discus. 10 (1), 405–448.

Le Toan, T., Quegan, S., Davidson, M.W.J., Balzter, H., Paillou, P., Papathanassiou, K., et al., 2011. The biomass mission: mapping global forest biomass to better understand the terrestrial carbon cycle. Remote Sens. Environ. 115, 2850–2860.

Leahey, A., 1947. Characteristics of soils adjacent to the MacKenzie river in the Northwest territories of Canada. Proc. Soil Sci. Soc. Am. 12, 458–461.

Lean, D.R., 1973. Phosphorus dynamics in lake water. Science 179 (4074), 678–680.

Lean, J., Warrilow, D.A., 1989. Simulation of the regional climatic impact of Amazon deforestation. Nature 342, 411–413.

LeBauer, D.S., Treseder, K.K., 2008. Nitrogen limitation of net primary productivity in terrestrial ecosystems is globally distributed. Ecology 89 (2), 371–379.

Leblans, N.I.W., Sigurdsson, B.D., Roefs, P., Thuys, R., Magnússon, B., Janssens, I.A., 2014. Effects of seabird nitrogen input on biomass and carbon accumulation after 50 years of primary succession on a Young Volcanic Island, Surtsey. Biogeosci. Discuss. 11 (5), 6269–6302.

Lebo, M.E., 1991. Particle-bound phosphorus along an urbanized coastal plain estuary. Mar. Chem. 34, 225–246.

Lechowicz, M.J., Bell, G., 1991. The ecology and genetics of fitness in forest plants. II. Microspatial heterogeneity of the edaphic environment. J. Ecol. 79, 687–696.

Leckie, S.E., Prescott, C.E., Grayston, S.J., Neufeld, J.D., Mohn, W.W., 2004. Comparison of chloroform fumigation-extraction, phospholipid fatty acid, and DNA methods to determine microbial biomass in forest humus. Soil Biol. Biochem. 36, 529–532.

Lecuyer, C., Simon, L., Guyot, F., 2000. Comparison of carbon, nitrogen and water budgets on venus and the earth. Earth Planet. Sci. Lett. 181, 33–40.

Ledwell, J.R., Watson, A.J., Law, C.S., 1993. Evidence for slow mixing across the pycnocline from an open-ocean tracer-release experiment. Nature 364, 701–703.

Lee, K., 2001. Global net community production estimated from the annual cycle of surface water total dissolved inorganic carbon. Limnol. Oceanogr. 46, 1287–1297.

Lee, C.-S., Fisher, N.S., 2019. Microbial generation of elemental mercury from dissolved methylmercury in seawater. Limnol. Oceanogr. 64 (2), 679–693.

Lee, J.A., Stewart, G.R., 1978. Ecological aspects of nitrogen assimilation. Adv. Bot. Res. 6, 1–43.

Lee, D.H., Granja, J.R., Martinez, J.A., Severin, K., Ghadiri, M.R., 1996. A self-replicating peptide. Nature 382 (6591), 525–528.

Lee, D.C., Halliday, A.N., Snyder, G.A., Taylor, L.A., 1997. Age and origin of the Moon. Science 278, 1098–1103.

Lee, S.J., Seo, Y.C., Jang, H.N., Park, K.S., Baek, J.I., An, H.S., Song, K.C., 2006. Speciation and mass distribution of mercury in a bituminous coal-fired power plant. Atmos. Environ. 40, 2215–2224.

Lee, K., Kim, T.-W., Byrne, R.H., Millero, F.J., Feely, R.A., Liu, Y.-M., 2010. The universal ratio of boron to chlorinity for the North Pacific and North Atlantic oceans. Geochim. Cosmochim. Acta 74 (6), 1801–1811.

Lee, C., Martin, R.V., van Donkelaar, A., Lee, H., Dickerson, R.R., Hains, J.C., Krotkov, N., Richter, A., Vinnikov, K., Schwab, J.J., 2011. SO_2 emissions and lifetimes: estimates from inverse modeling using in situ and global, space-based (SCIAMACHY and OMI) observations. J. Geophys. Res.-Atmos. 116.

Legendre, L., 1998. Flux of particulate organic material from the euphotic zone of oceans: estimation from phytoplankton biomass. J. Geophys. Res. Oceans 103, 2897–2903.

Legrand, M., Delmas, R.J., 1987. A 220-year continuous record of volcanic H_2SO_4 in the Antarctic ice sheet. Nature 327, 671–676.

Legrand, M., Fenietsaigne, C., Saltzman, E.S., Germain, C., Barkov, N.I., Petrov, V.N., 1991. Ice-core record of oceanic emissions of dimethylsulfide during the last climate cycle. Nature 350, 144–146.

Legrand, M., Deangelis, M., Staffelbach, T., Neftel, A., Stauffer, B., 1992. Large perturbations of ammonium and organic acids content in the summit Greenland ice core. Fingerprint from forest fires. Geophys. Res. Lett. 19, 473–475.

Lehman, J.T., 1980. Lower Food Web Dynamics. Cycling of P among Phytoplankton, Herbivorous Zooplankton, and the Lake Water. Special Report of the Great Lakes Research Division. University of Michigan Great Lakes and Marine Waters Center.

Lehman, J.T., 1988. Hypolimnetic metabolism in Lake Washington—relative effects of nutrinet load and food web structure on lake productivity. Limnol. Oceanogr. 33 (6), 1334–1347.

Lehmann, M.F., Sigman, D.M., McCorkle, D.C., Granger, J., Hoffmann, S., Cane, G., Brunelle, B.G., 2007. The distribution of nitrate $^{15}N/^{14}N$ in marine sediments and the impact of benthic nitrogen loss on the isotopic composition of oceanic nitrate. Geochim. Cosmochim. Acta 71, 5384–5404.

Lehner, B., Doll, P., 2004. Development and validation of a global database of lakes, reservoirs and wetlands. J. Hydrol. 296 (1–4), 1–22.

Lehner, B., Verdin, K., Jarvis, A., 2008. New global hydrography derived from spaceborne elevation data. Eos Trans. 89 (10), 93–94.

Lehner, B., Liermann, C.R., Revenga, C., Vorosmarty, C., Fekete, B., Crouzet, P., et al., 2011. High-resolution mapping of the world's reservoirs and dams for sustainable river-flow management. Front. Ecol. Environ. 9, 494–502.

Lehner, F., Wood, A.W., Vano, J.A., Lawrence, D.M., Clark, M.P., Mankin, J.S., 2019. The potential to reduce uncertainty in regional runoff projections from climate models. Nat. Clim. Chang. 9 (12), 926–933.

LeHouerou, H.N., 1984. Rain use efficiency: a unifying concept in arid-land ecology. J. Arid Environ. 7.

Leifer, I., Melton, C., Tratt, D.M., Buckland, K.N., Clarisse, L., Coheur, P., Frash, J., et al., 2017. Remote sensing and in situ measurements of methane and ammonia emissions from a megacity dairy complex: Chino, CA. Environ. Pollut. 221 (February), 37–51.

Leighton, R.B., Murray, B.C., 1966. Behavior of carbon dioxide and other volatiles on Mars. Science 153 (3732), 136–144.

Lein, A.Y., 1984. Anaerobic consumption of organic matter in modern marine sediments. Nature 312, 148–150.

Leininger, S., Urich, T., Schloter, M., Schwark, L., Qi, J., Nicol, G.W., Prosser, J.I., Schuster, S.C., Schleper, C., 2006. Archaea predominate among ammoniaoxidizing prokaryotes in soils. Nature 442, 806–809.

Lelieveld, J., Crutzen, P.J., 1990. Influences of cloud photochemical processes on tropospheric ozone. Nature 343, 227–233.

Lelieveld, J., Crutzen, P.J., Ramanathan, V., Andreae, M.O., Brenninkmeijer, C.M., Campos, T., Cass, G.R., et al., 2001. The Indian Ocean experiment: widespread air pollution from south and Southeast Asia. Science 291 (5506), 1031–1036.

Lelieveld, J., van Aardenne, J., Fischer, H., de Reus, M., Williams, J., Winkler, P., 2004. Increasing ozone over the Atlantic Ocean. Science 304 (5676), 1483–1487.

Lelieveld, J., Evans, J.S., Fnais, M., Giannadaki, D., Pozzer, A., 2015. The contribution of outdoor air pollution sources to premature mortality on a global scale. Nature 525 (7569), 367–371.

Leman, L., Orgel, L., Reza Ghadiri, M., 2004. Carbonyl sulfide-mediated prebiotic formation of peptides. Science 306 (5694), 283–286.

Lengke, M., Southam, G., 2006. Bioaccumulation of gold by sulfate-reducing bacteria cultured in the presence of gold (I)-thio sulfate complex. Geochim. Cosmochim. Acta 70, 3646–3661.

Lenhart, K., Althoff, F., Greule, M., Keppler, F., 2015. Technical note: methionine, a precursor of methane in living plants. Biogeosciences 12 (March), 1907–1914.

Lenton, T.M., 1998. Gaia and natural selection. Nature 394 (6692), 439–447.

Lenton, T.M., 2001. The role of land plants, phosphorus weathering and fire in the rise and regulation of atmospheric oxygen. Glob. Chang. Biol. 7, 613–629.

Leopold, L.B., Wolman, M.G., Miller, J.R., 1964. Fluvial Processes in Geomorphology. W.H. Freeman.

Leri, A.C., Myneni, S.C.B., 2010. Organochlorine turnover in forest ecosystems: the missing link in the terrestrial chlorine cycle. Glob. Biogeochem. Cycles 24.

Lerner, J., Matthews, E., Fung, I., 1988. Methane emission from animals: a global high-resolution data base. Glob. Biogeochem. Cycles 2, 139–156.

Lesack, L.F.W., Melack, J.M., 1991. The deposition, composition, and potential sources of major ionic solutes in rain of the Central Amazon Basin. Water Resour. Res. 27, 2953–2977.

Lesack, L.F.W., Melack, J.M., 1996. Mass balance of major solutes in a rainforest catchment in the Central Amazon: implications for nutrient budgets in tropical rainforests. Biogeochemistry 32, 115–142.

Leschine, S.B., Holwell, K., Canale-Parola, E., 1988. Nitrogen fixation by anaerobic cellulolytic bacteria. Science 242 (4882), 1157–1159.

Leshin, L.A., Mahaffy, P.R., Webster, C.R., Cabane, M., Coll, P., Conrad, P.G., Archer Jr., P.D., et al., 2013. Volatile, isotope, and organic analysis of Martian fines with the Mars curiosity rover. Science 341 (6153), 1238937.

Lesschen, J.P., Velthof, G.L., de Vries, W., Kros, J., 2011. Differentiation of nitrous oxide emission factors for agricultural soils. Environ. Pollut. 159 (11), 3215–3222.

Leuenberger, M., Siegenthaler, U., 1992. Ice age atmospheric concentration of nitrous oxide from an Antarctic ice core. Nature 360, 449–451.

Lever, M.A., Rouxel, O., Alt, J.C., Shimizu, N., Ono, S., Coggon, R.M., Shanks 3rd, W.C., et al., 2013. Evidence for microbial carbon and sulfur cycling in deeply buried ridge flank basalt. Science 339 (6125), 1305–1308.

Levesque, C., Limen, H., Juniper, S.K., 2005. Origin, composition and nutritional quality of particulate matter at deepsea hydrothermal vents on axial volcano, NE Pacific. Mar. Ecol.—Prog. Ser. 289, 43–52.

Levia, D.F., Germer, S., 2015. A review of stemflow generation dynamics and stemflow-environment interactions in forests and shrublands: stemflow review. Rev. Geophys. Pap. 349 53 (3), 673–714.

Levin, S.A., 1989. Challenges in the development of a theory of community and ecosystem structure and function. In: Roughgarden, R.M.M.J., Levin, S.A. (Eds.), Perspectives in Ecological Theory. Princeton University Press, pp. 242–255.

Levine, S.N., Stainton, M.P., Schindler, D.W., 1986. A radiotracer study of phosphorus cycling in a eutrophic Canadian shield lake, Lake-227, northwestern Ontario. Can. J. Fish. Aquat. Sci. 43, 366–378.

Levine, J.S., Cofer, W.R., Pinto, J.P., 1993. Biomass burning. In: Khalil, M.A.K. (Ed.), Atmospheric Methane: Sources, Sinks and Role in Global Change. Springer, pp. 299–313.

Levitus, S., Antonov, J.I., Wang, J., Delworth, T.L., Dixon, K.W., Broccoli, A.J., 2001. Anthropogenic warming of Earth's climate system. Science 292 (5515), 267–270.

Levitus, S., Antonov, J., Boyer, T., 2005. Warming of the world ocean, 1955–2003. Geophys. Res. Lett. 32.

Levy, H., Moxim, W., Klonecki, A., Kasibhatla, P.S., 1999. Simulated tropospheric NOx: its evaluation, global distribution and individual source contributions. J. Geophys. Res. 104, 26279–26306.

Lewicka-Szczebak, D., Well, R., Köster, J.R., Fuß, R., Senbayram, M., Dittert, K., Flessa, H., 2014. Experimental determinations of isotopic fractionation factors associated with N_2O production and reduction during denitrification in soils. Geochim. Cosmochim. Acta 134 (June), 55–73.

Lewis, W.M., 1974. Effects of fire on nutrient movement in a South Carolina pine forest. Ecology 55, 1120–1127.

Lewis, W.M., 1981. Precipitation chemistry and nutrient loading by precipitation in a tropical watershed. Water Resour. Res. 17, 169–181.

Lewis, W.M., 1986. Nitrogen and phosphorus runoff losses from a nutrient-poor tropical moist forest. Ecology 67, 1275–1282.

Lewis, W.M., 1988. Primary production in the Orinoco river. Ecology 69, 679–692.

Lewis, W.M., 2002. Yield of nitrogen from minimally disturbed watersheds of the United States. Biogeochemistry 57, 375–385.

Lewis, W.M., Grant, M.C., 1979. Relationships between stream discharge and yield of dissolved substances from a Colorado Mountain watershed. Soil Sci. 128, 353–363.

Lewis Jr., W.M., Wurtsbaugh, W.A., 2008. Control of lacustrine phytoplankton by nutrients: erosion of the phosphorus paradigm. Int. Rev. Hydrobiol. 93, 446–465.

Lewis, J.S., Prinn, R.G., 1984. Planets and their Atmospheres. Academic Press.

Lewis, R., Schwartz, E., 2004. Sea Salt Aerosol Production: Mechanisms, Methods, Measurements and Models—A Critical Review. vol. 152. Geophysical Monograph Series, American Geophysical Union, Washington, DC.

Lewis, M.R., Hebert, D., Harrison, W.G., Platt, T., Oakey, N.S., 1986. Vertical nitrate fluxes in the oligotrophic ocean. Science 234 (4778), 870–873.

Lewis, W.M., Hamilton, S.K., Jones, S.L., Runnels, D.D., 1987. Major element chemistry, weathering and element yields for the Caura River drainage, Venezuela. Biogeochemistry 4, 159–181.

Lewis, S.L., Lopez-Gonzalez, G., Sonké, B., Affum-Baffoe, K., Baker, T.R., Ojo, L.O., Phillips, O.L., et al., 2009. Increasing carbon storage in intact African tropical forests. Nature 457 (7232), 1003–1006.

Lewis, W.M., Wurtsbaugh, W.A., Paerl, H.W., 2011. Rationale for control of anthropogenic nitrogen and phosphorus to reduce eutrophication of inland waters. Environ. Sci. Technol. 45 (24), 10300–10305.

Li, Y.-H., 1972. Geochemical mass balance among lithosphere, hydrosphere, and atmosphere. Am. J. Sci. 272, 119–137.

Li, Y.H., 1981. Geochemical cycles of elements and human perturbation. Geochim. Cosmochim. Acta 45, 2073–2084.

Li, Z., Cassar, N., 2017. A mechanistic model of an upper bound on oceanic carbon export as a function of mixed layer depth and temperature. Biogeosciences 14 (22), 5015–5027.

Li, Y., Keppler, H., 2014. Nitrogen speciation in mantle and crustal fluids. Geochim. Cosmochim. Acta 129 (March), 13–32.

Li, Y.-H., Takahashi, T., Broecker, W.S., 1969. The degree of saturation of $CaCO_3$ in the oceans. J. Geophys. Res. 74, 5507–5525.

Li, W.K., Rao, D.V., Harrison, W.G., Smith, J.C., Cullen, J.J., Irwin, B., Platt, T., 1983. Autotrophic picoplankton in the tropical ocean. Science 219 (4582), 292–295.

Li, Z.P., Han, F.X., Su, Y., Zhang, T.L., Sun, B., Monts, D.L., Plodinec, M.J., 2007. Assessment of soil organic and carbonate carbon storage in China. Geoderma 138, 119–126.

Li, S., Matthews, J., Sinha, A., 2008. Atmospheric hydroxyl radical production from electronically excited NO_2 and H_2O. Science 319 (5870), 1657–1660.

Li, D., Niu, S., Luo, Y., 2012. Global patterns of the dynamics of soil carbon and nitrogen stocks following afforestation: a meta-analysis. New Phytol. 195 (1), 172–181.

Li, W., Czaja, A.D., Van Kranendonk, M.J., Beard, B.L., Roden, E.E., Johnson, C.M., 2013. An anoxic, Fe(II)-rich, U-poor ocean 3.46 billion years ago. Geochim. Cosmochim. Acta 120 (November), 65–79.

Li, P., Yang, Y., Han, W., Fang, J., 2014. Global patterns of soil microbial nitrogen and phosphorus stoichiometry in forest ecosystems: stoichiometric patterns in microbial system. Glob. Ecol. Biogeogr.: J. Macroecol. 23 (9), 979–987.

Li, Y., Schichtel, B.A., Walker, J.T., Schwede, D.B., Chen, X., Lehmann, C.M.B., Puchalski, M.A., Gay, D.A., Collett Jr., J.L., 2016a. Increasing importance of deposition of reduced nitrogen in the United States. Proc. Natl. Acad. Sci. U. S. A. 113 (21), 5874–5879.

Li, Y., Ma, C., Cheng, Z., Huang, R., Zheng, C., 2016b. Historical anthropogenic contributions to mercury accumulation recorded by a peat core from Dajiuhu Montane Mire, Central China. Environ. Pollut. 216 (September), 332–339.

Li, Z., Liu, L., Chen, J., Henry Teng, H., 2016c. Cellular dissolution at hypha- and spore-mineral interfaces revealing unrecognized mechanisms and scales of fungal weathering. Geology 44 (4), 319–322.

Li, W., Ciais, P., Peng, S., Yue, C., Wang, Y., Thurner, M., Saatchi, S.S., et al., 2017. Land-use and land-cover change carbon emissions between 1901 and 2012 constrained by biomass observations. Biogeosciences 14 (22), 5053–5067.

Li, S., Lucey, P.G., Milliken, R.E., Hayne, P.O., Fisher, E., Williams, J.-P., Hurley, D.M., Elphic, R.C., 2018a. Direct evidence of surface exposed water ice in the lunar polar regions. Proc. Natl. Acad. Sci. U. S. A. 115 (36), 8907–8912.

Li, X., Xiao, J., He, B., 2018b. Chlorophyll fluorescence observed by OCO-2 is strongly related to gross primary productivity estimated from flux towers in temperate forests. Remote Sens. Environ. 204 (January), 659–671.

Li, L., Ni, J., Chang, F., Yao, Y., Frolova, N., Magritsky, D., Borthwick, A., et al., 2019a. September. Global Trends in Water and Sediment Fluxes of the World's Large Rivers. Science Bulletin of the Faculty of Agriculture. Kyushu University.

Li, S., Li, W., Beard, B.L., Raymo, M.E., Wang, X., Yang, C., Chen, J., 2019b. K isotopes as a tracer for continental weathering and geological K cycling. Proc. Natl. Acad. Sci. U. S. A. 116 (18), 8740–8745.

Li, W., Zhang, H., Huang, G., Liu, R., Wu, H., Zhao, C., McDowell, N., 2019c. Effects of nitrogen enrichment on tree carbon allocation: a global synthesis. Glob. Ecol. Biogeogr.: J. Macroecol, 1–17.

Li, X., Ding, Y., Hood, E., Raiswell, R., Han, T., He, X., Kang, S., et al., 2019d. Dissolved iron supply from Asian glaciers: local controls and a regional perspective. Glob. Biogeochem. Cycles 33 (10), 1223–1237.

Li, Z., Tian, D., Wang, B., Wang, J., Wang, S., Chen, H.Y.H., Xu, X., Wang, C., He, N., Niu, S., 2019e. Microbes drive global soil nitrogen mineralization and availability. Glob. Chang. Biol. 25, 1078–1088.

Liang, J.Y., Horowitz, L.W., Jacob, D.J., Wang, Y.H., Fiore, A.M., Logan, J.A., Gardner, G.M., Munger, J.W., 1998. Seasonal budgets of reactive nitrogen species and ozone over the United States, and export fluxes to the global atmosphere. J. Geophys. Res. 103, 13435–13450.

Liang, Q., Newman, P.A., Daniel, J.S., Reimann, S., Hall, B.D., Dutton, G., Kuijpers, L.J.M., 2014. Constraining the carbon tetrachloride (CCl_4) budget using its global trend and inter-hemispheric gradient: the carbon tetrachloride budget. Geophys. Res. Lett. 41 (14), 5307–5315.

Liang, J., Crowther, T.W., Picard, N., Wiser, S., Zhou, M., Alberti, G., Schulze, E.-D., et al., 2016. Positive biodiversity-productivity relationship predominant in global forests. Science 354 (6309).

Liang, C., Amelung, W., Lehmann, J., Kaestner, M., 2019. Quantitative assessment of microbial necromass contribution to soil organic matter. Glob. Chang. Biol. 25 (July).

Lichter, J., 1998. Rates of weathering and chemical depletion in soils across a chronosequence of lake Michigan sand dunes. Geoderma 85, 255–282.

Lichter, J., Billings, S.A., Ziegler, S.E., Gaindh, D., Ryals, R., Finzi, A.C., Jackson, R.B., Stemmler, E.A., Schlesinger, W.H., 2008. Soil carbon sequestration in a pine forest after 9 years of atmospheric CO_2 enrichment. Glob. Chang. Biol. 14, 2910–2922.

Lieffers, V.J., Macdonald, S.E., 1990. Growth and foliar nutrient status of black spruce and tamarack in relation to depth of water table in some Alberta Peatlands. Can. J. Forest Res. 20, 805–809.

Liengen, T., 1999. Conversion factor between acetylene reduction and nitrogen fixation in free-living cyanobacteria from high Arctic habitats. Can. J. Microbiol. 45, 223–229.

Lieth, H., 1975. Modeling the primary productivity of the world. In: Lieth, H., Whittaker, R.H. (Eds.), Primary Productivity of the Biosphere. Springer, pp. 237–263.

Likens, G.E., 1975. Primary production of inland water ecosystems. In: Lieth, H., Whittaker, R.H. (Eds.), Primary Productivity of the Biosphere. Springer-Verlag, pp. 185–202.

Likens, G.E., 2013. Biogeochemistry of a Forested Ecosystem. Springer Science & Business Media.

Likens, G.E., Bormann, F.H., 1974a. Linkages between terrestrial and aquatic ecosystems. Bioscience 24 (8), 447–456.

Likens, G.E., Bormann, F.H., 1974b. Acid rain: a serious regional environmental problem. Science 184 (4142), 1176–1179.

Likens, G.E., Bormann, F.H., 1995. Biogeochemistry of a Forested Ecosystem, second ed. Springer.

Likens, G.E., Buso, D.C., 2012. Dilution and the elusive baseline. Environ. Sci. Technol. 46 (8), 4382–4387.

Likens, G.E., Bormann, F.H., Johnson, N.M., 1969. Nitrification: importance to nutrient losses from a cutover forested ecosystem. Science 163 (3872), 1205–1206.

Likens, G.E., Bormann, F.H., Johnson, N.M., Fisher, D.W., Pierce, R.S., 1970. Effects of forest cutting and herbicide treatment on nitrogen budgets in the Hubbard Brook watershed ecosystem. Ecol. Monogr. 40, 23–47.

Likens, G.E., Bormann, F.H., Johnson, N.M., Likens, G.E., 1981. Interactions between major biogeochemical cycles in terrestrial ecosystems. In: Some Perspectives of the Major Biogeochemical Cycles. Wiley, pp. 93–112.

Likens, G.E., Bormann, F.H., Pierce, R.S., Eaton, J.S., Munn, R.E., 1984. Long-term trends in precipitation chemistry at Hubbard Brook, New Hampshire. Atmos. Environ. 18, 2641–2647.

Likens, G.E., Bormann, F.H., Hedin, L.O., Driscoll, C.T., Eaton, J.S., 1990. Dry deposition of sulfur: a 23 year record for the Hubbard Brook Forest ecosystem. Tellus B Chem. Phys. Meteorol. 42, 319–329.

Likens, G.E., Driscoll, C.T., Buso, D.C., Siccama, T.G., Johnson, C.E., Lovett, G.M., Ryan, D.F., Fahey, T., Reiners, W.A., 1994. The biogeochemistry of potassium at Hubbard Brook. Biogeochemistry 25, 61–125.

Likens, G.E., Driscoll, C.T., Buso, D.C., 1996. Long-term effects of acid rain: response and recovery of a forest ecosystem. Science 272, 244–246.

Likens, G.E., Driscoll, C.T., Buso, D.C., Mitchell, M.J., Lovett, G.M., Bailey, S.W., Siccama, T.G., Reiners, W.A., Alewell, C., 2002. The biogeochemistry of sulfur at Hubbard Brook. Biogeochemistry 60, 235–316.

Likens, G.E., Buso, D.C., Butler, T.J., 2005. Long-term relationships between SO_2 and NO_x emissions and SO_4^2 and NO_3 concentration in bulk deposition at the Hubbard Brook experimental forest, NH. J. Environ. Monitor. 7, 964–968.

Limpens, J., Berendse, F., Blodau, C., Canadell, J.G., Freeman, C., Holden, J., Roulet, N., Rydin, H., Schaepman-Strub, G., 2008. Peatlands and the carbon cycle: from local processes to global implications—a synthesis. Biogeosci. Discuss. 5 (October), 1379–1419.

Lin, C.J., Pehkonen, S.O., 1999. The chemistry of atmospheric mercury: a review. Atmos. Environ. 33, 2067–2079.

Lin, B.L., Sakoda, A., Shibasaki, R., Goto, N., Suzuki, M., 2000. Modelling a global biogeochemical nitrogen cycle in terrestrial ecosystems. Ecol. Model. 135, 89–110.

Lin, Y.-P., Singer, P.C., Aiken, G.R., 2005. Inhibition of calcite precipitation by natural organic material: kinetics, mechanism, and thermodynamics. Environ. Sci. Technol. 39 (17), 6420–6428.

Lin, L.-H., Wang, P.-L., Rumble, D., Lippmann-Pipke, J., Boice, E., Pratt, L.M., Lollar, B.S., et al., 2006. Long-term sustainability of a high-energy, low-diversity crustal biome. Science 314 (5798), 479–482.

Lin, P., Chen, M., Guo, L., 2012. Speciation and transformation of phosphorus and its mixing behavior in the bay of St. Louis estuary in the northern Gulf of Mexico. Geochim. Cosmochim. Acta 87, 283–298.

Lindberg, S.E., Garten, C.T., 1988. Sources of sulfur in forest canopy throughfall. Nature 336, 148–151.

Lindberg, S.E., Lovett, G.M., 1985. Field measurements of particle dry deposition rates to foliage and inert surfaces in a forest canopy. Environ. Sci. Technol. 19 (3), 238–244.

Lindberg, S.E., Lovett, G.M., Richter, D.D., Johnson, D.W., 1986. Atmospheric deposition and canopy interactions of major ions in a forest. Science 231 (4734), 141–145.

Lindeboom, H.J., 1984. The nitrogen pathway in a penguin rookery. Ecology 65, 269–277.

Lindell, M.J., Graneli, H.W., Tranvik, L.J., 1996. Effects of sunlight on bacterial growth in lakes of different humic content. Aquatic Microb. Ecol.: Int. J. 11, 135–141.

Lindsay, W.L., 1979. Chemical Equilibria in Soils. Wiley.

Lindsay, W.L., Moreno, E.C., 1960. Phosphate phase equilibria in soils. Proc. Soil Sci. Soc. Am. 24, 177–182.

Lindsay, W.L., Vlek, P.L.G., 1977. Phosphate minerals. In: Dixon, J.B., Weed, S.B. (Eds.), Minerals in Soil Environments. Soil Science Society of America, Madison, Wisconsin, pp. 639–672.

Linkins, A.E., Sinsabaugh, R.L., McClaugherty, C.A., Melills, J.M., 1990. Cellulase activity on decomposing leaf litter in microcosms. Plant Soil 123, 17–25.

Lins, H.F., Slack, J.R., 1999. Streamflow trends in the United States. Geophys. Res. Lett. 26, 227–230.

Liss, P.S., Lovelock, J.E., 2008. Climate change: the effect of DMS emissions. Environ. Chem. 4 (6), 377–378.

Liss, P.S., Slater, P.G., 1974. Flux of gases across the air-sea interface. Nature 247, 181–184.

Litton, C.M., Raich, J.W., Ryan, M.G., 2007. Carbon allocation in forest ecosystems. Glob. Chang. Biol. 13, 2089–2109.

Liu, L., Greaver, T.L., 2009. A review of nitrogen enrichment effects on three biogenic GHGs: the CO_2 sink may be largely offset by stimulated N_2O and CH_4 emission. Ecol. Lett. 12 (10), 1103–1117.

Liu, L., Greaver, T.L., 2010. A global perspective on belowground carbon dynamics under nitrogen enrichment. Ecol. Lett. 13 (7), 819–828.

Liu, K.K., Kaplan, I.R., 1989. The eastern tropical Pacific as a source of ^{15}N-enriched nitrate in seawater off Southern California. Limnol. Oceanogr. 34, 820–830.

Liu, T.S., Gu, X.F., An, Z.S., Fan, Y.X., 1981. The dust fall in Beijing, China on April 18, 1980. In: Pewe, T.L. (Ed.), Desert Dust: Origin, Characteristics, and Effects on Man. Geological Society of America, Boulder, CO, pp. 149–157.

Liu, J., You, L., Amini, M., Obersteiner, M., Herrero, M., Zehnder, A.J.B., Yang, H., 2010. A high-resolution assessment on global nitrogen flows in cropland. Proc. Natl. Acad. Sci. U. S. A. 107 (17), 8035–8040.

Liu, X., Duan, L., Mo, J., Enzai, D., Shen, J., Lu, X., Zhang, Y., Zhou, X., He, C., Zhang, F., 2011. Nitrogen deposition and its ecological impact in China: an overview. Environ. Pollut. 159 (10), 2251–2264.

Liu, R., Li, Y., Wang, Q.-X., 2012. Variations in water and CO_2 fluxes over a saline desert in Western China. Hydrol. Process. 26, 513–522.

Liu, J., Chen, H., Zhu, Q., Shen, Y., Xue, W., Wang, M., Peng, C., 2015. A novel pathway of direct methane production and emission by eukaryotes including plants, animals and fungi: an overview. Atmos. Environ. 115 (August), 26–35.

Liu, S., Lin, F., Wu, S., Cheng, J., Sun, Y., Jin, Y., Li, S., Li, Z., Zou, J., 2017a. A meta-analysis of fertilizer-induced soil NO and combined NO^+ N_2O emissions. Glob. Chang. Biol. 23 (6), 2520–2532.

Liu, W., Yu, L., Zhang, T., Kang, R., Zhu, J., Mulder, J., Huang, Y., Duan, L., 2017b. In situ ^{15}N labeling experiment reveals different long-term responses to ammonium and nitrate inputs in N-saturated subtropical forest: [15]N tracing at TSP. J. Geophys. Res. Biogeosci. 122 (9), 2251–2264.

Liu, S., Huang, S., Xie, Y., Huang, Q., Wang, H., Leng, G., 2019a. Assessing the non-stationarity of low flows and their scale-dependent relationships with climate and human forcing. Sci. Total Environ. 687 (October), 244–256.

Liu, S., Wei, Y., Post, W.M., Cook, R.B., Schaefer, K., Thornton, M.M., 2013. The unified North American soil map and its implication on the soil organic carbon stock in North America. Biogeosciences 10, 2915–2930.

Liu, S., Xie, Z., Zeng, Y., Liu, B., Li, R., Wang, Y., Wang, L., Qin, P., Jia, B., Xie, J., 2019b. Effects of anthropogenic nitrogen discharge on dissolved inorganic nitrogen transport in global rivers. Glob. Change Biol. 25, 1493–1513.

Livingston, G.P., Vitousek, P.M., Matson, P.A., 1988. Nitrous oxide flux and nitrogen transformations across a landscape gradient in Amazonia. J. Geophys. Res. 93, 1593–1599.

Livingstone, D.A., 1963. Chemical composition of rivers and lakes. In: Fleischer, M. (Ed.), Data of Geochemistry. U.S. Geological Survey, Washington, DC.

Llácer, J.L., Fita, I., Rubio, V., 2008. Arginine and nitrogen storage. Curr. Opin. Struct. Biol. 18 (6), 673–681.

Llewellyn, M., 1975. The effects of the lime aphid (*Eucallipterus tiliae* L.) (Aphididae) on the growth of lime (*Tilia* X *vulgaris* Hayne). II. The primary production of saplings and mature trees, the energy drain imposed by the aphid population and revised standard deviations O. J. Appl. Ecol. 12, 15–23.

Lloyd, J., Farquhar, G.D., 1994. [13]C discrimination during CO_2 assimilation by the terrestrial biosphere. Oecologia 99 (3–4), 201–215.

Lloyd, J., Domingues, T.F., Schrodt, F., Ishida, F.Y., Feldpausch, T.R., Saiz, G., Quesada, C.A., et al., 2015. Edaphic, structural and physiological contrasts across Amazon basin forest-savanna ecotones suggest a role for potassium as a key modulator of tropical woody vegetation structure and function. Biogeosciences 12, 6529–6571.

Loaiciga, H.A., Valdes, J.B., Vogel, R., Garvey, J., Schwarz, H., 1996. Global warming and the hydrologic cycle. J. Hydrol. 174, 83–127.

Lobell, D.B., Schlenker, W., Costa-Roberts, J., 2011. Climate trends and global crop production since 1980. Science 333 (6042), 616–620.

Lobert, J.M., Scharffe, D.H., Hao, W.M., Crutzen, P.J., 1990. Importance of biomass burning in the atmospheric budgets of nitrogen-containing gases. Nature 346, 552–554.

Locarnini, R.A., Mishonov, A.V., Antonov, J.I., Boyer, T.P., Garcia, H.E., Baranova, O.K., Zweng, M.M., Paver, C.R., Reagan, J.R., Johnson, D.R., Hamilton, M., Seidov, D., 2013. Temperature. In: Levitus, S., A. Mishonov (Eds.), World Ocean Atlas 2013, Vol. 1. NOAA Atlas NESDIS 73, 40 pp.

Logan, J.A., 1985. Tropospheric ozone—seasonal behavior, trends, and anthropogenic influence. J. Geophys. Res. 90, 10463–10482.

Logan, J.A., Prather, M.J., Wofsy, S.C., McElroy, M.B., 1981. Tropospheric chemistry: a global perspective. J. Geophys. Res. D: Atmos. 86, 7210–7254.

Longanathan, P., Hedley, M.J., Grace, N.D., Lee, J., Cronin, S.J., Bolan, N.S., Zanders, J.M., 2003. Fertiliser contaminants in New Zealand grazed pasture with special reference to cadmium and fluorine: a review. Aust. J. Soil Res. 41, 501–532.

Loh, A.N., Bauer, J.E., 2000. Distribution, partitioning and fluxes of dissolved and particulate organic C, N and P in the eastern North Pacific and southern oceans. Deep-Sea Res. Part I—Oceanogr. Res. Pap. 47, 2287–2316.

Loh, A.N., Bauer, J.E., Druffel, E.R.M., 2004. Variable ageing and storage of dissolved organic components in the open ocean. Nature 430 (7002), 877–881.

Lohrenz, S.E., Fahnenstiel, G.L., Redalje, D.G., Lang, G.A., Dagg, M.J., Whitledge, T.E., Dortch, Q., 1999. Nutrients, irradiance, and mixing as factors regulating primary production in coastal waters impacted by the Mississippi river plume. Cont. Shelf Res. 19, 1113–1141.

Lohrer, A.M., Thrush, S.F., Gibbs, M.M., 2004. Bioturbators enhance ecosystem function through complex biogeochemical interactions. Nature 431 (7012), 1092–1095.

Lohrmann, R., Orgel, L.E., 1973. Prebiotic activation processes. Nature 244 (5416), 418–420.

Lohse, K.A., Brooks, P.D., McIntosh, J.C., Meixner, T., Huxman, T.E., 2009. Interactions between biogeochemistry and hydrologic systems. Annu. Rev. Environ. Resour. 34 (1), 65–96.

Loladze, I., Elser, J.J., 2011. The origins of the redfield nitrogen-to-phosphorus ratio are in a homoeostatic protein-to-rRNA ratio. Ecol. Lett. 14 (3), 244–250.

Lollar, B.S., Onstott, T.C., Lacrampe-Couloume, G., Ballentine, C.J., 2014. The contribution of the Precambrian continental lithosphere to global H_2 production. Nature 516 (7531), 379–382.

Long, S.P., Ainsworth, E.A., Leakey, A.D.B., Nösberger, J., Ort, D.R., 2006. Food for thought: lower-than-expected crop yield stimulation with rising CO_2 concentrations. Science 312 (5782), 1918–1921.

Longmore, M.E., Oleary, B.M., Rose, C.W., Chandica, A.L., 1983. Mapping soil erosion and accumulation with the fallout isotope Cesium-137. Aust. J. Soil Res. 21, 373–385.

Lonsdale, W.M., 1988. Predicting the amount of litterfall in forests of the world. Ann. Bot. 61, 319–324.

Lorenz, R.D., Mitchell, K.L., Kirk, R.L., Hayes, A.G., Aharonson, O., Zebker, H.A., Paillou, P., et al., 2008. Titan's inventory of organic surface materials. Geophys. Res. Lett. 35.

Lottig, N.R., Stanley, E.H., Hanson, P.C., Kratz, T.K., 2011. Comparison of regional stream and lake chemistry: differences, similarities, and potential drivers. Limnol. Oceanogr. 56 (5), 1551–1562.

Loulergue, L., Schilt, A., Spahni, R., Masson-Delmotte, V., Blunier, T., Lemieux, B., Barnola, J.-M., Raynaud, D., Stocker, T.F., Chappellaz, J., 2008. Orbital and millennial-scale features of atmospheric CH_4 over the past 800,000 years. Nature 453 (7193), 383–386.

Love, S.G., Brownlee, D.E., 1993. A direct measurement of the terrestrial mass accretion rate of cosmic dust. Science 262 (5133), 550–553.

Lovelock, J.E., 1979. Gaia: A New Look at Life on Earth. Oxford University Press.

Lovelock, J.E., Whitfield, M., 1982. Life span of the biosphere. Nature 296, 561–563.

Lovelock, J.E., Maggs, R.J., Rasmussen, R.A., 1972. Atmospheric dimethyl sulphide and the natural sulphur cycle. Nature 237, 452–453.

Lovett, G.M., 1994. Atmospheric deposition of Nutrients and pollutants in North America: An ecological perspective. Ecol. Appl. 4, 629–650.

Lovett, G.M., Goodale, C.L., 2011. A new conceptual model of nitrogen saturation based on experimental nitrogen addition to an oak forest. Ecosystems 14, 615–631.

Lovett, G.M., Lindberg, S.E., 1986. Dry deposition of nitrate to a deciduous forest. Biogeochemistry 2, 137–148.

Lovett, G.M., Lindberg, S.E., 1993. Atmospheric deposition and canopy interactions of nitrogen in forests. Can. J. Forest Res. 23, 1603–1616.

Lovett, G.M., Reiners, W.A., Olson, R.K., 1982. Cloud droplet deposition in subalpine balsam fir forests: hydrological and chemical inputs. Science 218 (4579), 1303–1304.

Lovett, G.M., Weathers, K.C., Sobczak, W.V., 2000. Nitrogen saturation and retention in forested watersheds of the Catskill Mountains, New York. Ecol. Appl. 10, 73–84.

Lovett, G.M., Weathers, K.C., Arthur, M.A., Schultz, J.C., 2004. Nitrogen cycling in a northern hardwood forest: Do species matter? Biogeochemistry 67, 289–308.

Lovett, G.M., Likens, G.E., Buso, D.C., Driscoll, C.T., Bailey, S.W., 2005. The biogeochemistry of chlorine at Hubbard Brook, New Hampshire. Biogeochemistry 72, 191–232.

Lovett, G.M., Cole, J.J., Pace, M.L., 2006. Is net ecosystem production equal to ecosystem carbon accumulation? Ecosystems 9, 152–155.

Lovett, G.M., Arthur, M.A., Weathers, K.C., Fitzhugh, R.D., Templer, P.H., 2013. Nitrogen addition increases carbon storage in soils, but not in trees, in an eastern U.S. deciduous forest. Ecosystems 16 (6), 980–1001.

Lovley, D.R., 1991. Dissimilatory Fe(III) and Mn (IV) reduction. Microbiol. Rev. 55, 259–287.

Lovley, D.R., Phillips, E.J.P., 1987. Competitive mechanisms for inhibition of sulfate reduction and methane production in the zone of ferric iron reduction in sediments. Appl. Environ. Microbiol. 53 (11), 2636–2641.

Lovley, D., Phillips, E., 1988a. Manganese inhibition of microbial iron reduction in anaerobic sediments. Geomicrobiol J. 6 (3), 145–155.

Lovley, D.R., Phillips, E.J.P., 1988b. Novel mode of microbial energy metabolism: organic carbon oxidation coupled to dissimilatory reduction of iron or manganese. Appl. Environ. Microbiol. 54 (6), 1472–1480.

Lovley, D.R., Giovannoni, S.J., White, D.C., Champine, J.E., Phillips, E.J.P., Gorby, Y.A., Goodwin, S., 1993. Geobacter metallireducens gen. Nov. Sp. Nov., a microorganism capable of coupling the complete oxidation of organic compounds to the reduction of iron and other metals. Arch. Microbiol. 159 (4), 336–344.

Lovley, D.R., Coates, J.D., Blunt-Harris, E.L., Phillips, E.J.P., Woodward, J.C., 1996. Humic substances as electron acceptors for microbial respiration. Nature 382, 445–448.

Lowry, B., Hebant, C., Lee, D., 1980. The origin of land plants: a new look at an old problem. Taxon 29, 183–197.

Lozier, M.S., 2012. Overturning in the North Atlantic. Annu. Rev. Mar. Sci. 4, 291–315.

Lozier, M.S., Dave, A.C., Palter, J.B., Gerber, L.M., Barber, R.T., 2011. On the relationship between stratification and primary productivity in the North Atlantic. Geophys. Res. Lett. 38.

Lozier, M.S., Gary, S.F., Bower, A.S., 2013. Simulated pathways of the overflow waters in the North Atlantic: subpolar to subtropical export. Deep-Sea Res. Part II, Top. Stud. Oceanogr. 85 (January), 147–153.

Lozier, M.S., Bacon, S., Bower, A.S., Cunningham, S.A., Femke De Jong, M., De Steur, L., De Young, B., et al., 2017. Overturning in the subpolar North Atlantic program: a new international ocean observing system. Bull. Am. Meteorol. Soc. 98 (4), 737–752.

Lu, M., Yang, Y., Luo, Y., Fang, C., Zhou, X., Chen, J., Yang, X., Li, B., 2011. Responses of ecosystem nitrogen cycle to nitrogen addition: a meta-analysis. New Phytol. 189 (4), 1040–1050.

Lu, M., Zhou, X., Yang, Q., Li, H., Luo, Y., Fang, C., Chen, J., Yang, X., Li, B., 2013. Responses of ecosystem carbon cycle to experimental warming: a meta-analysis. Ecology 94, 726–738.

Lu, X., Kicklighter, D.W., Melillo, J.M., Reilly, J.M., Liyi, X., 2015a. Land carbon sequestration within the conterminous United States: regional- and state-level analyses: regional U.S. land sinks and legacies. J. Geophys. Res. Biogeosci. 120 (2), 379–398.

Lu, X., Bottomley, P.J., Myrold, D.D., 2015b. Contributions of ammonia-oxidizing archaea and bacteria to nitrification in oregon forest soils. Soil Biol. Biochem. 85 (June), 54–62.

Lu, X., Taylor, A.E., Myrold, D.R., Neufeld, J.D., 2020. Expanding perspectives of soil nitrification to include ammonia-oxidizing archaea and comammox bacteria. Soil Sci. Soc. Am. J. 84, 287–302.

Lucas, Y., Luizao, F.J., Chauvel, A., Rouiller, J., Nahon, D., 1993. The relation between biological activity of the rain forest and mineral composition of soils. Science 260 (5107), 521–523.

Luce, C.H., Holden, Z.A., 2009. Declining annual streamflow distributions in the Pacific Northwest United States, 1948–2006. Geophys. Res. Lett. 36.

Ludwig, J., Meixner, F.X., Vogel, B., Forstner, J., 2001. Soil-air exchange of nitric oxide: an overview of processes, environmental factors, and modeling studies. Biogeochemistry 52, 225–257.

Luke, W.T., Dickerson, R.R., Ryan, W.F., Pickering, K.E., Nunnermacker, L.J., 1992. Tropospheric chemistry over the lower Great Plains of the United States 2. Trace gas profiles and distributions. J. Geophys. Res. 97, 20647–20670.

Lund, D.C., Lynch-Stieglitz, J., Curry, W.B., 2006. Gulf stream density structure and transport during the past millennium. Nature 444 (7119), 601–604.

Lundgren, D.G., Silver, M., 1980. Ore leaching by bacteria. Annu. Rev. Microbiol. 34, 263–283.

Lundstro, M.U.S., van Breemen, N., Bain, D., 2000. The Podzolization process. A review. Geoderma 94, 91–107.

Lundström, U.S., 1993. The role of organic acids in the soil solution chemistry of a podzolized soil. J. Soil Sci. 44 (1), 121–133.

Lundström, U.S., van Breemen, N., Bain, D.C., van Hees, P.A.W., Giesler, R., Gustafsson, J.P., Ilvesniemi, H., et al., 2000. Advances in understanding the podzolization process resulting from a multidisciplinary study of three coniferous forest soils in the Nordic countries. Geoderma 94 (2), 335–353.

Lunine, J.I., 1989. Origin and evolution of outer solar system atmospheres. Science 245 (4914), 141–147.

Luo, S., Ku, T.-L., 2003. Constraints on deep-water formation from the oceanic distributions of ^{10}Be. J. Geophys. Res. C: Oceans 108 (C5).

Luo, S., Ku, T.-L., 2004. On the importance of opal, carbonate, and lithogenic clays in scavenging and fractionating ^{230}Th, ^{231}Pa and ^{10}Be in the ocean. Earth Planet. Sci. Lett. 220 (1), 201–211.

Luo, Y., Su, B., Currie, W.S., Dukes, J.S., Finzi, A.C., Hartwig, U., Hungate, B., McMurtrie, R.E., Oren, R., Parton, W.J., Pataki, D.E., Shaw, M.R., Zak, D.R., Field, C.B., 2004. Progressive nitrogen limitation of ecosystem responses to rising atmospheric carbon dioxide. Bioscience 54, 731–739.

Luo, M., Shephard, M.W., Cady-Pereira, K.E., Henze, D.K., Zhu, L., Bash, J.O., Pinder, R.W., Capps, S.L., Walker, J.T., Jones, M.R., 2015. Satellite observations of tropospheric ammonia and carbon monoxide: global distributions, regional correlations and comparisons to model simulations. Atmos. Environ. 106 (April), 262–277.

Luo, X., Croft, H., Chen, J.M., He, L., Keenan, T.F., 2019. Improved estimates of global terrestrial photosynthesis using information on leaf chlorophyll content. Glob. Chang. Biol. 25 (7), 2499–2514.

Lupton, J.E., Craig, H., 1981. A major ^{3}He source at 15°S on the east Pacific rise. Science 214, 13–18.

Luterbacher, J., Dietrich, D., Xoplaki, E., Grosjean, M., Wanner, H., 2004. European seasonal and annual temperature variability, trends, and extremes since 1500. Science 303 (5663), 1499–1503.

Lüthi, D., Le Floch, M., Bereiter, B., Blunier, T., Barnola, J.-M., Siegenthaler, U., Raynaud, D., et al., 2008. High-resolution carbon dioxide concentration record 650,000–800,000 years before present. Nature 453 (7193), 379–382.

Lutz, R.A., Shank, T.M., Fornari, D.J., Haymon, R.M., Lilley, M.D., Vondamm, K.L., Desbruyeres, D., 1994. Rapid growth at deep sea vents. Nature 371, 663–664.

Lutz, B.D., Bernhardt, E.S., Roberts, B.J., 2011. Examining the coupling of carbon and nitrogen cycles in Appalachian streams: the role of dissolved organic nitrogen. Ecology 92 (3), 720–732.

Lutz, B.D., Bernhardt, E.S., Roberts, B.J., Cory, R.M., Mulholland, P.J., 2012a. Distinguishing dynamics of dissolved organic matter components in a forested stream using kinetic enrichments. Limnol. Oceanogr. 57, 76–89.

Lutz, B.D., Mulholland, P.J., Bernhardt, E.S., 2012b. Long-term data reveal patterns and controls on stream water chemistry in a forested stream: walker branch Tennessee. Ecol. Monogr. 82, 367–387.

Luxmoore, R.J., Grizzard, T., Strand, R.H., 1981. Nutrient translocation in the outer canopy and understory of an eastern deciduous forest. For. Sci. 27, 505–518.

Luyssaert, S., Inglima, I., Jung, M., Richardson, A.D., Reichsteins, M., Papale, D., Piao, S.L., et al., 2007. CO_2 balance of boreal, temperate, and tropical forests derived from a global database. Glob. Chang. Biol. 13, 2509–2537.

Luyssaert, S., Schulze, E.-D., Börner, A., Knohl, A., Hessenmöller, D., Law, B.E., Ciais, P., Grace, J., 2008. Old-growth forests as global carbon sinks. Nature 455 (7210), 213–215.

Luyssaert, S., Reichstein, M., Schulze, E.D., Janssens, I.A., Law, B.E., Papale, D., Dragoni, D., et al., 2009. Toward a consistency cross-check of eddy covariance flux-based and biometric estimates of ecosystem carbon balance. Glob. Biogeochem. Cycles 23.

Luz, B., Barkan, E., 2009. Net and gross oxygen production from O_2/Ar, $^{17}O/^{16}O$ and $^{18}O/^{16}O$ ratios. Aquat. Microb. Ecol.: Int. J. 56, 133–145.

Luz, B., Barkan, E., 2011. The isotopic composition of atmospheric oxygen. Glob. Biogeochem. Cycles. 25.

Lvovitch, M.I., 1973. The Global Water Balance. National Academy of Sciences, Washington, DC.

Lyons, T.W., Reinhard, C.T., Planavsky, N.J., 2014. The rise of oxygen in Earth's early ocean and atmosphere. Nature 506 (7488), 307–315.

Ma, Z., Bielenberg, D.G., Brown, K.M., Lynch, J.P., 2001. Regulation of root hair density by phosphorus availability in Arabidopsis Thaliana. Plant Cell Environ. 24, 459–467.

Ma, Z., Guo, D., Xu, X., Lu, M., Bardgett, R.D., Eissenstat, D.M., Luke McCormack, M., Hedin, L.O., 2018. Evolutionary history resolves global organization of root functional traits. Nature 555 (7694), 94–97.

Ma, T., Dai, G., Zhu, S., Chen, D., Chen, L., Lü, X., Wang, X., et al., 2019. Distribution and preservation of root- and shoot-derived carbon components in soils across the Chinese-Mongolian grasslands. J. Geophys. Res. Biogeosci. 124 (2), 420–431.

Maaroufi, N.I., Nordin, A., Hasselquist, N.J., Bach, L.H., Palmqvist, K., Gundale, M.J., 2015. Anthropogenic nitrogen deposition enhances carbon sequestration in boreal soils. Glob. Chang. Biol. 21 (8), 3169–3180.

Maathuis, F.J., Sanders, D., 1994. Mechanism of high-affinity potassium uptake in roots of Arabidopsis Thaliana. Proc. Natl. Acad. Sci. U. S. A. 91 (20), 9272–9276.

Maavara, T., Parsons, C.T., Ridenour, C., Stojanovic, S., Dürr, H.H., Powley, H.R., Van Cappellen, P., 2015. Global phosphorus retention by river damming. Proc. Natl. Acad. Sci. U. S. A. 112 (51), 15603–15608.

Mac Low, M.M., 2013. From gas to stars over cosmic time. Science 340 (6140), 1229229.

Macdonald, G.J., 1990. Role of methane clathrates in past and future climates. Clim. Chang. 16, 247–281.

Macdonald, N.W., Hart, J.B., 1990. Relating sulfate adsorption to soil properties in Michigan forest soils. Soil Sci. Soc. Am. J. 54, 238–245.

MacDonald, G.K., Bennett, E.M., Carpenter, S.R., 2012. Embodied phosphorus and the global connections of United States agriculture. Environ. Res. Lett. 7(4), 044024.

Macdonald, F.A., Swanson-Hysell, N.L., Park, Y., Lisiecki, L., Jagoutz, O., 2019. Arc-continent collisions in the tropics set Earth's climate state. Science 364 (6436), 181–184.

Mace, K.A., Duce, R.A., Tindale, N.W., 2003. Organic nitrogen in rain and aerosol at cape grim, Tasmania, Australia. J. Geophys. Res.-Atmos. 108.

Mach, D.L., Ramirez, A., Holland, H.D., 1987. Organic phosphorus and carbon in marine sediments. Am. J. Sci. 287, 429–441.

Macia, E., Hernandez, M.V., Oro, J., 1997. Primary sources of phosphorus and phosphates in chemical evolution. Origins Life Evol. Biosphere 27 (5–6), 459–480.

MacIntyre, F., 1974. The top millimeter of the ocean. Sci. Am. 230, 62–77.

Mack, M.C., Schuur, E.A.G., Bret-Harte, M.S., Shaver, G.R., Chapin, F.S., 2004. Ecosystem carbon storage in Arctic tundra reduced by long-term nutrient fertilization. Nature 431 (7007), 440–443.

Mack, M.C., Syndonia Bret-Harte, M., Hollingsworth, T.N., Jandt, R.R., Schuur, E.A.G., Shaver, G.R., Verbyla, D.L., 2011. Carbon loss from an unprecedented Arctic tundra wildfire. Nature 475 (7357), 489–492.

MacKay, N.A., Elser, J.J., 1998. Factors potentially preventing trophic cascades: food quality, invertebrate predation, and their interaction. Limnol. Oceanogr. 43, 339–347.

Mackenzie, F.T., Kump, L.R., 1995. Reverse weathering, clay mineral formation, and oceanic element cycles. Science 270 (5236), 586.

Mackenzie, F.T., Stoffyn, M., Wollast, R., 1978. Aluminum in seawater: control by biological activity. Science 199 (4329), 680–682.

Mackney, D., 1961. A Podzol development sequence in Oakwoods and heath in Central England. J. Soil Sci. 12, 23–40.

Madigan, M.T., Martinko, J.M., 2006. Microorganisms and microbiology. In: Brock Biology of Microorganisms, 11th ed. Pearson Prentice Hall, Upper Saddle River, New Jersey (NJ), pp. 1–20.

Madsen, H.B., Nornberg, P., 1995. Mineralogy of four sandy soils developed under heather, oak, spruce and grass in the same fluvioglacial deposit in Denmark. Geoderma 64, 233–256.

Maeda, M., Ihara, H., Ota, T., 2008. Deep-Soil adsorption of nitrate in a Japanese Andisol in response to different nitrogen sources. Soil Sci. Soc. Am. J. 72, 702–710.

Magnani, F., Mencuccini, M., Borghetti, M., Berbigier, P., Berninger, F., Delzon, S., Grelle, A., et al., 2007. The human footprint in the carbon cycle of temperate and boreal forests. Nature 447 (7146), 848–850.

Magnuson, J.J., Webster, K.E., Assel, R.A., Bowser, C.J., Dillon, P.J., Eaton, J.G., Evans, H.E., et al., 1997. Potential effects of climate changes on aquatic systems: Laurentian Great lakes and Precambrian shield region. Hydrol. Process. 11, 825–871.

Magnuson, J.J., Robertson, D.M., Benson, B.J., Wynne, R.H., Livingstone, D.M., Arai, T., Assel, R.A., et al., 2000. Historical trends in lake and river ice cover in the northern hemisphere. Science 289 (5485), 1743–1746.

Mahaffey, C., Williams, R.G., Wolff, G.A., Mahowald, N., Anderson, W., Woodward, M., 2003. Biogeochemical signatures of nitrogen fixation in the eastern North Atlantic. Geophys. Res. Lett. 30.

Mahaffy, P.R., Webster, C.R., Atreya, S.K., Franz, H., Wong, M., Conrad, P.G., Harpold, D., et al., 2013. Abundance and isotopic composition of gases in the Martian atmosphere from the curiosity rover. Science 341 (6143), 263–266.

Mahaffy, P.R., Webster, C.R., Stern, J.C., Brunner, A.E., Atreya, S.K., Conrad, P.G., Domagal-Goldman, S., et al., 2015. Mars atmosphere. The imprint of atmospheric evolution in the D/H of Hesperian clay minerals on Mars. Science 347 (6220), 412–414.

Mahall, B.E., Schlesinger, W.H., 1982. Effects of irradiance on growth, photosynthesis, and water use efficiency of seedlings of the chaparral shrub, *Ceanothus megacarpus*. Oecologia 54 (3), 291–299.

Mahdavi, S., Salehi, B., Granger, J., Amani, M., Brisco, B., Huang, W., 2018. Remote sensing for wetland classification: a comprehensive review. GISci. Remote Sens. 55 (5), 623–658.

Maher, K., 2010. The dependence of chemical weathering rates on fluid residence time. Earth Planet. Sci. Lett. 294, 101–110.

Maher, K., 2011. The role of fluid residence time and topographic scales in determining chemical fluxes from landscapes. Earth Planet. Sci. Lett. 312, 48–58.

Maher, K., Chamberlain, C.P., 2014. Hydrologic regulation of chemical weathering and the geologic carbon cycle. Science 343 (6178), 1502–1504.

Mahieu, N., Powlson, D.S., Randall, E.W., 1999. Statistical analysis of published carbon-13 CPMAS NMR spectra of soil organic matter. Soil Sci. Soc. Am. J. 63, 307–319.

Mahowald, N., 2011. Aerosol indirect effect on biogeochemical cycles and climate. Science 334 (6057), 794–796.

Mahowald, N.M., Artaxo, P., Baker, A.R., Jickells, T.D., Okin, G.S., Randerson, J.T., Townsend, A.R., 2005a. Impacts of biomass burning emissions and land use change on Amazonian atmospheric phosphorus cycling and deposition. Glob. Biogeochem. Cycles 19.

Mahowald, N.M., Baker, A.R., Bergametti, G., Brooks, N., Duce, R.A., Jickells, T.D., Kubilay, N., Prospero, J.M., Tegen, I., 2005b. Atmospheric global dust cycle and iron inputs to the ocean. Glob. Biogeochem. Cycles 19.

Mahowald, N., Jickells, T.D., Baker, A.R., Artaxo, P., Benitez-Nelson, C.R., Bergametti, G., Bond, T.C., et al., 2008. Global distribution of atmospheric phosphorus sources, concentrations and deposition rates, and anthropogenic impacts. Glob. Biogeochem. Cycles 22.

Maia, S.M.F., Ogle, S.M., Cerri, C.E.P., Cerri, C.C., 2010. Soil organic carbon stock change due to land use activity along the agricultural frontier of the southwestern Amazon, Brazil, between 1970 and 2002. Glob. Chang. Biol. 16, 2775–2788.

Makkonen, K., Helmisaari, H.S., 1999. Assessing fine-root biomass and production in a scots pine stand—comparison of soil core and root-ingrowth core methods. Plant Soil 210, 43–50.

Malaney, R.A., Fowler, W.A., 1988. The transformation of matter after the big bang. Am. Sci. 76, 472–477.

Malcolm, R.E., 1983. Assessment of phosphatase activity in soils. Soil Biol. Biochem. 15 (March), 403–408.

Malcolm, E.G., Schaefer, J.K., Ekstrom, E.B., Tuit, C.B., Jayakumar, A., Park, H., Ward, B.B., Morel, F.M.M., 2010. Mercury methylation in oxygen deficient zones of the oceans: no evidence for the predominance of anaerobes. Mar. Chem. 122, 11–19.

Malhi, Y., 2012. The productivity, metabolism and carbon cycle of tropical forest vegetation: carbon cycle of tropical forests. J. Ecol. 100 (1), 65–75.

Mallen-Cooper, M., Nakagawa, S., Eldridge, D.J., 2019. Global meta-analysis of soil-disturbing vertebrates reveals strong effects on ecosystem patterns and processes. Keith, S. (Ed.), Glob. Ecol. Biogeogr.: J. Macroecol. 28 (5), 661–679.

Mallin, M.A., Johnson, V.L., Ensign, S.H., MacPherson, T.A., 2006. Factors contributing to hypoxia in rivers, lakes, and streams. Limnol. Oceanogr. 51 (1), 690–701.

Malmer, N., 1975. Development of bog mires. In: Hassler, A.D. (Ed.), Coupling of Land and Water Systems. Springer, pp. 75–92.

Malone, M.J., Baker, P.A., Burns, S.J., 1994. Recrystallization of dolomite: evidence from the monterey formation (Miocene), California. Sedimentology 41, 1223–1239.

Malone, E.T., Abbott, B.W., Klaar, M.J., Kidd, C., Sebilo, M., Milner, A.M., Pinay, G., 2018. Decline in ecosystem δ^{13}C and mid-successional nitrogen loss in a two-century postglacial chronosequence. Ecosystems 21 (8), 1659–1675.

Manabe, S., Wetherald, R.T., 1980. Distribution of climate change resulting from an increase in the CO_2 content of the atmosphere. J. Atmos. Sci. 37, 99–118.

Manabe, S., Wetherald, R.T., 1986. Reduction in summer soil wetness induced by an increase in atmospheric carbon dioxide. Science 232 (4750), 626–628.

Mancinelli, R.L., McKay, C.P., 1988. The evolution of nitrogen cycling. Origins Life Evol. Biosphere 18, 311–325.

Mankin, W.G., Coffey, M.T., 1984. Increased stratospheric hydrogen chloride in the El Chichon cloud. Science 226 (4671), 170–172.

Mankin, W.G., Coffey, M.T., Goldman, A., 1992. Airborne observations of SO_2, HCl, and O_3 in the stratospheric plume of the Pinatubo volcano in July 1991. Geophys. Res. Lett. 19, 179–182.

Mann, M.E., Bradley, R.S., Hughes, M.K., 1999. Northern hemisphere temperatures during the past millennium: inferences, uncertainties, and lim itations. Geophys. Res. Lett. 26, 759–762.

Manney, G.L., Santee, M.L., Rex, M., Livesey, N.J., Pitts, M.C., Veefkind, P., Nash, E.R., et al., 2011. Unprecedented Arctic ozone loss in 2011. Nature 478 (7370), 469–475.

Manning, C.V., McKay, C.P., Zahnle, K.J., 2008. The nitrogen cycle on Mars: impact decomposition of near-surface nitrates as a source for a nitrogen steady state. Icarus 197, 60–64.

Mansy, S.S., Schrum, J.P., Krishnamurthy, M., Tobé, S., Treco, D.A., Szostak, J.W., 2008. Template-directed synthesis of a genetic polymer in a model protocell. Nature 454 (7200), 122–125.

Manzoni, S., Jackson, R.B., Trofymow, J.A., Porporato, A., 2008. The global stoichiometry of litter nitrogen mineralization. Science 321 (5889), 684–686.

Manzoni, S., Trofymow, J.A., Jackson, R.B., Porporato, A., 2010. Stoichiometric controls on carbon, nitrogen, and phosphorus dynamics in decomposing litter. Ecol. Monogr. 80, 89–106.

Manzoni, S., Čapek, P., Porada, P., Thurner, M., Winterdahl, M., Beer, C., Brüchert, V., et al., 2018. Reviews and syntheses: carbon use efficiency from organisms to ecosystems—definitions, theories, and empirical evidence. Biogeosciences 15 (19), 5929–5949.

Mao, J.-D., Johnson, R.L., Lehmann, J., Olk, D.C., Neves, E.G., Thompson, M.L., Schmidt-Rohr, K., 2012. Abundant and stable char residues in soils: implications for soil fertility and carbon sequestration. Environ. Sci. Technol. 46 (17), 9571–9576.

Mao, K.B., Ma, Y., Xia, L., Chen, W.Y., Shen, X.Y., He, T.J., Xu, T.R., 2014. Global aerosol change in the last decade: an analysis based on MODIS data. Atmos. Environ. 94 (September), 680–686.

Maranger, R., Bird, D.F., Price, N.M., 1998. Iron acquisition by photosynthetic marine phytoplankton from ingested bacteria. Nature 396, 248–251.

Maranger, R., Jones, S.E., Cotner, J.B., 2018. Stoichiometry of carbon, nitrogen, and phosphorus through the freshwater pipe. Limnol. Oceanogr. Lett. 3 (3), 89–101.

Marcano-Martinez, E., McBride, M.B., 1989. Calcium and sulfate retention by two oxisols of the Brazilian Cerrado. Soil Sci. Soc. Am. J. 53, 63–69.

Marchal, K., Vanderleyden, J., 2000. The "oxygen Paradox" of dinitrogen-fixing bacteria. Biol. Fertil. Soils 30, 363–373.

Marchese, A.J., Vaughn, T.L., Zimmerle, D.J., Martinez, D.M., Williams, L.L., Robinson, A.L., Mitchell, A.L., et al., 2015. Methane emissions from United States natural gas gathering and processing. Environ. Sci. Technol. 49 (17), 10718–10727.

Marenco, A., Gouget, H., Nedelec, P., Pages, J.P., Karcher, F., 1994. Evidence of a long-term increase in tropospheric ozone from PIC DU MIDI data series: consequences: positive radiative forcing. J. Geophys. Res. 99, 16617–16632.

Marino, R., Howarth, R.W., Chan, F., Cole, J.J., Likens, G.E., 2003. Sulfate inhibition of molybdenum-dependent nitrogen fixation by planktonic cyanobacteria under seawater conditions: a non-reversible effect. Hydrobiologia 500, 277–293.

Marino, R., Chan, F., Howarth, R.W., Pace, M.L., Likens, G.E., 2006. Ecological constraints on planktonic nitrogen fixation in saline estuaries. I. Nutrient and trophic controls. Mar. Ecol. Prog. Ser. 309, 25–39.

Marinos, R.E., Campbell, J.L., Driscoll, C.T., Likens, G.E., McDowell, W.H., Rosi, E.J., Rustad, L.E., Bernhardt, E.S., 2018. Give and take: a watershed acid rain mitigation experiment increases baseflow nitrogen retention but increases stormflow nitrogen export. Environ. Sci. Technol. 52 (22), 13155–13165.

Marion, G.M., Black, C.H., 1987. The effect of time and temperature on nitrogen mineralization in Arctic tundra soils. Soil Sci. Soc. Am. J. 51, 1501–1508.

Marion, G.M., Black, C.H., 1988. Potentially available nitrogen and phosphorus along a chaparral fire cycle chronosequence. Soil Sci. Soc. Am. J. 52, 1155–1162.

Marion, G.M., Schlesinger, W.H., Fonteyn, P.J., 1985. CALDEP: a regional model for soil $CaCO_3$ (caliche) deposition in southwestern deserts. Soil Sci. 139, 468–481.

Marion, G.M., Fritsen, C.H., Eicken, H., Payne, M.C., 2003. The search for life on Europa: limiting environmental factors, potential habitats, and earth analogues. Astrobiology 3 (4), 785–811.

Marion, G.M., Verburg, P.S.J., Stevenson, B., Arnone, J.A., 2008. Soluble element distributions in a Mojave desert soil. Soil Sci. Soc. Am. J. 72, 1815–1823.

Markewich, H.W., Pavich, M.J., 1991. Soil chronosequence studies in temperate to subtropical, low-latitude, low-relief terrain with data from the eastern United States. Geoderma 51 (1), 213–239.

Markewich, H.W., Pavich, M.J., Mausbach, M.J., Johnson, R.G., Gonzalez, V.M., 1989. A Guide for Using Soil and Weathering Profile Data in Chronosequence Studies of the Coastal Plain of the Eastern United States. U.S. Geological Survey.

Markewitz, D., Richter, D.D., 1998. The bio in aluminum and silicon geochemistry. Biogeochemistry 42, 235–252.

Markewitz, D., Richter, D.D., Allen, H.L., Urrego, J.B., 1998. Three decades of observed soil acidification in the Calhoun experimental forest: has acid rain made a difference? Soil Sci. Soc. Am. J. 62, 1428–1439.

Marklein, A.R., Houlton, B.Z., 2012. Nitrogen inputs accelerate phosphorus cycling rates across a wide variety of terrestrial ecosystems. New Phytol. 193 (3), 696–704.

Marklein, A.R., Winbourne, J.B., Enders, S.K., Gonzalez, D.J.X., van Huysen, T.L., Izquierdo, J.E., Light, D.R., et al., 2016. Mineralization ratios of nitrogen and phosphorus from decomposing litter in temperate versus tropical forests. Glob. Ecol. Biogeogr.: J. Macroecol. 25 (3), 335–346.

Marks, P.L., Bormann, F.H., 1972. Revegetation following forest cutting: mechanisms for return to steady-state nutrient cycling. Science 176 (4037), 914–915.

Marlon, J.R., Bartlein, P.J., Daniau, A.-L., Harrison, S.P., Maezumi, S.Y., Power, M.J., Tinner, W., Vanniére, B., 2013. Global biomass burning: a synthesis and review of Holocene Paleofire records and their controls. Quat. Sci. Rev. 65 (April), 5–25.

Marlow, J., Peckmann, J., Orphan, V., 2015. Autoendoliths: a distinct tpe of rock-hosted microbial life. Geobiology 13 (4), 303–307.

Marnette, E.C., Hordijk, C., Van Breeman, N., Cappenberg, T., 1992. Sulfate reduction and S-oxidation in a moorland pool sediment. Biogeochemistry 17, 123–143.

Maron, J.L., Estes, J.A., Croll, D.A., Danner, E.M., Elmendorf, S.C., Buckelew, S.L., 2006. An introduced predator alters Aleutian island plant communities by thwarting nutrient subsidies. Ecol. Monogr. 76, 3–24.

Maroulis, P.J., Bandy, A.R., 1977. Estimate of the contribution of biologically produced dimethyl sulfide to the global sulfur cycle. Science 196 (4290), 647–648.

Marquis, R.J., Whelan, C.J., 1994. Insectivorous birds increase growth of wite oak through consumption of chewing insects. Ecology 75, 2007–2014.

Marshall, J., Speer, K., 2012. Closure of the meridional overturning circulation through Southern Ocean upwelling. Nat. Geosci. 5 (February), 171.

Marshall, F., Reid, R.E.B., Goldstein, S., Storozum, M., Wreschnig, A., Hu, L., Kiura, P., Shahack-Gross, R., Ambrose, S.H., 2018. Ancient herders enriched and restructured African grasslands. Nature 561 (7723), 387–390.

Marteel, A., Boutron, C.F., Barbante, C., Gabrielli, P., Cozzi, G., Gaspari, V., Cescon, P., et al., 2008. Changes in atmospheric heavy metals and metalloids in dome C (East Antarctica) ice back to 672.0 Kyr BP (marine isotopic stages 16.2). Earth Planet. Sci. Lett. 272, 579–590.

Martens, C.S., Val Klump, J., 1984. Biogeochemical cycling in an organic-rich coastal marine basin. 4. An organic-carbon budget for sediments dominated by sulfate reduction and methanogenesis. Geochim. Cosmochim. Acta 48, 1987–2004.

Martens-Habbena, W., Berube, P.M., Urakawa, H., de la Torre, J.R., Stahl, D.A., 2009. Ammonia oxidation kinetics determine niche separation of nitrifying archaea and bacteria. Nature 461 (7266), 976–979.

Martin, J.H., 1990. Glacial-interglacial CO_2 change: the iron hypothesis. Paleoceanography 5, 1–13.

Martin, R.V., 2008. Satellite remote sensing of surface air quality. Atmos. Environ. 42 (March), 7823–7843.

Martin, M.E., Aber, J.D., 1997. High spectral resolution remote sensing of forest canopy lignin, nitrogen, and ecosystem processes. Ecol. Appl. 7, 431–443.

Martin, J.H., Fitzwater, S.E., 1992. Dissolved Organic Carbon in the Atlantic, southern and Pacific oceans. Nature 356, 699–700.

Martin, J.H., Gordon, R.M., 1988. Northeast Pacific iron distributions in relation to phytoplankton productivity. Deep-Sea Res. Part A—Oceanogr. Res. Pap. 35, 177–196.

Martin, C.W., Harr, R.D., 1988. Precipitation and sreamwater chemistry from undisturbed watersheds in the Cascade Mountains of Oregon. Water Air Soil Pollut. 42, 203–219.

Martin, J.-M., Meybeck, M., 1979. Elemental mass-balance of material carried by major world rivers. Mar. Chem. 7 (3), 173–206.

Martin, H.W., Sparks, D.L., 1985. On the behavior of nonexchangeable potassium in soils. Commun. Soil Sci. Plant Anal. 16, 133–162.

Martin, W.R., Bender, M., Leinen, M., Orchardo, J., 1991. Benthic organic carbon degradation and biogenic silica dissolution in the central equatorial Pacific. Deep-Sea Res. Part A—Oceanogr. Res. Pap. 38, 1481–1516.

Martin, J.H., Coale, K.H., Johnson, K.S., Fitzwater, S.E., Gordon, R.M., Tanner, S.J., Hunter, C.N., et al., 1994. Testing the iron hypothesis in ecosystems of the equatorial Pacific Ocean. Nature 371 (6493), 123–129.

Martin, R.V., Jacob, D.J., Chance, K., Kurosu, T.P., Palmer, P.I., Evans, M.J., 2003. Global inventory of nitrogen oxide emissions constrained by space-based observations of NO_2 columns. J. Geophys. Res. Atmos. 108, 1–12.

Martin, S.L., Hayes, D.B., Rutledge, D.T., Hyndman, D.W., 2011. The land-use legacy effect: adding temporal context to lake chemistry. Limnol. Oceanogr. 56, 2362–2370.

Martinelli, L.A., Piccolo, M.C., Townsend, A.R., Vitousek, P.M., Cuevas, E., McDowell, W., Robertson, G.P., Santos, O.C., Treseder, K., 1999. Nitrogen stable isotopic composition of leaves and soil: tropical versus temperate forests. Biogeochemistry 46, 45–65.

Martín-Español, A., Zammit-Mangion, A., Clarke, P.J., Flament, T., Helm, V., King, M.A., Luthcke, S.B., et al., 2016. Spatial and temporal Antarctic ice sheet mass trends, glacio-isostatic adjustment, and surface processes from a joint inversion of satellite altimeter, gravity, and GPS data. J. Geophys. Res. Earth Surf. 121 (2), 182–200.

Martinez-Cortizas, A., Pontevedra-Pombal, X., Garcia-Rodeja, E., Novoa-Munoz, J.C., Shotyk, W., 1999. Mercury in a Spanish peat bog: archive of climate change and atmospheric metal deposition. Science 284 (5416), 939–942.

Martínez-Garcia, A., Rosell-Melé, A., Jaccard, S.L., Geibert, W., Sigman, D.M., Haug, G.H., 2011. Southern ocean dust-climate coupling over the past four million years. Nature 476 (7360), 312–315.

Marty, B., 1995. Nitrogen content of the mantle inferred from N_2-Ar correlation in oceanic basalts. Nature 377, 326–329.

Marty, B., 2012. The origins and concentrations of water, carbon, nitrogen and noble gases on earth. Earth Planet. Sci. Lett. 313 (314), 56–66.

Marty, B., Chaussidon, M., Wiens, R.C., Jurewicz, A.J.G., Burnett, D.S., 2011. A ^{15}N-poor isotopic composition for the solar system as shown by genesis solar wind samples. Science 332 (6037), 1533–1536.

Marty, B., Zimmermann, L., Pujol, M., Burgess, R., Philippot, P., 2013. Nitrogen isotopic composition and density of the Archean atmosphere. Science 342 (6154), 101–104.

Marty, B., Altwegg, K., Balsiger, H., Bar-Nun, A., Bekaert, D.V., Berthelier, J.-J., Bieler, A., et al., 2017. Xenon isotopes in 67P/Churyumov-Gerasimenko show that comets contributed to Earth's atmosphere. Science 356 (6342), 1069–1072.

Marufu, L.T., Taubman, B.F., Bloomer, B., Piety, C.A., Doddridge, B.G., Stehr, J.W., Dickerson, R.R., 2004. The 2003 North American electrical blackout: an accidental experiment in atmospheric chemistry. J. Geophys. Res. 31.

Marumoto, T., Anderson, J.P.E., Domsch, K.H., 1982. Mineralization of nutrients from soil microbial biomass. Soil Biol. Biochem. 14, 469–475.

Marx, D.H., Hatch, A.B., Mendicino, J.F., 1977. High soil fertility decreases surcrose content and susceptibility of loblolly pine roots to ectomycorrrhizal infection by *Pisolithus tinctorius*. Can. J. Bot. 55, 1569–1574.

Masek, J.G., Cohen, W.B., Leckie, D., Wulder, M.A., Vargas, R., de Jong, B., Healey, S., et al., 2011. Recent rates of forest harvest and conversion in North America. J. Geophys. Res. 116.

Masion, A., Rose, J., Bottero, J.-Y., Tchoubar, D., Elmerich, P., 1997a. Nucleation and growth mechanisms of iron oxyhydroxides in the presence of PO_4 ions. 3. Speciation of Fe by small angle X-ray scattering. Langmuir: ACS J. Surf. Colloids 13 (14), 3882–3885.

Masion, A., Rose, J., Bottero, J.-Y., Tchoubar, D., Garcia, F., 1997b. Nucleation and growth mechanisms of iron oxyhydroxides in the presence of PO_4 ions. 4. Structure of the aggregates. Langmuir: ACS J. Surf. Colloids 13 (14), 3886–3889.

Mason, R.P., Fitzgerald, W.F., 1993. The distribution and biogeochemical cycling of mercury in the equatorial Pacific Ocean. Deep-Sea Res. Part I—Oceanogr. Res. Pap. 40, 1897–1924.

Mason, E., Edmonds, M., Turchyn, A.V., 2017. Remobilization of crustal carbon may dominate volcanic arc emissions. Science 357 (6348), 290–294.

Mast, M.A., Turk, J.T., Ingersoll, G.P., Clow, D.W., Kester, C.L., 2001. Use of stable sulfur isotopes to identify sources of sulfate in Rocky Mountain Snowpacks. Atmos. Environ. 35, 3303–3313.

Mast, M.A., Manthorne, D.J., Roth, D.A., 2010. Historical deposition of mercury and selected trace elements to high-elevation national parks in the Western US inferred from lake-sediment cores. Atmos. Environ. 44, 2577–2586.

Matamala, R., Gonzàlez-Meler, M.A., Jastrow, J.D., Norby, R.J., Schlesinger, W.H., 2003. Impacts of fine root turnover on forest NPP and soil C sequestration potential. Science 302 (5649), 1385–1387.

Mathieu, O., Leveque, J., Henault, C., Milloux, M.J., Bizouard, F., Andreux, F., 2006. Emissions and spatial variability of N_2O, N_2 and nitrous oxide mole fraction at the field scale, revealed with ^{15}N-isotopic techniques. Soil Biol. Biochem. 38, 941–951.

Matson, P.A., Vitousek, P.M., 1990. Ecosystem approach to a global nitrous oxide budget. Bioscience 40, 667–671.

Matson, P.A., Vitousek, P.M., Ewel, J.J., Mazzarino, M.J., Robertson, G.P., 1987. Nitrogen transformations following tropical forest felling and burning on a volcanic soil. Ecology 68, 491–502.

Matson, P.A., Johnson, L., Billow, C., Miller, J., Pu, R., 1994. Seasonal patterns and remote spectral estimation of canopy chemistry across the oregon transect. Ecol. Appl. 4, 280–298.

Matthews, E., 1997. Global litter production, pools, and turnover times: estimates from measurement data and regression models. J. Geophys. Res. D: Atmos. 102, 18771–18800.

Matthews, E., Fung, I., 1987. Methane emission from natural wetlands: global distribution, area, and environmental characteristics of sources. Glob. Biogeochem. Cycles 1, 61–86.

Matthews, E., Fung, I., Lerner, J., 1991. Methane emission from rice cultivation: geographic and seasonal distribution of cultivated areas and emissions. Glob. Biogeochem. Cycles 5, 3–24.

Mattson, W.J., 1980. Herbivory in relation to plant nitrogen content. Annu. Rev. Ecol. Syst. 11, 119–161.

Mattson, W.J., Addy, N.D., 1975. Phytophagous insects as regulators of forest primary production. Science 190, 515–522.

Mayer, L.M., 1994. Relationships between mineral surfaces and organic carbon concentrations in soils and sediments. Chem. Geol. 114, 347–363.

Mayer, B., Feger, K.H., Giesemann, A., Jager, H.J., 1995. Interpretation of sulfur cycling in two catchments in the black forest (Germany) using stable sulfur-and oxygen-isotope data. Biogeochemistry 30, 31–58.

Mayewski, P.A., Lyons, W.B., Spencer, M.J., Twickler, M., Dansgaard, W., Koci, B., Davidson, C.I., Honrath, R.E., 1986. Sulfate and nitrate concentrations from a South Greenland ice core. Science 232 (4753), 975–977.

Mayewski, P.A., Lyons, W.B., Spencer, M.J., Twickler, M.S., Buck, C.F., Whitlow, S., 1990. An ice-core record of atmospheric response to anthropogenic sulfate and nitrate. Nature 346, 554–556.

Maynard, J.J., O'Geen, A.T., Dahlgren, R.A., 2011. Sulfide induced mobilization of wetland phosphorus depends strongly on redox and iron geochemistry. Soil Sci. Soc. Am. J. 75(5).

Mayorga, E., Aufdenkampe, A.K., Masiello, C.A., Krusche, A.V., Hedges, J.I., Quay, P.D., Richey, J.E., Brown, T.A., 2005. Young organic matter as a source of carbon dioxide outgassing from Amazonian rivers. Nature 436 (7050), 538–541.

Mazumder, A., Dickman, M.D., 1989. Factors affecting the spatial and temporal distribution of phototrophic sulfur bacteria. Arch. Hydrobiol. 116, 209–226.

McCabe, P.J., 2009. Depositional environments of coal and coal-bearing strata. In: Sedimentology of Coal and Coal-Bearing Sequences. Blackwell Publishing Ltd, pp. 11–42.

McCabe, G.J., Wolock, D.M., 2002. A step increase in streamflow in the conterminous United States. Geophys. Res. Lett. 29.

McCalley, C.K., Woodcroft, B.J., Hodgkins, S.B., Wehr, R.A., Kim, E.-H., Mondav, R., Crill, P.M., et al., 2014. Methane dynamics regulated by microbial community response to permafrost thaw. Nature 514 (7523), 478–481.

McCarthy, J.J., Goldman, J.C., 1979. Nitrogenous nutrition of marine phytoplankton in nutrient-depleted waters. Science 203 (4381), 670–672.

McCarthy, M.D., Hedges, J.I., Benner, R., 1998. Major bacterial contribution to marine dissolved organic nitrogen. Science 281 (5374), 231–234.

McCarthy, H.R., Oren, R., Johnsen, K.H., Gallet-Budynek, A., Pritchard, S.G., Cook, C.W., LaDeau, S.L., Jackson, R.B., Finzi, A.C., 2010. Re-assessment of plant carbon dynamics at the duke free-air CO_2 enrichment site: interactions of atmospheric CO_2 with nitrogen and water availability over stand development. New Phytol. 185, 514–528.

McClelland, J.W., Dery, S.J., Peterson, B.J., Holmes, R.M., Wood, E.F., 2006. A Pan-Arctic evaluation of changes in river discharge during the latter half of the 20th century. Geophys. Res. Lett. 33.

McClintock, J., Ducklow, H., Fraser, W., 2008. Ecological responses to climate change on the Antarctic peninsula: the peninsula is an icy world that's warming faster than anywhere else on earth, threatening a rich but delicate biological community. Am. Sci. 96 (4), 302–310.

McClure, C.D., Jaffe, D.A., 2018. US particulate matter air quality improves except in wildfire-prone areas. Proc. Natl. Acad. Sci. U. S. A. 115 (31), 7901–7906.

McColl, J.G., Grigal, D.F., 1975. Forest fire—effects on phosphorus movement to lakes. Science 188, 1109–1111.

McConnell, J.R., Wilson, A.I., Stohl, A., Arienzo, M.M., Chellman, N.J., Eckhardt, S., Thompson, E.M., Mark Pollard, A., Steffensen, J.P., 2018. Lead pollution recorded in Greenland ice indicates European emissions tracked plagues, wars, and imperial expansion during antiquity. Proc. Natl. Acad. Sci. U. S. A. 115 (22), 5726–5731.

McCormack, M.L., Dickie, I.A., Eissenstat, D.M., Fahey, T.J., Fernandez, C.W., Guo, D., Helmisaari, H.-S., et al., 2015. Redefining fine roots improves understanding of below-ground contributions to terrestrial biosphere processes. New Phytol. 207 (3), 505–518.

McCormick, M.P., Thomason, L.W., Trepte, C.R., 1995. Atmospheric effects of the Mt. Pinatubo eruption. Nature 373, 399–404.

McCrackin, M.L., Elser, J.J., 2010. Atmospheric nitrogen deposition influences denitrification and nitrous oxide production in lakes. Ecology 91 (2), 528–539.

McCrackin, M.L., Elser, J.J., 2011. Greenhouse gas dynamics in lakes receiving atmospheric nitrogen deposition. Glob. Biogeochem. Cycles 25.

McCrackin, M.L., Elser, J.J., 2012. Denitrification kinetics and denitrifier abundances in sediments of lakes receiving atmospheric nitrogen deposition (Colorado, USA). Biogeochemistry 108 (1–3), 39–54.

McCrackin, M.L., Harrison, J.A., Compton, J.E., 2014. Factors influencing export of dissolved inorganic nitrogen by major rivers: a new, seasonal, spatially explicit, global model. Glob. Biogeochem. Cycles 28 (3), 269–285.

McCulloch, A., Aucott, M.L., Benkovitz, C.M., Graedel, T.E., Kleiman, G., Midgley, P.M., Li, Y.F., 1999. Global emissions of hydrogen chloride and chloromethane from coal combustion, incineration and industrial activities: reactive chlorine emissions inventory. J. Geophys. Res. 104, 8391–8403.

McCutchan, J.H., Lewis, W.M., 2002. Relative importance of carbon sources for macroinvertebrates in a Rocky Mountain stream. Limnol. Oceanogr. 47 (3), 742–752.

McDiffett, W.F., Beidler, A.W., Dominick, T.F., McCrea, K.D., 1989. Nutrient concentration-stream discharge relationships during storm events in a first-order stream. Hydrobiologia 179, 97–102.

McDonald, B.C., de Gouw, J.A., Gilman, J.B., Jathar, S.H., Akherati, A., Cappa, C.D., Jimenez, J.L., et al., 2018. Volatile chemical products emerging as largest petrochemical source of urban organic emissions. Science 359 (6377), 760–764.

McDowell, W.H., Asbury, C.E., 1994. Export of carbon, nitrogen, and major ions from three tropical montane watersheds. Limnol. Oceanogr. 39, 111–125.

McDowell, W.H., Likens, G.E., 1988. Origin, composition, and flux of dissolved organic carbon in the Hubbard Brook Valley. Ecol. Monogr. 58, 177–195.

McDowell, W.H., Wood, T., 1984. Podzolization: soil processes control dissolved organic carbon concentrations in stream water. Soil Sci. 137, 23–32.

McElroy, M.B., 1983. Marine biological controls on atmospheric CO_2 and climate. Nature 302, 328–329.

McElroy, M.B., 2002. The Atmospheric Environment. Princeton University Press.

McElroy, M.B., Wang, Y.X.X., 2005. Human and animal wastes: implications for atmospheric N_2O and NO_x. Glob. Biogeochem. Cycles 19.

McElroy, M.B., Yung, Y.L., Nier, A.O., 1976. Isotopic composition of nitrogen: implications for the past history of Mars' atmosphere. Science 194 (4260), 70–72.

McElroy, M.B., Prather, M.J., Rodriguez, J.M., 1982. Escape of hydrogen from Venus. Science 215 (4540), 1614–1615.

McGarvey, J.C., Thompson, J.R., Epstein, H.E., Shugart Jr., H.H., 2015. Carbon storage in old-growth forests of the mid-Atlantic: toward better understanding the eastern forest carbon sink. Ecology 96 (2), 311–317.

McGill, W.B., Cole, C.V., 1981. Comparative aspects of cycling of organic C, N, S and P through soil organic matter. Geoderma 26, 267–286.

McGill, G.E., Warner, J.L., Malin, M.C., Arvidson, R.E., Eliason, E., Nozette, S., Reasenberg, R.D., 1983. Topography, surface properties, and tectonic evolution. In: Hunten, D.M., Colin, L., Donahue, T.M., Moroz, V.I. (Eds.), Venus. University of Arizona Press, pp. 69–130.

McGillicuddy, D.J., Robinson, A.R., Siegel, D.A., Jannasch, H.W., Johnson, R., Dickeys, T., McNeil, J., Michaels, A.F., Knap, A.H., 1998. Influence of mesoscale eddies on new production in the Sargasso sea. Nature 394, 263–266.

McGroddy, M.E., Daufresne, T., Hedin, L.O., 2004. Scaling of C:N:P stoichiometry in forests worldwide: implications of terrestrial redfield-type ratios. Ecology 85, 2390–2401.

McGuire, A.D., Melillo, J.M., Joyce, L.A., Kicklighter, D.W., Grace, A.L., Moore, B., Vorosmarty, C.J., 1992. Interactions between carbon and nitrogen dynamics in estimating net primary production for potential vegetation in North America. Glob. Biogeochem. Cycles 6, 101–124.

McGuire, K.J., McDonnell, J.J., Weiler, M., Kendall, C., McGlynn, B.L., Welker, J.M., Seibert, J., 2005. The role of topography on catchment-scale water residence time. Water Resour. Res. 41, 14.

McGuire, A.D., Macdonald, R.W., Schuur, E.A., Harden, J.W., Kuhry, P., Hayes, D.J., Christensen, T.R., Heimann, M., 2010. The carbon budget of the northern cryosphere region. Curr. Opin. Environ. Sustain. 2 (4), 231–236.

McKay, C.P., 2014. Requirements and limits for life in the context of exoplanets. Proc. Natl. Acad. Sci. U. S. A. 111 (35), 12628–12633.

McKay, C.P., Borucki, W.J., 1997. Organic synthesis in experimental impact shocks. Science 276 (5311), 390–392.

McKay, C.P., Toon, O.B., Kasting, J.F., 1991. Making Mars habitable. Nature 352 (August), 489–496.

McKay, D.S., Gibson Jr., E.K., Thomas-Keprta, K.L., Vali, H., Romanek, C.S., Clemett, S.J., Chillier, X.D., Maechling, C.R., Zare, R.N., 1996. Search for past life on Mars: possible relic biogenic activity in Martian meteorite ALH84001. Science 273 (5277), 924–930.

McKelvey, V.E., 1980. Seabed minerals and the law of the sea. Science 209, 464–472.

McKenney, D.J., Drury, C.F., Findlay, W.I., Mutus, B., McDonnell, T., Gajda, C., 1994. Kinetics of denitrification by *Pseudomonas fluorescens*: oxygen effects. Soil Biol. Biochem. 26, 901–908.

McKenzie, R., Connor, B., Bodeker, G., 1999. Increased summertime UV radiation in New Zealand in response to ozone loss. Science 285 (5434), 1709–1711.

McLain, J.E.T., Kepler, T.B., Ahmann, D.M., 2002. Belowground factors mediating changes in methane consumption in a forest soil under elevated CO_2. Glob. Biogeochem. Cycles 16.

McLatchey, G.P., Reddy, K.R., 1998. Regulation of organic matter decomposition and nutrient release in a wetland soil. J. Environ. Qual. 27 (5), 1268–1274.

McLauchlan, K.K., Hobbie, S.E., Post, W.M., 2006. Conversion from agriculture to grassland builds soil organic matter on decadal timescales. Ecol. Appl. 16 (1), 143–153.

McLaughlin, S.B., Taylor, G.E., 1981. Relative humidity: important modifier of pollutant uptake by plants. Science 211 (4478), 167–169.

McMahon, W.J., Davies, N.S., 2018. Evolution of alluvial Mudrock forced by early land plants. Science 359 (6379), 1022–1024.

McMurtrie, R.E., Näsholm, T., 2018. Quantifying the contribution of mass flow to nitrogen acquisition by an individual plant root. New Phytol. 218 (1), 119–130.

McNaughton, S.J., 1988. Mineral nutrition and spatial concentrations of African ungulates. Nature 334 (6180), 343–345.

McNaughton, S.J., 1990. Mineral nutrition and seasonal movements of African migratory ungulates. Nature 345, 613–615.

McNaughton, S.J., Chapin, F.S., 1985. Effects of phosphorus nutrition and defoliation on C-4 Graminoids from the Serengeti plains. Ecology 66, 1617–1629.

McNaughton, S.J., Oesterheld, M., Frank, D.A., Williams, K.J., 1989. Ecosystem-level patterns of primary productivity and herbivory in terrestrial habitats. Nature 341 (6238), 142–144.

McNaughton, S.J., Banyikwa, F.F., McNaughton, M.M., 1997. Promotion of the cycling of diet-enhancing nutrients by African grazers. Science 278 (5344), 1798–1800.

McNaughton, S.J., Stronach, N.R.H., Georgiadis, N.J., 1998. Combustion in natural fires and global emissions budgets. Ecol. Appl. 8, 464–468.

McNeil, B.I., Matear, R.J., Key, R.M., Bullister, J.L., Sarmiento, J.L., 2003. Anthropogenic CO_2 uptake by the ocean based on the global chlorofluorocarbon data set. Science 299 (5604), 235–239.

McNorton, J., Chipperfield, M.P., Gloor, M., Wilson, C., Feng, W., Hayman, G.D., Rigby, M., et al., 2016. Role of OH variability in the stalling of the global atmospheric CH_4 growth rate from 1999 to 2006. Atmos. Chem. Phys. 16 (12), 7943–7956.

McNulty, S.G., Boggs, J., Aber, J.D., Rustad, L., Magill, A., 2005. Red spruce ecosystem level changes following 14 years of chronic N fertilization. For. Ecol. Manag. 219, 279–291.

McSween, H.Y., 1989. Chondritic meteorites and the formation of planets. Am. Sci. 77, 146–153.

Meade, R.H., Dunne, T., Richey, J.E., Santos, U.D.E.M., Salati, E., 1985. Storage and remobilization of suspended sediment in the lower Amazon River of Brazil. Science 228 (4698), 488–490.

Meade, C., Reffner, J.A., Ito, E., 1994. Synchrotron infrared absorbency measurements of hydrogen in $MgSiO_3$ perovskite. Science 264, 1558–1560.

Meador, J., Jeffrey, W.H., Kase, J.P., Dean Pakulski, J., Chiarello, S., Mitchell, D.L., 2002. Seasonal fluctuation of DNA photodamage in marine plankton assemblages at Palmer Station, Antarctica. Photochem. Photobiol. 75 (3), 266–271.

Measures, C.I., Vink, S., 2000. On the use of dissolved aluminum in surface waters to estimate dust deposition to the ocean. Glob. Biogeochem. Cycles 14, 317–327.

Medina-Elizalde, M., Rohling, E.J., 2012. Collapse of classic Maya civilization related to modest reduction in precipitation. Science 335 (6071), 956–959.

Medinets, S., Skiba, U., Rennenberg, H., Butterbach-Bahl, K., 2015. A review of soil NO transformation: associated processes and possible physiological significance on organisms. Soil Biol. Biochem. 80 (January), 92–117.

Medlyn, B.E., De Kauwe, M.G., Lin, Y.-S., Knauer, J., Duursma, R.A., Williams, C.A., Arneth, A., et al., 2017. How do leaf and ecosystem measures of water-use efficiency compare? New Phytol. 216 (3), 758–770.

Meentemeyer, V., 1978a. Climatic regulation of decomposition rates of organic matter in terrestrial ecosystems. In: Adriano, D.C., Brisbin, I.L. (Eds.), Environmental Chemistry and Cycling Processes. National Technical Information Service, Springfield, Virginia, pp. 779–789.

Meentemeyer, V., 1978b. Macroclimate and lignin control of litter decomposition rates. Ecology 59, 465–472.

Meentemeyer, V., Box, E.O., Thompson, R., 1982. World patterns and amounts of terrestrial plant litter production. Bioscience 32, 125–128.

Megonigal, J.P., Guenther, A.B., 2008. Methane emissions from upland forest soils and vegetation. Tree Physiol. 28 (4), 491–498.

Megonigal, J.P., Schlesinger, W.H., 1997. Enhanced CH_4 emissions from a wetland soil exposed to elevated CO_2. Biogeochemistry 37 (1), 77–88.

Megonigal, J.P., Schlesinger, W.H., 2002. Methane-limited methanotrophy in tidal freshwater swamps. Glob. Biogeochem. Cycles 16.

Megonigal, J.P., Patrick, W.H., Faulkner, S.P., 1993. Wetland identification in seasonally flooded forest soils—soil morphology and redox dyanmics. Soil Sci. Soc. Am. J. 57 (1), 140–149.

Megonigal, J.P., Conner, W.H., Kroeger, S., Sharitz, R.R., 1997. Aboveground production in southeastern floodplain forests: a test of the subsidy-stress hypothesis. Ecology 78, 370–384.

Megonigal, J.P., Hines, M.E., Visscher, P.T., 2003a. Anaerobic metabolism: linkages to trace gases and aerobic processes. In: Schlesinger, W.H. (Ed.), Biogeochemistry. Elsevier-Pergamon, pp. 317–324.

Megonigal, J.P., Hines, M.E., Visser, 2003b. Anaerobic metabolism: linkages to trace gases and aerobic processes. Treat. Geochem. 8, 317–424.

Mehner, T., Diekmann, M., Gonsiorczyk, T., Kasprzak, P., Koschel, R., Krienitz, L., Rumpf, M., Schulz, M., Wauer, G., 2008. Rapid recovery from eutrophication of a stratified lake by disruption of internal nutrient load. Ecosystems 11 (7), 1142–1156.

Meier, R., Owen, T.C., Matthews, H.E., Jewitt, D.C., Bockelée-Morvan, D., Biver, N., Crovisier, J., Gautier, D., 1998. A determination of the HDO/H_2O ratio in comet C/1995 O1 (Hale-Bopp). Science 279 (5352), 842–844.

Meier, M.F., Dyurgerov, M.B., Rick, U.K., O'Neel, S., Pfeffer, W.T., Anderson, R.S., Anderson, S.P., Glazovsky, A.F., 2007. Glaciers dominate eustatic sealevel rise in the 21st century. Science 317, 1064–1067.

Melack, J.M., 1976. Primary productivity and fish yields in tropical lakes. Trans. Am. Fish. Soc. 105, 575–580.

Melillo, J.M., Aber, J.D., Muratore, J.F., 1982. Nitrogen and lignin control of hardwood leaf litter decomposition dynamics. Ecology 63, 621–626.

Melillo, J.M., Steudler, P.A., Feigl, B.J., Neill, C., Garcia, D., Piccolo, M.C., Cerri, C.C., Tian, H., 2001. Nitrous oxide emissions from forests and pastures of various ages in the Brazilian Amazon. J. Geophys. Res. 106, 34179–34188.

Melillo, J.M., Steudler, P.A., Aber, J.D., Newkirk, K., Lux, H., Bowles, F.P., Catricala, C., Magill, A., Ahrens, T., Morrisseau, S., 2002. Soil warming and Carbon-cycle feedbacks to the climate system. Science 298 (5601), 2173–2176.

Melillo, J.M., Reilly, J.M., Kicklighter, D.W., Gurgel, A.C., Cronin, T.W., Paltsev, S., Felzer, B.S., Wang, X., Sokolov, A.P., Adam Schlosser, C., 2009. Indirect emissions from biofuels: how important? Science 326 (5958), 1397–1399.

Melillo, J.M., Butler, S., Johnson, J., Mohan, J., Steudler, P., Lux, H., Burrows, E., et al., 2011. Soil warming, carbon-nitrogen interactions, and forest carbon budgets. Proc. Natl. Acad. Sci. U. S. A. 108 (23), 9508–9512.

Melillo, J.M., Frey, S.D., DeAngelis, K.M., Werner, W.J., Bernard, M.J., Bowles, F.P., Pold, G., Knorr, M.A., Grandy, A.S., 2017. Long-term pattern and magnitude of soil carbon feedback to the climate system in a warming world. Science 358 (6359), 101–105.

Melosh, H.J., Vickery, A.M., 1989. Impact erosion of the primordial atmosphere of Mars. Nature 338 (6215), 487–489.

Mendelssohn, I.A., Kleiss, B.A., Wakeley, J.S., 1995. Factors controlling the formation of oxidized root channels: a review. Wetlands 15 (1), 37–46.

Mendez, J., Guieu, C., Adkins, J., 2010. Atmospheric input of manganese and Iron to the ocean: seawater dissolution experiments with Saharan and North American dusts. Mar. Chem. 120 (1–4), 34–43.

Mendonça, R., Müller, R.A., Clow, D., Verpoorter, C., Raymond, P., Tranvik, L.J., Sobek, S., 2017. Organic carbon burial in global lakes and reservoirs. Nat. Commun. 8 (1), 1694.

Ménez, B., Pisapia, C., Andreani, M., Jamme, F., Vanbellingen, Q.P., Brunelle, A., Richard, L., Dumas, P., Réfrégiers, M., 2018. Abiotic synthesis of amino acids in the recesses of the oceanic lithosphere. Nature 564 (7734), 59–63.

Menge, D.N.L., Batterman, S.A., Hedin, L.O., Liao, W., Pacala, S.W., Taylor, B.N., 2017. Why are nitrogen-fixing trees rare at higher compared to lower latitudes? Ecology 98 (12), 3127–3140.

Mengis, M., Gachter, R., Wehrli, B., 1997. Sources and sinks of nitrous oxide (N_2O) in deep lakes. Biogeochemistry 38, 281–301.

Mensing, S.A., Michaelsen, J., Byrne, R., 1999. A 560-year record of Santa Ana Fires reconstructed from charcoal deposited in the Santa Barbara Basin, California. Quat. Res. 51, 295–305.

Merrifield, M.A., Merrifield, S.T., Mitchum, G.T., 2009. An anomalous recent acceleration of global sea level rise. J. Clim. 22, 5772–5781.

Metcalfe, D.B., Asner, G.P., Martin, R.E., Silva Espejo, J.E., Huasco, W.H., Farfán Amézquita, F.F., Carranza-Jimenez, L., et al., 2014. Herbivory makes major contributions to ecosystem carbon and nutrient cycling in tropical forests. Ecol. Lett. 17 (3), 324–332.

Metson, G.S., Bennett, E.M., Elser, J.J., 2012a. The role of diet in phosphorus demand. Environ. Res. Lett. 7 (4), 044043.

Metson, G.S., Hale, R.L., Iwaniec, D.M., Cook, E.M., Corman, J.R., Galletti, C.S., Childers, D.L., 2012b. Phosphorus in Phoenix: a budget and spatial representation of phosphorus in an urban ecosystem. Ecol. Appl. 22 (2), 705–721.

Meure, C.M., Etheridge, D., Trudinger, C., Steele, P., Langenfelds, R., van Ommen, T., Smith, A., Elkins, J., 2006. Law Dome CO_2, CH_4 and N_2O ice-core records extended to 2000 years BP. Geophys. Res. Lett. 33.

Meybeck, M., 1977. Dissolved and suspended matter carried by rivers: composition, time and space variations, and world balance. In: Golterman, H.L. (Ed.), Interactions between Sediments and Fresh Water. W. Junk, pp. 25–32.

Meybeck, M., 1979. Major elements contents of river waters and dissolved inputs to the oceans. Rev. Geol. Dyn. Geogr. Phys. 21, 215–246.

Meybeck, M., 1982. Carbon, nitrogen, and phosphorus transport by world rivers. Am. J. Sci. 282, 401–450.

Meybeck, M., 1987. Global chemical weathering of surficial rocks estimated from river dissolved loads. Am. J. Sci. 287 (5), 401–428.

Meybeck, M., 1993. Riverine transport of atmospheric carbon—sources, global typology and budget. Water Air Soil Pollut. 70, 443–463.

Meybeck, M., 2003. Global occurrence of major elements in rivers. Treat. Geochem. 5–9(605).

Meyer, J.L., 1979. Role of sediments and bryophytes in phosphorus dynamics in a headwater stream ecosystem. Limnol. Oceanogr. 24, 365–375.

Meyer, J.L., 1980. Dynamics of phosphorus and organic matter during leaf decomposition in a forest stream. Oikos 34, 44–53.

Meyer, C., 1985. Ore metals through geologic history. Science 227 (4693), 1421–1428.

Meyer, O., 1994. Functional groups of microorganisms. In: Schulze, E.-D., Mooney, H.A. (Eds.), Biodiversity and Ecosystem Function. Springer-Verlag, pp. 67–96.

Meyer, J.L., Likens, G.E., 1979. Transport and transformation of phosphorus in a forest stream ecosystem. Ecology 60, 1255–1269.

Meyer, J.L., Likens, G.E., Sloane, J., 1981. Phosphorus, nitrogen and organic carbon flux in a headwater stream. Arch. Hydrobiol. 91, 28–44.

Meyer, J.L., McDowell, W.H., Bott, T.L., Elwood, J.W., Ishizaki, C., Melack, J.M., Peckarsky, B.L., Peterson, B.J., Rublee, P.A., 1988. Elemental dynamics in streams. J. N. Am. Benthol. Soc. 7, 410–432.

Meyer, J.L., Benke, A.C., Edwards, R.T., Wallace, J.B., 1997. Organic matter dynamics in the Ogeechee river, a Black-water river in Georgia. J. N. Am. Benthol. Soc. 16, 82–87.

Meyer, J.B., Wallaca, J.L., Eggert, S.L., 1998. Leaf litter as a source of dissolved organic carbon to streams. Ecosystems 1 (3), 240–249.

Meyers-Schulte, K.J., Hedges, J.I., 1986. Molecular evidence for a terrestrial component of organic matter dissolved in ocean water. Nature 321 (6065), 61–63.

Meysman, F.J.R., Risgaard-Petersen, N., Malkin, S.Y., Nielsen, L.P., 2015. The geochemical fingerprint of microbial long-distance eectron transport in the seafloor. Geochim. Cosmochim. Acta 152 (March), 122–142.

Michaelis, W., Seifert, R., Nauhaus, K., Treude, T., Thiel, V., Blumenberg, M., Knittel, K., et al., 2002. Microbial reefs in the Black Sea fueled by anaerobic oxidation of methane. Science 297 (5583), 1013–1015.

Michaels, A.F., Bates, N.R., Buesseler, K.O., Carlson, C.A., Knap, A.H., 1994. Carbon cycle imbalances in the Sargasso Sea. Nature 372, 537–540.

Michalopoulos, P., Aller, R.C., 1995. Rapid clay mineral formation in Amazon Delta sediments: reverse weathering and oceanic elemental cycles. Science 270 (5236), 614–617.

Michalski, G., Bohlke, J.K., Thiemens, M., 2004a. Long term atmospheric deposition as the source of nitrate and other salts in the Atacama Desert, Chile: new evidence from mass-independent oxygen isotopic compositions. Geochim. Cosmochim. Acta 68, 4023–4038.

Michalski, G., Meixner, T., Fenn, M., Hernandez, L., Sirulnik, A., Allen, E., Thiemens, M., 2004b. Tracing atmospheric nitrate deposition in a complex semiarid ecosystem using Delta ^{17}O. Environ. Sci. Technol. 38, 2175–2181.

Michel, R.L., Suess, H.E., 1975. Bomb tritium in the Pacific Ocean. J. Geophys. Res. 80 (30), 4139–4152.

Middag, R., de Baar, H.J.W., Bruland, K.W., 2019. The relationships between dissolved zinc and major nutrients phosphate and silicate along the GEOTRACES GA02 transect in the West Atlantic Ocean. Glob. Biogeochem. Cycles 33 (1), 63–84.

Middelburg, J.J., 2011. Chemoautotrophy in the ocean. Geophys. Res. Lett. 38.

Middelburg, J.J., 2017. Reviews and synthesis: to the bottom of carbon processing at the seafloor. Biogeosci. Discuss.

Middelburg, J.J., 2018. Review and syntheses: to the bottom of carbon processing at the seafloor. Biogeosciences 15, 413–427.

Middleton, K.R., Smith, G.S., 1979. Comparison of ammoniacal and nitrate nutrition of prerennial ryegrass through a thermodynamic model. Plant Soil 53, 487–504.

Midolo, G., Alkemade, R., Schipper, A.M., Benítez-López, A., Perring, M.P., De Vries, W., 2018. Impacts of nitrogen addition on plant species richness and abundance: a global meta-analysis. Glob. Ecol. Biogeogr.: J. Macroecol. 12 (December), 374.

Mieville, A., Granier, C., Liousse, C., Guillaume, B., Mouillot, F., Lamarque, J.F., Gregoire, J.M., Petron, G., 2010. Emissions of gases and particles from biomass burning during the 20th century using satellite data and an historical reconstruction. Atmos. Environ. 44, 1469–1477.

Migdisov, A.A., Ronov, A.B., Grinenko, V.A., 1983. The sulphur cycle in the lithosphere. In: Ivanov, M.V., Freney, J.R. (Eds.), The Global Biogeochemical Sulphur Cycle. Wiley, pp. 25–127.

Mikhail, S., Berry, P.H., Sverjensky, D.A., 2017. The relationship between mantle pH and the deep nitrogen cycle. Geochim. Cosmochim. Acta 209, 149–160.

Mikhailov, V.N., 2010. Water and sediment runoff at the Amazon river mouth. Water Resour. 37, 145–159.

Mikola, J., Silfver, T., Paaso, U., Possen, B.J.M.H., Rousi, M., 2018. Leaf N resorption efficiency and litter N mineralization rate have a genotypic tradeoff in a silver birch population. Ecology 99 (5), 1227–1235.

Mikutta, R., Kleber, M., Torn, M.S., Jahn, R., 2006. Stabilization of soil organic matter: association with minerals or chemical recalcitrance? Biogeochemistry 77, 25–56.

Milesi, C., Elvidge, C.D., Nemani, R.R., Running, S.W., 2003. Assessing the impact of urban land development on net primary productivity in the southeastern United States. Remote Sens. Environ. 86, 401–410.

Miller, S.L., 1953. A production of amino acids under possible primitive earth conditions. Science 117 (3046), 528–529.

Miller, S.L., 1957. The formation of organic compounds on the primitive earth. Ann. N. Y. Acad. Sci. 69 (2), 260–275.

Miller, L., Douglas, B.C., 2004. Mass and volume contributions to twentieth-century global sea level rise. Nature 428 (6981), 406–409.

Miller, W.R., Drever, J.I., 1977. Chemical weathering and related controls on surface water chemistry in Absaroka Mountains, Wyoming. Geochim. Cosmochim. Acta 41, 1693–1702.

Miller, J.R., Russell, G.L., 1992. The impact of global warming on river runoff. J. Geophys. Res. 97, 2757–2764.

Miller, H.G., Cooper, J.M., Miller, J.D., 1976. Effect of nitrogen supply on nutrients in litter fall and crown leaching in a stand of Corsican pine. J. Appl. Ecol. 13, 233–248.

Miller, E.K., Blum, J.D., Friedland, A.J., 1993. Determination of soil exchangeable cation loss and weathering rates using Sr isotopes. Nature 362, 438–441.

Miller, M.N., Dandie, C.E., Zebarth, B.J., Burton, D.L., Goyer, C., Trevors, J.T., 2012. Influence of carbon amendments on soil denitrifier abundance in soil microcosms. Geoderma 170, 48–55.

Millero, F.J., Feistel, R., Wright, D.G., McDougall, T.J., 2008. The composition of standard seawater and the definition of the reference-composition salinity scale. Deep-Sea Res. Part I, Oceanogr. Res. Pap. 55, 50–72.

Millett, J., Godbold, D., Smith, A.R., Grant, H., 2012. N_2 fixation and cycling in *Alnus glutinosa*, *Betula pendula* and *Fagus sylvatica* woodland exposed to free air CO_2 enrichment. Oecologia 169 (2), 541–552.

Milliman, J.D., 1993. Production and accumulation of calcium carbonate in the ocean: budget of a nonsteady state. Glob. Biogeochem. Cycles 7, 927–957.

Milliman, J.D., Meade, R.H., 1983. World-wide delivery of river sediment to the oceans. J. Geol. 91, 1–21.

Milliman, J.D., Syvitski, J.P.M., 1992. Geomorphic tectonic control of sediment discharge to the ocean: the importance of small mountainous rivers. J. Geol. 100, 525–544.

Milliman, J.D., Troy, P.J., Balch, W.M., Adams, A.K., Li, Y.H., Mackenzie, F.T., 1999. Biologically mediated dissolution of calcium carbonate above the chemical lysocline? Deep-Sea Res. Part I—Oceanogr. Res. Pap. 46, 1653–1669.

Mills, M.M., Ridame, C., Davey, M., La Roche, J., Geider, R.J., 2004. Iron and phosphorus co-limit nitrogen fixation in the eastern tropical North Atlantic. Nature 429 (6989), 292–294.

Milly, P.C.D., Wetherald, R.T., Dunne, K.A., Delworth, T.L., 2002. Increasing risk of great floods in a changing climate. Nature 415 (6871), 514–517.

Milly, P.C.D., Dunne, K.A., Vecchia, A.V., 2005. Global pattern of trends in streamflow and water availability in a changing climate. Nature 438 (7066), 347–350.

Milly, P.C.D., Betancourt, J., Falkenmark, M., Hirsch, R.M., Kundzewicz, Z.W., Lettenmaier, D.P., Stouffer, R.J., 2008. Stationarity is dead: Whither water management? Science 319, 573–574.

Mimmo, T., Del Buono, D., Terzano, R., Tomasi, N., Vigani, G., Crecchio, C., Pinton, R., Zocchi, G., Cesco, S., 2014. Rhizospheric organic compounds in the soil-microorganism-plant system: their role in iron availability. Eur. J. Soil Sci. SSSA Book Ser. n°1 65 (5), 629–642.

Min, S.-K., Zhang, X., Zwiers, F.W., Hegerl, G.C., 2011. Human contribution to more-intense precipitation extremes. Nature 470 (7334), 378–381.

Minderman, G., 1968. Addition, decomposition and accumulation of organic matter in forests. J. Ecol. 56, 355–362.

Minoura, H., Iwasaka, Y., 1996. Rapid change in nitrate and sulfate concentrations observed in early stage of precipitation and their deposition processes. J. Atmos. Chem. 24, 39–55.

Minschwaner, K., Salawitch, R.J., McElroy, M.B., 1993. Absorption of solar radiation by O_2: implications for O_3 and lifetimes of N_2O, $CFCl_3$, and CF_2Cl_2. J. Geophys. Res. 98, 10543–10561.

Minshall, G.W., Petersen, R.C., Cummins, K.W., Bott, T.L., Sedell, J.R., Cushing, C.E., Vannote, R.L., 1983. Interbiome comparison of stream ecosystem dynamics. Ecol. Monogr. 53, 1–25.

Miranda, M.L., Hastings, D.A., Aldy, J.E., Schlesinger, W.H., 2011. The environmental justice dimensions of climate change. Environ. Justice 4, 17–25.

Mispagel, M.E., 1978. Ecology and bioenergetics of the Acridid grasshopper, *Bootettix punctatus*, on Creosotebush, *Larrea tridentata*, in the northern Mojave Desert. Ecology 59, 779–788.

Misra, S., Froelich, P.N., 2012. Lithium isotope history of cenozoic seawater: changes in silicate weathering and reverse weathering. Science 335 (6070), 818–823.

Mitchard, E.T.A., 2018. The tropical forest carbon cycle and climate change. Nature 559 (7715), 527–534.

Mitchard, E.T.A., Feldpausch, T.R., Brienen, R.J.W., Lopez-Gonzalez, G., Monteagudo, A., Baker, T.R., Lewis, S.L., et al., 2014. Markedly divergent estimates of Amazon forest carbon density from ground plots and satellites. Glob. Ecol. Biogeogr.: J. Macroecol. 23 (8), 935–946.

Mitchell, J.F.B., 1983. The hydrological cycle as simulated by an atmospheric general circulation model. In: Street-Perrott, A., Beran, M. (Eds.), Variations in the Global Water Budget. Reidel, pp. 429–446.

Mitchell, M.J., David, M.B., Maynard, D.G., Telang, S.A., 1986. Sulfur constituents in soils and streams of a watershed in the Rocky Mountains of Alberta. Can. J. Forest Res. 16, 315–320.

Mitchell, M.J., Driscoll, C.T., Fuller, R.D., David, M.B., Likens, G.E., 1989. Effect of whole-tree harvesting on the sulfur dynamics of a forest soil. Soil Sci. Soc. Am. J. 53, 933–940.

Mitchell, M.J., Foster, N.W., Shepard, J.P., Morrison, I.K., 1992. Nutrient cycling in Huntington forest and Turkey lakes deciduous stands: nitrogen and sulfur. Can. J. Forest Res. 22, 457–464.

Mitchell, J.F.B., Johns, T.C., Gregory, J.M., Tett, S.F.B., 1995. Climate response to increasing levels of greenhouse gases and sulfate aerosols. Nature 376, 501–504.

Mitchell, L.E., Brook, E.J., Sowers, T., McConnell, J.R., Taylor, K., 2011a. Multidecadal variability of atmospheric methane, 1000–1800 CE. J. Geophys. Res. 116.

Mitchell, M.J., Lovett, G., Bailey, S., Beall, F., Burns, D., Buso, D., Clair, T.A., et al., 2011b. Comparisons of watershed sulfur budgets in southeast Canada and northeast US: new approaches and implications. Biogeochemistry 103, 181–207.

Mitchell, L., Brook, E., Lee, J.E., Buizert, C., Sowers, T., 2013. Constraints on the late Holocene anthropogenic contribution to the atmospheric methane budget. Science 342 (6161), 964–966.

Mitra, S., Wassmann, R., Vlek, P.L.G., 2005. An appraisal of global wetland area and its organic carbon stock. Curr. Sci. 88, 25–35.

Mitsch, W.J., Gosselink, J.G., 2007. Wetlands, fourth ed. John Wiley and Sons, Inc., New York.

Mitsch, W.J., Dorage, C.L., Wiemhoff, J.R., 1979. Ecosystem dynamics and a phosphorus budget of an alluvial cypress swamp in southern Illinois. Ecology 60 (6), 1116.

Mitsch, W.J., Nahlik, A., Wolski, P., Bernal, B., Zhang, L., Ramberg, L., 2010. Tropical wetlands: seasonal hydrologic pulsing, carbon sequestration, and methane emissions. Wetl. Ecol. Manag. 18, 573–586.

Mizutani, H., Wada, E., 1988. Nitrogen and carbon isotope ratios in seabird rookeries and their ecological implications. Ecology 69, 340–349.

Mizutani, H., Hasegawa, H., Wada, E., 1986. High nitrogen isotope ratio for soils of seabird rookeries. Biogeochemistry 2, 221–247.

Mochizuki, Y., Koba, K., Muneoki, Y., 2012. Strong inhibitory effect of nitrate on atmospheric methane oxidation in forest soils. Soil Biol. Biochem. 50, 164–166.

Mohammed, G.H., Colombo, R., Middleton, E.M., Rascher, U., van der Tol, C., Nedbal, L., Goulas, Y., et al., 2019. Remote sensing of solar-induced chlorophyll fluorescence (SIF) in vegetation: 50 years of progress. Remote Sens. Environ. 231 (September), 111177.

Mohren, G.M.J., Vandenburg, J., Burger, F.W., 1986. Phosphorus deficiency induced by nitrogen input in Douglas fir in the Netherlands. Plant Soil 95, 191–200.

Mojzsis, S.J., Arrhenius, G., McKeegan, K.D., Harrison, T.M., Nutman, A.P., Friend, C.R., 1996. Evidence for life on earth before 3,800 million years ago. Nature 384 (6604), 55–59.

Mojzsis, S.J., Harrison, T.M., Pidgeon, R.T., 2001. Oxygen-isotope evidence from ancient zircons for liquid water at the Earth's surface 4,300 Myr ago. Nature 409 (6817), 178–181.

Mokany, K., Raison, R.J., Prokushkin, A.S., 2006. Critical analysis of root: shoot ratios in terrestrial biomes. Glob. Chang. Biol. 12, 84–96.

Moldowan, P.D., Alex Smith, M., Baldwin, T., Bartley, T., Rollinson, N., Wynen, H., 2019. Nature's pitfall trap: Salamanders as rich prey for carnivorous plants in a nutrient-poor northern bog ecosystem. Ecology 100 (10), e02770.

Molina, M.J., Rowland, F.S., 1974. Stratospheric sink for chlorofluoromethanes: chlorine atom-catalyzed destruction of ozone. Nature 249, 810–812.

Molina, M.J., Tso, T.L., Molina, L.T., Wang, F.C., 1987. Antarctic stratospheric chemistry of chlorine nitrate, hydrogen chloride, and ice: release of active chlorine. Science 238 (4831), 1253–1257.

Möller, D., 1990. The Na/Cl ratio in rainwater and the seasalt chloride cycle. Tellus B Chem. Phys. Meteorol. 42 (3), 254–262.

Molles, M.C., Dahm, C.N., 1990. A perspective on El Niño and La Nina—global implications for stream ecology. J. N. Am. Benthol. Soc. 9, 68–76.

Monger, H.C., Daugherty, L.A., Lindemann, W.C., Liddell, C.M., 1991. Microbial precipitation of Pedogenic calcite. Geology 19, 997–1000.

Monk, C.D., 1966. An ecological significance to evergreenness. Ecology 47, 504–505.

Monson, R.K., Holland, E.A., 2001. Biospheric trace gas fluxes and their control over tropospheric chemistry. Annu. Rev. Ecol. Syst. 32, 547–576.

Monteiro, L.R., Furness, R.W., 1997. Accelerated increase in mercury contamination in North Atlantic mesopelagic food chains as indicated by time series of seabird feathers. Environ. Toxicol. Chem./SETAC 16, 2489–2493.

Monteiro, F.M., Bach, L.T., Brownlee, C., Bown, P., Rickaby, R.E.M., Poulton, A.J., Tyrrell, T., et al., 2016. Why marine phytoplankton calcify. Sci. Adv. 2 (7), e1501822.

Montgomery, D.R., Dietrich, W.E., 1988. Where do channels begin? Nature 336, 232–234.

Montzka, S.A., Butler, J.H., Myers, R.C., Thompson, T.M., Swanson, T.H., Clarke, A.D., Lock, L.T., Elkins, J.W., 1996. Decline in the tropospheric abundance of halogen from halocarbons: implications for stratospheric ozone depletion. Science 272 (5266), 1318–1322.

Montzka, S.A., Aydin, M., Battle, M., Butler, J.H., Saltzman, E.S., Hall, B.D., Clarke, A.D., Mondeel, D., Elkins, J.W., 2004. A 350-year atmospheric history for carbonyl sulfide inferred from Antarctic firn air and air trapped in ice. J. Geophys. Res. 109.

Montzka, S.A., Calvert, P., Hall, B.D., Elkins, J.W., Conway, T.J., Tans, P.P., Sweeney, C., 2007. On the global distribution, seasonality, and budget of atmospheric carbonyl sulfide (COS) and some similarities to CO_2. J. Geophys. Res. D: Atmos. 112.

Montzka, S.A., Dlugokencky, E.J., Butler, J.H., 2011a. Non-CO_2 greenhouse gases and climate change. Nature 476 (7358), 43–50.

Montzka, S.A., Krol, M., Dlugokencky, E., Hall, B., Jöckel, P., Lelieveld, J., 2011b. Small interannual variability of global atmospheric hydroxyl. Science 331 (6013), 67–69.

Montzka, S.A., Dutton, G.S., Yu, P., Ray, E., Portmann, R.W., Daniel, J.S., Kuijpers, L., et al., 2018. An unexpected and persistent increase in global emissions of ozone-depleting CFC-11. Nature 557 (7705), 413–417.

Moon, S., Chamberlain, C.P., Hilley, G.E., 2014. New estimates of silicate weathering rates and their uncertainties in global rivers. Geochim. Cosmochim. Acta 134 (June), 257–274.

Mooney, H.A., Billings, W.D., 1961. Comparative physiological ecology of Arctic and alpine populations of *Oxyria digyna*. Ecol. Monogr. 31, 1–29.

Moore, W.S., 2010. The effect of submarine groundwater discharge on the ocean. Annu. Rev. Mar. Sci. 2, 59–88.

Moore, T.R., Dalva, M., 1993. The influence of temperature and water-table position on carbon dioxide and methane emissions from laboratory columns of peatland soils. J. Soil Sci. 44, 651–664.

Moore, J.K., Doney, S.C., 2007. Iron availability limits the ocean nitrogen inventory stabilizing feedbacks between marine denitrification and nitrogen fixation. Glob. Biogeochem. Cycles 21.

Moore, T.R., Knowles, R., 1989. The influence of water-table levels on Methane and Carbon dioxide emissions from peatland Soils. Can. J. Soil Sci. 69 (1), 33–38.

Moore, T.R., Roulet, N.T., 1993. Methane flux—water table relations in northern wetlands. Geophys. Res. Lett. 20 (7), 587–590.

Moore, J.K., Doney, S.C., Glover, D.M., Fung, I.Y., 2001. Iron cycling and nutrient-limitation patterns in surface waters of the world ocean. Deep-Sea Res. Part II Top. Stud. Oceanogr. 49 (1), 463–507.

Moore, T., Blodau, C., Turunen, J., Roulet, N., Richard, P.J.H., 2005. Patterns of nitrogen and sulfur accumulation and retention in ombrotrophic bogs, eastern Canada. Glob. Chang. Biol. 11, 356–367.

Moore, C.M., Mills, M.M., Achterberg, E.P., Geider, R.J., LaRoche, J., Lucas, M.I., McDonagh, E.L., et al., 2009. Large-scale distribution of Atlantic nitrogen fixation controlled by iron availability. Nat. Geosci. 2 (12), 867–871.

Moorhead, D.L., Reynolds, J.F., Fonteyn, P.J., 1989. Patterns of stratified soil-water loss in a Chihuahuan desert community. Soil Sci. 148, 244–249.

Moosdorf, N., Hartmann, J., Lauerwald, R., Hagedorn, B., Kempe, S., 2011. Atmospheric CO_2 consumption by chemical weathering in North America. Geochim. Cosmochim. Acta 75, 7829–7854.

Mooshammer, M., Wanek, W., Zechmeister-Boltenstern, S., Richter, A., 2014. Stoichiometric imbalances between terrestrial decomposer communities and their resources: mechanisms and implications of microbial adaptations to their resources. Front. Microbiol. 5 (February), 22.

Mora, C.I., Driese, S.G., Colarusso, L.A., 1996. Middle to late paleozoic atmospheric CO_2 levels from soil carbonate and organic matter. Science 271, 1105–1107.

Moran, S.B., Moore, R.M., 1988. Evidence from Mesocosm studies for biological removal of dissolved aluminum from seawater. Nature 335, 706–708.

Moran, M.A., Zepp, R.G., 1997. Role of photoreactions in the formation of biologically labile compounds from dissolved organic matter. Limnol. Oceanogr. 42 (6), 1307–1316.

Morée, A.L., Beusen, A.H.W., Bouwman, A.F., Willems, W.J., 2013. Exploring global nitrogen and phosphorus flows in urban wastes during the twentieth century: twentieth century urban N and P flows. Glob. Biogeochem. Cycles 27 (3), 836–846.

Morel, F.M.M., Hudson, R.J.M., Price, N.M., 1991. Limitation of productivity by trace metals in the sea. Limnol. Oceanogr. 36 (8), 1742–1755.

Morel, F.M.M., Reinfelder, J.R., Roberts, S.B., Chamberlain, C.P., Lee, J.G., Yee, D., 1994. Zinc and carbon co-limitation of marine phytoplankton. Nature 369 (6483), 740–742.

Morford, S.L., Houlton, B.Z., Dahlgren, R.A., 2011. Increased forest ecosystem carbon and nitrogen storage from nitrogen rich bedrock. Nature 477 (7362), 78–81.

Morgan, C.G., Allen, M., Liang, M.C., Shia, R.L., Blake, G.A., Yung, Y.L., 2004. Isotopic fractionation of nitrous oxide in the stratosphere: comparison between model and observations. J. Geophys. Res. 109.

Morison, J., Kwok, R., Peralta-Ferriz, C., Alkire, M., Rigor, I., Andersen, R., Steele, M., 2012. Changing Arctic ocean freshwater pathways. Nature 481 (7379), 66–70.

Morley, N., Baggs, E.M., 2010. Carbon and oxygen controls on N_2O and N_2 production during nitrate reduction. Soil Biol. Biochem. 42, 1864–1871.

Morner, N.A., Etiope, G., 2002. Carbon degassing from the lithosphere. Glob. Planet. Chang. 33, 185–203.

Morowitz, H.J., 1968. Energy Flow in Biology: Biological Organization as a Problem in Thermal Physics. Academic Press.

Morris, R.V., Ruff, S.W., Gellert, R., Ming, D.W., Arvidson, R.E., Clark, B.C., Golden, D.C., et al., 2010. Identification of carbonate-rich outcrops on Mars by the spirit rover. Science 329 (5990), 421–424.

Morris, J.L., Puttick, M.N., Clark, J.W., Edwards, D., Kenrick, P., Pressel, S., Wellman, C.H., Yang, Z., Schneider, H., Donoghue, P.C.J., 2018. The timescale of early land plant evolution. Proc. Natl. Acad. Sci. U. S. A. 115 (10), E2274–E2283.

Morrow, P.A., Lamarche, V.C., 1978. Tree-ring evidence for chronic insect suppression of productivity in *Sulalpine eucalyptus*. Science 201, 1244–1246.

Morse, J.W., Mackenzie, F.T., 1998. Hadean ocean carbonate geochemistry. Aquat. Geochem. 4, 301–319.

Morse, J.W., Cornwell, J.C., Arakaki, T., Lin, S., Huertadiaz, M., 1992. Iron sulfide and carbonate mineral diagenesis in Baffin Bay, Texas. J. Sediment. Petrol. 62, 671–680.

Morse, J.L., Ardón, M., Bernhardt, E.S., 2012. Greenhouse gas fluxes in southeastern U.S. coastal plain wetlands under contrasting Land uses. Ecol. Appl. 22 (1), 264–280.

Morse, J.L., Durán, J., Beall, F., Enanga, E.M., Creed, I.F., Fernandez, I., Groffman, P.M., 2015. Soil denitrification fluxes from three northeastern North American forests across a range of nitrogen deposition. Oecologia 177 (1), 17–27.

Mortimer, C.H., 1941. The exchange of dissolved substances between mud and water in lakes. J. Ecol. 29 (2), 280–329.

Morton, S.D., Lee, T.H., 1974. Algal blooms. Possible effects of iron. Environmental Science & Technology 8, 673–674.

Moses, M.E., Brown, J.H., 2003. Allometry of human fertility and energy use. Ecol. Lett. 6, 295–300.

Mosier, A., Schimel, D., Valentine, D., Bronson, K., Parton, W., 1991. Methane and nitrous oxide fluxes in native, fertilized and cultivated grasslands. Nature 350, 330–332.

Mosier, A., Kroeze, C., Nevison, C., Oenema, O., Seitzinger, S., van Cleemput, O., 1998. Closing the global N_2O budget: nitrous oxide emissions through the agricultural nitrogen cycle. Nutr. Cycl. Agroecosyst. 52, 225–248.

Mouginot, J., Rignot, E., Bjørk, A.A., van den Broeke, M., Millan, R., Morlighem, M., Noël, B., Scheuchl, B., Wood, M., 2019. Forty-six years of Greenland ice sheet mass balance from 1972 to 2018. Proc. Natl. Acad. Sci. U. S. A. 116 (19), 9239–9244.

Mouillot, F., Field, C.B., 2005. Fire history and the global carbon budget: a 1o X 1o fire-history reconstruction for the 20th century. Glob. Chang. Biol. 11, 398–420.

Mouillot, F., Narasimha, A., Balkanski, Y., Lamarque, J.F., Field, C.B., 2006. Global carbon emissions from biomass burning in the 20th century. Geophys. Res. Lett. 33.

Moulton, K.L., West, J., Berner, R.A., 2000. Solute flux and mineral mass balance approaches to the quantification of plant effects on silicate weathering. Am. J. Sci. 300, 539–570.

Mount, G.H., 1992. The measurement of tropospheric OH by long-path absorption. 1. Instrumentation. J. Geophys. Res. 97, 2427–2444.

Mount, G.H., Brault, J.W., Johnston, P.V., Marovich, E., Jakoubek, R.O., Volpe, C.J., Harder, J., Olson, J., 1997. Measurement of tropospheric OH by long-path laser absorption at Fritz peak observatory, Colorado, during the OH photochemistry experiment, fall 1993. J. Geophys. Res. 102, 6393–6413.

Mowbray, T., Schlesinger, W.H., 1988. The buffer capacity of organic soils of the bluff mountain fen, North Carolina. Soil Sci. 146 (2), 73–79.

Moyers, J.L., Ranweiler, L.E., Hopf, S.B., Korte, N.E., 1977. Evaluation of particulate trace species in the Southwest Desert atmosphere. Environ. Sci. Technol. 11, 789–795.

Moyes, A.B., Kueppers, L.M., Pett-Ridge, J., Carper, D.L., Vandehey, N., O'Neil, J., Carolin Frank, A., 2016. Evidence for foliar endophytic nitrogen fixation in a widely distributed subalpine conifer. New Phytol. 210 (2), 657–668.

Mu, Q., Zhao, M., Running, S.W., 2011. Improvements to a MODIS global terrestrial evapotranspiration algorithm. Remote Sens. Environ. 115, 1781–1800.

Muhs, D.R., Bettis, E.A., Been, J., McGeehin, J.P., 2001. Impact of climate and parent material on chemical weathering in loess-derived soils of the Mississippi river valley. Soil Sci. Soc. Am. J. 65, 1761–1777.

Mulder, J., Stein, A., 1994. The solubility of aluminum in acidic forest soils: long-term changes due to acid deposition. Geochim. Cosmochim. Acta 58, 85–94.

Mulder, A., Vandegraaf, A.A., Robertson, L.A., Kuenen, J.G., 1995. Anaerobic ammonium oxidation discovered in a denitrifying fluidized-bed reactor. FEMS Microbiol. Ecol. 16, 177–183.

Mulholland, P.J., 1993. Hydrometric and stream chemistry evidence of three storm flowpaths in Walker branch watershed. J. Hydrol. 151, 129–316.

Mulholland, P.J., 2004. The importance of in-stream uptake for regulating stream concentrations and outputs of N and P from a forested watershed: evidence from long-term chemistry records for Walker Branch Watershed. Biogeochemistry 70 (3), 403–426.

Mulholland, P.J., Elwood, J.W., 1982. The role of lake and reservoir sediments as sinks in the perturbed global carbon cycle. Tellus 34, 490–499.

Mulholland, P.J., Newbold, J.D., Elwood, J.W., Ferren, L.A., Webster, J.R., 1985. Phosphorus spiraling in a woodland stream—seasonal variations. Ecology 66 (3), 1012–1023.

Mulholland, P.J., Helton, A.M., Poole, G.C., Hall, R.O., Hamilton, S.K., Peterson, B.J., Tank, J.L., et al., 2008. Stream denitrification across biomes and its response to anthropogenic nitrate loading. Nature 452 (7184), 202–205.

Mulholland, P.J., Hall, O., Sobota, D.J., Dodds, W.K., Findlay, S.E.G., Grimm, N.B., Hamilton, S.K., et al., 2009. Nitrate removal in stream ecosystems measured by ^{15}N-addition experiments: denitrification. Limnol. Oceanogr. 54, 666–680.

Mulitza, S., Heslop, D., Pittauerova, D., Fischer, H.W., Meyer, I., Stuut, J.-B., Zabel, M., et al., 2010. Increase in African dust flux at the onset of commercial agriculture in the Sahel region. Nature 466 (7303), 226–228.

Muller, P.J., Suess, E., 1979. Productivity, sedimentation rate, and sedimentary organic matter in the oceans. 1. Organic carbon preservation. Deep-Sea Res. Part A—Oceanogr. Res. Pap. 26, 1347–1362.

Muller, B., Maerki, M., Schmid, M., Vologina, E.G., Wehrli, B., Wuest, A., Sturm, M., 2005. Internal carbon and nutrient cycling in Lake Baikal: sedimentation, upwelling, and early diagenesis. Glob. Planet. Chang. 46, 101–124.

Muller, R.D., Sdrolias, M., Gaina, C., Roest, W.R., 2008. Age, spreading rates, and spreading asymmetry of the World's Ocean crust. Geochem. Geophys. Geosyst. 9.

Müller, R.D., Sdrolias, M., Gaina, C., Roest, W.R., 2008. Age, spreading rates, and spreading asymmetry of the World's Ocean crust. Geochem. Geophys. Geosyst. 9(4).

Muller, A.L., Hardy, S.P., Mamet, S.D., Ota, M., Lamb, E.G., Siciliano, S.D., 2017. Salix Arctica changes root distribution and nutrient uptake in response to subsurface nutrients in high Arctic deserts. Ecology 98 (8), 2158–2169.

Mummey, D.L., Smith, J.L., Bolton, H., 1994. Nitrous oxide flux from a shrub-steppe ecosystem: sources and regulation. Soil Biol. Biochem. 26, 279–286.

Munger, J.W., Eisenreich, S.J., 1983. Continental-scale variations in precipitation chemistry. Environ. Sci. Technol. 17, 32–42.

Munger, J.W., Fan, S.M., Bakwin, P.S., Goulden, M.L., Goldstein, A.H., Colman, A.S., Wofsy, S.C., 1998. Regional budgets for nitrogen oxides from continental sources: variations of rates for oxidation and deposition with season and distance from source regions. J. Geophys. Res. D: Atmos. 103, 8355–8368.

Munson, K.M., Lamborg, C.H., Boiteau, R.M., Saito, M.A., 2018. Dynamic mercury methylation and demethylation in oligotrophic marine water. Biogeosciences 15, 6451–6460.

Murakami, M., Hirose, K., Yurimoto, H., Nakashima, S., Takafuji, N., 2002. Water in Earth's lower mantle. Science 295 (5561), 1885–1887.

Murphy, T.P., Lean, D.R., Nalewajko, C., 1976. Blue-Green algae: their excretion of iron-selective chelators enables them to dominate other algae. Science 192 (4242), 900–902.

Murphy, D.M., Thomson, D.S., Mahoney, M.J., 1998. In situ measurements of organics, meteoritic material, mercury, and other elements in aerosols at 5 to 19 kilometers. Science 282 (5394), 1664–1669.

Murray, J.W., Grundmanis, V., 1980. Oxygen consumption in pelagic marine sediments. Science 209 (4464), 1527–1530.

Murray, J.W., Kuivila, K.M., 1990. Organic matter diagenesis in the Northeast Pacific: transition from aerobic red clay to suboxic hemipelagic sediments. Deep-Sea Res. Part A—Oceanogr. Res. Pap. 37, 59–80.

Murray, J.W., Barber, R.T., Roman, M.R., Bacon, M.P., Feely, R.A., 1994. Physical and biological controls on Carbon cycling in the equatorial Pacific. Science 266 (5182), 58–65.

Murray, J.W., Johnson, E., Garside, C., 1995. A U.S. JGOFS process study in the Equatorial Pacific (EqPac): introduction. Deep Sea Res. Part II: Top. Stud. Oceanogr. 42, 275–293.

Murty, S.V.S., Mohapatra, R.K., 1997. Nitrogen and heavy noble gases in ALH 84001: signatures of ancient Martian atmosphere. Geochim. Cosmochim. Acta 61, 5417–5428.

Mushinski, R.M., Phillips, R.P., Payne, Z.C., Abney, R.B., Jo, I., Fei, S., Pusede, S.E., White, J.R., Rusch, D.B., Raff, J.D., 2019. Microbial mechanisms and ecosystem flux estimation for aerobic NO_y emissions from deciduous forest soils. Proc. Natl. Acad. Sci. U. S. A. 116 (6), 2138–2145.

Mustard, J.F., Cooper, C.D., Rifkin, M.K., 2001. Evidence for recent climate change on Mars from the identification of youthful near-surface ground ice. Nature 412 (6845), 411–414.

Myneni, R.B., Keeling, C.D., Tucker, C.J., Asrar, G., Nemani, R.R., 1997. Increased plant growth in the northern high latitudes from 1981 to 1991. Nature 386, 698–702.

Myneni, S.C., Brown, J.T., Martinez, G.A., Meyer-Ilse, W., 1999. Imaging of humic substance macromolecular structures in water and soils. Science 286 (5443), 1335–1337.

Myneni, R.B., Dong, J., Tucker, C.J., Kaufmann, R.K., Kauppi, P.E., Liski, J., Zhou, L., Alexeyev, V., Hughes, M.K., 2001. A large carbon sink in the woody biomass of northern forests. Proc. Natl. Acad. Sci. U. S. A. 98 (26), 14784–14789.

Myriokefalitakis, S., Nenes, A., Baker, A.R., Mihalopoulos, N., Kanakidou, M., 2016. Bioavailable atmospheric phosphorous supply to the global ocean: a 3-D global modeling study. Biogeosciences 13 (24), 6519–6543.

Myrold, D.D., Matson, P.A., Peterson, D.L., 1989. Relationships between soil microbial properties and above-ground stand characteristics of conifer forests in Oregon. Biogeochemistry 8, 265–281.

Nacry, P., Bouguyon, E., Gojon, A., 2013. Nitrogen acquisition by roots: physiological and developmental mechanisms ensuring plant adaptation to a fluctuating resource. Plant Soil 370 (1), 1–29.

Nadelhoffer, K.F., Fry, B., 1988. Controls on natural ^{15}N and ^{13}C abundances in forest soil organic matter. Soil Sci. Soc. Am J. 52, 1633–1640.

Nadelhoffer, K.J., Raich, J.W., 1992. Fine root production estimates and belowground carbon allocation in forest ecosystems. Ecology 73, 1139–1147.

Nadelhoffer, K.J., Aber, J.D., Melillo, J.M., 1984. Seasonal patterns of ammonium and nitrate uptake in nine temperate forest ecosystems. Plant Soil 80, 321–335.

Nadelhoffer, K.J., Colman, B.P., Currie, W.S., Magill, A., Aber, J.D., 2004. Decadal-scale fates of ^{15}N tracers added to oak and pine stands under ambient and elevated N inputs at the Harvard Forest (USA). For. Ecol. Manag. 196, 89–107.

Naeem, S., Thompson, L.J., Lawler, S.P., Lawton, J.H., Woodfin, R.M., 1994. Declining biodiversity can Alter the performance of ecosystems. Nature 368, 734–737.

Najjar, R.G., Keeling, R.F., 1997. Analysis of the mean annual cycle of the dissolved oxygen anomaly in the world ocean. J. Mar. Res. 55, 117–151.

Najjar, R.G., Erickson, D.J., Madronich, S., 1995. Modeling the air-sea fluxes of gases formed from the decomposition of dissolved organic matter: carbonyl sulfide and carbon monoxide. In: Zepp, R.G., Sonntag, C. (Eds.), The Role of Nonliving Organic Matter in the Earth's Carbon Cycle. Wiley, pp. 107–132.

Nakagawa, T., Spiegelman, M.W., 2017. Global-scale water circulation in the Earth's mantle: implications for the mantle water budget in the early earth. Earth Planet. Sci. Lett. 464 (April), 189–199.

Nakagawa, F., Suzuki, A., Daita, S., Ohyama, T., Komatsu, D.D., Tsunogai, U., 2013. Tracing atmospheric nitrate in groundwater using triple oxygen isotopes: evaluation based on bottled drinking water. Biogeosciences 10 (6), 3547–3558.

Nance, J.D., Hobbs, P.V., Radke, L.F., 1993. Airborne measurements of gases and particles from an Alaskan wildfire. J. Geophys. Res.—Atmos. 98, 14873–14882.

Napieralski, S.A., Buss, H.L., Brantley, S.L., Lee, S., Xu, H., Roden, E.E., 2019. Microbial Chemolithotrophy Mediates Oxidative Weathering of Granitic Bedrock. Proceedings of the National Academy of Sciences of the United States of America, December.

Naqvi, S.W., Jayakumar, D.A., Narvekar, P.V., Naik, H., Sarma, V.V., D'Souza, W., Joseph, S., George, M.D., 2000. Increased marine production of N_2O due to intensifying anoxia on the Indian continental shelf. Nature 408 (6810), 346–349.

Nasholm, T., Ekblad, A., Nordin, A., Reiner, G., Hogberg, M., Hogberg, P., Näsholm, T., Högberg, M., Högberg, P., 1998. Boreal forest plants take up organic nitrogen. Nature 392 (6679), 914–916.

Nassar, R., Hill, T.G., McLinden, C.A., Wunch, D., Jones, D.B.A., Crisp, D., 2017. Quantifying CO_2 emissions from individual power plants from space. Geophys. Res. Lett. 44 (19), 38.

Nasto, M.K., Alvarez-Clare, S., Lekberg, Y., Sullivan, B.W., Townsend, A.R., Cleveland, C.C., 2014. Interactions among nitrogen fixation and soil phosphorus acquisition strategies in lowland tropical rain forests. Ecol. Lett. 17 (10), 1282–1289.

Nasto, M.K., Winter, K., Turner, B.L., Cleveland, C.C., 2019. Nutrient acquisition strategies augment growth in tropical N_2-fixing trees in nutrient-poor soil and under elevated CO_2. Ecology 100 (4), e02646.

Nault, B.A., Laughner, J.L., Wooldridge, P.J., Crounse, J.D., Dibb, J., Diskin, G., Peischl, J., et al., 2017. Lightning NO_X emissions: reconciling measured and modeled estimates with updated NO_X chemistry: LNO$_X$ emission with updated NO_X chemistry. Geophys. Res. Lett. 44 (18), 9479–9488.

Navarre-Sitchler, A., Thyne, G., 2007. Effects of carbon dioxide on mineral weathering rates at earth surface conditions. Chem. Geol. 243, 53–63.

Nave, L.E., Vance, E.D., Swanston, C.W., Curtis, P.S., 2009. Impacts of elevated N inputs on North temperate Forest Soil C Storage, C/N, and net N-mineralization. Geoderma 153, 231–240.

Neef, L., van Weele, M., van Velthoven, P., 2010. Optimal estimation of the present-day global methane budget. Glob. Biogeochem. Cycles 24.

Neff, J.C., Holland, E.A., Dentener, F.J., McDowell, W.H., Russell, K.M., 2002a. The origin, composition and rates of organic nitrogen deposition: a missing piece of the nitrogen cycle? Biogeochemistry 57, 99–136.

Neff, J.C., Townsend, A.R., Gleixner, G., Lehman, S.J., Turnbull, J., Bowman, W.D., 2002b. Variable effects of nitrogen additions on the stability and turnover of soil carbon. Nature 419 (6910), 915–917.

Neill, C., 1992. Comparison of soil coring and ingrowth methods for measuring belowground production. Ecology 73, 1918–1921.

Neilson, R.P., Marks, D., 1994. A global perspective of regional vegetation and hydrologic sensitivities from climatic change. J. Veg. Sci. 5, 715–730.

Nelson, D.W., 1982. Gaseous losses of nitrogen other than through denitrification. In: Stevenson, F.J. (Ed.), Nitrogen in Agricultural Soils. American Society of Agronomy, Madison, Wisconsin, pp. 327–363.

Nelson, D.M., Treguer, P., Brzezinski, M.A., Leynaert, A., Queguiner, B., 1995. Production and dissolution of biogenic silica in the ocean: revised global estimates, comparison with regional data and relationship to biogenic sedimentation. Glob. Biogeochem. Cycles 9, 359–372.

Nelson, D.M., Anderson, R.F., Barber, R.T., Brzezinski, M.A., Buesseler, K.O., Chase, Z., Collier, R.W., et al., 2002. Vertical budgets for organic carbon and biogenic silica in the Pacific sector of the Southern Ocean, 1996–1998. Deep-Sea Res. Part II-Top. Stud. Oceanogr. 49, 1645–1674.

Nelson, R., Gobakken, T., Næsset, E., Gregoire, T.G., Ståhl, G., Holm, S., Flewelling, J., 2012. Lidar sampling—using an airborne profiler to estimate forest biomass in Hedmark County, Norway. Remote Sens. Environ. 123 (August), 563–578.

Nemani, R.R., Keeling, C.D., Hashimoto, H., Jolly, W.M., Piper, S.C., Tucker, C.J., Myneni, R.B., Running, S.W., 2003. Climate-driven increases in global terrestrial net primary production from 1982 to 1999. Science 300 (5625), 1560–1563.

Nepstad, D.C., Decarvalho, C.R., Davidson, E.A., Jipp, P.H., Lefebvre, P.A., Negreiros, G.H., Dasilva, E.D., Stone, T.A., Trumbore, S.E., Vieira, S., 1994. The role of deep roots in the hydrological and carbon cycles of Amazonian forests and pastures. Nature 372, 666–669.

Nerem, R.S., Beckley, B.D., Fasullo, J.T., Hamlington, B.D., Masters, D., Mitchum, G.T., 2018. Climate-change-driven accelerated sea-level rise detected in the altimeter era. Proc. Natl. Acad. Sci. U. S. A. 115 (9), 2022–2025.

Nesme, T., Metson, G.S., Bennett, E.M., 2018. Global phosphorus flows through agricultural trade. Glob. Environ. Change: Hum. Policy Dimens. 50 (May), 133–141.

Nettleton, W.D., Peterson, F.F., 2011. Landform, soil, and plant relationships to nitrate accumulation, central Nevada. Geoderma 160, 265–270.

Neubauer, S.C., Givler, K., Valentine, S., Megonigal, J.P., 2005. Seasonal patterns and plant-mediated controls of subsurface wetland biogeochemistry. Ecology 86 (12), 3334–3344.

Neue, H.U., Wassmann, R., Kludze, H.K., Wang, B., Lantin, R.S., 1997. Factors and processes controlling methane emissions from rice fields. Nutr. Cycl. Agroecosyst. 49 (1–3), 111–117.

Neukum, G., 1977. Lunar cratering. Philos. Trans. R. Soc. Lond. Ser. A—Math. Phys. Eng. Sci. 285, 267–272.

Neukum, G., Jaumann, R., Hoffmann, H., Hauber, E., Head, J.W., Basilevsky, A.T., Ivanov, B.A., et al., 2004. Recent and episodic volcanic and glacial activity on Mars revealed by the high resolution stereo camera. Nature 432 (7020), 971–979.

Neumann, M., Ukonmaanaho, L., Johnson, J., Benham, S., Vesterdal, L., Novotný, R., Verstraeten, A., et al., 2018. Quantifying carbon and nutrient input from Litterfall in European forests using field observations and modeling. Glob. Biogeochem. Cycles 32 (5), 784–798.

Nevison, C.D., Lueker, T.J., Weiss, R.F., 2004. Quantifying the nitrous oxide source from coastal upwelling. Glob. Biogeochem. Cycles. 18.

Nevison, C., Andrews, A., Thoning, K., Dlugokencky, E., Sweeney, C., Miller, S., Saikawa, E., et al., 2018. Nitrous oxide emissions estimated with the CarbonTracker-Lagrange North American regional inversion framework. Glob. Biogeochem. Cycles 32 (3), 463–485.

Newbold, J.D., 1992. Cycles and spirals of nutrients. In: Calow, P., Petts, G. (Eds.), Rivers Handbook. Blackwell Publishing Ltd, pp. 379–408.

Newbold, J.D., Elwood, J.W., O'Neill, R.V., VanWinkle, W., 1981. Measuring nutrient spiraling in streams. Can. J. Fish. Aquat. Sci. 38, 860–863.

Newman, E.I., 1995. Phosphorus inputs to terrestrial ecosystems. J. Ecol. 83, 713–726.

Newman, E.I., Andrews, R.E., 1973. Uptake of phosphorus and potassium in relation to root growth and density. Plant Soil 38, 49–69.

Newman, D.K., Banfield, J.F., 2002. Geomicrobiology: how molecular-scale interactions underpin biogeochemical systems. Science 296 (5570), 1071–1077.

Newsom, H.E., Sims, K.W., 1991. Core formation during early accretion of the earth. Science 252 (5008), 926–933.

Nezat, C.A., Blum, J.D., Klaue, A., Johnson, C.E., Siccama, T.G., 2004. Influence of landscape position and vegetation on long-term weathering rates at the Hubbard Brook experimental forest, New Hampshire, USA. Geochim. Cosmochim. Acta 68, 3065–3078.

Nguyen, B.C., Mihalopoulos, N., Putaud, J.P., Bonsang, B., 1995. Carbonyl sulfide emissions from biomass burning in the tropics. J. Atmos. Chem. 22, 55–65.

Ni, X., Groffman, P.M., 2018. Declines in methane uptake in forest soils. Proc. Natl. Acad. Sci. U. S. A. 115 (34), 8587–8590.

Nicewonger, M.R., Aydin, M., Prather, M.J., Saltzman, E.S., 2018. Large changes in biomass burning over the last millennium inferred from paleoatmospheric ethane in polar ice cores. Proc. Natl. Acad. Sci. U. S. A. 115 (49), 12413–12418.

Nicholls, R.J., 2004. Coastal flooding and wetland loss in the 21st century: changes under the SRES climate and socio-economic scenarios. Glob. Environ. Change—Hum. Policy Dimens. 14, 69–86.

Nielsen, H., 1974. Isotopic composition of the major contributors to atmospheric sulfur. Tellus 26, 213–221.

Nielsen, D.R., Vangenuchten, M.T., Biggar, J.W., 1986. Water flow and solute transport processes in the unsaturated zone. Water Resour. Res. 22, S89–108.

Niemann, H.B., Atreya, S.K., Carignan, G.R., Donahue, T.M., Haberman, J.A., Harpold, D.N., Hartle, R.E., et al., 1996. The Galileo probe mass spectrometer: composition of Jupiter's atmosphere. Science 272 (5263), 846–849.

Niemann, H.B., Atreya, S.K., Bauer, S.J., Carignan, G.R., Demick, J.E., Frost, R.L., Gautier, D., et al., 2005. The abundances of constituents of Titan's atmosphere from the GCMS instrument on the Huygens probe. Nature 438 (7069), 779–784.

Niklas, K.J., Enquist, B.J., 2002. On the vegetative biomass partitioning of seed plant leaves, stems, and roots. Am. Nat. 159 (5), 482–497.

Niles, P.B., Boynton, W.V., Hoffman, J.H., Ming, D.W., Hamara, D., 2010. Stable isotope measurements of martian atmospheric CO_2 at the Phoenix landing site. Science 329 (5997), 1334–1337.

Nilsson, C., Reidy, C.A., Dynesius, M., Revenga, C., 2005. Fragmentation and flow regulation of the world's large river systems. Science 308 (5720), 405–408.

Nimmo, F., Pappalardo, R.T., 2016. Ocean worlds in the outer solar system. J. Geophys. Res. Planets 121 (8), 1378–1399.

Nisbet, E.G., Ingham, B., 1995. Methane output from natural and quasinatural sources: a review of the potential for change and for biotic and abiotic feedbacks. In: Woodwell, G.M., MacKenzie, F.T. (Eds.), Biotic Feedbacks in the Global Climatic System. Oxford University Press, pp. 189–218.

Nisbet, R.E.R., Fisher, R., Nimmo, R.H., Bendall, D.S., Crill, P.M., Gallego-Sala, A.V., Hornibrook, E.R.C., et al., 2009. Emission of methane from plants. Proc. Biol. Sci. 276 (1660), 1347–1354.

Nisbet, E.G., Dlugokencky, E.J., Manning, M.R., Lowry, D., Fisher, R.E., France, J.L., Michel, S.E., et al., 2016. Rising atmospheric methane: 2007–2014 growth and isotopic shift: rising methane 2007–2014. Glob. Biogeochem. Cycles 30 (9), 1356–1370.

Nisbet, E.G., Manning, M.R., Dlugokencky, E.J., Fisher, R.E., Lowry, D., Michel, S.E., Lund Myhre, C., et al., 2019. Very strong atmospheric methane growth in the 4 years 2014–2017: implications for the Paris agreement. Glob. Biogeochem. Cycles 33 (3), 318–342.

Nixon, S.W., 1980. Between coastal marshes and coastal waters—a review of twenty years of speculation and research on the role of salt marshes in estuarine productivity and water chemistry. In: Hamilton, P.B., MacDonald, K.B. (Eds.), Estuarine and Wetland Processes. Plenum, pp. 437–525.

Nixon, S.W., 2003. Replacing the Nile: are anthropogenic nutrients providing the fertility once brought to the Mediterranean by a Great River? Ambio 32 (1), 30–39.

Nixon, S.W., Granger, S.L., Nowicki, B.L., 1995. An assessment of the annual mass balance of carbon, nitrogen, and phosphorus in Narragansett Bay. Biogeochemistry 31, 15–61.

Nixon, S.W., Ammerman, J.W., Atkinson, L.P., Berounsky, V.M., Billen, G., Boicourt, W.C., Boynton, W.R., et al., 1996. The fate of nitrogen and phosphorus at the land sea margin of the North Atlantic Ocean. Biogeochemistry 35, 141–180.

Nodvin, S.C., Driscoll, C.T., Likens, G.E., 1988. Soil processes and sulfate loss at the Hubbard Brook experimental forest. Biogeochemistry 5, 185–199.

Nogueira, E.M., Fearnside, P.M., Nelson, B.W., Barbosa, R.I., Keizer, E.W.H., 2008. Estimates of forest biomass in the Brazilian Amazon: new allometric equations and adjustments to biomass from wood-volume inventories. For. Ecol. Manag. 256, 1853–1867.

Nogueira, E.M., Yanai, A.M., Fonseca, F.O.R., Fearnside, P.M., 2015. Carbon stock loss from deforestation through 2013 in Brazilian Amazonia. Glob. Chang. Biol. 21 (3), 1271–1292.

Nolan, B.T., Hitt, K.J., Ruddy, B.C., 2002. Probability of nitrate contamination of recently recharged groundwaters in the conterminous United States. Environ. Sci. Technol. 36 (10), 2138–2145.

Norby, R.J., Iversen, C.M., 2006. Nitrogen uptake, distribution, turnover, and efficiency of use in a CO_2-enriched sweetgum forest. Ecology 87 (1), 5–14.

Norby, R.J., Zak, D.R., 2011. Ecological lessons from free-air CO_2 enrichment (FACE) experiments. Annu. Rev. Ecol. Evol. Syst. 42 (1), 181–203.

Norby, R.J., Delucia, E.H., Gielen, B., Calfapietra, C., Giardina, C.P., King, J.S., Ledford, J., et al., 2005. Forest response to elevated CO_2 is conserved across a broad range of productivity. Proc. Natl. Acad. Sci. U. S. A. 102 (50), 18052–18056.

Norby, R.J., Warren, J.M., Iversen, C.M., Medlyn, B.E., McMurtrie, R.E., 2010. CO_2 enhancement of forest productivity constrained by limited nitrogen availability. Proc. Natl. Acad. Sci. U. S. A. 107 (45), 19368–19373.

Nordin, A., Schmidt, I.K., Shaver, G.R., 2004. Nitrogen uptake by Arctic soil microbes and plants in relation to soil nitrogen supply. Ecology 85, 955–962.

Norman, J.S., Lin, L., Barrett, J.E., 2015. Paired carbon and nitrogen metabolism by ammonia-oxidizing bacteria and archaea in temperate forest soils. Ecosphere. 6, https://doi.org/10.1890/ES14-00299.1.

Norris, C.A., Wood, B.J., 2017. Earth's volatile contents established by melting and vaporization. Nature 549 (7673), 507–510.

Norton, A.J., Rayner, P.J., Koffi, E.N., Scholze, M., Silver, J.D., Wang, Y.-P., 2019. Estimating global gross primary productivity using chlorophyll fluorescence and a data assimilation system with the BETHY-SCOPE model. Biogeosciences 16 (15), 3069–3093.

Norval, M., Cullen, A.P., de Gruijl, F.R., Longstreth, J., Takizawa, Y., Lucas, R.M., Noonan, F.P., van der Leun, J.C., 2007. The effects on human health from stratospheric ozone depletion and its interactions with climate change. Photochem. Photobiol. Sci.: Off. J. Eur. Photochem. Assoc. Eur. Soc. Photobiol. 6 (3), 232–251.

Notholt, J., Kuang, Z., Rinsland, C.P., Toon, G.C., Rex, M., Jones, N., Albrecht, T., et al., 2003. Enhanced upper tropical tropospheric COS: impact on the stratospheric aerosol layer. Science 300 (5617), 307–310.

Notz, D., Stroeve, J., 2016. Observed Arctic sea-ice loss directly follows anthropogenic CO_2 emission. Science 354 (6313), 747–750.

Novak, M., Kirchner, J.W., Fottova, D., Prechova, E., Jackova, I., Kram, P., Hruska, J., 2005. Isotopic evidence for processes of sulfur retention/release in 13 forested catchments spanning a strong pollution gradient (Czech Republic, Central Europe). Glob. Biogeochem. Cycles 19.

Novelli, P.C., Masarie, K.A., Tans, P.P., Lang, P.M., 1994. Recent changes in atmospheric carbon monoxide. Science 263 (5153), 1587–1590.

Novelli, P.C., Masarie, K.A., Lang, P.M., Hall, B.D., Myers, R.C., Elkins, J.W., 2003. Reanalysis of tropospheric CO trends: effects of the 1997–1998 wildfires. J. Geophys. Res.-Atmos. 108.

Nowak, D.J., Greenfield, E.J., Hoehn, R.E., Lapoint, E., 2013. Carbon Storage and sequestration by trees in urban and community areas of the United States. Environ. Pollut. 178 (July), 229–236.

Nowinski, N.S., Taneva, L., Trumbore, S.E., Welker, J.M., 2010. Decomposition of old organic matter as a result of deeper active layers in a snow depth manipulation experiment. Oecologia 163 (3), 785–792.

Nowlan, C.R., Martin, R.V., Philip, S., Lamsal, L.N., Krotkov, N.A., Marais, E.A., Wang, S., Zhang, Q., 2014. Global dry deposition of nitrogen dioxide and sulfur dioxide inferred from space-based measurements. Glob. Biogeochem. Cycles 28 (10), 1025–1043.

Nozette, S., Lewis, J.S., 1982. Venus: chemical weathering of igneous rocks and buffering of atmospheric composition. Science 216 (4542), 181–183.

Nriagu, J.O., 1989. A global assessment of natural sources of atmospheric trace metals. Nature 338, 47–49.

Nriagu, J.O., Coker, R.D., 1978. Isotopic composition of sulfur in precipitation within the Great Lakes Basin. Tellus 30, 365–375.

Nriagu, J.O., Holdway, D.A., Coker, R.D., 1987. Biogenic sulfur and the acidity of rainfall in remote areas of Canada. Science 237 (4819), 1189–1192.

Nriagu, J.O., Coker, R.D., Barrie, L.A., 1991. Origin of sulfur in Canadian Arctic haze from isotope measurements. Nature 349, 142–145.

Nugent, M.A., Brantley, S.L., Pantano, C.G., Maurice, P.A., 1998. The influence of natural mineral coatings on feldspar weathering. Nature 395, 588–591.

Nye, P.H., 1977. The rate-limiting step in plant nutrient absorption from soil. Soil Sci. 123, 292–297.

Nye, P.H., 1981. Changes of pH across the rhizosphere induced by roots. Plant Soil 61, 7–26.

O'D Alexander, C.M., Bowden, R., Fogel, M.L., Howard, K.T., Herd, C.D.K., Nittler, L.R., 2012. The provenances of asteroids, and their contributions to the volatile inventories of the terrestrial planets. Science 337 (6095), 721–723.

O'Hara, G.W., Boonkerd, N., Dilworth, M.J., 1988. Mineral constraints to nitrogen fixation. Plant Soil 108 (1), 93–110.

O'Leary, M.H., 1988. Carbon isotopes in photosynthesis. Bioscience 38, 328–336.

O'Neil, J., Carlson, R.W., Francis, D., Stevenson, R.K., 2008. Neodymium-142 evidence for Hadean Mafic Crust. Science 321, 1828–1831.

O'Neill, R.V., De Angelis, D.L., 1981. Comparative productivity and biomass relations of forest ecosystems. In: Reichle, D.E. (Ed.), Dynamic Properties of Forest Ecosystems. Cambridge University Press, pp. 411–449.

Oades, J.M., 1988. The retention of organic matter in soils. Biogeochemistry 5, 35–70.

Oaks, A., 1992. A reevaluation of nitrogen assimilation in roots. Bioscience 42, 103–111.

Oaks, A., 1994. Primary nitrogen assimilation in higher plants and its regulation. Can. J. Bot. 72, 739–750.

Oberg, G., Holm, M., Sanden, P., Svensson, T., Parikka, M., 2005. The role of organic-matter-bound chlorine in the chlorine cycle: a case study of the Stubbetorp catchment, Sweden. Biogeochemistry 75, 241–269.

Öberg, K.I., Guzmán, V.V., Furuya, K., Qi, C., Aikawa, Y., Andrews, S.M., Loomis, R., Wilner, D.J., 2015. The comet-like composition of a protoplanetary disk as revealed by complex cyanides. Nature 520 (7546), 198–201.

Oberhummer, H., Csótó, A., Schlattl, H., 2000. Stellar production rates of carbon and its abundance in the universe. Science 289 (5476), 88–90.

Obrist, D., Agnan, Y., Jiskra, M., Olson, C.L., Colegrove, D.P., Hueber, J., Moore, C.W., Sonke, J.E., Helmig, D., 2017. Tundra uptake of atmospheric elemental mercury drives Arctic mercury pollution. Nature 547 (7662), 201–204.

Oczkowski, A.J., Nixon, S.W., Granger, S.L., El-sayed, A.-F.M., Mckinney, R.A., 2009. Anthropogenic enhancement of Egypt ' S Mediterranean fishery. Proc. Natl. Acad. Sci. U. S. A. 106, 10–13.

Oda, T., Green, M.B., Urakawa, R., Scanlon, T.M., Sebestyen, S.D., McGuire, K.J., Katsuyama, M., Fukuzawa, K., Adams, M.B., Ohte, N., 2018. Stream runoff and nitrate recovery times after forest disturbance in the USA and Japan. Water Resour. Res. 54 (9), 6042–6054.

Odada, E.O., 1992. Growth rates of ferromanganese encrustations on rocks from the Romanche fracture zone, equatorial Atlantic. Deep-Sea Res. Part A—Oceanogr. Res. Pap. 39, 235–244.

Odasz-Albrigtsen, A.M., Tommervik, H., Murphy, P., 2000. Decreased photosynthetic efficiency in plant species exposed to multiple airborne pollutants along the Russian-Norwegian border. Can. J. Bot. 78, 1021–1033.

Odum, H.T., 1956. Primary production in flowing waters. Limnol. Oceanogr. 1 (2), 102–117.

Odum, H.T., 1957. Trophic structure and productivity of Silver Springs. Ecol. Monogr. 27 (1), 55–112.

Odum, H.T., Hoskin, C.M., 1958. Comparative studies on the metabolism of marine waters. Publ. Inst. Mar. Sci. 5, 16–46.

Odum, E.P., Finn, J.T., Franz, E.H., 1979. Subsidy-stress. Bioscience 29 (6), 349–352.

Oechel, W.C., Laskowski, C.A., Burba, G., Gioli, B., Kalhori, A.A.M., 2014. Annual patterns and budget of CO_2 flux in an Arctic tussock tundra ecosystem. J. Geophys. Res. Biogeosci. 119 (3), 323–339.

Oesterheld, M., Sala, O.E., McNaughton, S.J., 1992. Effect of animal husbandry on herbivore-carrying capacity at a regional scale. Nature 356 (6366), 234–236.

Officer, C.B., Ryther, J.H., 1980. The possible importance of silicon in marine eutrophication. Mar. Ecol. Prog. Ser. 3 (1), 83–91.

Ogawa, H., Amagai, Y., Koike, I., Kaiser, K., Benner, R., 2001. Production of refractory dissolved organic matter by bacteria. Science 292 (5518), 917–920.

Oh, N.H., Richter, D.D., 2005. Elemental translocation and loss from three highly weathered Soil-bedrock profiles in the southeastern United States. Geoderma 126, 5–25.

Ohmoto, H., Kakegawa, T., Lowe, D.R., 1993. 3.4-billion-year-old biogenic pyrites from Barberton, South Africa: sulfur isotope evidence. Science 262 (October), 555–557.

Ohte, N., Tokuchi, N., 1999. Geographical variation of the acid buffering of vegetated catchments: factors determining the bicarbonate leaching. Glob. Biogeochem. Cycles 13, 969–996.

Oki, T., Kanae, S., 2006. Global hydrological cycles and world water resources. Science 313 (5790), 1068–1072.

Okin, G.S., Mahowald, N., Chadwick, O.A., Artaxo, P., 2004. Impact of desert dust on the biogeochemistry of phosphorus in terrestrial ecosystems. Glob. Biogeochem. Cycles 18.

Olefeldt, D., Roulet, N.T., 2014. Permafrost conditions in Peatlands regulate magnitude, timing, and chemical composition of catchment dissolved organic carbon export. Glob. Chang. Biol. 20 (10), 3122–3136.

Ollinger, S.V., 2011. Sources of variability in canopy reflectance and the convergent properties of plants. New Phytol. 189 (2), 375–394.

Ollinger, S.V., Smith, M.L., 2005. Net primary production and canopy nitrogen in a temperate forest landscape: an analysis using imaging spectroscopy, modeling and field data. Ecosystems 8, 760–778.

Ollinger, S.V., Aber, J.D., Lovett, G.M., Millham, S.E., Lathrop, R.G., Ellis, J.M., 1993. A spatial model of atmospheric deposition for the northeastern United States. Ecol. Appl. 3, 459–472.

Ollinger, S.V., Smith, M.L., Martin, M.E., Hallett, R.A., Goodale, C.L., Aber, J.D., 2002. Regional variation in foliar chemistry and N cycling among forests of diverse history and composition. Ecology 83, 339–355.

Ollinger, S.V., Richardson, A.D., Martin, M.E., Hollinger, D.Y., Frolking, S.E., Reich, P.B., Plourde, L.C., et al., 2008. Canopy nitrogen, carbon assimilation, and albedo in temperate and boreal forests: functional relations and potential climate feedbacks. Proc. Natl. Acad. Sci. U. S. A. 105 (49), 19336–19341.

Olsen, S.R., Cole, C.V., Watanabe, F.S., Dean, L.A., 1954. Estimation of Available Phosphorus in Soils by Extraction with Sodium Bicarbonate. U.S. Department of Agriculture.

Olson, J.S., 1963. Energy storage and the balance of producers and decomposers in ecological systems. Ecology 44, 322–331.

Olson, R.K., Reiners, W.A., 1983. Nitrification in subalpine balsam fir soils: tests for inhibitory factors. Soil Biol. Biochem. 15, 413–418.

Olson, J.S., Watts, J.A., Allison, L.J., 1983. Carbon in Live Vegetation of Major World Ecosystems. National Technical Information Service, Springfield, Virginia.

Olson, J.S., Garrels, R.M., Berner, R.A., Armentano, T.V., Dyer, M.I., Yaalon, D.H., 1985. The natural carbon cycle. In: Trabalka, J.R. (Ed.), Atmospheric Carbon Dioxide and the Global Carbon Cycle. U.S. Department of Energy, pp. 175–213.

Olson, S.L., Reinhard, C.T., Lyons, T.W., 2016. Cyanobacterial diazotrophy and Earth's delayed oxygenation. Front. Microbiol. 7 (September), 1526.

Olson, C., Jiskra, M., Biester, H., Chow, J., Obrist, D., 2018. Mercury in active-layer tundra soils of Alaska: concentrations, pools, origins, and spatial distribution. Glob. Biogeochem. Cycles 32 (7), 1058–1073.

Olsson, M., Melkerud, P.A., 1989. Chemical and mineralogical changes during genesis of a Podzol from till in southern Sweden. Geoderma 45, 267–287.

Olsson, M.T., Melkerud, P.A., 2000. Weathering in three Podzolized Pedons on glacial deposits in northern Sweden and Central Finland. Geoderma 94, 149–161.

Oltmans, S.J., Hofmann, D.J., 1995. Increase in lower stratospheric water vapor at a mid-latitude northern hemisphere site from 1981 to 1994. Nature 374, 146–149.

Oltmans, S.J., Levy, H., 1992. Seasonal cycle of surface ozone over the Western North Atlantic. Nature 358, 392–394.

Oort, A.H., Anderson, L.A., Peixoto, J.P., 1994. Estimates of the energy cycle of the oceans. J. Geophys. Res. 99, 7665–7688.

Oppenheimer, C., 2003. Volcanic degassing. Treat. Geochem.

Oppenheimer, M., Epstein, C.B., Yuhnke, R.E., 1985. Acid deposition, smelter emissions, and the linearity issue in the Western United States. Science 229 (4716), 859–862.

Opsahl, S., Benner, R., 1997. Distribution and cycling of terrigenous dissolved organic matter in the ocean. Nature 386, 480–482.

Ordonez, J.C., van Bodegom, P.M., Witte, J.P.M., Wright, I.J., Reich, P.B., Aerts, R., 2009. A global study of relationships between leaf traits, climate and soil measures of nutrient fertility. Glob. Ecol. Biogeogr.: J. Macroecol. 18, 137–149.

Oren, R., Werk, K.S., Schulze, E.-D., Meyer, J., Schneider, B.U., Schramel, P., 1988. Performance of two *Picea abies* (L.) karst. Stands at different stages of decline: VI. Nutrient concentration. Oecologia 77 (2), 151–162.

Orgel, L.E., 1992. Molecular replication. Nature 358 (6383), 203–209.

Orgel, L.E., 1994. The origin of life on the earth. Sci. Am. 271, 77–83.

Orians, K.J., Bruland, K.W., 1986. The biogeochemistry of aluminum in the Pacific Ocean. Earth Planet. Sci. Lett. 78, 397–410.

Orians, K.J., Boyle, E.A., Bruland, K.W., 1990. Dissolved titanium in the open ocean. Nature 348, 322–325.

Orihel, D.M., Bird, D.F., Brylinsky, M., Chen, H., Donald, D.B., Huang, D.Y., Giani, A., et al., 2012. High microcystin concentrations occur only at low nitrogen-to-phosphorus ratios in nutrient-rich Canadian lakes. Can. J. Fish. Aquat. Sci. 69 (9), 1457–1462.

Orosei, R., Lauro, S.E., Pettinelli, E., Cicchetti, A., Coradini, M., Cosciotti, B., Di Paolo, F., et al., 2018. Radar evidence of subglacial liquid water on Mars. Science 361 (6401), 490–493.

Orr, J.C., Fabry, V.J., Aumont, O., Bopp, L., Doney, S.C., Feely, R.A., Gnanadesikan, A., et al., 2005. Anthropogenic ocean acidification over the twenty-first century and its impact on calcifying organisms. Nature 437 (7059), 681–686.

Orsi, W.D., Edgcomb, V.P., Christman, G.D., Biddle, J.F., 2013. Gene expression in the deep biosphere. Nature 499 (7457), 205–208.

Ortega, J., Helmig, D., Daly, R.W., Tanner, D.M., Guenther, A.B., Herrick, J.D., 2008. Approaches for quantifying reactive and low-volatility biogenic organic compound emissions by vegetation enclosure techniques—part B: applications. Chemosphere 72 (3), 365–380.

Osborn, T.J., 2006. Recent variations in the Winter North Atlantic oscillation. Wea 61 (12), 353–355.

Osborne, B.B., Nasto, M.K., Asner, G.P., Balzotti, C.S., Cleveland, C.C., Sullivan, B.W., Taylor, P.G., Townsend, A.R., Porder, S., 2017. Climate, topography, and canopy chemistry exert hierarchical control over soil N cycling in a neotropical lowland forest. Ecosystems 20 (6), 1089–1103.

Oschlies, A., Garcon, V., 1998. Eddy-induced enhancement of primary production in a model of the North Atlantic Ocean. Nature 394, 266–269.

Osmond, C.B., Winter, K., Ziegler, H., 1982. Functional significance of different pathways of CO_2 fixation in photosynthesis. In: Person, A., Zimmerman, M.H. (Eds.), Encyclopedia of Plant Physiology. Springer, pp. 479–547.

Ostendorf, B., Reynolds, J.F., 1993. Relationships between a terrain-based hydrologic model and patch-scale vegetation patterns in an Arctic tundra landscape. Landsc. Ecol. 8, 229–237.

Ostlund, H.G., 1983. Tritium and Radiocarbon, TTO Western North Atlantic Section, GEOSECS Reoccupation. Rosentiel School of Marine and Atmospheric Sciences, University of Miami.

Ostman, N.L., Weaver, G.T., 1982. Autumnal nutrient transfers by re-translocation, leaching, and litter fall in a chestnut oak Forest in southern Illinois. Can. J. Forest Res. 12, 40–51.

Ostrofsky, M.L., Stolarski, A.G., Dagen, K.A., 2018. Export of total, particulate, and apatite phosphorus from forested and agricultural watersheds. J. Environ. Qual. 47 (1), 106–112.

Oudot, C., Andrie, C., Montel, Y., 1990. Nitrous oxide production in the tropical Atlantic ocean. Deep-Sea Res. Part A—Oceanogr. Res. Pap. 37, 183–202.

Ouimet, R., Moore, J.-D., 2015. Effects of fertilization and liming on tree growth, vitality and nutrient status in boreal balsam fir stands. For. Ecol. Manag. 345 (June), 39–49.

Outridge, P.M., Mason, R.P., Wang, F., Guerrero, S., Heimbürger-Boavida, L.E., 2018. Updated global and oceanic mercury budgets for the United Nations Global mercury assessment 2018. Environ. Sci. Technol. 52 (20), 11466–11477.

Ovenden, L., 1990. Peat accumulation in northern wetlands. Quat. Res. 33, 377–386.

Owen, T., Biemann, K., 1976. Composition of the atmosphere at the surface of Mars: detection of Argon-36 and preliminary analysis. Science 193 (4255), 801–803.

Owen, R.M., Rea, D.K., 1985. Sea-floor hydrothermal activity links climate to tectonics: the eocene carbon dioxide greenhouse. Science 227 (4683), 166–169.

Owen, D.F., Wiegert, R.G., 1976. Do consumers maximize plant fitness? Oikos 27, 488–492.

Owen, T., Maillard, J.P., Debergh, C., Lutz, B.L., 1988. Deuterium on Mars: the abundance of HDO and the value of D/H. Science 240, 1767–1770.

Oyama, V.I., Carle, G.C., Woeller, F., Pollack, J.B., 1979. Venus lower atmospheric composition: analysis by gas chromatography. Science 203 (4382), 802–805.

Oyewole, O.A., Inselsbacher, E., Näsholm, T., 2014. Direct estimation of mass flow and diffusion of nitrogen compounds in solution and soil. New Phytol. 201 (3), 1056–1064.

Pabian, S.E., Brittingham, M.C., 2007. Terrestrial liming benefits birds in an acidified forest in the northeast. Ecol. Appl. 17 (8), 2184–2194.

Pace, M.L., Cole, J.J., Carpenter, S.R., Kitchell, J.F., Hodgson, J.R., Van de Bogert, M.C., Bade, D.L., Kritzberg, E.S., Bastviken, D., 2004. Whole-lake carbon-13 additions reveal terrestrial support of aquatic food webs. Nature 427 (6971), 240–243.

Pace, M.L., Carpenter, S.R., Cole, J.J., Coloso, J.J., Kitchell, J.F., Hodgson, J.R., Middelburg, J.J., Preston, N.D., Solomon, C.T., Weidel, B.C., 2007. Does terrestrial organic carbon subsidize the planktonic food web in a clearwater lake? Limnol. Oceanogr. 52 (5), 2177–2189.

Pacyna, E.G., Pacyna, J.M., Steenhuisen, F., Wilson, S., 2006. Global anthropogenic mercury emission inventory for 2000. Atmos. Environ. 40, 4048–4063.

Paerl, H.W., 1995. Coastal eutrophication in relation to atmospheric nitrogen deposition: current perspectives. Ophelia 41, 237–259.

Paerl, H.W., 1996. A comparison of cyanobacterial bloom dynamics in freshwater, estuarine and marine environments. Phycologia 35, 25–35.

Paerl, H.W., 2018. Why does N-limitation persist in the World's marine waters? Mar. Chem. 206 (October), 1–6.

Paerl, H.W., Willey, J.D., Go, M., Peierls, B.L., Pinckney, J.L., Fogel, M.L., 1999. Rainfall stimulation of primary production in Western Atlantic ocean waters: Roles of different nitrogen sources and co-limiting nutrients. Mar. Ecol.—Prog. Ser. 176, 205–214.

Paerl, H.W., Bales, J.D., Ausley, L.W., Buzzelli, C.P., Crowder, L.B., Eby, L.A., Fear, J.M., et al., 2001. Ecosystem impacts of three sequential hurricanes (Dennis, Floyd, and Irene) on the United States' largest Lagoonal estuary, Pamlico sound, NC. Proc. Natl. Acad. Sci. U. S. A. 98 (10), 5655–5660.

Paerl, H.W., Scott, J.T., McCarthy, M.J., Newell, S.E., Gardner, W.S., Havens, K.E., Hoffman, D.K., Wilhelm, S.W., Wurtsbaugh, W.A., 2016. It takes two to tango: when and where dual nutrient (N & P) reductions are needed to protect lakes and downstream ecosystems. Environ. Sci. Technol. 50 (20), 10805–10813.

Paerl, H.W., Otten, T.G., Kudela, R., 2018. Mitigating the expansion of harmful algal blooms across the freshwater-to-marine continuum. Environ. Sci. Technol. 52 (10), 5519–5529.

Pagani, M., Zachos, J.C., Freeman, K.H., Tipple, B., Bohaty, S., 2005. Marked decline in atmospheric carbon dioxide concentrations during the paleogene. Science 309 (5734), 600–603.

Page, S.E., Siegert, F., Rieley, J.O., Boehm, H.-D.V., Jaya, A., Limin, S., 2002. The amount of carbon released from peat and forest fires in Indonesia during 1997. Nature 420 (6911), 61–65.

Pagel, B.E., 1993. Abundances of light elements. Proc. Natl. Acad. Sci. U. S. A. 90 (11), 4789–4792.

Paine, R.T., 1971. The measurement and application of the calorie to ecological problems. Annu. Rev. Ecol. Syst. 2, 145–164.

Palit, S., Sharma, A., Talukder, G., 1994. Effects of cobalt on plants. Bot. Rev.; Interpr. Bot. Prog. 60, 149–181.

Palle, E., Goode, P.R., Montañés-Rodríguez, P., Shumko, A., Gonzalez-Merino, B., Martinez Lombilla, C., Jimenez-Ibarra, F., et al., 2016. Earth's albedo variations 1998–2014 as measured from ground-based earthshine observations. Geophys. Res. Lett. 43 (9), 4531–4538.

Palmer, S.M., Driscoll, C.T., 2002. Acidic deposition: Decline in mobilization of toxic Aluminium. Nature 417 (6886), 242–243.

Palmer, T.N., Räisänen, J., 2002. Quantifying the risk of extreme seasonal precipitation events in a changing climate. Nature 415 (6871), 512–514.

Palmer, S.M., Driscoll, C.T., Johnson, C.E., 2004. Long-term trends in soil solution and stream water chemistry at the Hubbard Brook experimental forest: relationship with landscape position. Biogeochemistry 68, 51–70.

Palter, J.B., Susan Lozier, M., Barber, R.T., 2005. The effect of advection on the nutrient reservoir in the North Atlantic subtropical gyre. Nature 437 (7059), 687–692.

Pan, Y., Birdsey, R., Hom, J., McCullough, K., Clark, K., 2006. Improved estimates of net primary productivity from Modis satellite data at regional and local scales. Ecol. Appl. 16 (1), 125–132.

Pan, Y., Birdsey, R.A., Fang, J., Houghton, R., Kauppi, P.E., Kurz, W.A., Phillips, O.L., et al., 2011. A large and persistent carbon sink in the world's forests. Science 333 (6045), 988–993.

Pangala, S.R., Enrich-Prast, A., Basso, L.S., Peixoto, R.B., Bastviken, D., Hornibrook, E.R.C., Gatti, L.V., et al., 2017. Large emissions from floodplain trees close the Amazon methane budget. Nature 552 (7684), 230–234.

Panichev, A.M., Golokhvast, K.S., Gulkov, A.N., Chekryzhov, I.Y., 2013. Geophagy in animals and geology of Kudurs (mineral licks): a review of Russian publications. Environ. Geochem. Health 35 (1), 133–152.

Paolini, J., 1995. Particulate organic carbon and nitrogen in the Orinoco river (Venezuela). Biogeochemistry 29, 59–70.

Papineau, D., Mojzsis, S.J., Karhu, J.A., Marty, B., 2005. Nitrogen isotopic composition of ammoniated phyllosilicates: case studies from precambrian metamorphosed sedimentary rocks. Chem. Geol. 216, 37–58.

Pardo, L.H., Hemond, H.F., Montoya, J.P., Fahey, T.J., Siccama, T.G., 2002. Response of the natural abundance of N-15 in Forest Soils and foliage to high nitrate loss following clear cutting. Can. J. Forest Res. 32, 1126–1136.

Pardo, L.H., McNulty, S.G., Boggs, J.L., Duke, S., 2007. Regional patterns in foliar ^{15}N across a gradient of nitrogen deposition in the northeastern US. Environ. Pollut. 149, 293–302.

Pardo, L.H., Fenn, M.E., Goodale, C.L., Geiser, L.H., Driscoll, C.T., Allen, E.B., Baron, J.S., et al., 2011. Effects of nitrogen deposition and empirical nitrogen critical loads for ecoregions of the United States. Ecol. Appl. 21, 3049–3082.

Parfitt, R.L., Smart, R.S.C., 1978. Mechanism of sulfate adsorption on iron oxides. Soil Sci. Soc. Am. J. 42, 48–50.

Park, H., Schlesinger, W.H., 2002. Global biogeochemical cycle of boron. Glob. Biogeochem. Cycles 16.

Park, H., McGinn, P.J., Morel, F.M.M., 2008. Expression of cadmium carbonic anhydrase of diatoms in seawater. Aquat. Microb. Ecol.: Int. J. 51, 183–193.

Park, S., Croteau, P., Boering, K.A., Etheridge, D.M., Ferretti, D., Fraser, P.J., Kim, K.-R., et al., 2012. Trends and seasonal cycles in the isotopic composition of nitrous oxide since 1940. Nat. Geosci. 5, 262–265.

Park, J.-H., Goldstein, A.H., Timkovsky, J., Fares, S., Weber, R., Karlik, J., Holzinger, R., 2013. Active atmosphere-ecosystem exchange of the vast majority of detected volatile organic compounds. Science 341 (6146), 643–647.

Parker, G.G., 1983. Throughfall and stemflow in the forest nutrient cycle. Adv. Ecol. Res. 13, 57–133.

Parker, J.L., Newstead, S., 2014. Molecular basis of nitrate uptake by the plant nitrate transporter NRT1.1. Nature 507 (7490), 68–72.

Parker, R.S., Troutman, B.M., 1989. Frequency distribution for suspended sediment loads. Water Resour. Res. 25, 1567–1574.

Parkes, D., Marzeion, B., 2018. Twentieth-century contribution to sea-level rise from uncharted glaciers. Nature 563 (7732), 551–554.

Parkes, R.J., Cragg, B.A., Bale, S.J., Getliff, J.M., Goodman, K., Rochelle, P.A., Fry, J.C., Weightman, A.J., Harvey, S.M., 1994. Deep bacterial biosphere in Pacific Ocean sediments. Nature 371, 410–413.

Parkin, T.B., 1987. Soil microsites as a source of denitrification variability. Soil Sci. Soc. Am. J. 51, 1194–1199.

Parkin, T.B., Sexstone, A.J., Tiedje, J.M., 1985. Comparison of field denitrification rates determined by acetylene-based soil core and ^{15}N methods. Soil Sci. Soc. Am. J. 49, 94–99.

Parkinson, C.L., 2006. Earth's cryosphere: current state and recent changes. Annu. Rev. Environ. Resour. 31, 33–60.

Parkinson, C.L., 2019. A 40-Y record reveals gradual Antarctic Sea ice increases followed by decreases at rates far exceeding the rates seen in the Arctic. Proc. Natl. Acad. Sci. U. S. A. 116 (29), 14414–14423.

Parkinson, C.L., Cavalieri, D.J., 2008. Arctic Sea ice variability and trends, 1979–2006. J. Geophys. Res. 113.

Parks, J.M., Johs, A., Podar, M., Bridou, R., Hurt Jr., R.A., Smith, S.D., Tomanicek, S.J., et al., 2013. The genetic basis for bacterial mercury methylation. Science 339 (6125), 1332–1335.

Parnell, J., Boyce, A.J., Mark, D., Bowden, S., Spinks, S., 2010. Early oxygenation of the terrestrial environment during the mesoproterozoic. Nature 468 (7321), 290–293.

Parrenin, F., Masson-Delmotte, V., Köhler, P., Raynaud, D., Paillard, D., Schwander, J., Barbante, C., Landais, A., Wegner, A., Jouzel, J., 2013. Synchronous change of atmospheric CO_2 and Antarctic temperature during the last deglacial warming. Science 339 (6123), 1060–1063.

Parrington, J.R., Zoller, W.H., Aras, N.K., 1983. Asian dust: seasonal transport to the Hawaiian islands. Science 220 (4593), 195–197.

Parrish, D.D., Holloway, J.S., Trainer, M., Murphy, P.C., Fehsenfeld, F.C., Forbes, G.L., 1993. Export of North American ozone pollution to the North Atlantic Ocean. Science 259 (5100), 1436–1439.

Parton, W.J., Schimel, D.S., Cole, C.V., Ojima, D.S., 1987. Analysis of factors controlling soil organic matter levels in Great Plains grasslands. Soil Sci. Soc. Am. J. 51, 1173–1179.

Parton, W.J., Stewart, J.W.B., Cole, C.V., 1988. Dynamics of C, N, P and S in grassland soils: a model. Biogeochemistry 5, 109–131.

Parton, W.J., Holland, E.A., Del Grosso, S.J., Hartman, M.D., Martin, R.E., Mosier, A.R., Ojima, D.S., Schimel, D.S., 2001. Generalized model for NO_X and N_2O emissions from soils. J. Geophys. Res. SCOPE 106 (D15), 17403–17419.

Parton, W., Silver, W.L., Burke, I.C., Grassens, L., Harmon, M.E., Currie, W.S., King, J.Y., et al., 2007. Global-scale similarities in nitrogen release patterns during long-term decomposition. Science 315 (5810), 361–364.

Parton, W. J., J. Neff, and P. M. Vitousek. n.d. "Modelling phosphorus, carbon and nitrogen dynamics in terrestrial ecosystems." Org. Phosp. Environ.

Pasek, M., Block, K., 2009. Lightning-induced reduction of phosphorus oxidation state. Nat. Geosci. 2, 553–556.

Pasek, M.A., Harnmeijer, J.P., Buick, R., Gull, M., Atlas, Z., 2013. Evidence for reactive reduced phosphorus species in the early Archean ocean. Proc. Natl. Acad. Sci. U. S. A. 110 (25), 10089–10094.

Pastor, J., Bridgham, S.D., 1999. Nutrient efficiency along nutrient availability gradients. Oecologia 118 (1), 50–58.

Pastor, J., Aber, J.D., McClaugherty, C.A., Melillo, J.M., 1984. Above-ground production and N and P cycling along a nitrogen mineralization gradient on Blackhawk Island, Wisconsin. Ecology 65, 256–268.

Pastore, M.A., Megonigal, J.P., Adam Langley, J., 2016. Elevated CO_2 promotes long-term nitrogen accumulation only in combination with nitrogen addition. Glob. Chang. Biol. 22 (1), 391–403.

Patterson, D.B., Farley, K.A., Norman, M.D., 1999. 4He as a tracer of continental dust: a 1.9-million-year record of Aeolian flux to the West equatorial Pacific Ocean. Geochim. Cosmochim. Acta 63, 615–625.

Paulot, F., Jacob, D.J., Johnson, M.T., Bell, T.G., Baker, A.R., Keene, W.C., Lima, I.D., Doney, S.C., Stock, C.A., 2015. Global oceanic emission of ammonia: constraints from seawater and atmospheric observations. Glob. Biogeochem. Cycles 29 (8), 1165–1178.

Paulot, F., Ginoux, P., Cooke, W.F., Donner, L.J., Fan, S., Lin, M.-Y., Mao, J., Naik, V., Horowitz, L.W., 2016. Sensitivity of nitrate aerosols to ammonia emissions and to nitrate chemistry: implications for present and future nitrate optical depth. Atmos. Chem. Phys. 16 (3), 1459–1477.

Pauly, D., Christensen, V., 1995. Primary production required to sustain global fisheries. Nature 374, 255–257.

Pausch, J., Kuzyakov, Y., 2018. Carbon input by roots into the soil: quantification of rhizodeposition from root to ecosystem scale. Glob. Chang. Biol. 24 (1), 1–12.

Paytan, A., Kastner, M., Chavez, F.P., 1996. Glacial to interglacial fluctuations in productivity in the equatorial Pacific as indicated by marine barite. Science 274 (5291), 1355–1357.

Paytan, A., Kastner, M., Campbell, D., Thiemens, M.H., 2004. Seawater sulfur isotope fluctuations in the cretaceous. Science 304 (5677), 1663–1665.

Pearson, P.N., Palmer, M.R., 2000. Atmospheric carbon dioxide concentrations over the past 60 million years. Nature 406 (6797), 695–699.

Pearson, J.A., Knight, D.H., Fahey, T.J., 1987. Biomass and nutrient accumulation during stand development in Wyoming Lodgepole pine forests. Ecology 68 (6), 1966–1973.

Pearson, L.K., Hendy, C.H., Hamilton, D.P., Pickett, R.C., 2010. Natural and anthropogenic lead in sediments of the Rotorua Lakes, New Zealand. Earth Planet. Sci. Lett. 297, 536–544.

Pedersen, H., Dunkin, K.A., Firestone, M.K., 1999. The relative importance of autotrophic and heterotrophic nitrification in a conifer forest soil as measured by ^{15}N-tracer and pool-dilution techniques. Biogeochemistry 44, 135–150.

Pedersen, L.L., Smets, B.F., Dechesne, A., 2015. Measuring biogeochemical heterogeneity at the micro scale in soils and sediments. Soil Biol. Biochem. 90 (November), 122–138.

Pedro, G., Jamagne, M., Begon, J.C., 1978. Two routes in genesis of strongly differentiated acid soils under humid, cool-temperate conditions. Geoderma 20, 173–189.

Peel, M.C., McMahon, T.A., Finlayson, B.L., 2010. Vegetation impact on mean annual evapotranspiration at a global catchment scale. Water Resour. Res. 46.

Peizhen, Z., Molnar, P., Downs, W.R., 2001. Increased sedimentation rates and grain sizes 2–4 Myr ago due to the influence of climate change on erosion rates. Nature.

Pekel, J.-F., Cottam, A., Gorelick, N., Belward, A.S., 2016. High-resolution mapping of global surface water and its long-term changes. Nature 540 (7633), 418–422.

Pellegrini, A.F.A., Ahlström, A., Hobbie, S.E., Reich, P.B., Nieradzik, L.P., Carla Staver, A., Scharenbroch, B.C., et al., 2018. Fire frequency drives decadal changes in soil carbon and nitrogen and ecosystem productivity. Nature 553 (7687), 194–198.

Pellerin, B.A., Wollheim, W.M., Hopkinson, C.S., McDowell, W.H., Williams, M.R., Vorosmarty, C.J., Daley, M.L., Vörösmarty, C.J., 2004. Role of wetlands and developed land use on dissolved organic nitrogen concentrations and DON/TDN in northeastern US rivers and streams. Limnol. Oceanogr. 49 (4), 910–918.

Pellerin, B.A., Kaushal, S.S., McDowell, W.H., 2006. Does anthropogenic nitrogen enrichment increase organic nitrogen concentrations in runoff from forested and human-dominated watersheds? Ecosystems 9 (5), 852–864.

Peng, T.H., Wanninkhof, R., Bullister, J.L., Feely, R.A., Takahashi, T., 1998. Quantification of decadal anthropogenic CO_2 uptake in the ocean based on dissolved inorganic carbon measurements. Nature 396, 560–563.

Pennell, K.D., Allen, H.L., Jackson, W.A., 1990. Phosphorus uptake capacity of a 14-year-old loblolly pine as indicated by a ^{32}P root bioassay. For. Sci. 36, 358–366.

Penner, J.E., Dickinson, R.E., Oneill, C.A., 1992. Effects of aerosols from biomass burning on the global radiation budget. Science 256, 1432–1434.

Penuelas, J., Azcon-Bieto, J., 1992. Changes in leaf delta13C of herbarium plant species during the last 3 centuries of CO_2 increase. Plant Cell Environ.

Peñuelas, J., Filella, I., 2001. Herbaria century record of increasing eutrophication in Spanish terrestrial ecosystems. Glob. Chang. Biol. 7, 427–433.

Peñuelas, J., Matamala, R., 1990. Changes in N and S leaf content, stomatal density and specific leaf area of 14 plant species during the last three centuries of CO_2 increase. J. Exp. Bot.

Peñuelas, J., Sardans, J., Rivas-ubach, A., Janssens, I.A., 2012. The human-induced imbalance between C, N and P in Earth's life system. Glob. Chang. Biol. 18 (1), 3–6.

Peñuelas, J., Jannssens, I., Ciais, P., Obersteiner, M., Sardans, J., 2020. Anthropogenic global shifts in biospheric N and P concentrations and ratios and their impacts on biodiversity, ecosystem productivity, food security, and human health. Glob. Change Biol. https://doi.org/10.1111/gcb.14981.

Penzias, A.A., 1979. Origin of the elements. Science 205, 549–554.

Pepin, R.O., 2006. Atmospheres on the terrestrial planets: clues to origin and evolution. Earth Planet. Sci. Lett. 252, 1–14.

Perakis, S.S., Hedin, L.O., 2002. Nitrogen loss from unpolluted South American forests mainly via dissolved organic compounds. Nature 415 (6870), 416–419.

Perakis, S.S., Pett-Ridge, J.C., 2019. Nitrogen-fixing red alder trees tap rock-derived nutrients. Proc. Natl. Acad. Sci. U. S. A. 116 (11), 5009–5014.

Perakis, S.S., Sinkhorn, E.R., 2011. Biogeochemistry of a temperate forest nitrogen gradient. Ecology 92 (7), 1481–1491.

Perdue, E.M., Beck, K.C., Reuter, J.H., 1976. Organic complexes of Iron and aluminum in natural waters. Nature 260, 418–420.

Perez-Fodich, A., Derry, L.A., 2019. Organic acids and high soil CO_2 drive intense chemical weathering of Hawaiian basalts: insights from reactive transport models. Geochim. Cosmochim. Acta 249 (March), 173–198.

Pérez-Rodríguez, M., Biester, H., Aboal, J.R., Toro, M., Cortizas, A.M., 2019. Thawing of snow and ice caused extraordinary high and fast mercury fluxes to lake sediments in Antarctica. Geochim. Cosmochim. Acta 248 (March), 109–122.

Peterjohn, W.T., Correll, D.L., 1984. Nutrient dynamics in an agricultural watershed: observations on the role of a riparian forest. Ecology 65, 1466–1475.

Peterjohn, W.T., Schlesinger, W.H., 1991. Factorscontrolling denitrification in a Chihuahuan Desert ecosystem. Soil Sci. Soc. Am. J. 55, 1694–1701.

Peterjohn, W.T., Melillo, J.M., Steudler, P.A., Newkirk, K.M., Bowles, F.P., Aber, J.D., 1994. Responses of trace gas fluxes and N availability to experimentally-elevated soil temperatures. Ecol. Appl. 4, 617–625.

Peterjohn, W.T., Adams, M.B., Gilliam, F.S., 1996. Symptoms of nitrogen saturation in two central Appalachian hardwood forest ecosystems. Biogeochemistry 35, 507–522.

Peterjohn, W.T., McGervey, R.J., Sexstone, A.J., Christ, M.J., Foster, C.J., Adams, M.B., 1998. Nitrous oxide production in two forested watersheds exhibiting symptoms of nitrogen saturation. Can. J. Forest Res. 28, 1723–1732.

Petersen, J.M., Zielinski, F.U., Pape, T., Seifert, R., Moraru, C., Amann, R., Hourdez, S., et al., 2011. Hydrogen is an energy source for hydrothermal vent symbioses. Nature 476 (7359), 176–180.

Peterson, B.J., 1980. Aquatic primary productivity and the ^{14}C-CO$_2$ method—a history of the productivity problem. Annu. Rev. Ecol. Syst. 11, 359–385.

Peterson, L.C., Haug, G.H., 2005. Climate and the collapse of Maya civilization. Am. Sci. 93, 322–329.

Peterson, B.J., Howarth, R.W., 1987. Sulfur, carbon, and nitrogen isotopes used to trace organic matter flow in the salt marsh estuaries of Sapelo Island, Georgia. Limnol. Oceanogr. 32, 1195–1213.

Peterson, B.J., Melillo, J.M., 1985. The potential storage of carbon caused by eutrophication of the biosphere. Tellus B Chem. Phys. Meteorol. 37, 117–127.

Peterson, B.J., Howarth, R.W., Garritt, R.H., 1985. Multiple stable isotopes used to trace the flow of organic matter in estuarine food webs. Science 227 (4692), 1361–1363.

Peterson, B.J., Howarth, R.W., Garritt, R.H., 1986. Sulfur and carbon isotopes as tracers of salt marsh organic matter flow. Ecology 67, 865–874.

Peterson, D.L., Spanner, M.A., Running, S.W., Teuber, K.B., 1987. Relationship of thematic mapper simulator data to leaf-area index of temperate coniferous forests. Remote Sens. Environ. 22, 323–341.

Peterson, B.J., Wollheim, W.M., Mulholland, P.J., Webster, J.R., Meyer, J.L., Tank, J.L., Marti, E., et al., 2001. Control of nitrogen export from watersheds by headwater streams. Science 292 (5514), 86–90.

Peterson, B.J., Holmes, R.M., McClelland, J.W., Vörösmarty, C.J., Lammers, R.B., Shiklomanov, A.I., Shiklomanov, I.A., Rahmstorf, S., 2002. Increasing river discharge to the Arctic ocean. Science 298 (5601), 2171–2173.

Peterson, B.J., McClelland, J., Curry, R., Holmes, R.M., Walsh, J.E., Aagaard, K., 2006. Trajectory shifts in the Arctic and subarctic freshwater cycle. Science 313 (5790), 1061–1066.

Petkov, B.H., Vitale, V., Tomasi, C., Siani, A.M., Seckmeyer, G., Webb, A.R., Smedley, A.R.D., et al., 2014. Response of the ozone column over Europe to the 2011 Arctic ozone depletion event according to ground-based observations and assessment of the consequent variations in surface UV irradiance. Atmos. Environ. 85 (March), 169–178.

Petrenko, V.V., Smith, A.M., Brook, E.J., Lowe, D., Riedel, K., Brailsford, G., Hua, Q., et al., 2009. (CH$_4$)-^{14}C measurements in Greenland ice: investigating last glacial termination CH$_4$ sources. Science 324, 506–508.

Petrenko, V.V., Smith, A.M., Schaefer, H., Riedel, K., Brook, E., Baggenstos, D., Harth, C., et al., 2017. Minimal geological methane emissions during the younger Dryas-Preboreal abrupt warming event. Nature 548 (7668), 443–446.

Petroff, A., Mailliat, A., Amielh, M., Anselmet, F., 2008. Aerosol dry deposition on vegetative canopies. Part I: Review of present knowledge. Atmos. Environ. 42, 3625–3653.

Petsch, S.T., Eglington, T.I., Edwards, K.J., 2001. ^{14}C-dead living biomass: evidence for microbial assimilation of ancient organic carbon during shale weathering. Science 292 (5519), 1127–1131.

Petty, G.W., 1995. The status of satellite-based rainfall estimation over land. Remote Sens. Environ. 51, 125–137.

Pfeiffer, W.C., Malm, O., Souza, C.M.M., Delacerda, L.D., Silveira, E.G., Bastos, W.R., 1991. Mercury in the Madeira river ecosystem, Rondonia, Brazil. For. Ecol. Manag. 38, 239–245.

Philander, S.G., 1989. El Nino, La Nina, and the Southern Oscillation. Academic Press.

Phillips, R., Milo, R., 2009. A feeling for the numbers in biology. Proc. Natl. Acad. Sci.

Phillips, F.M., Mattick, J.L., Duval, T.A., Elmore, D., Kubik, P.W., 1988. Chlorine-36 and tritium from nuclear weapons fallout as tracers for long-term liquid and vapor movement in desert soils. Water Resour. Res. 24, 1877–1891.

Phillips, R.L., Whalen, S.C., Schlesinger, W.H., 2001. Influence of atmospheric CO$_2$ enrichment on methane consumption in a temperate forest soil. Glob. Chang. Biol. 7, 557–563.

Phillips, O.L., Aragão, L.E.O.C., Lewis, S.L., Fisher, J.B., Lloyd, J., López-González, G., Malhi, Y., et al., 2009. Drought sensitivity of the Amazon rainforest. Science 323 (5919), 1344–1347.

Phillips, R.J., Davis, B.J., Tanaka, K.L., Byrne, S., Mellon, M.T., Putzig, N.E., Haberle, R.M., et al., 2011a. Massive CO$_2$ ice deposits sequestered in the south polar layered deposits of Mars. Science 332, 838–841.

Phillips, R.P., Finzi, A.C., Bernhardt, E.S., 2011b. Enhanced root exudation induces microbial feedbacks to N cycling in a pine forest under long-term CO$_2$ fumigation. Ecol. Lett. 14 (2), 187–194.

Phillips, R.P., Meier, I.C., Bernhardt, E.S., Stuart Grandy, A., Wickings, K., Finzi, A.C., 2012. Roots and fungi accelerate carbon and nitrogen cycling in forests exposed to elevated CO$_2$. Ecol. Lett. 15 (9), 1042–1049.

Phillips, R.P., Brzostek, E., Midgley, M.G., 2013. The mycorrhizal-associated nutrient economy: a new framework for predicting carbon-nutrient couplings in temperate forests. New Phytol. 199 (1), 41–51.

Phoenix, V.R., Bennett, P.C., Engel, A.S., Tyler, S.W., Ferris, F.G., 2006. Chilean high-altitude hot-spring sinters: a model system for UV screening mechanisms by early Precambrian cyanobacteria. Geobiology 4, 15–28.

Piao, S., Ciais, P., Friedlingstein, P., Peylin, P., Reichstein, M., Luyssaert, S., Margolis, H., et al., 2008. Net carbon dioxide losses of northern ecosystems in response to autumn warming. Nature 451 (7174), 49–52.

Piao, S., Fang, J., Ciais, P., Peylin, P., Huang, Y., Sitch, S., Wang, T., 2009. The carbon balance of terrestrial ecosystems in China. Nature 458 (7241), 1009–1013.

Piao, S., Luyssaert, S., Ciais, P., Janssens, I.A., Chen, A., Cao, C., Fang, J., Friedlingstein, P., Luo, Y., Wang, S., 2010. Forest annual carbon cost: a global-scale analysis of autotrophic respiration. Ecology 91 (3), 652–661.

Piao, S.L., Wang, X.H., Ciais, P., Zhu, B., Wang, T., Liu, J., 2011. Changes in satellite-derived vegetation growth trend in temperate and boreal Eurasia from 1982 to 2006. Glob. Chang. Biol. 17, 3228–3239.

Piao, S., Liu, Z., Wang, Y., Ciais, P., Yao, Y., Peng, S., Chevallier, F., et al., 2018. On the causes of trends in the seasonal amplitude of atmospheric CO_2. Glob. Chang. Biol. 24 (2), 608–616.

Piccolo, M.C., Neill, C., Melillo, J.M., Cerri, C.C., Steudler, P.A., 1996. ^{15}N natural abundance in forest and pasture soils of the Brazilian Amazon Basin. Plant Soil 182, 249–258.

Piccot, S.D., Watson, J.J., Jones, J.W., 1992. A global inventory of volatile organic compound emissions from anthropogenic sources. J. Geophys. Res. 97, 9897–9912.

Plecuch, C.G., Huybers, P., Hay, C.C., Kemp, A.C., Little, C.M., Mitrovica, J.X., Ponte, R.M., Tingley, M.P., 2018. Origin of spatial variation in US East Coast sea-level trends during 1900–2017. Nature 564 (7736), 400–404.

Pierre, S., Hewson, I., Sparks, J.P., Litton, C.M., Giardina, C., Groffman, P.M., Fahey, T.J., 2017. Ammonia oxidizer populations vary with nitrogen cycling across a tropical montane mean annual temperature gradient. Ecology 98 (7), 1896–1907.

Pietruczuk, A., Krzyscin, J.W., Jaroslawski, J., Podgorski, J., Sobolewski, P., Wink, J., 2010. Eyjafjallajokull volcano ash observed over Belsk (52°N, 21°E), Poland, in April 2010. Int. J. Remote Sens. 31, 3981–3986.

Pigott, C.D., Taylor, K., 1964. The distribution of some woodland herbs in relation to the supply of nitrogen and phosphorus in the soil. J. Ecol. 52, 175–185.

Piña-Ochoa, E., Álvarez-Cobelas, M., 2006. Denitrification in aquatic environments: a cross-system analysis. Biogeochemistry 81 (1), 111–130.

Pinay, G., Roques, L., Fabre, A., 1993. Spatial and temporal patterns of denitrification in a riparian forest. J. Appl. Ecol. 30, 581–591.

Piñero, E., Marquardt, M., Hensen, C., Haeckel, M., Wallmann, K., 2013. Estimation of the global inventory of methane hydrates in marine sediments using transfer functions. Biogeosciences 10 (2), 959–975.

Pingitore, N.E., Eastman, M.P., 1986. The coprecipitation of Sr2þ with calcite at 25°C and 1-Atm. Geochim. Cosmochim. Acta 50, 2195–2203.

Pinker, R.T., Zhang, B., Dutton, E.G., 2005. Do satellites detect trends in surface solar radiation? Science 308 (5723), 850–854.

Pinol, J., Alcaniz, J.M., Roda, F., 1995. Carbon dioxide efflux and pCO_2 in soils of three Quercus-Ilex montane forests. Biogeochemistry 30, 191–215.

Pinto, J.P., Gladstone, G.R., Yung, Y.L., 1980. Photochemical production of formaldehyde in Earth's primitive atmosphere. Science 210 (4466), 183–185.

Pinto-Tomás, A.A., Anderson, M.A., Suen, G., Stevenson, D.M., Chu, F.S.T., Cleland, W.W., Weimer, P.J., Currie, C.R., 2009. Symbiotic nitrogen fixation in the fungus gardens of leaf-cutter ants. Science 326 (5956), 1120–1123.

Pirozynski, K.A., Malloch, D.W., 1975. Origin of land plants: a matter of mycotropism. Bio Systems 6, 153–164.

Pirrone, N., Cinnirella, S., Feng, X., Finkelman, R.B., Friedli, H.R., Leaner, J., Mason, R., et al., 2010. Global mercury emissions to the atmosphere from anthropogenic and natural sources. Atmos. Chem. Phys. 10, 5951–5964.

Pistone, K., Eisenman, I., Ramanathan, V., 2014. Observational determination of albedo decrease caused by vanishing Arctic sea ice. Proc. Natl. Acad. Sci. U. S. A. 111 (9), 3322–3326.

Pitz, S., Megonigal, J.P., 2017. Temperate forest methane sink diminished by tree emissions. New Phytol. 214 (4), 1432–1439.

Pizzarello, S., Weber, A.L., 2004. Prebiotic amino acids as asymmetric catalysts. Science 303 (5661), 1151.

Planavsky, N.J., Asael, D., Hofmann, A., Reinhard, C.T., Lalonde, S.V., Knudsen, A., Wang, X., et al., 2014a. Evidence for oxygenic photosynthesis half a billion years before the great oxidation event. Nat. Geosci. 7 (4), 283–286.

Planavsky, N.J., Reinhard, C.T., Wang, X., Thomson, D., McGoldrick, P., Rainbird, R.H., Johnson, T., Fischer, W.W., Lyons, T.W., 2014b. Earth history. Low mid-Proterozoic atmospheric oxygen levels and the delayed rise of animals. Science 346 (6209), 635–638.

Plank, T., 2014. The chemical composition of subducting sediments. In: Turekian, K.K., Holland, H.D. (Eds.), Treatise on Geochemistry. second ed. Elsevier, Amsterdam, pp. 607–629.

Plank, T., Langmuir, C.H., 1998. The chemical composition of subducting sediment and its consequences for the crust and mantle. Chem. Geol. 145 (3–4), 325–394.

Plank, T., Manning, C.E., 2019. Subducting carbon. Nature 574, 343–352.

Plant, G., Kort, E.A., Floerchinger, C., Gvakharia, A., Vimont, I., Sweeney, C., 2019. Large fugitive mthane emissions from urban centers along the U.S. east coast. Geophys. Res. Lett. 46 (14), 8500–8507.

Platt, T., Sathyendranath, S., 1988. Oceanic primary production: estimation by remote sensing at local and regional scales. Science 241 (4873), 1613–1620.

Platt, U., Rateike, M., Junkermann, W., Rudolph, J., Ehhalt, D.H., 1988. New tropospheric OH measurements. J. Geophys. Res. D: Atmos. 93, 5159–5166.

Pletscher, D.H., Bormann, F.H., Miller, R.S., 1989. Importance of deer compared to other vertebrates in nutrient cycling and energy flow in a northern hardwood ecosystem. Am. Midl. Nat. 121, 302–311.

Pocklington, R., Tan, F.C., 1987. Seasonal and annual variations in the organic matter contributed by the St Lawrence River to the Gulf of St. Lawrence. Geochim. Cosmochim. Acta 51 (9), 2579–2586.

Pohlker, C., Wiedemann, K.T., Sinha, B., Shiraiwa, M., Gunthe, S.S., Smith, M., Su, H., et al., 2012. Biogenic potassium salt particles as seeds for secondary organic aerosol in the Amazon. Science 337, 1075–1078.

Pöhlker, C., Wiedemann, K.T., Sinha, B., Shiraiwa, M., Gunthe, S.S., Smith, M., Hang, S., et al., 2012. Biogenic potassium salt particles as seeds for secondary organic aerosol in the Amazon. Science 337 (6098), 1075–1078.

Pokhrel, Y.N., Hanasaki, N., Yeh, P.J.-F., Yamada, T.J., Kanae, S., Oki, T., 2012. Model estimates of sea-level change due to anthropogenic impacts on terrestrial water storage. Nat. Geosci. 5, 389–392.

Polag, D., Keppler, F., 2019. Global methane emissions from the human body: past, present and future. Atmos. Environ. 214 (July), 116823.

Polglase, P.J., Attiwill, P.M., Adams, M.A., 1992. Nitrogen and phosphorus cycling in relation to stand age of *Eucalyptus regnans* F-Muell. III. Labile inorganic and organic P, phosphatase activity and P availability. Plant Soil 142, 177–185.

Pollack, J.B., Black, D.C., 1982. Noble gases in planetary atmospheres: implications for the origin and evolution of atmospheres. Icarus 51, 169–198.

Pollack, J.B., Kasting, J.F., Richardson, S.M., Poliakoff, K., 1987. The case for a wet, warm climate on early Mars. Icarus 71, 203–224.

Pollard, R.T., Salter, I., Sanders, R.J., Lucas, M.I., Moore, C.M., Mills, R.A., Statham, P.J., et al., 2009. Southern ocean deep-water carbon export enhanced by natural iron fertilization. Nature 457 (7229), 577–580.

Polovina, J.J., Howell, E.A., Abecassis, M., 2008. Ocean's least productive waters are expanding. Geophys. Res. Lett. 35.

Polubesova, T.A., Chorover, J., Sposito, G., 1995. Surface-charge characteristics of podzolized soil. Soil Sci. Soc. Am. J. 59, 772–777.

Pomeroy, L.R., Deibel, D., 1986. Temperature regulation of bacterial activity during the spring bloom in new foundland coastal waters. Science 233 (4761), 359–361.

Pommier, M., McLinden, C.A., Deeter, M., 2013. Relative changes in CO emissions over megacities based on observations from space. Geophys. Res. Lett. 40 (14), 3766–3771.

Ponnamperuma, F.N., Tianco, E.M., Loy, T., 1967. Redox equilibria in flooded soils: I. The iron hydroxide systems. Soil Sci. 103 (6), 374.

Poole, G.C., O'Daniel, S.J., Jones, K.L., Woessner, W.W., Bernhardt, E.S., Helton, A.M., Stanford, J.A., Boer, B.R., Beechie, T.J., 2008. Hydrologic spiralling: the role of multiple interactive flow paths in stream ecosystems. River Res. Appl. 24 (7), 1018–1031.

Poorter, H., 1993. Interspecific variation in the growth response of plants to an elevated ambient CO_2 concentration. Vegetatio 104, 77–97.

Poorter, H., Pérez-Soba, M., 2001. The growth response of plants to elevated CO_2 under non-optimal environmental conditions. Oecologia 129 (1), 1–20.

Poorter, H., Niklas, K.J., Reich, P.B., Oleksyn, J., Poot, P., Mommer, L., 2012. Biomass allocation to leaves, stems and roots: meta-analyses of interspecific variation and environmental control. New Phytol. 193 (1), 30–50.

Poorter, L., Frans, B., Mitchell Aide, T., Almeyda, A.M., Zambrano, P.B., Becknell, J.M., Boukili, V., et al., 2016. Biomass resilience of neotropical secondary forests. Nature 530 (7589), 211–214.

Pope 3rd, C.A., Ezzati, M., Dockery, D.W., 2009. Fine-particulate air pollution and life expectancy in the United States. N. Engl. J. Med. 360 (4), 376–386.

Pope, E.C., Bird, D.K., Rosing, M.T., 2012. Isotope composition and volume of Earth's early oceans. Proc. Natl. Acad. Sci. U. S. A. 109 (12), 4371–4376.

Porder, S., Chadwick, O.A., 2009. Climate and soilage constraints on nutrient uplift and retention by plants. Ecology 90, 623–636.

Porder, S., Clark, D.A., Vitousek, P.M., 2006. Persistence of rock-derived nutrients in the wet tropical forests of La Selva, Costa Rica. Ecology 87 (3), 594–602.

Porder, S., Vitousek, P.M., Chadwick, O.A., Page Chamberlain, C., Hilley, G.E., 2007. Uplift, erosion, and phosphorus limitation in terrestrial ecosystems. Ecosystems 10 (1), 159–171.

Porras, R.C., Hicks Pries, C.E., McFarlane, K.J., Hanson, P.J., Torn, M.S., 2017. Association with pedogenic iron and aluminum: effects on soil organic carbon storage and stability in four temperate forest soils. Biogeochemistry 133 (3), 333–345.

Porter, K.G., 1976. Enhancement of algal growth and productivity by grazing zooplankton. Science 192 (4246), 1332–1334.

Portillo-Estrada, M., Pihlatie, M., Korhonen, J.F.J., Levula, J., Frumau, A.K.F., Ibrom, A., Lembrechts, J.J., et al., 2016. Climatic controls on leaf litter decomposition across European forests and grasslands revealed by reciprocal litter transplantation experiments. Biogeosciences 13 (5), 1621–1633.

Post, W.M., Kwon, K.C., 2000. Soil carbon sequestration and land-use change: processes and potential. Glob. Chang. Biol. 6, 317–327.

Post, W.M., Pastor, J., Zinke, P.J., Stangenberger, A.G., 1985. Global patterns of soil nitrogen storage. Nature 317, 613–616.

Postberg, F., Schmidt, J., Hillier, J., Kempf, S., Srama, R., 2011. A salt-water reservoir as the source of a compositionally stratified plume on *Enceladus*. Nature 474 (7353), 620–622.

Postma, D., Jakobsen, R., 1996. Redox zonation: equilibrium constraints on the $Fe(III)/SO_4$-reduction interface. Geochim. Cosmochim. Acta 60 (17), 3169–3175.

Potter, C.S., Matson, P.A., Vitousek, P.M., Davidson, E.A., 1996. Process modeling of controls on nitrogen trace gas emissions from soils worldwide. J. Geophys. Res. 101, 1361–1377.

Potter, C., Klooster, S., Hiatt, C., Genovese, V., Castilla-Rubio, J.C., 2011. Changes in the carbon cycle of Amazon ecosystems during the 2010 drought. Environ. Res. Lett. 6.

Potts, M.J., 1978. Deposition of air-borne salt on *Pinus radiata* and underlying soil. J. Appl. Ecol. 15, 543–550.

Poulton, S.W., Raiswell, R., 2002. The low-temperature geochemical cycle of iron: from continental fluxes to marine sediment deposition. Am. J. Sci. 302 (9), 774–805.

Pouyat, R.V., Yesilonis, I.D., Golubiewski, N.E., 2009. A comparison of soil organic stocks between residential turf grass and native soil. Urban Ecosyst. 12, 49–62.

Powell, T.M., Cloern, J.E., Huzzey, L.M., 1989. Spatial and temporal variability in South San Francisco Bay (USA). I. Horizontal distributions of salinity, suspended sediments, and phytoplankton biomass and productivity. Estuar. Coast. Shelf Sci. 28, 583–597.

Powers, J.S., Veldkamp, E., 2005. Regional variation in soil carbon and Delta ^{13}C in forests and pastures of northeastern Costa Rica. Biogeochemistry 72, 315–336.

Powers, J.S., Treseder, K.K., Lerdau, M.T., 2005. Fine roots, arbuscular mycorrhizal hyphae and soil nutrients in four neotropical rain forests: patterns across large geographic distances. New Phytol. 165 (3), 913–921.

Powers, J.S., Montgomery, R.A., Adair, E.C., Brearley, F.Q., DeWalt, S.J., Castanho, C.T., Chave, J., et al., 2009. Decomposition in tropical forests: a pan-tropical study of the effects of litter type, litter placement and mesofaunal exclusion across a precipitation gradient. J. Ecol. 97, 801–811.

Powlson, D.S., Stirling, C.M., Jat, M.L., Gerard, B.G., Palm, C.A., Sanchez, P.A., Cassman, K.G., 2014. Limited potential of no-till agriculture for climate change mitigation. Nat. Clim. Chang. 4 (8), 678–683.

Powner, M.W., Gerland, B., Sutherland, J.D., 2009. Synthesis of activated pyrimidine ribonucleotides in prebiotically plausible conditions. Nature 459 (7244), 239–242.

Prahl, F.G., Ertel, J.R., Goni, M.A., Sparrow, M.A., Eversmeyer, B., 1994. Terrestrial organic carbon contributions to sediments on the Washington margin. Geochim. Cosmochim. Acta 58, 3035–3048.

Prairie, Y.T., Duarte, C.M., 2006. Direct and indirect metabolic CO2 release by humanity. Biogeosci. Discuss. 3, 1781–1789.

Prairie, Y.T., Alm, J., Beaulieu, J., Barros, N., Battin, T., Cole, J., del Giorgio, P., et al., 2018. Greenhouse gas emissions from freshwater reservoirs: what does the atmosphere see? Ecosystems.

Prather, M.J., 1985. Continental sources of halocarbons and nitrous oxide. Nature 317, 221–225.

Prather, M.J., Hsu, J., 2008. NF3, the greenhouse gas missing from Kyoto. Geophys. Res. Lett. 35.

Prather, M.J., Derwent, R., Ehhalt, D., Fraser, P., Sanhueza, E., Zhou, X., 1995. Other trace gases and atmospheric chemistry. In: Houghton, J.T., Filho, L.G.M., Bruce, J., Lee, H., Callender, B.A., Haites, E., Harris, N., Maskell, K. (Eds.), Climate Change 1994. Cambridge University Press, pp. 73–126.

Prather, M.J., Hsu, J., DeLuca, N.M., Jackman, C.H., Oman, L.D., Douglass, A.R., Fleming, E.L., et al., 2015. Measuring and modeling the lifetime of nitrous oxide including its variability. J. Geophys. Res. D: Atmos. 120 (11), 5693–5705.

Prather, C.M., Laws, A.N., Cuellar, J.F., Reihart, R.W., Gawkins, K.M., Pennings, S.C., 2018. Seeking salt: herbivorous Prairie insects can be co-limited by macronutrients and sodium. Ecol. Lett. 21 (10), 1467–1476.

Pregitzer, K.S., Euskirchen, E.S., 2004. Carbon cycling and storage in world forests: biome patterns related to forest age. Glob. Chang. Biol. 10, 2052–2077.

Pregitzer, K.S., Burton, A.J., Zak, D.R., Talhelm, A.F., 2008. Simulated chronic nitrogen deposition increases carbon storage in northern temperate forests. Glob. Chang. Biol. 14, 142–153.

Pregitzer, K.S., Zak, D.R., Talhelm, A.F., Burton, A.J., Eikenberry, J.R., 2010. Nitrogen turnover in the leaf litter and fine roots of sugar maple. Ecology 91 (12), 3456–3462 (discussion 3503–3514).

Press, F., Siever, R., 1986. Earth. W.H. Freeman.

Prestbo, E.M., Gay, D.A., 2009. Wet deposition of mercury in the US and Canada, 1996–2005: results and analysis of the NADP mercury deposition network (MDN). Atmos. Environ. 43, 4223–4233.

Preunkert, S., Legrand, M., Wagenbach, D., 2001. Sulfate trends in a col Du dome (French Alps) ice core: a record of anthropogenic sulfate levels in the European midtroposphere over the twentieth century. J. Geophys. Res. 106, 31991–31994.

Preunkert, S., Wagenbach, D., Legrand, M., 2003. A seasonally resolved alpine ice core record of nitrate: comparison with anthropogenic inventories and estimation of preindustrial emissions of NO in Europe. J. Geophys. Res. 108.

Preunkert, S., McConnell, J.R., Hoffmann, H., Legrand, M., Wilson, A.I., Eckhardt, S., Stohl, A., Chellman, N.J., Arienzo, M.M., Friedrich, R., 2019. Lead and antimony in basal ice from col Du dome (*French alps*) dated with radiocarbon: a record of pollution during antiquity. Geophys. Res. Lett. 46 (9), 4953–4961.

Prézelin, B.B., Boczar, B.A., 1986. Molecular bases of cell absorption and fluorescence in phytoplankton: potential applications to studies in optical oceanography. Prog. Phycol. Res. 4, 349–464.

Price, N.M., Morel, F.M.M., 1990. Cadmium and cobalt substitution for zinc in a marine diatom. Nature 344, 658–660.

Price, S.J., Sherlock, R.R., Kelliher, F.M., McSeveny, T.M., Tate, K.R., Condron, L.M., 2004. Pristine New Zealand forest soil is a strong methane sink. Glob. Chang. Biol. 10, 16–26.

Prietzel, J., Zimmermann, L., Schubert, A., Christophel, D., 2016. Organic matter losses in German Alps forest soils since the 1970s most likely caused by warming. Nat. Geosci. 9 (7), 543–548.

Prince, S.D., De Colstoun, E.B., Kravitz, L.L., 1998. Evidence from rain-use efficiencies does not indicate extensive Sahelian desertification. Glob. Chang. Biol. 4, 359–374.

Pringle, C.M., 1987. Effects of water and substratum nutrient supplies on lotic periphyton growth: an integrated bioassay. Can. J. Fish. Aquat. Sci. 44, 619–629.

Prinn, R.G., 2003. The cleansing capacity of the atmosphere. Annu. Rev. Environ. Resour. 28, 29–57.

Prinn, R.G., Weiss, R.F., Miller, B.R., Huang, J., Alyea, F.N., Cunnold, D.M., Fraser, P.J., Hartley, D.E., Simmonds, P.G., 1995. Atmospheric trends and lifetime of CH_3CCl_3 and global OH concentrations. Science 269, 187–192.

Prinn, R.G., Huang, J., Weiss, R.F., Cunnold, D.M., Fraser, P.J., Simmonds, P.G., McCulloch, A., et al., 2005. Evidence for variability of atmospheric hydroxyl radicals over the past quarter century. Geophys. Res. Lett. 32.

Priscu, J.C., Fritsen, C.H., Adams, E.E., Giovannoni, S.J., Paerl, H.W., McKay, C.P., Doran, P.T., Gordon, D.A., Lanoil, B.D., Pinckney, J.L., 1998. Perennial Antarctic lake ice: an oasis for life in a polar desert. Science 280 (5372), 2095–2098.

Pritchard, S.G., Strand, A.E., McCormack, M.L., Davis, M.A., Finzi, A.C., Jackson, R.B., Matamala, R., Rogers, H.H., Oren, R., 2008a. Fine root dynamics in a loblolly pine forest are influenced by free-air-CO_2-enrichment: a six-year-Minirhizotron study. Glob. Chang. Biol. 14, 588–602.

Pritchard, S.G., Strand, A.E., McCormack, M.L., Davis, M.A., Oren, R., 2008b. Mycorrhizal and rhizomorph dynamics in a loblolly pine forest during 5 years of free-air-CO_2-enrichment. Glob. Chang. Biol. 14, 1252–1264.

Pritchard, H.D., Arthern, R.J., Vaughan, D.G., Edwards, L.A., 2009. Extensive dynamic thinning on the margins of the Greenland and Antarctic ice sheets. Nature 461 (7266), 971–975.

Probst, J.L., Tardy, Y., 1987. Long range streamflow and world continental runoff fluctuations since the beginning of this century. J. Hydrol. 94, 289–311.

Prokopenko, M.G., Sigman, D.M., Berelson, W.M., Hammond, D.E., Barnett, B., Chong, L., Townsend-Small, A., 2011. Denitrification in anoxic sediments supported by biological nitrate transport. Geochim. Cosmochim. Acta 75, 7180–7199.

Prospero, J.M., Savoie, D.L., Saltzman, E.S., Larsen, R., 1991. Impact of oceanic sources of biogenic sulfur on sulfate aerosol concentrations at Mawson, Antarctica. Nature 350, 221–223.

Protz, R., Ross, G.J., Martini, I.P., Terasmae, J., 1984. Rate of podzolic soil formation near Hudson Bay, Ontario. Can. J. Soil Sci. 64, 31–49.

Protz, R., Ross, G.J., Shipitalo, M.J., Terasmae, J., 1988. Podzolic soil development in the southern James Bay lowlands, Ontario. Can. J. Soil Sci. 68, 287–305.

Pryor, S.C., Barthelmie, R.J., Sorensen, L.L., Jensen, B., 2001. Ammonia concentrations and fluxes over a forest in the Midwestern USA. Atmos. Environ. 35, 5645–5656.

Puente-Sánchez, F., Arce-Rodríguez, A., Oggerin, M., García-Villadangos, M., Moreno-Paz, M., Blanco, Y., Rodríguez, N., et al., 2018. Viable cyanobacteria in the deep continental subsurface. Proc. Natl. Acad. Sci. U. S. A. 115 (42), 10702–10707.

Puig, R., Avila, A., Soler, A., 2008. Sulphur isotopes as tracers of the influence of a coal-fired power plant on a scots pine forest in Catalonia (NE Spain). Atmos. Environ. 42, 733–745.

Pujol, M., Marty, B., Burgess, R., Turner, G., Philippot, P., 2013. Argon isotopic composition of archaean atmosphere probes early earth geodynamics. Nature 498 (7452), 87–90.

Putaud, J.P., Belviso, S., Nguyen, B.C., Mihalopoulos, N., 1993. Dimethylsulfide, aerosols, and condensation nuclei over the tropical northeastern Atlantic ocean. J. Geophys. Res. 98, 14863–14871.

Pye, K., 1987. Eolian Dust and Dust Deposits. Academic Press, London.

Pyle, D.M., Mather, T.A., 2003. The importance of volcanic emissions for the global atmospheric mercury cycle. Atmos. Environ. 37, 5115–5124.

Qin, S., Pang, Y., Clough, T., Wrage-Mönnig, N., Hu, C., Zhang, Y., Zhou, S., Fang, Y., 2017. N_2 production via aerobic pathways may play a significant role in nitrogen cycling in upland soils. Soil Biol. Biochem. 108 (May), 36–40.

Quack, B., Wallace, D.W.R., 2003. Air-sea flux of bromoform: controls, rates, and implications. Glob. Biogeochem. Cycles 17.

Quade, J., Cerling, T.E., Bowman, J.R., 1989. Development of Asian monsoon revealed by marked ecological shift during the latest Miocene in northern Pakistan. Nature 342, 163–166.

Qualls, R.G., Haines, B.L., 1991. Fluxes of dissolved organic nutrients and humic substances in a deciduous forest. Ecology 72, 254–266.

Qualls, R.G., Haines, B.L., 1992. Biodegradability of dissolved organic matter in forest throughfall, soil solution, and stream water. Soil Sci. Soc. Am. J. 56, 578–586.

Quay, P.D., 1988. Isotopic composition of methane released from wetlands: implications for the increase in atmospheric methane. Glob. Biogeochem. Cycles 2, 385–397.

Quay, P.D., King, S.L., Stutsman, J., Wilbur, D.O., Steele, L.P., Fung, I., Gammon, R.H., et al., 1991. Carbon isotopic composition of atmospheric CH_4: fossil and biomass burning source strengths. Glob. Biogeochem. Cycles 5, 25–47.

Quay, P., Stutsman, J., Wilbur, D., Snover, A., Dlugokencky, E., Brown, T., 1999. The isotopic composition of atmospheric methane. Glob. Biogeochem. Cycles 13, 445–461.

Quay, P., Sonnerup, R., Westby, T., Stutsman, J., McNichol, A., 2003. Changes in the $^{13}C/^{12}C$ of dissolved inorganic carbon in the ocean as a tracer of anthropogenic CO_2 uptake. Glob. Biogeochem. Cycles 17.

Quay, P.D., Peacock, C., Björkman, K., Karl, D.M., 2010. Measuring primary production rates in the ocean: enigmatic results between incubation and non-incubation methods at station ALOHA. Glob. Biogeochem. Cycles 24 (3).

Quegan, S., Le Toan, T., Chave, J., Dall, J., Exbrayat, J.-F., Minh, D.H.T., Lomas, M., et al., 2019. The European space agency BIOMASS mission: measuring forest above-ground biomass from space. Remote Sens. Environ. 227 (June), 44–60.

Quinn, P.K., Bates, T.S., 2011. The case against climate regulation via oceanic phytoplankton sulphur emissions. Nature 480 (7375), 51–56.

Quinn, P.K., Charlson, R.J., Bates, T.S., 1988. Simultaneous observations of ammonia in the atmosphere and ocean. Nature 335, 336–338.

Quinn, T.P., Helfield, J.M., Austin, C.S., Hovel, R.A., Bunn, A.G., 2018. A multidecade experiment shows that fertilization by Salmon carcasses enhanced tree growth in the riparian zone. Ecology 99 (11), 2433–2441.

Quintana, E.V., Barclay, T., Raymond, S.N., Rowe, J.F., Bolmont, E., Caldwell, D.A., Howell, S.B., et al., 2014. An earth-sized planet in the habitable zone of a cool star. Science 344 (6181), 277–280.

Quist, M.E., Nasholm, T., Lindeberg, J., Johannisson, C., Hogbom, L., Hogberg, P., 1999. Responses of a nitrogen-saturated forest to a sharp decrease in nitrogen input. J. Environ. Qual. 28, 1970–1977.

Raaimakers, D., Boot, R.G.A., Dijkstra, P., Pot, S., Pons, T., 1995. Photosynthetic rates in relation to leaf phosphorus content in Pioneer versus climax tropical rain forest trees. Oecologia 102, 120–125.

Rabalais, N.N., Eugene Turner, R., Dubravko, J.Ć., Dortch, Q., Wiseman, W.J., Sen, B.K., Gupta, 1996. Nutrient changes in the Mississippi River and system responses on the adjacent continental shelf. Estuaries 19 (2), 386–407.

Rabalais, N.N., Turner, R.E., Wiseman, W.J., 2002. Gulf of Mexico hypoxia, Aka "the dead zone". Annu. Rev. Ecol. Syst. 33, 235–263.

Rabenhorst, M.C., Haering, K.C., 1989. Soil micromorphology of a Chesapeake Bay tidal marsh—implications for sulfur accumulation. Soil Sci. 147, 339–347.

Rabenhorst, M.C., Hively, W.D., James, B.R., 2009. Measurements of soil redox potential. Soil Sci. Soc. Am. J. 73 (2), 668.

Rabinowitz, J., Flores, J., Kresbach, R., Rogers, G., 1969. Peptide formation in the presence of linear or cyclic polyphosphates. Nature 224 (5221), 795–796.

Rachmilevitch, S., Cousins, A.B., Bloom, A.J., 2004. Nitrate assimilation in plant shoots depends on photorespiration. Proc. Natl. Acad. Sci. U. S. A. 101 (31), 11506–11510.

Raciti, S.M., Groffman, P.M., Jenkins, J.C., Pouyat, R.V., Fahey, T.J., Pickett, S.T.A., Cadenasso, M.L., 2011. Accumulation of carbon and nitrogen in residential soils with different land-use histories. Ecosystems 14, 287–297.

Rae, J.W.B., Burke, A., Robinson, L.F., Adkins, J.F., Chen, T., Cole, C., Greenop, R., et al., 2018. CO_2 storage and release in the deep Southern Ocean on millennial to centennial timescales. Nature 562 (7728), 569–573.

Raes, F., Van Dingenen, R., Vignati, E., Wilson, J., Putaud, J.P., Seinfeld, J.H., Adams, P., 2000. Formation and cycling of aerosols in the Global troposphere. Atmos. Environ. 34, 4215–4240.

Raghoebarsing, A.A., Pol, A., van de Pas-Schoonen, K.T., Smolders, A.J.P., Ettwig, K.F., Rijpstra, W.I.C., Schouten, S., et al., 2006. A microbial consortium couples anaerobic methane oxidation to denitrification. Nature 440 (7086), 918–921.

Raghubanshi, A.S., 1992. Effect of topography on selected soil properties and nitrogen mineralization in a dry tropical forest. Soil Biol. Biochem. 24, 145–150.

Ragueneau, O., Treguer, P., Leynaert, A., Anderson, R.F., Brzezinski, M.A., DeMaster, D.J., Dugdale, R.C., et al., 2000. A review of the Si cycle in the modem ocean: recent progress and missing gaps in the application of biogenic opal as a paleoproductivity proxy. Glob. Planet. Chang. 26, 317–365.

Rahman, S., Aller, R.C., Cochran, J.K., 2017. The missing silica sink: revisiting the marine sedimentary Si cycle using cosmogenic ^{32}Si. Glob. Biogeochem. Cycles 31 (10), 1559–1578.

Rahn, K.A., Lowenthal, D.H., 1984. Elemental tracers of distant regional pollution aerosols. Science 223 (4632), 132–139.

Raich, J.W., Nadelhoffer, K.J., 1989. Belowground carbon allocation in forest ecosystems: global trends. Ecology 70, 1346–1354.

Raich, J.W., Schlesinger, W.H., 1992. The global carbon dioxide flux in soil respiration and its relationship to vegetation and climate. Tellus Ser. B Chem. Phys. Meteorol. 44, 81–99.

Raich, J.W., Tufekcioglu, A., 2000. Vegetation and soil respiration: correlations and controls. Biogeochemistry 48, 71–90.

Raich, J.W., Potter, C.S., Bhagawati, D., 2002. Interannual variability in global soil respiration, 1980–94. Glob. Chang. Biol. 8, 800–812.

Raison, R.J., 1979. Modification of the soil environment by vegetation fires, with particular reference to nitrogen transformations: review. Plant Soil 51, 73–108.

Raison, R.J., Khanna, P.K., Woods, P.V., 1985. Mechanisms of element transfer to the atmosphere during vegetation fires. Can. J. Forest Res. 15, 132–140.

Raison, R.J., Connell, M.J., Khanna, P.K., 1987. Methodology for studying fluxes of soil mineral N in situ. Soil Biol. Biochem. 19, 521–530.

Raiswell, R., Berner, R.A., 1986. Pyrite and organic matter in phanerozoic normal marine shales. Geochim. Cosmochim. Acta 50, 1967–1976.

Raiswell, R., Canfield, D.E., 2012. The iron biogeochemical cycle past and present. Geochem. Perspect. 1 (1), 1–2.

Ralph, B.J., 1979. Oxidative reactions in the sulfur cycle. In: Trudinger, P.A., Swaine, D.J. (Eds.), Biogeochemical Cycling of Mineral-Forming Elements. Elsevier, pp. 369–400.

Ramanathan, V., 1988. The greenhouse theory of climate change: a test by an inadvertent global experiment. Science 240 (4850), 293–299.

Ramankutty, N., Foley, J.A., 1998. Characterizing patterns of global land use: an analysis of global croplands data. Glob. Biogeochem. Cycles 12, 667–685.

Ramirez, A.J., Rose, A.W., 1992. Analytical geochemistry of organic phosphorus and its correlation with organic carbon in marine and fluvial sediments and soils. Am. J. Sci. 292, 421–454.

Randerson, J.T., Chapin, F.S., Harden, J.W., Neff, J.C., Harmon, M.E., 2002. Net ecosystem production: a comprehensive measure of net carbon accumulation by ecosystems. Ecol. Appl. 12, 937–947.

Randerson, J.T., Chen, Y., van der Werf, G.R., Rogers, B.M., Morton, D.C., 2012. Global burned area and biomass burning emissions from small fires: burned area from small fires. J. Geophys. Res. 117 (G4).

Randlett, D.L., Zak, D.R., Macdonald, N.W., 1992. Sulfate adsorption and microbial immobilization in northern hardwood forests along an atmospheric deposition gradient. Can. J. Forest Res. 22, 1843–1850.

Rao, D.V.S., 1981. Effect of boron on primary production of nanoplankton. Can. J. Fish. Aquat. Sci. 38 (1), 52–58.

Raper, C.D., Osmond, D.L., Wann, M., Weeks, W.W., 1978. Interdependence of root and shoot activities in determining nitrogen uptake rate of roots. Bot. Gaz. 139, 289–294.

Rasmussen, B., 1996. Early-diagenetic REE-phosphate minerals (florencite, gorceixite, crandallite, and xenotime) in marine sandstones: a major sink for oceanic phosphorus. Am. J. Sci. 296, 601–632.

Rasmussen, B., 2000. Filamentous microfossils in a 3,235-million-year-old volcanogenic massive sulphide deposit. Nature 405 (6787), 676–679.

Rasmussen, C., Southard, R.J., Horwath, W.R., 2006. Mineral control of organic carbon mineralization in a range of temperate conifer forest soils. Glob. Chang. Biol. 12, 834–847.

Rastetter, E.B., McKane, R.B., Shaver, G.R., Melillo, J.M., 1992. Changes in C-storage by terrestrial ecosystems: how C-N interactions restrict responses to CO_2 and temperature. Water Air Soil Pollut. 64, 327–344.

Rauch, J.N., Pacyna, J.M., 2009. Earth's global Ag, Al, Cr, Cu, Fe, Ni, Pb, and Zn cycles: global metal cycles. Glob. Biogeochem. Cycles Geophys. Monogr. Ser. 23 (2).

Raupach, M.R., Gloor, M., Sarmiento, J.L., Canadell, J.G., Frölicher, T.L., Gasser, T., Houghton, R.A., Le Quéré, C., Trudinger, C.M., 2014. The declining uptake rate of atmospheric CO_2 by land and ocean sinks. Biogeosciences 11 (13), 3453–3475.

Raval, A., Ramanathan, V., 1989. Observational determination of the greenhouse effect. Nature 342, 758–761.

Raven, J.A., Cockell, C.S., 2006. Influence on photosynthesis of starlight, moonlight, planetlight, and light pollution (reflections on photosynthetically active radiation in the universe). Astrobiology 6 (4), 668–675.

Raven, J.A., Falkowski, P.G., 1999. Oceanic sinks for atmospheric CO_2. Plant Cell Environ. 22, 741–755.

Raven, J.A., Franco, A.A., Dejesus, E.L., Jacobneto, J., 1990. H^+ extrusion and organic acid synthesis in N_2-fixing symbioses involving vascular plants. New Phytol. 114, 369–389.

Raven, J.A., Wollenweber, B., Handley, L.L., 1992. A comparison of ammonium and nitrate as nitrogen sources for photolithotrophs. New Phytol. 121, 19–32.

Raven, J.A., Wollenweber, B., Handley, L.L., 1993. The quantitative role of ammonia/ammonium transport and metabolism by plants in the global nitrogen cycle. Physiol. Plant. 89, 512–518.

Raven, J.A., Lambers, H., Smith, S.E., Westoby, M., 2018. Costs of acquiring phosphorus by vascular land plants: patterns and implications for plant coexistence. New Phytol. 217 (4), 1420–1427.

Ravi, S., D'Odorico, P., Breshears, D.D., Field, J.P., Goudie, A.S., Huxman, T.E., Li, J., et al., 2011. Aeolian processes and the biosphere. Rev. Geophys. 49.

Ravishankara, A.R., 1997. Heterogeneous and multiphase chemistry in the troposphere. Science 276, 1058–1065.

Ravishankara, A.R., Daniel, J.S., Portmann, R.W., 2009. Nitrous oxide (N_2O): the dominant ozone-depleting substance emitted in the 21st century. Science 326 (5949), 123–125.

Raymo, M.E., Mitrovica, J.X., 2012. Collapse of polar ice sheets during the stage 11 interglacial. Nature 483 (7390), 453–456.

Raymond, P.A., 2017. Temperature versus hydrologic controls of chemical weathering fluxes from United States forests. Chem. Geol. 458 (May), 1–13.

Raymond, P.A., Bauer, J.E., 2001a. Riverine export of aged terrestrial organic matter to the North Atlantic ocean. Nature 409 (6819), 497–500.

Raymond, P.A., Bauer, J.E., 2001b. Use of ^{14}C and ^{13}C natural abundances for evaluating riverine, estuarine, and coastal DOC and POC sources and cycling: a review and synthesis. Org. Geochem. 32 (4), 469–485.

Raymond, P.A., Cole, J.J., 2003. Increase in the export of alkalinity from North America's largest river. Science 301 (5629), 88–91.

Raymond, P.A., Hamilton, S.K., 2018. Anthropogenic influences on riverine fluxes of dissolved inorganic carbon to the oceans: riverine fluxes of inorganic carbon to the oceans. Limnol. Oceanogr. 3 (3), 143–155.

Raymond, P.A., Neung-Hwan, O., 2009. Long term changes of chemical weathering products in rivers heavily impacted from acid mine drainage: insights on the impact of coal mining on regional and global carbon and sulfur budgets. Earth Planet. Sci. Lett. 284 (1–2), 50–56.

Raymond, P.A., Saiers, J.E., 2010. Event controlled DOC export from forested watersheds. Biogeochemistry 100 (1), 197–209.

Raymond, J., Segrè, D., 2006. The effect of oxygen on biochemical networks and the evolution of complex life. Science 311 (5768), 1764–1767.

Raymond, P.A., Caraco, N.F., Cole, J.J., 1997. Carbon dioxide concentration and atmospheric flux in the Hudson River. Estuaries 20 (2), 381–390.

Raymond, P.A., Oh, N.H., 2009. Long term changes of chemical weathering products in rivers heavily impacted from acid mine drainage: insights on the impact of coal mining on regional and global carbon and sulfur budgets. Earth Planet. Sci. Lett. 284, 50–56.

Raymond, P.A., Bauer, J.E., Cole, J.J., 2000. Atmospheric CO_2 evasion, dissolved inorganic carbon production, and net heterotrophy in the York River estuary. Limnol. Oceanogr. 45, 1707–1717.

Raymond, P.A., Neung-Hwan, O., Turner, R.E., Broussard, W., 2008. Anthropogenically enhanced fluxes of water and carbon from the Mississippi river. Nature 451 (7177), 449–452.

Raymond, P.A., Hartmann, J., Lauerwald, R., Sobek, S., McDonald, C., Hoover, M., Butman, D., et al., 2013. Global carbon dioxide emissions from inland waters. Nature 503 (7476), 355–359.

Raymond, P.A., Saiers, J.E., Sobczak, W.V., 2016. Hydrological and biogeochemical controls on watershed dissolved organic matter transport: pulse-shunt concept. Ecology 97 (1), 5–16.

Rea, D.K., Ruff, L.J., 1996. Composition and mass flux of sediment entering the world's subduction zones: implications for global sediment budgets, great earthquakes, and volcanism. Earth Planet. Sci. Lett. 140 (1), 1–12.

Reader, R.J., Stewart, J.M., 1972. Relationship between net primary production and accumulation for a peatland in southeastern Manitoba. Ecology 53, 1024–1037.

Reager, J.T., Gardner, A.S., Famiglietti, J.S., Wiese, D.N., Eicker, A., Lo, M.-H., 2016. A decade of sea level rise slowed by climate-driven hydrology. Science 351 (6274), 699–703.

Real, R., Báez, J.C., 2013. The North Atlantic oscillation affects the quality of cava (Spanish sparkling wine). Int. J. Biometeorol. 57 (3), 493–496.

Reay, D.S., Dentener, F., Smith, P., Grace, J., Feely, R.A., 2008. Global nitrogen deposition and carbon sinks. Nat. Geosci. 1, 430–437.

Reddy, K.R., DeLaune, R.D., 2008. Biogeochemistry of Wetlands: Science and Applications. Taylor and Francis.

Reddy, K.R., Graetz, D.A., 1988. Carbon and nitrogen dynamics in wetland soils. In: Hook, D.D., McKee, W.H., Smith, H.K., James, G., Burrell, V.G., Richard DeVoe, M., Sojka, R.E., et al. (Eds.), The Ecology and Management of Wetlands: Volume 1: Ecology of Wetlands. Springer US, New York, NY, pp. 307–318.

Reddy, K.J., Lindsay, W.L., Workman, S.M., Drever, J.I., 1990. Measurement of calcite ion activity products in soils. Soil Sci. Soc. Am. J. 54, 67–71.

Reddy, K.R., DeLaune, R.D., Debusk, W.F., Koch, M.S., 1993. Long-term nutrient accumulation rates in the everglades. Soil Sci. Soc. Am. J. 57, 1147–1155.

Reddy, K.R., Kadlec, R.H., Flaig, E., Gale, P.M., 1999. Phosphorus retention in streams and wetlands: a review. Crit. Rev. Environ. Sci. Technol. 29 (1), 83–146.

Redfield, A.C., 1958. The biological control of chemical factors in the environment. Am. Sci. 46, 205–222.

Redfield, A.C., 1963. The influence of organisms on the composition of seawater. Sea 2, 26–77.

Redmond, K.T., Koch, R.W., 1991. Surface climate and streamflow variability in the Western United States and their relationship to large-scale circulation indexes. Water Resour. Res. 27, 2381–2399.

Reeburgh, W.S., 1983. Rates of biogeochemical processes in anoxic sediments. Annu. Rev. Earth Planet. Sci. 11 (1), 269–298.

Reeburgh, W.S., 2007. Oceanic methane biogeochemistry. Chem. Rev. 107 (2), 486–513.

Reeburgh, W.S., Whalen, S.C., Alperin, M.J., 1993. The role of methylotrophy in the global methane budget. In: Microbial Growth on C1 Compounds, pp. 1–14.

Reed, S.C., Townsend, A.R., Cleveland, C.C., Nemergut, D.R., 2010. Microbial community shifts influence patterns in tropical forest nitrogen fixation. Oecologia 164 (2), 521–531.

Reed, S.C., Cleveland, C.C., Townsend, A.R., 2011. Functional ecology of free-living nitrogen fixation: a contemporary perspective. Ann. Rev. Ecol Evol. Syst. 42, 489–512.

Reed, S.C., Townsend, A.R., Davidson, E.A., Cleveland, C.C., 2012. Stoichiometric patterns in foliar nutrient resorption across multiple scales. New Phytol. 196 (1), 173–180.

Rees, C.E., Jenkins, W.J., Monster, J., 1978. Sulfur isotopic composition of ocean water sulfate. Geochim. Cosmochim. Acta 42, 377–381.

Reeves, H., 1994. On the origin of the light elements. Rev. Mod. Phys. 66, 193–216.

Reeves, R.D., Baker, A.J.M., Jaffré, T., Erskine, P.D., Echevarria, G., van der Ent, A., 2018. A global database for plants that hyperaccumulate metal and metalloid trace elements. New Phytol. 218 (2), 407–411.

Regnier, P., Friedlingstein, P., Ciais, P., Mackenzie, F.T., Gruber, N., Janssens, I.A., Laruelle, G.G., et al., 2013. Anthropogenic perturbation of the carbon fluxes from land to ocean. Nat. Geosci. 6 (8), 597–607.

Reheis, M.C., Kihl, R., 1995. Dust deposition in southern Nevada and California, 1984–1989: relations to climate, source area, and source lithology. J. Geophys. Res.-Atmos. 100, 8893–8918.

Reich, P.B., Oleksyn, J., 2004. Global patterns of plant leaf N and P in relation to temperature and latitude. Proc. Natl. Acad. Sci. U. S. A. 101 (30), 11001–11006.

Reich, P.B., Schoettle, A.W., 1988. Role of phosphorus and nitrogen in photosynthetic and whole plant carbon gain and nutrient use efficiency in eastern white pine. Oecologia 77 (1), 25–33.

Reich, P.B., Walters, M.B., Ellsworth, D.S., 1992. Leaf life span in relation to leaf, plant, and stand characteristics among diverse ecosystems. Ecol. Monogr. 62, 365–392.

Reich, P.B., Walters, M.B., Ellsworth, D.S., Uhl, C., 1994. Photosynthesis-nitrogen relations in Amazonian tree species: I. Patterns among species and communities. Oecologia 97 (1), 62–72.

Reich, P.B., Walters, M.B., Kloeppel, B.D., Ellsworth, D.S., 1995. Different photosynthesis-nitrogen relations in deciduous hardwood and evergreen coniferous tree species. Oecologia 104 (1), 24–30.

Reich, P.B., Grigal, D.F., Aber, J.D., Gower, S.T., 1997. Nitrogen mineralization and productivity in 50 hardwood and conifer stands on diverse soils. Ecology 78, 335–347.

Reich, P.B., Ellsworth, D.S., Walters, M.B., Vose, J.M., Gresham, C., Volin, J.C., Bowman, W.D., 1999. Generality of leaf trait relationships: a test across six biomes. Ecology 80, 1955–1969.

Reich, P.B., Oleksyn, J., Modrzynski, J., Mrozinski, P., Hobbie, S.E., Eissenstat, D.M., Chorover, J., Chadwick, O.A., Hale, C.M., Tjoelker, M.G., 2005. Linking litter calcium, earthworms and soil properties: acommon garden test with 14 tree species. Ecol. Lett. 8, 811–818.

Reich, P.B., Tjoelker, M.G., Machado, J.-L., Oleksyn, J., 2006. Universal scaling of respiratory metabolism, size and nitrogen in plants. Nature 439 (7075), 457–461.

Reich, P.B., Oleksyn, J., Wright, I.J., Niklas, K.J., Hedin, L., Elser, J.J., 2010. Evidence of a general 2/3-power law of scaling leaf nitrogen to phosphorus among major plant groups and biomes. Proc. Biol. Sci. 277 (1683), 877–883.

Reich, P.B., Sendall, K.M., Stefanski, A., Wei, X., Rich, R.L., Montgomery, R.A., 2016. Boreal and temperate trees show strong acclimation of respiration to warming. Nature 531 (7596), 633–636.

Reichle, D.E., Dinger, B.E., Edwards, N.T., Harris, W.F., Sollins, P., 1973a. Carbon flow and storage in a forest ecosystem. In: Woodwell, G.M., Pecan, E.V. (Eds.), Carbon and the Biosphere. National Technical Information Service, pp. 345–365.

Reichle, D.E., Goldstein, R.A., Van Hook, R.I., Dodson, C.J., 1973b. Analysis of insect consumption in a forest canopy. Ecology 54, 1076–1084.

Reid, C.D., Fiscus, E.L., 1998. Effects of elevated CO_2 and/or ozone on limitations to CO_2 assimilation in soybean (glycine max). J. Exp. Bot. 49, 885–895.

Reiners, W.A., 1972. Structure and energetics of three Minnesota forests. Ecol. Monogr. 42, 71–94.

Reiners, W.A., 1986. Complementary models for ecosystems. Am. Nat. 127, 59–73.

Reiners, W.A., 1992. Twenty years of ecosystem reorganization following experimental deforestation and regrowth suppression. Ecol. Monogr. 62, 503–523.

Reiners, P.W., Turchyn, A.V., 2018. Extraterrestrial dust, the marine lithologic record, and global biogeochemical cycles. Geology 46 (10), 863–866.

Reiners, P.W., Ehlers, T.A., Mitchell, S.G., Montgomery, D.R., 2003. Coupled spatial variations in precipitation and long-term erosion rates across the Washington cascades. Nature 426 (6967), 645–647.

Reinfelder, J.R., Fisher, N.S., 1991. The assimilation of elements ingested by marine copepods. Science 251 (4995), 794–796.

Reinhard, C.T., Raiswell, R., Scott, C., Anbar, A.D., Lyons, T.W., 2009. A late Archaen sulfidic sea simulated by early oxidative weathering of the continents. Science 326, 713–716.

Reinhard, C.T., Planavsky, N.J., Olson, S.L., Lyons, T.W., Erwin, D.H., 2016. Earth's oxygen cycle and the evolution of animal life. Proc. Natl. Acad. Sci. U. S. A. 113 (32), 8933–8938.

Reinhard, C.T., Planavsky, N.J., Gill, B.C., Ozaki, K., Robbins, L.J., Lyons, T.W., Fischer, W.W., Wang, C., Cole, D.B., Konhauser, K.O., 2017. Evolution of the global phosphorus cycle. Nature 541 (7637), 386–389.

Reinhold, A.M., Poole, G.C., Izurieta, C., Helton, A.M., Bernhardt, E.S., 2019. Constraint-based simulation of multiple interactive elemental cycles in biogeochemical systems. Ecol. Inform. 50 (March), 102–121.

Remde, A., Conrad, R., 1991. Role of nitrification and denitrifciation for NO metabolism in soil. Biogeochemistry 12, 189–205.

Ren, H., Sigman, D.M., Meckler, A.N., Plessen, B., Robinson, R.S., Rosenthal, Y., Haug, G.H., 2009. Foraminiferal isotope evidence of reduced nitrogen fixation in the ice age Atlantic Ocean. Science 323 (5911), 244–248.

Ren, H., Chen, Y.-C., Wang, X.T., Wong, G.T.F., Cohen, A.L., DeCarlo, T.M., Weigand, M.A., Mii, H.-S., Sigman, D.M., 2017a. 21st-century rise in anthropogenic nitrogen deposition on a remote coral reef. Science 356 (6339), 749–752.

Ren, H., Sigman, D.M., Martínez-García, A., Anderson, R.F., Chen, M.-T., Ravelo, A.C., Straub, M., Wong, G.T.F., Haug, G.H., 2017b. Impact of glacial/interglacial sea level change on the ocean nitrogen cycle. Proc. Natl. Acad. Sci. U. S. A. 114 (33), E6759–E6766.

Ren, X., Hall, D.L., Vinciguerra, T., Benish, S.E., Stratton, P.R., Ahn, D., Hansford, J.R., et al., 2019. Methane emissions from the Marcellus shale in southwestern Pennsylvania and northern West Virginia based on airborne measurements. J. Geophys. Res. D: Atmos. 124 (3), 1862–1878.

Renberg, I., Korsman, T., Birks, H.J.B., 1993. Prehistoric increases in the pH of acid-sensitive Swedish Lakes caused by land use changes. Nature 362 (6423), 824–827.

Renforth, P., Henderson, G., 2017. Assessing ocean alkalinity for carbon sequestration: ocean alkalinity for C sequestration. Rev. Geophys. Nato ASI Ser. 55 (3), 636–674.

Rennenberg, H., 1991. The significance of higher plants in the emission of sulfur compounds from terrestrial ecosystems. In: Sharkey, T.D., Holland, E.A., Mooney, H.A. (Eds.), Trace Gas Emissions by Plants. Academic Press, pp. 217–260.

Retallack, G.J., 1997. Early forest soils and their role in devonian global change. Science 276 (5312), 583–585.

Reuss, J.O., 1980. Simulation of soil nutrient losses resulting from rainfall acidity. Ecol. Model. 11, 15–38.

Reuss, J.O., Johnson, D.W., 1986. Acid Deposition and the Acidification of Soils and Waters. Springer.

Reuss, J.O., Cosby, B.J., Wright, R.F., 1987. Chemical processes governing soil and water acidification. Nature 329, 27–32.

Reynolds, R.C., 1978. Polyphenol inhibition of calcite precipitation in Lake Powell. Limnol. Oceanogr. 23, 585–597.

Reynolds, C.S., Davies, P.S., 2001. Sources and bioavailability of phosphorus fractions in freshwaters: a British perspective. Biol. Rev. Camb. Philos. Soc. 76 (1), 27–64.

Reynolds-Vargas, J.S., Richter, D.D., Bornemisza, E., 1994. Environmental impacts of nitrification and nitrate adsorption in fertilized andisols in the Valle central of Costa Rica. Soil Sci. 157, 289–299.

Rhein, M., Rintoul, S., Aoki, S., Campos, E., Chambers, D., Feely, R., Wang, F., 2013. Observations: ocean. In: Climate Change 2013: The Physical Science Basis. Contribution of Working Group I to the Fifth Assessment Report of the Intergovernmental Panel on Climate Change. Fifth Assessment Report of the Intergovernmental Panel on Climate Change, January, pp. 255–315.

Rhew, R.C., Whelan, M.E., Min, D.-H., 2014. Large methyl halide emissions from South Texas salt marshes. Biogeosciences 11 (22), 6427–6434.

Riccobono, F., Schobesberger, S., Scott, C.E., Dommen, J., Ortega, I.K., Rondo, L., Almeida, J., et al., 2014. Oxidation products of biogenic emissions contribute to nucleation of atmospheric particles. Science 344 (6185), 717–721.

Rice, A.L., Butenhoff, C.L., Teama, D.G., Röger, F.H., Aslam, M., Khalil, K., Rasmussen, R.A., 2016. Atmospheric methane isotopic record favors fossil sources flat in 1980s and 1990s with recent increase. Proc. Natl. Acad. Sci. U. S. A. 113 (39), 10791–10796.

Rich, P.H., Wetzel, R.G., 1978. Detritus in the lake ecosystem. Am. Nat. 112, 57–71.

Richardson, C.J., 1985. Mechanisms controlling phosphorus retention capacity in freshwater wetlands. Science 228 (4706), 1424–1427.

Richardson, C.J., Sasek, T.W., Fendick, E.A., 1992. Implications of physiological responses to chronic air pollution for forest decline in the southeastern United States. Environ. Toxicol. Chem. 11, 1105–1114.

Richardson, S.J., Peltzer, D.A., Allen, R.B., McGlone, M.S., Parfitt, R.L., 2004. Rapid development of phosphorus limitation in temperate rainforest along the Franz Josef soil chronosequence. Oecologia 139 (2), 267–276.

Richey, J.E., Wissmar, R.C., Devol, A.H., Likens, G.E., Eaton, J.S., Wetzel, R.G., Odum, W.E., et al., 1978. Carbon flow in four lake ecosystems—structural approach. Science 202 (4373), 1183–1186.

Richey, J.E., Melack, J.M., Aufdenkampe, A.K., Ballester, V.M., Hess, L.L., 2002. Outgassing from Amazonian rivers and wetlands as a large tropical source of atmospheric CO_2. Nature 416 (6881), 617–620.

Richmond, E.K., Rosi, E.J., Walters, D.M., Fick, J., Hamilton, S.K., Brodin, T., Sundelin, A., Grace, M.R., 2018. A diverse suite of pharmaceuticals contaminates stream and riparian food webs. Nat. Commun. 9 (1), 4491.

Richter, D.D., 2007. Humanity's transformation of Earth's soil: pedology's new frontier. Soil Sci. 172, 957–967.

Richter, D.D., Babbar, L.I., 1991. Soil diversity in the tropics. Adv. Ecol. Res. 21, 315–389.

Richter, D.D., Markewitz, D., 1995. How deep is soil? Bioscience 45, 600–609.

Richter, D.D., Ralston, C.W., Harms, W.R., 1982. Prescribed fire: effects on water quality and forest nutrient cycling. Science 215 (4533), 661–663.

Richter, F.M., Rowley, D.B., Depaolo, D.J., 1992. Sr-isotope evolution of seawater: the role of tectonics. Earth Planet. Sci. Lett. 109, 11–23.

Richter, D.D., Markewitz, D., Wells, C.G., Allen, H.L., April, R., Heine, P.R., Urrego, B., 1994. Soil chemical change during three decades in an old-field loblolly pine (*Pinus taeda* L.) ecosystem. Ecology 75, 1463–1473.

Richter, A., Burrows, J.P., Nüss, H., Granier, C., Niemeier, U., 2005. Increase in tropospheric nitrogen dioxide over China observed from space. Nature 437 (7055), 129–132.

Richter, D.D., Lee Allen, H., Li, J., Markewitz, D., Raikes, J., 2006. Bioavailability of slowly cycling soil phosphorus: major restructuring of soil P fractions over four decades in an aggrading forest. Oecologia 150 (2), 259–271.

Riddick, S.N., Blackall, T.D., Dragosits, U., Daunt, F., Newell, M., Braban, C.F., Tang, Y.S., et al., 2016. Measurement of ammonia emissions from temperate and sub-polar seabird colonies. Atmos. Environ. 134 (June), 40–50.

Riebe, C.S., Kirchner, J.W., Finkel, R.C., 2003. Long-term rates of chemical weathering and physical erosion from cosmogenic nuclides and geochemical mass balance. Geochim. Cosmochim. Acta 67, 4411–4427.

Riebesell, U., Zondervan, I., Rost, B., Tortell, P.D., Zeebe, R.E., Morel, F.M., 2000. Reduced calcification of marine plankton in response to increased atmospheric CO_2. Nature 407 (6802), 364–367.

Riebesell, U., Körtzinger, A., Oschlies, A., 2009. Sensitivities of marine carbon fluxes to ocean change. Proc. Natl. Acad. Sci. U. S. A. 106 (49), 20602–20609.

Riedel, T., Zak, D., Biester, H., Dittmar, T., 2013. Iron traps terrestrially derived dissolved organic matter at redox interfaces. Proc. Natl. Acad. Sci. U. S. A. 110 (25), 10101–10105.

Rigby, M., Prinn, R.G., O'Doherty, S., Miller, B.R., Ivy, D., Mühle, J., Harth, C.M., et al., 2014. Recent and future trends in synthetic greenhouse gas radiative forcing: synthetic greenhouse gas trends. Geophys. Res. Lett. 41 (7), 2623–2630.

Rigby, M., Montzka, S.A., Prinn, R.G., White, J.W.C., Young, D., O'Doherty, S., Lunt, M.F., et al., 2017. Role of atmospheric oxidation in recent methane growth. Proc. Natl. Acad. Sci. U. S. A. 114 (21), 5373–5377.

Rigby, M., Park, S., Saito, T., Western, L.M., Redington, A.L., Fang, X., Henne, S., et al., 2019. Increase in CFC-11 emissions from eastern China based on atmospheric observations. Nature 569 (7757), 546–550.

Rignot, E., Bamber, J.L., Van Den Broeke, M.R., Davis, C., Li, Y.H., Van De Berg, W.J., Van Meijgaard, E., 2008. Recent Antarctic ice mass loss from radar interferometry and regional climate modelling. Nat. Geosci. 1, 106–110.

Rignot, E., Jacobs, S., Mouginot, J., Scheuchl, B., 2013. Ice-shelf melting around Antarctica. Science 341 (6143), 266–270.

Riley, W.J., Gaudinski, J.B., Torn, M.S., Joslin, J.D., Hanson, P.J., 2009. Fine-root mortality rates in a temperate forest: estimates using radiocarbon data and numerical modeling. New Phytol. 184 (2), 387–398.

Rind, D., Goldberg, R., Hansen, J., Rosenzweig, C., Ruedy, R., 1990. Potential evapotranspiration and the likelihood of future drought. J. Geophys. Res.-Atmos. 95, 9983–10004.

Rind, D., Chiou, E.W., Chu, W., Larsen, J., Oltmans, S., Lerner, J., McCormick, M.P., McMaster, L., 1991. Positive water-vapor feedback in climate models confirmed by satellite data. Nature 349, 500–503.

Ringeval, B., Ducoudré, N.D.N., Ciais, P., Bousquet, P., Prigent, C., Papa, F., Rossow, W.B., 2010. An attempt to quantify the impact of changes in wetland extent on methane emissions on the seasonal and interannual time scales. Glob. Biogeochem. Cycles 24, 1–12.

Rinne, J., Hakola, H., Laurila, T., Rannik, U., 2000. Canopy-scale monoterpene emissions of *Pinus sylvestris* dominated forests. Atmos. Environ. 34, 1099–1107.

Rinne, K.T., Rajala, T., Peltoniemi, K., Chen, J., Smolander, A., Mäkipää, R., 2017. Accumulation rates and sources of external nitrogen in decaying wood in a Norway spruce dominated forest. Treseder, K. (Ed.), Funct. Ecol. 31 (2), 530–541.

Rinsland, C.P., Goldman, A., Mahieu, E., Zander, R., Notholt, J., Jones, N.B., Griffith, D.W.T., Stephen, T.M., Chiou, L.S., 2002. Ground-based infrared spectroscopic measurements of carbonyl sulfide: free tropospheric trends from a 24-year time series of solar absorption measurements. J. Geophys. Res.—Atmos. 107.

Rippey, B., Anderson, N.J., 1996. Reconstruction of lake phosphorus loading and dynamics using the sedimentary record. Environ. Sci. Technol. 30, 1786–1788.

Rippey, B., McSorley, C., 2009. Oxygen depletion in lake hypolimnia. Limnol. Oceanogr. 54, 905–916.

Risley, L.S., Crossley, D.A., 1988. Herbivore-caused greenfall in the southern Appalachians. Ecology 69, 1118–1127.

Rivkin, R.B., Legendre, L., 2001. Biogenic carbon cycling in the upper ocean: effects of microbial respiration. Science 291 (5512), 2398–2400.

Roberts, B.J., Mulholland, P.J., 2007. In-stream biotic control on nutrient biogeochemistry in a forested stream, West Fork of Walker Branch. JGR Biogeosci. 112.

Roberts, T.L., Stewart, J.W.B., Bettany, J.R., 1985. The influence of topography on the distribution of organic and inorganic soil phosphorus across a narrow environmental gradient. Can. J. Soil Sci. 65, 651–665.

Roberts, B.J., Mulholland, P.J., Hill, W.R., 2007. Multiple scales of temporal variability in ecosystem metabolism rates: results from 2 years of continuous monitoring in a forested headwater stream. Ecosystems 10 (4), 588–606.

Roberts, K.G., Gloy, B.A., Joseph, S., Scott, N.R., Lehmann, J., 2010. Life cycle assessment of biochar systems: estimating the energetic, economic, and climate change potential. Environ. Sci. Technol. 44 (2), 827–833.

Roberts, K., Defforey, D., Turner, B.L., Condron, L.M., Peek, S., Silva, S., Kendall, C., Paytan, A., 2015. Oxygen isotopes of phosphate and soil phosphorus cycling across a 6500 year chronosequence under lowland temperate rainforest. Geoderma 257–258 (November), 14–21.

Robertson, G.P., 1982. Nitrification in forested ecosystems. Philos. Trans. R. Soc. Lond. Ser. B—Biol. Sci. 296, 445–457.

Robertson, M.P., Miller, S.L., 1995. An efficient prebiotic synthesis of cytosine and uracil. Nature 375 (6534), 772–774.

Robertson, G.P., Vitousek, P.M., 1981. Nitrification potentials in primary and secondary succession. Ecology 62, 376–386.

Robertson, G.P., Vitousek, P.M., 2009. Nitrogen in agriculture: balancing the cost of an essential resource. Annu. Rev. Environ. Resour. 34, 97–125.

Robertson, G.P., Huston, M.A., Evans, F.C., Tiedje, J.M., 1988. Spatial variability in a successional plant community: patterns of nitrogen availability. Ecology 69, 1517–1524.

Robertson, J.E., Robinson, C., Turner, D.R., Holligan, P., Watson, A.J., Boyd, P., Fernandez, E., Finch, M., 1994. The impact of a coccolithophore bloom on oceanic carbon uptake in the Northeast Atlantic during Summer 1991. Deep-Sea Res. Part I Oceanogr. Res. Pap. 41, 297–314.

Robertson, D.M., Garn, H.S., Rose, W.J., 2007. Response of calcareous Nagawicka Lake, Wisconsin, to changes in phosphorus loading. Lake Reser. Manag. 23, 298–312.

Robinson, D., 1986. Limits to nutrient inflow rates in roots and root systems. Physiol. Plant. 68, 551–559.

Robinson, D., 1994. The responses of plants to nonuniform supplies of nutrients. New Phytol. 127, 635–674.

Robinson, D., 2001. Delta ^{15}N as an integrator of the nitrogen cycle. Trends Ecol. Evol. 16, 153–162.

Robock, A., Marquardt, A., Kravitz, B., Stenchikov, G., 2009. Benefits, risks, and costs of stratospheric geoengineering. Geophysical Research Letters 36.

Rockmann, T., Levin, I., 2005. High-precision determination of the changing isotopic composition of atmospheric N_2O from 1990 to 2002. J. Geophys. Res. 110.

Rockström, J., Steffen, W., Noone, K., Asa, P., Stuart Chapin 3rd, F., Lambin, E.F., Lenton, T.M., et al., 2009. A safe operating space for humanity. Nature 461 (7263), 472–475.

Rodbell, D.T., Seltzer, G.O., Anderson, D.M., Abbott, M.B., Enfield, D.B., Newman, J.H., 1999. An 15,000-year record of El Niño-driven alluviation in Southwestern Ecuador. Science 283 (5401), 516–520.

Rodell, M., Velicogna, I., Famiglietti, J.S., 2009. Satellite-based estimates of groundwater depletion in India. Nature 460 (7258), 999–1002.

Rodell, M., Famiglietti, J.S., Wiese, D.N., Reager, J.T., Beaudoing, H.K., Landerer, F.W., Lo, M.-H., 2018. Emerging trends in global freshwater availability. Nature 557 (7707), 651–659.

Rödenbeck, C., Zaehle, S., Keeling, R., Heimann, M., 2018. History of El Niño impacts on the global carbon cycle 1957–2017: a quantification from atmospheric CO_2 data. Philos. Trans. R. Soc. Lond. Ser. B Biol. Sci. 373 (1760), 20170303.

Rodhe, H., 1981. Formation of sulfuric and nitric acid in the atmosphere during long-range transport. Tell'Us 33, 132–141.

Rodhe, H., 1999. Human impact on the atmospheric sulfur balance. Tellus A Dyn. Meteorol. Oceanogr. 51, 110–122.

Rodriguez, A., Duran, J., Covelo, F., Maria Fernandez-Palacios, J., Gallardo, A., 2011. Spatial pattern and variability in soil N and P availability under the influence of two dominant species in a pine forest. Plant Soil 345, 211–221.

Rodriguez, F., Lillington, J., Johnson, S., Timmel, C.R., Lea, S.M., Berks, B.C., 2014. Crystal structure of the *Bacillus subtilis* phosphodiesterase PhoD reveals an iron and calcium-containing active site. J. Biol. Chem. 289 (45), 30889–30899.

Roehm, C.L., Prairie, Y.T., del Giorgio, P.A., 2009. The pCO_2 dynamics in lakes in the boreal region of northern Quebec, Canada. Glob. Biogeochem. Cycles Canada 23.

Roelandt, C., van Wesemael, B., Rounsevell, M., 2005. Estimating anual N_2O emissions from agricultural soils in temperate climates. Glob. Change Biol. 11, 1701–1711.

Roelle, P., Aneja, V.P., O'Connor, J., Robarge, W., Kim, D.S., Levine, J.S., 1999. Measurement of nitrogen oxide emissions from an agricultural soil with a dynamic chamber system. J. Geophys. Res. D: Atmos. 104, 1609–1619.

Rogers, H.H., Runion, G.B., Krupa, S.V., 1994. Plant responses to atmospheric CO_2 enrichment with emphasis on roots and the rhizosphere. Environ. Pollut. 83 (1–2), 155–189.

Rogers, J.R., Bennett, P.C., Choi, W.J., 1998. Feldspars as a source of nutrients for microorganisms. Am. Mineral. 83, 1532–1540.

Rohrer, F., Berresheim, H., 2006. Strong correlation between levels of tropospheric hydroxyl radicals and solar ultraviolet radiation. Nature 442 (7099), 184–187.

Roman, J., McCarthy, J.J., 2010. The whale pump: marine mammals enhance primary productivity in a coastal basin. PLoS ONE 5 (10), e13255.

Román, M.O., Wang, Z., Sun, Q., Kalb, V., Miller, S.D., Molthan, A., Schultz, L., et al., 2018. NASA's black marble nighttime lights product suite. Remote Sens. Environ. 210 (June), 113–143.

Romell, L.G., 1935. Ecological Problems in the Humus Layer in the Forest. Cornell University, Agricultural Experiment Station.

Rondon, A., Granat, L., 1994. Studies on the dry deposition of NO_2 to *Coniferous* species at low NO_2 concentrations. Tellus B Chem. Phys. Meteorol. 46 (5), 339–352.

Ronen, D., Magaritz, M., Almon, E., 1988. Contaminated aquifers are a forgotten component of the global N_2O budget. Nature 335, 57–59.

Roos-Barraclough, F., Martinez-Cortizas, A., Garcia-Rodeja, E., Shotyk, W., 2002. A 14,500-year record of the accumulation of atmospheric mercury in peat: volcanic signals, anthropogenic influences and a correlation to bromine accumulation. Earth Planet. Sci. Lett. 202, 435–451.

Rose, S.L., Youngberg, C.T., 1981. Tripartite associations in snowbrush (*Ceanothus velutinus*): effect of vesicular-arbuscular mycorrhizae on growth, nodulation, and nitrogen fixation. Can. J. Bot. 59, 34–39.

Rosemond, P.J.M.A.D., Elwood, J.W., 1993. Top-down and bottom-up control of stream periphyton: effects of nutrients and herbivores. Ecology 74 (4), 1264–1280.

Rosemond, A.D., Benstead, J.P., Bumpers, P.M., Gulis, V., Kominoski, J.S., Manning, D.W.P., Suberkropp, K., Wallace, J.B., 2015. Experimental nutrient additions accelerate terrestrial carbon loss from stream ecosystems. Science 347, 1142–1145.

Rosen, M.R., Turner, J.V., Coshell, L., Gailitis, V., 1995. The effects of water temperature, stratification, and biological activity on the stable isotopic composition and timing of carbonate precipitation in a hypersaline lake. Geochim. Cosmochim. Acta 59, 979–990.

Rosenzweig, M.L., 1968. Net primary productivity of terrestrial communities: prediction from climatological data. Am. Natur. 102, 67–74.

Rosi-Marshall, E.J., Bernhardt, E.S., Buso, D.C., Driscoll, C.T., Likens, G.E., 2016. Acid rain mitigation experiment shifts a forested watershed from a net sink to a net source of nitrogen. Proc. Natl. Acad. Sci. U. S. A. 113 (27), 7580–7583.

Rosing, M.T., 1999. ^{13}C-depleted carbon microparticles in >3700-Ma sea-floor sedimentary rocks from west Greenland. Science 283, 674–676.

Roskoski, J.P., 1980. Nitrogen fixation in hardwood forests of the northeastern United States. Plant Soil 54, 33–44.

Rosling, A., Suttle, K.B., Johansson, E., Van Hees, P.A.W., Banfield, J.F., 2007. Phosphorous availability influences the dissolution of apatite by soil fungi. Geobiology 5 (3), 265–280.

Ross, D.S., Bartlett, R.J., 1996. Field-extracted spodosol solutions and soils: aluminum, organic carbon, and pH interrelationships. Soil Sci. Soc. Am. J. 60, 589–595.

Ross, A.N., Wooster, M.J., Boesch, H., Parker, R., 2013. First satellite measurements of carbon dioxide and methane emission ratios in wildfire plumes: gosat measure of CO_2:CH_4 emission ratio. Geophys. Res. Lett. 40 (15), 4098–4102.

Ross, M.R.V., Nippgen, F., Hassett, B.A., McGlynn, B.L., Bernhardt, E.S., 2018. Pyrite oxidation drives exceptionally high weathering rates and geologic CO_2 release in mountaintop-mined landscapes. Glob. Biogeochem. Cycles 32 (8), 1182–1194.

Rosswall, T., 1982. Microbiological regulation of the biogeochemical nitrogen cycle. Plant Soil 67, 15–34.

Rost, S., Gerten, D., Bondeau, A., Lucht, W., Rohwer, J., Schaphoff, S., 2008. Agricultural green and blue water consumption and its influence on the global water system. Water Resour. Res. 44.

Rothman, D.H., 2002. Atmospheric carbon dioxide levels for the last 500 million years. Proc. Natl. Acad. Sci. U. S. A. 99 (7), 4167–4171.

Rothman, J.M., Van Soest, P.J., Pell, A.N., 2006. Decaying wood is a sodium source for mountain Gorillas. Biol. Lett. 2 (3), 321–324.

Rothschild, L.J., Mancinelli, R.L., 2001. Life in extreme environments. Nature 409 (6823), 1092–1101.

Roulet, N.T., 2000. Peatlands, carbon storage, greenhouse gases, and the Kyoto protocol: prospects and significance for Canada. Wetlands 20 (4), 605–615.

Roulet, N., Moore, T., Bubier, J., Lafleur, P., 1992. Northern fens—methane flux and climatic change. Tellus B-Chem. Phys. Meteorol. 44 (2), 100–105.

Roulet, N.T., Lafleur, P.M., Richard, P.J.H., Moore, T.R., Humphreys, E.R., Bubier, J., 2007. Contemporary carbon balance and late holocene carbon accumulation in a northern peatland. Glob. Biol. 13 (2), 397–411.

Rousk, J., Bååth, E., 2011. Growth of saprotrophic fungi and bacteria in soil. FEMS Microbiol. Ecol. 78 (1), 17–30.

Roussel, E.G., Bonavita, M.-A.C., Querellou, J., Cragg, B.A., Webster, G., Prieur, D., John Parkes, R., 2008. Extending the sub-sea-floor biosphere. Science 320 (5879), 1046.

Routson, R.C., Wildung, R.E., Garland, T.R., 1977. Mineral weathering in an arid watershed containing soil developed from mixed Basaltic-Felsic parent materials. Soil Sci. 124, 303–308.

Routson, C.C., Arcusa, S.H., McKay, N.P., Overpeck, J.T., 2019. A 4,500-year-long record of southern rocky mountain dust deposition. Geophys. Res. Lett. 46 (14), 8281–8288.

Rowland, F.S., 1989. Chlorofluorocarbons and the depletion of stratospheric ozone. Am. Sci. 77, 36–45.

Rowland, F.S., 1991. Stratospheric ozone depletion. Ann. Rev. Phys. Chem. 42, 731–768.

Rowley, M.C., Grand, S., Verrecchia, É.P., 2018. Calcium-mediated stabilisation of soil organic carbon. Biogeochemistry 137 (1), 27–49.

Røy, H., Kallmeyer, J., Adhikari, R.R., Pockalny, R., Jørgensen, B.B., D'Hondt, S., 2012. Aerobic microbial respiration in 86-million-year-old deep-sea red clay. Science 336 (6083), 922–925.

Roy, E.D., Richards, P.D., Martinelli, L.A., Coletta, L.D., Lins, S.R.M., Vazquez, F.F., Willig, E., Spera, S.A., VanWey, L.K., Porder, S., 2016. The phosphorus cost of agricultural intensification in the tropics. Nat. Plants 2 (5), 16043.

Royer, D.L., Berner, R.A., Beerling, D.J., 2001. Phanerozoic atmospheric CO_2 change: evaluating geochemical and paleobiological approaches. Earth-Sci. Rev. 54, 349–392.

Rudaz, A.O., Davidson, E.A., Firestone, M.K., 1991. Sources of nitrous oxide production following wetting of dry soil. FEMS Microbiol. Ecol. 85, 117–124.

Rudd, J.W.M., Hamilton, R.D., 1978. Methane cycling in a eutrophic shield lake and its effects on whole lake metabolism. Limnol. Oceanogr. 23, 337–348.

Rudd, J.W.M., Kelly, C.A., Furutani, A., 1986. The role of sulfate reduction in long-term accumulation of organic and inorganic sulfur in lake sediments. Limnol. Oceanogr. 31, 1281–1291.

Rue, E.L., Bruland, K.W., 1997. The role of organic complexation on ambient iron chemistry in the equatorial pacific ocean and the response of a mesoscale iron addition experiment. Limnol. Oceanogr. 42, 901–910.

Ruess, R.W., Seagle, S.W., 1994. Landscape patterns in soil microbial processes in the Serengeti National Park, Tanzania. Ecology 75, 892–904.

Ruhe, R.V., 1984. The soil-climate system across the Prairies in the midwestern USA. Geoderma 34 (201–219), 143–158.

Ruiz-Mirazo, K., Briones, C., de la Escosura, A., 2014. Prebiotic systems chemistry: new perspectives for the origins of life. Chem. Rev. 114 (1), 285–366.

Rumbold, D.G., Lange, T.R., Richard, D., DelPizzo, G., Hass, N., 2018. Mercury biomagnification through food webs along a salinity gradient down-estuary from a biological hotspot. Estuar. Coast. Shelf Sci. 200 (January), 116–125.

Rumpel, C., Kogel-Knabner, I., 2011. Deep soil organic matter—a key but poorly understood component of terrestrial C cycle. Plant Soil 338, 143–158.

Running, S.W., Nemani, R.R., Peterson, D.L., Band, L.E., Potts, D.F., Pierce, L.L., Spanner, M.A., 1989. Mapping regional forest evapotranspiration and photosynthesis by coupling satellite data with ecosystem simulation. Ecology 70, 1090–1101.

Running, S.W., Nemani, R.R., Heinsch, F.A., Zhao, M.S., Reeves, M., Hashimoto, H., 2004. A continuous satellite-derived measure of global terrestrial primary production. Bioscience 54, 547–560.

Runyon, J., Waring, R.H., Goward, S.N., Welles, J.M., 1994. Environmental limits on net primary production and light-use efficiency across the oregon transect. Ecol. Appl. 4, 226–237.

Rupert, M.G., 2008. Decadal-scale changes of nitrate in ground water of the United States, 1988–2004. J. Environ. Qual. 37 (5 Suppl), S240–S248.

Russell, M., 2006. First life. Am. Sci. 94, 32–39.

Russell, G.L., Miller, J.R., 1990. Global river runoff calculated from a global atmospheric general circulation model. J. Hydrol. 117, 241–254.

Russell, J.M., Luo, M.Z., Cicerone, R.J., Deaver, L.E., 1996. Satellite confirmation of the dominance of chlorofluorocarbons in the global stratospheric chlorine budget. Nature 379, 526–529.

Russell, A.E., Laird, D.A., Parkin, T.B., Mallarino, A.P., 2005. Impact of nitrogen fertilization and cropping system on carbon sequestration in midwestern Mollisols. Soil Sci. Soc. Am. J. 69, 413–422.

Rustad, L.E., 1994. Element dynamics along a decay continuum in a red spruce ecosystem in maine, USA. Ecology 75, 867–879.

Rustad, L., Campbell, J., Marion, G., Norby, R., Mitchell, M., Hartley, A., Cornelissen, J., Gurevitch, J., GCTE-NEWS, 2001. A meta-analysis of the response of soil respiration, net nitrogen mineralization, and aboveground plant growth to experimental ecosystem warming. Oecologia 126 (4), 543–562.

Ruttenberg, K.C., 1993. Reassessment of the oceanic residence time of phosphorus. Chem. Geol. 107, 405–409.

Ruttenberg, K.C., Berner, R.A., 1993. Authigenic apatite formation and burial in sediments from non-upwelling, continental marine environments. Geochim. Cosmochim. Acta 57, 991–1007.

Ruttenberg, K.C., Sulak, D.J., 2011. Sorption and desorption of dissolved organic phosphorus onto iron (oxyhydr)oxides in seawater. Geochim. Cosmochim. Acta 75, 4095–4112.

Rutting, T., Huygens, D., Ller, C.M., Cleemput, O., Godoy, R., Boeckx, P., 2008. Functional role of DNRA and nitrite reduction in a pristine south Chilean Nothofagus forest. Biogeochemistry 90, 243–258.

Ryan, M.G., 1991. Effects of climate change on plant respiration. Ecol. Appl. 1 (2), 157–167.

Ryan, M.G., 1995. Foliar maintenance respiration of subalpine and boreal trees and shrubs in relation to nitrogen content. Plant Cell Environ. 18, 765–772.

Ryan, D.F., Bormann, F.H., 1982. Nutrient resorption in northern hardwood frests. Bioscience 32, 29–32.

Ryan, M.G., Hubbard, R.M., Clark, D.A., Sanford, R.L., 1994. Woody tissue respiration for *Simarouba amara* and *Minqyartia guianensis*, two tropical wet forest trees with different growth habits. Oecologia 100, 213–220.

Ryan, M.G., Gower, S.T., Hubbard, R.M., Waring, R.H., Gholz, H.L., Cropper Jr., W.P., Running, S.W., 1995. Woody tissue maintenance respiration of four conifers in contrasting climates. Oecologia 101 (2), 133–140.

Ryan, M.C., Graham, G.R., Rudolph, D.L., 2001. Contrasting nitrate adsorption in andisols of two coffee plantations in costa rica. J. Environ.Qual. 30 (5), 1848–1852.

Ryan, M.H., Tibbett, M., Edmonds-Tibbett, T., Suriyagoda, L.D.B., Lambers, H., Cawthray, G.R., Pang, J., 2012. Carbon trading for phosphorus gain: the balance between rhizosphere carboxylates and arbuscular mycorrhizal symbiosis in plant phosphorus acquisition. Plant Cell Environ. 35 (12), 2170–2180.

Rychert, R., Skujins, J., Sorensen, D., Porcella, D., 1978. Nitrogen fixation by *Lichens* and free-living microorganisms in deserts. In: West, N.E., Skujins, J. (Eds.), Nitrogen in Desert Ecosystems. Dowden, Hutchinson and Ross, pp. 20–30.

Rye, R., Kuo, P.H., Holland, H.D., 1995. Atmospheric carbon dioxide concentrations before 2.2 billion years ago. Nature 378 (6557), 603–605.

Rygiewicz, P.T., Andersen, C.P., 1994. Mycorrhizae alter quality and quantity of carbon allocated below ground. Nature 369, 58–60.

Rysgaard, S., Risgaardpetersen, N., Sloth, N.P., Jensen, K., Nielsen, L.P., 1994. Oxygen regulation of nitrification and denitrification in sediments. Limnol. Oceanogr. 39, 1643–1652.

Ryu, Y., Baldocchi, D.D., Kobayashi, H., van Ingen, C., Li, J., Black, T.A., Beringer, J., et al., 2011. Integration of MODIS land and atmosphere products with acoupled-process model to estimate gross primary productivity and evapotranspiration from 1 km to global scales. Glob. Biogeochem. Cycles 25.

Ryu, Y., Berry, J.A., Baldocchi, D.D., 2019. What is global photosynthesis? History, uncertainties and opportunities. Remote Sens. Environ. 223 (March), 95–114.

Saa, A., Trasarcepeda, M.C., Gilsotres, F., Carballas, T., 1993. Changes in soil phosphorus and acid phosphatase activity immediately following forest fires. Soil Biol. Biochem. 25, 1223–1230.

Saá, A., Trasar-Cepeda, M.C., Soto, B., Gil-Sotres, F., Díaz-Fierros, F., 1994. Forms of phosphorus in sediments eroded from burnt soils. J. Environ. Qual. 23, 739–746.

Sabine, C.L., Feely, R.A., Gruber, N., Key, R.M., Lee, K., Bullister, J.L., Wanninkhof, R., et al., 2004. The oceanic sink for anthropogenic CO_2. Science 305 (5682), 367–371.

Sabo, J.L., Power, M.E., 2002. River–watershed exchange: effects of riverine subsidies on riparian lizards and their terrestrial prey. Ecology 83 (7), 1860–1869.

Sabo, J.L., Sinha, T., Bowling, L.C., Schoups, G.H.W., Wallender, W.W., Campana, M.E., Cherkauer, K.A., et al., 2010. Reclaiming freshwater sustainability in the Cadillac desert. Proc. Natl. Acad. Sci. U. S. A. 107 (50), 21263–21270.

Sabo, R.D., Nelson, D.M., Eshleman, K.N., 2016. Episodic, seasonal, and annual export of atmospheric and microbial nitrate from a temperate forest. Geophys. Res. Lett. 43 (2), 683–691.

Sadler, P.M., 1981. Sediment accumulation rates and the completeness of stratigraphic sections. J. Geol. 89, 569–584.

Sadro, S., Holtgrieve, G.W., Solomon, C.T., Koch, G.R., 2014. Widespread variability in overnight patterns of ecosystem respiration linked to gradients in dissolved organic matter, residence time, and productivity in a global set of lakes. Limnol. Oceanogr. 59 (5), 1666–1678.

Sagan, C., Thompson, W.R., Carlson, R., Gurnett, D., Hord, C., 1993. A search for life on earth from the Galileo Spaccecraft. Nature 365, 715–721.

Saggar, S., Bettany, J.R., Stewart, J.W.B., 1981. Sulfur transformations in relation to carbon and nitrogen in incubated soils. Soil Biol. Biochem. 13, 499–511.

Sahagian, D.L., Schwartz, F.W., Jacobs, D.K., 1994. Direct anthropogenic contributions to sea-level rise in the twentieth century. Nature 367, 54–57.

Sahai, N., 2000. Estimating adsorption enthalpies and affinity sequences of monovalent electrolyte ions on oxide surfaces in aqueous solution. Geochim. Cosmochim. Acta 64, 3629–3641.

Saito, M.A., Goepfert, T.J., Ritt, J.T., 2008. Some thoughts on the concept of colimitation: three definitions and the importance of bioavailability. Limnol. Oceanogr. 53 (1), 276–290.

Sala, O.E., Parton, W.J., Joyce, L.A., Lauenroth, W.K., 1988. Primary production of the central grassland region of the United States. Ecology 69, 40–45.

Salati, E., Vose, P.B., 1984. Amazon basin: a system in equilibrium. Science 225 (4658), 129–138.

Saleska, S.R., Harte, J., Torn, M.S., 1999. The effect of experimental ecosystem warming on CO_2 fluxes in a Montane Meadow. Glob. Change Biol. 5, 125–141.

Salifu, K.F., Timmer, V.R., 2003. Nitrogen retranslocation response of young *Picea mariana* to nitrogen-15 supply. Soil Sci. Soc. Am. J. 67, 309–317.

Saltzman, E.S., Aydin, M., Tatum, C., Williams, M.B., 2008. 2,000-Year record of atmospheric methyl bromide from a south pole ice core. J. Geophys. Res.—Atmos. 113.

Saltzman, E.S., Aydin, M., Williams, M.B., Verhulst, K.R., Gun, B., 2009. Methyl chloride in a deep ice core from Siple Dome, Antarctica. Geophys. Res. Lett. 36.

Sambrotto, R.N., Savidge, G., Robinson, C., Boyd, P., Takahashi, T., Karl, D.M., Langdon, C., Chipman, D., Marrs, J., Codispoti, L., 1993. Elevated consumption of carbon relative to nitrogen in the surface ocean. Nature 363, 248–250.

Samet, J.M., Dominici, F., Curriero, F.C., Coursac, I., Zeger, S.L., 2000. Fine particulate air pollution and mortality in 20 U.S. cities, 1987–1994. N. Engl. J. Med. 343 (24), 1742–1749.

Samset, B.H., Sand, M., Smith, C.J., Bauer, S.E., Forster, P.M., Fuglestvedt, J.S., Osprey, S., Schleussner, C.-F., 2018. Climate impacts from a removal of Anthropogenic aerosol emissions. Geophys. Res. Lett. 45 (2), 1020–1029.

Sanborn, P.T., Ballard, T.M., 1991. Combustion losses of sulfur from conifer foliage: implications of chemical form and soil nitrogen status. Biogeochemistry 12, 129–134.

Sanchez, P.A., Bandy, D.E., Villachica, J.H., Nicholaides, J.J., 1982a. Amazon basin soils: management for continuous crop production. Science 216 (4548), 821–827.

Sanchez, P.A., Gichuru, M.P., Katz, L.B., 1982b. Organic matter in major soils of the tropical and temperate regions. Int. Congr. Soil Sci. 12, 99–114.

SanClements, M.D., Fernandez, I.J., Norton, S.A., 2010. Phosphorus in soils of temperate forests: linkages to acidity and aluminum. Soil Sci. Soc. Am. J. 74, 2175–2186.

Sander, S.G., Koschinsky, A., 2011. Metal flux from hydrothermal vents increased by organic complexation. Nat. Geosci. 4 (3), 145–150.

Sanderman, J., Hengl, T., Fiske, G.J., 2017. Soil carbon debt of 12,000 years of human land use. Proc. Natl. Acad. Sci. U. S. A. 114 (36), 9575–9580.

Sanderson, M.G., 1996. Biomass of termites and their emissions of methane and carbon dioxide: a global database. Glob. Biogeochem. Cycles 10, 543–557.

Sand-Jensen, K., Staehr, P.A., 2009. Net heterotrophy in small Danish lakes: a widespread feature over gradients in trophic status and land cover. Ecosystems 12, 336–348.

Sandor, J.A., Norton, J.B., Homburg, J.A., Muenchrath, D.A., White, C.S., Williams, S.E., Havener, C.I., Stahl, P.D., 2007. Biogeochemical studies of a native American runoff agroecosystem. Geoarchaeol.—Int. J. 22, 359–386.

Sanford, R.A., Wagner, D.D., Wu, Q., Chee-Sanford, J.C., Thomas, S.H., Cruz-García, C., Rodríguez, G., et al., 2012. Unexpected nondenitrifier nitrous oxide reductase gene diversity and abundance in soils. Proc. Natl. Acad. Sci. U. S. A. 109 (48), 19709–19714.

Sanhueza, E., Cardenas, L., Donoso, L., Santana, M., 1994. Effect of plowing on CO_2, CO, CH_4, N_2O, and NO fluxes from tropical savanna soils. J. Geophys. Res.-Atmos. 99, 16429–16434.

Sanhueza, E., Dong, Y., Scharffe, D., Lobert, J.M., Crutzen, P.J., 1998. Carbon monoxide uptake by temperate forest soils: the effects of leaves and humus layers. Tellus B—Chem. Phys. Meteorol. 50, 51–58.

Sano, Y., Takahata, N., Nishio, Y., Fischer, T.P., Williams, S.N., 2001. Volcanic flux of nitrogen from the earth. Chem. Geol. 171, 263–271.

Sansone, F.J., Martens, C.S., 1981. Methane production from acetate and associated methane fluxes from anoxic coastal sediments. Science 211 (4483), 707–709.

Santi, C., Bogusz, D., Franche, C., 2013. Biological nitrogen fixation in non-legume plants. Ann. Bot. 111 (5), 743–767.

Santiago, L.S., Joseph Wright, S., Harms, K.E., Yavitt, J.B., Korine, C., Garcia, M.N., Turner, B.L., 2012. Tropical tree seedling growth responses to nitrogen, phosphorus and potassium addition: tropical tree seedling N-P-K responses. J. Ecol. 100 (2), 309–316.

Santín, C., Doerr, S.H., Kane, E.S., Masiello, C.A., Ohlson, M., de la Rosa, J.M., Preston, C.M., Dittmar, T., 2016. Towards a global assessment of pyrogenic carbon from vegetation fires. Glob. Chang. Biol.

Santner, J., Smolders, E., Wenzel, W.W., Degryse, F., 2012. First observation of diffusion-limited plant root phosphorus uptake from nutrient solution. Plant Cell Environ. 35 (9), 1558–1566.

Santoro, A.E., Buchwald, C., McIlvin, M.R., Casciotti, K.L., 2011. Isotopic signature of N_2O produced by marine ammonia-oxidizing archaea. Science 1282 (July), 1282–1285.

Sanyal, A., Hemming, N.G., Hanson, G.N., Broecker, W.S., 1995. Evidence for a higher pH in the glacial ocean from boron isotopes in foraminifera. Nature 373, 234–236.

Sapart, C.J., Monteil, G., Prokopiou, M., van de Wal, R.S.W., Kaplan, J.O., Sperlich, P., Krumhardt, K.M., et al., 2012. Natural and anthropogenic variations in methane sources during the past two millennia. Nature 490 (7418), 85–88.

Sarafian, A.R., Nielsen, S.G., Marschall, H.R., McCubbin, F.M., Monteleone, B.D., 2014. Early solar system. Early accretion of water in the inner solar system from a carbonaceous chondrite-like source. Science 346 (6209), 623–626.

Sarbu, S.M., Kane, T.C., Kinkle, B.K., 1996. A chemoautotrophically based cave ecosystem. Science 272 (5270), 1953–1955.

Sardans, J., Peñuelas, J., 2015. Potassium: a neglected nutrient in global change: potassium stoichiometry and global change. Glob. Ecol. Biogeogr.: J. Macroecol. 24 (3), 261–275.

Sarmiento, J.L., Gruber, N., 2006. Ocean Biogeochemical Dynamics. Princeton University Press.

Sarmiento, J.L., Sundquist, E.T., 1992. Revised budget for the oceanic uptake of anthropogenic carbon dioxide. Nature 356, 589–593.

Sarmiento, J.L., Gruber, N., Brzezinski, M.A., Dunne, J.P., 2004. High-latitude controls of thermocline nutrients and low latitude biological productivity. Nature 427 (6969), 56–60.

Sass, R.L., Fisher, F.M., 1997. Methane emissions from rice paddies: a process study summary. Nutr. Cycl. Agroecosyst. 49, 119–127.

Sathyendranath, S., Platt, T., Horne, E.P.W., Harrison, W.G., Ulloa, O., Outerbridge, R., Hoepffner, N., 1991. Estimation of new production in the ocean by compound remote sensing. Nature 353 (6340), 129–133.

Saugier, B., Roy, J., Mooney, H.A., 2001. Estimations of global terrestrial productivity: converging toward a single number? In: Roy, B.S.J., Mooney, H.A. (Eds.), Terrestrial Global Productivity. Academic Press, pp. 543–557.

Saunders, M.J., Jones, M.B., Kansiime, F., 2007. Carbon and water cycles in tropical papyrus wetlands. Wetl. Ecol. Manag. 15, 489–498.

Saunois, M., Bousquet, P., Poulter, B., Peregon, A., Ciais, P., Canadell, J.G., Dlugokencky, E.J., et al., 2016. The global methane budget 2000-2012. Earth Syst. Sci. Data 8 (2), 697–751.

Saurer, M., Siegwolf, R.T.W., Schweingruber, F.H., 2004. Carbon isotope discrimination indicates improving water-use efficiency of trees in northern Eurasia over the last 100 years. Glob. Chang. Biol. 10, 2109–2120.

Savarino, J., Legrand, M., 1998. High northern latitude forest fires and vegetation emissions over the last millennium inferred from the chemistry of a Central Greenland ice Core. J. Geophys. Res.-Atmos. 103, 8267–8279.

Savoie, D.L., Prospero, J.M., 1989. Comparison of oceanic and continental sources of Non-Sea-salt sulfate over the Pacific Ocean. Nature 339, 685–687.

Savoie, D.L., Prospero, J.M., Nees, R.T., 1987. Nitrate, Non-Sea-salt sulfate, and mineral aerosol over the northwestern Indian Ocean. J. Geophys. Res. 92, 933–942.

Savoie, D.I., Prospero, J.M., Larsen, R.J., Huang, F., Izaguirre, M.A., Huang, T., Snowdon, T.H., Custals, L., Sanderson, C.G., 1993. Nitrogen and sulfur species in Antarctic aerosols at Mawson, Palmer Station, and Marsh (King George Island). J. Atmos. Chem. 17, 95–122.

Savoy, P., Appling, A.P., Heffernan, J.B., Stets, E.G., Read, J.S., Harvey, J.W., Bernhardt, E.S., 2019. Metabolic rhythms in flowing waters: an approach for classifying river productivity regimes. Limnol. Oceanogr. 81 (March), AAAI Technical Report WS-94-03. 345.

Savva, Y., Berninger, F., 2010. Sulphur deposition causes a large-scale growth decline in boreal forests in Eurasia. Glob. Biogeochem. Cycles 24.

Sawyer, A.H., David, C.H., Famiglietti, J.S., 2016. Continental patterns of submarine groundwater discharge reveal coastal vulnerabilities. Science 353 (6300), 705–707.

Sayavedra-Soto, L.A., Arp, D.J., 2011. Ammonia-oxidizing bacteria: their biochemistry and molecular biology. In: Klotz, M.G., Ward, B.B., Arp, D.J. (Eds.), Nitrification. American Society of Microbiology, pp. 11–37.

Sayles, F.L., Mangelsdorf, P.C., 1977. The equilibration of clay minerals with sea water: exchange reactions. Geochim. Cosmochim. Acta 41 (7), 951–960.

Sayles, F.L., Martin, W.R., Deuser, W.G., 1994. Response of benthic oxygen demand to particulate organic carbon supply in the Deep Sea near Bermuda. Nature 371, 686–689.

Scanlon, B.R., Keese, K.E., Flint, A.L., Flint, L.E., Gaye, C.B., Edmunds, W.M., Simmers, I., 2006. Global synthesis of groundwater recharge in semiarid and arid regions. Hydrol. Process. 20, 3335–3370.

Scanlon, B.R., Gates, J.B., Reedy, R.C., Jackson, W.A., Bordovsky, J.P., 2010. Effects of irrigated agroecosystems: 2. Quality of soil water and groundwater in the Southern High Plains, Texas. Water Resour. Res. 46.

Schaefer, M., 1990. The soil Fauna of a beech forest on limestone: trophic structure and energy budget. Oecologia 82 (1), 128–136.

Schaefer, L., Fegley, B., 2010. Chemistry of atmospheres formed during accretion of the earth and other terrestrial planets. Icarus 208, 438–448.

Schaefer, D.A., Reiners, W.A., 1989. Throughfall chemistry and canopy processing mechanisms. In: Lindberg, S.E., Page, A.L., Norton, S.A. (Eds.), Acid Precipitation: Sources, Deposition and Canopy Interactions. Springer-Verlag, pp. 241–284.

Schaefer, D.A., Whitford, W.G., 1981. Nutrient cycling by the subterranean termite *Gnathamitermes tubiformans* in a Chihuahuan Desert ecosystem. Oecologia 48 (2), 277–283.

Schaefer, K., Zhang, T., Bruhwiler, L., Barrett, A.P., 2011. Amount and timing of permafrost carbon release in response to climate warming. Tellus B—Chem. Phys. Meteorol. 63, 165–180.

Schaefer, H., Mikaloff, S.E., Fletcher, C.V., Lassey, K.R., Brailsford, G.W., Bromley, T.M., Dlugokencky, E.J., et al., 2016. A 21st-century shift from fossil-fuel to biogenic methane emissions indicated by $^{13}CH\square$. Science 352 (6281), 80–84.

Schaller, M., Blum, J.D., Hamburg, S.P., Vadeboncoeur, M.A., 2010. Spatial variability of long-term chemical weathering rates in the White Mountains, New Hampshire. Geoderma 154, 294–301.

Scheffer, M., Bascompte, J., Brock, W.A., Brovkin, V., Carpenter, S.R., Dakos, V., Held, H., van Nes, E.H., Rietkerk, M., Sugihara, G., 2009. Early-warning signals for critical transitions. Nature 461 (7260), 53–59.

Schelske, C.L., Stoermer, E.F., 1971. Eutrophication, silica depletion, and predicted changes in algal quality in Lake Michigan. Science 173 (3995), 423–424.

Schidlowski, M., 1980. The atmosphere. In: Hutzinger, O. (Ed.), The Handbook of Environmental Chemistry. Volume 1, Part A. The Natural Environment and the Biogeochemical Cycles. Springer-Verlag, pp. 1–16.

Schidlowski, M., 1983. Evolution of photoautotrophy and early atmospheric oxygen levels. Precambrian Res. 20, 319–335.

Schidlowski, M., 2001. Carbon isotopes as biogeochemical recorders of life over 3.8 Ga of earth history: evolution of a concept. Precambrian Res. 106, 117–134.

Schiff, S., Aravena, R., Mewhinney, E., Elgood, R., Warner, B., Dillon, P., Trumbore, S., 1998. Precambrian shield wetlands: hydrologic control of the sources and export of dissolved organic matter. Clim. Chang. 40 (2), 167–188.

Schillawski, S., Petsch, S., 2008. Release of biodegradable dissolved organic matter from ancient sedimentary rocks. Glob. Biogeochem. Cycles 22.

Schilling, K.E., Chan, K.-S., Liu, H., Zhang, Y.-K., 2010. Quantifying the effect of land use-land cover change on increasing discharge in the upper Mississippi River. J. Hydrol. 387, 343–345.

Schilt, A., Baumgartner, M., Schwander, J., Buiron, D., Capron, E., Chappellaz, J., Loulergue, L., et al., 2010. Atmospheric nitrous oxide during the last 140,000 years. Earth Planet. Sci. Lett. 300, 33–43.

Schilt, A., Brook, E.J., Bauska, T.K., Baggenstos, D., Fischer, H., Joos, F., Petrenko, V.V., et al., 2014. Isotopic constraints on marine and terrestrial N_2O emissions during the last deglaciation. Nature 516 (7530), 234–237.

Schimel, J.P., Bennett, J., 2004. Nitrogen mineralization: challenges of a changing paradigm. Ecology 85, 591–602.

Schimel, J.P., Chapin, F.S., 1996. Tundra plant uptake of amino acid and NH_4^+ nitrogen in situ: plants compete well for amino acid N. Ecology 77, 2142–2147.

Schimel, J.P., Firestone, M.K., 1989. Nitrogen incorporation and flow through a coniferous forest soil profile. Soil Sci. Soc. Am. J. 53, 779–784.

Schimel, J.P., Firestone, M.K., Killham, K.S., 1984. Identification of heterotrophic nitrification in a Sierran forest soil. Appl. Environ. Microbiol. 48 (4), 802–806.

Schimel, D., Stillwell, M.A., Woodmansee, R.G., 1985. Biogeochemistry of C, N, and P in a soil catena of the shortgrass steppe. Ecology 66, 276–282.

Schindlbacher, A., Schnecker, J., Takriti, M., Borken, W., Wanek, W., 2015. Microbial physiology and soil CO_2 efflux after 9 years of soil warming in a temperate forest—no indications for thermal adaptations. Glob. Chang. Biol. 21 (11), 4265–4277.

Schindler, D.W., 1974. Eutrophication and recovery in experimental lakes: implications for lake management. Science 184 (4139), 897–899.

Schindler, D.W., 1977. Evolution of phosphorus limitation in lakes. Science 195, 260–262.

Schindler, D.W., 1978. Factors regulating phytoplankton production and standing crop in the world's freshwaters. Limnol. Oceanogr. 23, 478–486.

Schindler, D.W., 1986. The significance of in-lake production of alkalinity. Water Air Soil Pollut. 30, 931–944.

Schindler, D.W., 1988. Effects of acid rain on freshwater ecosystems. Science 239 (4836), 149–157.

Schindler, S.C., Mitchell, M.J., Scott, T.J., Fuller, R.D., Driscoll, C.T., 1986. Incorporation of 35S-sulfate into inorganic and organic constituents of two forest soils. Soil Sci. Soc. Am. J. 50, 457–462.

Schindler, D.W., Hecky, R.E., Findlay, D.L., Stainton, M.P., Parker, B.R., Paterson, M.J., Beaty, K.G., Lyng, M., Kasian, S.E.M., 2008. Eutrophication of lakes cannot be controlled by reducing nitrogen input: results of a 37-year whole-ecosystem experiment. Proc. Natl. Acad. Sci. U. S. A. 105 (32), 11254–11258.

Schipanski, M.E., Bennett, E.M., 2012. The influence of agricultural trade and livestock production on the global phosphorus cycle. Ecosystems 15, 256–268.

Schippers, P., Lurling, M., Scheffer, M., 2004. Increase of atmospheric CO_2 promotes phytoplankton productivity. Ecol. Lett. 7, 446–451.

Schippers, A., Neretin, L.N., Kallmeyer, J., Ferdelman, T.G., Cragg, B.A., John Parkes, R., Jørgensen, B.B., 2005. Prokaryotic cells of the deep sub-seafloor biosphere identified as living bacteria. Nature 433 (7028), 861–864.

Schlesinger, W.H., 1977. Carbon balance in terrestrial detritus. Annu. Rev. Ecol. Syst. 8, 51–81.

Schlesinger, W.H., 1978. Community structure, dyanamics and nutrient cycling in Okefenokee cypress swamp forest. Ecol. Monogr. 48 (1), 43–65.

Schlesinger, W.H., 1984. Soil organic matter: a source of atmospheric CO_2. In: Woodwell, G.M. (Ed.), The Role of Terrestrial Vegetation in the Global Carbon Cycle. Wiley, pp. 111–127.

Schlesinger, W.H., 1985. The formation of caliche in soils of the Mojave Desert, California. Geochim. Cosmochim. Acta 49, 57–66.

Schlesinger, W.H., 1986. Changes in soil carbon storage and associated properties with disturbance and recovery. In: Trabalka, J.R., Reichle, D.E. (Eds.), The Changing Carbon Cycle: A Global Analysis. Springer, pp. 194–220.

Schlesinger, W.H., 1990. Evidence from cronosequence studies for a low carbon-storage potential of soils. Nature 348, 232–234.

Schlesinger, W.H., 2000. Carbon sequestration in soils: some cautions amidst optimism. Agric. Ecosyst. Environ. 82, 121–127.

Schlesinger, W.H., 2009. On the fate of anthropogenic nitrogen. Proc. Natl. Acad. Sci. U. S. A. 106 (1), 203–208.

Schlesinger, W.H., 2013. An estimate of the global sink for nitrous oxide in soils. Glob. Chang. Biol. 19 (10), 2929–2931.

Schlesinger, W.H., 2017. An evaluation of abiotic carbon sinks in deserts. Glob. Chang. Biol. 23 (1), 25–27.

Schlesinger, W.H., Amundson, R., 2019. Managing for soil carbon sequestration: let's get realistic. Glob. Chang. Biol. 25 (2), 386–389.

Schlesinger, W.H., Hartley, A.E., 1992. A global budget for atmospheric NH_3. Biogeochemistry 15, 191–211.

Schlesinger, W.H., Jasechko, S., 2014. Transpiration in the global water cycle. Agric. For. Meteorol. 189–190 (June), 115–117.

Schlesinger, W.H., Melack, J.M., 1981. Transport of organic carbon in the world's rivers. Tellus 33, 172–187.

Schlesinger, M.E., Ramankutty, N., 1994. An oscillation in the global climate system of period 65–70 years. Nature 367 (6465), 723.

Schlesinger, W.H., Vengosh, A., 2016. Global boron cycle in the Anthropocene. Glob. Biogeochem. Cycles CCC/203. 30 (2), 219–230.

Schlesinger, W.H., Gray, J.T., Gilliam, F.S., 1982. Atmospheric deposition processes and their importance as sources of nutrients in a chaparral ecosystem of Southern California. Water Resour. Res. 18, 623–629.

Schlesinger, W.H., Delucia, E.H., Billings, W.D., 1989. Nutrient-use efficiency of Woody plants on contrasting soils in the Western Great Basin, Nevada. Ecology 70, 105–113.

Schlesinger, W.H., Raikes, J.A., Hartley, A.E., Cross, A.E., 1996. On the spatial pattern of soil nutrients in desert ecosystems. Ecology 77, 364–374.

Schlesinger, W.H., Bruijnzeel, L.A., Bush, M.B., Klein, E.M., Mace, K.A., Raikes, J.A., Whittaker, R.J., 1998. The biogeochemistry of phosphorus after the first century of soil development on Rakata Island, Krakatau, Indonesia. Biogeochemistry 40, 37–55.

Schlesinger, M.E., Ramankutty, N., Andronova, N., Margolis, M., Kerr, R.A., 2000a. Temperature oscillations in the North Atlantic. Science 289 (5479), 547b–548b.

Schlesinger, W.H., Ward, T.J., Anderson, J., 2000b. Nutrient losses in runoff from grassland and srubland habitats in southern New Mexico: II. Field plots. Biogeochemistry 49, 69–86.

Schlesinger, W.H., Pippen, J.S., Wallenstein, M.D., Hofmockel, K.S., Klepeis, D.M., Mahall, B.E., 2003. Community composition and photosynthesis by photoautotrophs under quartz pebbles, southern Mojave Desert. Ecology 84, 3222–3231.

Schlesinger, W.H., Tartowski, S.L., Schmidt, S.M., 2006. Nutrient cycling within an arid ecosystem. In: Havstad, K.M., Huenneke, L.F., Schlesinger, W.H. (Eds.), Structure and Function of a Chihuahuan Desert Ecosystem. Oxford University Press, pp. 133–149.

Schlesinger, W.H., Cole, J.J., Finzi, A.C., Holland, E.A., 2011. Introduction to coupled biogeochemical cycles. Front. Ecol. Environ. 9, 5–8.

Schlesinger, W.H., Dietze, M.C., Jackson, R.B., Phillips, R.P., Rhoades, C.C., Rustad, L.E., Vose, J.M., 2016. Forest biogeochemistry in response to drought. Glob. Chang. Biol. 22 (7), 2318–2328.

Schlesinger, W.H., Klein, E.M., Vengosh, A., 2017. Global biogeochemical cycle of vanadium. Proc. Natl. Acad. Sci. U. S. A. 114 (52), E11092–E11100.

Schlesinger, W.H., Klein, E.M., Vengosh, A., 2020. Global biogeochemical cycle of fluorine. Glob. Biogeochem. Cycles (in prep).

Schmidt, I., Sliekers, O., Schmid, M., Cirpus, I., Strous, M., Eberhard, B., Gijs Kuenen, J., Jetten, M.S.M., 2002. Aerobic and anaerobic ammonia oxidizing bacteria—competitors or natural partners? FEMS Microbiol. Ecol. 39 (3), 175–181.

Schmidt, R., Schwintzer, P., Flechtner, F., Reigber, C., Guntner, A., Doll, P., Ramillien, G., et al., 2006. GRACE observations of changes in continental water storage. Glob. Planet. Chang. 50, 112–126.

Schmidt, G.A., Ruedy, R.A., Miller, R.L., Lacis, A.A., 2010a. Attribution of the present-day total green-house effect. J. Geophys. Res.-Atmos. 115.

Schmidt, H., Eickhorst, T., Tippkötter, R., 2010b. Monitoring of root growth and redox conditions in paddy soil Rhizotrons by redox electrodes and image analysis. Plant Soil 341 (1–2), 221–232.

Schmidt, M.W.I., Torn, M.S., Abiven, S., Dittmar, T., Guggenberger, G., Janssens, I.A., Kleber, M., et al., 2011. Persistence of soil organic matter as an ecosystem property. Nature 478 (7367), 49–56.

Schmitt, J., Schneider, R., Elsig, J., Leuenberger, D., Lourantou, A., Chappellaz, J., Koehler, P., et al., 2012. Carbon isotope constraints on the deglacial CO_2 rise from ice cores. Science 336, 711–714.

Schmittner, A., 2005. Decline of the marine ecosystem caused by a reduction in the Atlantic overturning circulation. Nature 434 (7033), 628–633.

Schmitz Jr., W.J., 1995. On the Interbasin-scale thermohaline circulation. Rev. Geophys. 33 (2), 151–173.

Schmitz, O.J., Hawlena, D., Trussell, G.C., 2010. Predator control of ecosystem nutrient dynamics. Ecol. Lett. 13 (10), 1199–1209.

Schmitz, O.J., Wilmers, C.C., Leroux, S.J., Doughty, C.E., Atwood, T.B., Galetti, M., Davies, A.B., Goetz, S.J., 2018. Animals and the zoogeochemistry of the carbon cycle. Science 362 (6419), eaar3213.

Schneider, P., R. J. van der A., 2012. A global single-sensor analysis of 2002–2011 tropospheric nitrogen dioxide trends observed from space: global No 2 trends observed from space. J. Geophys. Res. 117 (D16).

Schneider, B., Schlitzer, R., Fischer, G., Nothig, E.M., 2003. Depth-dependent elemental compositions of particulate organic matter (POM) in the ocean. Glob. Biogeochem. Cycles 17.

Schnell, S., King, G.M., 1994. Mechanistic analysis of ammonium inhibition of atmospheric methane consumption in forest soils. Appl. Environ. Microbiol. 60 (10), 3514–3521.

Schoenau, J.J., Bettany, J.R., 1987. Organic matter leaching as a component of carbon, nitrogen, phosphorus, and sulfur cycles in a forest, grassland, and gleyed soil. Soil Sci. Soc. Am. J. 51, 646–651.

Schoepfer, V.A., Bernhardt, E.S., Burgin, A.J., 2014. Iron clad wetlands: soil iron-sulfur buffering determines coastal wetland response to salt water incursion. J. Geophys. Res.: Biogeosci. 119 (12).

Scholander, P.F., Van Dam, L., Scholander, S.I., 1955. Gas exchange in the roots of mangroves. Am. J. Bot. 42 (1), 92–98.

Scholefield, D., Hawkins, J.M.B., Jackson, S.M., 1997. Development of a helium atmosphere soil incubation technique for direct measurement of nitrous oxide and dinitrogen fluxes during denitrification. Soil Biol. Biochem. 29, 1345–1352.

Schönheit, P., Kristjansson, J.K., Thauer, R.K., 1982. Kinetic mechanism for the ability of sulfate reducers to out-compete methanogens for acetate. Arch. Microbiol. 132 (3), 285–288.

Schöttelndreier, M., Falkengren-Grerup, U., 1999. Plant induced alteration in the rhizosphere and the utilisation of soil heterogeneity. Plant Soil 209, 297–309.

Schrenk, M.O., Edwards, K.J., Goodman, R.M., Hamers, R.J., Banfield, J.F., 1998. Distribution of Thiobacillus ferrooxidans and Leptospirillum ferrooxidans: implications for generation of acid mine drainage. Science 279 (5356), 1519–1522.

Schroth, A.W., Bostick, B.C., Graham, M., Kaste, J.M., Mitchell, M.J., Friedland, A.J., 2007. Sulfur species behavior in soil organic matter during decomposition. J. Geophys. Res.-Biogeosci. 112, 1–10.

Schrumpf, M., Kaiser, K., Guggenberger, G., Persson, T., Kögel-Knabner, I., Schulze, E.-D., 2013. Storage and stability of organic carbon in soils as related to depth, occlusion within aggregates, and attachment to minerals. Biogeosciences 10 (3), 1675–1691.

Schuerch, M., Spencer, T., Temmerman, S., Kirwan, M.L., Wolff, C., Lincke, D., McOwen, C.J., et al., 2018. Future response of global coastal wetlands to sea-level rise. Nature 561 (7722), 231–234.

Schuessler, J.A., von Blanckenburg, F., Bouchez, J., Uhlig, D., Hewawasam, T., 2018. Nutrient cycling in a tropical montane rainforest under a supply-limited weathering regime traced by elemental mass balances and Mg stable isotopes. Chem. Geol. 497 (October), 74–87.

Schuffert, J.D., Jahnke, R.A., Kastner, M., Leather, J., Sturz, A., Wing, M.R., 1994. Rates of formation of modern phosphorite off Western Mexico. Geochim. Cosmochim. Acta 58, 5001–5010.

Schulte-Uebbing, L., de Vries, W., 2018. Global-scale impacts of nitrogen deposition on tree carbon sequestration in tropical, temperate, and boreal forests: a meta-analysis. Glob. Chang. Biol. 24 (2), e416–e431.

Schulthess, C.P., Huang, C.P., 1991. Humic and fulvic acid adsorption by silicon and aluminum oxide surfaces on clay minerals. Soil Sci. Soc. Am. J. 55, 34–42.

Schultz, R.C., Kormanik, P.P., Bryan, W.C., Brister, G.H., 1979. Vesicular-arbuscular mycorrhiza influence growth but not mineral concentrations in seedlings of eight sweetgum families. Can. J. Forest Res. 9, 218–223.

Schultz, M.G., Heil, A., Hoelzemann, J.J., Spessa, A., Thonicke, K., Goldammer, J.G., Held, A.C., Pereira, J.M.C., van het Bolscher, M., 2008. Global wildland fire emissions from 1960 to 2000. Glob. Biogeochem. Cycles 22.

Schulze, E.D., 1989. Air pollution and forest decline in a spruce (*Picea abies*) forest. Science 244 (4906), 776–783.

Schulze, E.-D., Wirth, C., Heimann, M., 2000. Managing forests after Kyoto. Science 289 (5487), 2058–2059.

Schulze, E.-D., Ciais, P., Luyssaert, S., Schrumpf, M., Janssens, I.A., Thiruchittampalam, B., Theloke, J., et al., 2010. The European carbon balance. Part 4: Integration of carbon and other trace-gas fluxes. Glob. Chang. Biol. 16, 1451–1469.

Schuster, P.F., Schaefer, K.M., Aiken, G.R., Antweiler, R.C., Dewild, J.F., Gryziec, J.D., Gusmeroli, A., et al., 2018. Permafrost stores a globally significant amount of mercury. Geophys. Res. Lett. 45 (3), 1463–1471.

Schutz, H., Schroder, P., Rennenberg, H., 1991. Role of plants in regulating the methane flux to the atmosphere. In: Sharkey, T.D., Holland, E.A., Mooney, H.A. (Eds.), Trace Gas Emissions by Plants. Academic Press, San Diego, pp. 29–63.

Schütz, H., Schröder, P., Rennenberg, H., 1991. Role of plants in regulating the methane flux to the atmosphere. In: Sharkey, T.D., Holland, E.A., Mooney, H.A. (Eds.), Trace Gas Emissions by Plants. Academic Press, San Diego, pp. 29–63.

Schuur, E.A.G., 2003. Productivity and global climate revisited: the sensitivity of tropical frest growth to precipitation. Ecology 84, 1165–1170.

Schuur, E.A.G., Bockheim, J., Canadell, J.G., Eugenie, E., Christopher, B., Goryachkin, S.V., Hagemann, S., et al., 2008. Vulnerability of permafrost carbon to climate change: implications for the global carbon cycle. Bioscience 58 (8), 701–714.

Schuur, E.A.G., Vogel, J.G., Crummer, K.G., Lee, H., Sickman, J.O., Osterkamp, T.E., 2009. The effect of permafrost thaw on old carbon release and net carbon exchange from tundra. Nature 459 (7246), 556–559.

Schuur, E.A.G., McGuire, A.D., Schädel, C., Grosse, G., Harden, J.W., Hayes, D.J., Hugelius, G., et al., 2015. Climate change and the permafrost carbon feedback. Nature 520 (7546), 171–179.

Schwander, T., von Borzyskowski, L.S., Burgener, S., Cortina, N.S., Erb, T.J., 2016. A synthetic pathway for the fixation of carbon dioxide in vitro. Science 354 (6314), 900–904.

Schwartz, S.E., 1988. Are global cloud albedo and climate controlled by marine phytoplankton? Nature 336, 441–445.

Schwartz, S.E., 1989. Acid deposition: unraveling a regional phenomenon. Science 243 (4892), 753–763.

Schwartzman, D.W., Volk, T., 1989. Biotic enhancement of weathering and the habitability of earth. Nature 340, 457–460.

Schwertmann, U., 1966. Inhibitory effect of soil organic matter on the crystallization of amorphous ferric hydroxide. Nature 212, 645–646.

Schwietzke, S., Michael Griffin, W., Scott Matthews, H., Bruhwiler, L.M.P., 2014. Natural gas fugitive emissions rates constrained by global atmospheric methane and ethane. Environ. Sci. Technol. 48 (14), 7714–7722.

Schwintzer, C.R., 1983. Non-symbiotic and symbiotic nitrogen fixation in a weakly minerotrophic peatland. Am. J. Bot. 70 (7), 1071–1078.

Schwintzer, C.R., Tjepkema, J.D., 1994. Factors affecting the acetylene to $^{15}N_2$ conversion ratio in root nodules of *Myrica gale* L. Plant Physiol. 106 (3), 1041–1047.

Scott, R.M., 1962. Exchangeable bases of mature, well-drained soils in relation to rainfall in East Africa. J. Soil Sci. 13, 1–9.

Scott, R.L., Biederman, J.A., 2017. Partitioning evapotranspiration using long-term carbon dioxide and water vapor fluxes: new approach to ET partitioning. Geophys. Res. Lett. 44 (13), 6833–6840.

Scott, N.A., Binkley, D., 1997. Foliage litter quality and annual net N mineralization: comparison across north American forest sites. Oecologia 111 (2), 151–159.

Scott, A.C., Glasspool, I.J., 2006. The diversification of Paleozoic fire systems and fluctuations in atmospheric oxygen concentration. Proc. Natl. Acad. Sci. U. S. A. 103 (29), 10861–10865.

Scott, D., Harvey, J., Alexander, R., Schwarz, G., 2007. Dominance of organic nitrogen from headwater streams to large rivers across the conterminous United States. Glob. Biogeochem. Cycles 21.

Scott, J.T., McCarthy, M.J., Gardner, W.S., Doyle, R.D., 2008. Denitrification, dissimilatory nitrate reduction to ammonium, and nitrogen fixation along a nitrate concentration gradient in a created freshwater Wetland. Biogeochemistry 87 (1), 99–111.

Scriber, J.M., 1977. Limiting effects of low leaf water content on nitrogen-utilization, energy budget, and larval growth of *Hyalophora cecropia* (Lepidoptera Saturniidae). Oecologia 28, 269–287.

Scudder, B.C., Chasar, L.C., Wentz, D.A., Bauch, N.J., Brigham, M.E., Moran, P.W., Krabbenhoft, D.P., 2009. Mercury in Fish, Bed Sediment, and Water From Streams Across the United States, 1998–2005. U.S. Geological Survey.

Scurlock, J.M.O., Olson, R.J., 2002. Terrestrial net primary productivity—a brief history and a new worldwide database. Environ. Res. 10, 91–109.

Seager, R., Ting, M., Held, I., Kushnir, Y., Lu, J., Vecchi, G., Huang, H.-P., et al., 2007. Model projections of an imminent transition to a more arid climate in southwestern North America. Science 316 (5828), 1181–1184.

Sears, S.O., Langmuir, D., 1982. Sorption and mineral equilibria controls on moisture chemistry in a C-horizon soil. J. Hydrol. 56, 287–308.

Seastedt, T.R., 1985. Maximization of primary and secondary productivity by grazers. Am. Nat. 126, 559–564.

Seastedt, T.R., 1988. Mass, nitrogen, and phosphorus dynamics in foliage and root detritus of tallgrass prairie. Ecology 69, 59–65.

Seastedt, T.R., Crossley, D.A., 1980. Effects of microarthropods on the seasonal dynamics of nutrients in forest litter. Soil Biol. Biochem. 12, 337–342.

Seastedt, T.R., Hayes, D.C., 1988. Factors influencing nitrogen concentrations in soil water in a north American tallgrass prairie. Soil Biol. Biochem. 20, 725–729.

Seastedt, T.R., Tate, C.M., 1981. Decomposition rates and nutrient contents of arthropod remains in forest litter. Ecology 62, 13–19.

Sebacher, D.I., Harriss, R.C., Bartlett, K.B., 1985. Methane emissions to the atmosphere through aquatic plants. J. Environ. Qual.

Sebacher, D.I., Harriss, R.C., Bartlett, K.B., Sebacher, S.M., Grice, S.S., 1986. Atmospheric methane sources: Alaskan Tundra Bogs, an Alpine Fen, and a Subarctic Boreal Marsh. Tellus B Chem. Phys. Meteorol. 38 (1), 1–10.

Sebastian, M., Ammerman, J.W., 2009. The alkaline phosphatase PhoX is more widely distributed in marine bcteria than the classical PhoA. ISME J. 3 (5), 563–572.

Sebilo, M., Mayer, B., Nicolardot, B., Pinay, G., Mariotti, A., 2013. Long-term fate of nitrate fertilizer in agricultural soils. Proc. Natl. Acad. Sci. U. S. A. 110 (45), 18185–18189.

See, C.R., McCormack, M.L., Hobbie, S.E., 2019. Global patterns in fine root decomposition, climate, chemistry, mycorrhizal association and woodiness. Ecol. Lett. 22, 946–953.

Seemann, J.R., Sharkey, T.D., Wang, J., Osmond, C.B., 1987. Environmental effects on photosynthesis, nitrogen-use efficiency, and metabolite pools in leaves of sun and shade plants. Plant Physiol. 84 (3), 796–802.

Segura, A., Navarro-Gonzalez, R., 2005. Nitrogen fixation on early Mars by volcanic lightning and other sources. Geophys. Res. Lett. 32.

Sehmel, G.A., 1980. Particle and gas dry deposition: a review. Atmos. Environ. 14, 983–1011.

Seiler, W., Conrad, R., 1987. Contribution of tropical ecosystems to the global budgets of trace gases, especially CH_4 H_2, CO, and N_2O. In: Dickinson, R.E. (Ed.), The Geophysiology of Amazonia. Wiley, pp. 133–162.

Seinfeld, J.H., 1989. Urban air pollution: state of the science. Science 243 (4892), 745–752.

Seiter, K., Hensen, C., Zabel, M., 2005. Benthic carbon mineralization on a global scale. Glob. Biogeochem. Cycles. 19.

Seitzinger, S.P., 1988. Denitrification in freshwater and coastal marine ecosystems: ecological and geochemical significance. Limnol. Oceanogr. 33 (4), 702–724.

Seitzinger, S.P., 1991. The effect of pH on the release of phosphorus from potomac estuary sediments—implications for Blue-Green algal blooms. Estuar. Coast. Shelf Sci. 33 (4), 409–418.

Seitzinger, S., Nixon, S., Pilson, M.E.Q., Burke, S., 1980. Denitrification and N_2O production in near-shore marine sediments. Geochim. Cosmochim. Acta 44 (11), 1853–1860.

Seitzinger, S.P., Nixon, S.W., Pilson, M.E.Q., 1984. Denitrification and nitrous oxide production in a coastal marine ecosystem. Limnol. Oceanogr. 29, 73–83.

Seitzinger, S.P., Nielsen, L.P., Caffrey, J., Christensen, P.B., 1993. Denitrification measurements in aquatic sediments—a comparison of three methods. Biogeochemistry 23 (3), 147–167.

Seitzinger, S.P., Styles, R.V., Boyer, E.W., Alexander, R.B., Billen, G., Howarth, R.W., Mayer, B., Van Breemen, N., 2002. Nitrogen retention in rivers: model development and application to watersheds in the northeastern United States. Biogeochemistry 57, 199–237.

Seitzinger, S.P., Harrison, J.A., Dumont, E., Beusen, A.H.W., Bouwman, A.F., 2005. Sources and delivery of carbon, nitrogen, and phosphorus to the coastal zone: an overview of global nutrient export from watersheds (NEWS) models and their application. Glob. Biogeochem. Cycles. 19. https://doi.org/10.1029/2005GB002606.

Seitzinger, S., Harrison, J.A., Böhlke, J.K., Bouwman, A.F., Lowrance, R., Peterson, B., Tobias, C., Van Drecht, G., 2006. Denitrification across landscapes and waterscapes: a synthesis. Ecol. Appl. 16 (6), 2064–2090.

Seitzinger, S.P., Mayorga, E., Bouwman, A.F., Kroeze, C., Beusen, A.H.W., Billen, G., Van Drecht, G., et al., 2010. Global river nutrient export: a scenario analysis of past and future trends. Glob. Biogeochem. Cycles 24 (May), GB0A08.

Self, S., Blake, S., Sharma, K., Widdowson, M., Sephton, S., 2008. Sulfur and chlorine in late cretaceous Deccan magmas and eruptive gas release. Science 319 (5870), 1654–1657.

Selin, N.E., 2009. Global biogeochemical cycling of mercury: a review. Annu. Rev. Environ. Resour. 34, 43–63.

Selin, N.E., Jacob, D.J., 2008. Seasonal and spatial patterns of mercury wet deposition in the United States: constraints on the contribution from north American anthropogenic sources. Atmos. Environ. 42 (21), 5193–5204.

Selin, N.E., Jacob, D.J., Yantosca, R.M., Strode, S., Jaegle, L., Sunderland, E.M., 2008. Global 3-D land-ocean-atmosphere model for mercury: presentday versus preindustrial cycles and anthropogenic enrichment factors for deposition. Glob. Biogeochem. Cycles 22.

Sella, G.F., Stein, S., Dixon, T.H., Craymer, M., James, T.S., Mazzotti, S., Dokka, R.K., 2007. Observation of glacial isostatic adjustment in "stable" North America with GPS. Geophys. Res. Lett. 34.

Sellers, P., Kelly, C.A., Rudd, J.W.M., MacHutchon, A.R., 1996. Photodegradation of methylmercury in lakes. Nature 380, 694–697.

Selmants, P.C., Hart, S.C., 2010. Phosphorus and soil development: does the Walker and Syers model apply to semi-arid ecosystems? Ecology 91 (2), 474–484.

Sen, I.S., Peucker-Ehrenbrink, B., 2012. Anthropogenic disturbance of element cycles at the Earth's surface. Environ. Sci. Technol. 46 (16), 8601–8609.

Seo, K.H., Bowman, K.P., 2002. Lagrangian estimate of global stratosphere-troposphere mass exchange. J. Geophys. Res. 107.

Sequeira, R., 1993. On the large-scale impact of arid dust on precipitation chemistry of the continental northern hemisphere. Atmos. Environ. A—Gen. Top. 27, 1553–1565.

Serbin, S.P., Singh, A., McNeil, B.E., Kingdon, C.C., Townsend, P.A., 2014. Spectroscopic determination of leaf morphological and biochemical traits for northern temperate and boreal tree species. Ecol. Appl. 24 (7), 1651–1669.

Serca, D., Guenther, A., Klinger, L., Helmig, D., Hereid, D., Zimmerman, P., 1998. Methyl bromide deposition to soils. Atmos. Environ. 32, 1581–1586.

Serna-Chavez, H.M., Fierer, N., van Bodegom, P.M., 2013. Global drivers and patterns of microbial abundance in soil: global patterns of soil microbial biomass. Glob. Ecol. Biogeogr.: J. Macroecol. 22 (10), 1162–1172.

Serrasolsas, I., Khanna, P.K., 1995. Changes in heated and autoclaved forest soils of SE Australia. II. Phosphorus and phosphatase activity. Biogeochemistry 29, 25–41.

Serreze, M.C., Holland, M.M., Stroeve, J., 2007. Perspectives on the Arctic's shrinking sea-ice cover. Science 315 (5818), 1533–1536.

Servant, J., 1986. The burden of the sulphate layer of the stratosphere during volcanic "quiescent" periods. Tellus B Chem. Phys. Meteorol. 38, 74–79.

Servant, J., 1994. The continental carbon cycle during the last glacial maximum. Atmos. Res. 31, 253–268.

Sessions, A.L., Doughty, D.M., Welander, P.V., Summons, R.E., Newman, D.K., 2009. The continuing puzzle of the great oxidation event. Curr. Biol. 19 (14), R567–R574.

Severinghaus, J.P., Brook, E.J., 1999. Abrupt climate change at the end of the last glacial period inferred from trapped air in polar ice. Science 286 (5441), 930–934.

Sexstone, A.J., Parkin, T.B., Tiedje, J.M., 1985a. Temporal response of soil denitrification rates to rainfall and irrigation. Soil Sci. Soc. Am. J. 49, 99–103.

Sexstone, A.J., Revsbech, N.P., Parkin, T.B., Tiedje, J.M., 1985b. Direct measurement of oxygen profiles and denitrification rates in soil aggregates. Soil Sci. Soc. Am. J. 49, 645–651.

Sgouridis, F., Ullah, S., 2015. Relative magnitude and controls of in situ N_2 and N_2O fluxes due to denitrification in natural and seminatural terrestrial ecosystems using (15)N tracers. Environ. Sci. Technol. 49 (24), 14110–14119.

Shaffer, G., 1993. Effects of the marine biota on global carbon cycling. In: Heimann, M. (Ed.), The Global Carbon Cycle. Springer-Verlag, pp. 431–455.

Shaked, Y., Xu, Y., Leblanc, K., Morel, F.M.M., 2006. Zinc availability and alkaline phosphatase activity in *Emiliania huxleyi*: implications for Zn-P co-limitation in the ocean. Limnol. Oceanogr. 51 (1), 299–309.

Shakun, J.D., Clark, P.U., He, F., Marcott, S.A., Mix, A.C., Liu, Z., Otto-Bliesner, B., Schmittner, A., Bard, E., 2012. Global warming preceded by increasing carbon dioxide concentrations during the last deglaciation. Nature 484 (7392), 49–54.

Shanks, A.L., Trent, J.D., 1979. Marine snow: microscale nutrient patches. Limnol. Oceanogr. 24, 850–854.

Shannon, R.D., White, J.R., 1994. Three-year study of controls on methane emissions from two Michigan Peatlands. Biogeochemistry 27 (1), 35–60.

Sharkey, T.D., 1985. Photosynthesis in intact leaves of C-3 plants: physics, physiology and rate limitations. Bot. Rev. Interpret. Bot. Prog. 51, 53–105.

Sharkey, T.D., 1988. Estimating the rate of photorespiration in leaves. Physiol. Plant. 73, 147–152.

Sharples, J., Middelburg, J.J., Fennel, K., Jickells, T.D., 2017. What proportion of riverine nutrients reaches the open ocean? Glob. Biogeochem. Cycles 31 (1), 39–58.

Sharpley, A.N., 1985. The selective erosion of plant nutrients in runoff. Soil Sci. Soc. Am. J. 49, 1527–1534.

Sharpley, A.N., Tiessen, H., Cole, C.V., 1987. Soil phosphorus forms extracted in soil tests as a function of pedogenesis. Soil Sci. Soc. Am. J. 51, 362–365.

Sharpley, A., Daniel, T.C., Sims, J.T., Pote, D.H., 1996. Determining environmentally sound soil phosphorus levels. J. Soil Water Conserv. 51, 160–166.

Shaver, G.R., Chapin III, F.S., 1986. Effect of fertilizer on production and biomass of Tussock Tundra, Alaska. USA. Arct. Alp. Res. 18 (3), 261–268.

Shaver, G.R., Melillo, J.M., 1984. Nutrient budgets of Marsh plants: efficiency concepts and relation to availability. Ecology 65, 1491–1510.

Shaver, G.R., Giblin, A.E., Nadelhoffer, K.J., Thieler, K.K., Downs, M.R., Laundre, J.A., Rastetter, E.B., 2006. Carbon turnover in Alaskan tundra soils: effects of organic matter quality, temperature, moisture and fertilizer. J. Ecol. 94, 740–753.

Shaver, G.R., Billings, W.D., Stuart Chapin, F., Giblin, A.E., Nadelhoffer, K.J., Oechel, W.C., Rastetter, E.B., 2011. Global change and carbon balance of Arctic ecosystems changes in global terrestrial carbon cycling. Bioscience 42 (6), 433–441.

Shaw, G.E., 1983. Bio-controlled thermostasis involving the sulfur cycle. Clim. Chang. 5, 297–303.

Shaw, R.W., 1987. Air pollution by particles. Sci. Am. 257, 96–103.

Shaw, M.R., Harte, J., 2001. Response of nitrogen cycling to simulated climate change: differential responses along a subalpine ecotone. Glob. Chang. Biol. 7, 193–210.

Shcherbak, I., Millar, N., Philip Robertson, G., 2014. Global metaanalysis of the nonlinear response of soil nitrous oxide (N_2O) emissions to fertilizer nitrogen. Proc. Natl. Acad. Sci. U. S. A. 111 (25), 9199–9204.

Shearer, G., Kohl, D.H., 1988. Natural ^{15}N abundance as a method of estimating the contribution of biologically-fixed nitrogen to N_2-fixing systems: potential for non-legumes. Plant Soil 110, 317–327.

Shearer, G., Kohl, D.H., 1989. Estimates of N_2 fixation in ecosystems: the need for and basis of the ^{15}N natural abundance method. In: Rundel, P.W., Ehleringer, J.R., Nagy, K.A. (Eds.), Stable Isotopes in Ecological Research. Springer, pp. 342–374.

Shearer, G., Kohl, D.H., Virginia, R.A., Bryan, B.A., Skeeters, J.L., Nilsen, E.T., Sharifi, M.R., Rundel, P.W., 1983. Estimates of N_2-fixation from variation in the natural abundance of ^{15}N in Sonoran desert ecosystems. Oecologia 56 (2–3), 365–373.

Shelley, R.U., Sedwick, P.N., Bibby, T.S., Cabedo-Sanz, P., Church, T.M., Johnson, R.J., Macey, A.I., et al., 2012. Controls on dissolved cobalt in surface waters of the Sargasso Sea: comparisons with iron and aluminum. Glob. Biogeochem. Cycles 26.

Shen, Y., Benner, R., 2018. Mixing it up in the ocean carbon cycle and the removal of refractory dissolved organic carbon. Sci. Rep. 8 (1), 2542.

Shen, Y., Buick, R., Canfield, D.E., 2001. Isotopic evidence for microbial sulphate reduction in the early Archaean era. Nature 410 (6824), 77–81.

Shen, G., Dongmei Chen, Y.W., Lu, L., Liu, C., 2019. Spatial patterns and estimates of global forest litterfall. Ecosphere 10 (2), e02587.

Sheng, J.-X., Weisenstein, D.K., Luo, B.-P., Rozanov, E., Stenke, A., Anet, J., Bingemer, H., Peter, T., 2015. Global atmospheric sulfur budget under volcanically quiescent conditions: aerosol-chemistry-climate model predictions and validation. J. Geophys. Res. D: Atmos. 120 (1), 256–276.

Shepherd, M.F., Barzetti, S., Hastie, D.R., 1991. The production of atmospheric Nox and N_2O from a fertilized agricultural soil. Atmos. Environ. A—Gen. Top. 25, 1961–1969.

Shepherd, A., Ivins, E.R., Geruo, A., Barletta, V.R., Bentley, M.J., Bettadpur, S., Briggs, K.H., et al., 2012. A reconciled estimate of ice-sheet mass balance. Science 338 (6111), 1183–1189.

Shepherd, A., Lin, G., Muir, A.S., Konrad, H., McMillan, M., Slater, T., Briggs, K.H., Sundal, A.V., Hogg, A.E., Engdahl, M.E., 2019. Trends in Antarctic ice sheet elevation and mass. Geophys. Res. Lett. 46 (14), 8174–8183.

Shi, L., Dong, H., Reguera, G., Beyenal, H., Lu, A., Liu, J., Yu, H.-Q., Fredrickson, J.K., 2016a. Extracellular electron transfer mechanisms between microorganisms and minerals. Nat. Rev. Microbiol. 14 (10), 651–662.

Shi, M., Fisher, J.B., Brzostek, E.R., Phillips, R.P., 2016b. Carbon cost of plant nitrogen acquisition: global carbon cycle impact from an improved plant nitrogen cycle in the community land model. Glob. Chang. Biol. 22 (3), 1299–1314.

Shields, C.A., Band, L.E., Law, N., Groffman, P.M., Kaushal, S.S., Savvas, K., Fisher, G.T., Belt, K.T., 2008. Streamflow distribution of non-point source nitrogen export from urban-rural catchments in the Chesapeake Bay watershed. Water Resour. Res. 44.

Shiller, A.M., 1997. Manganese in surface waters of the Atlantic ocean. Geophys. Res. Lett. 24, 1495–1498.

Shimshock, J.P., de Pena, R.G., 1989. Below-cloud scavenging of tropospheric ammonia. Tellus B Chem. Phys. Meteorol. 41, 296–304.

Shindell, D.T., Faluvegi, G., Koch, D.M., Schmidt, G.A., Unger, N., Bauer, S.E., 2009. Improved attribution of climate forcing to emissions. Science 326 (5953), 716–718.

Shindell, D., Kuylenstierna, J.C.I., Vignati, E., van Dingenen, R., Amann, M., Klimont, Z., Anenberg, S.C., et al., 2012. Simultaneously mitigating near-term climate change and improving human health and food security. Science 335 (6065), 183–189.

Shoji, S., Nanzyo, M., Shirato, Y., Ito, T., 1993. Chemical kinetics of weathering in young andisols from northeastern Japan using soil age normalized to 10 C. Soil Sci. 155, 53–60.

Sholkovitz, E.R., 1976. Flocculation of dissolved organic and inorganic matter during mixing of river water and seawater. Geochim. Cosmochim. Acta 40, 831–845.

Shon, Z.H., Davis, D., Chen, G., Grodzinsky, G., Bandy, A., Thornton, D., Sandholm, S., et al., 2001. Evaluation of the DMS flux and its conversion to SO_2 over the Southern Ocean. Atmos. Environ. 35, 159–172.

Shorter, J.H., Kolb, C.E., Crill, P.M., Kerwin, R.A., Talbot, R.W., Hines, M.E., Harriss, R.C., 1995. Rapid degradation of atmospheric methyl bromide in soils. Nature 377, 717–719.

Shortle, W.C., Smith, K.T., 1988. Aluminum-induced calcium deficiency syndrome in declining red spruce. Science 240 (4855), 1017–1018.

Shotyk, W., Goodsite, M.E., Roos-Barraclough, F., Frei, R., Heinemeier, J., Asmund, G., Lohse, C., Hansen, T.S., 2003. Anthropogenic contributions to atmospheric Hg, Pb and as accumulation recorded by peat cores from southern Greenland and Denmark, dated using the ^{14}C "bomb pulse curve". Geochim. Cosmochim. Acta 67, 3991–4011.

Shugart, H.H., Saatchi, S., Hall, F.G., 2010. Importance of structure and its measurement in quantifying function of forest ecosystems. J. Geophys. Res. 115.

Shukla, S.P., Sharma, M., 2010. Neutralization of rainwater acidity at Kanpur, India. Tellus B—Chem. Phys. Meteorol. 62, 172–180.

Shukla, J., Nobre, C., Sellers, P., 1990. Amazon deforestation and climate change. Science 247 (4948), 1322–1325.

Shuttleworth, W.J., 1988. Evaporation from Anazonian rainforest. Proc. R. Soc. Lond. Ser. A Math. Phys. Sci. 233, 321–346.

Siddall, M., Rohling, E.J., Almogi-Labin, A., Ch, H., Meischner, D., Schmelzer, I., Smeed, D.A., 2003. Sea-level fluctuations during the last glacial cycle. Nature 423 (6942), 853–858.

Sidle, R.C., Tsuboyama, Y., Noguchi, S., Fujieda, M., Hosoda, I., Shimizu, T., 2000. Stormflow generation in steep forested headwaters: a linked hydrogeomorphic paradigm. Hydrol. Process. 14, 369–385.

Siebert, S., Burke, J., Faures, J.M., Frenken, K., Hoogeveen, J., Doell, P., Portmann, F.T., 2010. Groundwater use for irrigation—a global inventory. Hydrol. Earth Syst. Sci. 14, 1863–1880.

Siegel, D.A., McGillicuddy, D.J., Fields, E.A., 1999. Mesoscale eddies, satellite altimetry, and new production in the Sargasso Sea. J. Geophys. Res. 104, 13359–13379.

Siever, R., 1974. The steady state of the Earth's crust, atmosphere, and oceans. Sci. Am. 230, 72–79.

Sigg, A., Neftel, A., 1991. Evidence for a 50-percent increase in H_2O_2 over the past 200 years from a Greenland ice Core. Nature 351, 557–559.

Sigl, M., Winstrup, M., McConnell, J.R., Welten, K.C., Plunkett, G., Ludlow, F., Büntgen, U., et al., 2015. Timing and climate forcing of volcanic eruptions for the past 2,500 years. Nature 523 (7562), 543–549.

Sigler, J.M., Lee, X., 2006. Recent trends in anthropogenic mercury emission in the Northeast United States. J. Geophys. Res. 111.

Sigman, D.M., Altabet, M.A., McCorkle, D.C., Francois, R., Fischer, G., 2000. The Delta ^{15}N of nitrate in the southern ocean: nitrogen cycling and circulation in the ocean interior. J. Geophys. Res. 105, 19599–19614.

Sigman, D.M., Hain, M.P., Haug, G.H., 2010. The polar ocean and glacial cycles in atmospheric CO_2 concentration. Nature 466, 47–55.

Sillen, L.G., 1966. Regulation of O_2, N_2 and CO_2 in the atmosphere; thoughts of a laboratory chemist. Tellus 18 (2–3), 198–206.

Sillitoe, R.H., Folk, R.L., Saric, N., 1996. Bacteria as mediators of copper sulfide enrichment during weathering. Science 272 (5265), 1153–1155.

Sillman, S., 1999. The relation between ozone, NOx and hydrocarbons in Urban and polluted rural environments. Atmos. Environ. 33, 1821–1845.

Silva, L.C.R., Anand, M., 2013. Probing for the influence of atmospheric CO_2 and climate change on forest ecosystems across biomes. Glob. Ecol. Biogeogr.: J. Macroecol. 22 (1), 83–92.

Silver, W.L., 1994. Is nutrient availability related to plant nutrient use in humid tropical forests? Oecologia 98 (3–4), 336–343.

Silver, W.L., Miya, R.K., 2001. Global patterns in root decomposition: comparisons of climate and litter quality effects. Oecologia 129 (3), 407–419.

Silver, W.L., Herman, D.J., Firestone, M.K., 2001. Dissimilatory nitrate reduction to ammonium in upland tropical forest soils. Ecology 82 (9), 2410–2416.

Silvester, W.B., 1989. Molybdenum limitation of asymbiotic nitrogen fixation in forests of the Pacific northwest America. Soil Biol. Biochem. 21, 283–289.

Silvester, W.B., Sollins, P., Verhoeven, T., Cline, S.P., 1982. Nitrogen fixation and acetylene reduction in decaying conifer boles: effects of incubation time, aeration, and moisture content. Can. J. Forest Res. 12, 646–652.

Sim, M.S., Bosak, T., Ono, S., 2011. Large sulfur isotope fractionation does not require disproportionation. Science 333 (6038), 74–77.

Simas, F.N.B., Schaefer, C., Melo, V.F., Albuquerque-Filho, M.R., Michel, R.F.M., Pereira, V.V., Gomes, M.R.M., da Costa, L.M., 2007. Ornithogenic cryosols from maritime Antarctica: phosphatization as a soil-forming process. Geoderma 138, 191–203.

Simmons, J.S., Klemedtsson, L., Hultberg, H., Hines, M.E., 1999. Consumption of atmospheric carbonyl sulfide by coniferous boreal forest soils. J. Geophys. Res. 104, 11569–11576.

Simon, N.S., 1988. Nitrogen cycling between sediment and the shallow water column in the transition zone of the Potomac River and estuary. 1. Nitrate and ammonium fluxes. Estuar. Coast. Shelf Sci. 26 (5), 483–497.

Simon, N.S., 1989. Nitrogen cycling between sediment and the shallow water column in the transition zone of the Potomac river and estuary. 2. The role of wind-driven resuspension and adsorbed ammonium. Estuar. Coast. Shelf Sci. 28 (5), 531–547.

Simon, L., Bousquet, J., Levesque, R.C., Lalonde, M., 1993. Origin and diversification of endomycorrhizal fungi and coincidence with vascular land plants. Nature 363, 67–69.

Simonich, S.L., Hites, R.A., 1994. Importance of vegetation in removing polycyclic aromatic hydrocarbons from the atmosphere. Nature 370, 49–51.

Simonich, S.L., Hites, R.A., 1995. Global distribution of persistent organochlorine compounds. Science 269 (5232), 1851–1854.

Simonson, R.W., 1995. Airborne dust and its significance to soils. Geoderma 65, 1–43.

Šímová, I., Sandel, B., Enquist, B.J., Michaletz, S.T., Kattge, J., Violle, C., McGill, B.J., et al., 2019. The relationship of woody plant size and leaf nutrient content to large-scale productivity for forests across the Americas. Hector, A. (Ed.), J. Ecol. 107 (5), 2278–2290.

Simpson, I.J., Sulbaek, M.P., Andersen, S.M., Bruhwiler, L., Blake, N.J., Detlev, H., Sherwood Rowland, F., Blake, D.R., 2012. Long-term decline of global atmospheric ethane concentrations and implications for methane. Nature 488 (7412), 490–494.

Sinclair, S.A., Senger, T., Talke, I.N., Cobbett, C.S., Haydon, M.J., Krämer, U., 2018. Systemic upregulation of MTP2- and HMA2-mediated Zn partitioning to the shoot supplements local Zn deficiency responses. Plant Cell 30 (10), 2463–2479.

Singer, A., 1989. Illite in the hot aridic soil environment. Soil Sci. 147, 126–133.

Singer, G.A., Fasching, C., Wilhelm, L., Niggemann, J., Steier, P., Dittmar, T., Battin, T.J., 2012. Biogeochemically diverse organic matter in alpine glaciers and its downstream fate. Nat. Geosci. 5 (10), 710–714.

Singh, J.S., Gupta, S.R., 1977. Plant decomposition and soil respiration in terrestrial ecosystems. Bot. Rev. Interpret. Bot. Progr. 43, 499–528.

Singh, J.S., Lauenroth, W.K., Steinhorst, R.K., 1975. Review and assessment of various techniques for estimating net primary production in grasslands from harvest data. Bot. Rev. Interpret. Bot. Progr. 41, 181–232.

Singh, A.P., Singh, R., Mina, U., Singh, M.P., Varshney, C.K., 2011. Emissions of monoterpene from tropical Indian plant species and assessment of VOC emission from the forest of Haryana state. Atmos. Pollut. Res. 2 (1), 72–79.

Singleton, G.A., Lavkulich, L.M., 1987a. A soil chronosequence on beach sands, Vancouver Island, British Columbia. Can. J. Soil Sci. 67, 795–810.

Singleton, G.A., Lavkulich, L.M., 1987b. Phosphorus transformations in a soil chronosequence, Vancouver Island, British Columbia. Can. J. Soil Sci. 67, 787–793.

Singsaas, E.L., Ort, D.R., DeLucia, E.H., 2001. Variation in measured values of photosynthetic quantum yield in eco-physiological studies. Oecologia 128 (1), 15–23.

Sinsabaugh, R.L., Antibus, R.K., Linkins, A.E., McClaugherty, C.A., Rayburn, L., Repert, D., Weiland, T., 1993. Wood decomposition: nitrogen and phosphorus dynamics in relation to extracellular enzyme activity. Ecology 74, 1586–1593.

Sinsabaugh, R.L., Carreiro, M.M., Repert, D.A., 2002. Allocation of extracellular enzymatic activity in relation to litter composition, N deposition, and mass loss. Biogeochemistry 60, 1–24.

Sinsabaugh, R.L., Lauber, C.L., Weintraub, M.N., Ahmed, B., Allison, S.D., Crenshaw, C., Contosta, A.R., et al., 2008. Stoichiometry of soil enzyme activity at global scale. Ecol. Lett. 11 (11), 1252–1264.

Sinsabaugh, R.L., Hill, B.H., Follstad Shah, J.J., 2009. Ecoenzymatic stoichiometry of microbial organic nutrient acqui-sition in soil and sediment. Nature 462 (7274), 795–798.

Sinsabaugh, R.L., Turner, B.L., Talbot, J.M., Waring, B.G., Powers, J.S., Kuske, C.R., Moorhead, D.L., Follstad Shah, J.J., 2016. Stoichiometry of microbial carbon use efficiency in soils. Ecol. Monogr. 86 (2), 172–189.

Sistla, S.A., Moore, J.C., Simpson, R.T., Gough, L., Shaver, G.R., Schimel, J.P., 2013. Long-term warming restructures Arctic tundra without changing net soil carbon storage. Nature 497 (7451), 615–618.

Sivan, O., Schrag, D.P., Murray, R.W., 2007. Rates of methanogenesis and methanotrophy in deep-sea sediments. Geobiology 5, 141–151.

Sjöberg, A., 1976. Phosphate analysis of anthropic soils. J. Field Archaeol. 3 (4), 447–454.

Skiba, U., Smith, K.A., Fowler, D., 1993. Nitrification and denitrificaiton as sources of nitric oxide and nitrous oxide in a sandy loam soil. Soil Biol. Biochem. 25, 1527–1536.

Skinner, L.C., Primeau, F., Freeman, E., de la Fuente, M., Goodwin, P.A., Gottschalk, J., Huang, E., McCave, I.N., Noble, T.L., Scrivner, A.E., 2017. Radiocarbon constraints on the glacial ocean circulation and its impact on atmo-spheric CO_2. Nat. Commun. 8 (July), 16010.

Skyring, G.W., 1987. Sulfate reduction in coastal ecosystems. Geomicrobiol J. 5, 295–374.

Sleep, N.H., 2018. Geological and geochemical constraints on the origin and evolution of life. Astrobiology 18 (9), 1199–1219.

Sleep, N.H., Zahnle, K.J., Kasting, J.F., Morowitz, H.J., 1989. Annihilation of ecosystems by large asteroid impacts on the early earth. Nature 342 (6246), 139–142.

Slessarev, E.W., Lin, Y., Bingham, N.L., Johnson, J.E., Dai, Y., Schimel, J.P., Chadwick, O.A., 2016. Water balance cre-ates a threshold in soil pH at the global scale. Nature 540 (7634), 567–569.

Slinn, W.G.N., 1988. A simple model for Junge's relationship between concentration fluctuations and residence times for tropospheric trace gases. Tellus B Chem. Phys. Meteorol. 40, 229–232.

Sloan, G.C., Matsuura, M., Zijlstra, A.A., Lagadec, E., Groenewegen, M.A.T., Wood, P.R., Szyszka, C., Bernard-Salas, J., van Loon, J.T., 2009. Dust formation in a galaxy with primitive abundances. Science 323 (5912), 353–355.

Slutsky, A.H., Yen, B.C., 1997. A macro-scale natural hydrologic cycle water availability model. J. Hydrol. 201, 329–347.

Smeck, N.E., 1985. Phosphorus dynamics in soils and landscapes. Geoderma 36, 185–199.

Smedley, S.R., Eisner, T., 1995. Sodium uptake by puddling in a moth. Science 270 (5243), 1816–1818.

Smeed, D.A., McCarthy, G., Cunningham, S.A., Frajka-Williams, E., Rayner, D., Johns, W.E., Meinen, C.S., et al., 2014. Observed decline of the Atlantic meridional overturning circulation 2004–2012. Ocean Sci. 10 (1), 29–38.

Smejda, L., Hejcman, M., Horak, J., Shai, I., 2017. Ancient settlement activities as important sources of nutrients (P, K, S, Zn and Cu) in eastern Mediterranean ecosystems—the case of biblical Tel Burna, Israel. Catena 156 (September), 62–73.

Smemo, K.A., Yavitt, J.B., 2011. Anaerobic oxidation of methane: an underappreciated aspect of methane cycling in peatland ecosystems? Biogeosciences 8 (3), 779–793.

Smetacek, V., 1998. Biological oceanography: diatoms and the silicate factor. Nature 391 (6664), 224.

Smil, V., 2000. Phosphorus in the environment: natural flows and human interferences. Annu. Rev. Energy Environ. 25, 53–88.

Smil, V., 2001. Enriching the Earth. MIT Press.

Smirnoff, N., Todd, P., Stewart, G.R., 1984. The occurrence of nitrate reduction in the leaves of woody plants. Ann. Bot. 54, 363–374.

Smith, W.H., 1976. Character and significance of forest tree root exudates. Ecology 57, 324–331.

Smith, S.V., 1981. Marine macrophytes as a global carbon sink. Science 211 (4484), 838–840.

Smith, V.H., 1982. The nitrogen and phosphorus dependence of algal biomass in lakes: an empirical and theoretical analysis. Limnol. Oceanogr. 27, 1101–1112.

Smith, V.H., 1983. Low nitrogen to phosphorus ratios favor dominance by blue-Green algae in lake phytoplankton. Science 221 (4611), 669–671.

Smith, W.H., 1990. The atmosphere and the rhizosphere: linkages with potential significance for forest tree health. In: Lucier, A.A., Haines, S.G. (Eds.), Mechanisms of Forest Response to Acidic Deposition. Springer-Verlag, pp. 188–241.

Smith, K.L., 1992. Benthic boundary layer communities and carbon cycling at abyssal depths in the central North Pacific. Limnol. Oceanogr. 37, 1034–1056.

Smith, V.H., Schindler, D.W., 2009. Eutrophication science: where do we go from here? Trends Ecol. Evol. 24, 201–207.

Smith, T.M., Shugart, H.H., 1993. The transient response of terrestrial carbon storage to a perturbed climate. Nature 361, 523–526.

Smith, W.H., Siccama, T.G., 1981. The Hubbard Brook ecosystem study: biogeochemistry of lead in the northern hardwood forest. J. Environ. Qual. 10, 323–332.

Smith, M.S., Tiedje, J.M., 1979. Phases of denitrification following oxygen depletion in soil. Soil Biol. Biochem. 11, 261–267.

Smith, R.B., Waring, R.H., Perry, D.A., 1981. Interpreting foliar analyses from Douglas fir as weight per unit of leaf area. Can. J. Forest Res. 11, 593–598.

Smith, R.A., Alexander, R.B., Wolman, M.G., 1987. Water-quality trends in the nation's rivers. Science 235 (4796), 1607–1615.

Smith, R.C., Prézelin, B.B., Baker, K.S., Bidigare, R.R., Boucher, N.P., Coley, T., Karentz, D., MacIntyre, S., Matlick, H.A., Menzies, D., 1992a. Ozone depletion: ultraviolet radiation and phytoplankton biology in Antarctic Waters. Science 255 (5047), 952–959.

Smith, T.M., Leemans, R., Shugart, H.H., 1992b. Sensitivity of terrestrial carbon storage to CO_2-induced climate change: comparison of four scenarios based on general-circulation models. Clim. Chang. 21, 367–384.

Smith, S.J., Power, J.F., Kemper, W.D., 1994. Fixed ammonium and nitrogen availability indexes. Soil Sci. 158, 132–140.

Smith, K.A., Dobbie, K.E., Ball, B.C., Bakken, L.R., Sitaula, B.K., Hansen, S., Brumme, R., et al., 2000. Oxidation of atmospheric methane in northern European soils, comparison with other ecosystems, and uncertainties in the global terrestrial sink. Glob. Chang. Biol. 6, 791–803.

Smith, M.L., Ollinger, S.V., Martin, M.E., Aber, J.D., Hallett, R.A., Goodale, C.L., 2002. Direct estimation of aboveground forest productivity through hyperspectral remote sensing of canopy nitrogen. Ecol. Appl. 12, 1286–1302.

Smith, L.C., MacDonald, G.M., Velichko, A.A., Beilman, D.W., Borisova, O.K., Frey, K.E., Kremenetski, K.V., Sheng, Y., 2004. Siberian Peatlands a net carbon sink and global methane source since the early Holocene. Science 303 (5656), 353–356.

Smith, T.M., Yin, X.G., Gruber, A., 2006. Variations in annual global precipitation (1979–2004), based on the global precipitation climatology project 2.5 degrees analysis. Geophys. Res. Lett. 33.

Smith, P.H., Tamppari, L.K., Arvidson, R.E., Bass, D., Blaney, D., Boynton, W.V., Carswell, A., et al., 2009. H_2O at the phoenix landing site. Science 325 (5936), 58–61.

Smith, S.J., van Aardenne, J., Klimont, Z., Andres, R.J., Volke, A., Delgado Arias, S., 2011. Anthropogenic sulfur dioxide emissions: 1850–2005. Atmos. Chem. Phys. 11 (3), 1101–1116.

Smith, W.K., Cleveland, C.C., Reed, S.C., Running, S.W., 2014. Agricultural conversion without external water and nutrient inputs reduces terrestrial vegetation productivity: agricultural conversion reduces NPP. Geophys. Res. Lett. 41 (2), 449–455.

Smith, F.A., Hammond, J.I., Balk, M.A., Elliott, S.M., Kathleen Lyons, S., Pardi, M.I., Tomé, C.P., Wagner, P.J., Westover, M.L., 2016. Exploring the influence of ancient and historic Megaherbivore extirpations on the global methane budget. Proc. Natl. Acad. Sci. U. S. A. 113 (4), 874–879.

Smith, W.K., Biederman, J.A., Scott, R.L., Moore, D.J.P., He, M., Kimball, J.S., Yan, D., et al., 2018. Chlorophyll fluorescence better captures seasonal and interannual Gross primary productivity dynamics across dryland ecosystems of southwestern North America. Geophys. Res. Lett. 45 (2), 748–757.

Smolders, A.J.P., Lamers, L.P.M., Lucassen, E.C.H.E.T., Van Der Velde, G., Roelofs, J.G.M., 2006. Internal eutrophication: how it works and what to do about it—a review. Chem. Ecol. 22, 93–111.

Smrekar, S.E., Stofan, E.R., Mueller, N., Treiman, A., Elkins-Tanton, L., Helbert, J., Piccioni, G., Drossart, P., 2010. Recent hotspot volcanism on Venus from VIRTIS emissivity data. Science 328 (5978), 605–608.

Snaydon, R.W., 1962. Micro-distribution of *Trifolium repens* L. and its relation to soil factors. J. Ecol. 50, 133–143.

Snider, D.M., Schiff, S.L., Spoelstra, J., 2009. $^{15}N/^{14}N$ and $^{18}O/^{16}O$ stable isotope ratios of nitrous oxide produced during denitrification in temperate forest soils. Geochim. Cosmochim. Acta 73, 877–888.

Sobczak, W.V., Findlay, S., Dye, S., 2003. Relationships between DOC bioavailability and nitrate removal in an upland stream: an experimental approach. Biogeochemistry 62, 309–327.

Sobek, S., DelSontro, T., Wongfun, N., Wehrli, B., 2012. Extreme organic carbon burial fuels intense methane bubbling in a temperate reservoir. Geophys. Res. Lett. 39, 4.

Soden, B.J., Jackson, D.L., Ramaswamy, V., Schwarzkopf, M.D., Huang, X., 2005. The radiative signature of upper tropospheric moistening. Science 310 (5749), 841–844.

Sokol, N.W., Bradford, M.A., 2019. Microbial formation of stable soil carbon is more efficient from belowground than aboveground input. Nat. Geosci. 12 (1), 46–53.

Sollins, P., Spycher, G., Glassman, C.A., 1984. Net nitrogen mineralization from light- and heavy-fraction forest soil organic matter. Soil Biol. Biochem. 16, 31–37.

Sollins, P., Robertson, G.P., Uehara, G., 1988. Nutrient mobility in variable-charge and permanent-charge soils. Biogeochemistry 6, 181–199.

Solomon, S., 1990. Progress towards a quantitative understanding of Antarctic ozone depletion. Nature 347, 347–354.

Solomon, D.K., Cerling, T.E., 1987. The annual carbon dioxide cycle in a montane soil: observations, modeling, and implications for weathering. Water Resour. Res. 23, 2257–2265.

Solomon, S., Garcia, R.R., Rowland, F.S., Wuebbles, D.J., 1986. On the depletion of Antarctic ozone. Nature 321, 755–758.

Solomon, P., Barrett, J., Mooney, T., Connor, B., Parrish, A., Siskind, D.E., 2006. Rise and decline of active chlorine in the stratosphere. Geophys. Res. Lett. 33.

Solomon, S., Portmann, R.W., Thompson, D.W.J., 2007. Contrasts between Antarctic and Arctic ozone depletion. Proc. Natl. Acad. Sci. U. S. A. 104 (2), 445–449.

Solomon, C.T., Bruesewitz, D.A., Richardson, D.C., Rose, K.C., Van de Bogert, M.C., Hanson, P.C., Kratz, T.K., et al., 2013. Ecosystem respiration: drivers of daily variability and background respiration in lakes around the globe. Limnol. Oceanogr. 58 (3), 849–866.

Solomon, C.T., Jones, S.E., Weidel, B.C., Buffam, I., Fork, M.L., Karlsson, J., Larsen, S., et al., 2015. Ecosystem consequences of changing inputs of terrestrial dissolved organic matter to lakes: current knowledge and future challenges. Ecosystems 18 (3), 376–389.

Solomon, S., Ivy, D.J., Kinnison, D., Mills, M.J., Neely 3rd, R.R., Schmidt, A., 2016. Emergence of healing in the Antarctic ozone layer. Science 353 (6296), 269–274.

Somes, C.J., Landolfi, A., Koeve, W., Oschlies, A., 2016. Limited impact of atmospheric nitrogen deposition on marine productivity due to biogeochemical feedbacks in a global ocean model. Geophys. Res. Lett. 43 (9), 4500–4509.

Sommer, M., Jochheim, H., Höhn, A., Breuer, J., Zagorski, Z., Busse, J., Barkusky, D., et al., 2013. Si cycling in a forest biogeosystem—the importance of transient state biogenic Si pools. Biogeosciences 10 (7), 4991–5007.

Son, Y., 2001. Non-symbiotic nitrogen fixation in forest ecosystems. Ecol. Res. 16, 183–196.

Soo, R.M., Hemp, J., Parks, D.H., Fischer, W.W., Hugenholtz, P., 2017. On the origins of oxygenic photosynthesis and aerobic respiration in cyanobacteria. Science 355 (6332), 1436–1440.

Soper, F.M., Chamberlain, S.D., Crumsey, J.M., Gregor, S., Derry, L.A., Sparks, J.P., 2018a. Biological cycling of mineral nutrients in a temperate forested shale catchment. J. Geophys. Res. Biogeosci. 123 (10), 3204–3215.

Soper, F.M., Sullivan, B.W., Nasto, M.K., Osborne, B.B., Bru, D., Balzotti, C.S., Taylor, P.G., et al., 2018b. Remotely sensed canopy nitrogen correlates with nitrous oxide emissions in a lowland tropical rainforest. Ecology 99 (9), 2080–2089.

Sorooshian, S., Li, J.L., Hsu, K.L., Gao, X.G., 2011. How significant is the impact of irrigation on the local hydroclimate in California's Central Valley? Comparison of model results with ground and remote-sensing data. J. Geophys. Res. 116.

Soulé, P.T., Knapp, P.A., 2006. Radial growth rate increases in naturally occurring ponderosa pine trees: a late-20th century CO_2 fertilization effect? New Phytol. 171 (2), 379–390.

South, P.F., Cavanagh, A.P., Liu, H.W., Ort, D.R., 2019. Synthetic glycolate metabolism pathways stimulate crop growth and productivity in the field. Science 363 (6422).

Sowers, T., 2006. Late quaternary atmospheric CH_4 isotope record suggests marine clathrates are stable. Science 311 (5762), 838–840.

Sowers, T., Alley, R.B., Jubenville, J., 2003. Ice core records of atmospheric N_2O covering the last 106,000 years. Science 301 (5635), 945–948.

Spalding, R.F., Exner, M.E., 1993. Occurrence of nitrate in groundwater—a review. J. Environ. Qual. 22, 392–402.

Sparks, D.L., 2002. Environmental Soil Chemistry, second ed. Academic/Elsevier.

Sparks, J.P., 2009. Ecological ramifications of the direct foliar uptake of nitrogen. Oecologia 159 (1), 1–13.

Sparks, J.P., Roberts, J.M., Monson, R.K., 2003. The uptake of gaseous organic nitrogen by leaves: a significant global nitrogen transfer process. Geophys. Res. Lett. 30.

Speidel, D.H., Agnew, A.F., 1982. The Natural Geochemistry of Our Environment. Westview Press.

Sperling, E.A., Frieder, C.A., Raman, A.V., Girguis, P.R., Levin, L.A., Knoll, A.H., 2013. Oxygen, ecology, and the Cambrian radiation of animals. Proc. Natl. Acad. Sci. U. S. A. 110 (33), 13446–13451.

Sperling, E.A., Wolock, C.J., Morgan, A.S., Gill, B.C., Kunzmann, M., Halverson, G.P., Macdonald, F.A., Knoll, A.H., Johnston, D.T., 2015. Statistical analysis of Iron geochemical data suggests limited late proterozoic oxygenation. Nature 523 (7561), 451–454.

Spichtinger, N., Wenig, M., James, P., Wagner, T., Platt, U., Stohl, A., 2001. Satellite detection of a continental-scale plume of nitrogen oxides from boreal forest fires. Geophys. Res. Lett. 28, 4579–4582.

Spielmann, F.M., Wohlfahrt, G., Hammerle, A., Kitz, F., Migliavacca, M., Alberti, G., Ibrom, A., et al., 2019. Gross primary productivity of four European ecosystems constrained by joint CO_2 and COS flux measurements. Geophys. Res. Lett. 46 (10), 5284–5293.

Spiro, P.A., Jacob, D.J., Logan, J.A., 1992. Global inventory of sulfur emissions with 1o X 1o resolution. J. Geophys. Res. 97, 6023–6036.

Spivack, A.J., You, C.F., Smith, H.J., 1993. Foraminiferal boron isotope ratios as a proxy for surface ocean pH over the past 21-Myr. Nature 363, 149–151.

Spoelstra, J., Schiff, S.L., Hazlett, P.W., Jeffries, D.S., Semkin, R.G., 2007. The isotopic composition of nitrate produced from nitrification in a hardwood Forest floor. Geochim. Cosmochim. Acta 71, 3757–3771.

Spohn, M., Sierra, C.A., 2018. How long do elements cycle in terrestrial ecosystems? Biogeochemistry 139 (1), 69–83.

Spohn, M., Müller, K., Höschen, C., Mueller, C.W., Marhan, S., 2019. Dark microbial CO_2 fixation in temperate forest soils increases with CO_2 concentration. Global Change Biology.

Sposito, G., 1989. The Surface Chemistry of Natural Particles. Oxford University Press.

Spracklen, D.V., Arnold, S.R., Taylor, C.M., 2012. Observations of increased tropical rainfall preceded by air passage over forests. Nature 489 (7415), 282–285.

Spratt, H.G., Morgan, M.D., 1990. Sulfur cycling in a cedar-dominated, fresh-water wetland. Limnol. Oceanogr. 35 (7), 1586–1593.

Springer, C.J., DeLucia, E.H., Thomas, R.B., 2005. Relationships between net photosynthesis and foliar nitrogen concentrations in a loblolly pine forest ecosystem grown in elevated atmospheric carbon dioxide. Tree Physiol. 25 (4), 385–394.

Spycher, G., Sollins, P., Rose, S., 1983. Carbon and nitrogen in the light fraction of a forest soil: vertical distribution and seasonal patterns. Soil Sci. 135, 79–87.

Squyres, S.W., Arvidson, R.E., Ruff, S., Gellert, R., Morris, R.V., Ming, D.W., Crumpler, L., et al., 2008. Detection of silica-rich deposits on Mars. Science 320 (5879), 1063–1067.

Srivastava, S.C., Singh, J.S., 1988. Carbon and phosphorus in the soil biomass of some tropical soils of India. Soil Biol. Biochem. 20, 743–747.

St Clair, J., Moon, S., Holbrook, W.S., Perron, J.T., Riebe, C.S., Martel, S.J., Carr, B., Harman, C., Singha, K., Deb Richter, D., 2015. Geophysical imaging reveals topographic stress control of bedrock weathering. Science 350 (6260), 534–538.

St. Louis, V.L., Kelly, C.A., Duchemin, É., Rudd, J.W.M., Rosenberg, D.M., 2000. Reservoir surfaces as sources of greenhouse gases to the atmosphere: a global estimate. Bioscience 50 (9), 766.

Staaf, H., Berg, B., 1982. Accumulation and release of plant nutrients in decomposing scots pine needle litter. Long-term decomposition in a scots pine forest. II. Can. J. Bot. 60, 1561–1568.

Staccone, A., Liao, W., Perakis, S., Compton, J., Clark, C., Menge, D., 2020. A spatially explicit, empirical estimate of tree-based biological nitrogen fixation in forests of the United States. Glob. Biogeochem. Cycles. https://doi.org/10.1029/2019GB006241.

Staehr, P.A., Sand-Jensen, K., Raun, A.L., Nilsson, B., Kidmose, J., 2010a. Drivers of metabolism and net heterotrophy in contrasting lakes. Limnol. Oceanogr. 55, 817–830.

Staehr, P.A., Bade, D., Koch, G.R., Williamson, C., Hanson, P., Cole, J.J., Kratz, T., 2010b. Lake metabolism and the diel oxygen technique: state of the science. Limnol. Oceanogr.: Meth., 628–644.

Staffelbach, T., Neftel, A., Stauffer, B., Jacob, D., 1991. A record of the atmospheric methane sink from formaldehyde in polar ice cores. Nature 349, 603–605.

Stallard, R.F., 1998. Terrestrial sedimentation and the carbon cycle: coupling weathering and erosion to carbon burial. Glob. Biogeochem. Cycles 12, 231–257.

Stallard, R.F., Edmond, J.M., 1981. Geochemistry of the Amazon. 1. Precipitation chemistry and the marine contribution to the dissolved load at the time of peak discharge. J. Geophys. Res. D: Atmos. 86, 9844–9858.

Stallard, R.F., Edmond, J.M., 1983. Geochemistry of the Amazon. 2. The influence of geology and weathering environment on the dissolved load. J. Geophys. Res. D: Atmos. 88, 9671–9688.

Stanley, S.R., Ciolkosz, E.J., 1981. Classification and genesis of spodosols in the Central Appalachians. Soil Sci. Soc. Am. J. 45, 912–917.

Stanley, D.W., Hobbie, J.E., 1981. Nitrogen recycling in a North Carolina Coastal River. Limnol. Oceanogr. 26, 30–42.

Stanley, E.H., Maxted, J.T., 2008. Changes in the dissolved nitrogen pool across land cover gradients in Wisconsin Streams. Ecol. Appl. 18 (7), 1579–1590.

Stanley, E.H., Fisher, S.G., Grimm, N.B., 1997. Ecosystem expansion and contraction in streams. Bioscience 47 (7), 427–435.

Stanley, E.H., Powers, S.M., Lottig, N.R., Buffam, I., Crawford, J.T., 2012. Contemporary changes in dissolved organic carbon (DOC) in human-dominated rivers: is there a role for DOC management? Freshw. Biol. 57, 26–42.

Stanley, E.H., Casson, N.J., Christel, S.T., Crawford, J.T., Loken, L.C., Oliver, S.K., 2016. The ecology of methane in streams and rivers: patterns, controls, and global significance. Ecol. Monogr. 86 (2), 146–171.

Stap, L.B., de Boer, B., Ziegler, M., Bintanja, R., Lourens, L.J., van de Wal, R.S.W., 2016. CO_2 over the past 5 million years: continuous simulation and new $\delta^{11}B$-based proxy data. Earth Planet. Sci. Lett. 439 (April), 1–10.

Stark, J.M., Hart, S.C., 1997. High rates of nitrification and nitrate turnover in undisturbed coniferous forests. Nature 385, 61–64.

Starr, G., Oberbauer, S.F., 2003. Photosynthesis of Arctic evergreens under snow: implications for tundra ecosystem carbon balance. Ecology 84, 1415–1420.

Staubes, R., Georgii, H.-W., Ockelmann, G., 1989. Flux of COS, DMS, and CS2 from various soils in Germany. Tellus B Chem. Phys. Meteorol. 41, 305–313.

Steele, L.P., Fraser, P.J., Rasmussen, R.A., Khalil, M.A.K., Conway, T.J., Crawford, A.J., Gammon, R.H., Masarie, K.A., Thoning, K.W., 1987. The global distribution of methane in the troposphere. J. Atmos. Chem. 5, 125–171.

Steffen, W., Richardson, K., Rockström, J., Cornell, S.E., Fetzer, I., Bennett, E.M., Biggs, R., et al., 2015. Sustainability. planetary boundaries: guiding human development on a changing planet. Science 347 (6223), 1259855.

Stehfest, E., Bouwman, L., 2006. N_2O and NO emission from agricultural fields and soils under natural vegetation: summarizing available measurement data and modeling of global annual emissions. Nutr. Cycl. Agroecosyst. 74, 207–228.

Steidinger, B.S., Crowther, T.W., Liang, J., Van Nuland, M.E., Werner, G.D.A., Reich, P.B., Nabuurs, G.J., et al., 2019. Climatic controls of decomposition drive the global biogeography of Forest-tree symbioses. Nature 569 (7756), 404–408.

Steinbacher, M., Bingemer, H.G., Schmidt, U., 2004. Measurements of the exchange of carbonyl sulfide (OCS) and carbon disulfide (CS_2) between soil and atmosphere in a spruce Forest in Central Germany. Atmos. Environ. 38, 6043–6052.

Steinke, M., Malin, G., Archer, S.D., Burkill, P.H., Liss, P.S., 2002. DMS production in a coccolithophorid bloom: evidence for the importance of dinoflagellate DMSP lyases. Aquat. Microb. Ecol.: Int. J. 26, 259–270.

Stelzer, R.S., Bartsch, L.A., Richardson, W.B., Strauss, E.A., 2011. The dark side of the hyporheic zone: depth profiles of nitrogen and its processing in stream sediments. Freshw. Biol 56, 2021–2033.

Stemmler, I., Hense, I., Quack, B., Maier-Reimer, E., 2014. Methyl iodide production in the open ocean. Biogeosciences 11 (16), 4459–4476.

Stemmler, I., Hense, I., Quack, B., 2015. Marine sources of bromoform in the global open ocean—global patterns and emissions. Biogeosciences 12 (6), 1967–1981.

Stendahl, J., Berg, B., Lindahl, B.D., 2017. Manganese availability is negatively associated with carbon storage in northern coniferous forest humus layers. Sci. Rep. 7 (1), 15487.

Stepanauskas, R., Leonardson, L., Tranvik, L.J., 1999. Bioavailability of wetland-derived DON to freshwater and marine bacterioplankton. Limnol. Oceanogr. 44 (6), 1477–1485.

Stephen, K., Aneja, V.P., 2008. Trends in agricultural ammonia emissions and ammonium concentrations in precipitation over the southeast and midwest United States. Atmos. Environ. 42 (14), 3238–3252.

Stephens, J.C., Stewart, E.H., 1976. Effect of climate on organic soil subsidence. In: Proceedings of the 2nd International Symposium on Land Subsidence. vol. 121. International Association of Hydrological Sciences, Publication, Anaheim, CA, pp. 649–655.

Stephens, G.L., Li, J., Wild, M., Clayson, C.A., Loeb, N., Kato, S., L'Ecuyer, T., Stackhouse, P.W., Lebsock, M., Andrews, T., 2012. An update on Earth's energy balance in light of the latest global observations. Nat. Geosci. 5 (10), 691–696.

Sterling, S., Ducharne, A., 2008. Comprehensive data set of global land cover change for land surface model applications. Glob. Biogeochem. Cycles 22.

Stern, D.I., 2006. Reversal of the trend in global anthropogenic sulfur emissions. Glob. Environ. Change-Hum. Policy Dimens. 16, 207–220.

Stern, J.C., Sutter, B., Freissinet, C., Navarro-González, R., McKay, C.P., Douglas Archer Jr., P., Buch, A., et al., 2015. Evidence for indigenous nitrogen in sedimentary and Aeolian deposits from the curiosity rover investigations at Gale Crater, Mars. Proc. Natl. Acad. Sci. U. S. A. 112 (14), 4245–4250.

Sternberg, E., Tang, D.G., Ho, T.Y., Jeandel, C., Morel, F.M.M., 2005. Barium uptake and adsorption in diatoms. Geochim. Cosmochim. Acta 69, 2745–2752.

Sterner, R.W., 1989. Resource competition during seasonal succession toward dominance by cyanobacteria. Ecology 70, 229–245.

Sterner, R.W., Elser, J.J., 2002. Ecological Stoichiometry. Princeton University Press.

Sterner, R.W., Elser, J.J., Hessen, D.O., 1992. Stoichiometric relationships among producers, consumers and nutrient cycling in pelagic ecosystems. Biogeochemistry 17, 49–67.

Sterner, R.W., Smutka, T.M., McKay, R.M.L., Qin, X.M., Brown, E.T., Sherrell, R.M., 2004. Phosphorus and trace metal limitation of algae and bacteria in lake superior. Limnol. Oceanogr. 49, 495–507.

Stets, E.G., Butman, D., McDonald, C.P., Stackpoole, S.M., DeGrandpre, M.D., Striegl, R.G., 2017. Carbonate buffering and metabolic controls on carbon dioxide in rivers. Glob. Biogeochem. Cycles 31 (4), 663–677.

Stetter, K.O., Lauerer, G., Thomm, M., Neuner, A., 1987. Isolation of extremely thermophilic sulfate reducers: evidence for a novel branch of Archaebacteria. Science 236 (4803), 822–824.

Steudler, P.A., Peterson, B.J., 1985. Annual cycle of gaseous sulfur emissions from a New England Spartina Alterniflora Marsh. Atmos. Environ. 19 (9), 1411–1416.

Steudler, P.A., Bowden, R.D., Melillo, J.M., Aber, J.D., 1989. Influence of nitrogen fertilization on methane uptake in temperate forest soils. Nature 341, 314–316.

Stevens, T.O., McKinley, J.P., 1995. Lithoautotrophic microbial ecosystems in deep basalt aquifers. Science 270, 450–454.

Stevens, C.J., Dise, N.B., Gowing, D.J.G., Mountford, J.O., 2006. Loss of forb diversity in relation to nitrogen deposition in the UK: regional trends and potential controls. Glob. Chang. Biol. 12, 1823–1833.

Stevens, C.J., Duprè, C., Dorland, E., Gaudnik, C., Gowing, D.J.G., Bleeker, A., Diekmann, M., et al., 2010. Nitrogen deposition threatens species richness of grasslands across Europe. Environ. Pollut. 158 (9), 2940–2945.

Stevens, C.J., Manning, P., van den Berg, L.J.L., de Graaf, M.C.C., Wamelink, G.W.W., Boxman, A.W., Bleeker, A., et al., 2011. Ecosystem responses to reduced and oxidised nitrogen inputs in European terrestrial habitats. Environ. Pollut. 159 (3), 665–676.

Stevens, C.J., Lind, E.M., Hautier, Y., Harpole, W.S., Borer, E.T., Hobbie, S., Seabloom, E.W., et al., 2015. Anthropogenic Nitrogen Deposition Predicts Local Grassland Primary Production Worldwide.

Stevenson, F.J., 1982. Origin and distribution of nitrogen in soil. In: Stevenson, F.J. (Ed.), Nitrogen in Agricultural Soils. Soil Science Society of America, pp. 1–42.

Stevenson, D.J., 1983. The nature of the Earth prior to the oldest known rock record: the Hadean Earth. In: Schopf, J.E. (Ed.), Earth's Earliest Biosphere. Princeton University Press, pp. 32–40.

Stevenson, F.J., 1986. Cycles of Soil. Wiley.

Stevenson, D.J., 2008. A planetary perspective on the deep earth. Nature 451 (7176), 261–265.

Stewart, A.J., Wetzel, R.G., 1982. Influence of dissolved humic materials on carbon assimilation and alkaline-phosphatase activity in natural algal bacterial assemblages. Freshw. Biol. 12, 369–380.

Still, C.J., Berry, J.A., Collatz, G.J., DeFries, R.S., 2003. Global distribution of C-3 and C-4 vegetation: carbon cycle implications. Glob. Biogeochem. Cycles 17.

Stinecipher, J.R., Cameron-Smith, P.J., Blake, N.J., Kuai, L., Lejeune, B., Mahieu, E., Simpson, I.J., Campbell, J.E., 2019. Biomass burning unlikely to account for missing source of carbonyl sulfide. Geophys. Res. Lett. 46, 14912–14920.

St-Laurent, P., Friedrichs, M.A.M., Najjar, R.G., Martins, D.K., Herrmann, M., Miller, S.K., Wilkin, J., 2017. Impacts of atmospheric nitrogen deposition on surface waters of the western North Atlantic mitigated by multiple feedbacks. J. Geophys. Res. Oceans 122, 8406–8426.

Stock, J.B., Stock, A.M., Mottonen, J.M., 1990. Signal transduction in bacteria. Nature 344 (6265), 395–400.

Stocker T.F., D. Qin, G-K Plattner, M. Tignor, S. K. Allen, J. Boschung, A. Nauels, Y. Xia, V. Bex, and P. M. Midgley. n. d. IPCC, 2013: Climate Change 2013: The Physical Science Basis. Contribution of Working Group I to the Fifth Assessment Report of the Intergovernmental Panel on Climate Change. Cambridge, United Kingdom/New York, NY, USA: Cambridge University Press.

Stockmann, U., Minasny, B., McBratney, A.B., 2014. How fast does soil grow? Geoderma 216, 48–61. 2014 v.216.

Stockner, J.G., Antia, N.J., 1986. Algal picoplankton from marine and freshwater ecosystems: a multidisciplinary perspective. Can. J. Fish. Aquat. Sci. 43, 2472–2503.

Stoddard, J.L., et al., 1999. Regional trends in aquatic recovery from acidification in North America and Europe. Nature 401, 575–576.

Stofan, E.R., Elachi, C., Lunine, J.I., Lorenz, R.D., Stiles, B., Mitchell, K.L., Ostro, S., et al., 2007. The lakes of titan. Nature 445 (7123), 61–64.

Stoken, O.M., Riscassi, A.L., Scanlon, T.M., 2016. Association of dissolved mercury with dissolved organic carbon in U.S. rivers and streams: the role of watershed soil organic carbon: association of HgD and DOC in U.S. streams: the role of SOC. Water Resour. Res. 52 (4), 3040–3051.

Stolper, D.A., Brenhin Keller, C., 2018. A record of deep-ocean dissolved O_2 from the oxidation state of Iron in submarine basalts. Nature 553 (7688), 323–327.

Stolper, D.A., Bender, M.L., Dreyfus, G.B., Yan, Y., Higgins, J.A., 2016. A pleistocene ice core record of atmospheric O_2 concentrations. Science 353 (6306), 1427–1430.

Stolt, M.H., Baker, J.C., Simpson, T.W., 1992. Characterization and genesis of Saprolite derived from gneissic rocks of Virginia. Soil Sci. Soc. Am. J. 56, 531–539.

Stone, E.L., Boonkird, S., 1963. Calcium accumulation in the bark of *Terminalia* spp. in Thailand. Ecology 44, 586–588.

Stone, E.L., Kszystyniak, R., 1977. Conservation of potassium in the *Pinus resinosa* ecosystem. Science 198 (4313), 192–194.

Stott, P.A., Tett, S.F., Jones, G.S., Allen, M.R., Mitchell, J.F., Jenkins, G.J., 2000. External control of 20th century temperature by natural and anthropogenic forcings. Science 290 (5499), 2133–2137.

Strahm, B.D., Harrison, R.B., 2006. Nitrate sorption in a variable-charge forest soil of the Pacific northwest. Soil Sci. 171, 313–321.

Strahm, B.D., Harrison, R.B., 2007. Mineral and organic matter controls on the sorption of macronutrient anions in variable-charge soils. Soil Sci. Soc. Am. J. 71, 1926–1933.

Strand, A.E., Pritchard, S.G., Luke McCormack, M., Davis, M.A., Oren, R., 2008. Irreconcilable differences: fine-root life spans and soil carbon persistence. Science 319 (5862), 456–458.

Straub, M., Sigman, D.M., Ren, H., Alfredo, M.-G., Nele Meckler, A., Hain, M.P., Haug, G.H., 2013. Changes in North Atlantic nitrogen fixation controlled by ocean circulation. Nature 501 (7466), 200–203.

Strauss, E.A., Lamberti, G.A., 2000. Regulation of nitrification in aquatic sediments by organic carbon. Limnol. Oceanogr. 45, 1854–1859.

Strauss, E.A., Lamberti, G.A., 2002. Effect of dissolved organic carbon quality on microbial decomposition and nitrification rates in stream sediments. Freshw. Biol. 47, 65–74.

Strauss, E.A., Mitchell, N.L., Lamberti, G.A., 2002. Factors regulating nitrification in aquatic sediments: effects of organic carbon, nitrogen availability, and pH. Can. J. Fish. Aquat. Sci. 59, 554–563.

Street, G., Mcnickle, G., 2019. A global estimate of terrestrial net secondary production of primary consumers. Glob. Ecol. Biogeogr. 23, 1763–1773.

Streets, D.G., Yu, C., Wu, Y., Chin, M., Zhao, Z., Hayasaka, T., Shi, G., 2008. Aerosol trends over China, 1980–2000. Atmos. Res. 88, 174–182.

Streets, D.G., Devane, M.K., Lu, Z., Bond, T.C., Sunderland, E.M., Jacob, D.J., 2011. All-time releases of mercury to the atmosphere from human activities. Environ. Sci. Technol. 45 (24), 10485–10491.

Streets, D.G., Horowitz, H.M., Lu, Z., Levin, L., Thackray, C.P., Sunderland, E.M., 2019. Global and regional trends in mercury emissions and concentrations, 2010–2015. Atmos. Environ. 201 (March), 417–427.

Striegl, R.G., McConnaughey, T.A., Thorstenson, D.C., Weeks, E.P., Woodward, J.C., 1992. Consumption of atmospheric methane by desert soils. Nature 357, 145–147.

Strobel, B.W., 2001. Influence of vegetation on low-molecular-weight carboxylic acids in soil solution—a review. Geoderma 99, 169–198.

Strojan, C.L., Randall, D.C., Turner, F.B., 1987. Relationship of leaf litter decomposition rates to rainfall in the Mojave Desert. Ecology 68, 741–744.

Strother, P.K., Battison, L., Brasier, M.D., Wellman, C.H., 2011. Earth's earliest non-marine eukaryotes. Nature 473 (7348), 505–509.

Strous, M., Fuerst, J.A., Kramer, E.H., Logemann, S., Muyzer, G., van de Pas-Schoonen, K.T., Webb, R., Kuenen, J.G., Jetten, M.S., 1999. Missing lithotroph identified as new planctomycete. Nature 400 (6743), 446–449.

Stüeken, E.E., Buick, R., Guy, B.M., Koehler, M.C., 2015. Isotopic evidence for biological nitrogen fixation by molybdenum-nitrogenase from 3.2 Gyr. Nature 520 (7549), 666–669.

Stuiver, M., 1980. ^{14}C distribution in the Atlantic Ocean. J. Geophys. Res. D: Atmos. 85, 2711–2718.

Stuiver, M., Quay, P.D., Ostlund, H.G., 1983. Abyssal water carbon-14 distribution and the age of the world oceans. Science 219 (4586), 849–851.

Stumm, W., Morgan, J.J., 1996. Aquatic Chemistry Chemical Equilibria and Rates in Natural Waters, third ed. Wiley and Sons, New York.

Sturges, W.T., Penkett, S.A., Barnola, J.M., Chappellaz, J., Atlas, E., Stroud, V., 2001. A long-term record of carbonyl sulfide (COS) in two hemispheres from firn air measurements. Geophys. Res. Lett. 28, 4095–4098.

Stutter, M.I., Demars, B.O.L., Langan, S.J., 2010. River phosphorus cycling: separating biotic and abiotic uptake during short-term changes in sewage effluent loading. Water Res. 44 (15), 4425–4436.

Su, H., Cheng, Y., Oswald, R., Behrendt, T., Trebs, I., Meixner, F.X., Andreae, M.O., Cheng, P., Zhang, Y., Pöschl, U., 2011. Soil nitrite as a source of atmospheric HONO and OH radicals. Science 333 (6049), 1616–1618.

Suarez, D.L., Wood, J.D., Ibrahim, I., 1992. Reevaluation of calcite supersaturation in soils. Soil Sci. Soc. Am. J. 56, 1776–1784.

Subalusky, A.L., Dutton, C.L., Njoroge, L., Rosi, E.J., Post, D.M., 2018. Organic matter and nutrient inputs from large wildlife influence ecosystem function in the Mara River, Africa. Ecology 99 (11), 2558–2574.

Subbarao, G.V., Nakahara, K., Hurtado, M.P., Ono, H., Moreta, D.E., Salcedo, A.F., Yoshihashi, A.T., et al., 2009. Evidence for biological nitrification inhibition in *Brachiaria pastures*. Proc. Natl. Acad. Sci. U. S. A. 106 (41), 17302–17307.

Suberkropp, K., Godshalk, G.L., Klug, M.J., 1976. Changes in the chemical composition of leaves during processing in a woodland stream. Ecology 57, 720–727.

Subke, J.A., Inglima, I., Cotrufo, M.F., 2006. Trends and methodological impacts in soil CO_2 efflux partitioning: a metaanalytical review. Glob. Chang. Biol. 12, 921–943.

Subke, J.A., Voke, N.R., Leronni, V., Garnett, M.H., Ineson, P., 2011. Dynamics and pathways of autotrophic and heterotrophic soil CO_2 efflux revealed by forest girdling. J. Ecol. 99, 186–193.

Suchet, P.A., Probst, J.L., Ludwig, W., 2003. Worldwide distribution of continental rock lithology: implications for the atmospheric/soil CO_2 uptake by continental weathering and alkalinity river transport to the oceans. Glob. Biogeochem. Cycles 17.

Sudduth, E.B., Perakis, S.S., Bernhardt, E.S., 2013. Nitrate in watersheds: straight from soils to streams? J. Geophys. Res. Biogeosci. 118 (1), 291–302.

Suga, M., Akita, F., Yamashita, K., Nakajima, Y., Ueno, G., Li, H., Yamane, T., et al., 2019. An oxyl/oxo mechanism for oxygen-oxygen coupling in PSII revealed by an X-ray free-electron laser. Science 366 (6463), 334–338.

Suh, S., Yee, S., 2011. Phosphorus use-efficiency of agriculture and food system in the US. Chemosphere 84 (6), 806–813.

Sulpis, O., Boudreau, B.P., Mucci, A., Jenkins, C., Trossman, D.S., Arbic, B.K., Key, R.M., 2018. Current $CaCO_3$ dissolution at the seafloor caused by anthropogenic CO_2. Proc. Natl. Acad. Sci. 115, 11700–11705.

Sumaila, U.R., Cheung, W.W.L., Lam, V.W.Y., Pauly, D., Herrick, S., 2011. Climate change impacts on the biophysics and economics of world fisheries. Nat. Clim. Chang. 1 (9), 449–456.

Sun, J., Bankston, J.R., Payandeh, J., Hinds, T.R., Zagotta, W.N., Zheng, N., 2014. Crystal structure of the plant dual-affinity nitrate transporter NRT1.1. Nature 507 (7490), 73–77.

Sun, Y., Piao, S., Huang, M., Ciais, P., Zeng, Z., Cheng, L., Li, X., et al., 2016. Global patterns and climate drivers of water-use efficiency in terrestrial ecosystems deduced from satellite-based datasets and carbon cycle models. Glob. Ecol. Biogeogr. 25, 311–323.

Sun, Y., Frankenberg, C., Wood, J.D., Schimel, D.S., Jung, M., Guanter, L., Drewry, D.T., et al., 2017. OCO-2 advances photosynthesis observation from space via solar-induced chlorophyll fluorescence. Science 358 (6360).

Sun, R., Jiskra, M., Amos, H.M., Zhang, Y., Sunderland, E.M., Sonke, J.E., 2019. Modelling the mercury stable isotope distribution of earth surface reservoirs: implications for global Hg cycling. Geochim. Cosmochim. Acta 246 (February), 156–173.

Sunda, W.G., Huntsman, S.A., 1992. Feedback interactions between zinc and phytoplankton in seawater. Limnol. Oceanogr. 37, 25–40.

Sundby, B., Gobeil, C., Silverberg, N., Mucci, A., 1992. The phosphorus cycle in coastal marine sediments. Limnol. Oceanogr. 37, 1129–1145.

Sundquist, E.T., Visser, K., 2005. The geologic history of the carbon cycle. In: Schlesinger, W.H. (Ed.), Biogeochemistry. Elsevier, pp. 425–472.

Sutherland, R.A., Vankessel, C., Farrell, R.E., Pennock, D.J., 1993. Landscape-scale variations in plant and soil ^{15}N natural abundance. Soil Sci. Soc. Am. J. 57, 169–178.

Sutka, R.L., Ostrom, N.E., Ostrom, P.H., Breznak, J.A., Gandhi, H., Pitt, A.J., Li, F., 2006. Distinguishing nitrous oxide production from nitrification and denitrification on the basis of isotopomer abundances. Appl. Environ. Microbiol. 72 (1), 638–644.

Sutton, R., Sposito, G., 2005. Molecular structure in soil humic substances: the new view. Environ. Sci. Technol. 39 (23), 9009–9015.

Sutton, M.A., Pitcairn, C.E.R., Fowler, D., 1993. The exchange of ammonia between the atmosphere and plant communities. Adv. Ecol. Res. 24, 301–393.

Sutton-Grier, A.E., Keller, J.K., Koch, R., Gilmour, C., Megonigal, J.P., 2011. Electron donors and acceptors influence anaerobic soil organic matter mineralization in tidal marshes. Soil Biol. Biochem. 43 (7), 1576–1583.

Svedrup, H.U., Johnson, M.W., Fleming, R.H., 1942. The Oceans. Prentice Hall.

Svensson, T., Lovett, G.M., Likens, G.E., 2012. Is chloride a conservative ion in forest ecosystems? Biogeochemistry 107, 125–134.

Swain, E.B., Engstrom, D.R., Brigham, M.E., Henning, T.A., Brezonik, P.L., 1992. Increasing rates of atmospheric mercury deposition in Midcontinental North America. Science 257 (5071), 784–787.

Swain, M.R., Vasisht, G., Tinetti, G., 2008. The presence of methane in the atmosphere of an extrasolar planet. Nature 452, 329–331.

Swank, W.T., Douglass, J.E., 1977. Nutrient budgets for undisturbed and manipulated hardwood forest ecosystems in the mountains of North Carolina. In: Correll, D.L. (Ed.), Watershed Research in Eastern North America. Smithsonian Institution, pp. 343–362.

Swank, W.T., Henderson, G.S., 1976. Atmospheric input of some cations and anions to forest ecosystems in North Carolina and Tennessee. Water Resour. Res. 12, 541–546.

Swank, W.T., Waide, J.B., Crossley Jr., D.A., Todd, R.L., 1981. Insect defoliation enhances nitrate export from forest ecosystems. Oecologia 51 (3), 297–299.

Swank, W.T., Fitzgerald, J.W., Ash, J.T., 1984. Microbial transformation of sulfate in forest soils. Science 223 (4632), 182–184.

Swanson, F.J., Fredriksen, F.L., McCorison, F.M., 1982. Material transfer in a Western Oregon forested watershed. In: Edmonds, R.L. (Ed.), Analysis of Coniferous Forest Ecosystems in the Western United States. Dowden, Hutchinson and Ross, pp. 233–266.

Swap, R., Garstang, M., Greco, S., Talbot, R., Kallberg, P., 1992. Saharan dust in the Amazon Basin. Tellus B Chem. Phys. Meteorol. 44, 133–149.

Sweeney, C., Gloor, E., Jacobson, A.R., Key, R.M., McKinley, G., Sarmiento, J.L., Wanninkhof, R., 2007. Constraining global air-sea gas exchange for CO_2 with recent bomb ^{14}C measurements. Glob. Biogeochem. Cycles 21.

Sweet, W., Horton, R., Kopp, R., Romanou, A., 2017a. Sea Level Rise. Agencies and Staff of the U.S. Department of Commerce.

Sweet, W.V., Horton, R., Kopp, R.E., LeGrande, A.N., Romanou, A., 2017b. Sea level rise. In: Wuebbles, D.J., Fahey, D.W., Hibbard, K.A., Dokken, D.J., Stewart, B.C., Maycock, T.K. (Eds.), Climate Science Special Report: Fourth National Climate Assessment. In: vol. 1. U.S. Global Change Research Program, Washington, DC, pp. 333–363.

Swetnam, T.W., Betancourt, J.L., 1990. Fire-southern oscillation relations in the southwestern United States. Science 249, 1017–1020.

Swift, M.J., Heal, O.W., Anderson, J.M., 1979. Decomposition in Terrestrial Ecosystems. University of California Press.

Syakila, A., Kroeze, C., 2011. The global nitrous oxide budget revisited. Greenh. Gas Measur. Manag. 1, 17–26.

Syed, T.H., Famiglietti, J.S., Rodell, M., Chen, J., Wilson, C.R., 2008. Analysis of terrestrial water storage changes from GRACE and GLDAS. Water Resour. Res. 44.

Syed, T.H., Famiglietti, J.S., Chambers, D.P., Willis, J.K., Hilburn, K., 2010. Satellite-based global-ocean mass balance estimates of interannual variability and emerging trends in continental freshwater discharge. Proc. Natl. Acad. Sci. U. S. A. 107 (42), 17916–17921.

Syvitski, J.P.M., Saito, Y., 2007. Morphodynamics of deltas under the influence of humans. Glob. Planet. Chang. 57, 261–282.

Syvitski, J.P.M., Vorosmarty, C.J., Kettner, A.J., Green, P., 2005. Green impact of humans on the flux of terrestrial sediment to the global coastal ocean. Science 308, 376–380.

Syvitski, J.P.M., Kettner, A.J., Overeem, I., Hutton, E.W.H., Hannon, M.T., Brakenridge, G.R., Day, J., et al., 2009. Sinking deltas due to human activities. Nat. Geosci. 2, 681–686.

Szikszay, M., Kimmelmann, A.A., Hypolito, R., Figueira, R.M., Sameshima, R.H., 1990. Evolution of the chemical composition of water passing through the unsaturated zone to groundwater at an experimental site at the University of Sao Paulo, Brazil. J. Hydrol. 118, 175–190.

Szilagyi, J., Katul, G.G., Parlange, M.B., 2001. Evapotranspiration intensifies over the conterminous United States. J. Water Resour. Plan. Manag.-ASCE 127, 354–362.

Tabatabai, M.A., Dick, W.A., 1979. Distribution and stability of pyrophosphatase in soils. Soil Biol. Biochem. 11, 655–659.

Tabazadeh, A., Turco, R.P., 1993. Stratospheric chlorine injection by volcanic eruptions: HCl scavenging and implications for ozone. Science 260, 1082–1086.

Tagliabue, A., Bopp, L., Dutay, J.-C., Bowie, A.R., Chever, F., Jean-Baptiste, P., Bucciarelli, E., et al., 2010. Hydrothermal contribution to the oceanic dissolved iron inventory. Nat. Geosci. 3 (4), 252–256.

Tagliabue, A., Bowie, A.R., Boyd, P.W., Buck, K.N., Johnson, K.S., Saito, M.A., 2017. The integral role of Iron in ocean biogeochemistry. Nature 543 (7643), 51–59.

Tajika, E., 1998. Mantle degassing of major and minor volatile elements during the Earth's history. Geophys. Res. Lett. 25, 3991–3994.

Takahashi, T., Broecker, W.S., Langer, S., 1985. Redfield ratio based on chemical data from isopycnal surfaces. J. Geophys. Res. Oceans 90, 6907–6924.

Takahashi, T., Sutherland, S.C., Feely, R.A., Wanninkhof, R., 2006. Decadal change of the surface water pCO_2 in the North Pacific: a synthesis of 35 years of observations. J. Geophys. Res. Oceans 111.

Talbot, R.W., Harriss, R.C., Browell, E.V., Gregory, G.L., Sebacher, D.I., Beck, S.M., 1986. Distribution and geochemistry of aerosols in the tropical North Atlantic troposphere: relationship to Saharan dust. J. Geophys. Res.-Atmos. 91, 5173–5182.

Talukdar, R.K., Mellouki, A., Schmoltner, A.M., Watson, T., Montzka, S., Ravishankara, A.R., 1992. Kinetics of the OH reaction with methyl chloroform and its atmospheric implications. Science 257 (5067), 227–230.

Tamrazyan, G.P., 1989. Global peculiarities and tendencies in river discharge and wash-down of the suspended sediments—the earth as a whole. J. Hydrol. 107, 113–131.

Tan, K.H., 1980. The release of silicon, aluminum, and potassium during decomposition of soil minerals by humic acid. Soil Sci. 129, 5–11.

Tan, K.H., Troth, P.S., 1982. Silica-sesquioxide ratios as aids in characterization of some temperate region and tropical soil clays. Soil Sci. Soc. Am. J. 46, 1109–1114.

Tanaka, N., Turekian, K.K., 1991. Use of cosmogenic ^{35}S to determine the rates of removal of atmospheric SO_2. Nature 352, 226–228.

Tanaka, N., Turekian, K.K., 1995. Determination of the dry deposition flux of SO_2 using cosmogenic ^{35}S and ^7Be measurements. J. Geophys. Res.-Atmos. 100, 2841–2848.

Tanentzap, A.J., Kielstra, B.W., Wilkinson, G.M., Berggren, M., Craig, N., Del Giorgio, P.A., Grey, J., et al., 2017. Terrestrial support of lake food webs: synthesis reveals controls over cross-ecosystem resource use. Sci. Adv. 3(3), e1601765.

Tang, G., Wen, Y., Gao, J., Long, D., Ma, Y., Wan, W., Yang, H., 2017. Similarities and differences between three coexisting Spaceborne Radars in global rainfall and snowfall estimation: comparing space precipitation radars. Water Resour. Res. 53 (5), 3835–3853.

Tang, X., Zhao, X., Bai, Y., Tang, Z., Wang, W., Zhao, Y., Wan, H., et al., 2018. Carbon pools in China's terrestrial ecosystems: new estimates based on an intensive field survey. Proc. Natl. Acad. Sci. U. S. A. 115 (16), 4021–4026.

Tang, W., Li, Z., Cassar, N., 2019. Machine learning estimates of global marine nitrogen fixation. J. Geophys. Res.: Biogeosci. 124 (3), 717–730.

Tanhua, T., Peter Jones, E., Jeansson, E., Jutterström, S., Smethie Jr., W.M., Wallace, D.W.R., Anderson, L.G., 2009. Ventilation of the Arctic ocean: mean ages and inventories of anthropogenic CO_2 and CFC-11. J. Geophys. Res. C: Oceans 114 (C1).

Tanre, D., Breon, F.M., Deuze, J.L., Herman, M., Goloub, P., Nadal, F., Marchand, A., 2001. Global observation of anthropogenic aerosols from satellite. Geophys. Res. Lett. 28, 4555–4558.

Tapia-Hernández, A., Bustillos-Cristales, M.R., Jiménez-Salgado, T., Caballero-Mellado, J., Fuentes-Ramírez, L.E., 2000. Natural endophytic occurrence of *Acetobacter diazotrophicus* in pineapple plants. Microb. Ecol. 39 (1), 49–55.

Tarnocai, C., Canadell, J.G., Schuur, E.A.G., Kuhry, P., Mazhitova, G., Zimov, S., 2009. Soil organic carbon pools in the northern circumpolar permafrost region. Glob. Biogeochem. Cycles 23 (2), 1–11.

Tarrasón, L., Turner, S., Fløisand, I., 1995. Estimation of seasonal dimethyl sulphide fluxes over the North Atlantic Ocean and their contribution to European pollution levels. J. Geophys. Res. 100, 11623–11639.

Tashiro, T., Ishida, A., Hori, M., Igisu, M., Koike, M., Méjean, P., Takahata, N., Sano, Y., Komiya, T., 2017. Early trace of life from 3.95 Ga sedimentary rocks in Labrador, Canada. Nature 549 (7673), 516–518.

Tavares, P., Pereira, A.S., Moura, J.J.G., Moura, I., 2006. Metalloenzymes of the denitrification pathway. J. Inorg. Biochem. 100 (12), 2087–2100.

Taylor, K., 1999. Rapid climate change. Am. Sci. 87, 320–327.

Taylor, A.B., Velbel, M.A., 1991. Geochemical mass balances and weathering rates in forested watersheds of the southern blue Ridge. II. Effects of botanical uptake terms. Geoderma 51, 29–50.

Taylor, A.H., Watson, A.J., Robertson, J.E., 1992. The influence of the spring phytoplankton bloom on carbon dioxide and oxygen concentrations in the surface Waters of the Northeast Atlantic during 1989. Deep-Sea Res. Part A—Oceanogr. Res. Pap. 39, 137–152.

Taylor, S., Lever, J.H., Harvey, R.P., 1998. Accretion rate of cosmic spherules measured at the south pole. Nature 392 (6679), 899–903.

Taylor, B.W., Flecker, A.S., Hall, R.O., 2006. Loss of a harvested fish species disrupts carbon flow in a diverse tropical river. Science 313 (5788), 833–836.

Taylor, P.G., Wieder, W.R., Weintraub, S., Cohen, S., Cleveland, C.C., Townsend, A.R., 2015. Organic forms dominate hydrologic nitrogen export from a lowland tropical watershed. Ecology 96 (5), 1229–1241.

Taylor, P.G., Cleveland, C.C., Wieder, W.R., Sullivan, B.W., Doughty, C.E., Dobrowski, S.Z., Townsend, A.R., 2017. Temperature and rainfall interact to control carbon cycling in tropical forests. Ecol. Lett. 20 (6), 779–788.

Taylor, M.A., Celis, G., Ledman, J.D., Bracho, R., Schuur, E.A.G., 2018. Methane efflux measured by Eddy covariance in Alaskan Upland tundra undergoing permafrost degradation. J. Geophys. Res. Biogeosci. 123 (9), 2695–2710.

Taylor, B.N., Chazdon, R.L., Menge, D.N.L., 2019. Successional dynamics of nitrogen fixation and forest growth in regenerating Costa Rican rainforests. Ecology 100 (4), e02637.

Tcherkez, G., Gauthier, P., Buckley, T.N., Busch, F.A., Barbour, M.M., Bruhn, D., Heskel, M.A., et al., 2017. Leaf day respiration: low CO_2 flux but high significance for metabolism and carbon balance. New Phytol. 216 (4), 986–1001.

Tegen, I., Werner, M., Harrison, S.P., Kohfeld, K.E., 2004. Relative importance of climate and land use in determining present and future global soil dust emission. Geophys. Res. Lett. 31.

Teh, Y.A., Diem, T., Jones, S., Huaraca Quispe, L.P., Baggs, E., Morley, N., Richards, M., Smith, P., Meir, P., 2014. Methane and nitrous oxide fluxes across an elevation gradient in the tropical *Peruvian andes*. Biogeosciences 11 (8), 2325–2339.

Temple, K.L., Colmer, A.R., 1951. The autotrophic oxidation of iron by a new bacterium, *Thiobacillus ferrooxidans*. J. Bacteriol. 62 (5), 605–611.

Templer, P.H., Arthur, M.A., Lovett, G.M., Weathers, K.C., 2007. Plant and soil natural abundance delta [15]N: indicators of relative rates of nitrogen cycling in temperate forest ecosystems. Oecologia 153, 399–406.

Templer, P.H., Silver, W.L., Pett-Ridge, J., DeAngelis, K.M., Firestone, M.K., 2008. Plant and microbial controls on nitrogen retention and loss in a humid tropical forest. Ecology 89 (11), 3030–3040.

Templer, P.H., Mack, M.C., Chapin 3rd, F.S., Christenson, L.M., Compton, J.E., Crook, H.D., Currie, W.S., et al., 2012. Sinks for nitrogen inputs in terrestrial ecosystems: a meta-analysis of ^{15}N tracer field studies. Ecology 93 (8), 1816–1829.

Templer, P.H., Weathers, K.C., Ewing, H.A., Dawson, T.E., Mambelli, S., Lindsey, A.M., Webb, J., Boukili, V.K., Firestone, M.K., 2015. Fog as a source of nitrogen for redwood trees: evidence from fluxes and stable isotopes. Austin, A. (Ed.), J. Ecol. 103 (6), 1397–1407.

Teodoro, G.S., Lambers, H., Nascimento, D.L., de Britto Costa, P., Flores-Borges, D.N.A., Abrahão, A., Mayer, J.L.S., et al., 2019. Specialized roots of Velloziaceae weather quartzite rock while mobilizing phosphorus using carboxylates. Power, S. (Ed.), Funct. Ecol. 33 (5), 762–773.

Teodoru, C., Wehrli, B., 2005. Retention of sediments and nutrients in the iron gate I reservoir on the Danube river. Biogeochemistry 76, 539–565.

Teodoru, C.R., Bastien, J., Bonneville, M.C., del Giorgio, P.A., Demarty, M., Garneau, M., Helie, J.F., et al., 2012. The net carbon footprint of a newly created boreal hydroelectric reservoir. Glob. Biogeochem. Cycles 26.

Teramoto, M., Liang, N., Ishida, S., Zeng, J., 2018. Long-term stimulatory warming effect on soil heterotrophic respiration in a cool-temperate broad-leaved deciduous forest in northern Japan. J. Geophys. Res. Biogeosci. 123 (4), 1161–1177.

Terao, Y., Mukai, H., Nojiri, Y., Machida, T., Tohjima, Y., Saeki, T., Maksyutov, S., 2011. Interannual variability and trends in atmospheric methane over the Western Pacific from 1994 to 2010. J. Geophys. Res. 116.

Terman, G.L., 1977. Quantitative relationships among nutrients leached from soils. Soil Sci. Soc. Am. J. 41, 935–940.

Terman, G.L., 1979. Volatilization losses of nitrogen as ammonia from surface-applied fertilizers, organic amendments, and crop residues. Adv. Agron. 31, 189–223.

Terrer, C., Vicca, S., Hungate, B.A., Phillips, R.P., Colin Prentice, I., 2016. Mycorrhizal association as a primary control of the CO_2 fertilization effect. Science 353 (6294), 72–74.

Teste, F.P., Laliberté, E., 2019. Plasticity in root symbioses following shifts in soil nutrient availability during long-term ecosystem development. Chang, C. (Ed.), J. Ecol. 107 (2), 633–649.

Textor, C., Graf, H.F., Herzog, M., Oberhuber, J.M., 2003. Injection of gases into the stratosphere by explosive volcanic eruptions. J. Geophys. Res. 108.

Tezuka, Y., 1990. Bacterial regeneration of ammonium and phosphate as affected by the carbon:nitrogen:phosphorus ratio of organic substrates. Microb. Ecol. 19 (3), 227–238.

Thamdrup, B., Fossing, H., Jorgensen, B.B., 1994. Manganese, iron, and sulfur cycling in a coastal marine sediment, Aarhus Bay, Denmark. Geochim. Cosmochim. Acta 58, 5115–5129.

Thiel, J., Byrne, J.M., Kappler, A., Schink, B., Pester, M., 2019. Pyrite formation from FeS and H_2S is mediated through microbial redox activity. Proc. Natl. Acad. Sci. U. S. A. 116 (14), 6897–6902.

Thiemens, M.H., Trogler, W.C., 1991. Nylon production: an unknown source of atmospheric nitrous oxide. Science 251 (4996), 932–934.

Thienemann, A., 1921. Seetypen. Die Naturwissenschaften 18, 1–3.

Thöle, L.M., Eri Amsler, H., Moretti, S., Auderset, A., Gilgannon, J., Lippold, J., Vogel, H., et al., 2019. Glacial-interglacial dust and export production records from the southern Indian Ocean. Earth Planet. Sci. Lett. 525. https://doi.org/10.1016/j.epsl.2019.115716.

Thomas, G.E., Olivero, J.J., Jensen, E.J., Schroeder, W., Toon, O.B., 1989. Relation between increasing methane and the presence of ice clouds at the Mesopause. Nature 338, 490–492.

Thomas, R.B., Lewis, J.D., Strain, B.R., 1994. Effects of leaf nutrient status on photosynthetic capacity in loblolly pine (*Pinus taeda* L.) seedlings grown in elevated atmospheric CO_2. Tree Physiol. 14, 947–960.

Thomas, C.D., Cameron, A., Green, R.E., Bakkenes, M., Beaumont, L.J., Collingham, Y.C., Erasmus, B.F.N., et al., 2004a. Extinction risk from climate change. Nature 427 (6970), 145–148.

Thomas, H., Bozec, Y., Elkalay, K., de Baar, H.J.W., 2004b. Enhanced open ocean storage of CO_2 from shelf sea pumping. Science 304, 1005–1008.

Thomas, R.Q., Canham, C.D., Weathers, K.C., Goodale, C.L., 2010. Increased tree carbon storage in response to nitrogen deposition in the US. Nat. Geosci. 3, 13–17.

Thomas-Keprta, K.L., Clemett, S.J., Messenger, S., Ross, D.K., Le, L., Rahman, Z., McKay, D.S., Gibson, E.K., Gonzalez, C., Peabody, W., 2014. Organic matter on the Earth's Moon. Geochim. Cosmochim. Acta 134 (June), 1–15.

Thomazo, C., Ader, M., Philippot, P., 2011. Extreme ^{15}N-enrichments in 2.72-Gyr-old sediments: evidence for a turning point in the nitrogen cycle. Geobiology 9 (2), 107–120.

Thompson, A.M., 1992. The oxidizing capacity of the Earth's atmosphere: probable past and future changes. Science 256 (5060), 1157–1165.

Thompson, S.K., Cotner, J.B., 2018. Bioavailability of dissolved organic phosphorus in temperate lakes. Front. Environ. Sci. Eng. China 6, 62.

Thompson, L.G., Yao, T., Mosley-Thompson, E., Davis, M.E., Henderson, K.A., Lin, P., 2000. A high-resolution millennial record of the South Asian monsoon from Himalayan ice cores. Science 289 (5486), 1916–1920.

Thompson, A.E., Anderson, R.S., Rudolph, J., Huang, L., 2002. Stable carbon isotope signatures of background tropospheric chloromethane and CFC113. Biogeochemistry 60, 191–211.

Thomson, J., Higgs, N.C., Croudace, I.W., Colley, S., Hydes, D.J., 1993. Redox zonation of elements at an oxic post-oxic boundary in deep sea sediments. Geochim. Cosmochim. Acta 57, 579–595.

Thorneloe, S.A., Barlaz, M.A., Peer, R., Huff, L.C., Davis, L., Mangino, J., 1993. Waste management. In: Khalil, M.A.K. (Ed.), Atmospheric Methane: Sources, Sinks and Role in Global Change. Springer, pp. 362–398.

Thorpe, S.A., 1985. Small-scale processes in the upper ocean boundary layer. Nature 318, 519–522.

Thurman, E.M., 1985. Organic Geochemistry of Natural Waters. Springer Science & Business Media, Dordrecht.

Ti, C., Gao, B., Luo, Y., Wang, X., Wang, S., Yan, X., 2018. Isotopic characterization of NHx-N in deposition and major emission sources. Biogeochemistry 138 (1), 85–102.

Tian, H.Q., Melillo, J.M., Kicklighter, D.W., McGuire, A.D., Helfrich, J.V.K., Moore, B., Vorosmarty, C.J., 1998. Effect of interannual climate variability on carbon storage in Amazonian ecosystems. Nature 396, 664–667.

Tian, F., Toon, O.B., Pavlov, A.A., De Sterck, H., 2005. A hydrogen-rich early earth atmosphere. Science 308 (5724), 1014–1017.

Tian, D., Enzai, D., Jiang, L., Ma, S., Zeng, W., Zou, A., Feng, C., et al., 2018. Responses of forest ecosystems to increasing N deposition in China: a critical review. Environ. Pollut. 243 (Pt A), 75–86.

Tian, D., Kattge, J., Chen, Y., Han, W., Luo, Y., He, J., Hu, H., et al., 2019. A global database of paired leaf nitrogen and phosphorus concentrations of terrestrial plants. Ecology 100 (9), e02812.

Tian, H., Xu, R., Pan, S., Yao, Y., Bian, Z., Cai, W.-J., Hopkinson, C.S., Justic, D., Lohrenz, S., Lu, C., Ren, W., Yang, J., 2020. Long-term trajectory of nitrogen loading and delivery from Mississippi River basin to the Gulf of Mexico. Glob. Biogeochem. Cycles. https://doi.org/10.1029/2019/GB006475.

Tice, M.M., Lowe, D.R., 2004. Photosynthetic microbial mats in the 3,416-Myr-old ocean. Nature 431 (7008), 549–552.

Tice, M.M., Lowe, D.R., 2006. Hydrogen-based carbon fixation in the earliest known photosynthetic organisms. Geology 34, 37–40.

Tiedje, J.M., 1988. Ecology of denitrification and dissimilatory nitrate reduction to ammonium. In: Zehnder, A.J.B. (Ed.), Biology of Anaerobic Microorganisms. John Wiley and Sons, New York, pp. 179–244.

Tiedje, J.M., Sexstone, A.J., Parkin, T.B., Revsbech, N.P., Shelton, D.R., 1984. Anaerobic processes in soil. Plant Soil 76, 197–212.

Tiedje, J.M., Simkins, S., Groffman, P.M., 1989. Perspectives on measurement of denitrification in the field including recommended protocols for acetylene-based methods. Plant Soil 115, 261–284.

Tiessen, H., Stewart, J.W.B., 1983. Particle-size fractions and their use in studies of soil organic matter: II. Cultivation effects on organic matter composition in size fractions. Soil Sci. Soc. Am. J. 47, 509–514.

Tiessen, H., Stewart, J.W.B., 1988. Light and electron microscopy of stained microaggregates-the role of organic matter and microbes in soil aggregation. Biogeochemistry 5, 312–322.

Tiessen, H., Stewart, J.W.B., Bettany, J.R., 1982. Cultivation effects on the amounts and concentration of carbon, nitrogen, and phosphorus in grassland soils. Agron. J. 74, 831–835.

Tiessen, H., Stewart, J.W.B., Cole, C.V., 1984. Pathways of phosphorus transformations in soils of differing pedogenesis. Soil Sci. Soc. Am. J. 48, 853–858.

Tilman, D., 1985. The resource-ratio hypothesis of plant succession. Am. Nat. 125, 827–852.

Tilman, D., Kiesling, R., Sterner, R., Kilham, S.S., Johnson, F.A., 1986. Green, bluegreen and diatom algae: taxonomic differences in competitive ability for phosphorus, silicaon and nitrogen. Arch. Hydrobiol. 106, 473–485.

Tilton, D.L., 1978. Comparative growth and foliar element concentrations of *Larix laricina* over a range of wetland types in Minnesota. J. Ecol. 66 (2), 499–512.

Timmer, V.R., Stone, E.L., 1978. Comparative foliar analysis of young balsam fir fertilized with nitrogen, phosphorus, potassium, and lime. Soil Sci. Soc. Am. J. 42, 125–130.

Tipping, E., Chamberlain, P.M., Bryant, C.L., Buckingham, S., 2010. Soil organic matter turnover in British deciduous woodlands, quantified with radiocarbon. Geoderma 155, 10–18.

Tipping, E., Somerville, C.J., Luster, J., 2016. The C:N:P:S stoichiometry of soil organic matter. Biogeochemistry 130 (1), 117–131.

Tischner, R., 2000. Nitrate uptake and reduction in higher and lower plants. Plant Cell Environ. 23, 1005–1024.

Tisdall, J.M., Oades, J.M., 1982. Organic matter and water-stable aggregates in soils. J. Soil Sci. 33, 141–163.

Tissue, D.T., Megonigal, J.P., Thomas, R.B., 1997. Nitrogenase activity and N_2 fixation are stimulated by elevated CO_2 in a tropical N_2-fixing tree. Oecologia.

Titman, D., 1976. Ecological competition between algae: experimental confirmation of resource-based competition theory. Science 192 (4238), 463–465.

Titus, J.H., Nowak, R.S., Smith, S.D., 2002. Soil resource heterogeneity in the Mojave desert. J. Arid Environ. 52, 269–292.

Titus, T.N., Kieffer, H.H., Christensen, P.R., 2003. Exposed water ice discovered near the South pole of Mars. Science 299 (5609), 1048–1051.

Toggweiler, J.R., Samuels, B., 1993. New radiocarbon constraints on the upwelling of abyssal water to the ocean's surface. In: Heimann, M. (Ed.), The Global Carbon Cycle. Springer, pp. 333–366.

Tolbert, N.E., Benker, C., Beck, E., 1995. The oxygen and carbon dioxide compensation points of C-3 plants: possible role in regulating atmospheric oxygen. Proc. Natl. Acad. Sci. 92, 11230–11233.

Toner, B.M., Fakra, S.C., Manganini, S.J., Santelli, C.M., Marcus, M.A., Moffett, J.W., Rouxel, O., German, C.R., Edwards, K.J., 2009. Preservation of iron(II) by carbon-Rich matrices in a hydrothermal plume. Nat. Geosci. 2 (3), 197–201.

Tonini, D., Saveyn, H.G.M., Huygens, D., 2019. Environmental and health co-benefits for advanced phosphorus recovery. Nat. Sustain. 2 (11), 1051–1061.

Toon, O.B., Kasting, J.F., Turco, R.P., Liu, M.S., 1987. The sulfur cycle in the marine atmosphere. J. Geophys. Res. 92, 943–963.

Torn, M.S., Trumbore, S.E., Chadwick, O.A., Vitousek, P.M., Hendricks, D.M., 1997. Mineral control of soil organic carbon storage and turnover. Nature 389, 170–173.

Torn, M.S., Lapenis, A.G., Timofeev, A., Fischer, M.L., Babikov, B.V., Harden, J.W., 2002. Organic carbon and carbon isotopes in modern and 100-year-old-soil archives of the Russian steppe. Glob. Chang. Biol. 8, 941–953.

Torn, M.S., Kleber, M., Zavaleta, E.S., Zhu, B., Field, C.B., Trumbore, S.E., 2013. A dual isotope approach to isolate soil carbon pools of different turnover times. Biogeosciences 10 (12), 8067–8081.

Tor-ngern, P., Oren, R., Ward, E.J., Palmroth, S., McCarthy, H.R., Domec, J.-C., 2015. Increases in atmospheric CO_2 have little influence on transpiration of a temperate Forest canopy. New Phytol. 205 (2), 518–525.

Tornqvist, T.E., Wallace, D.J., Storms, J.E.A., Wallinga, J., Van Dam, R.L., Blaauw, M., Derksen, M.S., Klerks, C.J.W., Meijneken, C., Snijders, E.M.A., 2008. Mississippi Delta subsidence primarily caused by compaction of Holocene strata. Nat. Geosci. 1, 173–176.

Tortell, P.D., Maldonado, M.T., Price, N.M., 1996. The role of heterotrophic bacteria in iron-limited ocean ecosystems. Nature 383, 330–332.

Tostevin, R., Turchyn, A.V., Farquhar, J., Johnston, D.T., Eldridge, D.L., Bishop, J.K.B., McIlvin, M., 2014. Multiple sulfur isotope constraints on the modern sulfur cycle. Earth Planet. Sci. Lett. 396 (June), 14–21.

Townsend, A.R., Braswell, B.H., Holland, E.A., Penner, J.E., 1996. Spatial and temporal patterns in terrestrial carbon storage due to deposition of fossil fuel nitrogen. Ecol. Appl. 6, 806–814.

Townsend, A.R., Howarth, R.W., Bazzaz, F.A., Booth, M.S., Cleveland, C.C., Collinge, S.K., Dobson, A.P., et al., 2003. Human health effects of a changing global nitrogen cycle. Front. Ecol. Environ. 1, 240–246.

Townsend-Small, A., Czimczik, C.I., 2010. Carbon sequestration and greenhouse gas emissions in Urban turf. Geophys. Res. Lett. 37.

Toyoda, S., Yano, M., Nishimura, S., Akiyama, H., Hayakawa, A., Koba, K., Sudo, S., et al., 2011. Characterization and production and consumption processes of N_2O emitted from temperate agricultural soils determined via isotopomer ratio analysis. Glob. Biogeochem. Cycles 25.

Trail, D., Bruce Watson, E., Tailby, N.D., 2011. The oxidation state of hadean magmas and implications for early Earth's atmosphere. Nature 480 (7375), 79–82.

Tranvik, L.J., Downing, J.A., Cotner, J.B., Loiselle, S.A., Striegl, R.G., Ballatore, T.J., Dillon, P., et al., 2009. Lakes and reservoirs as regulators of carbon cycling and climate. Limnol. Oceanogr. 54 (1), 2298–2314.

Travis, C.C., Etnier, E.L., 1981. A survey of sorption relationships for reactive solutes in soil. J. Environ. Qual. 10, 8–17.

Treat, C.C., Kleinen, T., Broothaerts, N., Dalton, A.S., Dommain, R., Douglas, T.A., Drexler, J.Z., et al., 2019. Widespread global peatland establishment and persistence over the last 130,000 Y. Proc. Natl. Acad. Sci. U. S. A. 116 (11), 4822–4827.

Trefry, J.H., Metz, S., 1989. Role of hydrothermal precipitates in the geochemical cycling of vanadium. Nature 342, 531–533.

Trefry, J.H., Metz, S., Trocine, R.P., Nelsen, T.A., 1985. A decline in lead transport by the Mississippi River. Science 230 (4724), 439–441.

Tréguer, P.J., De La Rocha, C.L., 2013. The world ocean silica cycle. Annu. Rev. Mar. Sci. 5, 477–501.

Tréguer, P., Bowler, C., Moriceau, B., Dutkiewicz, S., Gehlen, M., Aumont, O., Bittner, L., et al., 2017. Influence of diatom diversity on the ocean biological carbon pump. Nat. Geosci. 11 (1), 27–37.

Trenberth, K.E., 1998. Atmospheric moisture residence times and cycling implications from rainfall rate and climate change. Clim. Chang. 39, 667–694.

Trenberth, K.E., Caron, J.M., 2001. Estimates of meridional atmosphere and ocean heat transports. J. Clim. 14 (16), 3433–3443.

Trenberth, K.E., Guillemot, C.J., 1994. The total mass of the atmosphere. J. Geophys. Res. 99, 23079–23088.

Trenberth, K.E., Smith, L., Qian, T., Dai, A., Fasullo, J., 2007. Estimates of the global water budget and its annual cycle using observational and model data. J. Hydrometeorol. 8 (4), 758–769.

Treseder, K.K., 2004. A meta-analysis of mycorrhizal responses to nitrogen, phosphorus, and atmospheric CO_2 in field studies. New Phytol. 164 (2), 347–355.

Treseder, K.K., 2008. Nitrogen additions and microbial biomass: a meta-analysis of ecosystem studies. Ecol. Lett. 11 (10), 1111–1120.

Treseder, K.K., Vitousek, P.M., 2001. Effects of soil nutrient availability on investment in acquisition of N and P in Hawaiian rain forests. Ecology 82, 946–954.

Treuhaft, R.N., Law, B.E., Asner, G.P., 2004. Forest attributes from radar interferometric structure and its fusion with optical remote sensing. Bioscience 54, 561–571.

Treuhaft, R.N., Goncalves, F.G., Drake, J.B., Chapman, B.D., dos Santos, J.R., Dutra, L.V., Graca, P., Purcell, G.H., 2010. Biomass estimation in a tropical wet forest using Fourier transforms of profiles from Lidar or interferometric SAR. Geophys. Res. Lett. 37.

Trimble, S.W., 1977. Fallacy of stream equilibrium in contemporary denudation studies. Am. J. Sci. 277, 876–887.

Trimble, V., 1997. Origin of the biologically important elements. Origins Life Evol. Biosph. 27 (1–3), 3–21.

Tripati, A.K., Roberts, C.D., Eagle, R.A., 2009. Coupling of CO_2 and ice sheet stability over major climate transitions of the last 20 million years. Science 326 (5958), 1394–1397.

Tripler, C.E., Kaushal, S.S., Likens, G.E., Todd Walter, M., 2006. Patterns in potassium dynamics in forest ecosystems. Ecol. Lett. 9 (4), 451–466.

Triska, F.J., Sedell, J.R., Cromack, K., Gregory, S.V., McCorison, F.M., Michael McCorison, F.J.F., 1984. Nitrogen budget for a small coniferous forest stream. Ecol. Monogr. 54 (1), 119–140.

Trouwborst, R.E., Clement, B.G., Tebo, B.M., Glazer, B.T., Luther 3rd, G.W., 2006. Soluble Mn(III) in suboxic zones. Science 313 (5795), 1955–1957.

Trouwborst, R.E., Johnston, A., Koch, G., Luther, G.W., Pierson, B.K., 2007. Biogeochemistry of Fe(II) oxidation in a photosynthetic microbial mat: implications for Precambrian Fe(II) oxidation. Geochim. Cosmochim. Acta 71, 4629–4643.

Trumbore, S.E., 1993. Comparison of carbon dynamics in tropical and temperate soils Uding radiocarbon measurements. Glob. Biogeochem. Cycles 7, 275–290.

Trumbore, S.E., 1997. Potential responses of soil organic carbon to global environmental change. Proc. Natl. Acad. Sci. U. S. A. 94 (16), 8284–8291.

Tsigaridis, K., Daskalakis, N., Kanakidou, M., Adams, P.J., Artaxo, P., Bahadur, R., Balkanski, Y., et al., 2014. The AeroCom evaluation and intercomparison of organic aerosol in global models. Atmos. Chem. Phys. 14 (19), 10845–10895.

Tsujii, Y., Onoda, Y., Kitayama, K., 2017. Phosphorus and nitrogen resorption from different chemical fractions in senescing leaves of tropical tree species on mount Kinabalu, Borneo. Oecologia 185 (2), 171–180.

Tukey, H.B., 1970. The leaching of substances from plants. Annu. Rev. Plant Physiol. 21, 305–324.

Tully, K., Gedan, K., Epanchin-Niell, R., Strong, A., Bernhardt, E.S., BenDor, T., Mitchell, M., et al., 2019. The invisible flood: the chemistry, ecology, and social implications of coastal saltwater intrusion. Bioscience 69 (5), 368–378.

Turchin, P., 2009. Long-term population cycles in human societies. Ann. N. Y. Acad. Sci. 1162 (April), 1–17.

Turco, R.P., Whitten, R.C., Toon, O.B., Pollack, J.B., Hamill, P., 1980. OCS, stratospheric aerosols and climate. Nature 283, 283–286.

Turekian, K.K., 1977. Fate of metals in the oceans. Geochim. Cosmochim. Acta 41, 1139–1144.

Turetsky, M.R., Donahue, W.F., Benscoter, B.W., 2011. Experimental drying intensifies burning and carbon losses in a northern peatland. Nat. Commun. 2 (November), 514.

Turner, J., 1982. The mass-flow component of nutrient supply in three Western Washington Forest types. Acta Oecol.-Oecol. Plantar. 3, 323–329.

Turner, R.E., 2004. Coastal wetland subsidence arising from local hydrologic manipulations. Estuaries 27 (2), 266–272.

Turner, B.L., Engelbrecht, B.M.J., 2011. Soil organic phosphorus in lowland tropical rain forests. Biogeochemistry 103, 297–315.

Turner, A., Millward, G.E., 2002. Suspended particles: their role in estuarine biogeochemical cycles. Estuar. Coast. Shelf Sci. 55, 857–883.

Turner, J., Olson, P.R., 1976. Nitrogen relations in a Douglas fir plantation. Ann. Bot. 40, 1185–1193.

Turner, R.E., Rabalais, N.N., 1994. Coastal eutrophication near the Mississippi River Delta. Nature 368, 619–621.

Turner, J., Johnson, D.W., Lambert, M.J., 1980. Sulfur cycling in a Douglas fir forest and its modification by nitrogen applications. Acta Oecol.-Oecol. Plantar. 1, 27–35.

Turner, D.P., Koerper, G.J., Harmon, M.E., Lee, J.J., 1995. A carbon budget for forests of the conterminous United States. Ecol. Appl. 5, 421–436.

Turner, S.M., Nightingale, P.D., Spokes, L.J., Liddicoat, M.I., Liss, P.S., 1996. Increased dimethyl sulphide concentrations in sea water from in situ Iron enrichment. Nature 383, 513–517.

Turner, B.L., Condron, L.M., Richardson, S.J., Peltzer, D.A., Allison, V.J., 2007. Soil organic phosphorus transformations during Pedogenesis. Ecosystems 10, 1166–1181.

Turner, B.L., Wells, A., Condron, L.M., 2014. Soil organic phosphorus transformations along a coastal dune chronosequence under New Zealand temperate rain forest. Biogeochemistry 121, 595–611.

Turner, A.J., Frankenberg, C., Kort, E.A., 2019. Interpreting contemporary trends in atmospheric methane. Proc. Natl. Acad. Sci. U. S. A. 116 (8), 2805–2813.

Turnipseed, A.A., Huey, L.G., Nemitz, E., Stickel, R., Higgs, J., Tanner, D.J., Slusher, D.L., Sparks, J.P., Flocke, F., Guenther, A., 2006. Eddy covariance fluxes of peroxyacetyl nitrates (PANs) and NOy to a coniferous Forest. J. Geophys. Res. 111.

Turpault, M.-P., Calvaruso, C., Kirchen, G., Redon, P.-O., Cochet, C., 2018. Contribution of fine tree roots to the silicon cycle in a temperate forest ecosystem developed on three soil types. Biogeosciences 15 (7), 2231–2249.

Turtle, E.P., Perry, J.E., Hayes, A.G., Lorenz, R.D., Barnes, J.W., McEwen, A.S., West, R.A., et al., 2011. Rapid and extensive surface changes near Titan's equator: evidence of April showers. Science 331 (6023), 1414–1417.

Turunen, J., Tahvanainen, T., Tolonen, K., Pitkanen, A., 2001. Carbon accumulation in West Siberian mires, Russia. Glob. Biogeochem. Cycles 15, 285–296.

Turunen, J., Tomppo, E., Tolonen, K., Reinikainen, A., 2002. Estimating carbon accumulation rates of undrained mires in Finland—application to boreal and subarctic regions. The Holocene 12, 69–80.

Turunen, J., Roulet, N.T., Moore, T.R., 2004. Nitrogen deposition and increased carbon accumulation in ombrotrophic Peatlands in eastern Canada. Glob. Biogeochem. Cycles 18 (3), 1–12.

Twilley, R.R., Chen, R.H., Hargis, T., 1992. Carbon sinks in mangroves and their implications to the carbon budget of tropical coastal ecosystems. Water Air Soil Pollut. 64, 265–288.

Tyler, G., 1994. A new approach to understanding the calcifuge habit of plants. Ann. Bot. 73, 327–330.

Tyler, G., Ström, L., 1995. Differing organic acid exudation pattern explains calcifuge and acidifuge behaviour of plants. Ann. Bot. 75 (1), 75–78.

Tyrrell, T., 2013. On Gaia: A Critical Investigation of the Relationship Between Life and Earth. Princeton University Press.

Uehara, G., Gillman, G., 1981. The Mineralogy, Chemistry, and Physics of Tropical Soils With Variable Charge Clays. Westview Press.

Uematsu, M., Wang, Z.F., Uno, I., 2003. Atmospheric input of mineral dust to the western North Pacific region based on direct measurements and a regional chemical transport model. Geophys. Res. Lett. 30.

Ueno, Y., Yamada, K., Yoshida, N., Maruyama, S., Isozaki, Y., 2006. Evidence from fluid inclusions for microbial methanogenesis in the early Archaean era. Nature 440 (7083), 516–519.

Ugolini, F.C., Sletten, R.S., 1991. The role of proton donors in pedogenesis as revealed by soil solution studies. Soil Sci. 151, 59–75.

Ugolini, F.C., Dawson, H., Zachara, J., 1977a. Direct evidence of particle migration in the soil solution of a Podzol. Science 198 (4317), 603–605.

Ugolini, F.C., Minden, R., Dawson, H., Zachara, J., 1977b. Example of soil processes in the Abies Amabilis zone of the central cascades, Washington. Soil Sci. 124, 291–302.

Ugolini, F.C., Stoner, M.G., Marrett, D.J., 1987. Arctic Pedogenesis: I. Evidence for contemporary Podzolization. Soil Sci. 144, 90–100.

Uhl, C., Jordan, C.F., 1984. Succession and nutrient dynamics following forest cutting and burning in Amazonia. Ecology 65, 1476–1490.

Ullah, S., Moore, T.R., 2011. Biogeochemical controls on methane, nitrous oxide, and carbon dioxide fluxes from deciduous Forest soils in eastern Canada. J. Geophys. Res.—Biogeosci. 116.

UN Environment, 2019. Global Mercury Assessment 2018. UN Environment Programme. Chemicals and Health Branch, Geneva, Switzerland. ISBN: 978-92-807-3744-8.

UNEP (United Nations Environment Programme), 2013. Global Mercury Assessment 2013—Sources, Emissions, Releases and Environmental Transport. UNEP, Chemicals Branch, Geneva, Switzerland.

United Nations, 2010. World population prospects: the 2008 revision, volume 1: comprehensive tables and United Nations, world population prospects: the 2008 revision, highlights. Popul. Dev. Rev. 36 (4), 854–855.

Uno, I., Eguchi, K., Yumimoto, K., Takemura, T., Shimizu, A., Uematsu, M., Liu, Z.Y., Wang, Z.F., Hara, Y., Sugimoto, N., 2009. Asian dust transported one full circuit around the globe. Nat. Geosci. 2, 557–560.

Unrau, P.J., Bartel, D.P., 1998. RNA-catalysed nucleotide synthesis. Nature 395 (6699), 260–263.

Updegraff, K., Bridgham, S.D., Pastor, J., Weishampel, P., Harth, C., 2001. Response of CO_2 and CH_4 emissions from Peatlands to warming and water table manipulation. Ecol. Appl. 11 (2), 311–326.

Urban, N.R., Bayley, S.E., Eisenreich, S.J., 1989a. Export of dissolved organic carbon and acidity from Peatlands. Water Resour. Res. 25 (7), 1619–1628.

Urban, N.R., Eisenreich, S.J., Grigal, D.F., 1989b. Sulfur cycling in a forested Sphagnum bog in northern Minnesota. Biogeochemistry 7, 81–109.

Urban, N.R., Brezonik, P.L., Baker, L.A., Sherman, L.A., 1994. Sulfate reduction and diffusion in sediments of Little Rock Lake, Wisconsin. Limnol. Oceanogr. 39, 797–815.

Ushio, M., Fujiki, Y., Hidaka, A., Kitayama, K., 2015. Linkage of root physiology and morphology as an adaptation to soil phosphorus impoverishment in tropical Montane forests. Poorter, L. (Ed.), Funct. Ecol. 29 (9), 1235–1245.

Uusitalo, M., Kitunen, V., Smolander, A., 2008. Response of C and N transformations in birch soil to coniferous resin volatiles. Soil Biol. Biochem. 40 (10), 2643–2649.

Uz, B.M., Yoder, J.A., Osychny, V., 2001. Pumping of nutrients to ocean surface waters by the action of propagating planetary waves. Nature 409 (6820), 597–600.

Vadas, R.L., Wright, W.A., Beal, B.F., 2004. Biomass and productivity of intertidal rockweeds (*Ascophyllum nodosum* LeJolis) in Cobscook Bay. Northeast. Nat. 11, 123–142.

Vadstein, O., Olsen, Y., Reinertsen, H., Jensen, A., 1993. The role of planktonic bacteria in phosphorus cycling in lakes—sink and link. Limnol. Oceanogr. 38, 1539–1544.

Vahtera, E., Conley, D.J., Gustafsson, B.G., Kuosa, H., Pitkänen, H., Savchuk, O.P., Tamminen, T., et al., 2007. Internal ecosystem feedbacks enhance nitrogen-fixing cyanobacteria blooms and complicate management in the Baltic Sea. Ambio 36 (2–3), 186–194.

Valentine, D.W., Holland, E.A., Schimel, D.S., 1994. Ecosystem and physiological controls over methane production in Northern Wetlands. J. Geophys. Res. Agri. Handb. 99 (D1), 1563.

Valentini, R., Matteucci, G., Dolman, A.J., Schulze, E.D., Rebmann, C., Moors, E.J., Granier, A., et al., 2000. Respiration as the main determinant of carbon balance in European forests. Nature 404 (6780), 861–865.

Vallina, S.M., Simó, R., 2007. Strong relationship between DMS and the solar radiation dose over the global Surface Ocean. Science 315 (5811), 506–508.

Van Bodegom, P.M., Broekman, R., Van Dijk, J., Bakker, C., Aerts, R., 2005. Ferrous iron stimulates phenol oxidase activity and organic matter decomposition in waterlogged wetlands. Biogeochemistry 76 (1), 69–83.

van Breemen, N., 1995. How sphagnum bogs down other plants. Trends Ecol. Evol. 10 (7).

van Breemen, N., Burrough, P.A., Velthorst, E.J., Vandobben, H.F., Dewit, T., Ridder, T.B., Reijnders, H.F.R., 1982. Soil acidification from atmospheric ammonium sulfate in forest canopy throughfall. Nature 299, 548–550.

van Breemen, N., Finlay, R., Lundstrom, U., Jongmans, A.G., Giesler, R., Olsson, M., 2000. Mycorrhizal weathering: a true case of mineral plant nutrition? Biogeochemistry 49, 53–67.

van Breemen, N., Boyer, E.W., Goodale, C.L., Jaworski, N.A., Paustian, K., Seitzinger, S.P., Lajtha, K., et al., 2002. Where did all the nitrogen go? Fate of nitrogen inputs to large watersheds in the Northeastern USA. Biogeochemistry 57, 267–293.

Van Cappellen, P., Ingall, E.D., 1994. Benthic phosphorus regeneration, net primary production, and ocean anoxia—a model of the coupled marine biogeochemical cycles of carbon and phosphorus. Paleoceanography 9 (5), 677–692.

Van Cappellen, P., Ingall, E.D., 1996. Redox stabilization of the atmosphere and oceans by phosphorus-limited marine productivity. Science 271 (January), 493–496.

Van Cappellen, P., Dixit, S., van Beusekom, J., 2002. Biogenic silica dissolution in the oceans: reconciling experimental and field-based dissolution rates. Glob. Biogeochem. Cycles 16.

Van Cleve, K., White, R., 1980. Forest-floor nitrogen dynamics in a 60-year old paper birch ecosystem in interior Alaska. Plant Soil 54, 359–381.

Van Cleve, K., Barney, R., Schlentner, R., 1981. Evidence of temperature control of production and nutrient cycling in two interior Alaska black spruce ecosystems. Can. J. Forest Res. 11, 258–273.

Van Cleve, K., Oechel, W.C., Hom, J.L., 1990. Response of black spruce (*Picea mariana*) ecosystems to soil temperature modification in interior Alaska. Can. J. Forest Res. 20, 1530–1535.

Van Damme, M., Erisman, J.W., Clarisse, L., Dammers, E., Whitburn, S., Clerbaux, C., Dolman, A.J., Coheur, P.-F., 2015. Worldwide spatiotemporal atmospheric ammonia (NH_3) columns variability revealed by satellite: NH_3 spatiotemporal variability. Geophys. Res. Lett. 42 (20), 8660–8668.

Van Damme, M., Clarisse, L., Whitburn, S., Hadji-Lazaro, J., Hurtmans, D., Clerbaux, C., Coheur, P.-F., 2018. Industrial and agricultural ammonia point sources exposed. Nature 564 (7734), 99–103.

Van de Water, P.K., Leavitt, S.W., Betancourt, J.L., 1994. Trends in stomatal density and $^{13}C/^{12}C$ ratios of *Pinus flexilis* needles during last glacial-interglacial cycle. Science 264 (5156), 239–243.

van den Broeke, M., Bamber, J., Ettema, J., Rignot, E., Schrama, E., van de Berg, W.J., van Meijgaard, E., Velicogna, I., Wouters, B., 2009. Partitioning recent greenland mass loss. Science 326 (5955), 984–986.

Van den Driessche, R., 1974. Prediction of mineral nutrient status of trees by foliar analysis. Bot. Rev. Interpr. Bot. Progr. 40 (3), 347–394.

van der Ent, R.J., Savenije, H.H.G., Schaefli, R., Steele, S.C.D., 2010. Origin and fate of atmospheric moisture over continents. Water Resour. Res. 46.

Van der Hoven, S.J., Quade, J., 2002. Tracing spatial and temporal variations in the sources of calcium in pedogenic carbonates in a semiarid environment. Geoderma 108 (3), 259–276.

Van Devender, T.R., Spaulding, W.G., 1979. Development of vegetation and climate in the southwestern United States. Science 204 (4394), 701–710.

Van Dijk, A., Dolman, A.J., 2004. Estimates of CO_2 uptake and release among European forests based on Eddy covariance data. Glob. Chang. Biol. 10, 1445–1459.

Van Dijk, S.M., Duyzer, J.H., 1999. Nitric oxide emissions from forest soils. J. Geophys. Res. 104, 15955–15961.

van Groenigen, K.-J., Six, J., Hungate, B.A., de Graaff, M.-A., van Breemen, N., van Kessel, C., 2006. Element interactions limit soil carbon storage. Proc. Natl. Acad. Sci. U. S. A. 103 (17), 6571–6574.

van Groenigen, K.J., Osenberg, C.W., Hungate, B.A., 2011. Increased soil emissions of potent greenhouse gases under increased atmospheric CO_2. Nature 475 (7355), 214–216.

van Hees, P.A.W., Jones, D.L., Finlay, R., Godbold, D.L., Lundstomd, U.S., 2005. The carbon we do not see—the impact of low molecular weight compounds on carbon dynamics and respiration in forest soils: a review. Soil Biol. Biochem. 37, 1–13.

Van Houten, F.B., 1973. Origin of red beds: a review. Annu. Rev. Earth Planet. Sci. 1, 39–61.

van Keken, P.E., Hacker, B.R., Syracuse, E.M., Abers, G.A., 2011. Subduction factory: 4. Depth-dependent flux of H_2O from subducting slabs worldwide. J. Geophys. Res. Subduct.: Top Bot. 116 (B1), 387.

van Kessel, C., Farrell, R.E., Roskoski, J.P., Keane, K.M., 1994. Recycling of the naturally occurring ^{15}N in an established stand of Leucaena Leucocephala. Soil Biol. Biochem. 26, 757–762.

van Kessel, M.A.H.J., Speth, D.R., Albertsen, M., Nielsen, P.H., Op den Camp, H.J.M., Kartal, B., Jetten, M.S.M., Lücker, S., 2015. Complete nitrification by a single microorganism. Nature 528 (7583), 555–559.

van Lent, J., Hergoualc'h, K., Verchot, L.V., 2015. Reviews and syntheses: soil N_2O and NO emissions from land use and land-use change in the tropics and subtropics: a meta-analysis. Biogeosciences 12 (23), 7299–7313.

van Loon, H., Rogers, J.C., 1978. The seesaw in winter temperatures between Greenland and northern Europe. Part I: General description. Mon. Weather Rev. 106 (3), 296–310.

Van Meter, K.J., Van Cappellen, P., Basu, N.B., 2018. Legacy nitrogen may prevent achievement of water quality goals in the Gulf of Mexico. Science 360 (6387), 427–430.

Van Oost, K., Quine, T.A., Govers, G., De Gryze, S., Six, J., Harden, J.W., Ritchie, J.C., et al., 2007. The impact of agricultural soil erosion on the global carbon cycle. Science 318 (5850), 626–629.

van Scholl, L., Kuyper, T.W., Smits, M.M., Landeweert, R., Hoffland, E., van Breemen, N., 2008. Rock-eating mycorrhizas: their role in plant nutrition and biogeochemical cycles. Plant Soil 303, 35–47.

Van Sickle, J., 1981. Long-term distributions of annual sediment yields from small watersheds. Water Resour. Res. 17, 659–663.

Van Trump, J.E., Miller, S.L., 1972. Prebiotic synthesis of methionine. Science 178 (4063), 859–860.

van Vliet, J., Eitelberg, D.A., Verburg, P.H., 2017. A global analysis of land take in cropland areas and production displacement from urbanization. Glob. Environ. Change: Hum. Policy Dimens. 43 (March), 107–115.

van Wijngaarden, W.A., Syed, A., 2015. Changes in annual precipitation over the Earth's land mass excluding Antarctica from the 18th century to 2013. J. Hydrol. 531 (December), 1020–1027.

van Zuilen, M.A., Lepland, A., Arrhenius, G., 2002. Reassessing the evidence for the earliest traces of life. Nature 418 (6898), 627–630.

Vance, G.F., David, M.B., 1991. Chemical characteristics and acidity of soluble organic substances from a northern hardwood Forest floor, Central Maine, USA. Geochim. Cosmochim. Acta 55, 3611–3625.

Vance, D., Teagle, D.A.H., Foster, G.L., 2009. Variable quaternary chemical weathering fluxes and imbalances in marine geochemical budgets. Nature 458 (7237), 493–496.

Vandaele, A.C., Korablev, O., Belyaev, D., Chamberlain, S., Evdokimova, D., Th, E., Esposito, L., et al., 2017. Sulfur dioxide in the Venus atmosphere: I. Vertical distribution and variability. Icarus 295 (October), 16–33.

Vandal, G.M., Fitzgerald, W.F., Boutron, C.F., Candelone, J.P., 1993. Variations in mercury deposition to Antarctica over the past 34,000 years. Nature 362, 621–623.

Vandenberg, J.J., Knoerr, K.R., 1985. Comparison of surrogate surface techniques for estimation of sulfate dry deposition. Atmos. Environ. 19, 627–635.

Van Drecht, G., Bouwman, A.F., Knoop, J.M., Beusen, A.H.W., Meinardi, C.R., 2003. Global modeling of the fate of nitrogen from point and non-point sources in soils, groundwater and surface water. Glob. Biogeochem. Cycles 17, 1115.

Vanmaercke, M., Poesen, J., Govers, G., Verstraeten, G., 2015. Quantifying human impacts on catchment sediment yield: a continental approach. Glob. Planet. Chang. 130 (July), 22–36.

Vann, C.D., Megonigal, J.P., 2003. Elevated CO_2 and water-depth regulation of methane emissions: comparison of woody and non-woody wetland plant species. Biogeochemistry 63, 117–134.

Vanni, M.J., 2002. Nutrient cycling by animals in freshwater ecosystems. Annu. Rev. Ecol. Syst. 33, 341–370.

Vanni, M.J., Boros, G., McIntyre, P.B., 2013. When are fish sources vs. sinks of nutrients in lake ecosystems? Ecology 94 (10), 2195–2206.

Vannote, R.L., et al., Minshall, G.W., Cummins, K.W., Sedell, J.R., Cushing, C.E., 1980. The river continuum concept. Can. J. Fish. Aquat. Sci. 37 (1), 130–137.

Varshney, C.K., Singh, A.P., 2003. Isoprene emission from Indian trees. J. Geophys. Res. 108 (D24).

Vasconcelos, C., McKenzie, J.A., Bernasconi, S., Grujic, D., Tien, A.J., 1995. Microbial mediation as a possible mechanism for natural dolomite formation at low temperatures. Nature 377, 220–222.

Vasileva, A.V., Moiseenko, K.B., Mayer, J.C., Jurgens, N., Panov, A., Heimann, M., Andreae, M.O., 2011. Assessment of the regional atmospheric impact of wildfire emissions based on CO observations at the ZOTTO tall tower station in Central Siberia. J. Geophys. Res. 116.

Velbel, M.A., 1990. Mechanisms of saprolitization, isovolumetric weathering, and pseudomorphous replacement during rock weathering—a review. Chem. Geol. 84, 17–18.

Velbel, M.A., 1992. Geochemical mass balances and weathering rates in forested watersheds of the southern blue Ridge. III. Cation budgets and the weathering rate of amphibole. Am. J. Sci. 292, 58–78.

Velders, G.J.M., Snijder, A., Hoogerbrugge, R., 2011. Recent decreases in observed atmospheric concentrations of SO_2 in the Netherlands in line with emission reductions. Atmos. Environ. 45, 5647–5651.

Ven, A., Verlinden, M., Verbruggen, E., Vicca, S., 2019. Experimental evidence that phosphorus fertilization and arbuscular mycorrhizal symbiosis can reduce the carbon cost of phosphorus uptake. Funct. Ecol. 33, 2215–2225.

Venkatesan, A.K., Hamdan, A.-H.M., Chavez, V.M., Brown, J.D., Halden, R.U., 2016. Mass balance model for sustainable phosphorus recovery in a US wastewater treatment plant. J. Environ. Qual. 45 (1), 84–89.

Venterea, R.T., Rolston, D.E., 2000. Mechanisms and kinetics of nitric and nitrous oxide production during nitrification in agricultural soil. Glob. Chang. Biol. 6, 303–316.

Venterea, R.T., Groffman, P.M., Verchot, L.V., Magill, A.H., Aber, J.D., Steudler, P.A., 2003. Nitrogen oxide gas emissions from temperate forest soils receiving long-term nitrogen inputs. Glob. Chang. Biol. 9, 346–357.

Venterink, H.O., Van der Vliet, R.E., Wassen, M.J., 2001. Nutrient limitation along a productivity gradient in wet meadows. Plant Soil, 171–179.

Verchot, L.V., Davidson, E.A., Cattanio, J.H., Ackerman, I.L., Erickson, H.E., Keller, M., 1999. Land use change and biogeochemical controls of nitrogen oxide emissions from soils in eastern Amazonia. Glob. Biogeochem. Cycles 13, 31–46.

Vergutz, L., Manzoni, S., Porporato, A., Novais, R.F., Jackson, R.B., 2012. Global resorption efficiencies and concentrations of carbon and nutrients in leaves of terrestrial plants. Ecol. Monogr. 82, 205–220.

Verhoeven, J.T.A., Arheimer, B., Yin, C.Q., Hefting, M.M., 2006. Regional and global concerns over wetlands and water quality. Trends Ecol. Evol. 21 (2), 96–103.

Verhulst, K.R., Aydin, M., Saltzman, E.S., 2013. Methyl chloride variability in the Taylor Dome ice core during the Holocene: Taylor Dome methyl chloride. J. Geophys. Res. D: Atmos. 118 (21), 12218–12228.

Vernadsky, V., 1998. The Biosphere. Copernicus.

Verpoorter, C., Kutser, T., Seekell, D.A., 2014. A global inventory of lakes based on high-resolution satellite imagery. Geophys. Res. Lett.

Verstraten, J.M., Dopheide, J.C.R., Duysings, J., Tietema, A., Bouten, W., 1990. The proton cycle of a deciduous Forest ecosystem in the Netherlands and its implications for soil acidification. Plant Soil 127, 61–69.

Vico, G., Way, D.A., Hurry, V., Manzoni, S., 2019. Can leaf net photosynthesis acclimate to rising and more variable temperatures? Plant Cell Environ. 42 (6), 1913–1928.

Vidon, P., Hill, A.R., 2004. Denitrification and patterns of electron donors and acceptors in eight riparian zones with contrasting hydrogeology. Biogeochemistry 1997, 259–283.

Viereck, L.A., 1966. Plant succession and soil development on gravel outwash of the Muldrow glacier, Alaska. Ecol. Monogr. 36, 181–199.

Viers, J., Dupré, B., Gaillardet, J., 2009. Chemical composition of suspended sediments in world rivers: new insights from a new database. Sci. Total Environ. 407 (2), 853–868.

Vile, M.A., Bridgham, S.D., Kelman Wieder, R., Novak, M., 2003. Atmospheric sulfur deposition alters pathways of gaseous carbon production in Peatlands. Glob. Biogeochem. Cycles 17 (2), 1–7.

Villanueva, G.L., Mumma, M.J., Novak, R.E., Radeva, Y.L., Käufl, H.U., Smette, A., Tokunaga, A., Khayat, A., Encrenaz, T., Hartogh, P., 2013. A sensitive search for organics (CH_4, CH_3OH, H_2CO, C_2H_6, C_2H_2, C_2H_4), hydroperoxyl (HO_2), nitrogen compounds (N_2O, NH_3, HCN) and chlorine species (HCl, CH_3Cl) on Mars using ground-based high-resolution infrared spectroscopy. Icarus 223 (1), 11–27.

Villanueva, G.L., Mumma, M.J., Novak, R.E., Käufl, H.U., Hartogh, P., Encrenaz, T., Tokunaga, A., Khayat, A., Smith, M.D., 2015. Strong water isotopic anomalies in the Martian atmosphere: probing current and ancient reservoirs. Science 348 (6231), 218–221.

Villareal, T.A., Pilskaln, C., Brzezinski, M., Lipschultz, F., Dennett, M., Gardner, G.B., 1999. Upward transport of oceanic nitrate by migrating diatom mats. Nature 397 (6718), 423–425.

Vilmin, L., Mogollón, J.M., Beusen, A.H.W., Bouwman, A.F., 2018. Forms and subannual variability of nitrogen and phosphorus loading to global river networks over the 20th century. Glob. Planet. Chang. 163 (April), 67–85.

Virginia, R.A., Delwiche, C.C., 1982. Natural ^{15}N abundance of presumed N_2-fixing and non-N_2-fixing plants from selected ecosystems. Oecologia 54 (3), 317–325.

Virginia, R.A., Jarrell, W.M., 1983. Soil properties in a mesquite-dominated Sonoran Desert ecosystem. Soil Sci. Soc. Am. J. 47, 138–144.

Virginia, R.A., Jarrell, W.M., Francovizcaino, E., 1982. Direct measurement of denitrification in a *Prosopis* (mesquite)-dominated Sororan Desert ecosystem. Oecologia 53, 120–122.

Vitousek, P.M., 1977. Regulation of element concentrations in mountain streams in the northeastern United States. Ecol. Monogr. 47, 65–87.

Vitousek, P.M., 1994. Beyond global warming: ecology and global change. Ecology 75, 1861–1876.

Vitousek, P.M., 2004. Nutrient Cycling and Limitation. Princeton University Press.

Vitousek, P.M., Howarth, R.W., 1991. Nitrogen limitation on land and in the sea: how can it occur? Biogeochemistry 13, 87–115.

Vitousek, P.M., Matson, P.A., 1984. Mechanisms of nitrogen retention in forest ecosystems: a field experiment. Science 225 (4657), 51–52.

Vitousek, P.M., Matson, P.A., 1988. Nitrogen transformations in a range of tropical forest soils. Soil Biol. Biochem. 20, 361–367.

Vitousek, P.M., Melillo, J.M., 1979. Nitrate losses from disturbed forests: patterns and mechanisms. For. Sci. 25, 605–619.

Vitousek, P.M., Reiners, W.A., 1975. Ecosystem succession and nutrient retention: a hypothesis. Bioscience 25, 376–381.

Vitousek, P.M., Sanford, R.L., 1986. Nutrient cycling in moist tropical forest. Annu. Rev. Ecol. Syst. 17, 137–167.

Vitousek, P.M., Gosz, J.R., Grier, C.C., Melillo, J.M., Reiners, W.A., 1982. A comparative analysis of potential nitrification and nitrate mobility in forest ecosystems. Ecol. Monogr. 52, 155–177.

Vitousek, P.M., Ehrlich, P.R., Ehrlich, A.H., Matson, P.A., 1986. Human appropriation of the products of photosynthesis. Bioscience 36, 368–373.

Vitousek, P.M., Walker, L.R., Whiteaker, L.D., Mueller-Dombois, D., Matson, P.A., 1987. Biological invasion by *Myrica faya* alters ecosystem development in Hawaii. Science 238 (4828), 802–804.

Vitousek, P.M., Fahey, T., Johnson, D.W., Swift, M.J., 1988. Element interactions in forest ecosystems: succession, allometry and input-output budgets. Biogeochemistry 5, 7–34.

Vitousek, P.M., Aber, J.D., Howarth, R.W., Likens, G.E., Matson, P.A., Schindler, D.W., Schlesinger, W.H., Tilman, D.G., 1997. Human alteration of the global nitrogen cycle: sources and consequences. Ecol. Appl. 7, 737–750.

Vitousek, P.M., Ladefoged, T.N., Kirch, P.V., Hartshorn, A.S., Graves, M.W., Hotchkiss, S.C., Tuljapurkar, S., Chadwick, O.A., 2004. Soils, agriculture, and society in Precontact Hawai. Science 304, 1665–1669.

Vitousek, P.M., Naylor, R., Crews, T., David, M.B., Drinkwater, L.E., Holland, E., Johnes, P.J., et al., 2009. Agriculture. Nutrient imbalances in agricultural development. Science 324 (5934), 1519–1520.

Vitousek, P.M., Menge, D.N.L., Reed, S.C., Cleveland, C.C., 2013. Biological nitrogen fixation: rates, patterns and ecological controls in terrestrial ecosystems. Philos. Trans. R. Soc. Lond. Ser. B Biol. Sci. 368 (1621), 20130119.

Vitt, D.H., Halsey, L.A., Bauer, I.E., Campbell, C., 2000. Spatial and temporal trends in carbon storage of Peatlands of continental Western Canada through the Holocene. Can. J. Earth Sci. 37, 683–693.

Vo, A.-T.E., Bank, M.S., Shine, J.P., Edwards, S.V., 2011. Temporal increase in organic mercury in an endangered pelagic seabird assessed by century-old museum specimens. Proc. Natl. Acad. Sci. U. S. A. 108 (18), 7466–7471.

Vogt, K.A., Grier, C.C., Meier, C.E., Edmonds, R.L., 1982. Mycorrhizal role in net primary production and nutrient cycling in Abies amabilis ecosystems in western Washington. Ecology 63, 370–380.

Vogt, K.A., Grier, C.C., Meier, C.E., Keyes, M.R., 1983. Organic matter and nutrient dynamics in forest floors of young and mature *Abies amabilis* stands in Western Washington, as affected by fine root input. Ecol. Monogr. 53, 139–157.

Vogt, K.A., Grier, C.C., Vogt, D.J., 1986. Production, turnover, and nutrient dynamics of aboveground and belowground detritus of world forests. Adv. Ecol. Res. 15, 303–377.

Vogt, K.A., Vogt, D.J., Palmiotto, P.A., Boon, P., Ohara, J., Asbjornsen, H., 1996. Review of root dynamics in forest ecosystems grouped by climate, climatic forest type and species. Plant Soil 187, 159–219.

Vogt, K.A., Vogt, D.J., Bloomfield, J., 1998. Analysis of some direct and indirect methods for estimating root biomass and production of forests at an ecosystem level. Plant Soil 200, 71–89.

Voigt, C., Marushchak, M.E., Mastepanov, M., Lamprecht, R.E., Christensen, T.R., Dorodnikov, M., Jackowicz-Korczyński, M., et al., 2019. Ecosystem carbon response of an Arctic peatland to simulated permafrost thaw. Glob. Chang. Biol. 25 (5), 1746–1764.

Volk, T., 1998. Gaia's Body. Springer/Copernicus.

Volk, T., Hoffert, M.I., 1985. Ocean carbon pumps: analysis of relative strengths and efficiencies in ocean-driven atmospheric CO_2 changes. Carbon Cycle Atmos. CO_2: Nat. Variat. Archean Present 32, 99–110.

Vollenweider, R.A., 1976. Advances in defining critical loading levels for phosphorus in Lake eutrophication. Memorie dell'Istituto Italiano Di Idrobiologia 33, 53–83.

Volz, A., Kley, D., 1988. Evaluation of the Montsouris series of ozone measurements made in the 19th-century. Nature 332, 240–242.

von Fischer, J.C., Rhew, R.C., Ames, G.M., Fosdick, B.K., von Fischer, P.E., 2010. Vegetation height and other controls of spatial variability in methane emissions from the arctic coastal tundra at Barrow, Alaska. J. Geophys. Res. 115 (September), 1–11.

Voroney, R.P., Paul, E.A., 1984. Determination of kc and Kn in situ for calibration of the chloroform fumigation-incubation method. Soil Biol. Biochem. 16, 9–14.

Vorosmarty, C.J., Sahagian, D., 2000. Anthropogenic disturbance of the terrestrial water cycle. Bioscience 50, 753–765.

Vörösmarty, C.J., Berrien Moore, I.I.I., Grace, A.L., Patricia Gildea, M., Melillo, J.M., Peterson, B.J., Rastetter, E.B., Steudler, P.A., 1989. Continental scale models of water balance and fluvial transport: an application to South America. Glob. Biogeochem. Cycles Monogr. 3 (3), 241–265.

Vorosmarty, C.J., Sharma, K.P., Fekete, B.M., Copeland, A.H., Holden, J., Marble, J., Lough, J.A., 1997. The storage and aging of continental runoff in large reservoir systems of the world. Ambio 26 (4), 210–219.

Vörösmarty, C.J., Green, P., Salisbury, J., Lammers, R.B., 2000. Global water resources: vulnerability from climate change and population growth. Science 289 (5477), 284–288.

Vorosmarty, C.J., Green, P., Salisbury, J., Lammers, R.B., Vörösmarty, C.J., 2000. Global water resources: vulnerability from climate change acid population growth. Science 289 (5477), 284–288.

Vörösmarty, C.J., Meybeck, M., Fekete, B., Sharma, K., Green, P., Syvitski, J.P.M., 2003. Anthropogenic sediment retention: major global impact from registered river impoundments. Glob. Planet. Chang. 39 (1), 169–190.

Vose, J.M., Ryan, M.G., 2002. Seasonal respiration of foliage, fine roots, and woody tissues in relation to growth, tissue N, and photosynthesis. Glob. Chang. Biol. 8, 182–193.

Voss, M., Dippner, J.W., Montoya, J.P., 2001. Nitrogen isotope patterns in the oxygen-deficient waters of the eastern tropical North Pacific Ocean. Deep-Sea Res. Part I—Oceanogr. Res. Pap. 48, 1905–1921.

Voss, M., Bange, H.W., Dippner, J.W., Middelburg, J.J., Montoya, J.P., Ward, B., 2013. The marine nitrogen cycle: recent discoveries, uncertainties and the potential relevance of climate change. Philos. Trans. R. Soc. Lond. Ser. B Biol. Sci. 368 (1621), 20130121.

Vossbrinck, C.R., Coleman, D.C., Woolley, T.A., 1979. Abiotic and biotic factors in litter decomposition in a semiarid grassland. Ecology 60, 265–271.

Vrede, T., Tranvik, L.J., 2006. Iron constraints on planktonic primary production in oligotrophic lakes. Ecosystems 9, 1094–1105.

Vuichard, N., Ciais, P., Belelli, L., Smith, P., Valentini, R., 2008. Carbon sequestration due to the abandonment of agriculture in the former USSR since 1990. Glob. Biogeochem. Cycles. 22.

Wackett, L.P., Dodge, A.G., Ellis, L.B.M., 2004. Microbial genomics and the periodic table. Appl. Environ. Microbiol. 70 (2), 647–655.

Wada, Y., van Beek, L.P.H., van Kempen, C.M., Reckman, J., Vasak, S., Bierkens, M.F.P., 2010. Global depletion of groundwater resources. Geophys. Res. Lett. 37.

Wada, Y., van Beek, L.P.H., Bierkens, M.F.P., 2012. Nonsustainable groundwater sustaining irrigation: a global assessment. Water Resour. Res. 48.

Wade, J., Dyck, B., Palin, R.M., Moore, J.D.P., Smye, A.J., 2017. The divergent fates of primitive hydrospheric water on earth and Mars. Nature 552 (7685), 391–394.

Wade, A.M., Richter, D.D., Medjibe, V.P., Bacon, A.R., Heine, P.R., White, L.J.T., Poulsen, J.R., 2019. Estimates and determinants of stocks of deep soil carbon in Gabon, Central Africa. Geoderma 341 (May), 236–248.

Wadhams, P., Holfort, J., Hansen, E., Wilkinson, J.P., 2002. A deep convective chimney in the winter Greenland Sea. Geophys. Res. Lett. 29.

Wagener, S.M., Oswood, M.W., Schimel, J.P., 1998. Rivers and soils: parallels in carbon and nutrient processing. Bioscience 48, 104–108.

Wahlen, M., Tanaka, N., Henry, R., Deck, B., Zeglen, J., Vogel, J.S., Southon, J., Shemesh, A., Fairbanks, R., Broecker, W., 1989. ^{14}C in methane sources and in atmospheric methane: the contribution from fossil carbon. Science 245, 286–290.

Waite, T.D., 2001. Thermodynamics of the iron system in seawater. In: The Biogeochemistry of Iron in Seawater.

Waite Jr., J.H., Combi, M.R., Ip, W.-H., Cravens, T.E., McNutt Jr., R.L., Kasprzak, W., Yelle, R., et al., 2006. Cassini ion and neutral mass spectrometer: *Enceladus plume* composition and structure. Science 311 (5766), 1419–1422.

Waite, J.H., Glein, C.R., Perryman, R.S., Teolis, B.D., Magee, B.A., Miller, G., Grimes, J., et al., 2017. Cassini finds molecular hydrogen in the *Enceladus plume*: evidence for hydrothermal processes. Science 356 (6334), 155–159.

Wakatsuki, T., Rasyidin, A., 1992. Rates of weathering and soil formation. Geoderma 52, 251–263.

Wakefield, A.E., Gotelli, N.J., Wittman, S.E., Ellison, A.M., 2005. Prey addition alters nutrient stoichiometry of the carnivorous plant *Sarracenia purpurea*. Ecology 86, 1737–1743.

Walbridge, M.R., Vitousek, P.M., 1987. Phosphorus mineralization potentials in acid organic soils—processes affecting PO$_4^3$ ^{32}P isotope-dilution measurements. Soil Biol. Biochem. 19, 709–717.

Walbridge, M.R., Richardson, C.J., Swank, W.T., 1991. Vertical-distribution of biological and geochemical phosphorus subcycles in two southern Appalachian forest soils. Biogeochemistry 13, 61–85.

Waldman, J.M., Munger, J.W., Jacob, D.J., Flagan, R.C., Morgan, J.J., Hoffmann, M.R., 1982. Chemical composition of acid fog. Science 218 (4573), 677–680.

Waldman, J.M., Munger, J.W., Jacob, D.J., Hoffmann, M.R., 1985. Chemical characterization of stratus cloudwater and its role as a vector for pollutant deposition in a Los Angeles pine forest. Tellus B—Chem. Phys. Meteorol. 37, 91–108.

Waldrop, M.P., Wickland, K.P., White, R., Berhe, A.A., Harden, J.W., Romanovsky, V.E., 2010. Molecular investigations into a globally important carbon pool: permafrost-protected carbon in Alaskan soils. Glob. Chang. Biol. 16, 2543–2554.

Walker, J.C.G., 1977. Evolution of the Atmosphere. Macmillan.

Walker, J.C.G., 1980. The oxygen cycle. In: Hutzinger, O. (Ed.), Handbook of Environmental Chemistry. The Natural Environment and the Biogeochemical Cycles, Part A. vol. 1. Springer-Verlag, pp. 87–104.

Walker, J.C.G., 1983. Possible limits on the composition of the Archean Ocean. Nature 302, 518–520.

Walker, J.C., 1984. How life affects the atmosphere. Bioscience 34 (8), 486–491.

Walker, J.C., 1985. Carbon dioxide on the early earth. Origins Life Evol. Biosph. 16, 117–127.

Walker, T.W., Adams, A.F.R., 1958. Studies on soil organic matter: I. Influence of phosphorus content of parent materials on accumulations of carbon, nitrogen, sulfur, and organic phosphorus in grassland soils. Soil Sci. 85, 307–318.

Walker, J.C., Brimblecombe, P., 1985. Iron and sulfur in the pre-biologic ocean. Precambrian Res. 28, 205–222.

Walker, T.W., Syers, J.K., 1976. Fate of phosphorus during pedogenesis. Geoderma 15, 1–19.

Walker, J.C.G., Klein, C., Schopf, S.M.J.W., Stevenson, D.J., Walter, M.R., 1983. Environmental evolution of the Archean-early proterozoic earth. In: Schopf, J.W. (Ed.), Earth's Earliest Biosphere. Princeton University Press, pp. 260–290.

Walker, T.W.N., Kaiser, C., Strasser, F., Herbold, C.W., Leblans, N.I.W., Woebken, D., Janssens, I.A., Sigurdsson, B.D., Richter, A., 2018. Microbial temperature sensitivity and biomass change explain soil carbon loss with warming. Nat. Clim. Chang. 8 (10), 885–889.

Walker, X.J., Baltzer, J.L., Cumming, S.G., Day, N.J., Ebert, C., Goetz, S., Johnstone, J.F., et al., 2019. Increasing wildfires threaten historic carbon sink of boreal forest soils. Nature 572 (7770), 520–523.

Wallace, J.B., Webster, J.R., 1996. The role of macroinvertebrates in stream ecosystem function. Annu. Rev. Entomol. 41, 115–139.

Wallace, A., Bamburg, S.A., Cha, J.W., 1974. Quantitative studies of roots of perennial plants in the Mojave Desert. Ecology 55, 1160–1162.

Wallace, J.B., Eggert, S.L., Meyer, J.L., Webster, J.R., 1997. Multiple trophic levels of a forest stream linked to terrestrial litter inputs. Science 277 (5322), 102–104.

Wallace, J.B., Eggert, S.L., Meyer, J.L., Webster, J.R., 1999. Effects of resource limitation on a detrital-based ecosystem. Ecol. Monogr. 69, 409–442.

Wallenstein, M.D., McNulty, S., Fernandez, I.J., Boggs, J., Schlesinger, W.H., 2006. Nitrogen fertilization decreases forest soil fungal and bacterial biomass in three long-term experiments. For. Ecol. Manag. 222, 459–468.

Wallerstein, G., 1988. Mixing in stars. Science 240, 1743–1750.

Walling, D.E., Fang, D., 2003. Recent trends in the suspended sediment loads of the World's Rivers. Glob. Planet. Chang. 39, 111–126.

Wallington, T.J., Wiesen, P., 2014. N_2O emissions from global transportation. Atmos. Environ. 94 (September), 258–263.

Wallmann, K., 2001. The geological water cycle and the evolution of Marine Delta ^{18}O values. Geochim. Cosmochim. Acta 65, 2469–2485.

Wallmann, K., 2010. Phosphorus imbalance in the global ocean? Glob. Biogeochem. Cycles 24.

Walsh, J.J., 1984. The role of ocean biota in accelerated ecological cycles: a temporal view. Bioscience 34, 499–507.

Walsh, J.J., 1991. Importance of continental margins in the marine biogeochemical cycling of carbon and nitrogen. Nature 350, 53–55.

Walsh, C.J., Roy, A.H., Feminella, J.W., Cottingham, P.D., Groffman, P.M., Morgan, R.P., 2005. The urban stream syndrome: current knowledge and the search for a cure. J. N. Am. Benthol. Soc. 24, 706–723.

Walter, M.T., Wilks, D.S., Parlange, J.Y., Schneider, R.L., 2004a. Increasing evapotranspiration from the conterminous United States. J. Hydrometeorol. 5, 405–408.

Walter, S., Bange, H.W., Wallace, D.W.R., 2004b. Nitrous oxide in the surface layer of the tropical North Atlantic Ocean along a west to east transect. Geophys. Res. Lett. 31, L23S07.

Walter, K.M., Zimov, S.A., Chanton, J.P., Verbyla, D., Chapin 3rd., F.S., 2006. Methane bubbling from Siberian thaw lakes as a positive feedback to climate warming. Nature 443 (7107), 71–75.

Walter, K.M., Edwards, M.E., Grosse, G., Zimov, S.A., Chapin 3rd, F.S., 2007. Thermokarst lakes as a source of atmospheric CH_4 during the last deglaciation. Science 318 (5850), 633–636.

Walter, M.J., Kohn, S.C., Araujo, D., Bulanova, G.P., Smith, C.B., Gaillou, E., Wang, J., Steele, A., Shirey, S.B., 2011. Deep mantle cycling of oceanic crust: evidence from diamonds and their mineral inclusions. Science 334 (6052), 54–57.

Walther, S., Voigt, M., Thum, T., Gonsamo, A., Zhang, Y., Köhler, P., Jung, M., Varlagin, A., Guanter, L., 2016. Satellite chlorophyll fluorescence measurements reveal large-scale decoupling of photosynthesis and greenness dynamics in boreal evergreen forests. Glob. Chang. Biol. 22 (9), 2979–2996.

Walvoord, M.A., Phillips, F.M., Stonestrom, D.A., Evans, R.D., Hartsough, P.C., Newman, B.D., Striegl, R.G., 2003. A reservoir of nitrate beneath desert soils. Science 302, 1021–1024.

Wan, S.Q., Hui, D.F., Luo, Y.Q., 2001. Fire effects on nitrogen pools and dynamics in terrestrial ecosystems: a meta-analysis. Ecol. Appl. 11, 1349–1365.

Wanek, W., Zotz, G., 2011. Are vascular epiphytes nitrogen or phosphorus limited? A study of plant (15)N fractionation and foliar N:P stoichiometry with the tank bromeliad *Vriesea sanguinolenta*. New Phytol. 192 (2), 462–470.

Wanek, W., Mooshammer, M., Bloechl, A., Hanreich, A., Richter, A., 2010. Determination of gross rates of amino acid production and immobilization in decomposing leaf litter by a novel ^{15}N-isotope pool dilution technique. Soil Biol. Biochem. 42, 1293–1302.

Wanek, W., Zezula, D., Wasner, D., Mooshammer, M., Prommer, J., 2019. A novel isotope pool dilution approach to quantify gross rates of key abiotic and biological processes in the soil phosphorus cycle. Biogeosciences 16 (15), 3047–3068.

Wang, Z.H.A., Cai, W.J., 2004. Carbon dioxide degassing and inorganic carbon export from a Marsh-dominated estuary (the Duplin River): a Marsh CO_2 pump. Limnol. Oceanogr. 49, 341–354.

Wang, F.Y., Chapman, P.M., 1999. Biological implications of sulfide in sediment—a review focusing on sediment toxicity. Environ. Toxicol. Chem. 18 (11), 2526–2532.

Wang, D.B., Hejazi, M., 2011. Quantifying the relative contribution of the climate and direct human impacts on mean annual streamflow in the contiguous United States. Water Resour. Res. 47, 16.

Wang, Y.P., Houlton, B.Z., 2009. Nitrogen constraints on terrestrial carbon uptake: implications for the global carbon-climate feedback. Geophys. Res. Lett. 36.

Wang, L., Macko, S.A., 2011. Constrained preferences in nitrogen uptake across plant species and environments. Plant Cell Environ. 34 (3), 525–534.

Wang, J., Yan, X., 2016. Denitrification in upland of China: magnitude and influencing factors: denitrification in Chinese upland soils. J. Geophys. Res. Biogeosci. Iss. Ecol. 121 (12), 3060–3071.

Wang, J.S., Logan, J.A., McElroy, M.B., Duncan, B.N., Megretskaia, I.A., Yantosca, R.M., 2004. A 3-D model analysis of the slowdown and interannual variability in the methane growth rate from 1988 to 1997. Glob. Biogeochem. Cycles 18.

Wang, Y., He, Y., Zhang, H., Schroder, J., Li, C., Zhou, D., 2008. Phosphate mobilization by citric, tartaric, and oxalic acids in a clay loam ultisol. Soil Sci. Soc. Am. J. 72, 1263–1268.

Wang, Y.P., Law, R.M., Pak, B., 2010a. A global model of carbon, nitrogen and phosphorus cycles for the terrestrial biosphere. Biogeosciences 7 (7), 2261–2282.

Wang, Z., Chappellaz, J., Park, K., Mak, J.E., 2010b. Large variations in southern hemisphere biomass burning during the last 650 years. Science 330 (6011), 1663–1666.

Wang, F., Sims, J.T., Ma, L., Ma, W., Dou, Z., Zhang, F., 2011. The phosphorus footprint of China's food chain: implications for food security, natural resource management, and environmental quality. J. Environ. Qual. 40 (4), 1081–1089.

Wang, D., Heckathorn, S.A., Wang, X., Philpott, S.M., 2012. A meta-analysis of plant physiological and growth responses to temperature and elevated CO_2. Oecologia 169, 1–13.

Wang, L., Good, S.P., Caylor, K.K., 2014. Global synthesis of vegetation control on evapotranspiration partitioning: vegetation and et partitioning. Geophys. Res. Lett. 41 (19), 6753–6757.

Wang, H., Richardson, C.J., Ho, M., 2015. Dual controls on carbon loss during drought in Peatlands. Nat. Clim. Chang. 5 (6), 584–587.

Wang, Z.-P., Qian, G., Deng, F.-D., Jian-Hui, H., Megonigal, J.P., Yu, Q., Lü, X.-T., et al., 2016. Methane emissions from the trunks of living trees on upland soils. New Phytol. 211 (2), 429–439.

Wang, H., Weiss, B.P., Bai, X.-N., Downey, B.G., Wang, J., Wang, J., Suavet, C., Fu, R.R., Zucolotto, M.E., 2017a. Lifetime of the solar nebula constrained by meteorite paleomagnetism. Science 355 (6325), 623–627.

Wang, Z.-P., Han, S.-J., Li, H.-L., Deng, F.-D., Zheng, Y.-H., Liu, H.-F., Han, X.-G., 2017b. Methane production explained largely by water content in the heartwood of living trees in upland forests: methane in heartwood. J. Geophys. Res. Biogeosci. 122 (10), 2479–2489.

Wang, J., Sun, J., Xia, J., He, N., Li, M., Niu, S., 2018a. Soil and vegetation carbon turnover times from tropical to boreal forests. Funct. Ecol. 32 (1), 71–82.

Wang, X.T., Cohen, A.L., Luu, V., Ren, H., Zhan, S., Haug, G.H., Sigman, D.M., 2018b. Natural forcing of the North Atlantic nitrogen cycle in the Anthropocene. Proc. Natl. Acad. Sci. U. S. A. 115 (42), 10606–10611.

Wang, Y., Guo, J., Vogt, R.D., Mulder, J., Wang, J., Zhang, X., 2018c. Soil pH as the chief modifier for regional nitrous oxide emissions: new evidence and implications for global estimates and mitigation. Glob. Chang. Biol. 24 (2), e617–e626.

Wang, H., River, M., Richardson, C.J., 2019a. Does an 'iron gate' carbon preservation mechanism exist in organic-rich wetlands? Soil Biol. Biochem. 135, 48–50.

Wang, J., Kan, J., Qian, G., Chen, J., Xia, Z., Zhang, X., Liu, H., Sun, J., 2019b. Denitrification and anammox: understanding nitrogen loss from Yangtze estuary to the East China Sea (ECS). Environ. Pollut. 252 (Pt B), 1659–1670.

Wang, J.-J., Bowden, R.D., Lajtha, K., Washko, S.E., Wurzbacher, S.J., Simpson, M.J., 2019c. Long-term nitrogen addition suppresses microbial degradation, enhances soil carbon storage, and alters the molecular composition of soil organic matter. Biogeochemistry 142 (2), 299–313.

Wang, L., Amelung, W., Prietzel, J., Willbold, S., 2019d. Transformation of organic phosphorus compounds during 1500 years of organic soil formation in Bavarian alpine forests—a ^{31}P NMR study. Geoderma 340 (April), 192–205.

Wang, Q., Zhao, X., Chen, L., Yang, Q., Chen, S., Zhang, W., 2019e. Global synthesis of temperature sensitivity of soil organic carbon decomposition: latitudinal patterns and mechanisms. Funct. Ecol. 33 (3), 514–523.

Wang, W.-L., Keith Moore, J., Martiny, A.C., Primeau, F.W., 2019f. Convergent estimates of marine nitrogen fixation. Nature 566 (7743), 205–211.

Wanninkhof, R., Mulholland, P.J., Elwood, J.W., 1990. Gas-exchange rates for a 1st-order stream determined with deliberate and natural tracers. Water Resour. Res. 26 (7), 1621–1630.

Warby, R.A.F., Johnson, C.E., Driscoll, C.T., 2009. Continuing acidification of organic soils across the northeastern USA: 1984–2001. Soil Sci. Soc. Am. J. 73, 274–284.

Ward, R.C., 1967. Principles of Hydrology. McGraw-Hill.

Ward, B.B., Devol, A.H., Rich, J.J., Chang, B.X., Bulow, S.E., Naik, H., Pratihary, A., Jayakumar, A., 2009. Denitrification as the dominant nitrogen loss process in the Arabian Sea. Nature 461 (7260), 78–81.

Ward, D.M., Nislow, K.H., Folt, C.L., 2010. Bioaccumulation syndrome: identifying factors that make some stream food webs prone to elevated mercury bioaccumulation. Ann. N. Y. Acad. Sci. 1195 (May), 62–83.

Ward, N.D., Bianchi, T.S., Sawakuchi, H.O., 2016. The reactivity of plant-derived organic matter and the potential importance of priming effects along the lower Amazon river. J. Geophys. Res.: Biogeosci. 121, 1522–1539.

Ward, D.S., Shevliakova, E., Malyshev, S., Rabin, S., 2018. Trends and variability of global fire emissions due to historical anthropogenic activities. Glob. Biogeochem. Cycles 32 (1), 122–142.

Wardle, D.A., 1992. A comparative assessment of factors which influence microbial biomass carbon and nitrogen levels in soil. Biol. Rev. Camb. Philos. Soc. 67, 321–358.

Wardle, D.A., Walker, L.R., Bardgett, R.D., 2004. Ecosystem properties and forest decline in contrasting long-term chronosequences. Science 305 (5683), 509–513.

Ware, D.M., Thomson, R.E., 2005. Bottom-up ecosystem trophic dynamics determine fish production in the Northeast Pacific. Science 308 (5726), 1280–1284.

Warembourg, F.R., Paul, E.A., 1977. Seasonal transfers of assimilated ^{14}C in grassland: plant production and turnover, soil and plant respiration. Soil Biol. Biochem. 9, 295–301.

Waring, R.H., Schlesinger, W.H., 1985. Forest Ecosystems. Academic.

Waring, R.H., Rogers, J.J., Swank, W.T., 1981. Water relations and hydrologic cycles. In: Reichle, D.E. (Ed.), Dynamic Properties of Forest Ecosystems. Cambridge University Press, pp. 205–264.

Waring, R.H., Landsberg, J.J., Williams, M., 1998. Net primary production of forests: a constant fraction of gross primary production? Tree Physiol. 18 (2), 129–134.

Waring, B.G., Álvarez-Cansino, L., Barry, K.E., Becklund, K.K., Dale, S., Gei, M.G., Keller, A.B., et al., 2015. Pervasive and strong effects of plants on soil chemistry: a meta-analysis of individual plant 'Zinke' effects. Proc. R. Soc. B Biol. Sci. 282 (1812), 20151001.

Warneck, P., 2000. Chemistry of the Natural Atmosphere, second ed. Academic Press.

Warner, J.X., Dickerson, R.R., Wei, Z., Strow, L.L., Wang, Y., Liang, Q., 2017. Increased atmospheric ammonia over the world's major agricultural areas detected from space. Geophys. Res. Lett. 44 (6), 2875–2884.

Warner, D.L., Bond-Lamberty, B., Jian, J., Stell, E., Vargas, R., 2019. Spatial predictions and associated uncertainty of annual soil respiration at the global scale. Glob. Biogeochem. Cycles 33, 1733–1745.

Warren, D.R., Bernhardt, E.S., Hall, R.O., Likens, G.E., 2007. Forest age, wood and nutrient dynamics in headwater streams of the Hubbard Brook experimental forest, NH. Earth Surf. Process. Landf. 32, 1154–1163.

Watanabe, S., Yamamoto, H., Tsunogai, S., 1995. Relation between the concentrations of DMS in surface seawater and air in the temperate North Pacific region. J. Atmos. Chem. 22, 271–283.

Watanabe, Y., Martini, J.E., Ohmoto, H., 2000. Geochemical evidence for terrestrial ecosystems 2.6 billion years ago. Nature 408 (6812), 574–578.

Watkins, N.D., Sparks, R.S.J., Sigurdsson, H., Huang, T.C., Federman, A., Carey, S., Ninkovich, D., 1978. Volume and extent of Minoan tephra from Santorini volcano: new evidence from deep sea sediment cores. Nature 271, 122–126.

Watmough, S.A., McNeely, R., Lafleur, P.M., 2001. Changes in wood and foliar delta 13C in sugar maple at Gatineau Park, Quebec, Canada. Glob. Chang. Biol. 7, 955–960.

Watson, A.J., Lovelock, J.E., 1983. Biological homeostasis of the global environment: the parable of daisyworld. Tellus B Chem. Phys. Meteorol. 35, 284–289.

Watson, A., Lovelock, J.E., Margulis, L., 1978. Methanogenesis, fires and the regulation of atmospheric oxygen. Bio Systems 10 (4), 293–298.

Watson, A.J., Bakker, D.C., Ridgwell, A.J., Boyd, P.W., Law, C.S., 2000. Effect of iron supply on Southern Ocean CO_2 uptake and implications for glacial atmospheric CO_2. Nature 407 (6805), 730–733.

Watson, D., Hansen, C.J., Selsing, J., Koch, A., Malesani, D.B., Andersen, A.C., Fynbo, J.P.U., et al., 2019. Identification of strontium in the merger of two neutron stars. Nature 574 (7779), 497–500.

Watts, S.F., 2000. The mass budgets of carbonyl sulfide, dimethyl sulfide, carbon disulfide, and hydrogen sulfide. Atmos. Environ. 34, 761–779.

Watts, R.D., Compton, R.W., McCammon, J.H., Rich, C.L., Wright, S.M., Owens, T., Ouren, D.S., 2007. Roadless space of the conterminous United States. Science 316 (5825), 736–738.

Watwood, M.E., Fitzgerald, J.W., 1988. Sulfur transformations in forest litter and soil: results of laboratory and field incubations. Soil Sci. Soc. Am. J. 52, 1478–1483.

Waugh, D.W., Hall, T.M., 2002. Age of stratospheric air: theory, observations, and models. Rev. Geophys. 40.

Waugh, D.W., Primeau, F., Devries, T., Holzer, M., 2013. Recent changes in the ventilation of the southern oceans. Science 339 (6119), 568–570.

Waughman, G.J., Bellamy, D.J., 1980. Nitrogen fixation and the nitrogen balance in peatland ecosystems. Ecology 61 (5), 1185–1198.

Wayne, R.P., 1991. Chemistry of Atmospheres: An Introduction to the Chemistry of the Atmospheres of Earth, the Planets, and Their Satellites, second ed. Oxford University Press.

Weatherall, P., Marks, K.M., Jakobsson, M., Schmitt, T., Tani, S., Arndt, J.E., Rovere, M., Chayes, D., Ferrini, V., Wigley, R., 2015. A new digital bathymetric model of the world's oceans. Life Supp. Biosp. Sci.: Int. J. Earth Space, U.S. Geol. Surv. Water Supply Pap., 968-C 2 (8), 331–345.

Weathers, K.C., Likens, G.E., Bormann, F.H., Eaton, J.S., Bowden, W.B., Andersen, J.L., Cass, D.A., et al., 1986. A regional acidic cloud fog water event in the eastern United States. Nature 319, 657–658.

Weathers, K.C., Lovett, G.M., Likens, G.E., Lathrop, R., 2000. The effect of landscape features on deposition to Hunter Mountain, Catskill Mountains, New York. Ecol. Appl. 10, 528–540.

Webb, W., Szarek, S., Lauenroth, W., Kinerson, R., Smith, M., 1978. Primary productivity and Wateruse in native forest, grassland, and desert ecosytsems. Ecology 59, 1239–1247.

Webb, W.L., Lauenroth, W.K., Szarek, S.R., Kinerson, R.S., 1983. Primary production and abiotic controls in forests, grasslands, and desert ecosystems in the United States. Ecology 64, 134–151.

Webb, R.S., Rosenzweig, C.E., Levine, E.R., 1993. Specifying land surface characteristics in general circulation models: soil profile data set and derived water-holding capacities. Glob. Biogeochem. Cycles 7, 97–108.

Weber, T.S., Deutsch, C., 2010. Ocean nutrient ratios governed by plankton biogeography. Nature 467, 550–554.

Weber, K.A., Achenbach, L.A., Coates, J.D., 2006. Microorganisms pumping iron: anaerobic microbial iron oxidation and reduction. Nat. Rev. Microbiol. 4 (10), 752–764.

Weber, B., Wu, D., Tamm, A., Ruckteschler, N., Rodríguez-Caballero, E., Steinkamp, J., Meusel, H., et al., 2015. Biological soil crusts accelerate the nitrogen cycle through large NO and HONO emissions in drylands. Proc. Natl. Acad. Sci. U. S. A. 112 (50), 15384–15389.

Webster, J.R., Ehrman, T.P., 1996. Solute dynamics. In: Hauer, F.R., Lamberti, G.A. (Eds.), Methods in Stream Ecology. Elsevier, pp. 145–160.

Webster, J.R., Meyer, J.L., 1997. Organic matter budgets for streams: a synthesis. J. N. Am. Benthol. Soc. 16, 141–161.

Webster, J.R., Patten, B.C., 1979. Effects of watershed perturbation on stream potassium and calcium dynamics. Ecol. Monogr. 19, 51–52.

Webster, J.R., Benfield, E.F., Ehrman, T.P., Schaeffer, M.A., Tank, J.L., Hutchens, J.J., D'Angelo, D.J., 1999. What happens to allochthonous material that falls into Streams? A synthesis of new and published information from Coweeta. Freshw. Biol. 41, 687–705.

Webster, J.R., Mulholland, P.J., Tank, J.L., Valett, H.M., Dodds, W.K., Peterson, B.J., Bowden, W.B., et al., 2003. Factors affecting ammonium uptake in Streams—an inter-biome perspective. Freshw. Biol. 48, 1329–1352.

Webster, A.J.R., Meyer, J.L., Journal, S., American, N., Society, B., No, M., Edwards, R.T., Melack, J.M., 2012. Organic matter budgets for streams: a synthesis. J. N. Am. Benthol. Soc. 16 (1), 141–161.

Webster, C.R., Mahaffy, P.R., Atreya, S.K., Moores, J.E., Flesch, G.J., Malespin, C., McKay, C.P., et al., 2018. Background levels of methane in Mars' atmosphere show strong seasonal variations. Science 360 (6393), 1093–1096.

Wedepohl, K.H., 1995. The composition of the continental crust. Geochim. Cosmochim. Acta 59, 1217–1232.

Wehr, R., Munger, J.W., McManus, J.B., Nelson, D.D., Zahniser, M.S., Davidson, E.A., Wofsy, S.C., Saleska, S.R., 2016. Seasonality of temperate forest photosynthesis and daytime respiration. Nature 534 (7609), 680–683.

Wehr, R., Róisín, C., William Munger, J., Barry McManus, J., Nelson, D.D., Zahniser, M.S., Saleska, S.R., Wofsy, S.C., 2017. Dynamics of canopy stomatal conductance, transpiration, and evaporation in a temperate deciduous forest, validated by carbonyl sulfide uptake. Biogeosciences 14 (2), 389–401.

Wei, J., Dirmeyer, P., 2019. Sensitivity of land precipitation to surface evapotranspiration: a nonlocal perspective based on water vapor transport. Geophys. Res. Lett. 46 (November).

Wei, Z., Yoshimura, K., Wang, L., Miralles, D.G., Jasechko, S., Lee, X., 2017. Revisiting the contribution of transpiration to global terrestrial evapotranspiration: revisiting global ET partitioning. Geophys. Res. Lett. 44 (6), 2792–2801.

Wei, X., Hayes, D.J., Fraver, S., Chen, G., 2018. Global pyrogenic carbon production during recent decades has created the potential for a large, long-term sink of atmospheric CO_2. J. Geophys. Res. Biogeosci. 123 (12), 3682–3696.

Weier, K.L., Gilliam, J.W., 1986. Effect of acidity on denitrification and nitrous oxide evolution from Atlantic coastal plain soils. Soil Sci. Soc. Am. J. 50, 1202–1206.

Weier, K.L., Doran, J.W., Power, J.F., Walters, D.T., 1993. Denitrification and the dinitrogen/nitrous oxide ratio as affected by soil water, available carbon, and nitrate. Soil Sci. Soc. Am. J. 57, 66–72.

Weihrauch, C., Opp, C., 2018. Ecologically relevant phosphorus pools in soils and their dynamics: the story so far. Geoderma 325 (September), 183–194.

Weiner, T., Gross, A., Moreno, G., Migliavacca, M., Schrumpf, M., Reichstein, M., Hilman, B., Carrara, A., Angert, A., 2018. Following the turnover of soil bioavailable phosphate in mediterranean savanna by oxygen stable isotopes. J. Geophys. Res. Biogeosci. Landsc. Ser. 123 (6), 1850–1862.

Weir, J.S., 1972. Spatial distribution of elephants in an African National Park in relation to environmental sodium. Oikos 23, 1–13.

Weirich, S.W., da Silva, R.F., Perrando, E.R., Da Ros, C.O., Dellai, A., Scheid, D.L., Trombeta, H.W., 2018. Influence of ectomycorrhizae on the growth of seedlings of *Eucalyptus grandis*, *Corymbia citriodora*, *Eucalyptus saligna* and *Eucalyptus dunnii*. Ciencia Florestal 28 (2), 765–777.

Weiss, H.V., Koide, M., Goldberg, E.D., 1971. Mercury in a Greeland ice sheet: evidence of recent input by man. Science 174, 692–694.

Weiss, H., Courty, M.A., Wetterstrom, W., Guichard, F., Senior, L., Meadow, R., Curnow, A., 1993. The genesis and collapse of third millennium north Mesopotamian civilization. Science 261 (5124), 995–1004.

Weiss, P.S., Johnson, J.E., Gammon, R.H., Bates, T.S., 1995. Reevaluation of the open ocean source of carbonyl sulfide to the atmosphere. J. Geophys. Res. 100, 23083–23092.

Weiss, J.V., Emerson, D., Megonigal, J.P., 2005. Rhizosphere iron(III) deposition and reduction in a *Juncus effusus* L.-dominated wetland. Soil Sci. Soc. Am. J. 69 (6), 1861.

Weiss, R.F., Muehle, J., Salameh, P.K., Harth, C.M., 2008. Nitrogen trifluoride in the global atmosphere. Geophys. Res. Lett. 35.

Weiss-Penzias, P.S., Gustin, M.S., Lyman, S.N., 2011. Sources of gaseous oxidized mercury and mercury dry deposition at two southeastern US sites. Atmos. Environ. 45, 4569–4579.

Weitz, A.M., Veldkamp, E., Keller, M., Neff, J., Crill, P.M., 1998. Nitrous oxide, nitric oxide, and methane fluxes from soils following clearing and burning of tropical secondary forest. J. Geophys. Res. 103, 28047–28058.

Welch, S.A., Ullman, W.J., 1993. The effect of organic acids on plagioclase dissolution rates and stoichiometry. Geochim. Cosmochim. Acta 57, 2725–2736.

Welch, S.A., Taunton, A.E., Banfield, J.F., 2002. Effect of microorganisms and microbial metabolites on apatite dissolution. Geomicrobiol J. 19, 343–367.

Well, R., Eschenbach, W., Flessa, H., von der Heide, C., Weymann, D., 2012. Are dual isotope and Isotopomer ratios of N_2O useful indicators for N_2O turnover during denitrification in nitrate-contaminated aquifers? Geochim. Cosmochim. Acta 90 (August), 265–282.

Wells, P.V., 1983. Paleobiogeography of Montane Islands in the Great Basin since the last Glaciopluvial. Ecol. Monogr. 53, 341–382.

Welti, E.A.R., Sanders, N.J., de Beurs, K.M., Kaspari, M., 2019. A distributed experiment demonstrates widespread sodium limitation in grassland food webs. Ecology 100 (3), e02600.

Weng, F.Z., Ferraro, R.R., Grody, N.C., 1994. Global Precipitation estimations using defense meteorological satellite program F10 and F11 special sensor microwave imager data. J. Geophys. Res. 99, 14493–14502.

Wennberg, P.O., Cohen, R.C., Stimpfle, R.M., Koplow, J.P., Anderson, J.G., Salawitch, R.J., Fahey, D.W., et al., 1994. Removal of stratospheric O_3 by radicals: in situ measurements of OH, HO_2, NO, NO_2, ClO, and BrO. Science 266 (5184), 398–404.

Wentworth, G.R., Murphy, J.G., Gregoire, P.K., Cheyne, C.A.L., Tevlin, A.G., Hems, R., 2014. Soil–atmosphere exchange of ammonia in a non-fertilized grassland: measured emission potentials and inferred fluxes. Biogeosciences 11 (20), 5675–5686.

Wentz, F.J., Ricciardulli, L., Hilburn, K., Mears, C., 2007. How much more rain will global warming bring? Science 317 (5835), 233–235.

Werner, F., de la Haye, T.R., Spielvogel, S., Prietzel, J., 2017. Small-scale spatial distribution of phosphorus fractions in soils from silicate parent material with different degree of Podzolization. Geoderma 302 (September), 52–65.

Wesely, M.L., Hicks, B.B., 2000. A review of the current status of knowledge on dry deposition. Atmos. Environ. 34, 2261–2282.

Wessman, C.A., Aber, J.D., Peterson, D.L., Melillo, J.M., 1988. Remote sensing of canopy chemistry and nitrogen cycling in temperate forest ecosystems. Nature 335, 154–156.

West, T.O., Marland, G., 2002. A synthesis of carbon sequestration, carbon emissions, and net carbon flux in agriculture: comparing tillage practices in the United States. Agric. Ecosyst. Environ. 91, 217–232.

West, A.J., Galy, A., Bickle, M., 2005. Tectonic and climatic controls on silicate weathering. Earth Planet. Sci. Lett. 235, 211–228.

West, T.O., Marland, G., Singh, N., Bhaduri, B.L., Roddy, A.B., 2009. The human carbon budget: an estimate of the spatial distribution of metabolic carbon consumption and release in the United States. Biogeochemistry 94 (1), 29–41.

West, T.O., Brandt, C.C., Baskaran, L.M., Hellwinckel, C.M., Mueller, R., Bernacchi, C.J., Bandaru, V., et al., 2010. Cropland carbon fluxes in the United States: increasing geospatial resolution of inventory-based carbon accounting. Ecol. Appl. 20 (4), 1074–1086.

Westheimer, F.H., 1987. Why nature chose phosphates. Science 235 (4793), 1173–1178.

Westman, C.J., Laiho, R., 2003. Nutrient dynamics of drained peatland forests. Biogeochemistry 63, 269–298.

Weston, N.B., Dixon, R.E., Joye, S.B., 2006. Ramifications of increased salinity in tidal freshwater sediments: geochemistry and microbial pathways of organic matter mineralization. J. Geophys. Res. 111 (G1), G01009.

Wetherill, G.W., 1985. Occurrence of giant impacts during the growth of the terrestrial planets. Science 228 (4701), 877–879.

Wetherill, G.W., 1994. Provenance of the terrestrial planets. Geochim. Cosmochim. Acta 58 (20), 4513–4520.

Wetselaar, R., 1968. Soil organic nitrogen mineralization as affected by low soil water potentials. Plant Soil 29, 9–17.

Wetzel, R.G., 1992. Gradient-dominated ecosystems - sources and regulatory functions of dissolved organic matter in freshwater ecosystems. Hydrobiologia 229, 181–198.

Wetzel, R.G., 2001. Limnology, third ed. Academic Press.

Wetzel, R.G., Likens, G.E., 2000. Limnological Analyses, third ed. Springer-Verlag.

Weyl, P.K., 1978. Micro-paleontology and ocean surface climate. Science 202, 475–481.

Whalen, S.C., 2005. Biogeochemistry of methane exchange between natural wetlands and the atmosphere. Environ. Eng. Sci. 22 (1), 73–94.

Whalen, S.C., Reeburgh, W.S., 1990. Consumption of atmospheric methane by tundra soils. Nature 346, 160–162.

Whalen, S.C., Reeburgh, W.S., Sandbeck, K.A., 1990. Rapid methane oxidation in a landfill cover soil. Appl. Environ. Microbiol. 56 (11), 3405–3411.

Whalen, S.C., Reeburgh, W.S., Kizer, K.S., 1991. Methane consumption and emission by taiga. Glob. Biogeochem. Cycles 5, 261–273.

Wheat, C.G., McManus, J., Mottl, M.J., Giambalvo, E., 2003. Oceanic phosphorus imbalance: magnitude of the mid-ocean ridge flank hydrothermal sink. Geophys. Res. Lett. 30.

Whelan, M.E., Rhew, R.C., 2016. Reduced sulfur trace gas exchange between a seasonally dry grassland and the atmosphere. Biogeochemistry 128 (3), 267–280.

Whelan, M.E., Lennartz, S.T., Gimeno, T.E., Wehr, R., Wohlfahrt, G., Wang, Y., Kooijmans, L.M.J., et al., 2018. Reviews and syntheses: carbonyl sulfide as a multi-scale tracer for carbon and water cycles. Biogeosciences 15 (12), 3625–3657.

Whelpdale, D.M., Galloway, J.N., 1994. Sulfur and reactive nitrogen oxide fluxes in the North Atlantic atmosphere. Glob. Biogeochem. Cycles 8, 481–493.

Whitburn, S., Van Damme, M., Kaiser, J.W., van der Werf, G.R., Turquety, S., Hurtmans, D., Clarisse, L., Clerbaux, C., Coheur, P.-F., 2015. Ammonia emissions in tropical biomass burning regions: comparison between satellite-derived emissions and bottom-up fire inventories. Atmos. Environ. 121 (November), 42–54.

White, T.C.R., 1984. The abundance of invertebrate herbivores in relation to the availability of nitrogen in stressed food plants. Oecologia 63 (1), 90–105.

White, C.S., 1988. Nitrification inhibition by monoterpenoids: theoretical mode of action based on molecular structures. Ecology 69, 1631–1633.

White, A.F., Blum, A.E., 1995. Effects of climate on chemical weathering in watersheds. Geochim. Cosmochim. Acta 59, 1729–1747.

White, P.J., Broadley, M.R., 2001. Chloride in soils and its uptake and movement within the plant: a review. Ann. Bot. 88, 967–988.

White, E.J., Turner, F., 1970. A method of estimating income of nutrients in catch of airborne particles by a woodland canopy. J. Appl. Ecol. 7, 441–461.

White, A.F., Blum, A.E., Schulz, M.S., Bullen, T.D., Harden, J.W., Peterson, M.L., 1996. Chemical weathering rates of a soil chronosequence on granitic alluvium: I. Quantification of mineralogical and surface area changes and calculation of primary silicate reaction rates. Geochim. Cosmochim. Acta 60, 2533–2550.

White, A.F., Blum, A.E., Schulz, M.S., Vivit, D.V., Stonestrom, D.A., Larsen, M., Murphy, S.F., Eberl, D., 1998. Chemical weathering in a tropical watershed, Luquillo Mountains, Puerto Rico: I. Long-term versus short-term weathering fluxes. Geochim. Cosmochim. Acta 62, 209–226.

White, A.F., Bullen, T.D., Schulz, M.S., Blum, A.E., Huntington, T.G., Peters, N.E., 2001. Differential rates of feldspar weathering in granitic Regoliths. Geochim. Cosmochim. Acta 65, 847–869.

Whitehead, D.C., Lockyer, D.R., Raistrick, N., 1988. The volatilization of ammonia from perennial ryegrass during decomposition, drying and induced senescence. Ann. Bot. 61, 567–571.

Whiteside, M.D., Treseder, K.K., Atsatt, P.R., 2009. The brighter side of soils: quantum dots track organic nitrogen through fungi and plants. Ecology 90 (1), 100–108.

Whitfield, P.H., Schreier, H., 1981. Hysteresis in relationships between discharge and water chemistry in the Fraser River basin, British Columbia. Limnol. Oceanogr. 26, 1179–1182.

Whitfield, M., Turner, D.R., 1979. Water–rock partition coefficients and the composition of seawater and river water. Nature 278 (5700), 132–137.

Whitfield, C.J., Aherne, J., Baulch, H.M., 2011. Controls on greenhouse gas concentrations in polymictic headwater lakes in Ireland. Sci. Total Environ. 410-411 (December), 217–225.

Whiticar, M.J., Faber, E., Schoell, M., 1986. Biogenic methane formation in marine and freshwater environments—CO_2 reduction vs acetate fermentation isotope evidence. Geochim. Cosmochim. Acta 50 (5), 693–709.

Whiting, G.J., Chanton, J.P., 1993. Primary production control of methane emission from wetlands. Nature 364, 794–795.

Whiting, B.G.J., Chanton, J.P., 2001. Greenhouse carbon balance of wetlands: methane emission versus carbon sequestration. Tellus 521–528.

Whitlow, S., Mayewski, P., Dibb, J., Holdsworth, G., Twickler, M., 1994. An ice-core based record of biomass burning in the arctic and subarctic, 1750–1980. Tellus B Chem. Phys. Meteorol. 46, 234–242.

Whitney, F.A., Freeland, H.J., Robert, M., 2007. Persistently declining oxygen levels in the interior waters of the eastern subarctic Pacific. Prog. Oceanogr. 75, 179–199.

Whittaker, R.H., 1975. Communities and Ecosystems, second ed. Macmillian.

Whittaker, R.H., Likens, G.E., 1973. Carbon in the biota. In: Woodwell, G.M., Pecan, E.V. (Eds.), Carbon and the Biosphere. National Technical Information Service, pp. 281–302.

Whittaker, R.H., Marks, P.L., 1975. Methods of assessing terrestrial productivity. In: Lieth, H., Whittaker, R.H. (Eds.), Primary Productivity of the Biosphere. Springer, pp. 55–118.

Whittaker, R.H., Niering, W.A., 1975. Vegetation of the Santa Catalina Mountains, Arizona. V. Biomass, production and diversity along the elevation gradient. Ecology 56, 771–790.

Whittaker, R.H., Woodwell, G.M., 1968. Dimension and production relations of trees and shrubs in the Brookhaven Forest. J. Ecol. 56, 1–25.

Whittaker, R.H., Bormann, F.H., Likens, G.E., Siccama, T.G., 1974. The Hubbard Brook ecosystem study: forest biomass and production. Ecol. Monogr. 44, 233–254.

Widdel, F., Schnell, S., Heising, S., Ehrenreich, A., Assmus, B., Schink, B., 1993. Ferrous iron oxidation by anoxygenic phototrophic Bacteria. Nature 362, 834–836.

Widmer, F., Shaffer, B.T., Porteous, L.A., Seidler, R.J., 1999. Analysis of nifH Gene pool complexity in soil and litter at a Douglas fir forest site in the Oregon Cascade Mountain range. Appl. Environ. Microbiol. 65 (2), 374–380.

Wieder, R.K., Lang, G.E., 1988. Cycling of inorganic and organic sulfur in peat from big run bog, West Virginia. Biogeochemistry 5 (2), 221–242.

Wieder, R.K., Yavitt, J.B., Lang, G.E., 1990. Methane production and sulfate reduction in two Appalachian Peatlands. Biogeochemistry 10 (2), 81–104.

Wieder, W.R., Cleveland, C.C., Townsend, A.R., 2009. Controls over leaf litter decomposition in wet tropical forests. Ecology 90 (12), 3333–3341.

Wieder, W.R., Cleveland, C.C., Kolby Smith, W., Todd-Brown, K., 2015. Future productivity and carbon storage limited by terrestrial nutrient availability. Nat. Geosci. 8 (6), 441–444.

Wiegert, R.G., Evans, F.C., 1964. Primary production and the disappearance of dead vegetation on an old field in southeastern Michigan. Ecology 45, 49–63.

Wiegner, T.N., Seitzinger, S.P., Glibert, P.M., Bronk, D.A., 2006. Bioavailability of dissolved organic nitrogen and carbon from nine rivers in the eastern United States. Aquat. Microb. Ecol.: Int. J. 43 (3), 277–287.

Wigley, T.M.L., 1989. Possible climate change due to SO_2-derived cloud condensation nuclei. Nature 339, 365–367.

Wilcke, W., Velescu, A., Leimer, S., Bigalke, M., Boy, J., Valarezo, C., 2019. Temporal trends of phosphorus cycling in a tropical montane forest in Ecuador during 14 years. J. Geophys. Res. Biogeosci. 124 (5), 1370–1386.

Wilde, S.A., Valley, J.W., Peck, W.H., Graham, C.M., 2001. Evidence from detrital zircons for the existence of continental crust and oceans on the earth 4.4 Gyr ago. Nature 409 (6817), 175–178.

Wilhelm, S.W., Maxwell, D.P., Trick, C.G., 1996. Growth, iron requirements, and siderophore production in iron-limited *Synechococcus* PCC 7002. Limnol. Oceanogr. 41, 89–97.

Wilkinson, B.H., McElroy, B.J., 2007. The impact of humans on continental erosion and sedimentation. Geol. Soc. Am. Bull. 119, 140–156.

Wilkinson, D.M., Nisbet, E.G., Ruxton, G.D., 2012. Could methane produced by sauropod dinosaurs have helped drive mesozoic climate warmth? Curr. Biol. 22 (9), R292–R293.

Wilkinson, G.M., Pace, M.L., Cole, J.J., 2013. Terrestrial dominance of organic matter in north temperate lakes. Glob. Biogeochem. Cycles 27 (1), 43–51.

Willenbring, J.K., Codilean, A.T., McElroy, B., 2013. Earth is (mostly) flat: apportionment of the flux of continental sediment over millennial time scales. Geology 41 (3), 343–346.

Willett, K.M., Gillett, N.P., Jones, P.D., Thorne, P.W., 2007. Attribution of observed surface humidity changes to human influence. Nature 449 (7163), 710–712.

Williams, G.R., 1996. The Molecular Biology of Gaia. Columbia University Press.

Williams, P.M., Druffel, E.R.M., 1987. Radiocarbon in dissolved organic matter in the central North Pacific Ocean. Nature 330 (6145), 246–248.

Williams, L.E., Miller, A.J., 2001. Transporters responsible for the uptake and partitioning of nitrogenous solutes. Annu. Rev. Plant Physiol. Plant Mol. Biol. 52 (June), 659–688.

Williams, C.D., Mukhopadhyay, S., 2019. Capture of nebular gases during Earth's accretion is preserved in deep-mantle neon. Nature 565 (7737), 78–81.

Williams, E.J., Hutchinson, G.L., Fehsenfeld, F.C., 1992. NO_x and N_2O emissions from soil. Glob. Biogeochem. Cycles 6, 351–388.

Williams, E.L., Walter, L.M., Ku, T.C.W., Kling, G.W., Zak, D.R., 2003. Effects of CO_2 and nutrient availability on mineral weathering in controlled tree growth experiments. Glob. Biogeochem. Cycles. 17.

Williams, R.M.E., Grotzinger, J.P., Dietrich, W.E., Gupta, S., Sumner, D.Y., Wiens, R.C., Mangold, N., et al., 2013. Martian fluvial conglomerates at Gale crater. Science 340 (6136), 1068–1072.

Williams, M., Zalasiewicz, J., Waters, C.N., Edgeworth, M., Bennett, C., Barnosky, A.D., Ellis, E.C., et al., 2016. The Anthropocene: a conspicuous stratigraphical signal of anthropogenic changes in production and consumption across the biosphere: anthropocene biosphere evolution. Earth's Future 4 (3), 34–53.

Willison, T.W., Goulding, K.W.T., Powlson, D.S., 1995. Effect of land-use change and methane mixing ratio on methane uptake from United Kingdom soil. Glob. Chang. Biol. 1, 209–212.

Wilson, A.T., 1978. Pioneer agriculture explosion and CO_2 levels in the atmosphere. Nature 273, 40–41.

Windolf, J., Jeppesen, E., Jensen, J.P., Kristensen, P., 1996. Modelling of seasonal variation in nitrogen retention and in-lake concentration: a four-year mass balance study in 16 shallow Danish lakes. Biogeochemistry 33, 25–44.

Winner, W.E., Smith, C.L., Koch, G.W., Mooney, H.A., Bewley, J.D., Krouse, H.R., 1981. Rates of emission of H_2S from plants and patterns of stable sulfur isotope fractionation. Nature 289, 672–673.

Winslow, L.A., Zwart, J.A., Batt, R.D., Dugan, H.A., Iestyn Woolway, R., Corman, J.R., Hanson, P.C., Read, J.S., 2016. LakeMetabolizer: an R package for estimating lake metabolism from free-water oxygen using diverse statistical models. Inland Waters 6 (4), 622–636.

Wittmann, H., von Blanckenburg, F., Maurice, L., Guyot, J.L., Filizola, N., Kubik, P.W., 2011. Sediment production and delivery in the Amazon River basin quantified by in situ-produced cosmogenic nuclides and recent river loads. Geol. Soc. Am. Bull. 123, 934–950.

Wlostowski, A.N., Gooseff, M.N., McKnight, D.M., Lyons, W.B., 2018. Transit times and rapid chemical equilibrium explain chemostasis in glacial meltwater streams in the McMurdo dry valleys, Antarctica. Geophys. Res. Lett. 45 (24), 13322–13331.

Wogelius, R.A., Walther, J.V., 1991. Olivine dissolution at 25 C—Effects of pH, CO_2, and organic acids. Geochim. Cosmochim. Acta 55, 943–954.

Wohl, E., Dwire, K., Sutfin, N., Polvi, L., Bazan, R., 2012. Mechanisms of carbon storage in mountainous headwater rivers. Nat. Commun. 3, 1263–1268.

Wolf, I., Russow, R., 2000. Different pathways of formation of N_2O, N_2 and NO in black earth soil. Soil Biol. Biochem. 32, 229–239.

Wolf, A.A., Drake, B.G., Erickson, J.E., Megonigal, J.P., 2007. An oxygen-mediated positive feedback between elevated carbon dioxide and soil organic matter decomposition in a simulated anaerobic Wetland. Glob. Chang. Biol. 13 (9), 2036–2044.

Wolf, J., West, T.O., Le Page, Y., Page Kyle, G., Xuesong, Z., James Collatz, G., Imhoff, M.L., 2015. Biogenic carbon fluxes from global agricultural production and consumption: global agricultural carbon fluxes. Glob. Biogeochem. Cycles Int. Trade Stat. 2011 29 (10), 1617–1639.

Wolf, J., Asrar, G.R., West, T.O., 2017. Revised methane emissions factors and spatially distributed annual carbon fluxes for global livestock. Carbon Balance Manag. 12 (1), 16.

Wolfe, D.W., 2001. Tales from the Underground. Perseus.

Wolfe, G.V., Steinke, M., Kirst, G.O., 1997. Grazing-activated chemical defence in a unicellular marine alga. Nature 387, 894–897.

Wolfe, G.M., Nicely, J.M., St Clair, J.M., Hanisco, T.F., Liao, J., Oman, L.D., Brune, W.B., et al., 2019. Mapping hydroxyl variability throughout the global remote troposphere via synthesis of airborne and satellite formaldehyde observations. Proc. Natl. Acad. Sci. U. S. A. 116 (23), 11171–11180.

Wolff, E.W., Fischer, H., Fundel, F., Ruth, U., Twarloh, B., Littot, G.C., Mulvaney, R., et al., 2006. Southern ocean sea-ice extent, productivity and iron flux over the past eight glacial cycles. Nature 440 (7083), 491–496.

Wolin, M.J., Miller, T.L., 1987. Bioconversion of organic carbon to CH_4 and CO_2. Geomicrobiol J. 5, 239–259.

Wollast, R., 1981. Interactions between major biogeochemical cycles in marine ecosystems. In: Likens, G.E. (Ed.), Some Perspectives of the Major Biogeochemical Cycles. Wiley, pp. 125–142.

Wollast, R., 1993. Interactions of carbon and nitrogen cycles in the coastal zone. In: Wollast, F.T.M.R., Chou, L. (Eds.), Interactions of C, N, P, and S Biogeochemical Cycles and Global Change. Springer, pp. 195–210.

Wollheim, W.M., Vorosmarty, C.J., Peterson, B.J., Seitzinger, S.P., Hopkinson, C.S., 2006. Relationship between river size and nutrient removal. Geophys. Res. Lett. 33.

Wollheim, W.M., Vorosmarty, C.J., Bouwman, A.F., Green, P., Harrison, J., Linder, E., Peterson, B.J., Seitzinger, S.P., Syvitski, J.P.M., 2008. Global N removal by freshwater aquatic systems using a spatially distributed, within-basin approach. Glob. Biogeochem. Cycles 22.

Woltemate, I., Whiticar, M.J., Schoell, M., 1984. Carbon and hydrogen isotopic composition of bacterial methane in a shallow freshwater lake. Limonol. Oceanogr. 29 (5), 985–992.

Wong, M.T.F., Hughes, R., Rowell, D.L., 1990. Retarded leaching of nitrate in acid soils from the tropics: measurement of the effective anion exchange capacity. J. Soil Sci. 41, 655–663.

Wong, M.H., Atreya, S.K., Mahaffy, P.N., Franz, H.B., Malespin, C., Trainer, M.G., Stern, J.C., et al., 2013. Isotopes of nitrogen on Mars: atmospheric measurements by curiosity's mass spectrometer. Geophys. Res. Lett. 40 (23), 6033–6037.

Wood, W.W., Petraitis, M.J., 1984. Origin and distribution of carbon dioxide in the unsaturated zone of the southern high plains of Texas. Water Resour. Res. 20, 1193–1208.

Wood, T., Bormann, F.H., Voigt, G.K., 1984. Phosphorus cycling in a northern hardwood forest: biological and chemical control. Science 223 (4634), 391–393.

Wood, B.J., Walter, M.J., Wade, J., 2006. Accretion of the earth and segregation of its core. Nature 441, 825–833.

Woodbury, P.B., Smith, J.E., Heath, L.S., 2007. Car-bon sequestration in the US forest sector from 1990 to 2010. For. Ecol. Manag. 241, 14–27.

Woodmansee, R.G., Dodd, J.L., Bowman, R.A., Clark, F.E., Dickinson, C.E., 1978. Nitrogen budget of a shortgrass prairie ecosystem. Oecologia 34 (3), 363–376.

Woodward, F.I., 1987. Stomatal numbers are sensitive to increases in CO_2 from preindustrial levels. Nature 327, 617–618.

Woodward, F.I., 1993. Plant responses to past concentrations of CO_2. Veg. Hist. Archaeobotany 104, 145–155.

Woodward, G., Gessner, M.O., Giller, P.S., Gulis, V., Hladyz, S., Lecerf, A., Malmqvist, B., et al., 2012. Continental-scale effects of nutrient pollution on stream ecosystem functioning. Science 336 (6087), 1438–1440.

Woodwell, G.M., 1974. Variation in the nutrient content of leaves of *Quercus alba*, *Quercus coccinea*, and *Pinus rigida* in the Brookhaven forest from bud-break to abscission. Am. J. Bot. 61, 749–753.

Woosley, S.E., 1986. Nucleosynthesis and Stellar evolution. In: Chiosi, C., Audouze, J., Woosley, S.E. (Eds.), Nucleosynthesis and Chemical Evolution. The Geneva Observatory, pp. 1–195.

Woosley, S.E., Phillips, M.M., 1988. Supernova 1987A! Science 240 (4853), 750–759.

Woosley, S.E., Weaver, T.A., 1995. The evolution and explosion of massive stars. II. Explosive hydrodynamics and nucleosynthesis. Astrophys. J. Suppl. Ser. 101 (November), 181.

Workshop, S.S., 1990. Concepts and methods for assessing solute dynamics in stream ecosystems. J. N. Am. Benthol. Soc. 9, 95–119.

Worm, B., Barbier, E.B., Nicola, B., Emmett Duffy, J., Folke, C., Halpern, B.S., Jackson, J.B.C., et al., 2006. Impacts of biodiversity loss on ocean ecosystem services. Science 314 (5800), 787–790.

Worm, B., Hilborn, R., Baum, J.K., Branch, T.A., Collie, J.S., Costello, C., Fogarty, M.J., et al., 2009. Rebuilding global fisheries. Science 325 (5940), 578–585.

Worman, S.L., Pratson, L.F., Karson, J.A., Klein, E.M., 2016. Global rate and distribution of H_2 gas produced by serpentinization within oceanic lithosphere: H_2 formation in ocean lithosphere. Geophys. Res. Lett. 43 (12), 6435–6443.

Worrall, F., Clay, G.D., Masiello, C.A., Mynheer, G., 2013. Estimating the oxidative ratio of the global terrestrial biosphere carbon. Biogeochemistry 115 (1), 23–32.

Worsley, T.R., Davies, T.A., 1979. Sea-level fluctuations and deep-sea sedimentation rates. Science 203 (4379), 455–456.

Wortmann, U.G., Paytan, A., 2012. Rapid variability of seawater chemistry over the past 130 million years. Science 337 (6092), 334–336.

Wotawa, G., Trainer, M., 2000. The influence of Canadian forest fires on pollutant concentrations in the United States. Science 288 (5464), 324–328.

Wotawa, G., Novelli, P.C., Trainer, M., Granier, C., 2001. Inter-annual variability of summertime CO concentrations in the northern hemisphere explained by boreal Forest fires in North America and Russia. Geophys. Res. Lett. 28, 4575–4578.

Wrage, N., Velthof, G.L., van Beusichem, M.L., Oenema, O., 2001. Role of nitrifier denitrification in the production of nitrous oxide. Soil Biol. Biochem. 33, 1723–1732.

Wrage-Mönnig, N., Horn, M.A., Well, R., Müller, C., Velthof, G., Oenema, O., 2018. The role of nitrifier denitrification in the production of nitrous oxide revisited. Soil Biol. Biochem. 123 (August), A3–16.

Wright, R.F., 1976. Impact of forest fire on nutrient influxes to small lakes in northeastern Minnesota. Ecology 57, 649–663.

Wright, S.J., 2019. Plant responses to nutrient addition experiments conducted in tropical forests. Ecol. Monogr. 89 (4), 1581.

Wright, J.R., Leahey, A., Rice, H.M., 1959. Chemical, morphological and mineralogical characteristics of a chronosequence of soils on alluvial deposits in the Northwest Territories. Can. J. Soil Sci. 39, 32–43.

Wright, R.F., Lotse, E., Semb, A., 1994. Experimental acidification of alpine catchments at Sogndal, Norway: results after 8 years. Water Air Soil Pollut. 72, 297–315.

Wright, S.J., Yavitt, J.B., Wurzburger, N., Turner, B.L., Tanner, E.V.J., Sayer, E.J., Santiago, L.S., et al., 2011. Potassium, phosphorus, or nitrogen limit root allocation, tree growth, or litter production in a lowland tropical forest. Ecology 92 (8), 1616–1625.

Wright, L.P., Zhang, L., Marsik, F.J., 2016. Overview of mercury dry deposition, Litterfall, and throughfall studies. Atmos. Chem. Phys. 16 (21), 13399–13416.

Wu, J., 1981. Evidence of sea spray produced by bursting bubbles. Science 212 (4492), 324–326.

Wu, J., Odonnell, A.G., Syers, J.K., 1995. Influences of glucose, nitrogen and plant residues on the immobilization of sulfate-S in soil. Soil Biol. Biochem. 27, 1363–1370.

Wu, J.F., Sunda, W., Boyle, E.A., Karl, D.M., 2000. Phosphate depletion in the Western North Atlantic Ocean. Science 289 (5480), 759–762.

Wu, Z.T., Dijkstra, P., Koch, G.W., Penuelas, J., Hungate, B.A., 2011. Responses of terrestrial ecosystems to temperature and precipitation change: a meta-analysis of experimental manipulation. Glob. Chang. Biol. 17, 927–942.

Wuebbles, D.J., Hayhoe, K., 2002. Atmospheric methane and global change. Earth Sci. Rev. 57, 177–210.

Wuebbles, D.J., Tamaresis, J.S., 1993. The role of methane in the global environment. In: Khalil, M.A.K. (Ed.), Atmospheric Methane: Sources, Sinks, and Role in Global Climate. Springer, pp. 469–513.

Wuebbles, D.J., Dong, W., Patten, K.O., Olsen, S.C., 2013. Analyses of new short-lived replacements for HFCs with large GWPs. Geophys. Res. Lett. 40 (17), 4767–4771.

Wujeska-Klause, A., Crous, K.Y., Ghannoum, O., Ellsworth, D.S., 2019. Lower photorespiration in elevated CO_2 reduces leaf N concentrations in mature *Eucalyptus* trees in the field. Glob. Chang. Biol. 25 (February).

Wullstein, L.H., Pratt, S.A., 1981. Scanning electron microscopy of rhizosheaths of *Oryzopsis hymenoides*. Am. J. Bot. 68, 408–419.

Xi, F., Davis, S.J., Ciais, P., Crawford-Brown, D., Guan, D., Pade, C., Shi, T., et al., 2016. Substantial global carbon uptake by cement carbonation. Nat. Geosci. 9 (12), 880–883.

Xia, J., Wan, S., 2008. Global response patterns of terrestrial plant species to nitrogen addition. New Phytol. 179 (2), 428–439.

Xia, M., Talhelm, A.F., Pregitzer, K.S., 2017. Chronic nitrogen deposition influences the chemical dynamics of leaf litter and fine roots during decomposition. Soil Biol. Biochem. 112, 24–34.

Xiao, J.F., Zhuang, Q.L., Law, B.E., Chen, J.Q., Baldocchi, D.D., Cook, D.R., Oren, R., et al., 2010. A continuous measure of gross primary production for the conterminous United States derived from MODIS and AmeriFlux data. Remote Sens. Environ. 114, 576–591.

Xiao, J., Li, X., He, B., Altaf Arain, M., Beringer, J., Desai, A.R., Emmel, C., et al., 2019. Solar-induced chlorophyll fluorescence exhibits a universal relationship with gross primary productivity across a wide variety of biomes. Glob. Chang. Biol. 25 (4), e4–e6.

Xie, Z.-Q., Sun, L.-G., Wang, J.-J., Liu, B.-Z., 2002. A potential source of atmospheric sulfur from penguin colony emissions. J. Geophys. Res. 107.

Xie, R., Zhao, J., Lu, L., Ge, J., Brown, P., Wei, S., Wang, R., Qiao, Y., Webb, S., Tian, S., 2019. Efficient phloem remobilisation of Zn protects apple trees during the early stages of Zn deficiency. Plant Cell Environ. 42 (July).

Xiong, J., Fischer, W.M., Inoue, K., Nakahara, M., Bauer, C.E., 2000. Molecular evidence for the early evolution of photosynthesis. Science 289 (5485), 1724–1730.

Xu, X., Liu, W., 2017. The global distribution of Earth's critical zone and its controlling factors. Geophys. Res. Lett. 44 (7), 3201–3208.

Xu, X.F., Tian, H.Q., Hui, D.F., 2008a. Convergence in the relationship of CO_2 and N_2O exchanges between soil and atmosphere within terrestrial ecosystems. Glob. Chang. Biol. 14, 1651–1660.

Xu, Y., Liang, F., Jeffrey, P.D., Shi, Y., Morel, F.M.M., 2008b. Structure and metal exchange in the cadmium carbonic anhydrase of marine diatoms. Nature 452 (7183), 56–61.

Xu, X., Thornton, P.E., Post, W.M., 2013. A global analysis of soil microbial biomass carbon, nitrogen and phosphorus in terrestrial ecosystems. Glob. Ecol. Biogeogr.: J. Macroecol. 22 (6), 737–749.

Xu, H., Vandecasteele, B., Zavattaro, L., Sacco, D., Wendland, M., Boeckx, P., Haesaert, G., Sleutel, S., 2019a. Maize root-derived C in soil and the role of physical protection on its relative stability over shoot-derived C. Eur. J. Soil Sci. 70 (5), 1489.

Xu, R., Tian, H., Pan, S., Prior, S.A., Feng, Y., Batchelor, W.D., Chen, J., Yang, J., 2019b. Global ammonia emissions from synthetic nitrogen fertilizer applications in agricultural systems: empirical and process-based estimates and uncertainty. Glob. Chang. Biol. 25 (1), 314–326.

Xu-Ri, I.C.P., Spahni, R., Niu, H.S., 2012. Modelling terrestrial nitrous oxide emissions and implications for climate feedback. New Phytol. 196 (2), 472–488.

Yaalon, D.H., 1965. Downward movement and distribution of anions in soil profiles with limited wetting. In: Hallsworth, E.D., Crawford, D.V. (Eds.), Experimental Pedology. Butterworth, pp. 157–164.

Yagi, K., Williams, J., Wang, N.Y., Cicerone, R.J., 1995. Atmospheric methyl bromide (CH_3Br) from agricultural soil fumigations. Science 267 (5206), 1979–1981.

Yamada, A., Inoue, T., Wiwatwitaya, D., Ohkuma, M., Kudo, T., Sugimoto, A., 2006. Nitrogen fixation by termites in tropical forests, Thailand. Ecosystems 9, 75–83.

Yamagata, Y., Watanabe, H., Saitoh, M., Namba, T., 1991. Volcanic production of polyphosphates and its relevance to prebiotic evolution. Nature 352 (6335), 516–519.

Yanagawa, H., Ogawa, Y., Kojima, K., Ito, M., 1988. Construction of Protocellular structures under stimulated primitive earth conditions. Origins Life Evol. Biosph. 18, 179–207.

Yanai, R.D., 1992. Phosphorus budget of a 70-year-old northern hardwood forest. Biogeochemistry 17, 1–22.

Yanai, R.D., Vadeboncoeur, M.A., Hamburg, S.P., Arthur, M.A., Fuss, C.B., Groffman, P.M., Siccama, T.G., Driscoll, C.T., 2013. From missing source to missing sink: long-term changes in the nitrogen budget of a northern hardwood forest. Environ. Sci. Technol. 47 (20), 11440–11448.

Yang, L.H., 2004. Periodical cicadas as resource pulses in north American forests. Science 306 (5701), 1565–1567.

Yang, X., Post, W.M., 2011. Phosphorus transformations as a function of pedogenesis: a synthesis of soil phosphorus data using Hedley fractionation method. Biogeosciences 8, 2907–2916.

Yang, E.S., Gupta, P., Christopher, S.A., 2009. Net radiative effect of dust aerosols from satellite measurements over Sahara. Geophys. Res. Lett. 36.

Yang, K., Dickerson, R.R., Carn, S.A., Ge, C., Wang, J., 2013a. First observations of SO_2 from the satellite Suomi NPP OMPS: widespread air pollution events over China. Geophys. Res. Lett. 40 (18), 4957–4962.

Yang, X., Post, W.M., Thornton, P.E., Jain, A., 2013b. The distribution of soil phosphorus for global biogeochemical modeling. Biogeosciences 10 (4), 2525–2537.

Yang, K., Carn, S.A., Ge, C., Wang, J., Dickerson, R.R., 2014. Advancing measurements of tropospheric NO_2 from space: new algorithm and first global results from OMPS. Geophys. Res. Lett. 41 (13), 4777–4786.

Yang, Y., Li, P., He, H., Zhao, X., Datta, A., Ma, W., Zhang, Y., et al., 2015. Long-term changes in soil pH across major Forest ecosystems in China: soil pH dynamics in forests. Geophys. Res. Lett. 42 (3), 933–940.

Yang, S., Gruber, N., Long, M.C., Vogt, M., 2017. ENSO-driven variability of denitrification and Suboxia in the eastern tropical Pacific Ocean. Glob. Biogeochem. Cycles 31 (10), 1470–1487.

Yang, H., Liu, Y., Liu, J., Meng, J., Hu, X., Tao, S., 2019. Improving the imbalanced global supply chain of phosphorus fertilizers. Earth's Future 7 (6), 638–651.

Yano, J., Kern, J., Sauer, K., Latimer, M.J., Pushkar, Y., Biesiadka, J., Loll, B., et al., 2006. Where water is oxidized to dioxygen: structure of the photosynthetic Mn_4Ca cluster. Science 314 (5800), 821–825.

Yavitt, J.B., Fahey, T.J., 1993. Production of methane and nitrous oxide by organic soils within a Northern Hardwood forest ecosystem. In: Oremland, R.S. (Ed.), Biogeochemistry and Global Change. Chapman and Hall, pp. 261–277.

Yavitt, J.B., Knapp, A.K., 1995. Methane emission to the atmosphere through emergent cattail (*Typha latifolia* L.) plants. Tellus B—Chem. Phys. Meteorol. 47 (5), 521–534.

Yavitt, J.B., Harms, K.E., Garcia, M.N., Mirabello, M.J., Joseph Wright, S., 2011. Soil fertility and fine root dynamics in response to 4 years of nutrient (N, P, K) fertilization in a lowland tropical moist forest, Panama: tropical forest soil and fine roots. Aust. Ecol. 36 (4), 433–445.

Yeung, L.Y., Murray, L.T., Martinerie, P., Witrant, E., Hu, H., Banerjee, A., Orsi, A., Chappellaz, J., 2019. Isotopic constraint on the twentieth-century increase in tropospheric ozone. Nature 570 (7760), 224–227.

Yi, Z.G., Wang, X.M., Ouyang, M.G., Zhang, D.Q., Zhou, G.Y., 2010. Air-soil exchange of dimethyl sulfide, carbon disulfide, and dimethyl disulfide in three subtropical forests in South China. J. Geophys. Res.-Atmos. 115.

Yi-Balan, S.A., Amundson, R., Buss, H.L., 2014. Decoupling of sulfur and nitrogen cycling due to biotic processes in a tropical rainforest. Geochim. Cosmochim. Acta 142 (October), 411–428.

Yin, X.W., 1993. Variation in foliar nitrogen concentration by forest type and climatic gradients in North America. Can. J. Forest Res. 23, 1587–1602.

Yin, Q., Jacobsen, S.B., Yamashita, K., Blichert-Toft, J., Télouk, P., Albarède, F., 2002. A short timescale for terrestrial planet formation from Hf-W chronometry of meteorites. Nature 418 (6901), 949–952.

Yoh, M., Terai, H., Saijo, Y., 1983. Accumulation of nitrous oxide in the oxygen-deficient layer of freshwater lakes. Nature 301, 327–329.

Yoh, M., Terai, H., Saijo, Y., 1988. Nitrous oxide in freshwater lakes. Arch. Hydrobiol. 113, 273–294.

Yokouchi, Y., Nojiri, Y., Toom-Sauntry, D., Fraser, P., Inuzuka, Y., Tanimoto, H., Nara, H., Murakami, R., Mukai, H., 2012. Long-term variation of atmospheric methyl iodide and its link to global environmental change: long-term change of methyl iodide. Geophys. Res. Lett. 39 (23).

Yokoyama, Y., Lambeck, K., De Deckker, P., Johnston, P., Fifield, L.K., 2000. Timing of the last glacial maximum from observed sea-level minima. Nature 406 (6797), 713–716.

Yoneyama, T., Muraoka, T., Murakami, T., Boonkerd, N., 1993. Natural abundance of ^{15}N in tropical plants with emphasis on tree legumes. Plant Soil 153, 295–304.

Yong, S.C., Roversi, P., Lillington, J., Rodriguez, F., Krehenbrink, M., Zeldin, O.B., Garman, E.F., Lea, S.M., Berks, B.C., 2014. A complex iron-calcium cofactor catalyzing phosphotransfer chemistry. Science 345 (6201), 1170–1173.

Yool, A., Martin, A.P., Fernández, C., Clark, D.R., 2007. The significance of nitrification for oceanic new production. Nature 447 (7147), 999–1002.

Yoshida, N., Toyoda, S., 2000. Constraining the atmospheric N_2O budget from intramolecular site preference in N_2O isotopomers. Nature 405 (6784), 330–334.

Yoshinari, T., Knowles, R., 1976. Acetylene inhibition of nitrous oxide reduction by denitrifying bacteria. Biochem. Biophys. Res. Commun. 69 (3), 705–710.

Young, J.R., Ellis, E.C., Hidy, G.M., 1988. Deposition of air-borne acidifiers in the Western environment. J. Environ. Qual. 17, 1–26.

Youngberg, C.T., Wollum, A.G., 1976. Nitrogen accretion in developing *Ceanothus velutinus* stands. Soil Sci. Soc. Am. J. 40, 109–112.

Yu, Z.C., 2012. Northern peatland carbon stocks and dynamics: a review. Biogeosciences 9 (10), 4071–4085.

Yu, Z., Zhang, Q., Kraus, T.E.C., Dahlgren, R.A., Anastasio, C., Zasoski, R.J., 2002. Contribution of amino compounds to dissolved organic nitrogen in forest soils. Biogeochemistry 61, 173–198.

Yu, Z., Campbell, I.D., Campbell, C., Vitt, D.H., Bond, G.C., Apps, M.J., 2003. Carbon sequestration in Western Canadian peat highly sensitive to Holocene wet-dry climate cycles at millennial timescales. The Holocene 13 (6), 801–808.

Yu, J., Broecker, W.S., Elderfield, H., Jin, Z., McManus, J., Zhang, F., 2010a. Loss of carbon from the deep sea since the last glacial maximum. Science 330 (6007), 1084–1087.

Yu, Z., Loisel, J., Brosseau, D.P., Beilman, D.W., Hunt, S.J., 2010b. Global peatland dynamics since the last glacial maximum: global peatlands since the LGM. Geophys. Res. Lett. Geophys. Monogr. Ser. 37 (13).

Yu, H., Remer, L.A., Chin, M., Bian, H., Tan, Q., Yuan, T., Zhang, Y., 2012. Aerosols from overseas rival domestic emissions over North America. Science 337 (6094), 566–569.

Yu, H., Chin, M., Yuan, T., Bian, H., Remer, L.A., Prospero, J.M., Omar, A., et al., 2015. The fertilizing role of African dust in the Amazon rainforest: a first multiyear assessment based on data from cloud-aerosol Lidar and infrared pathfinder satellite observations: dust deposition in the Amazon rainforest. Geophys. Res. Lett. 42 (6), 1984–1991.

Yu, L., Zhu, J., Mulder, J., Dörsch, P., 2016. Multiyear dual nitrate isotope signatures suggest that N-saturated subtropical forested catchments can act as robust N sinks. Glob. Chang. Biol. 22 (11), 3662–3674.

Yu, L., Jin, X., Josey, S.A., Lee, T., Kumar, A., Wen, C., Xue, Y., 2017. The global ocean water cycle in atmospheric reanalysis, satellite, and ocean salinity. J. Clim. 30 (10), 3829–3852.

Yu, L., Zhu, J., Zhang, X., Wang, Z., 2019a. Humid subtropical forests constitute a net methane source: a catchment-scale study. J. Geophys. Res. Biogeosci. 124, 2927–2942.

Yu, L., Mulder, J., Zhu, J., Zhang, X., Wang, Z., Dörsch, P., 2019b. Denitrification as a major regional nitrogen sink in subtropical forest catchments: evidence from multi-site dual nitrate isotopes. Glob. Chang. Biol. 25 (5), 1765–1778.

Yu, M., Wang, Y., Jiang, J., Wang, C., Zhou, G., Yan, J., 2019c. Soil organic carbon stabilization in the three subtropical forests: importance of clay and metal oxides. J. Geophys. Res.: Biogeosci. 124 (August).

Yuan, Z.Y., Chen, H.Y.H., 2015. Negative effects of fertilization on plant nutrient resorption. Ecology 96 (2), 373–380.

Yuan, G., Lavkulich, L.M., 1994. Phosphate sorption in relation to extractable iron and aluminum in Spodosols. Soil Sci. Soc. Am. J. 58, 343–346.

Yuan, T.L., Gammon, N., Leighty, R.G., 1967. Relative contribution of organic and clay fractions to cation-exchange capacity of sandy soils from several groups. Soil Sci. 104, 123–128.

Yuan, X., Xiao, S., Taylor, T.N., 2005. Lichen-like symbiosis 600 million years ago. Science 308 (5724), 1017–1020.

Yuan, Z., Sun, L., Bi, J., Wu, H., Zhang, L., 2011. Phosphorus flow analysis of the socioeconomic ecosystem of Shucheng County, China. Ecol. Appl. 21 (7), 2822–2832.

Yung, Y.L., DeMore, W.B., 1999. Photochemistry of Planetary Atmospheres. Oxford University Press.

Yung, Y.L., Wen, J.S., Moses, J.I., Landry, B.M., Allen, M., Hsu, K.J., 1989. Hydrogen and deuterium loss from the terrestrial atmosphere: a quantitative assessment of nonthermal escape fluxes. J. Geophys. Res. 94 (D12), 14971–14989.

Yung, Y.L., Lee, T., Wang, C.H., Shieh, Y.T., 1996. Dust: a diagnostic of the hydrologic cycle during the last glacial maximum. Science 271 (5251), 962–963.

Yvon-Lewis, S.A., Saltzman, E.S., Montzka, S.A., 2009. Recent trends in atmospheric methyl bromide: analysis of post-montreal protocol variability. Atmos. Chem. Phys. 9, 5963–5974.

Zachos, J.C., Röhl, U., Schellenberg, S.A., Sluijs, A., Hodell, D.A., Kelly, D.C., Thomas, E., et al., 2005. Rapid acidification of the ocean during the Paleocene-Eocene thermal maximum. Science 308 (5728), 1611–1615.

Zachos, J.C., Dickens, G.R., Zeebe, R.E., 2008. An early cenozoic perspective on greenhouse warming and carbon-cycle dynamics. Nature 451 (7176), 279–283.

Zaferani, S., Pérez-Rodríguez, M., Biester, H., 2018. Diatom ooze-A large marine mercury sink. Science 361 (6404), 797–800.

Zagal, E., Persson, J., 1994. Immobilization and remineralization of nitrate during glucose decomposition at four rates of nitrogen addition. Soil Biol. Biochem. 26, 1313–1321.

Zak, D.R., Host, G.E., Pregitzer, K.S., 1989. Regional variability in nitrogen mineralization, nitrification, and overstory biomass in northern lower Michigan. Can. J. Forest Res. 19, 1521–1526.

Zak, D.R., Tilman, D., Parmenter, R.R., Rice, C.W., Fisher, F.M., Vose, J., Milchunas, D., Martin, C.W., 1994. Plant production and soil microorganisms in late successional ecosystems: a continental-scale study. Ecology 75, 2333–2347.

Zak, D.R., Argiroff, W.A., Freedman, Z.B., Upchurch, R.A., Entwistle, E.M., Romanowicz, K.J., 2019. Anthropogenic N deposition, fungal gene expression, and an increasing soil carbon sink in the northern hemisphere. Ecology 100 (10), e02804.

Zamin, T.J., Syndonia Bret-Harte, M., Grogan, P., 2014. Evergreen shrubs dominate responses to experimental summer warming and fertilization in Canadian Mesic Low Arctic Tundra. Aerts, R. (Ed.), J. Ecol. 102 (3), 749–766.

Zamora, L.M., Oschlies, A., 2014. Surface nitrification: a major uncertainty in marine N_2O emissions. Geophys. Res. Lett. 41 (12), 4247–4253.

Zbieranowski, A.L., Aherne, J., 2011. Long-term trends in atmospheric reactive nitrogen across Canada: 1988–2007. Atmos. Environ. 45, 5853–5862.

Zechmeister-Boltenstern, S., Keiblinger, K.M., Mooshammer, M., Peñuelas, J., Richter, A., Sardans, J., Wanek, W., 2015. The Application of Ecological Stoichiometry to Plant–Microbial–Soil Organic Matter Transformations.

Zedler, J.B., Kercher, S., 2005. Wetland resources: status, trends, ecosystem services, and restorability. Annu. Rev. Environ. Resour. 30 (1), 39–74.

Zeebe, R.E., Archer, D., 2005. Feasibility of ocean fertilization and its impact on future atmospheric CO_2 levels. Geophys. Res. Lett. 32.

Zeebe, R.E., Zachos, J.C., Caldeira, K., Tyrrell, T., 2008. Oceans: carbon emissions and acidification. Science 321, 51–52.

Zehetner, F., 2010. Does organic carbon sequestration in volcanic soils offset volcanic CO_2 emissions? Quat. Sci. Rev. 29, 1313–1316.

Zehnder, A.J., Stumm, W., 1988. Geochemistry and biogeochemistry of anaerobic habitats. In: Zehnder, A.J. (Ed.), Biology of Anaerobic Microorganisms. Wiley and Sons, New York, NY, pp. 1–38.

Zehr, J.P., Ward, B.B., 2002. Minireview nitrogen cycling in the ocean: new perspectives on processes and paradigms. Appl. Environ. Microbiol. 68 (3), 1015–1024.

Zeikus, J.G., Ward, J.C., 1974. Methane formation in living trees: a microbial origin. Science 184 (4142), 1181–1183.

Zektser, I.S., Loaiciga, H.A., 1993. Groundwater fluxes in the global hydrologic cycle: past, present and future. J. Hydrol. 144, 405–427.

Zelles, L., 1999. Fatty acid patterns of phospholipids and lipopolysaccharides in the characterisation of microbial communities in soil: a review. Biol. Fertil. Soils 29 (2), 111–129.

Zemp, M., Huss, M., Thibert, E., Eckert, N., McNabb, R., Huber, J., Barandun, M., et al., 2019. Global glacier mass changes and their contributions to sea-level rise from 1961 to 2016. Nature 568 (7752), 382–386.

Zender, C.S., Miller, R.L., Tegen, I., 2004. Supplemental material to "quantifying mineral dust mass budgets: systematic terminology, constraints, and current estimates". Eos Trans. Aguiaine/Societe D'etudes Folkloriques Du Centre-Ouest 85, 509.

Zerefos, C.S., Tetsis, P., Kazantzidis, A., Amiridis, V., Zerefos, S.C., Luterbacher, J., Eleftheratos, K., Gerasopoulos, E., Kazadzis, S., Papayannis, A., 2014. Further evidence of important environmental information content in red-to-green ratios as depicted in paintings by great masters. Atmos. Chem. Phys. 14 (6), 2987–3015.

Zerihun, A., McKenzie, B.A., Morton, J.D., 1998. Photosynthate costs associated with the utilization of different nitrogen-forms: Influence on the carbon balance of plants and shoot-root biomass partitioning. New Phytol. 138, 1–11.

Zerkle, A.L., Poulton, S.W., Newton, R.J., Mettam, C., Claire, M.W., Bekker, A., Junium, C.K., 2017. Onset of the aerobic nitrogen cycle during the great oxidation event. Nature 542 (7642), 465–467.

Zhan, X., Yu, G., He, N., Jia, B., Zhou, M., Wang, C., Zhang, J., et al., 2015. Inorganic nitrogen wet deposition: evidence from the north-south transect of eastern China. Environ. Pollut. 204 (September), 1–8.

Zhang, H., Bloom, P.R., 1999. Dissolution kinetics of hornblende in organic acid solutions. Soil Sci. Soc. Am. J. 63, 815–822.

Zhang, H., Forde, B.G., 1998. An Arabidopsis MADS box gene that controls nutrient-induced changes in root architecture. Science 279 (5349), 407–409.

Zhang, X., Kondragunta, S., 2006. Estimating forest biomass in the USA using generalized allometric models and MODIS land products. Geophys. Res. Lett. 33 (9), L10501.

Zhang, Y., Zindler, A., 1993. Distribution and evolution of carbon and nitrogen in earth. Earth Planet. Sci. Lett. 117 (3–4), 331–345.

Zhang, X., Zwiers, F.W., Hegerl, G.C., Hugo Lambert, F., Gillett, N.P., Solomon, S., Stott, P.A., Nozawa, T., 2007. Detection of human influence on twentieth-century precipitation trends. Nature 448 (7152), 461–465.

Zhang, K., Kimball, J.S., Nemani, R.R., Running, S.W., 2010. A continuous satellite-derived global record of land surface evapotranspiration from 1983 to 2006. Water Resour. Re. 46.

Zhang, D.D., Lee, H.F., Wang, C., Li, B., Zhang, J., Pei, Q., Chen, J., 2011a. Climate change and large-scale human population collapses in the pre-industrial era. Glob. Ecol. Biogeogr.: J. Macroecol. 20, 520–531.

Zhang, X., Hester, K.C., Ussler, W., Walz, P.M., Peltzer, E.T., Brewer, P.G., 2011b. In situ raman-based measurements of high dissolved methane concentrations in hydrate-rich ocean sediments. Geophys. Res. Lett. 38.

Zhang, F., Chen, J.M., Pan, Y., Birdsey, R.A., Shen, S., Weimin, J., He, L., 2012a. Attributing carbon changes in conterminous U.S. forests to disturbance and non-disturbance factors from 1901 to 2010. J. Geophys. Res.: Biogeosci.

Zhang, T., Kim, B., Levard, C., Reinsch, B.C., Lowry, G.V., Deshusses, M.A., Hsu-Kim, H., 2012b. Methylation of mercury by bacteria exposed to dissolved, nano-particulate and microparticulate mercuric sulfides. Environ. Sci. Technol. 46, 6950–6958.

Zhang, Y., Song, L., Liu, X.J., Li, W.Q., Lu, S.H., Zheng, L.X., Bai, Z.C., Cai, G.Y., Zhang, F.S., 2012c. Atmospheric organic nitrogen deposition in China. Atmos. Environ. 46, 195–204.

Zhang, Y., Jaeglé, L., Thompson, L., Streets, D.G., 2014. Six centuries of changing oceanic mercury: anthropogenic mercury in ocean. Glob. Biogeochem. Cycles 28 (11), 1251–1261.

Zhang, X., Davidson, E.A., Mauzerall, D.L., Searchinger, T.D., Dumas, P., Shen, Y., 2015a. Managing nitrogen for sustainable development. Nature 528 (7580), 51–59.

Zhang, Y., Mahowald, N., Scanza, R.A., Journet, E., Desboeufs, K., Albani, S., Kok, J.F., et al., 2015b. Modeling the global emission, transport and deposition of trace elements associated with mineral dust. Biogeosciences 12 (19), 5771–5792.

Zhang, B., Tian, H., Ren, W., Tao, B., Lu, C., Yang, J., Banger, K., Pan, S., 2016a. Methane emissions from global rice fields: magnitude, spatiotemporal patterns, and environmental controls. Glob. Biogeochem. Cycles 30 (9), 1246–1263.

Zhang, X., McRose, D.L., Romain, D., Bellenger, J.P., Morel, F.M.M., Kraepiel, A.M.L., 2016b. Alternative nitrogenase activity in the environment and nitrogen cycle implications. Biogeochemistry 127 (2), 189–198.

Zhang, B., Tian, H., Lu, C., Chen, G., Pan, S., Anderson, C., Poulter, B., 2017a. Methane emissions from global wetlands: an assessment of the uncertainty associated with various Wetland extent data sets. Atmos. Environ. 165 (September), 310–321.

Zhang, Y.G., Pagani, M., Henderiks, J., Ren, H., 2017b. A long history of equatorial deep-water upwelling in the Pacific Ocean. Earth Planet. Sci. Lett. 467, 1–9.

Zhang, H.-Y., Lü, X.-T., Hartmann, H., Keller, A., Han, X.-G., Trumbore, S., Phillips, R.P., 2018a. Foliar nutrient resorption differs between arbuscular mycorrhizal and ectomycorrhizal trees at local and global scales. Glob. Ecol. Biogeogr.: J. Macroecol. 27 (7), 875–885.

Zhang, H., Yee, L.D., Lee, B.H., Curtis, M.P., Worton, D.R., Isaacman-VanWertz, G., Offenberg, J.H., et al., 2018b. Monoterpenes are the largest source of summertime organic aerosol in the southeastern United States. Proc. Natl. Acad. Sci. U. S. A. 115 (9), 2038–2043.

Zhang, K., Song, C., Zhang, Y., Dang, H., Cheng, X., Zhang, Q., 2018c. Global-scale patterns of nutrient density and partitioning in forests in relation to climate. Glob. Chang. Biol. 24 (1), 536–551.

Zhang, Y., Dannenberg, M., Hwang, T., Song, C., 2019. El Niño-southern oscillation-induced variability of terrestrial gross primary production during the satellite era. J. Geophys. Res.: Biogeosci. 124 (July).

Zhao, M., Running, S.W., 2010. Drought-induced reduction in global terrestrial net primary production from 2000 through 2009. Science 329 (5994), 940–943.

Zhao, J., Gao, Q., Liu, Q., Guo, F., 2020. Lake eutrophication recovery trajectories: some recent findings and challenges ahead. Ecol. Indic.. 110. https://doi.org/10.1016/j.ecolind.2019.105878.

Zheng, B., Chevallier, F., Yin, Y., Ciais, P., Fortems-Cheiney, A., Deeter, M.N., Parker, R.J., Wang, Y., Worden, H.M., Zhao, Y.H., 2019. Global atmospheric carbon monoxide budget 2000–2017 inferred from multispecies atmospheric inversions. Earth Syst. Sci. Data 21, 1411–1436.

Zhou, T., Luo, Y.Q., 2008. Spatial patterns of ecosystem carbon residence time and NPP-driven carbon uptake in the conterminous United States. Glob. Biogeochem. Cycles 22.

Zhou, G., Liu, S., Li, Z., Zhang, D., Tang, X., Zhou, C., Yan, J., Mo, J., 2006. Old-growth forests can accumulate carbon in soils. Science 314 (5804), 1417.

Zhou, L., Tian, Y., Myneni, R.B., Ciais, P., Saatchi, S., Liu, Y.Y., Piao, S., et al., 2014. Widespread decline of Congo rainforest greenness in the past decade. Nature 509 (7498), 86–90.

Zhou, J., Baojing, G., Schlesinger, W.H., Xiaotang, J., 2016. Significant accumulation of nitrate in Chinese semi-humid croplands. Sci. Rep. 6 (1), 25088.

Zhou, K., Barjenbruch, M., Kabbe, C., Inial, G., Remy, C., 2017. Phosphorus recovery from municipal and fertilizer wastewater: China's potential and perspective. J. Environ. Sci. 52 (February), 151–159.

Zhou, J., Bing, H., Wu, Y., Sun, H., Wang, J., 2018. Weathering of primary mineral phosphate in the early stages of ecosystem development in the Hailuogou glacier foreland chronosequence: weathering of P in initial stage of pedogenesis. Eur. J. Soil Sci. NATO ASI Ser. I4 69 (3), 450–461.

Zhou, Y., Sawyer, A.H., David, C.H., Famiglietti, J.S., 2019. Fresh submarine groundwater discharge to the near-global coast. Geophys. Res. Lett. 46 (11), 5855–5863.

Zhu, R., Glindemann, D., Kong, D., Sun, L., Geng, J., Wang, X., 2007. Phosphine in the marine atmosphere along a hemispheric course from China to Antarctica. Atmos. Environ. 41 (7), 1567–1573.

Zhu, J., Mulder, J., Bakken, L., Dörsch, P., 2013. The importance of denitrification for N_2O emissions from an N-saturated forest in SW China: results from in situ ^{15}N labeling experiments. Biogeochemistry 116 (1), 103–117.

Zhu, Q., Peng, C., Chen, H., Fang, X., Liu, J., Jiang, H., Yang, Y., Yang, G., 2015. Estimating global natural Wetland methane emissions using process modelling: spatio-temporal patterns and contributions to atmospheric methane fluctuations. Glob. Ecol. Biogeogr.: J. Macroecol. 24 (8), 959–972.

Zhu, L., Mickley, L.J., Jacob, D.J., Marais, E.A., Jianxiong Sheng, L.H., Abad, G.G., Chance, K., 2017. Long-term (2005–2014) trends in formaldehyde (HCHO) columns across North America as seen by the OMI satellite instrument: evidence of changing emissions of volatile organic compounds: HCHO trend across North America. Geophys. Res. Lett. 44 (13), 7079–7086.

Zhuang, Q., Melillo, J.M., Sarofim, M.C., Kicklighter, D.W., David McGuire, A., Felzer, B.S., Sokolov, A., Prinn, R.G., Steudler, P.A., Shaomin, H., 2006. CO_2 and CH_4 exchanges between land ecosystems and the atmosphere in northern high latitudes over the 21st century. Geophys. Res. Lett. 33 (17), 512.

Zhuang, Q., Lu, Y., Chen, M., 2012. An inventory of global N_2O emissions from the soils of natural terrestrial ecosystems. Atmos. Environ. 47 (February), 66–75.

Zhu-Barker, X., Cavazos, A.R., Ostrom, N.E., Horwath, W.R., Glass, J.B., 2015. The importance of abiotic reactions for nitrous oxide production. Biogeochemistry 126 (3), 251–267.

Ziebis, W., Forster, S., Huettel, M., Jorgensen, B.B., 1996. Complex burrows of the mud shrimp *Callianassa truncata* and their geochemical impact in the sea bed. Nature 382, 619–622.

Zimmerman, P.R., Greenberg, J.P., 1988. Measurements of atmospheric hydrocarbons and biogenic emission fluxes in the Amazon boundary layer. J. Geophys. Res. 93, 1407–1416.

Zimov, S.A., Voropaev, Y.V., Semiletov, I.P., Davidov, S.P., Prosiannikov, S.F., Chapin, F.S., Chapin, M.C., Trumbore, S., Tyler, S., 1997. North Siberian lakes: a methane source fueled by pleistocene carbon. Science 277 (5327), 800–802.

Zimov, S.A., Davydov, S.P., Zimova, G.M., Davydova, A.I., Schuur, E.A.G., Dutta, K., Chapin III, F.S., 2006. Permafrost carbon: stock and decomposability of a globally significant carbon pool. Geophys. Res. Lett. 33 (20), 151.

Zinke, P.J., 1962. The pattern of influence of individual forest trees on soil properties. Ecology 43 (1), 130–133.

Zinke, P.J., 1980. Influence of Chronic Air Pollution on Mineral Cycling in Forests. Pacific Southwest Forest and Range Experiment Station, U.S. Forest Service.

Zobel, D.B., Antos, J.A., 1991. 1980 tephra from Mount St. Helens: spatial and temporal variation beneath forest canopies. Biol. Fertil. Soils 12 (1), 60–66.

Zörb, C., Senbayram, M., Peiter, E., 2014. Potassium in agriculture—status and perspectives. J. Plant Physiol. 171, 656–669.

Zumkehr, A., Hilton, T.W., Whelan, M., Smith, S., Elliott Campbell, J., 2017. Gridded anthropogenic emissions inventory and atmospheric transport of carbonyl sulfide in the U.S. J. Geophys. Res. D: Atmos. 122 (4), 2169–2178.

Zumkehr, A., Hilton, T.W., Whelan, M., Smith, S., Kuai, L., Worden, J., Elliott Campbell, J., 2018. Global gridded anthropogenic emissions inventory of carbonyl sulfide. Atmos. Environ. 183 (June), 11–19.

Zwally, H.J., Comiso, J.C., Parkinson, C.L., Cavalieri, D.J., Gloersen, P., 2002. Variability of Antarctic sea ice 1979–1998. J. Geophys. Res. Oceans 107.

Zychowski, J., 2012. Impact of cemeteries on groundwater chemistry: a review. Catena 93, 29–37.

Zysset, M., Blaser, P., Luster, J., Gehring, A.U., 1999. Aluminum solubility control in different horizons of a Podzol. Soil Sci. Soc. Am. J. 63, 1106–1115.

Index

Note: Page numbers followed by *f* indicate figures, *t* indicate tables and *np* indicate footnotes.

735